Elementary Particles and Their Interactions

Springer-Verlag Berlin Heidelberg GmbH

Quang Ho-Kim   Xuan-Yem Pham

# Elementary Particles and Their Interactions

## Concepts and Phenomena

With 116 Figures, 36 Tables, Numerous Examples, and 102 Problems with Selected Solutions

Springer

Professor Quang Ho-Kim
Physics Department
Université Laval
Ste-Foy, QC Canada G1K 7P4
E-mail: qhokim@phy.ulaval.ca

Professor Xuan-Yem Pham
Directeur de recherche au CNRS
Universités Paris VI et VII
Laboratoire de Physique Théorique et Hautes Energies
Tour 16, 1er Etage, 4 Place Jussieu
F-75252 Paris Cedex 05, France
E-mail: pham@lpthe.jussieu.fr

DOI 10.1007/978-3-662-03712-6

Library of Congress Cataloging-in-Publication Data
Ho-Kim, Q. (Quang), 1938- Elementary particles and their interactions: concepts and phenomena / Quang Ho-Kim, Xuan-Yem Pham. p. cm. Includes bibliographical references and index.
1. Particles (Nuclear physics) 2. Nuclear reactions. I. Pham, Xuân-Yêm. II. Title.
QC793.2.H6 1998 539.7'2–dc21 98-4242 CIP


Originally published by Springer-Verlag Berlin Heidelberg New York in 1998
MyCopy version of the original edition 1998



Typesetting: Data conversion by Fa. Steingraeber, Heidelberg
Cover design: *design & production* GmbH, Heidelberg

SPIN 10540816 56/3144 – 5 4 3 2 1 0 – Printed on acid-free paper
www.springer.com/mycopy

*To our families*

# Preface

The last few decades have seen major advances in the physics of elementary particles. New generations of particle accelerators and detectors have come into operation, and have successfully contributed to improving the quantity and quality of data on diverse interaction processes and to the discoveries of whole new families of particles. At the same time, important new ideas have emerged in quantum field theory, culminating in the developments of theories for the weak and strong interactions to complement quantum electrodynamics, the theory of the electromagnetic force. The simplest of the new theories that are at the same time mathematically consistent and physically successful constitute what is known as the standard model of the fundamental interactions. This book is an attempt to present these remarkable advances at an elementary level, making them accessible to students familiar with quantum mechanics, special relativity, and classical electrodynamics.

The main content of the book is roughly divided into two parts; one on theories to lay the foundation and the other on further developments of concepts and descriptions of phenomena to prepare the student for more advanced work. After a brief overview of the subject and a presentation of some basic ideas, two chapters which deal mostly with relativistic one-body wave equations, quantization of fields, and Lorentz invariance follow. In the spirit of the practical approach taken in this book, a heuristic derivation of the Feynman rules is given in the fourth chapter, where the student is shown how to calculate cross-sections and decay rates at the lowest order. The following chapter contains a discussion on discrete symmetries and the concept of symmetry breaking. Isospin is introduced next as the simplest example of internal symmetries in order to ease the reader into the notion of unitary groups in general and of SU(3) in particular, which is discussed next together with the recent discoveries of new particles. The next two chapters present the standard model of the fundamental interactions. We make contact with experiments in subsequent chapters with detailed studies of some fundamental electroweak processes, such as the deep inelastic lepton–nucleon scattering, the CP violation in the neutral K mesons, the neutrino oscillations and the related problem of the solar neutrino deficit, and finally, the $\tau$ lepton decay, which touch upon many aspects of weak interactions. The very high precision of the data that is now attained in some of these processes requires a careful examination of higher-order effects. This leads to a detailed study of one-loop QCD corrections to weak interactions. The next chapter demonstrates the remarkable property of asymptotic freedom of quantum chromodynamics and introduces the powerful concept of the renormalization group which plays a central role in many phenomena. The heavy flavors of quarks, which pose new questions on several aspects of interactions and could open windows on the 'new' physics, form the subject of a separate chapter. We close with a review of the present status of the standard model and, briefly, of its extensions. Selected solutions to problems are given. Finally, important formulas are collected in an Appendix for convenient reference.

In writing this book we have constantly borne in mind the beginning student learning the subject for the first time. For this reason we have avoided a presentation of the formalism based either on canonical quantization or path integral methods. We have adopted instead a decidedly more practical approach based on perturbative field theory. Many particle phenomena may thus be described in detail early in the book, and the student, in turn, can carry out actual calculations. The importance of the physical point of view is further emphasized by the many examples found throughout the book. The first part of the book gives the student the basic (and some extra) material needed to follow the arguments leading to the standard model and to understand the physics that flows from it. The second part is an attempt to reflect recent advances in experimental particle physics (such as neutrino oscillations, B meson physics, and precision tests of electroweak processes). These topics are selected mainly on the strength of their lasting intrinsic value or because they bring out some novel physics. Whatever the motivations, we introduce all topics at an elementary level, work out the calculations in detail, and carry the development to the point where the reader can start deepening his or her own understanding through a meaningful independent study.

We owe thanks to our teachers, students, and colleagues for the physics they have taught us. Many have helped us in our present project. We are in particular grateful to Pierre Fayet, Michel Gourdin, Chi-Sing Lam, Serguey Petcov, and Pham Tri-Nang for reading parts of the book and for making judicious comments and suggestions. Thanks are also due to Dr. Hans Kölsch, our editor at Springer for a pleasant and fruitful collaboration. One of us (QHK) acknowledges with gratitude the financial support given by the Natural Sciences and Engineering Research Council of Canada and the gracious hospitality extended to him by the Laboratoire de Physique Théorique et Hautes Énergies (Université Paris VI et Université Paris VII) and the Laboratoire de Physique Théorique et Modélisation (Université Cergy-Pontoise). Finally, we are greatly indebted to our families, to whom this work is dedicated, for their support and encouragement throughout the writing of this book.

Paris 1998 *Q. Ho-Kim & X.-Y. Pham*

# Contents

# 1 Particles and Interactions: An Overview

In this introductory chapter, we shall get acquainted with the fundamental particles and their interactions, and have a first look at their characteristic properties which we shall study more fully later in this book. We shall also ponder on the crucial and pervasive role of the concept of symmetry, and close the chapter with considerations of the indispensable practical matter of physical units.

## 1.1 A Preview

The idea that a basic simplicity and regularity govern the apparent complexity and diversity of the universe seems to have always been an important aspect of natural philosophy. Less evident is the realization of that idea in terms of irreducible ultimate elements as the fundamental building blocks of all matter, because equally plausible is the notion of an indefinitely divisible matter, conserving all of its properties at all levels of fragmentation. It was probably the discovery of the atom and certainly the discovery of the electron and the proton that finally gave a decisive argument in favor of the concept of the fundamental constituents of matter or elementary particles. In any case, this constant search for order and simplicity has acted as a powerful driving force for progress in physics.

The history of the physics of the infinitely small is largely the history of the uncovering of successive layers of structure, each one a new microcosm existing within older, less fundamental worlds. The notion of what constitutes an elementary particle in fact is not static but evolves with time, changing in step with technological advances, or more precisely with the growth in the power of the sources of energy that become available to the experimenter. The higher the energy of the particle beam used to illuminate or probe the object under study is, the shorter are the wavelengths associated with the incoming particles and the finer the resolutions obtained in the measure. Thus, it is successively discovered that matter is built up from molecules; that the molecules are composed of atoms; the atoms of electrons and nuclei; and the nuclei of protons and neutrons. As the power of the modern particle accelerators keeps on increasing, it has become possible to accelerate particles to higher and higher velocities, to attain resolutions surpassing $10^{-16}$ centimeters and to observe more violent collisions between particles, which have revealed all the wonders of the subatomic universe, not only in the presence of ever finer structure levels, but also in the existence at every level of new particles of ever greater masses. Particle physics has now become synonymous with high-energy physics.

Considered not so long ago, along with the electron and its neutrino, as the fundamental elements of matter, the proton and the neutron have now lost their primary status, as have all particles that respond similarly to the strong interactions and that are generically called *hadrons*, to appear merely as composites of more fundamental objects called *quarks*. Such objects, designated by the symbols u and d, and the first of many to be postulated, replace the proton and neutron to form with the first known *leptons* (the electron e and its neutrino $\nu_e$) the basic components of the stable matter of the universe. The discovery of all kinds of unstable hadrons requires however the introduction of other types of quarks, forming with the more recently discovered leptons new *generations* of quarks and leptons, repeating the original pattern – (u, d; $\nu_e$, e), (c, s; $\nu_\mu$, $\mu$), (t, b; $\nu_\tau$, $\tau$). Since we have as yet no evidence for the existence of structure within quarks and leptons, these particles are considered to be pointlike. In the view of contemporary physics, matter is in large part composed of quarks and leptons, distinct by the fact that the latter, in contrast to the former, are utterly indifferent to strong interactions.

The study of the structure of matter is therefore inextricably tied to the study of the *fundamental forces*, which seeks to explain in every possible way and at every physical level of structure how particles interact. In spite of the wonderful diversity and the bewildering complexity of its multifarious manifestations, nature seems content to use with an admirable sense of economy only four basic forces. Of these, two – *gravitation* and *electromagnetism* – have been known for a long time and are historically the first to be studied; the former by Isaac Newton as early as 1666 and the latter by Charles Augustin Coulomb in 1776. They act over very large distances and are responsible for many familiar phenomena, such as the alternate rise and fall of the sea, the orbiting of the planets, the propagation of radiowaves and the colors of the rainbow. The two other forces, simply called the *strong interaction* and the *weak interaction*, cannot be directly experienced in our everyday life because they exert their influence over very short distances, about $10^{-13}$ centimeters in the first case and $10^{-16}$ centimeters in the second. The strong interaction (also known as the *hadronic interaction*) holds atomic nuclei together and, in another context, binds quarks within hadrons. It is then the force that ultimately ensures the stability of matter. The weak interaction triggers off the $\beta$-decay of some neutron-rich nuclei and, more generally, the slow decay of many particles; in spite of its apparent feebleness, it plays a crucial role in the evolution of the stars.

Up until recently each of these forces has been described by a different theory formulated by a few physicists of great genius – gravitation by Isaac Newton and Albert Einstein, electromagnetism by James Clerk Maxwell, the nuclear strong interaction by Hideki Yukawa, and the nuclear weak interaction by Enrico Fermi. But in the persistent pursuit of the physicist's dream of a unified theory that would include all known forces, remarkable progress has been achieved in the last few decades that closely parallels advances made in our understanding of the particles. Significant similarities between these forces begin to emerge, and now three of the basic interactions can be described by quantum theories that have the same mathematical form. Such theories, among the most beautiful in physics, are based on a symmetry postulate, *the principle of local gauge invariance*, which now appears to physicists to be fundamental. It is a technical term which means that theories formulated in this way, called *gauge theories*, must remain invariant to a certain class of transformations independently performed on all the *particle fields* at different points in space and at different instants in time. Among these theories, the simplest version that is at the same time physically realistic and mathematically complete is the so-called *standard model* (that is, of the strong, weak, and electromagnetic interactions). Without really achieving the long sought-after unification of all forces, this theory nevertheless treats the electromagnetic force, the weak interaction, and the strong interaction on one footing in the same mathematical formalism, and successfully describes all relevant experimental observations.

The main objectives of this book are first to discuss the essential concepts of elementary particle physics and the observed phenomena that have contributed to their developments, and second to explain at an introductory level how the standard model is formulated, how to use it to calculate physical quantities, and finally, how to determine the limits of its applicability.

Before plunging into the long exposition of the theory and the arduous task of complex and at times difficult calculations, and in order to give ourselves an overview of the situation and useful guide posts for the work to come, we describe in the rest of this chapter some general properties of particles and of their interactions, and discuss the importance of the role the symmetry concept plays in high-energy physics.

## 1.2 Particles

According to a widely held view in particle physics, there exist two main classes of particles: the *matter constituents*, which include quarks and leptons, and the *interaction quanta*, which include photons and other particles that mediate interactions. We will describe the first group in the next few paragraphs, leaving the other for the following section.

### 1.2.1 Leptons

Leptons are indivisible particles, apparently devoid of any structure and having in common the property of being completely unaffected by the strong interaction. They all have spin 1/2, obey the Fermi–Dirac statistics, and are therefore called *fermions*. There exist six distinct types of leptons distinguishable by their masses, electric charges, and interaction modes. Three leptons – the electron $e^-$, the muon $\mu^-$, and the tau $\tau^-$ – have a nonvanishing electric charge equal to $-1$ (the sign of which is fixed by convention and its value is given in units of charge $e > 0$); they differ however in the values of their masses. The other three leptons, the *neutrinos*, are all electrically neutral and have a mass either vanishing or very small (see Table 1.1). As a general rule, to every particle corresponds an *antiparticle* (which may or may not be distinct); a particle and its associated antiparticle have the same mass, spin, and lifetime; however their electric charge (as well as other characteristics of similar nature, called *generalized charges*) is the same in magnitude but differs in signs. Thus, the antielectron or, more commonly, the positron ($e^+$), the antimuon ($\mu^+$), and the antitau ($\tau^+$), all have a positive charge equal to 1, whereas the three antineutrinos($\bar{\nu}_e$, $\bar{\nu}_\mu$, $\bar{\nu}_\tau$) are electrically neutral. An example of generalized charge is the *leptonic number*, $L_\ell$, defined as the quantum number with value of $+1$ for the leptons, $-1$ for the antileptons, and 0 for any other particles. This number has been introduced to express the experimental fact that the net number of leptons (i.e. the number of all leptons in presence minus the number of all antileptons) is conserved, that is, unchanged in any reaction, exactly as the familiar electrical charge.

Each charged lepton is associated with a neutrino, the pair forming what one sometimes refers to as a *family* of leptons. There exist three such lepton families: ($\nu_e$, $e^-$), ($\nu_\mu$, $\mu^-$), and ($\nu_\tau$, $\tau^-$). By no means artificial, this classification reflects rather an observed physical property – namely, that lepton families are preserved in all processes – which is mathematically realized by introducing three other conserved generalized charges, the electronic number $L_e$, the muonic number $L_\mu$, and the tauic number $L_\tau$. Each of these numbers

**Table 1.1.** Leptons

| Flavor | Symbol | Mass[a] | Charge[b] |
|---|---|---|---|
| Electronic neutrino | $\nu_e$ | $< 15 \times 10^{-6}$ | 0 |
| Electron | $e^-$ | 0.5 | $-1$ |
| Muonic neutrino | $\nu_\mu$ | $< 0.17$ | 0 |
| Muon | $\mu^-$ | 105.7 | $-1$ |
| Tauonic neutrino | $\nu_\tau$ | $< 19$ | 0 |
| Tauon | $\tau^-$ | 1777 | $-1$ |

[a] In units of MeV/c$^2$; [b] In units of $e$.

is assigned the value +1 for the corresponding charged lepton and its neutrino, −1 for the corresponding antileptons, and 0 for every other particle. As of now, no profound reason for the existence of such rules is known.

### 1.2.2 Quarks

At present, six different types or *flavors* of quarks are known to exist, whimsically called *up* (u), *down* (d), *charm* (c), *strange* (s), *top* or *truth* (t), and *bottom* or *beauty* (b), and arranged into three families according to their main modes of interactions: (u, d), (c, s), and (t, b). The quarks in the first family constitute the basic components of existent matter, whereas the quarks of the other families, having a more fleeting life, are the main stuff of unstable particles. The quarks, just like the leptons, have spin $1/2$ and therefore exist in two spinorial states. But the similarities end there (see Table 1.2). First, all quarks have a fractional electrical charge: the u, c, and t quarks have a charge of $2/3$ while the d, s, and b quarks have a charge of $-1/3$ (always in units of charge $e > 0$). The corresponding antiquarks have charges of opposite signs, $-2/3$ for $\bar{u}$, $\bar{c}$, and $\bar{t}$, and $1/3$ for $\bar{d}$, $\bar{s}$, and $\bar{b}$. To keep track of another empirical conservation rule (conservation of 'matter'), yet another generalized charge has been introduced, the *baryonic number* $N_B$, defined as being $+1/3$ for the quarks, $-1/3$ for the antiquarks, and 0 for all leptons and antileptons. But what really differentiates quarks from leptons is the fact that they have a characteristic that the leptons do not, a quantum number called *color*. Each quark flavor can exist in one of three color states, that we may call, without any profound reasons, red, blue, and green, or 1, 2, and 3, or whatever we like. The color plays the role of the charge for strong interactions among quarks. As the leptons have no colors, they cannot respond to these forces.

If particles of fractional electric charges have never been observed as free particles, it is because quarks, unlike leptons, cannot exist in isolation, but always in clusters, such that the aggregate charge, given by the sum of all constituent charges, is a whole multiple of the unit charge $e$. All hadrons are thus composed, either by combining a quark of a given color and an antiquark of the opposite color, or by combining three quarks, each with a different basic color. It turns out that these structures have whole baryonic numbers and neutral color charges (in other words, are colorless). Besides their multiples, they are the only possible combinations to possess these properties and the only ones to have been observed.

As free quarks do not exist, the definition of their mass is somewhat problematic and not without ambiguities. One could, for example, take the quark mass as the average energy of the quark bound in a hadron in its ground state, or as the probable mass it would have were it to be free. This 'free mass' is the mass that appears in expressions describing quark currents. All these mass values are experimentally determined in one way or another; at present no one knows how to calculate them from first principles.

**Table 1.2.** Quarks

| Flavor | Symbol | Free mass[a] | Constituent mass[a] | Charge[b] |
|---|---|---|---|---|
| up | u | $(5.6 \pm 1.1) \times 10^{-3}$ | 0.33 | $2/3$ |
| down | d | $(9.9 \pm 1.1) \times 10^{-3}$ | 0.33 | $-1/3$ |
| charm | c | $1.35 \pm 0.05$ | 1.5 | $2/3$ |
| strange | s | $0.199 \pm 0.033$ | 0.5 | $-1/3$ |
| top | t | | 180 | $2/3$ |
| bottom | b | | 4.5 | $-1/3$ |

[a]In units of GeV/$c^2$; [b] In units of $e$.

### 1.2.3 Hadrons

As we have mentioned earlier, *hadrons* have an internal structure and are thus not elementary particles at all. It is a generic term used to designate *mesons*, *baryons*, and their antiparticles. Mesons are mainly composed of a quark and an antiquark (not necessarily of the same kind); they have a spin of 0 or 1, obey the Bose–Einstein statistics and for this reason are called *bosons.* Baryons are structures predominantly formed from three quarks, and are fermions of spin $1/2$ or $3/2$. Hundreds of hadrons have been produced, observed and identified, and their properties (mass, spin, charge, lifetime) determined. A small selection of such particles with relatively small masses are shown in Table 1.3. There will be ample time to get better acquainted with each of them as we progress. For now let us simply point out that, on the one hand, a couple of them, such as $\pi^+$, $\pi^-$, and $\pi^0$, have almost identical masses, and on the other hand, all particles of the same spin and parity have very similar masses. Could there be some deep relationships between these particles? Let us also note that particles which mainly decay through electromagnetic interactions, signaled by the production of photons, have a mean lifetime in the range $10^{-20}$–$10^{-16}$ s, whereas particles that decay through weak forces have a mean lifetime generally superior to $10^{-10}$ s. A careful study of such properties and other data on reactions involving hadrons could reveal the underlying dynamics as well as the physical behavior of the constituents, which would serve to guide thinking and test ideas.

## 1.3 Interactions

Table 1.4 exhibits the four fundamental forces together with their coupling strengths, ranges, and typical interaction times. Also shown are the masses of the interaction quanta or the particles that carry the forces. These particles are also known as the *gauge bosons* because they have integral spins and because their existence and physical behavior are predicted and studied by gauge theories.

**Table 1.3.** Low-lying hadrons

| Hadrons | $I(J^P)^a$ | Mass$^b$ | Mean life (s) | Decay modes |
|---|---|---|---|---|
| Mesons | | | | |
| $\pi^\pm$ | $1(0^-)$ | 139.6 | $2.6 \times 10^{-8}$ | $\mu^\pm\nu$ |
| $\pi^0$ | $1(0^-)$ | 135.0 | $0.8 \times 10^{-16}$ | $\gamma\gamma$ |
| $\eta$ | $0(0^-)$ | 548.8 | $0.8 \times 10^{-18}$ | $\gamma\gamma$, $3\pi^0, \pi^+\pi^-\pi^0$ |
| $\mathrm{K}^\pm$ | $1/2(0^-)$ | 493.7 | $1.2 \times 10^{-8}$ | $\mu^\pm\nu$, $\pi^\pm\pi^0$ |
| $\mathrm{K}^0$, $\bar{\mathrm{K}}^0$ | $1/2(0^-)$ | 497.7 | | 50% $\mathrm{K}^0_\mathrm{S}$, 50% $\mathrm{K}^0_\mathrm{L}$ |
| $\mathrm{K}^0_\mathrm{S}$ | | | $0.9 \times 10^{-10}$ | $\pi^+\pi^-$, $2\pi^0$ |
| $\mathrm{K}^0_\mathrm{L}$ | | | $5.2 \times 10^{-8}$ | $3\pi^0$, $\pi^+\pi^-\pi^0$, $\pi^\pm\mathrm{e}^\mp\nu$, $\pi^\pm\mu^\mp\nu$ |
| Baryons | | | | |
| p | $1/2(1/2^+)$ | 938.3 | $> 10^{31}$yrs | stable |
| n | $1/2(1/2^+)$ | 939.6 | 917 | $\mathrm{pe}^-\bar{\nu}$ |
| $\Lambda^0$ | $0(1/2^+)$ | 1115.6 | $2.6 \times 10^{-10}$ | $\mathrm{p}\pi^-$, $\mathrm{n}\pi^0$ |
| $\Sigma^+$ | $1(1/2^+)$ | 1189.4 | $0.8 \times 10^{-10}$ | $\mathrm{p}\pi^0$, $\mathrm{n}\pi^+$ |
| $\Sigma^0$ | $1(1/2^+)$ | 1192.5 | $7 \times 10^{-20}$ | $\Lambda\gamma$ |
| $\Sigma^-$ | $1(1/2^+)$ | 1197.4 | $1.5 \times 10^{-10}$ | $\mathrm{n}\pi^-$ |
| $\Xi^0$ | $1/2(1/2^+)$ | 1314.9 | $2.9 \times 10^{-10}$ | $\Lambda\pi^0$ |
| $\Xi^-$ | $1/2(1/2^+)$ | 1321.3 | $1.6 \times 10^{-10}$ | $\Lambda\pi^-$ |

[a] Isospin, Spin, Parity; [b] In units of MeV/c$^2$.

From *Review of Particle Properties.* Phys. Rev. **D54** (1996) 1

The strength of a force is measured by its coupling constant. We may note that of all the forces, gravitation is by far the weakest. Although it exerts its influence on all objects at all distances and produces a tremendously powerful force on the cosmological scale, it is so feeble on the microscopic scale when compared with the other forces present on this scale that its effects are insignificant in short distance phenomena normally observed in particle physics, and so it can be completely neglected (which does not however exclude the possibility that it may recover its importance, and even dominance, at the extreme end of the short distance scale of the order of $10^{-33}$ cm, called the *Planck scale*).

We have introduced earlier, rather casually, the term 'particle field'. It is an important concept that comes naturally from relativity and quantum mechanics, and is used to convey the idea that particles can spread their effects

**Table 1.4.** Fundamental interactions

| Interaction | Effective coupling | Boson | Mass[a] | Range[b] | Typical time[c] |
|---|---|---|---|---|---|
| Gravitation | $10^{-39}$ | graviton | 0 | $\infty$ | – |
| Electromagnetism | 1/137 | photon | 0 | $\infty$ | $10^{-20}$ |
| Weak force | $10^{-5}$ | $W^{\pm}$, $Z^0$ | 80–90 | $10^{-16}$ | $10^{-10}$ |
| Strong force | 1 | gluons | 0 | $< 10^{-13}$[d] | $10^{-23}$ |

[a]In units of GeV/c$^2$; [b]In cm; [c] In seconds.

[d]This is the range of the nuclear force, not that of the quark–quark force.

over entire space and time. Consider a particle that receives a sudden push in some way: it cannot produce in turn an instantaneous change in another particle nearby, because, naturally, no signals can travel faster than light. To have the extra energy transferred to the second particle, conservation of energy and momentum at all points in space and time requires that the excited particle emits a quantum, or field, that carries the additional energy and momentum over to the second particle. Thus, two particles separated by some distance can still have an effect on each other by the exchange of this intermediary field. In a quantum-mechanical context, the field concept represents the existence of a given particle everywhere in space and at every instant in time in terms of discrete energy quanta. A theory based on this concept of particle is called a *quantum field theory*. It predicts in particular that interactions between particles are induced by an exchange of energy quanta, which have all the attributes of ordinary matter particles (Fig. 1.1). Now, paraphrasing an argument due to Yukawa, if two particles interact by exchanging a virtual particle of mass $m$, then the maximum distance over which effects of this exchange are felt is given by $\hbar/mc$, where $\hbar$ is the Planck constant and $c$ the speed of light. Indeed, the emission of a quantum of mass $m$ by one of the particles in interaction causes the energy of the system to change by an amount $\Delta E = mc^2$, a violation of energy conservation, which is nevertheless allowed by the Heisenberg uncertainty principle provided the energy fluctuation lasts no longer than $\Delta t = \hbar/\Delta E$. During this time interval the quantum must reach the second particle and be absorbed by it for an interaction between the two particles to effectively take place. The maximum distance traveled by the quantum, called the force range, is then given by $c\Delta t = \hbar c/mc^2$.

As gravitation and electromagnetism are known from experience to have a very long range, the mass of the exchanged bosons, the graviton and the photon, must be correspondingly very small. In fact, the theories devoted to the study of their properties – Einstein's gravitational theory and quantum electrodynamics – demand their masses to be exactly zero. On the other hand, ever since the nuclear $\beta$-decay was discovered and studied (in the early 1930s), it has been realized that the range of the weak interaction is extremely short, about $10^{-16}$ cm, which would imply a large mass for the interaction quantum. We now know that there are actually three gauge bosons associated with the weak forces: two, $W^+$ and $W^-$, are electrically charged and bear equal masses of $80\,\mathrm{GeV}/c^2$, and the other, $Z^0$, is electrically neutral with a mass of $91\,\mathrm{GeV}/c^2$. How is it that the gauge bosons in this case can have a nonvanishing mass? The most likely correct answer to this question, based on one of the most beautiful ideas in modern physics, will be discussed in detail later on. We will see then, to have a complete theory of particles and fields, one has to introduce yet another class of spin-0 particles called the *Higgs bosons.* The existence of these, however, has not yet been confirmed by experiment.

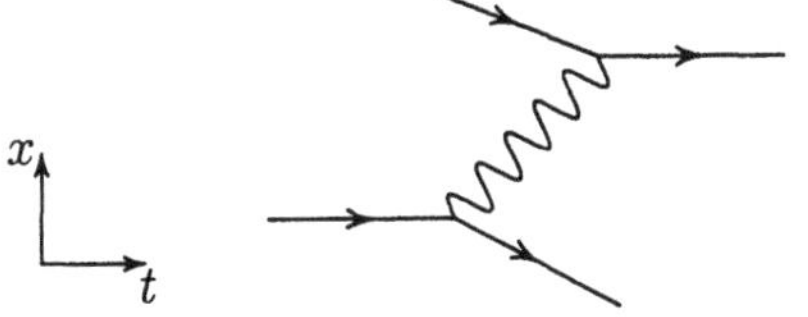


**Fig. 1.1.** Space-time representation of the basic interaction between two particles by a quantum exchange

The range of the nuclear force is at most of the same magnitude as the size of the lightest bound atomic nucleus, the deuteron, which is of the order of $10^{-13}$ cm, corresponding to a mass of $200\,\mathrm{MeV}/c^2$ for the exchanged particle. Nuclear physics tells us that interactions between protons and neutrons arise from exchanges of mesons, whose masses range from 140 MeV to 700 MeV, and even beyond. These interactions are not simple. Nor are they universal because they do not apply to other hadrons. The obvious reason, of course, is that interactions between hadrons are not of a fundamental nature. They are the complex result of the basic interactions between the quark constituents of the hadrons, exactly in the same way that the atomic force between two atoms is the global manifestation of the electromagnetic forces among the electrons and protons that make up those atoms. The elementary strong interaction between the colored quarks acts through eight kinds of bosons, known as *gluons*, which are themselves colored, each carrying at the same time a color and an anticolor. This unique property of gluons gives the quark interaction a distinctive behavior: it increases in strength with the interquark separations to preclude the appearance of isolated quarks, but decreases sufficiently at distances less than $10^{-13}$ cm to make the quarks relatively free within the hadrons in which they evolve. The quantum field theory devoted to the study of the interaction between the color charges is another gauge theory, known as *quantum chromodynamics* or QCD.

## 1.4 Symmetries

The recent history of physics gives us several examples that illustrate the importance of the symmetry considerations in explaining empirical observations or in developing new ideas. Thus, the intriguing regularities found in the atomic periodic table can be naturally explained as resulting from the rotational symmetry that characterizes atoms in their ground states; similarly, the relativity theory owes the clarity and the elegance of its formulation to its guiding principle, Lorentz invariance. However, more than any other field, particle physics, perhaps because of the very nature of the subject or because of the absence of relevant macroscopic analogies or useful classical correspondences, has by necessity conferred upon the symmetry concept a key role that has become essential in formulating new theories. The existence of the $\Omega^-$ particle and the reality of quarks are two outstanding demonstrations of the power of this line of reasoning, but no less impressive is the prediction of the existence of the electronic neutrino by Wolfgang Pauli back in 1930 solely on the basis of the conservation of energy, momentum, and angular momentum, the validity of which was still in doubt at the time. Pauli took a road 'less traveled by' and opened up a whole new world. The prominent place taken by the symmetry considerations throughout this book only reflects their importance in particle physics. In this section, we will sketch a general picture of the idea, and briefly define various symmetry operations.

As we have seen above, every particle is identified by a set of quantum numbers. These numbers summarize the intrinsic properties of the particle and, for this reason, are called the *internal* quantum numbers, meaning that they have nothing to do with the kinetic state of the particle, which is described by other conserved quantities that depend on the state the particle is in, such as the energy, momentum, or angular momentum.

The existence of a quantum number in a system always arises from the invariance of the system under a *global* geometrical transformation, that is, one that does not depend on the coordinates of the space-time point where it is applied. A simple example suffices to illustrate the general situation. Consider two particles in a reference frame in which their interaction energy depends only on the relative distance of the particles. It follows then, first, that a displacement of the origin of the coordinates by an arbitrary distance produces no measurable physical effects on the system, and second, that the total momentum of the system remains constant in time because its rate of change, given by the total gradient of the interaction energy, is strictly zero. So, generally, if we have a physical system in which the absolute positions are not observable (its energy depending on the relative distance rather than individual particle positions) and if we apply on it a geometrical transformation (spatial translation), then we obtain as direct consequences the invariance of the system to the applied transformation (translational invariance) and the existence of a conservation rule (momentum conservation). These are, in short, the interdependent aspects found in every symmetry principle.

**Table 1.5.** Examples of symmetries

| Transformations | Conservation laws | Nonobservables |
|---|---|---|
| *Continuous transformations in space-time:* | | |
| Spatial translation | Momentum | Absolute position |
| Translation in time | Energy | Absolute time |
| Rotation | Angular momentum | Absolute orientation |
| Lorentz transformation | Group generators | Absolute velocity |
| *Discrete transformations:* | | |
| Spatial inversion | Parity | Left–right distinction |
| Time inversion | Invariance to time inversion | Absolute time direction |
| Charge conjugation | Charge parity | Absolute sign of charge |
| *Phase transformations:* | | |
| $\psi \rightarrow \mathrm{e}^{i\alpha N}\psi$ | Generalized charge | Relative phase-angle between states of different charges |
| Transformations between admixtures of proton and neutron | Isospin | Distinction between coherent admixtures of proton and neutron |

Table 1.5 gives a summary of the properties of some of the symmetries of relevance to particle physics that will be discussed in this book. There exist three main types:

1. *Continuous symmetries in space-time.* The corresponding quantum numbers are additive, that is, the quantum number associated with a given symmetry of a composite system is obtained by adding together (algebraically or vectorially) the corresponding quantum numbers of all the components of the system.
2. *Discrete symmetries.* The quantum numbers are multiplicative in this case: such a quantum number in a composite system is given by the product of the quantum numbers of all the constituents.
3. *Unitary symmetries.* They can be considered as arising from phase transformations of fields, or from generalized rotations in the internal space of the system. They are related, for example, to the conservation of a generalized charge (such as the electric charge, the baryonic number, or the leptonic number) or the conservation of isospin, flavors, or colors. The associated quantum numbers are additive.

This list would not be complete without mentioning the *permutation symmetry* in systems of identical particles, a symmetry that arises from the indistinguishability of identical quantum particles. There is a general result in quantum field theory (known as the *spin–statistics connection*) which states that identical particles of half-integral spins obey the *Fermi–Dirac statistics* such that their wave function is antisymmetric in the permutation of any two particles, whereas identical particles of integral spins obey the *Bose–Einstein statistics* such that their wave function is completely symmetric in the variables of all particles.

There is no doubt that exact symmetry is important in the study of particles. It contributes to defining the identity of a new particle produced in a reaction when the identities of all other particles involved are known. It tells us which reactions can proceed and which are inhibited. More remarkable still is a relation that exists between symmetries of a dynamical model and conservation laws, a relation known as *Noether's theorem.* According to this theorem, invariance of a physical system to a class of continuous symmetry transformations always gives rise to some conserved quantity. In other words, symmetries of a dynamical model and conserved quantities are intimately related. The significance of this important result is to be fully realized in the building of physically acceptable models.

However, many symmetry laws in particle physics are not exact, they are only approximate. A symmetry is said to be violated or broken if a quantity, presumed nonobservable by symmetry, turns out to be actually observable under some circumstances. One could think that a study of such symmetries is unproductive. On the contrary, it can be very fruitful because a symmetry breaking in physical systems is always orderly and systematic, leaving many a trace of its presence, many a clue to its behavior for the physicist to discover and exploit.

Finally, another important facet of the symmetry concept is that any continuous transformation may be made *local*, that is, dependent on the coordinates of the space-time point where it is applied. The corresponding symmetry, called *local symmetry*, changes completely in nature to take on the attributes of a dynamical law. However, only a few of such symmetries are endowed with the remarkable property of generating fundamental observable forces. Such exceptional symmetries (as far as we now know) are: invariance to the general space-time transformations, symmetries in the electric charge space, in the (weak) isospin space, and in the color charge space. The latter three, which act on the internal space, are usually referred to as *gauge symmetries*, and even though of a different origin, they have close but yet undefined relation with the first. Are there other symmetries of this kind? For example, does the local symmetry associated with the baryonic quantum number lead to some as yet unobserved force in nature? These are very deep questions which have at present no answers.

## 1.5 Physical Units

In the familiar cgs unit system, the basic physical units are the centimeter (cm) for length, the gram (g) for weight, and the second (s) for time. However, in the realm of high energies and short distances of direct interest to particle physics, it is better to adopt more suitable units, for example, one million electron volts (MeV= $10^6$ eV) or even one billion electron volts (GeV= $10^9$ eV) for energy, and the femtometer (1 fm = $10^{-13}$ cm) for length. In these units, the values of two important universal physical constants, the Planck constant ($\hbar = h/2\pi$) and the speed of light ($c$), and their product $\hbar c$ are given by

$$\hbar = 6.582 \times 10^{-22}\ \text{MeV s}\ , \tag{1.1}$$

$$c = 3 \times 10^{23}\ \text{fm s}^{-1}\ , \tag{1.2}$$

$$\hbar c = 197.33\ \text{MeV fm}\ . \tag{1.3}$$

As formulas in particle physics frequently contain these constants, it is very useful to make a systematic simplification by using a system of units in which the action function (energy multiplied by time) is measured in $\hbar$, and velocity (length divided by time) is measured in $c$. These units are referred to as the *natural units*. In any practical calculation, one may set

$$\hbar = c = 1 \tag{1.4}$$

throughout. At the very end of the calculation, one may recover, if one so wishes, the formulas in the conventional units by inserting the correct powers of $\hbar$ and $c$ at the right places via a dimensional analysis and with the help[1] of (1)–(3).

Setting $c = 1$ means that length and time are equivalent dimensions, $[L] = [T]$. With the usual relativistic relation between energy and momentum $E^2 = p^2c^2 + m^2c^4$, it is seen that energy, momentum, and mass are all equivalent in this sense. The additional choice $\hbar = 1$ implies the dimensional equivalence of energy and inverse length, $[E] = [L]^{-1}$. It is then possible to use a single independent dimension in the system of natural units. Conversion to any other dimensions is readily effected via the equivalence relations

$$1\ \text{MeV} = 1.52 \times 10^{21}\ \text{s}^{-1}\ , \tag{1.5}$$

$$1\ \text{s} = 3 \times 10^{23}\ \text{fm}\ , \tag{1.6}$$

$$1\ \text{fm} = 5.07 \times 10^{-3}\ \text{MeV}^{-1}\ . \tag{1.7}$$

---

[1] In this book, the formulas that are enumerated are identified by the chapter number followed by the formula number in the chapter. When reference is made to a formula defined in the same chapter, the chapter number is omitted; but when the formula comes from another chapter, the full identification number is given.

When natural units are used, the symbol $\boldsymbol{p}$ may mean not only the momentum but also the wave vector $\boldsymbol{k} = \boldsymbol{p}/\hbar$; the symbol $\omega$ may mean either a frequency or an energy $\hbar\omega$; the symbol $m$ may mean not only a mass but also an energy $mc^2$, a reciprocal length $mc/\hbar$, or a reciprocal time $mc^2/\hbar$. The conversion factors let us convert a resonance width $\Gamma$ given in MeV to the equivalent lifetime $\tau = \hbar/\Gamma$ in s, and the range of a force $R$ given in fm to the equivalent energy transmitted $\hbar c/R$ in MeV. Thus, the Planck length $L_{\mathrm{P}} = 1.6 \times 10^{-33}$ cm or $1.6 \times 10^{-20}$ fm is equivalent to the reciprocal Planck energy of

$$1.6 \times 10^{-20} \times 5 \times 10^{-3}\ \mathrm{MeV}^{-1} = 8 \times 10^{-23}\ \mathrm{MeV}^{-1}\,,$$

or the Planck energy $E_{\mathrm{P}} = 1.25 \times 10^{19}$ GeV. Similarly, saying that a certain resonance $\omega$ has a full width of 8.43 MeV is equivalent to saying that it has a reciprocal lifetime of

$$8.43 \times 1.52 \times 10^{21}\ \mathrm{s}^{-1} = 1.28 \times 10^{22}\ \mathrm{s}^{-1}\,,$$

or a lifetime $\tau = 0.78 \times 10^{-22}$ s.

The Compton wavelength of a particle of mass $m$ is defined in natural units by $\lambda = 1/m$. As in the usual units $mc^2$ has the dimension of energy and $\hbar c$ the dimension of length multiplied by energy, $\lambda$ can be expressed in units of length by inserting the appropriate factors of $\hbar$ and $c$:

$$\lambda = \hbar c/mc^2 = \hbar/mc\,.$$

For example, the mass of the $\pi$ meson being 140 MeV/$c^2$, its Compton wavelength is:

$$\lambda\,(\mathrm{fm}) = \frac{1}{m}\,\mathrm{MeV}^{-1} = \frac{1\,\mathrm{fm}}{140 \times 5 \times 10^{-3}} = 1.42\,\mathrm{fm}.$$

As another example, consider the force couplings. The electromagnetic coupling constant is given by the dimensionless fine structure constant, which in natural units is simply

$$\alpha = \frac{e^2}{4\pi} = \frac{1}{137.036}\,. \tag{1.8}$$

On the other hand, the Fermi coupling constant of the weak interaction is not dimensionless, being given in various equivalent units by

$$G_{\mathrm{F}} = 1.166 \times 10^{-5}\,\mathrm{GeV}^{-2} = 0.878 \times 10^{-4}\,\mathrm{MeV\ fm}^3\,. \tag{1.9}$$

For comparison with other coupling strengths, it is useful to define a dimensionless effective coupling constant by multiplying $G_{\mathrm{F}}$ by the proton squared mass, $G_{\mathrm{F}}M_{\mathrm{p}}^2 \approx 10^{-5}$.

## Problems

**1.1 Dimension of wave function.** (a) Let $[L]$ be the dimension of length, $[E]$ the dimension of energy and so on. What is the dimension of a wave function $\phi_c(\boldsymbol{x})$ of a particle the norm of which is given by $\int d^3x\, \phi_c^*(\boldsymbol{x})\phi_c(\boldsymbol{x})$?
(b) The transition rate for i→f is given by Fermi's golden rule

$$w_{fi} = 2\pi |\langle \phi_{cf} \,|\, H_{int} \,|\, \phi_{ci} \rangle|^2 \rho\,,$$

where $H_{int}$ is the interaction Hamiltonian, $\rho$ the number of final states per unit of energy, and $\phi_{ci}$, $\phi_{cf}$ are the wave function of the initial and final states. Restore the appropriate factors of $\hbar$ and $c$ to have $w$ in numbers of events per second.
(c) As in quantum mechanics, the Hamiltonian is the energy operator. It is equal to the space integral of the Hamiltonian density $\mathcal{H}$, so that in natural units the dimension of $\mathcal{H}$ is $[M]^4$. Given that the Hamiltonian density for a boson field $\phi$ contains terms such as $(\partial\phi/\partial x_\mu)^2$, $(mc/\hbar)^2\phi^2$, find the dimension of $\phi$. Similarly for a fermion field, $\mathcal{H}$ contains terms like $mc^2\bar{\psi}\psi$, find the dimension of the fermion field $\psi$.

**1.2 Natural units, conventional units.** Rewrite in conventional units the following expressions, given in natural units:
(a) The differential cross-section of a nonrelativistic electron by a point nucleus:

$$\frac{d\sigma}{d\Omega} = 4m^2(Z\alpha)^2 q^{-4},$$

where $m$ is the electron mass (MeV), $q$ the momentum transferred to the nucleus ($fm^{-1}$), and $\alpha = e^2/4\pi \approx 1/137$ ;
(b) The mean lifetime of the muon of mass $m_\mu$

$$\tau_\mu = 192\pi^3/G_F^2\, m_\mu^5\,,$$

where the coupling $G_F$ is given in $MeV^{-2}$, and $m$ in MeV.

**1.3 Estimations of order magnitudes.** To guide the physical sense, it is often useful to have rough estimates of physical quantities. Such an approximate calculation is based on simple physical considerations and a dimensional analysis. We consider in this problem the total cross-sections for some processes in the limit of very high energies, where only the coupling constant and the reaction energy are relevant. Give in each case an estimate of the cross-section in GeV or in barn ($1b = 10^{-24}\,cm^2$ ).
(a) The total cross-section for proton–proton elastic scattering;
(b) The total cross-section for the electromagnetic annihilation process $e^+e^- \to \mu^+\mu^-$;
(c) The weak interaction scattering $\nu_e$+proton→ $\nu_e$+proton.

**1.4 The Bohr radius.** (a) Make an estimate of the radius of the hydrogen (the Bohr radius), assuming known the electron mass $m_e = 0.51$ MeV and the fine structure constant $\alpha = 1/137$.
(b) Make an estimate of the Bohr radius for a 'gravitational atom' composed of two neutrons bound only by their gravitational attraction at the ground state level.

## Suggestions for Further Reading

*The history of particle physics is the subject of many recent excellent books. In particular,*

Close, F., *The Particle Explosion.* Oxford U. Press, New York 1994
Ezhela, V. V. et al, *Particle Physics: One Hundred Years of Discoveries: An Annotated Chronological Bibliography.* AIP Press, New York 1996
Pais, A., *Inward Bound.* Oxford U. Press, New York 1986

*For reviews with little or no mathematics, see*

Davies, P. C. W., *The New Physics.* Cambridge U. Press, Cambridge 1989
Georgi, H., *A Unified Theory of Elementary Particles and Forces.* Scientific American **244** (April 1981) 48
Ho-Kim, Q., Kumar, N. and Lam, C.S., *Invitation to Contemporary Physics,* World Scientific, Singapore 1991; Chap. VI
't Hooft, G., *Gauge Theories of the Forces between Elementary Particles.* Scientific American **242** (June 1980) 104
Quigg, C., *Elementary Particles and Forces.* Scientific American **252** (April 1985) 84
Ramond, P., *Gauge Theories and their Unification.* Ann. Rev. Nucl. Part. Sci. **41** (1983) 31

*Symmetry has always fascinated philosophers and artists alike. Some examples are*

Brack, A., *et al., La symétrie d'aujourd'hui.* Eds. du Seuil, Paris 1989
Hargittai, I. and Hargittai, M., *Symmetry through the Eyes of a Chemist.* VCH Publishers, New York 1987
MacGillavry, C. H., *Symmetry Aspects of M. C. Escher's Periodic Drawings.* Bohn, Scheltema, and Holkema, Utrecht 1976
Schröder, E., *Dürer. Kunst und Geometrie.* Akademie Verlag, Berlin 1980
Shubnikov, A. V. and Koptsik, V. A., *Symmetry in Science and Art.* Plenum Press, New York 1974
Weyl, H., *Symmetry.* Princeton U. Press, Princeton 1973
Wigner, E., *Symmetries and Reflections.* Indiana U. Press, Bloomington 1967
Yang, C. N., *Elementary Particles.* Princeton U. Press, Princeton 1962

# 2 Boson Fields

Nonrelativistic quantum mechanics, useful as it is in the formulation of all fields of modern physics and in their applications, nevertheless has limitations. In particular, it is not generally applicable to the study of elementary particles because it cannot properly predict the dynamic behavior of systems evolving at high velocities and, in its usual formulation, cannot account for the phenomena of creation and annihilation of particles that regularly occur at high energies. Two concepts – special relativity and field – have crucially contributed to the creation of relativistic quantum field theory, which has unquestionably become the foundation of particle physics. It is then possible, for example, to have a natural explanation for the existence of spins of particles, to make the connection between particles and interactions, and to associate to each particle a charge conjugate particle.

In nonrelativistic quantum mechanics, the wave function $\phi(t, \boldsymbol{x})$ of a particle of mass $m$ in the absence of any interaction obeys the Schrödinger equation, written in natural units with $\hbar = c = 1$,

$$\mathrm{i}\frac{\partial}{\partial t}\phi(t, \boldsymbol{x}) = -\frac{1}{2m}\nabla^2 \phi(t, \boldsymbol{x})\,. \tag{2.1}$$

Comparison with the energy-momentum relation for a nonrelativistic particle

$$E = \boldsymbol{p}^2/2m \tag{2.2}$$

suggests the correspondence rules

$$E \to \mathrm{i}\frac{\partial}{\partial t}\,; \qquad \boldsymbol{p} \to -\mathrm{i}\nabla = -\mathrm{i}\left(\frac{\partial}{\partial x}, \frac{\partial}{\partial y}, \frac{\partial}{\partial z}\right)\,. \tag{2.3}$$

An equation that generalizes (1) to the relativistic regime must have at least a homogeneous coordinate dependence and be symmetric in space and time, a requirement that suggests two possibilities; one, involving only second derivatives and of the general form

$$a\frac{\partial^2}{\partial t^2}\phi(t, \boldsymbol{x}) = \left(\sum_{i=1}^{3} b^i \frac{\partial^2}{\partial x^{i2}} + m^2\right)\phi(t, \boldsymbol{x})\,, \tag{2.4}$$

and the other, involving only first derivatives,

$$\mathrm{i}\gamma^0 \frac{\partial}{\partial t}\phi(t,\boldsymbol{x}) = \left(-\mathrm{i}\sum_{i=1}^{3}\gamma^i \frac{\partial}{\partial x^i} + m\right)\phi(t,\boldsymbol{x}) \,. \tag{2.5}$$

In these equations, $a$, $b^i$, $\gamma^0$, $\gamma^i$, and $m$ are *a priori* complex constants to be determined in accordance with the relativistic energy-momentum relation for a free particle,

$$E^2 = \boldsymbol{p}^2 + m^2 \,. \tag{2.6}$$

So it becomes apparent that relativity will be called on to play a key role in the solution to this problem and justifies a careful examination of Lorentz invariance. It forms the subject of the following section. In the next four sections we discuss the second-quantized scalar and vector field solutions to (4), called the Klein–Gordon equation. We close the chapter with considerations of the action function and of Noether's theorem.

## 2.1 Lorentz Symmetry

In Newtonian physics, physical laws are stated in terms of equations that preserve their forms when coordinates are changed by a Galilean transformation (one that independently changes the position vector and the time parameter by constant amounts). This invariance, called the *Galilean invariance*, characterizes Newtonian mechanics. We now know, since Einstein, that nature is endowed with a higher symmetry, including Galilean symmetry as a special case. The transformations that define this symmetry are called the Lorentz transformations; they leave invariant (unchanged in magnitude) the speed of light, and covariant (unchanged in form) the Maxwell equations and, in general, all physical equations. Therefore, it is important to have a detailed look at what constitutes the fundamental principle of special relativity and indeed of all contemporary physics.

### 2.1.1 Lorentz Transformations

As the coordinates of space and time are to be treated on the same footing, it is convenient to gather them into a single four-component object which behaves by its transformation properties as a vector in a four-dimensional space that one may call *space-time:*

$$x^\mu = (x^0, x^1, x^2, x^3) = (t, \boldsymbol{x}) \,. \tag{2.7}$$

In this notation, $\mu$ and other Greek indices take on the values $0, 1, 2, 3$; the coordinate $x^0$ stands for the time parameter $t$, and $x^1$, $x^2$, $x^3$ (or $x$, $y$, $z$) are the Cartesian components of the usual position vector $\boldsymbol{x}$. We will use Latin indices $i, j, \ldots$, restricted to values $1, 2, 3$, to indicate space components.

A Lorentz transformation is a transformation of some coordinate system $\{x^\mu\}$ into another system $\{x'^\mu\}$ such that

$$\begin{aligned} x'^\mu &= a^\mu{}_0\, x^0 + a^\mu{}_1\, x^1 + a^\mu{}_2\, x^2 + a^\mu{}_3\, x^3 \\ &= a^\mu{}_\nu\, x^\nu \,. \end{aligned} \tag{2.8}$$

We follow here the usual convention that a summation over repeated indices (taking all admissible values) is understood. The parameters of the transformation, $a^\mu{}_\nu$, are *real constants* that specify the relative spatial orientations and the relative velocities of the two reference frames. One can recognize in (8) a generalization to four dimensions of the familiar rotation laws relating to cartesian coordinates. Let us consider some examples.

**Example 2.1 Space Rotation**

A rotation of a coordinate system about the 3 or $z$ axis by a positive angle $\theta$ in the *counterclockwise* direction is defined by the real coefficients $a^\mu{}_\nu$ arranged as the elements of a $4 \times 4$ matrix, where the first index $\mu = 0, 1, 2, 3$ labels the columns and the second index $\nu = 0, 1, 2, 3$ labels the rows,

$$a^\mu{}_\nu = \begin{pmatrix} 1 & 0 & 0 & 0 \\ 0 & \cos\theta & \sin\theta & 0 \\ 0 & -\sin\theta & \cos\theta & 0 \\ 0 & 0 & 0 & 1 \end{pmatrix}, \tag{2.9}$$

where $0 \le \theta < 2\pi$. Note that the matrix is unimodular, $\det a = 1$. As with any other space rotation, this particular rotation mixes space coordinates, leaving untouched the time component:

$$\begin{aligned} x'^0 &= x^0 \,, \\ x'^1 &= x^1 \cos\theta + x^2 \sin\theta \,, \\ x'^2 &= -x^1 \sin\theta + x^2 \cos\theta \,, \\ x'^3 &= x^3 \,. \end{aligned} \tag{2.10}$$

Occasionally, it is useful to define the complex combinations

$$x^{(\pm)} = \mp \frac{1}{\sqrt{2}} (x^1 \pm \mathrm{i} x^2) \tag{2.11}$$

(called the circular or spherical components), which transform as

$$x'^{(\pm)} = \mathrm{e}^{\mp \mathrm{i}\theta}\, x^{(\pm)} \,. \tag{2.12}$$

When $\theta$ is very small we may use a linear approximation in which $a^\mu{}_\nu = \delta^\mu{}_\nu + \epsilon^\mu{}_\nu$ for

$$\epsilon^\mu{}_\nu = \theta \begin{pmatrix} 0 & 0 & 0 & 0 \\ 0 & 0 & 1 & 0 \\ 0 & -1 & 0 & 0 \\ 0 & 0 & 0 & 0 \end{pmatrix}. \tag{2.13}$$

As defined in (10), $a^\mu{}_\nu$ describes a *passive rotation*, which leaves the physical system unchanged. In an *active rotation*, where the physical system (rather than the reference system) is rotated (clockwise about the $z$ axis), the coordinates of the transformed object and of the original object in the same reference system are still related by (10) but with $\theta$ replaced by $-\theta$ (see Fig. 2.1).

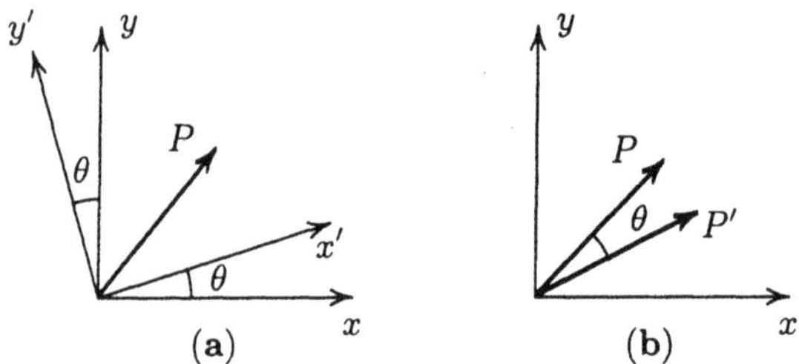


**Fig. 2.1.** Equivalent rotations about the $\hat{z}$ axis: (**a**) reference axes are rotated; (**b**) physical system $P$ is rotated

### Example 2.2 Pure Lorentz Transformation

A pure Lorentz transformation (Lorentz boost) relates two reference frames which differ only by a uniform relative motion of velocity $v$. When the motion is in the positive $x$ direction, the transformation is given by the matrix

$$a^\mu{}_\nu = \begin{pmatrix} \cosh\omega & -\sinh\omega & 0 & 0 \\ -\sinh\omega & \cosh\omega & 0 & 0 \\ 0 & 0 & 1 & 0 \\ 0 & 0 & 0 & 1 \end{pmatrix}, \tag{2.14}$$

having determinant $\det a = 1$. It is a kind of rotation that mixes space coordinates with the time parameter:

$$\begin{aligned} x'^0 &= \cosh\omega\,(\,x^0 - x^1\tanh\omega) = \gamma\,(\,x^0 - vx^1)\,, \\ x'^1 &= \cosh\omega\,(-x^0\tanh\omega + x^1) = \gamma\,(-vx^0 + x^1)\,, \\ x'^2 &= x^2\,, \\ x'^3 &= x^3\,, \end{aligned} \tag{2.15}$$

where $-\infty < \omega < \infty$, and $\cosh\omega = \gamma = 1/\sqrt{1-v^2}$, or $\tanh\omega = v$.

If an *active* transformation is considered, where a particle of mass $m$ at rest is boosted to velocity $v$ in the positive $x$ direction, the coordinates of the two particle states, all measured in the same, unchanged reference, are related by

$$\begin{aligned} x'^0 &= x^0\cosh\omega + x^1\sinh\omega\,, \\ x'^1 &= x^0\sinh\omega + x^1\cosh\omega\,, \\ x'^2 &= x^2\,, \\ x'^3 &= x^3\,. \end{aligned} \tag{2.16}$$

It will be seen below that the energy and momentum form a four-vector $p^\mu = (E, \boldsymbol{p})$ which transforms as $x^\mu$, so that the two vectors $(m, 0, 0, 0)$ and $(E, p, 0, 0)$ are related by

$$\begin{aligned} E &= m \cosh \omega \,, \\ p &= m \sinh \omega \,, \end{aligned} \tag{2.17}$$

from which $\tanh \omega = p/E = v$, $\cosh \omega = 1/\sqrt{1-v^2}$, and $\sinh \omega = v/\sqrt{1-v^2}$. A particle of mass $m$ at rest acquires through a Lorentz boost an energy $E = m \cosh \omega$ and momentum $p = m \sinh \omega$. The parameter $\omega$ is called the particle *rapidity*.

**Example 2.3 Space Inversion**
An inversion in space is defined by the matrix

$$a^\mu{}_\nu = \begin{pmatrix} 1 & 0 & 0 & 0 \\ 0 & -1 & 0 & 0 \\ 0 & 0 & -1 & 0 \\ 0 & 0 & 0 & -1 \end{pmatrix} . \tag{2.18}$$

Note that in this case $\det a = -1$. The coordinates then transform as $x'^0 = x^0$, $x'^1 = -x^1$, $x'^2 = -x^2$, and $x'^3 = -x^3$. ■

The distinctive property of Lorentz transformations is that they leave invariant the proper time interval $\mathrm{d}\tau$, defined by

$$\mathrm{d}\tau^2 \equiv \mathrm{d}t^2 - \mathrm{d}\boldsymbol{x}^2 = g_{\mu\nu}\, \mathrm{d}x^\mu\, \mathrm{d}x^\nu \,. \tag{2.19}$$

The symbol $g_{\mu\nu}$ stands for a tensor, called the space-time metric, with components $g_{00} = 1$, $g_{11} = g_{22} = g_{33} = -1$, and $g_{\mu\nu} = 0$ for $\mu \neq \nu$, which can be represented by a matrix,

$$g_{\mu\nu} = \begin{pmatrix} 1 & 0 & 0 & 0 \\ 0 & -1 & 0 & 0 \\ 0 & 0 & -1 & 0 \\ 0 & 0 & 0 & -1 \end{pmatrix} . \tag{2.20}$$

It is numerically equal to its inverse, $g^{\mu\nu}$, which carries upper indices.

In another Lorentz frame, the infinitesimal elements of the coordinates are given, according to (8) with constant $a^\mu_\nu$, by

$$\mathrm{d}x'^\mu = a^\mu{}_\nu\, \mathrm{d}x^\nu \,, \tag{2.21}$$

and the proper time interval is given by

$$\mathrm{d}\tau'^2 = g_{\mu\nu}\, \mathrm{d}x'^\mu\, \mathrm{d}x'^\nu = g_{\mu\nu}\, a^\mu{}_\rho\, a^\nu{}_\sigma\, \mathrm{d}x^\rho\, \mathrm{d}x^\sigma \,. \tag{2.22}$$

Invariance of proper time,

$$d\tau'^2 = d\tau^2 , \tag{2.23}$$

which expresses the experimental observation that the speed of light in the vacuum is the same in all inertial frames, imposes a condition on the Lorentz transformation matrix similar to the orthogonality relation for the space rotation matrix,

$$g_{\mu\nu}\, a^{\mu}{}_{\lambda}\, a^{\nu}{}_{\kappa} = g_{\lambda\kappa} \,. \tag{2.24}$$

We limit ourselves to *real* Lorentz transformations. This is in fact the case of physical transformations, which map real (coordinate) space into real space; but complex extensions are also possible. From (24), two conditions on $a^{\mu}_{\nu}$ may be written down:

$$(\det a)^2 = 1 ; \tag{2.25}$$

$$a_{00}^2 - \sum_i a_{ii}^2 = 1 . \tag{2.26}$$

They divide the real Lorentz transformations into four classes, namely,

| | | | | |
|---|---|---|---|---|
| (1) $L_{+}^{\uparrow}$ | $\det a = +1$ | $a_{00} \geq 1$ | proper orthochronous | (1) , |
| (2) $L_{+}^{\downarrow}$ | $\det a = +1$ | $a_{00} \leq -1$ | proper nonorthochronous | (TP) , |
| (3) $L_{-}^{\uparrow}$ | $\det a = -1$ | $a_{00} \geq 1$ | improper orthochronous | (P) , |
| (4) $L_{-}^{\downarrow}$ | $\det a = -1$ | $a_{00} \leq -1$ | improper nonorthochronous | (T) . |

These four classes are disconnected because neither $a_{00}$ nor det $a$ can be changed continuously from a value less than 1 to a value greater than 1. But a transformation in each class can be continuously deformed into any other transformation of that class and in particular to the basic transformation characteristic of the class, namely, 1, $P$, $T$, or $TP$, where $P$ is space inversion (parity) and $T$ is time inversion.

*Remarks.* The set of all Lorentz transformations constitute an algebraic structure, called the *Lorentz group*. It has the three key defining properties of a group; namely,

(a) there exists an identity transformation (which effects no changes at all);

(b) to each transformation there corresponds an inverse which is also a member of the set;

(c) two transformations successively applied are equivalent to some element of the set.

In particular, the subset of proper orthochronous Lorentz transformations form a subgroup of the Lorentz group, called the special orthogonal group SO(3,1), where the notation reflects the condition $\det a = +1$ and the asymmetry between space and time as manifest in the metric. In general, it is this specific group one refers to when one speaks of 'Lorentz invariance', and it is to this group that we will limit our discussion for the rest of the chapter.

### 2.1.2 Tensor Algebra

Any vector that transforms as $x^\mu$ according to (8) is said to be a *contravariant* (Lorentz) vector,

$$V^\mu \to V'^\mu = a^\mu{}_\nu V^\nu . \tag{2.27}$$

A *covariant* vector $U_\mu$ is one that transforms as

$$U_\mu \to U'_\mu = a_\mu{}^\nu U_\nu , \tag{2.28}$$

that is, with the matrix inverse of $a^\mu{}_\nu$:

$$a_\mu{}^\nu \equiv (a^{-1})^\nu{}_\mu = g_{\mu\lambda}\, g^{\nu\kappa}\, a^\lambda{}_\kappa . \tag{2.29}$$

From these definitions and (24) follow several useful relations. First, to each contravariant vector corresponds a covariant vector, and inversely,

$$V_\mu \equiv g_{\mu\nu} V^\nu , \qquad U^\mu \equiv g^{\mu\nu} U_\nu . \tag{2.30}$$

Note in particular that the sign of the space components changes when indices change positions, reflecting the presence of both signs in $g_{\mu\nu}$:

$$V_0 = +V^0, \;\; V_1 = -V^1, \;\; V_2 = -V^2, \;\; V_3 = -V^3 . \tag{2.31}$$

Moreover, the scalar product of a covariant vector and a contravariant vector

$$U_\nu V^\nu = U^0 V^0 - \boldsymbol{U} \cdot \boldsymbol{V} \tag{2.32}$$

is a Lorentz-invariant scalar:

$$U'_\mu V'^\mu = a_\mu{}^\lambda\, a^\mu{}_\nu\, U_\lambda V^\nu = U_\nu V^\nu . \tag{2.33}$$

There exist objects, neither vectors nor scalars, that also transform in a well-defined though complicated manner and that may carry several upper or lower indices. They are called (Lorentz) *tensors*. For example, a rank-3 mixed tensor transforms according to

$$T^\lambda{}_{\mu\nu} \to T'^\lambda{}_{\mu\nu} = a^\lambda{}_\alpha\, T^\alpha{}_{\beta\gamma}\, a_\mu{}^\beta\, a_\nu{}^\gamma . \tag{2.34}$$

This rule can be readily extended to tensors of any rank.

The four-gradient, $\partial_\mu \equiv \partial/\partial x^\mu$, is a covariant vector:

$$\frac{\partial}{\partial x^\mu} \to \frac{\partial}{\partial x'^\mu} = \frac{\partial x^\nu}{\partial x'^\mu}\frac{\partial}{\partial x^\nu} , \tag{2.35}$$

where $\partial x^\nu/\partial x'^\mu = a_\mu{}^\nu$, from (28). One can similarly prove that $\partial^\mu \equiv \partial/\partial x_\mu$ is a contravariant vector. It is important to note the sign difference in the following two formulas:

$$\partial_\mu \equiv \frac{\partial}{\partial x^\mu} = \left(\frac{\partial}{\partial t}, \nabla\right), \qquad \partial^\mu \equiv \frac{\partial}{\partial x_\mu} = \left(\frac{\partial}{\partial t}, -\nabla\right). \tag{2.36}$$

It follows, for example, that the divergence of an arbitrary contravariant vector, $\partial V^\mu/\partial x^\mu$, is invariant, as are also the d'Alembertian $\Box = \partial^\mu \partial_\mu$ and any operator of the form $V^\mu \partial_\mu$:

$$V^\mu \partial_\mu = V^0 \frac{\partial}{\partial t} + \boldsymbol{V} \cdot \nabla, \tag{2.37}$$

$$\partial^\mu \partial_\mu = \frac{\partial^2}{\partial t^2} - \nabla^2. \tag{2.38}$$

Another example of physical interest is the energy-momentum vector of a particle with mass $m$, defined by

$$p^\mu \equiv m \frac{\mathrm{d}x^\mu}{\mathrm{d}\tau}. \tag{2.39}$$

It is evidently a vector because $\mathrm{d}x^\mu$ is one while $m$ and $\mathrm{d}\tau$ are both invariants. The particle energy and momentum can then be identified with the time and space components of $p^\mu$:

$$p^0 = E = m\gamma, \tag{2.40}$$

$$\boldsymbol{p} = m\gamma\boldsymbol{v}, \tag{2.41}$$

where

$$\boldsymbol{v} \equiv \frac{\mathrm{d}\boldsymbol{x}}{\mathrm{d}t}, \qquad \gamma \equiv \frac{\mathrm{d}t}{\mathrm{d}\tau} = (1 - \boldsymbol{v}^2)^{-1/2}. \tag{2.42}$$

If the total energy-momentum is conserved in a reaction $\mathrm{i} \to \mathrm{f}$ in some reference frame (i.e. $P_\mathrm{i}^\mu = P_\mathrm{f}^\mu$, where $P_\mathrm{i}^\mu$ and $P_\mathrm{f}^\mu$ denote the total energy-momentum in the initial and final states), it is also conserved in any other frame related to the first by a Lorentz transformation (i.e. $P_\mathrm{i}'^\mu = P_\mathrm{f}'^\mu$). This simple yet significant result follows directly from (8) because a four-vector that vanishes in a given frame necessarily vanishes in any other Lorentz frame.

### 2.1.3 Tensor Fields

Consider two observers $\mathcal{O}$ and $\mathcal{O}'$ moving in two different inertial reference frames related by a Lorentz transformation. If observer $\mathcal{O}$ describes a field by a certain function $\varphi(x) = \varphi(t, \boldsymbol{x})$ using the coordinates of her own frame, then observer $\mathcal{O}'$ will describe the same field by another function $\varphi'(x') = \varphi'(t', \boldsymbol{x}')$ in terms of the transformed coordinates $x'^\mu = a^\mu{}_\nu x^\nu$. The question is, how are $\varphi(x)$ and $\varphi'(x')$ related? A theory consistent with relativistic principles can only contain fields that have well-defined transformation properties. They include

(a) *scalar fields*, that remain invariant in every Lorentz transformation:
$\phi'(x') = \phi(x)$;

(b) *vector fields*, such as the electromagnetic field; a Lorentz transformation acts on both the field and its arguments such that
$A'^{\mu}(x') = a^{\mu}{}_{\nu} A^{\nu}(x)$;

(c) *tensor fields*, such as the gravitational field or the electromagnetic field tensor, for which the transformation rule is
$F'^{\mu\nu}(x') = a^{\mu}{}_{\rho} a^{\nu}{}_{\sigma} F^{\rho\sigma}(x)$.

One characterizes these rules by saying that $\phi$, $A^{\mu}$, and $F^{\mu\nu}$ belong to different *representations* of the Lorentz group. There also exist many other representations that cannot be constructed in such a simple manner. For example, the *spinor representation*, to which belong fields of spin-$1/2$ particles in four-dimensional space-time, also transforms with a $4 \times 4$ matrix with elements given by nontrivial functions of $a^{\mu}{}_{\nu}$. We will examine this case in the next chapter. The rest of this chapter is devoted to the study of the two simplest representations of the Lorentz group, the scalar and the vector fields.

## 2.2 Scalar Fields

The observed mesons $\pi$, K, and $\eta$, and the postulated Higgs bosons of the standard model are all spin-0 particles, described by scalar fields generically represented by the symbol $\phi$. Suppose an observer $\mathcal{O}$ constructs in her frame the coordinates $x^{\mu}$ and describes a certain scalar field by the space-time function $\phi(x)$ and the state vector of the observed physical system by $\Phi_A$. Suppose also a second observer $\mathcal{O}'$, with similarly constructed coordinates $x'^{\mu}$, describes the same field by $\phi'(x')$ and the same physical state by the vector $\Phi'_{A'}$. How are the corresponding objects in the two frames related? What observables do such relations imply?

### 2.2.1 Space-Time Translation of a Scalar Field

Let us illustrate the way symmetry arguments are applied to fields by the simple example of space-time translation, which is defined by

$$x^{\mu} \to x'^{\mu} = x^{\mu} - a^{\mu}, \qquad (2.43)$$

where $a^{\mu}$ is the constant displacement parameter. As translation is a continuous transformation and can be constructed by a succession of small translations, it suffices to consider an infinitesimal transformation,

$$x^{\mu} \to x'^{\mu} = x^{\mu} - \delta a^{\mu}, \qquad (2.44)$$

where $\delta a^\mu$ is a very small constant vector. If the *total variation* of the field is defined as

$$\delta\phi(x) = \phi'(x') - \phi(x)\,, \tag{2.45}$$

then $\delta\phi(x) = 0$, since $\phi(x)$ is by definition a scalar, invariant field. The question then is, if $\mathcal{O}'$ is given $\phi(x)$, how will he obtain from this information the field in his own frame but written in terms of $x$? What he wants is an expression of the form

$$\phi'(x) = \phi(x) + \delta_0\phi(x)\,, \tag{2.46}$$

where $\delta_0\phi$ defines the variation of the field alone, keeping the argument fixed. Setting $\delta\phi = 0$ in (45) and calling $x'$ simply $x$, observer $\mathcal{O}'$ will obtain by expanding $\phi(x + \delta a)$ up to terms linear in $\delta a^\mu$

$$\phi'(x) = \phi(x^\mu + \delta a^\mu) \approx (1 + \delta a^\mu \partial_\mu)\phi(x)\,. \tag{2.47}$$

Therefore,

$$\delta_0\phi = \mathrm{i}\delta a^\mu(-\mathrm{i}\partial_\mu\phi(x))\,, \tag{2.48}$$

where $-\mathrm{i}\partial_\mu$ induces an infinitesimal variation of the field and is called, for this reason, the *generator* of infinitesimal translations. A finite translation is obtained by replication,

$$\phi'(x) = \mathrm{limit}_{\,n\to\infty}\left(1 + \frac{a^\mu}{n}\partial_\mu\right)^n \phi(x) = \exp(a^\mu\partial_\mu)\,\phi(x)\,. \tag{2.49}$$

The operator $U(a) \equiv \exp(a^\mu\partial_\mu)$ is unitary: by Hermitian conjugation one has $\partial_\mu^\dagger = -\partial_\mu$, hence $U^\dagger(a) = \exp(-a^\mu\partial_\mu) = U^{-1}(a) = U(-a)$, which can also be written as $U^\dagger U = 1$. The set of all $U(a)$ form a group, the *translation group*. This group is *Abelian*, meaning that the result of two successive translations does not depend on the order in which they are applied:

$$U(a)U(a') = U(a')U(a)\,. \tag{2.50}$$

This property is equivalent to commutativity of the generators,

$$\partial_\mu\partial_\nu - \partial_\nu\partial_\mu \equiv [\partial_\mu, \partial_\nu] = 0\,, \tag{2.51}$$

and $\{\partial_\mu\}$ is said to form an *Abelian algebra*.

At this point it is useful to introduce a Hermitian operator, $P_\mu = P_\mu^\dagger$, realized in $x$ space by $-\mathrm{i}\partial_\mu$, and interpreted, just as in quantum mechanics, as the total energy-momentum operator of the system. The transformation operator, similarly abstracted, is then given by $U(a) = \exp(\mathrm{i}a^\mu P_\mu)$. Two vectors in the Hilbert space constructed respectively by the two observers to describe the same physical state are then related by

$$\Phi'_{A'} = U(a)\,\Phi_A\,, \tag{2.52}$$

and the operators representing the same observable are similarly related:

$$X' = U(a)XU^{-1}(a)\,. \tag{2.53}$$

These transformation rules are motivated by the general physical condition that a scalar product in the Hilbert space, interpreted as usual as a probability amplitude, remains unchanged in any symmetry transformation,

$$\langle \Phi'_{B'}\,|\,X'\,|\,\Phi'_{A'}\rangle = \langle \Phi_B\,|\,X\,|\,\Phi_A\rangle\,. \tag{2.54}$$

Now, $X$ is said to be invariant to the transformation if

$$X' = X \quad \text{or} \quad UX = XU\,; \tag{2.55}$$

that is, it commutes with all the generators of the transformation group,

$$[X, P_\mu] = 0\,, \qquad \mu = 0,1,2,3\,. \tag{2.56}$$

In particular, the physical system described by a Hamiltonian $H$ is invariant to translations if $[H, P_\mu] = 0$, for $\mu = 0,1,2,3$. Considered as a Heisenberg operator, $P_\mu$ is a constant of the motion; that is, the total energy and momentum of the system are conserved. This conservation law, which is a direct consequence of the invariance under constant displacements of space-time coordinates, is valid whenever cosmological effects are negligible. The energy-momentum conservation law turns out to be among the most useful tools in particle physics.

*Remarks.* (a) Relation (54) can be understood as follows. Let $a$, $b$ be complex numbers, and $|\varphi_i\rangle$ arbitrary vectors. An operator $U$ is said to be *unitary* if it is *linear*,

$$U(a\,|\varphi_1\rangle + b\,|\varphi_2\rangle) = a U\,|\varphi_1\rangle + b U\,|\varphi_2\rangle\,, \tag{2.57}$$

and *preserves the norm* of every vector

$$\langle U\varphi\,|U\varphi\rangle = \langle \varphi\,|\varphi\rangle\,. \tag{2.58}$$

It immediately follows that the scalar product is also preserved:

$$\langle U\varphi_1\,|U\varphi_2\rangle = \langle \varphi_1\,|\varphi_2\rangle\,. \tag{2.59}$$

(b) In quantum mechanics, a physical system is most often described by a wave function which evolves according to the Schrödinger equation which, for a time-independent Hamiltonian, can be formally solved to yield

$$\phi(t, \boldsymbol{x}) = \mathrm{e}^{-\mathrm{i}Ht}\phi(0, \boldsymbol{x}) . \tag{2.60}$$

The matrix element of an arbitrary operator $A$ (assumed for simplicity to be time independent) may be written as

$$\int \mathrm{d}^3x\, \phi_f^*(t, \boldsymbol{x}) A\, \phi_i(t, \boldsymbol{x}) = \int \mathrm{d}^3x\, \phi_f^*(0, \boldsymbol{x})\, A(t)\, \phi_i(0, \boldsymbol{x}) . \tag{2.61}$$

The time-dependent operator thus defined,

$$A(t) \equiv \mathrm{e}^{\mathrm{i}Ht} A\, \mathrm{e}^{-\mathrm{i}Ht} , \tag{2.62}$$

satisfies what is known as the Heisenberg equation:

$$\mathrm{i}\, \frac{\mathrm{d}A(t)}{\mathrm{d}t} = [A(t),\, H] . \tag{2.63}$$

Thus, a quantum system may be described either by time-dependent wave functions of the Schrödinger representation or, alternatively, by fixed state vectors and time-varying operators, like $A(t)$, of the Heisenberg representation. When $[A(t), H] = 0$, the operator $A(t)$ does not depend on time and, according to (61), its expectation value in an arbitrary state is constant. We shall return to this subject in Chap. 4.

### 2.2.2 Lorentz Transformation of a Scalar Field

We proceed now to the study of the Lorentz transformation properties of a scalar field. For this purpose, it suffices to consider infinitesimal transformations, defined by

$$x^\mu \to x'^\mu = a^\mu{}_\nu\, x^\nu \approx (\delta^\mu{}_\nu + \epsilon^\mu{}_\nu)\, x^\nu , \tag{2.64}$$

where $\epsilon_{\mu\nu} = g_{\mu\lambda}\epsilon^\lambda{}_\nu$ are all very small constants. The basic condition (24) on the transformation matrix tells us that $\epsilon_{\mu\nu}$ is antisymmetric: $\epsilon_{\mu\nu} = -\epsilon_{\nu\mu}$. And so in the examples considered above the infinitesimal rotation $(\hat{z}, \delta\theta)$ is defined by the nonvanishing matrix elements $\epsilon_{12} = -\epsilon_{21} = -\delta\theta$, while the Lorentz boost $(\hat{x}, \delta\omega)$ is defined by the nonzero elements $\epsilon_{01} = -\epsilon_{10} = -\delta\omega$.

Now, from the defining property of a scalar field, namely

$$\phi'(x') = \phi(x) = \phi(a^{-1}x') , \tag{2.65}$$

one obtains to the first-order terms in the variation

$$\begin{aligned} \phi'(x) &= \phi(a^{-1}x) = \phi(x^\mu - \epsilon^\mu{}_\nu\, x^\nu) \\ &\approx (1 - \epsilon^\mu{}_\nu\, x^\nu \partial_\mu)\, \phi(x) . \end{aligned} \tag{2.66}$$

The intrinsic variations of the field immediately follow:

$$\begin{aligned}\delta_0\phi &= \phi'(x) - \phi(x) \\ &= \tfrac{1}{2}\epsilon^{\mu\nu}\left(x_\mu\partial_\nu - x_\nu\partial_\mu\right)\phi(x) \equiv -\tfrac{\mathrm{i}}{2}\epsilon^{\mu\nu}\, L_{\mu\nu}\,\phi(x)\,.\end{aligned} \tag{2.67}$$

This result shows that the infinitesimal Lorentz transformations of scalar fields are essentially described by the operators

$$L_{\mu\nu} = \mathrm{i}(x_\mu\partial_\nu - x_\nu\partial_\mu)\,, \tag{2.68}$$

called the generators of infinitesimal transformations. They are Hermitian, $L^\dagger{}_{\mu\nu} = L_{\mu\nu}$, antisymmetric, $L_{\mu\nu} = -L_{\nu\mu}$, and form a complete set in the sense that

$$[L_{\mu\nu}, L_{\rho\sigma}] = -\mathrm{i}g_{\mu\rho}\, L_{\nu\sigma} - \mathrm{i}g_{\nu\sigma}\, L_{\mu\rho} + \mathrm{i}g_{\mu\sigma}\, L_{\nu\rho} + \mathrm{i}g_{\nu\rho}\, L_{\mu\sigma}\,. \tag{2.69}$$

The lack in commutativity of these generators reflects the fact that the order of successive applications of two arbitrary Lorentz transformations is important. The result of the application of a Lorentz boost followed by a rotation, for example, is not the same as that obtained when the order of applications is reversed. Hence the Lorentz group and the corresponding algebra of generators are *non-Abelian.*

An observable $X$ is said to be invariant to the Lorentz group when it commutes with all generators $L_{\mu\nu}$. In particular, a physical system is said to be invariant if the Hamiltonian that describes it satisfies

$$[H, L_{\mu\nu}] = 0 \quad \text{for all } \mu,\nu\,. \tag{2.70}$$

Space rotations are generated by the operators

$$\begin{aligned}&L^1 \equiv L_{23} = -\mathrm{i}\left(y\frac{\partial}{\partial z} - z\frac{\partial}{\partial y}\right), \quad L^2 \equiv L_{31} = -\mathrm{i}\left(z\frac{\partial}{\partial x} - x\frac{\partial}{\partial z}\right), \\ &L^3 \equiv L_{12} = -\mathrm{i}\left(x\frac{\partial}{\partial y} - y\frac{\partial}{\partial x}\right),\end{aligned} \tag{2.71}$$

which represent the total angular momentum of a scalar field. They obey the familiar commutation relations

$$[L^i,\, L^j] = \mathrm{i}\epsilon^{ijk}\, L^k\,, \tag{2.72}$$

where $\epsilon^{ijk}$ is the completely antisymmetric Levi-Cività tensor, such that $\epsilon^{123} = +1$. We learn from these results that the orbital angular momentum is the whole contribution to the total angular momentum. For a translational invariant field,

$$\phi(x) = \mathrm{e}^{\mathrm{i}x^\mu P_\mu}\,\phi(0) = \mathrm{e}^{\mathrm{i}P_0 t - \mathrm{i}\boldsymbol{P}\cdot\boldsymbol{x}}\,\phi(0)\,.$$

In its rest frame, where $\boldsymbol{P} = 0$, the field is independent of space coordinates and the total angular momentum of a particle at rest described by $\phi$ vanishes, $L^i\phi = 0$ for $i = 1,2,3$. The *intrinsic spin* of the scalar field is zero, $s = 0$.

## 2.3 Vector Fields

A vector field $A^\alpha(x)$ carries a Lorentz index and is defined by its characteristic Lorentz transformation property

$$A'^\alpha(x') = a^\alpha{}_\beta\, A^\beta(x) = a^\alpha{}_\beta\, A^\beta\left(a^{-1}x'\right) . \tag{2.73}$$

For an infinitesimal transformation, in which $a^\alpha{}_\beta = \delta^\alpha{}_\beta + \epsilon^\alpha{}_\beta$, we expand the right-hand side of the equation, keeping terms up to the first order in the transformation parameters, to get

$$\begin{aligned} A'^\alpha(x) &= a^\alpha{}_\beta\, A^\beta(a_\rho{}^\sigma\, x_\sigma) = (\delta^\alpha{}_\beta + \epsilon^\alpha{}_\beta)\left[A^\beta(x) - \epsilon^\mu{}_\nu\, x^\nu\, \partial_\mu A^\beta(x)\right] \\ &= \left[\delta^\alpha{}_\beta - \tfrac{\mathrm{i}}{2}\,\epsilon^{\mu\nu}(\Sigma_{\mu\nu})^\alpha{}_\beta\right]\left[A^\beta(x) - \tfrac{\mathrm{i}}{2}\,\epsilon^{\mu\nu}\, L_{\mu\nu}\, A^\beta(x)\right] \\ &= A^\alpha(x) - \tfrac{\mathrm{i}}{2}\,\epsilon^{\mu\nu}\,(L_{\mu\nu} + \Sigma_{\mu\nu})\, A^\alpha(x) . \end{aligned} \tag{2.74}$$

Here we have introduced the matrix operator

$$(\Sigma_{\mu\nu})^\alpha{}_\beta = \mathrm{i}(\delta^\alpha{}_\mu\, g_{\nu\beta} - \delta^\alpha{}_\nu\, g_{\beta\mu}) \tag{2.75}$$

and used the simplified notation $\Sigma_{\mu\nu}\, A^\alpha = (\Sigma_{\mu\nu})^\alpha{}_\beta\, A^\beta$. This operator $\Sigma$ acts on the vector label, i.e. generates the variations stemming from the vectorial character of the field, while $L_{\mu\nu}$, given by the same expression as in (68), generates rather the variations due to the functional field dependence. The antisymmetric tensor operator

$$J_{\mu\nu} = L_{\mu\nu} + \Sigma_{\mu\nu} \tag{2.76}$$

is the full generator of infinitesimal Lorentz transformations of vector fields:

$$\delta_0 A^\alpha(x) = -\tfrac{\mathrm{i}}{2}\,\epsilon^{\mu\nu}\, J_{\mu\nu}\, A^\alpha(x) . \tag{2.77}$$

The pure space components, $J_{ij}$ for $i, j = 1, 2, 3$, act nontrivially on the space field components to generate space rotations and represent the total angular momentum of the field. Let us denote the three independent space components by

$$J^i \equiv L^i + S^i \equiv \tfrac{1}{2}\,\epsilon^{ijk}\, J_{jk} \tag{2.78}$$

with

$$(L^i)^a_b = \tfrac{\mathrm{i}}{2}\,\epsilon^{ijk}\,(x_j\partial_k - x_k\partial_j)\,\delta^a_b\, , \tag{2.79}$$

$$(S^i)^a_b = -\mathrm{i}\epsilon^{iab} . \tag{2.80}$$

It can be checked that $L^i$, $S^i$ and $J^i$ satisfy the usual commutation relations of the algebra of angular momenta. For example,

$$[L^1, L^2] = \mathrm{i}L^3, \qquad [S^1, S^2] = \mathrm{i}S^3, \qquad [J^1, J^2] = \mathrm{i}J^3\, ; \tag{2.81}$$

together with relations obtained by permuting indices, and $[L^i, S^j] = 0$ for all $i$, $j$.

The square of the operator $\boldsymbol{S}$ is

$$(\boldsymbol{S}^2)^a_b = \sum_{k,c}(S^k)^a_c(S^k)^c_b = \sum_{k,c}\epsilon^{kca}\,\epsilon^{kcb} = 2\,\delta^a_b\, , \tag{2.82}$$

which simply shows that $\boldsymbol{S}^2 = s(s+1) = 1(1+1)$, or that the *intrinsic spin* of a vector field is equal to $s = 1$.

## 2.4 The Klein–Gordon Equation

Let us return now to the relativistic generalizations of the equation of motion for a free particle which we started to consider in the introduction. We shall focus for now on the first of the two possibilities,

$$a\,\frac{\partial^2}{\partial t^2}\phi(t,\boldsymbol{x}) = \left(\sum_{i=1}^{3} b^i\frac{\partial^2}{\partial x^{i2}} + m^2\right)\phi(t,\boldsymbol{x})\,. \tag{2.83}$$

To be consistent with the relativistic energy-momentum relation (6), the coefficients must have the values $a = b^i = -1$ for $i = 1,2,3$. From this choice follows the *Klein–Gordon equation* for a free particle:

$$\left(\frac{\partial^2}{\partial t^2} - \nabla^2 + m^2\right)\phi(x) = 0\,, \tag{2.84}$$

which in relativistic notations reads

$$(\Box + m^2)\,\phi(x) = 0\,. \tag{2.85}$$

As the operator $\Box = \partial^\mu\partial_\mu$, the complex function $\phi$, and the mass parameter $m$ are all Lorentz scalars, the equation (85) is manifestly invariant in the Lorentz group, retaining the same form in any Lorentz frame.

### 2.4.1 Free-Particle Solutions

For any given momentum $\boldsymbol{p}$, (85) admits a solution

$$\phi(x) = f_{\boldsymbol{p}}(\boldsymbol{x})\mathrm{e}^{-\mathrm{i}Wt} \tag{2.86}$$

provided $W^2 = \boldsymbol{p}^2 + m^2$. The simplest is a plane wave, $f_{\boldsymbol{p}}(\boldsymbol{x}) = C\exp(\mathrm{i}\boldsymbol{p}\cdot\boldsymbol{x})$, with normalization constant $C$. For a given momentum vector $\boldsymbol{p}$, two energy values are allowed; one, $W = E = \sqrt{\boldsymbol{p}^2 + m^2}$, associated with the *positive-energy solution*

$$\phi^{(+)}(x) = f_{\boldsymbol{p}}\mathrm{e}^{-\mathrm{i}Et}\,; \tag{2.87}$$

and the other, $W = -E = -\sqrt{\boldsymbol{p}^2 + m^2}$, with the *negative-energy solution*

$$\phi^{(-)}(x) = f_{\boldsymbol{p}}\mathrm{e}^{\mathrm{i}Et}\,. \tag{2.88}$$

Note that now, unlike in the nonrelativistic case, the particle energy may take all possible values, positive as well as negative, and the energy spectrum extends from $-\infty$ to $+\infty$ just like the momentum. Thus, the symmetry between energy and momentum is restored.

### 2.4.2 Particle Probability

To understand the physical meaning of the new, negative-energy solution (88), it is useful to examine its implications on the particle probability. It is recalled that in nonrelativistic quantum mechanics the probability density is defined by $\phi^*\phi$ and is related to a current density via a continuity equation. But now the presence of a second-order time derivative in (85) will force us to reconsider the definition of the probability density. To find out the required changes, let us look at the following equations:

$$\phi^*(\partial^\mu\partial_\mu + m^2)\phi = 0\,,$$
$$\phi(\partial^\mu\partial_\mu + m^2)\phi^* = 0\,.$$

Taking the difference of the two equations, we obtain a local conservation equation

$$\partial_\mu j^\mu = 0\,, \tag{2.89}$$

for the four-current

$$j^\mu \equiv \mathrm{i}[\phi^*\partial^\mu\phi - (\partial^\mu\phi^*)\phi]\,, \tag{2.90}$$

which has the components

$$j^0 = \mathrm{i}\left(\phi^*\frac{\partial\phi}{\partial t} - \frac{\partial\phi^*}{\partial t}\phi\right) \equiv \mathrm{i}\phi^*\,\frac{\overleftrightarrow{\partial}}{\partial t}\,\phi\,, \tag{2.91}$$

$$\boldsymbol{j} = -\mathrm{i}\,[\phi^*\nabla\phi - (\nabla\phi^*)\phi] \equiv -\mathrm{i}\phi^*\,\overleftrightarrow{\nabla}\,\phi\,. \tag{2.92}$$

Here, we have used the notation

$$\frac{\overleftrightarrow{\partial}}{\partial t} \equiv \frac{\overrightarrow{\partial}}{\partial t} - \frac{\overleftarrow{\partial}}{\partial t}\,, \qquad \overleftrightarrow{\nabla} \equiv \overrightarrow{\nabla} - \overleftarrow{\nabla}\,.$$

Now, in the nonrelativistic limit $E \approx m$, the positive-energy solution yields $j^0 \approx 2m|f_{\boldsymbol{p}}|^2$, which is indeed proportional to the expected nonrelativistic probability density. However, as we presently see, this naive interpretation is not justified. In fact, from (87)–(88) and (91)–(92), one deduces that

$$j^0 = \pm 2E\,|f_{\boldsymbol{p}}|^2\,, \qquad \boldsymbol{j} = 2\boldsymbol{p}\,|f_{\boldsymbol{p}}|^2\,, \tag{2.93}$$

where $\pm$ refer respectively to $\phi^{(+)}$ and $\phi^{(-)}$. It follows that the particle probability, given by the integral $\int \mathrm{d}^3x\, j^0$ up to a constant, is *positive* for the positive-energy solution, but *negative* for the negative-energy solution – a rather unsatisfactory result. To avoid this awkward situation, could one simply omit the embarrassing negative-energy solution? No, because the wave functions $\phi = f_{\boldsymbol{p}} \exp(-\mathrm{i}Wt)$, with $W > 0$, do not by themselves form a complete set, and a description of a physical system in an arbitrary state in terms of these functions alone is not possible. Both positive- and negative-energy solutions are indispensable. Thus, either the Klein–Gordon equation should be abandoned, as it had effectively been for a time following 1926, or the interpretation of $j^0$ as a probability density should be given up.

In 1934, six years after P. A. M. Dirac's success in discovering another relativistic equation which now carries his name and in giving the correct physical interpretation of its solutions, the interest for the Klein–Gordon equation was revived by W. Pauli and V. F. Weisskopf's original idea that $j^\nu$ should not be interpreted as a *current of probability density*, as nonrelativistic quantum mechanics would suggest, but rather as a *current of charge density*. The reinterpretation is crucial . The sign change of $j^0$ that occurs when $\phi^{(+)}$ is replaced by $\phi^{(-)}$ would come simply from the fact that the two solutions describe states of electric charges of opposite signs. In the light of this suggestion, let us again examine the two solutions (87)–(88). The function

$$\phi^{(+)}_{\boldsymbol{p}}(x) = C_{\boldsymbol{p}} \mathrm{e}^{\mathrm{i}(\boldsymbol{p}\cdot\boldsymbol{x}-Et)} = C_{\boldsymbol{p}} \mathrm{e}^{-\mathrm{i}p\cdot x} \tag{2.94}$$

describes a state of a particle moving in the normal time direction, whereas

$$\phi^{(-)}_{\boldsymbol{p}}(x) = C_{\boldsymbol{p}} \mathrm{e}^{\mathrm{i}[\boldsymbol{p}\cdot\boldsymbol{x}-E(-t)]} \tag{2.95}$$

can be viewed as a state of a particle moving in the reverse time direction. Now, if the charge density $j^0$ differs in sign for the two states because of their opposite charges, the current density $\boldsymbol{j}$ must, for the very same reason, carry opposite signs for the two cases; in other words, one ought to have $\boldsymbol{j} = \pm 2\boldsymbol{p}\,|f_{\pm\boldsymbol{p}}|^2$, which means that the momentum direction in the negative-energy solution should be reversed:

$$\phi^{(-)}_{-\boldsymbol{p}}(x) = C_{\boldsymbol{p}} \mathrm{e}^{\mathrm{i}(-\boldsymbol{p}\cdot\boldsymbol{x}+Et)} = C_{\boldsymbol{p}} \mathrm{e}^{\mathrm{i}p\cdot x}\,. \tag{2.96}$$

The interpretation of $\phi^{(-)}$ as a state of negative energy propagating in the reverse time direction is certainly counterintuitive and rather difficult to visualize. It would be conceptually clearer to replace it with the picture intuitively more familiar to our senses of a state of *antiparticle of positive energy* propagating in the *normal time direction*. By antiparticle, we mean a *charge conjugate* state to the positive-energy state, with the same mass, but with opposite charge and reversed momentum direction. In this interpretation, the antiparticle is treated as any other particle, with the appropriate quantum numbers. No more negative energies, no more reversed time arrow. This interpretation will become even more transparent when the wave function itself is quantized, that is, made into a quantum operator that depends on space-time parameters in a procedure called the *second quantization*.

The eigensolutions of the Klein–Gordon equation $\{\phi_{\boldsymbol{p}}^{(+)}\}$ and $\{\phi_{-\boldsymbol{p}}^{(-)}\}$ form a complete set. With normalization $C_{\boldsymbol{p}} = 1/\sqrt{(2\pi)^3 2E_p}$ , they satisfy the orthonormality relations

$$
\begin{aligned}
\mathrm{i}\int \mathrm{d}^3x\,\phi_{\boldsymbol{p}'}^{(+)*}\,\overleftrightarrow{\frac{\partial}{\partial t}}\,\phi_{\boldsymbol{p}}^{(-)} &= 0\,,\\
\mathrm{i}\int \mathrm{d}^3x\,\phi_{\boldsymbol{p}'}^{(+)*}\,\overleftrightarrow{\frac{\partial}{\partial t}}\,\phi_{\boldsymbol{p}}^{(+)} &= \delta(\boldsymbol{p}'-\boldsymbol{p})\,,\\
\mathrm{i}\int \mathrm{d}^3x\,\phi_{\boldsymbol{p}'}^{(-)*}\,\overleftrightarrow{\frac{\partial}{\partial t}}\,\phi_{\boldsymbol{p}}^{(-)} &= -\delta(\boldsymbol{p}'-\boldsymbol{p})\,.
\end{aligned}
\tag{2.97}
$$

It is important to note the sign on the right-hand side of the last equation. An arbitrary wave function of the particle can then be written in this basis:

$$
\phi(x) = \sum_p \left[a_{\boldsymbol{p}}\phi_{\boldsymbol{p}}^{(+)}(x) + b_{\boldsymbol{p}}^*\phi_{-\boldsymbol{p}}^{(-)}(x)\right]\,,
\tag{2.98}
$$

where $a_{\boldsymbol{p}}$ and $b_{\boldsymbol{p}}^*$ are the complex amplitudes of the eigenmodes of $\phi$. We have used here, as we often will in the following, the shorthand notation $\sum_p = \int \mathrm{d}^3p$ .

### 2.4.3 Second Quantization

In the formalism of quantum field theories, the function $\phi(x)$ becomes a quantum operator on the Hilbert space of state vectors. The algebra of this field operator and its conjugate momentum density operator is defined by a set of quantization rules. The operator $\phi(x)$ still obeys the same dynamic equation as the classical field, but the coefficients $a_{\boldsymbol{p}}$ and $b_{\boldsymbol{p}}$ in the series (98) now become quantum operators. Using the same notations as for the classical field (98), we have the following expressions for the scalar field operator and its Hermitian adjoint:

$$
\begin{aligned}
\phi(x) &= \sum_p \left[a_{\boldsymbol{p}}\phi_{\boldsymbol{p}}^{(+)}(x) + b_{\boldsymbol{p}}^\dagger\phi_{-\boldsymbol{p}}^{(-)}(x)\right]\,,\\
\phi^\dagger(x) &= \sum_p \left[a_{\boldsymbol{p}}^\dagger\phi_{\boldsymbol{p}}^{(+)*}(x) + b_{\boldsymbol{p}}\phi_{-\boldsymbol{p}}^{(-)*}(x)\right]\,.
\end{aligned}
\tag{2.99}
$$

We will show in what follows that

$a_{\boldsymbol{p}}$ destroys a particle of momentum $\boldsymbol{p}$ and positive (by convention) unit charge;
$a_{\boldsymbol{p}}^\dagger$ creates a particle of momentum $\boldsymbol{p}$ and positive unit charge;
$b_{\boldsymbol{p}}$ destroys an antiparticle of momentum $\boldsymbol{p}$ and negative unit charge;
$b_{\boldsymbol{p}}^\dagger$ creates a particle of momentum $\boldsymbol{p}$ and negative unit charge.

Thus, the field operator $\phi$ reduces the total charge of the system by one unit, either by destroying a (say, positive) charge with the operator $a_{\boldsymbol{p}}$ or by creating an antiparticle of opposite (negative) charge with $b_{\boldsymbol{p}}^{\dagger}$. Similarly, the operator $\phi^{\dagger}$ increases the total charge by one unit, either by creating a particle with $a_{\boldsymbol{p}}^{\dagger}$ or by destroying an antiparticle with $b_{\boldsymbol{p}}$. In general, the field operator $\phi$ can operate on a state vector containing an arbitrary number of particles and antiparticles, and so the Klein–Gordon equation which determines the evolution of the field $\phi$ is also an equation that governs the dynamics of the whole system of particles and antiparticles. A quantum field theory is, for this reason, necessarily a many-particle theory. Evidently, other more complex processes are possible. For example, the quadratic operator $\phi^{\dagger}(x)\phi(x)$ represents any one of several charge-conserving processes: the creation and the annihilation of a particle, or of an antiparticle, at space-time point $x$, or the creation or destruction of a particle–antiparticle pair. In any of these events, there are no changes in the net charge and the total charge of the system remains unchanged.

Finally, if the field is Hermitian, $\phi = \phi^{\dagger}$, then $a_{\boldsymbol{p}} = b_{\boldsymbol{p}}$, which means that the particle is identical to its antiparticle.

### 2.4.4 Operator Algebra

In this section, we give a physical motivation for the quantization rules of the scalar field operators. For this purpose, let us introduce (see Sect. 2.6 below) the Hamiltonian operator for a complex scalar field $\phi$,

$$H = \int \mathrm{d}^3x \left[ \frac{\partial \phi^{\dagger}}{\partial t} \frac{\partial \phi}{\partial t} + \partial^i \phi^{\dagger} \partial^i \phi + m^2 \phi^{\dagger} \phi \right] . \tag{2.100}$$

An integration by parts of the second term on the right-hand side leads, with the help of the Klein–Gordon equation (which $\phi$ obeys), to

$$H = - \int \mathrm{d}^3x \, \phi^{\dagger} \, \frac{\overleftrightarrow{\partial}}{\partial t} \left( \frac{\partial \phi}{\partial t} \right) . \tag{2.101}$$

A surface term has been dropped, which is justified by assuming $\phi$ to vanish on the integration surfaces. Now, using the expansion series (99) and the orthonormality relations (97), one can simplify the expression as follows, carefully keeping the relative order of all operators,

$$\begin{aligned} H = & \int \mathrm{d}^3x \sum_{\boldsymbol{p}'} \left[ a_{\boldsymbol{p}'}^{\dagger} \phi_{\boldsymbol{p}'}^{(+)*}(x) + b_{\boldsymbol{p}'} \phi_{-\boldsymbol{p}'}^{(-)*}(x) \right] \\ & \times \mathrm{i} \, \frac{\overleftrightarrow{\partial}}{\partial t} \sum_{\boldsymbol{p}} E_{\boldsymbol{p}} \left[ a_{\boldsymbol{p}} \phi_{\boldsymbol{p}}^{(+)}(x) - b_{\boldsymbol{p}}^{\dagger} \phi_{-\boldsymbol{p}}^{(-)}(x) \right] \\ = & \sum_{\boldsymbol{p}} E_{\boldsymbol{p}} \left( a_{\boldsymbol{p}}^{\dagger} a_{\boldsymbol{p}} + b_{\boldsymbol{p}} b_{\boldsymbol{p}}^{\dagger} \right) . \end{aligned} \tag{2.102}$$

The first term on the right-hand side, $E_p a_p^\dagger a_p \equiv E_p \mathcal{N}_p$, can be interpreted as the energy density of the particle, and $\mathcal{N}_p = a_p^\dagger a_p$, as the particle number operator. However, in the second term the order of the factors $b$ and $b^\dagger$ must be reversed before their product can be similarly interpreted as an antiparticle number operator, $\overline{\mathcal{N}}_p = b_p^\dagger b_p$.

The creation and annihilation operators and observables, such as $H$ and $\mathcal{N}$, operate on an abstract space characterized by a vacuum state $|0\rangle$ defined by

$$a_p\,|0\rangle = b_p\,|0\rangle = 0\,, \qquad \langle 0\,|0\rangle = 0\,. \tag{2.103}$$

The basis states of that space, called Fock space, are constructed by applying all positive powers of $a^\dagger$ and $b^\dagger$ on the vacuum, and Fock states are linear combinations of the basis states with definite numbers of excitation modes. In particular, $a^\dagger\,|0\rangle$ and $b^\dagger\,|0\rangle$ are one-particle states conventionally normalized:

$$\langle 0\,|\,a_p a_p^\dagger\,|\,0\rangle = 1\,, \qquad \langle 0\,|\,b_p b_p^\dagger\,|\,0\rangle = 1\,. \tag{2.104}$$

Therefore, when we reverse the order of $b$ and $b^\dagger$ in (102), we may write

$$b_p b_p^\dagger = 1 \pm b_p^\dagger b_p\,,$$

and obtain the expression for $H$ in the form

$$H = \sum_p E_p\,(a_p^\dagger a_p \pm b_p^\dagger b_p) + \sum_p E_p\,. \tag{2.105}$$

The last term on the right-hand side is a c-number quantity which can be interpreted as the (infinite) vacuum energy. In general, it may be dropped (because it is not observable) by redefining the energy of the system relatively to the vacuum, such that the rescaled vacuum energy vanishes:

$$H \to H - \sum_p E_p = \sum_p E_p\,(a_p^\dagger a_p \pm b_p^\dagger b_p)\,. \tag{2.106}$$

Clearly, the Hamiltonian without the vacuum energy is non-negative, provided the contributions from antiparticles are non-negative, that is, provided that one adopts the + sign, rather than the − sign, in the above expressions. This leads to

$$H = \sum_p E_p\,(a_p^\dagger a_p + b_p^\dagger b_p) = \sum_p E_p\,(\mathcal{N}_p + \overline{\mathcal{N}}_p)\,. \tag{2.107}$$

Thus, the condition that the total energy of the system be non-negative implies that the operators $a$ and $b$ obey *commutation relations.* This is consistent with a natural generalization of the quantum rules of coordinates and momenta in a discrete system,

$$[q_i,\, p_j] = \mathrm{i}\delta_{ij}\,, \quad [q_i,\, q_j] = [p_i,\, p_j] = 0\,,$$

to the quantization rules of a scalar field and its conjugate momentum density:

$$[\phi(t,\boldsymbol{x}),\, \pi(t,\boldsymbol{y})] = \mathrm{i}\delta^{(3)}(\boldsymbol{x}-\boldsymbol{y}),$$
$$[\phi(t,\boldsymbol{x}),\, \phi(t,\boldsymbol{y})] = [\pi(t,\boldsymbol{x}),\, \pi(t,\boldsymbol{y})] = 0, \tag{2.108}$$

where $\pi \equiv (\partial\phi^\dagger/\partial t)$. From this postulate, follow the commutation relations

$$[a_{\boldsymbol{p}'}, a^\dagger_{\boldsymbol{p}}] = \delta(\boldsymbol{p}'-\boldsymbol{p}); \quad [a_{\boldsymbol{p}'}, a_{\boldsymbol{p}}] = 0; \quad [a_{\boldsymbol{p}'}, b_{\boldsymbol{p}}] = 0;$$
$$[b_{\boldsymbol{p}'}, b^\dagger_{\boldsymbol{p}}] = \delta(\boldsymbol{p}'-\boldsymbol{p}); \quad [b_{\boldsymbol{p}'}, b_{\boldsymbol{p}}] = 0; \quad [a_{\boldsymbol{p}'}, b^\dagger_{\boldsymbol{p}}] = 0. \tag{2.109}$$

### 2.4.5 Physical Significance of the Fock Operators

The charge operator, which is given by

$$Q = \mathrm{i}\int \mathrm{d}^3x\, j^0(x) = \mathrm{i}\int \mathrm{d}^3x\, \phi^\dagger(x)\, \overleftrightarrow{\frac{\partial}{\partial t}}\, \phi(x), \tag{2.110}$$

may be rewritten in the following form by retracing the same steps as in the calculation of $H$,

$$\begin{aligned} Q &= \int \mathrm{d}^3x \sum_{p'} \left[a^\dagger_{\boldsymbol{p}'}\phi^{(+)*}_{\boldsymbol{p}'}(x) + b_{\boldsymbol{p}'}\phi^{(-)*}_{-\boldsymbol{p}'}(x)\right] \\ &\quad \times \mathrm{i}\, \overleftrightarrow{\frac{\partial}{\partial t}} \sum_{p} \left[a_{\boldsymbol{p}}\phi^{(+)}_{\boldsymbol{p}}(x) + b^\dagger_{\boldsymbol{p}}\phi^{(-)}_{-\boldsymbol{p}}(x)\right] \\ &= \sum_p (a^\dagger_{\boldsymbol{p}} a_{\boldsymbol{p}} - b_{\boldsymbol{p}} b^\dagger_{\boldsymbol{p}}). \end{aligned} \tag{2.111}$$

Using the commutation relations (109) and dropping the vacuum terms yields

$$Q = \sum_p (a^\dagger_{\boldsymbol{p}} a_{\boldsymbol{p}} - b^\dagger_{\boldsymbol{p}} b_{\boldsymbol{p}}) = \sum_p (\mathcal{N}_{\boldsymbol{p}} - \overline{\mathcal{N}}_{\boldsymbol{p}}). \tag{2.112}$$

Since $[\mathcal{N}_{\boldsymbol{p}}, \overline{\mathcal{N}}_{\boldsymbol{p}'}] = 0$, operators $H$ and $Q$ commute, which simply means that charge is conserved. Of course, when a particle is identical to its antiparticle, i.e. $\phi^\dagger = \phi$, its charge operator identically vanishes, $Q = 0$. A Hermitian field therefore represents a *neutral particle.*

As already mentioned, $a$, $b$, $a^\dagger$, $b^\dagger$, and the operators $H$ and $Q$ act on Fock states. For example, the Fock state of one particle of momentum $p$ is

$$|p\rangle = \sqrt{(2\pi)^3 2E_{\boldsymbol{p}}}\, a^\dagger_{\boldsymbol{p}}\, |0\rangle = C^{-1}_{\boldsymbol{p}}\, a^\dagger_{\boldsymbol{p}}\, |0\rangle, \tag{2.113}$$

normalized such that

$$\langle p'\, |p\rangle = (2\pi)^3\, 2E_{\boldsymbol{p}}\, \delta^3(\boldsymbol{p}'-\boldsymbol{p}). \tag{2.114}$$

With the help of (107), (109), and (112), one shows that it is an eigenstate of energy $E_{\boldsymbol{p}}$:

$$\begin{aligned} H\,|p\rangle &= C_{\boldsymbol{p}}^{-1} H a_{\boldsymbol{p}}^{\dagger}\,|0\rangle = C_{\boldsymbol{p}}^{-1}[H, a_{\boldsymbol{p}}^{\dagger}]\,|0\rangle \\ &= E_{\boldsymbol{p}}\,|p\rangle\,; \end{aligned} \tag{2.115}$$

and of one unit of charge:

$$\begin{aligned} Q\,|p\rangle &= C_{\boldsymbol{p}}^{-1} Q a_{\boldsymbol{p}}^{\dagger}\,|0\rangle = C_{\boldsymbol{p}}^{-1}[Q, a_{\boldsymbol{p}}^{\dagger}]\,|0\rangle \\ &= |p\rangle\,. \end{aligned} \tag{2.116}$$

Thus, the operator $a_{\boldsymbol{p}}^{\dagger}$ creates a discrete excitation of positive unit charge, of energy $E_{\boldsymbol{p}}$ related to momentum by the usual relation $E_{\boldsymbol{p}} = \sqrt{\boldsymbol{p}^2 + m^2}$. One may call such a state a *particle*. When applied on an arbitrary state describing the system, $a_{\boldsymbol{p}}^{\dagger}$ adds a particle to the system, increasing its energy by $E_{\boldsymbol{p}}$ and its charge by a unit of charge.

A similar analysis can be done for a one-antiparticle state, defined by

$$|\bar{p}\rangle = \sqrt{(2\pi)^3 2E_p}\, b_{\boldsymbol{p}}^{\dagger}\,|0\rangle = C_{\boldsymbol{p}}^{-1}\, b_{\boldsymbol{p}}^{\dagger}\,|0\rangle\,. \tag{2.117}$$

It will then be seen that $b_{\boldsymbol{p}}^{\dagger}$ adds an antiparticle to the system, thus increasing its energy by $E_{\boldsymbol{p}}$ and reducing its charge by a unit of charge:

$$[H,\, b_{\boldsymbol{p}}^{\dagger}] = E_{\boldsymbol{p}}\, b_{\boldsymbol{p}}^{\dagger}\,, \qquad [Q,\, b_{\boldsymbol{p}}^{\dagger}] = -b_{\boldsymbol{p}}^{\dagger}\,.$$

What is the connection between the classical wave function and the quantum field operators $\phi$, $\phi^{\dagger}$ ? For a plane-wave free-particle state (99), it is given by the relation

$$\langle 0\,|\,\phi(x)\,|\,p\rangle = \mathrm{e}^{-\mathrm{i}p\cdot x}\,. \tag{2.118}$$

The matrix element $\langle 0\,|\,\phi(x)\,|\,p\rangle$, which represents the amplitude for the annihilation at point $x$ of a particle carrying momentum $p$, is thus shown, with the help of (109) and (113), to be identical to the plane-wave function of a free particle having momentum $p$. Hence the following amplitudes describe the initial or final states of a particle or antiparticle:

| | | |
|---|---|---|
| $< 0\vert\phi(x)\vert p > = \mathrm{e}^{-\mathrm{i}p\cdot x}$ | →• | annihilation of a particle, |
| $< p\vert\phi^{\dagger}(x)\vert 0 > = \mathrm{e}^{\mathrm{i}p\cdot x}$ | •→ | creation of a particle, |
| $< \bar{p}\vert\phi(x)\vert 0 > = \mathrm{e}^{\mathrm{i}p\cdot x}$ | •← | creation of an antiparticle, |
| $< 0\vert\phi^{\dagger}(x)\vert\bar{p} > = \mathrm{e}^{-\mathrm{i}p\cdot x}$ | ←• | annihilation of an antiparticle. |

## 2.5 Quantized Vector Fields

It was already shown that a vector field has intrinsic spin equal to 1. Quantum mechanics tells us that a spin-1 particle generally has three spinorial (polarization) states, eigenstates of the spin projection along some quantization axis, normally chosen to be the $z$ axis. It is exactly the case of particles with nonvanishing masses, such as the vector mesons ($\rho$, $\phi$, $J/\psi$) or the weak interaction bosons ($W^{\pm}$, Z). However, the electromagnetic field does not have this property; it is described by a four-vector field which has just two independent components. The particle associated with it, the photon, is massless and has only two polarization states. We will describe in this section the nature of the vector fields, the differences between the two cases, and the role of gauge invariance and of relativistic invariance.

### 2.5.1 Massive Vector Fields

For a massive vector field in the absence of interactions, it is natural to assume in a relativistic theory that each of the four components of the field $A^{\mu}$ satisfies the Klein–Gordon equation

$$(\partial_{\nu}\partial^{\nu} + m^2)A^{\mu} = 0\,, \tag{2.119}$$

subject to the condition

$$\partial_{\nu}A^{\nu} = 0\,. \tag{2.120}$$

This manifestly Lorentz-invariant condition is needed to reduce the number of independent equations from four to three, which is precisely the number of the independent degrees of freedom of a spin-1 particle.

For a plane-wave solution

$$A^{\mu}(x) \approx a^{\mu}(k)\,\mathrm{e}^{-\mathrm{i}k\cdot x}\,, \tag{2.121}$$

the equation (119) implies a constraint on the four-momentum $k$:

$$\begin{aligned} k^2 &\equiv (k^0)^2 - \boldsymbol{k}^2 = m^2\,, \qquad \text{or} \\ \pm k^0 &= E_k = \sqrt{\boldsymbol{k}^2 + m^2}, \end{aligned} \tag{2.122}$$

which shows that $m$ is the mass of the particle associated with the field. On the other hand, (120) imposes a restriction on the vector $a^{\mu}(k)$ which describes the polarization states of the field:

$$k_{\mu}a^{\mu}(k) = 0\,. \tag{2.123}$$

It is not difficult to construct a set of three independent vectors satisfying this condition. For example, placing the $z$ axis in the direction of the momentum $\boldsymbol{k}$, so that $k^{\mu} = (E_k, 0, 0, |\boldsymbol{k}\,|)$, a possible choice of polarization vectors is

$$\begin{aligned} e^{\mu}(k,1) &= (0,1,0,0)\,, \\ e^{\mu}(k,2) &= (0,0,1,0)\,, \\ e^{\mu}(k,3) &= \frac{1}{m}\,(|\boldsymbol{k}\,|, 0, 0, E_k)\,. \end{aligned} \tag{2.124}$$

In this choice, the basis vectors are *real*, spacelike, normalized to $-1$, and reduce to $\hat{\boldsymbol{x}}$, $\hat{\boldsymbol{y}}$, and $\hat{\boldsymbol{z}}$ when the particle comes to rest. Together with $k^\mu$, they form a complete set in the sense that

$$\sum_{\lambda=1}^{3} e^\mu(k,\lambda)\, e^\nu(k,\lambda) = -g^{\mu\nu} + \frac{k^\mu k^\nu}{m^2}\,, \tag{2.125}$$

where $k^\mu \equiv (\sqrt{\boldsymbol{k}^2+m^2},\, \boldsymbol{k})$. The vectors $e^\mu(k,1)$ and $e^\mu(k,2)$ are orthogonal to $k^\mu$ as well as to $\boldsymbol{k}$, and describe the *transverse polarizations* of the field, while the *longitudinal polarization* vector $e^\mu(k,3)$ has its space components parallel to $\boldsymbol{k}$ and its time component determined by (123). Of course, other choices are possible, leading generally to complex polarization vectors, which may be more convenient for some purposes.

The expansion series of the massive vector field in terms of plane waves and of complex polarization basis vectors reads

$$A^\mu(x) = \sum_k \sum_{\lambda=1}^{3} \frac{1}{\sqrt{(2\pi)^3 2E_k}} \times \left[e^\mu(k,\lambda)\, a(\boldsymbol{k},\lambda)\, \mathrm{e}^{-\mathrm{i}k\cdot x} + e^{\mu *}(k,\lambda)\, b^\dagger(\boldsymbol{k},\lambda)\, \mathrm{e}^{\mathrm{i}k\cdot x}\right], \tag{2.126}$$

where $E_k$ and $e^\mu(k,\lambda)$ satisfy (122)–(123). Considered as a quantum field, $A^\mu$ is a superposition of eigenmodes of particles and antiparticles of momentum $k^\nu$ and polarization $\lambda$. These modes are described by the creation and annihilation operators $a^\dagger$, $b^\dagger$ and $a$, $b$, which obey the commutation relations

$$\begin{aligned}
[a(\boldsymbol{k}',\lambda'),\, a(\boldsymbol{k},\lambda)] &= [b(\boldsymbol{k}',\lambda'),\, b(\boldsymbol{k},\lambda)] = 0\,,\\
[a(\boldsymbol{k}',\lambda'),\, b(\boldsymbol{k},\lambda)] &= [a^\dagger(\boldsymbol{k}',\lambda'),\, b^\dagger(\boldsymbol{k},\lambda)] = 0\,,\\
[a(\boldsymbol{k}',\lambda'),\, a^\dagger(\boldsymbol{k},\lambda)] &= \delta_{\lambda'\lambda}\, \delta(\boldsymbol{k}'-\boldsymbol{k})\,,\\
[b(\boldsymbol{k}',\lambda'),\, b^\dagger(\boldsymbol{k},\lambda)] &= \delta_{\lambda'\lambda}\, \delta(\boldsymbol{k}'-\boldsymbol{k})\,.
\end{aligned} \tag{2.127}$$

### 2.5.2 The Maxwell Equations

A classical electromagnetic field is described by two three-vectors, the electric field $\boldsymbol{E}$ and the magnetic field $\boldsymbol{B}$, which obey a set of coupled differential equations discovered by James Clerk Maxwell in 1864. Equivalently, it can be represented by a four-vector that satisfies a Klein–Gordon equation for a particle of vanishing mass.

The Maxwell equations for a field in the presence of an electrical source $\rho$ and a current $\boldsymbol{j}$ are

$$\begin{aligned}
\nabla\cdot\boldsymbol{E} &= \rho\,,\\
\nabla\times\boldsymbol{B} - \frac{\partial\boldsymbol{E}}{\partial t} &= \boldsymbol{j}\,,\\
\nabla\cdot\boldsymbol{B} &= 0\,,\\
\nabla\times\boldsymbol{E} + \frac{\partial\boldsymbol{B}}{\partial t} &= 0\,.
\end{aligned} \tag{2.128}$$

They are written in the rationalized Heaviside–Lorentz units, with $c = \hbar = 1$, so that the fine structure constant is defined by $\alpha = e^2/4\pi \approx 1/137$.

Experience tells us that electric and magnetic phenomena are not independent, except in static situations, and that a certain symmetry exists between $\boldsymbol{E}$ and $\boldsymbol{B}$ in spite of the absence of observable magnetic charges and currents. It is thus quite possible that these apparently distinct forces are actually different aspects of the same kind of interaction. Therefore, it would be of advantage to combine the components of $\boldsymbol{E}$ and $\boldsymbol{B}$ into a single object, which must be an antisymmetric tensor of second rank as it has the correct number of (six) independent components,

$$F^{i0} = -F^{0i} = E^i, \qquad F^{ij} = -\epsilon^{ijk} B^k,$$

or

$$F^{\mu\nu} = \begin{pmatrix} 0 & -E^1 & -E^2 & -E^3 \\ E^1 & 0 & -B^3 & B^2 \\ E^2 & B^3 & 0 & -B^1 \\ E^3 & -B^2 & B^1 & 0 \end{pmatrix}. \tag{2.129}$$

In terms of this field tensor, the two nonhomogeneous equations, which are driven by the charge and current densities, take on a really simple form

$$\partial_\mu F^{\mu\nu} = j^\nu, \tag{2.130}$$

where $j^\mu = (\rho, \boldsymbol{j})$. The current conservation is contained in this equation and is a direct consequence of the antisymmetry of $F^{\mu\nu}$ and of the commutativity of $\partial_\mu$ and $\partial_\nu$:

$$\partial_\nu j^\nu = 0. \tag{2.131}$$

Now, if a *vector field* $A^\mu(x)$ is defined through

$$F^{\mu\nu} = \partial^\mu A^\nu - \partial^\nu A^\mu, \tag{2.132}$$

the homogeneous Maxwell equations are automatically satisfied because

$$\partial_\lambda F_{\mu\nu} + \partial_\mu F_{\nu\lambda} + \partial_\nu F_{\lambda\mu} = 0 \tag{2.133}$$

is an algebraic identity when considered in terms of $A^\mu$, and reproduces the two homogeneous Maxwell equations when considered in terms of $F_{\mu\nu}$ with indices $(\lambda, \mu, \nu) = (i, j, k)$ and $(0, i, j)$. Given $A^\mu$, the fields $\boldsymbol{E}$ and $\boldsymbol{B}$ may be calculated from (132):

$$\begin{aligned} \boldsymbol{E} &= -\frac{\partial \boldsymbol{A}}{\partial t} - \nabla A^0, \\ \boldsymbol{B} &= \nabla \times \boldsymbol{A}. \end{aligned} \tag{2.134}$$

To sum up, the Maxwell equations can be seen as the dynamic equations for the vector field $A^\mu(x)$:

$$\begin{aligned} F^{\mu\nu}(x) &= \partial^\mu A^\nu(x) - \partial^\nu A^\mu(x), \\ \partial_\mu F^{\mu\nu}(x) &= j^\nu(x). \end{aligned} \tag{2.135}$$

### 2.5.3 Quantization of the Electromagnetic Field

We now limit ourselves to a free field, in the absence of external sources. Then (135) reduces to

$$\partial_\mu F^{\mu\nu} = 0\,, \tag{2.136}$$

or, in terms of $A^\mu(x)$ now considered the fundamental variable, to

$$\Box A^\mu - \partial^\mu(\partial{\cdot}A) = 0\,. \tag{2.137}$$

This equation has a remarkable symmetry property, absent in the case of massive vector fields: it is invariant to transformations defined by

$$A^\mu(x) \to A'^\mu(x) = A^\mu(x) + \partial^\mu\Lambda(x)\,, \tag{2.138}$$

where $\Lambda(x)$ is an *arbitrary* function of $x$. Here the field argument $x$ is unchanged, but the magnitude of each component is modified by an additive quantity (a 'gauge') that depends on the space-time coordinates of the field. Such a transformation is called a *local gauge transformation.* Invariance of the field tensor can easily be proved:

$$\begin{aligned} F'^{\mu\nu} &= \partial^\mu A'^\nu - \partial^\nu A'^\mu = \partial^\mu(A^\nu + \partial^\nu\Lambda) - \partial^\nu(A^\mu + \partial^\mu\Lambda) \\ &= \partial^\mu A^\nu - \partial^\nu A^\mu = F^{\mu\nu}\,, \end{aligned} \tag{2.139}$$

from which it follows that (137) is invariant. Saying that (137) is invariant to transformations (138) means that whenever $A^\mu(x)$ is a solution to (137), $A'^\mu(x)$ is also a solution. Thus, the equation for $A^\mu$ does not have a single solution but rather an infinite class of solutions related by (138). The situation in fact is quite similar to what one finds in general relativity where Einstein's field equations, which are invariant to general coordinate transformations, cannot determine by themselves the solution; to make the solution unique, one has to specify the coordinate system in which the problem is to be solved. In the present case, it corresponds to specifying the function $\Lambda(x)$, or as one says, to *choosing a gauge*. Since this choice is not unique, the intermediate steps in calculations may differ, depending on the gauge chosen. However, since the physics described by the theory is invariant to gauge transformations, it must not depend on the gauge chosen for the calculation and the final expression of any observable correctly calculated must be independent of the function $\Lambda$.

To solve (137), one can proceed in two steps: first, a constraint is imposed on the field to reduce the number of independent solutions from four to three, and next, an appropriate function $\Lambda(x)$ is chosen to fix the gauge, thus reducing the number of independent solutions from three to two.

In the first step, a possible constraint is inspired by the condition (120) for massive vector fields, which has the advantage of being covariant. With this constraint (the *Lorentz condition*), the problem is reduced to solving

$$\begin{aligned} &\Box A^\mu = 0 \qquad \text{(field equation)}, \\ &\partial_\mu A^\mu = 0 \qquad \text{(Lorentz condition)}\,. \end{aligned} \tag{2.140}$$

Even then, the field equation still remains invariant under a residual class of gauge transformations defined by

$$A^\mu(x) \to A'^\mu(x) = A^\mu(x) + \partial^\mu \Lambda(x)\,, \qquad \Box\Lambda = 0\,. \tag{2.141}$$

This means, if $A_\mu$ is solution to (140), then all $A'^\mu = A^\mu + \partial^\mu\Lambda$ such that $\Box\Lambda = 0$ form a class of admissible solutions. The solution becomes unique only by fixing $\Lambda(x)$.

A plane-wave solution to (140) has the form

$$A_\mu(x) \approx a_\mu(k)\, \mathrm{e}^{-\mathrm{i}k\cdot x}\,, \tag{2.142}$$

together with the constraints

$$k^2 = {k_0}^2 - \boldsymbol{k}^2 = 0 \quad \to \quad \pm k_0 = |\boldsymbol{k}| = \omega\,, \qquad k^\mu\, a_\mu(k) = 0\,. \tag{2.143}$$

The condition $k^2 = 0$ tells us that the quantum (photon) mass is zero, while the second, $k{\cdot}a = 0$, lets us determine one component of $a^\mu$, say the time component, in terms of the others:

$$k^0 a_0(k) = \boldsymbol{k}{\cdot}\boldsymbol{a}(k) \;\to\; a_0(k) = \hat{\boldsymbol{k}}{\cdot}\boldsymbol{a}(k) \qquad \text{with } \hat{\boldsymbol{k}} \equiv \boldsymbol{k}/|\boldsymbol{k}|\,. \tag{2.144}$$

So $a_0(k)$ is equal to the length of the longitudinal component of $\boldsymbol{a}$:

$$\boldsymbol{a}_\parallel = (\boldsymbol{a}{\cdot}\hat{\boldsymbol{k}})\hat{\boldsymbol{k}}\,. \tag{2.145}$$

Suppose now that we have the solutions $A^\mu(x)$ of (140) and hence the coefficients $a^\mu(k)$. Let us consider a particular solution to $\Box\Lambda = 0$,

$$\Lambda(x) \approx \lambda(k)\, \mathrm{e}^{-\mathrm{i}k\cdot x}\,, \tag{2.146}$$

with the four-momentum $k^\mu$ satisfying the same constraint, $k^2 = 0$, as in $A_\mu(x)$. The residual gauge transformation (141) now appears as

$$a_\mu(k) \to a'_\mu(k) = a_\mu(k) - \mathrm{i}k_\mu\, \lambda(k)\,. \tag{2.147}$$

Now choose the gauge such that

$$\lambda(k) = a_0(k)/\mathrm{i}k_0\,, \tag{2.148}$$

which implies that the time component of the polarization vector, and hence also the longitudinal component, vanishes: $a'_0(k) = 0$, $\boldsymbol{a}'_\parallel = 0$. It follows that $a'_\mu(k)$ reduces to a spacelike vector lying in a plane perpendicular to $\boldsymbol{k}$;

that is, it reduces to a three-vector $\boldsymbol{a}' = \boldsymbol{a}'_\perp$ perpendicular to $\boldsymbol{k}$. Hence it can always be expressed in terms of two basis vectors spanning that plane:

$$\hat{\boldsymbol{k}}\cdot\boldsymbol{e}(k,\lambda) = 0, \qquad \lambda = 1, 2\,. \tag{2.149}$$

The polarization vector $\boldsymbol{a}(k)$ (with its prime accent suppressed) can now be written in this basis as

$$\boldsymbol{a}(k) = \sum_{\lambda=1}^{2} a(k,\lambda)\,\boldsymbol{e}(k,\lambda)\,,$$

where $a(k,1)$ and $a(k,2)$ are scalar coefficients.

The three vectors $\boldsymbol{e}(k,1)$, $\boldsymbol{e}(k,2)$, and $\hat{\boldsymbol{k}}$ form a complete, orthonormal basis in three-dimensional space, so that

$$\sum_{\lambda=1}^{2} e^{i*}(k,\lambda)\,e^{j}(k,\lambda) = \delta_{ij} - \frac{k^i k^j}{\boldsymbol{k}^2}\,. \tag{2.150}$$

The simplest choice consists of two real unit vectors orthogonal to each other as well as to the propagation direction $\boldsymbol{k}$,

$$\begin{aligned} \boldsymbol{e}(k,\lambda)\cdot\boldsymbol{e}(k,\lambda') &= \delta_{\lambda\lambda'}\,, \\ \boldsymbol{e}(k,1)\times\boldsymbol{e}(k,2) &= \hat{\boldsymbol{k}}\,. \end{aligned} \tag{2.151}$$

These vectors represent transverse *linear polarizations.* If the $z$ axis is chosen to coincide with $\boldsymbol{k}$, then $\boldsymbol{e}(k,1)$ and $\boldsymbol{e}(k,2)$ lie respectively along and in the same sense as $\hat{\boldsymbol{x}}$ and $\hat{\boldsymbol{y}}$ (see Fig. 2.2). In an infinitesimal rotation $(\hat{\boldsymbol{z}}, \delta\theta)$, they transform according to (77), with $L^i = 0$ and $(S^i)_{ab} = -\mathrm{i}\,\epsilon^{iab}$,

$$\begin{aligned} \delta\boldsymbol{e}(k,1) &= -\epsilon_{12}\,\boldsymbol{e}(k,2) = \phantom{-}\delta\theta\,\boldsymbol{e}(k,2)\,, \\ \delta\boldsymbol{e}(k,2) &= +\epsilon_{12}\,\boldsymbol{e}(k,1) = -\delta\theta\,\boldsymbol{e}(k,1)\,. \end{aligned} \tag{2.152}$$

Another possible basis may include the complex vectors

$$\boldsymbol{e}(k,\pm) = \mp\frac{1}{\sqrt{2}}\,[\boldsymbol{e}(k,1) \pm \mathrm{i}\boldsymbol{e}(k,2)]\,, \tag{2.153}$$

which transform without mixing as the spherical components of vectors:

$$\delta\boldsymbol{e}(k,\pm) = \mp\mathrm{i}\,\delta\theta\,\boldsymbol{e}(k,\pm)\,, \tag{2.154}$$

and therefore represent states of *circular polarizations.* They are associated with the $m = \pm 1$ spin components, with the quantization axis chosen to coincide with the propagation vector $\boldsymbol{k}$. Since the longitudinal polarization is absent, the $m = 0$ spin component is also absent. Thus, the massless

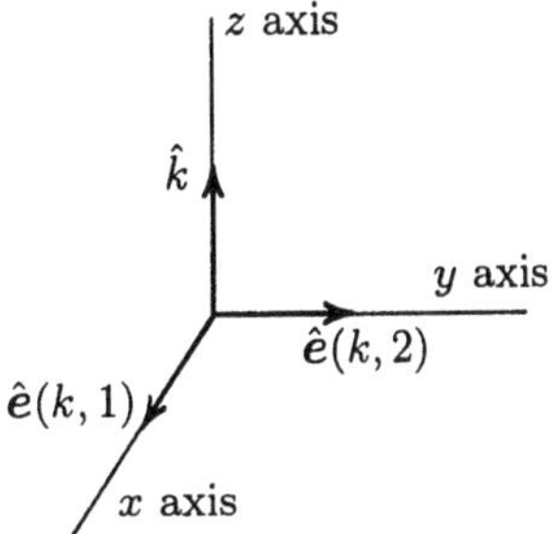


**Fig. 2.2.** Relative orientations of the linear polarization vectors and the propagation vector

photon has only two orientations, one parallel and the other antiparallel to the propagation vector.

The solution to (140) in the gauge thus chosen (called the *Coulomb gauge*),

$$A^0(x) = 0\,, \qquad \nabla\cdot\boldsymbol{A}(x) = 0\,, \tag{2.155}$$

is therefore uniquely determined. It reads, with $k^\mu = (\omega = |\boldsymbol{k}|,\, \boldsymbol{k})$,

$$\boldsymbol{A}(x) = \sum_k \sum_{\lambda=1}^{2} \frac{1}{\sqrt{(2\pi)^3\, 2\omega}} \times \left[\boldsymbol{e}(k,\lambda)\, a(k,\lambda)\, \mathrm{e}^{-\mathrm{i}k\cdot x} + \boldsymbol{e}^*(k,\lambda)\, a^\dagger(k,\lambda)\, \mathrm{e}^{\mathrm{i}k\cdot x}\right]. \tag{2.156}$$

As *quantum fields*, $A^i(x)$ and its conjugate momentum $\dot{A}^i(x)$ obey the canonical commutation relations at equal times

$$[A^i(t,\boldsymbol{x}),\, \dot{A}^j(t,\boldsymbol{y})] = \mathrm{i}\int \frac{\mathrm{d}^3k}{(2\pi)^3} \left(\delta_{ij} - \frac{k^i k^j}{\boldsymbol{k}^2}\right) \mathrm{e}^{\mathrm{i}\boldsymbol{k}\cdot(\boldsymbol{x}-\boldsymbol{y})}\,. \tag{2.157}$$

The transversality of the field explains the presence of a 'divergenceless' $\delta$-function on the right-hand side. $A^i(x)$ is a superposition of particle modes of different momenta with two allowed polarizations represented by the creation and annihilation operators, $a(k,\lambda)$ and $a^\dagger(k,\lambda)$, which obey commutation relations (127). The field is Hermitian, $\boldsymbol{A}^\dagger = \boldsymbol{A}$, because the particle associated with it, the photon, has all generalized charges equal to 0.

By quantizing in Coulomb gauge (where $A^0$ is constrained), we have lost manifest Lorentz invariance; but since the Maxwell theory is Lorentz- and gauge-invariant, the final physical results will be both Lorentz-covariant and independent of gauge. Alternatively, we can quantize in the Lorentz gauge ($\partial_\nu A^\nu = 0$) by modifying the field equation, adding $\partial_\mu(\partial\cdot A)$, and treating $A^0$ as a dynamical variable. The commutation relations for the fields will appear in a more familiar form, $[A^\mu(t,\boldsymbol{x}),\, \dot{A}^\nu(t,\boldsymbol{y})] = -\mathrm{i}g^{\mu\nu}\delta(\boldsymbol{x}-\boldsymbol{y})$, but the extra minus sign in $[A^0,\, \dot{A}^0]$ indicates that the $A^0$ generates a space of indefinite metric. It turns out that quantization with constraints can be more easily done in a functional integral formulation, especially in the case of non-Abelian fields, as we will see in Chap. 15.

### 2.5.4 Field Energy and Momentum

As an application, we may calculate the energy and momentum of the electromagnetic field, given by the classical expressions

$$H = \tfrac{1}{2}\int \mathrm{d}^3x\,(\boldsymbol{E}^2 + \boldsymbol{B}^2)\,,$$
$$\boldsymbol{P} = \int \mathrm{d}^3x\,(\boldsymbol{E}\times\boldsymbol{B})\,. \tag{2.158}$$

The same expressions remain valid in quantum mechanics, only with $\boldsymbol{E}$ and $\boldsymbol{B}$ interpreted as operators. In the gauge chosen, the fields are

$$\boldsymbol{B} = \nabla \times \boldsymbol{A} = \sum_{k,\lambda} C_{\boldsymbol{k}}(\mathrm{i}\boldsymbol{k}\times \boldsymbol{e}(k,\lambda))\,[a(\boldsymbol{k},\lambda)\mathrm{e}^{-\mathrm{i}k.x} - a^\dagger(\boldsymbol{k},\lambda)\mathrm{e}^{\mathrm{i}k\cdot x}]\,,$$
$$\boldsymbol{E} = -\frac{\partial \boldsymbol{A}}{\partial t} = \sum_{k,\lambda} C_{\boldsymbol{k}}\,\mathrm{i}\omega\,\boldsymbol{e}(k,\lambda)\,[a(\boldsymbol{k},\lambda)\mathrm{e}^{-\mathrm{i}k.x} - a^\dagger(\boldsymbol{k},\lambda)\mathrm{e}^{\mathrm{i}k\cdot x}]\,,$$

with normalization constant $C_{\boldsymbol{k}} = \sqrt{1/(2\pi)^3\,2\omega}$ and energy $\omega \equiv |\boldsymbol{k}|$. The vectors $\boldsymbol{e}(k,\lambda)$ have been assumed, without loss in generality, to be real. The final results are

$$H = \sum_{k,\lambda} \omega\, a^\dagger(\boldsymbol{k},\lambda)a(\boldsymbol{k},\lambda)\,; \tag{2.159}$$

$$\boldsymbol{P} = \sum_{k,\lambda} \boldsymbol{k}\, a^\dagger(\boldsymbol{k},\lambda)a(\boldsymbol{k},\lambda)\,. \tag{2.160}$$

It follows from (159) and (160) that the energy and momentum of the one-photon state are

$$Ha^\dagger(\boldsymbol{k},\lambda)\,|0\rangle = [H, a^\dagger(\boldsymbol{k},\lambda)]\,|0\rangle = \omega\, a^\dagger(\boldsymbol{k},\lambda)\,|0\rangle\,,$$
$$\boldsymbol{P}a^\dagger(\boldsymbol{k},\lambda)\,|0\rangle = [\boldsymbol{P}, a^\dagger(\boldsymbol{k},\lambda)]\,|0\rangle = \boldsymbol{k}\, a^\dagger(\boldsymbol{k},\lambda)\,|0\rangle\,.$$

Since $|\boldsymbol{k}|^2 = \omega^2$, the photon is massless: $(\text{mass})^2 = \omega^2 - |\boldsymbol{k}|^2 = 0$. The operator $a^\dagger(\boldsymbol{k},\lambda)$ creates a photon of energy $\omega$, momentum $\boldsymbol{k}$, and polarization $\lambda$. The wave functions for a photon absorbed and a photon emitted at point $x$ are given by

$$\langle 0|A_\mu(x)|k,\lambda\rangle = e_\mu(k,\lambda)e^{-ik.x} \qquad \xrightarrow{k,\lambda} \text{〰〰○}\,,$$

$$\langle k,\lambda|A_\mu(x)|0\rangle = e^*_\mu(k,\lambda)e^{ik.x} \qquad \text{○〰〰} \xrightarrow{k,\lambda}\,.$$

## 2.6 The Action

In particle physics, one of the most useful theoretical tools is the Lagrangian, more exactly the Lagrangian density. As it constitutes the complete definition of the model, from which the dynamic field equations are generated, one can incorporate into it from the beginning all the characteristics one wishes to see emerge from the model.

### 2.6.1 The Euler–Lagrange Equation

In the Hamiltonian formulation of classical mechanics, the equations of motion of a system of particles can be derived from a certain function of the generalized coordinates $q(t)$ and velocities $\dot{q}(t)$ of the particles, called the *action*,

$$S = \int_{t_1}^{t_2} \mathrm{d}t \, L(q(t), \dot{q}(t)) \,, \tag{2.161}$$

where $L$ is the Lagrange function. Hamilton's minimum action principle holds that, among all possible paths joining any two fixed points at times $t_1$ and $t_2$ $(t_2 > t_1)$, the path for which $S$ is minimum corresponds to the physical path that determines the actual motion of the particles. Thus, by varying $q \to q + \delta q$, subject to the constraint $\delta q(t_1) = \delta q(t_2) = 0$, one gets

$$S \to S + \delta S \,,$$

$$\delta S = \int_{t_1}^{t_2} \mathrm{d}t \left( \frac{\partial L}{\partial q} \delta q + \frac{\partial L}{\partial \dot{q}} \delta \dot{q} \right) . \tag{2.162}$$

After an integration by part of the second term on the right-hand side, this becomes

$$\delta S = \int_{t_1}^{t_2} \mathrm{d}t \left( \frac{\partial L}{\partial q} - \frac{\mathrm{d}}{\mathrm{d}t} \frac{\partial L}{\partial \dot{q}} \right) \delta q \, \mathrm{d}t + \int_{t_1}^{t_2} \mathrm{d}t \frac{\mathrm{d}}{\mathrm{d}t} \left( \frac{\partial L}{\partial \dot{q}} \delta q \right) . \tag{2.163}$$

The last term on the right-hand side vanishes because $\delta q(t_1) = \delta q(t_2) = 0$. The variation $\delta q$ being arbitrary, it follows that the minimization condition on $\delta S$ implies the Euler–Lagrange equation of classical mechanics

$$\frac{\partial L}{\partial q} - \frac{\mathrm{d}}{\mathrm{d}t} \frac{\partial L}{\partial \dot{q}} = 0 \qquad \text{(Euler–Lagrange equation)} . \tag{2.164}$$

One may then introduce momentum $p$ conjugate to coordinate $q$ via

$$p = \partial L / \partial \dot{q} \,, \tag{2.165}$$

a relation that can be inverted to express $\dot{q}$ in terms of $p$ and $q$. The Hamiltonian is now defined by a Lagrange transformation as a function of the dynamical variables $p$ and $q$:

$$H = p\dot{q} - L \,. \tag{2.166}$$

When the situation is extended to a continuous distribution, as in field theories, each generalized coordinate $q(t)$ is replaced by a field, generically called $\varphi(t, \boldsymbol{x})$, that represents a continuum of particles in space-time, and each corresponding velocity $\dot{q}(t)$ by the four-derivative $\partial_\mu \varphi$. In a local field theory the Lagrangian can be written as a spatial integral of a functional of the field $\varphi$ and its derivative $\partial_\mu \varphi$, called the *Lagrangian density*, $\mathcal{L}$,

$$L = \int \mathrm{d}^3 x \, \mathcal{L}(\varphi, \partial_\mu \varphi) \,, \tag{2.167}$$

and the action assumes the form

$$S = \int_{\tau_1}^{\tau_2} \mathrm{d}^4 x \, \mathcal{L}(\varphi, \partial_\mu \varphi) \,, \tag{2.168}$$

where $\tau_1$ and $\tau_2$ represent the limiting surfaces of integration. As in classical mechanics, $L$ has the dimension of energy and $S$ that of angular momentum. In natural units, $[S] = 1$ and $[\mathcal{L}] = [E]^4$. From now on we will refer to $\mathcal{L}$ simply as the 'Lagrangian', with little risk of confusing it with $L$ since the latter function will scarcely reappear.

Now, in an arbitrary variation of the field

$$\varphi(x) \to \varphi'(x) = \varphi(x) + \delta_0 \varphi(x) \,, \tag{2.169}$$

such that $\delta_0 \varphi = 0$ at the integration limits, the action changes $S \to S + \delta S$ by an amount

$$\begin{aligned} \delta S &= \int_{\tau_1}^{\tau_2} \mathrm{d}^4 x \left( \frac{\partial \mathcal{L}}{\partial \varphi} \delta_0 \varphi + \frac{\partial \mathcal{L}}{\partial(\partial_\mu \varphi)} \, \delta_0(\partial_\mu \varphi) \right) \\ &= \int_{\tau_1}^{\tau_2} \mathrm{d}^4 x \left( \frac{\partial \mathcal{L}}{\partial \varphi} - \partial_\mu \frac{\partial \mathcal{L}}{\partial(\partial_\mu \varphi)} \right) \delta_0 \varphi + \int_{\tau_1}^{\tau_2} \mathrm{d}^4 x \, \partial_\mu \left( \frac{\partial \mathcal{L}}{\partial(\partial_\mu \varphi)} \, \delta_0 \varphi \right). \end{aligned} \tag{2.170}$$

The second integral on the right-hand side vanishes because of the boundary condition, $\delta_0 \varphi = 0$ on the integration surfaces. Then the demand that $S$ be stationary for such an arbitrary variation $\delta_0 \varphi$ immediately implies the Euler–Lagrange equation for the field $\varphi$ :

$$\frac{\partial \mathcal{L}}{\partial \varphi} - \partial_\mu \frac{\partial \mathcal{L}}{\partial(\partial_\mu \varphi)} = 0 \qquad \text{(Euler–Lagrange equation).} \tag{2.171}$$

This equation can be readily generalized to cases with several fields of arbitrary tensor characters. For each independent field component $(\varphi_i)$, there is an equation of the form (171). As a functional of the fields $\varphi_i(x)$ and their conjugate momentum densities, $\pi_i(x)$, the Hamiltonian density is given by

$$\mathcal{H}(\pi_i, \varphi_i) = \sum_i \left( \pi_i \frac{\partial \varphi_i}{\partial t} \right) - \mathcal{L} \,, \quad \text{where} \quad \pi_i = \frac{\partial \mathcal{L}}{\partial \left( \partial \varphi_i / \partial t \right)} \,. \tag{2.172}$$

How to find $\mathcal{L}$? If the equations of motion for a system are known, it is always possible to find the corresponding Lagrangian. Let us consider the three cases we have studied so far, and for each let us write down the dynamic equations and the associated Lagrangian.

(a) *Real scalar field:*

$$(\Box + m^2)\phi = -\frac{\lambda}{6}\phi^3\,;$$
$$\mathcal{L} = \frac{1}{2}\partial_\mu\phi\,\partial^\mu\phi - \frac{1}{2}m^2\phi^2 - \frac{\lambda}{4!}\,\phi^4\,. \tag{2.173}$$

The expression for the Lagrangian contains, in order, a kinematic term, a mass term (for particle of mass $m$), and a self-interaction term (of coupling constant $\lambda$). It can be checked that the Euler–Lagrange equation derived from $\mathcal{L}$ coincides precisely with the dynamic equation given. As $\mathcal{L}$ has the dimension of $[E]^4$, the field $\phi$ has dimension $[E]$, and the coupling constant $\lambda$ is dimensionless. The momentum density conjugate to $\phi$ is $\pi \equiv \partial\phi/\partial t$.

(b) *Complex scalar field:*

$$(\Box + m^2)\phi = -2\lambda\phi(\phi^*\phi)\,, \qquad (\Box + m^2)\phi^* = -2\lambda\phi^*(\phi^*\phi)\,; \tag{2.174}$$

$$\mathcal{L} = \partial_\mu\phi^*\partial^\mu\phi - m^2\phi^*\phi - \lambda(\phi^*\phi)^2\,. \tag{2.175}$$

Note the absence of factors $1/2$ in the present case; but with $\phi = \frac{1}{\sqrt{2}}(\phi_1 + \mathrm{i}\phi_2)$, (175) will coincide with the Lagrangian for two real fields $\phi_1$ and $\phi_2$. The momentum density conjugate to $\phi$ is $\pi \equiv \partial\phi^*/\partial t$.

(c) *Real vector field:*

$$\partial_\mu F^{\mu\nu} + m^2 A^\nu = j^\nu\,, \tag{2.176}$$

$$\mathcal{L} = -\tfrac{1}{4}F_{\mu\nu}F^{\mu\nu} + \tfrac{1}{2}\,m^2 A_\mu A^\mu - j_\mu A^\mu\,. \tag{2.177}$$

The overall sign in $\mathcal{L}$ is conventional, and the sign of the mass term for the physical degrees of freedom (spacelike components $A^i$) agrees with that found in the scalar field case. The vector field has dimension $[A_\mu] = [E]$. For each component $A_\mu$ considered as an independent field, one gets

$$\frac{\partial\mathcal{L}}{\partial A_\nu} = m^2 A^\nu - j^\nu\,,$$

and then

$$\begin{aligned}
\partial_\mu\frac{\partial\mathcal{L}}{\partial(\partial_\mu A_\nu)} &= -\frac{1}{4}\partial_\mu\frac{\partial}{\partial(\partial_\mu A_\nu)}[(\partial_\rho A_\sigma - \partial_\sigma A_\rho)(\partial^\rho A^\sigma - \partial^\sigma A^\rho)] \\
&= -\frac{1}{4}\partial_\mu\frac{\partial}{\partial(\partial_\mu A_\nu)}(2\partial_\rho A_\sigma\partial^\rho A^\sigma - 2\partial_\rho A_\sigma\partial^\sigma A^\rho) \\
&= -\partial_\mu(\partial^\mu A^\nu - \partial^\nu A^\mu) \;=\; -\partial_\mu F^{\mu\nu}\,.
\end{aligned}$$

Hence the Euler–Lagrange equations for $A^\nu$ correctly reproduce the equations of motion of the vector field.

On the other hand, if the dynamic equations are unknown, which is the case most often encountered in practice and where the Lagrangian approach turns out to be the most fruitful, one can construct the Lagrangian for the model by imposing certain conditions, general enough for the model to be useful, yet restrictive enough to make it well defined, even unique. The most important conditions one may consider are

(a) $\mathcal{L}$ must be Hermitian, so that the Hamiltonian itself is Hermitian;
(b) it must be relativistically invariant, that is, invariant to space-time translations and Lorentz transformations;
(c) it should be invariant to other universal symmetries, but may break certain symmetries in some definite manner, as required by experiments;
(d) it could be limited so that it contains no space-time derivatives of fields higher than the first (so that the field equations are at most of second order), and so that it contains only local couplings built up from field quantities evaluated at the same space-time point.

### 2.6.2 Conserved Current

The advantages of using a formalism based on the Lagrangian density $\mathcal{L}$ become apparent when we want to study the implications of the symmetries that the model described by $\mathcal{L}$ may have. In particular, it can be shown that there is a close relationship between the *invariance of the action* in an arbitrary *continuous global transformation* and the existence of a *conserved current*. This important result is referred to as *Noether's theorem.*

The symmetries of a model may be uncovered by studying the changes in the Lagrangian or the action function that defines the model following transformations on space-time coordinates or on internal variables. Leaving the second class of symmetries for later chapters, we limit our discussion for now to coordinate transformations. To begin, consider the following infinitesimal coordinate transformations and their subsequent effects on the typical field:

$$\begin{aligned} x^\mu \to x'^\mu &= x^\mu + \delta x^\mu \,, \\ \varphi(x) \to \varphi'(x') &= \varphi(x) + \delta\varphi(x) \,. \end{aligned} \tag{2.178}$$

For example, for a translation, $\delta x^\mu = -a^\mu$, while for a Lorentz transformation, $\delta x^\mu = \epsilon^\mu{}_\nu x^\nu$. Generally, $\delta x^\mu$ may depend on coordinates even if the transformation parameters are themselves constant.

The total variation of the field, $\delta\varphi(x)$, may be determined by a series expansion of $\varphi'(x')$:

$$\begin{aligned} \delta\varphi(x) &= \varphi'(x') - \varphi(x) \\ &= (\varphi'(x) - \varphi(x)) + (\varphi'(x') - \varphi'(x)) \\ &\approx \delta_0\varphi(x) + \delta x^\mu\, \partial_\mu\varphi(x) \,. \end{aligned} \tag{2.179}$$

It consists in general of a variation of the field due to its coordinate dependence and of a variation specific to the functional structure of the field, $\delta_0\varphi$. This variation is left arbitrary and, in particular, need *not* vanish on the integration surfaces, in contrast to the situation where one wants to derive the Euler–Lagrange equation. The action function, defined for an arbitrary volume $V$, varies by an amount

$$\delta S = \int_V \delta(\mathrm{d}^4x)\mathcal{L} + \int_V \mathrm{d}^4x\, \delta\mathcal{L}\,, \tag{2.180}$$

in which the two terms arise from the variations of the integration volume element and from the variations of the Lagrangian itself.

To calculate the variation $\delta(\mathrm{d}^4x)$, consider the element of volume in the transformed coordinates

$$\begin{aligned} \mathrm{d}^4x' &= \left|\det\left(\frac{\partial x'^\mu}{\partial x^\nu}\right)\right| \mathrm{d}^4x = |\det[\delta^\mu{}_\nu + \partial_\nu(\delta x^\mu)]|\, \mathrm{d}^4x \\ &\approx [1 + \partial_\mu(\delta x^\mu)]\, \mathrm{d}^4x\,, \end{aligned} \tag{2.181}$$

which implies

$$\delta(\mathrm{d}^4x) = (\partial_\mu(\delta x^\mu))\, \mathrm{d}^4x\,. \tag{2.182}$$

The first term in (180) then becomes

$$\int_V \delta(\mathrm{d}^4x)\mathcal{L} = \int_V \mathrm{d}^4x\, (\partial_\mu(\delta x^\mu))\, \mathcal{L}\,. \tag{2.183}$$

In the second term of (180), the variation $\delta\mathcal{L}$ arises from both $\delta x$ and $\delta\varphi$, the latter coming from $\delta x$ and $\delta_0\varphi$. The $x$ variations of $\mathcal{L}$ as function of $x$ yield the quantity $\int \mathrm{d}x\, (\partial\mathcal{L}/\partial x^\mu)\delta x^\mu$, while contributions from $\delta_0\varphi$ are given by the same expression as in (170), which includes two terms. If $\varphi$ satisfies the equation of motion (171), which is being assumed, the first term vanishes, leaving only the second, nonzero term (which vanished there because of the imposed boundary conditions). Therefore,

$$\int_V \mathrm{d}^4x\, \delta\mathcal{L} = \int_V \mathrm{d}^4x \left[(\partial_\mu\mathcal{L})\delta x^\mu + \partial_\mu\left(\frac{\partial\mathcal{L}}{\partial(\partial_\mu\varphi)}\, \delta_0\varphi\right)\right]\,. \tag{2.184}$$

Adding together both terms (183) and (184), we get for the total variation of the action

$$\delta S = \int_V \mathrm{d}^4x\, \partial_\mu \left(\frac{\partial\mathcal{L}}{\partial(\partial_\mu\varphi)}\, \delta_0\varphi\, +\, \mathcal{L}\delta x^\mu\right)\,. \tag{2.185}$$

We are interested in *global transformations*, that is, those with constant parameters. The corresponding infinitesimal transformations are defined by the small constant parameters $\delta\omega_a$, with the values of the subscripts left undefined for now. Then invariance of $S$ to a symmetry transformation defined by $\delta\omega_a$ means that its variations in an arbitrary finite volume vanish:

$$\frac{\partial S}{\partial \omega_a} \equiv \int_V \mathrm{d}^4x\, \partial_\mu j_a^\mu = 0\,, \tag{2.186}$$

for every value of $a$. Since this is true for an arbitrary volume, the integrand itself vanishes, and so does the divergence:

$$\partial_\mu j_a^\mu = 0\,. \tag{2.187}$$

The four-vector current introduced is shorthand for

$$j_a^\mu = \frac{\partial \mathcal{L}}{\partial(\partial_\mu \varphi)} \frac{\delta_0 \varphi}{\delta \omega_a} + \mathcal{L} \frac{\delta x^\mu}{\delta \omega_a}\,. \tag{2.188}$$

Note that contributions from variations of the coordinates and of the field are well separated in the two terms. When $\mathcal{L}$ contains several fields $\varphi_i$ for $i = 1, 2, \ldots$, all fields contribute to the current

$$j_a^\mu = \frac{\partial \mathcal{L}}{\partial(\partial_\mu \varphi_i)} \frac{\delta_0 \varphi_i}{\delta \omega_a} + \mathcal{L} \frac{\delta x^\mu}{\delta \omega_a}\,. \tag{2.189}$$

The current conservation in (187) implies that the associated 'charge' defined by the space integral of the current time component

$$Q_a \equiv \int_V \mathrm{d}^3x\, j_a^0(t, \boldsymbol{x}) \tag{2.190}$$

is a *constant of the motion*, since

$$\begin{aligned} \frac{\mathrm{d}Q_a}{\mathrm{d}t} &= \int_V \mathrm{d}^3x\, \partial_0\, j_a^0 = \int_V \mathrm{d}^3x\, (\partial_\mu j_a^\mu - \nabla \cdot \boldsymbol{j}_a) \\ &= -\int_S \mathrm{d}\boldsymbol{S} \cdot \boldsymbol{j}_a = 0\,, \end{aligned} \tag{2.191}$$

where use has been made of (187) and of Gauss's theorem. In the last step it is assumed that $\boldsymbol{j} = 0$ on the surface of integration.

In summary, when the action function is invariant to a continuous symmetry transformation of the coordinates and of the fields involved in the model, a locally conserved density current and a conserved (constant in time) charge associated with the symmetry can be defined. In other words, if the fields or the different components of each of the tensor fields describing the model can be transformed among themselves without changing the physical content of the fields and their interactions, the system possesses a symmetry, and this symmetry implies well-defined conserved quantities. This result, due to

Emmy Noether, makes mathematically precise the relationship between the symmetries of a system and the existence of conserved quantum numbers, and therefore, between dynamics and conservation laws. In the examples that follow, we consider, for simplicity, free fields. But it must be kept in mind that Noether's theorem may be applied to general situations, classical or quantized, with or without interactions.

**Example 2.4 Translation of a Generic Field**
Consider the translation $x^\mu \to x'^\mu = x^\mu - \delta a^\mu$. It implies the following variations:

$$\delta x^\mu = -\delta a^\mu\,, \quad \delta_0\varphi = \delta a^\mu\,\partial_\mu\varphi\,;$$
$$\delta\omega_a \equiv \delta a^\mu\,,$$

which then give

$$\frac{\delta x^\mu}{\delta a^\nu} = -\delta^\mu{}_\nu, \quad \frac{\delta_0\varphi}{\delta a^\nu} = \partial_\nu\varphi\,.$$

If the action for a generic field $\varphi(x)$ is invariant to space-time translations, the conserved current and charge associated with this invariance are

$$j^\mu{}_\nu \equiv \mathcal{T}^\mu{}_\nu = \left(\frac{\partial\mathcal{L}}{\partial(\partial_\mu\varphi)}\right)\partial_\nu\varphi - \delta^\mu{}_\nu\,\mathcal{L}\,, \tag{2.192}$$

$$P_\nu \equiv T^0_\nu \quad = \int \mathrm{d}^3x\,\mathcal{T}^0{}_\nu\,. \tag{2.193}$$

It turns out that the current, $\mathcal{T}^\mu{}_\nu$, is the energy-momentum tensor of the system and the charge, $P_\nu$, is the energy-momentum vector. In particular, the energy of the field can be identified with $P_0$ :

$$P_0 = H = \int \mathrm{d}^3x\,(\pi\dot\varphi - \mathcal{L})\,, \tag{2.194}$$

where the velocity and momentum conjugates to the field are

$$\dot\varphi = \partial_0\varphi \quad \text{and} \quad \pi = \partial\mathcal{L}/\partial\dot\varphi\,,$$

in agreement with the classical relation (166).

**Example 2.5 Lorentz Transformation of a Real Scalar Field**
In this case, the variations are

$$\delta x^\mu = \epsilon^\mu{}_\nu\,x^\nu = \epsilon^{\mu\nu}\,x_\nu\,; \qquad \delta_0\phi = -\frac{\mathrm{i}}{2}\,\epsilon^{\mu\nu}\,L_{\mu\nu}\,\phi\,.$$

Letting $\delta\omega_a \equiv \epsilon^{\rho\sigma}$, we obtain the derivatives

$$\frac{\delta x^\mu}{\delta\epsilon^{\rho\sigma}} = \delta^\mu{}_\rho\,x_\sigma - \delta^\mu{}_\sigma\,x_\rho\,,$$
$$\frac{\delta_0\phi}{\delta\epsilon^{\rho\sigma}} = -\mathrm{i}L_{\rho\sigma}\phi = (x_\rho\,\partial_\sigma - x_\sigma\,\partial_\rho)\,\phi\,.$$

The Lagrangian for a scalar field being given Lorentz-invariant, we can immediately deduce the associated current (called the angular momentum density tensor)

$$j^\mu{}_{\rho\sigma} \equiv \mathcal{M}^\mu{}_{\rho\sigma} = x_\rho \mathcal{T}^\mu{}_\sigma - x_\sigma \mathcal{T}^\mu{}_\rho \,, \tag{2.195}$$

and the corresponding conserved charge (or angular momentum)

$$Q_{\rho\sigma} \equiv M_{\rho\sigma} = \int \mathrm{d}^3x \, \mathcal{M}^0{}_{\rho\sigma} \,. \tag{2.196}$$

The space components of $Q_{\rho\sigma}$ give the familiar angular momentum

$$M_{ij} = \int \mathrm{d}^3x \, (x_i \mathcal{P}_j - x_j \mathcal{P}_i) \,, \tag{2.197}$$

where $\mathcal{P}_i = \mathcal{T}^0{}_i$ is the momentum density of the field.

**Example 2.6 Internal Transformation of a Complex Scalar Field**
We consider here an example of transformation on internal space. In such a transformation, space-time coordinates are not affected, $\delta x^\mu = 0$, only the fields vary, $\delta\phi = \delta_0\phi \neq 0$. As an example of physical interest, consider the Lagrangian for a free complex scalar field (175). It is clearly invariant to the *phase transformations*

$$\begin{aligned} \phi &\to \phi' = \mathrm{e}^{-\mathrm{i}\alpha}\phi \approx \phi - \mathrm{i}\alpha\,\phi \,, \\ \phi^* &\to \phi'^* = \mathrm{e}^{\mathrm{i}\alpha}\phi^* \approx \phi^* + \mathrm{i}\alpha\phi^* \,, \end{aligned} \tag{2.198}$$

where $\alpha$ is a real constant. Posing $\delta\omega \equiv \delta\alpha$, one gets the derivatives

$$\frac{\delta_0\phi}{\delta\alpha} = -\mathrm{i}\phi \,, \qquad \frac{\delta_0\phi^*}{\delta\alpha} = \mathrm{i}\phi^* \,, \tag{2.199}$$

and subsequently the associated current and charge

$$j^\mu = \mathrm{i}\left(\frac{\partial\mathcal{L}}{\partial(\partial_\mu\phi^*)}\,\phi^* - \frac{\partial\mathcal{L}}{\partial(\partial_\mu\phi)}\,\phi\right), \tag{2.200}$$

$$Q = \mathrm{i}\int \mathrm{d}^3x \left(\frac{\partial\mathcal{L}}{\partial(\partial_0\phi^*)}\,\phi^* - \frac{\partial\mathcal{L}}{\partial(\partial_0\phi)}\,\phi\right). \tag{2.201}$$

Thus, the Noether current associated with the phase invariance is identical to the charge current density postulated in Sect. 2.4.2. ■

## Problems

**2.1 Spin of $\pi^0$ meson.** (a) Show that two real photons in the reference frame of their center-of-mass (where the total momentum is zero) cannot be in a state of angular momentum 1. Your proof will be based on rotational invariance, Bose statistics, the transversality of photons, and the superposition principle (which says that the state of two photons is a homogeneous linear function of their polarization vectors).
(b) The $\pi^0$ meson decays mainly through the channel $\pi^0 \to 2\gamma$. Show that if angular momentum is conserved, the $\pi^0$ spin cannot be 1 (it is known now that its spin is 0).
(c) Given that the $\pi^0$ spin is 0, show that in the rest frame of $\pi^0$, the two photons emitted in the decay have the same polarization.

**2.2 Spin of $\mathrm{K}^0$ meson.** The mode $\mathrm{K}^0 \to 2\pi^0$ accounts for 31% of all $\mathrm{K}^0$ meson decays. Given that $\pi^0$ spin is 0, show that the $\mathrm{K}^0$ spin is an even number (in $\hbar$ units). We now know that its spin is 0.

**2.3 Dilation.** Consider the Lagrangian for a real scalar field $\phi(x)$ in four-space-time,

$$\mathcal{L} = \frac{1}{2}(\partial\phi)^2 - \frac{1}{2}m^2\phi^2 - \frac{1}{4!}g\phi^4 .$$

Under a scale transformation, $x' = \lambda^{-1}x$ and $\phi'(x) = \exp[D\ln\lambda]\phi(\lambda x)$, (with $\lambda > 0$). Show that the action is invariant when $D = 1$ and $m = 0$, with the field satisfying the equation of motion.

**2.4 Angular momentum of the electromagnetic field.** Consider an electromagnetic field in a region where the sources are absent. In an infinitesimal Lorentz transformation (cf. Sect. 2.3) the variations are defined by

$$\begin{aligned}\delta x^\mu &= \epsilon^\mu{}_\nu x^\nu , \\ \delta_0 A^\alpha(x) &= -\frac{\mathrm{i}}{2}\epsilon^{\mu\nu}\left(L_{\mu\nu} + \Sigma_{\mu\nu}\right) A^\alpha(x) .\end{aligned}$$

(a) Show that the density of the energy-momentum tensor is given by

$$\mathcal{T}^\mu{}_\nu = -F^{\mu\lambda}\partial_\nu A_\lambda + \delta^\mu{}_\nu \frac{1}{4} F_{\rho\sigma}F^{\rho\sigma},$$

and that the density of the angular momentum tensor is given by

$$\mathcal{M}^\mu_{\rho\sigma} = x_\rho \mathcal{T}^\mu{}_\sigma - x_\sigma \mathcal{T}^\mu{}_\rho - F^{\mu\lambda}(g_{\lambda\rho}A_\sigma - g_{\lambda\sigma}A_\rho) .$$

(b) The intrinsic spin of the field is defined by

$$S^i = \frac{1}{2}\epsilon^{ijk} S_{jk} ,$$

where

$$S_{jk} = \int d^3x \, \mathcal{M}^0_{jk} = -\int d^3x \, (\dot{A}_j A_k - \dot{A}_k A_j) \, .$$

Prove that

$$\begin{aligned} \boldsymbol{S} &= -\mathrm{i} \sum_{\boldsymbol{k}} \hat{\boldsymbol{k}} \, [a^\dagger(\boldsymbol{k},1) a(\boldsymbol{k},2) - a^\dagger(\boldsymbol{k},2) a(\boldsymbol{k},1)] \\ &= \sum_{\boldsymbol{k}} \hat{\boldsymbol{k}} \, [a^\dagger(\boldsymbol{k},+) a(\boldsymbol{k},+) - a^\dagger(\boldsymbol{k},-) a(\boldsymbol{k},-)] \end{aligned}$$

and hence, with $S_\parallel \equiv \boldsymbol{S} \cdot \hat{\boldsymbol{k}}$,

$$[S_\parallel, a^\dagger(\boldsymbol{k},\pm)] = \pm a^\dagger(\boldsymbol{k},\pm) \, ,$$

where $a(\boldsymbol{k},\lambda)$ and $a^\dagger(\boldsymbol{k},\lambda)$ are defined in Sect. 2.5.3.

## Suggestions for Further Reading

*For their historical value:*

Gordon, W., Z. Phys. **40** (1926) 117, 121

Klein, O., Z. Phys. **37** (1926) 895

Pauli, W. and Weisskopf, V. F., Helv. Phys. Acta **7** (1934) 709

Proca, A., J. Phys. Radium **9** (1938) 61

*Detailed study of the Klein–Gordon equation:*

Feshbach, H. and Villars, F., Rev. Mod. Phys. **30** (1958) 24

*Discussion on Noether's theorem:*

Hill, E. L., Rev. Mod. Phys. **23** (1957) 253

Noether, E., Nachr. K. Geo. Wiss. Göttingen **37** (1918) 235

# 3 Fermion Fields

In the previous chapter, we have seen that the nonrelativistic equation of motion of a free particle can be generalized in a natural way to the relativistic regime by making homogeneous its dependence on space and time. There are two possibilities. The first, involving second-order derivatives and called the Klein–Gordon equation, governs the evolution of fields of integral spins which are associated with operators that obey commutation relations. The second, containing only first-order derivatives and discovered by P. A. M. Dirac in 1928 in his search for a relativistic formalism admitting a non-negative probability density, describes the dynamics of fields having spins $^1\!/_2$. These fields must then represent particles, such as the electron, the proton, or the quarks, that constitute the bulk of visible matter in the universe. They are the subject of the present chapter.

## 3.1 The Dirac Equation

The equation in question is of the form

$$(\mathrm{i}\gamma^\mu \partial_\mu - m)\psi(x) = 0\,, \tag{3.1}$$

where the parameter $m$, of the dimension of mass, can be chosen to be real (by redefining if necessary the phase of the complex function $\psi$), but the four quantities $\gamma^\mu = (\gamma^0, \gamma^1, \gamma^2, \gamma^3)$ remain in general complex and behave, by assumption, as the components of some Lorentz vector. In particular, $\gamma_\mu = g_{\mu\nu}\gamma^\nu$. Application of the operator $(\mathrm{i}\gamma^\mu\partial_\mu + m)$ on this equation gives

$$-(\gamma^\mu\gamma^\nu\,\partial_\mu\partial_\nu + m^2)\psi \;=\; -\left[\tfrac{1}{2}\,(\gamma^\mu\gamma^\nu + \gamma^\nu\gamma^\mu)\,\partial_\mu\partial_\nu + m^2\right]\psi \;=\; 0\,. \tag{3.2}$$

Comparing the operator on the left-hand side of this equation with the energy-momentum relation for a free particle of mass $m$,

$$p^\mu p_\mu - m^2 = 0\,,$$

considered as an operator by the correspondence $p^\mu \to \mathrm{i}\partial^\mu$, identifies $m$ in (1) as the particle mass and leads to the condition

$$\gamma_\mu\gamma_\nu + \gamma_\nu\gamma_\mu \equiv \{\gamma_\mu, \gamma_\nu\} \;=\; 2g_{\mu\nu}\,1\,. \tag{3.3}$$

The notation $\{a,b\} = ab+ba$ stands for an *anticommutator*, and 1 for a unit matrix of a certain order since $\gamma^\mu$ are not necessarily simple numbers. From (3) we see that, first $\gamma_0^2 = 1$ and $\gamma_i^2 = -1$ and hence the eigenvalues of $\gamma_0$ are $\pm 1$ and those of $\gamma_i$ are $\pm\sqrt{-1}$, and second, $\gamma_0 = \gamma_i\gamma_0\gamma_i$ and $\gamma_i = -\gamma_0\gamma_i\gamma_0$ for $i = 1,2,3$. Taking the traces of the last two identities, one gets

$$\begin{aligned} \mathrm{Tr}\gamma_0 &= \mathrm{Tr}\,(\gamma_i\gamma_0\gamma_i) = -\mathrm{Tr}\gamma_0\,, \\ \mathrm{Tr}\gamma_i &= -\mathrm{Tr}\,(\gamma_0\gamma_i\gamma_0) = -\mathrm{Tr}\gamma_i\,, \end{aligned}$$

where the cyclic property of trace has been used: $\mathrm{Tr}(abc) = \mathrm{Tr}(cab)$. So $\mathrm{Tr}\gamma_\mu = 0$. Since the trace of a matrix is equal to the sum of its eigenvalues, $\gamma_0$ must have as many eigenvalues equal to $+1$ as those equal to $-1$, and similarly for $\gamma_i$. It follows that the order of the matrix $\gamma_\mu$ must be an even number. The smallest possible order, $N = 2$, is not admissible because it has just enough room for the three Pauli matrices and the unit matrix. The next smallest order for which the $\gamma_\mu$ matrices can be realized according to (3) is $N = 4$, which can accommodate 16 independent matrices, and it is this case that interests us. In this representation (called the *spinor representation*), $\psi$ is a column vector with four components (called the *Dirac spinor*) and the $\gamma_\mu$ are $4 \times 4$ complex matrices. The mass $m$ will be assumed to be nonvanishing for now. The special case $m = 0$, for which the representation $N = 2$ is perfectly appropriate, will be reconsidered at the end of this chapter. Let us note in passing that the equality of the dimensions of the spinor representation and of space-time is a pure coincidence that occurs only in four-dimensional space-time.

Let us rewrite (1) in the form

$$\mathrm{i}\gamma^0 \frac{\partial}{\partial t}\psi = (-\mathrm{i}\boldsymbol{\gamma}\cdot\nabla + m)\psi\,.$$

Multiplying both sides by $\gamma^0$ yields a relativistic version of the Schrödinger equation for the system

$$\mathrm{i}\frac{\partial}{\partial t}\psi = (-\mathrm{i}\gamma^0\boldsymbol{\gamma}\cdot\nabla + m\gamma^0)\psi\,. \tag{3.4}$$

The operator on the right-hand side may be identified with the Hamiltonian of the Dirac particle

$$\hat{H} = (-\mathrm{i}\gamma^0\boldsymbol{\gamma}\cdot\nabla + m\gamma^0)\,, \tag{3.5}$$

the Hermitian conjugate of which is

$$\hat{H}^\dagger = (-\mathrm{i}\nabla\cdot\boldsymbol{\gamma}^\dagger\gamma^{0\dagger} + m\gamma^{0\dagger})\,. \tag{3.6}$$

Being an observable, $\hat{H}$ must be Hermitian, $\hat{H} = \hat{H}^\dagger$, which implies

$$\gamma_0^\dagger = \gamma_0\,, \qquad \gamma_i^\dagger = \gamma_0\gamma_i\gamma_0 = -\gamma_i\,,$$

or more concisely

$$\gamma_\mu^\dagger = \gamma_0 \gamma_\mu \gamma_0 \,. \tag{3.7}$$

The basic properties of the $\gamma_\mu$ given in (3) and (7) should suffice to define $\gamma_\mu$. Nevertheless, it is sometimes useful to have an explicit matrix representation. In the most popular representation (which therefore is called the *standard representation* or the *Dirac–Pauli representation*) $\gamma^0$ is diagonal:

$$\gamma^0 = \begin{pmatrix} 1 & 0 \\ 0 & -1 \end{pmatrix} , \qquad \gamma^i = \begin{pmatrix} 0 & \sigma^i \\ -\sigma^i & 0 \end{pmatrix} . \tag{3.8}$$

Here 1 is the $2 \times 2$ unit matrix and $\sigma^i$ the usual $2 \times 2$ Pauli matrices:

$$\sigma^1 = \begin{pmatrix} 0 & 1 \\ 1 & 0 \end{pmatrix} , \qquad \sigma^2 = \begin{pmatrix} 0 & -\mathrm{i} \\ \mathrm{i} & 0 \end{pmatrix} , \qquad \sigma^3 = \begin{pmatrix} 1 & 0 \\ 0 & -1 \end{pmatrix} . \tag{3.9}$$

Products of the form $\gamma^\mu A_\mu$, which occur frequently in calculations involving the Dirac particles, deserve a distinctive symbol: $\not{A} \equiv \gamma^\mu A_\mu$; for example,

$$\not{\partial} = \gamma^\mu \partial_\mu = \gamma^0 \frac{\partial}{\partial t} + \boldsymbol{\gamma} \cdot \nabla \,.$$

With this notation the Dirac equation assumes the form

$$(\mathrm{i}\not{\partial} - m)\psi(x) = 0 \,. \tag{3.10}$$

The Hermitian conjugation of this equation followed by an application of rule (7) yields

$$-(\mathrm{i}\,\partial_\mu \psi^\dagger \gamma^0 \gamma^\mu \gamma^0 + m\psi^\dagger) = 0 \,,$$

which reduces after multiplication from the right by $\gamma^0$ to a simpler form

$$\overline{\psi}(\mathrm{i}\overleftarrow{\not{\partial}} + m) = 0 \,. \tag{3.11}$$

Here $\overline{\psi}\overleftarrow{\not{\partial}} \equiv \partial_\mu \overline{\psi} \gamma^\mu$, and $\overline{\psi} \equiv \psi^\dagger \gamma_0$ is a row vector with four components (called the *Hermitian adjoint spinor*). If $\psi_1, \ldots, \psi_4$ are the components of the column vector $\psi$, the adjoint spinor $\overline{\psi}$ is given in the representation (8) of the $\gamma_\mu$ by the row vector $(\psi_1^*, \psi_2^*, -\psi_3^*, -\psi_4^*)$, with $\psi_a^*$ standing for complex conjugates of $\psi_a$.

Let us finally note that in the Dirac equation, just as in the Schrödinger equation, the time evolution is determined by a first-order time derivative. In both cases, if the wave function is known at time $t = 0$, it is also known at any later time $t > 0$. In contrast, to solve an equation of second-order time derivative, such as the Klein–Gordon equation, one needs to know both the wave function and its time derivative at the initial point.

## 3.2 Lorentz Symmetry

In this section, three important results will be derived from the Lorentz symmetry of the theory:

- the covariance of the Dirac equation, without which the theory would not be viable;
- the spin of the Dirac field;
- the bilinear covariants in $\psi$, a result essential to model building.

### 3.2.1 Covariance of the Dirac Equation

To say that the Dirac equation is covariant means that first, if an observer $\mathcal{O}$ provided with the coordinates $x$ describes a field by $\psi(x)$ as a solution of (10), then another Lorentz observer $\mathcal{O}'$ describes the same physical field by $\psi'(x')$ that satisfies an equation of the same form written in coordinates $x'$ of $\mathcal{O}'$, and second, that there is a well-defined relation between $\psi(x)$ and $\psi'(x')$.

A Lorentz transformation is defined by the real parameters $a^{\mu}{}_{\nu}$:

$$\begin{aligned} x^{\mu} \to x'^{\mu} &= a^{\mu}{}_{\nu}\, x^{\nu}\,, \\ a^{\mu}{}_{\rho}\, g_{\mu\nu}\, a^{\nu}{}_{\sigma} &= g_{\rho\sigma}\,. \end{aligned} \tag{3.12}$$

Since the Dirac equation and the Lorentz transformation are both linear relations, $\psi(x)$ and $\psi'(x')$ must also be connected by a linear relation, that is, each component $\psi'_a(x')$ $(a = 1, 2, 3, 4)$ can be written as a linear combination of the components $\psi_b(x)$ :

$$\psi'_a(x') = S_{ab}(a)\, \psi_b(x)\,, \tag{3.13}$$

[cf. the simpler transformation law for vector fields: $A'^{\mu}(x') = a^{\mu}{}_{\nu}\, A^{\nu}(x)$ ]. The $4 \times 4$ matrix $S$, which depends on the parameters $a^{\mu}{}_{\nu}$, is determined by requiring $\psi'(x')$ to be a solution to

$$(\mathrm{i}\gamma^{\mu}\partial'_{\mu} - m)\psi'(x') = 0\,. \tag{3.14}$$

Multiplying (10) from the left by $S$, considered as a linear operator, and inserting $S^{-1}S = 1$, one obtains

$$(\mathrm{i}S\gamma^{\mu}S^{-1}\partial_{\mu} - m)\psi'(x') = 0\,, \tag{3.15}$$

which exactly coincides with (14) provided that $S\gamma^{\mu}S^{-1}\partial_{\mu} = \gamma^{\mu}\partial'_{\mu}$. Since $\partial_{\mu} = a^{\nu}{}_{\mu}\partial'_{\nu}$, this condition means

$$S^{-1}(a)\gamma^{\mu}S(a) = a^{\mu}{}_{\nu}\gamma^{\nu}\,. \tag{3.16}$$

This is precisely what is meant when we say that $\gamma^{\mu}$ behaves as a Lorentz vector.

The relation (16) holds for any Lorentz transformation parameterized by $a^{\mu}{}_{\nu}$. Now we use it to determine $S(a)$ for a proper Lorentz transformation. In this case it suffices to consider an infinitesimal deviation from the identity

$$a^{\mu}{}_{\nu} = \delta^{\mu}{}_{\nu} + \epsilon^{\mu}{}_{\nu}\,, \quad \text{where} \quad \epsilon_{\mu\nu} = -\epsilon_{\nu\mu}\,. \tag{3.17}$$

To first order in $\epsilon_{\mu\nu}$, $S(a)$ must have the general form

$$S(a) \approx 1 - \frac{\mathrm{i}}{4}\epsilon^{\mu\nu}\sigma_{\mu\nu}\,, \tag{3.18}$$

where $\sigma_{\mu\nu}$ are $4\times4$ matrices antisymmetric in their Lorentz indices, $\sigma_{\mu\nu} = -\sigma_{\nu\mu}$, and the numerical factor $-\mathrm{i}/4$ has been introduced by convention. The condition (16) then becomes to first order in $\epsilon_{\mu\nu}$,

$$\epsilon^{\nu}{}_{\mu}\gamma^{\mu} = -\frac{\mathrm{i}}{4}\epsilon^{\kappa\lambda}(\gamma^{\nu}\sigma_{\kappa\lambda} - \sigma_{\kappa\lambda}\gamma^{\nu})\,,$$

which reduces to

$$2\mathrm{i}(\delta^{\nu}{}_{\kappa}\gamma_{\lambda} - \delta^{\nu}{}_{\lambda}\gamma_{\lambda}) = [\gamma^{\nu}\,,\,\sigma_{\kappa\lambda}]\,.$$

As solution to this equation, one finds with the help of (3)

$$\sigma_{\mu\nu} = \frac{\mathrm{i}}{2}\,[\gamma_{\mu}, \gamma_{\nu}]\,. \tag{3.19}$$

Its Hermitian conjugate is

$$\sigma^{\dagger}_{\mu\nu} = \gamma_0\sigma_{\mu\nu}\gamma_0\,, \tag{3.20}$$

or explicitly,

$$\sigma^{\dagger}_{ij} = \sigma_{ij}\,, \qquad \sigma^{\dagger}_{0i} = -\sigma_{0i}\,, \tag{3.21}$$

where the following identities have been used: $[\gamma_0, \sigma_{ij}] = 0; \{\gamma_0, \sigma_{0i}\} = 0$. The six independent matrices $\sigma^{\mu\nu}$ are $\sigma^{0j} = \mathrm{i}\gamma^0\gamma^j$ and $\sigma^{ij} = \mathrm{i}\gamma^i\gamma^j$ for $i \neq j$. Accordingly, we introduce the useful notations $\alpha^i$ and $\Sigma^i$, which are given in the standard representation by

$$\sigma^{0j} \equiv \mathrm{i}\alpha^j = \mathrm{i}\begin{pmatrix} 0 & \sigma^j \\ \sigma^j & 0 \end{pmatrix}\,,$$

$$\sigma^{ij} \equiv \epsilon^{ijk}\,\Sigma^k = \epsilon^{ijk}\begin{pmatrix} \sigma^k & 0 \\ 0 & \sigma^k \end{pmatrix}\,. \tag{3.22}$$

An important property of $S$ follows from (20), namely,

$$S^{\dagger} = \gamma_0\,S^{-1}\,\gamma_0\,, \tag{3.23}$$

which also shows that in general $S$ is not unitary. The Lorentz transforms of the Hermitian conjugate spinor and of the adjoint spinor are

$$\begin{aligned} \psi'(x')^\dagger &= \psi^\dagger(x) S^\dagger , \\ \overline{\psi}'(x')^\dagger &= \psi^\dagger(x) S^\dagger \gamma^0 = \overline{\psi}(x) S^{-1} . \end{aligned} \tag{3.24}$$

They show that bilinear products of the form $\overline{\psi}\Gamma\psi$ in general transform more simply than $\psi^\dagger\Gamma\psi$. As $S(a)$ will play a central role in what immediately follows, let us consider a few examples.

**Example 3.1 Rotation $(\hat{z}, \delta\theta)$**
An infinitesimal rotation about the $z$ axis through an angle $\delta\theta$ is defined by the parameter $\epsilon_{12} = -\delta\theta$, and the corresponding rotation matrix for the spinor is

$$S_\mathrm{R}(\hat{z}, \delta\theta) \approx 1 - \frac{\mathrm{i}}{4}\epsilon_{ij}\,\sigma^{ij} = 1 + \frac{\mathrm{i}}{2}\delta\theta\,\sigma^{12} . \tag{3.25}$$

More generally, for the finite rotation through an angle $\theta$ about an arbitrary axis $\hat{\boldsymbol{n}}$, the rotation matrix for the spinor is obtained by replicating (25):

$$\begin{aligned} S_\mathrm{R}(\hat{\boldsymbol{n}}, \theta) &= \exp\left(\frac{\mathrm{i}}{2}\,\theta\,\hat{\boldsymbol{n}}\cdot\boldsymbol{\Sigma}\right) \\ &= \cos\frac{\theta}{2} + \mathrm{i}\,\hat{\boldsymbol{n}}\cdot\boldsymbol{\Sigma}\,\sin\frac{\theta}{2} . \end{aligned} \tag{3.26}$$

To obtain the second line, the exponential has first been expanded in a series in powers of $\hat{\boldsymbol{n}}\cdot\boldsymbol{\Sigma}$, then even and odd powers have been summed up separately using $(\hat{\boldsymbol{n}}\cdot\boldsymbol{\Sigma})^2 = (\hat{\boldsymbol{n}})^2 = 1$. Hermiticity of $\sigma^{ij}$ (or $\Sigma^i$) then implies the unitarity of $S$ in this case: $S_\mathrm{R}^\dagger = S_\mathrm{R}^{-1}$.

**Example 3.2 Lorentz Boost**
An infinitesimal Lorentz boost in the $x$ direction is defined by the parameter $\epsilon_{01} = -\delta\omega$, and the corresponding transformation matrix for the spinor is

$$S_\mathrm{L}(\delta\omega) = 1 - \frac{1}{2}\,\delta\omega\,\alpha^1 .$$

For an arbitrary finite boost $\omega\hat{\boldsymbol{n}}$ the matrix reads

$$\begin{aligned} S_\mathrm{L}(\omega\hat{\boldsymbol{n}}) &= \exp\left(-\tfrac{1}{2}\,\omega\hat{\boldsymbol{n}}\cdot\boldsymbol{\alpha}\right) \\ &= \cosh\frac{\omega}{2} - \hat{\boldsymbol{n}}\cdot\boldsymbol{\alpha}\,\sinh\frac{\omega}{2} . \end{aligned}$$

To obtain the second line, it is useful to note $(\hat{\boldsymbol{n}}\cdot\boldsymbol{\alpha})^2 = (\hat{\boldsymbol{n}})^2 = 1$. As the matrices $\alpha^j = -\mathrm{i}\sigma^{0j}$ are Hermitian, $S_\mathrm{L}$ is also Hermitian, $S_\mathrm{L}^\dagger = S_\mathrm{L}$, rather than unitary. ■

### 3.2.2 Spin of the Dirac Field

Just as for a vector field (see Chap. 2), we proceed first by determining the total angular momentum of the spinor field through the examination of its property (13),

$$\psi'(x') = S(a)\psi(x)$$

or

$$\psi'(x) = S(a)\psi(a^{-1}x)\,. \tag{3.27}$$

To the first order of an infinitesimal transformation, $a^{\mu}{}_{\nu} \approx \delta^{\mu}{}_{\nu} + \epsilon^{\mu}{}_{\nu}$, and the transformed field is

$$\begin{aligned}\psi'(x) &= \left(1 - \tfrac{\mathrm{i}}{4}\epsilon^{\mu\nu}\sigma_{\mu\nu}\right)\,\psi(x^{\rho} - \epsilon^{\rho}{}_{\sigma}x^{\sigma}) \\ &= \left(1 - \tfrac{\mathrm{i}}{4}\epsilon^{\mu\nu}\sigma_{\mu\nu}\right)\left[\psi(x) - \tfrac{\mathrm{i}}{2}\epsilon^{\mu\nu}L_{\mu\nu}\,\psi(x)\right] \\ &= \psi(x) - \tfrac{\mathrm{i}}{2}\epsilon^{\mu\nu}\left(L_{\mu\nu} + \tfrac{1}{2}\,\sigma_{\mu\nu}\right)\,\psi(x)\,,\end{aligned}$$

which implies the field variation

$$\delta_0\psi(x) = -\frac{\mathrm{i}}{2}\epsilon^{\mu\nu}J_{\mu\nu}\,\psi(x)\,. \tag{3.28}$$

Let us write for once this relation with spinor labels explicitly shown:

$$\delta_0\psi_a(x) = -\frac{\mathrm{i}}{2}\epsilon^{\mu\nu}\left(J_{\mu\nu}\right)_{ab}\,\psi_b(x)\,. \tag{3.29}$$

Here $J_{\mu\nu}$ are the generators for infinitesimal Lorentz transformations,

$$\left(J_{\mu\nu}\right)_{ab} = L_{\mu\nu}\delta_{ab} + \tfrac{1}{2}\,\left(\sigma_{\mu\nu}\right)_{ab}\,, \tag{3.30}$$

where $\sigma_{\mu\nu}$, defined by (19), corresponds to the intrinsic part of the transformation, and $L_{\mu\nu}$, defined in the previous chapter, to the orbital contribution,

$$L_{\mu\nu} = \mathrm{i}(x_{\mu}\partial_{\nu} - x_{\nu}\partial_{\mu})\,.$$

When the transformation being applied is a pure rotation, the corresponding generators are of course simply the angular momentum components

$$J^k = \tfrac{1}{2}\,\epsilon^{ijk}\left(L_{ij} + \tfrac{1}{2}\,\sigma_{ij}\right) \equiv L^k + \tfrac{1}{2}\,\Sigma^k. \tag{3.31}$$

The square of the spin $\frac{1}{2}\boldsymbol{\Sigma}$ is

$$\tfrac{1}{2}\,\boldsymbol{\Sigma}\cdot\tfrac{1}{2}\,\boldsymbol{\Sigma} = \left(\tfrac{1}{2}\,\boldsymbol{\sigma}\right)\cdot\left(\tfrac{1}{2}\,\boldsymbol{\sigma}\right) = \tfrac{3}{4} = \tfrac{1}{2}\,(1 + \tfrac{1}{2})\,,$$

which shows that the field described by the Dirac equation has spin $s = 1/2$ .

### 3.2.3 Bilinear Covariants

The presence of half-angles in Lorentz rotations, such as in (26), implies that a rotation through an angle of $4\pi$ or a multiple of $4\pi$ is needed to bring a spinor $\psi(x)$ back to its initial value. Therefore, physical observables in the Dirac theory must involve combinations of even powers of $\psi(x)$, the simplest being of the second power.

A simple example is provided by the current density. It can be derived by multiplying (10) from the left by $\overline{\psi}$ and (11) from the right by $\psi$, and by summing up the resulting expressions:

$$\overline{\psi}(\mathrm{i}\gamma \cdot \partial + \mathrm{i}\gamma \cdot \overleftarrow{\partial})\psi = \mathrm{i}\partial_\mu(\overline{\psi}\gamma^\mu\psi) = 0 .$$

The result has the form of a conservation law

$$\partial_\mu j^\mu(x) = 0 , \tag{3.32}$$

and immediately suggests the definition of a current for the free Dirac field

$$j^\mu(x) = \overline{\psi}(x)\gamma^\mu\psi(x) . \tag{3.33}$$

The zero-component, $j^0 = \overline{\psi}(x)\gamma^0\psi(x) = \psi^\dagger\psi = |\psi_1|^2 + |\psi_2|^2 + |\psi_3|^2 + |\psi_4|^2$, can be interpreted as a probability density; it is real and positive, exactly the property required. The current density $j^\mu$ behaves as a Lorentz vector, because using (24),

$$\begin{aligned} j'^\mu(x') &= \overline{\psi}'(x')\gamma^\mu\psi'(x') \\ &= \overline{\psi}(x)S^{-1}\gamma^\mu S\psi(x) \\ &= a^\mu{}_\nu\overline{\psi}(x)\gamma^\nu\psi(x) . \end{aligned}$$

Then, since $j^\mu$ is a vector, the divergence $\partial_\mu j^\mu$ is a Lorentz-invariant and the continuity equation (32) itself is invariant.

In view of applications to come, it is useful to determine now all the basic bilinear covariants of the form $\overline{\psi}\Gamma\psi$, of which $j^\mu = \overline{\psi}\gamma^\mu\psi$ is but an example. Since the dimension of the spinor representation is 4, there must exist 16 linearly independent $4 \times 4$ matrices, which can be constructed from products of 0, 1, 2, 3, and 4 $\gamma$-matrices. They are

$$\begin{aligned} &\Gamma_\mathrm{S} = 1 , \qquad \Gamma^\mu_\mathrm{V} = \gamma^\mu , \qquad \Gamma^{\mu\nu}_\mathrm{T} = \sigma^{\mu\nu} , \\ &\Gamma^\mu_\mathrm{A} = \gamma_5\gamma^\mu , \qquad \Gamma_\mathrm{P} = \mathrm{i}\gamma_5 . \end{aligned} \tag{3.34}$$

Here we have introduced a new symbol

$$\begin{aligned} \gamma_5 = \gamma^5 &= \mathrm{i}\gamma^0\gamma^1\gamma^2\gamma^3 \\ &= \frac{\mathrm{i}}{4!}\,\epsilon_{\mu\nu\rho\sigma}\,\gamma^\mu\gamma^\nu\gamma^\rho\gamma^\sigma . \end{aligned} \tag{3.35}$$

In a general Lorentz transformation $\gamma_5$ obeys the relation

$$S^{-1}\gamma_5 S = S^{-1}(\mathrm{i}\gamma^0\gamma^1\gamma^2\gamma^3)S = (\det a)\,\gamma_5\,, \tag{3.36}$$

and so is invariant to a proper transformation ($\det a = +1$) but changes sign in an inversion or a reflection ($\det a = -1$). Such transformation properties are characteristic of *pseudoscalar* quantities. For similar reasons, $\gamma_5\gamma^\mu$ transforms as an *axial vector*. In the standard representation of the $\gamma$-matrices, $\gamma_5$ is given by

$$\gamma_5 = \begin{pmatrix} 0 & 1 \\ 1 & 0 \end{pmatrix}. \tag{3.37}$$

The bilinear covariants associated with the Γs represent various couplings the Dirac fields may have with themselves or with other fields. With the Γs so chosen, they are real (Problem 3.3), independent from one another, and have the characteristic Lorentz transformation properties shown in Table 3.1.

**Table 3.1.** Bilinear covariants

| Representations | Γ | Lorentz transformations |
|---|---|---|
| Scalar | $1$ | $\bar{\psi}'(x')\psi'(x') = \bar{\psi}(x)\psi(x)$ |
| Pseudoscalar | $\mathrm{i}\gamma_5$ | $\bar{\psi}'(x')\mathrm{i}\gamma_5\psi'(x') = \det[a]\,\bar{\psi}(x)\mathrm{i}\gamma_5\psi(x)$ |
| Vector | $\gamma^\mu$ | $\bar{\psi}'(x')\gamma^\mu\psi'(x') = a^\mu{}_\nu\,\bar{\psi}(x)\gamma^\nu\psi(x)$ |
| Axial vector | $\gamma_5\gamma^\mu$ | $\bar{\psi}'(x')\gamma_5\gamma^\mu\psi'(x') = \det[a]\,a^\mu{}_\nu\,\bar{\psi}(x)\gamma_5\gamma^\nu\psi(x)$ |
| Tensor | $\sigma^{\mu\nu}$ | $\bar{\psi}'(x')\sigma^{\mu\nu}\psi'(x') = a^\mu{}_\alpha\,a^\nu{}_\beta\,\bar{\psi}(x)\sigma^{\alpha\beta}\psi(x)$ |

## 3.3 Free-Particle Solutions

A plane-wave solution to the Dirac equation (10) is

$$\psi(x) = w(p)\,\mathrm{e}^{-\mathrm{i}p\cdot x}\,, \tag{3.38}$$

where the coefficient $w(p)$ is an $x$-independent spinor the four components of which satisfy a system of homogeneous equations

$$(\not{p} - m)_{ab}\,w_b(p) = 0\,. \tag{3.39}$$

For a nontrivial solution to the latter equations to exist, it is necessary that $\det(\not{p} - m) = 0$, which together with (8) gives $m^2 + \boldsymbol{p}^2 - p_0^2 = 0$. This characteristic equation yields two possible eigenvalues for the energy, $p^0 = \pm E$, with $E = \sqrt{\boldsymbol{p}^2 + m^2}$, to which correspond the wave functions

$$\psi_{\pm}(x) = \begin{cases} u(\boldsymbol{p})\mathrm{e}^{-\mathrm{i}p\cdot x} \\ v(\boldsymbol{p})\mathrm{e}^{+\mathrm{i}p\cdot x} \end{cases}, \tag{3.40}$$

where now $p^\mu = (E, \boldsymbol{p})$, with $p^0 = E > 0$ in both cases. The stationary spinors satisfy the equations

$$(\not{p} - m)\, u(\boldsymbol{p}) = 0\,, \tag{3.41}$$

$$(\not{p} + m)\, v(\boldsymbol{p}) = 0\,. \tag{3.42}$$

The spinor $u(\boldsymbol{p})$ will be referred to as the *positive-energy* solution, and $v(\boldsymbol{p})$ as the *negative-energy* solution. The corresponding adjoint spinors, $\bar{u} = u^\dagger\gamma^0$ and $\bar{v} = v^\dagger\gamma^0$, obey the equations

$$\bar{u}(\boldsymbol{p})(\not{p} - m) = 0\,, \qquad \bar{v}(\boldsymbol{p})(\not{p} + m) = 0\,. \tag{3.43}$$

As spinors play an essential role in the study of the Dirac particles, it is important to examine in detail their properties. To facilitate the arguments, it will be useful to adopt the standard representation of the $\gamma$-matrices. Therefore, some results, such as the explicit form of the spinors, depend on the specific representation chosen, but the final expressions for the observables should be independent of the representation.

### 3.3.1 Normalized Spinors

In the rest frame of the particle ($\boldsymbol{p} = 0$) equation (41) reads

$$m(\gamma^0 - 1)u(\mathbf{0}) = -2m \begin{pmatrix} 0 & 0 \\ 0 & 1 \end{pmatrix} u(\mathbf{0}) = 0\,.$$

Here the matrix elements 0 and 1 are themselves $2 \times 2$ matrices. Writing the Dirac four-component spinor $u$ in terms of two two-component Pauli spinors $\xi$ and $\eta$,

$$u = \begin{pmatrix} \xi \\ \eta \end{pmatrix},$$

the above equation yields $\eta = 0$ and two degenerate solutions for $\xi$.

For $\boldsymbol{p} \neq 0$ (41) may be cast in the form

$$(E - m)\xi - \boldsymbol{\sigma}\cdot\boldsymbol{p}\,\eta = 0\,,$$
$$\boldsymbol{\sigma}\cdot\boldsymbol{p}\,\xi - (E + m)\eta = 0\,.$$

The solution obtained for the Pauli spinor,

$$\eta = \frac{\boldsymbol{\sigma}\cdot\boldsymbol{p}}{E+m}\,\xi\,, \tag{3.44}$$

then leads to two independent solutions

$$u(\boldsymbol{p},s) = N\begin{pmatrix} \chi_s \\ \dfrac{\boldsymbol{\sigma}\cdot\boldsymbol{p}}{E+m}\chi_s \end{pmatrix}, \qquad (s=1,2), \tag{3.45}$$

where $N$ is a normalization factor. The Pauli spinors $\chi_s$, for $s = 1,2$, are linearly independent and may be normalized according to $\chi_s^\dagger \chi_{s'} = \delta_{ss'}$.

The solutions to (42) can be similarly found:

$$v(\boldsymbol{p},s) = N'\begin{pmatrix} \dfrac{\boldsymbol{\sigma}\cdot\boldsymbol{p}}{E+m}\eta_s \\ \eta_s \end{pmatrix}, \qquad (s=1,2), \tag{3.46}$$

where $\eta_s$ are two normalized Pauli spinors, $\eta_s^\dagger \eta_{s'} = \delta_{ss'}$.

The spinors $\chi_s$ and $\eta_s$ may be chosen for example as the eigenvectors of the spin operator $^1/_2\,\sigma^3$,

$$\chi_1 = \begin{pmatrix} 1 \\ 0 \end{pmatrix}, \quad \chi_2 = \begin{pmatrix} 0 \\ 1 \end{pmatrix}, \quad \eta_1 = \begin{pmatrix} 0 \\ 1 \end{pmatrix}, \quad \eta_2 = \begin{pmatrix} -1 \\ 0 \end{pmatrix}. \tag{3.47}$$

They are related by

$$\eta_s = -\mathrm{i}\sigma^2 \chi_s = (-)^{(1-2s)/2}\chi_{-s} \qquad \text{for } s = \pm\tfrac{1}{2}\,. \tag{3.48}$$

In this choice the subscripts $s$ correspond to the eigenvalues of $^1/_2\,\sigma^3$. The phases of $\eta_s$ are fixed so that the corresponding charge conjugate spinors (to be introduced in Chap. 5) can be more simply defined. It is sometimes useful to make the spin eigenvalues explicit in the labels according to the conversion rules

$$\chi_1 = \chi_{1/2}\,, \quad \chi_2 = \chi_{-1/2}\,, \qquad \eta_1 = \eta_{1/2}\,, \quad \eta_2 = \eta_{-1/2}\,. \tag{3.49}$$

They correspond up to a normalization constant to the Dirac spinors

$$u(\mathbf{0},1) = \begin{pmatrix} 1 \\ 0 \\ 0 \\ 0 \end{pmatrix}, \; u(\mathbf{0},2) = \begin{pmatrix} 0 \\ 1 \\ 0 \\ 0 \end{pmatrix}, \; v(\mathbf{0},1) = \begin{pmatrix} 0 \\ 0 \\ 0 \\ 1 \end{pmatrix}, \; v(\mathbf{0},2) = \begin{pmatrix} 0 \\ 0 \\ -1 \\ 0 \end{pmatrix}. \tag{3.50}$$

In conclusion, the Dirac equation has four independent solutions $u(\boldsymbol{p},s)$ and $v(\boldsymbol{p},s)$, with $s = 1,\,2$ representing positive-energy and negative-energy polarization states of a spin-$^1/_2$ field.

### 3.3.2 Completeness Relations

To fix the normalization constant of the spinors, let us first calculate the scalar product

$$u^\dagger(\boldsymbol{p},s)u(\boldsymbol{p},s') = N^2 \left(1 \quad \frac{\boldsymbol{\sigma}\cdot\boldsymbol{p}}{E+m}\right)\begin{pmatrix}1\\ \boldsymbol{\sigma}\cdot\boldsymbol{p}/(E+m)\end{pmatrix}\chi_s^\dagger\chi_{s'}$$
$$= N^2\left(1+\frac{(\boldsymbol{\sigma}\cdot\boldsymbol{p})^2}{(E+m)^2}\right)\delta_{ss'} = N^2\frac{2E}{E+m}\,\delta_{ss'}\,.$$

Since $u^\dagger u = \bar{u}\gamma^0 u$ behaves as the time component of a Lorentz vector, the normalization factor $N$ must be chosen so that the right-hand side of the last equation has the same property,

$$u^\dagger(\boldsymbol{p},s)\,u(\boldsymbol{p},s') = 2E\,\delta_{ss'}\,, \tag{3.51}$$

with a numerical factor fixed for convenience. This implies $N = \sqrt{E+m}$. By the same token, with the normalization $N' = \sqrt{E+m}$ the noncovariant norm of $v$ is given by

$$v^\dagger(\boldsymbol{p},s)\,v(\boldsymbol{p},s') = 2E\,\delta_{ss'}\,. \tag{3.52}$$

Note however that $v^\dagger(\boldsymbol{p},-s)u(\boldsymbol{p},s) \neq 0$ but in contrast $v^\dagger(-\boldsymbol{p},s')\,u(\boldsymbol{p},s) = 0$ and $u^\dagger(\boldsymbol{p},s)v(-\boldsymbol{p},s') = 0$. The reason is that $u(\boldsymbol{p},s)$ and $v(-\boldsymbol{p},s)$, rather than $v(\boldsymbol{p},s)$, are the eigenspinors of the Dirac Hamiltonian

$$H_{\boldsymbol{p}} = \gamma^0(\boldsymbol{\gamma}\cdot\boldsymbol{p}+m)\,, \tag{3.53}$$

with respective eigenvalues $E$ and $-E$:

$$\begin{aligned} H_{\boldsymbol{p}}u(\boldsymbol{p}) &= E\,u(\boldsymbol{p})\,,\\ H_{\boldsymbol{p}}v(-\boldsymbol{p}) &= -E\,v(-\boldsymbol{p})\,, \end{aligned} \tag{3.54}$$

and are therefore mutually orthogonal. The four spinors $u(\boldsymbol{p},s)$ and $v(-\boldsymbol{p},s)$ for $s=1,2$ form a complete set in the spinor representation for any given $\boldsymbol{p}$, which implies the closure relation

$$\sum_{s=1}^{2}\left[u_a(\boldsymbol{p},s)u_b^\dagger(\boldsymbol{p},s) + v_a(-\boldsymbol{p},s)v_b^\dagger(-\boldsymbol{p},s)\right] = 2E\,\delta_{ab}\,, \tag{3.55}$$

where $a$, $b = 1,\ldots,4$ label the spinor components.

The above results, written in terms of the Hermitian conjugates of spinors, would be better re-expressed in terms of the spinor adjoint conjugates because the latter are just those involved in bilinear covariants. In general, in order to make explicit calculations of bilinear covariants, one has to use

Dirac's equations (41)–(42). Let us for example evaluate the scalar product $\bar{u}(\boldsymbol{p},s)\,u(\boldsymbol{p},s')$. Multiplying (41) from the left by $u^\dagger(\boldsymbol{p},s)$ gives

$$u^\dagger(\boldsymbol{p},s)(\gamma\cdot p - m)\,u(\boldsymbol{p},s') = u^\dagger(\boldsymbol{p},s)(\gamma^0 E - \boldsymbol{\gamma}\cdot\boldsymbol{p} - m)\,u(\boldsymbol{p},s') = 0. \quad (3.56)$$

Its Hermitian conjugate, with the spin indices $s$, $s'$ interchanged, reads

$$u^\dagger(\boldsymbol{p},s)(\gamma^\dagger\cdot p - m)\,u(\boldsymbol{p},s') = u^\dagger(\boldsymbol{p},s)(\gamma^0 E + \boldsymbol{\gamma}\cdot\boldsymbol{p} - m)\,u(\boldsymbol{p},s') = 0. \quad (3.57)$$

Summing the last two equations gives

$$E u^\dagger(\boldsymbol{p},s)\gamma^0 u(\boldsymbol{p},s') = m u^\dagger(\boldsymbol{p},s)\,u(\boldsymbol{p},s') = 2Em\,\delta_{ss'}\,,$$

by making use of (51). The other three invariant products are calculated in the same way, with the help of (41)–(42), (51)–(52), and $\gamma_0^\dagger = \gamma_0$, $\gamma_i^\dagger = -\gamma_i$, leading to the results

$$\begin{aligned}
\bar{u}(\boldsymbol{p},s)\,u(\boldsymbol{p},s') &= \;\; 2m\,\delta_{ss'}\,,\\
\bar{v}(\boldsymbol{p},s)\,v(\boldsymbol{p},s') &= -2m\,\delta_{ss'}\,,\\
\bar{u}(\boldsymbol{p},s)\,v(\boldsymbol{p},s') &= \bar{v}(\boldsymbol{p},s)\,u(\boldsymbol{p},s') \;=\; 0\,.
\end{aligned} \quad (3.58)$$

The norms defined in this way are covariant. Note also that the norm of the negative-energy spinor $v$ is negative; the sign difference with the corresponding noncovariant norm can be traced to the minus sign in the $\gamma^0$ matrix that comes with $\bar{v} = v^\dagger\gamma^0$.

To rewrite the completeness relation (55) in terms of $\bar{u}$ and $\bar{v}$, first note that, with (54),

$$\begin{aligned}
(H_{\boldsymbol{p}} + E)\,u(\boldsymbol{p},s) &= \;\; 2E\,u(\boldsymbol{p},s)\,,\\
(H_{\boldsymbol{p}} - E)\,v(-\boldsymbol{p},s) &= -2E\,v(-\boldsymbol{p},s)\,;
\end{aligned} \quad (3.59)$$

that is, $H_{\boldsymbol{p}} \pm E$ act as projection operators: $H_{\boldsymbol{p}} - E$ cancels $u(\boldsymbol{p})$ but leaves $v(-\boldsymbol{p})$ essentially unchanged, whereas $H_{\boldsymbol{p}} + E$ cancels $v(-\boldsymbol{p})$ but leaves $u(\boldsymbol{p})$ unchanged. However, these operators are not in a covariant form, an inconvenience in an invariant theory, and so should be replaced by equivalent operators which are. For this purpose, it suffices to note that

$$\begin{aligned}
(\not{p} - m)\,u &= 0\,,\\
(\not{p} + m)\,u &= 2m\,u\,,
\end{aligned}$$

and also

$$\begin{aligned}
(\not{p} + m)\,v &= 0\,,\\
(\not{p} - m)\,v &= -2m\,v\,.
\end{aligned}$$

Then it is clear that $\not{p} \pm m$ are precisely the operators we are seeking. It is also useful to have them re-expressed in terms of spinors, which requires somewhat more work. When $H_{\boldsymbol{p}} + E$ is applied on both sides of (55), one gets $\sum uu^\dagger$ on the left-hand side and $H_{\boldsymbol{p}} + E$ on the right-hand side:

$$\sum_s u(\boldsymbol{p}, s)u^\dagger(\boldsymbol{p}, s) = H_{\boldsymbol{p}} + E = \gamma^0(\boldsymbol{\gamma} \cdot \boldsymbol{p} + m) + E\,. \tag{3.60}$$

The desired result is obtained by multiplying both sides from the right by $\gamma^0$:

$$\sum_s u(\boldsymbol{p}, s)\bar{u}(\boldsymbol{p}, s) = \gamma^0 E - \boldsymbol{\gamma} \cdot \boldsymbol{p} + m = \gamma_\mu p^\mu + m\,. \tag{3.61}$$

Proceeding in the same manner with $H_{\boldsymbol{p}} - E$, one gets

$$\sum_s v(-\boldsymbol{p}, s)\bar{v}(-\boldsymbol{p}, s) = \gamma^0 E + \boldsymbol{\gamma} \cdot \boldsymbol{p} - m\,, \tag{3.62}$$

which, after reversing the sign of $\boldsymbol{p}$ on both sides, leads to the result

$$\sum_s v(\boldsymbol{p}, s)\bar{v}(\boldsymbol{p}, s) = \gamma^0 E - \boldsymbol{\gamma} \cdot \boldsymbol{p} - m = \gamma_\mu p^\mu - m\,. \tag{3.63}$$

It is useful to introduce next the operators

$$\begin{aligned}
\Lambda_+(p) &\equiv \frac{\not{p} + m}{2m} = \frac{1}{2m}\sum_s u(\boldsymbol{p}, s)\bar{u}(\boldsymbol{p}, s)\,, \\
\Lambda_-(p) &\equiv \frac{-\not{p} + m}{2m} = -\frac{1}{2m}\sum_s v(\boldsymbol{p}, s)\bar{v}(\boldsymbol{p}, s)\,.
\end{aligned} \tag{3.64}$$

Applied on an arbitrary spinor the operator $\Lambda_+(p)$ gives the positive-energy components, while $\Lambda_-(p)$ projects out the negative-energy components. They are therefore the projection operators for the positive-energy solutions and the negative-energy solutions, respectively. Their basic properties are summarized in the following relations, valid for any given momentum,

$$\begin{aligned}
\Lambda_\pm^2 &= \Lambda_\pm\,, \\
\Lambda_+\Lambda_- = \Lambda_-\Lambda_+ &= 0\,, \\
\Lambda_+ + \Lambda_- &= 1\,.
\end{aligned} \tag{3.65}$$

The last of these relations is just the covariant form of the closure relation exactly equivalent to (55):

$$\sum_s [\,u(\boldsymbol{p}, s)\bar{u}(\boldsymbol{p}, s) - v(\boldsymbol{p}, s)\bar{v}(\boldsymbol{p}, s)\,] = 2m\,. \tag{3.66}$$

### 3.3.3 Helicities

In this subsection we examine the exact physical significance of the two degrees of freedom $s$ for each of the positive-energy or negative-energy solutions of a free particle. For a particle at rest, $\boldsymbol{p} = 0$, it is clear that the spinors $u(\mathbf{0}, s)$ and $v(\mathbf{0}, s)$ can be constructed as eigenspinors of the spin operator

$$\tfrac{1}{2}\,\Sigma^3 = \tfrac{1}{2}\begin{pmatrix}\sigma^3 & 0\\ 0 & \sigma^3\end{pmatrix},$$

with $s = 1, 2$ associated with the eigenvalues $\pm 1/2$. It suffices then to choose the spinors as in (50). However, in general, for any nonvanishing momentum vector $\boldsymbol{p}$ not lying in the $z$ direction, the free-particle solutions given in (45) and (46) are not eigenvectors of $\Sigma^3$. In other words, let $\chi_\lambda$ be the two-component spinors that are eigenfunctions of the Hermitian matrix $\boldsymbol{\sigma}\cdot\hat{\boldsymbol{p}}$,

$$\tfrac{1}{2}\boldsymbol{\sigma}\cdot\hat{\boldsymbol{p}}\,\chi_\lambda = \lambda\,\chi_\lambda\,. \tag{3.67}$$

Then the solutions to the Dirac equation for a free particle in (45) and (46), but with the Pauli spinors $\chi_s$ and $\eta_s$ both replaced by $\chi_\lambda$, are the eigenspinors of $\boldsymbol{\Sigma}\cdot\hat{\boldsymbol{p}}$ for $\hat{\boldsymbol{p}} = \boldsymbol{p}/|\boldsymbol{p}|$. This follows from the simple fact that $\boldsymbol{\sigma}\cdot\hat{\boldsymbol{p}}$ commutes with itself.

In two-component spinor space, it is even possible to diagonalize $\boldsymbol{\sigma}\cdot\hat{\boldsymbol{n}}$ for an *arbitrary* unit vector $\hat{\boldsymbol{n}}$. However, as this operator does not generally commute with $\boldsymbol{\sigma}\cdot\hat{\boldsymbol{p}}$ [which appears in the Dirac spinors (45) and (46) ], it is not possible to construct the solutions to the Dirac equation for a free particle as four-component eigenspinors of $\boldsymbol{\Sigma}\cdot\hat{\boldsymbol{n}}$ for an arbitrary $\hat{\boldsymbol{n}}$, unless $\hat{\boldsymbol{n}} = \pm\hat{\boldsymbol{p}}$ or $\hat{\boldsymbol{n}} = 0$. Indeed, with the Hamiltonian defined in (53) and $\boldsymbol{\Sigma}$ considered as a Heisenberg operator, the Heisenberg equation for $\boldsymbol{\Sigma}$ reads

$$\frac{\mathrm{d}\boldsymbol{\Sigma}}{\mathrm{d}t} = \mathrm{i}\,[H_{\boldsymbol{p}}, \boldsymbol{\Sigma}] = -2(\boldsymbol{\alpha}\times\boldsymbol{p})\,. \tag{3.68}$$

As in general $\boldsymbol{\alpha}\times\boldsymbol{p} \neq 0$, it follows that $\mathrm{d}\boldsymbol{\Sigma}/\mathrm{d}t \neq 0$ and the spin $\boldsymbol{\Sigma}$ is not a constant vector (although the total angular momentum $\boldsymbol{J} = \boldsymbol{L} + \tfrac{1}{2}\,\boldsymbol{\Sigma}$ of course is). Now forming the dot product of both sides of the above equation with some constant vector $\hat{\boldsymbol{n}}$,

$$\frac{\mathrm{d}\boldsymbol{\Sigma}\cdot\hat{\boldsymbol{n}}}{\mathrm{d}t} = -2(\boldsymbol{\alpha}\times\boldsymbol{p})\cdot\hat{\boldsymbol{n}}\,, \tag{3.69}$$

it is seen that $\mathrm{d}\boldsymbol{\Sigma}\cdot\hat{\boldsymbol{n}}/\mathrm{d}t \neq 0$, unless $\hat{\boldsymbol{n}} = 0$ or $\hat{\boldsymbol{n}} = \pm\hat{\boldsymbol{p}}$.

As $\boldsymbol{L}\cdot\hat{\boldsymbol{p}} = 0$, one has $\boldsymbol{J}\cdot\hat{\boldsymbol{p}} = \tfrac{1}{2}\,\boldsymbol{\Sigma}\cdot\hat{\boldsymbol{p}}$. The operator $\boldsymbol{J}\cdot\hat{\boldsymbol{p}}$ or $\tfrac{1}{2}\,\boldsymbol{\Sigma}\cdot\hat{\boldsymbol{p}}$ is called the *helicity operator* for a spin-$1/2$ particle. One refers to the eigenstates of helicity $h = +1/2$ as the *right-handed states* (with spin oriented in the

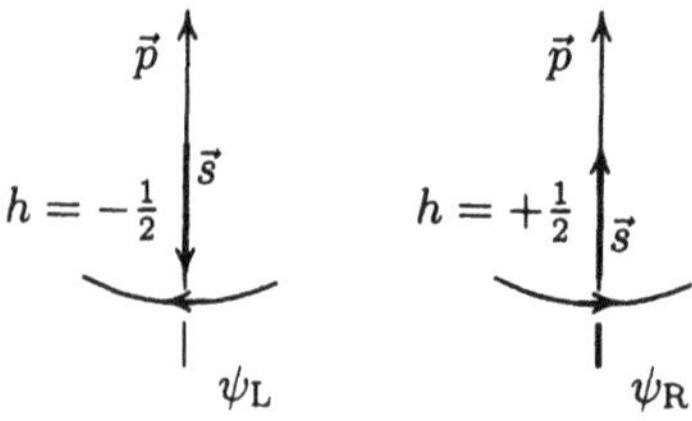


**Fig. 3.1.** Relative orientation of spin and momentum for left-handed and right-handed particles

direction of motion), and to those of helicity $h = -1/2$ as the *left-handed states* (with spin opposite to the direction of motion). See Fig. 3.1.

Given an arbitrary spinor, how can we extract its component having a specified (circular) polarization? We expect that the operators that perform this task are some covariant generalizations of an operator found in nonrelativistic quantum mechanics,

$$P(\hat{\boldsymbol{n}}) = \tfrac{1}{2}\,(1 + \boldsymbol{\sigma}\cdot\hat{\boldsymbol{n}})\,,$$

which projects out of a given Pauli spinor the component polarized in the $\hat{\boldsymbol{n}}$ direction. The operators we are seeking will separate states $s = 1$ and $s = 2$, just as $\Lambda_\pm$ separate the spinors $u$ and $v$. They should be orthonormalized, should have the correct nonrelativistic limit, and finally, should commute with both $\Lambda_\pm$. The operators that satisfy these conditions are

$$P(\pm n) = \tfrac{1}{2}\,(1 \pm \gamma_5\, \not{n})\,, \tag{3.70}$$

where $n^\mu$ is a normalized spacelike vector, $n_\mu n^\mu = -1$ (to satisfy the first two conditions), and is orthogonal to the particle momentum, $n_\mu p^\mu = 0$ (to satisfy the last condition).

For a system at rest, $\boldsymbol{p} = 0$, the condition $n_\mu p^\mu = 0$ implies $n^0 = 0$, and so to have $n\cdot n = -1$, it suffices to orient $n^\mu$ in the $z$ direction, so that $n^\mu = (0,0,0,1)$. Then, in the standard representation of the $\gamma$-matrices,

$$P(\pm n) = \tfrac{1}{2}\,(1 \mp \gamma_5\gamma^3) = \tfrac{1}{2}\begin{pmatrix} 1\pm\sigma^3 & 0 \\ 0 & 1\mp\sigma^3 \end{pmatrix}. \tag{3.71}$$

With the spinors $\chi_s$ and $\eta_s$ in (45) and (46) chosen as eigenspinors of $\sigma^3$, the operators $P(\pm n)$ perform the required tasks:

$$\begin{aligned} P(+n)\,u(\mathbf{0},1) &= u(\mathbf{0},1)\,,\\ P(-n)\,u(\mathbf{0},2) &= u(\mathbf{0},2)\,,\\ P(+n)\,v(\mathbf{0},1) &= v(\mathbf{0},1)\,,\\ P(-n)\,v(\mathbf{0},2) &= v(\mathbf{0},2)\,, \end{aligned} \tag{3.72}$$

and cancel the other three spinors in each case. The operator $P(+n)$ projects out the state with the polarization $1/2$ in the rest frame of the particle for a positive-energy solution (right-handed particle), and $-1/2$ for a negative-energy solution (left-handed antiparticle). Similarly $P(-n)$, *mutatis mutandis.* Since (70) is Lorentz-invariant and gives the correct solution in a particular Lorentz frame, it is the required operator valid in every Lorentz frame.

When $\boldsymbol{p} \neq \boldsymbol{0}$, one may choose $n = n_p$ such that $\boldsymbol{n}_p$ is parallel to $\boldsymbol{p}$:

$$n_p = \left( \frac{|\boldsymbol{p}|}{m}, \frac{p^0}{m} \hat{\boldsymbol{p}} \right) . \tag{3.73}$$

With this choice, the polarization becomes identical to the helicity. This can be seen for example by matrix multiplication, using the standard representation of the $\gamma$, as follows:

$$(1 + \gamma_5 \not{n}_p)(\pm \not{p} + m) = (1 \pm \boldsymbol{\Sigma} \cdot \hat{\boldsymbol{p}})(\pm \not{p} + m) ,$$
$$(1 - \gamma_5 \not{n}_p)(\pm \not{p} + m) = (1 \mp \boldsymbol{\Sigma} \cdot \hat{\boldsymbol{p}})(\pm \not{p} + m) ;$$

or alternatively,

$$\begin{aligned} P(n_p)\, \Lambda_\pm(p) &= \tfrac{1}{2} (1 \pm \boldsymbol{\Sigma} \cdot \hat{\boldsymbol{p}})\, \Lambda_\pm(p) , \\ P(-n_p)\, \Lambda_\pm(p) &= \tfrac{1}{2} (1 \mp \boldsymbol{\Sigma} \cdot \hat{\boldsymbol{p}})\, \Lambda_\pm(p) . \end{aligned} \tag{3.74}$$

Just as expected, $P(n_p)$ projects out the positive-helicity component from a positive-energy state and the negative-helicity component from a negative-energy state, and similarly, $P(-n_p)$ projects out the negative-helicity component from a positive-energy state and the positive-helicity component from a negative-energy state.

Projection operators are very useful in practice. In expressions where specific states are selectively considered, they make possible the use of closure relations and unnecessary explicit calculations of the spinors, replacing these by known spin matrices. For example, the probability of a certain process may be given by $\sum_i (\bar{u}_f \Gamma u_i)(\bar{u}_i \Gamma' u_f)$, where the summation is to be performed over the two positive-energy spinors $u_i$. Then the sum $\sum_i u_i \bar{u}_i$ can be replaced by $\not{p} + m = 2m\, \Lambda_+(p)$. On the other hand, if instead of summing over both spin states, one calculates rather the probability for some given polarization, then one just inserts the operator $P(n_p)$ to project out the appropriate spin component.

## 3.4 The Lagrangian for a Free Dirac Particle

As we have seen in the last chapter, a Lagrangian completely defines the dynamics of any given system and embodies all of its symmetries. The Lagrangian for a free Dirac field is

$$\mathcal{L} = \overline{\psi}(\mathrm{i}\gamma^\mu \partial_\mu - m)\psi . \tag{3.75}$$

Since $\mathcal{L}$ has dimension $[E]^4$, the Dirac field must have dimension $[E]^{3/2}$. The Euler–Lagrange equations for the field variables $\psi$ and $\overline{\psi}$,

$$\begin{aligned}\frac{\partial \mathcal{L}}{\partial \overline{\psi}} - \partial_\mu \frac{\partial \mathcal{L}}{\partial(\partial_\mu \overline{\psi})} &= (\mathrm{i}\gamma^\mu \partial_\mu - m)\psi = 0\,,\\ \frac{\partial \mathcal{L}}{\partial \psi} - \partial_\mu \frac{\partial \mathcal{L}}{\partial(\partial_\mu \psi)} &= -(\partial_\mu \overline{\psi}\,\mathrm{i}\gamma^\mu + m\overline{\psi}) = 0\,,\end{aligned}$$

correctly reproduce the Dirac equations (10) and (11). Noether's theorem introduced in the last chapter will now be applied to derive the conserved currents associated with the symmetries of the system.

It is clear that the Dirac Lagrangian (75) is invariant to any constant translation,

$$\begin{aligned}x^\mu \to \quad & x'^\mu = x^\mu - a^\mu\,,\\ \psi(x) \to & \psi'(x') = \psi(x)\,.\end{aligned}$$

The associated current is the energy-momentum tensor

$$\begin{aligned}\mathcal{T}^\mu{}_\nu &= \frac{\partial \mathcal{L}}{\partial(\partial_\mu \psi)}\,\partial_\nu \psi + \partial_\nu \overline{\psi}\,\frac{\partial \mathcal{L}}{\partial(\partial_\mu \overline{\psi})} - \delta^\mu{}_\nu\,\mathcal{L}\\ &= \overline{\psi}\,\mathrm{i}\gamma^\mu \partial_\nu \psi\,,\end{aligned} \tag{3.76}$$

where $\psi$ is a solution to the Dirac equation and, therefore, $\mathcal{L} = 0$. Since the current is conserved, $\partial_\mu \mathcal{T}^\mu{}_\nu = 0$, the corresponding 'charge' or momentum

$$P_\nu = \int \mathrm{d}^3x\,\mathcal{T}^0{}_\nu = \int \mathrm{d}^3x\,\overline{\psi}\,\mathrm{i}\gamma^0\,\partial_\nu \psi \tag{3.77}$$

is a constant of the motion. In particular, its zero-component defines the energy or Hamiltonian of the system

$$\begin{aligned}H = P_0 &= \int \mathrm{d}^3x\,\overline{\psi}\,\mathrm{i}\gamma^0\,\partial_0 \psi\\ &= \int \mathrm{d}^3x\,\psi^\dagger \gamma^0(-\mathrm{i}\boldsymbol{\gamma}\cdot\nabla + m)\psi\,,\end{aligned} \tag{3.78}$$

where we can recognize the Hamiltonian operator $\hat{H} = \gamma^0(-\mathrm{i}\boldsymbol{\gamma}\cdot\nabla + m)$.

From the Lorentz transformation properties of the bilinear covariants $\overline{\psi}\psi$ and $\overline{\psi}\gamma^\mu\psi$ (see Table 3.1), $\mathcal{L}$ is seen to be Lorentz-invariant. In an infinitesimal Lorentz transformation, parameterized by $\epsilon_{\mu\nu} = -\epsilon_{\nu\mu}$,

$$\begin{aligned}x^\mu \to x'^\mu &= x^\mu + \epsilon^\mu{}_\nu\, x^\nu\,,\\ \psi(x) \to \psi'(x') &= S(1+\epsilon)\,\psi(x)\,,\end{aligned}$$

the field variation is, according to (28),

$$\delta_0\psi(x) = -\frac{\mathrm{i}}{2}\,\epsilon^{\mu\nu} J_{\mu\nu}\,\psi(x)\,, \tag{3.79}$$

where $J_{\mu\nu}$ are the generators of the infinitesimal Lorentz transformation,

$$J_{\mu\nu} = L_{\mu\nu} + \tfrac{1}{2}\,\sigma_{\mu\nu}, \tag{3.80}$$

with $L_{\mu\nu} = \mathrm{i}(x_\mu\partial_\nu - x_\nu\partial_\mu)$ representing the orbital part and $\sigma_{\mu\nu} = \frac{\mathrm{i}}{2}[\gamma_\mu, \gamma_\nu]$ the intrinsic part. To evaluate the Noether current density associated with this symmetry, the general expression (2.194) becomes in this case

$$\mathcal{M}^{\mu}_{\ \rho\sigma} = \frac{\partial\mathcal{L}}{\partial(\partial_\mu\psi)}\,\frac{\delta_0\psi}{\delta\epsilon^{\rho\sigma}} + \frac{\delta_0\overline{\psi}}{\delta\epsilon^{\rho\sigma}}\,\frac{\partial\mathcal{L}}{\partial(\partial_\mu\overline{\psi})} + \mathcal{L}\,\frac{\delta x^\mu}{\delta\epsilon^{\rho\sigma}}\,. \tag{3.81}$$

As only the first term on the right-hand side is nonvanishing, using

$$\frac{\delta_0\psi}{\delta\epsilon^{\rho\sigma}} = -\mathrm{i}J_{\rho\sigma}\psi = (x_\rho\partial_\sigma - x_\sigma\partial_\rho)\psi - \frac{\mathrm{i}}{2}\sigma_{\rho\sigma}\psi \tag{3.82}$$

immediately carries (81) into the desired result

$$\mathcal{M}^{\mu}_{\ \rho\sigma} = x_\rho \mathcal{T}^\mu{}_\sigma - x_\sigma \mathcal{T}^\mu{}_\rho + \frac{1}{2}\overline{\psi}\,\gamma^\mu\sigma_{\rho\sigma}\,\psi\,. \tag{3.83}$$

The associated conserved 'charge' is the angular momentum tensor

$$\begin{aligned} M_{\rho\sigma} &= \int \mathrm{d}^3x\, \mathcal{M}^0_{\ \rho\sigma} \\ &= \int \mathrm{d}^3x\,(x_\rho\mathcal{T}^0{}_\sigma - x_\sigma\mathcal{T}^0{}_\rho) + \frac{1}{2}\int \mathrm{d}^3x\,\overline{\psi}\,\gamma^0\sigma_{\rho\sigma}\,\psi\,, \end{aligned} \tag{3.84}$$

where evidently the first integral on the last line represents the orbital component, and the second, the intrinsic component.

The Lagrangian (75) is also invariant to complex phase transformations of the fields,

$$\begin{aligned} \psi(x) &\to \psi'(x) = \mathrm{e}^{-\mathrm{i}\alpha}\,\psi(x) \approx \psi(x) - \mathrm{i}\alpha\,\psi(x), \\ \overline{\psi}(x) &\to \overline{\psi}'(x) = \mathrm{e}^{\mathrm{i}\alpha}\,\overline{\psi}(x) \approx \overline{\psi}(x) + \mathrm{i}\alpha\,\overline{\psi}(x), \end{aligned} \tag{3.85}$$

where $\alpha$ stands for a real constant. These internal transformations change only the fields, leaving untouched their coordinate arguments,

$$\frac{\delta_0\psi}{\delta\alpha} = -\mathrm{i}\psi\,, \qquad \frac{\delta_0\overline{\psi}}{\delta\alpha} = \mathrm{i}\overline{\psi}\,. \tag{3.86}$$

The associated conserved current density and charge are

$$j^\mu = \frac{\partial\mathcal{L}}{\partial(\partial_\mu\psi)}\,\frac{\delta_0\psi}{\delta\alpha} + \frac{\delta_0\overline{\psi}}{\delta\alpha}\,\frac{\partial\mathcal{L}}{\partial(\partial_\mu\overline{\psi})} = \overline{\psi}\gamma^\mu\psi\,, \tag{3.87}$$

$$Q = \int \mathrm{d}^3x\, j^0 = \int \mathrm{d}^3x\,\psi^\dagger\psi\,. \tag{3.88}$$

## 3.5 Quantization of the Dirac Field

In Dirac's theory, the probability density given by the time component of a conserved current, $j^0 = \psi^\dagger\psi$ for $\psi = \psi_+$ or $\psi = \psi_-$, is evidently positive-definite. This result, by avoiding one of the obstacles initially met by the Klein–Gordon equation, makes it possible to interpret the Dirac equation as the basic equation for a *one-particle* system. However, Dirac could not prevent the presence of negative-energy solutions, and it is a measure of his genius to be able to turn this apparent difficulty to his advantage, giving us at the same time the novel concept of *antiparticle.* As we have seen in the last chapter, the negative-energy solution to the Klein–Gordon equation can be interpreted as the wave function of an antiboson of electric charge opposite to that of the particle described by the positive-energy solution, and the current density of the theory must be considered not as a probability current density but rather as a charge current density. A similar interpretation applies to the present case as well, and should even emerge quite naturally when one considers processes in which particles are created or destroyed, such as $\mathrm{n} \to \mathrm{p} + \mathrm{e}^- + \bar{\nu}$ or $\gamma \to \mathrm{e}^+ + \mathrm{e}^-$, because, clearly, what is conserved then is not the probability of finding a given particle in the space volume but rather the total electric charge of the system. The problem becomes a *many-body* problem for which the quantum field theory is the most appropriate approach. The classical Dirac field is then treated as a field operator which describes the creation and annihilation of fermions and antifermions at all points in space-time, paralleling the role played by the quantized Klein–Gordon field for bosons and antibosons. However, there is a fundamental difference between the two cases that must be taken into account in the formulation, namely, the existence for fermions of a rule (the Pauli exclusion principle) that forbids the presence of more than one fermion of the same kind in the same state.

The four solutions to the Dirac equation for a free particle

$$\begin{aligned} \psi^{(+)}_{\boldsymbol{p},s}(x) &= C_{\boldsymbol{p}}\psi_+(x) = C_{\boldsymbol{p}}u(p,s)\,\mathrm{e}^{-\mathrm{i}p\cdot x}\,, \\ \psi^{(-)}_{-\boldsymbol{p},-s}(x) &= C_{\boldsymbol{p}}\psi_-(x) = C_{\boldsymbol{p}}v(p,s)\,\mathrm{e}^{\mathrm{i}p\cdot x}\,, \end{aligned} \tag{3.89}$$

are the eigenvectors of $\hat{H}$ with energies $E_p$ and $-E_p$ ($p_0 = E_{\boldsymbol{p}} = \sqrt{\boldsymbol{p}^2 + m^2}$), and of spin $s_z = \pm 1/2$. According to (66) they form a complete set in the spinor representation and, with normalization $C_{\boldsymbol{p}} = 1/\sqrt{(2\pi)^3\,2E_{\boldsymbol{p}}}$, satisfy the orthonormality relations

$$\begin{aligned} &\int \mathrm{d}^3x\,\psi^{(+)\dagger}_{\boldsymbol{p}',s'}(x)\,\psi^{(+)}_{\boldsymbol{p},s}(x) = \delta(\boldsymbol{p}'-\boldsymbol{p})\,\delta_{s's}\,, \\ &\int \mathrm{d}^3x\,\psi^{(-)\dagger}_{\boldsymbol{p}',s'}(x)\,\psi^{(-)}_{\boldsymbol{p},s}(x) = \delta(\boldsymbol{p}'-\boldsymbol{p})\,\delta_{s's}\,, \\ &\int \mathrm{d}^3x\,\psi^{(+)\dagger}_{\boldsymbol{p}',s'}(x)\,\psi^{(-)}_{-\boldsymbol{p},-s}(x) = \int \mathrm{d}^3x\,\psi^{(-)\dagger}_{-\boldsymbol{p}',-s'}(x)\,\psi^{(+)}_{\boldsymbol{p},s}(x) = 0\,. \end{aligned} \tag{3.90}$$

The wave function $\psi^{(+)}_{\boldsymbol{p},s}(x)$ is the solution for a positive-energy state of momentum $\boldsymbol{p}$ and polarization $s$, whereas $\psi^{(-)}_{-\boldsymbol{p},-s}(x)$ is the wave function for a negative-energy state of momentum $-\boldsymbol{p}$ and polarization $-s$, which is however more conveniently reinterpreted as describing a state of an antiparticle of positive energy, momentum $\boldsymbol{p}$, and polarization $s$. Note that, apart from the presence of the spin, the situation is exactly the same as in the case of the solutions to the Klein–Gordon equation and therefore, this reinterpretation can be similarly justified.

The formalism of the classical fields discussed in the previous section is converted into a quantum field theory by simply treating the Dirac field $\psi$ as a quantum operator. The fields $\psi$ and $\overline{\psi}$ are expanded over the complete set of eigenspinors $\psi^{(\pm)}_{\boldsymbol{p},s}$:

$$\begin{aligned} \psi(x) &= \sum_{\boldsymbol{p},s} \left[\psi^{(+)}_{\boldsymbol{p},s}(x)\, b(\boldsymbol{p},s) + \psi^{(-)}_{-\boldsymbol{p},-s}(x)\, d^\dagger(\boldsymbol{p},s)\right] , \\ \overline{\psi}(x) &= \sum_{\boldsymbol{p},s} \left[\overline{\psi}^{(+)}_{\boldsymbol{p},s}(x)\, b^\dagger(\boldsymbol{p},s) + \overline{\psi}^{(-)}_{-\boldsymbol{p},-s}(x)\, d(\boldsymbol{p},s)\right] \end{aligned} \tag{3.91}$$

(where $\sum_{\boldsymbol{p}} = \int \mathrm{d}^3p$), and the expansion coefficients are treated as operators of creation and destruction:

$b(\boldsymbol{p},s)$ destroys a particle of momentum $\boldsymbol{p}$ and polarization $s$,
$d(\boldsymbol{p},s)$ destroys an antiparticle of momentum $\boldsymbol{p}$ and polarization $s$,
$b^\dagger(\boldsymbol{p},s)$ creates a particle of momentum $\boldsymbol{p}$ and polarization $s$,
$d^\dagger(\boldsymbol{p},s)$ creates an antiparticle of momentum $\boldsymbol{p}$ and polarization $s$.

### 3.5.1 Spins and Statistics

The creation and annihilation operators applied on the ground state, the vacuum $|0\rangle$, produce states of one or several particles called the Fock states. Thus, for example, the state of one particle of momentum $p$ (suppressing spin for the moment) is given by

$$|\boldsymbol{p}\rangle = C^{-1}_{\boldsymbol{p}} b^\dagger_{\boldsymbol{p}}\, |0\rangle , \tag{3.92}$$

and the state of two identical particles of momenta $p$ and $p'$ is given by

$$|\boldsymbol{p},\boldsymbol{p}'\rangle = C^{-1}_{\boldsymbol{p}'} C^{-1}_{\boldsymbol{p}} b^\dagger_{\boldsymbol{p}'} b^\dagger_{\boldsymbol{p}}\, |0\rangle . \tag{3.93}$$

The probability for finding two particles of the same kind of momenta $p$ and $p'$ in an arbitrary physical state $\Psi$ is $|\langle \Psi | \boldsymbol{p},\boldsymbol{p}'\rangle|^2$. As the two particles are identical, they cannot be distinguished by any experiment; all that can be said is that one of them has momentum $p$ and the other momentum $p'$, a statement that can be translated into the equation

$$|\langle \Psi | \boldsymbol{p},\boldsymbol{p}'\rangle|^2 = |\langle \Psi | \boldsymbol{p}',\boldsymbol{p}\rangle|^2 , \tag{3.94}$$

which implies (for some real $\phi$)

$$\langle \Psi \,|\, \boldsymbol{p}, \boldsymbol{p}' \rangle = \mathrm{e}^{i\phi} \langle \Psi \,|\, \boldsymbol{p}', \boldsymbol{p} \rangle \; .$$

Since two successive permutations of the particles must return $|\boldsymbol{p}, \boldsymbol{p}'\rangle$ to the same state, this also means

$$\langle \Psi \,|\, \boldsymbol{p}, \boldsymbol{p}' \rangle = \pm \langle \Psi \,|\, \boldsymbol{p}', \boldsymbol{p} \rangle \; , \tag{3.95}$$

or, since $\Psi$ is arbitrary,

$$|\boldsymbol{p}, \boldsymbol{p}'\rangle = \pm \; |\boldsymbol{p}', \boldsymbol{p}\rangle \; . \tag{3.96}$$

The two solutions correspond to the two possible statistics for identical quantum particles: in the Bose–Einstein statistics the Fock states are *symmetric* under a permutation of any two particles, while in the Fermi–Dirac statistics they are *antisymmetric*. Particles obeying the Bose–Einstein statistics are referred to as bosons, and those obeying the Fermi–Dirac statistics as fermions. The creation and annihilation operators for a boson field satisfy *commutation relations*, whereas those for a fermion field satisfy *anticommutation relations.*

Let us assume from now on that $b$ and $b^\dagger$ are operators for a fermion field. The anticommutation relation

$$b_{\boldsymbol{p}'} b_{\boldsymbol{p}} + b_{\boldsymbol{p}} b_{\boldsymbol{p}'} \equiv \{b_{\boldsymbol{p}'}, b_{\boldsymbol{p}}\} = 0$$

implies that $b_{\boldsymbol{p}} b_{\boldsymbol{p}} = 0$, or that two identical fermions cannot occupy the same state. The operator for the occupation number of the individual state $p$ is

$$\mathcal{N}_b = b^\dagger_{\boldsymbol{p}} b_{\boldsymbol{p}} \, .$$

It follows from the anticommutation rules that $\mathcal{N}_b(1 - \mathcal{N}_b) = 0$, which means that the number of fermions of a given kind occupying a given individual state is either 0 or 1. This is the *Pauli exclusion principle.*

There exists in quantum field theory a general theorem giving a connection between spins and statistics. It states that for a Lorentz-invariant local field theory in four-dimensional space-time admitting a unique vacuum state, the fields of integral spins are quantized as Bose–Einstein fields and the fields of half-integral spins are quantized as Fermi–Dirac fields if the microcausality condition is satisfied. A local theory means the Lagrangian density describing the theory contains fields that refer to a single space-time point. The microcausality condition means the local density operators do not interfere, that is, they commute (or anticommute) for spacelike separations. The predictions of this fundamental theorem are in perfect agreement with experimental observations.

### 3.5.2 Dirac Field Observables

From the above arguments, the Dirac field operator $\psi$ and its canonical momentum, $i\psi^\dagger$, must satisfy the following *anticommutation* (rather than commutation) quantization rules:

$$\begin{aligned} &\{\psi_a(t,\boldsymbol{x}),\, \psi_b^\dagger(t,\boldsymbol{y})\} = \delta_{ab}\delta(\boldsymbol{x}-\boldsymbol{y})\,; \\ &\{\psi_a(t,\boldsymbol{x}),\, \psi_b(t,\boldsymbol{y})\} = 0\,; \end{aligned} \tag{3.97}$$

which lead to the corresponding algebra for the associated creation and annihilation operators:

$$\begin{aligned} \{b(\boldsymbol{p}',s'),\, b^\dagger(\boldsymbol{p},s)\} &= \delta_{ss'}\delta(\boldsymbol{p}'-\boldsymbol{p})\,; \\ \{d(\boldsymbol{p}',s'),\, d^\dagger(\boldsymbol{p},s)\} &= \delta_{ss'}\delta(\boldsymbol{p}'-\boldsymbol{p})\,; \\ \{b(\boldsymbol{p}',s'),\, b(\boldsymbol{p},s)\} &= \{b(\boldsymbol{p}',s'),\, d(\boldsymbol{p},s)\} = 0\,; \\ \{d(\boldsymbol{p}',s'),\, d(\boldsymbol{p},s)\} &= \{b(\boldsymbol{p}',s'),\, d^\dagger(\boldsymbol{p},s)\} = 0\,. \end{aligned} \tag{3.98}$$

The Hamiltonian $H$ can be expressed in terms of the static operators by substituting (91) in (78) and using (89):

$$\begin{aligned} H &= \int \mathrm{d}^3x\, \psi^\dagger(x) \sum_{\boldsymbol{p},s} E_{\boldsymbol{p}} \left[\psi^{(+)}_{\boldsymbol{p},s}(x)\, b(\boldsymbol{p},s) - \psi^{(-)}_{-\boldsymbol{p},-s}(x)\, d^\dagger(\boldsymbol{p},s)\right] \\ &= \sum_{\boldsymbol{p},s} E_{\boldsymbol{p}} \left[b^\dagger(\boldsymbol{p},s)\, b(\boldsymbol{p},s) - d(\boldsymbol{p},s)\, d^\dagger(\boldsymbol{p},s)\right]\,. \end{aligned} \tag{3.99}$$

If $\psi$ and $\overline{\psi}$ were classical fields, $b$ and $d$ would be c-number coefficients, the second term in (99) would be negative, and the field energy $H$ could not be positive-definite. Therefore a classical Dirac field cannot exist. On the other hand, if $b$ and $d$ are operators that commute as in the case of the boson fields, the energy again will not be positive-definite and will not have a lower bound. The only possible way to have a positive value for the second term in (99) is to make $d(\boldsymbol{p},s)d^\dagger(\boldsymbol{p},s)$ change signs when $d$ and $d^\dagger$ are interchanged, that is, to require that $d$ and $d^\dagger$ (and by extension $b$ and $b^\dagger$) obey the anticommutation relations (98). The Hamiltonian operator is then given by

$$H = \sum_{\boldsymbol{p},s} E_{\boldsymbol{p}}\, [b^\dagger(\boldsymbol{p},s)\, b(\boldsymbol{p},s) + d^\dagger(\boldsymbol{p},s)\, d(\boldsymbol{p},s)\,]\,. \tag{3.100}$$

As in the boson field case, an additive constant, interpreted as the vacuum energy, has been dropped. The total energy of the field appears then as a sum of positive-energy contributions from all different modes of fermions and antifermions.

The procedure leading to (100) can be summarized by the formula

$$\begin{aligned} H &= \int \mathrm{d}^3x\, :\, \overline{\psi} i\gamma^0 \partial_0 \psi\, : \\ &= \sum_{\boldsymbol{p},s} E_{\boldsymbol{p}} \left[b^\dagger(\boldsymbol{p},s)\, b(\boldsymbol{p},s) + d^\dagger(\boldsymbol{p},s)\, d(\boldsymbol{p},s)\right]\,. \end{aligned} \tag{3.101}$$

It consists in writing the products of the creation and annihilation operators in the *normal order* (symbolized by : :) by reordering the factors such that the creation operators are to the left of the destruction operators taking into account all sign changes arising from permutations of operators in accordance with their statistics. The final additive constant term, independent of operators, which results from these operations, is identified with the vacuum expectation value $\langle 0 | H | 0 \rangle$ and dropped. In what follows, the normal order of field products in the expressions for observables is always assumed, even though the notation : : may not be used explicitly.

The field momentum (77) can be similarly calculated by noting that $\nabla \psi^{(+)}_{\boldsymbol{p},s}(x) = \mathrm{i}\boldsymbol{p}\, \psi^{(+)}_{\boldsymbol{p},s}(x)$ and $\nabla \psi^{(-)}_{-\boldsymbol{p},-s}(x) = -\mathrm{i}\boldsymbol{p}\, \psi^{(-)}_{-\boldsymbol{p},-s}(x)$ and using the orthogonality properties of the basis functions:

$$\begin{aligned} \boldsymbol{P} &= -\mathrm{i} \int \mathrm{d}^3x \, : \psi^\dagger(x) \, \nabla \psi(x) : \\ &= \sum_{\boldsymbol{p},s} \boldsymbol{p} : \left[ b^\dagger(\boldsymbol{p}, s)\, b(\boldsymbol{p}, s) - d(\boldsymbol{p}, s)\, d^\dagger(\boldsymbol{p}, s) \right] : \\ &= \sum_{\boldsymbol{p},s} \boldsymbol{p} \left[ b^\dagger(\boldsymbol{p}, s)\, b(\boldsymbol{p}, s) + d^\dagger(\boldsymbol{p}, s)\, d(\boldsymbol{p}, s) \right] . \end{aligned} \tag{3.102}$$

Finally, the charge operator (88) is found to be

$$\begin{aligned} Q &= \int \mathrm{d}^3x \, : \psi^\dagger(x) \, \psi(x) : \\ &= \sum_{\boldsymbol{p},s} : \left[ b^\dagger(\boldsymbol{p}, s)\, b(\boldsymbol{p}, s) + d(\boldsymbol{p}, s)\, d^\dagger(\boldsymbol{p}, s) \right] : \\ &= \sum_{\boldsymbol{p},s} \left[ b^\dagger(\boldsymbol{p}, s)\, b(\boldsymbol{p}, s) - d^\dagger(\boldsymbol{p}, s)\, d(\boldsymbol{p}, s) \right] . \end{aligned} \tag{3.103}$$

### 3.5.3 Fock Space

To gain a better physical understanding of the formalism, let us study the observables for states in the Fock space and in particular for the one-fermion or one-antifermion states

$$|p, s\rangle = C_{\boldsymbol{p}}^{-1} b^\dagger(\boldsymbol{p}, s)\, |0\rangle \, , \qquad |\overline{p, s}\rangle = C_{\boldsymbol{p}}^{-1} d^\dagger(\boldsymbol{p}, s)\, |0\rangle \, . \tag{3.104}$$

First note the identity valid for any three arbitrary operators

$$[AB, C] = A\{B, C\} - \{A, C\}B \, . \tag{3.105}$$

The operator algebra (98) and the above expressions for $H$, $\boldsymbol{P}$, and $Q$ can be used to derive the following relations:

$$\begin{aligned} [H, b^\dagger(\boldsymbol{p}, s)] &= E_{\boldsymbol{p}} b^\dagger(\boldsymbol{p}, s) \, , & [H, d^\dagger(\boldsymbol{p}, s)] &= E_{\boldsymbol{p}} d^\dagger(\boldsymbol{p}, s) \, , \\ [\boldsymbol{P}, b^\dagger(\boldsymbol{p}, s)] &= \boldsymbol{p}\, b^\dagger(\boldsymbol{p}, s) \, , & [\boldsymbol{P}, d^\dagger(\boldsymbol{p}, s)] &= \boldsymbol{p}\, d^\dagger(\boldsymbol{p}, s) \, , \\ [Q, b^\dagger(\boldsymbol{p}, s)] &= b^\dagger(\boldsymbol{p}, s) \, , & [Q, d^\dagger(\boldsymbol{p}, s)] &= -d^\dagger(\boldsymbol{p}, s) \, . \end{aligned}$$

By taking their Hermitian conjugates while recalling the hermiticity of the operators $H$, $\boldsymbol{P}$, and $Q$, one obtains similar equations involving $b(\boldsymbol{p}, s)$ or $d(\boldsymbol{p}, s)$ in place of $b^\dagger(\boldsymbol{p}, s)$ or $d^\dagger(\boldsymbol{p}, s)$. Then the energies, momenta, and charges for one-particle states are given by

$$\begin{aligned}
H\ |p, s\rangle &= [H, b^\dagger(\boldsymbol{p}, s)]\ |0\rangle\ C_{\boldsymbol{p}}^{-1} = E_{\boldsymbol{p}}\ |p, s\rangle\ ,\\
\boldsymbol{P}\ |p, s\rangle &= [\boldsymbol{P}, b^\dagger(\boldsymbol{p}, s)]\ |0\rangle\ C_{\boldsymbol{p}}^{-1} = \ \boldsymbol{p}\ |p, s\rangle\ ,\\
Q\ |p, s\rangle &= [Q, b^\dagger(\boldsymbol{p}, s)]\ |0\rangle\ C_{\boldsymbol{p}}^{-1} = \ |p, s\rangle\ ,
\end{aligned}$$

and for one-antiparticle states by

$$\begin{aligned}
H\ |\overline{p, s}\rangle &= [H, d^\dagger(\boldsymbol{p}, s)]\ |0\rangle\ C_{\boldsymbol{p}}^{-1} = E_{\boldsymbol{p}}\ |\overline{p, s}\rangle\ ,\\
\boldsymbol{P}\ |\overline{p, s}\rangle &= [\boldsymbol{P}, d^\dagger(\boldsymbol{p}, s)]\ |0\rangle\ C_{\boldsymbol{p}}^{-1} = \ \boldsymbol{p}\ |\overline{p, s}\rangle\ ,\\
Q\ |\overline{p, s}\rangle &= [Q, d^\dagger(\boldsymbol{p}, s)]\ |0\rangle\ C_{\boldsymbol{p}}^{-1} = -\ |\overline{p, s}\rangle\ .
\end{aligned}$$

It is now clear that $b^\dagger(\boldsymbol{p}, s)$ increases the energy of the system by $E_{\boldsymbol{p}}$, its momentum by $\boldsymbol{p}$, and its charge by a unit of charge, while $d^\dagger(\boldsymbol{p}, s)$ also increases the energy of the system by $E_{\boldsymbol{p}}$, its momentum by $\boldsymbol{p}$, but reduces its charge by a unit of charge. One can similarly show that $b(\boldsymbol{p}, s)$ and $d(\boldsymbol{p}, s)$ both reduce the energy and momentum of the system by $E_{\boldsymbol{p}}$ and $\boldsymbol{p}$, but while $b(\boldsymbol{p}, s)$ reduces the charge by one unit, $d(\boldsymbol{p}, s)$ increases it by the same amount. In other words, $b^\dagger(\boldsymbol{p}, s)$ creates and $b(\boldsymbol{p}, s)$ destroys a particle of energy $E_{\boldsymbol{p}}$, momentum $\boldsymbol{p}$, and of unit charge, whereas $d^\dagger(\boldsymbol{p}, s)$ creates and $d(\boldsymbol{p}, s)$ destroys an antiparticle of energy $E_{\boldsymbol{p}}$, momentum $\boldsymbol{p}$, and of charge equal in magnitude but opposite in sign to the unit charge. Since $E_{\boldsymbol{p}}^2 - \boldsymbol{p}^2 = m^2$ in both cases, a particle and its conjugate antiparticle that are associated with the same field operator have equal masses.

The polarization states can be understood as follows. The intrinsic part of the angular momentum tensor, given by (84)

$$S_{ij} = \tfrac{1}{2} \int \mathrm{d}^3x\, \psi^\dagger \sigma_{ij} \psi\, , \tag{3.106}$$

leads to the definition for the spin operator

$$S^i = \mathrm{i}\epsilon^{ijk} S_{jk}\, . \tag{3.107}$$

To simplify we consider just its third component

$$\begin{aligned}
S^3 \equiv S_z = \sum_{\boldsymbol{p},s,s'} (2\pi)^3 C_{\boldsymbol{p}}^2 \Big[ & u^\dagger(\boldsymbol{p}, s')\tfrac{1}{2}\, \Sigma_z u(\boldsymbol{p}, s)\, b^\dagger(\boldsymbol{p}, s') b(\boldsymbol{p}, s) \\
& - v^\dagger(\boldsymbol{p}, s')\tfrac{1}{2}\, \Sigma_z v(\boldsymbol{p}, s)\, d^\dagger(\boldsymbol{p}, s) d(\boldsymbol{p}, s') \\
& + u^\dagger(\boldsymbol{p}, s')\tfrac{1}{2}\, \Sigma_z v(-\boldsymbol{p}, s)\, b^\dagger(\boldsymbol{p}, s') d^\dagger(-\boldsymbol{p}, s)\, \mathrm{e}^{2\mathrm{i}Et} \\
& + v^\dagger(-\boldsymbol{p}, s)\tfrac{1}{2}\, \Sigma_z u(\boldsymbol{p}, s')\, d(-\boldsymbol{p}, s) b(\boldsymbol{p}, s')\, \mathrm{e}^{-2\mathrm{i}Et} \Big]\, .
\end{aligned}$$

By definition $b(\boldsymbol{p})\,|0\rangle = d(\boldsymbol{p})\,|0\rangle = 0$, and by rotational invariance $S_z\,|0\rangle = 0$. Application of $S_z$ on a one-particle state and a one-antiparticle state yields

$$\begin{aligned} S_z b^\dagger(\boldsymbol{k},r)\,|0\rangle &= \left[S_z, b^\dagger(\boldsymbol{k},r)\right]\,|0\rangle \\ &= +(2\pi)^3 C_{\boldsymbol{k}}^2 \sum_s u^\dagger(\boldsymbol{k},s)\tfrac{1}{2}\,\Sigma_z u(\boldsymbol{k},r) b^\dagger(\boldsymbol{k},s)\,|0\rangle\,, \end{aligned} \tag{3.108}$$

$$\begin{aligned} S_z d^\dagger(\boldsymbol{k},r)\,|0\rangle &= \left[S_z, d^\dagger(\boldsymbol{k},r)\right]\,|0\rangle \\ &= -(2\pi)^3 C_{\boldsymbol{k}}^2 \sum_s v^\dagger(\boldsymbol{k},r)\tfrac{1}{2}\,\Sigma_z v(\boldsymbol{k},s) d^\dagger(\boldsymbol{k},s)\,|0\rangle\,. \end{aligned} \tag{3.109}$$

If the $z$ axis is chosen in the same direction as the momentum vector $\hat{\boldsymbol{k}} = \boldsymbol{k}/|\boldsymbol{k}|$, then $\frac{1}{2}\,\Sigma_z = \frac{1}{2}\boldsymbol{\Sigma}\cdot\hat{\boldsymbol{k}}$ represents the helicity. In the rest frame, where $\boldsymbol{k} = 0$, it is then convenient to choose the spinors $u(0,r)$ and $v(0,r)$ as eigenspinors of $\frac{1}{2}\,\Sigma_z$ with eigenvalues $+1/2$ and $-1/2$ for $r = 1, 2$. In a general frame where $\boldsymbol{k} = |\boldsymbol{k}|\hat{z} \neq 0$, the spinors $u(\boldsymbol{k},r)$ and $v(\boldsymbol{k},r)$ remain eigenspinors of $\frac{1}{2}\,\Sigma_z$ with eigenvalues $\lambda_r$ ($\lambda_1 = +1/2$, $\lambda_2 = -1/2$):

$$\tfrac{1}{2}\,\boldsymbol{\Sigma}\cdot\hat{\boldsymbol{k}}\,u(\boldsymbol{k},r) = \lambda_r u(\boldsymbol{k},r)\,, \qquad \tfrac{1}{2}\,\boldsymbol{\Sigma}\cdot\hat{\boldsymbol{k}}\,v(\boldsymbol{k},r) = \lambda_r v(\boldsymbol{k},r)\,.$$

Using the normalization (51) and (52), we immediately obtain

$$S_z b^\dagger(\boldsymbol{k},r)\,|0\rangle = \lambda_r\, b^\dagger(\boldsymbol{k},r)\,|0\rangle\,, \tag{3.110}$$

$$S_z d^\dagger(\boldsymbol{k},r)\,|0\rangle = -\lambda_r\, d^\dagger(\boldsymbol{k},r)\,|0\rangle\,. \tag{3.111}$$

These results tell us that $b^\dagger(\boldsymbol{k},1)$ and $d^\dagger(\boldsymbol{k},2)$ create states of helicity $+1/2$, while $b^\dagger(\boldsymbol{k},2)$ and $d^\dagger(\boldsymbol{k},1)$ create states of helicity $-1/2$.

Just as for the boson fields, the quantum fields $\psi(x)$ and $\overline{\psi}(x)$ are related to the c-valued wave functions $\psi^{(+)}_{\boldsymbol{p},s}(x)$ and $\psi^{(-)}_{-\boldsymbol{p},-s}(x)$. For example, $\psi^{(+)}_{\boldsymbol{p},s}(x)$ can be interpreted as the annihilation amplitude of a particle at point $x$, $\langle 0\,|\,\psi(x)\,|\,ps\rangle$, and the spinor $u(p,s)$ is associated with an incoming fermion of momentum $\boldsymbol{p}$ and polarization $s$. Similarly, $\psi^{(-)}_{-\boldsymbol{p},-s}(x)$ represents the creation amplitude of an antiparticle at $x$, and the spinor $v(p,s)$ is associated with an outgoing antifermion of momentum $p$ and polarization $s$. To summarize,

| | | |
|---|---|---|
| $\langle 0\vert\psi(x)\vert ps\rangle = u(p,s)\,\mathrm{e}^{-\mathrm{i}p\cdot x}$ | →• | annihilation of a particle, |
| $\langle ps\vert\overline{\psi}(x)\vert 0\rangle = \bar{u}(p,s)\,\mathrm{e}^{\mathrm{i}p\cdot x}$ | •→ | creation of a particle, |
| $\langle\overline{ps}\vert\psi(x)\vert 0\rangle = v(p,s)\,\mathrm{e}^{\mathrm{i}p\cdot x}$ | •← | creation of an antiparticle, |
| $\langle 0\vert\overline{\psi}(x)\vert\overline{ps}\rangle = \bar{v}(p,s)\,\mathrm{e}^{-\mathrm{i}p\cdot x}$ | ←• | annihilation of an antiparticle. |

## 3.6 Zero-Mass Fermions

When the field is massless, the Dirac theory may be formulated in terms of two-component spinors. This simplification proves to be quite useful in the study of neutrinos, which have very small masses, or of particles of non-vanishing masses at very high energies, where their masses can be neglected in comparison with their kinetic energies.

If $\psi$ is a solution to the Dirac equation

$$(\mathrm{i}\gamma^\mu\partial_\mu - m)\psi = 0\,, \tag{3.112}$$

$\gamma_5\psi$ obeys the equation

$$(\mathrm{i}\gamma^\mu\partial_\mu + m)\gamma_5\psi = 0\,. \tag{3.113}$$

The two equations are different for a nonvanishing mass. But when $m = 0$ they become identical and the spinors $\psi$ and $\gamma_5\psi$ are proportional to each other. In other words, for $m = 0$ the Dirac equation may be written as

$$\mathrm{i}\frac{\partial}{\partial t}\psi = \hat{H}\psi\,, \tag{3.114}$$

where $\hat{H} = -\mathrm{i}\gamma_0\boldsymbol{\gamma}\cdot\nabla$. Since $[\hat{H},\gamma_5] = 0$, the matrices $\hat{H}$ and $\gamma_5$ are simultaneously diagonalizable, and common eigenfunctions can be found for $\hat{H}$ and $\gamma_5$. As $\gamma_5^2 = 1$, the eigenvalues of $\gamma_5$ are $\pm 1$. An eigenspinor with eigenvalue $+1$ for $\gamma^5$ is said to have a *positive chirality*; when its eigenvalue for $\gamma^5$ is $-1$, it is said to have a *negative chirality*. We now proceed to describe these spinors.

For $m = 0$, (54) reduces to

$$\begin{aligned} \gamma_0\boldsymbol{\gamma}\cdot\boldsymbol{p}\,u(\boldsymbol{p},h) &= \;\;E\,u(\boldsymbol{p},h)\,,\\ \gamma_0\boldsymbol{\gamma}\cdot\boldsymbol{p}\,v(-\boldsymbol{p},h) &= -E\,v(-\boldsymbol{p},h)\,, \end{aligned} \tag{3.115}$$

where $E = |\boldsymbol{p}|$. As already mentioned, $u(\boldsymbol{p},h)$ and $v(\boldsymbol{p},h)$ can always be chosen as eigenspinors of the helicity operator:

$$\begin{aligned} \boldsymbol{\Sigma}\cdot\hat{\boldsymbol{p}}\,u(\boldsymbol{p},h) &= 2h\,u(\boldsymbol{p},h)\,,\\ \boldsymbol{\Sigma}\cdot\hat{\boldsymbol{p}}\,v(-\boldsymbol{p},h) &= 2h\,v(-\boldsymbol{p},h)\,. \end{aligned} \tag{3.116}$$

From the definition $\gamma_5 = \mathrm{i}\gamma^0\gamma^1\gamma^2\gamma^3$, one gets $\gamma_5\gamma^0 = -\mathrm{i}\,\gamma^1\gamma^2\gamma^3$, and so $\gamma_5\gamma^0\gamma^1 = \mathrm{i}\,\gamma^2\gamma^3 = \sigma^{23} = \Sigma^1$. In general, $\gamma_5\gamma^0\gamma^i = \Sigma^i$. Applying $\gamma_5$ from the left on both sides of (115) and using (116), one gets

$$\begin{aligned} \gamma_5\,u(\boldsymbol{p},h) &= \;\;\boldsymbol{\Sigma}\cdot\hat{\boldsymbol{p}}\,u(\boldsymbol{p},h) = \;\;2h\,u(\boldsymbol{p},h)\,,\\ \gamma_5\,v(-\boldsymbol{p},h) &= -\boldsymbol{\Sigma}\cdot\hat{\boldsymbol{p}}\,v(-\boldsymbol{p},h) = -2h\,v(-\boldsymbol{p},h)\,. \end{aligned} \tag{3.117}$$

This means in particular that for an arbitrary vector $\boldsymbol{p}$, the spinors $u(\boldsymbol{p},+1/2)$ and $v(\boldsymbol{p},-1/2)$ are eigenspinors of $\gamma^5$ of positive chirality $+1$, while $u(\boldsymbol{p},-1/2)$ and $v(\boldsymbol{p},+1/2)$ have negative chirality $-1$. Thus, for a zero-mass particle, chirality and helicity are equivalent and are Lorentz-invariant. On the other hand, for a particle with a nonvanishing mass, chirality is not well defined, but states of such a particle can still be identified by their helicities. As it is the scalar product of two three-vectors, $\boldsymbol{\Sigma}\cdot\hat{\boldsymbol{p}}$, the helicity is invariant to spatial rotations, which makes a description of states in terms of helicities often useful. However, a general Lorentz transformation will mix up states of different helicities. For example, a particle of nonvanishing mass with a positive helicity in a given inertial frame will have a negative helicity in a frame in which its direction of motion is reversed. Thus, a Dirac particle with a nonzero mass must occur in both helicity states.

Any spinor $\psi$, massive or not, can be decomposed into two components of well-defined chiralities, called the *Weyl spinors*,

$$\begin{aligned}
&\psi = \psi_{\rm R} + \psi_{\rm L}\,, \\
&\psi_{\rm R} = \tfrac{1}{2}\,(1+\gamma_5)\psi, \quad \gamma_5\psi_{\rm R} = +\psi_{\rm R}, \\
&\psi_{\rm L} = \tfrac{1}{2}\,(1-\gamma_5)\psi, \quad \gamma_5\psi_{\rm L} = -\psi_{\rm L}\,.
\end{aligned} \tag{3.118}$$

From (91) and (117), their Fourier series can be written as

$$\begin{aligned}
\psi_{\rm R} &= \sum_{\boldsymbol{p}} \left[\, \psi^{(+)}_{\boldsymbol{p},\,1/2}\, b(\boldsymbol{p},\,1/2) \;+ \psi^{(-)}_{-\boldsymbol{p},\,1/2}\, d^\dagger(\boldsymbol{p},-1/2)\right], \\
\psi_{\rm L} &= \sum_{\boldsymbol{p}} \left[\psi^{(+)}_{\boldsymbol{p},-1/2}\, b(\boldsymbol{p},-1/2) + \psi^{(-)}_{-\boldsymbol{p},-1/2}\, d^\dagger(\boldsymbol{p},\,1/2)\right].
\end{aligned} \tag{3.119}$$

According to (119), $\psi_{\rm R}$ destroys positive-helicity states of particle and creates negative-helicity states of antiparticle, whereas $\psi_{\rm L}$ destroys $h = -1/2$ states of particle and creates $h = +1/2$ states of antiparticle. Table 3.2 summarizes these results.

For a zero-mass particle, chirality is well defined and Lorentz-invariant, and so $\psi$ can exist either as a left-handed state, $\psi_{\rm L}$, or as a right-handed state, $\psi_{\rm R}$. For a given momentum $\boldsymbol{p}$, a massless particle can have its spin oriented parallel or antiparallel to its direction of motion, and each state can be described by a two-component spinor. Indeed, when $m = 0$, the Dirac Hamiltonian $\hat{H} = -{\rm i}\gamma_0\boldsymbol{\gamma}\cdot\nabla$ involves only three matrices, namely, $\gamma_0\gamma_i$ for $i = 1, 2, 3$, which satisfy the algebra

$$\{\gamma_0\gamma_i,\, \gamma_0\gamma_j\} = 2\,\delta_{ij}, \quad i, j = 1, 2, 3\,. \tag{3.120}$$

These relations can be satisfied by the $2\times 2$ Pauli matrices, obviating the need for a formulation in terms of four-component spinors. The situation becomes particularly transparent when $\gamma_5$ is diagonal, as in the Weyl or *chiral representation* of the $\gamma_\mu$ matrices defined by

$$\gamma^0 = \begin{pmatrix} 0 & -1 \\ -1 & 0 \end{pmatrix}, \quad \gamma^i = \begin{pmatrix} 0 & \sigma^i \\ -\sigma^i & 0 \end{pmatrix}, \quad \gamma_5 = \begin{pmatrix} 1 & 0 \\ 0 & -1 \end{pmatrix}. \tag{3.121}$$

In this representation the Weyl spinors of chiralities $\gamma_5 = +1$ and $\gamma_5 = -1$ take respectively the forms

$$\psi_{\rm R} = \begin{pmatrix} \chi_{\rm R} \\ 0 \end{pmatrix} \quad \text{and} \quad \psi_{\rm L} = \begin{pmatrix} 0 \\ \chi_{\rm L} \end{pmatrix}, \tag{3.122}$$

**Table 3.2.** Chirality and helicity

| Spinors | | Chirality $\gamma_5$ | Helicity $h$ |
|---|---|---|---|
| $\psi^{(+)}_{\boldsymbol{p},1/2}(x)$ | $u(\boldsymbol{p},+1/2)$ | $+1$ | $1/2$ |
| $\psi^{(-)}_{-\boldsymbol{p},1/2}(x)$ | $v(\boldsymbol{p},-1/2)$ | $+1$ | $-1/2$ |
| $\psi^{(+)}_{\boldsymbol{p},-1/2}(x)$ | $u(\boldsymbol{p},-1/2)$ | $-1$ | $-1/2$ |
| $\psi^{(-)}_{-\boldsymbol{p},-1/2}(x)$ | $v(\boldsymbol{p},+1/2)$ | $-1$ | $1/2$ |

where $\chi_{\mathrm{R}}$ and $\chi_{\mathrm{L}}$ are two-component spinors. Dirac's equation then becomes a system of two uncoupled equations for two-component spinors:

$$\begin{aligned} \mathrm{i}\partial_0\chi_{\mathrm{R}} &= -\mathrm{i}\boldsymbol{\sigma}\cdot\nabla\chi_{\mathrm{R}}\,, \\ \mathrm{i}\partial_0\chi_{\mathrm{L}} &= \phantom{-}\mathrm{i}\boldsymbol{\sigma}\cdot\nabla\chi_{\mathrm{L}}\,. \end{aligned} \tag{3.123}$$

The Lagrangian for a Dirac particle with mass $m \neq 0$, given by (75), may be rewritten in terms of Weyl spinors:

$$\mathcal{L} = \overline{\psi}_{\mathrm{R}}\,\mathrm{i}\gamma^\mu\partial_\mu\,\psi_{\mathrm{R}} + \overline{\psi}_{\mathrm{L}}\,\mathrm{i}\gamma^\mu\partial_\mu\,\psi_{\mathrm{L}} - m(\overline{\psi}_{\mathrm{R}}\,\psi_{\mathrm{L}} + \overline{\psi}_{\mathrm{L}}\,\psi_{\mathrm{R}})\,. \tag{3.124}$$

The components of opposite helicities are connected in the mass term and both therefore necessarily appear for a fermion of nonvanishing mass. However, when $m = 0$, the Lagrangian breaks up into two independent parts, one for each chirality,

$$\mathcal{L} = \chi^\dagger_{\mathrm{R}}\,\mathrm{i}\sigma^\mu\partial_\mu\chi_{\mathrm{R}} + \chi^\dagger_{\mathrm{L}}\,\mathrm{i}\bar{\sigma}^\mu\partial_\mu\chi_{\mathrm{L}}\,, \tag{3.125}$$

where

$$\sigma^\mu = (1,\boldsymbol{\sigma})\,, \qquad \bar{\sigma}^\mu = (1,-\boldsymbol{\sigma})\,. \tag{3.126}$$

This decomposition into left- and right-handed states not only simplifies the formalism but also turns out to be a necessity because it is now known that only left-handed neutrinos and right-handed antineutrinos exist and they can be described naturally in terms of the Weyl spinors. Right-handed neutrinos, even if they exist, are not observed in weak interaction reactions, are not coupled to known particles, and cannot acquire mass through interactions. Therefore, models of weak interactions will involve only left-handed neutrinos and right-handed antineutrinos.

## Problems

**3.1 Boosting a fermion from rest.** The Dirac spinor for a free particle of momentum $\boldsymbol{p}$ can be obtained from the corresponding solution for $\boldsymbol{p}=0$ by a Lorentz boost. As an example, calculate $u(\boldsymbol{p},s)=S_{\rm L}(\omega)\sqrt{2m}\,u(\boldsymbol{0},s)$, where $\omega_{\mu\nu}$ give the boost parameters.

**3.2 Γ matrices.** (a) Prove that $\Gamma_i$, $i=$S, V, T, A, P, satisfy the conjugation property, $\gamma_0\Gamma_i^\dagger\gamma_0=\Gamma_i$, and produce 16 linearly independent matrices. (b) Show that two sets $\gamma_\mu$ and $\gamma'_\mu$ satisfying the relation $\gamma_\mu\gamma_\nu+\gamma_\nu\gamma_\mu=2g_{\mu\nu}$ are related by $\gamma'_\mu=S\gamma_\mu S^{-1}$ for some $4\times 4$ matrix $S$ [for help, consult Good, R. H., Rev. Mod. Phys. **27** (1955) 187].

**3.3 Bilinear covariants.** Prove that the bilinear covariants given in Table 3.1 are Hermitian, and satisfy the Lorentz transformation properties shown in the table.

**3.4 Majorana and Weyl representations.** (a) Find the matrices $S$ that transform the Pauli–Dirac standard representation of the $\gamma$-matrices into their Majorana and the Weyl representations:

$$\begin{aligned}\gamma_\mu^{(\rm M)}&=S_{\rm M}\gamma_\mu S_{\rm M}^{-1}\,,\\ \gamma_\mu^{(\rm W)}&=S_{\rm W}\gamma_\mu S_{\rm W}^{-1}\,,\end{aligned}$$

where the Majorana representation is defined by

$$\gamma^1=\begin{pmatrix}\mathrm{i}\sigma^3&0\\0&\mathrm{i}\sigma^3\end{pmatrix},\quad \gamma^2=\begin{pmatrix}0&-\sigma^2\\\sigma^2&0\end{pmatrix},\quad \gamma^3=\begin{pmatrix}-\mathrm{i}\sigma^1&0\\0&-\mathrm{i}\sigma^1\end{pmatrix},$$
$$\gamma^0=\begin{pmatrix}0&\sigma^2\\\sigma^2&0\end{pmatrix},\quad \gamma_5=\begin{pmatrix}\sigma^2&0\\0&-\sigma^2\end{pmatrix};$$

and the Weyl representation by

$$\gamma^0=\begin{pmatrix}0&-1\\-1&0\end{pmatrix},\quad \gamma^i=\begin{pmatrix}0&\sigma^i\\-\sigma^i&0\end{pmatrix},\quad \gamma_5=\begin{pmatrix}1&0\\0&-1\end{pmatrix}.$$

(b) Find the analogs of the spinors (45) and (46) in the Majorana and Weyl representations of the $\gamma_\mu$ matrices.

**3.5 Orthogonality of spinors.** Prove the following relations:
(a) $v^\dagger(\boldsymbol{p},s)u(\boldsymbol{p},s')=u^\dagger(\boldsymbol{p},s)v(\boldsymbol{p},s')=\delta_{ss'}\,2\boldsymbol{\sigma}\cdot\boldsymbol{p}/(E+m)\,,$
(b) $v^\dagger(-\boldsymbol{p},s)u(\boldsymbol{p},s)=0.$

**3.6 Closure relation.** Let $\psi(\boldsymbol{p})$ be an arbitrary spinor of momentum $p_\mu$ given by the sum

$$\psi(\boldsymbol{p})=\sum_s\left[A_s u(\boldsymbol{p},s)+B_s v(-\boldsymbol{p},s)\right].$$

Calculate the coefficients $A_s$ and $B_s$, and prove the closure relation (55).

**3.7 Gordon identities.** Let $\psi_1$ and $\psi_2$ be solutions to the Dirac equations $(\mathrm{i}\gamma^\mu\partial_\mu - m_i)\psi_i(x) = 0$. Prove the following relations:

$$(m_1+m_2)\psi_2\gamma_\mu\psi_1 = \overline{\psi}_2\left(-\mathrm{i}\overleftarrow{\partial}_\mu + \mathrm{i}\overrightarrow{\partial}_\mu\right)\psi_1 + \partial^\nu\left(\overline{\psi}_2\sigma_{\mu\nu}\psi_1\right),$$

$$(m_1+m_2)\psi_2\gamma_\mu\gamma_5\psi_1 = (-\mathrm{i}\partial_\mu)\left(\overline{\psi}_2\gamma_5\psi_1\right) + \overline{\psi}_2\left(-\mathrm{i}\overleftarrow{\partial}_\nu + \mathrm{i}\overrightarrow{\partial}_\nu\right)\sigma_\mu{}^\nu\gamma_5\psi_1 .$$

**3.8 Fierz transformation.** As the 16 $\Gamma^i$ matrices in Problem 3.2, with $i$ =S, V, T, A, P, form a complete set of $N = 4$ matrices, any product of bilinear covariants of the form $(\bar{u}_1\Gamma^i u_2)(\bar{u}_3\Gamma^j u_4)$ can be expressed as a linear combination of similar products written with a different sequence of spinors

$$(\bar{u}_1\Gamma^i u_2)(\bar{u}_3\Gamma^j u_4) = \sum_{mn} C^{ij}_{mn}(\bar{u}_1\Gamma^m u_4)(\bar{u}_3\Gamma^n u_2).$$

In general, the spinors $u_i$ refer to different particles. Show that

$$C^{ij}_{mn} = \frac{1}{N_m N_n}\mathrm{Tr}\,(\Gamma^i\Gamma^n\Gamma^j\Gamma^m).$$

$\Gamma^i$ are assumed to be orthonormalized such that $\mathrm{Tr}\,(\Gamma^i\Gamma^j) = N_i\delta_{ij}$.

**3.9 The Dirac Lagrangian.** Show that the Lagrangian

$$\mathcal{L}_F = \overline{\psi}\left[\tfrac{1}{2}\mathrm{i}\gamma^\mu\overleftrightarrow{\partial}_\mu - m\right]\psi$$

differs from (75) only by a total derivative and therefore leads to the same equation of motion.

**3.10 Spin of hyperon Λ.** $\Lambda^0$ particles are produced in the reaction $\pi^- + \mathrm{p} \to \mathrm{K}^0 + \Lambda^0$ and are identified by their subsequent disintegrations $\Lambda \to \pi^- + \mathrm{p}$. Assuming that the proton spin ($1/2$) and the $\pi$, K spins (0) are known, we want to determine the spin of $\Lambda^0$. (a) For $\Lambda$ produced in the direction of the incoming beam (parallel to the $z$ axis chosen as the axis of quantization), what are the possible values of $S_z$, the $z$ component of the $\Lambda$ spin? (b) For polarized protons, show that in the $\Lambda$ rest frame the angular distributions of $\Lambda$ produced in the incident beam direction are given for different values of $\Lambda$ by

$S_\Lambda = 1/2$ isotropic,
$S_\Lambda = 3/2$ $(1 + 3\cos^2\theta)$,
$S_\Lambda = 5/2$ $(1 - 2\cos^2\theta + 5\cos^4\theta)$,

where $\theta$ is the relative angle of the $\Lambda$ disintegration products. (The observed distributions turn out to be isotropic, indicating that the $\Lambda$ spin is most likely equal to $1/2$.)

**3.11 Anticommutation relations between Dirac fields.** Let $\psi_a(x)$, with $a = 1, \ldots, 4$, be the components of the Dirac spinor of a free particle of mass $m$.
(a) Using the completeness of spinors and (98), show that for $x_0 = y_0$,

$$\{\psi_a(x), \psi_b^\dagger(y)\}|_{x^0=y^0} = \delta_{ab}\, \delta(\boldsymbol{x} - \boldsymbol{y})\,.$$

(b) Using projection operators and (98), show that for arbitrary $x$ and $y$,

$$\{\psi_a(x),\, \psi_b^\dagger(y)\} = -\mathrm{i} S_{ab}(x - y, m),$$

where

$$\begin{aligned} S(x,m) &= -(\mathrm{i}\gamma \cdot \partial + m)\, \Delta(x,m) \\ &= -(\mathrm{i}\gamma \cdot \partial + m)\, \frac{-\mathrm{i}}{(2\pi)^3} \int \mathrm{d}^4 p\, \mathrm{e}^{-\mathrm{i}p\cdot x} \delta(p^2 - m^2)\, \epsilon(p_0)\,. \end{aligned}$$

Here $\epsilon(x) = 1$ for $x \geq 0$ and $\epsilon(x) = 0$ for $x < 0$.

## Suggestions for Further Reading

*The classics:*
Dirac, P. A. M., Proc. Roy. Soc. (London) **A117** (1928) 610
Majorana, E., Nuovo Cimento **14** (1937) 171
*Connections between spins and statistics:*
Lüders, C., Ann. Phys. **2** (1957) 1
Pauli, W., Phys. Rev. **58** (1940) 716
*Nonrelativistic limit of Dirac's equation:*
Foldy, L. L. and Wouthuysen, S. A., Phys. Rev. **78** (1950) 29

# 4 Collisions and Decays

In the preceding two chapters we studied freely moving particles. However, a complete absence of interactions can hardly be considered realistic, for in general, particles do interact with one another or with external fields. After all, it is precisely through these interactions that the particles are observed. In this chapter, we will develop a practical method for introducing interactions among relativistic quantum fields and for calculating experimentally measurable quantities such as reaction cross-sections and decay rates.

In dealing with processes involving relativistic particles, it is crucial to have a fully covariant description, and it turns out that such a description is feasible even within the framework of the Hamiltonian formalism. The basic quantity to consider is the amplitude of transition; one can expand it in a power series of the interaction strength, and if the interaction may be regarded as a weak perturbation, only the first few lowest-order nontrivial terms suffice to produce physical quantities. We shall assume that it is the case of the processes discussed in this chapter.

We shall start with the simple example of interacting scalar particles ($\sigma$–$\pi$ system) and progress on to the technically more difficult cases of the electromagnetic field interacting with spin-$1/2$ fermions. In each case, the transition amplitude will be derived and the result graphically interpreted in terms of *Feynman diagrams.* This graphical representation will in turn suggest empirical rules of calculation, called the *Feynman rules*, many of which turn out to have a broader range of applicability than the way they are obtained would at first indicate. Although these rules are derived for certain second-order processes, it is assumed, without proof, that they can be generalized to more complicated situations. Such a general proof is far from trivial. Nevertheless, we will have acquired by the end of the chapter a practical method for calculating the lowest nontrivial order of the transition matrix and, in particular, the rules for spinor electrodynamics. We will still be missing the rules for higher-order diagrams and the rules associated with renormalization, some of which will be discussed in later chapters.

## 4.1 Interaction Representation

We described briefly in Chap. 2 two different formulations of quantum mechanics, referred to as the Schrödinger and the Heisenberg representations. There exists yet another, called the interaction representation, which proves to be very useful in particle physics because it allows a manifestly covariant calculation of the transition matrices.

### 4.1.1 The Three Pictures

Let us first recall that in the *Schrödinger representation* the time evolution of a physical system is governed by the Schrödinger equation

$$\mathrm{i}\frac{\partial}{\partial t}\Psi_{\mathrm{S}}(t) = H_{\mathrm{S}}\,\Psi_{\mathrm{S}}(t)\,, \tag{4.1}$$

where $H_{\mathrm{S}}$ is the Hamiltonian of the system. If $H_{\mathrm{S}}$ is not time dependent, (1) can be integrated over time to give the formal solution

$$\Psi_{\mathrm{S}}(t) = U_{\mathrm{S}}(t,t_0)\,\Psi_{\mathrm{S}}(t_0)\,, \tag{4.2}$$

where the time evolution operator $U_{\mathrm{S}}(t,t_0)$ is defined by

$$U_{\mathrm{S}}(t,t_0) = \mathrm{e}^{-\mathrm{i}H_{\mathrm{S}}(t-t_0)}\,. \tag{4.3}$$

It is unitary provided the Hamiltonian is Hermitian, $H_{\mathrm{S}}^{\dagger} = H_{\mathrm{S}}$. Given a solution to (1) at time $t_0$, the operator $U_{\mathrm{S}}(t,t_0)$ generates the corresponding solution at any later time $t$. In this picture, physical observables usually have no explicit time dependence (which for simplicity we shall assume to be the case), and thus the state vectors carry the entire time dependence.

In contrast, the state vectors in the *Heisenberg representation* are unchanging in time,

$$\mathrm{i}\frac{\partial}{\partial t}\Psi_{\mathrm{H}}(t) = 0\,. \tag{4.4}$$

It is clear from (1) and (4) that $\Psi_{\mathrm{S}}(t)$ and $\Psi_{\mathrm{H}}(t)$ are related by a unitary transformation

$$\Psi_{\mathrm{H}}(t) = \mathrm{e}^{\mathrm{i}H_{\mathrm{S}}t}\,\Psi_{\mathrm{S}}(t) \tag{4.5}$$

[in fact $\Psi_{\mathrm{H}}(t) = \Psi_{\mathrm{S}}(0)$], such that the matrix element of an arbitrary operator remains unchanged in passing from one picture to the other:

$$\langle\,\Psi_{\mathrm{H}}(b;t)\,|\,A_{\mathrm{H}}(t)\,|\,\Psi_{\mathrm{H}}(a;t)\rangle = \langle\,\Psi_{\mathrm{S}}(b;t)\,|\,A_{\mathrm{S}}\,|\,\Psi_{\mathrm{S}}(a;t)\rangle\,, \tag{4.6}$$

provided the corresponding operators are related by the transformation rule

$$A_{\mathrm{H}}(t) = \mathrm{e}^{\mathrm{i}H_{\mathrm{S}}t}A_{\mathrm{S}}\,\mathrm{e}^{-\mathrm{i}H_{\mathrm{S}}t}\,. \tag{4.7}$$

In other words, if an operator $A_S$ in the Schrödinger picture does not depend on time, the corresponding operator $A_H(t)$ in the Heisenberg picture varies with time in a nontrivial way according to the Heisenberg equation

$$\mathrm{i}\frac{\mathrm{d}}{\mathrm{d}t}A_{\mathrm{H}}(t) = [A_{\mathrm{H}}(t), H_{\mathrm{S}}]\,. \tag{4.8}$$

However, it will become *constant* in time if $A_S$ commutes with $H_S$, in which case $A_H(t) = A_S$. In particular, since $H_S$ commutes with itself, the Hamiltonians in the two representations are identical,

$$H_{\mathrm{H}}(t) = H_{\mathrm{S}}\,, \qquad \text{for all } t\,. \tag{4.9}$$

Finally, in the *interaction representation*, the time dependence of the system is shared by state vectors and observables according to the way in which the Hamiltonian of the system splits up into two terms,

$$H_{\mathrm{S}} = H_{\mathrm{S}}^{0} + H_{\mathrm{S}}'\,. \tag{4.10}$$

It is often the case that this splitting is physically motivated, for example in a two-particle collision, where $H_S^0$ may describe the motion of two particles far apart in space, as they are well before a collision, and $H_S'$ may describe how the particles interact when they come close enough to affect each other noticeably, which occurs, say, at time $t \approx 0$. But well before or well after that instant in time, the system is effectively described by $H_S^0$ alone through the equation

$$\mathrm{i}\frac{\partial}{\partial t}\Psi(t) = H_{\mathrm{S}}^{0}\,\Psi(t)\,, \qquad \text{for } t \ll 0 \quad \text{or} \quad t \gg 0\,, \tag{4.11}$$

which we know how to solve, as in (2). Thus, instead of suppressing all time dependence in the state vectors as in (5), the interaction picture lets $H_S'$ govern the time evolution of the state vectors, leaving $H_S^0$ to play the same role for the observables. In other words, the state vector $\Psi_I$ in this picture is obtained from $\Psi_S$, in analogy with (5), by the similarity transformation

$$\Psi_{\mathrm{I}}(t) = \mathrm{e}^{\mathrm{i}H_{\mathrm{S}}^{0}t}\,\Psi_{\mathrm{S}}(t)\,, \tag{4.12}$$

which, together with (1), implies

$$\mathrm{i}\frac{\partial}{\partial t}\Psi_{\mathrm{I}}(t) = H_{\mathrm{I}}'(t)\,\Psi_{\mathrm{I}}(t)\,, \tag{4.13}$$

provided the Hamiltonian $H_S'$ transforms as expected, that is,

$$H_{\mathrm{I}}'(t) = \mathrm{e}^{\mathrm{i}H_{\mathrm{S}}^{0}t}H_{\mathrm{S}}'\,\mathrm{e}^{-\mathrm{i}H_{\mathrm{S}}^{0}t}\,. \tag{4.14}$$

Of course, for scalar products to be invariant, all observables must similarly transform,

$$A_{\mathrm{I}}(t) = \mathrm{e}^{\mathrm{i}H_{\mathrm{S}}^{0}t} A_{\mathrm{S}}\, \mathrm{e}^{-\mathrm{i}H_{\mathrm{S}}^{0}t}, \tag{4.15}$$

which is equivalent to the evolution equation

$$\mathrm{i}\frac{\mathrm{d}}{\mathrm{d}t} A_{\mathrm{I}}(t) = [A_{\mathrm{I}}(t), H_{\mathrm{S}}^{0}]. \tag{4.16}$$

All three representations coincide at $t = 0$:

$$\begin{aligned} &\Psi_{\mathrm{S}}(0) = \Psi_{\mathrm{H}} = \Psi_{\mathrm{I}}(0), \\ &A_{\mathrm{S}} = A_{\mathrm{H}}(0) = A_{\mathrm{I}}(0). \end{aligned} \tag{4.17}$$

But as time evolves, they diverge. While in the Schrödinger representation the state vectors carry the entire time dependence, they are completely time independent in the Heisenberg representation. In contrast, in the interaction representation, the time evolution of the system is described by both state vectors and operators. It is this characteristic that makes the latter formulation relevant to quantum field theory because in the interaction representation, the field operator satisfies the free-field equation of motion (16) which involves only $H_{\mathrm{S}}^{0}$. Thus, even in the presence of an interaction, the field operator is completely known, being a linear combination of free-field solutions. The nontrivial time dependence of the system is contained in the state vectors and is determined by the interaction Hamiltonian $H'_{\mathrm{I}}$, which differs from the corresponding Hamiltonian in the Schrödinger picture $H'_{\mathrm{S}}$ in that the free-field operators vary in time. It is to this essential time dependence of the state vectors that we now turn our attention.

### 4.1.2 Time Evolution in the Interaction Picture

Just as in the Schrödinger picture where a solution at some time $t$ can be obtained from the corresponding solution at an earlier time $t_0$ by a time evolution operator $U_{\mathrm{S}}(t, t_0)$, so too can a state vector $\Psi_{\mathrm{I}}(t)$ in the interaction picture that satisfies (13) be similarly deduced from the corresponding state vector $\Psi_{\mathrm{I}}(t_0)$:

$$\Psi_{\mathrm{I}}(t) = U(t, t_0)\, \Psi_{\mathrm{I}}(t_0), \quad \text{such that} \quad U(t, t) = 1. \tag{4.18}$$

In terms of this time evolution operator, (13) is equivalent to

$$\mathrm{i}\frac{\mathrm{d}}{\mathrm{d}t} U(t, t_0) = H'_{\mathrm{I}}(t)\, U(t, t_0). \tag{4.19}$$

From (2) and (12),

$$\begin{aligned} \Psi_{\mathrm{I}}(t) &= \mathrm{e}^{\mathrm{i}H_{\mathrm{S}}^{0}t}\, \Psi_{\mathrm{S}}(t) \\ &= \mathrm{e}^{\mathrm{i}H_{\mathrm{S}}^{0}t}\, U_{\mathrm{S}}(t, t_0)\, \mathrm{e}^{-\mathrm{i}H_{\mathrm{S}}^{0}t_0}\, \Psi_{\mathrm{I}}(t_0), \end{aligned} \tag{4.20}$$

which gives a formal solution for $U$:

$$U(t,t_0) = \mathrm{e}^{\mathrm{i}H_{\mathrm{S}}^0 t}\, U_{\mathrm{S}}(t,t_0)\, \mathrm{e}^{-\mathrm{i}H_{\mathrm{S}}^0 t_0}\,. \tag{4.21}$$

$U(t,t_0)$ is a linear operator on the Hilbert space of state vectors. It is *unitary*

$$U^\dagger(t',t)\,U(t',t) = U(t',t)\,U^\dagger(t',t) = 1\,, \tag{4.22}$$

has the characteristic group composition property

$$U(t',t'')\,U(t'',t) = U(t',t)\,, \tag{4.23}$$

and has an inverse

$$U^{-1}(t,t') = U(t',t) = U^\dagger(t,t')\,. \tag{4.24}$$

Although an explicit formal solution for $U(t,t_0)$ is available, it proves more convenient for our purpose to have it in an implicit form as solution to an integral equation that is equivalent to the differential equation (19) subject to the boundary condition $U(t_0,t_0) = 1$,

$$U(t,t_0) = 1 - \mathrm{i}\int_{t_0}^{t} \mathrm{d}t'\, H_{\mathrm{I}}'(t')\,U(t',t_0)\,. \tag{4.25}$$

From here on, we simplify the notation by writing $H'(t)$ in place of $H_{\mathrm{I}}'(t)$. We can solve this equation by the iterative procedure, leading to a series in powers of the interaction:

$$\begin{aligned}
U(t,t_0) &= 1 - \mathrm{i}\int_{t_0}^{t} \mathrm{d}t_1\, H'(t_1)U(t_1,t_0) \\
&= 1 - \mathrm{i}\int_{t_0}^{t} \mathrm{d}t_1\, H'(t_1)\left[1 - \mathrm{i}\int_{t_0}^{t_1} \mathrm{d}t_2\, H'(t_2)\,U(t_2,t_0)\right] \\
&= 1 - \mathrm{i}\int_{t_0}^{t} \mathrm{d}t_1\, H'(t_1) + (-\mathrm{i})^2\int_{t_0}^{t} \mathrm{d}t_1 \int_{t_0}^{t_1} \mathrm{d}t_2\, H'(t_1)H'(t_2) + \dots \\
&\quad + (-\mathrm{i})^n \int_{t_0}^{t} \mathrm{d}t_1 \int_{t_0}^{t_1} \mathrm{d}t_2 \dots \int_{t_0}^{t_{n-1}} \mathrm{d}t_n\, H'(t_1)H'(t_2)\dots H'(t_n) + \dots\,.
\end{aligned} \tag{4.26}$$

Now, $U(t,t_0)$ can be interpreted as the operator that gives the probability, $|\langle \Phi_{\mathrm{f}}\,|\,U(t,t_0)\,|\,\Phi_{\mathrm{i}}\rangle|^2$, for finding the system in state f at some time $t$, if the system is known to be in state i at some earlier time $t_0$. Therefore, this series is useful in practice only if it converges rapidly enough so that only the first few terms suffice to give physically meaningful results. This will be the case if

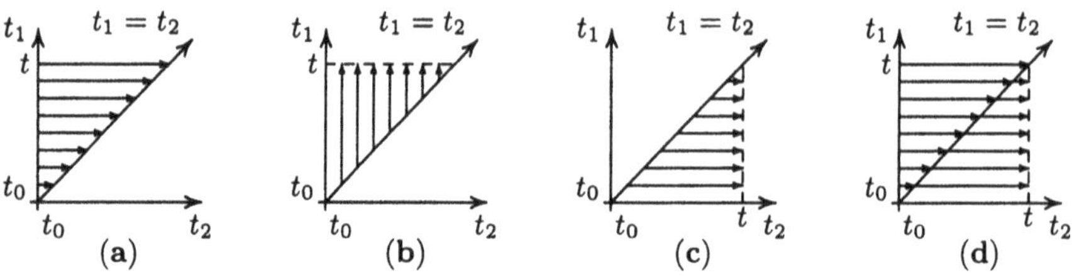


**Fig. 4.1a–d.** Transformations of the integrals in the second-order term

the interaction is weak enough, with a dimensionless coupling constant small compared to unity.

The above series can be rewritten in a more compact form by transforming the integrals so that they all have the same integration limits, $t_0$ and $t$. It suffices to show the procedure for the second-order term, $U^{(2)}$; the higher-order terms will be obtained by analogy. The second-order term is given, apart from a sign, by

$$\int_{t_0}^{t} \mathrm{d}t_1 \int_{t_0}^{t_1} \mathrm{d}t_2\, H'(t_1)H'(t_2)\,,$$

which we split equally into two; we next change the order of integrations in the second half,

$$\tfrac{1}{2} \int_{t_0}^{t} \mathrm{d}t_1 \int_{t_0}^{t_1} \mathrm{d}t_2\, H'(t_1)H'(t_2) + \tfrac{1}{2} \int_{t_0}^{t} \mathrm{d}t_2 \int_{t_2}^{t} \mathrm{d}t_1\, H'(t_1)H'(t_2)\,.$$

The integration to be performed first in each of the two terms is represented respectively by Fig. 4.1a–b. Next, in the second term, the integration labels are interchanged:

$$\tfrac{1}{2} \int_{t_0}^{t} \mathrm{d}t_2 \int_{t_2}^{t} \mathrm{d}t_1\, H'(t_1)H'(t_2) = \tfrac{1}{2} \int_{t_0}^{t} \mathrm{d}t_1 \int_{t_1}^{t} \mathrm{d}t_2\, H'(t_2)H'(t_1)\,.$$

Integration over $t_2$ is shown in Fig. 4.1c. Together with the first half, left unchanged, it gives the whole second-order term of the series in the form

$$\int_{t_0}^{t} \mathrm{d}t_1 \int_{t_0}^{t_1} \mathrm{d}t_2\, H'(t_1)H'(t_2)$$
$$= \tfrac{1}{2} \int_{t_0}^{t} \mathrm{d}t_1 \int_{t_0}^{t} \mathrm{d}t_2 \left[ H'(t_1)H'(t_2)\theta(t_1 - t_2) + H'(t_2)H'(t_1)\theta(t_2 - t_1) \right],$$

where $\theta(t)$ is the usual step function. This result is illustrated in Fig. 4.1d.

We now introduce the time-ordered product of factors $H'(t)$ at different times:

$$\mathrm{T}[H'(t_1)H'(t_2)\dots H'(t_n)] \equiv H'(t_{i_1})H'(t_{i_2})\dots H'(t_{i_n}) \quad \text{for} \quad t_{i_1} \geq t_{i_2} \geq \dots \geq t_{i_n} \tag{4.27}$$

for all possible permutations $t_{i_1}, t_{i_2}, \dots, t_{i_n}$ of the set $t_1, t_2, \dots, t_n$, and the relative orders of $H'(t_i)$ and $H'(t_j)$ are the same on both sides of the above equation whenever $t_i = t_j$. The series (26) may then be rewritten as

$$\begin{aligned} U(t, t_0) = 1 - \mathrm{i} \int_{t_0}^{t} \mathrm{d}t_1\, H'(t_1) + \frac{(-\mathrm{i})^2}{2!} \int_{t_0}^{t} \mathrm{d}t_1 \int_{t_0}^{t} \mathrm{d}t_2\, \mathrm{T}[H'(t_1)H'(t_2)] + \dots \\ + \frac{(-\mathrm{i})^n}{n!} \int_{t_0}^{t} \mathrm{d}t_1 \int_{t_0}^{t} \mathrm{d}t_2 \dots \int_{t_0}^{t} \mathrm{d}t_n\, \mathrm{T}[H'(t_1)H'(t_2)\dots H'(t_n)] + \dots . \end{aligned} \tag{4.28}$$

It can now be formally summed up, leading to the symbolic representation

$$\begin{aligned} U(t, t_0) &= \mathrm{T} \exp\Big[-\mathrm{i} \int_{t_0}^{t} \mathrm{d}t'\, H'(t')\Big] \\ &\equiv \mathrm{T} \exp\Big[-\mathrm{i} \int_{t_0}^{t} \mathrm{d}t' \int \mathrm{d}^3x\, \mathcal{H}'(t', \boldsymbol{x}')\Big], \end{aligned} \tag{4.29}$$

where $\mathcal{H}'(t', \boldsymbol{x}')$ is the interaction Hamiltonian density in the interaction picture. The series (28) is complicated because of the need for time ordering of the operators $H'(t)$ for different $t$, which in turn is dictated by the fact that $H'(t)$ at different times do not commute in general. When they do commute, $[H'(t),\, H'(t')] = 0$ for all $t$ and $t'$, the result becomes much simpler:

$$U(t, t_0) = \exp\Big[-\mathrm{i} \int_{t_0}^{t} \mathrm{d}t'\, H'(t')\Big]. \tag{4.30}$$

### 4.1.3 The $S$-Matrix

As mentioned above, $U$ has a definite physical meaning. Suppose a physical system is known to be in state $\Phi_{\mathrm{i}}$ at time $t = t_0$. When it goes through the interaction process described by the interaction Hamiltonian $H'$, its time evolution is given in the interaction representation by

$$\Psi_{\mathrm{I}}(\mathrm{i}; t) = U(t, t_0)\, \Phi_{\mathrm{i}}\,. \tag{4.31}$$

The amplitude of the probability for finding it in state $\Phi_{\mathrm{f}}$ in the distant future, long after the interaction has ceased to act, is

$$\langle\, \Phi_{\mathrm{f}} \,|\, \Psi_{\mathrm{I}}(\mathrm{i}; t)\,\rangle = \langle \Phi_{\mathrm{f}} \,|\, U(t, t_0) \,|\, \Phi_{\mathrm{i}} \rangle = U_{\mathrm{fi}}(t, t_0)\,. \tag{4.32}$$

The average rate per unit time for the transition i $\to$ f (with i $\neq$ f) is obtained by dividing the probability by the interaction time, $\Delta t = t - t_0$, with the result

$$\frac{1}{\Delta t} \left| U_{\rm fi}(t, t_0) - \delta_{\rm fi} \right|^2 . \tag{4.33}$$

As $t_0 \to -\infty$ and $t \to +\infty$, this expression has a well-defined limit which represents a measurable quantity. Therefore, it is useful to introduce

$$S \equiv U(\infty, -\infty) , \tag{4.34}$$

called the $S$-matrix, which one knows how to calculate from (29) in terms of the interaction Hamiltonian density:

$$\begin{aligned} S &= \mathrm{T} \exp\Big[-\mathrm{i} \int_{-\infty}^{\infty} \mathrm{d}t \int \mathrm{d}^3 x \, \mathcal{H}'(t, \boldsymbol{x})\Big] \\ &= \mathrm{T} \exp\Big[-\mathrm{i} \int \mathrm{d}^4 x \, \mathcal{H}'(x)\Big] . \end{aligned} \tag{4.35}$$

In considering any transition i $\to$ f, it is convenient to separate out the zeroth order (no interaction) term in (28) and define a transition matrix $T_{\rm fi}$:

$$S_{\rm fi} = \delta_{\rm fi} + \mathrm{i}\, T_{\rm fi} . \tag{4.36}$$

The energy-momentum conservation condition may be factored out, leaving the corresponding invariant reduced transition matrix $\mathcal{M}_{\rm fi}$:

$$T_{\rm fi} = (2\pi)^4 \, \delta^{(4)}(P_{\rm f} - P_{\rm i}) \, \mathcal{M}_{\rm fi} . \tag{4.37}$$

It is this matrix $\mathcal{M}_{\rm fi}$ that encapsulates the whole dynamic content of the transition and that directly produces, together with the relevant kinematic factors, its cross-section or decay rate.

## 4.2 Cross-Sections and Decay Rates

We first derive the general expression for the cross section of a two body reaction to any $n$-particle final state, then specialize it to the case of final two-particle states.

### 4.2.1 General Formulas

To simplify, we assume the particles to have no spins; but the final result we obtain can be readily modified to include spin effects. Thus, we consider a state with two distinct particles of momenta $\boldsymbol{p}_1$ and $\boldsymbol{p}_2$, and masses $m_1$ and $m_2$. At $t = -\infty$, it is given by

$$|\Phi_{\rm i}\rangle = \int \mathrm{d}^3 \tilde{p}_1' \int \mathrm{d}^3 \tilde{p}_2' \, f(\boldsymbol{p}_1') \, f(\boldsymbol{p}_2') \, |\boldsymbol{p}_1' \, \boldsymbol{p}_2'\rangle , \tag{4.38}$$

where

$$\mathrm{d}^3\tilde{p}' \equiv \mathrm{d}^3 p' / \left[(2\pi)^3 \, 2E_{\boldsymbol{p}'}\right] . \tag{4.39}$$

The particle states are normalized such that

$$\begin{aligned} \langle \alpha' \boldsymbol{p}' \,|\alpha \boldsymbol{p}\rangle &= (2\pi)^3 \, 2E_{\boldsymbol{p}} \, \delta(\boldsymbol{p}' - \boldsymbol{p}) \, \delta_{\alpha'\alpha} \, , \\ \int \mathrm{d}^3\tilde{p}' \sum_{\alpha} |\alpha \boldsymbol{p}'\rangle \, \langle \alpha \boldsymbol{p}'| &= 1 \, . \end{aligned} \tag{4.40}$$

Each wave packet $f(\boldsymbol{p}')$ attains its peak value at $\boldsymbol{p}' = \boldsymbol{p}$, and its Fourier transform in configuration space $f_{\boldsymbol{p}}(\boldsymbol{x})$ is given in the plane-wave limit by $\exp(-\mathrm{i}\boldsymbol{p} \cdot \boldsymbol{x})$ modulated by some slowly varying function $F(\boldsymbol{x})$. The corresponding current probability density reads

$$J_\mu = \mathrm{i} \, f^*(x) \overleftrightarrow{\partial}_\mu f(x) \approx 2p_\mu \, |f(x)|^2 . \tag{4.41}$$

The total number of particles contained in a space volume $V$,

$$N = 2E_{\boldsymbol{p}} \int_V \mathrm{d}^3 x \, |f(x)|^2 , \tag{4.42}$$

is just the space integral of the particle density $\varrho = J_0 = 2E_{\boldsymbol{p}} \, |f(x)|^2$.

To calculate the total transition probability for $\mathrm{i}(p_1, p_2) \to \mathrm{f}(P_\mathrm{f})$ $(\mathrm{i} \neq \mathrm{f})$

$$w_{\mathrm{fi}} \equiv |T_{\mathrm{fi}}|^2 = \left| (2\pi)^4 \delta^{(4)}(P_\mathrm{f} - p_1 - p_2) \, \langle \mathrm{f} \,|\, \mathcal{M} \,|\, \Phi_\mathrm{i} \rangle \right|^2 , \tag{4.43}$$

we use (38) for $\Phi_\mathrm{i}$ and

$$(2\pi)^4 \delta^{(4)}(p' - p) = \int \mathrm{d}^4 x \, \mathrm{e}^{\mathrm{i}(p - p') \cdot x}$$

for one of the $\delta$-factors. This will carry (43) into

$$w_{\mathrm{fi}} = (2\pi)^4 \delta^{(4)}(P_\mathrm{f} - p_1 - p_2) \, | \, \langle \mathrm{f} \,|\, \mathcal{M} \,|\, \mathrm{i} \rangle \, |^2 \int \mathrm{d}^4 x \; |f_1(x)|^2 \, |f_2(x)|^2 . \tag{4.44}$$

The result can be interpreted as a density integrated over the interaction volume and time, and the transition probability density itself (number of events per units of time and volume) is then given by

$$\frac{\mathrm{d}w_{\mathrm{fi}}}{\mathrm{d}V \, \mathrm{d}t} = (2\pi)^4 \delta^{(4)}(P_\mathrm{f} - p_1 - p_2) \, | \, \langle \mathrm{f} \,|\, \mathcal{M} \,|\, \mathrm{i} \rangle \, |^2 \; |f_1(x)|^2 \, |f_2(x)|^2 . \tag{4.45}$$

The initial scattering state is characterized by an incident flux (relative velocity times the particle density in the incoming beam)

$$I_\mathrm{i} \equiv |\boldsymbol{v}_1 - \boldsymbol{v}_2| \, \varrho_1 = |\boldsymbol{v}_1 - \boldsymbol{v}_2| \, 2E_1 \, |f_1(x)|^2 , \tag{4.46}$$

and the particle density in the target or in the second colliding beam

$$\varrho_2 \equiv 2E_2 \left|f_2(x)\right|^2 , \tag{4.47}$$

($\boldsymbol{v}_1$ and $\boldsymbol{v}_2$ are the velocities of the incoming particles, and $E_1 = E_{\boldsymbol{p}_1}$ and $E_2 = E_{\boldsymbol{p}_2}$ their energies). On the other hand, the final state is defined by the set of momenta of the collision products, which take values in a certain phase space volume consistent with energy-momentum conservation and represented by an integration measure called $\mathrm{d}\Phi_\mathrm{f}$.

A quantity independent of the initial wave functions, called the differential cross-section, is introduced:

$$\mathrm{d}\sigma = \frac{\mathrm{d}w_\mathrm{fi}}{\mathrm{d}V\,\mathrm{d}t}\,\frac{1}{I_\mathrm{i}\varrho_2}\,\mathrm{d}\Phi_\mathrm{f} \,. \tag{4.48}$$

The cross-section represents an effective area over which one incident particle interacts with one target particle. That it has the dimension of an area can be seen from the fact that the quantity

$$\mathrm{d}\sigma|\boldsymbol{v}_1 - \boldsymbol{v}_2|\,\mathrm{d}t = \frac{\mathrm{d}w_\mathrm{fi}}{\mathrm{d}V}\,\frac{1}{\varrho_1\varrho_2}\,\mathrm{d}\Phi_\mathrm{f} \tag{4.49}$$

has the dimension of spatial volume, and $|\boldsymbol{v}_{12}|\mathrm{d}t$ has the dimension of length (see Fig. 4.2).

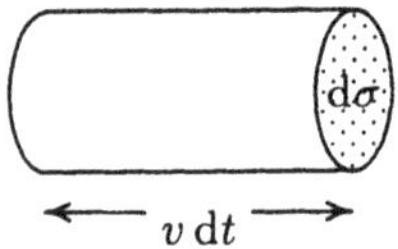


**Fig. 4.2.** Interpretation of the cross-section as an effective area

We will call the product $F = E_1\,E_2\,|\boldsymbol{v}_{12}|$, which arises from $I_\mathrm{i}\varrho_2$ in (48), the flux factor. It turns out that it is invariant to Lorentz transformations in the direction of motion of the incident beams, as shown by

$$F = E_1\,E_2\,|\boldsymbol{v}_{12}| = \left[(p_1 \cdot p_2)^2 - m_1^2\,m_2^2\right]^{1/2} . \tag{4.50}$$

This relation holds when $\boldsymbol{p}_2 = 0$, or when $\boldsymbol{p}_1$ and $\boldsymbol{p}_2$ are collinear. In a general situation, it should be replaced by

$$(p_1 \cdot p_2)^2 - m_1^2\,m_2^2 = (E_2\boldsymbol{p}_1 - E_1\boldsymbol{p}_2)^2 + (\boldsymbol{p}_1 \cdot \boldsymbol{p}_2)^2 - \boldsymbol{p}_1^2\,\boldsymbol{p}_2^2 \,. \tag{4.51}$$

In the laboratory (lab) system where particle 2 is at rest, and in the center-of-mass (cm) system where $\boldsymbol{p}_1 + \boldsymbol{p}_2 = 0$, the flux factor has the values

$$F_\mathrm{lab} = |\boldsymbol{p}_1|\,m_2 \,, \tag{4.52}$$

$$F_\mathrm{cm} = |\boldsymbol{p}_1|\,(E_1 + E_2) \,. \tag{4.53}$$

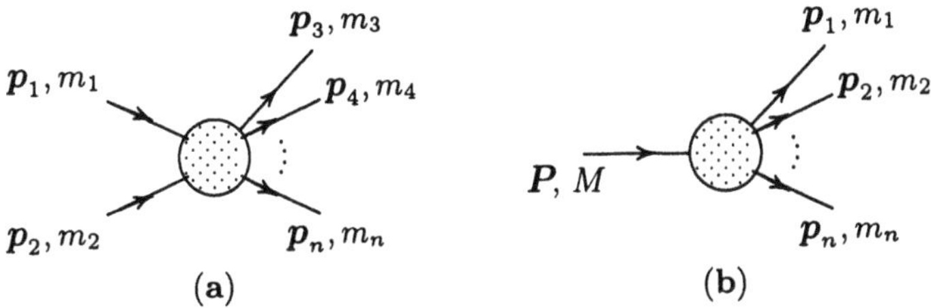


**Fig. 4.3.** (**a**) Two-body reaction to an $(n-2)$-particle final state; (**b**) decay to an $n$-particle final state

For the reaction $(p_1, p_2) \to (p_3, p_4, \ldots, p_n)$ illustrated in Fig. 4.3 a, the expression for the differential cross-section reads

$$d\sigma = \frac{|\mathcal{M}|^2}{4F}\, d\Phi_f\, \mathcal{S}\,. \tag{4.54}$$

Here, $\mathcal{M}$ is shorthand for $\langle f\,|\,\mathcal{M}\,|\,i\rangle = \langle p_3, \ldots, p_n\,|\,\mathcal{M}\,|\,p_1, p_2\rangle$ and $d\Phi_f$ stands for the phase space volume element of the final state

$$\begin{aligned} d\Phi_f(p_3, \ldots, p_n) &= (2\pi)^4\, \delta^{(4)}(p_3 + \ldots + p_n - P_i) \\ &\quad \times \frac{1}{(2\pi)^{3(n-2)}} \frac{d^3 p_3}{2E_3} \cdots \frac{d^3 p_n}{2E_n} \end{aligned} \tag{4.55}$$

$(P_i = p_1 + p_2)$, and finally, $\mathcal{S}$ is a combinatorial factor needed to avoid overcounting identical configurations whenever there are identical particles in the final state, and is given by $\mathcal{S} = \prod_a 1/\ell_a!$, where $\ell_a$ denotes the number of identical particles of type $a$ in the final state.

## 4.2.2 Two-Body Reaction to Two-Body Final States

To illustrate, consider now the scattering of two particles of momenta $p_1^\mu$, $p_2^\mu$ and masses $m_1$, $m_2$ leading to a final state of two distinct particles of momenta $p_3^\mu$, $p_4^\mu$ and masses $m_3$, $m_4$.

We are interested in the probability for observing a final product of the reaction emitted in a certain direction $(\varphi, \theta)$ within an element of the solid angle $d\Omega = d\varphi\, d\cos\theta$, that is,

$$d\sigma = \int_{d\Omega} \frac{|\mathcal{M}|^2}{4F}\, d\Phi_f\,. \tag{4.56}$$

At fixed $(\varphi, \theta)$ angles, the two-particle phase space integral yields, after integrations over $\boldsymbol{p}_4$ and then over $E_3$,

$$\begin{aligned} \int_{d\Omega} d\Phi_f(p_3, p_4) &= \int_{d\Omega} (2\pi)^4\, \delta^{(4)}(p_1 + p_2 - p_3 - p_4) \frac{1}{(2\pi)^6} \frac{d^3 p_3}{2E_3} \frac{d^3 p_4}{2E_4} \\ &= \frac{d\Omega\, \boldsymbol{p}_3^2}{16\pi^2\, E_3\, E_4} \frac{d|\boldsymbol{p}_3|}{d(E_3 + E_4)}\,, \end{aligned}$$

where

$$E_3^2 = \boldsymbol{p}_3^2 + m_3^2\,,$$
$$E_4^2 = (\boldsymbol{p}_1 + \boldsymbol{p}_2 - \boldsymbol{p}_3)^2 + m_4^2\,.$$

The differential cross-section is then given by

$$\frac{\mathrm{d}\sigma}{\mathrm{d}\Omega} = \frac{|\mathcal{M}|^2}{64\pi^2 F}\,\frac{\boldsymbol{p}_3^2}{E_3\,E_4}\,\frac{\mathrm{d}|\boldsymbol{p}_3|}{\mathrm{d}(E_3+E_4)} \qquad \text{(in general).} \tag{4.57}$$

For spinless particles, the matrix $\mathcal{M}$ depends on momenta only through the Lorentz-invariant variables $s$, $t$, and $u$ (called the Mandelstam variables), defined by

$$\begin{aligned} s &= (p_1^\mu + p_2^\mu)^2 = (p_3^\mu + p_4^\mu)^2\,, \\ t &= (p_1^\mu - p_3^\mu)^2 = (p_4^\mu - p_2^\mu)^2\,, \\ u &= (p_1^\mu - p_4^\mu)^2 = (p_3^\mu - p_2^\mu)^2\,, \end{aligned} \tag{4.58}$$

only two of which are independent because $s + t + u = m_1^2 + m_2^2 + m_3^2 + m_4^2$.

In the *center-of-mass* system (see Fig. 4.4) defined by the four-momenta

$$p_1^\mu = (E_1, \boldsymbol{p}), \quad p_2^\mu = (E_2, -\boldsymbol{p}), \quad p_3^\mu = (E_3, \boldsymbol{p}'), \quad p_4^\mu = (E_4, -\boldsymbol{p}'),$$

we have

$$E_3\,E_4\,\frac{\mathrm{d}(E_3+E_4)}{\mathrm{d}|\boldsymbol{p}_3|} = E_3\,E_4\left(\frac{|\boldsymbol{p}'|}{E_3} + \frac{|\boldsymbol{p}'|}{E_4}\right) = |\boldsymbol{p}'|\,(E_1+E_2)\,,$$

and

$$F = |\boldsymbol{p}|\,(E_1+E_2)\,,$$
$$s = (E_1+E_2)^2\,.$$

The differential cross-section (57) then becomes

$$\left(\frac{\mathrm{d}\sigma}{\mathrm{d}\Omega}\right)_{\mathrm{cm}} = \frac{|\mathcal{M}|^2}{64\pi^2 s}\,\frac{|\boldsymbol{p}'|}{|\boldsymbol{p}|} \qquad \text{(center-of-mass).} \tag{4.59}$$

It depends on a single variable, the squared energy $s$, because

$$\begin{aligned} |\boldsymbol{p}|^2 &= \frac{1}{4s}\lambda(s, m_1^2, m_2^2)\,, \\ |\boldsymbol{p}'|^2 &= \frac{1}{4s}\lambda(s, m_3^2, m_4^2)\,, \end{aligned} \tag{4.60}$$

where

$$\lambda(a,b,c) \equiv (a-b-c)^2 - 4bc = [a - (\sqrt{b}+\sqrt{c})^2]\,[a - (\sqrt{b}-\sqrt{c})^2]\,. \tag{4.61}$$

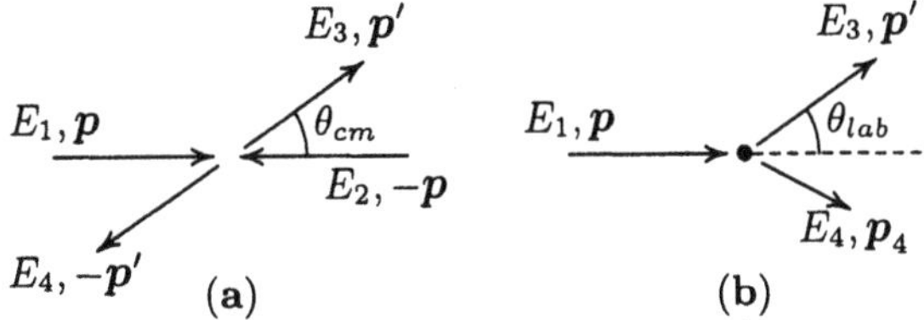


**Fig. 4.4.** Kinematic variables in two-body scattering: (**a**) in center-of-mass frame; (**b**) in laboratory frame

Alternatively, one can write the differential cross-section in terms of the Mandelstam variables $s$ and $t$:

$$\left(\frac{\mathrm{d}\sigma}{\mathrm{d}t}\right)_{\mathrm{cm}} = \frac{|\mathcal{M}|^2}{64\pi\, s\boldsymbol{p}^2} = \frac{|\mathcal{M}|^2}{16\pi\,\lambda(s, m_1^2, m_2^2)} \qquad \text{(center-of-mass)}\,, \tag{4.62}$$

where $t$ is related to $\theta_{\mathrm{cm}}$, the scattering angle, through

$$t = (p_1^\mu - p_3^\mu)^2 = m_1^2 + m_3^2 - 2E_1E_3 + 2|\boldsymbol{p}||\boldsymbol{p}'|\cos\theta_{\mathrm{cm}}\,,$$

so that $\mathrm{d}t = 2|\boldsymbol{p}||\boldsymbol{p}'|\,\mathrm{d}\cos\theta_{\mathrm{cm}}$. Corresponding to $0 \le \theta_{\mathrm{cm}} \le \pi$, we have the interval of definition for $t$:

$$t_\pi \le t \le t_0$$

$$t_0 \equiv t(\theta_{\mathrm{cm}} = 0) = (E_1 - E_3)^2 - (|\boldsymbol{p}| - |\boldsymbol{p}'|)^2\,,$$
$$t_\pi \equiv t(\theta_{\mathrm{cm}} = \pi) = (E_1 - E_3)^2 - (|\boldsymbol{p}| + |\boldsymbol{p}'|)^2\,.$$

In the *laboratory* system (see Fig. 4.4), where particle 2 is at rest, the kinematic variables are

$$p_1^\mu = (E_1, \boldsymbol{p}), \quad p_2^\mu = (m_2, \boldsymbol{0}), \quad p_3^\mu = (E_3, \boldsymbol{p}'), \quad p_4^\mu = (E_4, \boldsymbol{p}_4)\,,$$

where

$$E_4 = E_1 + m_2 - E_3\,,$$
$$\boldsymbol{p}_4^2 = (\boldsymbol{p} - \boldsymbol{p}')^2 = \boldsymbol{p}^2 + \boldsymbol{p}'^2 - 2|\boldsymbol{p}||\boldsymbol{p}'|\cos\theta_{\mathrm{lab}}\,.$$

For given angles $(\varphi, \theta)_{\mathrm{lab}}$, one has $|\boldsymbol{p}_4|\mathrm{d}|\boldsymbol{p}_4| = (|\boldsymbol{p}'| - |\boldsymbol{p}|\cos\theta_{\mathrm{lab}})\,\mathrm{d}|\boldsymbol{p}'|$, so that

$$E_3\,E_4\,\frac{\mathrm{d}(E_3 + E_4)}{\mathrm{d}|\boldsymbol{p}'|} = |\boldsymbol{p}'|(E_1 + m_2) - |\boldsymbol{p}|E_3\,\cos\theta_{\mathrm{lab}}\,. \tag{4.63}$$

The flux factor is $F = |\boldsymbol{p}|m_2$, and the squared energy, $s = m_1^2 + m_2^2 + 2E_1m_2$. Therefore, in the laboratory system the differential cross-section (57) becomes

$$\left(\frac{\mathrm{d}\sigma}{\mathrm{d}\Omega}\right)_{\mathrm{lab}} = \frac{|\mathcal{M}|^2}{64\pi^2\,m_2}\,\frac{|\boldsymbol{p}'|}{|\boldsymbol{p}|}\,\frac{1}{E_1 + m_2 - (|\boldsymbol{p}|/|\boldsymbol{p}'|)\,E_3\,\cos\theta_{\mathrm{lab}}} \qquad \text{(lab)}, \tag{4.64}$$

where $E_3 = \sqrt{\boldsymbol{p}'^2 + m_3^2}$ and, by energy conservation,

$$E_3(E_1 + m_2) - |\boldsymbol{p}||\boldsymbol{p}'| \cos\theta_{\text{lab}} = E_1 m_2 + \tfrac{1}{2}\left(m_1^2 + m_2^2 + m_3^2 - m_4^2\right)$$
$$= \tfrac{1}{2}\left(s + m_3^2 - m_4^2\right). \tag{4.65}$$

In the remaining of this subsection, we consider the case $m_3 = m_1$ and $m_4 = m_2$, still in the laboratory system. Then (65) reduces to

$$E_3(E_1 + m_2) - |\boldsymbol{p}||\boldsymbol{p}'| \cos\theta_{\text{lab}} = E_1 m_2 + m_1^2. \tag{4.66}$$

This gives, with the four-momentum transfer $q = p_3 - p_1$ related to various variables by $q^2 = (p_3 - p_1)^2 = t = 2m_2(E_3 - E_1)$, the relation

$$\frac{|\boldsymbol{p}|}{|\boldsymbol{p}'|} \cos\theta_{\text{lab}} = \frac{q^2}{2\boldsymbol{p}'^2} + \frac{E_1 E_3 - m_1^2}{\boldsymbol{p}'^2}.$$

The cross-section (64) then takes the form (with $m_1 = m_3$, $m_2 = m_4$)

$$\left(\frac{\mathrm{d}\sigma}{\mathrm{d}\Omega}\right)_{\text{lab}} = \frac{|\mathcal{M}|^2}{64\pi^2 m_2^2} \frac{|\boldsymbol{p}'|}{|\boldsymbol{p}|} \left[1 - \frac{q^2}{2m_2^2 |\boldsymbol{p}'|^2}\left(m_2 E_3 - m_1^2\right)\right]^{-1}. \tag{4.67}$$

In the limit of a static target, where $\boldsymbol{p}_4 = 0$, so that $|\boldsymbol{p}| = |\boldsymbol{p}'|$ and $E_1 = E_3$, and

$$E_1 + m_2 - \frac{|\boldsymbol{p}|}{|\boldsymbol{p}'|} E_3 \cos\theta_{\text{lab}} \approx m_2 + E_1(1 - \cos\theta_{\text{lab}}),$$

the cross-section may be approximated by

$$\left(\frac{\mathrm{d}\sigma}{\mathrm{d}\Omega}\right)_{\text{lab}} \approx \frac{|\mathcal{M}|^2}{64\pi^2 m_2^2} \frac{1}{1 + (E_1/m_2)(1 - \cos\theta_{\text{lab}})} \quad \text{(static target)}. \tag{4.68}$$

On the other hand, when one of the particles is extremely relativistic, $E_1 \approx |\boldsymbol{p}|$ and $E_3 \approx |\boldsymbol{p}'|$, so that

$$E_1 + m_2 - (|\boldsymbol{p}|/|\boldsymbol{p}'|)\, E_3 \cos\theta_{\text{lab}} \approx m_2 \left(1 + \frac{2E_1}{m_2} \sin^2\frac{\theta_{\text{lab}}}{2}\right)$$
$$\approx m_2 \left(\frac{E_1}{E_3} + \frac{m_1^2}{E_3 m_2}\right)$$

[the second line follows from (66)], the cross-section has the limiting value

$$\left(\frac{\mathrm{d}\sigma}{\mathrm{d}\Omega}\right)_{\text{lab}} \approx \frac{|\mathcal{M}|^2}{64\pi^2 m_2^2} \frac{E_3/E_1}{\left[1 + 2(E_1/m_2)\sin^2(\theta_{\text{lab}}/2)\right]}$$
$$\approx \frac{|\mathcal{M}|^2}{64\pi^2 m_2^2} \left(\frac{E_3}{E_1}\right)^2 \quad \text{(ultra-relativistic)}. \tag{4.69}$$

### 4.2.3 Decay Rates

Let us now consider a particle of energy $E_P$ and mass $M$ decaying into a final state of $n$ particles of momenta $p_1, \dots, p_n$ in the phase space volume element $\mathrm{d}^3p_1 \dots \mathrm{d}^3p_n$. We shall obtain the decay rate by merely adapting (48) to the present situation in which $\varrho_2$ should refer to the state density of the decaying particle, and $I_\mathrm{i} = 1$ since there are no other particles in the initial state. The differential decay rate can then be read off from (54):

$$\mathrm{d}\Gamma(P \to p_1 + \dots + p_n) = \frac{|\mathcal{M}|^2}{2E_P}\, \mathrm{d}\Phi_\mathrm{f}(p_1, \dots, p_n)\, \mathcal{S}\,, \tag{4.70}$$

where $\mathrm{d}\Phi_\mathrm{f}(p_1, \dots, p_n)$ and $\mathcal{S}$ are defined as in (54).

Consider for example a particle of mass $M$ decaying from rest through a channel leading to two distinct particles of masses $m_1$ and $m_2$ in the final state. The momentum of either of the two emitted particles, $\boldsymbol{p}$, has magnitude

$$\begin{aligned} |\boldsymbol{p}| &= \frac{1}{2M}\sqrt{\lambda(M^2, m_1^2, m_2^2)} \\ &= \frac{1}{2M}\left\{[M^2 - (m_1 + m_2)^2][M^2 - (m_1 - m_2)^2]\right\}^{1/2}. \end{aligned}$$

It is clear that no decays can occur unless $M \geq m_1 + m_2$. With the phase space volume in the solid angle $\mathrm{d}\Omega = \sin\theta_p\, \mathrm{d}\theta_p\, \mathrm{d}\varphi_p$ given by

$$\frac{\mathrm{d}\Omega}{16\pi^2}\frac{|\boldsymbol{p}|}{M}\,,$$

the differential decay rate reads

$$\mathrm{d}\Gamma(M \to p_1 + p_2) = \frac{|\mathcal{M}|^2\, |\boldsymbol{p}|}{32\pi^2\, M^2}\, \mathrm{d}\Omega\,. \tag{4.71}$$

The partial decay rate for the two-particle mode is obtained by integrating over all directions of emission,

$$\Gamma(M \to 1 + 2) = \frac{|\boldsymbol{p}|}{32\pi^2\, M^2}\int \mathrm{d}\Omega\, |\mathcal{M}|^2\,. \tag{4.72}$$

The total decay rate of a particle of mass $M$ is defined as the sum of the partial decay rates for all allowed modes to any numbers of particles:

$$\Gamma(M) = \sum_n \Gamma(M \to 1 + 2 + \dots + n)\,. \tag{4.73}$$

## 4.3 Interaction Models

In the expressions given above for the cross-sections and decay rates, the kinematic factors are completely defined, but the dynamic component, represented by the transition matrix $\mathcal{M}$, is still to be determined. This matrix $\mathcal{M}$ depends on the specific model for the physical phenomena being considered, but once given the Lagrangian of the model, it can be systematically calculated by applying a few simple rules, as will be shown by examples in the following sections.

Let us recall from Chap. 2 the general conditions the Lagrangian density $\mathcal{L}$ that describes any physical model must satisfy. For the theory to be relativistically invariant, $\mathcal{L}$ must be invariant to restricted Lorentz transformations. It must be Hermitian so that the Hamiltonian to which it is related is a physical observable. It may also have other invariance properties as justified by the physical situation. We may also require that it contains no space-time derivatives of fields of orders higher than the first.

The Lagrangian of the system generally includes a free-field part, $\mathcal{L}_0$, which represents the uncoupled fields and an interaction part, $\mathcal{L}'$, which describes the interactions between the fields. The simplest form of coupling between fields is *local*, which means the field quantities that enter the interaction terms refer to the same space-time point. When no derivatives of fields appear in the coupling, it is said to be direct or *nonderivative*, otherwise it is said to be a *derivative* coupling. For example, a relativistically invariant interaction between a pseudoscalar field $\phi$ and a spinor field $\psi$ can be constructed either by multiplying $\phi$ and the pseudoscalar bilinear product $\bar{\psi}(x)\gamma_5\psi(x)$ to have a nonderivative coupling, $\mathcal{L}' = g\bar{\psi}(x)\gamma_5\psi(x)\phi(x)$, or by forming the Lorentz scalar product of $\mathrm{i}\bar{\psi}(x)\gamma_5\gamma^\mu\psi(x)$ and $-\mathrm{i}\partial_\mu\phi(x)$ to get a derivative coupling, $\mathcal{L}'' = g'\bar{\psi}(x)\gamma_5\gamma^\mu\psi(x)\partial_\mu\phi(x)$. The multiplicative constants $g$ and $g'$ that appear in the interaction terms measure the strengths of the interactions, and are called the *coupling constants.* In natural units, the Lagrangian density has dimension $[\text{mass}]^4$, the field $\phi$ has the dimension of mass, and the field $\psi$ has dimension $[\text{mass}]^{3/2}$, so that $g$ is dimensionless, whereas $g'$ has the dimension of inverse mass.

The operator that appears in the expression for the $S$-matrix is the interaction Hamiltonian density $\mathcal{H}'$, rather than the interaction Lagrangian density. However, for any system, the total Hamiltonian density $\mathcal{H}$ can be obtained from the Lagrangian density $\mathcal{L} = \mathcal{L}(\varphi_i, \partial_\mu\varphi_i)$ by the relation

$$\mathcal{H} = \frac{\partial\mathcal{L}}{\partial\dot{\varphi}_i}\dot{\varphi}_i - \mathcal{L}\,. \tag{4.74}$$

In particular, if the interaction contains no time derivatives of field quantities, $\mathcal{H}'$ is simply

$$\mathcal{H}' = -\mathcal{L}'\,. \tag{4.75}$$

## 4.4 Decay Modes of Scalar Particles

Let us first construct a simple model for a system composed of a real scalar field $\sigma(x)$ of mass $\mu$ and three real scalar fields $\phi_a(x)$ for $a = 1, 2, 3$ of equal masses $m$. Let $\phi_1$ and $\phi_2$ combine to form a complex field

$$\varphi(x) = \frac{1}{\sqrt{2}}(\phi_1 - \mathrm{i}\phi_2)\,,$$

which is meant to represent a charged field, called $\pi^{\pm}$. The neutral fields $\phi \equiv \phi_3$ and $\sigma$ are to describe the $\pi^0$ and $\sigma$ particles, respectively. In the absence of interactions, the Lagrangian of the system is

$$\begin{aligned}\mathcal{L}_0 &= \left(\frac{1}{2}\partial_\mu\sigma\,\partial^\mu\sigma - \frac{1}{2}\mu^2\sigma^2\right) + \sum_a\left(\frac{1}{2}\partial_\mu\phi_a\,\partial^\mu\phi_a - \frac{1}{2}m^2\phi_a^2\right)\\ &= \left(\frac{1}{2}\partial_\mu\sigma\,\partial^\mu\sigma - \frac{1}{2}\mu^2\sigma^2\right) + \left(\frac{1}{2}\partial_\mu\phi\,\partial^\mu\phi - \frac{1}{2}m^2\phi^2\right)\\ &\quad + \left(\partial_\mu\varphi^\dagger\,\partial^\mu\varphi - m^2\varphi^\dagger\varphi\right).\end{aligned} \tag{4.76}$$

### 4.4.1 Neutral Decay Mode

Let us first consider the $\sigma$ and $\pi^0$ fields alone. Local nonderivative couplings may now be introduced by taking products of the field quantities, these products being normal ordered when the fields are quantized. Since both field quantities have the dimension of mass, interaction terms like $\phi^4$, $\sigma\phi^3$, $\sigma^2\phi^2$, $\sigma^3\phi$, or $\sigma^4$ have dimensionless coupling constants, whereas $\phi^3$, $\sigma\phi^2$, $\sigma^2\phi$, or $\sigma^3$ have coupling constants with the dimension of mass.

As discussed in the next chapter, an important symmetry operation is space inversion, defined by $\boldsymbol{x} \to -\boldsymbol{x}$. Let us assume for the sake of illustration that under this transformation the fields vary as

$$\begin{aligned}\mathcal{P}:\quad &\sigma(t,\boldsymbol{x}) \to \sigma(t,-\boldsymbol{x}) = +\sigma(t,\boldsymbol{x})\,,\\ &\phi(t,\boldsymbol{x}) \to \phi(t,-\boldsymbol{x}) = -\phi(t,\boldsymbol{x})\,.\end{aligned}$$

Then $\sigma(x)$ is said to be a *scalar* field, and $\phi(x)$ a *pseudoscalar* field. Now the free-field Lagrangian (76) is evidently invariant under space inversion; if the interacting system is to preserve this symmetry, the couplings must contain only even powers of $\phi$ but may have even or odd powers of $\sigma$. Thus, the interaction Lagrangian that is invariant under both space inversion and restricted Lorentz transformations has the general form

$$\mathcal{L}_1 = -g\phi^4 - g'\sigma^2\phi^2 - g''\sigma^4 - \lambda\sigma\phi^2 - \lambda'\sigma^3\,, \tag{4.77}$$

where the coupling constants $g$, $g'$, $g''$, $\lambda$ and $\lambda'$ are real numbers. Since the couplings are nonderivative, the interaction Hamiltonian is just $-\mathcal{L}_1$:

$$\mathcal{H}_1 = g\phi^4 + g'\sigma^2\phi^2 + g''\sigma^4 + \lambda\sigma\phi^2 + \lambda'\sigma^3\,. \tag{4.78}$$

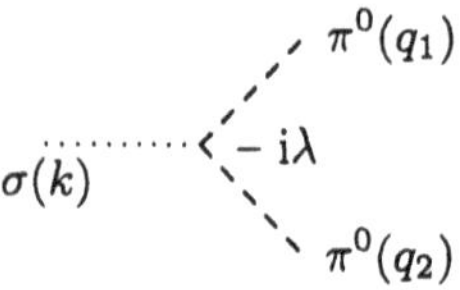


**Fig. 4.5.** Diagram representing the neutral decay mode of $\sigma$ described by the interaction Lagrangian (77). Time evolves from left to right

Let us now calculate the rate for the decay mode shown in Fig. 4.5,

$$\sigma(k) \to \pi^0(q_1) + \pi^0(q_2)\,, \tag{4.79}$$

where the kinematic variables of the system are given by $k = (E_k, \boldsymbol{k})$ and $q_i = (E_i, \boldsymbol{q}_i)$ for $i = 1, 2$, with the energies of the particles related to their momenta in the usual way, $E_k = \sqrt{\boldsymbol{k}^2 + \mu^2}$ and $E_i = \sqrt{\boldsymbol{q}_i^2 + m^2}$. The initial state is described in terms of the $\sigma$-particle creation operator $c_{\boldsymbol{k}}^\dagger$ by the ket

$$|k\rangle = c_{\boldsymbol{k}}^\dagger\; |0\rangle\, C_{\boldsymbol{k}}^{-1}\,, \tag{4.80}$$

whereas the final two-$\pi^0$ state is described in terms of the $\pi^0$ Fock operators $a_{\boldsymbol{q}_i}$ by the bra

$$\langle q_2\, q_1| = (C_{\boldsymbol{q}_2} C_{\boldsymbol{q}_1})^{-1}\, \langle 0|\; a_{\boldsymbol{q}_2} a_{\boldsymbol{q}_1}\,, \tag{4.81}$$

where the constants $C_{\boldsymbol{k}}$, $C_{\boldsymbol{q}_i}$ are chosen to be consistent with the normalization (40), i.e. $C_{\boldsymbol{k}}^{-1} = \sqrt{(2\pi)^3\, 2E_{\boldsymbol{k}}}$. The convention we use to define a bra for several particles is motivated by the conjugation relation between a ket

$$|1\rangle\; |2\rangle = |1, 2\rangle = a_1^\dagger\, a_2^\dagger\; |0\rangle$$

and the corresponding bra

$$(\,|1\rangle\; |2\rangle)^\dagger = \langle 2|\; \langle 1| \;= \langle 2, 1| \;= \langle 0|\; a_2\, a_1\,.$$

In this problem, $\mathcal{L}_0$ describes the unperturbed states, whereas $\mathcal{L}_1$ plays the role of a perturbation of the system. The quantized neutral fields $\sigma$ and $\phi$ have the expansion series

$$\sigma(x) = \sum_{\boldsymbol{k}} \left[ c_{\boldsymbol{k}} \phi_{\boldsymbol{k}}^{(+)}(x) + c_{\boldsymbol{k}}^\dagger \phi_{-\boldsymbol{k}}^{(-)}(x) \right]\,,$$

$$\phi(x) = \sum_{\boldsymbol{q}} \left[ a_{\boldsymbol{q}} \phi_{\boldsymbol{q}}^{(+)}(x) + a_{\boldsymbol{q}}^\dagger \phi_{-\boldsymbol{q}}^{(-)}(x) \right]\,, \tag{4.82}$$

where all Fock operators obey the usual commutation relations consistent with Bose statistics, and the wave amplitudes are

$$\phi_{\boldsymbol{p}}^{(+)}(x) = C_{\boldsymbol{p}}\, \mathrm{e}^{-\mathrm{i}p\cdot x}\,,$$

$$\phi_{-\boldsymbol{p}}^{(-)}(x) = C_{\boldsymbol{p}}\, \mathrm{e}^{\mathrm{i}p\cdot x}\,, \qquad (\,C_{\boldsymbol{p}} = [(2\pi)^3\, 2E_{\boldsymbol{p}}]^{-1/2}\,)\,. \tag{4.83}$$

In the lowest order of interaction, only the coupling $-\lambda\sigma\phi^2$ contributes to the process (79), so that the corresponding $S$-matrix element is given by

$$S_{\mathrm{fi}}^{(1)} = -\mathrm{i}\,\lambda\left\langle q_2\, q_1\left|\int \mathrm{d}^4x\,\sigma(x)\,\phi^2(x)\right|k\right\rangle . \tag{4.84}$$

Given the initial and final states of the system, it is clear that the only operators from $\sigma(x)\phi^2(x)$ that can contribute to the matrix element are of the type $a^\dagger_{\boldsymbol{q}_1'}\, a^\dagger_{\boldsymbol{q}_2'}\, c_{\boldsymbol{k}'}$, and therefore we may just consider

$$S_{\mathrm{fi}}^{(1)} = -\mathrm{i}\,\lambda\left\langle q_2\, q_1\left|\int \mathrm{d}^4x\,\Big(\sum_{\boldsymbol{q}_1'} a^\dagger_{\boldsymbol{q}_1'}\phi^{(-)}_{-\boldsymbol{q}_1'}\Big)\Big(\sum_{\boldsymbol{q}_2'} a^\dagger_{\boldsymbol{q}_2'}\phi^{(-)}_{-\boldsymbol{q}_2'}\Big)\Big(\sum_{\boldsymbol{k}'} c_{\boldsymbol{k}'}\phi^{(+)}_{\boldsymbol{k}'}\Big)\right|k\right\rangle .$$

The matrix element is calculated with the canonical commutation relations (2.109) which the Fock operators $a$, $a^\dagger$ and $c$, $c^\dagger$ must obey,

$$\left\langle 0\left| a_{\boldsymbol{q}_2}\, a_{\boldsymbol{q}_1}\, a^\dagger_{\boldsymbol{q}_1'}\, a^\dagger_{\boldsymbol{q}_2'}\, c_{\boldsymbol{k}'}\, c^\dagger_{\boldsymbol{k}}\right|0\right\rangle = \delta_{\boldsymbol{k}\boldsymbol{k}'}\big(\delta_{\boldsymbol{q}_1\boldsymbol{q}_1'}\delta_{\boldsymbol{q}_2\boldsymbol{q}_2'} + \delta_{\boldsymbol{q}_1\boldsymbol{q}_2'}\delta_{\boldsymbol{q}_2\boldsymbol{q}_1'}\big)\,,$$

where, for example, $\delta_{\boldsymbol{k}\boldsymbol{k}'} = \delta(\boldsymbol{k}-\boldsymbol{k}')$. All normalization factors exactly cancel out, and the first-order $S$-matrix for the decay mode (79) reduces to

$$\begin{aligned} S_{\mathrm{fi}}^{(1)} &= -\mathrm{i}\,\lambda\int \mathrm{d}^4x\,\left[\mathrm{e}^{-\mathrm{i}(k-q_1-q_2)\cdot x} + \mathrm{e}^{-\mathrm{i}(k-q_2-q_1)\cdot x}\right] \\ &= -2\mathrm{i}\,\lambda\,(2\pi)^4\delta^{(4)}(q_1+q_2-k)\,, \end{aligned} \tag{4.85}$$

from which the reduced transition amplitude is identified:

$$\mathrm{i}\mathcal{M} = -2\mathrm{i}\,\lambda\,. \tag{4.86}$$

This result is shown graphically in Fig. 4.5. The diagram represents a three-point vertex corresponding to the interaction Lagrangian $-\lambda\sigma\phi^2$, to which are attached the three external lines describing the particles in the initial and final states. A value of $-\mathrm{i}\,\lambda$ is assigned to the vertex. Since the final two particles are identical bosons, the final state should be *symmetrized*, which results in a numerical factor of 2, corresponding to the number of possible ways in which two identical mesons can be attached to the final two legs of the vertex, cf. (85). This symmetry factor is not indicated in the diagram. We have here a simple example of the Feynman diagrams and rules found so useful in the study of particle physics.

The rate of the $\sigma$ decay from rest via the process (79) can be calculated from formula (72), with the symmetry factor $\mathcal{S} = 1/2$. Since $\mathcal{M}$ is just a constant, the integration over the solid angle yields $4\pi$. That the invariant amplitude has no angular dependence follows from Lorentz invariance, since there is no preferred direction in which a spinless particle at rest decays into two spinless particles. The decay rate of $\sigma$ to a final state of two neutral spinless particles follows from (72), with momentum $|\boldsymbol{q}| = \frac{1}{2}\sqrt{\mu^2 - 4m^2}$,

$$\begin{aligned}\Gamma(\sigma \to \pi^0\pi^0) &= \frac{|\boldsymbol{q}|}{32\pi^2\mu^2}\frac{4\pi}{2}|\mathcal{M}|^2 \\ &= \frac{\lambda^2}{8\pi\mu^2}\sqrt{\mu^2 - 4m^2}\,. \end{aligned} \tag{4.87}$$

### 4.4.2 Charged Decay Mode

Turning now to the charged decay mode, $\sigma \to \pi^+\pi^-$, we will assume that $\pi^\pm$ are coupled to $\sigma$ in exactly the same way as is $\pi^0$, so that the interaction term describing the decay mode $\sigma \to \pi^+\pi^-$ is

$$\begin{aligned}\mathcal{L}_2 &= -\lambda\,\sigma\,(\phi_1^2 + \phi_2^2) + \ldots \\ &= -2\lambda\sigma\,\varphi^\dagger\varphi + \ldots\,, \end{aligned} \tag{4.88}$$

where the quantized complex fields $\varphi$ and $\varphi^\dagger$ are expressed as usual in terms of the Fock operators,

$$\begin{aligned}\varphi(x) &= \sum_{\boldsymbol{q}}\left[a_{\boldsymbol{q}}\phi_{\boldsymbol{q}}^{(+)}(x) + b_{\boldsymbol{q}}^\dagger\phi_{-\boldsymbol{q}}^{(-)}(x)\right]\,, \\ \varphi^\dagger(x) &= \sum_{\boldsymbol{q}}\left[a_{\boldsymbol{q}}^\dagger\phi_{\boldsymbol{q}}^{(+)*}(x) + b_{\boldsymbol{q}}\phi_{-\boldsymbol{q}}^{(-)*}(x)\right]\,. \end{aligned} \tag{4.89}$$

Here $a_{\boldsymbol{q}}^\dagger$ creates a $\pi^+$ and $b_{\boldsymbol{p}}^\dagger$ creates a $\pi^-$.

With the final $\pi^+\pi^-$ state represented by the bra

$$\langle q_2\, q_1| = (C_{\boldsymbol{q}_2}C_{\boldsymbol{q}_1})^{-1}\,\langle 0|\; b_{\boldsymbol{q}_2}a_{\boldsymbol{q}_1}\,, \tag{4.90}$$

the decay $\sigma \to \pi^+\pi^-$ is described to lowest order by

$$S_{\mathrm{fi}}^{(1)} = -2\mathrm{i}\,\lambda\left\langle q_2\, q_1\left|\int \mathrm{d}^4x\,\sigma(x)\,\varphi^\dagger(x)\varphi(x)\right|k\right\rangle\,. \tag{4.91}$$

Since only operator products of the type $a_{\boldsymbol{q}_1'}^\dagger\, b_{\boldsymbol{q}_2'}^\dagger\, c_{\boldsymbol{k}'}$ can contribute, the matrix element we wish to evaluate is

$$S_{\mathrm{fi}}^{(1)} = -2\mathrm{i}\,\lambda\left\langle q_2\, q_1\left|\int \mathrm{d}^4x \sum_{q_1'} a_{\boldsymbol{q}_1'}^\dagger\phi_{\boldsymbol{q}_1'}^{(+)*}\sum_{q_2'} b_{\boldsymbol{q}_2'}^\dagger\phi_{-\boldsymbol{q}_2'}^{(-)}\sum_{\boldsymbol{k}'} c_{\boldsymbol{k}'}\phi_{\boldsymbol{k}'}^{(+)}\right|k\right\rangle\,.$$

With the basic matrix element

$$\left\langle 0\left|b_{\boldsymbol{q}_2}\,a_{\boldsymbol{q}_1}\,a_{\boldsymbol{q}_1'}^\dagger\,b_{\boldsymbol{q}_2'}^\dagger\,c_{\boldsymbol{k}'}\,c_{\boldsymbol{k}}^\dagger\right|0\right\rangle = \delta_{\boldsymbol{k}\boldsymbol{k}'}\,\delta_{\boldsymbol{q}_1\boldsymbol{q}_1'}\,\delta_{\boldsymbol{q}_2\boldsymbol{q}_2'}\,,$$

the $S$-matrix turns out to be identical to that found for the neutral decay,

$$S_{\mathrm{fi}}^{(1)} = -2\mathrm{i}\,\lambda\,(2\pi)^4\delta^{(4)}(q_1 + q_2 - k)\,, \tag{4.92}$$

from which is extracted the invariant transition amplitude

$$\mathrm{i}\mathcal{M} = -2\mathrm{i}\,\lambda\,. \tag{4.93}$$

The Feynman rule for the calculation of this amplitude is very simple: to each vertex $\sigma\pi^+\pi^-$ described by the field coupling $-2\lambda\,\sigma\,\varphi^\dagger\varphi$, assign the value $-\mathrm{i}\,2\lambda$. There is no need for symmetrization of the final state since it contains only distinct particles. The corresponding Feynman diagram is shown in Fig. 4.6.

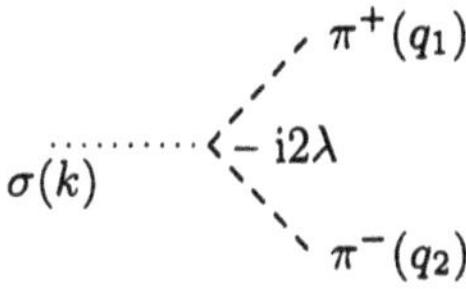


**Fig. 4.6.** Diagram representing the charged decay mode of $\sigma$ described by the interaction Lagrangian (88)

The decay rate is easily calculated from (72), yielding

$$\Gamma(\sigma \to \pi^+\pi^-) = 2\,\Gamma(\sigma \to \pi^0\pi^0)\,. \tag{4.94}$$

This result can be understood from an internal (isospin) symmetry residing in the model (see Chap. 6); it may be compared, for example, with the measured decay rates of the short-lived neutral meson $\mathrm{K}^0$:

$$\Gamma(\mathrm{K}_\mathrm{S}^0 \to \pi^+\pi^-) = 2.18\,\Gamma(\mathrm{K}_\mathrm{S}^0 \to \pi^0\pi^0)\,.$$

## 4.5 Pion Scattering

Although the basic interaction between fields may be local, particles separated by space-time distances can, and do, interact with each other. In relativistic quantum field theory such an interaction at a distance between two particles is explained by an exchange process, in which an interaction quantum is emitted by one particle and subsequently reabsorbed by the second particle. This exchange process may be represented by a probability amplitude for creating the quantum at one space-time point and annihilating it at another point at a later instant. This amplitude is referred to as a *propagator*.

### 4.5.1 The Scalar Boson Propagator

Let us consider a generic Hermitian field $\phi(x)$ of mass $m$, which one may write as a Fourier series, as in (82). The propagator for $\phi$, to be denoted by $\Delta$, is defined as the vacuum expectation value of the time-ordered product of a pair of field values at two different space-time points:

$$\begin{aligned} \mathrm{i}\,\Delta(x,y) &\equiv \langle 0\,|\,\mathrm{T}[\phi(x)\phi(y)]\,|\,0\rangle \\ &= \theta(x_0-y_0)\,\langle 0\,|\,\phi(x)\phi(y)\,|\,0\rangle + \theta(y_0-x_0)\,\langle 0\,|\,\phi(y)\phi(x)\,|\,0\rangle\,. \end{aligned} \tag{4.95}$$

Only operator products of the type $a_{\boldsymbol{k}}a^\dagger_{\boldsymbol{k}'}$ have nonvanishing vacuum-to-vacuum expectation values:

$$\begin{aligned} \mathrm{i}\,\Delta(x,y) &= \theta(x_0-y_0)\left\langle 0\left|\left(\sum_{\boldsymbol{k}}\phi^{(+)}_{\boldsymbol{k}}(x)a_{\boldsymbol{k}}\right)\left(\sum_{\boldsymbol{k}'}\phi^{(-)}_{-\boldsymbol{k}'}(y)a^\dagger_{\boldsymbol{k}'}\right)\right|0\right\rangle \\ &\quad + \theta(y_0-x_0)\left\langle 0\left|\left(\sum_{\boldsymbol{k}}\phi^{(+)}_{\boldsymbol{k}}(y)a_{\boldsymbol{k}}\right)\left(\sum_{\boldsymbol{k}'}\phi^{(-)}_{-\boldsymbol{k}'}(x)a^\dagger_{\boldsymbol{k}'}\right)\right|0\right\rangle \\ &= \theta(x_0-y_0)\sum_{\boldsymbol{k}}\phi^{(+)}_{\boldsymbol{k}}(x)\phi^{(-)}_{-\boldsymbol{k}}(y) + \theta(y_0-x_0)\sum_{\boldsymbol{k}}\phi^{(+)}_{\boldsymbol{k}}(y)\phi^{(-)}_{-\boldsymbol{k}}(x) \\ &= \int \mathrm{d}^3k\,C^2_{\boldsymbol{k}}\,\mathrm{e}^{-\mathrm{i}\boldsymbol{k}\cdot(\boldsymbol{x}-\boldsymbol{y})} \\ &\quad \times\left[\theta(x_0-y_0)\,\mathrm{e}^{-\mathrm{i}E(x_0-y_0)} + \theta(y_0-x_0)\,\mathrm{e}^{-\mathrm{i}E(y_0-x_0)}\right]. \end{aligned} \tag{4.96}$$

To obtain the same factor $\exp[\mathrm{i}\boldsymbol{k}\cdot(\boldsymbol{y}-\boldsymbol{x})]$ in both terms on the right-hand side, we have changed the sign of $\boldsymbol{k}$ in the second integral without affecting the integrand [which is an even function of $\boldsymbol{k}$ because $E=\sqrt{\boldsymbol{k}^2+m^2}$ and $C^2_{\boldsymbol{k}} = 1/(2\pi)^3\,2E$]. These terms can actually be summed up by using the integral representation of the step function:

$$\theta(t) = \frac{-1}{2\pi\mathrm{i}}\int_{-\infty}^{\infty}\mathrm{d}z\,\frac{\mathrm{e}^{-\mathrm{i}zt}}{z+\mathrm{i}\varepsilon}\,, \tag{4.97}$$

in which $\varepsilon \to 0^+$. The integral on the right-hand side can be evaluated by considering it as part of a contour integral in the complex $z$ plane. The integrand is analytic except for the single pole at $-\mathrm{i}\varepsilon$ in the lower half-plane. If $t>0$ the contour must be closed in the lower half-plane, while if $t<0$ it must be closed in the upper half-plane, so that in either case the exponential tends to zero and the half-circle at infinity makes no contributions. Therefore, the right-hand side gives 1 for $t>0$ and 0 for $t<0$. Using the representation of the step function (97), the expression inside the square brackets in (96), which will be called $X$, can be simplified as follows:

$$\begin{aligned} X &= \theta(x_0 - y_0)\, \mathrm{e}^{-\mathrm{i}E(x_0-y_0)} + \theta(y_0 - x_0)\, \mathrm{e}^{-\mathrm{i}E(y_0-x_0)} \\ &= \frac{1}{2\pi\mathrm{i}}\left[\int_{-\infty}^{\infty} \mathrm{d}k_0\, \frac{\mathrm{e}^{-\mathrm{i}k_0(x_0-y_0)}}{E-k_0-\mathrm{i}\varepsilon} + \int_{-\infty}^{\infty} \mathrm{d}k_0\, \frac{\mathrm{e}^{-\mathrm{i}k_0(y_0-x_0)}}{E-k_0-\mathrm{i}\varepsilon}\right] \\ &= \frac{1}{2\pi\mathrm{i}}\left[\int_{-\infty}^{\infty} \mathrm{d}k_0\, \frac{\mathrm{e}^{-\mathrm{i}k_0(x_0-y_0)}}{E-k_0-\mathrm{i}\varepsilon} + \int_{-\infty}^{\infty} \mathrm{d}k_0\, \frac{\mathrm{e}^{-\mathrm{i}k_0(x_0-y_0)}}{E+k_0-\mathrm{i}\varepsilon}\right] \\ &= \frac{2E}{2\pi\mathrm{i}} \int_{-\infty}^{\infty} \mathrm{d}k_0\, \frac{\mathrm{e}^{-\mathrm{i}k_0(x_0-y_0)}}{E^2-k_0^2-\mathrm{i}\varepsilon'}\,, \end{aligned} \tag{4.98}$$

where $\varepsilon' = 2E\varepsilon$ is another infinitesimal positive real quantity, completely equivalent to $\varepsilon$ for our purpose. It is interesting to note that the two time-ordered terms on the first line above, representing forward and backward propagations in time, are summed up to give a single term on the last line. The denominator on this line is manifestly covariant:

$$E^2 - k_0^2 - \mathrm{i}\varepsilon = m^2 - (k_0^2 - \boldsymbol{k}^2) - \mathrm{i}\varepsilon = m^2 - k^2 - \mathrm{i}\varepsilon\,,$$

with $k^2 = k_0^2 - \boldsymbol{k}^2 \neq m^2$. Therefore, inserting the result for $X$ back into (96), one obtains the propagator for a spin-0 particle as a covariant integral in four-momentum space,

$$\mathrm{i}\Delta(x-y) = \int \frac{\mathrm{d}^4k}{(2\pi)^4}\, \mathrm{e}^{-\mathrm{i}k\cdot(x-y)} \frac{\mathrm{i}}{k^2-m^2+\mathrm{i}\varepsilon}\,, \tag{4.99}$$

which is valid for both $x_0 > y_0$ and $x_0 < y_0$. In configuration space it depends only on the coordinate difference $x - y$, while in momentum space it is a function of the square of the particle four-momentum,

$$\Delta(k) = \frac{1}{k^2 - m^2 + \mathrm{i}\varepsilon}\,. \tag{4.100}$$

The physical meaning of the terms displayed in (96) is clear: the first term describes particles (positive-energy solutions) propagating forward in time, $x_0 > y_0$, while the second term corresponds to antiparticles representing negative-energy solutions propagating backward in time, $x_0 < y_0$. The beautifully simple result (100) takes care of both particle and antiparticle propagations.

As for the complex scalar field, its propagator is defined by the vacuum expectation value

$$\begin{aligned} &\langle 0\,|\,\mathrm{T}[\varphi(x)\varphi^\dagger(y)]\,|\,0\rangle \\ &\quad = \theta(x_0-y_0)\,\langle 0\,|\,\varphi(x)\varphi^\dagger(y)\,|\,0\rangle + \theta(y_0-x_0)\,\langle 0\,|\,\varphi(y)^\dagger\varphi(x)\,|\,0\rangle\,. \end{aligned} \tag{4.101}$$

When the fields $\varphi$ and $\varphi^\dagger$ are expressed in terms of the Fock operators through the expansion series (89), the nonvanishing contributions which arise from matrix elements of $a\,a^\dagger$ and $b\,b^\dagger$ yield the result

$$\begin{aligned} &\mathrm{i}\Delta_{\mathrm{c}}(x,y) \\ &\quad = \theta(x_0 - y_0)\sum_{\boldsymbol{k}} \phi^{(+)}_{\boldsymbol{k}}(x)\phi^{(+)*}_{\boldsymbol{k}}(y) + \theta(y_0 - x_0)\sum_{\boldsymbol{k}} \phi^{(-)}_{-\boldsymbol{k}}(x)\phi^{(-)*}_{\boldsymbol{k}}(y)\,. \end{aligned}$$

Thus, the propagator of a charged boson field is identical to the propagator of a neutral boson field having the same mass, $\Delta_{\mathrm{c}}(k,m) = \Delta(k,m)$.

### 4.5.2 Scattering Processes

The elastic scattering of two neutral $\pi^0$ can be described by the terms $-\lambda\sigma\phi^2$ and $-g\phi^4$ of the interaction Lagrangian (77). However, to study scattering of charged particles as well, these couplings must be modified to include the components $\phi_1$ and $\phi_2$ in a symmetric manner:

$$\begin{aligned} \mathcal{L}_3 &= -g(\phi_1^2 + \phi_2^2 + \phi_3^2)^2 - \lambda\sigma(\phi_1^2 + \phi_2^2 + \phi_3^2) \\ &= -g\phi^4 - \lambda\sigma\phi^2 - 4g(\varphi^\dagger\varphi)^2 - 4g\varphi^\dagger\varphi\phi^2 - 2\lambda\sigma\varphi^\dagger\varphi\,, \end{aligned} \tag{4.102}$$

where $\phi = \phi_3$ and $\varphi = (\phi_1 - \mathrm{i}\phi_2)/\sqrt{2}$, as before.

Let us consider first the scattering of charged particles:

$$\pi^+(q_1) + \pi^-(q_2) \to \pi^+(q_3) + \pi^-(q_4)\,. \tag{4.103}$$

The two-particle ket

$$|q_1\,q_2\rangle = (C_{\boldsymbol{q}_1}C_{\boldsymbol{q}_2})^{-1} a^\dagger_{\boldsymbol{q}_1} b^\dagger_{\boldsymbol{q}_2}\,|0\rangle \tag{4.104}$$

represents the initial state of the process, and the bra

$$\langle q_4\,q_3| = (C_{\boldsymbol{q}_4}C_{\boldsymbol{q}_3})^{-1}\,\langle 0|\,b_{\boldsymbol{q}_4} a_{\boldsymbol{q}_3} \tag{4.105}$$

the final state. The momenta of the $\pi^+$ are $q_1$ and $q_3$, while those of the $\pi^-$ are $q_2$ and $q_4$. We will exclude forward scattering, that is, we will assume $q_1 \neq q_3$ and $q_2 \neq q_4$.

To first order of interaction, the only contribution to the $S$-matrix comes from the four-field coupling $-4g(\varphi^\dagger\varphi)^2$:

$$\begin{aligned} S^{(1)}_{\mathrm{fi}} &= \left\langle q_4\,q_3 \left| \mathrm{i}\int \mathrm{d}^4x\,\mathcal{L}_3 \right| q_1\,q_2 \right\rangle \\ &= -\mathrm{i}4g \left\langle q_4\,q_3 \left| \int \mathrm{d}^4x\,(\varphi^\dagger\varphi)^2 \right| q_1\,q_2 \right\rangle\,. \end{aligned} \tag{4.106}$$

Since forward scattering is explicitly excluded, the matrix element is nonvanishing only if the coupling contributes exactly the types of operators needed to balance the operators $a^\dagger_{q_1}$ and $b^\dagger_{q_2}$ of the initial state and the operators $b_{q_4}$ and $a_{q_3}$ of the final state. The only terms that survive are two of the type $a^\dagger_i b_m$ coming from $(\varphi^\dagger)^2$, plus two others of the type $b^\dagger_k a_j$ coming from $\varphi^2$. Thus, there are four terms in all, each of the form

$$\left\langle 0 \left| b_{q_4} a_{q_3}\, a^\dagger_i\, a_j\, b^\dagger_k\, b_m\, a^\dagger_{q_1} b^\dagger_{q_2} \right| 0 \right\rangle = \delta_{i,q_3}\, \delta_{j,q_1}\, \delta_{k,q_4}\, \delta_{m,q_2}\,,$$

and each making an equal contribution to the integral

$$\left\langle q_4\, q_3 \left| \int \mathrm{d}^4x\, (\varphi^\dagger \varphi)^2 \right| q_1\, q_2 \right\rangle = 4 \times (2\pi)^4 \delta^{(4)}(q_3 + q_4 - q_1 - q_2)\,.$$

We thus have the $S$-matrix in momentum space

$$S^{(1)}_{\mathrm{fi}} = -16\,\mathrm{i}g(2\pi)^4\delta^{(4)}(q_3 + q_4 - q_1 - q_2)\,, \tag{4.107}$$

or equivalently, the invariant amplitude

$$\mathrm{i}\mathcal{M}^{(1)} = -16\,\mathrm{i}g\,. \tag{4.108}$$

This result can be interpreted as the product of two factors, a vertex $-\mathrm{i}\,4g$ which comes from the coupling $-4g(\varphi^\dagger\varphi)^2$ and a combinatorial factor of 4 which reflects the number of ways of attaching the incoming $\pi^+$ and $\pi^-$ to two points of the four-point vertex and the outgoing $\pi^+$ and $\pi^-$ to the other two points. This is illustrated in Fig. 4.7.

**Fig. 4.7.** Four-$\pi$ vertex

In second order the interaction term $-2\lambda\,\sigma\varphi^\dagger\varphi$ gives

$$S^{(2)}_{\mathrm{fi}} = \frac{(-2\mathrm{i}\lambda)^2}{2} \int \mathrm{d}^4x \int \mathrm{d}^4y\, \langle \mathrm{f} \,|\, \mathrm{T}[\,\sigma(x)\varphi^\dagger(x)\varphi(x)\, \sigma(y)\varphi^\dagger(y)\varphi(y)\,]\,|\, \mathrm{i}\rangle\,. \tag{4.109}$$

Again, we need interaction operators of the type $a^\dagger_i\, a_j\, b^\dagger_k\, b_m$ to correctly pair off the operators in the initial and final states. As before, $\varphi^\dagger(x)\varphi(x)\varphi^\dagger(y)\varphi(y)$ yields four such terms, but here one must carefully keep track of the coordinate dependence:

$$\begin{aligned} &\langle \mathrm{f} \,|\, \varphi^\dagger(x)\varphi(x)\, \varphi^\dagger(y)\varphi(y) \,|\, \mathrm{i}\rangle \\ &\quad = \exp[\mathrm{i}x\cdot(q_3 - q_1) + \mathrm{i}y\cdot(q_4 - q_2)] + \exp[\mathrm{i}x\cdot(q_3 + q_4) - \mathrm{i}y\cdot(q_1 + q_2)] \\ &\qquad + \exp[-\mathrm{i}x\cdot(q_1 + q_2) + \mathrm{i}y\cdot(q_3 + q_4)] + \exp[\mathrm{i}x\cdot(q_4 - q_2) + \mathrm{i}y\cdot(q_3 - q_1)]\,. \end{aligned}$$

The resulting $S$-matrix,

$$S_{\rm fi}^{(2)} = (-2{\rm i}\lambda)^2 \int {\rm d}^4x \int {\rm d}^4y \, \{\exp[{\rm i}x\cdot(q_3+q_4) - {\rm i}y\cdot(q_1+q_2)] + \exp[{\rm i}x\cdot(q_4-q_2) - {\rm i}y\cdot(q_1-q_3)]\} \, \langle 0 \,|\, {\rm T}[\sigma(x)\,\sigma(y)] \,|\, 0\rangle \,, \tag{4.110}$$

is proportional to the Fourier transforms of the expectation value of the time-ordered product of two $\sigma(x)$ field operators, that is, the propagator for a boson field of the kind we have just introduced. Inserting its Fourier transform

$$\langle 0 \,|\, {\rm T}[\sigma(x)\,\sigma(y)] \,|\, 0\rangle = {\rm i}\Delta(x-y) = \int \frac{{\rm d}^4k}{(2\pi)^4} \, {\rm e}^{-{\rm i}k\cdot(x-y)} \, {\rm i}\Delta(k) \,, \tag{4.111}$$

we arrive at the $S$-matrix

$$S_{\rm fi}^{(2)} = (-2{\rm i}\lambda)^2 \, (2\pi)^4 \delta^{(4)}(q_3+q_4-q_1-q_2) \, [\,{\rm i}\Delta(q_1-q_3) + {\rm i}\Delta(q_1+q_2)\,] \,,$$

and the corresponding reduced transition matrix

$${\rm i}\mathcal{M}^{(2)} = (-{\rm i}2\lambda)^2 \, [\,{\rm i}\Delta(q_2+q_1) + {\rm i}\Delta(q_3-q_1)\,] \,. \tag{4.112}$$

The first term may be understood as resulting from the annihilation of the initial $\pi^+\pi^-$ pair followed by the creation of a virtual $\sigma$ of momentum $q_1+q_2$ and its subsequent conversion into a final $\pi^+\pi^-$-pair, whereas the second term may be visualized as a direct scattering of $\pi^+\pi^-$ via the exchange of a virtual $\sigma$ of momentum $q_3-q_1$. They are represented in Fig. 4.8 a, b.

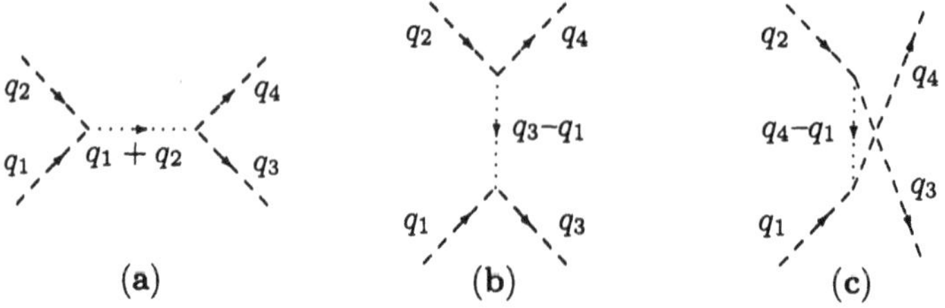


**Fig. 4.8a–c.** Tree diagrams representing the $\sigma$-exchange mechanism in $\pi$–$\pi$ scattering. Time evolves from left to right

In neutral $\pi^0$ scattering

$$\pi^0(q_1) + \pi^0(q_2) \to \pi^0(q_3) + \pi^0(q_4) \,, \tag{4.113}$$

from a state of a $\pi^0$-pair

$$|q_1\,q_2\rangle = a_{q_1}^\dagger a_{q_1}^\dagger \,|0\rangle \, (C_{q_1}C_{q_2})^{-1} \tag{4.114}$$

to a state of another $\pi^0$-pair

$$\langle q_4\, q_3| = \langle 0|\, a_{q_4} a_{q_3} (C_{q_3} C_{q_4})^{-1}\,, \tag{4.115}$$

the interaction processes at the tree-diagram level arise from the couplings $-g\phi^4$ and $-\lambda\sigma\phi^2$. Proceeding as before, we first consider

$$S_{\mathrm{fi}}^{(1)} = -\mathrm{i}g \left\langle q_4\, q_3 \left| \int \mathrm{d}^4x\, \phi^4(x) \right| q_1\, q_2 \right\rangle . \tag{4.116}$$

Since all the fields are Hermitian, capable of creating and destroying particles of the same kind, there are more possibilities here than in the previous case: the coupling $\phi^4(x)$ gives rise to six operators of the type $a^\dagger_{q'_3} a^\dagger_{q'_4} a_{q'_1} a_{q'_2}$ needed to yield nonvanishing matrix elements between $|q_1\, q_2\rangle$ and $\langle q_4\, q_3|$ . Each such operator leads to four possible terms:

$$\begin{aligned} &\left\langle 0 \left| a_{q_4} a_{q_3}\, a^\dagger_{q'_3} a^\dagger_{q'_4} a_{q'_1} a_{q'_2}\, a^\dagger_{q_1} a^\dagger_{q_1} \right| 0 \right\rangle \\ &\qquad = (\delta_{q'_1 q_1} \delta_{q'_2 q_2} + 1 \leftrightarrow 2)(\delta_{q'_3 q_3} \delta_{q'_4 q_4} + 3 \leftrightarrow 4)\,, \end{aligned}$$

resulting in

$$S_{\mathrm{fi}}^{(1)} = (2\pi)^4 \delta^{(4)}(q_3 + q_4 - q_1 - q_2)\, \mathrm{i}\mathcal{M}^{(1)}\,, \tag{4.117}$$

where

$$\mathrm{i}\mathcal{M}^{(1)} = -\mathrm{i}\, 24\, g\,. \tag{4.118}$$

This result suggests the following rule: the invariant amplitude is obtained by multiplying the vertex $-\mathrm{i}\, g$ representing the coupling $-g\phi^4$ by the factor $4! = 24$, which corresponds to the number of ways in which four external lines can be independently hooked onto a four-point vertex.

The coupling $-\lambda\sigma\phi^2$ contributes to the second-order $S$-matrix

$$S_{\mathrm{fi}}^{(2)} = \frac{(-\mathrm{i}\lambda)^2}{2} \int \mathrm{d}^4x \int \mathrm{d}^4y\, \langle q_4\, q_3 \,|\, \mathrm{T}[\sigma(x)\phi^2(x)\, \sigma(y)\phi^2(y)] \,|\, q_1\, q_2\rangle\,. \tag{4.119}$$

Compared to the charged particle scattering, there are again more terms. The field product $\phi^2(x)$ includes operators $aa$, $a^\dagger a^\dagger$, and $a^\dagger a$. The two annihilation operators of $\phi^2(x)$ can be paired with the two creation operators in the initial state, and its two creation operators can be paired with the two annihilation operators of the final state. Taking into account $\phi^2(y)$, there will be eight possible terms describing the pair annihilation processes (Fig. 4.8 a). The annihilation operator in $a^\dagger a$ of $\phi^2(x)$ can be paired with one or the other creation operator in the initial state, leaving the other factor, a creation operator, to balance either of the annihilation operators in the final state, so that a given incoming $\pi^0$ scatters from a second $\pi^0$ into one or the other final directions, $q_3$ or $q_4$, exchanging a $\sigma$-particle. The two possible terms are referred to as 'direct' and 'exchange', and are illustrated in Fig. 4.8 b, c. (The exchange term is of course nonexistent in scattering of distinct particles.) The resulting invariant amplitude for the $\sigma$-exchange mechanism turns out to be

$$\mathrm{i}\mathcal{M}^{(2)} = (-\mathrm{i}\lambda)^2\, 4\,[\,\mathrm{i}\Delta(q_2+q_1) + \mathrm{i}\Delta(q_3-q_1) + \mathrm{i}\Delta(q_4-q_1)\,]\,. \tag{4.120}$$

Each vertex from the coupling $-\lambda\sigma\phi^2$ contributes a factor $-\mathrm{i}\lambda$, and each $\sigma$-exchange a factor $\mathrm{i}\Delta(q)$, where $q$ is fixed by energy-momentum conservation at either vertex. The resulting expression is to be multiplied by 4, the number of ways in which two $\pi^0$ can be connected to a $\sigma\pi\pi$ vertex multiplied by the number of ways in which two other $\pi^0$ can be attached to a second such vertex.

Finally, to this approximation order, the total transition amplitude may be written as

$$\mathrm{i}\mathcal{M} = \mathrm{i}\mathcal{M}^{(1)} + \mathrm{i}\mathcal{M}^{(2)} = F(s) + F(t) + F(u)\,, \tag{4.121}$$

where $s$, $t$, and $u$ are the Mandelstam variables of the reaction, and

$$F(x) = 4\left[-\mathrm{i}\,2\,g + (-\mathrm{i}\lambda)^2 \frac{\mathrm{i}}{x-\mu^2+\mathrm{i}\varepsilon}\right]. \tag{4.122}$$

[The exchange scattering term, $F(u)$, should not appear in $\pi^+\pi^-$ scattering.] Let us suppose the pion to be massless and take the limit when the four-momentum of one of the $\pi$ vanishes (the soft-pion limit). Then $s$, $t$, $u \to 0$ and $\mathrm{i}\mathcal{M} = 3\,F(0) = 24\mathrm{i}[-g+\lambda^2/(2\mu^2)]$. Thus, when the couplings are related by $g = \lambda^2/2\mu^2$, the transition amplitude $\mathcal{M}$ vanishes, and pions do not scatter at all from each other.

## 4.5.3 Summary and Generalization

Even before introducing higher-order terms in the expansion series of $S$, there are many more diagrams, including the following, that can contribute up to second order to the pion scattering than we have considered so far:

Each diagram includes *external lines* representing the observable particles participating in the scattering process, and may include *internal lines* connecting *vertices* of interaction. In the above illustration, we have already met the first four diagrams: all are *tree diagrams* in which all lines have momenta fixed by the external momenta. These are the only tree diagrams that may contribute to the amplitude for pion–pion scattering.

The next two contain closed loops and are called *loop diagrams.* Not all their internal momenta are fixed by momentum conservation, those which are not must be integrated over. Each closed loop has one such undetermined four-momentum.

The next two diagrams have *disconnected* parts (one of which has a disconnected vacuum bubble unattached to any line). It turns out that this type of diagrams is irrelevant in the calculation of $\mathcal{M}$. The last two diagrams are examples of *amputable diagrams*, in which some external legs have loops all for themselves (as self-energy insertions). Every one of these decorated external legs is to be removed from the rest of the diagram by cutting a single propagator. The diagram thus amputated, with only undecorated external legs left, is retained for calculation. Therefore, the Feynman diagrams contributing to a physical transition amplitude are *fully connected, amputated diagrams*, which contain only undecorated external lines, all connected to one another.

To evaluate the covariant amplitude $\mathrm{i}\mathcal{M}$, one calculates all the connected, amputated diagrams with the following Feynman rules:

1. for each $\sigma$-propagator, include $\dfrac{\mathrm{i}}{p^2-\mu^2+\mathrm{i}\varepsilon}$;
2. for each $\pi_a$-propagator, include $\dfrac{\mathrm{i}}{p^2-m^2+\mathrm{i}\varepsilon}$;
3. for each external line, include 1;
4. at each vertex, impose momentum conservation;
5. for each vertex, include a combinatorial factor to account for the different ways external lines can be connected to the vertex;
6. to each interaction vertex, assign an appropriate value (see below);
7. integrate over each undetermined internal momentum: $\displaystyle\int\frac{\mathrm{d}^4p}{(2\pi)^4}$.

Rule 6 depends on the specific model. For the system of $\pi$–$\sigma$ under study, the interactions are described by the sum $\mathcal{L}'=\mathcal{L}_1+\mathcal{L}_2+\mathcal{L}_3$. In order to have concise rules for the vertex, it is best to write out $\mathcal{L}'$ symmetrically in terms of the Cartesian isospin components of $\boldsymbol{\phi}=(\phi_1,\phi_2,\phi_3)$:

$$\mathcal{L}'=-g\left(\boldsymbol{\phi}^2\right)^2-\lambda\,\sigma\boldsymbol{\phi}^2-g'\sigma^2\boldsymbol{\phi}^2-\lambda'\sigma^3-g''\sigma^4\,. \tag{4.123}$$

Thus, for the interaction model in (123), we have the following Feynman diagrams and rules for the vertices (a dotted line represents a $\sigma$; a dashed line a $\pi$; and $a$, $b=1$, 2, or 3).

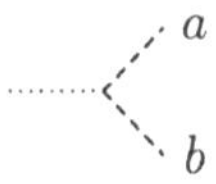

$$-\mathrm{i}\lambda\,\delta_{ab}$$

$$-\mathrm{i}\lambda'$$

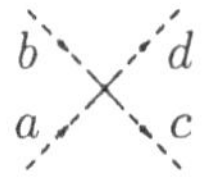

$$-\mathrm{i}g(\delta_{ab}\delta_{cd}+\delta_{ac}\delta_{bd}+\delta_{ad}\delta_{bc})$$

$$-\mathrm{i}g''$$

$$-\mathrm{i}g'\delta_{ab}$$

## 4.6 Electron–Proton Scattering

In this and the remaining sections of the chapter, we shall give a brief account of some practical aspects of quantum electrodynamics (QED), a theory that describes how charged particles interact with the electromagnetic field. QED is considered one of the most important theories in modern physics for its fundamental role in formulating atomic and molecular physics and many aspects of particle physics, and for its role as a prototype for gauge theories, the modern theories of interactions.

The first example of QED we shall study is the electron–proton elastic scattering. The particles involved are treated as pointlike particles interacting only through the electromagnetic field, which is an excellent approximation for the electron as this particle is known to be a structureless lepton, but is of a more restricted value for the proton because the latter is a bound state of quarks and gluons dominated by the strong interaction. Nevertheless, at low energies and as a first approximation, it is still useful to treat the proton as a pointlike particle, correcting for the effects of its complex structure and finite size at a later stage.

We will first introduce the Hamiltonian of the system, then proceed on to calculate the cross-section for electron–proton scattering in the lowest-order approximation. In this example, we will learn in particular how to deal with the propagator of the exchanged photon and the spinors of the external fermions, and how to perform summations over fermion spin states.

### 4.6.1 The Electromagnetic Interaction

From considerations in the previous two chapters, the Lagrangian for a system of interacting electron and proton may be written as

$$\mathcal{L} = \sum_{a=\mathrm{e,p}} \left[\overline{\psi}_a(\mathrm{i}\gamma^\mu\partial_\mu - m_a)\psi_a\right] - \frac{1}{4}F_{\mu\nu}F^{\mu\nu} - e(J^\mu_\mathrm{p} + J^\mu_\mathrm{e})A_\mu\,, \tag{4.124}$$

where $e > 0$ denotes the unit of charge, and

$$\begin{aligned} J^\mu_\mathrm{p}(x) &= \overline{\psi}_\mathrm{p}(x)\gamma^\mu\psi_\mathrm{p}(x)\,, \\ J^\mu_\mathrm{e}(x) &= -\overline{\psi}_\mathrm{e}(x)\gamma^\mu\psi_\mathrm{e}(x) \end{aligned} \tag{4.125}$$

are the proton and electron currents, both conserved by virtue of the Dirac equations they satisfy. Products of fields are understood as normal ordered. It is seen in Chap. 2 that this Lagrangian generates the correct Maxwell equations with $J^\mu(x) = J^\mu_\mathrm{p}(x) + J^\mu_\mathrm{e}(x)$ as the source of the electromagnetic field. Its interaction terms are in perfect accord with the minimal coupling, that is, with the substitution $\mathrm{i}\gamma^\mu\partial_\mu \to \mathrm{i}\gamma^\mu(\partial_\mu \pm eA_\mu)$ for charges $\mp e$, in both the free electron and free proton Dirac Lagrangians.

Following Chap. 2, we now impose the Coulomb (or radiation) gauge

$$\nabla\cdot\boldsymbol{A} = 0\,, \tag{4.126}$$

and solve the Maxwell equation for $A^0$ in the presence of a charge distribution to get

$$A^0(t,\boldsymbol{x}) = \frac{1}{4\pi}\int \mathrm{d}^3x'\,\frac{J^0(t,\boldsymbol{x}')}{|\boldsymbol{x}'-\boldsymbol{x}|}\,. \tag{4.127}$$

This expression gives the instantaneous Coulomb potential, in the sense that the potential field at time $t$ is produced instantaneously by the charge distribution at the same instant, in contrast to the retarded potential which solves the fully covariant equation $\Box A_0(x) = J_0$. It is then eliminated from the Lagrangian to give

$$\begin{aligned} \mathcal{L} = &\sum_{a=\mathrm{e,p}} \left[\overline{\psi}_a(\mathrm{i}\gamma^\mu\partial_\mu - m_a)\psi_a\right] + \frac{1}{2}(\boldsymbol{E}_\perp^2 - \boldsymbol{B}^2) \\ &- \frac{e^2}{8\pi}\int \mathrm{d}^3x'\,\frac{J^0(t,\boldsymbol{x})J^0(t,\boldsymbol{x}')}{|\boldsymbol{x}-\boldsymbol{x}'|} + e\,\boldsymbol{J}\cdot\boldsymbol{A}\,, \end{aligned} \tag{4.128}$$

where

$$\boldsymbol{E}_\perp = -\frac{\partial\boldsymbol{A}}{\partial t} \quad \text{and} \quad \boldsymbol{B} = \nabla\times\boldsymbol{A}\,. \tag{4.129}$$

With the electron and proton spinor fields and the spatial field components $A^i$, for $i = 1, 2, 3$, considered as independent quantities, one obtains the Hamiltonian for the system

$$H = H_0 + H' , \tag{4.130}$$

where the free-field part is

$$H_0 = \int \mathrm{d}^3x \sum_{a=\mathrm{e,p}} [\overline{\psi}_a(-\mathrm{i}\gamma\cdot\nabla + m_a)\psi_a] + \frac{1}{2}\int \mathrm{d}^3x\,(\boldsymbol{E}_\perp^2 + \boldsymbol{B}^2), \tag{4.131}$$

and the interaction part $H'$ is the sum of an *instantaneous* interaction term

$$H_{\text{inst}} = \frac{e^2}{4\pi}\int \mathrm{d}^3x \int \mathrm{d}^3y\, \frac{J^0_\mathrm{p}(t,\boldsymbol{x})\,J^0_\mathrm{e}(t,\boldsymbol{y})}{|\boldsymbol{x}-\boldsymbol{y}|}, \tag{4.132}$$

and a term describing the coupling of the *radiation field* to the matter currents

$$H_{\text{rad}} = \int \mathrm{d}^3x\,\mathcal{H}_{\text{rad}}(x) = -e\int \mathrm{d}^3x\,(\boldsymbol{J}_\mathrm{p} + \boldsymbol{J}_\mathrm{e})\cdot\boldsymbol{A}. \tag{4.133}$$

The self-interaction terms $J^0_\mathrm{p}J^0_\mathrm{p}$ and $J^0_\mathrm{e}J^0_\mathrm{e}$ were dropped from (132). The Hamiltonian $H = H_0 + H_{\text{inst}} + H_{\text{rad}}$ completely determines the dynamics of the structureless electron–proton system in an electromagnetic field. Although it is written in a specific gauge, the physics that it implies should be independent of this particular choice and therefore, the final result of any physical calculation should also reflect this fact.

### 4.6.2 Electron–Proton Scattering Cross-Section

We now consider the elastic scattering process

$$\mathrm{e}^-(p,s) + \mathrm{p}\,(P,S) \to \mathrm{e}^-(p',s') + \mathrm{p}\,(P',S'),$$

with the kinematics defined by the variables $p = (E_{\boldsymbol{p}},\boldsymbol{p})$, $p' = (E_{\boldsymbol{p}'},\boldsymbol{p}')$, $P = (E_{\boldsymbol{P}},\boldsymbol{P})$, and $P' = (E_{\boldsymbol{P}'},\boldsymbol{P}')$. The particle masses will be denoted by $m_\mathrm{e} = m$ and $m_\mathrm{p} = M$, and the initial and final states will be represented by the ket and bra

$$\begin{aligned} |\mathrm{i}\rangle &= |P\,p\rangle = c^\dagger_{\boldsymbol{P}} b^\dagger_{\boldsymbol{p}}\,|0\rangle\,(C_{\boldsymbol{P}}C_{\boldsymbol{p}})^{-1}, \\ \langle\mathrm{f}| &= \langle p'P'| = \langle 0|\,b_{\boldsymbol{p}'}c_{\boldsymbol{P}'}(C_{\boldsymbol{P}'}C_{\boldsymbol{p}'})^{-1}. \end{aligned} \tag{4.134}$$

$b^\dagger_{\boldsymbol{p}}$ and $b_{\boldsymbol{p}'}$ refer to the creation and annihilation operators for the electron field, while $c^\dagger_{\boldsymbol{P}}$ and $c_{\boldsymbol{P}'}$ play the same roles for the proton. In this problem, $H_0$ is treated as the unperturbed Hamiltonian, describing the free fermion fields, whereas $H'$ governs the interaction between fields. Since we are dealing with particle states in the present case, we may as well keep only positive-energy waves in the electron and proton fields,

$$\psi_{\mathrm{e}}(x) = \sum_{\boldsymbol{p},s} u(\boldsymbol{p},s)\, b(\boldsymbol{p},s)\, \phi_{\boldsymbol{p}}^{(+)}(x)\,,$$
$$\psi_{\mathrm{p}}(x) = \sum_{\boldsymbol{P},s} u(\boldsymbol{P},s)\, c(\boldsymbol{P},s)\, \phi_{\boldsymbol{P}}^{(+)}(x)\,. \tag{4.135}$$

The wave functions $\phi^{(+)}(x)$ are defined as in (83). To simplify, we will use the notations $u_1 = u(\boldsymbol{p},s)$, $u_2 = u(\boldsymbol{P},S)$, $u_3 = u(\boldsymbol{p}',s')$, and $u_4 = u(\boldsymbol{P}',S')$, and will calculate the electron–proton scattering amplitude in the lowest nontrivial order, which is proportional to $e^2$,

$$S_{\mathrm{fi}} = \left\langle \mathrm{f} \left| -\mathrm{i}\int_{-\infty}^{\infty} \mathrm{d}t\, H_{\mathrm{inst}} + \frac{(-\mathrm{i})^2}{2} \int \mathrm{d}^4x \int \mathrm{d}^4y\, \mathrm{T}[\mathcal{H}_{\mathrm{rad}}(x)\mathcal{H}_{\mathrm{rad}}(y)] \right| \mathrm{i} \right\rangle$$
$$\equiv S_{\mathrm{inst}} + S_{\mathrm{rad}}\,. \tag{4.136}$$

To evaluate the first term on the right-hand side, it suffices to remark that it consists essentially of the matrix element $\langle \mathrm{f} \,|\, J^0_{\mathrm{p}} J^0_{\mathrm{e}} \,|\, \mathrm{i} \rangle$ to which only operators of the form $(c^\dagger\gamma^0 c)(b^\dagger\gamma^0 b)$ contribute. The instantaneous interaction then produces

$$\begin{aligned} S_{\mathrm{inst}} &= \frac{\mathrm{i}e^2}{4\pi}(\bar{u}_3\gamma^0 u_1)(\bar{u}_4\gamma^0 u_2) \\ &\quad \times \int_{-\infty}^{\infty} \mathrm{d}t\; \mathrm{e}^{\mathrm{i}(E_{P'}+E_{p'}-E_P-E_p)t} \int \mathrm{d}^3x \int \mathrm{d}^3y\, \frac{1}{|\boldsymbol{x}-\boldsymbol{y}|}\, \mathrm{e}^{\mathrm{i}(\boldsymbol{P}-\boldsymbol{P}')\cdot\boldsymbol{x}+\mathrm{i}(\boldsymbol{p}-\boldsymbol{p}')\cdot\boldsymbol{y}} \\ &= (2\pi)^4\delta^{(4)}(P'+p'-P-p)\,(\bar{u}_3\gamma^0 u_1)\frac{\mathrm{i}e^2}{\boldsymbol{q}^2}(\bar{u}_4\gamma^0 u_2)\,, \end{aligned} \tag{4.137}$$

from which one identifies the reduced transition matrix

$$\mathrm{i}\mathcal{M}_{\mathrm{inst}} = (\bar{u}_3\gamma^0 u_1)\frac{\mathrm{i}e^2}{\boldsymbol{q}^2}(\bar{u}_4\gamma^0 u_2)\,. \tag{4.138}$$

Here, $\boldsymbol{q}$ stands for the space components of $q = p' - p = P - P'$. As we will soon see, this expression would be better written in an equivalent form by noting first that

$$\frac{1}{\boldsymbol{q}^2} = \frac{1}{q^2+\mathrm{i}\varepsilon}\,\frac{q_0^2-\boldsymbol{q}^2+\mathrm{i}\varepsilon}{\boldsymbol{q}^2}\,, \tag{4.139}$$

and secondly that

$$\bar{u}_3\gamma^\mu q_\mu u_1 = \bar{u}_3\gamma^\mu(p'_\mu - p_\mu)u_1 = 0\,,$$
$$\bar{u}_4\gamma^\mu q_\mu u_2 = \bar{u}_4\gamma^\mu(P_\mu - P'_\mu)u_2 = 0\,, \tag{4.140}$$

and therefore,

$$\bar{u}_3\gamma^0 q_0\, u_1 = \bar{u}_3\boldsymbol{\gamma}\cdot\boldsymbol{q}\, u_1\,; \qquad \bar{u}_4\gamma^0 q_0\, u_2 = \bar{u}_4\boldsymbol{\gamma}\cdot\boldsymbol{q}\, u_2\,. \tag{4.141}$$

It follows that we may trade a noncovariant function $1/\boldsymbol{q}^2$ for a covariant function $1/q^2$ in the scattering amplitude:

$$\mathrm{i}\mathcal{M}_{\text{inst}} = \frac{\mathrm{i}e^2}{q^2+\mathrm{i}\varepsilon}\left[-(\bar{u}_3\gamma^0 u_1)(\bar{u}_4\gamma^0 u_2) + (\bar{u}_3\boldsymbol{\gamma}\cdot\hat{\boldsymbol{q}}\, u_1)(\bar{u}_4\boldsymbol{\gamma}\cdot\hat{\boldsymbol{q}}\, u_2)\right]. \tag{4.142}$$

Turning now to the radiation field contribution,

$$\begin{aligned} S_{\text{rad}} &= \left\langle \mathrm{f} \left| \frac{(-\mathrm{i})^2}{2}\int \mathrm{d}^4x \int \mathrm{d}^4y\, \mathrm{T}[\mathcal{H}_{\text{rad}}(x)\mathcal{H}_{\text{rad}}(y)] \right| \mathrm{i} \right\rangle \\ &= \frac{-e^2}{2}\int \mathrm{d}^4x \int \mathrm{d}^4y\, \langle 0\,|\,\mathrm{T}[A^i(x)\,A^j(y)]\,|\,0\rangle \\ &\quad \times \left[\langle \mathrm{f}\,|\,J^i_{\mathrm{p}}(x)J^j_{\mathrm{e}}(y)\,|\,\mathrm{i}\rangle + \langle \mathrm{f}\,|\,J^i_{\mathrm{e}}(x)J^j_{\mathrm{p}}(y)\,|\,\mathrm{i}\rangle\right], \end{aligned} \tag{4.143}$$

we notice that its two terms are actually equal because

$$\mathrm{T}[A^i(x)\,A^j(y)] = \mathrm{T}[A^j(y)\,A^i(x)]\,.$$

The expectation value of this time-ordered product of two *transverse* photon fields can be calculated in the same way as in the case of scalar fields (see Problem 4.2), leading to the result

$$\begin{aligned} \mathrm{i}D^{ij}_{\mathrm{T}}(x-y) &\equiv \langle 0\,|\,\mathrm{T}[A^i(x)\,A^j(y)]\,|\,0\rangle \\ &= \int \frac{\mathrm{d}^4k}{(2\pi)^4}\, \mathrm{e}^{-\mathrm{i}k\cdot(x-y)}\, \mathrm{i}D^{ij}_{\mathrm{T}}(k)\,. \end{aligned} \tag{4.144}$$

In momentum space, the transverse photon propagator reads

$$D^{ij}_{\mathrm{T}}(k) = \frac{1}{k^2+\mathrm{i}\varepsilon}\left(\delta_{ij} - \frac{k^i k^j}{\boldsymbol{k}^2}\right). \tag{4.145}$$

In configuration space, the photon propagator, just like the scalar boson propagator, depends only on the difference of the coordinates of the space-time points at which the photon is created and subsequently destroyed.

The radiation part of the $S$-matrix is then given by

$$\begin{aligned} S_{\text{rad}} &= e^2\,(\bar{u}_3\gamma^i u_1)(\bar{u}_4\gamma^j u_2)\int \mathrm{d}^4x \int \mathrm{d}^4y\, \mathrm{i}D^{ij}_{\mathrm{T}}(x-y)\, \mathrm{e}^{\mathrm{i}(p'-p)\cdot x + \mathrm{i}(P'-P)\cdot y} \\ &= (2\pi)^4\delta^{(4)}(P'+p'-P-p)\, e^2\,(\bar{u}_3\gamma^i u_1)\, \mathrm{i}D^{ij}_{\mathrm{T}}(q)\,(\bar{u}_4\gamma^j u_2)\,, \end{aligned} \tag{4.146}$$

which corresponds to the reduced transition matrix

$$\mathrm{i}\mathcal{M}_{\text{rad}} = \frac{\mathrm{i}e^2}{q^2}\left[(\bar{u}_3\gamma^i u_1)(\bar{u}_4\gamma^i u_2) - (\bar{u}_3\boldsymbol{\gamma}\cdot\hat{\boldsymbol{q}}\, u_1)(\bar{u}_4\boldsymbol{\gamma}\cdot\hat{\boldsymbol{q}}\, u_2)\right]. \tag{4.147}$$

It is to be added to the instantaneous interaction part $\mathcal{M}_{\text{inst}}$ to give the full scattering amplitude to order $e^2$:

$$\begin{aligned}
\mathrm{i}\mathcal{M} &= \mathrm{i}\mathcal{M}_{\text{inst}} + \mathrm{i}\mathcal{M}_{\text{rad}} \\
&= \frac{\mathrm{i}e^2}{q^2}\left[-(\bar{u}_3\gamma^0 u_1)(\bar{u}_4\gamma^0 u_2) + (\bar{u}_3\gamma^i u_1)(\bar{u}_4\gamma^i u_2)\right] \\
&= (\bar{u}_3\gamma^\mu u_1)\,(\mathrm{i}e)\frac{-\mathrm{i}g_{\mu\nu}}{q^2+\mathrm{i}\varepsilon}\,(-\mathrm{i}e)(\bar{u}_4\gamma^\nu u_2)\,.
\end{aligned} \tag{4.148}$$

Introducing the covariant propagator $D_{\mu\nu}(q)$, we get a very simple expression for the invariant amplitude of the electron–proton elastic scattering via the exchange of a virtual photon:

$$\mathrm{i}\mathcal{M} = [\bar{u}(\boldsymbol{p}',s')\gamma^\mu u(\boldsymbol{p},s)]\,(\mathrm{i}e)\,\mathrm{i}D_{\mu\nu}(q)\,(-\mathrm{i}e)\,[\bar{u}(\boldsymbol{P}',S')\gamma^\nu u(\boldsymbol{P},S)]\,. \tag{4.149}$$

This comes as a somewhat surprising result because even though $\mathcal{M}_{\text{inst}}$ and $\mathcal{M}_{\text{rad}}$ taken separately are not relativistically invariant, their sum manifestly is. It corresponds to the Feynman diagram shown in Fig. 4.9 and can be calculated according to the following rules:

- assign a factor $-\mathrm{i}e\,\gamma^\mu$ to each vertex where a fermion with positive charge $e$ emits or absorbs a photon with polarization index $\mu$;
- assign to each internal photon carrying momentum $q$ and polarization indices $\mu$ and $\nu$ a propagator

$$\mathrm{i}D_{\mu\nu}(q) = \frac{-\mathrm{i}g_{\mu\nu}}{q^2+\mathrm{i}\varepsilon}\,;$$

- impose momentum conservation at each vertex;
- include an adjoint spinor $\bar{u}(\boldsymbol{p},s)$ for each outgoing fermion with momentum $\boldsymbol{p}$ and spin $s$, and a spinor $u(\boldsymbol{p},s)$ for each incoming fermion with momentum $\boldsymbol{p}$ and spin $s$; the relative order of the spinors in the matrix element, read from right to left, is the same as on a particle-oriented line from its beginning to its end so as to make the matrix element well defined, with $\bar{u}$ on the left and $u$ on the right of the relevant interaction operator.

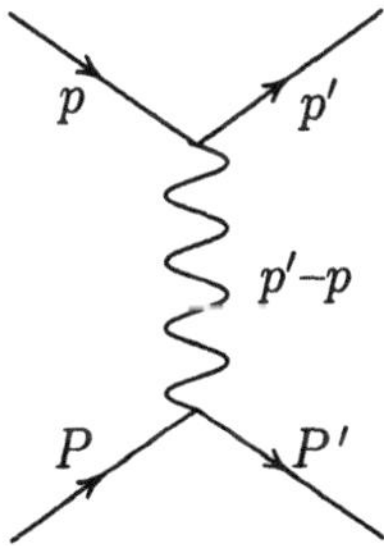


**Fig. 4.9.** First-order electron–proton scattering

The differential cross-section in the laboratory system can be calculated from the general result (67), modified to take into account the spin degrees of freedom. For unpolarized target and beam, and for undetected final spins, one sums over all final spin states and averages over all initial spin states. This spin averaging will introduce a factor of 1/4, since the electron and the proton have two possible spin states each. In the *laboratory frame* where the proton is initially at rest, $\boldsymbol{P}=0$, the differential cross-section for unpolarized electron–proton elastic scattering into the final electron momentum direction $\boldsymbol{p}'$ (such that $\hat{\boldsymbol{p}}\cdot\hat{\boldsymbol{p}}'=\cos\theta$) is then

$$\frac{\mathrm{d}\sigma}{\mathrm{d}\Omega}=\frac{e^4}{64\pi^2 M}\,\frac{|\boldsymbol{p}'|/|\boldsymbol{p}|}{[E+M-|\boldsymbol{p}|(E'/|\boldsymbol{p}'|)\cos\theta]}\,\frac{1}{4}|\overline{\mathcal{M}}|^2\,, \tag{4.150}$$

where we introduced the notation

$$|\overline{\mathcal{M}}|^2\equiv\frac{1}{(q^2)^2}\sum_{\text{spins}}|(\bar{u}_3\gamma^\mu u_1)\,(\bar{u}_4\gamma_\mu u_2)|^2\,. \tag{4.151}$$

The electron energies $E=E_{\boldsymbol{p}}=\sqrt{\boldsymbol{p}^2+m^2}$ and $E'=E_{\boldsymbol{p}'}=\sqrt{\boldsymbol{p}'^2+m^2}$ satisfy the (energy conservation) relation

$$E'(E+M)-|\boldsymbol{p}||\boldsymbol{p}'|\cos\theta=EM+m^2\,. \tag{4.152}$$

The calculation of $|\overline{\mathcal{M}}|^2$ will involve operations on spin states of the type

$$\begin{aligned}\sum_{s,s'}|\bar{u}(k',s')\Gamma u(k,s)|^2&=\sum_{s,s'}[\bar{u}(k',s')\Gamma u(k,s)]\,[\bar{u}(k',s')\Gamma u(k,s)]^*\\&=\sum_{s,s'}[\bar{u}(k',s')\Gamma u(k,s)]\,[u(k,s)^\dagger\Gamma^\dagger\gamma^0 u(k',s')]\\&=\sum_{s,s'}\bar{u}_i(k',s')\Gamma_{ij}u_j(k,s)\,\bar{u}_m(k,s)\bar{\Gamma}_{mn}u_n(k',s')\,,\end{aligned}$$

where $\bar{\Gamma}=\gamma^0\Gamma^\dagger\gamma^0$, and $i,j,\ldots$ indicate spinor components. Now, using a result found in the previous chapter,

$$\sum_s u_j(k,s)\,\bar{u}_m(k,s)=(\not{k}+m)_{jm}\,, \tag{4.153}$$

we get

$$\begin{aligned}\sum_{s,s'}|\bar{u}(k',s')\Gamma u(k,s)|^2&=(\not{k}'+m)_{ni}\Gamma_{ij}(\not{k}+m)_{jm}\bar{\Gamma}_{mn}\\&=\mathrm{Tr}\left[(\not{k}'+m)\Gamma(\not{k}+m)\bar{\Gamma}\right]\,.\end{aligned} \tag{4.154}$$

Thus, a summation over spins of a bilinear spinor product reduces to a calculation of the trace, or the sum of the diagonal elements, of a product of

$\gamma$-matrices. Such traces are invariant to unitarity transformations on the $\gamma$-matrices and are therefore independent of their representations. They can be found with the help of the following general results valid in four-dimensional space-time.

**Theorem 1.** *The trace of the product of an odd number of $\gamma$-matrices is zero:*

$$\mathrm{Tr}\,(\not{a}_1 \dots \not{a}_n) = 0, \qquad \text{for odd } n\,. \tag{4.155}$$

*Proof.* Since $\gamma_5^2 = 1$, one has for any $n$

$$\begin{aligned}\mathrm{Tr}\,(\not{a}_1 \dots \not{a}_n) &= \mathrm{Tr}\,(\not{a}_1 \dots \not{a}_n \gamma_5 \gamma_5) \\ &= \mathrm{Tr}\,(\gamma_5 \not{a}_1 \dots \not{a}_n \gamma_5) \\ &= (-)^n\, \mathrm{Tr}\,(\not{a}_1 \dots \not{a}_n \gamma_5 \gamma_5)\,,\end{aligned}$$

where on the second line we have used the cyclic property of the trace, i.e. $\mathrm{Tr}(abc) = \mathrm{Tr}(cab)$, and on the last line we have moved the first $\gamma_5$ to the right, using $\gamma_5 \gamma_\mu = -\gamma_\mu \gamma_5$. The presence of the resulting sign factor $(-)^n$ implies that the trace vanishes for odd $n$.

**Theorem 2.** *The traces of products of zero, two, and four $\gamma$-matrices are*

$$\begin{aligned}\mathrm{Tr}\,1 &= 4\,, \\ \mathrm{Tr}\,(\not{a}\not{b}) &= 4a{\cdot}b\,, \\ \mathrm{Tr}\,(\not{a}\not{b}\not{c}\not{d}) &= 4[(a{\cdot}b)(c{\cdot}d) + (a{\cdot}d)(b{\cdot}c) - (a{\cdot}c)(b{\cdot}d)]\,.\end{aligned} \tag{4.156}$$

*Proof.* For the power of 2,

$$\begin{aligned}\mathrm{Tr}\,(\not{a}\not{b}) &= \mathrm{Tr}\,(\not{b}\not{a}) = \tfrac{1}{2}\,\mathrm{Tr}(\not{a}\not{b} + \not{b}\not{a}) \\ &= (a{\cdot}b)\,\mathrm{Tr}\,1 = 4(a{\cdot}b)\,,\end{aligned}$$

while for the power of 4 (or any even power), we use $\not{a}\not{b} = -\not{b}\not{a} + 2a{\cdot}b$ to shift $\not{a}$ to the right of all the other factors and, at the end of the process, we move it back to the first position by using the cyclic property:

$$\begin{aligned}\mathrm{Tr}\,(\not{a}\not{b}\not{c}\not{d}) &= 2(a{\cdot}b)\,\mathrm{Tr}\,(\not{c}\not{d}) - \mathrm{Tr}\,(\not{b}\not{a}\not{c}\not{d}) \\ &= 8(a{\cdot}b)(c{\cdot}d) - 2(a{\cdot}c)\,\mathrm{Tr}\,(\not{b}\not{d}) + \mathrm{Tr}\,(\not{b}\not{c}\not{a}\not{d}) \\ &= 8(a{\cdot}b)(c{\cdot}d) - 8(a{\cdot}c)(b{\cdot}d) + 2(a{\cdot}d)\,\mathrm{Tr}\,(\not{b}\not{c}) - \mathrm{Tr}\,(\not{a}\not{b}\not{c}\not{d}) \\ &= 8[(a{\cdot}b)(c{\cdot}d) - (a{\cdot}c)(b{\cdot}d) + (a{\cdot}d)(b{\cdot}c)] - \mathrm{Tr}\,(\not{a}\not{b}\not{c}\not{d})\,.\end{aligned}$$

The announced result follows. One proceeds in the same way for a higher even power of $\gamma$ and expresses the result in terms of the traces of lower even powers of $\gamma$. ∎

Returning to our problem of electron–proton scattering, where $\Gamma = \gamma^\mu$ and therefore $\bar{\Gamma} = \gamma^0\gamma^{\mu\dagger}\gamma^0 = \gamma^\mu$, we now make use of the trace theorems to evaluate the required spin sums:

$$\begin{aligned}
|\overline{\mathcal{M}}|^2 &= \frac{1}{(q^2)^2}\mathrm{Tr}\left[(\not{p}' + m)\gamma^\mu(\not{p} + m)\gamma^\nu\right] \mathrm{Tr}\left[(\not{P}' + M)\gamma_\mu(\not{P} + M)\gamma_\nu\right] \\
&= \frac{16}{(q^2)^2}\left[p'^\mu p^\nu + p'^\nu p^\mu - g^{\mu\nu}(p'\cdot p - m^2)\right] \\
&\quad \times \left[P'_\mu P_\nu + P'_\nu P_\mu - g_{\mu\nu}(P'\cdot P - M^2)\right] \\
&= \frac{32}{(q^2)^2}\left[(p\cdot P)(p'\cdot P') + (p\cdot P')(p'\cdot P)\right. \\
&\quad \left. - m^2(P\cdot P') - M^2(p\cdot p') + 2m^2M^2\right] .
\end{aligned} \tag{4.157}$$

Inserting this expression for $|\overline{\mathcal{M}}|^2$ in (150) gives the exact differential cross-section for elastic scattering of structureless electron and proton in the laboratory system. We now consider its limiting values in two cases of interest.

For *nonrelativistic electrons* of energy $E \ll M$, energy conservation relation (152) implies that $E \approx E'$ and $|\boldsymbol{p}| \approx |\boldsymbol{p}'|$, so that

$$\begin{aligned}
|\overline{\mathcal{M}}|^2 &= \frac{32}{(q^2)^2}\left[2EE'M^2 - M^2(p\cdot p') + m^2M^2\right] \\
&= \frac{64M^2E^2}{(q^2)^2}\left(1 + \frac{q^2}{4E^2}\right) ,
\end{aligned}$$

which leads to the differential cross-section

$$\frac{\mathrm{d}\sigma}{\mathrm{d}\Omega} \approx \frac{e^4}{64\pi^2M^2}\,\frac{1}{4}|\overline{\mathcal{M}}|^2 = \frac{4\alpha^2E^2}{(q^2)^2}\left(1 + \frac{q^2}{4E^2}\right) , \qquad \text{for } \frac{E}{M} \ll 1 , \tag{4.158}$$

where, as usual, $\alpha = e^2/4\pi$ , and

$$q^2 = (p' - p)^2 \approx -4\boldsymbol{p}^2 \sin^2\frac{\theta}{2} .$$

Taking the limit $m \to 0$ gives the familiar formula for Mott cross-section

$$\sigma_{\mathrm{Mott}}(\theta) = \frac{\alpha^2}{4E^2}\frac{\cos^2(\theta/2)}{\sin^4(\theta/2)} . \tag{4.159}$$

When the proton recoil becomes important, the electron may be treated as *extremely relativistic* and its mass may be neglected, i.e. $m \ll E$, $E'$, so that $E \approx |\boldsymbol{p}|$ and $E' \approx |\boldsymbol{p}'|$. Then energy conservation (152) yields

$$M(E - E') \approx EE'(1 - \cos\theta) = 2EE' \sin^2(\theta/2) ,$$

and the momentum transfer factor becomes

$$q^2 = (p' - p)^2 \approx -2(p{\cdot}p') = -4EE'\sin^2(\theta/2)\,.$$

Inserting $P' = P + p - p'$ in (157), one obtains in the present situation

$$\begin{aligned}
|\overline{\mathcal{M}}|^2 &\approx \frac{32}{(q^2)^2}\left\{2(p{\cdot}P)(p'{\cdot}P) + (p{\cdot}p')[(p{\cdot}P) - (p'{\cdot}P) - M^2)]\right\} \\
&\approx \frac{32}{(q^2)^2}\left[2EE'M^2 + (p{\cdot}p')M(E - E' - M)\right] \\
&= \frac{64EE'M^2}{(q^2)^2}\left[\cos^2(\theta/2) - \frac{q^2}{2M^2}\sin^2(\theta/2)\right]\,. \qquad (4.160)
\end{aligned}$$

The differential cross-section is thus

$$\begin{aligned}
\frac{\mathrm{d}\sigma}{\mathrm{d}\Omega} &= \frac{e^4}{64\pi^2M^2}\left(\frac{E'}{E}\right)^2\frac{1}{4}|\overline{\mathcal{M}}|^2 \\
&= \sigma_{\text{Mott}}(\theta)\,\frac{E'}{E}\left[1 - \frac{q^2}{2M^2}\tan^2\frac{\theta}{2}\right] \\
&= \frac{\alpha^2}{4E^2}\frac{\cos^2(\theta/2)}{\sin^4(\theta/2)}\left(\frac{1 - (q^2/2M^2)\tan^2(\theta/2)}{1 + (2E/M)\sin^2(\theta/2)}\right)\,. \qquad (4.161)
\end{aligned}$$

To arrive at this formula, it was assumed that the proton is a structureless Dirac particle which behaves just like a heavy electron of mass $M$. The resulting cross-section is the Mott cross-section, corrected for recoil by the factor $E'/E$ and supplemented by a term proportional to $(q^2/2M^2)\tan^2(\theta/2)$ due to scattering from the *Dirac magnetic moment* of the proton. However, this description is still incomplete because it neglects the *structure* and the *anomalous magnetic moment* of the proton. It turns out that these effects, all due to strong interactions, can be taken into account at small momentum transfers without an explicit dynamic calculation by parameterizing them through two smooth functions of momentum, called *form factors*, in the proton electromagnetic current. We shall return to this point in Chap. 10.

## 4.7 Electron–Positron Annihilation

The $e^+e^-$ annihilation process is relatively simple to describe because, at energies lower than about 30 GeV, it is overwhelmingly dominated by the electromagnetic interaction. Yet it is physically interesting because, together with other lepton–lepton processes, it may be used to detect possible substructures of the leptons and to produce new types of leptons. At higher energies, hadrons will appear and while the same basic mechanism still prevails, strong interaction effects in the final state must be adequately included. The total cross-section for $e^+e^-$ annihilation into hadrons at very high energy

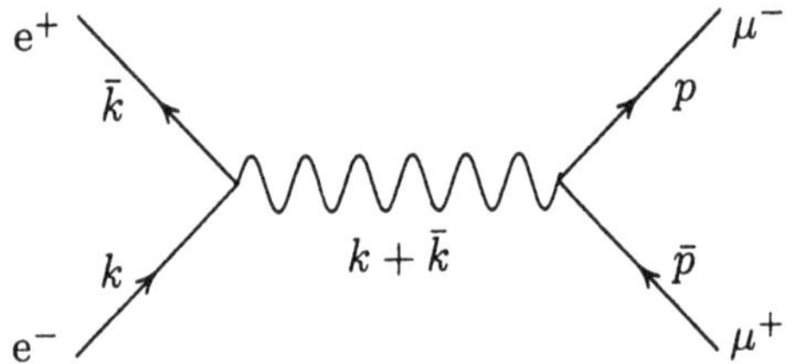


**Fig. 4.10.** Feynman diagram showing $\mu^+\mu^-$ pair production by $e^+e^-$ annihilation. Arrows indicate the directions of the particle (negative) charge flows; all momenta are in the direction of the time arrow, from left to right

has been instrumental in confirming the validity of quantum chromodynamics and the quark model.

We shall limit ourselves to a study of the $e^+e^- \to \mu^+\mu^-$ process in the lowest order of interaction, so that it may be treated as a virtual photon exchange mechanism, as in Fig. 4.10. The main purpose in the study of this example is to learn how to deal with antiparticle states.

The ket

$$|\mathrm{i}\rangle = |k\,\bar{k}\rangle = b^\dagger_{\boldsymbol{k}} d^\dagger_{\bar{\boldsymbol{k}}}\,|0\rangle\,(C_{\boldsymbol{k}}C_{\bar{\boldsymbol{k}}})^{-1} \tag{4.162}$$

represents an initial state composed of an electron and a positron with four-momenta $k$ and $\bar{k}$, such that $k^2 = \bar{k}^2 = m^2$, while the bra

$$\langle\mathrm{f}| = \langle\bar{p}\,p| = \langle 0|\, D_{\bar{\boldsymbol{p}}} B_{\boldsymbol{p}} (C_{\boldsymbol{p}} C_{\bar{\boldsymbol{p}}})^{-1} \tag{4.163}$$

describes a final state composed of $\mu^+$ and $\mu^-$ with momenta $p$ and $\bar{p}$, such that $p^2 = \bar{p}^2 = M^2$.

The approximate treatment we have in mind deals with the same expression for the $S$-matrix as in the previous section, i.e. (136), where $H_{\mathrm{inst}}$ and $H_{\mathrm{rad}}$ are given respectively by (132) and (133), but with the proton current replaced by the muon vector current

$$J^\nu_\mu(x) = -\overline{\psi}_\mu(x)\gamma^\nu\psi_\mu(x)\,. \tag{4.164}$$

Of course, in this process the full fields of both the electron and muon will participate in the reaction and so their Fourier series should be taken as

$$\psi_{\mathrm{e}}(x) = \sum_{\boldsymbol{p},s} \left[u(\boldsymbol{p},s)b(\boldsymbol{p},s)\phi^{(+)}_{\boldsymbol{p}}(x) + v(\boldsymbol{p},s)d^\dagger(\boldsymbol{p},s)\phi^{(-)}_{-\boldsymbol{p}}(x)\right], \tag{4.165}$$

$$\psi_\mu(x) = \sum_{\boldsymbol{p},s} \left[u(\boldsymbol{p},s)B(\boldsymbol{p},s)\phi^{(+)}_{\boldsymbol{p}}(x) + v(\boldsymbol{p},s)D^\dagger(\boldsymbol{p},s)\phi^{(-)}_{-\boldsymbol{p}}(x)\right]. \tag{4.166}$$

Proceeding as in the previous section, we obtain without difficulty the contribution from the instantaneous interaction part

$$\begin{aligned} S_{\mathrm{inst}} &= \left\langle \mathrm{f} \left| -\mathrm{i}\int_{-\infty}^{\infty} \mathrm{d}t\, H_{\mathrm{inst}} \right| \mathrm{i} \right\rangle \\ &= (2\pi)^4\delta^{(4)}(p+\bar{p}-k-\bar{k})\,[\bar{u}(p)\gamma^0 v(\bar{p})]\,\frac{-\mathrm{i}e^2}{\boldsymbol{K}^2}\,[\bar{v}(\bar{k})\gamma^0 u(k)]\,, \end{aligned} \tag{4.167}$$

where $K = k + \bar{k}$ and, to simplify, we have suppressed the spin arguments of the spinors. In a familiar step, we now replace the factor $1/\boldsymbol{K}^2$ with

$$\frac{1}{K^2 + \mathrm{i}\varepsilon} \frac{K_0^2 - \boldsymbol{K}^2}{\boldsymbol{K}^2}, \tag{4.168}$$

and use current conservation to rewrite the reduced matrix element in a more practical form

$$\begin{aligned} \mathrm{i}\mathcal{M}_{\text{inst}} &= [\bar{u}(p)\gamma^0 v(\bar{p})] \frac{-\mathrm{i}e^2}{\boldsymbol{K}^2} [\bar{v}(\bar{k})\gamma^0 u(k)] \\ &= \frac{\mathrm{i}e^2}{K^2} \left\{ [\bar{u}(p)\gamma^0 v(\bar{p})][\bar{v}(\bar{k})\gamma^0 u(k)] - [\bar{u}(p)\boldsymbol{\gamma}\cdot\hat{\boldsymbol{K}} v(\bar{p})][\bar{v}(\bar{k})\boldsymbol{\gamma}\cdot\hat{\boldsymbol{K}} u(k)] \right\}. \end{aligned} \tag{4.169}$$

Similarly, we calculate the radiation term

$$S_{\text{rad}} = \frac{(-\mathrm{i})^2}{2} \left\langle \mathrm{f} \left| \int \mathrm{d}^4x \int \mathrm{d}^4y \, \mathrm{T}[\mathcal{H}_{\text{rad}}(x)\mathcal{H}_{\text{rad}}(y)] \right| \mathrm{i} \right\rangle, \tag{4.170}$$

following the same steps as in the computation of the e–p transition matrix,

$$\begin{aligned} S_{\text{rad}} &= -e^2 \int \mathrm{d}^4x \int \mathrm{d}^4y \, \langle 0 | \mathrm{T}[A^i(x)A^j(y)] | 0 \rangle \langle \mathrm{f} | J^i_\mu(x) J^j_{\mathrm{e}}(y) | \mathrm{i} \rangle \\ &= -e^2 \int \mathrm{d}^3x \int \mathrm{d}^3y \, \mathrm{i}D^{ij}_{\mathrm{T}}(x-y) \langle \mathrm{f} | \overline{\psi}_\mu(x)\gamma^i\psi_\mu(x) \, \overline{\psi}_{\mathrm{e}}(y)\gamma^j\psi_{\mathrm{e}}(y) \, | \mathrm{i} \rangle \\ &= (2\pi)^4 \delta^{(4)}(p + \bar{p} - k - \bar{k})(-\mathrm{i}e^2) \, [\bar{u}(p)\gamma^i v(\bar{p})] \, D^{ij}_{\mathrm{T}}(K) \, [\bar{v}(\bar{k})\gamma^j u(k)], \end{aligned} \tag{4.171}$$

from which we extract the reduced matrix

$$\begin{aligned} \mathrm{i}\mathcal{M}_{\text{rad}} &= (-\mathrm{i}e^2) \, [\bar{u}(p)\gamma^i v(\bar{p})] \, D^{ij}_{\mathrm{T}}(K) \, [\bar{v}(\bar{k})\gamma^j u(k)] \\ &= \frac{-\mathrm{i}e^2}{K^2} \left\{ [\bar{u}(p)\gamma^i v(\bar{p})][\bar{v}(\bar{k})\gamma^i u(k)] - [\bar{u}(p)\gamma^i \hat{K}^i v(\bar{p})] \, [\bar{v}(\bar{k})\gamma^j \hat{K}^j u(k)] \right\}. \end{aligned} \tag{4.172}$$

We have made use of (145) for $D^{ij}_{\mathrm{T}}(K)$ with momentum transfer $K = k + \bar{k}$. Neither $\mathcal{M}_{\text{inst}}$ nor $\mathcal{M}_{\text{rad}}$ is separately relativistically covariant, but when they are added together, a cancelation of just the right terms miraculously occurs and correctly produces a covariant result, which is independent of the gauge initially selected for the calculations:

$$\begin{aligned} \mathrm{i}\mathcal{M} &= \mathrm{i}\mathcal{M}_{\text{inst}} + \mathrm{i}\mathcal{M}_{\text{rad}} \\ &= [\bar{u}(p)\gamma^\mu v(\bar{p})] \, (\mathrm{i}e) \, \frac{-\mathrm{i}g_{\mu\nu}}{K^2 + \mathrm{i}\varepsilon} \, (\mathrm{i}e) \, [\bar{v}(\bar{k})\gamma^\nu u(k)]. \end{aligned} \tag{4.173}$$

The Feynman rules suggested by this calculation are:

- a factor $-\mathrm{i}e\,\gamma^\mu$ at each vertex where a fermion with positive charge $e$ emits or absorbs a photon with polarization index $\mu$;
- a (covariant) propagator

$$\mathrm{i}D_{\mu\nu}(q) = \frac{-\mathrm{i}g_{\mu\nu}}{q^2 + \mathrm{i}\varepsilon}$$

  for each internal photon carrying momentum $q$ and polarization indices $\mu$ and $\nu$;
- momentum conservation at each vertex;
- matrix elements for fermions are formed with incoming fermion spinors, vertex operators and outgoing fermion spinors, such that the order of the factors read from right to left is the same as that found along each fermion line oriented by the negative charge flow, and
  - $u(k, s)$ for each incoming fermion of momentum $k$ and spin $s$;
  - $\bar{v}(\bar{k}, \bar{s})$ for each incoming antifermion of momentum $\bar{k}$ and spin $\bar{s}$;
  - $\bar{u}(p, S)$ for each outgoing fermion of momentum $p$ and spin $S$;
  - $v(\bar{p}, \bar{S})$ for each outgoing antifermion of momentum $\bar{p}$ and spin $\bar{S}$;

  in each matrix element, $u$ or $v$ stands to the right and $\bar{u}$ or $\bar{v}$ to the left of the vertex operator.

In recent years, $\mathrm{e}^+\mathrm{e}^-$ processes are often studied in colliding beam accelerators. In such experiments, the laboratory system coincides with the center-of-mass system, and the total energy available is considerably greater than in a fixed target experiment with the same beam energy ($s = 4E^2$ compared with $s \approx 2ME$, where $E$ is the beam energy and $M$ is the mass of the target particle).

The kinematic variables in the *center-of-mass system* are defined as

$$k = (E, \boldsymbol{k}), \quad \bar{k} = (E, -\boldsymbol{k}), \quad p = (E, \boldsymbol{p}), \quad \bar{p} = (E, -\boldsymbol{p}). \tag{4.174}$$

They are all related to the invariant $s$ through the relations

$$s = (k + \bar{k})^2 = K^2 = 4E^2, \tag{4.175}$$

$$|\boldsymbol{k}| = \frac{\sqrt{s}}{2}\sqrt{1 - \frac{4m^2}{s}}, \qquad |\boldsymbol{p}| = \frac{\sqrt{s}}{2}\sqrt{1 - \frac{4M^2}{s}}. \tag{4.176}$$

For an unpolarized experiment, where the incoming particles are unpolarized and the spin directions of the outgoing particles not measured, the cross-section in the center-of-mass system reads [cf. (59)]

$$\frac{\mathrm{d}\sigma}{\mathrm{d}\Omega} = \frac{e^4}{64\pi^2 s}\frac{|\boldsymbol{p}|}{|\boldsymbol{k}|}\frac{1}{4}|\overline{\mathcal{M}}|^2 = \frac{\alpha^2}{4s}\frac{|\boldsymbol{p}|}{|\boldsymbol{k}|}\frac{1}{4}|\overline{\mathcal{M}}|^2. \tag{4.177}$$

Here, as usual, $\alpha = e^2/4\pi$, and the sum over spins is

$$|\overline{\mathcal{M}}|^2 = \frac{1}{s^2} \sum_{\text{spins}} \left| [\bar{u}(p)\gamma^\mu v(\bar{p})]\, [\bar{v}(\bar{k})\gamma_\mu u(k)] \right|^2 . \tag{4.178}$$

In order to evaluate this sum, first note that $\gamma_0\gamma_\mu^\dagger\gamma_0 = \gamma_\mu$ and then recall the following results obtained in our discussion in Chap. 3 on the projection operators in spinor space:

$$\begin{aligned} \sum_s u(\boldsymbol{p}, s)\bar{u}(\boldsymbol{p}, s) &= \not{p} + m\,, \\ \sum_s v(\boldsymbol{p}, s)\bar{v}(\boldsymbol{p}, s) &= \not{p} - m\,, \qquad \text{where} \quad p = (E_{\boldsymbol{p}}, \boldsymbol{p})\,. \end{aligned} \tag{4.179}$$

Summations over electron and muon spins can now be converted without difficulties into traces of $\gamma$-matrix products:

$$|\overline{\mathcal{M}}|^2 = \frac{1}{s^2} \mathrm{Tr}[(\not{p} + M)\gamma^\mu(\bar{\not{p}} - M)\gamma^\nu]\; \mathrm{Tr}[(\not{k} + m)\gamma_\nu(\bar{\not{k}} - m)\gamma_\mu]\,.$$

It is the first step in a series of algebraic manipulations that, with the trace theorems helping, will finally lead to

$$\begin{aligned} |\overline{\mathcal{M}}|^2 &= \frac{16}{s^2}[p^\mu\bar{p}^\nu + p^\nu\bar{p}^\mu - g^{\mu\nu}(p{\cdot}\bar{p} + M^2)][k_\mu\bar{k}_\nu + k_\nu\bar{k}_\mu - g_{\mu\nu}(k{\cdot}\bar{k} + m^2)] \\ &= \frac{32}{s^2}[(p{\cdot}k)(\bar{p}{\cdot}\bar{k}) + (p{\cdot}\bar{k})(\bar{p}{\cdot}k) + M^2(k{\cdot}\bar{k}) + m^2(p{\cdot}\bar{p}) + 2M^2m^2]\,. \end{aligned}$$

It is then a simple matter to work out the kinematic relations to reach the final expression valid in the center-of-mass frame

$$|\overline{\mathcal{M}}|^2 = 4\left[1 + \frac{4}{s}(m^2 + M^2) + \left(1 - \frac{4m^2}{s}\right)\left(1 - \frac{4M^2}{s}\right)\cos^2\theta\right]. \tag{4.180}$$

The differential cross-section in the center-of-mass system for $\mathrm{e}^+\mathrm{e}^- \to \mu^+\mu^-$ depends on $s$ and $\theta$, the angle between $\boldsymbol{k}$ and $\boldsymbol{p}$:

$$\begin{aligned} \frac{\mathrm{d}\sigma}{\mathrm{d}\Omega} &= \frac{\alpha^2}{4s}\sqrt{\frac{1 - 4M^2/s}{1 - 4m^2/s}} \\ &\quad \times \left[1 + \frac{4}{s}(m^2 + M^2) + \left(1 - \frac{4m^2}{s}\right)\left(1 - \frac{4M^2}{s}\right)\cos^2\theta\right]. \end{aligned} \tag{4.181}$$

Upon integrating over the solid angle, one obtains the total cross-section

$$\sigma = \frac{4\pi\alpha^2}{3s}\sqrt{\frac{1 - 4M^2/s}{1 - 4m^2/s}}\left[1 + \frac{2(m^2 + M^2)}{s} + \frac{4m^2M^2}{s^2}\right]. \tag{4.182}$$

These results are exact, and are valid for any fermion–antifermion annihilation producing a different fermion–antifermion pair, $f\bar{f} \to F\bar{F}$, provided the particles involved have no structure. Several special limiting situations are of interest.

In the extremely relativistic limit where both masses are negligible, $m^2 \ll s$ and $M^2 \ll s$, the following approximate expressions may be used:

$$\frac{d\sigma}{d\Omega} = \frac{\alpha^2}{4s}(1+\cos^2\theta)\,, \tag{4.183}$$

$$\sigma = \frac{4\pi\alpha^2}{3s} = \frac{\pi\alpha^2}{3E^2}\,. \tag{4.184}$$

This result can be understood from simple scale arguments. The factor $\alpha^2$ follows from the order of the interaction (one quantum exchange, or two vertices). At high energy, where the masses are considered negligible, $1/\sqrt{s}$ remains as the only variable with the dimension of length, the cross-section must be proportional to $1/s$. In natural units, $1\ \mathrm{GeV}^{-2} = 390 \times 10^{-30}\,\mathrm{cm}^2$, and one may have, as a rough estimate,

$$\sigma \approx \frac{22\,\mathrm{nb}}{(E\text{ in GeV})^2} \qquad (1\,\mathrm{nb} = 10^{-33}\,\mathrm{cm}^2)\,. \tag{4.185}$$

This simple prediction that the total cross-section of pair production from pair annihilation through the electromagnetic interaction depends at very high energies only on the particle charges and the reaction energy, but not on any other parameters (e.g. masses), is in excellent agreement with data available on lepton pair productions, such as $e^+e^- \to \mu^+\mu^-$.

When the high-energy incoming beams produce heavy fermion pairs, such as in the production of heavy leptons $e^+e^- \to \tau^+\tau^-$, it is appropriate to consider $m^2/s \ll 1$ and $M/E \approx 1$, so that

$$\frac{|\boldsymbol{p}|}{|\boldsymbol{k}|} \approx \frac{|\boldsymbol{p}|}{E} \equiv \beta_\mathrm{f} \quad \text{and} \quad s = 4E^2 = 4(M^2+\boldsymbol{p}^2) \approx 4M^2(1+\beta_\mathrm{f}^2)\,. \tag{4.186}$$

The cross-sections may then be approximated by

$$\frac{d\sigma}{d\Omega} = \frac{\alpha^2}{4s}\beta_\mathrm{f}\left(2-\beta_\mathrm{f}^2\sin^2\theta\right)\,, \tag{4.187}$$

$$\sigma = \frac{2\pi\alpha^2}{s}\beta_\mathrm{f}\left(1-\frac{\beta_\mathrm{f}^2}{3}\right)\,. \tag{4.188}$$

Finally, if the situation calls for a heavy fermion pair in the initial state, such as in the leptonic decay of a heavy quark pair, the basic conditions are $M^2/s \ll 1$ and $m/E \approx 1$, so that

$$\frac{|\boldsymbol{p}|}{|\boldsymbol{k}|} \approx \frac{E}{|\boldsymbol{k}|} = \frac{1}{\beta_\mathrm{i}}, \quad \text{and} \quad s = 4E^2 = 4(m^2+\boldsymbol{k}^2) \approx 4m^2(1+\beta_\mathrm{i}^2)\,. \tag{4.189}$$

The cross-sections are then given by

$$\frac{\mathrm{d}\sigma}{\mathrm{d}\Omega} = \frac{\alpha^2}{4s}\frac{1}{\beta_\mathrm{i}}\left(2 - \beta_\mathrm{i}^2 \sin^2\theta\right), \tag{4.190}$$

$$\sigma = \frac{2\pi\alpha^2}{s\beta_\mathrm{i}}\left(1 - \frac{\beta_\mathrm{i}^2}{3}\right). \tag{4.191}$$

## 4.8 Compton Scattering

In this final example, which deals with the Compton effect, or the photon–electron scattering, we will learn how to describe incoming and outgoing states of real photons and the propagation of virtual spin-$1/2$ fermions.

We start by introducing the spin-$1/2$ fermion propagator, which is defined as a $4 \times 4$ matrix with elements given by the vacuum expectation values of the time-ordered products of a pair of fermion field operators:

$$\begin{aligned} \mathrm{i}S_{ij}(x,y) &= \langle 0 \,|\, \mathrm{T}[\psi_i(x)\overline{\psi}_j(y)] \,|\, 0\rangle \\ &= \theta(x_0 - y_0)\langle 0 \,|\, \psi_i(x)\overline{\psi}_j(y) \,|\, 0\rangle \\ &\quad - \theta(y_0 - x_0)\langle 0 \,|\, \overline{\psi}_j(y)\psi_i(x) \,|\, 0\rangle . \end{aligned} \tag{4.192}$$

The extra minus sign in the second term on the right-hand side arises from the interchange of two anticommuting operators. The fermion fields are operators with expansion series given in (166). Nonvanishing contributions to the vacuum-to-vacuum matrix elements in (192) can only come from products of operators of the types $bb^\dagger$ or $dd^\dagger$. For $x_0 > y_0$, the matrix element found in the first term on the right-hand side of (192) is evaluated on the basis of the spin sum of the positive-energy solutions in (179), with $E = \sqrt{\boldsymbol{p}^2 + m^2}$,

$$\begin{aligned} &\langle 0 \,|\, \psi_i(x)\overline{\psi}_j(y) \,|\, 0\rangle \\ &= \left\langle 0 \left| \sum_{\boldsymbol{p},s} u_i(\boldsymbol{p},s) b(\boldsymbol{p},s)\phi_{\boldsymbol{p}}^{(+)}(x) \sum_{\boldsymbol{p}',s'} \bar{u}_j(\boldsymbol{p}',s') b^\dagger(\boldsymbol{p}',s')\phi_{\boldsymbol{p}'}^{(+)*}(y) \right| 0 \right\rangle \\ &= \sum_{\boldsymbol{p}} \phi_{\boldsymbol{p}}^{(+)}(x)\phi_{\boldsymbol{p}}^{(+)*}(y) \sum_s u_i(\boldsymbol{p},s)\bar{u}_j(\boldsymbol{p},s) \\ &= \int \mathrm{d}^3p\, C_{\boldsymbol{p}}^2 \mathrm{e}^{-\mathrm{i}E(x_0-y_0)}\, \mathrm{e}^{\mathrm{i}\boldsymbol{p}\cdot(\boldsymbol{x}-\boldsymbol{y})}\,(m + E\gamma_0 - \boldsymbol{p}\cdot\boldsymbol{\gamma})_{ij} . \end{aligned} \tag{4.193}$$

For $y_0 > x_0$, we follow similar steps and get

$$\begin{aligned} &\langle 0 \,|\, \overline{\psi}_j(y)\psi_i(x) \,|\, 0\rangle \\ &= \left\langle 0 \left| \sum_{\boldsymbol{p},s} \bar{v}_j(\boldsymbol{p},s) d(\boldsymbol{p},s)\phi_{-\boldsymbol{p}}^{(-)*}(y) \sum_{\boldsymbol{p}',s'} v_i(\boldsymbol{p}',s') d^\dagger(\boldsymbol{p}',s')\phi_{-\boldsymbol{p}'}^{(-)}(x) \right| 0 \right\rangle \\ &= \sum_{\boldsymbol{p}} \phi_{-\boldsymbol{p}}^{(-)*}(y)\phi_{-\boldsymbol{p}}^{(-)}(x) \sum_s \bar{v}_j(\boldsymbol{p},s)v_i(\boldsymbol{p},s) \\ &= \int \mathrm{d}^3p\, C_{\boldsymbol{p}}^2 \mathrm{e}^{\mathrm{i}E(x_0-y_0)}\, \mathrm{e}^{\mathrm{i}\boldsymbol{p}\cdot(\boldsymbol{x}-\boldsymbol{y})}\,(-m + E\gamma_0 + \boldsymbol{p}\cdot\boldsymbol{\gamma})_{ij} , \end{aligned} \tag{4.194}$$

where we have used the projection on negative-energy solutions in (179), and changed the sign of the integration variable from $\boldsymbol{p}$ to $-\boldsymbol{p}$, so as to have the same exponential factor as in (193).

Putting these two results together, we have

$$\begin{aligned}
&\langle 0 \,|\, \mathrm{T}[\psi(x)\overline{\psi}(y)] \,|\, 0\rangle \\
&\quad = \int \mathrm{d}^3p\, C_{\boldsymbol{p}}^2 \mathrm{e}^{\mathrm{i}\boldsymbol{p}\cdot(\boldsymbol{x}-\boldsymbol{y})} \Big[\theta(x_0 - y_0)\, \mathrm{e}^{-\mathrm{i}E(x_0-y_0)}(m + E\gamma_0 - \boldsymbol{p}\cdot\boldsymbol{\gamma}) \\
&\qquad + \theta(y_0 - x_0)\, \mathrm{e}^{-\mathrm{i}E(y_0-x_0)}(m - E\gamma_0 - \boldsymbol{p}\cdot\boldsymbol{\gamma})\Big] . \qquad (4.195)
\end{aligned}$$

Let $Y$ stand for the expression inside the square brackets in (195), which combines the $x_0 > y_0$ contribution with the $y_0 > x_0$ contribution. In the next few steps, we merely repeat on $Y$ the same operations we performed on a similar expression found in the boson propagator (98), while making use of the integral representation of the step function (97). Thus,

$$\begin{aligned}
Y =& \frac{1}{2\pi\mathrm{i}} \Big[ \int_{-\infty}^{\infty} \mathrm{d}p_0 \, \frac{\mathrm{e}^{-\mathrm{i}p_0(x_0-y_0)}}{E - p_0 - \mathrm{i}\varepsilon} (m + E\gamma_0 - \boldsymbol{p}\cdot\boldsymbol{\gamma}) \\
&+ \int_{-\infty}^{\infty} \mathrm{d}p_0 \, \frac{\mathrm{e}^{-\mathrm{i}p_0(y_0-x_0)}}{E - p_0 - \mathrm{i}\varepsilon} (m - E\gamma_0 - \boldsymbol{p}\cdot\boldsymbol{\gamma}) \Big] \\
=& \frac{1}{2\pi\mathrm{i}} \int_{-\infty}^{\infty} \mathrm{d}p_0 \, \mathrm{e}^{-\mathrm{i}p_0(x_0-y_0)} \left( \frac{m + E\gamma_0 - \boldsymbol{p}\cdot\boldsymbol{\gamma}}{E - p_0 - \mathrm{i}\varepsilon} + \frac{m - E\gamma_0 - \boldsymbol{p}\cdot\boldsymbol{\gamma}}{E + p_0 - \mathrm{i}\varepsilon} \right) \\
=& \frac{2E}{2\pi\mathrm{i}} \int_{-\infty}^{\infty} \mathrm{d}p_0 \, \mathrm{e}^{-\mathrm{i}p_0(x_0-y_0)} \, \frac{m + \gamma_0 p_0 - \boldsymbol{\gamma}\cdot\boldsymbol{p}}{E^2 - p_0^2 - \mathrm{i}\varepsilon'} , \qquad (4.196)
\end{aligned}$$

where $\varepsilon' = 2E\varepsilon$ is an infinitesimal positive real quantity, to be simply called $\varepsilon$ from now on. Since

$$E^2 - p_0^2 - \mathrm{i}\varepsilon = m^2 - (p_0^2 - \boldsymbol{p}^2) - \mathrm{i}\varepsilon = m^2 - p^2 - \mathrm{i}\varepsilon ,$$

the denominator that appears on the last line is manifestly invariant. With this result inserted into (195), we obtain

$$\begin{aligned}
\mathrm{i}S(x - y) &= \langle 0 \,|\, \mathrm{T}[\psi(x)\overline{\psi}(y)] \,|\, 0\rangle \\
&= \int \frac{\mathrm{d}^4p}{(2\pi)^4} \, \mathrm{e}^{-\mathrm{i}p\cdot(x-y)} \, \frac{\mathrm{i}(\not{p} + m)}{p^2 - m^2 + \mathrm{i}\varepsilon} \\
&= \int \frac{\mathrm{d}^4p}{(2\pi)^4} \, \mathrm{e}^{-\mathrm{i}p\cdot(x-y)} \, \frac{\mathrm{i}}{\not{p} - m + \mathrm{i}\varepsilon} . \qquad (4.197)
\end{aligned}$$

This remarkable result, which holds for all times, regardless of whether $x_0$ is earlier or later than $y_0$, describes the propagation amplitude of a Dirac fermion. In coordinate space, it depends on the coordinate distance $x - y$, while in momentum space, it is

$$S(p) = \frac{(\not{p} + m)}{p^2 - m^2 + \mathrm{i}\varepsilon} = \frac{1}{\not{p} - m + \mathrm{i}\varepsilon} . \qquad (4.198)$$

Given this result, we are now ready to consider the Compton process

$$\gamma(k,\lambda) + \mathrm{e}^-(p,s) \to \gamma(k',\lambda') + \mathrm{e}^-(p',s') .$$

Here, $s$, $s'$, $\lambda$, and $\lambda'$ denote the spin states of the particles involved; and $k$, $k'$, $p$, and $p'$ stand for their four-momenta (with, in particular, the energy components $k_0 = \omega_k = |\boldsymbol{k}|$ and $k'_0 = \omega'_k = |\boldsymbol{k}'|$). The quantum system is described by an electron field $\psi(x)$ and an electromagnetic field $A_\mu(x)$, the latter driven by some external source that is not part of the system. The coupling of the field $A_\mu(x)$ to the electron current

$$J^\mu(x) = -\overline{\psi}(x)\gamma^\mu\psi(x) \tag{4.199}$$

is represented by the interaction Lagrangian

$$\mathcal{L}_{\mathrm{rad}} = -e\, J^\mu(x)\, A_\mu(x), \qquad (e > 0) . \tag{4.200}$$

Since the present problem involves only particle and not antiparticle states, it suffices to keep only the positive-energy components in the electron field

$$\psi(x) = \sum_{\boldsymbol{p},s} u(\boldsymbol{p},s) b(\boldsymbol{p},s) \phi^{(+)}_{\boldsymbol{p}}(x) . \tag{4.201}$$

As for the radiation field, the full field will participate in the process

$$A^\mu = \sum_{\boldsymbol{k},\lambda=1}^{2} \left[ e^\mu(k,\lambda)\, a(\boldsymbol{k},\lambda)\, \phi^{(+)}_{\boldsymbol{k}}(x) + e^{\mu *}(k,\lambda)\, a^\dagger(\boldsymbol{k},\lambda)\, \phi^{(-)}_{-\boldsymbol{k}}(x) \right] . \tag{4.202}$$

The wave functions $\phi^{(\pm)}(x)$ in the preceding two equations are defined as in (83). To simplify, the polarization vectors $e^\mu(k,\lambda)$ will be assumed real.

We start by writing down the initial and final state vectors:

$$|\mathrm{i}\rangle = \left|\gamma(k,\lambda), \mathrm{e}^-(p,s)\right\rangle = a^\dagger(\boldsymbol{k},\lambda) b^\dagger(\boldsymbol{p},s)\, |0\rangle\, (C_{\boldsymbol{k}} C_{\boldsymbol{p}})^{-1} , \tag{4.203}$$

$$\langle \mathrm{f}| = \left\langle \gamma(k',\lambda'), \mathrm{e}^-(p',s')\right| = (C_{\boldsymbol{k}'} C_{\boldsymbol{p}'})^{-1}\, \langle 0|\, a(\boldsymbol{k}',\lambda')\, b(\boldsymbol{p}',s') . \tag{4.204}$$

In the lowest nontrivial order of interaction, the Compton effect is described by the matrix element

$$S_{\mathrm{Com}} = \frac{(-\mathrm{i}e)^2}{2} \int \mathrm{d}^4x\, d^4y\, \langle \mathrm{f}\,|\,\mathrm{T}[J^\mu(x)A_\mu(x)\, J^\nu(y)A_\nu(y)]\,|\,\mathrm{i}\rangle . \tag{4.205}$$

We first calculate the contributions from the radiation field:

$$\begin{aligned} &\langle k'\lambda'\,|\,A_\mu(x)A_\nu(y)\,|\,k\lambda\rangle \\ &\quad = e_\mu(k,\lambda)e_\nu(k',\lambda')\,\mathrm{e}^{-\mathrm{i}k\cdot x+\mathrm{i}k'\cdot y} + e_\mu(k',\lambda')e_\nu(k,\lambda)\,\mathrm{e}^{\mathrm{i}k'\cdot x-\mathrm{i}k'\cdot y}, \end{aligned} \tag{4.206}$$

and find that the two terms which appear here actually result in equal contributions to the integrals over $x$ and $y$ in $S_{\text{Com}}$, so that

$$S_{\text{Com}} = (-\mathrm{i}e)^2 \int \mathrm{d}^4x\, \mathrm{d}^4y\, \langle \mathrm{f} \,|\, \mathrm{T}[J(x)\cdot e(k,\lambda)\, J(y)\cdot e(k'\lambda')] \,|\, \mathrm{i}\rangle\, \mathrm{e}^{-\mathrm{i}k\cdot x + \mathrm{i}k'\cdot y}\,. \tag{4.207}$$

To evaluate the matrix element in (207), which involves a product of four electron fields, we need an operator $b^\dagger(\boldsymbol{q})$ and an operator $b(\boldsymbol{q}')$ in the product $J^\mu(x)\, J^\nu(y)$ to annihilate the incident electron and create the outgoing electron. Two such possibilities exist in $J^\mu(x)\, J^\nu(y)$, leaving the other two Fock operators to create and annihilate a virtual electron or positron with the probability amplitude

$$\langle 0 \,|\, \mathrm{T}[\psi_i(x)\overline{\psi}_j(y)] \,|\, 0\rangle = \mathrm{i}\, S_{ij}(x-y)\,, \tag{4.208}$$

which is the electron propagator in configuration space. Therefore, if we discard the uninteresting forward scattering, the time-ordered product of the current operators in (207) may be effectively taken as

$$\begin{aligned} &\mathrm{T}[J^\mu(x)\, J^\nu(y)] \\ &= \overline{\psi}(x)\gamma^\mu\, \mathrm{i}S(x-y)\, \gamma^\nu\psi(y) + \overline{\psi}(y)\gamma^\nu\, \mathrm{i}S(y-x)\, \gamma^\mu\psi(x) + \dots \end{aligned}$$

(where ... represents terms that do not contribute), so that the matrix element produces

$$\begin{aligned} \langle p' \,|\, \mathrm{T}[J^\mu(x)J^\nu(y)] \,|\, p\rangle &= \bar{u}(\boldsymbol{p}',s')\gamma^\mu \mathrm{i}S(x-y)\gamma^\nu u(\boldsymbol{p},s)\, \mathrm{e}^{\mathrm{i}p'\cdot x - \mathrm{i}p\cdot y} \\ &+ \bar{u}(\boldsymbol{p}',s')\gamma^\nu \mathrm{i}S(y-x)\gamma^\mu u(\boldsymbol{p},s)\, \mathrm{e}^{\mathrm{i}p'\cdot y - \mathrm{i}p\cdot x}\,. \end{aligned} \tag{4.209}$$

Once this expression is substituted into (207) and the indicated integrations are performed, we can drop the four-momentum conservation factor $(2\pi)^4\delta^{(4)}(k'+p'-k-p)$ to have the reduced amplitude we are seeking:

$$\begin{aligned} \mathrm{i}\mathcal{M} &= (-\mathrm{i}e)^2\, [\bar{u}(p')\, \not{e}'\, \mathrm{i}S(p+k)\, \not{e}\, u(p) + \bar{u}(p')\, \not{e}\, \mathrm{i}S(p-k')\, \not{e}'\, u(p)] \\ &= (-\mathrm{i}e)^2 \bar{u}(p') \left[ \not{e}' \frac{\mathrm{i}(\not{p} + \not{k} + m)}{(p+k)^2 - m^2}\, \not{e} + \not{e} \frac{\mathrm{i}(\not{p} - \not{k}' + m)}{(p-k')^2 - m^2}\, \not{e}' \right] u(p)\,, \end{aligned} \tag{4.210}$$

where $\not{e} = \gamma_\mu e^\mu(k,\lambda)$ and $\not{e}' = \gamma_\mu e^\mu(k',\lambda')$. Notice that the substitution

$$k,\, e \leftrightarrow -k',\, e' \tag{4.211}$$

interchanges the two terms in (210) and so the Compton transition amplitude $\mathcal{M}$ in (210) is invariant under this transformation. This symmetry persists

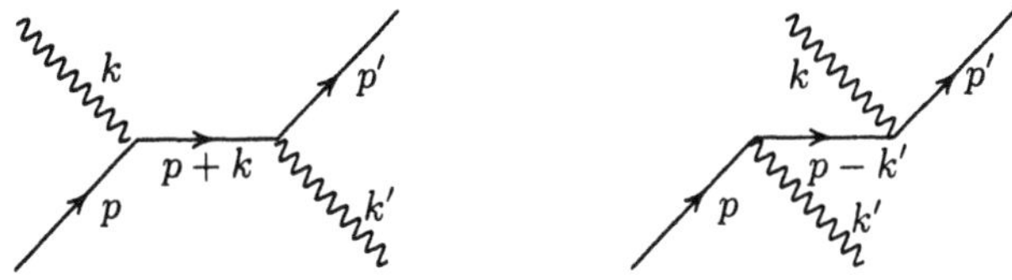


**Fig. 4.11.** Feynman diagrams for the Compton scattering

in all higher-order terms as an exact symmetry, which has come to be known in particle physics as the *crossing symmetry*.

The two terms written in (210) can be visualized by drawing the corresponding two Feynman diagrams shown in Fig. 4.11. They suggest the following calculation rules:

- to each vertex where a fermion with charge $e$ emits or absorbs a photon with polarization index $\mu$, assign a factor $-\mathrm{i}e\,\gamma^\mu$ ;
- to each internal fermion with momentum $p$ and Dirac indices $i$ and $j$ corresponds a factor

$$\mathrm{i}S_{ij}(p) = \left(\frac{\mathrm{i}}{\not{p} - m + \mathrm{i}\varepsilon}\right)_{ij} = \frac{\mathrm{i}(\not{p} + m)_{ij}}{p^2 - m^2 + \mathrm{i}\varepsilon} \,;$$

- impose four-momentum conservation at each vertex;
- matrix elements for fermions are formed as in the previous calculations;
- to a photon with momentum $k$ and polarization $\lambda$ absorbed at a vertex $-\mathrm{i}e\gamma^\mu$, assign a factor $e_\mu(k,\lambda)$; to a photon with momentum $k'$ and polarization $\lambda'$ emitted at a vertex $-\mathrm{i}e\gamma^\mu$, assign a factor $e^*_\mu(k',\lambda')$.

Before performing any further calculations on $\mathcal{M}$, it is useful to reduce it to its simplest form. The relations $k{\cdot}e = 0$, $k'{\cdot}e' = 0$, and $k^2 = k'^2 = 0$, $p^2 = p'^2 = m^2$ hold quite generally, and the spinors $u(p)$ and $u(p')$ satisfy free Dirac equations. Thus, the first term in (210) can be simplified with the relations

$$\begin{aligned}(p+k)^2 - m^2 &= k^2 + p^2 + 2k{\cdot}p - m^2 = 2k{\cdot}p\,,\\ \not{e}'(\not{p} + \not{k} + m)\,\not{e}\,u(p) &= \not{e}'[(2p{\cdot}e - \not{e}\,\not{p}) + (\not{k} + m)\,\not{e}\,]\,u(p)\\ &= \not{e}'(2p{\cdot}e + \not{k}\,\not{e})\,u(p)\,;\end{aligned}$$

and the second term, with

$$\begin{aligned}(p-k')^2 - m^2 &= k'^2 + p^2 - 2k'{\cdot}p - m^2 = -2k'{\cdot}p\,,\\ \not{e}(\not{p} - \not{k}' + m)\,\not{e}'u(p) &= \not{e}\,[(2p{\cdot}e' - \not{e}'\,\not{p}) + (-\not{k}' + m)\,\not{e}'\,]\,u(p)\\ &= \not{e}(2p{\cdot}e' - \not{k}'\,\not{e}')\,u(p)\,.\end{aligned}$$

These simplifications carry the reduced transition matrix into

$$\mathrm{i}\mathcal{M} = -\mathrm{i}e^2\bar{u}(p')\left(\frac{\not{e}'\,\not{k}\,\not{e} + 2\,\not{e}'p{\cdot}e}{2p{\cdot}k} + \frac{\not{e}\,\not{k}'\,\not{e}' - 2\,\not{e}p{\cdot}e'}{2p{\cdot}k'}\right)u(p)\,. \tag{4.212}$$

In a gauge where the initial and final photons are transversely polarized, their polarization vectors have the properties

$$
\begin{aligned}
e^\mu &= (0, \boldsymbol{e}) \qquad \text{with } \boldsymbol{k}\cdot\boldsymbol{e} = 0\,, \qquad \text{and } \boldsymbol{e}^2 = 1\,,\\
e'^\mu &= (0, \boldsymbol{e}') \qquad \text{with } \boldsymbol{k}'\cdot\boldsymbol{e}' = 0\,, \qquad \text{and } \boldsymbol{e}'^2 = 1\,.
\end{aligned}
$$

Now, if we choose the *laboratory* system with the initial electron at rest, $p = (m, \boldsymbol{0})$, in which to do further calculations, then the relations $p{\cdot}e = 0$ and $p{\cdot}e' = 0$, which hold in this reference frame, eliminate two terms in $\mathcal{M}$, reducing it to

$$
\mathrm{i}\mathcal{M} = -\mathrm{i}e^2 \bar{u}(p') \Big[ \frac{\not{e}'\,\not{k}\,\not{e}}{2p{\cdot}k} + \frac{\not{e}\,\not{k}'\,\not{e}'}{2p{\cdot}k'} \Big] u(p)\,. \tag{4.213}
$$

The cross-section in the laboratory frame has already been given in (64). In the present kinematics, it reads

$$
\begin{aligned}
\frac{\mathrm{d}\sigma}{\mathrm{d}\Omega} &= \frac{|\mathcal{M}|^2}{64\pi^2 m} \frac{|\boldsymbol{k}'|/|\boldsymbol{k}|}{|\boldsymbol{k}| + m - |\boldsymbol{k}|\cos\theta}\\
&= \frac{\omega'^2}{64\pi^2 m^2 \omega^2} |\mathcal{M}|^2\,,
\end{aligned} \tag{4.214}
$$

where we have used (66) to write

$$
\frac{\omega}{\omega'} = 1 + \frac{\omega}{m}(1 - \cos\theta)\,, \quad \omega = |\boldsymbol{k}|\,, \quad \text{and} \quad \omega' = |\boldsymbol{k}'|\,. \tag{4.215}
$$

Here $\theta$ is the photon scattering (lab) angle, $\cos\theta = \hat{\boldsymbol{k}}{\cdot}\hat{\boldsymbol{k}}'$.

We are interested in an unpolarized electron scattering, so that we have to average over the two spin states of the incoming electron, and sum over the final electron spin states. But the photon may have definite polarizations, $\lambda$ and $\lambda'$, in the initial and final states. Thus, the differential cross-section for the Compton scattering in the laboratory system is given by

$$
\frac{\mathrm{d}\sigma}{\mathrm{d}\Omega} = \Big( \frac{\alpha\omega'}{2m\omega} \Big)^2 \frac{1}{2} \overline{|\mathcal{M}|^2}\,, \tag{4.216}
$$

with the notation

$$
\overline{|\mathcal{M}|^2} = \sum_{ss'} \Big| \bar{u}(p's') \Big( \frac{\not{e}'\,\not{k}\,\not{e}}{2p{\cdot}k} + \frac{\not{e}\,\not{k}'\,\not{e}'}{2p{\cdot}k'} \Big) u(ps) \Big|^2\,. \tag{4.217}
$$

Now, since $\gamma_0 \gamma_\mu^\dagger \gamma_0 = \gamma_\mu$, the sum over spins can be reduced to a trace exactly in the same way as in the previous section:

$$
\begin{aligned}
\overline{|\mathcal{M}|^2} &= \mathrm{Tr}\Big[ (\not{p}' + m) \Big( \frac{\not{e}'\,\not{k}\,\not{e}}{2p{\cdot}k} + \frac{\not{e}\,\not{k}'\,\not{e}'}{2p{\cdot}k'} \Big) (\not{p} + m) \Big( \frac{\not{e}\,\not{k}\,\not{e}'}{2p{\cdot}k} + \frac{\not{e}'\,\not{k}'\,\not{e}}{2p{\cdot}k'} \Big) \Big]\\
&= \frac{1}{(2p{\cdot}k)^2} T_1 + \frac{1}{(2p{\cdot}k')^2} T_2 + \frac{1}{(2p{\cdot}k)(2p{\cdot}k')} (T_3 + T_4)\,,
\end{aligned} \tag{4.218}
$$

in which the final expression was divided into four terms, related under the substitution (211) by

$$T_1 \leftrightarrow T_2\,, \qquad T_3 \leftrightarrow T_4\,. \tag{4.219}$$

Therefore, it suffices to calculate, for example, just $T_1$ and $T_3$. The main technical difficulty in this problem resides in the calculation of these traces. The general approach is to use the trace Theorem 1 to eliminate the trivial terms and to shift factors in the remaining terms, using

$$\not{a}\not{b} = -\not{b}\not{a} + 2a{\cdot}b\,,$$

until one or another of the relations $\not{e}^2 = \not{e}'^2 = -1$, $\not{k}^2 = \not{k}'^2 = 0$, $\not{p}^2 = \not{p}'^2 = m^2$, and $p{\cdot}e = p{\cdot}e' = k{\cdot}e = k'{\cdot}e' = 0$ can be used to reduce the numbers of the $\gamma$-factors, and finally to apply the trace Theorem 2 for the final answer. In this way we proceed, first with $T_1$,

$$\begin{aligned}
T_1 &= \mathrm{Tr}[(\not{p}'+m)\;\not{e}'\;\not{k}\;\not{e}(\not{p}+m)\;\not{e}\;\not{k}\;\not{e}'] \\
&= \mathrm{Tr}(\not{p}'\;\not{e}'\;\not{k}\;\not{e}\;\not{p}\;\not{e}\;\not{k}\;\not{e}') + m^2\,\mathrm{Tr}(\not{e}'\;\not{k}\;\not{e}\;\not{e}\;\not{k}\;\not{e}') && \text{(even number of } \gamma) \\
&= \mathrm{Tr}(\not{p}'\;\not{e}'\;\not{k}\;\not{e}\;\not{p}\;\not{e}\;\not{k}\;\not{e}') && (e^2=-1,\ k^2=0) \\
&= \mathrm{Tr}(\not{p}'\;\not{e}'\;\not{k}\;\not{p}\;\not{k}\;\not{e}') && (\not{p}\;\not{e} = -\;\not{e}\;\not{p},\ e^2=-1) \\
&= 2p{\cdot}k\,\mathrm{Tr}(\not{p}'\;\not{e}'\;\not{k}\;\not{e}') && (\not{p}\;\not{k} = -\;\not{k}\;\not{p} + 2p{\cdot}k,\ k^2=0) \\
&= 8p{\cdot}k\,[2(k{\cdot}e')(p'{\cdot}e') - (p'{\cdot}k)e'^2] \\
&= 8p{\cdot}k\,[2(k{\cdot}e')^2 + p{\cdot}k'] && (p'{\cdot}e' = k{\cdot}e',\ p'{\cdot}k = p{\cdot}k')\,.
\end{aligned}$$

As for $T_3$, it proves convenient to substitute $p+k-k'$ for $p'$ and to split $T_3$ into two as follows:

$$\begin{aligned}
T_3 &= \mathrm{Tr}\,[(\not{p}'+m)\;\not{e}'\;\not{k}\;\not{e}(\not{p}+m)\;\not{e}'\;\not{k}'\;\not{e}] \\
&= \mathrm{Tr}\,[(\not{p}+m)\;\not{e}'\;\not{k}\;\not{e}(\not{p}+m)\;\not{e}'\;\not{k}'\;\not{e}] + \mathrm{Tr}\,[(\not{k}-\not{k}')\;\not{e}'\;\not{k}\;\not{e}\;\not{p}\;\not{e}'\;\not{k}'\;\not{e}] \\
&= T_{3a} + T_{3b}\,.
\end{aligned}$$

Let us start with $T_{3a}$:

$$\begin{aligned}
T_{3a} &= \mathrm{Tr}\,[(\not{p}\;\not{k}\;\not{p} + m^2\;\not{k})\;\not{e}\;\not{e}'\;\not{k}'\;\not{e}\;\not{e}'] && (\not{p}\;\not{e} = -\;\not{e}\;\not{p},\ \not{p}\;\not{e}' = -\;\not{e}'\;\not{p}) \\
&= 2(p{\cdot}k)\mathrm{Tr}(\not{p}\;\not{e}\;\not{e}'\;\not{k}'\;\not{e}\;\not{e}') && (\not{p}\;\not{k} = 2p{\cdot}k -\;\not{k}\;\not{p} \text{ and } p^2 = m^2)\,.
\end{aligned}$$

Next anticommute $\not{p}$ all the way through to the right and use $\mathrm{Tr}\,abc = \mathrm{Tr}\,cab$,

$$\begin{aligned}
T_{3a} &= 2p{\cdot}k\;\left[2p{\cdot}k'\mathrm{Tr}(\not{e}\;\not{e}'\;\not{e}\;\not{e}') - \mathrm{Tr}(\not{p}\;\not{e}\;\not{e}'\;\not{k}'\;\not{e}\;\not{e}')\right] \\
&= 2(p{\cdot}k)(p{\cdot}k')\,\mathrm{Tr}(\not{e}\;\not{e}'\;\not{e}\;\not{e}') \\
&= 8(p{\cdot}k)(p{\cdot}k')[2(e{\cdot}e')^2 - 1]\,.
\end{aligned}$$

In the second term, $T_{3b}$, we use the relations $\not k \not e' \not k = 2(k{\cdot}e') \not k$ and $\not k' \not e \not k' = 2(k'{\cdot}e) \not k'$ to get

$$\begin{aligned} T_{3b} &= 2k{\cdot}e' \operatorname{Tr}(\not k' \not e \not p \not e' \not k' \not e) - 2k'{\cdot}e \operatorname{Tr}(\not k' \not e' \not k \not e \not p \not e') \\ &= 2k{\cdot}e' \operatorname{Tr}(\not k' \not k \not p \not e') - 2k'{\cdot}e \operatorname{Tr}(\not k' \not k \not e \not p) \\ &= -8p{\cdot}k'(k{\cdot}e')^2 + 8p{\cdot}k(k'{\cdot}e)^2 \,. \end{aligned}$$

Summing up $T_{3a}$ and $T_{3b}$ yields

$$T_3 = 8(p{\cdot}k)(p{\cdot}k')[2(e{\cdot}e')^2 - 1] - 8(p{\cdot}k')(k{\cdot}e')^2 + 8(p{\cdot}k)(k'{\cdot}e)^2 \,.$$

Finally, by the crossing substitution (211),

$$\begin{aligned} T_2 &= -8\,(p{\cdot}k')\,[2(k'{\cdot}e')^2 - (p{\cdot}k)] \,, \\ T_4 &= T_3 \,. \end{aligned}$$

Putting all these partial results together in (216), we find the famous Klein–Nishina formula for the Compton scattering

$$\frac{\mathrm{d}\sigma}{\mathrm{d}\Omega} = \frac{\alpha^2}{4m^2}\left(\frac{\omega'}{\omega}\right)^2\left[\frac{\omega'}{\omega} + \frac{\omega}{\omega'} + 4(e{\cdot}e')^2 - 2\right] \qquad \text{(Klein–Nishina)}. \quad (4.220)$$

In the low-energy limit, when $\omega \to 0$ and, by (215), $\omega'/\omega \to 1$, the differential cross-section becomes

$$\frac{\mathrm{d}\sigma}{\mathrm{d}\Omega} \approx \left(\frac{\alpha}{m}\right)^2 (e{\cdot}e')^2 \,.$$

Note that it is proportional to the square of the classical radius of the electron:

$$\frac{\alpha}{m} = \frac{e^2}{4\pi mc^2} = 2.8 \times 10^{-13}\,\mathrm{cm} \,.$$

Finally, when the initial photons are unpolarized and the final photon polarizations are not observed, we average over the two possible $\lambda$ states and sum over the two possible $\lambda'$ states, so that

$$\begin{aligned} \frac{1}{2}\sum_{\lambda\lambda'} (e{\cdot}e')^2 &= \frac{1}{2}\sum_{\lambda\lambda'} \left[e^i(k,\lambda)\, e^i(k',\lambda')\right]^2 \\ &= \frac{1}{2}\left(\delta_{ij} - \frac{k^i k^j}{\boldsymbol{k}^2}\right)\left(\delta_{ij} - \frac{k'^i\, k'^j}{\boldsymbol{k}'^2}\right) \\ &= \frac{1}{2}(1 + \cos^2\theta) \,. \end{aligned}$$

The unpolarized Compton cross-section in the lab frame then comes out to be

$$\frac{\mathrm{d}\bar\sigma}{\mathrm{d}\Omega} = \frac{\alpha^2}{2m^2}\left(\frac{\omega'}{\omega}\right)^2\left[\frac{\omega'}{\omega} + \frac{\omega}{\omega'} - \sin^2\theta\right] . \qquad (4.221)$$

At low energies, as $\omega \to 0$ and $\omega'/\omega \to 1$, it reduces to

$$\frac{d\bar{\sigma}}{d\Omega} = \frac{\alpha^2}{2m^2}(1+\cos^2\theta)\,, \tag{4.222}$$

and when integrated over the solid angle, it gives the famous Thomson cross-section formula

$$\bar{\sigma} = \frac{8\pi}{3}\frac{\alpha^2}{m^2} \qquad \text{(Thomson)}. \tag{4.223}$$

This result, apart from the numerical factor, can be understood from purely dimensional considerations. Except for $m$, there is no other constant with the dimension of length. Therefore, the cross-section (an effective area) for this purely electrodynamics process, which is of order $e^2$ in the amplitude, must be proportional to the square the classical radius of the electron.

## Problems

**4.1 Three-particle decay mode.** Consider the decay of a spinless particle of mass $M$ into three spinless particles of masses $m_i$ and momenta $\boldsymbol{p}_i$, with $i = 1, 2, 3$. The final state consistent with energy-momentum conservation is determined by five independent variables, which may be chosen as the energies of two of the particles, $E_1$ and $E_2$, two angles that fix the direction of $\boldsymbol{p}_1$, and finally one angle that defines the rotations of the system $(\boldsymbol{p}_2, \boldsymbol{p}_3)$ about $\boldsymbol{p}_1$. Calculate the phase space volume. Compare the rate for the three-particle decay with the rate for the two-particle decay in the limit of massless final particles.

**4.2 Transverse photon propagator.** For the electromagnetic field satisfying the Coulomb gauge, prove that the propagator in momentum space is given by

$$D_{\mathrm{T}}^{ij}(k) = \frac{1}{k^2 + \mathrm{i}\varepsilon}\left(\delta_{ij} - \frac{k^i k^j}{\boldsymbol{k}^2}\right).$$

**4.3 Propagator of massive vector field.** Consider the Stueckelberg Lagrangian for a massive vector particle of the form

$$\mathcal{L} = -\frac{1}{4}F_{\mu\nu}F^{\mu\nu} + \frac{1}{2}\mu^2 A_\nu A^\nu - \frac{1}{2}\lambda(\partial\cdot A)^2\,,$$

where $\mu$ is the mass and the last term is an auxiliary term introduced so that the limit $\mu \to 0$ is not singular for $\lambda \neq 0$. Calculate the propagator of the field.

**4.4 The Lagrangian for electrodynamics in the Coulomb gauge.** The Lagrangian for the electromagnetic field coupled to a conserved current is given by

$$\mathcal{L} = -\frac{1}{4} F_{\mu\nu} F^{\mu\nu} - J^\mu A_\mu \, .$$

It is clearly invariant under the local gauge transformation $A_\mu \to A_\mu - \partial_\mu \Lambda$. Therefore, to have a definite solution for the Maxwell equations, the arbitrariness associated with this invariance must be removed by choosing a gauge for $A_\mu$. Further, since the time derivative of $A_0$ is absent from $\mathcal{L}$, its conjugate field vanishes, and one is free to remove $A_0$.
(a) Show that it is always possible to choose the (Coulomb) gauge $\nabla \cdot \boldsymbol{A} = 0$ at some given time $t$, and that the gauge condition holds at any later time.
(b) Find $A_0(x)$.
(c) Eliminate $A_0$ from $\mathcal{L}$ and prove that

$$\mathcal{L} = \frac{1}{2}(\boldsymbol{E}_\perp^2 - \boldsymbol{B}^2) - \frac{e^2}{8\pi} \int \mathrm{d}^3 x' \, \frac{J^0(t, \boldsymbol{x}) J^0(t, \boldsymbol{x}')}{|\boldsymbol{x} - \boldsymbol{x}'|} + e \, \boldsymbol{J} \cdot \boldsymbol{A} \, ,$$

where $\boldsymbol{E}_\perp = -(\partial \boldsymbol{A}/\partial t)$ and $\boldsymbol{B} = \nabla \times \boldsymbol{A}$.

## Suggestions for Further Reading

*A good discussion of the formulation of quantum field theory in the interaction and the Heisenberg representations can be found in*

Schweber, S. S., *An Introduction to Relativistic Quantum Field Theory.* Row, Peterson and Co., Evanston, IL 1961

*More systematic treatments of the perturbative theory are found in*

Itzykson, C. and Zuber, J.-B., *Quantum Field Theory.* McGraw-Hill, New York 1980

Peskin, M. E. and Schroeder, D. V., *Quantum Field Theory.* Addison-Wesley, Reading, MA 1995

*The reader will find other examples of physical processes in*

Bjorken, J. D. and Drell, S. D., *Relativistic Quantum Mechanics.* McGraw-Hill, New York 1964

Gross, F., *Relativistic Quantum Mechanics and Field Theory.* Wiley-Interscience, New York 1993

Halzen, F. and Martin, A. D. *Quarks and Leptons: An Introductory Course in Modern Particle Physics.* Wiley, New York 1984

Nachtmann, O., *Elementary Particle Physics, Concepts and Phenomena.* Springer, Berlin, Heidelberg 1990

*For further study of quantum electrodynamics, the reader may refer to*

Feynman, R. P., *Quantum Electrodynamics.* Benjamin, New York 1961

Feynman, R. P., *The Theory of Fundamental Processes.* Benjamin, New York 1962

Schwinger, J., *Selected Papers on Quantum Electrodynamics.* Dover, New York 1958

# 5 Discrete Symmetries

The group of all Lorentz transformations includes the proper continuous transformations already studied in previous chapters and the discrete transformations to be treated in this chapter. The latter class of transformations deals with *space* and *time inversions* as well as all operations formed by successive applications of a space or time inversion and a proper continuous transformation.

Invariance of physical systems with respect to the proper Lorentz group is one of the best-established properties, so much so that it is universally accepted as a fundamental principle of contemporary physics. It is then natural and, from the esthetic viewpoint, desirable to expect all physical phenomena to be invariant to the inversion operations as well: left–right symmetry and past–future symmetry. After all, the dynamic equations of classical mechanics appear unchanged in these transformations. What a surprise when it was discovered that the symmetry under space reflections was violated by the weak interactions. It then seems quite possible that the reversal of the time direction is not a universal symmetry either.

Related to these inversion operations is the *charge conjugation*, which acts not on space-time but rather on internal space. It reverses the signs of the electric charges of fields and all of their other additive quantum numbers (also called generalized charges) without changing any of their kinematic attributes, thus converting particles into antiparticles. There exists in fact a close relationship between these three discrete transformations: a successive application of all three transformations in any order constitutes a symmetry operation for all quantum field theories that satisfy very general conditions, even in cases where individual transformations may be violated in some interactions.

In this chapter we shall discuss applications of the inversion operations in quantum field theories. In view of model building, it is just as important to study the implications of invariance of physical systems to these transformations so as to discover how and in what circumstances these symmetries are violated.

## 5.1 Parity

The elementary discrete space transformation is the reflection in a spatial plane. However, as a reflection in a plane is equivalent to a rotation through an angle $\pi$ about an axis perpendicular to that plane followed by an inversion with respect to the intersection of that axis with the plane, it suffices to consider without any loss of generality just the inversion. This operation,

$$\mathcal{P}: \boldsymbol{x} \to \boldsymbol{x}' = -\boldsymbol{x}, \quad t \to t' = t, \tag{5.1}$$

is the basic *improper orthochronous* Lorentz transformation defined by

$$a^{\mu}{}_{\nu} = \begin{pmatrix} 1 & 0 & 0 & 0 \\ 0 & -1 & 0 & 0 \\ 0 & 0 & -1 & 0 \\ 0 & 0 & 0 & -1 \end{pmatrix}. \tag{5.2}$$

It is also often referred to as the *parity* operation. Invariance of a physical system to inversion means that the system cannot distinguish left from right in any interactions. On the other hand, the detection in the system of some physical quantity with a left–right asymmetry is a clear signal that the symmetry is broken. In the next few paragraphs, we will generalize the notion of space inversion of quantum mechanics to field theories, in particular defining the transformation rules for observables and introducing the concept of intrinsic parity, and briefly discuss the behavior of the fundamental interactions under the parity operation.

### 5.1.1 Parity in Quantum Mechanics

The inversion transforms the momentum $\boldsymbol{p}$ into $-\boldsymbol{p}$, as is evident from the operational form $\boldsymbol{p} = -\mathrm{i}\nabla$. The orbital angular momentum $\boldsymbol{L}$ remains unchanged since $\boldsymbol{L} = \boldsymbol{x} \times \boldsymbol{p}$. The generalized angular momentum $\boldsymbol{J}$ is also unchanged since space inversion commutes with all space rotations (see Fig. 5.1). A three-vector that changes sign in inversion (for example, $\boldsymbol{x}$ or $\boldsymbol{p}$) is called a *polar vector*; if it remains unchanged (for example, $\boldsymbol{x} \times \boldsymbol{p}$ or $\boldsymbol{J}$), it is called an *axial vector*. A scalar quantity that remains unchanged is a *scalar* (e.g. $\boldsymbol{p}^2$), but if it changes sign under inversion, it is a *pseudoscalar* (e.g. $\boldsymbol{p}\cdot\boldsymbol{J}$).

We assume there exists a linear operator $\mathcal{P}$ that performs space inversions on the Hilbert space and relates a given state vector to the transformed state vector. It is chosen to be *unitary* to preserve the normalization and orthogonality of states. Since $\mathcal{P}^2$, when acting on a state, brings it back to the original state, the phases can be fixed such that $\mathcal{P}^2 = 1$ provided we ignore (for the moment) spin degrees of freedom. With $\mathcal{P}^2 = 1$ and $\mathcal{P}^\dagger\mathcal{P} = 1$, the operator $\mathcal{P}$ is Hermitian ($\mathcal{P}^\dagger = \mathcal{P}^{-1} = \mathcal{P}$) and therefore is an *observable.*

If a system described by the Hamiltonian $H$ is invariant to inversion,

$$\mathcal{P}H\mathcal{P}^{-1} = H \quad \text{or} \quad [\mathcal{P}, H] = 0, \tag{5.3}$$

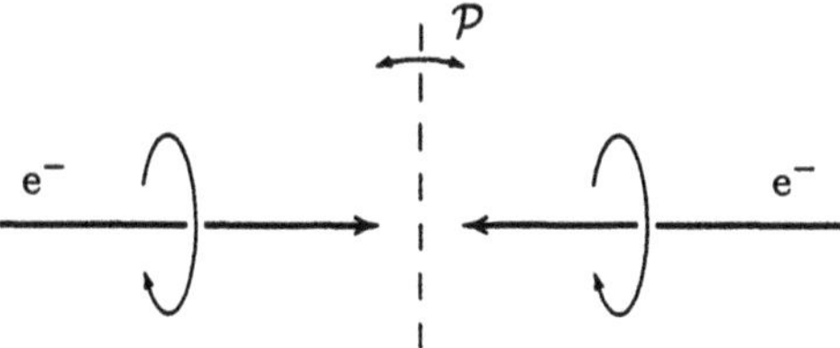


**Fig. 5.1.** $\mathcal{P}$ reverses the momentum of a particle without flipping its spin

$\mathcal{P}$ is a constant of the motion, and simultaneous eigenvectors of $H$ and $\mathcal{P}$ can be found. The corresponding eigenvalue of $\mathcal{P}$ for the state is called the *parity* of the state, $\eta = +1$ or $\eta = -1$. Therefore, parity is a multiplicative quantum number; that is, the parity of a compound system is equal to the product of the parities of its individual components.

Consider for example a particle in an orbit of angular momentum $\ell$. The angular part of its wave function is given by the spherical harmonics $Y_{\ell m}(\theta,\varphi)$. In inversion, $\theta \to \pi - \theta$ and $\varphi \to \varphi + \pi$, and $Y_{\ell m}(\theta,\varphi)$ changes into $Y_{\ell m}(\pi - \theta, \varphi + \pi) = (-)^\ell Y_{\ell m}(\theta,\varphi)$. It follows that

$$\mathcal{P}\,|\ell m\rangle = (-)^\ell\;|\ell m\rangle\,. \tag{5.4}$$

Thus, an eigenvector of orbital angular momentum $\ell$ also has a well-defined parity, which is $(-1)^\ell$. We have also verified in passing that the parity and the orbital angular momentum are simultaneously good quantum numbers, i.e. $[\mathcal{P}, \boldsymbol{L}] = 0$. In contrast, since $\mathcal{P}\boldsymbol{p}\mathcal{P}^\dagger = -\boldsymbol{p}$, an eigenvector of momentum does not have a well-defined parity and, conversely, an eigenvector of parity does not have a well-defined momentum. The plane wave of a spinless particle, $\langle \boldsymbol{x}|\,\boldsymbol{p}\rangle = \exp(-\mathrm{i}Et + \mathrm{i}\boldsymbol{p}\cdot\boldsymbol{x})$, becomes after inversion a plane wave propagating in the reversed direction:

$$\langle \boldsymbol{x}'\;|\boldsymbol{p}\rangle = \langle \mathcal{P}\boldsymbol{x}\;|\boldsymbol{p}\rangle = \exp[-\mathrm{i}Et + \mathrm{i}(-\boldsymbol{p})\cdot\boldsymbol{x}\,]\,.$$

Let $\mathcal{O}_+$ be an *even* operator under inversion, $\mathcal{P}\mathcal{O}_+\mathcal{P}^\dagger = \mathcal{O}_+$, and $\mathcal{O}_-$ an *odd* operator, $\mathcal{P}\mathcal{O}_-\mathcal{P}^\dagger = -\mathcal{O}_-$. Their matrix elements between states of well-defined parities are given by

$$\begin{aligned}\langle \eta'|\;\mathcal{O}_+\;|\eta''\rangle &= \langle \eta'|\;\mathcal{P}^\dagger\mathcal{P}\mathcal{O}_+\mathcal{P}^\dagger\mathcal{P}\;|\eta''\rangle = \eta'\eta''\,\langle \eta'|\;\mathcal{O}_+\;|\eta''\rangle\,,\\ \langle \eta'|\;\mathcal{O}_-\;|\eta''\rangle &= \langle \eta'|\;\mathcal{P}^\dagger\mathcal{P}\mathcal{O}_-\mathcal{P}^\dagger\mathcal{P}\;|\eta''\rangle = -\eta'\eta''\,\langle \eta'|\;\mathcal{O}_-\;|\eta''\rangle\,.\end{aligned} \tag{5.5}$$

These results show that an *even observable* has vanishing matrix elements between states of opposite parities, whereas an *odd observable* has vanishing matrix elements between states of equal parities. This selection rule is useful in studies of nuclear and electromagnetic transitions.

We have ignored up to now the notion of *intrinsic parity*. In fact, the parity of a state arises from both the relative motion of all the particles composing the system and the intrinsic parity of every particle. The intrinsic parity of a nuclear system can be inferred once we know the angular momentum couplings of individual particles and define the intrinsic parity of the nucleon. Take for example the deuteron, which is known to be mainly in a $^3S_1$ state. (In spectroscopic studies, states are often labeled by $^{2S+1}\ell_J$, where $\ell$, $S$, and $J$ denote the orbital angular momentum, intrinsic spin, and total angular momentum; $\ell = 0, 1, 2, \ldots$ are labeled by the letters $S, P, D, \ldots$. So $^3S_1$ means $\ell = 0$, $S = 1$, and $J = 1$.) In its center-of-mass, the deuteron has orbital angular momentum $\ell = 0$ and hence parity $\eta = +1$, provided the relative parity of the neutron and proton in their center-of-mass is defined as $+1$. If we then treat the deuteron as a particle, we may define its intrinsic parity to be $\eta_{\rm d} = +1$.

Consider now the $\pi^-$-capture reaction by a deuteron d, $\pi^- + {\rm d} \to {\rm n} + {\rm n}$. The neutron and the meson are taken for now as elementary particles. If $\ell$ and $\ell'$ stand for the relative orbital angular momenta of the particles respectively in the initial and final states, then the assumed parity conservation implies

$$\eta_\pi \eta_{\rm d} (-)^\ell = \eta_{\rm n}\eta_{\rm n} (-)^{\ell'} = (-)^{\ell'} .$$

In the capture process, the meson $\pi^-$ is slowed down and captured in an atomic s-state ($\ell = 0$) of the deuteron, which means that the parity of the initial state is simply $\eta_\pi$ and the total angular momentum is that of the deuteron, $J_{\rm i} = 1$. The final total angular momentum is, by conservation, $J_{\rm f} = 1$, and so the final two-neutron state must be one of the four configurations allowed by the rules of angular momentum couplings, namely $^3S_1$, $^3P_1$, $^1P_1$, $^3D_1$. However, since the final state is composed of two identical fermions, it must be antisymmetric under a permutation of the two neutrons (that is, $\ell$ and $S$ must be both even or both odd numbers), which rules out all possibilities except $^3P_1$, evidently of negative parity. Thus, conservation of parity requires the existence of an intrinsic parity for $\pi^-$ of value $\eta_\pi = -1$.

### 5.1.2 Parity in Field Theories

We now proceed to define the parities of boson and fermion fields and of their associated Fock operators. We shall discover, in particular, that the relative parity of a conjugate boson–antiboson pair is positive while that of a conjugate fermion–antifermion pair is negative.

**Scalar and Pseudoscalar Fields.** Let $\phi(t, \boldsymbol{x})$ be the operator that represents a Bose field of spin 0, and $\phi^\dagger(t, \boldsymbol{x})$ its Hermitian conjugate. Their transformations under inversion are defined by

$$\begin{aligned} \mathcal{P}\,\phi(t,\boldsymbol{x})\,\mathcal{P}^{-1} &= \eta_{\rm B}\phi(t,-\boldsymbol{x})\,, \\ \mathcal{P}\,\phi^\dagger(t,\boldsymbol{x})\,\mathcal{P}^{-1} &= \eta_{\rm B}\phi^\dagger(t,-\boldsymbol{x})\,, \end{aligned} \tag{5.6}$$

where $\eta_{\rm B} = +1$ or $-1$. Although $\phi$ behaves as a scalar field under proper Lorentz transformations, the inversion operation differentiates a *scalar field* with parity $\eta_{\mathrm{B}} = +1$ from a *pseudoscalar field* with $\eta_{\mathrm{B}} = -1$. This quantum number is determined by experiment. Considered as elementary entities, the meson $f_0(980\ \mathrm{MeV})$ is a scalar particle, and the mesons $\pi^0$, $\pi^\pm$, $\mathrm{K}^0$, and $\mathrm{K}^\pm$ are pseudoscalar particles.

In order to study the transformation properties of Fock states, we substitute into (6) the expansion series (2.99) of $\phi$ in terms of the operators $a_{\boldsymbol{p}}$ and $b_{\boldsymbol{p}}$, recalling that $\mathcal{P}$, as a linear operator in the Hilbert space, does not act on c-number quantities. We thus have, on the one hand,

$$\mathcal{P}\,\phi(x)\,\mathcal{P}^{-1} = \sum_{\boldsymbol{p}} C_{\boldsymbol{p}} \left[\mathcal{P}\,a_{\boldsymbol{p}}\,\mathcal{P}^{-1}\mathrm{e}^{-\mathrm{i}(Et-\boldsymbol{p}\cdot\boldsymbol{x})} + \mathcal{P}\,b^\dagger_{\boldsymbol{p}}\,\mathcal{P}^{-1}\mathrm{e}^{\mathrm{i}(Et-\boldsymbol{p}\cdot\boldsymbol{x})}\right] , \tag{5.7}$$

($C_{\boldsymbol{p}}$ being the usual normalization of the field), and on the other hand,

$$\begin{aligned}\eta_{\mathrm{B}}\,\phi(t,-\boldsymbol{x}) &= \eta_{\mathrm{B}} \sum_{\boldsymbol{p}} C_{\boldsymbol{p}} \left[a_{\boldsymbol{p}}\,\mathrm{e}^{-\mathrm{i}(Et+\boldsymbol{p}\cdot\boldsymbol{x})} + b^\dagger_{\boldsymbol{p}}\,\mathrm{e}^{\mathrm{i}(Et+\boldsymbol{p}\cdot\boldsymbol{x})}\right] \\ &= \eta_{\mathrm{B}} \sum_{\boldsymbol{p}} C_{\boldsymbol{p}} \left[a_{-\boldsymbol{p}}\,\mathrm{e}^{-\mathrm{i}(Et-\boldsymbol{p}\cdot\boldsymbol{x})} + b^\dagger_{-\boldsymbol{p}}\,\mathrm{e}^{\mathrm{i}(Et-\boldsymbol{p}\cdot\boldsymbol{x})}\right] .\end{aligned} \tag{5.8}$$

Together with similar relations for $\phi^\dagger$, one obtains the basic properties

$$\mathcal{P}\,a_{\boldsymbol{p}}\,\mathcal{P}^{-1} = \eta_{\mathrm{B}}\,a_{-\boldsymbol{p}}\,, \qquad \mathcal{P}\,a^\dagger_{\boldsymbol{p}}\,\mathcal{P}^{-1} = \eta_{\mathrm{B}}\,a^\dagger_{-\boldsymbol{p}}\,, \tag{5.9}$$

$$\mathcal{P}\,b_{\boldsymbol{p}}\,\mathcal{P}^{-1} = \eta_{\mathrm{B}}\,b_{-\boldsymbol{p}}\,, \qquad \mathcal{P}\,b^\dagger_{\boldsymbol{p}}\,\mathcal{P}^{-1} = \eta_{\mathrm{B}}\,b^\dagger_{-\boldsymbol{p}}\,. \tag{5.10}$$

The parity of a one-boson state of momentum $\boldsymbol{p}$ is therefore given by

$$\begin{aligned}\mathcal{P}\,|\boldsymbol{p}\rangle &= \mathcal{P}\,a^\dagger_{\boldsymbol{p}}\,|0\rangle = \mathcal{P}a^\dagger_{\boldsymbol{p}}\mathcal{P}^{-1}\mathcal{P}\,|0\rangle \\ &= \eta_{\mathrm{B}}a^\dagger_{-\boldsymbol{p}}\,|0\rangle = \eta_{\mathrm{B}}\,|-\boldsymbol{p}\rangle\,,\end{aligned} \tag{5.11}$$

where the parity of the vacuum is fixed by convention, $\mathcal{P}\,|0\rangle = +\,|0\rangle$. This result (11) restates the simple fact that momentum changes sign under inversion and that a state of a free boson of well-defined momentum is not an eigenstate of $\mathcal{P}$, exactly as found in the first quantization formalism. However, in the rest frame of the particle, where $\boldsymbol{p} = 0$,

$$\mathcal{P}\,|\boldsymbol{p}=0\rangle = \eta_{\mathrm{B}}\,|\boldsymbol{p}=0\rangle\,; \tag{5.12}$$

that is, $|\boldsymbol{p}=0\rangle$ is an eigenstate of $\mathcal{P}$. A particle at rest has a well-defined parity which is by definition its *intrinsic parity*, $\eta_{\mathrm{B}} = +1$ for a scalar particle and $\eta_{\mathrm{B}} = -1$ for a pseudoscalar particle. A similar analysis, starting from (10), shows that the corresponding antiparticle in a state of equal orbital angular momentum has equal parity. Hence the general result: a boson and its conjugate antiboson have *equal intrinsic parities*. It follows, for instance, that a $\pi^+\pi^-$ system in relative orbital angular momentum $\ell$ has parity $(-)^\ell$.

From the field properties (6), the transformation rules for dynamical variables can be found. For example, the current density for a boson field,

$$j^\mu(x) = \mathrm{i}\left[\phi^\dagger(x)\partial^\mu\phi(x) - (\partial^\mu\phi^\dagger(x))\phi(x)\right], \tag{5.13}$$

transforms according to (6) as

$$\mathcal{P}\, j^0(t,\boldsymbol{x})\,\mathcal{P}^{-1} = +j^0(t,-\boldsymbol{x}), \qquad \mathcal{P}\, j^i(t,\boldsymbol{x})\,\mathcal{P}^{-1} = -j^i(t,-\boldsymbol{x}). \tag{5.14}$$

These transformation laws state that $j^0$ behaves as a scalar field, and $\boldsymbol{j}$ as a polar vector field under space inversion. With the transformation matrix $a^\mu{}_\nu$ defined in (2), the above results can be expressed concisely:

$$\mathcal{P}\, j^\mu(t,\boldsymbol{x})\,\mathcal{P}^{-1} = a^\mu{}_\nu\, j^\nu(t,-\boldsymbol{x}). \tag{5.15}$$

**Electromagnetic Field.** As the electromagnetic field is a Lorentz vector, one expects that

$$\mathcal{P}\, A^\mu(t,\boldsymbol{x})\,\mathcal{P}^{-1} = \eta_A\, a^\mu{}_\nu\, A^\nu(t,-\boldsymbol{x}). \tag{5.16}$$

Given (15) and the experimental observation that the electromagnetic interaction, $H_{\rm em} = qj^\mu A_\mu$, is invariant to space inversion, one may infer the value of the phase factor $\eta_A = +1$. In particular, the space components of the field transform according to

$$\mathcal{P}\, \boldsymbol{A}(t,\boldsymbol{x})\,\mathcal{P}^{-1} = -\boldsymbol{A}(t,-\boldsymbol{x}). \tag{5.17}$$

The transformation properties of the electromagnetic field operators in Fock space can be obtained by substituting into (17) the expansion series of the transverse $\boldsymbol{A}$ given in (2.156). Thus, the right-hand side of (17) reads

$$\begin{aligned}
-\boldsymbol{A}(t,-\boldsymbol{x}) &= -\sum_{k\lambda} C_k \left[\boldsymbol{\epsilon}(\boldsymbol{k},\lambda)\, a(\boldsymbol{k},\lambda) \mathrm{e}^{-\mathrm{i}\omega t-\mathrm{i}\boldsymbol{k}\cdot\boldsymbol{x}} + \boldsymbol{\epsilon}^*(\boldsymbol{k},\lambda)\, a^\dagger(\boldsymbol{k},\lambda)\mathrm{e}^{\mathrm{i}\omega t+\mathrm{i}\boldsymbol{k}\cdot\boldsymbol{x}}\right] \\
&= -\sum_{k\lambda} C_k \left[\boldsymbol{\epsilon}(-\boldsymbol{k},\lambda)\, a(-\boldsymbol{k},\lambda) \mathrm{e}^{-\mathrm{i}k\cdot x} + \boldsymbol{\epsilon}^*(-\boldsymbol{k},\lambda)\, a^\dagger(-\boldsymbol{k},\lambda)\mathrm{e}^{\mathrm{i}k\cdot x}\right].
\end{aligned}$$

If the $z$ axis is chosen to coincide with the propagation vector $\boldsymbol{k}$, the polarization vectors given in (2.153) become $\boldsymbol{\epsilon}(\hat{\boldsymbol{z}},\pm) = \mp\frac{1}{\sqrt{2}}(\hat{\boldsymbol{x}} \pm \mathrm{i}\hat{\boldsymbol{y}})$. A rotation through 180° about the $y$ axis brings $\hat{\boldsymbol{x}}$ to $-\hat{\boldsymbol{x}}$ and $\hat{\boldsymbol{z}}$ to $-\hat{\boldsymbol{z}}$, and the polarization vectors to

$$\boldsymbol{\epsilon}(-\hat{\boldsymbol{z}},\pm) = \pm\frac{1}{\sqrt{2}}(\hat{\boldsymbol{x}} \mp \mathrm{i}\hat{\boldsymbol{y}}) = \boldsymbol{\epsilon}(\hat{\boldsymbol{z}},\mp) = -\boldsymbol{\epsilon}^*(\hat{\boldsymbol{z}},\pm), \tag{5.18}$$

or more generally, $\epsilon(\hat{\boldsymbol{k}},\lambda) = \epsilon(-\hat{\boldsymbol{k}},-\lambda) = -\epsilon^*(-\hat{\boldsymbol{k}},\lambda)$. With this property taken into account, (17) becomes

$$\mathcal{P}\,\boldsymbol{A}(x)\,\mathcal{P}^{-1} = -\sum_{k\lambda} C_k\Big[\epsilon(\boldsymbol{k},\lambda)\,a(-\boldsymbol{k},-\lambda)\,\mathrm{e}^{-\mathrm{i}k\cdot x} + \epsilon^*(\boldsymbol{k},\lambda)\,a^\dagger(-\boldsymbol{k},-\lambda)\,\mathrm{e}^{\mathrm{i}k\cdot x}\Big], \tag{5.19}$$

which implies the basic transformation rule for the photon Fock operator

$$\mathcal{P}\,a(\boldsymbol{k},\lambda)\,\mathcal{P}^{-1} = -a(-\boldsymbol{k},-\lambda)\,. \tag{5.20}$$

Thus, the photon has a *negative* intrinsic parity, $\eta_\gamma = -1$. Both its momentum and helicity change signs under the parity operation.

**Dirac Fermion Field.** Covariance of the Dirac equation requires the Dirac wave function $\psi(x)$ to transform as

$$\begin{aligned} \mathcal{P}\ :\ \psi(x) \to \psi'(t,-\boldsymbol{x}) &= S(a)\psi(x) \\ \text{or} \qquad \psi'(x) &= S(a)\psi(t,-\boldsymbol{x})\,. \end{aligned}$$

Here, $S(a)$ is defined by (3.13) and, with $a$ as in (2), it becomes

$$S(a) = \eta_{\mathrm{F}}\,\gamma_0, \qquad \eta_{\mathrm{F}} = \pm 1\,, \tag{5.21}$$

which holds for any representation of $\gamma_0$. Here, $\eta_{\mathrm{F}}$ is the intrinsic parity of the Dirac particle to be determined by experiment.

In analogy with the classical wave function, the Dirac field operator transforms according to

$$\mathcal{P}\,\psi(x)\,\mathcal{P}^{-1} = \eta_{\mathrm{F}}\,\gamma_0\,\psi(t,-\boldsymbol{x})\,. \tag{5.22}$$

We first recall the expansion series for $\psi$ given in (3.91):

$$\psi(x) = \sum_{\boldsymbol{p},s} C_p\left[b(\boldsymbol{p},s)u(\boldsymbol{p},s)\,\mathrm{e}^{-\mathrm{i}p\cdot x} + d^\dagger(\boldsymbol{p},s)v(\boldsymbol{p},s)\,\mathrm{e}^{\mathrm{i}p\cdot x}\right], \tag{5.23}$$

and also note that the free-particle spinors have the following properties, which can be proved by using their explicit expressions (3.45) and (3.46),

$$\begin{aligned} \gamma_0 u(-\boldsymbol{p},s) &= u(\boldsymbol{p},s)\,, \\ \gamma_0 v(-\boldsymbol{p},s) &= -v(\boldsymbol{p},s)\,. \end{aligned}$$

Then the right-hand side of (22) may be written as

$$\eta_{\mathrm{F}}\gamma_0\psi(t,-\boldsymbol{x}) = \eta_{\mathrm{F}}\sum_{\boldsymbol{p},s} C_p\left[b(-\boldsymbol{p},s)u(\boldsymbol{p},s)\,\mathrm{e}^{-\mathrm{i}p\cdot x} - d^\dagger(-\boldsymbol{p},s)v(\boldsymbol{p},s)\,\mathrm{e}^{\mathrm{i}p\cdot x}\right],$$

which leads to the transformation properties of the Fock operators

$$\mathcal{P}\, b(\boldsymbol{p}, s)\, \mathcal{P}^{-1} = \eta_{\mathrm{F}}\, b(-\boldsymbol{p}, s), \qquad \mathcal{P}\, b^{\dagger}(\boldsymbol{p}, s)\, \mathcal{P}^{-1} = \eta_{\mathrm{F}}\, b^{\dagger}(-\boldsymbol{p}, s), \tag{5.24}$$

$$\mathcal{P}\, d(\boldsymbol{p}, s)\, \mathcal{P}^{-1} = -\eta_{\mathrm{F}}\, d(-\boldsymbol{p}, s), \qquad \mathcal{P}\, d^{\dagger}(\boldsymbol{p}, s)\, \mathcal{P}^{-1} = -\eta_{\mathrm{F}}\, d^{\dagger}(-\boldsymbol{p}, s)\,. \tag{5.25}$$

Thus, the one-fermion state $b^{\dagger}(\boldsymbol{p}, s)\,|0\rangle$ transforms into $b^{\dagger}(-\boldsymbol{p}, s)\,|0\rangle$, and the one-antifermion state $d^{\dagger}(\boldsymbol{p}, s)\,|0\rangle$ into $-d^{\dagger}(-\boldsymbol{p}, s)\,|0\rangle$, with their spin orientations unchanged. However, since momentum reverses direction, $\boldsymbol{J}{\cdot}\hat{\boldsymbol{p}}$ changes sign, and helicity states are not invariant to $\mathcal{P}$.

The negative sign on the right-hand sides of (25) means that the intrinsic parity of an antifermion is opposite in sign to that of the corresponding fermion. As a result, the parity of an electron–positron system in a relative s-state ($\ell = 0$) is necessarily odd:

$$\mathcal{P} b^{\dagger}(\boldsymbol{0}, s) d^{\dagger}(\boldsymbol{0}, s')\,|0\rangle = -b^{\dagger}(\boldsymbol{0}, s) d^{\dagger}(\boldsymbol{0}, s')\,|0\rangle \tag{5.26}$$

(to be contrasted with a boson–antiboson pair, such as $\pi^{+}\pi^{-}$). Thus, in general, the relative intrinsic parity is *even* for a self-conjugate boson–antiboson system and *negative* for a self-conjugate fermion–antifermion system.

Let us finally remark that the transformation rule (15) is also valid for the current density $j_{\mu}(x)$ of a Dirac field, as can be seen by applying the rules $\psi \to \eta\gamma_0\psi$ and $\overline{\psi} \to \eta\overline{\psi}\gamma_0$ to the expression $\overline{\psi}\gamma_{\mu}\psi$. More generally, a bilinear covariant transforms according to

$$\mathcal{P}\ \overline{\psi}(x)\Gamma\psi(x)\,\mathcal{P}^{-1} = \overline{\psi}(t, -\boldsymbol{x})\,\gamma_0\Gamma\gamma_0\,\psi(t, -\boldsymbol{x})\,. \tag{5.27}$$

For pseudoscalar, vector, and axial-vector operators, one needs

$$\begin{aligned}
\gamma_0\gamma_5\gamma_0 &= -\gamma_5;\\
\gamma_0\gamma_{\mu}\gamma_0 &= \begin{cases} +\gamma_{\mu}, & \text{if } \mu = 0;\\ -\gamma_{\mu}, & \text{if } \mu = 1,\ 2,\ 3;\end{cases}\\
\gamma_0\gamma_{\mu}\gamma_5\gamma_0 &= \begin{cases} -\gamma_{\mu}\gamma_5, & \text{if } \mu = 0;\\ +\gamma_{\mu}\gamma_5, & \text{if } \mu = 1,\ 2,\ 3.\end{cases}
\end{aligned}$$

### 5.1.3 Parity and Interactions

In the mid-1950s, it was discovered that while parity was conserved to a high degree of precision in strong and electromagnetic interactions, it was badly broken in weak interactions. Experiments were devised and carried out to map these irregularities, which have eventually led to a deeper understanding of the dynamics of particles.

**Intrinsic Parity.** If we set the phase of the vacuum state of some Hilbert space to 1, the absolute phase of any state vector in the space is defined as its phase relative to the vacuum. In discrete symmetries, there always exists an ambiguity in defining the phases of transformed states. Take for example a state of electric charge $|q\rangle$ also assumed to have good parity, $\mathcal{P}\,|q\rangle = \eta\,|q\rangle$. If now the parity operator is redefined as $\mathcal{P}' \equiv \mathcal{P}\exp(\mathrm{i}\alpha Q)$, where $\alpha$ is some real constant and $Q$ the charge operator, then the parity of the state becomes $\eta' = \eta\exp(\mathrm{i}\alpha q)$ without causing observable physical effects on the system. Thus parity is defined only up to a phase factor. Its definition becomes unambiguous only if the charge vanishes, or is equal to the vacuum charge. More generally, the absolute parity is well defined only for *completely neutral particles* – particles that have all their generalized charges identically equal to zeros, such as the photon or the $\pi^0$ meson.

As we have seen, invariance of the electromagnetic interaction to inversion implies that the photon is *odd*, i.e. $\eta_\gamma = -1$. The parity of $\pi^0$ is also *negative*, a result that can be inferred from the following arguments. The meson $\pi^0$ has mean lifetime $\tau = 8\times 10^{-17}$ s, and decays in 99% of all cases through the channel $\pi^0 \to 2\gamma$. In the meson rest frame the initial angular momentum is $J_\mathrm{i} = 0$. By conservation, the final angular momentum is also $J_\mathrm{f} = 0$. The wave function of the two photons in the final state must contain the polarization vectors $\boldsymbol{\epsilon}_1$, $\boldsymbol{\epsilon}_2$ and the relative momentum $\boldsymbol{k}$, which obey the transversality conditions $\boldsymbol{k}\cdot\boldsymbol{\epsilon}_1 = \boldsymbol{k}\cdot\boldsymbol{\epsilon}_2 = 0$. It must be a scalar function ($J_\mathrm{f} = 0$), linear in $\boldsymbol{\epsilon}_1$ and $\boldsymbol{\epsilon}_2$, and symmetric under permutation of the two photons, i.e. in the simultaneous exchanges $\boldsymbol{\epsilon}_1 \leftrightarrow \boldsymbol{\epsilon}_2$ and $\boldsymbol{k} \leftrightarrow -\boldsymbol{k}$. There are two possibilities consistent with these conditions:

(i) $\boldsymbol{\epsilon}_1\cdot\boldsymbol{\epsilon}_2$, even under inversion, $\eta = +1$, and
(ii) $\boldsymbol{k}\cdot(\boldsymbol{\epsilon}_1 \times \boldsymbol{\epsilon}_2)$, odd under inversion, $\eta = -1$.

With $\varphi$ denoting the angle between $\boldsymbol{\epsilon}_1$ and $\boldsymbol{\epsilon}_2$, the corresponding angular distributions are

(i) $|\boldsymbol{\epsilon}_1\cdot\boldsymbol{\epsilon}_2|^2 \propto \cos^2\varphi$, $\quad \eta = +1$,
(ii) $|\boldsymbol{\epsilon}_1 \times \boldsymbol{\epsilon}_2|^2 \propto \sin^2\varphi$, $\quad \eta = -1$.

Parity conservation says that the intrinsic parity of $\pi^0$ must be equal to $\eta_{\pi^0} = \eta\,\eta_\gamma^2 = \eta$. To determine $\eta_{\pi^0}$, it suffices to measure the photon polarizations in the final state. If the photons are found with predominant parallel linear polarizations ($\varphi = 0$), $\pi^0$ is a scalar particle; if on the contrary they are seen emitted with perpendicular polarizations ($\varphi = \pi/2$), $\pi^0$ is a pseudoscalar meson. Experiments show a clear preference for the second possibility: $\pi^0$ is a pseudoscalar meson with $\eta_\pi = -1$.

As for particles having nonvanishing additive quantum numbers, it is necessary to fix first the parities of a minimum number of reference particles. The relative parities of all other particles, whenever they can be defined, are determined from arguments based on parity conservation in parity-conserving reactions. Thus, one must define *at least* the parities of the neutron, of the proton (for processes that conserve electric charge and baryon number), and of $\Lambda^0$ (for reactions that conserve *strangeness*, a number characteristic of a class of unstable particles). The conventional choice is

$$\eta_\mathrm{n} = \eta_\mathrm{p} = \eta_\Lambda = +1. \tag{5.28}$$

**Tests of Parity.** For *electromagnetic interactions*, a category of tests of parity conservation consists in detecting transitions forbidden by the symmetry. Such tests can be made relatively simpler by concentrating on atomic states where the stronger hadronic effects are absent. For example, transitions between two atomic states of equal spins and equal parities, $J_{\rm i}^P = 1^+ \to J_{\rm f}^P = 1^+$, may proceed via the electric quadrupole and magnetic dipole modes (described by even parity operators in both cases), but are forbidden for the (odd parity) electric dipole mode, which would otherwise be the kinematically favored mode. The fact that transitions between these two states have not been observed indicates that if parity is broken at all in electromagnetic interactions, such a symmetry violation must be a very small effect.

Parity conservation in *strong interactions* can be similarly verified. A typical experiment consists in observing the $\alpha$-decay of $^{20}$Ne through a channel forbidden by conservation of parity, namely $J_{\rm i} = 1^+ \overset{\alpha}{\nrightarrow} J_{\rm f} = 0^+$. The measured branching ratio for this mode is very small, again indicating that parity is indeed a symmetry of the strong interaction.

Parity conservation in a system demands its Lagrangian to obey

$$\mathcal{P}\,\mathcal{L}(t,\boldsymbol{x})\,\mathcal{P}^{-1} = \mathcal{L}(t,-\boldsymbol{x})\,. \tag{5.29}$$

As already mentioned, the electromagnetic interaction of a Dirac particle with an electromagnetic field obtained via the traditional *minimal coupling*, obtained by making the substitution $\mathrm{i}\partial_\mu \to \mathrm{i}\partial_\mu - qA_\mu$ ($q$ being the particle charge) in the particle kinetic term,

$$q\bar{\psi}(x)\gamma_\mu\psi(x)\,A^\mu(x)\,, \tag{5.30}$$

is clearly parity conserving. For the strong couplings of fermions to mesons, two possibilities consistent with (29) are $g_1\,\bar{\psi}(x)\psi(x)\,\varphi(x)$ for a scalar meson $\varphi$ and $\mathrm{i}g_2\,\bar{\psi}(x)\gamma_5\psi(x)\,\phi(x)$ for a pseudoscalar meson $\phi$. Here $g_1$ and $g_2$ stand for dimensionless coupling constants.

In the mid-1950s, there was a persistent problem referred to as the $\tau$–$\theta$ puzzle, that resisted any satisfactory solution for a long time. The so-called $\tau$ and $\theta$ particles have equal masses (494 MeV) and equal mean lifetimes ($1.23\times10^{-8}$ s), but decay through channels of opposite parities:

$$\begin{aligned}\theta^+ &\to \pi^+ + \pi^0\,,\\ \tau^+ &\to \pi^+ + \pi^+ + \pi^-\,.\end{aligned}$$

The $\theta$ mode is observed in 21% and the $\tau$ mode in 6% of all disintegrations. The values of their masses and lifetimes being identical, it is plausible that $\tau$ and $\theta$ are different decay modes of the same particle. However, this seemingly natural explanation has but one difficulty in that it runs counter to the accepted tenets of the time. *If parity is a conserved quantum number* in these decay processes, then, as $\tau$ and $\theta$ have opposite parities, they must be different particles in spite of their identical masses and mean lifetimes. On the other hand, *if parity is not a conserved quantum number*, the above argument does not hold and a given particle may decay through nonconserving interactions into two or three pions. Lee and Yang (1956) systematically re-examined the whole question and came to the conclusion that while parity was conserved in hadronic and electromagnetic interactions, there existed no firm experimental data verifying the validity of this symmetry in weak interactions. They suggested several ways to check the conservation or violation of parity in weak interactions. A series of experiments were subsequently performed and proved that the weak interactions indeed broke the parity symmetry. In particular, it was showed that $\tau$ and $\theta$ were in fact different manifestations of the same particle, now called the K meson.

The first observed weak process was the $\beta$-decay of neutron-rich nuclei, in which a bound neutron disintegrates into a proton, an electron and a neutrino. The same process also occurs with free neutrons. E. Fermi described $\beta$-decay by a local interaction involving the four fermions, which was later generalized to the form

$$\mathcal{H}_\beta = \sum_i C_i \left(\overline{\psi}_{\mathrm{p}}\Gamma_i\psi_{\mathrm{n}}\right) \left(\overline{\psi}_{\mathrm{e}}\Gamma^i\psi_\nu\right) . \tag{5.31}$$

Here $C_i = C_{\mathrm{S}}, C_{\mathrm{V}}, C_{\mathrm{T}}, C_{\mathrm{A}}, C_{\mathrm{P}}$ are real or complex coupling constants of dimensions $[\mathrm{mass}]^{-2}$, and the matrices

$$\Gamma^i = 1,\ \gamma^\mu,\ \sigma^{\mu\nu}/\sqrt{2},\ \gamma^\mu\gamma_5,\ \mathrm{i}\gamma_5\,,$$
$$\Gamma_i = 1,\ \gamma_\mu,\ \sigma_{\mu\nu}/\sqrt{2},\ \gamma_\mu\gamma_5,\ -\mathrm{i}\gamma_5$$

represent all possible couplings. If parity is *not* a symmetry, an even more general expression may be postulated:

$$\mathcal{H}_\beta = \sum_i C_i \left(\overline{\psi}_{\mathrm{p}}\Gamma_i\psi_{\mathrm{n}}\right) \left[\overline{\psi}_{\mathrm{e}} \left(1 + \alpha_i\gamma_5\right) \Gamma^i\psi_\nu\right] , \tag{5.32}$$

where $\alpha_i$ are dimensionless complex constants. Since the couplings $\overline{\psi}_{\mathrm{e}}\,\Gamma^i\psi_\nu$ and $\overline{\psi}_{\mathrm{e}}\gamma_5\,\Gamma^i\psi_\nu$ have opposite parities (cf. Table 3.1 or Table 5.3) the presence of both terms in the interaction breaks parity. If the momentum dependence in the neutron and proton spinors is neglected, the transition amplitude obtained from $\mathcal{H}_\beta$ can be written more simply in terms of the Pauli spinors:

$$\begin{aligned}\mathcal{M} \approx (\chi_{\mathrm{p}}^\dagger\chi_{\mathrm{n}})\Big[C_{\mathrm{S}}\bar{u}_{\mathrm{e}}(p_{\mathrm{e}})(1+\alpha_{\mathrm{S}}\gamma_5)v_{\bar{\nu}}(p_\nu) + C_{\mathrm{V}}\bar{u}_{\mathrm{e}}(p_{\mathrm{e}})(1+\alpha_{\mathrm{V}}\gamma_5)\gamma^0 v_{\bar{\nu}}(p_\nu)\Big] \\ + (\chi_{\mathrm{p}}^\dagger\boldsymbol{\sigma}\chi_{\mathrm{n}})\cdot\Big[C_{\mathrm{T}}\bar{u}_{\mathrm{e}}(p_{\mathrm{e}})(1+\alpha_{\mathrm{T}}\gamma_5)\boldsymbol{\sigma}\, v_{\bar{\nu}}(p_\nu) \\ + C_{\mathrm{A}}\bar{u}_{\mathrm{e}}(p_{\mathrm{e}})(1+\alpha_{\mathrm{A}}\gamma_5)\gamma_5\boldsymbol{\gamma} v_{\bar{\nu}}(p_\nu)\Big] .\end{aligned} \tag{5.33}$$

The S and V terms on the first line are responsible for the allowed Fermi transitions in nuclei, while the A and T terms produce the allowed Gamow–Teller transitions. All the constants $C_i$ and $\alpha_i$ have been measured.

The constants $\alpha_i$ are first determined by measuring the longitudinal polarization of the emitted electron. A method consists in measuring the left–right asymmetry of atomic scattering of the electrons emitted in $\beta$-decay by observing the helicities of the electrons (or positrons) emitted in Fermi or Gamow–Teller nuclear transitions, in the decay of free neutrons, in inverse $\beta$-decay ($\mathrm{p} \to \mathrm{n e^+ \nu_e}$), or in muon decay ($\mu^\pm \to \mathrm{e}^\pm \nu \bar{\nu}$). The results conclusively demonstrate that there exists a clear left–right asymmetry and thus confirm there is parity violation. Electrons emitted in $\beta$-decay are polarized in the direction opposite to their motion, whereas positrons are polarized in the direction of their motion:

$$\begin{aligned} \langle \mathrm{e}^- \,|\, \boldsymbol{\Sigma}\cdot\hat{\boldsymbol{p}} \,|\, \mathrm{e}^- \rangle &= -\,v, \quad \text{electrons}\,, \\ \langle \mathrm{e}^+ \,|\, \boldsymbol{\Sigma}\cdot\hat{\boldsymbol{p}} \,|\, \mathrm{e}^+ \rangle &= +\,v, \quad \text{positrons}\,. \end{aligned} \tag{5.34}$$

In the ultra-relativistic limit where velocity $v \to c = 1$, the helicities of emitted electrons and positrons go to $-1$ and $+1$ respectively. In the same limit, the projection for a left-handed electron becomes $(1-\gamma_5)/2$, and the interaction terms must involve only the electron left chiral components. In order to reproduce this limiting result, all constants $\alpha_i$ must be real and identical to $+1$, and (33) becomes

$$\begin{aligned} \mathcal{M} \approx (\chi_\mathrm{p}^\dagger \chi_\mathrm{n}) \Big[ C_\mathrm{S} \bar{u}_\mathrm{e}(p_\mathrm{e})(1+\gamma_5) v_{\bar{\nu}} + C_\mathrm{V} \bar{u}_\mathrm{e}(p_\mathrm{e})(1+\gamma_5)\gamma^0 v_{\bar{\nu}} \Big] + (\chi_\mathrm{p}^\dagger \boldsymbol{\sigma} \chi_\mathrm{n}) \\ \times \Big[ C_\mathrm{T} \bar{u}_\mathrm{e}(p_\mathrm{e})(1+\gamma_5)\boldsymbol{\sigma}\, v_{\bar{\nu}} + C_\mathrm{A} \bar{u}_\mathrm{e}(p_\mathrm{e})(1+\gamma_5)\gamma_5 \boldsymbol{\gamma} v_{\bar{\nu}} \Big]\,. \end{aligned} \tag{5.35}$$

It remains to determine $C_i$. In transitions $J_\mathrm{i}^P = 0^+ \to J_\mathrm{f}^P = 0^+$, as in $\beta^+$-decay $^{14}\mathrm{O}(0^+) \to {}^{14}\mathrm{N}^*(0^+, 2.31\,\mathrm{MeV})$, only the Fermi couplings $C_\mathrm{S}$ and $C_\mathrm{V}$ are allowed since $\chi_\mathrm{p}^\dagger \boldsymbol{\sigma} \chi_\mathrm{n} = 0$. (Here $\chi_\mathrm{p}$ and $\chi_\mathrm{n}$ denote the Pauli spinors of bound nucleons.) In the angular distribution of the antineutrinos relative to the electron direction, the contributions from the scalar S coupling vary as $(1 - v_\mathrm{e} \cos\theta)$, and those from the vector V term as $(1 + v_\mathrm{e} \cos\theta)$. Comparisons with experimental observations confirm that Fermi transitions are of the V type, i.e. $C_\mathrm{S} = 0$. A similar analysis for Gamow–Teller transitions, as in $\beta^-$-decays $^6\mathrm{He}(0^+) \to {}^6\mathrm{Li}(1^+)$ or $^{60}\mathrm{Co}(5^+) \to {}^{60}\mathrm{Ni}^*(4^+, 2.51\,\mathrm{MeV})$, shows that the angular distribution of the antineutrinos will be proportional to $(1 - {}^1\!/_3\, v_\mathrm{e} \cos\theta)$ for an axial-vector coupling and to $(1 + {}^1\!/_3\, v_\mathrm{e} \cos\theta)$ for a tensor coupling. Experiments are consistent with a small tensor coupling, showing that $C_\mathrm{T} \ll C_\mathrm{A}$.

Therefore, the amplitude for $\beta$-transitions is

$$\begin{aligned} \mathcal{M} &= (\chi_\mathrm{p}^\dagger \chi_\mathrm{n})\, C_\mathrm{V} \bar{u}_\mathrm{e}(p_\mathrm{e})(1+\gamma_5)\gamma^0 v_{\bar{\nu}} + (\chi_\mathrm{p}^\dagger \boldsymbol{\sigma} \chi_\mathrm{n}) \cdot C_\mathrm{A} \bar{u}_\mathrm{e}(p_\mathrm{e})(1+\gamma_5)\gamma_5 \boldsymbol{\gamma} v_{\bar{\nu}} \\ &= (\chi_\mathrm{p}^\dagger \chi_\mathrm{n})\, C_\mathrm{V} \bar{u}_\mathrm{e}(p_\mathrm{e})\gamma^0(1-\gamma_5) v_{\bar{\nu}} + (\chi_\mathrm{p}^\dagger \boldsymbol{\sigma} \chi_\mathrm{n}) \cdot C_\mathrm{A} \bar{u}_\mathrm{e}(p_\mathrm{e})\boldsymbol{\gamma}(1-\gamma_5) v_{\bar{\nu}}\,. \end{aligned}$$

The magnitudes of the remaining coupling constants, $C_\mathrm{V}$ and $C_\mathrm{A}$, are determined by the $\beta$-decay rates of the neutron and the pure Fermi transition in

$^{14}$O. Their relative phase is inferred from the electron angular distributions relative to the neutron spin in $\beta$-decay of polarized neutrons. This leads to

$$\begin{aligned} G_{\mathrm{F}} &\equiv \sqrt{2}\, C_{\mathrm{V}} = (1.14730 \pm 0.0006)\, 10^{-5}\, \mathrm{GeV}^{-2}\,, \\ \alpha &\equiv \frac{C_{\mathrm{A}}}{C_{\mathrm{V}}} = (1.2573 \pm 0.0028)\,. \end{aligned} \tag{5.36}$$

In summary, the nucleon $\beta$-decay may be described by the Lagrangian

$$\begin{aligned} \mathcal{L}_\beta(x) &= -\frac{G_{\mathrm{F}}}{\sqrt{2}}\, J_\mu(x) j^\mu(x) + \mathrm{h.c.} \\ &= -\frac{G_{\mathrm{F}}}{\sqrt{2}} \Big\{ \left[\bar{\psi}_{\mathrm{p}}(x)\gamma_\mu(1-\alpha\gamma_5)\psi_{\mathrm{n}}(x)\right] \left[\bar{\psi}_{\mathrm{e}}(x)\gamma^\mu(1-\gamma_5)\psi_\nu(x)\right] \\ &\quad + \left[\bar{\psi}_{\mathrm{n}}(x)\gamma_\mu(1-\alpha\gamma_5)\psi_{\mathrm{p}}(x)\right] \left[\bar{\psi}_\nu(x)\gamma^\mu(1-\gamma_5)\psi_{\mathrm{e}}(x)\right] \Big\}\,. \end{aligned} \tag{5.37}$$

The first term on the right-hand side describes the $\beta$-decay itself, in which a *right-handed antineutrino* is emitted. Its Hermitian conjugate (h.c.), given in the second term, makes the whole Lagrangian Hermitian; it represents the inverse $\beta$-decay processes, $\bar{\mathrm{n}} \to \bar{\mathrm{p}} + \mathrm{e}^+ + \nu$ or $\mathrm{p} \to \mathrm{n} + \mathrm{e}^+ + \nu$, in which a *left-handed neutrino* appears. Thus, this weak interaction involves only left-handed leptons and right-handed antileptons.

## 5.2 Time Inversion

The time inversion operator $\mathcal{T}$ reverses the sign of the time parameter,

$$\mathcal{T}\ :\ x = (t, \boldsymbol{x}) \to x' = (-t, \boldsymbol{x})\,, \tag{5.38}$$

and changes physical variables accordingly,

$$\begin{aligned} \boldsymbol{p} &= m\,\frac{\mathrm{d}\boldsymbol{x}}{\mathrm{d}t} \to \boldsymbol{p}' = -\boldsymbol{p}\,, \\ \boldsymbol{L} &= \boldsymbol{x} \times \boldsymbol{p} \to \boldsymbol{L}' = -\boldsymbol{L} \end{aligned}$$

(see Fig. 5.2). Newton's equation for a particle acted on by a nondissipative and time-independent force is invariant to this transformation, so that if $\boldsymbol{x}(t)$ is an allowed trajectory, then $\boldsymbol{x}(-t)$ is equally allowed. Classical mechanics cannot determine the time arrow. On the other hand, assuming classical electromagnetism to be invariant as well, one sees that the electric and magnetic fields transform as $\boldsymbol{E} \to +\boldsymbol{E}$ and $\boldsymbol{B} \to -\boldsymbol{B}$, since the electric charge is unchanged but the electric current (the product of charge and velocity) changes sign under time inversion.

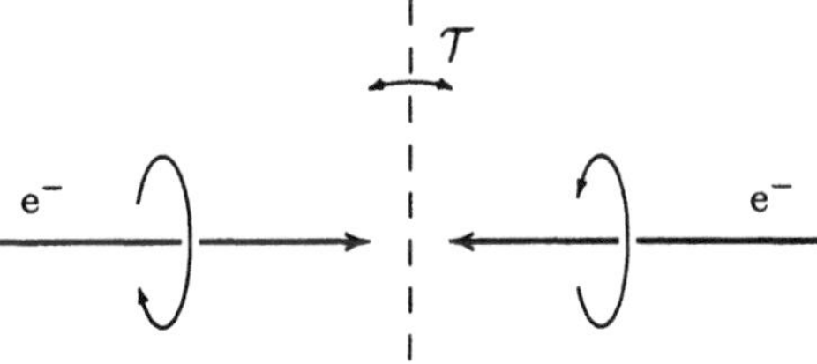


**Fig. 5.2.** $\mathcal{T}$ reverses the momentum of a particle and flips its spin

### 5.2.1 Time Inversion in Quantum Mechanics

In the Schrödinger representation, the state function satisfies the equation

$$\mathrm{i}\,\frac{\partial}{\partial t}\phi(t,\boldsymbol{x}) = H\phi(t,\boldsymbol{x})\,. \tag{5.39}$$

Invariance requires the transformed wave function $\mathcal{T}\phi(t,\boldsymbol{x})$ to satisfy the same equation with $t$ replaced by $t' = -t$. The question is, how is $\mathcal{T}\phi(t,\boldsymbol{x})$ related to $\phi(t,\boldsymbol{x})$?

The simplest possible postulate, $\mathcal{T}\phi(t,\boldsymbol{x}) = \phi(-t,\boldsymbol{x})$, leads to

$$\mathrm{i}\,\frac{\partial}{\partial(-t)}\phi(-t,\boldsymbol{x}) = -H\phi(-t,\boldsymbol{x})\,, \tag{5.40}$$

so that, for example, the wave function of a free particle, $\phi(t,\boldsymbol{x}) = \exp(\mathrm{i}\boldsymbol{p}\cdot\boldsymbol{x} - \mathrm{i}Et)$, will become after time inversion $\phi(-t,\boldsymbol{x}) = \exp(\mathrm{i}\boldsymbol{p}\cdot\boldsymbol{x} + \mathrm{i}Et)$. For the dynamic equation to preserve its form, the Hamiltonian must be modified to $H' = -H$, which implies that to each state of positive energy before the transformation, there corresponds a state of negative energy after the transformation. States of negative energies are unstable and would sink to a state of infinitely large negative energy. The assumption $\mathcal{T}\phi(t,\boldsymbol{x}) = \phi(-t,\boldsymbol{x})$ is thus unacceptable for an invariant theory with positive energies before as well as after a time reversal.

The solution to this difficulty was given by Wigner in 1932. It is first assumed that there exists a *unitary* operator $\mathcal{U}$ such that $\mathcal{U}H^*\mathcal{U}^\dagger = H$ (where * means complex conjugation). Applying $\mathcal{U}$ from the left on both sides of the complex conjugate of (39) yields

$$\mathrm{i}\,\frac{\partial}{\partial(-t)}\mathcal{U}\phi^*(t,\boldsymbol{x}) = \mathcal{U}H^*\phi^*(t,\boldsymbol{x}) = H\mathcal{U}\phi^*(t,\boldsymbol{x})\,,$$

or, changing the sign of $t$,

$$\mathrm{i}\,\frac{\partial}{\partial t}\mathcal{U}\phi^*(-t,\boldsymbol{x}) = H\mathcal{U}\phi^*(-t,\boldsymbol{x})\,. \tag{5.41}$$

Thus, if $\phi(t,\boldsymbol{x})$ is a solution to (39), so too is the transformed wave function

$$\phi'(t',\boldsymbol{x}') = \mathcal{U}\phi^*(-t,\boldsymbol{x})\,. \tag{5.42}$$

In particular, if $H$ is real, invariance to $\mathcal{T}$ means that if $\phi_E(\boldsymbol{x})$ represents a stationary wave function of energy $E$, the function $\phi_E^*(\boldsymbol{x})$ is also an energy eigenfunction with the same energy. This implies that if $E$ is nondegenerate, then $\phi_E(\boldsymbol{x}) \propto \phi_E^*(\boldsymbol{x})$ and thus can be chosen real.

The time inversion operator on state vector space is thus the product of two operators: a unitary transformation $\mathcal{U}$, which replaces the state vector on which it operates with a time-reversed state vector, and a complex conjugation $K$ of all the coefficients that may come with the state vector,

$$\mathcal{T}\,\alpha\psi(t) = \mathcal{U}K\,\alpha\psi(-t) = \alpha^*\,\mathcal{U}\psi^*(-t)\,. \tag{5.43}$$

The presence of a nontrivial $\mathcal{U}$ is necessary except in the most trivial cases. Since $K^2 = 1$, one has $\mathcal{T}^{-1} = K\mathcal{U}^\dagger$. Moreover, $\mathcal{T}$ has several properties worth noting:

(P1) *antilinearity*: $\mathcal{T}(a\,|\phi\rangle + b\,|\psi\rangle) = a^*\mathcal{T}\,|\phi\rangle + b^*\mathcal{T}\,|\psi\rangle$, where $a$ and $b$ are complex constants;
(P2) *antiunitarity*: $\langle \mathcal{T}\phi(t)\,|\mathcal{T}\psi(t)\,\rangle = \langle \phi(-t)\,|\psi(-t)\,\rangle^*$, a distinctive property of $\mathcal{T}$; but the norms of vectors and the probabilities remain invariant, just as in the case of unitary operators;
(P3) *operator transformation*: $\mathcal{O}' = \mathcal{T}\mathcal{O}\mathcal{T}^{-1} = \mathcal{U}\mathcal{O}^*\mathcal{U}^{-1}$ for an arbitrary operator $\mathcal{O}$, so that $\langle \mathcal{T}\psi(t)\,|\,\mathcal{O}'\,|\,\mathcal{T}\phi(t)\rangle = \langle \psi(-t)\,|\,\mathcal{O}\,|\,\phi(-t)\rangle^*$.

**Example 1. Spin-0 Particle**
For a spinless particle, $\mathcal{T}$ is merely complex conjugation $K$. Thus, its wave function in the $x$ representation $\langle x\,|p\rangle = \exp(\mathrm{i}\boldsymbol{p}\cdot\boldsymbol{x} - \mathrm{i}Et)$ becomes $\langle x'\,|p\rangle^* = \exp(-\mathrm{i}\boldsymbol{p}\cdot\boldsymbol{x} - \mathrm{i}Et)$. Also since $\langle Kx\,|p\rangle^* = \langle x\,|\,K\,|\,p\rangle$, the ket of a particle of momentum $\boldsymbol{p}$ transforms into

$$\mathcal{T}\,|\boldsymbol{p}\rangle = |-\boldsymbol{p}\rangle\,. \tag{5.44}$$

The radial part of a state having a well-defined angular momentum remains unchanged under $\mathcal{T}$, but its angular part $\langle x\,|\ell\, m\rangle = \mathrm{i}^\ell Y_{\ell m}(\theta,\varphi)$ transforms into

$$\langle x'\,|\ell\, m\rangle^* = (-\mathrm{i})^\ell\, Y^*_{\ell m}(\theta,\varphi) = (-)^{\ell-m}\,\mathrm{i}^\ell\, Y_{\ell,-m}(\theta,\varphi)\,,$$

which we may identify with $\langle x\,|\,\mathcal{T}\,|\,\ell\, m\rangle$ to get

$$\mathcal{T}\,|\ell\, m\rangle = (-)^{\ell-m}\,|\ell,-m\rangle\,. \tag{5.45}$$

This result agrees with the expected transformation of the angular momentum, $\boldsymbol{L} \to -\boldsymbol{L}$.

**Example 2. Spin-1/2 Particle**
By extension of the transformation rule for orbital angular momentum, it is assumed that the spin transforms as $\mathcal{T}\boldsymbol{S}\mathcal{T}^{-1} = -\boldsymbol{S}$, where $\mathcal{T} = \mathcal{U}K$. For spin 1/2, $\boldsymbol{S} = \boldsymbol{\sigma}/2$. In the standard representation, where $\sigma_x$ and $\sigma_z$ are real and $\sigma_y$ is imaginary, we have

$$\begin{aligned}
\mathcal{T}\sigma_x\mathcal{T}^{-1} &= \phantom{-}\mathcal{U}\sigma_x\mathcal{U}^{-1} = -\sigma_x\,,\\
\mathcal{T}\sigma_y\mathcal{T}^{-1} &= -\mathcal{U}\sigma_y\mathcal{U}^{-1} = -\sigma_y\,,\\
\mathcal{T}\sigma_z\mathcal{T}^{-1} &= \phantom{-}\mathcal{U}\sigma_z\mathcal{U}^{-1} = -\sigma_z\,.
\end{aligned}$$

These equations admit as solution $\mathcal{U} = \eta\sigma_y$, with $\eta$ an arbitrary unimodular phase factor, $|\eta| = 1$. One may choose for example $\eta = -\mathrm{i}$ to reproduce the conventional phase of angular momentum eigenvectors. Thus, for the basis vectors

$$\chi_+ = \begin{pmatrix} 1 \\ 0 \end{pmatrix}, \qquad \chi_- = \begin{pmatrix} 0 \\ 1 \end{pmatrix},$$

one gets $-i\sigma_y \chi_m = (-)^{1/2-m} \chi_{-m}$, where $m = \pm 1/2$. Therefore, the state vector of a spin-$1/2$ particle of momentum $\boldsymbol{p}$ and polarization $m$ transforms under $\mathcal{T}$ into

$$\mathcal{T} \,|\boldsymbol{p}, m\rangle = (-)^{1/2-m} \,|-\boldsymbol{p}, -m\rangle \,. \tag{5.46}$$

More generally, a vector of total angular momentum $j$, $j_z = m$ obeys the relation

$$\mathcal{T} \,|\alpha, j\, m\rangle = (-)^{j-m} \,|\alpha_{\mathrm{T}}, j, -m\rangle \,, \tag{5.47}$$

where $\alpha_{\mathrm{T}}$ stands for the time-reversed quantum numbers corresponding to $\alpha$. Note that $\mathcal{T}^2 \,|j\, m\rangle = (-)^{2j} \,|j\, m\rangle$, and hence

$$\begin{aligned} \mathcal{T}^2 &= +1 \quad \text{for an integral spin particle, and} \\ \mathcal{T}^2 &= -1 \quad \text{for a half-integral spin particle}\,. \end{aligned}$$

This is a general result which depends neither on the phase convention nor on the representation of state vectors. In particular, for a system of $N$ fermions, $\mathcal{T}^2 = (-)^N$. ■

### 5.2.2 Time Inversion in Field Theories

We begin by studying the behavior of classical c-numbered fields under time inversion and generalize the results to the corresponding field operators.

**Scalar Fields.** The Klein–Gordon equation for a classical scalar field $\phi_{\mathrm{c}}$ is invariant to time inversion whether the transformation rule is $\phi'_{\mathrm{c}}(t') = \phi_{\mathrm{c}}(-t)$ or $\phi'_{\mathrm{c}}(t') = \phi^*_{\mathrm{c}}(-t)$. However, for consistency we adopt the same rule as for the Schrödinger equation, up to an arbitrary phase,

$$\mathcal{T} \,: \phi_{\mathrm{c}}(t, \boldsymbol{x}) \to \phi'_{\mathrm{c}}(t', \boldsymbol{x}') = \zeta_{\mathrm{B}} \phi^*_{\mathrm{c}}(-t, \boldsymbol{x}), \qquad \text{such that } |\zeta_{\mathrm{B}}| = 1\,. \tag{5.48}$$

As seen in Chap. 2, this c-numbered function is related to the corresponding quantized field $\phi(x)$ by

$$\phi_{\mathrm{c}}(t, \boldsymbol{x}) = \langle 0 \,|\, \phi(t, \boldsymbol{x}) \,|\, \boldsymbol{p}\rangle \,, \tag{5.49}$$

which, upon application of $\mathcal{T}$ on both sides, results in

$$\zeta_{\mathrm{B}} \phi^*_{\mathrm{c}}(-t, \boldsymbol{x}) = \langle \mathcal{T} 0 \,|\, \phi(t, \boldsymbol{x}) \,|\, \mathcal{T} \boldsymbol{p}\rangle \,. \tag{5.50}$$

By property P3, the left-hand side is $\zeta_B \langle \mathcal{T}0 | \mathcal{T}\phi(-t,\boldsymbol{x})\mathcal{T}^{-1} | \mathcal{T}\boldsymbol{p}\rangle$, which leads to

$$\mathcal{T}\phi(t,\boldsymbol{x})\mathcal{T}^{-1} = \zeta_B\phi(-t,\boldsymbol{x}), \qquad (\zeta_B = \pm 1). \tag{5.51}$$

Note the transformed quantized field is *not* complex conjugated.

To obtain the transformation rules for Fock operators we substitute the plane-wave expansion of $\phi$ (2.99) into (51) to obtain for its left-hand side

$$\mathcal{T}\phi(x)\mathcal{T}^{-1} = \sum_p C_{\boldsymbol{p}} \left[\mathcal{T}a_{\boldsymbol{p}}\mathcal{T}^{-1}\mathrm{e}^{\mathrm{i}(Et-\boldsymbol{p}\cdot\boldsymbol{x})} + \mathcal{T}b_{\boldsymbol{p}}^\dagger\mathcal{T}^{-1}\mathrm{e}^{-\mathrm{i}(Et-\boldsymbol{p}\cdot\boldsymbol{x})}\right] \tag{5.52}$$

(the exponentials having been complex conjugated by antilinearity), and for its right-hand side

$$\begin{aligned}\zeta_B\,\phi(-t,\boldsymbol{x}) &= \zeta_B \sum_p C_{\boldsymbol{p}} \left[a_{\boldsymbol{p}}\mathrm{e}^{\mathrm{i}(Et+\boldsymbol{p}\cdot\boldsymbol{x})} + b_{\boldsymbol{p}}^\dagger\mathrm{e}^{-\mathrm{i}(Et+\boldsymbol{p}\cdot\boldsymbol{x})}\right] \\ &= \zeta_B \sum_p C_{\boldsymbol{p}} \left[a_{-\boldsymbol{p}}\mathrm{e}^{\mathrm{i}(Et-\boldsymbol{p}\cdot\boldsymbol{x})} + b_{-\boldsymbol{p}}^\dagger\mathrm{e}^{-\mathrm{i}(Et-\boldsymbol{p}\cdot\boldsymbol{x})}\right]\end{aligned} \tag{5.53}$$

(changing $t \to -t$ and $\boldsymbol{p} \to -\boldsymbol{p}$ in the sum). Identifying the right-hand sides of the two resulting equations, one gets (with $\zeta_B = \pm 1$)

$$\begin{aligned}\mathcal{T}a_{\boldsymbol{p}}\mathcal{T}^{-1} &= \zeta_B a_{-\boldsymbol{p}}\,, \\ \mathcal{T}b_{\boldsymbol{p}}^\dagger\mathcal{T}^{-1} &= \zeta_B b_{-\boldsymbol{p}}^\dagger\,.\end{aligned} \tag{5.54}$$

To illustrate, consider the probability amplitude for the presence of a boson of momentum $\boldsymbol{p}$ in an arbitrary state $\psi$,

$$\begin{aligned}\langle\psi|\boldsymbol{p}\rangle &= \langle\psi|a_{\boldsymbol{p}}^\dagger|0\rangle = \langle\mathcal{T}\psi|\mathcal{T}a_{\boldsymbol{p}}^\dagger\mathcal{T}^{-1}|\mathcal{T}0\rangle^* \\ &= \zeta\langle 0|a_{-\boldsymbol{p}}|\mathcal{T}\psi\rangle = \zeta\langle -\boldsymbol{p}|\mathcal{T}\psi\rangle\,.\end{aligned}$$

In the time-reversed amplitude the particle reverses its direction of motion, with the initial state found in the bra.

The current density for a scalar field (13) transforms into

$$\mathcal{T}j^0(t,\boldsymbol{x})\mathcal{T}^{-1} = +j^0(-t,\boldsymbol{x}), \qquad \mathcal{T}j^i(t,\boldsymbol{x})\mathcal{T}^{-1} = -j^i(-t,\boldsymbol{x}), \tag{5.55}$$

just as anticipated from classical arguments.

**Electromagnetic Field.** If the electromagnetic interaction is T-invariant as shown by observations, and if it may be described by $H_{\mathrm{em}} = qj^\mu A_\mu$, then given the property (55) of the current, $A^\mu(x)$ must satisfy

$$\mathcal{T}A^\mu(t,\boldsymbol{x})\mathcal{T}^{-1} = (A^0(-t,\boldsymbol{x}),\ -A^i(-t,\boldsymbol{x}))\,. \tag{5.56}$$

The two sides of this equation may be explicitly written out, making use of the plane-wave expansion series (2.156) for the transverse field $\boldsymbol{A}$,

$$\mathcal{T}\boldsymbol{A}(t,\boldsymbol{x})\mathcal{T}^{-1} = \sum_{k\lambda} C_{\boldsymbol{k}}\Big[\boldsymbol{\epsilon}^*(\boldsymbol{k},\lambda)\,\mathcal{T}a(\boldsymbol{k},\lambda)\,\mathcal{T}^{-1}\mathrm{e}^{\mathrm{i}k\cdot x} + \boldsymbol{\epsilon}(\boldsymbol{k},\lambda)\,\mathcal{T}a^\dagger(\boldsymbol{k},\lambda)\,\mathcal{T}^{-1}\mathrm{e}^{-\mathrm{i}k\cdot x}\Big]\,; \tag{5.57}$$

$$\begin{aligned} -\boldsymbol{A}(-t,\boldsymbol{x}) &= -\sum_{k\lambda} C_{\boldsymbol{k}}\,\left[\boldsymbol{\epsilon}(-\boldsymbol{k},\lambda)\,a(-\boldsymbol{k},\lambda)\mathrm{e}^{\mathrm{i}\omega t-\mathrm{i}\boldsymbol{k}\cdot\boldsymbol{x}} + \boldsymbol{\epsilon}^*(-\boldsymbol{k},\lambda)\,a^\dagger(-\boldsymbol{k},\lambda)\mathrm{e}^{-\mathrm{i}\omega t+\mathrm{i}\boldsymbol{k}\cdot\boldsymbol{x}}\right] \\ &= +\sum_{k\lambda} C_{\boldsymbol{k}}\,\left[\boldsymbol{\epsilon}^*(\boldsymbol{k},\lambda)\,a(-\boldsymbol{k},\lambda)\,\mathrm{e}^{\mathrm{i}k\cdot x} + \boldsymbol{\epsilon}(\boldsymbol{k},\lambda)\,a^\dagger(-\boldsymbol{k},\lambda)\,\mathrm{e}^{-\mathrm{i}k\cdot x}\right]. \end{aligned} \tag{5.58}$$

In the last step we have used (18). Identifying the right-hand sides of the two resulting equations leads to the transformation rules for the photon Fock operators:

$$\begin{aligned} \mathcal{T}a(\boldsymbol{p},\lambda)\,\mathcal{T}^{-1} &= a(-\boldsymbol{p},\lambda)\,, \\ \mathcal{T}a^\dagger(\boldsymbol{p},\lambda)\,\mathcal{T}^{-1} &= a^\dagger(-\boldsymbol{p},\lambda)\,. \end{aligned} \tag{5.59}$$

The result is consistent with a phase equal to +1. The photon helicity is unchanged by $\mathcal{T}$ because both its spin and momentum change signs.

**Dirac Field.** It is assumed, just as before, that the c-numbered Dirac wave function transforms under time inversion as

$$\mathcal{T}\;:\psi_{\mathrm{c}}(t,\boldsymbol{x}) \to \psi'_{\mathrm{c}}(t',\boldsymbol{x}') = \zeta_{\mathrm{F}}\,A\psi^*_{\mathrm{c}}(-t,\boldsymbol{x}), \qquad |\zeta_{\mathrm{F}}| = 1. \tag{5.60}$$

Here $A$ is a $4\times 4$ unitary matrix in terms of which any component of the transformed spinor is expressed as a linear combination of different components of the original spinor. However, in quantum field theory, if the rule $\psi \to A\psi^\dagger$ were adopted, a fermion would transform into an antifermion, which would not be physically acceptable. To find the correct transformation rules, one follows the same arguments as for the boson field and obtains

$$\begin{aligned} \mathcal{T}\psi(t,\boldsymbol{x})\,\mathcal{T}^{-1} &= \zeta_{\mathrm{F}}\,A\,\psi(-t,\boldsymbol{x})\,, \\ \mathcal{T}\psi^\dagger(t,\boldsymbol{x})\,\mathcal{T}^{-1} &= \zeta^*_{\mathrm{F}}\,\psi^\dagger(-t,\boldsymbol{x})A^\dagger\,, \\ \mathcal{T}\overline{\psi}(t,\boldsymbol{x})\,\mathcal{T}^{-1} &= \zeta^*_{\mathrm{F}}\,\overline{\psi}(-t,\boldsymbol{x})A^\dagger\,. \end{aligned} \tag{5.61}$$

The matrix $A$ cannot be calculated by the same relation which has served to determine $S(a)$ for parity, because time inversion is antiunitary whereas parity is unitary. Rather, one proceeds by imposing T-invariance on the Dirac Lagrangian,

$$\mathcal{T}\mathcal{L}_{\mathrm{F}}(x)\,\mathcal{T}^{-1} = \mathcal{L}_{\mathrm{F}}(x')\,, \qquad \text{where } x' = (-t,\boldsymbol{x})\,. \tag{5.62}$$

From (61) and properties P1–P3 of $\mathcal{T}$, one gets for the right-hand side

$$\mathcal{L}_{\mathrm{F}}(x') = \overline{\psi}(x')(\mathrm{i}\gamma^\mu \partial'_\mu - m)\psi(x'),$$

and for the left-hand side

$$\begin{aligned}\mathcal{T}\,\mathcal{L}_{\mathrm{F}}(x)\,\mathcal{T}^{-1} &= \mathcal{T}\,\psi^\dagger(x)\,\mathcal{T}^{-1}\left[-\mathrm{i}(\gamma^0\gamma^\mu)^*\partial_\mu - m\gamma^{0*}\right]\mathcal{T}\,\psi(x)\,\mathcal{T}^{-1}\\ &= \overline{\psi}(x')\gamma^0 A^\dagger\left(\mathrm{i}\gamma^{0\mathrm{T}}\gamma^{\mu\mathrm{T}}\partial'_\mu - m\gamma^{0\mathrm{T}}\right)A\psi(x'),\end{aligned}$$

where use has been made of the matrix properties $\gamma^{\mu\mathrm{T}} = (\gamma^0\gamma^\mu\gamma^0)^*$, or $\gamma^{0\mathrm{T}} = \gamma^{0*}$ and $\gamma^{i\mathrm{T}} = -\gamma^{i*}$, which follow from their Hermitian conjugation property, $\gamma^{\mu\dagger} = \gamma^0\gamma^\mu\gamma^0$. From these relations one can immediately infer the defining property of $A$:

$$A\gamma^\mu A^\dagger = \gamma^{\mu\mathrm{T}} \qquad \text{(in arbitrary representation).} \tag{5.63}$$

To find an explicit expression for $A$, a concrete representation of $\gamma_\mu$ is needed. In the standard representation where only $\gamma^2$ is imaginary, $A$ commutes with $\gamma^0$ and $\gamma^2$, and anticommutes with both $\gamma^1$ and $\gamma^3$ :

$$\begin{aligned}&A\gamma^0 = \gamma^0 A, \quad A\gamma^1 = -\gamma^1 A,\\ &A\gamma^2 = \gamma^2 A, \quad A\gamma^3 = -\gamma^3 A.\end{aligned}$$

These conditions hold provided that

$$A = \lambda\gamma^1\gamma^3, \quad |\lambda|^2 = 1 \qquad \text{(in standard representation)}\,. \tag{5.64}$$

As an application of this result, consider the fermion current $j_\mu(x) = \overline{\psi}(x)\gamma_\mu\psi(x)$, which transforms into

$$\begin{aligned}\mathcal{T}\,j_\mu(x)\,\mathcal{T}^{-1} &= |\zeta_{\mathrm{F}}|^2\,\psi^\dagger(x')A^\dagger(\gamma_0\gamma_\mu)^* A\psi(x')\\ &= \overline{\psi}(x')A^\dagger\gamma_\mu^* A\psi(x') = \overline{\psi}(x')\gamma_\mu^\dagger\psi(x')\,.\end{aligned}$$

This reduces to (55) on using (63) for $A$. Transformations of other bilinear covariants, which may describe various interaction models, can be similarly found. Thus, for example,

$$\begin{aligned}\mathcal{T}\,\overline{\psi}(x)\psi(x)\,\mathcal{T}^{-1} &= \overline{\psi}(x')A^\dagger A\psi(x') = \overline{\psi}(x')\psi(x)\,;\\ \mathcal{T}\,\overline{\psi}(x)\gamma^\mu\psi(x)\,\mathcal{T}^{-1} &= \overline{\psi}(x')A^\dagger(\gamma^\mu)^* A\psi(x') = \overline{\psi}(x')\gamma^{\mu\dagger}\psi(x)\,;\\ \mathcal{T}\,\overline{\psi}(x)\mathrm{i}\gamma_5\psi(x)\,\mathcal{T}^{-1} &= \overline{\psi}(x')A^\dagger(\mathrm{i}\gamma_5)^* A\psi(x') = -\overline{\psi}(x')\mathrm{i}\gamma_5\psi(x)\,;\\ \mathcal{T}\,\overline{\psi}(x)\gamma^\mu\gamma_5\psi(x)\,\mathcal{T}^{-1} &= \overline{\psi}(x')A^\dagger\gamma^{\mu*}\gamma_5^* A\psi(x') = \overline{\psi}(x')\gamma^{\mu\dagger}\gamma_5\psi(x)\,.\end{aligned}$$

Let us now examine the action of $A$ on spinors of polarizations $s = \pm 1/2$. With the phase conventions adopted in (3.45) and (3.46), we have

$$\begin{aligned}Au(p,s) &= -\lambda\,(-)^{1/2-s}u^*(-\boldsymbol{p},-s)\,,\\ Av(p,s) &= -\lambda\,(-)^{1/2-s}v^*(-\boldsymbol{p},-s)\,.\end{aligned} \tag{5.65}$$

In order to determine the transformations of the Fock operators for fermions, we write out the expansion series of (61). We have, on the one hand,

$$\begin{aligned} \mathcal{T}\,\psi(x)\,\mathcal{T}^{-1} = & \sum_{\boldsymbol{p},s} C_{\boldsymbol{p}}\big[\mathcal{T}\,b(\boldsymbol{p},s)\,\mathcal{T}^{-1}u^*(\boldsymbol{p},s)\,\mathrm{e}^{\mathrm{i}p\cdot x} \\ & + \mathcal{T}\,d^\dagger(\boldsymbol{p},s)\,\mathcal{T}^{-1}v^*(\boldsymbol{p},s)\,\mathrm{e}^{-\mathrm{i}p\cdot x}\big] \end{aligned} \tag{5.66}$$

(using antilinearity of $\mathcal{T}$), and on the other hand,

$$\begin{aligned} A\psi(-t,\boldsymbol{x}) = \lambda & \sum_{\boldsymbol{p},s} (-)^{1/2-s} C_{\boldsymbol{p}}[b(-\boldsymbol{p},-s)u^*(-\boldsymbol{p},-s)\mathrm{e}^{\mathrm{i}p\cdot x} \\ & + d^\dagger(-\boldsymbol{p},-s)v^*(\boldsymbol{p},s)\,\mathrm{e}^{-\mathrm{i}p\cdot x}\,]\,, \end{aligned} \tag{5.67}$$

where we have used (65) and changed the signs of $\boldsymbol{p}$ and $s$ in the sums; since $s$ is half-integral, $(-)^{1/2+s} = -(-)^{1/2-s}$. From these results follow the relations

$$\mathcal{T}\,b(\boldsymbol{p},s)\,\mathcal{T}^{-1} = (-)^{1/2-s}\zeta_{\mathrm{F}}\,b(-\boldsymbol{p},-s)\,, \tag{5.68}$$

$$\mathcal{T}\,d^\dagger(\boldsymbol{p},s)\,\mathcal{T}^{-1} = (-)^{1/2-s}\zeta_{\mathrm{F}}\,d^\dagger(-\boldsymbol{p},-s)\,, \tag{5.69}$$

where $\lambda \equiv +1$. This choice is adopted simply to be in accord with (46),

$$\mathcal{T}\,|\boldsymbol{p},s\rangle = (-)^{1/2-s}\,|-\boldsymbol{p},-s\rangle\,. \tag{5.70}$$

### 5.2.3 $\mathcal{T}$ and Interactions

The interaction Hamiltonian $H_{\mathrm{em}} = qj_\mu A^\mu$ describes electromagnetic phenomena. As we have seen, it is invariant to time inversion. An example of noninvariant interaction is that of the electric dipole moment which is described in the nonrelativistic limit by $H_{\mathrm{d}} = -\mu^{(\mathrm{el})}\boldsymbol{\sigma}\cdot\boldsymbol{E}$. Since, under time inversion, the electric field remains unchanged while the angular momentum reverses its direction, $H_{\mathrm{d}}$ is odd with respect to $\mathcal{T}$ (as it is also with respect to $\mathcal{P}$). Hermiticity of the Hamiltonian requires $\mu^{(\mathrm{el})}$ to be real. Experiments have proved that this interaction is negligible, as shown by the extremely small values of the measured electric dipole moments of typical fermions:

| | |
|---|---|
| e | $(-0.3 \pm 0.8)\times 10^{-26}\ e\,\mathrm{cm}$, |
| $\mu$ | $(3.7 \pm 3.4)\times 10^{-19}\ e\,\mathrm{cm}$, |
| p | $(-4 \pm 6)\times 10^{-23}\ e\,\mathrm{cm}$, |
| n | $< 1.1\times 10^{-25}\ e\,\mathrm{cm}$. |

Just as with other symmetries, T-invariance imposes restrictions on physical models. Consider for example a possible candidate for the interaction model of fermionic fields with a neutral scalar or pseudoscalar field. Nonderivative Yukawa couplings may take the general form

$$\mathcal{H}_{\mathrm{int}} = (g_{\mathrm{s}}\overline{\psi}\psi + \mathrm{i}g_{\mathrm{p}}\overline{\psi}\gamma_5\psi)\phi\,. \tag{5.71}$$

As $\mathcal{H}_{\text{int}}$ is Hermitian, the coupling constants $g_{\text{s}}$ and $g_{\text{p}}$ must be real. The time-reversed Hamiltonian is

$$\mathcal{T}\mathcal{H}_{\text{int}}\mathcal{T}^{-1} = (g_{\text{s}}\overline{\psi}\psi - \text{i}g_{\text{p}}\overline{\psi}\gamma_5\psi)\zeta_\phi\phi\,. \tag{5.72}$$

To have a T-invariant model, clearly one of the two terms must be absent. In this example, invariance of the model to time inversion also implies its invariance to space inversion.

An arbitrary quantum state is normally a very complex quantity, and only the state vectors of a stable particle are simple enough for the time-reversed vectors to be explicitly known. In general, the time reverse of an arbitrary state is complicated and its actual observation, unlikely. For this reason a direct verification of T-invariance of a dynamic equation is difficult. More often, the symmetry can only be indirectly tested, for example by the confirmation or invalidation of certain predicted phase relations. Let us consider again the $\beta$-decay Hamiltonian (37), where both $C_{\text{V}}$ and $C_{\text{A}}$ are now assumed to be *a priori* complex. Under time inversion,

$$\begin{aligned}&\mathcal{T}\left[\overline{\psi}_{\text{p}}\gamma_\mu(C_{\text{V}} - C_{\text{A}}\gamma_5)\psi_{\text{n}}\right]\left[\overline{\psi}_{\text{e}}\gamma^\mu(1-\gamma_5)\psi_\nu\right]\mathcal{T}^{-1}\\&\quad= \zeta_{\text{p}}^*\zeta_{\text{e}}^*\zeta_{\text{n}}\zeta_\nu\left[\overline{\psi}_{\text{p}}\gamma_\mu(C_{\text{V}}^* - C_{\text{A}}^*\gamma_5)\psi_{\text{n}}\right]\left[\overline{\psi}_{\text{e}}\gamma^\mu(1-\gamma_5)\psi_\nu\right].\end{aligned} \tag{5.73}$$

So the interaction (37) is T-invariant provided the constants $C_{\text{A}}$ and $C_{\text{V}}$ are relatively real. This condition can be experimentally checked. Available data

$$C_{\text{A}}/C_{\text{V}} \equiv |C_{\text{A}}/C_{\text{V}}|\exp(\text{i}\phi_{\text{AV}}), \quad \phi_{\text{AV}}^{\text{exp}} = (180.07 \pm 0.18)^\circ\,, \tag{5.74}$$

demonstrate without ambiguities that the $\beta$-decay is time-reversal-invariant. However, not all weak interactions have this property. In particular, a small but unmistakable violation of the T-symmetry has been detected in the $\text{K}^0$–$\overline{\text{K}}^0$ system (see Chap. 11).

## 5.3 Charge Conjugation

We have considered up to now symmetries in space-time and their associated quantum numbers. In addition to these numbers of kinematic origins, particles may have other attributes, called *internal*, such as the electric charge, the baryon number, the lepton number, and the strangeness. These numbers may also obey conservation rules that arise from the fact that the phases of non-Hermitian fields are nonobservable or, equivalently, from the invariance of the theory to such phase or gauge transformations. Each internal quantum number is associated with an abstract internal space in which symmetry operations, such as rotations or reflections, can be defined in analogy with similar operations in ordinary space. In this section, we will first introduce some such internal quantum numbers (also called generalized charges) and then discuss the *charge conjugation*, which is a discrete transformation that reverses the signs of all nonzero generalized charges of a given particle, converting it into the corresponding antiparticle.

### 5.3.1 Additive Quantum Numbers

We have seen in Chap. 2 that invariance of a theory to a global gauge transformation independent of space-time coordinates gives rise to a local conserved current and an associated charge, which is constant in time. Such a transformation effects a phase change on the wave function of the first-quantized formalism,

$$\varphi \to \exp(-iq\alpha)\varphi\,,$$

or on the field operator in the second-quantized formalism,

$$U\varphi U^{-1} = \mathrm{e}^{-\mathrm{i}q\alpha}\varphi\,, \quad U = \mathrm{e}^{\mathrm{i}Q\alpha}\,. \tag{5.75}$$

Here $\alpha$ is a real arbitrary constant and $q$ the electric charge of the field. The operator $Q$ which generates infinitesimal gauge transformations is identified with the electric charge operator and is, for this reason, Hermitian. Invariance of the theory means that the system remains unchanged on application of $U$ and that the total charge is conserved. This may be expressed for example in terms of the invariance of the $S$-matrix with respect to arbitrary phase transformations of the charged states, so that a typical $S$-matrix element for an allowed transition varies as

$$\langle \mathrm{f}\,|\,S\,|\,\mathrm{i}\rangle \to [1 + \mathrm{i}(Q_\mathrm{i} - Q_\mathrm{f})\alpha]\,\langle \mathrm{f}\,|\,S\,|\,\mathrm{i}\rangle\,,$$

where $Q_\mathrm{i}$ ($Q_\mathrm{f}$) is the total charge of the initial (final) state. Invariance implies that $S$ is unchanged, and hence that $Q_\mathrm{i} = Q_\mathrm{f}$. If the system is composed of particles of charges $q_1, q_2, \ldots$, then $U \to \exp[-\mathrm{i}\alpha(q_1 + q_2 + \ldots)]$ when applied on the state vector of the system, and the total charge is the algebraic sum of all individual charges $(q_1 + q_2 + \ldots)$. To the $Q$ operator corresponds then an *additive* quantum number, as is the general case of any Hermitian operator that generates a unitary transformation by exponentiation.

The evidence for the *electric charge conservation* is strong. If this conservation rule were broken, the lightest charged particle, the electron, would decay into lighter neutral particles in processes such as $\mathrm{e} \to \nu\gamma$ or $\mathrm{e} \to \nu\nu\bar{\nu}$. Data show that these processes rarely occur, if at all. The current value of the mean lifetime of the electron is $\tau_\mathrm{e} > 2.7 \times 10^{23}$ years (to be compared with the estimated age of the universe, $t_0 \approx 10^{10}$ years). In other words, the electron appears to be stable.

A remarkable property of the electric charge is its quantization: every *observable* particle carries an electric charge which is a whole multiple of the unit charge, $|q| = Ne$, where $N = 0, 1, 2, \ldots$. In particular, the neutron charge must be exactly zero and the charges of the electron and proton must be equal in magnitudes and opposite in signs. Data confirm this expectation:

$$q_\mathrm{n} = (-0.4 \pm 1.1) \times 10^{-21}\,e\,,$$
$$|q_\mathrm{p} + q_\mathrm{e}| < 1.0 \times 10^{-21}\,e\,.$$

As we have seen before, a free neutron decays into a proton, an electron, and an antineutrino. In this process as well as in any other reactions involving nucleons, the number of nucleons is always conserved regardless of the interactions involved. Although free neutrons decay, free protons are believed to be stable. This general property of matter is encoded in a new additive conserved quantum number, called the *baryon number* $N_B$, which takes value $N_B = +1$ for n, p, $\Lambda^0$, $\Sigma^{\pm,0}$, and $\Xi^{-0}$, and $N_B = -1$ for the corresponding antiparticles. Quarks and antiquarks are assigned fractional baryon numbers. Finally, the photon, mesons, and leptons all have vanishing baryon numbers.

Conservation of the baryon number implies that an antibaryon cannot be produced alone from baryons, but always in association with another baryon. For example,

$$\begin{aligned} \mathrm{p}+\mathrm{p} &\to \mathrm{p}+\mathrm{p}+\mathrm{p}+\bar{\mathrm{p}}\,, \\ \pi^- +\mathrm{p} &\to \mathrm{p}+\bar{\mathrm{p}}+\Lambda^0+\mathrm{K}^0\,. \end{aligned}$$

Just as for the electric charge, conservation of the baryon number is practically absolute, which implies that no matrix elements of a physical operator can exist between states of different baryon numbers. As far as we know, matter is stable and, in particular, the lightest baryon, the proton, is believed to be stable. Estimates of the mean lifetime of the proton vary with the assumed decay modes,

$$\begin{aligned} \tau_p &> 1.6\times 10^{25}\,\text{years} \qquad (\text{mode independent})\,, \\ &> 10^{31} - 5\times 10^{32}\,\text{years} \quad (\text{mode dependent})\,. \end{aligned}$$

But, regardless of the decay modes considered, these limits are always by far greater than the age of the universe.

The observability of processes involving leptons can be similarly encoded in various *lepton numbers* $L_\ell$, $L_e$, $L_\mu$, and $L_\tau$, whose values (+1 for each type of lepton and −1 for each antilepton) are assigned to different particles according to Table 5.1 .

Conservation of the lepton numbers implies that leptons are created or destroyed in charge-conjugate pairs. Thus, in the neutron $\beta$-decay process $\mathrm{n}\to\mathrm{p}+\mathrm{e}^-+\bar{\nu}_e$, an electronic antineutrino appears together with an electron while in the inverse $\beta$-decay of a bound proton, $\mathrm{p}\to\mathrm{n}+\mathrm{e}^++\nu_e$ (for example in the nuclear transition $^{14}\mathrm{O}\to{}^{14}\mathrm{N}^*+\mathrm{e}^++\nu_e$), an electronic neutrino and a positron are emitted. It is also possible, when enough energy is available, to observe an $L_e$-conserving double-$\beta$-decay in which two (bound) neutrons decay,

$$\begin{array}{lccccc} & 2\mathrm{n} & \to & 2\mathrm{p} & +2\mathrm{e}^- & +2\bar{\nu}_e\,, \\ L_e: & 0 & & 0 & +2 & -2 \end{array}$$

(as in the nuclear transition $^{48}\mathrm{Ca}\to{}^{48}\mathrm{Ti}+2\mathrm{e}^-+2\bar{\nu}_e$), giving rise to two antineutrinos in the final state. These elusive particles would be completely

**Table 5.1.** Lepton numbers [a]

| | $e^-, \nu_e$ | $e^+, \bar{\nu}_e$ | $\mu^-, \nu_\mu$ | $\mu^+, \bar{\nu}_\mu$ | $\tau^-, \nu_\tau$ | $\tau^+, \bar{\nu}_\tau$ |
|---|---|---|---|---|---|---|
| $L_e$ | +1 | −1 | 0 | 0 | 0 | 0 |
| $L_\mu$ | 0 | 0 | +1 | −1 | 0 | 0 |
| $L_\tau$ | 0 | 0 | 0 | 0 | +1 | −1 |
| $L_\ell$ | +1 | −1 | +1 | −1 | +1 | −1 |

[a] All other particles have zero lepton numbers.

absent if the lepton number $L_e$ were not conserved, because then an $L_e$-violating decay could first take place via $n \to p + e^- + \nu_e$ (in which $\nu_e$ rather than $\bar{\nu}_e$ is produced), to be followed by an $L_e$-conserving reaction, $\nu_e + n \to p + e^-$, which gives the net result

$$\begin{array}{lccc} & 2n & \to & 2p \quad +2e^- \\ L_e: & 0 & & 0 \quad\;\; +2\,. \end{array}$$

Experiments show that neutrinoless double-$\beta$-decays are by far less probable than the corresponding neutrino-emitting decays of the same nuclei.

Historically, the hypothesis of the existence of an additional lepton number, $L_\mu \neq L_e$, was introduced to explain the suppression of the decay mode

$$\begin{array}{lcccc} & \mu^+ & \to & e^+ & \gamma \\ L_e: & 0 & & -1 & 0\,, \\ L_\mu: & -1 & & 0 & 0\,, \\ L_\ell: & -1 & & -1 & 0\,, \end{array}$$

a process which would otherwise not be forbidden by the conservation rules of the generic lepton number $L_\ell$ and of any other known quantum number. The experimentally observed suppression of the process,

$$\frac{\tau(\mu \to e\,\gamma)}{\tau(\mu \to e\,\bar{\nu}\,\nu)} < 2 \times 10^{-8}\,, \tag{5.76}$$

implies that if a reaction is initiated by a $\mu$-neutrino, it must produce a muon as in $\nu_\mu + n \to \mu^- + p$, rather than an electron, as it would be the case in $\nu_\mu + n \to e^- + p$. This hypothesis is strongly supported by the much higher probability of observing the decay mode $\pi^+ \to \mu^+\nu_\mu$ (99.98% of all modes) compared to $\pi^+ \to \mu^+\nu_e$ ($8.0 \times 10^{-3}$ of all modes). The existence of the muonic neutrino distinct from the electronic neutrino is thus confirmed and motivates the introduction of a new additive quantum number $L_\mu \neq L_e$.

The $\tau$ lepton was discovered in 1975 in the reaction $e^+ + e^- \to \tau^+ + \tau^-$, followed by the decays $\tau^- \to \ell^- + \bar{\nu}_\ell + \nu_\tau$ and $\tau^+ \to \ell^+ + \nu_\ell + \bar{\nu}_\tau$, where $\ell = e, \mu$. As experiments show that the process $\nu_\mu + n \to \tau^- + p$ is highly improbable, it is concluded that $\nu_\mu \neq \nu_\tau$ and, by the same token, also $\nu_e \neq \nu_\tau$. If that is the case, there must exist a $\tau$-lepton number $L_\tau$.

While all quantum numbers discussed in this section are additively conserved, the electric charge is outstanding in its remarkable particularity of playing the double role of being an additive quantum number by its presence in the *global gauge* transformation $U = \exp(\mathrm{i}Q\alpha)$, and of being a coupling constant of an interaction by its presence in the *local gauge* transformation $U(x) = \exp[\mathrm{i}Q\alpha(x)]$ (as we shall see in detail in Chap. 8). In contrast, neither the baryon number nor the lepton numbers seem to be associated with detectable interactions. It follows that the electric charge alone can be expressed in terms of a measurable physical unit, while the baryon number and the lepton numbers have arbitrary units. The profound implication of this difference is that the conservation of electric charge is exact, being anchored by a principle considered as fundamental – the local gauge invariance – whereas the conservations of the baryon and lepton numbers may be approximate. Thus, in principle, one may not exclude, for example, processes like $\mathrm{p} \to \mathrm{e}^+\pi^0$, $\mathrm{p} \to \mathrm{e}^+\gamma$, or $\nu_e \leftrightarrow \nu_\mu$, $\nu_\mu \leftrightarrow \nu_\tau$.

Finally, leaving aside for the moment quantum numbers of more recent origins (charm, topness, bottomness), we now consider briefly another additive quantum number called *strangeness*. It is a quantum number assigned to some hadrons (baryons or mesons) that have apparently contradictory properties. These particles are copiously produced in nucleon–nucleus collisions and other hadronic reactions with total production cross-sections of the order of a millibarn, comparable in magnitude to cross-sections for other strong interaction processes, such as pion–nucleon reactions. However, their rather long lifetimes, typically $\tau \approx 10^{-10}\,\mathrm{s}$, suggest that their decays arise from weak interactions.

The hypothesis of strangeness $S$ gives a simple solution to this dilemma by assuming that $S$ is conserved in strong interactions, responsible for productions, but is nonconserved in weak interactions, responsible for decays. Strangeness is thus associated with an imperfect symmetry, in contrast to the other additive quantum numbers studied in this section. Given this basic assumption, the values of $S$ for all hadrons can be determined relatively to a few selected reference particles:

$$\begin{aligned} S &= 0 \qquad \text{for } \pi, \text{ nucleons}\,; \\ S &= +1 \qquad \text{for } \mathrm{K}^+\,. \end{aligned}$$

Mesons $\mathrm{K}^+$, among the first strange particles observed, are produced in

$$\mathrm{p} + \mathrm{p} \to \mathrm{p} + \mathrm{K}^+ + \Lambda^0\,.$$

Assuming conservation of strangeness in this production reaction, one obtains $S = -1$ for $\Lambda^0$. The quantum number $S$ for other hadrons is determined in a similar fashion:

$$
\begin{array}{ll}
\pi^- \mathrm{p} \to \mathrm{K}^0 \Lambda^0 & S(\mathrm{K}^0) = +1\,, \\
\pi^- \mathrm{p} \to \mathrm{n}\,\mathrm{K}^+ \mathrm{K}^- & S(\mathrm{K}^-) = -1\,, \\
\pi^- \mathrm{p} \to \mathrm{K}^+ \Sigma^- & S(\Sigma^-) = -1\,, \\
\mathrm{p}\mathrm{p} \to \mathrm{n}\,\mathrm{K}^+ \Sigma^+ & S(\Sigma^+) = -1\,, \\
\mathrm{p}\mathrm{p} \to \mathrm{p}\,\mathrm{K}^+ \Sigma^0 & S(\Sigma^0) = -1\,, \\
\pi^- \mathrm{p} \to \mathrm{K}^+ \mathrm{K}^0\, \Xi^- & S(\Xi^-) = -2\,, \\
\pi^+ \mathrm{p} \to \mathrm{K}^+ \mathrm{K}^+ \Xi^0 & S(\Xi^0) = -2\,.
\end{array}
$$

Note that the baryons $\Sigma^+$, $\Sigma^-$ and $\Sigma^0$ have the same value $S = -1$ as well as the same mass (see Table 1.3). This fact points to the existence of some new symmetry (which will be identified as the isospin symmetry in the following chapter). As for the mesons K, the situation is somewhat different. Mesons $\mathrm{K}^+$ and $\mathrm{K}^-$ have equal masses but electric charges and strangeness numbers of opposite signs, which indicates that they are charge conjugates to each other. It is then plausible that $\mathrm{K}^0$ must similarly have a charge conjugate of strangeness $S = -1$ and equal mass. It is possible to detect such a particle, called $\bar{\mathrm{K}}^0$, for example in

$$\pi^+ \mathrm{p} \to \mathrm{p}\,\bar{\mathrm{K}}^0\, \mathrm{K}^+\,.$$

Observations indicate that there must exist two doublets of mesons of $S = \pm 1$ conjugate to each other, $(\mathrm{K}^+, \mathrm{K}^0)$ and $(\mathrm{K}^-, \bar{\mathrm{K}}^0)$. Similarly, the doubly strange baryons $\Xi^-$ and $\Xi^0$ form a doublet to which corresponds a distinct antidoublet. Finally, there exists a particle, $\Omega^-$ (1672 MeV), with $S = -3$.

Electromagnetically induced reactions, such as

$$
\begin{array}{ll}
\gamma\, \mathrm{p} \to \Sigma^0 \mathrm{K}^+, & (S_\mathrm{i} = S_\mathrm{f} = 0)\,, \\
\gamma\, \mathrm{p} \to \Sigma^+ \mathrm{K}^0, & (S_\mathrm{i} = S_\mathrm{f} = 0)\,,
\end{array}
$$

have been observed but not, for example,

$$
\begin{array}{ll}
\gamma\, \mathrm{p} \to \Lambda^0 \pi^+, & (S_\mathrm{i} = 0,\ S_\mathrm{f} = -1)\,, \\
\gamma\, \mathrm{n} \to \Sigma^+ \mathrm{K}^-, & (S_\mathrm{i} = 0,\ S_\mathrm{f} = -2)
\end{array}
$$

(where the photon and leptons are assumed to have zero strangeness). These data indicate that strangeness is a symmetry in electromagnetic interactions, a conclusion reinforced by the observation of the strangeness-conserving decay

$$\Sigma^0 \to \Lambda^0 \gamma \qquad (\Delta S = S_\mathrm{f} - S_\mathrm{i} = 0)\,.$$

The baryon $\Sigma^0$ has mean lifetime $\tau = 7 \times 10^{-20}$ s.

As already mentioned, the relatively long lifetimes of strange particles indicate that, except for $\Sigma^0$, they decay via weak interactions by breaking strangeness symmetry:

$$
\begin{array}{ll}
\Lambda^0 \to \mathrm{p}\,\pi^-\,, & \Sigma^- \to \mathrm{n}\,\pi^-\,, \\
\mathrm{K}^0 \to \pi^+ \pi^-\,, & \Xi^0 \to \Lambda^0 \pi^0\,, \\
\mathrm{K}^+ \to \mu^+ \nu_\mu\,, & \Omega^- \to \Lambda^0 \mathrm{K}^-\,.
\end{array}
$$

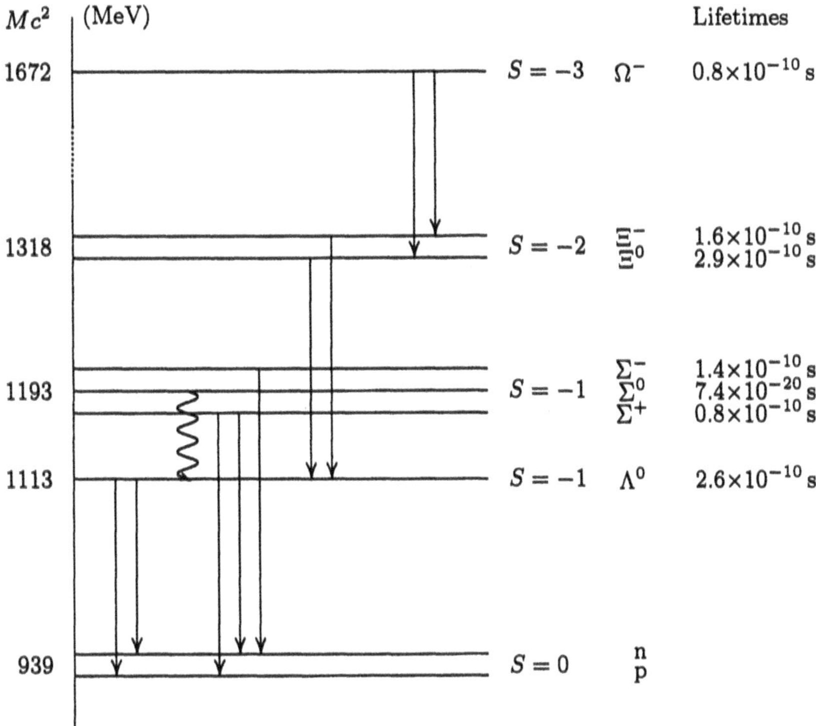


**Fig. 5.3.** Decay modes and lifetimes of some strange particles (a straight line represents a $\pi$ meson; a wavy line, a photon)

Decays of low-lying strange particles by emission of a photon or a pion are shown in Fig. 5.3. Note that symmetry breaking obeys in all cases an extremely accurate selection rule $|\Delta S| = 1$. Transitions $|\Delta S| \geq 2$, even in a phase-space-favored decay mode such as $\Xi^- \to n\,\pi^-$ for which $|\Delta S| = 2$, are either forbidden or very improbable.

To summarize, every particle is characterized by additive quantum numbers – electric charge, baryon number, lepton numbers, strangeness, as well as charm, topness (truth), bottomness (beauty) to be introduced later. The corresponding antiparticle has the same quantum numbers, but with reversed signs, and is therefore distinct from its conjugate, unless it is completely neutral. Such are the cases of the mesons $\pi^0$ (135 MeV) and $\eta^0$ (547 MeV).

### 5.3.2 Charge Conjugation in Field Theories

The notion of antiparticle originates from Dirac's theory of the electron. This theory predicted the existence of a particle identical to the electron except for having an electric charge with the opposite sign. This idea was substantiated by subsequent detections of the positron and other particles having the same masses and lifetimes as certain known particles, differing from them only in the signs of their respective additive quantum numbers. To relate the two types of particles it proves convenient to introduce a *unitary* operator $\mathcal{C}$ that reverses the signs of all the generalized charges of particles without affecting their spatial properties. Specifically, its action on a particle of momentum $\boldsymbol{p}$, spin $s$, and generalized charges, collectively represented by the symbol $Q$, is given by

$$\mathcal{C}\;|\boldsymbol{p},s,Q\rangle = \xi\;|\boldsymbol{p},s,-Q\rangle\;, \tag{5.77}$$

where $\xi$ is a unimodular phase factor. It is not necessarily true that this operation of *charge conjugation* is identical to *field conjugation*, which replaces particles with their antiparticles. However, it turns out just to be the case for all physical particles. Therefore, $\mathcal{C}$ will be taken to represent effectively the field conjugation for all particles as well.

It follows from (77) that if $Q \neq 0$, then $[\mathcal{C},Q] \neq 0$. In other words, a state of nonvanishing charge cannot be an eigenstate of $\mathcal{C}$. Nevertheless, the notion of charge conjugation remains useful even in these cases because of the physical consequences that follow from invariance of the system when this invariance holds. On the other hand, for a particle or system of particles completely neutral, $\mathcal{C}$ commutes with all the generalized charge generators and therefore may have common eigenstates with these operators. For such states, since $\mathcal{C}^2 = 1$, the eigenvalues of $\mathcal{C}$ are $\pm 1$.

As the antiparticle concept is unknown in nonrelativistic quantum mechanics, the language of relativistic quantum field theories is the only one suited to its study.

**Scalar Field.** We may begin by defining the field conjugation operator $\mathcal{C}$ by its action on a complex scalar field

$$\mathcal{C}\,\phi(x)\,\mathcal{C}^{-1} = \xi_{\mathrm{B}}\phi^\dagger(x); \qquad \mathcal{C}^\dagger\mathcal{C} = 1, \quad |\xi_{\mathrm{B}}|^2 = 1\,. \tag{5.78}$$

We will prove that it implies (77).

It is evident that the Lagrangian for the noninteracting complex scalar field

$$\mathcal{L}_{\mathrm{B}}(x) = \partial_\mu\phi^\dagger\partial^\mu\phi - m^2\phi^\dagger\phi \tag{5.79}$$

(where normal-ordered products are understood) is invariant to $\mathcal{C}$ because

$$\mathcal{C}\,\mathcal{L}_{\mathrm{B}}(x)\,\mathcal{C}^{-1} = |\xi_{\mathrm{B}}|^2(\partial_\mu\phi\partial^\mu\phi^\dagger - m^2\phi\phi^\dagger) = \mathcal{L}_{\mathrm{B}}(x)\,. \tag{5.80}$$

In the last step, the implicit convention of normal-ordered products has been used in permuting the field operators $\phi$ and $\phi^\dagger$.

The action of $\mathcal{C}$ on the Fock operators for the charged boson can be determined from (78) with $\phi$ replaced by its Fourier series

$$\sum_p C_{\boldsymbol{p}}[\,\mathcal{C}\,a_{\boldsymbol{p}}\,\mathcal{C}^{-1}\mathrm{e}^{-\mathrm{i}p\cdot x} + \mathcal{C}\,b^\dagger_{\boldsymbol{p}}\,\mathcal{C}^{-1}\mathrm{e}^{\mathrm{i}p\cdot x}\,] = \xi_{\mathrm{B}}\sum_p C_{\boldsymbol{p}}[\,a^\dagger_{\boldsymbol{p}}\mathrm{e}^{\mathrm{i}p\cdot x} + b_{\boldsymbol{p}}\mathrm{e}^{-\mathrm{i}p\cdot x}\,]\,.$$

Identifying the corresponding coefficients on both sides, one obtains

$$\mathcal{C}\, a_{\boldsymbol{p}}\, \mathcal{C}^{-1} = \xi_{\mathrm{B}}\, b_{\boldsymbol{p}}\,,$$
$$\mathcal{C}\, b_{\boldsymbol{p}}\, \mathcal{C}^{-1} = \xi_{\mathrm{B}}^{*} a_{\boldsymbol{p}}\,. \tag{5.81}$$

For a one-particle state one gets, assuming a C-invariant vacuum, $\mathcal{C}\,|0\rangle = |0\rangle$,

$$\mathcal{C} a_{\boldsymbol{p}}^{\dagger}\,|0\rangle = \mathcal{C}\, a_{\boldsymbol{p}}^{\dagger}\, \mathcal{C}^{-1}\mathcal{C}\,|0\rangle = \xi_{\mathrm{B}}^{*} b_{\boldsymbol{p}}^{\dagger}\,|0\rangle\,,$$
$$\mathcal{C} b_{\boldsymbol{p}}^{\dagger}\,|0\rangle = \mathcal{C}\, b_{\boldsymbol{p}}^{\dagger}\, \mathcal{C}^{-1}\mathcal{C}\,|0\rangle = \xi_{\mathrm{B}}\, a_{\boldsymbol{p}}^{\dagger}\,|0\rangle\,.$$

Thus, as defined, $\mathcal{C}$ transforms a particle into its antiparticle, and vice versa, without changing their momenta.

The Noether current associated with global gauge transformations of the Lagrangian (79) is given by an expression of the form (13). The action of $\mathcal{C}$ on this current is

$$\mathcal{C}\, j_{\mu}(x)\, \mathcal{C}^{-1} = \mathrm{i}\,[\phi\partial_{\mu}\phi^{\dagger} - (\partial_{\mu}\phi)\phi^{\dagger}]$$
$$= -\, j_{\mu}(x)\,. \tag{5.82}$$

The generalized charge $Q$ defined by the space integral of $j^{0}(x)$ is evidently conserved. It changes sign under the C-conjugation, $\mathcal{C}\,Q\,\mathcal{C}^{-1} = -Q$. Therefore, the operation $\mathcal{C}$ defined by (78) changes the signs of electric charge, baryon number etc., as expected from the definition of charge conjugation. If the particle is completely neutral, the associated field is Hermitian, $\phi^{\dagger} = \phi$, and therefore $a_{\boldsymbol{p}} = b_{\boldsymbol{p}}$, and the phase becomes real, $\xi_{\mathrm{B}} = \xi_{\mathrm{B}}^{*}$. Since $\mathcal{C}^{2} = 1$ and $\mathcal{C}^{\dagger}\mathcal{C} = 1$, it follows that $\xi_{\mathrm{B}} = \pm 1$. This *multiplicative* quantum number, when it can be defined, is called the *charge (conjugation) parity.*

**Electromagnetic Field.** If we assume that the transformation rule for the current density (82) is generally valid (as it proves indeed to be the case), the Maxwell field $A_{\mu}$ must transform according to

$$\mathcal{C}\, A_{\mu}(x)\, \mathcal{C}^{-1} = -A_{\mu}(x) \tag{5.83}$$

to generate an electromagnetic interaction invariant with respect to $\mathcal{C}$. It follows that

$$\mathcal{C}\, a(\boldsymbol{k},\lambda)\, \mathcal{C}^{-1} = -a(\boldsymbol{k},\lambda)\,. \tag{5.84}$$

Therefore, a photon state

$$\mathcal{C}\,|\boldsymbol{k},\lambda\rangle = -\,|\boldsymbol{k},\lambda\rangle \tag{5.85}$$

is also an eigenstate of $\mathcal{C}$ of eigenvalue $\xi_{\gamma} = -1$. The photon is *odd* under charge conjugation.

**Dirac Field.** Just as for a charged boson, the charge conjugate of a Dirac field must be proportional to its complex conjugate $\psi^*$, which suggests the following definition of the operator $\mathcal{C}$ on the Hilbert space for fermions:

$$\mathcal{C}\,\psi(x)\,\mathcal{C}^{-1} = \xi_{\mathrm{F}} B\psi^*(x), \quad |\xi_{\mathrm{F}}| = 1\,, \tag{5.86}$$

where $B$ is a $4 \times 4$ unitary matrix on the spinor representation. Since $\overline{\psi}$ rather than $\psi^*$ appears frequently in formulas, it is more practical to state the rule in the equivalent form

$$\mathcal{C}\,\psi(x)\,\mathcal{C}^{-1} = \xi_{\mathrm{F}} C\overline{\psi}^{\mathrm{T}}(x)\,, \quad |\xi_{\mathrm{F}}| = 1\,. \tag{5.87}$$

We have used the relation $\overline{\psi}^{\mathrm{T}} = \gamma_0^*\psi^*$ and introduced another $4 \times 4$ matrix $C = B\gamma_0^*$, which is also unitary $C^\dagger C = 1$. Note that in (87) the transposition T applies only to the spinor, not to the Fock operators. To find $C$ it is required that Dirac's equation for $\psi$ be covariant, or equivalently, the corresponding Lagrangian be invariant to charge conjugation. In the latter viewpoint the condition reads

$$\mathcal{C}\,\mathcal{L}_{\mathrm{F}}(x)\,\mathcal{C}^{-1} = \mathcal{L}_{\mathrm{F}}(x)\,. \tag{5.88}$$

It is then convenient to use the explicitly Hermitian version of $\mathcal{L}_{\mathrm{F}}$,

$$\begin{aligned}\mathcal{L}_{\mathrm{F}} &\equiv \mathcal{L}_1 + \mathcal{L}_1^\dagger \\ &= \frac{1}{2}\overline{\psi}\left[\mathrm{i}\gamma^\mu \overrightarrow{\partial}_\mu - m\right]\psi + \frac{1}{2}\overline{\psi}\left[-\mathrm{i}\gamma^\mu \overleftarrow{\partial}_\mu - m\right]\psi\,.\end{aligned} \tag{5.89}$$

Noting that $\mathcal{C}\,\psi^\dagger\,\mathcal{C}^{-1} = \xi_{\mathrm{F}}^*\psi^{\mathrm{T}}\gamma_0^* C^\dagger$, one gets for $\mathcal{L}_1$,

$$\mathcal{C}\,\mathcal{L}_1\,\mathcal{C}^{-1} = \frac{1}{2}\psi^{\mathrm{T}}\gamma_0^* C^\dagger(\mathrm{i}\gamma^0\gamma^\mu\partial_\mu - \gamma_0 m)C\overline{\psi}^{\mathrm{T}}\,. \tag{5.90}$$

Since the expression on the right-hand side is a scalar, it may be equivalently replaced by its transpose in spinor space. A permutation of the anticommuting operators $\psi$ and $\overline{\psi}$ has the effect of introducing an additional minus sign plus a c-number term given by their anticommutation rules. This c-number term drops out because $\mathcal{L}_{\mathrm{F}}$ is implicitly normal ordered, thus leaving

$$\mathcal{C}\,\mathcal{L}_1\,\mathcal{C}^{-1} = \frac{1}{2}\overline{\psi}C^{\mathrm{T}}\left(-\mathrm{i}\gamma^{\mu\mathrm{T}}\gamma^{0\mathrm{T}}\overleftarrow{\partial}_\mu + \gamma_0^{\mathrm{T}} m\right)C^*\gamma_0\psi\,. \tag{5.91}$$

A similar calculation applies to $\mathcal{C}\,\mathcal{L}_1^\dagger\,\mathcal{C}^{-1}$. To satisfy (88) it suffices to require

$$\mathcal{C}\,\mathcal{L}_1\,\mathcal{C}^{-1} = \mathcal{L}_1^\dagger\,, \tag{5.92}$$

which implies

$$C^\dagger\gamma_\mu C = -\gamma_\mu^{\mathrm{T}} \qquad \text{(in arbitrary representation of } \gamma_\mu)\,. \tag{5.93}$$

To obtain an explicit expression for the matrix $C$ it is useful to adopt a specific representation for the $\gamma_\mu$. In the standard representation, the basic condition (93) becomes

$$C^\dagger \gamma_\mu C = -\gamma_\mu \quad (\mu = 0, 2)\,,$$
$$C^\dagger \gamma_\mu C = +\gamma_\mu \quad (\mu = 1, 3)\,,$$

which admits the solution

$$C = \lambda \gamma^2 \gamma^0\,, \quad |\lambda| = 1 \quad (\text{in standard representation of } \gamma_\mu)\,. \tag{5.94}$$

With (93) the charge conjugate of the adjoint $\overline{\psi}$ can be easily found:

$$\begin{aligned} \mathcal{C}\,\overline{\psi}\,\mathcal{C}^{-1} &= \mathcal{C}\,\psi^\dagger \gamma_0\,\mathcal{C}^{-1} = \mathcal{C}\,\psi^\dagger\,\mathcal{C}^{-1}\gamma_0 \\ &= \xi_{\mathrm{F}}^* \psi^{\mathrm{T}} \gamma_0 C^\dagger \gamma_0 = -\xi_{\mathrm{F}}^* \psi^{\mathrm{T}} \gamma_0 \gamma_0 C^\dagger \\ &= -\xi_{\mathrm{F}}^*\, \psi^{\mathrm{T}} C^\dagger\,. \end{aligned} \tag{5.95}$$

It follows that an arbitrary bilinear covariant of field operators has the transformation property

$$\mathcal{C}\,\overline{\psi}\Gamma\psi\,\mathcal{C}^{-1} = \overline{\psi}\,C\Gamma^{\mathrm{T}} C^\dagger\,\psi\,. \tag{5.96}$$

In particular,

$$\begin{aligned} C\gamma_\mu^{\mathrm{T}} C^\dagger &= -\gamma_\mu\,; \\ C\gamma_5^{\mathrm{T}} C^\dagger &= \gamma_5\,; \\ C(\gamma_\mu\gamma_5)^{\mathrm{T}} C^\dagger &= \gamma_\mu\gamma_5\,. \end{aligned}$$

Thus, the current for the Dirac particle, $j_\mu = \overline{\psi}\gamma_\mu\psi$, and the associated charge, obey the expected transformation rules

$$\begin{aligned} \mathcal{C}\,j_\mu\,\mathcal{C}^{-1} &= \mathcal{C}\,\overline{\psi}\,\mathcal{C}^{-1}\gamma_\mu \mathcal{C}\,\psi\,\mathcal{C}^{-1} = -j_\mu, \\ \mathcal{C}\,Q\,\mathcal{C}^{-1} &= -Q\,. \end{aligned} \tag{5.97}$$

Next, we examine how the Fock operators and hence particle or antiparticle states transform. First, note that in the standard representation of the $\gamma_\mu$ and with the choice $\lambda = \mathrm{i}$, the spinors $u$ and $v$, explicitly given in (3.45) and (3.46), are related through the $C$-matrix by

$$\begin{aligned} C\bar{u}^{\mathrm{T}}(\boldsymbol{p}, s) &= v(\boldsymbol{p}, s)\,, \\ C\bar{v}^{\mathrm{T}}(\boldsymbol{p}, s) &= u(\boldsymbol{p}, s)\,, \qquad C = \mathrm{i}\gamma^2\gamma^0\,. \end{aligned} \tag{5.98}$$

The two sides of (87) then become in terms of the Fock operators

$$\mathcal{C}\,\psi\,\mathcal{C}^{-1} = \sum_{\boldsymbol{p},s} C_{\boldsymbol{p}} \left[\mathcal{C}\,b(\boldsymbol{p}, s)\,\mathcal{C}^{-1} u(\boldsymbol{p}, s)\mathrm{e}^{-\mathrm{i}p\cdot x} + \mathcal{C}\,d^\dagger(\boldsymbol{p}, s)\,\mathcal{C}^{-1} v(\boldsymbol{p}, s)\mathrm{e}^{\mathrm{i}p\cdot x}\right],$$

$$\xi_{\mathrm{F}} C\overline{\psi}^{\mathrm{T}} = \xi_{\mathrm{F}} \sum_{\boldsymbol{p},s} C_{\boldsymbol{p}} \left[b^\dagger(\boldsymbol{p}, s) v(\boldsymbol{p}, s)\,\mathrm{e}^{\mathrm{i}p\cdot x} + d(\boldsymbol{p}, s) u(\boldsymbol{p}, s)\,\mathrm{e}^{-\mathrm{i}p\cdot x}\right].$$

The transformation rules for the Fock operators immediately follow:

$$\begin{aligned} C\, b(\boldsymbol{p}, s)\, C^{-1} &= \xi_{\mathrm{F}}\, d(\boldsymbol{p}, s)\,, \\ C\, d(\boldsymbol{p}, s)\, C^{-1} &= \xi_{\mathrm{F}}^{*}\, b(\boldsymbol{p}, s)\,. \end{aligned} \tag{5.99}$$

They confirm a known result: field conjugation (87) converts a particle state $b^{\dagger}(\boldsymbol{p}, s)\,|0\rangle$ into the corresponding antiparticle state $\xi_{\mathrm{F}}^{*} d^{\dagger}(\boldsymbol{p}, s)\,|0\rangle$ without changing its spin or momentum. All charges however change signs according to (97). This result is illustrated in Fig. 5.4.

Finally, let us recall that a fermion and its conjugate partner have opposite parities, opposite chiralities, but equal helicities (see Problem 5.9).

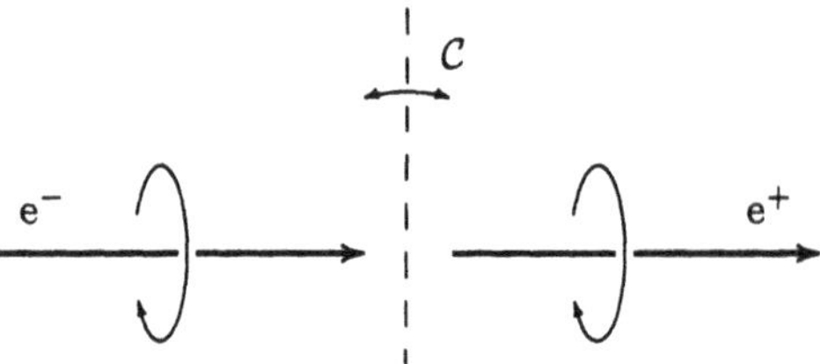


**Fig. 5.4.** $C$ flips the sign of the charge of a particle without changing its spin or momentum

### 5.3.3 Interactions

For completely neutral particles, one may define a multiplicative quantum number associated with charge conjugation symmetry called the *charge (conjugation) parity*, $\xi$, which takes values $+1$ or $-1$. Examples where this concept applies are the photon, the mesons $\pi^0$ and $\eta^0$, and the self-conjugate pairs, such as $e^-e^+$, $p\,\bar{p}$, and $\pi^-\pi^+$.

Assuming that the **electromagnetic interaction** is C-invariant, the photon is odd ($\xi_\gamma = -1$) and an $n$-photon system has charge parity $(-)^n$: states of an even number of photons are even and states of an odd number of photons are odd with respect to charge conjugation, regardless of their spatial configurations. *Furry's theorem* immediately follows from this result. It says that the matrix element of an operator invariant to charge conjugation vanishes if the number of external photons is odd and if there are no other particles. This is because if there are $n_{\mathrm{i}}$ and $n_{\mathrm{f}}$ photons in the initial and final states of some given process, the total number of external photons for the process is $N = n_{\mathrm{i}} + n_{\mathrm{f}}$. Since the interaction operator is assumed to be even, conservation of charge parity requires $(-)^{n_{\mathrm{i}}} = (-)^{n_{\mathrm{f}}}$, which implies $N$ an even integer.

From the observed two-gamma decay modes of $\pi^0$ and $\eta^0$

$$\begin{aligned} &\pi^0 \to 2\gamma \qquad \text{(branching ratio: 98.8\%)}\,, \\ &\eta^0 \to 2\gamma \qquad \text{(branching ratio: 38.8\%)}\,, \end{aligned}$$

one infers their charge parities, $\xi_{\pi^0} = 1$ and $\xi_{\eta^0} = 1$. On the other hand, the very small rates of their three-photon decay modes confirm that $C$ is indeed a symmetry for electromagnetic interactions:

$$\begin{aligned} \pi^0 &\to 3\gamma \qquad (3 \times 10^{-8}), \\ \eta^0 &\to 3\gamma \qquad (5 \times 10^{-4}). \end{aligned}$$

The charge parity $\xi$ also serves to identify states of particle–antiparticle systems. Consider first a state of *scalar–antiscalar* particles of relative orbital angular momentum $\ell$:

$$|\phi_\ell(\mathrm{s\bar{s}})\rangle = \int \mathrm{d}^3 p\, F_\ell(\boldsymbol{p}) a^\dagger_{\boldsymbol{p}} b^\dagger_{-\boldsymbol{p}}\, |0\rangle \tag{5.100}$$

($a^\dagger$, $b^\dagger$ create a boson and a conjugate antiboson, respectively). Application of $C$ on both sides leads to

$$C\, |\phi_\ell(\mathrm{s\bar{s}})\rangle = \int \mathrm{d}^3 p\, F_\ell(\boldsymbol{p}) b^\dagger_{\boldsymbol{p}} a^\dagger_{-\boldsymbol{p}}\, |0\rangle\ . \tag{5.101}$$

After permuting the positions of the two operators (without introducing any additional signs) and changing the sign of $\boldsymbol{p}$ in the integral, one gets

$$C\, |\phi_\ell(\mathrm{s\bar{s}})\rangle = \int \mathrm{d}^3 p\, F_\ell(-\boldsymbol{p}) a^\dagger_{\boldsymbol{p}} b^\dagger_{-\boldsymbol{p}}\, |0\rangle\ . \tag{5.102}$$

$F_\ell(-\boldsymbol{p})$ is an eigenstate of angular momentum $\ell$, and so $F_\ell(-\boldsymbol{p}) = (-)^\ell F_\ell(\boldsymbol{p})$ from (4). It follows that

$$C\, |\phi_\ell(\mathrm{s\bar{s}})\rangle = (-)^\ell\, |\phi_\ell(\mathrm{s\bar{s}})\rangle\,, \tag{5.103}$$

and the charge parity of the system is $\xi(\mathrm{s\bar{s}}) = (-)^\ell$. Thus for example, the $\pi^+\pi^-$ system is even or odd with respect to $C$ according to the parity of its orbital angular momentum. If in addition the boson is self-conjugate, i.e. identical to its antiparticle, $\ell$ may only be an even integer by Bose statistics, and $\xi(\mathrm{s\bar{s}}) = +1$ always. This is the case of a $\pi^0\pi^0$ pair, for example.

Let us consider now a pair of *fermion–antifermion* of individual spins $1/2$. Assuming the situation to be nonrelativistic, one may neglect the effects of spin-orbit coupling and of virtual fields. A state of relative orbital angular momentum $\ell$ and total spin $S = 0$ or 1 can then be described by

$$|\psi_{\ell,S}(\mathrm{f\bar{f}})\rangle = \int \mathrm{d}^3 p \sum_{s,s'} F^{ss'}_{\ell S}(\boldsymbol{p}) b^\dagger(\boldsymbol{p}, s) d^\dagger(-\boldsymbol{p}, s')\, |0\rangle \tag{5.104}$$

($b^\dagger$ creates a fermion, $d^\dagger$ an antifermion), and its charge conjugate is

$$C\, |\psi_{\ell,S}(\mathrm{f\bar{f}})\rangle = \int \mathrm{d}^3 p \sum_{s,s'} F^{ss'}_{\ell S}(\boldsymbol{p}) d^\dagger(\boldsymbol{p}, s) b^\dagger(-\boldsymbol{p}, s')\, |0\rangle\ . \tag{5.105}$$

After permuting the two Fock operators, which introduces an additional minus sign to account for their anticommutation, changing the sign of momentum integration variable, and exchanging the two spin variables, one gets

$$\begin{aligned} \mathcal{C} \left|\psi_{\ell,S}(\mathrm{f\bar{f}})\right\rangle &= -\int \mathrm{d}^3p \sum_{s,s'} F_{\ell S}^{s's}(-\boldsymbol{p}) b^\dagger(\boldsymbol{p},s) d^\dagger(-\boldsymbol{p},s') \left|0\right\rangle \\ &= -(-)^{\ell+S+1} \int \mathrm{d}^3p \sum_{s,s'} F_{\ell S}^{ss'}(\boldsymbol{p}) b^\dagger(\boldsymbol{p},s) d^\dagger(-\boldsymbol{p},s') \left|0\right\rangle , \end{aligned} \quad (5.106)$$

where on the last line $F_{\ell S}^{s's}(-\boldsymbol{p}) = (-)^{\ell+S+1} F_{\ell S}^{ss'}(\boldsymbol{p})$, whose sign arises from exchanging the particle space and spin variables. Therefore, the charge parity of a self-conjugate fermion–antifermion pair is given by

$$\mathcal{C} \left|\psi_{\ell,S}(\mathrm{f\bar{f}})\right\rangle = (-)^{\ell+S} \left|\psi_{\ell,S}(\mathrm{f\bar{f}})\right\rangle . \quad (5.107)$$

In Table 5.2 the charge parity is given together with the ordinary parity (defined by space inversion) for an arbitrary state $n\,^{2S+1}\ell_J$ with principal quantum number $n$.

**Table 5.2.** Parities of pion–pion and fermion–antifermion systems

| States | P | C | CP |
|---|---|---|---|
| $\pi^+\pi^-(\ell)$ | $(-)^\ell$ | $(-)^\ell$ | 1 |
| $\pi^0\pi^0(\ell)$ | 1 | 1 | 1 |
| $\mathrm{f\bar{f}}(n\,^{2S+1}\ell_J)$ | $(-)^{\ell+1}$ | $(-)^{\ell+S}$ | $(-)^{S+1}$ |

As a first example, take the *positronium* – a bound positron–electron system – which has an energy spectrum similar to that of the hydrogen atom but with energy spacings approximately halved. The ground state is a $1^1S_0$ ($\ell = 0,\ S = 0$) level, separated from the first excited level $1^3S_1$ ($\ell = 0,\ S = 1$) by $8.4 \times 10^{-4}$ eV. The charge parities are, from Table 5.2, $\xi(^1S_0) = +1$ and $\xi(^3S_1) = -1$. Since $\xi$ is conserved in electromagnetic interaction, the singlet state $^1S_0$ decays into an even number of photons, while the triplet state $^3S_1$ decays into 3, 5, ... photons (the one-real-photon mode being forbidden by energy-momentum conservation). Experiments show that the ground state $1^1S_0$ decays indeed into two real photons and the excited state $1^3S_1$ into three photons (at a rate reduced by a factor $\alpha = 1/137$ and, additionally, by a smaller phase space volume in the final state).

As another example, take the *mesons* treated not as elementary entities, but as bound quark–antiquark states. Neglecting the presence of gluons, we can apply the above results directly to this case. The quark and antiquark spins being 1/2, the total spin of the quark–antiquark pair is either $S = 0$

or $S = 1$. Assuming that the lowest energy state has the most symmetric spatial configuration ($\ell = 0$), one can expect the existence of *pseudoscalar* mesons ${}^1S_0$ of parity $P = -1$ and charge parity $C = +1$, and of *vector* mesons ${}^3S_1$ with $P = -1$ and $C = -1$. We have already noted above the electromagnetic decay modes of neutral pseudoscalar mesons $\pi^0$ and $\eta^0$:

$$\text{quark–antiquark } {}^1S_0\ (C = 1) \quad \to \quad \gamma\gamma\ (C = 1)\,.$$

But their three-photon modes are strongly suppressed:

$$\text{quark–antiquark } {}^1S_0\ (C = 1) \quad \to \quad \gamma\gamma\gamma\ (C = -1)\,,$$

which confirms the charge conjugation symmetry in electromagnetic interactions at this level as well. Decays of neutral vector mesons $\rho$ (770 MeV), $\omega$ (782 MeV), or $\phi$ (1020 MeV) into photons alone have not been observed. But with such high energies available, it is unlikely that photons produced by the decays could exist for long without being converted into pairs of leptons or hadrons. For instance, the $e^+e^-$ pair production observed in decays of any of these three mesons can be viewed as resulting from the quark–antiquark pair annihilation into one *virtual* photon which in turn creates a pair of relativistic positron–electron:

$$\text{quark–antiquark } {}^3S_1\ (C = -1) \quad \to \quad \text{virtual photon} \quad \to \quad e^+e^-\,.$$

**Hadronic interactions** are invariant to charge conjugation, as can be verified by comparing the energy and momentum distributions of the charge conjugate reactions like

$$\pi^+ p \to \pi^+ p \quad \text{and} \quad \pi^- \bar{p} \to \pi^- \bar{p}\,,$$

or

$$p\bar{p} \to \pi^+ + a + b \quad \text{and} \quad p\bar{p} \to \pi^- + \bar{a} + \bar{b}\,.$$

However, **weak interactions** break charge conjugation symmetry, as it was already clear from the first experiments performed in 1957 to detect parity violations. It was then established in particular that the helicities of electrons and positrons emitted in weak interactions have opposite signs. It turned out that negative and positive muons produced in weak processes also have opposite helicities. Now, the helicity operator, $\boldsymbol{\Sigma}\cdot\boldsymbol{p}$, is invariant to charge conjugation. Therefore, if charge conjugation is a conserving transformation, one should have

$$\begin{aligned}\langle e^- \,|\, \boldsymbol{\Sigma}\cdot\boldsymbol{p} \,|\, e^- \rangle &= \langle e^- \,|\, \mathcal{C}^{-1}\mathcal{C}\, \boldsymbol{\Sigma}\cdot\boldsymbol{p}\, \mathcal{C}^{-1}\mathcal{C} \,|\, e^- \rangle \\ &= \langle e^+ \,|\, \boldsymbol{\Sigma}\cdot\boldsymbol{p} \,|\, e^+ \rangle\,,\end{aligned}$$

which is contrary to observations, cf. (34).

The breakdown of charge conjugation symmetry is implicit in the empirical Hamiltonian for $\beta$-decay (37). It has already been experimentally established that the four coupling parameters contained in this Hamiltonian,

$C_{\rm V}$, $C_{\rm A}$, $C'_{\rm V} = \alpha_{\rm V} C_{\rm V}$, and $C'_{\rm A} = \alpha_{\rm A} C_{\rm A}$ are all relatively real. Now charge conjugation invariance of this Hamiltonian would imply $C_{\rm V}$ and $C_{\rm A}$ are relatively real, as are $C'_{\rm V}$ and $C'_{\rm A}$, but that $C_{\rm V}$ and $C'_{\rm V}$ differ by a relative phase of 90°, as do $C_{\rm A}$ and $C'_{\rm A}$. This contradicts observations.

Let us examine in more detail the decay Hamiltonian $\mathcal{H}_\beta$ given in (32), keeping only the V and A terms. To simplify notations, we denote $\psi_{\rm p}$ by $p$ and so on, and write the $\beta$-decay Hamiltonian as

$$\begin{aligned} \mathcal{H}_\beta = & C_{\rm V}(\bar{p}\gamma_\mu n)(\bar{e}\gamma^\mu \nu) + C'_{\rm V}(\bar{p}\gamma_\mu n)(\bar{e}\gamma_5\gamma^\mu \nu) \\ & + C_{\rm A}(\bar{p}\gamma_\mu\gamma_5 n)(\bar{e}\gamma^\mu\gamma_5 \nu) - C'_{\rm A}(\bar{p}\gamma_\mu\gamma_5 n)(\bar{e}\gamma^\mu \nu)\,, \end{aligned} \tag{5.108}$$

and obtain its Hermitian conjugate

$$\begin{aligned} \mathcal{H}^\dagger_\beta = & C^*_{\rm V}(\bar{n}\gamma_\mu p)(\bar{\nu}\gamma^\mu e) + C'^*_{\rm V}(\bar{n}\gamma_\mu p)(\bar{\nu}\gamma_5\gamma^\mu e) \\ & + C^*_{\rm A}(\bar{n}\gamma_\mu\gamma_5 p)(\bar{\nu}\gamma^\mu\gamma_5 e) - C'^*_{\rm A}(\bar{n}\gamma_\mu\gamma_5 p)(\bar{\nu}\gamma^\mu e)\,, \end{aligned} \tag{5.109}$$

and its charge conjugate

$$\begin{aligned} \mathcal{C}\,\mathcal{H}_\beta\,\mathcal{C}^{-1} = & C_{\rm V}(\bar{n}\gamma_\mu p)(\bar{\nu}\gamma^\mu e) - C'_{\rm V}(\bar{n}\gamma_\mu p)(\bar{\nu}\gamma_5\gamma^\mu e) \\ & + C_{\rm A}(\bar{n}\gamma_\mu\gamma_5 p)(\bar{\nu}\gamma^\mu\gamma_5 e) + C'_{\rm A}(\bar{n}\gamma_\mu\gamma_5 p)(\bar{\nu}\gamma^\mu e)\,. \end{aligned} \tag{5.110}$$

Invariance of the complete Hamiltonian $\mathcal{H}_\beta + \mathcal{H}^\dagger_\beta$ implies $\mathcal{C}\,\mathcal{H}_\beta\,\mathcal{C}^{-1} = \mathcal{H}^\dagger_\beta$, or the condition that $C_{\rm V}$ and $C_{\rm A}$ be both real, and $C'_{\rm V}$ and $C'_{\rm A}$ be both imaginary. Therefore, charge conjugation symmetry is violated. It is interesting to note however that if one now applies the parity operation $\mathcal{P}$ on both sides of (110), the only effect on the right-hand side is to flip the signs of the $C'_{\rm V}$ and $C'_{\rm A}$ terms. Therefore, comparison with $\mathcal{H}^\dagger_\beta$ tells us invariance of the full Hamiltonian under combined $\mathcal{P}$ and $\mathcal{C}$ requires all $C_i$, $C'_i$ for $i =$ V, A to be relatively real, in agreement with observations.

To summarize the results of this and previous sections, the weak processes described by the Hamiltonian $\mathcal{H}_\beta + \mathcal{H}^\dagger_\beta$ break parity and charge conjugation symmetries separately, but are invariant to $\mathcal{T}$, to the combined operation $\mathcal{CP}$, and evidently also to $\mathcal{CPT}$. This example illustrates the general situation to be treated in the following section.

## 5.4 The CPT Theorem

This theorem, due to Lüders, Pauli, and Schwinger, states that the product of the transformations $\mathcal{C}$, $\mathcal{P}$, $\mathcal{T}$ applied in any order is always a symmetry of a quantum theory if the Lagrangian that defines it is Hermitian, invariant under proper Lorentz transformations, is built up from normal-ordered products of fields, and if the fields are quantized in accord with the usual spin–statistics connection.

The operation $\Theta \equiv \mathcal{CPT}$ is defined as the product of the transformations $\mathcal{C}$, $\mathcal{P}$, and $\mathcal{T}$ performed in any order. From the definitions of $\mathcal{C}$, $\mathcal{P}$, and $\mathcal{T}$, the action of $\Theta$ on various variables can be found. Namely:

- all complex numbers are replaced by their complex conjugates;
- space-time coordinates $x^\mu$ are replaced by $x^{\mu\prime} = -x^\mu$;
- scalar fields are transformed as $\Theta\,\phi(x)\,\Theta^{-1} = \omega_{\rm B}\phi^\dagger(-x)$,

  where the phase should be *fixed* at $\omega_{\rm B} = +1$ in order to realize invariance of interactions involving scalar fields;
- the electromagnetic field transforms as $\Theta\,A_\mu(x)\,\Theta^{-1} = -A_\mu(-x)$;
- fermion fields transform as $\Theta\,\psi(x)\,\Theta^{-1} = \omega_{\rm F}\gamma_5\gamma_0\overline{\psi}^{\rm T}(-x)$, $\quad|\omega_{\rm F}| = 1$;
- bilinear products of fermion fields of the form $\overline{\psi}\Gamma\psi$ may be grouped according to their transformations under $\Theta$: $\Gamma_+ = \{1,\, \sigma^{\mu\nu}\,, {\rm i}\gamma_5\}$ are even (do not change signs), and $\Gamma_- = \{\gamma^\mu\,, \gamma^\mu\gamma_5\}$ are odd (change signs).

A Lorentz-invariant quantity is constructed for example by multiplying a $\Gamma_-$ by another $\Gamma_-$, $A^\mu(x)$, or $\partial/\partial_\mu$, or by multiplying $\Gamma_+$ by a factor having the same even number of Lorentz indices, and contracting all repeated Lorentz indices. In a $\Theta$-transformation, individual factors involving fermion fields may change their signs or phases, but their products do not change signs or phases. For example, in the familiar $\beta$-decay,

$$\Theta\,\left[\mathcal{H}_\beta(x) + \mathcal{H}_\beta^\dagger(x)\right]\Theta^{-1} = \mathcal{H}_\beta(-x) + \mathcal{H}_\beta^\dagger(-x)\,.$$

We have seen that a product of quantum fields representing a physical quantity should always be normal-ordered. When it is necessary to restore the field factors to such an order, no sign change is needed when permuting Bose fields, but a minus sign is introduced for each permutation of two Fermi fields. This is why the usual connection between spins and statistics is required in the CPT theorem. When the theory contains interactions between bosons, as in $\lambda\phi^3 + \lambda^*\phi^{\dagger 3}$, or interactions between a boson field and a fermion field, as in ${\rm i}g\overline{\psi}\gamma_5\psi\phi$ or in $g\overline{\psi}\gamma_5\gamma^\mu\psi\partial_\mu\phi$, the transformation rules given above, with the selected phase $\omega_{\rm B} = +1$, guarantee invariance of these interactions to $\Theta$.

In summary, the CPT theorem requires that the Lagrangian density of any physical quantum field theory transforms as

$$\Theta\,\mathcal{L}(x)\,\Theta^{-1} = \mathcal{L}(-x)^\dagger\,. \tag{5.111}$$

Since $\mathcal{L}$ is a Hermitian operator and the action function is given by the space-time integral $\int {\rm d}^4x\,\mathcal{L}$, the theorem guarantees invariance of the action and hence that of the theory itself. The validity of the theorem is based on the invariance to the group of continuous Lorentz transformations, the usual spin–statistics connection and the locality of the theory. It is not affected by whether $\mathcal{C}$, $\mathcal{P}$, and $\mathcal{T}$ separately are symmetries or not.

### 5.4.1 Implications of CPT Invariance

Using the same method as in the preceding sections, one can show that $\Theta$ transforms a one-particle state into an antiparticle state (up to a phase factor)

$$\Theta\,|a\rangle = |\bar{a}\rangle\, \mathrm{e}^{\mathrm{i}\vartheta}\,. \tag{5.112}$$

Therefore, in an invariant theory,

$$\langle a\,|\,H\,|\,a\rangle = \langle a\,\big|\,\Theta^{-1}H\Theta\,\big|\,a\rangle = \langle \bar{a}\,|\,H\,|\,\bar{a}\rangle\;. \tag{5.113}$$

Here $H$ is the total Hamiltonian. The result says that the energy spectra in the original and the transformed systems are identical. In particular, in the absence of interactions, $\langle a\,|\,H\,|\,a\rangle$ gives essentially the mass of the particle, and therefore CPT invariance implies the equality between the masses of the particle and the corresponding antiparticle. This result is experimentally well verified. For example, the proton and the antiproton differ in mass by

$$(m_{\mathrm{p}} - m_{\bar{\mathrm{p}}})/m_{\mathrm{p}} \approx 2 \times 10^{-11}\,.$$

The best test of CPT invariance comes from comparing the $\mathrm{K}^0$ and $\bar{K}^0$ masses. We will see in Chap. 11 that the $\mathrm{K}^0$–$\bar{\mathrm{K}}^0$ mass difference is related to certain CP violation parameters. The best available values for these yield

$$|(m_{\bar{\mathrm{K}}^0} - m_{\mathrm{K}^0})/m_{\mathrm{K}^0}| \leq 9 \times 10^{-19}. \tag{5.114}$$

The transition rate from state $|a\rangle$ to state $|b\rangle$ due to a weak interaction $H_{\mathrm{F}}$ is given by

$$w_{ba} = 2\pi |\,\langle b\,|\,H_{\mathrm{F}}\,|\,a\rangle\,|^2 \rho_b, \tag{5.115}$$

where $\rho_b$ is the final state density. The total transition rate, obtained by summing over all possible final channels, $w_a = \sum_b w_{ba}$, is related to the total lifetime of $|a\rangle$ by $\tau_a = 1/w_a$. From invariance of $H_{\mathrm{F}}$ to $\Theta$ it follows that the total lifetime is also invariant:

$$\tau_a = \tau_{\bar{a}}\,. \tag{5.116}$$

A particle and its antiparticle stable to strong and electromagnetic interactions have equal lifetimes. This agrees with observations. For example,

$$(\tau_{\pi^+} - \tau_{\pi^-})/\tfrac{1}{2}\,(\tau_{\pi^+} + \tau_{\pi^-}) = (6 \pm 7) \times 10^{-4}\,,$$
$$(\tau_{\mathrm{K}^+} - \tau_{\mathrm{K}^-})/\tfrac{1}{2}\,(\tau_{\mathrm{K}^+} + \tau_{\mathrm{K}^-}) = (0.11 \pm 0.09) \times 10^{-2}\,.$$

Finally, $\Theta$-invariance being assumed, if for some interaction one of the transformations $\mathcal{C}$, $\mathcal{P}$, or $\mathcal{T}$ is nonconserving, at least another is also nonconserving. We have already seen that in $\beta$-decay, neither $\mathcal{P}$ nor $\mathcal{C}$ is a symmetry, but $\mathcal{T}$ and $\mathcal{PC}$ are both symmetries. On the other hand, if for example neither $\mathcal{T}$ nor $\mathcal{CP}$ is a symmetry for a certain interaction, then either $\mathcal{P}$, but not $\mathcal{C}$, is conserving; or $\mathcal{C}$ is, but not $\mathcal{P}$. An interaction that breaks CP-invariance also breaks T-invariance. An example of this situation is the decay of neutral kaons to be studied in Chap. 11.

### 5.4.2 C, P, T, and CPT

For reference, we summarize in these paragraphs the conjugation rules for Dirac fermion fields and their bilinear products. We use the standard representation of the matrices $\gamma_\mu$.

(a) Hermitian conjugation

$$\left(\overline{\psi}_1(x)\Gamma\psi_2(x)\right)^\dagger = \overline{\psi}_2(x)\gamma_0\Gamma^\dagger\gamma_0\psi_1(x)\,. \tag{5.117}$$

(b) Parity $\mathcal{P}$

$$\mathcal{P}\,\psi(x)\,\mathcal{P}^{-1} = \eta\gamma_0\psi(\tilde{x}), \quad \tilde{x} = (t, -\boldsymbol{x}), \tag{5.118}$$

$$\mathcal{P}\,\overline{\psi}(x)\,\mathcal{P}^{-1} = \eta^*\overline{\psi}(\tilde{x})\gamma_0, \tag{5.119}$$

$$\mathcal{P}\,\overline{\psi}_1(x)\Gamma\psi_2(x)\,\mathcal{P}^{-1} = \eta_1^*\eta_2\overline{\psi}_1(\tilde{x})\,\gamma_0\Gamma\gamma_0\,\psi_2(\tilde{x})\,. \tag{5.120}$$

(c) Charge conjugation $\mathcal{C}$ $\;(C \equiv \mathrm{i}\gamma^2\gamma^0)$

$$\mathcal{C}\,\psi(x)\,\mathcal{C}^{-1} = \xi C\overline{\psi}^{\mathrm{T}}(x),$$

$$\mathcal{C}\,\overline{\psi}(x)\,\mathcal{C}^{-1} = -\xi^*\psi^{\mathrm{T}}(x)C^\dagger, \tag{5.121}$$

$$\mathcal{C}\,\overline{\psi}_1(x)\Gamma\psi_2(x)\,\mathcal{C}^{-1} = \xi_1^*\xi_2\overline{\psi}_2(x)\,C\Gamma^{\mathrm{T}}C^\dagger\,\psi_1(x)\,. \tag{5.122}$$

(d) Time reversal $\mathcal{T}$ $\;(A \equiv \gamma^1\gamma^3)$

$$\mathcal{T}\,\psi(x)\,\mathcal{T}^{-1} = \zeta A\psi(x'), \quad x' = (-t, \boldsymbol{x})\,, \tag{5.123}$$

$$\mathcal{T}\,\overline{\psi}(x)\,\mathcal{T}^{-1} = \zeta^*\overline{\psi}(x')\gamma_0 A^\dagger\gamma_0\,, \tag{5.124}$$

$$\mathcal{T}\,\overline{\psi}_1(x)\Gamma\psi_2(x)\,\mathcal{T}^{-1} = \zeta_1^*\zeta_2\overline{\psi}_1(x')\,A\Gamma^*A^\dagger\,\psi_(x')\,. \tag{5.125}$$

(e) $\mathcal{CP}$ transformation

$$\mathcal{CP}\psi(x)\mathcal{P}^{-1}\mathcal{C}^{-1} = \xi\eta C\gamma_0\overline{\psi}^{\mathrm{T}}(\tilde{x})\,, \tag{5.126}$$

$$\mathcal{CP}\overline{\psi}(x)\mathcal{P}^{-1}\mathcal{C}^{-1} = -\xi^*\eta^*\psi^{\mathrm{T}}(\tilde{x})C^\dagger\gamma_0\,, \tag{5.127}$$

$$\mathcal{CP}\overline{\psi}_1(x)\Gamma\psi_2(x)\mathcal{P}^{-1}\mathcal{C}^{-1} = -\xi_1^*\xi_2\eta_1^*\eta_2\overline{\psi}_2(\tilde{x})\,\gamma^2\Gamma^{\mathrm{T}}\gamma^2\,\psi_1(\tilde{x})\,. \tag{5.128}$$

(f) $\Theta = \mathcal{CPT}$

$$\Theta\,\psi(x)\,\Theta^{-1} = \omega\gamma_5\gamma_0\overline{\psi}^{\mathrm{T}}(-x), \quad -x = (-t, -\boldsymbol{x})\,, \tag{5.129}$$

$$\Theta\,\overline{\psi}(x)\,\Theta^{-1} = \omega^*\psi^{\mathrm{T}}(-x)\gamma_5\gamma_0\,, \tag{5.130}$$

$$\Theta\,\overline{\psi}_1(x)\Gamma\psi_2(x)\,\Theta^{-1} = \omega_1^*\omega_2\overline{\psi}_2(-x)\,\gamma_5\Gamma\gamma_5\,\psi_1(-x)\,. \tag{5.131}$$

Finally, Table 5.3 gives a list of transformation properties of the basic spinor operators in the discrete symmetries discussed in this chapter.

**Table 5.3.** Transformations of the Γs in discrete symmetries

| | | $\mathcal{P}$ | $\mathcal{C}$ | $\mathcal{T}$ | $\mathcal{CP}$ | $\mathcal{CPT}$ |
|---|---|---|---|---|---|---|
| | $\Gamma$ | $\gamma_0\Gamma\gamma_0$ | $C\Gamma^{\mathrm{T}}C^{\dagger}$ | $A\Gamma^{*}A^{\dagger}$ | $-\gamma^2\Gamma^{\mathrm{T}}\gamma^2$ | $\gamma_5\Gamma\gamma_5$ |
| S | 1 | 1 | 1 | 1 | 1 | 1 |
| P | $\mathrm{i}\gamma_5$ | $-\mathrm{i}\gamma_5$ | $\mathrm{i}\gamma_5$ | $-\mathrm{i}\gamma_5$ | $-\mathrm{i}\gamma_5$ | $\mathrm{i}\gamma_5$ |
| V | $\gamma^{\mu}$ | $\gamma_{\mu}$ | $-\gamma^{\mu}$ | $\gamma_{\mu}$ | $-\gamma_{\mu}$ | $-\gamma^{\mu}$ |
| A | $\gamma^{\mu}\gamma_5$ | $-\gamma_{\mu}\gamma_5$ | $\gamma^{\mu}\gamma_5$ | $\gamma_{\mu}\gamma_5$ | $-\gamma_{\mu}\gamma_5$ | $-\gamma^{\mu}\gamma_5$ |
| T | $\sigma^{\mu\nu}$ | $\sigma_{\mu\nu}$ | $-\sigma^{\mu\nu}$ | $-\sigma_{\mu\nu}$ | $-\sigma_{\mu\nu}$ | $\sigma^{\mu\nu}$ |

*Remark*: The position of the Lorentz index is important, e.g. $\gamma_{\mu} = (\gamma^0, -\gamma^i)$.

## Problems

**5.1 Symmetries in quantum mechanics.** (a) Consider $\pi$–p scattering or more generally scattering of a spin-0 particle by a spin-1/2 particle. The relevant variables are $\boldsymbol{p}_{\mathrm{i}}$(relative initial momentum), $\boldsymbol{p}_{\mathrm{f}}$ (relative final momentum), $\boldsymbol{n} = \boldsymbol{p}_{\mathrm{i}} \times \boldsymbol{p}_{\mathrm{f}}/|\boldsymbol{p}_{\mathrm{i}} \times \boldsymbol{p}_{\mathrm{f}}|$ and $\boldsymbol{\sigma}$ (fermion spin). What is the general rotational invariant form of the transition amplitude? What are the restrictions when $\mathcal{P}$ or $\mathcal{T}$ invariances are imposed?
(b) Repeat the analysis for the scattering of two spin-1/2 particles.

**5.2 Lepton decay.** The coupling constant for $\mu \to \mathrm{e}\nu\bar{\nu}$ is $G_{\mathrm{F}} = 1.166 \times 10^{-5}\,\mathrm{GeV}^{-2}$. From dimensional analysis the decay rate is proportional to $G_{\mathrm{F}}^2 m_{\mu}^5$. Derive its exact formula

$$\Gamma(\mu \to \mathrm{e}\nu\bar{\nu}) = \frac{G_{\mathrm{F}}^2 m_{\mu}^5}{192\pi^3}.$$

Give an estimate of the muon mean lifetime. In 1975 the $\tau$ lepton of mass 1.78 GeV was detected. Give estimates of the decay rates for $\tau \to \mathrm{e}\nu\bar{\nu}$ and $\tau \to \mu\nu\bar{\nu}$ and the corresponding branching ratios.

**5.3 $\pi^{\pm}$ decays.** The weak decays $\pi \to \mathrm{e}\nu$ and $\pi \to \mu\nu$ may be described by the V–A interactions

$$\mathcal{H}_{\mathrm{VA}}(x) = \frac{G_{\mathrm{F}}}{\sqrt{2}} J^{\alpha}_{\mathrm{hadr}}(x)\,[\bar{\ell}\gamma_{\alpha}(1-\gamma_5)\nu] + \mathrm{h.c.}\,, \quad \ell = \mathrm{e}, \mu\,.$$

The hadronic matrix $\langle 0 |\, J^{\alpha}_{\mathrm{hadr}} \,| \pi \rangle$ can depend only on the four-vector $p_{\pi} = p_{\ell} + p_{\nu}$, and by virtue of the (Dirac) equations of motion for the leptons, the decay amplitude is proportional to the charged lepton mass. Show that

$$\frac{\Gamma(\pi \to \mu\nu)}{\Gamma(\pi \to \mathrm{e}\nu)} = P\,\frac{m_{\mu}^2}{m_{\mathrm{e}}^2}, \quad \text{where } P = (m_{\pi}^2 - m_{\mathrm{e}}^2)^2/(m_{\pi}^2 - m_{\mu}^2)^2\,.$$

**5.4 The $\tau-\theta$ puzzle.** (a) Consider the decay mode $\theta \to \pi^+\pi^0$. Assuming parity invariance and 0 for the spin of $\theta$, find the parity of $\theta$.
(b) Now consider the decay process $\tau \to \pi^+\pi^+\pi^-$. (This $\tau$ is an old symbol for the K meson.) Let $\ell$ be the orbital angular momentum of $\pi^+\pi^+$, and $\ell'$ the orbital angular momentum of $\pi^-$ relative to the center-of-mass of $\pi^+\pi^+$. Assuming parity invariance and the spin of $\tau$ equal to 0, find its parity.

**5.5 $\Lambda^0$ decay.** The weak decay $\Lambda^0 \to \mathrm{p} + \pi^-$ can be described by the Hamiltonian

$$H_{\mathrm{int}} = \int_0^t \mathrm{d}t' \int \mathrm{d}^3x\, \phi^\dagger \overline{\psi}_{\mathrm{p}}(g + g'\gamma_5)\psi_\Lambda + \mathrm{h.c.}\,,$$

where $\psi_\Lambda$ destroys a $\Lambda$ and creates a $\bar{\Lambda}$, $\overline{\psi}_p$ creates a proton and destroys an antiproton, and $\phi^\dagger$ creates a $\pi^-$ and destroys a $\pi^+$. The h.c. term makes $\mathcal{H}_{\mathrm{int}}$ Hermitian. The initial and final states in the process $\Lambda^0 \to \mathrm{p} + \pi^-$ are given by $|\mathrm{i}\rangle = b_\Lambda^\dagger(\boldsymbol{p}, s)\,|0\rangle$, and $|\mathrm{f}\rangle = a^\dagger(\boldsymbol{k}) b_{\mathrm{p}}^\dagger(\boldsymbol{p}', s')\,|0\rangle$.
(a) Calculate the transition amplitude $\langle \mathrm{f}\,|\, H_{\mathrm{int}}\,|\,\mathrm{i}\rangle$ for $\Lambda$ at rest.
(b) Suppose that $\Lambda$ is polarized with spin oriented in the positive $z$ direction, show that the relative probabilities for observing protons produced with polarizations $\pm 1/2$ are

$$\begin{aligned} |a_{\mathrm{s}} + a_{\mathrm{p}} \cos\theta|^2 &\qquad \text{for spin } + 1/2\,, \\ |a_{\mathrm{p}} \sin\theta|^2 &\qquad \text{for spin } - 1/2\,, \end{aligned}$$

where $\cos\theta = p'_z/|\boldsymbol{p}'|$. Calculate $a_{\mathrm{s}}$, $a_{\mathrm{p}}$.

**5.6 Lepton number.** Discuss how to set up an experiment to decide whether the lepton number is additive or multiplicative.

**5.7 Weyl representation.** Find the time inversion matrix $A$ and the charge conjugation matrix $C$ in the Weyl representation of the $\gamma_\mu$ matrices.

**5.8 Quantum numbers for an antifermion.** Let $\psi$ be a Dirac wave function. Its parity $\eta$, helicity $h$, and chirality $\lambda$ (in the ultra-relativistic limit) are defined respectively by $\mathcal{P}\psi = \eta\gamma_0\psi$, $\boldsymbol{\Sigma}\cdot\hat{\boldsymbol{p}}\,\psi = h\psi$, and $\gamma_5\psi = \lambda\psi$. Show that the corresponding quantum numbers for the charge conjugate field $\psi^{\mathrm{c}} = \xi C\overline{\psi}^{\mathrm{T}}$ are $\eta^{\mathrm{c}} = -\eta$, $h^{\mathrm{c}} = h$, $\lambda^{\mathrm{c}} = -\lambda$.

**5.9 Invariance of the electromagnetic interaction.** Let $A_\mu$ and $F_{\mu\nu}$ be the electromagnetic field and field tensor. Study the transformation properties of the following interaction terms under the operations $\mathcal{P}$, $\mathcal{C}$, $\mathcal{T}$, and $\mathcal{PCT}$ : (a) $e\overline{\psi}\gamma_\mu\psi\, A^\mu$; (b) $1/2\mu_{\mathrm{m}} F_{\mu\nu}\overline{\psi}\sigma^{\mu\nu}\psi$; (c) $1/2\mu_{\mathrm{e}} F_{\mu\nu}\overline{\psi}\gamma_5\sigma^{\mu\nu}\psi$.

## Suggestions for Further Reading

*Analysis and demonstrations of nonconservation of parity:*

Friedman, J. I. and Telegdi, V. L., Phys. Rev. **105** (1957) 1681

Garwin, R. L., Lederman, L. M. and Weinrich, M., Phys. Rev. **105** (1957) 1415

Lee, T. D. and Yang, C. N., Phys. Rev. **104** (1956) 254

Wu, C. S., Ambler, E., Hayward, R. W., Hoppes, D. and Hudson, R. P., Phys. Rev. **105** (1957) 1413

*Time reversal:*

Cohen-Tannoudji, G. and Jacob, M. *Le temps et sa flèche* (ed. by Klein, E. and Spiro, M.). Editions Frontières, Gif-sur-Yvette 1994

Schwinger, J., Phys. Rev. **82** (1951) 914

Wigner, E., Nachr. Akad. Wiss. Göttingen **32** (1932) 35; *Group Theory and its Applications to Quantum Mechanics of Atomic Spectra*. Academic Press, New York 1959

*Charge conjugation:*

Furry, W. H., Phys. Rev. **51** (1937) 125

Kramers, H. A., Proc. Amst. Akad. Sci. **40** (1937) 814

Pauli, W., Annales de l'Inst. Henri Poincaré **6** (1936) 137

*CPT theorem:*

Lüders, G., Kgl. Danske Vidensk. Selsk. Mat.-Fys. Medd. **28** (1954) no. 5; Ann. of Phys. **2** (1957) 1

Pauli, W., *Niels Bohr and the Development of Physics*. McGraw-Hill, New York 1955; Nuovo Cimento **6** (1957) 204

Schwinger, J., Phys. Rev. **82** (1951) 914; **91** (1951) 713

Streater, R. F. and Wightman, A. S., *PCT, Spin, Statistics, and All That*. Benjamin, New York 1968

*A collection of reports on experiments with background introductions:*

Cahn, R. N. and Goldhaber, G., *The Experimental Foundations of Particle Physics*. Cambridge U. Press, Cambridge 1989

*Data from*

Review of Particle Properties, Phys. Rev. **D54** (1996) 1

# 6 Hadrons and Isospin

The six known *leptons* and their charge conjugates fall naturally into a simple pattern of classification suggestive of an underlying symmetry that may eventually lead to an uncovering of their dynamical laws. In contrast, the situation is vastly more complex with the *hadrons* because of their larger number and greater diversity. Nevertheless, similarities and relationships do exist among mesons and baryons, which have gradually come to light through both experimental and theoretical efforts. The experiments carried out and the ideas put forth during an effervescent period of over thirty years – roughly from 1932, when Werner Heisenberg introduced the concept of *isospin*, to the early 1960s, when Murray Gell-Mann and Yuval Ne'eman proposed the notion of the *eightfold way* – have contributed significantly to shaping our present-day view of the particles and their interactions. They form the subject matter of the present and the next chapters.

In this chapter, we introduce the concept of isospin and show how it is used in quantum field theories, especially in situations involving nucleons and pions. We also define the G-parity for 'unflavored' hadrons and the hypercharge for strange particles. Isospin is conserved in the strong interaction, but not in the electromagnetic and weak interactions. We give a brief discussion of how and where the symmetry is violated; the missing 'why' should be found in a future interaction model.

## 6.1 Charge Symmetry and Charge Independence

There exists ample evidence for *charge symmetry* in the physics of strong interactions. This principle holds that, apart from electromagnetic and weak effects, mesons and baryons behave in exactly the same way as their charge symmetric counterparts. As examples of manifestations of charge symmetry, we have the near equality of the proton and neutron masses and the small difference in the nuclear binding energies of $^3$H and $^3$He (0.8 MeV out of 8 MeV); we may also point out that spectra in mirror nuclei (such as $^7$Li and $^7$Be, or $^{11}$B and $^{11}$C) show levels of identical angular momenta and parities at approximately the same relative energies. These regularities must reflect some sort of symmetry – the symmetry of nuclear systems under interchanging neutrons and protons, which necessarily implies the equality of neutron–neutron and proton–proton nuclear forces.

More remarkable still, the stronger hypothesis of *charge independence* also appears to be generally valid, provided again that electromagnetic and weak effects can be neglected. In particular, the nuclear forces in any pairs of neutrons and protons in the same orbital angular momentum and spin states are expected to be the same. Evidence can be found in comparing similar (isobaric) states in certain groups of nuclei, such as $^{14}$C, $^{14}$N, and $^{14}$O, or $^{21}$F, $^{21}$Ne, $^{21}$Na, and $^{21}$Mg, in which nuclei differ from one another only in their last two or three neutrons or protons, and hence in the presence of different pair interactions, nn, pp, or pn. Further support comes from the near equality of the masses of certain mesons and hyperons, as shown in Table 6.1, as well as from the excellent agreement between the measures and calculations of the relative rates of production of neutral and charged pions from the collisions of protons with deuterons.

**Table 6.1.** Comparison of some meson and hyperon masses

| Mass difference $\Delta$ [a] | | $M$ (average) [a] | $\Delta/M$ |
|---|---|---|---|
| $\pi^\pm - \pi^0$ | 4.59 | 137 | .033 |
| $K^0 - K^\pm$ | 4.02 | 495 | .008 |
| $\Sigma^- - \Sigma^+$ | 8.07 | 1193 | .007 |
| $\Sigma^- - \Sigma^0$ | 4.89 | 1193 | .004 |

[a] In MeV/$c^2$.

Charge symmetry of a nuclear system means the physical properties of the system remain unchanged upon interchanging all protons with all neutrons, whereas charge independence implies invariance even when the proton and neutron states are replaced with *any* orthonormal, real or complex, admixtures of them. Just as there is no possible distinction between the spin-up and spin-down states of an electron in the absence of magnetic fields, no observable effects can discriminate protons from neutrons in the absence of electromagnetic interactions. This fact points to the advantages of introducing an abstract space spanned by vectors, whose discrete projections on a certain axis correspond to charge states. The *nucleon* is one such vector, with its two projections identified with the neutron and the proton. Charge independence can then be interpreted in geometrical terms as rotational invariance in charge space, and rotations in this space can be generated, as in ordinary space, by an operator, called the *isobaric spin* or simply *isospin*, with the same algebraic properties as the ordinary rotation operator, the angular momentum of coordinate space. And just as the conservation of angular momentum follows from rotational invariance in ordinary space, so too does the conservation of isospin from rotational invariance in isospin space.

The usefulness of the isospin concept stems from the fact that it can be generalized to all hadrons and that isospin is a conserved quantity in strong interactions. It follows that mesons and baryons can be classified into multiplets characterized by an isospin, and their strong interactions are rotationally invariant in isospin space, so that several general results can be derived without a detailed knowledge of the interaction or the labors of dynamical calculations. On a deeper level, as hadrons are regarded as quark composites, *isospin invariance* arises from the (still unexplained) near equality of the u and d quark masses. Although only approximate (since the electromagnetic interaction of nature introduces a preferred direction in isospin space), isospin invariance is expected to be valid to an accuracy of the order of the ratio of the electromagnetic coupling to the strong coupling, i.e. a few percent. The success of this symmetry in relating a large number of particles and in predicting many phenomena may lead one to wonder whether different isospin multiplets could be further regrouped, by virtue of some common properties, under some higher symmetry. Even the very fact that it is violated, which it is, not in a haphazard but rather systematic way by the electromagnetic and weak interactions, confers upon it a significant role in any model building.

## 6.2 Nucleon Field in Isospin Space

In this section we study the nucleon field as the simplest isospin representation. Although it is known that the proton and the neutron are composite particles, they will be treated for now as elementary, a perfectly valid point of view on phenomenological grounds. Rotations on the charge space will be described by unitary operators and generated by isospin operators. This treatment is generalized to other particles in later sections.

The nucleon field is represented by an eight-component column vector

$$\psi(x) = \begin{pmatrix} \psi_{\mathrm{p}}(x) \\ \psi_{\mathrm{n}}(x) \end{pmatrix}, \tag{6.1}$$

formed from the proton field $\psi_{\mathrm{p}}$ and the neutron field $\psi_{\mathrm{n}}$, assumed to have the same mass, $m$. Each of these is a quantized Dirac four-component spinor operator, which may be expressed as a Fourier series

$$\begin{aligned} \psi_{\mathrm{p}}(x) &= \sum_{\boldsymbol{k},s} C_{\boldsymbol{k}} \left[ u(\boldsymbol{k},s)\, \mathrm{e}^{-\mathrm{i}k\cdot x} b_{\mathrm{p}}(\boldsymbol{k},s) + v(\boldsymbol{k},s)\, \mathrm{e}^{\mathrm{i}k\cdot x}\, d^{\dagger}_{\mathrm{p}}(\boldsymbol{k},s) \right], \\ \psi_{\mathrm{n}}(x) &= \sum_{\boldsymbol{k},s} C_{\boldsymbol{k}} \left[ u(\boldsymbol{k},s)\, \mathrm{e}^{-\mathrm{i}k\cdot x} b_{\mathrm{n}}(\boldsymbol{k},s) + v(\boldsymbol{k},s)\, \mathrm{e}^{\mathrm{i}k\cdot x}\, d^{\dagger}_{\mathrm{n}}(\boldsymbol{k},s) \right], \end{aligned} \tag{6.2}$$

[$C_{\boldsymbol{k}} = 1/\sqrt{(2\pi)^3\, 2E_{\boldsymbol{k}}}$, $k_0 = E_{\boldsymbol{k}} = \sqrt{\boldsymbol{k}^2 + m^2}$], so that the components of the nucleon field are: $\psi_A = \psi_{\mathrm{p},i}$ for $A = i = 1, \ldots, 4$, and $\psi_A = \psi_{\mathrm{n},i}$ for $A = i+4$, $i = 1, \ldots, 4$.

To find the isospin operator for the nucleon field, it is best to start from what we already know, namely, the baryon number operator, $N_{\mathrm{B}}$, and the charge operator, $Q$. A slight generalization of results found in Chap. 3 yields

$$\begin{aligned} N_{\mathrm{B}} &= \int \mathrm{d}^3x\, \psi^\dagger \psi \\ &= \int \mathrm{d}^3x\, \psi_{\mathrm{p}}^\dagger \psi_{\mathrm{p}} + \int \mathrm{d}^3x\, \psi_{\mathrm{n}}^\dagger \psi_{\mathrm{n}} \\ &= (N_{\mathrm{p}} + N_{\mathrm{n}}) - (N_{\bar{\mathrm{p}}} + N_{\bar{\mathrm{n}}}) \end{aligned} \tag{6.3}$$

for the baryon number operator and

$$\begin{aligned} Q &= \int \mathrm{d}^3x\, \psi_{\mathrm{p}}^\dagger \psi_{\mathrm{p}} \\ &= N_{\mathrm{p}} - N_{\bar{\mathrm{p}}} \end{aligned} \tag{6.4}$$

for the charge operator (in units of $e > 0$). Here the number operators for protons, neutrons, antiprotons, and antineutrons are given respectively by

$$\begin{aligned} N_{\mathrm{p}} &= \sum_{\boldsymbol{k},s} b_{\mathrm{p}}^\dagger(\boldsymbol{k},s) b_{\mathrm{p}}(\boldsymbol{k},s), \qquad & N_{\mathrm{n}} &= \sum_{\boldsymbol{k},s} b_{\mathrm{n}}^\dagger(\boldsymbol{k},s) b_{\mathrm{n}}(\boldsymbol{k},s)\,, \\ N_{\bar{\mathrm{p}}} &= \sum_{\boldsymbol{k},s} d_{\mathrm{p}}^\dagger(\boldsymbol{k},s) d_{\mathrm{p}}(\boldsymbol{k},s), \qquad & N_{\bar{\mathrm{n}}} &= \sum_{\boldsymbol{k},s} d_{\mathrm{n}}^\dagger(\boldsymbol{k},s) d_{\mathrm{n}}(\boldsymbol{k},s)\,. \end{aligned} \tag{6.5}$$

When applied on a one-proton or a one-neutron state, they give, for example,

$$\begin{aligned} N_{\mathrm{p}} |\mathrm{p}\rangle &= |\mathrm{p}\rangle\,, \qquad & N_{\mathrm{n}} |\mathrm{p}\rangle &= 0, \\ N_{\mathrm{n}} |\mathrm{n}\rangle &= |\mathrm{n}\rangle\,, \qquad & N_{\mathrm{p}} |\mathrm{n}\rangle &= 0. \end{aligned}$$

So that neither $N_{\mathrm{B}}$ nor $Q$ may be considered to be an isospin operator for the nucleon field, but the linear combination

$$\begin{aligned} I_3 &= Q - \tfrac{1}{2} N_{\mathrm{B}} \\ &= \tfrac{1}{2}(N_{\mathrm{p}} - N_{\bar{\mathrm{p}}}) - \tfrac{1}{2}(N_{\mathrm{n}} - N_{\bar{\mathrm{n}}}) \end{aligned} \tag{6.6}$$

has the expected property when applied on a nucleon state: it gives $+1/2$ if the nucleon is in a proton state, and $-1/2$ if the nucleon is in a neutron state (and $-1/2$ for an antiproton state, $+1/2$ for an antineutron state). When written in the form

$$I_3 = \tfrac{1}{2} \int \mathrm{d}^3x\, \psi^\dagger \begin{pmatrix} 1 & 0 \\ 0 & -1 \end{pmatrix} \psi, \tag{6.7}$$

it suggests identification with a field version of the third component of the Pauli matrix in isospin space, and hence generalization to all components:

$$I_i = \tfrac{1}{2} \int \mathrm{d}^3x\, \psi^\dagger \tau_i \psi \quad \text{for } i = 1, 2, 3, \tag{6.8}$$

where

$$\tau_1 = \begin{pmatrix} 0 & 1 \\ 1 & 0 \end{pmatrix}, \quad \tau_2 = \begin{pmatrix} 0 & -\mathrm{i} \\ \mathrm{i} & 0 \end{pmatrix}, \quad \text{and} \quad \tau_3 = \begin{pmatrix} 1 & 0 \\ 0 & -1 \end{pmatrix} \tag{6.9}$$

are understood as tensor products of the ordinary $2 \times 2$ Pauli spin matrices and the $4 \times 4$ identity matrix in the Dirac isospin space. As in the case of the Pauli matrices, $\tau_i$ are Hermitian and satisfy the basic property

$$\tau_i \tau_j = \delta_{ij} + \mathrm{i}\epsilon_{ijk}\tau_k \quad \text{for } i, j = 1,\, 2,\, 3. \tag{6.10}$$

The Hermitian operators $I_i$ will be considered as the components of a three-component vector in isospin space, $\boldsymbol{I}$. Isospin invariance means that the hadronic (or strong) part of the Hamiltonian, $H_h$, is such that it commutes with each $I_i$:

$$[I_i,\, H_h] = 0 \quad \text{for } i = 1,\, 2,\, 3\,. \tag{6.11}$$

In other words, $\boldsymbol{I}$ is a conserved vector, constant in time, which implies, among other things, that the time parameter of the fields in (8) can be arbitrarily chosen.

We will show next that $I_i$ generate rotations of the field $\psi$ in isopin space. First, we calculate the commutation relation between $I_i$ and an arbitrary field component $\psi_A(x)$:

$$[I_i,\, \psi_A(x)] = \tfrac{1}{2} \int \mathrm{d}^3 y \, \left[\psi_B^\dagger(y) \tau_{iBC} \psi_C(y),\, \psi_A(x)\right] . \tag{6.12}$$

Since time is arbitrary under the integral, we may set $y_0 = x_0$ and use the equal times anticommutation relations

$$\begin{aligned} \left\{\psi_A^\dagger(t, \boldsymbol{x}),\, \psi_B(t, \boldsymbol{y})\right\} &= \delta_{AB}\, \delta(\boldsymbol{x} - \boldsymbol{y})\,, \\ \left\{\psi_A(t, \boldsymbol{x}),\, \psi_B(t, \boldsymbol{y})\right\} &= \left\{\psi_A^\dagger(t, \boldsymbol{x}),\, \psi_B^\dagger(t, \boldsymbol{y})\right\} = 0 \end{aligned} \tag{6.13}$$

in (12) to obtain

$$\begin{aligned} [I_i,\, \psi_A(x)] &= -\tfrac{1}{2}\, \tau_{iBC} \int \mathrm{d}^3 y \, \left\{\psi_B^\dagger(y),\, \psi_A(x)\right\} \psi_C(y) \Big|_{y_0 = x_0} \\ &= -\tfrac{1}{2}\, \tau_{iAC}\, \psi_C(x)\,. \end{aligned}$$

This relation and its Hermitian conjugation describes the action of the isospin operators on the nucleon field:

$$\begin{aligned} [I_i, \psi_A(x)] &= -\tfrac{1}{2}\, \tau_{iAB}\, \psi_B(x), \\ [I_i,\, \psi_A^\dagger(x)] &= +\tfrac{1}{2}\, \tau_{iBA}\, \psi_B^\dagger(x)\,. \end{aligned} \tag{6.14}$$

Note that since a proton or neutron state is created by $\psi^\dagger$ and destroyed by $\psi$, the signs in (14) come out correctly, in good accord with (6).

Next, to obtain the transformation rule for $\psi$, let $\theta$ be a small real parameter; $\hat{\boldsymbol{n}} = (n_1, n_2, n_3)$ a unit vector in isospin space; and define $\boldsymbol{\theta} = \theta\hat{\boldsymbol{n}}$. We may then rewrite the first of (14), exact up to terms linear in $\theta$ as follows:

$$\psi + \mathrm{i}\,[\boldsymbol{\theta}\cdot\boldsymbol{I},\, \psi] = \psi - \frac{\mathrm{i}}{2}\boldsymbol{\theta}\cdot\boldsymbol{\tau}\,\psi\,,$$

$$(1 + \mathrm{i}\boldsymbol{\theta}\cdot\boldsymbol{I})\psi(1 - \mathrm{i}\boldsymbol{\theta}\cdot\boldsymbol{I}) = (1 - \frac{\mathrm{i}}{2}\boldsymbol{\theta}\cdot\boldsymbol{\tau})\,\psi\,.$$

This result may then be cast into a transformation law,

$$\mathcal{U}\,\psi(x)\,\mathcal{U}^{-1} = S(\boldsymbol{\theta})\,\psi(x)\,, \tag{6.15}$$

where

$$\mathcal{U} \approx 1 + \mathrm{i}\,\boldsymbol{\theta}\cdot\boldsymbol{I} \tag{6.16}$$

is the infinitesimal *rotation operator* on the Hilbert space, and

$$S(\boldsymbol{\theta}) \approx 1 - \frac{\mathrm{i}}{2}\boldsymbol{\theta}\cdot\boldsymbol{\tau} \tag{6.17}$$

its *matrix representation* in the Dirac space. Since $I_i$ are Hermitian, $\mathcal{U}$ is unitary to the first order in $\theta$.

For finite $\theta$, we find, by integration or by repeated applications of infinitesimal rotations, just as in ordinary space [cf. (3.26)], the general *rotation operator*

$$\mathcal{U} = \mathrm{e}^{\mathrm{i}\boldsymbol{\theta}\cdot\boldsymbol{I}}\,, \qquad \mathcal{U}^\dagger\mathcal{U} = 1\,, \tag{6.18}$$

and its *matrix representation*

$$\begin{aligned} S(\boldsymbol{\theta}) = \mathrm{e}^{-\frac{\mathrm{i}}{2}\boldsymbol{\theta}\cdot\boldsymbol{\tau}} &= \cos\tfrac{\theta}{2} - \mathrm{i}\hat{\boldsymbol{n}}\cdot\boldsymbol{\tau}\sin\tfrac{\theta}{2} \\ &= \begin{pmatrix} \cos\frac{\theta}{2} - \mathrm{i}n_3\sin\frac{\theta}{2} & -\mathrm{i}(n_1 - \mathrm{i}n_2)\sin\frac{\theta}{2} \\ -\mathrm{i}(n_1 + \mathrm{i}n_2)\sin\frac{\theta}{2} & \cos\frac{\theta}{2} + \mathrm{i}n_3\sin\frac{\theta}{2} \end{pmatrix}. \end{aligned} \tag{6.19}$$

To resum the power series that defines $\exp(-{}^{\mathrm{i}}\!/_2\,\boldsymbol{\theta}\cdot\boldsymbol{\tau})$, we have used the identity $(\hat{\boldsymbol{n}}\cdot\boldsymbol{\tau})^2 = 1$, which follows from (10). As a special case, the operator

$$\mathcal{U}_2 = \mathrm{e}^{\mathrm{i}\pi I_2} \tag{6.20}$$

performs a rotation through $\pi$ about the isospin $y$ axis, and acts as the *charge-exchange operator* for the nucleon, interchanging the proton and neutron states:

$$S(\pi\hat{\boldsymbol{n}}_2)\psi = \mathrm{e}^{-\frac{\mathrm{i}}{2}\pi\tau_2}\psi = -\mathrm{i}\tau_2\begin{pmatrix}\psi_{\mathrm{p}}\\ \psi_{\mathrm{n}}\end{pmatrix} = \begin{pmatrix}-\psi_{\mathrm{n}}\\ \psi_{\mathrm{p}}\end{pmatrix}. \tag{6.21}$$

The commutation relations for the generators of infinitesimal isospin rotations can be determined from (10) and (14):

$$\begin{aligned}
[I_i, I_j] &= \frac{1}{2}\tau_{jCD} \int \mathrm{d}^3y \left[ I_i, \psi_C^\dagger(y)\psi_D(y) \right] \\
&= \frac{1}{2}\tau_{jCD} \int \mathrm{d}^3y \left\{ [I_i, \psi_C^\dagger(y)]\psi_D(y) + \psi_C^\dagger(y) [I_i, \psi_D(y)] \right\} \\
&= \frac{1}{4} \int \mathrm{d}^3y\, \psi_C^\dagger(y)\psi_B(y) \left[ \tau_{iCD}\, \tau_{jDB} - \tau_{jCD}\, \tau_{iDB} \right] \\
&= \mathrm{i}\epsilon_{ijk} \frac{1}{2} \int \mathrm{d}^3y\, \psi_C^\dagger\, \tau_{kCB}\, \psi_B\, .
\end{aligned}$$

The resulting algebra of the isospin operators

$$[I_i, I_j] = \mathrm{i}\epsilon_{ijk} I_k \quad \text{for } i, j, k = 1,\, 2,\, 3 \tag{6.22}$$

is identical to the usual angular momentum algebra, and so the familiar results can be readily adapted to this case. In particular, as the Hamiltonian for nucleons $H_h$ is isospin-invariant, a common basis for it and $\boldsymbol{I}^2$ and $I_3$ can be found. The eigenequations are

$$\begin{aligned}
\boldsymbol{I}^2 \phi_{jm} &= j(j+1)\phi_{jm}\, , \\
I_3 \phi_{jm} &= m\phi_{jm}\, ,
\end{aligned} \tag{6.23}$$

with $j = \frac{1}{2}$ and $m = \pm\frac{1}{2}$. With the standard phase convention, one defines the step-up and step-down operations, $I_\pm \equiv I_1 \pm \mathrm{i} I_2$, on $\phi_{jm}$,

$$I_\pm\, \phi_{j,m} = [(j \mp m)(j \pm m + 1)]^{\frac{1}{2}}\; \phi_{j,m\pm 1}\, , \tag{6.24}$$

which give, in particular,

$$I_- \phi_{\frac{1}{2},-\frac{1}{2}} = 0\, , \qquad I_+ \phi_{\frac{1}{2},-\frac{1}{2}} = \phi_{\frac{1}{2},\frac{1}{2}}\, . \tag{6.25}$$

This means, if we let the neutron state be $|\mathrm{n}\rangle = \phi_{\frac{1}{2},-\frac{1}{2}}$, we can identify immediately $\phi_{\frac{1}{2},\frac{1}{2}}$ with the proton state $|\mathrm{p}\rangle \equiv \phi_{\frac{1}{2},\frac{1}{2}}$, with its phase fixed by (14).

How does the charge conjugate nucleon field behave under isospin rotations? If the isospin formalism is to be useful in quantum field theory, equally applicable to both the nucleon and the antinucleon fields, it is essential that these fields transform identically in isospin space. Now, since the nucleon field and its conjugate adjoint transform as

$$\begin{aligned}
\psi \to \psi' &= S\psi\, , \qquad S = \mathrm{e}^{-\frac{\mathrm{i}}{2}\boldsymbol{\theta}\cdot\boldsymbol{\tau}}\, , \\
\overline{\psi} \to \overline{\psi}' &= S^*\overline{\psi}\, , \quad S^* = \mathrm{e}^{\frac{\mathrm{i}}{2}\boldsymbol{\theta}\cdot\boldsymbol{\tau}^*}\, ,
\end{aligned} \tag{6.26}$$

the spinor formed by simply stacking up the charge conjugate proton and neutron fields, $\psi_{\mathrm{p}}^{\mathrm{c}} = C\overline{\psi}_{\mathrm{p}}^{\mathrm{T}}$ and $\psi_{\mathrm{n}}^{\mathrm{c}} = C\overline{\psi}_{\mathrm{n}}^{\mathrm{T}}$ ($C$ being the charge conjugation matrix defined in Sect. 5.3) transforms differently from the nucleon field:

$$\begin{pmatrix} \psi_{\mathrm{p}}^{\mathrm{c}} \\ \psi_{\mathrm{n}}^{\mathrm{c}} \end{pmatrix} \rightarrow S^* \begin{pmatrix} \psi_{\mathrm{p}}^{\mathrm{c}} \\ \psi_{\mathrm{n}}^{\mathrm{c}} \end{pmatrix}, \tag{6.27}$$

and so is not admissible. We should rather seek a spinor in the form

$$\psi^{\mathrm{c}} \equiv AC\overline{\psi}^{\mathrm{T}} \tag{6.28}$$

that transforms exactly as $\psi$ in (26), that is,

$$\psi^{\mathrm{c}} \rightarrow \psi^{\mathrm{c}\prime} = S\psi^{\mathrm{c}}, \tag{6.29}$$

which means that the matrix $A$ must have the property

$$AS^*A^{-1} = S. \tag{6.30}$$

With $\tau_i$ defined in (9), consistent with the standard sign conventions, one has $\tau_2\tau_i\tau_2 = -\tau_i^*$ for $i = 1, 2, 3$. Therefore, $A = \eta\tau_2$, with $|\eta| = 1$. Choosing $\eta = \mathrm{i}$, one finds the correct charge conjugate nucleon isospinor

$$\psi^{\mathrm{c}} = \mathrm{i}\tau_2 C\overline{\psi}^{\mathrm{T}} = \mathrm{i}\tau_2 \begin{pmatrix} \psi_{\mathrm{p}}^{\mathrm{c}} \\ \psi_{\mathrm{n}}^{\mathrm{c}} \end{pmatrix} = \begin{pmatrix} \psi_{\mathrm{n}}^{\mathrm{c}} \\ -\psi_{\mathrm{p}}^{\mathrm{c}} \end{pmatrix}, \tag{6.31}$$

that is, $\psi_{\mathrm{n}}^{\mathrm{c}}$ is an isospin-up state and $-\psi_{\mathrm{p}}^{\mathrm{c}}$, with its important minus sign, an isospin-down state. This result tells us that the two doublets $\psi$ and $\psi^{\mathrm{c}}$ are *equivalent* representations, transforming in the same way under isospin rotations. The two sets of transformation matrices $\{S\}$ and $\{S^*\}$ coincide after rearranging their elements through (30). Since $A = \exp(\mathrm{i}\pi\tau_2/2)$ also belongs to the set $\{S\}$, (30) is just an expression of the composition law of the transformation group.

A system of several nucleons is described by higher-dimensional representations, which may be found by applying the usual rules of angular momentum addition. Two cases will be of special interest: a pair of nucleons, and a nucleon–antinucleon system. Vector addition of two isospins $1/2$ gives two possible total isospins, $I = 0$ and $I = 1$. Composition of the corresponding states is shown in Table 6.2. A nucleon state is denoted simply by $p$ (rather than $|\mathrm{p}\rangle$) if $I_3 = 1/2$, and $n$ (rather than $|\mathrm{n}\rangle$) if $I_3 = -1/2$. Similarly, an antinucleon state is denoted by $\bar{n}$ if $I_3 = 1/2$, and $-\bar{p}$ if $I_3 = -1/2$. For two-particle states, the notation $p(-\bar{p})$, for example, means that particle 1 is

**Table 6.2.** Two-nucleon isospin states

| $I$ | $I_3$ | nucleon–nucleon | nucleon–antinucleon |
|---|---|---|---|
| | $+1$ | $pp$ | $p\bar{n}$ |
| 1 | 0 | $\frac{1}{\sqrt{2}}(pn+np)$ | $\frac{1}{\sqrt{2}}[p(-\bar{p})+n\bar{n}]$ |
| | $-1$ | $nn$ | $n(-\bar{p})$ |
| 0 | 0 | $\frac{1}{\sqrt{2}}(pn-np)$ | $\frac{1}{\sqrt{2}}[p(-\bar{p})-n\bar{n}]$ |

in an isospin-up state (a proton) and particle 2 in an isospin-down state (an antiproton).

The two-nucleon charge wave functions have a definite symmetry under a permutation of the particle isospin labels: the $I=1$ (charge triplet) functions are symmetric, while the $I=0$ (charge singlet) function is antisymmetric. Since $n$ and $p$ are to be regarded as two different states of the same particle, the total wave function of two nucleons, which is the product of a space coordinate function, a spin function, and an isospin function, must be antisymmetric, that is, must change sign under the simultaneous interchange of the particle labels. As we interchange the spatial coordinate labels, the spatial wave function of relative orbital momentum $\ell$ changes by a sign $(-)^{\ell}$; similarly, when the spin labels are interchanged, the spin wave function of total spin $S$ changes by a sign $(-)^{S+1}$, and finally, when the isospin labels are interchanged, the isospin part of total isospin $I$ contributes a sign $(-)^{I+1}$. Thus, the *generalized Pauli principle* requires a two-nucleon state $(\ell, S, J, I)$, or in spectroscopic notation ${}^{2S+1}\ell_J$, to satisfy

$$(-1)^{\ell+S+I} = -1 \quad \text{for any admissible } J\,. \tag{6.32}$$

Physical two-nucleon states are then

$$\begin{aligned} &I=1\ (\mathrm{pp,\ pn,\ nn}) &&: {}^1S_0,\ \ {}^3P_{0,1,2},\ \ {}^1D_2,\ \ {}^3F_{2,3,4}\,,\ldots\,, \\ &I=0\ (\mathrm{pn}) &&: {}^3S_1,\ \ {}^1P_1,\ \ {}^3D_{1,2,3},\ \ {}^1F_3\,,\ldots\ . \end{aligned} \tag{6.33}$$

## 6.3 Pion Field in Isospin Space

Around 140 MeV the particle mass spectrum shows three, and only three, mesons, called $\pi^+$, $\pi^-$, and $\pi^0$. It is consistent with all existing data to assume that they form a self-conjugate charge triplet: $\pi^+$ and $\pi^-$ are charge conjugates to each other, and $\pi^0$ is the charge conjugate to itself. Therefore, $\pi^+$ and $\pi^-$ must have exactly the same mass by CPT invariance, and the $\pi^0$ mass differs from the $\pi^\pm$ mass only by isospin symmetry-breaking effects. The charged $\pi$ mesons, introduced by Yukawa (1935) as agents of the nuclear forces, were discovered in cosmic rays in 1947. The existence of the neutral

$\pi$ meson, predicted by Kemmer (1938) on the basis of the concept of isospin, was confirmed by experiment in 1950.

In the isopin formalism, we regard $\pi^+$, $\pi^-$, and $\pi^0$ as three states of a certain isospin vector field, denoted by $\boldsymbol{\phi}$, which must be Hermitian to represent a self-conjugate multiplet. It may be decomposed into either three real components,

$$\phi_1\,,\ \phi_2\,,\ \text{and}\ \phi_3\,, \tag{6.34}$$

or one complex component $\varphi$ plus one real component $\phi_3$,

$$\varphi = \frac{1}{\sqrt{2}}(\phi_1 - \mathrm{i}\phi_2)\ \text{and}\ \phi_3\,. \tag{6.35}$$

Since $\phi_3$ is real, the charge operator comes from the complex field alone [cf. (2.110)]:

$$Q = \mathrm{i}\int \mathrm{d}^3x\,\left(\varphi^\dagger\dot{\varphi} - \dot{\varphi}^\dagger\varphi\right) \tag{6.36}$$

($\dot{\varphi}$ denotes a time derivative). Rewritten in terms of the real fields, it is

$$Q = \int \mathrm{d}^3x\,\left(\phi_1\dot{\phi}_2 - \phi_2\dot{\phi}_1\right) = \int \mathrm{d}^3x\,\left(\boldsymbol{\phi}\times\dot{\boldsymbol{\phi}}\right)_3\,, \tag{6.37}$$

where the final expression is meant to read as the third component of an isovector. Since by (2.112) $Q$ equals (in units of charge $e$) the difference between the particle number and the antiparticle number, it gives $+1$, $-1$, and 0 when applied respectively on a one-particle, one-antiparticle, and one neutral particle state, exactly the results expected of an isospin operator. Hence, it is natural to identify without further ado $Q$ with $I_3$, the third component of an isospin vector $\boldsymbol{I}$, whose components we postulate to be

$$\begin{aligned} I_i &= \epsilon_{ijk}\int \mathrm{d}^3x\,\phi_j\dot{\phi}_k \\ &= -\mathrm{i}\int \mathrm{d}^3x\,\dot{\boldsymbol{\phi}}^{\mathrm{T}}\boldsymbol{t}_i\boldsymbol{\phi}\,, \quad i = 1,\,2,\,3, \end{aligned} \tag{6.38}$$

where $\boldsymbol{\phi}^{\mathrm{T}}$ stands for a transposed (i.e. row) vector, and

$$(\boldsymbol{t}_i)_{jk} = -\mathrm{i}\epsilon_{ijk}\,, \quad i,j,k = 1,\,2,\,3 \tag{6.39}$$

are the elements of three Hermitian $3\times3$ matrices, with the algebraic property characteristic of representations of angular momentum operators

$$[\boldsymbol{t}_i,\,\boldsymbol{t}_j] = \mathrm{i}\epsilon_{ijk}\boldsymbol{t}_k\,. \tag{6.40}$$

The coefficients $\epsilon_{ijk}$ are referred to as the *structure constants* of the angular momentum algebra. The isovector representation, whose dimension is equal to the number of generators, is called the *adjoint* representation of the isospin group; as shown in (39) it is always possible to construct its generators such that their matrix elements are given by the structure constants. Note also the following relations, which can be verified by direct calculation,

$$(\boldsymbol{t}_i{}^2)_{jk} = \delta_{jk}(1 - \delta_{ij}), \tag{6.41}$$

$$\boldsymbol{t}_i{}^3 = \boldsymbol{t}_i \quad \text{for } i = 1,\, 2,\, 3\,. \tag{6.42}$$

Isospin invariance being assumed, all $I_i$ are conserved, constant in time, and their commutation relations with the field components $\phi_j$ can be calculated using the canonical commutation relations of fields:

$$\begin{aligned} [\dot{\phi}_i(t,\boldsymbol{x}),\, \phi_j(t,\boldsymbol{y})\,] &= -\mathrm{i}\delta_{ij}\delta(\boldsymbol{x}-\boldsymbol{y})\,, \\ [\phi_i(t,\boldsymbol{x}),\, \phi_j(t,\boldsymbol{y})\,] &= [\phi_i^\dagger(t,\boldsymbol{x}),\, \phi_j^\dagger(t,\boldsymbol{y})\,] = 0\,. \end{aligned} \tag{6.43}$$

The result turns out to be

$$[I_i,\, \phi_j(x)\,] = -(\boldsymbol{t}_i)_{jk}\phi_k(x) \ = \mathrm{i}\epsilon_{ijk}\phi_k(x)\,. \tag{6.44}$$

This gives, for example,

$$\begin{aligned} [I_1 + \mathrm{i}I_2,\, \phi_3\,] &= -(\phi_1 + \mathrm{i}\phi_2) = -\sqrt{2}\varphi^\dagger\,, \\ [I_1 - \mathrm{i}I_2,\, \phi_3\,] &= +(\phi_1 - \mathrm{i}\phi_2) = +\sqrt{2}\varphi\,. \end{aligned} \tag{6.45}$$

Using (40) and (44), and proceeding in exactly the same way as for the nucleon field studied above, one obtains the commutation relations

$$[I_i,\, I_j\,] = \mathrm{i}\epsilon_{ijk}I_k\,, \tag{6.46}$$

and proves that $I_i$ are the generators of infinitesimal rotations on the pion isospin space. The general isospin rotation rule for the field $\boldsymbol{\phi}$ is given by

$$\mathcal{U}\boldsymbol{\phi}(x)\mathcal{U}^{-1} = S(\theta)\boldsymbol{\phi}(x)\,, \tag{6.47}$$

where $\mathcal{U}$ is a unitary operator,

$$\mathcal{U} = \mathrm{e}^{\mathrm{i}\boldsymbol{\theta}\cdot\boldsymbol{I}}\,, \tag{6.48}$$

and $S$ its representation in three-dimensional isospin space,

$$S(\boldsymbol{\theta}) = \mathrm{e}^{-\mathrm{i}\boldsymbol{\theta}\cdot\boldsymbol{t}}\,. \tag{6.49}$$

This result, or its infinitesimal form (44), proves that the pion field $\boldsymbol{\phi}(x)$ transforms indeed as a vector under isospin rotations.

A rotation through an angle $\theta$ about isospin $y$ axis is brought about by the $3 \times 3$ matrix $\mathrm{e}^{-\mathrm{i}\theta \boldsymbol{t}_2}$, which can be explicitly calculated by developing the exponential in power series and selectively resumming terms with the help of $\boldsymbol{t}_2{}^3 = \boldsymbol{t}_2$, as follows:

$$\begin{aligned}
\mathrm{e}^{-\mathrm{i}\theta \boldsymbol{t}_2} &= 1 + \sum_{n=1,3,\ldots}^{\infty} \frac{(-\mathrm{i}\theta \boldsymbol{t}_2)^n}{n!} + \sum_{n=2,4,\ldots}^{\infty} \frac{(-\mathrm{i}\theta \boldsymbol{t}_2)^n}{n!} \\
&= 1 - \mathrm{i}\boldsymbol{t}_2 \sum_{n=1}^{\infty} (-)^{n-1} \frac{(\theta)^{2n-1}}{(2n-1)!} + (\boldsymbol{t}_2)^2 \sum_{n=1}^{\infty} (-)^n \frac{(\theta)^{2n}}{(2n)!} \\
&= 1 - \mathrm{i}\boldsymbol{t}_2 \sin\theta - (1 - \cos\theta)(\boldsymbol{t}_2)^2 .
\end{aligned} \tag{6.50}$$

In particular, the matrix for a rotation through an angle $\theta = \pi$ about isospin $y$ axis is given by

$$S(\pi \hat{\boldsymbol{n}}_2) = 1 - 2(\boldsymbol{t}_2)^2 = \begin{pmatrix} -1 & 0 & 0 \\ 0 & 1 & 0 \\ 0 & 0 & -1 \end{pmatrix} . \tag{6.51}$$

It flips the signs of $\phi_1$ and $\phi_3$, leaving $\phi_2$ unchanged, or equivalently, it interchanges $\varphi$ and $-\varphi^\dagger$. The corresponding operator

$$\mathcal{U}_2 = \mathrm{e}^{\mathrm{i}\pi I_2} \tag{6.52}$$

can thus be interpreted as the *charge exchange operator* for the meson field.

As we have seen in Chap. 2, the operator $\varphi^\dagger$ creates a $\pi^+$ and destroys a $\pi^-$, while $\varphi$ creates a $\pi^-$ and destroys a $\pi^+$, and the Hermitian operator $\phi_3$ creates and destroys a $\pi^0$. It remains, however, to find their exact isospin character. For this purpose, recall relation (24), which defines the standard phase convention of isospin states and which reads, in the present case,

$$\begin{aligned}
I_+ \phi_{10} &= \sqrt{2}\, \phi_{11} , \\
I_- \phi_{10} &= \sqrt{2}\, \phi_{1,-1} .
\end{aligned} \tag{6.53}$$

Identifying one-$\pi^0$ state with the isospin state $\phi_{10}$, obtained by having the field operator $\phi_3$ act on the vacuum,

$$\phi_{10} = \left|\pi^0\right\rangle = \phi_3 \left|0\right\rangle , \tag{6.54}$$

and comparing (53) with (45), which reads

$$[I_\pm , \phi_3] = \mp\sqrt{2}\, \frac{(\phi_1 \pm \mathrm{i}\phi_2)}{\sqrt{2}} , \tag{6.55}$$

we arrive at the conclusion that $\phi_{11}$ is produced by $-\varphi^\dagger$, and $\phi_{1,-1}$ by $+\varphi$:

$$\begin{aligned}
\phi_{11} &= - \left|\pi^+\right\rangle = -\varphi^\dagger \left|0\right\rangle , \\
\phi_{1,-1} &= \phantom{-} \left|\pi^-\right\rangle = \phantom{-}\varphi \left|0\right\rangle .
\end{aligned} \tag{6.56}$$

The presence of the correct signs in the above definitions is essential if one is to make use of the standard coefficients of angular momentum coupling.

As a simple application, consider a system composed of a pion and a nucleon. It may exist in states with isospin $I = 3/2$ or $I = 1/2$. With the phase factors for the nucleon wave functions and the meson wave functions defined above, the eigenvectors $\Phi_{II_3}(\pi N)$ of $\boldsymbol{I}^2$ and $I_3$ can be readily calculated and are given in Table 6.3 for various $(I, I_3)$.

**Table 6.3.** Pion–nucleon isospin states

| $I$ | $I_3$ | $\Phi_{II_3}(\pi\text{–nucleon})$ |
|---|---|---|
| $\frac{3}{2}$ | $\frac{3}{2}$ | $-\pi^+p$ |
| $\frac{3}{2}$ | $\frac{1}{2}$ | $\sqrt{\frac{2}{3}}\,\pi^0p-\sqrt{\frac{1}{3}}\,\pi^+n$ |
| $\frac{3}{2}$ | $-\frac{1}{2}$ | $\sqrt{\frac{1}{3}}\,\pi^-p+\sqrt{\frac{2}{3}}\,\pi^0n$ |
| $\frac{3}{2}$ | $-\frac{3}{2}$ | $\pi^-n$ |
| $\frac{1}{2}$ | $\frac{1}{2}$ | $-\sqrt{\frac{1}{3}}\,\pi^0p-\sqrt{\frac{2}{3}}\,\pi^+n$ |
| $\frac{1}{2}$ | $-\frac{1}{2}$ | $-\sqrt{\frac{2}{3}}\,\pi^-p+\sqrt{\frac{1}{3}}\,\pi^0n$ |

For charge independence to be valid, the Lagrangian for a model of strong interaction must be invariant to isospin rotations. If the nucleon and the pion are treated as fundamental fields, they can be described, respectively, by the isospinor $\psi$ and the isovector $\boldsymbol{\phi}$. Then, as the operators $\overline{\psi}\gamma_5\boldsymbol{\tau}\psi$, $\overline{\psi}\gamma_5\gamma_\mu\boldsymbol{\tau}\psi$, $\boldsymbol{\phi}$, and $\partial_\mu\boldsymbol{\phi}$ all transform as isovectors, various isoscalar products, which must also be Lorentz scalar, can be formed from them to modelize the pion–nucleon interaction. One such possibility is the nonderivative pseudoscalar coupling

$$\mathcal{L}_{\text{int}} = \mathrm{i}g\,\overline{\psi}(x)\gamma_5\boldsymbol{\tau}\psi(x)\cdot\boldsymbol{\phi}(x)\,. \tag{6.57}$$

It is instructive to expand it out in terms of the charge states. We have

$$\begin{aligned}\mathcal{L}_{\text{int}} &= \mathrm{i}g\,(\overline{\psi}_{\mathrm{p}}\quad \overline{\psi}_{\mathrm{n}})\,\gamma_5\begin{pmatrix}\phi_3 & \phi_1-\mathrm{i}\phi_2\\ \phi_1+\mathrm{i}\phi_2 & -\phi_3\end{pmatrix}\begin{pmatrix}\psi_{\mathrm{p}}\\ \psi_{\mathrm{n}}\end{pmatrix}\\ &= \mathrm{i}g\sqrt{2}\,(\overline{\psi}_{\mathrm{p}}\gamma_5\psi_{\mathrm{n}}\varphi+\overline{\psi}_{\mathrm{n}}\gamma_5\psi_{\mathrm{p}}\varphi^\dagger)+\mathrm{i}g(\overline{\psi}_{\mathrm{p}}\gamma_5\psi_{\mathrm{p}}-\overline{\psi}_{\mathrm{n}}\gamma_5\psi_{\mathrm{n}})\phi_3\,. \end{aligned}\tag{6.58}$$

Each term represents a basic interaction process. For example, the first term, $\mathrm{i}g\sqrt{2}\,\overline{\psi}_{\mathrm{p}}\gamma_5\psi_{\mathrm{n}}\varphi$, describes an incoming neutron transmuting into a proton while emitting a $\pi^-$ (or absorbing a $\pi^+$); its coupling strength is given by $\mathrm{i}g\sqrt{2}$. Figure 6.1 shows various possible basic coupling vertices.

More complex interactions can be constructed from these (vertex) diagrams. For example, the nucleon–nucleon forces mediated by the exchange of a single pion are represented by the diagrams shown in Fig. 6.2. With the signs and numerical factors given in (58), one can verify, to the lowest order in $g$, the charge independence of the nuclear forces, i.e. the equality of the pp, pn, and nn forces for nucleons in the same spatial configurations.

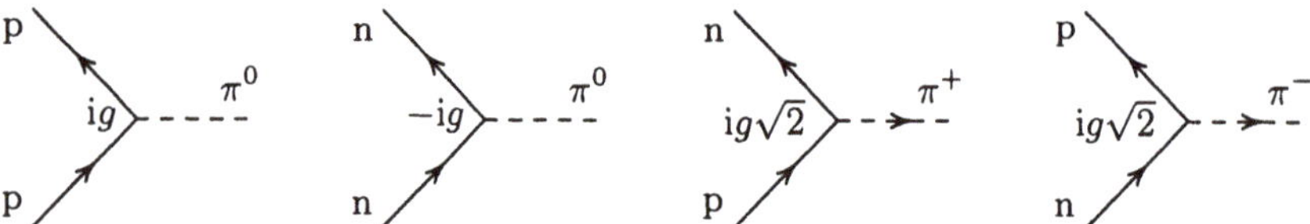


**Fig. 6.1.** Basic pion–nucleon couplings

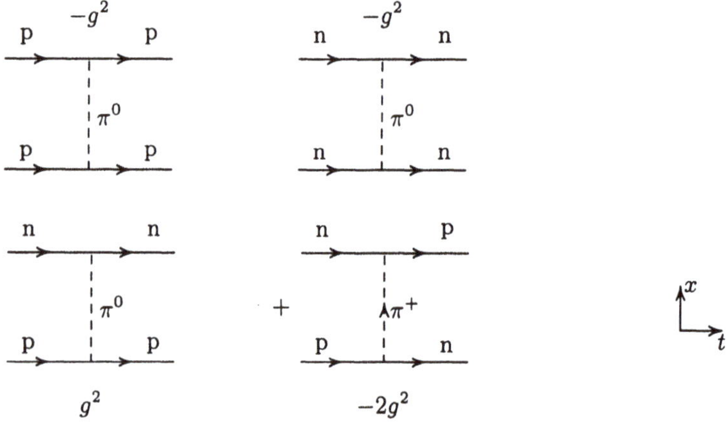


**Fig. 6.2.** Charge independence of the pp, nn, and pn forces

## 6.4 G-Parity

All hadrons have an isospin, but only for *unflavored mesons*, such as the familiar light mesons, $\pi(138)$, $\eta(547)$, $\rho(770)$, ..., or the more exotic, heavier mesons, $\eta_c(2980)$, $J/\psi(3097)$, $\Upsilon(9460)$, ..., can one define, in addition, a related quantum number called *G-parity.* By its interplay with the isospin, the G-parity gives rise to several useful selection rules in strong interaction (or equivalently hadronic) processes.

It is experimentally known that the strong interaction is invariant to isospin rotations as well as to charge conjugation. However, these two symmetry operations on charge space do not commute, and so charge states, which are eigenstates of $I_3$, are not eigenstates of the charge conjugation operator $\mathcal{C}$. It turns out, however, that the product of $\mathcal{C}$ and the charge exchange operator is isospin-invariant and hence is an observable in experiments involving charged (but unflavored) particles.

### 6.4.1 Nucleon and Pion Fields

As in the preceding sections, we start out with the nucleon and the pion fields. Let us first recall the effects of the charge conjugation $\mathcal{C}$, and the charge exchange operation $\mathcal{U}_2$. From Sect. 5.3 we have, for the nucleon field,

$$\mathcal{C}\psi_{\mathrm{p}}\mathcal{C}^{-1} = \psi_{\mathrm{p}}^{\mathrm{c}}, \qquad \mathcal{C}\psi_{\mathrm{n}}\mathcal{C}^{-1} = \psi_{\mathrm{n}}^{\mathrm{c}}, \tag{6.59}$$

and, for the pion field,

$$\begin{aligned} \mathcal{C}\varphi\mathcal{C}^{-1} &= \varphi^{\dagger} \\ \mathcal{C}\phi_3\mathcal{C}^{-1} &= \phi_3 \end{aligned} \qquad \text{or} \qquad \mathcal{C}\begin{pmatrix}\phi_1\\ \phi_2\\ \phi_3\end{pmatrix}\mathcal{C}^{-1} = \begin{pmatrix}\phi_1\\ -\phi_2\\ \phi_3\end{pmatrix}. \tag{6.60}$$

Note that we have chosen the phases such that $\xi_F = 1$ for both the neutron and the proton, and $\xi_B = 1$ for both $\pi^{\pm}$ and $\pi^0$. On the other hand, the action of the charge exchange operation or, equivalently, the rotation by 180° about the isospin $y$ axis, $\mathcal{U}_2 = \exp(\mathrm{i}\pi I_2)$, yields

$$\mathcal{U}_2\psi\mathcal{U}_2^{-1} = \mathrm{e}^{-\frac{1}{2}\pi\tau_2}\begin{pmatrix}\psi_{\mathrm{p}}\\ \psi_{\mathrm{n}}\end{pmatrix} = \begin{pmatrix}-\psi_{\mathrm{n}}\\ \psi_{\mathrm{p}}\end{pmatrix} \tag{6.61}$$

and

$$\mathcal{U}_2\phi\mathcal{U}_2^{-1} = \mathrm{e}^{-\mathrm{i}\pi t_2}\begin{pmatrix}\phi_1\\ \phi_2\\ \phi_3\end{pmatrix} = \begin{pmatrix}-\phi_1\\ \phi_2\\ -\phi_3\end{pmatrix}. \tag{6.62}$$

Now, we introduce the *G-conjugation* operator

$$\mathcal{G} \equiv \mathcal{C}\mathcal{U}_2 = \mathcal{C}\mathrm{e}^{\mathrm{i}\pi I_2}. \tag{6.63}$$

Its effects on fields are the results of the combined action of $\mathcal{C}$ and $\mathcal{U}_2$ that we have just described. Therefore,

$$\mathcal{G}\begin{pmatrix}\psi_{\mathrm{p}}\\ \psi_{\mathrm{n}}\end{pmatrix}\mathcal{G}^{-1} = \begin{pmatrix}-\psi_{\mathrm{n}}^{\mathrm{c}}\\ \psi_{\mathrm{p}}^{\mathrm{c}}\end{pmatrix} \qquad \text{or} \qquad \mathcal{G}\psi\mathcal{G}^{-1} = -\psi^{\mathrm{c}}, \tag{6.64}$$

$$\mathcal{G}\begin{pmatrix}\phi_1\\ \phi_2\\ \phi_3\end{pmatrix}\mathcal{G}^{-1} = \begin{pmatrix}-\phi_1\\ -\phi_2\\ -\phi_3\end{pmatrix} \qquad \text{or} \qquad \mathcal{G}\phi\mathcal{G}^{-1} = -\phi. \tag{6.65}$$

This shows that an isospin representation acted on by $\mathcal{G}$ results in (minus) its own charge conjugate, which has the same isospin but, in general, different additive quantum numbers. As these numbers (except $Q$) vanish for the

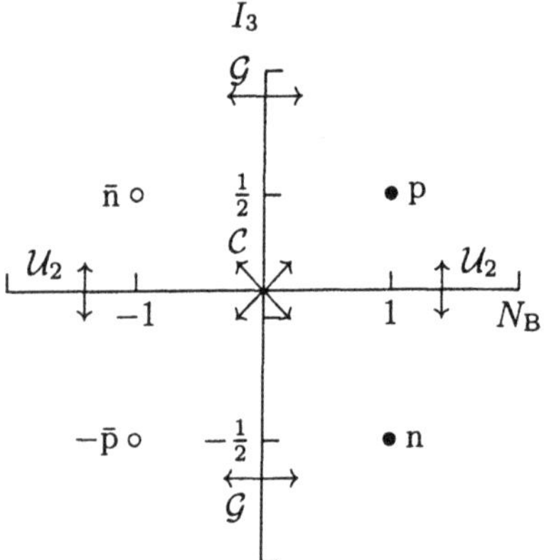


**Fig. 6.3.** Action of the conjugation operators $\mathcal{C}$, $\mathcal{U}_2$, and $\mathcal{G}$ on the nucleon field

pion, for which $\phi = \phi^c$, the action of $\mathcal{G}$ on the pion field appears merely as an inversion operation in isospin space (cf. Figs. 6.3 and 6.4).

Next, we seek the transformation rule for isospin under $\mathcal{G}$. The explicit expression for the nucleon isospin is given in (8) :

$$I_i^{(N)} = \tfrac{1}{2} \int d^3x \, \psi^\dagger \tau_i \psi \quad \text{for } i = 1, 2, 3 . \tag{6.66}$$

From the charge conjugation transforms of bilinear covariants listed in Table 5.3 and from the properties of the Pauli matrices,

$$\begin{aligned} \tau_i^T &= s_i \tau_i , \qquad \tau_2 \tau_i \tau_2 = -\tau_i^T , \\ s_i &= +1, \ i = 1, 3, \quad \text{and} \quad s_2 = -1 \end{aligned} \tag{6.67}$$

($\tau^T$ means transpose of $\tau$), we obtain

$$\mathcal{C} I_i^{(N)} \mathcal{C}^{-1} = -s_i I_i^{(N)} , \tag{6.68}$$

$$\mathcal{U}_2 I_i^{(N)} \mathcal{U}_2^{-1} = -s_i I_i^{(N)} . \tag{6.69}$$

Therefore,

$$\mathcal{G} I_i^{(N)} \mathcal{G}^{-1} = I_i^{(N)} \quad \Rightarrow \quad [\mathcal{G}, \boldsymbol{I}^{(N)}] = 0 . \tag{6.70}$$

$\mathcal{G}$ commutes with *all* components of the isospin vector, but $\mathcal{C}$ commutes only with its $y$ component, $I_2$, and so also with $\mathcal{U}_2$.

As for the pion field, we start from (38),

$$I_i^{(\pi)} = \epsilon_{ijk} \int d^3x \, \phi_j \dot{\phi}_k , \quad i = 1, 2, 3 , \tag{6.71}$$

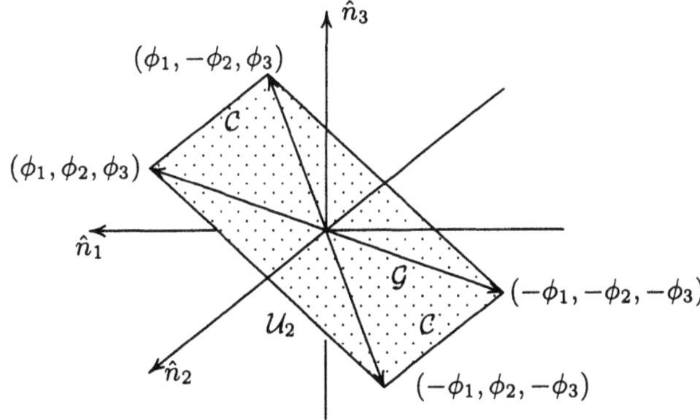


**Fig. 6.4.** Action of the conjugation operators $\mathcal{C}$, $\mathcal{U}_2$, and $\mathcal{G}$ on the pion field

and, making use of the transformation properties of the pion field, we can derive results similar to those for the nucleon field, namely, that $\boldsymbol{I}^{(\pi)}$ commutes with $\mathcal{G}$, but not with $\mathcal{C}$ nor with $\mathcal{U}_2$. Thus, provided the system consists only of nucleons, antinucleons, and pions, $\mathcal{G}$ is isospin-invariant:

$$[\mathcal{G}, \boldsymbol{I}] = 0 \tag{6.72}$$

for $\boldsymbol{I} = \boldsymbol{I}^{(\mathrm{N})} + \boldsymbol{I}^{(\pi)}$. This result must also hold for any field that can be reached by any strong interaction reaction from an initial state composed only of nucleons, antinucleons, and pions, because the strong interaction, being both isospin-invariant and C-invariant, and thus G-invariant, must preserve the validity of (72) all through the reaction. And so we will assume (72) to hold in general, for an isospin $\boldsymbol{I}$ that includes any hadrons.

Will $\mathcal{G}$ lead then to a conserved quantum number? Not necessarily. As seen above, the nucleon field transforms as

$$\mathcal{G}\psi\mathcal{G}^{-1} = -\psi^{\mathrm{c}} = -AC\overline{\psi}^{\mathrm{T}}, \tag{6.73}$$

and so $\mathcal{G}$ changes the baryon number. More generally, noting that the isospinor adjoint has the property

$$\mathcal{G}\overline{\psi}\mathcal{G}^{-1} = \psi^{\mathrm{T}}C^{\dagger}A^{\dagger}, \tag{6.74}$$

we have for any operator $\Omega$ the transformation rule

$$\mathcal{G}\overline{\psi}_1\Omega\psi_2\mathcal{G}^{-1} = \overline{\psi}_2 CA\Omega^{\mathrm{T}}A^{\dagger}C^{\dagger}\psi_1 . \tag{6.75}$$

In particular, for the frequently encountered cases of $\Omega = \Gamma$ and $\Omega = \Gamma\tau_j$, the bilinear covariants of isospinors transform under $\mathcal{G}$ as follows:

$$\overline{\psi}_1\Gamma^i\psi_2 \quad\to\quad \xi_c\overline{\psi}_2\Gamma^i\psi_1 \qquad \text{(isoscalar)}, \tag{6.76}$$

$$\overline{\psi}_1\Gamma^i\boldsymbol{\tau}\psi_2 \quad\to\quad -\xi_c\overline{\psi}_2\Gamma^i\boldsymbol{\tau}\psi_1 \qquad \text{(isovector)}, \tag{6.77}$$

$$\xi_c = +1 \quad \text{for } i = \mathrm{S, P, A}; \qquad \xi_c = -1 \quad \text{for } i = \mathrm{V, T}.$$

It follows that an *isovector current* commutes with $\mathcal{G}$,

$$\boldsymbol{J}_\mu^{(1)}(x) = \tfrac{1}{2}\overline{\psi}(x)\gamma_\mu \boldsymbol{\tau}\psi(x), \qquad \mathcal{G}\boldsymbol{J}_\mu^{(1)}\mathcal{G}^{-1} = \boldsymbol{J}_\mu^{(1)}, \tag{6.78}$$

but an *isoscalar current* does not,

$$J_\mu^{(0)}(x) = \tfrac{1}{2}\overline{\psi}(x)\gamma_\mu \psi(x), \qquad \mathcal{G}J_\mu^{(0)}\mathcal{G}^{-1} = -J_\mu^{(0)}. \tag{6.79}$$

As the electromagnetic current normally consists of an isoscalar and an isovector component, it is not invariant to $\mathcal{G}$: the G-operation is not a symmetry of the electromagnetic interaction. By the same token, neither the associated electric charge,

$$Q = \frac{1}{2}\int \mathrm{d}^3x\, \overline{\psi}\gamma^0(1+\tau_3)\psi\,, \tag{6.80}$$

nor any purely isoscalar charge, such as the baryon number,

$$N_\mathrm{B} = \int \mathrm{d}^3x\, \overline{\psi}\gamma^0\psi\,, \tag{6.81}$$

commutes with $\mathcal{G}$. This is because, as noted above, $\mathcal{G}$ interchanges nucleon and antinucleon states. It follows that an arbitrary system of hadrons and antihadrons does not admit eigenstates of $\mathcal{G}$. The notable exception is when all of the net additive quantum numbers of the system, apart from $Q$, vanish. The electric charge $Q$ may be nonvanishing because in this situation it exactly coincides with $I_3$, which always commutes with $\mathcal{G}$. An eigenvalue of $\mathcal{G}$, when it exists, is referred to as the *G-parity*. Unflavored (that is, nonstrange, with no charm nor beauty) mesons ($N_\mathrm{B} = 0$) may thus have a G-parity; for instance, the pion has an odd G-parity. We may regard the notion of G-conjugation parity as a generalization to charged but unflavored systems of the notion of charge conjugation parity, which is relevant only to completely neutral states. But, unlike the $C$ operation, $\mathcal{G}$ is not a symmetry of the electromagnetic interaction, just one of the strong interactions.

A simple example of states with well-defined G-parity is a system of pions. The G-parity of one pion being $-1$, that of a system of $n$ pions in an arbitrary spatial configuration must be

$$G(n\,\pi) = (-)^n \quad \text{for an } n\,\pi \text{ state}. \tag{6.82}$$

Consider now a system of nucleon–antinucleon. Its nonrelativistic wave function of relative orbital angular momentum $\ell$, spin $S$, and isospin $I$ may be written, in analogy with (5.104),

$$|\psi_{\ell,S,I}(\mathrm{N\bar{N}})\rangle = \int \mathrm{d}^3p \sum_{sts't'} F_{\ell SI}^{sts't'}(\boldsymbol{p})\, b^\dagger(\boldsymbol{p},s,t) d^\dagger(-\boldsymbol{p},s',t')\, |0\rangle\,, \tag{6.83}$$

where $s$ and $t$ are eigenvalues of the third components of individual spins and isospins, respectively, and

$b^\dagger(t = \ {}^1\!/_2)$ creates a proton (p),
$b^\dagger(t = -{}^1\!/_2)$ creates a neutron (n),
$d^\dagger(t = \ {}^1\!/_2)$ creates an antineutron ($\bar{\text{n}}$),
$d^\dagger(t = -{}^1\!/_2)$ creates an antiproton ($-\bar{\text{p}}$).

Being an unflavored state, a nucleon–antinucleon pair has a well-defined G-parity. The calculation of this number proceeds as in Chap. 5 for a similar calculation of the C-parity. What is new is the presence of the signs in

$$\begin{aligned} \mathcal{G}\psi\mathcal{G}^{-1} &= -\psi^{c} \quad \Rightarrow \quad \mathcal{G}b^\dagger(t)\mathcal{G}^{-1} = -d^\dagger(t), \\ \mathcal{G}\psi^{c}\mathcal{G}^{-1} &= \ \psi \quad \Rightarrow \quad \mathcal{G}d^\dagger(t)\mathcal{G}^{-1} = +b^\dagger(t). \end{aligned} \tag{6.84}$$

Applied on $\psi_{\ell,S,I}$ the G-conjugation produces the following sign factors:

$(-)^1$ from anticommuting two fermion operators,
$(-)^1$ from G-transform, given by (84),
$(-)^{\ell+S+1}$ from exchanging the particle space–spin labels,
$(-)^{I+1}$ from exchanging the particle isospin labels.

Given these facts, one readily finds the G-parity of a nucleon–antinucleon (or, in fact, of any isodoublet–anti-isodoublet) system

$$G(\text{N}\bar{\text{N}}) = (-)^{\ell+S+I} = C(-)^I. \tag{6.85}$$

For illustration, first take the multipion (strong) production reaction

$$\pi + \pi \to n\pi. \tag{6.86}$$

It follows from (82) that $n$ must be an even number. More generally, any Feynman diagram with an odd number of external pion lines and no other external particle lines must vanish. This is analogous to Furry's theorem for photons, discussed in the previous chapter.

The G selection rule for the strong production reaction $\text{N} + \bar{\text{N}} \to n\pi$ is $(-)^{\ell+S+I} = (-)^n$. States with even $\ell + S + I$ can decay only into an even number of pions, and states with odd $\ell + S + I$ can decay only into an odd number of pions. For example, consider a $\bar{\text{p}}$n system, which has a nonvanishing net charge, and thus no C-parity, but may have a G-parity. It is necessarily an isotriplet (since $I_3 = -1$). If it is in a ${}^1S_0$ state, $\ell+S+I = 1$ is odd and its G-parity is negative; so it cannot decay into $2\pi$, $4\pi, \ldots$. On the other hand, if it is in a ${}^3S_0$ state, $\ell + S + I = 2$ is even and its G-parity is positive; hence decays into $3\pi$, $5\pi, \ldots$ are all forbidden.

### 6.4.2 Other Unflavored Hadrons

The particle $\eta^0$ (548 MeV) decays mainly through the following modes:

$$\begin{aligned}\eta^0 &\to 2\gamma && (39\%)\,,\\ &\to 3\pi^0 && (32\%)\,,\\ &\to \pi^+\pi^-\pi^0 && (23\%)\,.\end{aligned}$$

The $\eta^0 \to 2\gamma$ decay is evidently electromagnetic. Since the other two modes have comparable rates, they too must be electromagnetic, involving virtual photons. The observed distribution of events in the $\eta^0 \to \pi^+\pi^-\pi^0$ is consistent with the $\eta^0$ spin being 0; and the $\eta^0 \to 2\gamma$ decay tells us that, just as in the case of $\pi^0$, its P-parity is negative and its C-parity positive. In other words, $\eta^0$ is a *pseudoscalar meson*, $J^{PC} = 0^{-+}$. Since no charged partners have been detected, it has to be an isosinglet (which makes it essentially different from $\pi^0$). Since $I = 0$, its G-parity is identical to its C-parity, which is +1. The notation $I^G = 0^+$ encodes these data. G-invariance would then allow hadronic decay to a state composed of an even number of pions. In particular, $\eta^0 \to 2\pi$ would be the predominant mode, favored by kinematics. However, this decay could not conserve both angular momentum and ordinary parity: the initial state being $J^P = 0^-$, angular momentum conservation requires the orbital angular momentum of the final pion state to be $\ell = 0$, which means a final parity +1, rather than −1 as demanded by parity conservation. This strong interaction mode is thus forbidden. The strong decay of $\eta$ must then proceed through an emission of four pions. But to have $J^P = 0^-$, the pions in the final state must have some orbital angular momentum, which is hardly kinematically possible, given the very small momentum available to the decay. Therefore, the next most likely decay channels must be to final $3\pi$ states of negative G-parity, which, although forbidden to the strong interaction, are open to the *G-parity-breaking* electromagnetic interaction. This is consistent with the observed partial widths being of the same magnitude as the full width. In principle, $\eta \to \pi\pi$ may take place through the P-parity-breaking weak interaction and, being strangeness-conserving, it is even more favored than the analogous weak decay $K^0 \to \pi\pi$; however, that decay mode is strongly suppressed (branching$< 1.5 \times 10^{-3}$) because it violates not only P-parity but also CP symmetry (see Table 5.2).

*Vector mesons* can be produced in pion–nucleon or proton–antiproton reactions as resonating multipion states, which rapidly decay into real pions. This is the case, for example, of the $\rho$ mesons, $\rho^\pm$ and $\rho^0$, which have approximately the same mass (770 MeV) and therefore must form an isospin triplet. They mainly decay into two pions. The large value of the observed width ($\Gamma = 151$ MeV) indicates that the $\rho \to \pi\pi$ decay proceeds via strong interactions. Since strong interactions are G-invariant, the G-parity of the $\rho$ mesons is well defined and must be even, $G = +1$. As an $I = 1$, $I_3 = 0$ state changes sign under $\mathcal{U}_2$, the $\rho^0$ meson must be odd under charge conjugation. (Needless to say, its charged partners have no C-parities.) On the other hand, the vector meson $\omega^0$ (782 MeV) is an isosinglet. Its main decay mode $\omega^0 \to \pi^+\pi^-\pi^0$ (branching ratio: 89%) occurs via strong interactions, so its G-parity is odd. Since $I = 0$, its C-parity must be $C = G = -1$.

The results of this section are summarized in Table 6.4.

**Table 6.4.** G-parities of unflavored systems

| | $\pi$ | $\eta$ | $\rho$ | $\omega$ | $n\pi$ | $\mathrm{N\bar{N}}$ |
|---|---|---|---|---|---|---|
| $G$ | $-1$ | $+1$ | $+1$ | $-1$ | $(-)^n$ | $(-)^{\ell+S+I}$ |

## 6.5 Isospin of Strange Particles

For particles with a net flavor, neither the C-parity nor the G-parity can be defined. We limit our discussion for now to strange particles, leaving the study of other flavors to later chapters. In the absence of any known charged partners, $\Lambda^0$ is taken to be an isosinglet, $I = 0$. It is produced together with the meson $\mathrm{K}^0$ in the strong reaction

$$\begin{array}{rccccccc} & \pi^- & + & \mathrm{p} & \to & \Lambda^0 & + & \mathrm{K}^0 \;, \\ I: & 1 & & 1/2 & & 0 & & 1/2,\, 3/2 \end{array}$$

where the particle isospins are indicated on the second line. By isospin conservation, the isospin of the kaon must be either $1/2$ or $3/2$. The four observed nearly degenerate states $\mathrm{K}^+$, $\mathrm{K}^0$, $\mathrm{K}^-$, and $\bar{\mathrm{K}}^0$ may form either two isospin doublets or one isospin quadruplet. But since the pairs $\mathrm{K}^+$, $\mathrm{K}^0$ and $\mathrm{K}^-$, $\bar{\mathrm{K}}^0$ have the strangeness quantum numbers opposite in signs, they must form two distinct isospin doublets ($I = 1/2$),

$$\begin{pmatrix} \mathrm{K}^+ \\ \mathrm{K}^0 \end{pmatrix} \quad \text{and} \quad \begin{pmatrix} \bar{\mathrm{K}}^0 \\ -\mathrm{K}^- \end{pmatrix}, \tag{6.87}$$

which transform under isospin rotations like the nucleon doublet $\psi$ and the antinucleon doublet $\psi^c$, respectively.

We have seen from the above that the charge $Q$ of a nucleon is related to its isospin $I_3$ by

$$Q = I_3 + \tfrac{1}{2}\, N_\mathrm{B}\,, \tag{6.88}$$

where $N_\mathrm{B} = 1$. This relation, which also holds for mesons provided $N_\mathrm{B} = 0$, is not valid for $\Lambda^0$, or the kaons, or any other strange particles, and therefore is not complete. But the modified relation

$$Q = I_3 + \tfrac{1}{2}\,(N_\mathrm{B} + S) \quad \text{(Gell-Mann–Nishijima relation)} \tag{6.89}$$

holds for both strange and nonstrange particles. It is evident from this formula that any interaction that conserves $N_\mathrm{B}$, $Q$, and $I_3$ also conserves $S$.

Consider now the following strong reaction:

| | $\pi^\pm$ | + | p | $\to$ | $\Sigma^\pm$ | + | $\mathrm{K}^+$. |
|---|---|---|---|---|---|---|---|
| $I_3$ : | $\pm 1$ | | $1/2$ | | $\pm 1$ | | $1/2$ |
| $S$ : | 0 | | 0 | | $-1$ | | 1 |
| $N_\mathrm{B}$ : | 0 | | 1 | | 1 | | 0 |
| $Q$ : | $\pm 1$ | | 1 | | $\pm 1$ | | 1 |

From the known quantum numbers of p, $\pi^\pm$, and $\mathrm{K}^+$, one deduces that $\Sigma^\pm$ has baryon number $N_\mathrm{B} = 1$, strangeness $S = -1$ and isospin $I$ either equal to 1 or 2. Since $\Sigma^\pm$ and $\Sigma^0$ are the only degenerate states (with a mass difference less than 0.5%) around 1190 MeV with such quantum numbers, they must be viewed as forming an isotriplet, distinct from the triplet of the corresponding antiparticles, $\bar{\Sigma}^\pm$ and $\bar{\Sigma}^0$, which have strangeness $S = 1$.

The production reaction

$$\mathrm{K}^- \; + \; \mathrm{p} \; \to \; \Xi^- \; + \; \mathrm{K}^+$$

is useful in identifying the quantum numbers of $\Xi$ from the characteristics of the other known particles. Thus, $\Xi^-$ must have $N_\mathrm{B} = 1$, $S = -2$, and isospin either $1/2$ or $3/2$. Observations are consistent with the assignment $I = 1/2$, its isospin-up neutral partner $I_3 = 1/2$ being $\Xi^0\,(1315)$. They differ in mass by $m_{\Xi^-} - m_{\Xi^0} = 6.4$ MeV. Together with their antiparticles they form two isospin doublets, charge conjugate to each other,

$$\begin{pmatrix} \Xi^0 \\ \Xi^- \end{pmatrix} \quad \text{and} \quad \begin{pmatrix} \bar{\Xi}^- \\ -\bar{\Xi}^0 \end{pmatrix} . \tag{6.90}$$

As a final example, consider

$$\mathrm{K}^- + \mathrm{p} \to \Omega^- + \mathrm{K}^+ + \mathrm{K}^0 .$$

Here the unknown particle $\Omega^-$ must have $N_\mathrm{B} = 1$, $S = -3$. Since there are no other particles with such characteristics around 1672 MeV, where it is observed, it must have isospin $I = 0$.

It has been found useful in discussions involving strange particles to introduce the *hypercharge*

$$Y = N_\mathrm{B} + S = 2(Q - I_3) . \tag{6.91}$$

Whenever $N_\mathrm{B}$ and $S$ or $Q$ and $I_3$ are conserved, $Y$ is also conserved. It is the simplest way to combine charge conservation with isospin invariance, and we shall see later that generalized $Y$ plays a fundamental role in local symmetry considerations. In Table 6.5, we summarize the values of the isospins and hypercharges of the light baryons $J^P = 1/2^+$, $N_\mathrm{B} = 1$ and light pseudoscalar mesons $J^P = 0^-$, $N_\mathrm{B} = 0$. The remarkable parallelism between the two sets of particles and the approximate mass degeneracies within each of the two groups strongly suggests that there must exist some common underlying symmetry among these particles.

**Table 6.5.** Low-lying states of baryons and pseudoscalar mesons

| $Y$ | $I$ | | | | | | Average masses |
|---|---|---|---|---|---|---|---|
| | $I_3$ : | $-1$ | $-1/2$ | $0$ | $1/2$ | $1$ | (MeV) |
| 1 | $1/2$ | | n | | p | | 938 |
| 0 | 1 | $\Sigma^-$ | | $\Sigma^0$ | | $\Sigma^+$ | 1193 |
| 0 | 0 | | | $\Lambda^0$ | | | 1115 |
| $-1$ | $1/2$ | | $\Xi^-$ | | $\Xi^0$ | | 1318 |
| 1 | $1/2$ | | $K^0$ | | $K^+$ | | 496 |
| 0 | 1 | $\pi^-$ | | $\pi^0$ | | $\pi^+$ | 138 |
| 0 | 0 | | | $\eta^0$ | | | 548 |
| $-1$ | $1/2$ | | $K^-$ | | $\bar{K}^0$ | | 496 |

## 6.6 Isospin Violations

Although a well-established symmetry of the strong interaction, isospin is not conserved in electromagnetic and weak interactions. Nevertheless, the systematic way in which the symmetry is broken has led to several empirical rules which may serve to check observations and to put constraints on models.

### 6.6.1 Electromagnetic Interactions

We know from the beginning that the electromagnetic interaction is not exactly charge independent, not even exactly charge symmetric, and therefore not invariant under any isospin rotation generated by $I_1$ or $I_2$. However, since reactions such as $\mathrm{p} \leftrightarrow \mathrm{p}+\gamma$ can take place, $I_3$ is possibly conserved, which turns out indeed to be the case. In what follows, the symbol $\Delta A$, where $A$ is any quantum number, always means the change in the total $A$ *of the hadrons.* Thus, in electromagnetic processes, $\Delta I_3 = 0$ and therefore, since $\Delta Q = \Delta N_{\mathrm{B}} = 0$, strangeness is also conserved, $\Delta S = 0$, by Gell-Mann–Nishijima's relation. Furthermore, the Lagrangian for the free-nucleon field and the free-pion field is invariant to the global gauge transformations of the isospin fields

$$\begin{aligned} \psi &\to \psi - \frac{\mathrm{i}}{2}\varepsilon(1+\tau_3)\psi\,, \\ \boldsymbol{\phi} &\to -\varepsilon(\boldsymbol{\phi}\times\hat{\boldsymbol{n}}_3)\,, \end{aligned} \tag{6.92}$$

where $\hat{\boldsymbol{n}}_3$ is a unit vector in isospin $z$ direction. The associated Noether current is the conserved electromagnetic current (cf. Problem 6.4):

$$J_\mu(x) = J_\mu^{(0)}(x) + J_\mu^{(1)}(x)\,, \tag{6.93}$$

$$J_\mu^{(0)}(x) = \tfrac{1}{2}\,\overline{\psi}(x)\gamma_\mu\psi(x)\,, \tag{6.94}$$

$$J_\mu^{(1)}(x) = \tfrac{1}{2}\,\overline{\psi}(x)\gamma_\mu\tau_3\psi(x) + (\boldsymbol{\phi}\times\partial_\mu\boldsymbol{\phi})\cdot\hat{\boldsymbol{n}}_3\,. \tag{6.95}$$

Its coupling to the electromagnetic field, $J_\mu(x)A^\mu(x)$, is correspondingly given by the sum of a term, $H^{(0)}$, invariant to isospin rotations, and a term, $H^{(1)}$, which transforms as the third component of an isovector. It follows that $\Delta I_3 = 0$, and that $H^{(0)}$ leaves the isospin of the system unchanged while $H^{(1)}$ changes the isospin by $\Delta I = 0, \pm 1$. (By definition, $\Delta I = I_\mathrm{f} - I_\mathrm{i}$, where $I_\mathrm{i}$ and $I_\mathrm{f}$ are the total isospins of the hadrons in the initial and final states, respectively.) Reactions $\mathrm{p} \to \mathrm{p} + \gamma$ and $\eta \to 2\gamma$ are examples of $\Delta I = 0$ isospin-conserving transitions, whereas $\pi^0 \to 2\gamma$ and $\Sigma^0 \to \Lambda^0 + \gamma$ are examples of $|\Delta I| = 1$ isospin-changing processes. (The isospin of $\gamma$ is conventionally set to 0.) In contrast, $|\Delta I_3| = 1/2$ processes, e.g. $\Sigma^+ \to \mathrm{p} + \gamma$ and $\Lambda \to n + \gamma$, are forbidden. In summary, the selection rules for the first-order electromagnetic transitions are

$$\Delta I_3 = 0\,, \quad \Delta I = 0, \pm 1 \qquad \text{(electromagnetic transitions)}. \tag{6.96}$$

### 6.6.2 Weak Interactions

It is conventional to refer to weak interaction events as being *leptonic, semi-leptonic*, or *nonleptonic* depending on whether they involve leptons only, both leptons and hadrons, or hadrons only, as shown by examples in Table 6.6. As we have seen in Chap. 5, they all obey, to first order of the weak couplings, the selection rule $\Delta S = 0, \pm 1$. Moreover, as indicated by the occurrence of the neutron $\beta$-decay, isospin invariance is clearly violated in weak processes. How is isospin conservation violated? Are there well-defined rules?

We have seen in the previous chapter that $\beta$-decay may be described with great accuracy by an interaction of the form

$$\mathcal{H}_\beta = \frac{G_\mathrm{F}}{\sqrt{2}}\left[\overline{\psi}_\mathrm{n}\gamma_\lambda(1-\alpha\gamma_5)\psi_\mathrm{p}\right]^\dagger \overline{\psi}_\mathrm{e}\gamma^\lambda(1-\gamma_5)\psi_\nu + \mathrm{h.c.}\,. \tag{6.97}$$

From this and similar results from analyses of other weak decays (e.g. $\mu$ and $\pi$ decays), it is tempting to suppose that all weak interactions might be described by a 'universal' interaction Hamiltonian of the form

$$H_\mathrm{weak} = \frac{G}{\sqrt{2}}\int \mathrm{d}^3x\, J_\mu^\dagger J^\mu\,. \tag{6.98}$$

Here $J_\lambda \equiv L_\lambda + H_\lambda$ is the full weak current, arising from leptons and hadrons. The Hamiltonian (98) represents all known weak interactions at low energies: $L_\lambda^\dagger L^\lambda$ describes the leptonic processes; $L_\lambda^\dagger H^\lambda + H_\lambda^\dagger L^\lambda$, the semileptonic

**Table 6.6.** Classification of weak interaction processes

| Processes | $\Delta S = 0$ | $\lvert\Delta S\rvert = 1$ |
|---|---|---|
| Leptonic | | |
| | $\mu^+ \to e^+ \nu_e \bar{\nu}_\mu$ | |
| | $\nu_e e^- \to \nu_e e^-$ | |
| Semileptonic | | |
| | $n \to p e^- \bar{\nu}_e$ | $K^+ \to \mu^+ \nu_\mu,\ e^+ \nu_e$ |
| | $\pi^+ \to \mu^+ \nu_\mu,\ e^+ \nu_e$ | $K^+ \to \pi^0 e^+ \nu_e$ |
| | $\pi^+ \to \pi^0 e^+ \nu_e$ | $K^0 \to \pi^- e^+ \nu_e$ |
| | $\nu p \to \nu p \pi^0$ | $\Lambda^0 \to p e^- \bar{\nu}_e$ |
| Nonleptonic | | |
| | | $K^+ \to \pi^+ \pi^0, \pi^+ \pi^+ \pi^-$ |
| | | $\Lambda^0 \to p \pi^-, n \pi^0$ |
| | | $\Sigma^+ \to p \pi^0, n \pi^+$ |
| | | $\Xi^- \to \Lambda \pi^-$ |
| | | $\Omega^- \to \Lambda K^-,\ \Xi^0 \pi^-$ |

processes; and $H_\lambda^\dagger H^\lambda$, the nonleptonic processes. But whereas the leptonic current is experimentally well established to be of the V–A form:

$$L_\lambda = \overline{\psi}_e \gamma_\lambda (1 - \gamma_5) \psi_{\nu_e} + \overline{\psi}_\mu \gamma_\lambda (1 - \gamma_5) \psi_{\nu_\mu} + \text{other lepton types}\,,$$

the weak current $H_\lambda$ for hadrons is not well known, except in that it contains strangeness-conserving and strangeness-changing terms, both of which have the V–A form:

$$\begin{aligned} H_\lambda &= H_\lambda^{\Delta S=0} + H_\lambda^{\Delta S=1} \\ &= V_\lambda^{\Delta S=0} - A_\lambda^{\Delta S=0} + V_\lambda^{\Delta S=1} - A_\lambda^{\Delta S=1}\,. \end{aligned}$$

All four terms are needed because the neutron $\beta$-decay rate is determined by the matrix element $\langle n \,|\, V_\lambda^{\Delta S=0} - A_\lambda^{\Delta S=0} \,|\, p \rangle$, while the $K^+ \to \mu^+ + \nu$ decay amplitude involves $\langle 0 \,|\, A_\lambda^{\Delta S=1} \,|\, K \rangle$ and the $K^+ \to \pi^0 + e^+ + \nu$ decay amplitude involves $\langle \pi \,|\, V_\lambda^{\Delta S=1} \,|\, K \rangle$. In the neutron $\beta$-decay, isospin is violated by $|\Delta I_3| = 1$, so that the $H_\lambda^{\Delta S=0}$ current behaves as an *isovector*, $\Delta I = 1$ (assuming $\Delta I \geq 2$ to be suppressed). On the other hand, the hadronic matrix elements of $\Delta S = 1$ decays indicate that $\Delta I_3 = \frac{1}{2}$, so that the $H_\lambda^{\Delta S=1}$ current could be characterized by $\Delta I = \frac{1}{2}$ or $\frac{3}{2}$. It is postulated (by Gell-Mann and Feynman) that the currents with $\Delta I_3 = \frac{1}{2}$ and $\Delta I = \frac{3}{2}$ do not exist. The strangeness-changing current for hadrons, $H_\lambda^{\Delta S=1}$, should behave as an isospin-$1/2$ operator.

**Nonleptonic weak decays** involve only hadrons and so it is always true that $\Delta N_{\mathrm{B}} = 0$ and $\Delta Q = 0$.

The $\Delta S = 0$ weak processes are described by

$$H_{\lambda}^{(\Delta S=0)\dagger} H^{\lambda(\Delta S=0)} + H_{\lambda}^{(\Delta S=1)\dagger} H^{\lambda(\Delta S=1)}.$$

They include weak interaction contributions to nuclear forces and are very difficult to detect, because they are masked by the ordinary nuclear effects.

The $\Delta S = 1$ weak processes, which have more relevance to the present discussion, are described by

$$H_{\lambda}^{(\Delta S=0)\dagger} H^{\lambda(\Delta S=1)} + H_{\lambda}^{(\Delta S=1)\dagger} H^{\lambda(\Delta S=0)}.$$

We have seen in the above discussion that $H_{\lambda}^{(\Delta S=0)}$ behaves as an isovector and $H_{\lambda}^{(\Delta S=1)}$ behaves as an isospinor. So their product gives rise to isospins $1/2$ and $3/2$. It is experimentally observed that transitions $|\Delta I| = 3/2$ are suppressed. We will refer to this empirical fact as the $|\Delta I| = 1/2$ rule. It is worth emphasizing that this selection rule – that first-order nonleptonic strangeness-changing weak transitions occur via $|\Delta I| = 1/2$ – is distinct from Gell-Mann–Feynman's postulate that the strangeness-changing current for hadrons transforms as an isospin-$1/2$ operator. The latter can now be understood in the context of the standard model, whereas the former as yet has no satisfactory explanations (see however Chap. 16). Let us illustrate the situation by a few examples.

• $\mathrm{K}^0$ *Decay.* Consider first the neutral kaon decay

$$\mathrm{K}^0 \to \pi^+\pi^-, \quad \pi^0\pi^0.$$

In both transitions, conservation of angular momentum requires that the relative orbital angular momentum of the final state be $\ell = 0$, corresponding to a symmetric spatial configuration. Bose statistics then requires that the isospin wave function be symmetric as well, which rules out $I = 1$. (A simple argument runs as follows: Since the pion is an isospin vector, the only vector obtained from two arbitrary unit vectors is $\hat{t}_1 \times \hat{t}_2$, which is antisymmetric. Alternatively, look at the composition of isospin states $I = 0$, 1, 2 and $I_3 = 0$ from $\pi^+\pi^-$ and $\pi^0\pi^0$.) So the isospin of the final state must be either 0 or 2. Now, $\mathrm{K}^0$ has isospin $I = 1/2$. The $|\Delta I| = \frac{1}{2}$ rule then favors $I = 0$ but not $I = 2$. From the usual angular momentum coupling rules, the isospin wave function reads

$$|I = 0, I_3 = 0\rangle = \tfrac{1}{\sqrt{3}} \left( |\pi_1^+\pi_2^-\rangle - |\pi^0\pi^0\rangle + |\pi_1^-\pi_2^+\rangle \right), \tag{6.99}$$

from which one deduces the ratio of the two decay rates

$$\frac{\Gamma(\mathrm{K}^0 \to \pi^+\pi^-)}{\Gamma(\mathrm{K}^0 \to \pi^0\pi^0)} = 2, \tag{6.100}$$

or, equivalently, the branching ratio

$$\frac{\Gamma(\mathrm{K}^0 \to \pi^+\pi^-)}{\Gamma(\mathrm{K}^0 \to \text{all})} = \frac{2}{3} = 0.67\,. \tag{6.101}$$

Of course, the same results apply to $\bar{\mathrm{K}}^0$ decays as well. Experimentally, one measures the decay of the combination $\mathrm{K}^0_\mathrm{S} = \frac{1}{\sqrt{2}}(\mathrm{K}^0 + \bar{\mathrm{K}}^0)$ with the result $\Gamma(\mathrm{K}^0_\mathrm{S} \to \pi^+\pi^-)/\Gamma(\mathrm{K}^0_\mathrm{S} \to \text{all}) = 68.61\%$, which is in excellent agreement with the theoretical prediction.

• $\mathrm{K}^+ \to \pi^+\pi^0$ *Decay.* As in the previous case, conservation of angular momentum and Bose statistics require the final state to have $\ell = 0$ and $I = 0$ or 2. But $\pi^+\pi^0$, being in the charge state $I_3 = 1$, rules out $I = 0$, leaving $I = 2$ as the only possibility. As K has isospin $1/2$, the $|\Delta I| = 1/2$ rule forbids this decay. It is in fact strongly suppressed, which explains in part why $\mathrm{K}^+$ is much more stable than $\mathrm{K}^0$:

$$\frac{\Gamma(\mathrm{K}^\pm \to \pi^\pm\pi^0)}{\Gamma(\mathrm{K}^0_\mathrm{S} \to 2\pi)} = 1.5 \times 10^{-3}\,. \tag{6.102}$$

Nevertheless, deviations from the heuristic $|\Delta I| = 1/2$ rule in certain circumstances are, of course, to be expected. Radiative corrections, ever present, can cause the charge-dependent strong interaction to bring in $|\Delta I| = 3/2$ or even $|\Delta I| = 5/2$ transitions. Such admixtures are unmistakably present in the following example.

• $\Omega^-$ *Decay.* Besides the dominant $\Omega^- \to \Lambda^0\,\mathrm{K}^-$ branch, $\Omega^-$ is also observed to decay through two other channels,

$$\begin{aligned} \Omega^- &\to \Xi^0\,\pi^- \qquad (23.6\%), \\ &\to \Xi^-\,\pi^0 \qquad (\ 8.6\%), \end{aligned} \tag{6.103}$$

which can shed light on the nature of the isospin violation of hadronic weak decays. The $\Omega$ isospin is 0, whereas the final $\Xi\,\pi$ state may have $I = 1/2,\ 3/2$. The $|\Delta I| = 1/2$ rule fixes isospin $I = 1/2$ for the final state, leading to the prediction (using the coefficients in Table 6.3)

$$\frac{\Gamma(\Omega^- \to \Xi^0\pi^-)}{\Gamma(\Omega^- \to \Xi^-\pi^0)} = \left|\frac{\sqrt{2/3}}{\sqrt{1/3}}\right|^2 = 2\,, \tag{6.104}$$

to be compared with the value of 2.74, deduced from the measured branching ratios (103). The difference between the measured and predicted ratios is small but undeniable. The $|\Delta I| = 1/2$ rule is broken in this situation, and the final states are not pure isospin eigenstates, but rather mixed states. It can be checked that a small admixture (7.3%) of $I = 3/2$ suffices to explain the deviation from the data.

Turning now to the **semileptonic weak decays** of hadrons, which are described by $L_\lambda^\dagger H^\lambda + H_\lambda^\dagger L^\lambda$, we still have conservation of the baryon number but, in the presence of leptons, not necessarily conservation of electric charge for hadrons. Two situations, $\Delta Q = 0$ and $|\Delta Q| = 1$, may occur.

- $\Delta Q = 0$. Examples of reactions involving a single $\pi$ in the final states are

$$\nu\,\mathrm{p} \to \nu\,\mathrm{p}\,\pi^0\,,\quad \nu\,\mathrm{n} \to \nu\,\mathrm{n}\,\pi^0\,,$$
$$\nu\,\mathrm{n} \to \nu\,\mathrm{p}\,\pi^-\,,\quad \nu\,\mathrm{p} \to \nu\,\mathrm{n}\,\pi^+\,.$$

The currents of hadrons responsible for these processes do not change charges and are called *neutral currents.* They are found to combine an isoscalar component with an isovector $z$ component (terms with $I \geq 2$ are however found suppressed). They conserve strangeness to a very high degree, with the limit of violation of the order of $[\Gamma(\mathrm{K}^+ \to \pi^+\nu\bar{\nu})/\Gamma(\mathrm{K}^+ \to \mathrm{all})] < 2.4 \times 10^{-9}$. Thus, the rule appears to be that neutral-current weak interactions do not change strangeness (or, more generally, flavor).

- $|\Delta Q| = 1$. Here we deal with charged currents. They may conserve or change strangeness; the magnitudes of change are determined by Gell-Mann–Nishijima's relation, $\Delta S = 2(\Delta Q - \Delta I_3)$.

When $\Delta S = 0$, the isospin is violated by $\Delta I_3 = \Delta Q = \pm 1$, and therefore $|\Delta I| = 1$, assuming violations $|\Delta I| \geq 2$ to be suppressed. These charged currents behave as the charge-raising or charge-lowering components of an isovector current analogous to (78). Several examples can be found in Table 6.6. For an application, see Problem 6.4.

When $|\Delta S| = 1$, Gell-Mann–Nishijima's relation allows $\Delta I_3 = \Delta Q - \tfrac{1}{2}\Delta S = \pm\tfrac{1}{2}, \tfrac{3}{2}$, and so $|\Delta I| = \tfrac{1}{2}, \tfrac{3}{2}, \ldots$, of which only $|\Delta I| = \tfrac{1}{2}$ should remain. This rule seems to hold rather well, as indicated for example by the relative rate of the three-body decays of the charged and neutral kaons (see Problem 6.5):

$$\frac{\Gamma\left(\mathrm{K}^0 \to \pi^- + \ell^+ + \nu_\ell\right)}{\Gamma\left(\mathrm{K}^+ \to \pi^0 + \ell^+ + \nu_\ell\right)} = 2\,,$$

which agrees with data. Let us note that the rule $|\Delta I| = \tfrac{1}{2}$ also implies

$$\Delta Q = \Delta S\,. \tag{6.105}$$

Indeed, if $|\Delta I| = \tfrac{1}{2}$, then $\Delta I_3 = \pm\tfrac{1}{2}$. The relation $|\Delta Q - \tfrac{1}{2}\Delta S| = |\Delta I_3|$ can be satisfied for $|\Delta I_3| = \tfrac{1}{2}$ only if $\Delta Q = \Delta S$. On the other hand, if $\Delta Q = -\Delta S$, it would imply $|\Delta I_3| = \tfrac{3}{2}$ and hence the presence of a $|\Delta I| = \tfrac{3}{2}$ admixture in the interaction. However, such impurities have been shown by experiment to be consistently small and should be viewed as a result of second-order weak processes. Other tests come from limits on decay rates such as $\Gamma(\Sigma^+ \to \mathrm{n}e^+\nu)/\Gamma(\Sigma^+ \to \mathrm{all}) < 5 \times 10^{-6}$.

Rule (105) states that the change in the strangeness quantum number of hadrons in a $|\Delta S| = 1$ transition equals the change in charge. Analogous rules also exist for charm and bottomness:

$$\Delta Q = \Delta C \qquad \text{and} \qquad \Delta Q = \Delta B\,. \tag{6.106}$$

In conclusion, while isospin invariance is a good symmetry of the strong interactions, it is broken by the electromagnetic and the weak interactions to varying degrees, but in a specific and systematic manner. These features constitute rather stringent constraints that must be accounted for by any interaction model.

## Problems

**6.1 Isospin eigenstates.** (a) Derive the isospin states given in Table 6.2. Starting from the definition $\Phi_{1,-1} = nn$ and using the step-up operator $I_+$, obtain $\Phi_{1,0}$ and $\Phi_{1,1}$. To get $\Phi_{00}$, find the normalized combination of $pn$ and $np$ orthogonal to $\Phi_{1,0}$. (b) Derive the pion–nucleon isospin states given in Table 6.3, starting from $\Phi_{3/2,-3/2}$ and using the same approach as in (a).

**6.2 Pion–nucleon scattering.** By inverting the wave functions given in Table 6.3, express the observable pion–nucleon states in terms of isospin states. Assuming isospin invariance, express the physical amplitudes $R_+$, $R_0$, and $R_-$ in terms of the isospin amplitudes, and find a linear condition relating the three observable amplitudes at the same energies and angles. Near the center-of-mass energy $m_{\mathrm{N}} + m_\pi + 159\,\mathrm{MeV}$, there is a resonance $\pi + \mathrm{N} \to \mathrm{N}^* \to \mathrm{N} + \pi$. Assuming that $\mathrm{N}^*$ is a pure isospin state, either $I = 1/2$ or $I = 3/2$, calculate the ratios of the cross-sections of the three observed reactions.

**6.3 p–d reactions.** Consider the reactions

$$\mathrm{p} + \mathrm{d} \to \pi^+ + {}^3\mathrm{H}, \qquad \mathrm{p} + \mathrm{d} \to \pi^0 + {}^3\mathrm{He}\,.$$

Since the deuteron is in a ${}^3S_1$ state, it must be an isospin singlet. Therefore, the initial state $\mathrm{p} + \mathrm{d}$ is a pure $I = 1/2$ state. Given that ${}^3\mathrm{H}$ and ${}^3\mathrm{He}$ form an isodoublet, write down the isospin decomposition of the final states, and from this, the ratio of the two cross-sections.

**6.4 Conserved currents in a pion–nucleon system.** A model Lagrangian for the interacting nucleon ($\psi$) and pion ($\boldsymbol{\phi}$) fields is taken to be

$$\mathcal{L}_{\mathrm{N}\pi} = \overline{\psi}(\mathrm{i}\not{\partial} - m)\psi + \tfrac{1}{2}\left(\partial_\mu \boldsymbol{\phi} \cdot \partial^\mu \boldsymbol{\phi} - \mu^2 \boldsymbol{\phi} \cdot \boldsymbol{\phi}\right) + \mathrm{i}g\overline{\psi}\gamma_5 \boldsymbol{\tau}\psi \cdot \boldsymbol{\phi}\,,$$

where $m$ is the nucleon mass and $\mu$ the pion mass. Consider the following global gauge transformations of the fields and, in each case, verify that the

Lagrangian is invariant, calculate the conserved current and the associated conserved charge.

$$\begin{array}{llllll} \text{(a)} & \psi \to & \psi - \mathrm{i}\varepsilon\psi, & \boldsymbol{\phi} & \to & \boldsymbol{\phi}; \\ \text{(b)} & \psi \to & \psi - \mathrm{i}\varepsilon\frac{1}{2}(1+\tau_3)\psi, & \boldsymbol{\phi} & \to & \boldsymbol{\phi} - \varepsilon(\boldsymbol{\phi}\times\hat{\boldsymbol{n}}_3); \\ \text{(c)} & \psi \to & \psi - \mathrm{i}\varepsilon\frac{1}{2}\boldsymbol{\tau}\cdot\hat{\boldsymbol{n}}\,\psi, & \boldsymbol{\phi} & \to & \boldsymbol{\phi} - \mathrm{i}\varepsilon\boldsymbol{t}\cdot\hat{\boldsymbol{n}}\,\boldsymbol{\phi}. \end{array}$$

In these equations, $\varepsilon$ is a very small real constant, $\hat{\boldsymbol{n}}$ an arbitrary unit vector in isospin space and $\hat{\boldsymbol{n}}_3$ a unit vector in isospin $z$ direction.

**6.5** $\mathbf{K_{\ell 3}}$ **decays.** Calculate the ratio of the decay rates for $K^0 \to \pi^- e^+ \nu_e$, $\pi^- \mu^+ \nu_\mu$, and $K^- \to \pi^0 e^- \bar{\nu}_e$, $\pi^0 \mu^- \bar{\nu}_\mu$.

**6.6 Gell-Mann–Nishijima relation.** This relation is given in (89). Apply it to the strange K meson and the $\Sigma$ baryon. After reading Chap. 7, apply it to charmed D, bottom-flavored B mesons, charm-flavored $\Lambda_c$ and bottom-flavored $\Lambda_b$ baryons.

**6.7 Particle production by strong interaction.** Explain why the processes $\pi^- + p \to \pi^+ + \Sigma^-$, $\pi^- + p \to K^0 + n$, $\pi^- + p \to \Sigma^+ + K^-$ cannot be observed.

## Suggestions for Further Reading

*The idea of proton and neutron as two states of the nucleon and the isospin concept:*

Cassen, B. and Condon, E. U., Phys. Rev. **50** (1936) 846

Heisenberg, W., Z. Phys. **77** (1932) 1

*Extension of the isospin concept to π mesons and experimental observations:*

Bjorkland, R. et al., Phys. Rev. **77** (1950) 213

Carlson, A. G. et al., Phil. Mag. **41** (1950) 701

Kemmer, N., Proc. Cambridge Phil. Soc. **34** (1938) 354

Lattes, C. M. G., Muirhead, H., Powell, C. F. and Occhialini, G. P., Nature **159** (1947) 694

*Extension of the isospin concept to strange particles:*

Gell-Mann, M., Phys. Rev. **92** (1953) 833

Gell-Mann, M. and Pais, A., Phys. Rev. **97** (1955) 1387

Nishijima, K., Progr. Theor. Phys. **12** (1954) 107; *ibid.* **13** (1955) 285

*G-parity:*

Lee, T. D. and Yang, C. N., Nuovo Cimento **3** (1956) 749

Michel, L., Nuovo Cimento **10** (1953) 319

Pais, A. and Jost, R., Phys. Rev. **87** (1952) 871

*Weak interactions in general:*

Marshak, R. E., Riazzuddin and Ryan, C. P., *Theory of Weak Interactions in Particle Physics.* Wiley-Interscience, New York 1969

*Neutral weak currents:*

Hung, P. Q. and Sakurai, J. J., *The structure of neutral currents.* Ann. Rev. Nucl. Part. Sci. **31** (1981) 375

# 7 Quarks and SU(3) Symmetry

By 1960 a great number of particles (which decay weakly) and resonances (which decay strongly) had been discovered. Some are seen in *production reactions*, where they are produced along with other final-state particles (such as the $\omega$ meson in $p\bar{p} \to \pi^+\pi^-\omega$), others in *formation reactions*, where they are the only products of collisions between the incident particles (such as the isobar resonance $\Delta$ in $\pi p \to \Delta$). This proliferation of particles and resonances calls for an organizing scheme more powerful than the Gell-Mann–Nishijima relation – in fact, a model that could embody the main features of known symmetry principles, establish or suggest relationships among particles, and provide a good basis for an eventual dynamic approach.

The precursor of the modern particle models is the Fermi–Yang model (1949) based on the fundamental set of the proton and neutron; nonstrange mesons are then built up from combinations of a nucleon and an antinucleon. Sakata (1956) added to this (p, n) pair the isosinglet hyperon $\Lambda$ of strangeness $-1$ and succeeded in giving a completely uniform treatment of all mesons, strange and nonstrange. But this model met with serious difficulties in dealing with baryons: their predicted mass spectrum is not as observed and their spins and parities are not correctly related. Nevertheless, it inspired later models. In terms of group theory, the Fermi–Yang model is based on the symmetry of the unitary group SU(2) and the Sakata model on that of the SU(3) group.

In a further extension, Gell-Mann and Ne'eman (1961) proposed the *eightfold-way* model in which the basic unit is an eight-member multiplet, or octet, of SU(3), not a triplet as in the Sakata model. The lowest-mass baryons of spin $1/2$ would then belong to an octet, and the pseudoscalar mesons $0^-$ to another analogous octet. All other particles and resonances would fall into octets or multiplets that could be made from the basic octets. This model, though remarkably successful in many practical aspects, lacks a fundamental basis. A much deeper understanding of the physical nature of SU(3) emerged when Gell-Mann and Zweig (1964) put forth a simple but drastic idea that hadrons are built from three basic constituents called *quarks*. The idea is simple because it retains the triplet as the basic building block for all hadrons, and drastic because the quarks are not only novel but would also have rather surprising properties.

Of course, even this SU(3) model is not final. But as with any good model, it is rich in implications and ramifications, and opens the way to further developments. The concept of *color* will be introduced, and new kinds of particles will be discovered. Still, no quark is seen. Yet the concept of quark endures, giving us the most elegant model of particles we have, and laying the groundwork for a theory of the fundamental interactions. These developments in particle spectroscopy up to the detections of the $\tau$ lepton and the c, b, and t quarks form the main topic of the present chapter.

## 7.1 Isospin: SU(2) Symmetry

In this section, we briefly review some of the concepts introduced in Chap. 6, rephrasing them in a language more readily generalizable to higher-order symmetries. In particular, we will introduce a description of particle multiplets by means of tensorial techniques frequently used in other fields of physics.

The conservation of baryons observed in particle physics may be thought of as a consequence of the invariance of the theory to arbitrary phase transformations of the baryon states. Taken as a typical baryon field, the neutron field transforms as

$$\psi_{\mathrm{n}} \to \mathrm{e}^{\mathrm{i}\alpha}\psi_{\mathrm{n}}\,, \tag{7.1}$$

for any arbitrary real constant $\alpha$. The set of all unitary transformations $\{\exp \mathrm{i}\alpha\}$ acting on the spinor $\psi_{\mathrm{n}}$, considered for the present purpose as a one-component object, forms a one-parameter *unitary group*, called U(1). It is not a very interesting group because it cannot lead to any relations between different fields. For this reason we seek higher-order symmetries.

Charge independence suggests that the proton (p) and neutron (n) are in some sense interchangeable states and should be considered as parts of a two-component spinor

$$\psi = \begin{pmatrix} \psi_{\mathrm{p}} \\ \psi_{\mathrm{n}} \end{pmatrix}, \tag{7.2}$$

with (contravariant) components $\psi^1 = \psi_{\mathrm{p}}$ and $\psi^2 = \psi_{\mathrm{n}}$. These spinors may be interpreted either as state vectors or as field operators that annihilate the proton or the neutron. The most general linear transformation of $\psi$

$$\psi^a \to \psi'^a = U^a{}_b\,\psi^b \qquad (a,\, b = 1,\, 2), \tag{7.3}$$

(summing over the repeated index as usual) is defined by a $2 \times 2$ complex matrix $U$. If it is required as usual that the scalar product in this vector space be invariant, $U$ must satisfy the *unitarity condition*

$$U^\dagger U = UU^\dagger = 1\,, \tag{7.4}$$

and can be parameterized by four real constants. All such transformations form a representation of the unitary group U(2). Unitarity (4) implies

$|\det U|^2 = 1$, so that $\det U = \exp i\alpha$ for an arbitrary real constant $\alpha$. This means that in general we can factor out the complex phase,

$$U = e^{i\alpha} S, \tag{7.5}$$

and treat it separately as an element of a one-parameter gauge group representing baryon conservation, as seen above. From now on, we shall limit ourselves to the *unitary, unimodular* transformations $S$, for which

$$S^\dagger S = 1 \quad \text{and} \quad \det S = 1. \tag{7.6}$$

They form the Lie group SU(2), the group of unitary $2 \times 2$ matrices of determinant equal to one. The unimodular condition reduces the number of independent real parameters to three (which defines the *dimension* of the group), so that the most general such transformation may be expressed as

$$S = \exp[-\tfrac{i}{2}(\alpha_1\tau_1 + \alpha_2\tau_2 + \alpha_3\tau_3)], \tag{7.7}$$

where $\alpha_i$ are real constants, and $\tau_i$ are $2 \times 2$ matrices [with elements $(\tau_i)^a{}_b$, for $a, b = 1$ or 2], which must be Hermitian and traceless, as a consequence respectively of the unitarity and unimodular conditions on $S$. The usual Pauli matrices satisfy these conditions. The matrices $I_i = \tau_i/2$, called the *generators* of the infinitesimal transformations of the group, form a closed algebra, i.e. the commutator of any two of them is again a member of the set,

$$[I_i, I_j] = i\epsilon_{ijk} I_k, \quad (i, j, k = 1, 2, \text{ or } 3), \tag{7.8}$$

where $\epsilon_{ijk}$ are the components of the totally antisymmetric Levi-Cività tensor, with $\epsilon_{123} = +1$.

Even though these relations are obtained from the $2 \times 2$ matrices $\tau_i/2$, they actually hold for any representation of the generators of SU(2) and define the Lie algebra associated with the Lie group SU(2) and characterized by the *structure constants* $\epsilon_{ijk}$. This algebra allows only one diagonal operator, conventionally taken to be $I_3$. In the two-dimensional representation, $I_3$ has diagonal elements $1/2$ and $-1/2$, corresponding to its eigenvalues for $\psi^1$ and $\psi^2$, respectively. We express this fact by saying that SU(2) has *rank* one. In general, the rank of a group is the number of generators that can be simultaneously diagonalized; it gives the number of independent additive quantum numbers whose conservation is implied by the invariance of the theory under the transformations of the group. The three generators $I_1$, $I_2$ and $I_3$ may be taken as the components of a vector called the isobaric spin (or *isospin*). The expectation value of its square is written as $\boldsymbol{I}^2 = I(I+1)$. For the nucleon multiplet, $I = 1/2$.

We are most interested in other multiplets of particles which, just like the proton and neutron, transform among themselves, and thus must have the same spin and parity and, at least roughly, the same mass. They constitute the basis vectors of *irreducible representations* of the group.

Besides the trivial one-dimensional representation, the simplest is the fundamental, or defining, representation formed by the set of transformations $\{\exp(-i\boldsymbol{\alpha}\cdot\boldsymbol{\tau}/2)\}$, as defined above, which act on a (carrier) space of dimension two, whose basis vectors transform under SU(2) as

$$\psi^a \to \psi'^a = S^a{}_b\, \psi^b\,, \qquad (a = 1, 2)\,. \tag{7.9}$$

In order to introduce the scalar product in this vector space, it is necessary to define the dual basis vectors, labeled by covariant (lower) indices, such that the scalar product remains invariant to SU(2) transformations:

$$\phi'_a\, \psi'^a = \phi'_a\, S^a{}_b\, \psi^b = \phi_b\, \psi^b\,. \tag{7.10}$$

Therefore, $\phi_a$ must transform as

$$\phi'_a = \left(S^{-1}\right)^b{}_a\, \phi_b = \left(S^\dagger\right)^b{}_a\, \phi_b = \left(S^a{}_b\right)^* \phi_b\,, \tag{7.11}$$

that is, exactly as $(\psi^a)^*$, and hence give the basis of the *conjugate* fundamental representation.

All higher representations of a group are constructed from its fundamental representations by tensor multiplication. The tensors that define the basis of their respective spaces transform by composition, that is, the upper indices, contravariantly, and the lower indices, covariantly. For example,

$$T^{ab}_{cd} \to T'^{ab}_{cd} = S^a{}_{a'}\, S^b{}_{b'}\, T^{a'b'}_{c'd'}\, (S^\dagger)^{c'}{}_c\, (S^\dagger)^{d'}{}_d\,. \tag{7.12}$$

In words, components of any tensor transform into combinations of themselves and nothing else, according to rule (12).

In fact, the fundamental representation of SU(2) and its conjugate turn out to be *equivalent* (meaning the two sets of transformations $\{S\}$ and $\{S^*\}$ are identical), as can be seen from the simple fact that $\tau_i^* = -\tau_2\tau_i\tau_2$ for $i = 1, 2, 3$. To put it another way, a covariant vector can always be expressed in a contravariant basis, as we will now show.

Let $\epsilon_{ab}$ be the antisymmetric symbol defined by $\epsilon_{11} = \epsilon_{22} = 0$ and $\epsilon_{12} = -\epsilon_{21} = 1$. Its inverse $\epsilon^{ab}$ ($\epsilon^{12} = -1$) satisfies

$$\epsilon_{ac}\, \epsilon^{cb} = \delta_a{}^b\,. \tag{7.13}$$

The tensors with components $\delta^a{}_b$, $\epsilon_{ab}$, and $\epsilon^{ab}$ are invariant, in the sense that each component transforms into itself, without mixing. Now, let $\chi^a$ be a contravariant basis, and define

$$\theta_a = \epsilon_{ab}\, \chi^b\,. \tag{7.14}$$

Since by assumption $\chi$ transforms as a contravariant vector, the transformation for $\theta$ is

$$\begin{aligned}\theta'_a &= \epsilon_{ab}\,\chi'^b = \epsilon_{ab}\,S^b{}_c\,\chi^c \\ &= \epsilon_{ab}\,\epsilon^{dm}\,\epsilon_{mc}\,S^b{}_d\,\chi^c = \epsilon_{ab}\,\theta_m\,\epsilon^{dm}\,S^b{}_d \\ &= \theta_m(S^{-1})^m{}_a\,,\end{aligned}$$

where, in the last step, we have made use of the unimodular condition

$$\det S = -\tfrac{1}{2}\,\epsilon_{ab}\,\epsilon^{cd}\,S^a{}_c\,S^b{}_d = 1\,. \tag{7.15}$$

The result shows that $\epsilon_{ab}\,\chi^b$, and hence $\theta_a$, transforms as a covariant vector.

As the conjugate representation of SU(2) is equivalent to the ordinary representation, there is no real need for tensors with covariant indices; contravariant tensors $T^{a\ldots d}$ alone would do for the bases of linear representations of the group. Since $\epsilon_{ab}$ is an invariant tensor under SU(2), a contraction of a rank-$n$ tensor $T^{a_1\ldots a_n}$,

$$\epsilon_{a_i a_j}\,T^{a_1\ldots a_n}\,, \quad \text{for } 1 \le i,j \le n\,, \tag{7.16}$$

may yield a nonvanishing tensor of rank $n-2$ for some $i$, $j$, in which case the tensor $T^{a_1\ldots a_n}$ is said to be *reducible.* If (16) vanishes for every possible contraction, it is said to be *irreducible*. Thus, an irreducible tensor of rank $n$ is a contravariant tensor totally symmetric in its $n$ indices. Because of this symmetry, it has only $n+1$ linearly independent components. If $n+1$ such components are selected to form a column vector $\phi$, then just as $\psi^a$ transforms with the matrix $S$, so too will $\phi$ transform according to

$$\phi \to \phi' = V\phi\,. \tag{7.17}$$

The $(n+1)\times(n+1)$ matrix $V$, given by a product of matrices $S$, defines an *irreducible representation* of the group having dimension $d=n+1$. A particularly interesting representation is the *adjoint*, or *regular*, representation, the dimension of which is identical to that of the group, i.e. $d = \dim \mathrm{SU}(2) = 3$, and the elements of the generator matrices are given by the structure constants, $(I_i)_{jk} = -\mathrm{i}\epsilon_{ijk}$.

Each independent component $T^{a_1\ldots a_n}$ of an irreducible rank-$n$ tensor corresponds to a member of an isospin multiplet, with eigenvalues

$$I_3 = \tfrac{1}{2}\,(n_1 - n_2)\,, \tag{7.18}$$

where $n_1$ and $n_2 = n - n_1$ are the numbers of indices $a_i$ with values 1 and 2, respectively. Since $I_3$ has $n_1/2 = n/2$ as its largest value and $-n_2/2 = -n/2$ as its smallest, the multiplet has isospin $I = n/2$ and multiplicity (number of its members) $d = n+1 = 2I+1$.

For illustration, let us take a totally symmetric tensor of rank $n = 2$ as defining a basis for the adjoint representation and call it $\pi^{ab}$. Its three independent components span a three-dimensional space and correspond to the three isospin states $I = 1$, $I_3 = 0, \pm 1$. The contravariant tensor $\pi^{ab}$ is equivalent to the traceless *mixed tensor* of second rank $\pi^a{}_b = \epsilon_{bc}\,\pi^{ac}$. The latter can be expanded in terms of the three Pauli matrices $(\tau_i)^a{}_b$ with real coefficients $\phi_i$:

$$\pi^a{}_b = \frac{1}{\sqrt{2}}\sum_{i=1}^{3}(\tau_i)^a{}_b\,\phi_i = \frac{1}{\sqrt{2}}\begin{pmatrix}\phi_3 & \phi_1 - \mathrm{i}\phi_2\\ \phi_1 + \mathrm{i}\phi_2 & -\phi_3\end{pmatrix}. \tag{7.19}$$

The normalization has been chosen such that

$$\pi^a{}_b\,\pi^b{}_a = \tfrac{1}{2}\sum \phi_i\,\phi_j \mathrm{Tr}(\tau_i\tau_j) = \sum \phi_i^2\,. \tag{7.20}$$

From (18), different elements $\pi^a{}_b$ have the following $I_3$-assignments:

$$\begin{aligned} I_3 &= \phantom{-}0: & \pi^1{}_1 &= -\pi^2{}_2 = \pi^{12}\,,\\ I_3 &= \phantom{-}1: & \pi^1{}_2 &= -\pi^{11}\,,\\ I_3 &= -1: & \pi^2{}_1 &= \phantom{-}\pi^{22}\,. \end{aligned} \tag{7.21}$$

Therefore, the elements will be designated by the particles having the appropriate quantum numbers:

$$\pi^a{}_b = \begin{pmatrix}\frac{1}{\sqrt{2}}\pi^0 & \pi^+\\ \pi^- & \frac{-1}{\sqrt{2}}\pi^0\end{pmatrix}. \tag{7.22}$$

As quantum operators, the fields

$$\pi^0 = \phi_3\,, \qquad \pi^+ = \tfrac{1}{\sqrt{2}}(\phi_1 - \mathrm{i}\phi_2)\,, \qquad \text{and} \quad \pi^- = \tfrac{1}{\sqrt{2}}(\phi_1 + \mathrm{i}\phi_2) \tag{7.23}$$

*annihilate* the particles whose symbols they carry.

Let us now pretend that particles of higher multiplets may be constructed as bound states of the basic doublets $\psi^a = (p,\ n)$ and $\psi_a = (\bar{p},\ \bar{n})$. Their compositions can then be determined from appropriate tensor products. For example, states composed of a 'nucleon' and an 'antinucleon' are obtained by rewriting the product $\psi^a\psi_b$, in a simple process called *reduction*:

$$\psi^a\,\psi_b = \tfrac{1}{2}\,(\psi^c\,\psi_c)\,\delta^a{}_b + \psi^a\,\psi_b - \tfrac{1}{2}\,(\psi^c\,\psi_c)\,\delta^a{}_b\,, \tag{7.24}$$

in terms of two irreducible tensors, one identified with the *isosinglet* and the other with the *isotriplet* $\pi^a{}_b$ defined above:

$$\sigma \equiv \tfrac{1}{\sqrt{2}}(\psi^c\,\psi_c) = \tfrac{1}{\sqrt{2}}(p\bar{p} + n\bar{n})\,, \tag{7.25}$$

$$\pi^a{}_b \equiv \psi^a \psi_b - \tfrac{1}{2}(\psi^c \psi_c)\,\delta^a{}_b = \begin{pmatrix} \frac{1}{2}(p\bar{p} - n\bar{n}) & p\bar{n} \\ n\bar{p} & -\frac{1}{2}(p\bar{p} - n\bar{n}) \end{pmatrix}. \tag{7.26}$$

Comparing the last equation with (22), one may deduce the structure of the composite fields

$$\pi^0 = \tfrac{1}{\sqrt{2}}(p\bar{p} - n\bar{n}), \qquad \pi^+ = p\bar{n}, \qquad \pi^- = n\bar{p}. \tag{7.27}$$

These are essentially the same as the corresponding relations listed in Table 6.2 obtained by other means.

Finally, let us construct states of three 'nucleons', an exercise as instructive as it is useful for later considerations. One begins by reducing the product of two spinors

$$\begin{aligned} \psi^a\psi^b &= \tfrac{1}{2}(\psi^a\psi^b - \psi^b\psi^a) + \tfrac{1}{2}(\psi^a\psi^b + \psi^b\psi^a) \\ &= -\tfrac{1}{2}\epsilon^{ab}A + \tfrac{1}{2}S^{ab}, \end{aligned} \tag{7.28}$$

where $A \equiv \epsilon_{cd}\psi^c\psi^d$ and $S^{ab} \equiv \psi^a\psi^b + \psi^b\psi^a$. In more familiar terms, this superposition of an invariant and a rank-2 tensor may be seen as the result of the coupling of two isospin-$1/2$ states, leading to an antisymmetric $I = 0$ state, represented by $A$, and a symmetric $I = 1$ state, represented by $S^{ab}$ (cf. Table 6.2). Now adding one more spinor to the system yields

$$\begin{aligned} \psi^a\psi^b\psi^c &= -\tfrac{1}{2}\epsilon^{ab}A\psi^c + \tfrac{1}{2}S^{ab}\psi^c \\ &= -\tfrac{1}{2}\epsilon^{ab}A\psi^c + \tfrac{1}{6}\epsilon_{ed}(\epsilon^{ca}S^{eb} + \epsilon^{cb}S^{ea})\psi^d \\ &\quad + \tfrac{1}{6}(S^{ab}\psi^c + S^{bc}\psi^a + S^{ca}\psi^b) \\ &= -\tfrac{1}{\sqrt{2}}\epsilon^{ab}\chi^c_{\mathrm{A}} + \tfrac{1}{\sqrt{6}}(\epsilon^{ca}\chi^b_{\mathrm{S}} + \epsilon^{cb}\chi^a_{\mathrm{S}}) + \chi^{abc}_{\mathrm{Q}}, \end{aligned} \tag{7.29}$$

where three irreducible tensors, two of rank 1 and one of rank 3, have been introduced:

$$\begin{aligned} \chi^a_{\mathrm{A}} &\equiv \tfrac{1}{\sqrt{2}}(\psi^1\psi^2 - \psi^2\psi^1)\psi^a, \\ \chi^a_{\mathrm{S}} &\equiv \tfrac{1}{\sqrt{6}}\epsilon_{bc}S^{ab}\psi^c, \\ \chi^{abc}_{\mathrm{Q}} &\equiv \tfrac{1}{6}(S^{ab}\psi^c + S^{bc}\psi^a + S^{ca}\psi^b). \end{aligned} \tag{7.30}$$

Thus, coupling three isospin-$1/2$ states produces one completely symmetric $I = 3/2$ (quadruplet) state, given by $\chi^{abc}_{\mathrm{Q}}$, and two $I = 1/2$ (doublet) states of mixed symmetry, given by $\chi^a_{\mathrm{A}}$ and $\chi^a_{\mathrm{S}}$, which are formed by coupling the third spinor in two different ways, either to $I = 0$ (antisymmetric) or to $I = 1$ (symmetric) two-particle states. Note in particular the absence of a totally antisymmetric combination of three two-component spinors.

## 7.2 Hypercharge: SU(3) Symmetry

We have seen in the previous chapter that the strong interaction conserves the electric and baryonic charges as well as isospin and strangeness. When conservation of strangeness is combined with conservation of the baryon number by introducing the notion of hypercharge $Y$,

$$Y = N_{\mathrm{B}} + S = 2\,(Q - I_3)\,, \tag{7.31}$$

it becomes apparent that the low-lying hadrons that have the same baryon numbers and the same spins and parities fall into regular patterns associating together several isospin multiplets, each characterized by some value of $Y$ (cf. Table 6.5). This suggests a symmetry beyond isospin.

### 7.2.1 The Fundamental Representation

In order to incorporate conservation of both isospin and hypercharge into a single group structure, the group must be at least of the *second rank*, with one diagonalized generator related to $I_3$ and the other to $Y$. In this minimal extension, the basic spinor defining the fundamental representation must have three components, each chosen to correspond to one of the three characteristics of the group, namely, *upness* ($I_3 = 1/2$), *downness* ($I_3 = -1/2$) and *strangeness* ($S = -1$). The up ($u$) and the down ($d$) states, both with $S = 0$, are assumed to form an isodoublet, while the strangeness state ($s$) is an isosinglet. Accordingly, we use the following notation for the basic three-component spinor:

$$q^a = \begin{pmatrix} u \\ d \\ s \end{pmatrix}\,. \tag{7.32}$$

In a linear transformation, it obeys the rule

$$q^a \to q'^a = U^a{}_b\, q^b\,. \tag{7.33}$$

From our successful experience with SU(2), we again restrict considerations to transformations defined by *unitary, unimodular* complex $3 \times 3$ matrices. The set of all such matrices form the Lie group SU(3). The one unimodular and nine unitarity constraints together reduce the 18 real transformation parameters to 8, which is the *dimension* of the group, $d(\mathrm{G}) = 8$. An arbitrary element of the group may thus be expanded in terms of 8 real constants $\alpha_i$:

$$S = \exp\left(-\frac{\mathrm{i}}{2}\sum_{i=1}^{8} \alpha_i\,\lambda_i\right)\,. \tag{7.34}$$

The $3 \times 3$ matrices $\lambda_i$ are generalizations of the Pauli matrices $\tau_i$. From the unitary and unimodular conditions on $S$, they must be Hermitian and traceless. They may also be made orthonormal:

$$\mathrm{Tr}\left(\frac{\lambda_i}{2}\frac{\lambda_j}{2}\right) = \frac{1}{2}\,\delta_{ij}\,. \tag{7.35}$$

Table 7.1 displays the explicit expressions for the $\lambda_i$ due to Gell-Mann consistent with these conditions. Two of the matrices are diagonalized (the group is of *rank* 2). It is evident that $\lambda_3/2$ gives the $I_3$ eigenvalues for $q^a$, but the physical interpretation for the other diagonal matrix, $\lambda_8$, remains for now undetermined, although the objective is to relate it to the hypercharge.

**Table 7.1.** Gell-Mann matrices

$$\lambda_1 = \begin{pmatrix} 0 & 1 & 0 \\ 1 & 0 & 0 \\ 0 & 0 & 0 \end{pmatrix} \quad \lambda_2 = \begin{pmatrix} 0 & -\mathrm{i} & 0 \\ \mathrm{i} & 0 & 0 \\ 0 & 0 & 0 \end{pmatrix} \quad \lambda_3 = \begin{pmatrix} 1 & 0 & 0 \\ 0 & -1 & 0 \\ 0 & 0 & 0 \end{pmatrix}$$

$$\lambda_4 = \begin{pmatrix} 0 & 0 & 1 \\ 0 & 0 & 0 \\ 1 & 0 & 0 \end{pmatrix} \quad \lambda_5 = \begin{pmatrix} 0 & 0 & -\mathrm{i} \\ 0 & 0 & 0 \\ \mathrm{i} & 0 & 0 \end{pmatrix}$$

$$\lambda_6 = \begin{pmatrix} 0 & 0 & 0 \\ 0 & 0 & 1 \\ 0 & 1 & 0 \end{pmatrix} \quad \lambda_7 = \begin{pmatrix} 0 & 0 & 0 \\ 0 & 0 & -\mathrm{i} \\ 0 & \mathrm{i} & 0 \end{pmatrix} \quad \lambda_8 = \frac{1}{\sqrt{3}} \begin{pmatrix} 1 & 0 & 0 \\ 0 & 1 & 0 \\ 0 & 0 & -2 \end{pmatrix}$$

The generators of the infinitesimal transformations in the fundamental representation, $\lambda_i/2$, satisfy the commutation relations

$$\left[\frac{\lambda_i}{2}, \frac{\lambda_j}{2}\right] = \mathrm{i} f_{ijk} \frac{\lambda_k}{2}, \qquad (i,\, j,\, k = 1, \ldots, 8)\,. \tag{7.36}$$

The coefficients $f_{ijk}$ are, of course, the structure constants of the group and can be calculated from the explicit $\lambda_i$. By definition $f_{ijk}$ is antisymmetric in the exchange of the first two indices $i$ and $j$, and, for the particular choice (35), is totally antisymmetric in all three indices, since then

$$f_{ijk} = -\frac{\mathrm{i}}{4} \mathrm{Tr}\left([\lambda_i, \lambda_j]\, \lambda_k\right)\,. \tag{7.37}$$

The explicit expressions for $\lambda_1$, $\lambda_2$, and $\lambda_3$ indicate that they have with each other the same commutation relations as the Pauli matrices $\tau_1$, $\tau_2$, and $\tau_3$. This means that $\lambda_1/2$, $\lambda_2/2$, and $\lambda_3/2$ form a subalgebra of (36) and generate a subgroup having the same structure as the isospin group. There exist two other SU(2) subgroups of SU(3) (called the V-spin subgroup and the U-spin subgroup) generated by $\lambda_4/2$, $\lambda_5/2$, and $(\sqrt{3}\,\lambda_8 + \lambda_3)/4$, and by $\lambda_6/2$, $\lambda_7/2$, and $(\sqrt{3}\,\lambda_8 - \lambda_3)/4$, respectively. Just as isospin invariance implies the existence of multiplets of different charges, e.g. $(\mathrm{K}^0, \mathrm{K}^+)$ or $(\pi^-, \pi^0, \pi^+)$, so does U-spin invariance the existence of multiplets of equal charges, e.g. $(\mathrm{K}^+, \pi^+)$ or $(\bar{\mathrm{K}}^0, \pi^0, \mathrm{K}^0)$. Many physical implications of SU(3) invariance may be deduced from considerations of U-spin.

### 7.2.2 Higher-Dimensional Representations

We now describe higher-dimensional representations of SU(3), defining their generators, their quadratic Casimir operator, and finally the basis vectors of their space. The generators in a given representation labeled by R will be denoted by $F_i(\mathrm{R})$, with $i = 1, \dots, 8$. The dimension of representation R will be called $d(\mathrm{R})$, whereas the dimension of SU(3) is $d(\mathrm{G}) = 8$.

**Adjoint Representation.** A very special representation is the *adjoint*, or *regular*, representation (A), in which the generators are the $8 \times 8$ matrices defined by

$$(F_i(\mathrm{A}))_{jk} \equiv -\mathrm{i} f_{ijk} \,. \tag{7.38}$$

With the help of (36) and (38), the Jacobi identity for $\lambda_i$,

$$[\lambda_i,\, [\lambda_j,\, \lambda_k]] + [\lambda_j,\, [\lambda_k,\, \lambda_i]] + [\lambda_k,\, [\lambda_i,\, \lambda_j]] = 0 \,,$$

can be cast in the form

$$-(F_j F_i)_{km} + (F_i F_j)_{km} = \mathrm{i} f_{ijn}\, (F_n)_{km} \,.$$

So $F_i(\mathrm{A})$ satisfy the same commutation relations (36) as $\lambda_i/2$:

$$[F_i(\mathrm{A}),\, F_j(\mathrm{A})] = \mathrm{i} f_{ijk} F_k(\mathrm{A}) \,. \tag{7.39}$$

Once the normalization of the fundamental representation is fixed by (35), the normalization of $F_i(\mathrm{A})$ is not free, being determined by the trace

$$\mathrm{Tr}\,[F_i(\mathrm{A}) F_j(\mathrm{A})] = (F_i(\mathrm{A}))_{km} (F_i(\mathrm{A}))_{mk} = f_{ikm}\, f_{jkm} \,, \tag{7.40}$$

which yields, with the values of the coefficients listed in Problem 7.3,

$$\mathrm{Tr}\,[F_i(\mathrm{A}) F_j(\mathrm{A})] = 3\, \delta_{ij} \,. \tag{7.41}$$

More generally, the algebra of the generators of SU(3) in any representation $R$ is defined by the commutation relations

$$[F_i(\mathrm{R}),\, F_j(\mathrm{R})] = \mathrm{i} f_{ijk} F_k(\mathrm{R}) \,, \tag{7.42}$$

together with the normalization condition

$$\mathrm{Tr}\,[F_i(\mathrm{R}) F_j(\mathrm{R})] = C(\mathrm{R})\, \delta_{ij} \,, \tag{7.43}$$

where $C(\mathrm{R})$ is a constant for each representation R. From previous results, we have $C(\mathrm{f}) = 1/2$ for the fundamental representation and $C(\mathrm{A}) = 3$ for the adjoint representation.

**Quadratic Casimir Operator.** We have seen that representations in SU(2) are characterized by the eigenvalues of the total isospin $\boldsymbol{I}^2 = I_i I_i$. Similarly, we may define for any representation of SU(3) (or of any simple Lie algebra) the operator

$$\boldsymbol{F}^2 = F_i F_i \,. \tag{7.44}$$

It is called the *quadratic Casimir operator.* It commutes with every generator of the group because

$$[F_i F_i,\, F_j] = F_i[F_i,\, F_j] + [F_i,\, F_j]F_i = \mathrm{i}f_{ijk}\{F_i,\, F_k\}$$

vanishes by the antisymmetry of $f_{ijk}$. Therefore $\boldsymbol{F}^2(\mathrm{R})$ is proportional to the identity matrix $\mathbf{1}_\mathrm{R}$ of representation R

$$\boldsymbol{F}^2(\mathrm{R}) = C_2(\mathrm{R})\mathbf{1}_\mathrm{R}\,, \tag{7.45}$$

where $C_2(\mathrm{R})$ is a constant characteristic of the representation. It is related to the normalization constant $C(\mathrm{R})$ defined in (43) by a simple relation, which may be obtained by contracting the two free labels in (43):

$$\mathrm{Tr}\,\boldsymbol{F}^2(\mathrm{R}) = C(\mathrm{R})\delta_{ij}\delta_{ij} = C(\mathrm{R})\,d(\mathrm{G})\,,$$

and by taking the trace of (45)

$$\mathrm{Tr}\,\boldsymbol{F}^2(\mathrm{R}) = C_2(\mathrm{R})\,\mathrm{Tr}\,\mathbf{1}_\mathrm{R} = C_2(\mathrm{R})\,d(\mathrm{R})\,.$$

Therefore $C_2(\mathrm{R})$ and $C(\mathrm{R})$ are related through

$$C_2(\mathrm{R})\,d(\mathrm{R}) = C(\mathrm{R})\,d(\mathrm{G})\,. \tag{7.46}$$

In the fundamental representation, $d(\mathrm{f}) = 3$ and $C(\mathrm{f}) = 1/2$, and so the quadratic Casimir operator is $C_2(\mathrm{f}) = 4/3$. In the adjoint representation, $d(\mathrm{A}) = d(\mathrm{G}) = 8$ and $C(\mathrm{A}) = 3$, and so $C_2(\mathrm{A}) = C(\mathrm{A}) = 3$.

**Vector Spaces of Representations.** We turn now to the study of the vector spaces that define representations $R$. Let $\psi^a$, $\phi^a, \ldots$ denote various triplets that transform as the basis of the fundamental representation:

$$\psi^a \to \psi'^a = S^a{}_b\,\psi^b\,, \qquad (a,\, b = 1,\, 2,\, 3)\,, \tag{7.47}$$

under a transformation of SU(3) with matrix elements $S^a{}_b$. Covariant spinors spanning the carrier space of the *conjugate* representation, which have components labeled by lower indices, must transform so as to make the inner product $\theta_a\psi^a$ invariant:

$$\theta_a \to \theta'_a = (S^a{}_b)^*\theta = \theta_b\,(S^\dagger)^b{}_a\,. \tag{7.48}$$

Spinors of the types $\psi^a$ and $\theta_a$ are the simplest nontrivial examples of tensors. Generally, tensors are objects whose components, carrying both upper and lower indices, transform among themselves, with the upper indices transforming contravariantly and the lower, covariantly. If $n$ stands for the number of upper indices, and $m$, the number of lower indices, the tensor will be denoted by $T(n,m)$ and its *rank* is defined by the ordered set of integers $(n,m)$. For instance, the mixed tensor of rank $(1,2)$ transforms as

$$T^a_{bc} \to T'^a_{bc} = S^a{}_{a'}\, T^{a'}_{b'c'}\, (S^\dagger)^{b'}{}_b\, (S^\dagger)^{c'}{}_c\,, \tag{7.49}$$

while the tensor of rank $(2,1)$ transforms as

$$T^{ab}_c \to T'^{ab}_c = S^a{}_{a'}\, S^b{}_{b'}\, T^{a'b'}_{c'}\, (S^\dagger)^{c'}{}_c\,. \tag{7.50}$$

The interest in these examples is that $T^a{}_{bc}$ obeys the same transformation rule as $(T^{bc}_a)^*$, a result which generalizes to tensors of arbitrary ranks: a tensor $T(n,m)$ transforms exactly as $T^*(m,n)$.

There exist three *invariant tensors*, whose components are unchanged under the transformations of SU(3). They are the Kronecker delta

$$\delta^a{}_b = \begin{cases} 1, & \text{if } a = b\,, \\ 0, & \text{otherwise;} \end{cases} \tag{7.51}$$

the totally antisymmetric covariant Levi-Cività symbol

$$\epsilon_{abc} = \begin{cases} 1, & \text{if } a,b,c \text{ is an even permutation of } 1,2,3\,, \\ -1, & \text{if } a,b,c \text{ is an odd permutation of } 1,2,3\,, \\ 0, & \text{otherwise;} \end{cases} \tag{7.52}$$

and the contravariant Levi-Cività symbol $\epsilon^{abc}$, which is numerically equal to $\epsilon_{abc}$, so that $\epsilon^{123} = +1$ and

$$\epsilon_{abm}\, \epsilon^{mcd} = \delta^c{}_a\, \delta^d{}_b - \delta^d{}_a\, \delta^c{}_b\,. \tag{7.53}$$

Invariance of $\delta^a{}_b$ follows immediately from (49) and the unitarity of $S$, while that of $\epsilon_{abc}$ (and similarly of $\epsilon^{abc}$) may be inferred from the unimodular condition on $S$ written in the form

$$\epsilon^{ijk}\, S^a{}_i\, S^b{}_j\, S^c{}_k = \epsilon^{abc}\,. \tag{7.54}$$

In SU(3) the fundamental representation is not equivalent to its conjugate. This means that the set of transformations $\{S = \exp(-\mathrm{i}\alpha_i\lambda_i/2)\}$ is different from the set $\{S^*\}$, even after rearranging the order of its elements (see Problem 7.2). Therefore, a covariant spinor $\theta_a$ is not linearly related to a contravariant spinor, and vice versa. It is rather related to an antisymmetric second-rank contravariant tensor,

$$\theta_a = \epsilon_{abc}\, \psi^b \phi^c\,. \tag{7.55}$$

That $\theta_a$ is indeed a first-rank covariant tensor follows from the invariance of $\epsilon_{abc}$ and from (54).

Therefore a lower index cannot be made equivalent to an upper index, as in SU(2), and there must exist mixed tensors carrying indices of both types. They may be reduced to tensors of lower ranks by contracting their indices with one or another invariant tensor. Thus, starting from a tensor $T^{a_1\dots a_n}_{b_1\dots b_m}$ of rank $(n, m)$, we may construct tensors of ranks $(n-1, m-1)$, $(n-2, m+1)$, and $(n+1, m-2)$ by contractions:

$$\begin{aligned} &\delta^{b_j}{}_{a_i}\, T^{a_1\dots a_n}_{b_1\dots b_m}\,, \quad \text{for } 1 \le i \le n,\ 1 \le j \le m\,; \\ &\epsilon_{a_i a_j b_{m+1}}\, T^{a_1\dots a_n}_{b_1\dots b_m}\,, \quad \text{for } 1 \le i, j \le n\,; \\ &\epsilon^{b_i b_j a_{n+1}}\, T^{a_1\dots a_n}_{b_1\dots b_m}\,, \quad \text{for } 1 \le i, j \le m\,. \end{aligned} \tag{7.56}$$

When no nonvanishing tensors can be obtained in this way, the tensor $T(n, m)$ is said to be *irreducible*. Thus, irreducible tensors of SU(3) are *traceless* and *totally symmetric* in indices of the same type. Because of these restrictions, not all components of an irreducible tensor are linearly independent. The number of independent components can be easily calculated by subtracting the number of independent conditions from the total number of components. The general formula is

$$d(\mathrm{R}) = \tfrac{1}{2}\,(n+1)(m+1)(n+m+2)\,. \tag{7.57}$$

As discussed in the previous section, the independent components of a tensor form a basis of the vector space of the corresponding *irreducible representation*, which will be denoted by $\mathrm{R}(n, m)$, the dimension of which is just $\dim \mathrm{R}(n, m) = d(\mathrm{R})$. If the representation conjugate to R is denoted by $\mathrm{R}^*$, then $\mathrm{R}^*(n, m) = \mathrm{R}(m, n)$, which follows from the equivalence of $T^*(n, m)$ and $T(m, n)$. Of course, representations $\mathrm{R}(n, n)$ are self-conjugate; in particular, $\mathrm{R}(1, 1)$ corresponds to the adjoint representation. An irreducible representation is usually labeled by its dimension, but this convenient shorthand notation is not without ambiguities, since, for example, $\dim \mathrm{R}(4, 0) = \dim \mathrm{R}(2, 1) = 15$. Some of the lower-dimensional irreducible tensors and representations are given in the following table:

| | | |
|---|---|---|
| 1 | (0,0) | **1** |
| $T^a$ | (1,0) | **3** |
| $T_a$ | (0,1) | **3*** |
| $T^a{}_b$ | (1,1) | **8** |
| $T^{ab}$ | (2,0) | **6** |
| $T_{ab}$ | (0,2) | **6*** |
| $T^{abc}$ | (3,0) | **10** |
| $T_{abc}$ | (0,3) | **10*** |
| $T^{ab}{}_{cd}$ | (2,2) | **27** |

### 7.2.3 Physical Significance of $F_3$ and $F_8$

What is the physical significance of the diagonalized matrices $F_3 = \lambda_3/2$ and $F_8 = \lambda_8/2$? All we know at this point is that two components of the basic spinor form an isodoublet ($I = 1/2$) and one, an isosinglet ($I = 0$), with eigenvalues of $F_3$ and $F_8$ for the triplet $q^a = (q^1, q^2, q^3)$ given by

$$F_3 = (\tfrac{1}{2}, -\tfrac{1}{2}, 0)\,,$$
$$F_8 = (\tfrac{1}{2\sqrt{3}}, \tfrac{1}{2\sqrt{3}}, -\tfrac{1}{\sqrt{3}})\,. \tag{7.58}$$

As the basic spinor transforms with $\exp(\mathrm{i}\alpha F_3)$ and $\exp(\mathrm{i}\beta F_8)$, so does its conjugate with $\exp(-\mathrm{i}\alpha F_3)$ and $\exp(-\mathrm{i}\beta F_8)$. Hence, the conjugate triplet $q_a$ has the same values of $F_3$ and $F_8$ as does $q^a$ but with reversed signs (see Fig. 7.1). Once we know this precise correspondence between the indices of the spinor and the elements of the diagonal matrices, it is easy to calculate the additive quantum numbers that are $F_3$ and $F_8$ for any components of any given irreducible tensor. We thus have for irreducible tensor $T(n, m)$:

$$F_3 = \tfrac{1}{2}(n_1 - n_2) - \tfrac{1}{2}(m_1 - m_2)\,, \tag{7.59}$$

$$\begin{aligned} F_8 &= \tfrac{1}{2\sqrt{3}}(n_1 + n_2 - m_1 - m_2) + \tfrac{1}{\sqrt{3}}(-n_3 + m_3) \\ &= \tfrac{\sqrt{3}}{2}(-n_3 + m_3) + \tfrac{1}{2\sqrt{3}}(n - m)\,, \end{aligned} \tag{7.60}$$

where $n_i$ denotes the number of *upper* indices with values $i$, and $m_j$, the number of *lower* indices with values $j$.

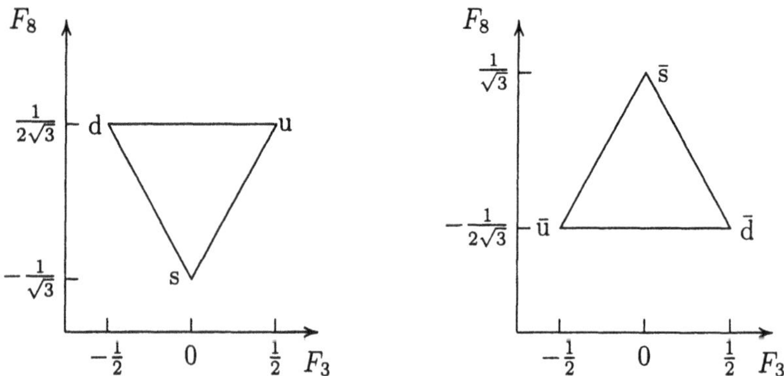


**Fig. 7.1.** Ordinary and conjugate fundamental representations of SU(3)

Since charge conservation is to be made part of the group structure, we ought to express the charge operator $Q$ in terms of $F_3$, $F_8$, and $N_B$. Since there should not be any distinction in group properties between meson and baryon multiplets of the same dimensions, $Q$ cannot vary with the baryon number and hence should depend only on $F_3$ and $F_8$, i.e. should transform as some component of an SU(3) octet:

$$Q = aF_3 + bF_8\,. \tag{7.61}$$

Now, from the discussion in Sect. 7.1 we know that the pions belong to the adjoint representation of SU(2), and from data we know that the pions are almost degenerate with the kaons ($K^{\pm}$, $K^{0}$). We may therefore reasonably surmise that the pions and the kaons belong to the adjoint representation of SU(3), i.e. to an octet, denoted by the tensor $M^{a}{}_{b}$. And just as $\pi^{1}{}_{2}$ of SU(2) is identified with $\pi^{+}$, so too is $M^{1}{}_{2}$ of SU(3) with that same particle and, by extension, $M^{1}{}_{3}$ with $K^{+}$. With these assignments, we obtain from (60), $F_3 = 1$, $F_8 = 0$ for $\pi^{+}$ and $F_3 = 1/2$, $F_8 = \sqrt{3}/2$ for $K^{+}$, which allows us to deduce that $a = 1$ and $b = 1/\sqrt{3}$, and thus

$$Q = F_3 + \frac{F_8}{\sqrt{3}} . \tag{7.62}$$

Comparing this relation with $Q = I_3 + Y/2$, it now becomes clear how $F_3$ and $F_8$ should be interpreted, namely,

$$F_3 = I_3 , \tag{7.63}$$

$$F_8 = \tfrac{\sqrt{3}}{2} Y . \tag{7.64}$$

For reference, we give the explicit expressions for eigenvalues of $I_3$, $Y$, and $Q$ in terms of the characteristics of components of an irreducible tensor of rank $(n, m)$:

$$\begin{aligned} I_3 &= \tfrac{1}{2}\,(n_1 - n_2 - m_1 + m_2) , \\ Y &= -n_3 + m_3 + \tfrac{1}{3}(n - m) , \\ Q &= n_1 - m_1 - \tfrac{1}{3}(n - m) , \end{aligned} \tag{7.65}$$

where $n_i$ and $m_i$ have the same meaning as in (60).

In Table 7.2 we show the values of the familiar quantum numbers for members of the fundamental triplet; the corresponding values for the conjugate triplet are obtained by reversing all signs. Thus, SU(3) symmetry makes the very surprising prediction that members of the fundamental representation have *fractional* charges, hypercharges, and baryon numbers. Nor is it the only representation with such unusual properties, as (65) shows. Therefore, assuming SU(3) symmetry valid in hadronic physics and admitting as observable only particles with *integral* charges and hypercharges, we must conclude that the only possible physical representations are those with a zero *triality number*, that is, with

$$t \equiv n - m \,(\text{mod } 3) = 0 . \tag{7.66}$$

Admissible candidates of the lower ranks are R(0, 0), R(1, 1), R(3, 0), and R(0, 3). In the context of SU(3), the physical particles are constructed from a triplet of fundamental particles (called *quark*) and its conjugate (*antiquark*), belonging respectively to the fundamental representations R(1, 0) and R(0, 1)

**Table 7.2.** Quantum numbers of the **SU(3)** fundamental triplet

| | $I_3$ | $S$ | $Y$ | $N_\mathrm{B}$ | $Q$ |
|---|---|---|---|---|---|
| $q^1$ | $\frac{1}{2}$ | 0 | $\frac{1}{3}$ | $\frac{1}{3}$ | $\frac{2}{3}$ |
| $q^2$ | $-\frac{1}{2}$ | 0 | $\frac{1}{3}$ | $\frac{1}{3}$ | $-\frac{1}{3}$ |
| $q^3$ | 0 | $-1$ | $-\frac{2}{3}$ | $\frac{1}{3}$ | $-\frac{1}{3}$ |

of SU(3). Since the triality of a triplet or antitriplet is 1, the physical particles should be bound states of a quark–antiquark pair, or three quarks, or a multiple of these. As a quark has $N_\mathrm{B} = 1/3$ and an antiquark $N_\mathrm{B} = -1/3$, mesons must be made of a quark–antiquark pair so that $N_\mathrm{B} = 0$, and baryons, made of three quarks so that $N_\mathrm{B} = 1$.

Finally, if quarks were spinless, one would expect to find scalar mesons lying lowest, just below the p-wave vector mesons, in the meson mass spectrum. This is not what is observed. On the other hand, if quarks and antiquarks are assumed to have spin $1/2$, the empirical meson spectrum can be understood in a natural way. Similarly, the simplest way to account for the existence of baryons of spin $1/2$ is to set the spins of all quarks and antiquarks to $1/2$. Hence, quarks and antiquarks will be taken as spin-$1/2$ fermions.

### 7.2.4 $3 \times 3^*$ Equal Mesons

Group theory provides us with powerful tools to generate particle multiplets from a quark–antiquark pair or from three quarks.

Let us begin with the first case. A quark–antiquark pair is represented symbolically by $\mathbf{3} \times \mathbf{3}^*$, or more explicitly by $q^a\, q_b$. This tensor product is reducible as it may be rewritten as

$$q^a\, q_b = \tfrac{1}{3}\delta^a{}_b\,(q^c q_c) + q^a\, q_b - \tfrac{1}{3}\delta^a{}_b\,(q^c q_c)\,. \tag{7.67}$$

It is clear that

$$S = \tfrac{1}{\sqrt{3}}\,(q^c q_c) \tag{7.68}$$

is invariant and corresponds to the representation R(0, 0), whereas

$$M^a{}_b = q^a\, q_b - \tfrac{1}{3}\delta^a{}_b\,(q^c q_c) \tag{7.69}$$

transforms under SU(3) as a tensor of rank (1, 1) and thus corresponds to the irreducible self-conjugate representation R(1, 1). The decomposition (67) is written symbolically as

$$\mathbf{3} \times \mathbf{3}^* = \mathbf{1} + \mathbf{8}\,. \tag{7.70}$$

Therefore, in as much as they can be viewed as quark–antiquark pairs, mesons fall into SU(3) singlets or octets, regardless of the spatial configurations of their constituents. Let us use the conventional symbols u, d, and s for different types, or *flavors*, of quarks, and $(q^1, q^2, q^3) = (u, d, s)$ and $(q_1, q_2, q_3) = (\bar{u}, \bar{d}, \bar{s})$ for their respective states. Then the quark contents of **1** and **8** are

$$S = \frac{1}{\sqrt{3}}(u\bar{u} + d\bar{d} + s\bar{s}), \tag{7.71}$$

$$M^a{}_b = \begin{pmatrix} \frac{1}{3}(2u\bar{u} - d\bar{d} - s\bar{s}) & u\bar{d} & u\bar{s} \\ d\bar{u} & \frac{1}{3}(-u\bar{u} + 2d\bar{d} - s\bar{s}) & d\bar{s} \\ s\bar{u} & s\bar{d} & \frac{1}{3}(-u\bar{u} - d\bar{d} + 2s\bar{s}) \end{pmatrix}. \tag{7.72}$$

The kinds of physical fields the tensors $S$ and $M^a{}_b$ can accommodate are determined by the quantum numbers assigned to each component. While the meson singlet must be completely neutral, the meson octet has the following values for $(I_3, Y, Q)$, as calculated from (65):

$$\begin{matrix}(I_3, Y, Q) \\ \text{of } M^a{}_b\end{matrix} = \begin{pmatrix} (0,0,0) & (1,0,1) & (\frac{1}{2},1,1) \\ (-1,0,-1) & (0,0,0) & (-\frac{1}{2},1,0) \\ (-\frac{1}{2},-1,-1) & (\frac{1}{2},-1,0) & (0,0,0) \end{pmatrix}. \tag{7.73}$$

Members of the same multiplet must of course have the same spins and parities (conserved in strong interactions), and approximately equal masses. To better identify the physical fields associated with tensor components, we must have some general idea about the internal structure of the mesons. For the lower mass particles, it is a reasonable approximation to assume that each quark is in the same s-orbit of a common potential and that there is no interquark interaction. This assumption implies, first, that the total angular momentum of a meson comes just from coupling two quark intrinsic spins $1/2$, leading to $J = 0$ or $J = 1$, and, second, that its parity is odd, as it should be for a fermion–antifermion system in an s-state. So, the low-lying mesons must be either *pseudoscalar* ($J^P = 0^-$) or *vectorial* ($J^P = 1^-$). The set of pseudoscalar mesons shown in Table 6.5, with masses well below the mass of the nucleon, is a clear candidate for an SU(3) octet, together with the pseudoscalar meson $\eta'(958)$ identified with the associated singlet, as in (70). Another candidate is offered by the vector mesons, with masses around the nucleon mass, listed in Table 7.3. These circumstantial associations of SU(3) multiplets and spins–parities look reasonable but need further and firmer justifications.

**Table 7.3.** Vector meson nonet

| | $I$ | $Y$ | Mass (MeV) |
|---|---|---|---|
| $\rho$ | 1 | 0 | 776 |
| K* | $\frac{1}{2}$ | $\pm 1$ | 872 |
| $\omega$ | 0 | 0 | 783 |
| $\phi$ | 0 | 0 | 1019 |

To be specific, let us focus on the pseudoscalar nonet, composed of a singlet and an octet. The singlet is realized by a pseudoscalar field, $\phi_0$, which forms the greater part of a field that annihilates an $\eta'(958)$. We call it the singlet-$\eta$, or $\eta_1$,

$$S = \phi_0 = \eta_1 \,. \tag{7.74}$$

Its quark content is given in (71).

As for the octet, its basis, $M^a{}_b$, is a $3 \times 3$ traceless tensor, and thus may be expanded in terms of the generators of the fundamental representation, $\lambda_i$, with real field coefficients, $\phi_i$ for $i = 1, \ldots, 8$,

$$M^a{}_b = \frac{1}{\sqrt{2}} \sum_{i=1}^{8} (\lambda_i)^a{}_b \, \phi_i \,, \tag{7.75}$$

with the normalization chosen such that

$$M^a{}_b \, M^b{}_a = \frac{1}{2} \sum \mathrm{Tr}(\lambda_i \lambda_j) \, \phi_i \phi_j = \phi_1^2 + \ldots + \phi_8^2 \,. \tag{7.76}$$

Using the known expressions for $\lambda_i$, one gets

$$M^a{}_b = \begin{pmatrix} \frac{1}{\sqrt{2}}\phi_3 + \frac{1}{\sqrt{6}}\phi_8 & \frac{1}{\sqrt{2}}(\phi_1 - \mathrm{i}\phi_2) & \frac{1}{\sqrt{2}}(\phi_4 - \mathrm{i}\phi_5) \\ \frac{1}{\sqrt{2}}(\phi_1 + \mathrm{i}\phi_2) & -\frac{1}{\sqrt{2}}\phi_3 + \frac{1}{\sqrt{6}}\phi_8 & \frac{1}{\sqrt{2}}(\phi_6 - \mathrm{i}\phi_7) \\ \frac{1}{\sqrt{2}}(\phi_4 + \mathrm{i}\phi_5) & \frac{1}{\sqrt{2}}(\phi_6 + \mathrm{i}\phi_7) & -\frac{2}{\sqrt{6}}\phi_8 \end{pmatrix} . \tag{7.77}$$

Following the field definitions in (22) for the SU(2) case and making use of the assigned quantum numbers in (73), we identify the physical fields with the symbols of the particles they *annihilate*, and relate them to the eight cartesian fields $\{\phi_i\}$ as follows:

$$\begin{aligned} \pi^0 &= \phi_3 \,, \\ \pi^\pm &= \frac{1}{\sqrt{2}}(\phi_1 \mp \mathrm{i}\phi_2) \,, \\ K^\pm &= \frac{1}{\sqrt{2}}(\phi_4 \mp \mathrm{i}\phi_5) \,, \\ K^0 &= \frac{1}{\sqrt{2}}(\phi_6 - \mathrm{i}\phi_7) \,, \\ \bar{K}^0 &= \frac{1}{\sqrt{2}}(\phi_6 + \mathrm{i}\phi_7) \,, \\ \eta_8 &= \phi_8 \,. \end{aligned} \tag{7.78}$$

The octet-$\eta$, or $\eta_8$, field overlaps but does not completely coincide with the physical meson field $\eta(547)$, as we shall see shortly.

Similar considerations apply to the vector mesons as well. The quark flavor contents of both octets can be obtained from (72) and are given as

$$
\begin{aligned}
\mathrm{K}^+\,,\quad & \mathrm{K}^{*+} && = u\bar{s}\,,\\
\mathrm{K}^0\,,\quad & \mathrm{K}^{*0} && = d\bar{s}\,,\\
\mathrm{K}^-\,,\quad & \mathrm{K}^{*-} && = s\bar{u}\,,\\
\bar{\mathrm{K}}^0\,,\quad & \bar{\mathrm{K}}^{*0} && = s\bar{d}\,,\\
\pi^+\,,\quad & \rho^+ && = u\bar{d}\,,\\
\pi^-\,,\quad & \rho^- && = d\bar{u}\,,\\
\pi^0\,,\quad & \rho^0 && = \tfrac{1}{\sqrt{2}}(u\bar{u} - d\bar{d})\,,\\
\eta_8\,,\quad & \omega_8 && = \tfrac{1}{\sqrt{6}}(u\bar{u} + d\bar{d} - 2s\bar{s})\,,\\
\eta_1\,,\quad & \omega_1 && = \tfrac{1}{\sqrt{3}}(u\bar{u} + d\bar{d} + s\bar{s})\,.
\end{aligned}
$$

Pseudoscalar and vector mesons (shown in $Y$–$I_3$ plots in Fig. 7.2a–b) have identical quark–antiquark compositions and thus may be differentiated only by the properties that lie outside SU(3).

### 7.2.5 3 × 3 × 3 Equal Baryons

Three quarks give $N_\mathrm{B} = 1$ and thus make baryons. The product of three quark fields can be reduced in two steps to a linear combination of irreducible tensors by simple tensor calculus,

$$3 \times 3 \times 3 = (\mathbf{3}^* + \mathbf{6}) \times \mathbf{3} = \mathbf{1} + \mathbf{8} + \mathbf{8} + \mathbf{10}\,. \tag{7.79}$$

In the first step, a product of two quarks $\mathbf{3} \times \mathbf{3}$ yields

$$
\begin{aligned}
q^a q^b &= \tfrac{1}{2}(q^a q^b - q^b q^a) + \tfrac{1}{2}(q^a q^b + q^b q^a)\\
&= \tfrac{1}{2}\epsilon^{abk}\theta_k + \tfrac{1}{2}S^{ab}\,,
\end{aligned}
$$

which is a combination of a $\mathbf{3}^*$ tensor and a $\mathbf{6}$ tensor:

$$\theta_k = \epsilon_{kmn} q^m q^n\,, \qquad S^{ab} = q^a q^b + q^b q^a\,. \tag{7.80}$$

In the second step, the products $\mathbf{3}^* \times \mathbf{3}$ and $\mathbf{6} \times \mathbf{3}$ are decomposed into the sums $\mathbf{1} + \mathbf{8}$ and $\mathbf{8} + \mathbf{10}$, respectively,

$$
\begin{aligned}
\theta_k q^c &= \tfrac{1}{3}\delta^c{}_k(\theta_m q^m) + [\theta_k q^c - \tfrac{1}{3}\delta^c{}_k(\theta_m q^m)]\,,\\
S^{ab} q^c &= \tfrac{1}{3}\left(\epsilon^{ack} S^{bm} + \epsilon^{bck} S^{am}\right) q^n\,\epsilon_{mnk} + \tfrac{1}{3}(S^{ab} q^c + S^{bc} q^a + S^{ca} q^b)\,.
\end{aligned}
$$

Thus, the full product $q^a q^b q^c$ reduces to the sum

$$
\begin{aligned}
q^a q^b q^c &= \tfrac{1}{2}\epsilon^{abk}\theta_k q^c + \tfrac{1}{2}S^{ab} q^c\,,\\
&= \tfrac{1}{\sqrt{6}}\epsilon^{abc}\, S + \tfrac{1}{\sqrt{2}}\epsilon^{abd}\, B^c{}_d + \tfrac{1}{\sqrt{6}}\left(\epsilon^{acd}\, N^b{}_d + \epsilon^{bcd}\, N^a{}_d\right) + D^{abc}\,,
\end{aligned}
\tag{7.81}
$$

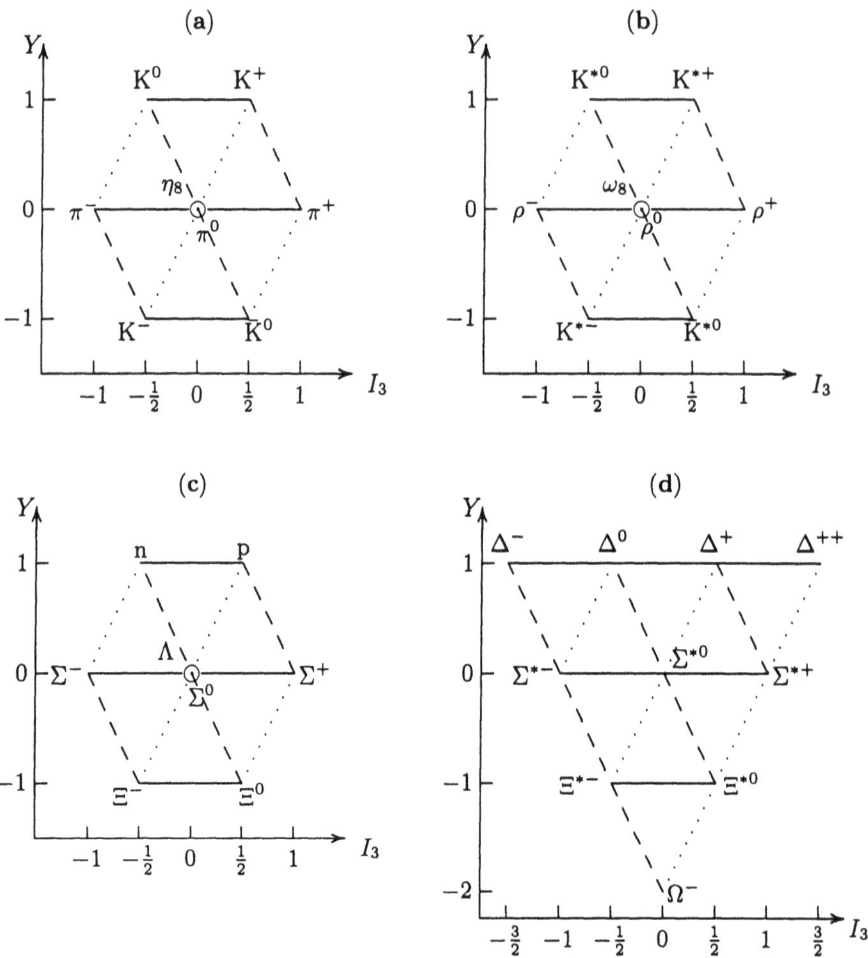


**Fig. 7.2.** (**a**) $0^-$ mesons; (**b**) $1^-$ mesons; (**c**) $\frac{1}{2}^+$ baryons; (**d**) $\frac{3}{2}^+$ baryons. Solid lines join members of I-spin multiplets, dashed lines join members of U-spin multiplets, and dotted lines join members of V-spin multiplets

where the irreducible tensors of ranks (0,0), (1,1), (1,1), and (3,0) are defined as follows:

$$
\begin{aligned}
S &= \tfrac{1}{\sqrt{6}}\theta_a q^a ,\\
B^a{}_b &= \tfrac{1}{\sqrt{2}}\left[\theta_b q^a - \tfrac{1}{3}\delta^a{}_b(\theta_c q^c)\right] ,\\
N^a{}_b &= \tfrac{1}{\sqrt{6}}\,\epsilon_{bcd}\, S^{ac}\, q^d ,\\
D^{abc} &= \tfrac{1}{6}\left(S^{ab}q^c + S^{bc}q^a + S^{ca}q^b\right) .
\end{aligned} \tag{7.82}
$$

Note that the relative orders of the quark factors are important, since the position of each factor is meant to correspond to some specific attributes other than flavors – coordinate or kinematic variables – of the first, second, or third quark. For example,

$$\begin{aligned} B^1{}_2 &= \tfrac{1}{\sqrt{2}}\theta_2 q^1 = \tfrac{1}{\sqrt{2}}\,(q^3q^1 - q^1q^3)q^1 \\ &= \tfrac{1}{\sqrt{2}}\,[s(1)\,u(2) - u(1)\,s(2)]\,u(3)\,, \end{aligned} \tag{7.83}$$

$$\begin{aligned} B^1{}_3 &= \tfrac{1}{\sqrt{2}}\theta_3 q^1 = \tfrac{1}{\sqrt{2}}\,(q^1q^2 - q^2q^1)q^1 \\ &= \tfrac{1}{\sqrt{2}}\,[u(1)\,d(2) - d(1)\,u(2)]\,u(3)\,, \end{aligned} \tag{7.84}$$

$$\begin{aligned} N^1{}_2 &= \tfrac{1}{\sqrt{6}}\epsilon_{2cd}\,S^{1c}q^d = \tfrac{1}{\sqrt{6}}\,\left[(q^1q^3 + q^3q^1)q^1 - 2q^1q^1q^3\right] \\ &= \tfrac{1}{\sqrt{6}}\,\{[u(1)\,s(2) + s(1)\,u(2)]\,u(3) - 2u(1)u(2)s(3)\}\,, \end{aligned} \tag{7.85}$$

$$\begin{aligned} N^1{}_3 &= \tfrac{1}{\sqrt{6}}\epsilon_{3cd}\,S^{1c}q^d = \tfrac{1}{\sqrt{6}}\,\left[2q^1q^1q^2 - (q^1q^2 + q^2q^1)q^1\right] \\ &= \tfrac{1}{\sqrt{6}}\,\{2u(1)u(2)d(3) - [u(1)\,d(2) + d(1)\,u(2)]\,u(3)\}\,. \end{aligned} \tag{7.86}$$

Even though the two eight-dimensional representations are completely indistinguishable under SU(3) transformations, they can be distinguished from their properties outside the group, which, in this case, are symmetries under permutations of individual quarks. Thus, the octet $B^a{}_b$ is antisymmetric under interchanges of the first two quarks while the octet $N^a{}_b$ is symmetric. To know which one is to be assigned to which physically observed baryons, one needs additional assumptions or information beyond SU(3) symmetry.

Ordinary baryons, which are made up of u, d, and s quarks, should occur as SU(3) singlets, octets and decuplets (Fig. 7.2c–d). The lightest baryons with $J^P = {}^1\!/_2{}^+$, which include the neutron and the proton (Table 6.5), form an octet. The lightest $J^P = {}^3\!/_2{}^+$, which include the $\Delta$-resonance and the $\Omega^-$ (see Table 7.4), form a decuplet.

**Table 7.4.** Baryon states with $J^P = {}^3\!/_2{}^+$, with average masses given in the last column

| $Y$ | $I$ | | | | | | | | Masses |
|---|---|---|---|---|---|---|---|---|---|
| | $I_3$ : | $-3/2$ | $-1$ | $-1/2$ | $0$ | $1/2$ | $1$ | $3/2$ | (MeV) |
| 1 | $3/2$ | $\Delta^-$ | | $\Delta^0$ | | $\Delta^+$ | | $\Delta^{++}$ | 1232 |
| 0 | 1 | | $\Sigma^{*-}$ | | $\Sigma^{*0}$ | | $\Sigma^{*+}$ | | 1385 |
| −1 | $1/2$ | | | $\Xi^{*-}$ | | $\Xi^{*0}$ | | | 1530 |
| −2 | 0 | | | | $\Omega^-$ | | | | 1672 |

Members of the SU(3) multiplets are identified with physical baryons by their quantum numbers, calculated from relations (65). The SU(3) singlet **1** is a uds-state ($\Lambda_1$), similar in content to the $\Lambda$ found in the octet; although it may occur at a higher energy, it is forbidden in the ground state multiplet by

Fermi–Dirac statistics (see below), and therefore cannot mix with the octet-$\Lambda$. Nor can mixing occur among the $1/2^+$ and $3/2^+$ multiplets since no two states have the same quantum numbers $I$, $J$, and $Y$.

The *baryon octet* $B^a{}_b$ (or $N^a{}_b$) is symbolically represented in terms of the physical fields by

$$B^a{}_b = \begin{pmatrix} \frac{1}{\sqrt{2}}\Sigma^0 + \frac{1}{\sqrt{6}}\Lambda^0 & \Sigma^+ & p \\ \Sigma^- & -\frac{1}{\sqrt{2}}\Sigma^0 + \frac{1}{\sqrt{6}}\Lambda^0 & n \\ -\Xi^- & \Xi^0 & -\frac{2}{\sqrt{6}}\Lambda^0 \end{pmatrix}, \tag{7.87}$$

where $\Xi^-$ comes with a minus sign to conform to a sign convention for ladder-operators of the isospin subgroup. The quark contents of the baryon fields can be inferred from (82), and their normalization is fixed by requiring

$$\begin{aligned}(B^a{}_b)^* B^a{}_b = \Sigma^{+*}\Sigma^+ + \Sigma^{0*}\Sigma^0 + \Sigma^{-*}\Sigma^- \\ + p^*p + n^*n + \Xi^{-*}\Xi^- + \Xi^{0*}\Xi^0 + \Lambda^{0*}\Lambda^0 ,\end{aligned} \tag{7.88}$$

where (*) means complex conjugation.

The *decuplet* contains the isospin multiplets $I = 0,\ 1/2,\ 1$, and $3/2$ corresponding respectively to the tensor components $D^{333}$, $D^{i33}$, $D^{ij3}$, and $D^{ijk}$, for $i, j, k = 1, 2$ . These are assigned to the lowest excited baryon states:

$$\begin{aligned}
D^{111} &= \Delta^{++}, & D^{112} &= \tfrac{1}{\sqrt{3}}\Delta^+, & D^{122} &= \Delta^0, & D^{222} &= \Delta^-, \\
D^{113} &= \tfrac{1}{\sqrt{3}}\Sigma^{*+}, & D^{123} &= \tfrac{1}{\sqrt{6}}\Sigma^{*0}, & D^{223} &= \tfrac{1}{\sqrt{3}}\Sigma^{*-}, \\
D^{133} &= \tfrac{1}{\sqrt{3}}\Xi^{*0}, & D^{233} &= \tfrac{1}{\sqrt{3}}\Xi^{*-}, \\
D^{333} &= \Omega^- .
\end{aligned} \tag{7.89}$$

Here (*) refers to an excited state. Again, the numerical factors come from the normalization condition

$$(D^{abc})^* D^{abc} = \sum_{i=1}^{10} \psi_i^* \psi_i , \tag{7.90}$$

where $\psi_i$ stand for the decuplet states $\Delta$, $\Sigma^*$, $\Xi^*$, and $\Omega$.

## 7.3 Mass Splitting of the Hadron Multiplets

In the final analysis, the validity of the SU(3) symmetry discussed in the previous section rests on the presumed symmetry among the three quark flavors u, d, and s, that is, on the flavor independence of the strong forces and on the equality of the quark masses:

$$m_\mathrm{u} = m_\mathrm{d} = m_\mathrm{s} . \tag{7.91}$$

However, these mass relations cannot be exact because data show that mesons or baryons in a given SU(3) multiplet, degenerate though they may be within the same I-spin multiplets, differ in mass by a few hundred MeV for different values of hypercharges. From the measured masses and from the predicted quark compositions of hadrons, it seems more realistic to assume rather that the SU(3) *symmetry of flavor is broken* but in such a way as to leave the isospin symmetry intact. This can be realized by requiring

$$m_{\mathrm{u}} = m_{\mathrm{d}} < m_{\mathrm{s}} . \tag{7.92}$$

Therefore, the hadronic Hamiltonian may take the form

$$H_{\mathrm{st}} = H_0 + H_8 ,$$

where $H_0$ is SU(3)-invariant, and $H_8$ symmetry breaking. In the following, it is assumed that $H_8$ is weaker than $H_0$ to the extent that it may be regarded as a perturbation. (Hadron masses indicate symmetry violations of the order of 1/10; cf. Table 7.4.) To zeroth order in $H_8$, the theory is SU(3)-invariant and the SU(3) multiplets are degenerate in mass values. If one assumes further that the symmetry-breaking term transforms in a definite way under SU(3) transformations, one may derive relations among the masses of the particles belonging to a particular SU(3) multiplet. The assumption that the part of $H_8$ responsible for the mass splitting of hadron multiplets transforms precisely as the ${T^3}_3$ component of an octet tensor, so that the baryon number, charge and isospin are all conserved, has led to the famous Gell-Mann–Okubo (GMO) mass formula,

$$M(I,Y) = a + bY + c\,[I(I+1) - \tfrac{1}{4}Y^2] . \tag{7.93}$$

This relation proves to be in reasonably good agreement with experiment.

The underlying assumption of the GMO formula – that the symmetry-breaking term transforms as ${T^3}_3$ – could be plausibly argued as a mere extension of the case of noninteracting quarks whose masses satisfy (92). In this simpler situation, the mass term in the Lagrangian is given by

$$m_{\mathrm{u}}(q_1q^1 + q_2q^2) + m_{\mathrm{s}}q_3q^3 = m_0\, q_aq^a + m_1\, {M^3}_3 , \tag{7.94}$$

where $m_0 = (2m_{\mathrm{u}} + m_{\mathrm{s}})/3$ and $m_1 = m_{\mathrm{s}} - m_{\mathrm{u}}$. Thus, (94) is composed of an SU(3)-invariant term, $q_aq^a$, and a symmetry-breaking term, ${M^3}_3 = q_3q^3 - (q_aq^a/3)$, which is a component of an octet tensor. To simplify, we are ignoring Dirac $\gamma$-matrices in writing down these expressions. In what follows, we will derive mass relations for specific multiplets from effective symmetry-breaking mass terms, which are bilinear in the state functions and which transform as ${M^3}_3$ under SU(3).

### 7.3.1 Baryons

The effective mass terms in the baryon Lagrangian are quadratic in the baryon fields. To begin, let us consider members of the nucleon octet, and therefore the bilinear product $\bar{B}^a{}_b B^c{}_d$, where $\bar{B} = B^\dagger$. The decomposition of the product $\mathbf{8} \times \mathbf{8}$ according to

$$\mathbf{8} \times \mathbf{8} = \mathbf{1} + \mathbf{8}_\mathrm{f} + \mathbf{8}_\mathrm{d} + \mathbf{10} + \mathbf{10}^* + \mathbf{27} \tag{7.95}$$

produces two types of octet [cf. (82)]: an antisymmetric, or f-type, and a symmetric, or d-type, defined by

$$\begin{aligned}(\overline{B}B_\mathrm{f})^a_b &= \overline{B}^a_c B^c_b - \overline{B}^c_b B^a_c\,,\\ (\overline{B}B_\mathrm{d})^a_b &= \overline{B}^a_c B^c_b + \overline{B}^c_b B^a_c - \tfrac{2}{3}\delta^a_b\,\overline{B}^c_d B^d_c\,.\end{aligned}$$

Their components of interest are

$$\begin{aligned}(\overline{B}B_\mathrm{f})^3{}_3 &= \bar{p}p + \bar{n}n - (\overline{\Xi}^-\Xi^- + \overline{\Xi}^0\Xi^0)\,,\\ (\overline{B}B_\mathrm{d})^3{}_3 &= \tfrac{1}{3}(\bar{p}p + \bar{n}n + \overline{\Xi}^-\Xi^- + \overline{\Xi}^0\Xi^0)\\ &\quad + \tfrac{2}{3}\bar{\Lambda}\Lambda - \tfrac{2}{3}(\overline{\Sigma}^+\Sigma^+ + \overline{\Sigma}^-\Sigma^- + \overline{\Sigma}^0\Sigma^0)\,,\end{aligned}$$

where $\overline{\psi}$ means $\psi^\dagger$. The most general mass terms for baryons that include symmetry breaking of the kind $T^3{}_3$ are given by the linear combination

$$m_0 \overline{B}^a_b B^b_a + m_\mathrm{d}(\overline{B}B_\mathrm{d})^3{}_3 + m_\mathrm{f}(\overline{B}B_\mathrm{f})^3{}_3\,, \tag{7.96}$$

where $m_0$, $m_\mathrm{d}$, and $m_\mathrm{f}$ are free parameters, in terms of which one expresses the masses of the members of the baryon octet:

$$\begin{aligned}M_\mathrm{N} &= m_0 + \tfrac{1}{3}m_\mathrm{d} + m_\mathrm{f}, & M_\Sigma &= m_0 - \tfrac{2}{3}m_\mathrm{d}\,,\\ M_\Xi &= m_0 + \tfrac{1}{3}m_\mathrm{d} - m_\mathrm{f}, & M_\Lambda &= m_0 + \tfrac{2}{3}m_\mathrm{d}\,.\end{aligned}$$

This yields the GMO mass relation for the baryon octet

$$\tfrac{1}{2}\,(M_\mathrm{N} + M_\Xi) = \tfrac{1}{4}\,(3M_\Lambda + M_\Sigma)\,. \tag{7.97}$$

The corresponding experimental values are 1129 MeV and 1135 MeV.

To construct the effective mass term for the decuplet isobars, one considers the product reduction

$$\mathbf{10}^* \times \mathbf{10} = \mathbf{1} + \mathbf{8} + \mathbf{27} + \mathbf{64}\,, \tag{7.98}$$

which contains a single octet; call it $H$. Thus, the only possible effective mass term transforming like $T^3{}_3$ is

$$\begin{aligned}H^3{}_3 &= \overline{D}_{3cd}D^{3cd} - \tfrac{1}{3}\,\left(\overline{D}_{ecd}D^{ecd}\right)\\ &= \tfrac{2}{3}\,\overline{\Omega}\Omega + \tfrac{1}{3}(\overline{\Xi}^{*-}\Xi^{*-} + \overline{\Xi}^{*0}\Xi^{*0})\\ &\quad - \tfrac{1}{3}\,(\bar{\Delta}^{++}\Delta^{++} + \bar{\Delta}^+\Delta^+ + \bar{\Delta}^0\Delta^0 + \bar{\Delta}^-\Delta^-)\,,\end{aligned}$$

where $\overline{D}_{abc} = (D^{abc})^\dagger$ and $\overline{\psi}_A = \psi^\dagger_A$. The most general mass term, which is

$$m_0\,\overline{D}_{abc}D^{abc} + m_1 H^3{}_3\,, \tag{7.99}$$

immediately yields the mass relations

$$M_\Omega - M_{\Xi^*} = M_{\Xi^*} - M_{\Sigma^*} = M_{\Sigma^*} - M_\Delta\,, \tag{7.100}$$

to be compared with the corresponding measured mass differences: 142, 145, and 153 MeV. This result – equal mass spacings in the decuplet – follows from $Y = 2(I-1)$ which holds for this multiplet, so that the general GMO relation reduces to a linear function of $Y$ in this case.

### 7.3.2 Mesons

The meson octets can be treated in the same way as the baryon octets, with two differences. First, since the boson mass parameters enter the Lagrangian quadratically, the mass relations will involve squared masses rather than simply masses. Secondly, since the meson multiplets are self-conjugate, the antisymmetric bilinear is absent so that the mass term arises from pure d-type coupling, $\overline{M}M_{\rm d}$. The resulting mass formulas are, for the pseudoscalar mesons,

$$\begin{array}{cc} 4M^2_{\rm K} & = 3M^2_{\eta 8} + M^2_\pi\,, \\ 0.984\,{\rm GeV}^2 & 0.916\,{\rm GeV}^2 \end{array} \tag{7.101}$$

and for the vector mesons,

$$\begin{array}{cc} 4M^2_{\rm K^*} & = 3M^2_{\omega 8} + M^2_\rho\,. \\ 3.18\,{\rm GeV}^2 & 3.71\,{\rm GeV}^2 \end{array} \tag{7.102}$$

For comparison, experimental data are also shown, setting $\eta_8 = \eta(547)$ and $\omega_8 = \phi(1019)$. The relations are not very well satisfied, especially for the vector mesons. This signals the presence of another important source of symmetry breaking. The singlet and the octet $I = 0$ components, while belonging to different representations, have the same quantum numbers and therefore could be mixed to give rise to the states that are actually observed.

To be specific, let us consider pseudoscalar mesons. The physically observed $\eta$ and $\eta'$ particles are linear combinations of the pure singlet and octet $I = 0$ states, $\eta_1$ and $\eta_8$, or inversely,

$$\begin{aligned} \eta_1 &= \eta\sin\theta + \eta'\cos\theta\,, \\ \eta_8 &= -\eta\cos\theta + \eta'\sin\theta\,, \end{aligned} \tag{7.103}$$

where $\theta$ is the mixing angle to be determined. The mass term in the effective Lagrangian is

$$m_0^2\, M^a{}_b M^b{}_a + m_1^2\,\eta_1^2 + m_{\rm d}^2\,(\overline{M}M_{\rm d})^3{}_3 + m_{18}\,\eta_1\,M^3{}_3\,. \tag{7.104}$$

Therefore we must have in the $\eta_1$–$\eta_8$ sector

$$m_1^2\,\eta_1^2 + M_8^2\,\eta_8^2 - 2\lambda\,\eta_1\eta_8 = M_\eta^2\,\eta^2 + M_{\eta'}^2\,\eta'^2\,, \tag{7.105}$$

where $\lambda = m_{18}/\sqrt{6}$. Substituting (103) into this equation, we obtain

$$\begin{aligned} M_\eta^2 &= m_1^2 \sin^2\theta + M_8^2 \cos^2\theta + 2\lambda\,\cos\theta\sin\theta\,, \\ M_{\eta'}^2 &= m_1^2 \cos^2\theta + M_8^2 \sin^2\theta - 2\lambda\,\cos\theta\sin\theta\,, \\ 0 &= (m_1^2 - M_8^2)\cos\theta\,\sin\theta + \lambda\,(\cos^2\theta - \sin^2\theta)\,. \end{aligned} \tag{7.106}$$

Eliminating $m_1$ and $\lambda$ from these equations leads to

$$\tan^2\theta = \left(M_\eta^2 - M_8^2\right) / \left(M_8^2 - M_{\eta'}^2\right)\,, \tag{7.107}$$

where $M_8 = M_{\eta 8}$ is found in (101).

Similarly for the vector mesons, the physical $\phi$ and $\omega$ are given in terms of the pure singlet and octet states, $\omega_1$ and $\omega_8$, by

$$\begin{aligned} \omega_1 &= \phi\sin\theta + \omega\cos\theta\,, \\ \omega_8 &= -\phi\cos\theta + \omega\sin\theta\,, \end{aligned} \tag{7.108}$$

where the mixing angle is to be determined by

$$\tan^2\theta = \left(M_\phi^2 - M_8^2\right) / \left(M_8^2 - M_\omega^2\right)\,, \tag{7.109}$$

for $M_8 = M_{\omega_8}$ as in (102).

The observed mass values yield $\theta \approx 11^\circ$ for the pseudoscalar mesons, and $\theta \approx 47^\circ$ for the vector mesons. Thus, while the $\eta_1$–$\eta_8$ mixing is relatively small, there is a sizable $\omega_1$–$\omega_8$ mixing. To see what this large mixing may imply, let us approximate it, in the latter case by $\theta \approx 35^\circ$, in order to have a round number for $\sin\theta = 1/\sqrt{3}$. Then, we get from (108)

$$\begin{aligned} \phi &= \tfrac{1}{\sqrt{3}}(\omega_1 - \sqrt{2}\omega_8) = s\bar{s}\,, \\ \omega &= \tfrac{1}{\sqrt{3}}(\omega_8 + \sqrt{2}\omega_1) = \tfrac{1}{\sqrt{2}}(u\bar{u} + d\bar{d})\,. \end{aligned} \tag{7.110}$$

In this 'ideal mixing', $\phi$ is made up entirely of strange quarks, and $\omega$, of nonstrange quarks. This may explain why $\omega$ has nearly the same mass as $\rho$ while $\phi$ is more massive than any other member of the vector meson octet.

A large $\omega_1$–$\omega_8$ mixing is also consistent with another observed fact. As $\phi$ and $\omega$ have the same quantum numbers, one would expect that they have similar strong interaction properties and, in particular, comparable strong decay widths. Experiment shows otherwise. While $\omega$ has a width typical of hadrons ($\Gamma_\omega = 8.4$ MeV) and decays predominantly as expected into the $\pi^+\pi^-\pi^0$ channel, $\phi$ has a significantly *smaller width* ($\Gamma_\phi = 4.4$ MeV) and decays predominantly via $\overline{\mathrm{K}}^0\,\mathrm{K}^0$ and $\mathrm{K}^+\,\mathrm{K}^-$ modes, rather than via the energetically more favorable $3\pi$ channel. To explain this and other similar data, Okubo (1963), Zweig (1964), and Iizuka (1966) independently suggested that strong processes in which the final states can only be reached through $q\bar{q}$ annihilations are suppressed (the OZI rule). This is illustrated by the quark flow diagrams in Fig. 7.3, in which individual constituent quarks are viewed as flowing from one hadron to another through interactions. Diagrams a and b, which involve only internally unbroken quark lines, are allowed, whereas diagram c, which involves internally disconnected quark lines, is suppressed by the phenomenological OZI rule. In the quantum chromodynamics language, this suppression is explained in terms of multigluon exchanges in the intermediate states and plays an especially important role in our understanding of the very narrow widths of the more massive mesons composed of quark–antiquark pairs of heavy flavors.

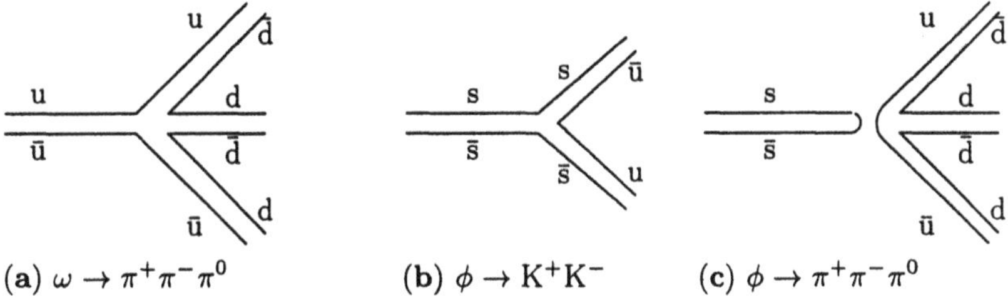


**Fig. 7.3.** Quark flow diagrams for $\omega$ and $\phi$ strong decays: (**a**) and (**b**) are allowed; (**c**) is forbidden by the OZI rule

## 7.4 Including Spin: SU(6)

To make the picture of hadrons in terms of quarks more precise, one has to introduce space-time degrees of freedom. In a nonrelativistic treatment, it means placing quarks in a common potential of some kind and letting them occupy the single-particle levels in this well, coupling their orbital angular momenta to a relative $\boldsymbol{L}$ and their spins to a total $\boldsymbol{S}$. The coupling $\boldsymbol{L}+\boldsymbol{S}=\boldsymbol{J}$ gives the total angular momentum of the system which is identified with the spin of the hadron thus formed. Disregarding for now particle statistics, the lowest hadronic states should be generated by quarks and antiquarks sitting on 1s levels, so that they have $L=0$. Higher in the mass spectrum will see states with $L=1,2,\ldots$ appearing. We shall limit our discussion to the $L=0$ states and, therefore, may concentrate on the properties that are spin and flavor dependent. It was shown by Gürsey and Radicati (1964) and by Sakita (1964), following an earlier work by Wigner (1937), that it is possible to determine, through group-theoretic relations between the internal properties and the space-time properties, the spin–parity values of the SU(3) multiplets. Wigner found that if the nuclear force is independent of both charge and spin, an interesting connection exists between the spin and isospin properties in nuclei. This assumption implies that the four possible degrees of freedom of the nucleon, two charge and two spin states – $p\uparrow$, $p\downarrow$, $n\uparrow$, and $n\downarrow$ – are interchangeable in transformations of a unitary group, the SU(4) group, and thus together form the fundamental representation of the group. The tensor products of the spin and isospin matrices,

$$\boldsymbol{I}_2' \otimes \boldsymbol{\sigma}, \quad \boldsymbol{\tau} \otimes \boldsymbol{I}_2, \quad \boldsymbol{\tau} \otimes \boldsymbol{\sigma},$$

where $\boldsymbol{I}_2$ and $\boldsymbol{I}_2'$ are $2 \times 2$ unit matrices in spin and isospin spaces, define the 15 generators of the infinitesimal transformations of the group in the basic representation. The adjoint representation is a multiplet of 15 states characterized by $I = 1$ and $S = 0$ (the $\pi$ of particle physics), by $I = 0$ and $S = 1$ (or $\omega$), and by $I = 1$ and $S = 1$ (or $\rho$). It thus provides a precise connection between internal and external quantum numbers.

The generalization of the above idea to SU(3) will lead to SU(6). It consists in replacing the three $2 \times 2$ isospin matrices $\tau_i$ with the eight $3 \times 3$ matrices $\lambda_i$ in the definition of the generators:

$$\lambda_0 \otimes \sigma_m, \quad \lambda_i \otimes \boldsymbol{I}_2, \quad \lambda_i \otimes \sigma_m, \qquad (i = 1, \ldots, 8;\ m = 1, 2, 3); \tag{7.111}$$

where $\lambda_0 = \frac{2}{\sqrt{6}} \boldsymbol{I}_3$ ($\boldsymbol{I}_3$ is the $3 \times 3$ unit matrix). These are the 35 Hermitian traceless matrices which generate the transformations in the fundamental representation of SU(6). They act on the six-component quark, which spans the six-dimensional representation **6**,

$$q^A = \begin{pmatrix} q^{1\uparrow} \\ q^{1\downarrow} \\ q^{2\uparrow} \\ q^{2\downarrow} \\ q^{3\uparrow} \\ q^{3\downarrow} \end{pmatrix} = \begin{pmatrix} u^\uparrow \\ u^\downarrow \\ d^\uparrow \\ d^\downarrow \\ s^\uparrow \\ s^\downarrow \end{pmatrix}; \qquad \begin{array}{l} A = a, \alpha; \\ (a = 1,\ 2,\ 3 \quad \alpha = 1,\ 2); \end{array} \tag{7.112}$$

and its conjugate $q_A$, which defines the **6**$^*$ representation. If the forces responsible for the quark binding depend weakly on SU(2) and SU(3) spins, hadrons made up of quarks and antiquarks may be considered as SU(6)-invariant and their states can be identified with irreducible representations of this larger group. This group contains SU(3) $\otimes$ SU(2) as a subgroup, so that the flavor and spin content of any SU(6) representation can be seen from its reduction to representations of the subgroup SU(3) $\otimes$ SU(2). This is how, by a symmetry principle, spins and parities pair up with various flavor multiplets.

### 7.4.1 Mesons

First, consider the $q\bar{q}$ system, represented by $q^A q_B$. This tensor product may be rewritten as

$$q^A q_B = \tfrac{1}{6}\delta^A{}_B\, q^C q_C + (q^A q_B - \tfrac{1}{6}\delta^A{}_B\, q^C q_C)\,, \tag{7.113}$$

or symbolically as $\mathbf{6}\times\mathbf{6}^* = \mathbf{1}+\mathbf{35}$. That is, it reduces to the sum of a singlet,

$$S = \frac{1}{\sqrt{6}}\, q^A q_A\,, \tag{7.114}$$

and a **35** (or adjoint) representation,

$$M^A{}_B = q^A q_B - \tfrac{1}{6}\delta^A{}_B\, q^C q_C\,. \tag{7.115}$$

Whereas the singlet is clearly a scalar in both SU(3) and SU(2), the adjoint representation acts variously under SU(3) and SU(2) as shown by its decomposition

$$\begin{aligned} M^A{}_B \equiv M^{a\alpha}{}_{b\beta} = {} & \tfrac{1}{3}\delta^a{}_b\, M^{c\alpha}{}_{c\beta} + \tfrac{1}{2}\delta^\alpha_\beta\, M^{a\gamma}{}_{b\gamma} \\ & + \left[M^{a\alpha}{}_{b\beta} - \tfrac{1}{3}\delta^a{}_b\, M^{c\alpha}{}_{c\beta} - \tfrac{1}{2}\delta^\alpha_\beta\, M^{a\gamma}{}_{b\gamma}\right]\,, \end{aligned} \tag{7.116}$$

which is a sum of tensor products of representations of SU(3) and SU(2):

$$\mathbf{35} = (\mathbf{1},\mathbf{3}) + (\mathbf{8},\mathbf{1}) + (\mathbf{8},\mathbf{3})\,.$$

As for any other adjoint representation, **35** may be expanded in terms of the group generators,

$$\begin{aligned} M^A{}_B &= \frac{1}{2}\left\{\delta^\alpha_\beta\,(\lambda_i)^a{}_b\, P_i + (\sigma_m)^\alpha{}_\beta\left[(\lambda_0)^a{}_b S_m + (\lambda_i)^a{}_b\, V_{im}\right]\right\} \\ &= \frac{1}{\sqrt{2}}\delta^\alpha_\beta P^a{}_b + \frac{1}{\sqrt{3}}\delta^a_b\, S^\alpha{}_\beta + V^{a\alpha}{}_{b\beta}\,. \end{aligned} \tag{7.117}$$

The fields are normalized such that

$$M^A{}_B\, M^B{}_A = \sum_{i=1}^{8}(P_i)^2 + \sum_{m=1}^{3}(S_m)^2 + \sum_{i=1}^{8}\sum_{m=1}^{3}(V_{im})^2\,. \tag{7.118}$$

This result tells us that the lowest-mass $q\bar{q}$ states include, in addition to the singlet $S$, an SU(3) octet of spin 0, an octet of spin 1, and a singlet of spin 1. As $L = 0$ by assumption, the parity is negative in all cases. Therefore, the **35** multiplet gathers in a single supermultiplet all the known low-lying boson states with the correct combinations of SU(3) quantum numbers, spins, and parities.

Vector mesons differ from pseudoscalar mesons by more than just their spins. As seen in the previous chapters, $\rho^0$ and $\phi$ have negative charge conjugation parities, in contrast to $\pi^0$ and $\eta$; and similarly, $\rho^{0,\pm}$ have positive G-parities, while $\phi$ has a negative G-parity, opposite in sign to those of $\pi^{0,\pm}$ and $\eta$, respectively. Such characteristics can be readily incorporated into the quark wave functions. Under C-conjugation,

$$\begin{aligned} \mathcal{C}: \quad & u \to \bar{u}\,, \quad d \to \bar{d}\,, \\ & u\bar{u} \to \bar{u}u\,, \quad d\bar{d} \to \bar{d}d\,, \quad s\bar{s} \to \bar{s}s\,; \end{aligned}$$

while under G-conjugation,

$$\begin{aligned} \mathcal{G}: \quad & u \to -\bar{d}\,, \quad d \to \bar{u}\,, \quad \bar{u} \to -d\,, \quad \bar{d} \to u\,, \\ & u\bar{u} \to \bar{d}d\,, \quad d\bar{d} \to \bar{u}u\,, \quad s\bar{s} \to \bar{s}s\,, \\ & u\bar{d} \to -\bar{d}u\,, \quad d\bar{u} \to -\bar{u}d\,. \end{aligned}$$

Therefore, symmetric and antisymmetric combinations of $q^a\bar{q}^b$ and $\bar{q}^b q^a$ have opposite C-parities and opposite G-parities. The $q\bar{q}$ combinations with well-defined C- and G-parities are found in Table 7.5.

**Table 7.5.** Quark–antiquark combinations of good C- and G-parities

| $0^-$ | $1^-$ | | C | G |
|---|---|---|---|---|
| $K^+$ | $K^{*+}$ | $\frac{1}{\sqrt{2}}(u\bar{s} \pm \bar{s}u)$ | | |
| $K^0$ | $K^{*0}$ | $\frac{1}{\sqrt{2}}(d\bar{s} \pm \bar{s}d)$ | | |
| $K^-$ | $K^{*-}$ | $\frac{1}{\sqrt{2}}(s\bar{u} \pm \bar{u}s)$ | | |
| $\bar{K}^0$ | $\bar{K}^{*0}$ | $\frac{1}{\sqrt{2}}(s\bar{d} \pm \bar{d}s)$ | | |
| $\pi^+$ | $\rho^+$ | $\frac{1}{\sqrt{2}}(u\bar{d} \pm \bar{d}u)$ | | $\mp$ |
| $\pi^-$ | $\rho^-$ | $\frac{1}{\sqrt{2}}(d\bar{u} \pm \bar{u}d)$ | | $\mp$ |
| $\pi^0$ | $\rho^0$ | $\frac{1}{2}[(u\bar{u} - d\bar{d}) \pm (\bar{u}u - \bar{d}d)]$ | $\pm$ | $\mp$ |
| $\eta_8$ | $\omega_8$ | $\frac{1}{2\sqrt{3}}[(uu + d\bar{d} - 2s\bar{s}) \pm (\bar{u}u + dd - 2\bar{s}s)]$ | $\pm$ | $\pm$ |
| $\eta_1$ | $\omega_1$ | $\frac{1}{\sqrt{6}}[(u\bar{u} + d\bar{d} + s\bar{s}) \pm (\bar{u}u + \bar{d}d + \bar{s}s)]$ | $\pm$ | $\pm$ |

More generally, if $L$ is the relative orbital angular momentum of the quark–antiquark pair, the parity of the meson is $P = (-1)^{L+1}$ and its spin $J = |L-S|, \ldots, L+S$, where $S = 0$ or 1. A state composed of a quark and its own antiquark is also an eigenstate of charge conjugation, with $C = (-1)^{L+S}$. When new flavors of quarks (see below) are included, nearly all known mesons can be described as bound states of a quark and an antiquark. These new possibilities all occur in the upper parts of the mass spectrum and will not be considered here.

### 7.4.2 Baryons

Let us turn now to the tensor product $q^A q^B q^C$ for $A, B, C = 1, \ldots, 6$. It can be shown (see Problem 7.7) that it reduces to one completely antisymmetric tensor of dimension 20, one completely symmetric tensor of dimension 56, and two tensors of mixed symmetry, both of dimension 70:

$$\mathbf{6} \times \mathbf{6} \times \mathbf{6} = \mathbf{20} + \mathbf{70} + \mathbf{70} + \mathbf{56}\,, \tag{7.119}$$

to be compared with the reduction $\mathbf{3} \times \mathbf{3} \times \mathbf{3} = \mathbf{1} + \mathbf{8} + \mathbf{8} + \mathbf{10}$ in SU(3).

At first, one would expect the antisymmetric **20** representation to give a good description of the lowest baryons as s-wave three-quark bound states. Its reduction under SU(6) $\to$ SU(3)$\times$SU(2) is $\mathbf{20} = (\mathbf{8},\mathbf{2}) + (\mathbf{1},\mathbf{4})$. It contains a spin-$1/2$ octet of baryons but just one spin-$3/2$ state, not enough to account for the rich $\frac{3}{2}^+$ spectrum that is observed. Besides, the magnetic moments for the $(\mathbf{8},\mathbf{2})$ are completely wrong, not only in magnitude but also in sign, i.e. they point in the wrong direction. As seen in previous sections, the totally symmetric representation **56** should be more interesting for the baryon ground states and thus deserves closer scrutiny. In reducing it to irreducible representations of flavor and spin, it is perhaps evident that only three kinds of combination of SU(3) representations with SU(2) representations can lead to a completely symmetric representation of SU(6): a completely symmetric representation with a completely symmetric representation, a representation of mixed symmetry with a representation of similarly mixed symmetry, and finally, a completely antisymmetric representation with a completely antisymmetric representation. Since no SU(2) singlet can be built from three two-component spinors, the latter possibility is closed, and the reduction must be $\mathbf{56} = (\mathbf{10},\mathbf{4}) + (\mathbf{8},\mathbf{2})$. Algebraically, it is equivalent to writing the completely symmetric tensor as

$$\begin{aligned}\Psi^{ABC} &\equiv \Psi^{a\alpha,b\beta,c\gamma} = D^{abc,\alpha\beta\gamma} \\ &+ \frac{1}{3\sqrt{2}}\left[\epsilon^{abd}\epsilon^{\alpha\beta}\,N_d^{c,\gamma} + \epsilon^{bcd}\epsilon^{\beta\gamma}\,N_d^{a,\alpha} + \epsilon^{cad}\epsilon^{\gamma\alpha}\,N_d^{b,\beta}\right].\end{aligned} \tag{7.120}$$

The tensor $D^{abc,\alpha\beta\gamma}$, symmetric in SU(3) labels $a$, $b$, $c$ and equally in spin indices $\alpha$, $\beta$, $\gamma$, has ten components $D^{abc}$ appearing in four spin states; it is the $(\mathbf{10},\mathbf{4})$. The $N_b^{a,\alpha}$ is an SU(3) octet appearing in two spin components $\alpha = 1, 2$; it is the $(\mathbf{8},\mathbf{2})$. In terms of the SU(2) and SU(3) tensors defined in Sects. 7.1–7.2, they read

$$D^{abc,\alpha\beta\gamma} = D^{abc}\,\chi_Q^{\alpha\beta\gamma}\,, \tag{7.121}$$

$$N_b^{a,\alpha} = \tfrac{1}{\sqrt{2}}\left(N^a{}_b\,\chi_S^\alpha + B^a{}_b\,\chi_A^\alpha\right). \tag{7.122}$$

They represent a **10** of spin $3/2$ and an **8** of spin $1/2$, both of positive parities. Note that now both $N^a{}_b$ and $B^a{}_b$ come together in the wave functions of the spin-$1/2$ baryons. Thus, the two familiar SU(3) baryon multiplets fit neatly into a single SU(6) supermultiplet with the correct spin and parity assignments.

To end this discussion, a few brief remarks on higher-energy states. In the spirit of the potential model, excited states of baryons are obtained by letting one or several quarks occupy higher and higher single-particle levels; their angular momenta are determined by the usual vector coupling, and their parities, by the values of their orbital angular momenta. Thus, for example, one expects the excited states lying just above the ground states to have the following configurations:

$$\begin{aligned}
&(1\mathrm{s})^2\,(1\mathrm{p}) \quad (J^P = \tfrac{1}{2}^-, \tfrac{3}{2}^-, \tfrac{5}{2}^-)\,; \\
&(1\mathrm{s})^2\,(1\mathrm{d}) \quad (J^P = \tfrac{1}{2}^+, \tfrac{3}{2}^+, \tfrac{5}{2}^+, \tfrac{7}{2}^+)\,; \\
&(1\mathrm{s})^2\,(2\mathrm{s}) \quad (J^P = \tfrac{1}{2}^+, \tfrac{3}{2}^+)\,; \\
&(1\mathrm{s})\,(1\mathrm{p})^2 \quad (J^P = \tfrac{1}{2}^+, \tfrac{3}{2}^+, \tfrac{5}{2}^+, \tfrac{7}{2}^+)\,.
\end{aligned}$$

To narrow down the possibilities, one assumes further that each baryon composed of three quarks must be in a state *completely symmetric* in the combined space and SU(6) degrees of freedom, i.e. it must belong to a symmetric SU(6) × O(3) representation, O(3) being the orthogonal rotational group of coordinate space. This assumption means that the SU(6) and O(3) substates have matching permutation symmetries; it is the basis of the *symmetric quark model*. States assigned to the different SU(6) representations of three quarks, **56**, **70**, and **20**, emerge with assigned spins and parities. Most of the observed excited states can be identified with one or another predicted state. For example, the negative-parity states around 1.5–1.7 GeV would belong to a **70**, while many positive parity resonances around 1.9–2.0 GeV would be candidates for slots in an excited **56**.

### 7.4.3 Application: Magnetic Moments of Hadrons

As a simple application, let us calculate the magnetic moments of the proton and the neutron. The proton wave function with spin $S_z = +1/2$ is given by (122):

$$|\mathrm{p}\uparrow\rangle = N_3^{11} = \tfrac{1}{\sqrt{2}}\,({B^1}_3\,\chi_A^1 + {N^1}_3\,\chi_S^1)\,, \tag{7.123}$$

where, it is recalled,

$$\begin{aligned}
\chi_A^1 &= \tfrac{1}{\sqrt{2}}(\uparrow\downarrow - \downarrow\uparrow)\uparrow\,, & \chi_S^1 &= \tfrac{1}{\sqrt{6}}\,[2\uparrow\uparrow\downarrow - (\uparrow\downarrow + \downarrow\uparrow)\uparrow]\,, \\
{B^1}_3 &= \tfrac{1}{\sqrt{2}}(ud - du)u\,, & {N^1}_3 &= \tfrac{1}{\sqrt{6}}[2uud - (ud + du)u]\,.
\end{aligned}$$

Assuming $L = 0$, the spatial part may be suppressed in the wave function as well as in the magnetic moment operator. This operator therefore includes just contributions from the spins:

$$[\mu_z]_{\mathrm{op}} = \mu_0 \sum_{i=1}^{3} e_i\,\sigma_z^{(i)}\,, \tag{7.124}$$

where the sum is over the three quarks on which it operates; $e_i$ is the charge number of quark $i$, and $\mu_0$ a constant. By symmetry, the magnetic moment of the proton is

$$\begin{aligned}\mu_\mathrm{p} &= \left\langle \mathrm{p}\uparrow \middle| [\mu_z]_\mathrm{op} \middle| \mathrm{p}\uparrow \right\rangle = 3\mu_0 \left\langle \mathrm{p}\uparrow \middle| e_3\, \sigma_z^{(3)} \middle| \mathrm{p}\uparrow \right\rangle \\ &= \frac{3}{2}\mu_0 \left[ \left\langle B^1{}_3\, \chi_A^{\frac{1}{2}} \middle| e_3\, \sigma_z^{(3)} \middle| B^1{}_3\, \chi_A^{\frac{1}{2}} \right\rangle + \left\langle N^1{}_3\, \chi_S^{\frac{1}{2}} \middle| e_3\, \sigma_z^{(3)} \middle| N^1{}_3\, \chi_S^{\frac{1}{2}} \right\rangle \right].\end{aligned}$$

The matrix elements needed are

$$\left\langle \chi_A^{\frac{1}{2}} \middle| \sigma_z^{(3)} \middle| \chi_A^{\frac{1}{2}} \right\rangle = 1\,, \qquad \left\langle \chi_S^{\frac{1}{2}} \middle| \sigma_z^{(3)} \middle| \chi_S^{\frac{1}{2}} \right\rangle = -\frac{1}{3}\,,$$

$$\langle B^1{}_3 \,|\, e_3 \,|\, B^1{}_3 \rangle = e_\mathrm{u}\,, \qquad \langle N^1{}_3 \,|\, e_3 \,|\, N^1{}_3 \rangle = \frac{1}{3}(e_\mathrm{u} + 2e_\mathrm{d})\,.$$

Therefore, the magnetic moment of the proton is

$$\mu_\mathrm{p} = \mu_0\, \tfrac{1}{3}(4e_\mathrm{u} - e_\mathrm{d})\,. \tag{7.125}$$

For the neutron, it suffices to interchange u and d quarks to get

$$\mu_\mathrm{n} = \mu_0\, \tfrac{1}{3}(4e_\mathrm{d} - e_\mathrm{u})\,. \tag{7.126}$$

Since $e_\mathrm{u} = 2/3$ and $e_\mathrm{d} = -1/3$, the ratio of the two nuclear magnetic moments is $\mu_\mathrm{p}/\mu_\mathrm{n} = -3/2$, which is in excellent agreement in magnitude and sign with data, $\mu_\mathrm{p}/\mu_\mathrm{n} = 2.79/(-1.91) = -1.46$.

The individual nuclear magnetic moments agree with data as well if each of the three constituent quarks contributes an effective magnetic moment

$$\mu_i = 2.79\, \frac{e\hbar}{2M_\mathrm{p}c}\, e_i\,, \quad i = \mathrm{u},\, \mathrm{d}, \tag{7.127}$$

so that $\mu_\mathrm{p} = 2\mu_\mathrm{u} + \mu_\mathrm{d}$. A quark would have such a magnetic moment if it acts as a Dirac particle with an effective mass

$$m_\mathrm{q} = M_\mathrm{p}/2.79 \approx 336\ \mathrm{MeV}/c^2\,. \tag{7.128}$$

Thus, each constituent u or d quark would effectively have about one third of the proton mass.

These results give indirect but convincing evidence that $L = 0$ in proton and neutron. The orbital three-quark wave functions of the nucleon and other ground state baryons are symmetric. So too are their spin and flavor wave functions. Therefore, the total (space, spin, and flavor) wave functions of ground state baryons are completely symmetric under the interchange of any two quarks. This symmetry persists in excited states as well. On the other hand, experiments show complete accord with spin-1/2 quarks. We are thus faced with the paradox of having particles of half-integral spins described by symmetric wave functions, in violation of the Fermi–Dirac statistics. So, either the quarks have unusual particle statistics, or they obey conventional statistics but have some unsuspected degrees of freedom alongside the existing ones. Of these two possibilities, the more conservative solution would be simply to introduce a new degree of freedom without giving up well-established statistics.

## 7.5 The Color of Quarks

The new degree of freedom is called '*color*'. If we assume that each flavor of quark comes in different colors, then three quarks in, say, the proton may be in the same single-particle state without violating the generalized Pauli principle as long as they all have different colors. This requires at least three colors. But if there are *exactly three colors*, more can be said. We can then assume that nature is *invariant under color change*, defined by a color-generated SU(3) group and labeled $\mathrm{SU_c(3)}$, which acts on color space. Exact $\mathrm{SU_c(3)}$ implies in particular degeneracy in energy for different colors of quark, to be contrasted with the symmetry breaking in flavor. In addition, to agree with current observations, we may also adopt a working rule that *only color-neutral, or* $\mathrm{SU_c(3)}$*-singlet, states are physically observable.* This means in particular that no isolated quarks can be seen. Nor can colored hadrons. Color nonsinglets, if they exist, must have very high or infinite masses. The ultimate objective is to deduce this rule from a gauge theory based on color – quantum chromodynamics (QCD).

The space of the color fundamental representation is spanned by the quark spinor $q^{Ak}$, which carries, along with the usual flavor–spin index $A$, the additional color index, $k$ = 1, 2, 3, or R (red), B (blue), and G (green). The quark index $A$ transforms according to SU(6) [or SU(6) × O(3)] and index $k$ according to $\mathrm{SU_c(3)}$. The antiquark $q_{Ak}$ carrying negative color generates the conjugate representation.

Baryons therefore must belong to completely antisymmetric representations of SU(6) × O(3) × $\mathrm{SU_c(3)}$ if the generalized Pauli principle is to be satisfied. With the flavor–spin state function symmetric, the three-color state function must be an antisymmetric color-singlet. For example, the complete wave function for the proton is obtained by multiplying (123) by a space s-wave part $\psi_s(r_1, r_2, r_3)$ and a color-singlet part given by

$$\frac{1}{\sqrt{6}}\epsilon_{klm}\chi^k\chi^l\chi^m = \frac{1}{\sqrt{6}}\Big[(\chi^{\mathrm{R}}\chi^{\mathrm{B}} - \chi^{\mathrm{B}}\chi^{\mathrm{R}})\chi^{\mathrm{G}} + (\chi^{\mathrm{B}}\chi^{\mathrm{G}} - \chi^{\mathrm{G}}\chi^{\mathrm{B}})\chi^{\mathrm{R}} + (\chi^{\mathrm{G}}\chi^{\mathrm{R}} - \chi^{\mathrm{R}}\chi^{\mathrm{G}})\chi^{\mathrm{B}}\Big], \quad (7.129)$$

where $\chi^k$ is a three-component color spinor.

As for mesons, their complete wave functions are obtained by multiplying their SU(6) × O(3) wave functions by the color-singlet combination

$$\frac{1}{\sqrt{3}}\left(\chi^{\mathrm{R}}\chi_{\mathrm{R}} + \chi^{\mathrm{G}}\chi_{\mathrm{G}} + \chi^{\mathrm{B}}\chi_{\mathrm{B}}\right) . \quad (7.130)$$

If mesons were not color-neutral, there would be nine possible varieties of mesons, corresponding to nine possible color combinations of the same quark–antiquark flavors. The failure to see nine colored pions makes the color-singlet assignment for physical hadrons all the more credible.

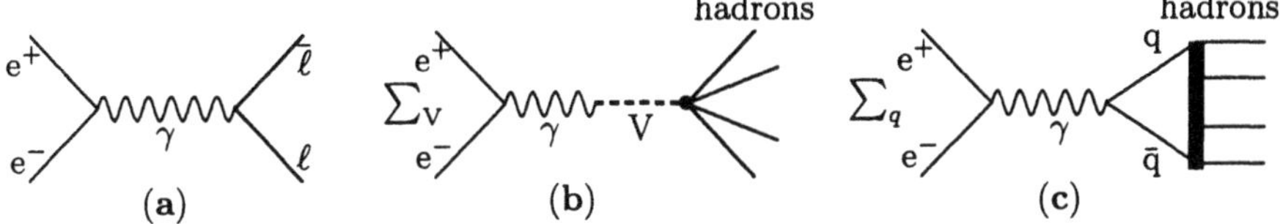


**Fig. 7.4.** (**a**) One-photon exchange diagram for $e^+e^- \to e^+e^-$, $\mu^+\mu^-$; hadronic productions from $e^+e^-$ annihilations can proceed through (**b**) vector meson resonances V= $\rho$, $\omega$, $\phi$, $\psi$, ..., or (**c**) quark–antiquark pair pointlike production followed by fragmentation

Experimental data definitely supports the idea that quarks exist in three colors. The most direct evidence comes from measurements of the total cross-sections for the annihilation of electron–positron pairs in colliding beam experiments. The reactions occur predominantly through the annihilation of the incoming pair into a single virtual photon, which then materializes as a lepton or quark pair (see Fig. 7.4). At high energy, the total cross-section for each channel depends on the channel only through the square of the electric charge of particle being pair-produced, and thus is directly related to the cross-section for $e^+e^- \to \mu^+\mu^-$ (see Sect. 4.7):

$$\sigma(e^+e^- \to \mu^+\mu^-) = \frac{4\pi\alpha^2}{3s},$$

where $s = E^2_{\text{c.m.}}$. If hadrons are considered composites of pointlike quarks, we expect that, again at large $s$, processes $e^+e^- \to$ hadrons take place via $e^+e^- \to q\bar{q}$ and $e^+e^- \to$ heavy leptons (if any exist), and these reaction products subsequently emerge through some unspecified mechanism as the observed hadrons:

$$\begin{aligned}\sigma(e^+e^- \to \text{hadrons}) &= \sum_q (e^+e^- \to q\bar{q}) + \sum_L (e^+e^- \to L\bar{L}) \\ &= \sigma(e^+e^- \to \mu^+\mu^-) \sum_i e_i^2 \\ &= \frac{4\pi\alpha^2}{3s} \sum_i e_i^2 .\end{aligned} \tag{7.131}$$

Therefore, the all-inclusive cross-section, subtracted of resonances and final-state interactions, should be proportional to the sum of squared charges of all the leptons and quarks that can be energetically pair-produced. The result can be expressed as the ratio

$$R \equiv \frac{\sigma(e^+e^- \to \text{hadrons})}{\sigma(e^+e^- \to \mu^+\mu^-)} = \sum_i e_i^2 . \tag{7.132}$$

The $e^-$ and $\mu^-$ charges are not included in the sum, but heavier lepton final states, if any, are, provided they can decay into hadrons. At each production threshold, when a pair of new quarks or leptons appears, $R$ increases by the squared charge of the new particle. In between thresholds, $R$ remains constant with increasing energy, and its magnitude directly measures the sum of the squared charges of the fundamental fermion fields present at that energy. Below the center-of-mass energy of 3 GeV, only u, d, and s quarks contribute, and one would expect

$$R = \left(\frac{2}{3}\right)^2 + \left(-\frac{1}{3}\right)^2 + \left(-\frac{1}{3}\right)^2 = \frac{2}{3} \qquad \text{for uncolored u, d, s}. \tag{7.133}$$

The measured ratio turns out to be constant, as expected of pointlike quarks, but at a value larger than predicted, rather closer to 2, within errors. The value of $R = 2$ is precisely what would come out of a three-flavor quark model with quark charges of $2/3$, $-1/3$, and $-1/3$, each flavor present in three colors. Once this is accepted, any sudden increase in $R$ at an energy beyond 3 GeV should be viewed as the opening up of a new degree of freedom. By 1973, data accumulated at various laboratories indicated a jump in $R$ to a value of about 4.5 at energy of 4 GeV, an unmistakable signal of the appearance of a new quark or a new lepton.

## 7.6 The New Particles

In November 1974, a resonance with a mass of 3.1 GeV and a very narrow width was observed. It was immediately understood that it involved a new type of quark. This remarkable discovery launched an exciting period in the history of particle physics marked by the opening of a new generation of particle accelerators and detectors and the observations and identifications of a whole new generation of fundamental particles.

### 7.6.1 J/$\psi$ and Charm

The resonance at 3.1 GeV (dubbed J/$\psi$ or simply $\psi$) was observed in two different experiments. In one, performed at the Brookhaven National Laboratory's alternating-gradient synchrotron (Aubert et al., 1974), it appeared as a narrow enhancement in the electron–positron pair invariant mass distribution in collisions of 28 GeV protons on a beryllium target,

$$\mathrm{p} + \mathrm{Be} \to \mathrm{e}^+\mathrm{e}^- + \text{anything}. \tag{7.134}$$

In the other, performed at the Stanford Linear Accelerator Center's electron–positron storage ring SPEAR (Augustin et al., 1974), it showed up as a very sharp peak in the total cross-section for

$$\begin{aligned} \mathrm{e}^+\mathrm{e}^- &\to \text{hadrons} \\ &\to \mathrm{e}^+\mathrm{e}^-,\ \mu^+\mu^-. \end{aligned} \tag{7.135}$$

In addition to $\psi$, a second resonance $\psi'$ at energy 3.7 GeV, almost as narrow, was also observed in this first series of experiments at SPEAR. In both cases, the extremely small widths, already remarkable in themselves, become even more intriguing in view of the very high energy involved.

The reported experimental width of about 2 MeV at 3.1 GeV, which results from the spread in energy of the electron and positron beams or of the secondary electrons, cannot be identified with the $\psi$-resonance width itself. The true width, which is much smaller, can however be determined from the total reaction rate and the leptonic branching ratio, both of which have been measured. The resonance cross-section for $\psi$ production in $e^+e^-$ annihilations at center-of-mass energy $E$, leading to any final observed state f, is given by the Breit–Wigner formula

$$\sigma(e^+e^- \to \psi \to f; E) = \frac{2J+1}{(2s_1+1)(2s_2+1)} \frac{\pi}{p^2} \frac{\Gamma_{in}\,\Gamma_{out}}{[(E-E_R)^2+\Gamma^2/4]}, \quad (7.136)$$

where the incoming particles have spins $s_1 = s_2 = 1/2$ and momentum $p$ in the center-of-mass system, with $p \approx E_R/2$ near resonance. The resonance has spin $J$, energy $E_R$, and full width $\Gamma$. The in-channel width, $\Gamma_{in} = \Gamma_{ee}$, is the electronic width for $e^+e^- \to \psi$, and $\Gamma_{out} = \Gamma_{\psi\to f}$ is the partial width for the decay of $\psi$ to some final state f. The total reaction rate is given by the area under the cross-section curve (assuming $\Gamma \ll E_R$):

$$\text{Area} = \int_0^\infty \sigma(E)\mathrm{d}E = \frac{2J+1}{(2s_1+1)(2s_2+1)} \frac{2\pi^2\,\Gamma_{in}\,\Gamma_{out}}{p^2\,\Gamma}. \quad (7.137)$$

Assuming that the observed resonance has spin $J = 1$ and energy $E_R = M_\psi$, we get

$$\text{Area} = \frac{6\pi^2}{M_\psi^2} \frac{\Gamma_{in}}{\Gamma} \frac{\Gamma_{out}}{\Gamma} \Gamma. \quad (7.138)$$

The area under the resonance curve for $e^+e^- \to$ hadrons measured at SPEAR is about 10 nb GeV. The branching ratio for $e^+e^-$ decay has been measured with the result

$$\frac{\Gamma_{ee}}{\Gamma} = (6.02 \pm 0.19)\%. \quad (7.139)$$

Assuming $\Gamma_{\mu\mu} = \Gamma_{ee}$, the ratio for decays to hadrons should be $\approx 88\%$. From (138), the full width of the $\psi$ resonance then comes out to be $\Gamma = 78$ keV (close to the current value of $87 \pm 5$ keV), an astonishingly small value for a particle weighing as much as 3098 MeV, and about a thousand times smaller than what one would have *a priori* anticipated for a hadronic resonance of that mass. For comparison, the isobar resonance $\Delta(1232)$ has a width of 115

MeV. In contrast, the electronic width of the resonance, $\Gamma_{ee} = 5.2$ keV, is typical of vector mesons (cf. Table 7.7).

That $\psi$ is indeed a vector meson, as we have assumed, can be established by observing the shape of the resonance curve for $e^+e^- \to \mu^+\mu^-$, which displays a pattern characteristic of the interference of two processes: one involving the $\psi$ resonance in the intermediate state and the other, the exchange of a virtual photon. The interference of these two processes means that $\psi$ must have the same quantum numbers as the photon, i.e. $J^{PC} = 1^{--}$. Its G-parity is odd because it decays predominantly to states with odd numbers of pions. Since isospin is related to G-parity and C-parity through $G = C\,(-1)^I$, $\psi$ must have an even isospin. The specific assignment $I = 0$ is established from the equal branching ratios ($4.2 \times 10^{-3}$) of $\psi$ decays to various charge states, $\psi \to \rho^+\pi^-$, $\rho^0\pi^0$, $\rho^-\pi^+$, and from the coupling coefficients of two isospins $\boldsymbol{I} = 1$ to $\boldsymbol{I} = 0$. In summary, the quantum numbers of $\psi$ are $I^G = 0^-$ and $J^{PC} = 1^{--}$.

An extremely narrow width for such a massive state as $\psi$ is very difficult to understand in terms of the known light (u, d, and s) quarks alone, without invoking new quantum numbers or new selection rules. Some years before, Glashow, Iliopoulos, and Maiani (1970) introduced a new type of quark to explain the very puzzling fact that the rate of $K^+ \to \pi^+e^+e^-$ was $10^{-5}$ times smaller than the rate of the similar decay $K^+ \to \pi^0e^+\nu$, which is an example of the general observed suppression of the charge-preserving but strangeness-changing weak processes (cf. Sects. 6.6 and 9.4). This new flavor, called c for *charmed* quark, carries a new quantum number, $C$ for *charm*, which, like strangeness, would be conserved in strong and electromagnetic interactions. The charmed quark, invented in response to a problem in weak interactions, will play an essential role in the structure of the unified electroweak interaction and will make a considerable impact on strong interaction physics. (It is hoped that the context will tell the reader what $C$ refers to, charm or charge conjugation parity.) Just before the discovery of $\psi$, Appelquist and Politzer (1975) were investigating the binding of a charmed and anticharmed quark in the framework of QCD. They found that – just as in the positronium but at a completely different energy scale – there would be a series of bound states with very small widths. Thus, there would exist spectroscopic levels $n\,^{2S+1}L_J$ at various energies corresponding to pseudoscalar states with $J^P = 0^-$, vector states with $J^P = 1^-$, and so on.

The $c\bar{c}$ (*charmonium*) bound states $1\,^3S_1$ and $2\,^3S_1$ were immediately considered as the leading candidates for the observed resonances $\psi$ and $\psi'$. Following an intensive spectroscopic search, other resonances were seen as well and identified with other $c\bar{c}$ bound states, some with the pseudoscalar states $\eta_c$ or vector states $\psi$, the others with the p-wave states $\chi_c$ (see Table 7.6). Specially noteworthy are the $\psi$ states above 3750 MeV, which are no more exceptionally narrow, a sign that they now have access to some new strong decay modes. These results give us a lot of confidence in the validity

**Table 7.6.** Some of the observed $c\bar{c}$ bound states

| | Name | Mass(MeV/$c^2$) | $\Gamma$(MeV) | $\Gamma_{ee}$(keV) |
|---|---|---|---|---|
| $^1S_0(0^{--})$ : | | | | |
| $1\,^1S_0$ | $\eta_c$ | $2979.8 \pm 2.1$ | $13.2 \pm 3.8$ | |
| $^3S_1(1^{--})$ : | | | | |
| $1\,^3S_1$ | $\psi$ | $3096.88 \pm 0.04$ | $0.087 \pm 0.005$ | $5.26 \pm 0.37$ |
| $2\,^3S_1$ | $\psi'$ | $3686.00 \pm 0.09$ | $0.277 \pm 0.031$ | $2.14 \pm 0.21$ |
| $3\,^3S_1$ | $\psi$ | $4040. \pm 10$ | $52 \pm 10$ | $0.75 \pm 0.15$ |
| $1\,^3P_J(J^{++})$ : | | | | |
| $^3P_0$ | $\chi_0$ | $3415.1 \pm 1.0$ | $14 \pm 5$ | |
| $^3P_1$ | $\chi_1$ | $3510.53 \pm 0.12$ | $0.88 \pm 0.14$ | |
| $^3P_2$ | $\chi_2$ | $3556.17 \pm 1.0$ | $2.00 \pm 0.18$ | |
| $^3D_1(1^{--})$ : | | | | |
| $1\,^3D_1$ | $\psi$ | $3769.9 \pm 2.5$ | $23.6 \pm 2.7$ | $0.26 \pm 0.04$ |

of the potential quark model. If this picture is indeed valid, the c quark mass must be of the order of $\frac{1}{2}M_\psi = 1.5$ GeV.

In addition to the uncharmed u, d, and s, we now have a fourth quark with isospin and strangeness zero and carrying one unit of charm. If we ignore, for the sake of the argument, the mass differences between flavors, the four quarks u, d, s, and c form the fundamental quartet representation **4** of the flavor symmetry group $SU_f(4)$, illustrated by the weight diagram in Fig. 7.5. Exactly as in $SU_f(3)$, the mesons are built up from $q\bar{q}$, i.e. from the tensor product

$$\mathbf{4} \times \bar{\mathbf{4}} = \mathbf{1} + \mathbf{15}\,, \tag{7.140}$$

leading to seven new states, which are
– three states with charm $C = 1$, forming a $\bar{\mathbf{3}}$ of subgroup SU(3):

$$c\bar{u} \quad (D^0 \text{ or } D^{*0}), \qquad c\bar{d} \quad (D^+ \text{ or } D^{*+}), \qquad c\bar{s} \quad (D_s^+ \text{ or } D_s^{*+}); \tag{7.141}$$

– three states with charm $C = -1$, forming a $\mathbf{3}$ of subgroup SU(3):

$$\bar{c}u \quad (\bar{D}^0 \text{ or } \bar{D}^{*0}), \qquad \bar{c}d \quad (D^- \text{ or } D^{*-}), \qquad \bar{c}s \quad (D_s^- \text{ or } D_s^{*-}); \tag{7.142}$$

– a state with charm zero, $C = 0$, which is a singlet of subgroup SU(3):

$$c\bar{c} \quad (\eta_c \text{ or } \psi)\,. \tag{7.143}$$

The pseudoscalar mesons are shown in Fig. 7.5b.

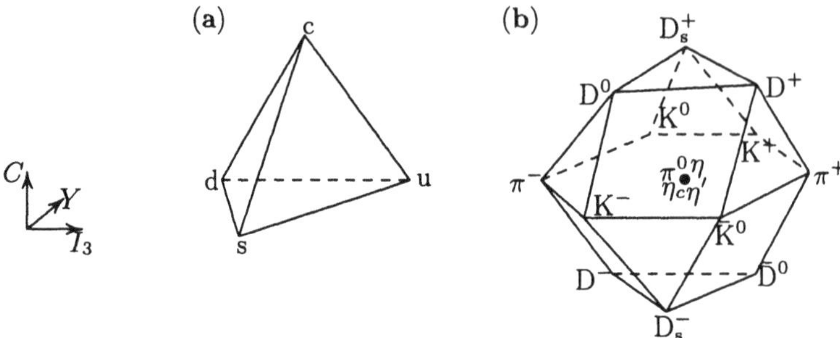


**Fig. 7.5.** $SU_f(4)$ group: (**a**) the fundamental representation made up of quarks u, d, s, and c; and (**b**) the 16-plet for the pseudoscalar mesons

In $SU_f(3)$ the qqq baryons form multiplets with dimensions

$$\mathbf{3} \times \mathbf{3} \times \mathbf{3} = \mathbf{1} + \mathbf{8} + \mathbf{8} + \mathbf{10}\,.$$

A parallel construction in $SU_f(4)$ leads to four multiplets,

$$\mathbf{4} \times \mathbf{4} \times \mathbf{4} = \mathbf{4} + \mathbf{20} + \mathbf{20} + \mathbf{20}\,.$$

There are now four possible antisymmetric states forming **4** in place of the single one in $SU_f(3)$. Each of the old octets gains twelve charmed members while the symmetric decuplet admits another ten. Figure 7.6a shows the **20** containing the familiar baryon octet and multiplets with $C = 1$ and $C = 2$. Figure 7.6b shows the $SU_f(4)$ **20** baryon multiplet that has as its ground level the $SU_f(3)$ decuplet containing the $\Delta(1232)$ and succeeding higher levels with $C = 1$, 2, and 3.

Charmed pseudoscalar mesons D and their antiparticles, charmed vector mesons $D^*$, mesons with both charm and strangeness $D_s$, and even charmed baryons that contain only one charmed quark each, all have been discovered. Of the charmed mesons, the lightest is the $D^0$ at 1865 MeV, followed by the $D^\pm$ at 1879 MeV. If the positively charged $D^+$ is indeed a combination of $c\bar{d}$, as the $SU_f(4)$ classification indicates, the c quark must have electric charge $e_c = 2/3$. This assignment is certainly consistent with the electronic width of $\psi$, $\Gamma_{ee} = 5.2\,\text{keV}$, which can be seen as follows.

In the nonrelativistic quark model, $\psi$ is described by a charm–anticharm s-state, with radial wave function $\varphi(r)$. The partial width $\Gamma(\psi \to e^+e^-)$ is given by $\Gamma = \sigma \varrho v$, where $\sigma$ is the total cross-section for $c\bar{c} \to e^+e^-$, $v$ is the relative velocity of $c\bar{c}$, and $\varrho$ the target density, $\varrho = |\varphi(0)|^2$. We may assume that the quarks are nonrelativistic and take the low energy limit of the cross-section for $e^+e^- \to \mu^+\mu^-$, derived in Sect. 4.7 and appropriately modified for this case:

$$\sigma(c\bar{c} \to e^+e^-) = 3 \times \frac{2\pi\alpha^2 e_c^2}{s\beta}\,, \tag{7.144}$$

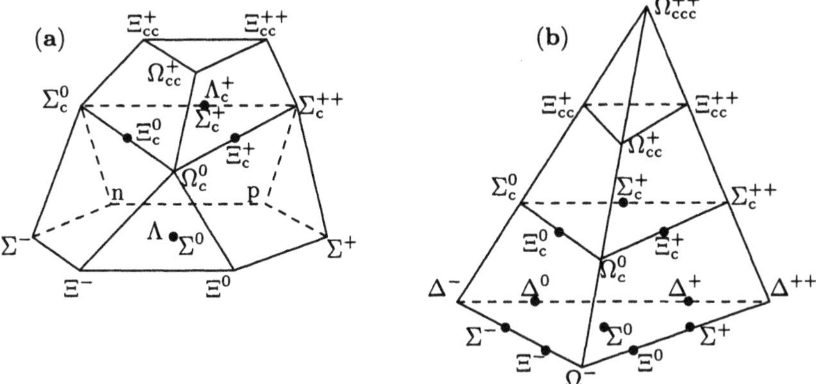


**Fig. 7.6.** $SU_f(4)$ multiplets of baryons: (**a**) the 20-plet with an $SU_f(3)$ octet, and (**b**) the 20-plet built on an $SU_f(3)$ decuplet

where $\beta = \frac{1}{2}v$ is the velocity of the quark or antiquark in the center-of-mass frame; $s$ the square of the CM energy; and $e_c$ the electric charge number of c. The factor 3 accounts for the three colors. The cross-section (144) already includes the averaging over the four possible quark spin states. Particle $\psi$ is a spin triplet, $S = 1$, and so may annihilate through a virtual photon to $e^+e^-$. The cross-section for its decay is therefore just 4/3 of the averaged cross-section. The final expression for the electronic width reads

$$\Gamma(\psi \to e^+e^-) = \frac{16\pi\alpha^2 e_c^2}{M_\psi^2} \, |\varphi(0)|^2 \,. \tag{7.145}$$

By a mild modification, one obtains a more general formula for the leptonic decay width of any vector meson decaying to a lepton pair via the exchange of a single virtual photon:

$$\Gamma(V \to \ell^+\ell^-) = \frac{16\pi\alpha^2 Q^2}{M_V^2} \, |\varphi(0)|^2 \,. \tag{7.146}$$

The assumption here is that the quark and antiquark in the meson are in a relative s-state, possibly in different flavor mixtures $a_i$, so that the squared sum of the charges of the quarks in the meson V is $Q^2 = |\sum a_i e_i|^2$.

The $q\bar{q}$ pair must come together to annihilate into the virtual photon, and the probability for this to occur is given by $|\varphi(0)|^2$. The radial wave function at the origin $\varphi(0)$ cannot be calculated in perturbation theory because the mean radius of the $q\bar{q}$ system is in the order of one fermi, a distance scale at which the effective coupling for the binding has already become large. Therefore $\varphi(0)$ must be determined by the nonperturbative, long-range part of the $q\bar{q}$ potential. From various potential model studies of the lower-mass vector

**Table 7.7.** Leptonic widths of vector mesons

| | $\rho$ | $\omega$ | $\phi$ | $\psi$ | $\Upsilon$ |
|---|---|---|---|---|---|
| | $\frac{1}{\sqrt{2}}(\mathrm{u\bar{u}-d\bar{d}})$ | $\frac{1}{\sqrt{2}}(\mathrm{u\bar{u}+d\bar{d}})$ | $\mathrm{s\bar{s}}$ | $\mathrm{c\bar{c}}$ | $\mathrm{b\bar{b}}$ |
| $\Gamma_{\mathrm{tot}}$(MeV)[a] | 150 | 8.43 | 4.43 | 0.087 | 0.052 |
| $\Gamma_{\mathrm{ee}}$(keV) | 6.77 | 0.60 | 1.37 | 5.26 | 1.32 |
| $Q^2$ | $\frac{1}{2}$ | $\frac{1}{18}$ | $\frac{1}{9}$ | $\frac{4}{9}$ | $\frac{1}{9}$ |
| $\Gamma_{\mathrm{ee}}/Q^2$ | 13.54 | 10.8 | 12.33 | 11.83 | 11.88 |

[a] The total widths are given for reference.

mesons, $\rho^0(770)$, $\omega(782)$, and $\phi(1020)$, it appears that the ratio $|\varphi(0)|^2/M_V^2$ has similar values for these mesons, as indicated by the rather good agreement of data with the predicted relative leptonic branching ratios

$$\Gamma(\rho \to \mathrm{e^+e^-}) \,:\, \Gamma(\omega \to \mathrm{e^+e^-}) \,:\, \Gamma(\phi \to \mathrm{e^+e^-}) = 9 : 1 : 2 \,.$$

If this observation may apply to other, heavier s-wave mesons as well, we can get an indication of the charges of their constituent quarks. With the measured values of $\Gamma_{\mathrm{ee}}$ and the assumed quark contents of the vector mesons, the ratios $\Gamma_{\mathrm{ee}}/Q^2$ are calculated with results shown in Table 7.7. We have considered $\psi(\mathrm{c\bar{c}})$, assuming $e_{\mathrm{c}} = 2/3$, and an even more recent meson $\Upsilon(\mathrm{b\bar{b}})$, assuming $e_{\mathrm{b}} = -1/3$ (of which more will be said below). To put it in another way, assuming known $x \equiv 16\pi\alpha^2|\varphi|^2/M_V^2 \approx 12.2$ from averaging its values for $\rho$, $\omega$, and $\phi$, one can estimate the squared charges of c quark and b quark:

$$Q^2(\psi) = \frac{\Gamma_{\mathrm{ee}}(\psi)}{12.2} \approx 0.43 \,, \qquad \text{compared with} \quad \left(\frac{2}{3}\right)^2 \approx 0.44 \,;$$

$$Q^2(\Upsilon) = \frac{\Gamma_{\mathrm{ee}}(\Upsilon)}{12.2} \approx 0.11 \,, \qquad \text{compared with} \quad \left(\frac{1}{3}\right)^2 \approx 0.11 \,.$$

Now that the c quark is shown to exist with charge 2/3, isospin and strangeness both zero, and charm $C = 1$, the extreme narrowness of the resonance $\psi(3100)$ can be readily understood in terms of the OZI rule. According to this rule, the decay $\psi \to \mathrm{D^+D^-}$ is preferred but not allowed by energy conservation, leaving the decay to uncharmed mesons energetically allowed but strongly suppressed by the OZI rule. This is also the case of all charmonium states below threshold $2M_{\mathrm{D}} = 3730\,\mathrm{MeV}$. Higher-energy resonances, such as $\psi(3770)$ and $\psi(4040)$, can decay into $\mathrm{D\bar{D}}$ and hence have much larger widths.

The empirical OZI rule can best be explained in terms of QCD, the theory based on the interaction between the quark colors via *gluons*, much like QED

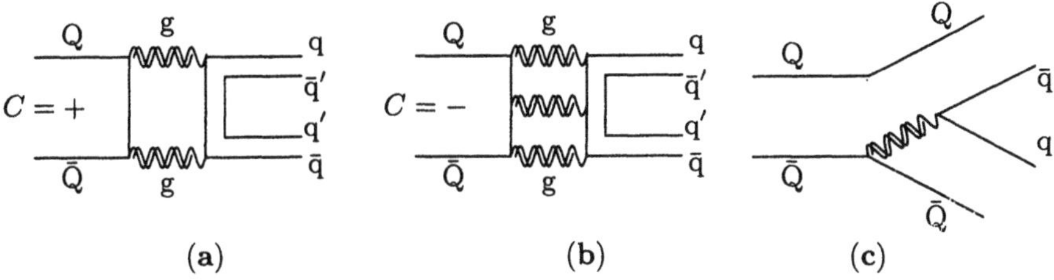


**Fig. 7.7.** OZI-suppressed hadronic decay of a quarkonium below flavor threshold: (**a**) for $C = +$ $(Q\overline{Q})$ states; (**b**) for $C = -$ $(Q\overline{Q})$ states; (**c**) decay of quarkonium above flavor threshold into mesons carrying flavor of heavy quark Q

is based on the interaction between electron charges via the photon. Whereas the photon is a single vector field, gluons come as a set of eight Lorentz vectors forming an $SU_c(3)$ octet. The hadronic decay of a meson can thus proceed by an exchange of gluons between the quark constituents. Now, consider the decay of $\psi(3100)$, a color-singlet with charge conjugation parity $-1$. Since charge conjugation is a symmetry of strong interactions, its $(C)$ parity is preserved throughout the process and, therefore, the number of gluons exchanged must be odd. The least number of gluons that can form a color-singlet state with odd charge conjugation parity is three. Therefore the simplest nonelectromagnetic annihilation process that can lead $\psi$ into hadrons is through an exchange of three gluons (Fig. 7.7b). The rate for the decay of a vector meson to hadrons through three gluons is similar to that for the decay of (ortho)positronium into three photons and is given to lowest order by

$$\Gamma(\mathrm{V} \to \mathrm{ggg} \to \text{hadrons}) = \frac{160}{81}(\pi^2 - 9)\left[\alpha_s(M_V^2)\right]^3 \frac{|\varphi(0)|^2}{M_V^2}, \qquad (7.147)$$

where the quark–gluon coupling strength $\alpha_s$ is defined at the meson mass. We want to make rough estimates of $\alpha_s$ from the expression

$$\alpha_s^3 = \frac{81\,\pi\,\alpha^2\,Q^2}{10(\pi^2-9)}\,\frac{\Gamma(\mathrm{V} \to \mathrm{ggg} \to \text{hadrons})}{\Gamma(\mathrm{V} \to \mathrm{ee})}. \qquad (7.148)$$

Results for $\phi$, $\psi$, and $\Upsilon$ are listed in Table 7.8. In the case of $\phi$, the electronic branching is calculated relative to the OZI-forbidden decays $\rho\pi$ and $\pi\pi\pi$ only (which together amount to 15.6% of all decays). Note that $\alpha_s$ is not a constant, it decreases with increasing energy, reflecting a general result of QCD (see Chap. 15). The OZI suppression is therefore due to the exchange of three 'hard' gluons associated with high-momentum transfer and a weak coupling parameter; this also explains why state $\psi$ is much more narrow than $\phi$. In contrast, the OZI-allowed decay mode $\psi(3770) \to \mathrm{D\overline{D}}$, which involves the exchange of a single 'soft' gluon carrying low momentum (Fig. 7.7c), proceeds far more intensely and has a much larger width.

**Table 7.8.** Quark–gluon couplings in vector mesons

| Meson | $Q^2$ | $\Gamma_{ee}/\Gamma_{total}$ | $\alpha_s$ |
|---|---|---|---|
| $\phi(1020)$ | 1/9 | $2 \times 10^{-3}$ | 0.44 |
| $\psi(3100)$ | 4/9 | $6 \times 10^{-2}$ | 0.21 |
| $\Upsilon(9460)$ | 1/9 | $2.5 \times 10^{-2}$ | 0.18 |

### 7.6.2 The Tau Lepton

Shortly after the discovery of charmonium states, an important new and unexpected particle was added to the list of known particles. The first evidence for the existence of the $\tau$ lepton was obtained (Perl et al., 1975) at the SPEAR $e^+e^-$ storage ring. The observation was confirmed soon afterward by experiments again at SPEAR and at the DORIS $e^+e^-$ storage ring of the Deutsches Elektronen Synchrotron (DESY). Examining the massive amount of data from the annihilation $e^+e^-$ process, Perl and collaborators found reactions of the form $e^+ + e^- \to e^\pm + \mu^\mp +$ missing energy, in which no other charged particles or photons were detected. Most of these events occurred at or above 4 GeV. They concluded that these events arose from the production and subsequent decay of a pair of unidentified objects, each having a mass in the range of 1.6 to 2 GeV (which coincidentally includes the D mesons). The production and decay sequence they had in mind was

$$\begin{aligned} &e^+ + e^- \to \gamma \to \tau^+ + \tau^- ; \\ &\qquad \tau^+ \to e^+ + \nu_e + \bar{\nu}_\tau , \ \mu^+ + \nu_\mu + \bar{\nu}_\tau , \\ &\qquad \tau^- \to e^- + \bar{\nu}_e + \nu_\tau , \ \mu^- + \bar{\nu}_\mu + \nu_\tau . \end{aligned} \tag{7.149}$$

The $e^\pm\mu^\mp$ pairs were the only detected particles in the events and, with their characteristic two-pronged signature and total zero charge, led to the discovery of $\tau$.

The mass of the new particle can be determined by measuring the threshold for $e^+ + e^- \to \tau^+ + \tau^-$. The current value is

$$m_\tau = 1777 \pm 0.30 \text{ MeV} .$$

To qualify as a lepton, a particle should not interact through the strong interaction but only through the weak interaction and, if charged, through the electromagnetic interaction; in addition, it should have no evidently observable structure and preferably should have spin 1/2. Experiments have established that the $\tau$ lepton has all these basic properties.

For spin 0 or 1/2, the electromagnetic production process for pointlike particles (149) is well understood from quantum electrodynamics; its cross-section at the Born level (cf. Sect. 4.7) is

$$\sigma(\mathrm{e}^+\mathrm{e}^- \to \tau^+\tau^-) = \frac{\pi\alpha^2\beta^3}{3s} \qquad \text{for spin } 0\,, \tag{7.150}$$

$$\sigma(\mathrm{e}^+\mathrm{e}^- \to \tau^+\tau^-) = \frac{4\pi\alpha^2}{3s}\,\frac{\beta(3-\beta^2)}{2} \qquad \text{for spin } 1/2\,, \tag{7.151}$$

where $\beta = p/E$ and $E$ are the velocity and energy of the $\tau^-$ or $\tau^+$, and $s = E_{\mathrm{cm}}^2 = 4E^2$. It has become customary to remove the $1/s$ dependence of cross-section by defining the ratio

$$R_\tau \equiv \sigma(\mathrm{e}^+\mathrm{e}^- \to \tau^+\tau^-)\,/\,\sigma(\mathrm{e}^+\mathrm{e}^- \to \mu^+\mu^-)\,. \tag{7.152}$$

Then

$$R_\tau = \tfrac{1}{4}\beta^3 \qquad \text{for spin } 0\,, \tag{7.153}$$

$$R_\tau = \tfrac{1}{2}\beta(3-\beta^2) \qquad \text{for spin } 1/2\,. \tag{7.154}$$

For spin 1 and higher integral spins, $R_\tau$ has a $\beta^3$-dependence. As $E$ rises above the $\tau$ threshold, $\beta \to 1$ and $R_\tau$ tends to a simple (upper) limit whose value depends on the $\tau$ spin:

$$R_\tau \to 1/4 \quad \text{for spin } 0\,; \qquad R_\tau \to 1 \quad \text{for spin } 1/2\,. \tag{7.155}$$

If $\tau$ has internal structure, (151) is modified by a form factor $F(s)$ to

$$\sigma(\mathrm{e}^+\mathrm{e}^- \to \tau^+\tau^-) = \frac{4\pi\alpha^2}{3s}\,\frac{\beta(3-\beta^2)}{2}\,|F(s)|^2\,. \tag{7.156}$$

A pointlike $\tau$ has $F(s) = 1$ for all $s$, but the presence of an internal structure will cause $F(s)$ and the cross-section to fall quickly to zero when $E_{\mathrm{cm}} \gg 2m_\tau$, just as it happens in pair production of hadrons such as $\mathrm{e}^+\mathrm{e}^- \to \mathrm{p}\bar{\mathrm{p}}$.

The measured value of $R_\tau$ has a maximum of 1, hence spin 0 is excluded. The observed behavior of $\sigma(\mathrm{e}^+\mathrm{e}^- \to \tau^+\tau^-)$ near threshold excludes a $\beta^3$ behavior and hence spin 1 or higher integral spins. While spin 3/2 or higher half-integral spins cannot be ruled out at the present time if $\tau$ has a structure, all observed behavior of $\tau$ seems consistent with spin 1/2 and no structure. Therefore, spin 1/2 is the currently accepted assignment.

$\tau$ is the only lepton heavy enough to decay through a hadronic mode. Besides the two leptonic modes shown in (149), it also decays through quark emission, $\tau^- \to \mathrm{d} + \bar{\mathrm{u}} + \nu_\tau$ (in 3 quark-color states), so that the hadronic decay branching is

$$\frac{\Gamma(\tau \to \text{hadrons})}{\Gamma(\tau \to \text{all})} = \frac{3}{5}\,,$$

which is in good agreement with the measured ratio, 0.64.

From experiments, one infers that the associated neutrino, $\nu_\tau$, has spin $1/2$ and a vanishingly small mass less than 19 MeV, even though the particle itself has not yet been directly detected. The $\tau$–$\nu_\tau$ current is of the V–A type (see Chap. 5). In parallel with the decay rate for $\mu^- \to \nu_\mu + \mathrm{e}^- + \bar{\nu}_\mathrm{e}$

$$\Gamma(\nu_\mu \mathrm{e}^- \bar{\nu}_\mathrm{e}) = \frac{G_\mathrm{F}^2 m_\mu^5}{192\pi^3}, \tag{7.157}$$

the decay rate for $\tau^- \to \nu_\tau + \mathrm{e}^- + \bar{\nu}_\mathrm{e}$ can be calculated as

$$\Gamma(\nu_\tau \mathrm{e}^- \bar{\nu}_\mathrm{e}) = \frac{G_\mathrm{e} G_\tau m_\tau^5}{192\pi^3}. \tag{7.158}$$

The electron mass is neglected in both cases; $G_\mathrm{F} = G_\mathrm{e} = G_\mu$ is the Fermi coupling, and $G_\tau$ is the corresponding $\tau$–$\nu_\tau$ coupling. Taking into account its branching, this decay rate yields the $\tau$ lifetime:

$$T_\tau = \frac{B(\nu_\tau \mathrm{e}^- \bar{\nu}_\mathrm{e})}{\Gamma(\nu_\tau \mathrm{e}^- \bar{\nu}_\mathrm{e})} = B(\nu_\tau \mathrm{e}^- \bar{\nu}_\mathrm{e}) \frac{G_\mathrm{e}}{G_\tau} \frac{m_\mu^5}{m_\tau^5} T_\mu, \tag{7.159}$$

where $B(\nu_\tau \mathrm{e}^- \bar{\nu}_\mathrm{e})$ is the branching ratio for $\tau^- \to \nu_\tau + \mathrm{e}^- + \bar{\nu}_\mathrm{e}$ and $T_\mu$ is the $\mu$ lifetime. Using the data

$$T_\mu = 2.2 \times 10^{-6}\ \mathrm{s} \qquad \text{and} \qquad B(\nu_\tau \mathrm{e}^- \bar{\nu}_\mathrm{e}) = 0.18, \tag{7.160}$$

the expression (159) predicts that

$$T_\tau = 2.7 \times 10^{-13}\ \mathrm{s} \quad \text{if} \quad G_\tau = G_\mathrm{e}, \tag{7.161}$$

which agrees quite well with the current measured value of $2.91 \times 10^{-13}$ s. This result therefore confirms that $G_\tau = G_\mathrm{e}$ and hence e–$\mu$–$\tau$ *universality* (more details in Chap. 13). More recent values of the $\tau$ mass, lifetime, and electronic branching fraction allow a test of the weak interaction universality with improved precision; the ratios of the couplings found are

$$\frac{G_\tau}{G_\mu} = 0.981 \pm 0.014, \qquad \frac{G_\tau}{G_\mathrm{e}} = 0.992 \pm 0.018. \tag{7.162}$$

### 7.6.3 From Bottom to Top

The newly found charmed quark completed a set of fermions ($\nu_\mu$, $\mu$; c, s), replicating the pattern of the first, ($\nu_\mathrm{e}$, e; u, d). The discovery of the $\tau$ lepton and its neutrino seemed to signal the formation of a new generation, and immediately intensified the search for its prospective new quark members. The search was carried out using again $\mathrm{e}^+\mathrm{e}^-$ annihilation and hadronic production of lepton pairs. These efforts were rewarded in mid-1977 with the discovery by Lederman and co-workers of two narrow resonances at 9.44 and 10.17 GeV (called respectively $\Upsilon$ and $\Upsilon'$) in the cross-section of collisions of 400 GeV protons on nuclear targets, p + (Cu, Pt) $\to \mu^+\mu^- + \dots$.

These events were confirmed a year later by two groups at the DORIS $e^+e^-$ storage ring at DESY. The mass of the resonance was determined with improved precision, $M_\Upsilon = 9.46 \pm 0.01$ GeV, and the partial width for $\Upsilon \to e^+e^-$ was calculated, as for $\psi$, from the area under the resonance curve with the result $\Gamma_{ee} = 1.32 \pm 0.03$ keV. We have already seen that it is possible to infer from $\Gamma_{ee}$ the electric charge of the new quark, which turns out to be $-1/3$ (see Table 7.7). The new quark is called the b or *bottom* quark, a name that reflects its charge of $-1/3$ and its place 'beneath' an anticipated quark of charge $2/3$ (which therefore would be called *top*). Soon after, other resonances were seen at 10.35, 10.58, 10.86, and 11.02 GeV. They are identified with other $b\bar{b}$ bound states, exactly as the $\psi$-family was with a set of $c\bar{c}$ levels. The $\Upsilon(9460)$, $\Upsilon'(10\,023)$, and $\Upsilon''(10\,355)$ have extremely narrow widths, which implies that their decays to the already established particles are energetically possible but OZI-suppressed, and therefore that the b quark must have a new quantum number, called *bottom* or *beauty*, $B = -1$. (The sign is conventional, as it is also in $I_3 = -1/2$ for the d quark and in $S = -1$ for the s quark.) On the other hand, the $\Upsilon'''(10\,580)$ resonance is much broader and so must lie above the threshold for bottom-flavored meson pair production, B $\bar{\text{B}}$. A bottom-flavored meson B is composed of a $\bar{b}$ quark and a u or d quark; its mass, $M_B$ is such that $M_{\Upsilon''} \leq 2\,M_B \leq M_{\Upsilon'''}$. The families of pseudoscalar and vector mesons with $B = \pm1$ and $S = 0$, and many other mesons with $B = \pm1$ but $S = \pm1$ have been identified.

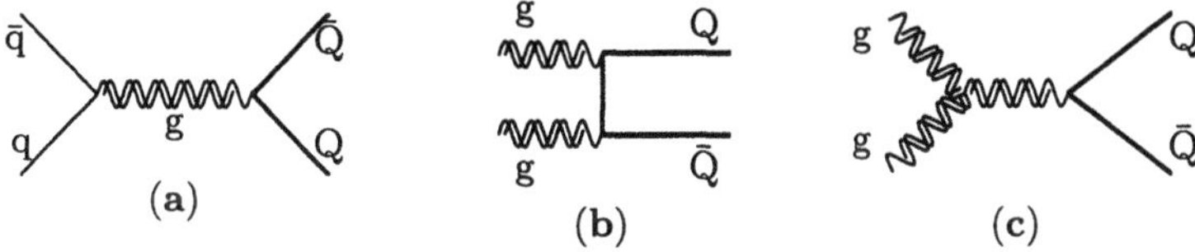


**Fig. 7.8.** Lowest-order mechanisms for hadronic production of heavy quark–antiquark pairs: (**a**) $q\bar{q}$ annihilation; (**b**)–(**c**) gluon–gluon fusion

Meanwhile the standard model of particles has been fully developed and has enjoyed an outstanding success for two decades. Yet one of its key predictions, the existence of the *top* quark, remained unfulfilled. The top quark is considered in the standard model as the (weak isospin) partner of the b quark, and is required to account for the absence of flavor-changing, charge-preserving weak decays of b. As a member of the third family of quark, it provides the simplest explanation for CP violation by the weak interaction. Its existence is crucial lest quantum corrections break the symmetries of the theory, leaving it internally inconsistent. The standard model does not predict its mass, and the experimental lower limit kept climbing with the years, from 23 to 50 GeV, then to 91, 131, and 175 GeV.

The long search finally bore fruit: evidence for the elusive quark was reported in 1994 by two groups working at the Fermilab Tevatron $p\bar{p}$ collider, in which a beam of 900 GeV protons collides with a beam of 900 GeV antiprotons. In $p\bar{p}$ collisions, top–antitop pairs are expected to appear following gluon–gluon fusion and $q\bar{q}$ annihilation, the latter mode being dominant for a top mass above 100 GeV (see Fig. 7.8). In the standard model, the top quark decays into a bottom quark and a W boson (which is the charged gauge boson of the weak interaction). The b quark's lifetime, about 1.5 picoseconds, is long enough for it to find an ordinary quark and together form a B meson. The W meson lives for a fraction of $10^{-24}$ s. Two-thirds of the time, it decays into a quark–antiquark pair, which manifests itself in two jets, or narrow spurts of hadrons moving nearly in the directions of the quarks. The remaining third of W decays produce a lepton pair, a high-energy charged lepton accompanied by an invisible neutrino.

The search for the t quark was carried out in channels where both W bosons decayed leptonically (e$\mu$+jets, ee+jets, and $\mu\mu$+jets) and in channels where just one W boson decayed leptonically (e+jets and $\mu$+jets) – see Fig. 7.9. The data established the existence of $t\bar{t}$ production in $p\bar{p}$ collisions at $\sqrt{s} = 1.8$ TeV, with the top quark mass of $m_t = 180 \pm 12$ GeV. Thus, a single top weighs as much as an atom of gold and more than 37 times a bottom quark, the heaviest among the other five quark varieties.

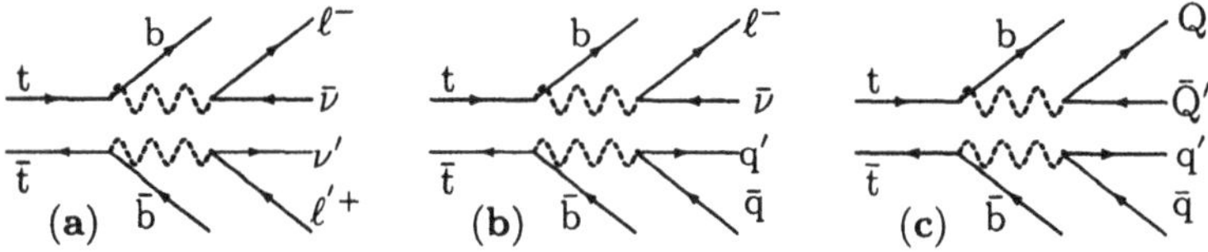


**Fig. 7.9.** Signature events in the top–antitop production: (**a**) dilepton mode; (**b**) lepton-plus-jets mode; (**c**) 4-jets mode

Implications of this astoundingly large mass are many and are still being explored. First, the large mass of t makes it a very short-lived particle. Assuming the dominant decay mode to be the signature decay $t \to b + W$ and neglecting QCD corrections, we can calculate the decay width (Problem 7.8)

$$\Gamma(t \to bW^+) \approx \frac{G_F}{8\sqrt{2}\pi} m_t^3 \left(1 - 3\,\frac{M_W^4}{m_t^4} + 2\,\frac{M_W^6}{m_t^6}\right) . \tag{7.163}$$

Note its strong dependence on the top mass. With $G_F \approx 10^{-5}\,\mathrm{GeV}^{-2}$, $m_t = 175\,\mathrm{GeV}$, and $M_W = 80\,\mathrm{GeV}$, we find $\Gamma(t \to bW^+) \approx 1$ GeV, which corresponds to a t quark's lifetime of $\approx 0.6 \times 10^{-24}$ s. Comparing this with the lifetime of the b quark, which is $1.5 \times 10^{-12}$ s, and that of the $\tau$ lepton, which is $0.3 \times 10^{-12}$ s, one realizes how special top is. Because the top quark decays before it can be hadronized, there are no bound $t\bar{t}$ states (toponia)

**Table 7.9.** Fundamental fermions

| Quark flavor | Mass (GeV) | Charge | Nonzero additive quantum number |
|---|---|---|---|
| u | 0.008 | $2/3$ | $I_3 = +1/2$ |
| d | 0.015 | $-1/3$ | $I_3 = -1/2$ |
| c | 1.5 | $2/3$ | $C = +1$ |
| s | 0.3 | $-1/3$ | $S = -1$ |
| t | 180.0 | $2/3$ | $T = +1$ |
| b | 4.5 | $-1/3$ | $B = -1$ |
| **Lepton flavor** | **Mass (MeV)** | **Charge** | **Nonzero lepton number** |
| $\nu_e$ | $15 \times 10^{-6}$ | 0 | $L_e = 1$ |
| e | 0.51 | $-1$ | $L_e = 1$ |
| $\nu_\mu$ | 0.17 | 0 | $L_\mu = 1$ |
| $\mu$ | 105.66 | $-1$ | $L_\mu = 1$ |
| $\nu_\tau$ | 19 | 0 | $L_\tau = 1$ |
| $\tau$ | 1777.0 | $-1$ | $L_\tau = 1$ |

and no top-flavored mesons or baryons, unlike the situation with the other, lighter flavors. Further, top's very large mass opens up many new decay channels, which might lead to productions of highly exotic particles. Finally, it is believed that a very heavy top, being so distinct from the other quarks, could hold the key to many questions about particle physics still unanswered in the framework of the standard model.

In just over twenty years, many crucial discoveries were made and the remaining half of all the fundamental constituents of matter were observed and identified, thus completing the basic bricks required for the foundation of the standard model. Table 7.9 summarizes the present situation.

## Problems

**7.1 Group and algebra.** (a) Show that the generators of the infinitesimal unitary and unimodular transformations are Hermitian and traceless. (b) Show that the group property $g_1 g_2 = g_3$, which says that the product of any two group elements is some group element, implies (for a Lie group) closed commutation relations among the generators.

**7.2 Nonequivalence of 3 and 3* of SU(3).** Let $\{S\}$ and $\{S^*\}$ be the ordinary and the conjugate fundamental representations of SU(3). To say that they are equivalent is to say that for every $S$ there is an $S^*$ such that $S^* = S_0 S S_0^\dagger$ for some fixed element $S_0$ of the group. $S^*$ is the ordinary complex conjugate of $S$. This relation may also be written as $\lambda_i^* = -S_0 \lambda_i S_0^\dagger$. Show that it cannot be satisfied for all $i = 1, \ldots, 8$.

**7.3 Structure constants of SU(3).** From the known Gell-Mann matrices, calculate the structure constants of SU(3) given in the following table. Verify also the values of the completely symmetric coefficients $d_{ijk}$ of the algebra.

| $[\lambda_i, \lambda_j] = 2\mathrm{i} f_{ijk} \lambda_k$ | | $\{\lambda_i, \lambda_j\} = \frac{4}{3}\delta_{ij} + 2 d_{ijk}\lambda_k$ | | | |
|---|---|---|---|---|---|
| $ijk$ | $f_{ijk}$ (antisymmetric) | $ijk$ | $d_{ijk}$ (symmetric) | $ijk$ | $d_{ijk}$ (symmetric) |
| 123 | 1 | 118 | $1/\sqrt{3}$ | 366 | $-1/2$ |
| 147 | $1/2$ | 146 | $1/2$ | 377 | $-1/2$ |
| 156 | $-1/2$ | 157 | $1/2$ | 448 | $-1/(2\sqrt{3})$ |
| 246 | $1/2$ | 228 | $1/\sqrt{3}$ | 558 | $-1/(2\sqrt{3})$ |
| 257 | $1/2$ | 247 | $-1/2$ | 668 | $-1/(2\sqrt{3})$ |
| 345 | $1/2$ | 256 | $1/2$ | 778 | $-1/(2\sqrt{3})$ |
| 367 | $-1/2$ | 338 | $1/\sqrt{3}$ | 888 | $-1/\sqrt{3}$ |
| 458 | $\sqrt{3}/2$ | 334 | $1/2$ | | |
| 678 | $\sqrt{3}/2$ | 355 | $1/2$ | | |

**7.4 Applications of U-spin.** (a) Show that $U_\pm = F_6 \pm \mathrm{i} F_7$ and $U_3 = (\sqrt{3}F_8 - F_3)/2$ satisfy the SU(2) algebra

$$[U_3, U_\pm] = \pm U_\pm, \quad [U_+, U_-] = 2U_3 .$$

(b) Show that the charge operator $Q = F_3 + F_8/\sqrt{3}$ is a U-scalar, that is, it has U-spin $U = 0$, or $[Q, U_i] = 0$ for $i = \pm, 3$. Write down the electromagnetic current operator. (c) Show that for the meson octet, the ($U_3 = 0$)-component of the U-triplet is $\pi_\mathrm{u}^0 = (-\pi^0 + \sqrt{3}\eta)/2$, and the U-singlet is $\eta_\mathrm{u}^0 = (\sqrt{3}\pi^0 + \eta)/2$. Since $\pi_\mathrm{u}^0$ is a U-spin vector component it cannot couple to an electromagnetic current. Show that for the $2\gamma$ decay mode, $\langle \pi^0 | 2\gamma \rangle = \sqrt{3} \langle \eta | 2\gamma \rangle$.

**7.5 Gell-Mann–Okubo mass formula.** The mass symmetry-breaking interaction for an isospin multiplet is proportional to the three-component of the isospin operator, $I_3$. Similarly, the symmetry-breaking Hamiltonian $H_8$ of SU(3) for the octet baryons is given by the eight-component of the octet operator $F_8 = \lambda_8/2$. In this case a further set of octet matrices can be formed, that is, $D_8 = d_{8ij} F_i F_j$, where $d_{ijk}$ are the completely symmetric coefficients of the group. Therefore, the most general symmetry-breaking interaction is of the form $H_8 = a F_8 + b D_8$. Derive the GMO mass formula.

**7.6 Reduction to irreducible representations in SU(3).** (a) Prove the reduction formula $\mathbf{8} \times \mathbf{8} = \mathbf{1} + \mathbf{8} + \mathbf{8} + \mathbf{10} + \mathbf{10^*} + \mathbf{27}$ of SU(3) by explicitly constructing the irreducible tensors from the two irreducible tensors $M^a{}_b$ and $N^a{}_b$. (b) Prove the formula $\mathbf{10^*} \times \mathbf{10} = \mathbf{1} + \mathbf{8} + \mathbf{27} + \mathbf{64}$ without an explicit construction of the irreducible tensors, e.g. using the prescription by S. Coleman, J. Math. Phys. **5** (1964) 1343–1344.

**7.7 Reduction to irreducible representations in SU(6).** Prove by direct construction the following reduction formula in SU(6):

$$\mathbf{6} \times \mathbf{6} \times \mathbf{6} = \mathbf{20} + \mathbf{70} + \mathbf{70} + \mathbf{56}\,.$$

**7.8 Top decay rate.** Assume that the amplitude for the decay of quark Q into quark q plus a massive vector boson W is given by

$$\mathcal{M} = \mathrm{i} f\, \bar{u}_{\mathrm{q}}(p)(\gamma^\mu - \alpha\gamma^\mu\gamma_5)u_{\mathrm{Q}}(P)\,\varepsilon^*_\mu(k)\,.$$

Here $P$, $p$, and $k$ are momenta; $f$ and $\alpha$ are constants; $\varepsilon_\mu$ is the polarization vector of the vector boson.

(a) Show that upon summing over spins and polarizations one has

$$\sum_{\text{spins}}\sum_{\text{pol}} |\mathcal{M}|^2 = 4f^2(1+\alpha^2)[p{\cdot}P + \frac{2}{M_{\mathrm{W}}^2}(k{\cdot}p)(k{\cdot}P)\,] - 12f^2(1-\alpha^2)m_{\mathrm{q}}m_{\mathrm{Q}}\,,$$

where $k^2 = M_{\mathrm{W}}^2$, $p^2 = m_{\mathrm{q}}^2$, and $P^2 = m_{\mathrm{Q}}^2$.

(b) Using formula (4.74) of Chap. 4, show that the rate of decay from rest, with $\alpha = 1$, is given by

$$\Gamma(\mathrm{Q} \to \mathrm{q} + \mathrm{W}) = \frac{f^2}{4\pi M_{\mathrm{W}}^2}\,|\boldsymbol{p}|m_{\mathrm{Q}}^2\,[1 + x - 2x^2 - y(2 - x - y)]\,,$$

where $|\boldsymbol{p}| = (m_{\mathrm{Q}}/2)\{[1-(\sqrt{x}+\sqrt{y})^2][1-(\sqrt{x}-\sqrt{y})^2]\}^{1/2}$, $x = (M_{\mathrm{W}}/m_{\mathrm{Q}})^2$, and $y = (m_{\mathrm{q}}/m_{\mathrm{Q}})^2$. See also I. Bigi et al., Phys. Lett. **181B** (1986) 157.

## Suggestions for Further Reading

*General references on group theory:*

Georgi, H., *Lie Algebras in Particle Physics.* Benjamin, Reading, MA 1982

Gilmore, R., *Lie Groups, Lie Algebras, and Some of Their Applications.* Wiley, New York 1974

Lichtenberg, D., *Unitary Symmetry and Elementary Particles.* Academic Press, New York 1978

Tung, Wu-Ki, *Group Theory in Physics.* World Scientific, Singapore 1985

*Early models:*

Fermi, E. and Yang, C. N., Phys. Rev. **76** (1949) 1739

Sakata, S., Progr. Theor. Phys. **16** (1956) 686

*Introduction of the* SU(3) *octet structure of mesons and baryons:*

Gell-Mann, M., Phys. Rev. **125** (1962) 1067

Gell-Mann, M. and Ne'eman, Y., *The Eightfold Way: A Review – With Collection of Reprints*. Benjamin, New York 1964

Ne'eman, Y., Nucl. Phys. **20** (1961) 222

*Introduction of quarks as fundamental building blocks for hadrons:*

Gell-Mann, M., Phys. Lett. **8** (1964) 214

Zweig, G., CERN-8419-TH-412 (1964). Reprinted in *Development in the Quark Theory of Hadrons*, (ed. by Lichtenberg, D. B. and Rosen, S. P.). Hadronic Press, Monamtum, MA 1980

*Introduction of the* SU(6) *classification of hadrons:*

Gürsey, F. and Radicati, L. A., Phys. Rev. Lett. **13** (1964) 173

Sakita, B., Phys. Rev. **136** (1964) B1756

*The quark potential model:*

Appelquist, T. and Politzer, H., Phys. Rev. Lett. **34** (1975) 43; Appelquist, T., Barnett, R. M. and Lane K., Ann. Rev. Nucl. Part. Sci. **28** (1978) 387

Close, F. E., *An Introduction to Quarks and Partons*. Academic Press, New York 1979

*Introduction of the color quantum number:*

Greenberg, O. W., Phys. Rev. Lett. **13** (1964) 598

Han, M. Y. and Nambu, Y., Phys. Rev. **139** (1965) B1006

*Discoveries of* c, b, t, *and* $\tau$*:*

Abachi, S. et al., Phys. Rev. Lett. **74** (1995) 2632 (top)

Abe, F. et al., Phys. Rev. **D50** (1994) 2966 (top)

Abe, F. et al., Phys. Rev. Lett. **74** (1995) 2626 (top)

Aubert, J. J. et al., Phys. Rev. Lett. **33** (1974) 1404 (charm)

Augustin, J. E. et al., Phys. Rev. Lett. **33** (1974) 1406 (charm)

Herb, S. W. et al., Phys. Rev. Lett. **39** (1977) 252 (bottom)

Perl, M. L. et al., Phys. Rev. Lett. **35** (1975) 1489 ($\tau$ lepton)

*Additional references may be found in*

Ezhela, V. V. et al, *Particle Physics: One Hundred Years of Discoveries: An Annotated Chronological Bibliography*. AIP Press, New York 1996

*Data are quoted from*

Review of Particle Properties, Phys. Rev. **D54** (1996) 1

# 8 Gauge Field Theories

All the current successful theories of the fundamental forces start from the premise of invariance of the physical laws to certain coordinate-dependent transformations. In particular, the quantum field theories of the electromagnetic, weak, and strong interactions of the fundamental particles all belong to the class of *local gauge theories*, so called because they are invariant to coordinate-dependent transformations on internal space of the particles. We start this chapter by describing the general relation between symmetries and interactions. Next, we take up the study of invariance under the Abelian gauge group U(1), the group of space-time-dependent phase transformations on charged fields; the resulting gauge theory is electrodynamics. The following section is devoted to theories for which the gauge group is non-Abelian. The results see immediate applications to quantum chromodynamics, a theory based on the color SU(3) group. The last two sections of the chapter contain a discussion on the mechanism of *spontaneous symmetry breaking*, which is an indispensable ingredient in the formulation of the standard theory of the electroweak interaction, the subject of the following chapter.

## 8.1 Symmetries and Interactions

In previous chapters, we have studied some of the implications of the conservation or violation of *global* symmetries that a theory may have. Under a symmetry transformation of this kind, fields are changed by an identical amount that remains fixed throughout space and time, and invariance of the theory to such changes implies the existence of a conserved quantity.

Generally, when this symmetry is made *local*, whereby all the particle fields are altered by an amount that varies with each space-time point, invariance may be preserved provided a set of vector fields (or higher-rank tensor fields) defined over all space-time is introduced into the theory to cancel the long-range effects of the vector gradient of the transformation parameter and to restore the symmetry. In particular, transformations in which this parameter is the phase-angle of the particle fields are called *gauge transformations* and considered as *internal*, for they act on the labels of the particles rather than on their space-time coordinates. If global, they give rise to conserved charges, of which the electric charge is an example. If local, they may lead to an observable force. Three of the four existing fundamental interactions are believed to be explicable in this fashion: the electric current is the source of the electromagnetic force; the weak isospin and the weak hypercharge are the sources of the unified electromagnetic and weak forces; and finally, the quark colors, the sources of the strong interactions between quarks. It is not understood why only those and no other conserved charges can produce observable dynamical effects. As for the fourth fundamental force, gravitation, it may be similarly viewed as arising from a local invariance, with the difference that the transformations that leave the theory invariant act on space-time coordinates themselves, and therefore the resulting force field, generated by the conserved energy-momentum tensor, is tensorial rather than vectorial. Otherwise, gravitation and gauge theories have close similarities with one another.

In any nontrivial quantum field theory, divergent integrals may appear in higher orders of the perturbation expansion of the transition amplitudes. Renormalization is a procedure of removing these ultraviolet divergences by adding extra terms, called counterterms, to the Lagrangian of the theory. A theory is said to be *renormalizable* when all the counterterms induced by this procedure are of the same form as terms in the original Lagrangian. A theory with an interaction of mass dimension greater than four is nonrenormalizable, although not all theories involving only interactions of mass dimension four or less are necessarily renormalizable. All three gauge theories mentioned above respect this simple but highly constraining demand of renormalizability. We shall return to this topic in Chap. 15.

As mentioned above, the vector gauge fields introduced into the theory to enforce gauge invariance have an infinite range or, equivalently, have no mass. The photon, which is supposed to mediate the electromagnetic interaction, is in effect observed to be massless ($m_\gamma < 6 \times 10^{-16}$ eV). Experiment is also consistent with the assumption that the gluons, the gauge fields of the fundamental strong force, have vanishing masses. However, the weak forces have always been known since their discovery to have a very short range. So the gauge fields associated with the conserved weak charges in a gauge-invariant theory cannot be immediately identified with the observed weak forces. They must first acquire mass. But we are not allowed to introduce artificially a mass term in the theory since this would break gauge invariance, which would in turn make the theory divergent and thus nonpredictive. The solution to this difficulty is to hide part of the gauge group, that is, to arrange so that, while remaining exact in the underlying field equations, the gauge symmetries are not realized in physical states. This *spontaneous symmetry breakdown* is similar to the loss of symmetry during certain phase transitions observed in condensed matter, such as the loss of translational symmetry when liquid water turns into an ice crystal lattice below 0° C, or the loss of rotational symmetry when a very large sample of ferromagnet acquires a net magnetization below the Curie temperature.

## 8.2 Abelian Gauge Invariance

In quantum electrodynamics (QED), the interaction of a charged particle with an electromagnetic field is obtained by coupling the field with the electromagnetic current for the particle, an empirical rule known in classical physics as the *minimal coupling postulate.* This rule can be better understood in terms of a general symmetry principle, called the *principle of gauge invariance*, susceptible of generalizations.

Consider, as a representative example of matter, a fermion field. The Lagrangian density for a free Dirac field of mass $m$ is

$$\mathcal{L}_0 = \overline{\psi}(\mathrm{i}\gamma^\mu \partial_\mu - m)\psi\,. \tag{8.1}$$

It is invariant to the *global* phase transformation

$$\psi(x) \to \psi'(x) = \mathrm{e}^{-\mathrm{i}q\omega}\,\psi(x)\,, \tag{8.2}$$

where $\omega$ is the transformation parameter, an arbitrary constant number, independent of $x$. Another constant $q$ has been inserted at this point to accord with common usage; it will take on the meaning of the particle electric charge in the present context. All operations (2) form a representation of the single-parameter Abelian group $\mathrm{U}(1)$, which is sometimes written with a subscript, such as in $\mathrm{U_Q}(1)$, to emphasize its association with a conserved quantum number. It is crucial for the invariance of $\mathcal{L}_0(\psi, \partial_\mu\psi)$ to the global symmetry transformation (2) that the field gradient transforms exactly like the field itself:

$$\partial_\mu\psi(x) \to \partial_\mu\psi'(x) = \mathrm{e}^{-\mathrm{i}q\omega}\,\partial_\mu\psi(x)\,. \tag{8.3}$$

As discussed in Sect. 2.7, this invariance implies the existence of a locally conserved current,

$$j^\mu(x) = q\,\overline{\psi}\gamma^\mu\psi\,. \tag{8.4}$$

The global transformation (2) can be generalized to the *local* transformation

$$\psi(x) \to \psi'(x) = \mathrm{e}^{-\mathrm{i}q\omega(x)}\,\psi(x)\,, \tag{8.5}$$

where $\omega$ is now a real function of $x$, i.e. $\omega(x)$ defines an independent phase transformation at each space-time point. However, the Lagrangian (1) and, in general, any free-field Lagrangian cannot be invariant under this local transformation because the transformation rule for the field gradient differs from that for the field,

$$\partial_\mu\psi(x) \to \partial_\mu\psi'(x) = \mathrm{e}^{-\mathrm{i}q\omega}\,[\partial_\mu\psi - \mathrm{i}q(\partial_\mu\omega)\,\psi]\,, \tag{8.6}$$

so that the transformed free-field Lagrangian acquires an extra term which spoils the invariance:

$$\mathcal{L}_0 \to \mathcal{L}_0' = \mathcal{L}_0 + q\,\overline{\psi}\gamma^\mu\psi\,\partial_\mu\omega = \mathcal{L}_0 + j^\mu\partial_\mu\omega\,. \tag{8.7}$$

The presence of a symmetry-violating term in (7) suggests that if we wish to make the theory invariant under (5), it is necessary to introduce a vector field $A_\mu$ that couples to the particle current so that this coupling when transformed may cancel $j^\mu\partial_\mu\omega$. The modified Lagrangian

$$\mathcal{L}_1 = \mathcal{L}_0 - j^\mu A_\mu \tag{8.8}$$

transforms under (5) as

$$\mathcal{L}_1 \to \mathcal{L}_1' = \mathcal{L}_0' - j'^\mu A_\mu' = \mathcal{L}_0 + j^\mu\,\partial_\mu\omega - j^\mu\,A_\mu' \tag{8.9}$$

(since $j_\mu' = q\,\overline{\psi}'\gamma_\mu\psi' = j_\mu$). Invariance of $\mathcal{L}_1$ then requires the vector field to have the property

$$A_\mu \to A_\mu' = A_\mu + \partial_\mu\omega \tag{8.10}$$

under the transformation that acts on $\psi$ according to (5). The 'scale' of the vector field has changed. Thus, the quantum theory of electrically charged particles is said to have local phase-angle independence – referring to the change of matter field in (5) – or more currently, local gauge invariance – emphasizing the scale change of the force field in (10). The field $A_\mu$ is accordingly called a *gauge field.* Rewriting the Lagrangian as

$$\mathcal{L}_1 = \overline{\psi}(\mathrm{i}\gamma^\mu\partial_\mu - m)\psi - q\,\overline{\psi}\gamma^\mu\psi A_\mu = \overline{\psi}(\mathrm{i}\gamma^\mu D_\mu - m)\psi\,, \tag{8.11}$$

where

$$D_\mu \equiv \partial_\mu + \mathrm{i}qA_\mu\,, \tag{8.12}$$

we observe that the gauge invariance of $\mathcal{L}_1$ has been realized by making the field gradient transform covariantly, that is,

$$D_\mu\psi \to D_\mu'\psi' = \mathrm{e}^{\mathrm{i}q\omega}\,D_\mu\psi\,, \tag{8.13}$$

and, for this reason, $D_\mu$ is called a *covariant derivative* for this gauge group. To make the vector field an integral part of the dynamic system, it is necessary to introduce gauge-invariant terms built up from $A_\mu$ and its derivatives. The combination

$$F_{\mu\nu} = \partial_\mu A_\nu - \partial_\nu A_\mu$$

is invariant under (10), whereas

$$\partial_\mu A_\nu + \partial_\nu A_\mu$$

is not. Therefore, the Lorentz scalar $-\frac{1}{4}F_{\mu\nu}F^{\mu\nu}$ (with a conventional normalization factor) may be added to the Lagrangian. The mass term $A_\mu A^\mu$ is not allowed since it is not invariant under (10). So the final gauge-invariant Lagrangian looks like

$$\mathcal{L}_1 = \overline{\psi}(\mathrm{i}\gamma^\mu D_\mu - m)\psi - \frac{1}{4}F_{\mu\nu}F^{\mu\nu}\,. \tag{8.14}$$

The term $\epsilon^{\mu\nu\rho\sigma}F_{\mu\nu}F_{\rho\sigma}$, which is equally gauge-invariant and of dimension four, need not be included because it may be rewritten as the divergence of a current, $\partial_\mu K^\mu$, and therefore contributes only as a surface term to the action. Under the usual assumption that fields vanish at infinity, it may be discarded. Other higher-dimensional gauge-invariant couplings, such as $\overline{\psi}\sigma_{\mu\nu}\psi F^{\mu\nu}$, are not allowed by the requirement of renormalizability. When the fields that appear here are reinterpreted as quantum fields, (14) is just the familiar form of the QED Lagrangian for a Dirac particle of charge $q$ interacting with the electromagnetic field. It is the most general U(1)-gauge-invariant dimension-four renormalizable Lagrangian, and it is in extremely good agreement with experiment.

Thus, we have shown that, when a free-field theory has an exact global phase symmetry, it may have the corresponding local phase invariance only upon becoming an interacting field theory involving a massless vector field (the photon) which interacts with the charged particle in a well-defined manner. In Abelian theories such as this one, there are no restrictions on the coupling strength between the gauge field and matter fields; the electron has charge $q = -e$ while another particle may carry any other charge $q = ze$. But the interaction appears in the same form regardless of the nature of the particle, be it lepton, quark, or hadron. That the interaction derived from imposing renormalizability and some kind of gauge invariance on the theory is *unique* and *universal* is precisely what has made this symmetry condition – the *gauge invariance principle* – so powerful that it has now become the guiding principle in the search for the theories of interactions in particle physics.

## 8.3 Non-Abelian Gauge Invariance

As particles usually come in multiplets, we might wonder what kind of gauge fields and interactions the principle of gauge invariance would imply in general. Suppose, for example, we have a number of Dirac spinor fields $\psi^a$ (with $a = 1, \ldots, n$ describing some internal degree of freedom, such as isospin or color) that form such a multiplet, $\psi$; that is to say, they have equal masses, $m_a = m$, and transform into one another by the rule

$$\psi^a \to \psi'^a = U^a{}_b\,\psi^b\,, \tag{8.15}$$

where $U$ is a *unitary* $n \times n$ matrix. In the following, we will further limit ourselves to *unimodular* matrices, so that $\det U = 1$. All such matrices define some representation of a Lie group, G. To simplify, we also assume that G is a simple group and $\psi$ belongs to its fundamental representation.

Unitary unimodular matrices $U$ may be parameterized by $N = n^2 - 1$ real phase-angles $\omega_i$ in the form

$$U = \exp(-\mathrm{i}gT_i\omega_i)\,, \tag{8.16}$$

where, as usual, a sum over repeated indices is implied. The real constant factor $g$, common to all terms in the sum, will turn out to be a coupling constant. Transformations very near the identity are given by $1 - \mathrm{i}gT_i\omega_i$, and for this reason the matrices $T_i$ are called the generators of infinitesimal transformations. They are Hermitian and traceless, $T_i^\dagger = T_i$ and $\mathrm{Tr}\, T_i = 0$, as respective consequences of the unitarity and unimodularity of $U$. They constitute a basis of a Lie algebra, and must satisfy commutation relations of the form

$$[T_i, T_j] = \mathrm{i}f_{ijk}\, T_k\,, \qquad \text{for} \quad i,\, j,\, k = 1, \ldots, N\,. \tag{8.17}$$

When not all the structure constants $f_{ijk}$ vanish, these relations define a *non-Abelian* algebra. It is convenient to normalize the generators such that

$$\mathrm{Tr}\,(T_iT_j) = \tfrac{1}{2}\,\delta_{ij}\,. \tag{8.18}$$

For G=SU(2), the generators in the fundamental representation are given by the familiar $2 \times 2$ Pauli matrices, $T_i = \frac{1}{2}\,\tau_i$, with $i = 1,\, 2,\, 3$, while for G=SU(3), $T_i = \frac{1}{2}\,\lambda_i$, with $i = 1, \ldots, 8$, are the $3 \times 3$ Gell-Mann matrices.

The free-field Lagrangian, assumed independent of the internal degree of freedom, is given by

$$\mathcal{L}_0 = \overline{\psi}_a\,(\mathrm{i}\gamma^\mu\partial_\mu - m)\,\psi^a \;= \overline{\psi}\;(\mathrm{i}\gamma^\mu\partial_\mu - m)\;\psi\,. \tag{8.19}$$

In the second equation, the operator contains an implicit unit matrix defined on the $n$-dimensional space of the group representation.

The free-field Lagrangian is invariant under the gauge transformation (15) provided it is a *global* transformation, independent of space-time coordinates $x$. The conserved fermion currents that follow from this invariance are given by Noether's theorem:

$$j_i^\mu = g\,\overline{\psi}\,\gamma^\mu T_i\,\psi\,. \tag{8.20}$$

For the isospin group SU(2), they are the conserved isospin currents.

On the other hand, if $U$ represents a *local* transformation, which depends on the space-time point where it acts, $U = U(x)$, then the free-field Lagrangian will not be invariant in general, but will rather vary as

$$\mathcal{L}_0 \to \mathcal{L}_0' = \mathcal{L}_0 + \overline{\psi}\,\mathrm{i}\gamma^\mu(U^\dagger\partial_\mu U)\,\psi\,, \tag{8.21}$$

where the symmetry-violating term arises from differences in the transformation rules for the field and the field gradient:

$$\psi \to \psi' = U\psi, \tag{8.22}$$

$$\partial_\mu \psi \to \partial_\mu \psi' = U\,\partial_\mu \psi + (\partial_\mu U)\psi\,. \tag{8.23}$$

This suggests that we must introduce extra fields with couplings to the Noether currents similar in form to the second term on the right-hand side of (21) to compensate for this unwanted term. The modified Lagrangian

$$\mathcal{L}_1 = \mathcal{L}_0 - g\overline{\psi}\gamma^\mu \boldsymbol{A}_\mu \psi\,, \tag{8.24}$$

where $\boldsymbol{A}_\mu$ is an $n \times n$ Hermitian traceless matrix whose elements are vector fields, transforms as

$$\begin{aligned}\mathcal{L}_1 \to \mathcal{L}_1' &= \mathcal{L}_0' - g\overline{\psi}'\gamma^\mu \boldsymbol{A}_\mu' \psi' \\ &= \mathcal{L}_0 + \overline{\psi}\,\mathrm{i}\gamma^\mu (U^\dagger \partial_\mu U)\psi - g\overline{\psi}\,\gamma^\mu\, U^\dagger \boldsymbol{A}_\mu' U\,\psi\,.\end{aligned} \tag{8.25}$$

The demand that $\mathcal{L}_1$ be invariant in this operation requires

$$\overline{\psi}\,\mathrm{i}\gamma^\mu (U^\dagger \partial_\mu U)\psi - g\overline{\psi}\gamma^\mu\, U^\dagger \boldsymbol{A}_\mu' U\,\psi = -g\overline{\psi}\gamma^\mu \boldsymbol{A}_\mu \psi\,.$$

Thus, in a gauge transformation that acts on $\psi$ according to (15), the vector field has the transformation property

$$\boldsymbol{A}_\mu \to \boldsymbol{A}_\mu' = \frac{\mathrm{i}}{g}(\partial_\mu U)U^\dagger + U\boldsymbol{A}_\mu U^\dagger\,. \tag{8.26}$$

For most practical purposes it suffices to restrict $\omega_i(x)$ to infinitesimal values so that, to first order,

$$U \approx 1 - \mathrm{i}g\,\boldsymbol{\omega}\,, \tag{8.27}$$

where $\boldsymbol{\omega} = \omega_j T_j$. To this order, the fermion field transforms as

$$\begin{aligned}\psi' &= U\psi \approx \psi + \delta\psi\,, \\ \delta\psi &= -\,\mathrm{i}g\,\boldsymbol{\omega}\,\psi\,,\end{aligned} \tag{8.28}$$

or in components

$$\delta\psi^a = -\mathrm{i}g\,\omega_j\,(T_j)^a{}_b\,\psi^b \qquad \text{for } a = 1, \ldots, n\,. \tag{8.29}$$

There are $(n^2 - 1)$ local gauge fields $A_{j\mu}$, which are independent of the representation of the particle fields $\psi$ and which form the elements of the matrices $\boldsymbol{A}_\mu$. They are chosen so that $\boldsymbol{A}_\mu = A_{j\mu}T_j$. To first order, the transformation rule (26) for the gauge field matrix becomes

$$\begin{aligned} \boldsymbol{A}'_\mu &= \frac{\mathrm{i}}{g}(\partial_\mu U)U^\dagger + U\boldsymbol{A}_\mu U^\dagger \approx \boldsymbol{A}_\mu + \delta\boldsymbol{A}_\mu\,, \\ \delta\boldsymbol{A}_\mu &= \partial_\mu\boldsymbol{\omega} + \mathrm{i}g\,[\boldsymbol{A}_\mu, \boldsymbol{\omega}]\,, \end{aligned} \tag{8.30}$$

or in components

$$\delta A_{i\mu} = \partial_\mu\omega_i - g\, f_{ijk}\, A_{j\mu}\, \omega_k \qquad \text{for } i = 1, \ldots, n^2 - 1\,. \tag{8.31}$$

If G is Abelian, (31) reduces to

$$\delta A_{i\mu} = \partial_\mu\omega_i \qquad \text{(Abelian group)}\,.$$

It corresponds to the first, inhomogeneous term in (31) and implies that the vector fields $A_{i\mu}$ have as sources the currents $j_i^\mu$, just as the transformation rule for the electromagnetic field identifies the electric current as its source. If, on the other hand, G is a non-Abelian group of global symmetry, the transformation rule becomes

$$\delta A_{i\mu} = -g\, f_{ijk}\, A_{j\mu}\, \omega_k \qquad \text{(global symmetry)}\,.$$

The right-hand side of this equation is the same in form as the right-hand side of (29) with $(T_j)^a{}_b$ replaced by $-\mathrm{i}f_{jab}$, which indicates that the gauge fields $A_{i\mu}$ belong to the adjoint representation of the group; that is, for example, they transform as an isovector in SU(2) and as an octet in SU(3).

In terms of the covariant derivative for the non-Abelian gauge group G

$$\boldsymbol{D}_\mu = \partial_\mu + \mathrm{i}g\boldsymbol{A}_\mu\,, \tag{8.32}$$

which obeys the relation

$$\boldsymbol{D}'_\mu U\psi = U\,\boldsymbol{D}_\mu\psi\,, \tag{8.33}$$

the Lagrangian (24) takes the form

$$\mathcal{L}_1 = \overline{\psi}\,\mathrm{i}\gamma^\mu\,(\partial_\mu + \mathrm{i}g\boldsymbol{A}_\mu)\psi - m\,\overline{\psi}\psi = \overline{\psi}\,(\mathrm{i}\gamma^\mu\boldsymbol{D}_\mu - m)\,\psi\,. \tag{8.34}$$

To this Lagrangian must be added contributions from the gauge fields themselves. In analogy with the identity

$$[D_\mu, D_\nu]\,\psi = \mathrm{i}q\,F_{\mu\nu}\psi\,, \tag{8.35}$$

which is satisfied by the electromagnetic field strength, we may define the field tensor in the non-Abelian case by the generalized relation

$$[\boldsymbol{D}_\mu, \boldsymbol{D}_\nu]\,\psi \equiv \mathrm{i}g\,\boldsymbol{F}_{\mu\nu}\psi\,. \tag{8.36}$$

Under the gauge transformation $U$, the left-hand side of (36) gives

$$[\boldsymbol{D}'_\mu,\ \boldsymbol{D}'_\nu]\,U\psi = U[\boldsymbol{D}_\mu,\ \boldsymbol{D}_\nu]\,\psi = \mathrm{i}g\,U\boldsymbol{F}_{\mu\nu}\psi\,,$$

where (33) has been used, while the right-hand side becomes

$$\mathrm{i}g\,\boldsymbol{F}'_{\mu\nu}\,\psi' = \mathrm{i}g\,\boldsymbol{F}'_{\mu\nu}\,U\psi\,.$$

Therefore, identifying the right-hand sides of the last two equations yields

$$\boldsymbol{F}'_{\mu\nu} = U\boldsymbol{F}_{\mu\nu}U^\dagger \approx \boldsymbol{F}_{\mu\nu} + \mathrm{i}g\,[\boldsymbol{F}_{\mu\nu},\,\boldsymbol{\omega}]\,. \tag{8.37}$$

Thus, we have learned that the non-Abelian field strength is not invariant, merely covariant; it transforms under (26) as an adjoint multiplet, just like $\boldsymbol{A}_\mu$ but without an inhomogeneous term. Since $\boldsymbol{F}_{\mu\nu}$ is an $n \times n$ matrix, it decomposes as

$$\boldsymbol{F}_{\mu\nu} = F^i_{\mu\nu}\,T_i\,, \tag{8.38}$$

where the expression

$$F^i_{\mu\nu} = \partial_\mu A^i_\nu - \partial_\nu A^i_\mu - g\,f_{ijk}\,A^j_\mu A^k_\nu \tag{8.39}$$

shows that the field strengths are independent of the fermion representation chosen in the defining relation (36). The kinetic term in the electromagnetic Lagrangian admits as non-Abelian generalization

$$\mathrm{Tr}\,(\boldsymbol{F}_{\mu\nu}\boldsymbol{F}^{\mu\nu})\,, \tag{8.40}$$

which is both Lorentz- and gauge-invariant, as it should be. With the help of the orthonormality relation (18) for $T_i$, it may also be rewritten as

$$\mathrm{Tr}\,(\boldsymbol{F}_{\mu\nu}\boldsymbol{F}^{\mu\nu}) = F^i_{\mu\nu}F^{\mu\nu}_j\,\mathrm{Tr}(T_iT_j) = \tfrac{1}{2}\,F^i_{\mu\nu}F^{\mu\nu}_i\,. \tag{8.41}$$

To summarize, the free-field Lagrangian (19) is invariant in the global non-Abelian symmetry group (15), but not in the corresponding local gauge group. Application of the principle of gauge invariance turns it into an interacting field theory when one introduces vector gauge fields, as many fields as there are generators in the gauge group and appropriately coupled to the conserved vector currents (20). The first theory of this type [for the case of the isospin SU(2) group] was constructed by C. N. Yang and R. L. Mills; for this reason a theory invariant under a local non-Abelian gauge group is frequently referred to as a Yang–Mills theory.

The full gauge-invariant Lagrangian for Dirac spinor fields interacting with vector gauge fields is

$$\mathcal{L} = \mathcal{L}_\mathrm{I} + \mathcal{L}_\mathrm{G}\,, \tag{8.42}$$

where

$$\begin{aligned}\mathcal{L}_1 &= \overline{\psi}\,(\mathrm{i}\gamma^\mu \boldsymbol{D}_\mu - m)\psi \\ &= \overline{\psi}\,(\mathrm{i}\gamma^\mu \partial_\mu - m)\psi - g\,A_{i\mu}\,\overline{\psi}\,\gamma^\mu T_i \psi\,;\end{aligned} \tag{8.43}$$

and

$$\begin{aligned}\mathcal{L}_\mathrm{G} &= -\frac{1}{2}\,\mathrm{Tr}\,(\boldsymbol{F}_{\mu\nu}\boldsymbol{F}^{\mu\nu}) \\ &= -\frac{1}{4}B^i_{\mu\nu}B^{\mu\nu}_i + \frac{g}{2}\,f_{ijk}\,B^i_{\mu\nu}A^\mu_j A^\nu_k - \frac{g^2}{4}\,f_{ijk}f_{i\ell m}\,A_{j\mu}A_{k\nu}A^\mu_\ell A^\nu_m\,,\end{aligned} \tag{8.44}$$

together with the definition

$$B^i_{\mu\nu} \equiv \partial_\mu A^i_\nu - \partial_\nu A^i_\mu\,. \tag{8.45}$$

The spinor fields transform as some representation of the gauge group, but the gauge fields must belong to the adjoint representation. Since it is not possible to construct gauge-invariant mass terms, the gauge fields are necessarily massless, just as in the Abelian case. However, in contrast to the Abelian case, the part of the Lagrangian that describes the gauge fields, $\mathcal{L}_\mathrm{G}$, constitutes by itself a nontrivial interacting theory (pure Yang–Mills theory): besides the expected kinetic terms, it includes self-couplings stemming from the nonlinear expression (39), with coupling strengths that depend on the single constant $g$. The physical reason for the presence of these couplings can be easily understood: each non-Abelian gauge field $A^\mu_i$ carries a charge characteristic of the group and labeled by the index $i$, and so it must couple to every field carrying any such charge, including itself and other members of the gauge multiplet. Exactly for the same reason, gravitation is also an inherently nonlinear theory because the gravitational field interacts with everything that has energy density, including itself.

We have considered so far a *simple* Lie group as the gauge group: in this case, the generators of the group transform irreducibly under the action of the group and therefore must have the same coupling constant $g$, regardless of the representation. Basically, $g$ cannot be arbitrarily scaled, its normalization being fixed by the commutation relations characteristic of the group. This is in sharp contrast with the $\mathrm{U_Q}(1)$ case, where there are no such constraints on the coupling constant $q$, which may assume different values for different representations. If the gauge group is a direct product of simple group factors, e.g. $\mathrm{SU}(m) \times \mathrm{SU}(n)$, generators of the different factors do not mix under the action of the group, and an independent gauge coupling constant comes with each factor in the product group. Finally, let us note that we may add, in a simple generalization of the above discussion, any other matter fields belonging to any other representations of the gauge group with appropriate matrices $T_i$. In particular, it is possible to have the left- and the right-handed components of Dirac fields transforming independently as different representations of the gauge group.

## 8.4 Quantum Chromodynamics

Although historically the non-Abelian gauge principle was first used to formulate a unified theory of weak and electromagnetic interactions, the theory of strong interactions of quarks is the more obvious extension of quantum electrodynamics because the gauge symmetry on which it is based is a simple Lie group and also because the symmetry remains manifestly intact throughout. This theory marks the culmination of significant progress made over many years on two levels in the study of the physics of elementary particles.

On the one hand, major quantitative advances were achieved by the quark model in correlating detailed data in hadron spectroscopy (masses, decay rates, etc.) and by the parton model in describing the scaling phenomenon as observed in deeply inelastic, large momentum-transfer processes (such as ep $\to$ e+X and $e^+e^- \to$ hadrons). Here *parton* is the generic name given by Feynman to an independently moving constituent within a hadron, of which the quark is but an example, and *scaling* refers to the property, predicted by J. D. Bjorken, that the structure functions appearing in the cross-sections of deeply inelastic, hard processes depend only on a certain dimensionless combination of energy variables. (The structure functions give the probability of finding a parton inside a hadron carrying a certain fraction of the hadron's momentum.) The success of the quark-parton model implies that the hadron, when viewed in a frame in which its momentum is very large, is composed of almost-free constituents; in other words, quarks can interact weakly at short distances (see Chaps. 10, 12). Another key result drawn from these studies is that quarks have a three-valued quantum number, called *color*. Observations require exact color symmetry and the absence of isolated color multiplets other than singlets; this suggests that the forces between the colored quarks must be color dependent or, equivalently, they must carry 'color charges'.

On the other hand, important new ideas emerged from developments in quantum field theory. These ideas revolve around the demand of renormalizability of physical theories and the notion of energy-dependent coupling strengths. To make sense, a quantum field theory must be finite or can be made finite (renormalized) by introducing a finite number of counterterms into the original Lagrangian without changing its basic form. Renormalizability in theories involving vector quanta can be ensured by gauge invariance. In a renormalization procedure, the kinematic point at which the physical parameters, such as the mass and the coupling constant, are defined is arbitrary. However, since the physical content of the theory should remain invariant under a mere change of the normalization condition, there must be relations between physical quantities taken at different reference points. The coupling constant, for example, should be regarded as a function of the reference point and, in this sense, is energy and momentum dependent. When this effective coupling constant decreases as the relevant energy scale grows (or, equivalently, as distances shrink), the theory is said to be *asymptotically free*. Asymptotic freedom offers a possible explanation for Bjorken's scaling and is part of the reason why quarks and other hadronic constituents behave as if weakly bound inside a target nucleon, yet are not produced as free particles in final states of deep inelastic scatterings. This suggests that the field theory of strong interactions must be asymptotically free. We now know that all pure Yang–Mills theories based on groups without Abelian factors are asymptotically free, and theories of non-Abelian gauge fields and fermion multiplets are asymptotically free only if the theory does not have too many fermions. This means, for example, if the gauge group is SU(3) the number of fermion triplets is limited to sixteen or less. Another known result is that a renormalizable field theory cannot be asymptotically free unless it involves non-Abelian gauge fields. (A more detailed discussion is found in Chap. 15.)

**The QCD Lagrangian.** All this leads to the conviction that the strong interactions should be described by non-Abelian gauge fields and that it is the color symmetry that should be gauged. The resulting theory is a Yang–Mills theory based on the color SU(3) group, containing eight vector gauge bosons called *gluons*, together with different flavors of quarks, each transforming as the fundamental triplet representation. It is assumed in addition that the color gauge invariance remains exact, unbroken by any mechanism, so that the gluons remain massless. The theory, called *quantum chromodynamics* (Gross and Wilczek 1973; Fritzsch, Gell-Mann, and Leutwyler 1973; Weinberg 1973), has a Lagrangian of the form

$$\mathcal{L}_{\text{QCD}} = -\frac{1}{4}\, F^i_{\mu\nu} F^{\mu\nu}_i + \sum_{A=1}^{N_\text{f}} \bar{\psi}^A\, (\mathrm{i}\gamma^\mu \boldsymbol{D}_\mu - m_A)\psi^A\,, \tag{8.46}$$

where

$$\begin{aligned} F^i_{\mu\nu} &= \partial_\mu G^i_\nu - \partial_\nu G^i_\mu - g_\text{s}\, f_{ijk}\, G^j_\mu G^k_\nu\,, \\ \boldsymbol{D}_\mu \psi^A_a &= \partial_\mu \psi^A_a + \tfrac{\mathrm{i}}{2} g_\text{s}\, G_{i\mu}\, (\lambda_i)_{ab}\, \psi^A_b\,. \end{aligned} \tag{8.47}$$

The matrices $\lambda_i$, with $i = 1, \ldots, 8$ as internal symmetry index, are the usual Gell-Mann matrices that satisfy the SU(3) Lie algebra

$$[\lambda_i, \lambda_j] = 2\,\mathrm{i}\, f_{ijk} \lambda_k\,. \tag{8.48}$$

The $f_{ijk}$ are the SU(3) structure constants. There are $3^2 - 1 = 8$ gluon fields, $G^i_\mu$, and $N_\text{f} = 6$ quark color-triplets, $\psi^A_a$ with $A = 1, \ldots, N_\text{f}$ denoting the flavors and $a = 1, 2, 3$ denoting the colors. The complete quark content of the model is

$$\psi^A_a : \quad \begin{pmatrix} u_1 \\ u_2 \\ u_3 \end{pmatrix}, \begin{pmatrix} d_1 \\ d_2 \\ d_3 \end{pmatrix}, \begin{pmatrix} c_1 \\ c_2 \\ c_3 \end{pmatrix}, \begin{pmatrix} s_1 \\ s_2 \\ s_3 \end{pmatrix}, \begin{pmatrix} t_1 \\ t_2 \\ t_3 \end{pmatrix}, \begin{pmatrix} b_1 \\ b_2 \\ b_3 \end{pmatrix}. \tag{8.49}$$

Since the gluons are flavor-neutral, that is, u, d, s, c, t, and b quarks have exactly the same strong interactions, the QCD Lagrangian (46) has all the flavor symmetries of a free-quark model, which are only broken by a lack of degeneracy in the quark masses. In particular, it conserves strangeness, charm, etc.. It also clearly has all the well-known strong interaction symmetries, such as invariance under charge conjugation and space inversion.

**Approaches to Solutions.** The Lagrangian (46) contains $N_\mathrm{f} + 1$ parameters: the quark masses, one for each flavor, plus one dimensionless coupling constant, $g_\mathrm{s}$. (Actually there is another parameter hidden, a vacuum angle related to the possibility of strong CP violation, which is however experimentally found consistent with zero.) With the fields second quantized, (46) forms the basis for a quantum description of the quark dynamics, and should in principle describe all the world of strong interactions. This description separates naturally into two regions: the short-distance (large invariant momentum transfer) regime in which the effective coupling strength is weak, and quarks and gluons may be treated as if they were free particles; and the large-distance (small invariant momentum transfer) regime in which the full force of the strong coupling comes into play.

In the short-distance regime, asymptotic freedom makes QCD calculable by *perturbative methods* under the right circumstances, and that is when the long-distance effects are irrelevant or can be factored out. Processes amenable to this kind of treatment include, but are not restricted to, deep inelastic lepton–hadron scattering ($\mathrm{e} + \mathrm{p} \to \mathrm{e}$+hadrons), electron–positron annihilation ($\mathrm{e}^+\mathrm{e}^- \to$ hadrons), large invariant mass lepton-pair production ($\mathrm{p}+\mathrm{p} \to \mu^+\mu^-+$ hadrons) and jet phenomena. The successes of perturbative QCD in calculating strong interaction corrections beyond the leading order make quantitative analyses of these processes possible, and contribute to reinforcing the general belief that QCD is an essentially correct theory of the strong interaction (see Chaps. 14, 16).

The situation is much more complex in the large-distance domain. If the gluons are massless, as they are assumed to be in QCD, why have long-range strong interactions never been detected? If the strong interactions are color dependent, why are color singlets only ever observed? This is the famous outstanding problem of *color confinement.* Many methods have been devised to deal with this aspect of the strong interaction, among which the most promising consists in formulating the gauge theory on a *lattice*, in which the space-time continuum is discretized. The lattice spacing thus introduced provides a natural cutoff for momenta and allows for a natural regularization scheme in the study of the long-range properties of QCD. The gauge fields appear there as gauge-invariant dynamical variables associated with links joining adjacent lattice points; because of gauge invariance, link variables must either form closed loops or begin and end on color sources (see Fig. 8.1).

A formulation on the lattice makes feasible expansions independent of the perturbation theory; it turns out that it is in fact simpler to perform an expansion in powers of $1/g_\mathrm{s}^2$, so that the lattice gauge theory can be treated as a perturbation in the strong coupling limit. It is then possible to study various physical quantities by computer simulation based on the Monte Carlo method in which loop configurations are sampled rather than summed over. Considerable progress has been made to the point where true precision hadron mass calculations can be performed for heavy quarkonium systems and heavy–light quark systems (although the light hadron spectroscopy still eludes concerted efforts). We will not study the lattice gauge theory or any other *nonperturbative methods* of gauge theory in this book, but rather refer the reader to the series of 'Lattice' Conference Proceedings for more recent developments.

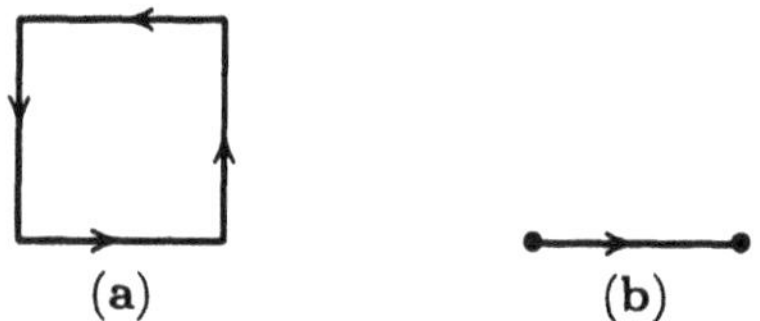


**Fig. 8.1.** (**a**) Simplest gluonic bound state; (**b**) simplest q$\bar{\text{q}}$ bound state in lattice gauge theory

**Feynman Rules for QCD.** We will end this section by giving a derivation of the Feynman rules for the tree diagrams in perturbative QCD. The rules are written in momentum space, where any field is represented by a Fourier transform of itself in space-time:

$$A(p) = \int \mathrm{d}^4x \; \mathrm{e}^{\mathrm{i}p\cdot x} A(x)\,. \tag{8.50}$$

The propagator for a quark is similar to that for the electron found in Chap. 4,

$$\mathrm{i}(S_{\mathrm{F}}(p))_{\beta\alpha}\delta_{ba}\delta_{BA} = \left(\frac{\mathrm{i}}{\not p - m_A + \mathrm{i}\varepsilon}\right)_{\beta\alpha} \delta_{ba}\delta_{BA}\,. \tag{8.51}$$

Each quark line is associated with three indices: $A$ for family, $a$ for color, and $\alpha$ for spinor. The quark–gluon coupling contained in (46)

$$\mathcal{L}_{\mathrm{h}}^0 = -g_{\mathrm{s}}\,\bar{\psi}(x)\gamma^\mu \frac{\lambda_i}{2}\psi(x)\,G_{i\mu}(x) \tag{8.52}$$

contributes $\mathrm{i}\int \mathrm{d}^4x\,\mathcal{L}_{\mathrm{h}}^0$ to the action, which leads to the quark–gluon vertex

$$-\mathrm{i}g_{\mathrm{s}}\,(\gamma^\mu)_{\beta\alpha}\,\frac{(\lambda_i)_{ba}}{2}\,\delta_{BA}\,. \tag{8.53}$$

The pure gauge part of the Lagrangian (46) is

$$\begin{aligned}\mathcal{L}_{\mathrm{G}} &= -\frac{1}{4}\,F^i_{\mu\nu}F_i^{\mu\nu} \\ &= -\frac{1}{4}G^i_{\mu\nu}G_i^{\mu\nu} + \frac{g_{\mathrm{s}}}{2}\,f_{ijk}\,G^i_{\mu\nu}G^\mu_j G^\nu_k - \frac{g_{\mathrm{s}}^2}{4}\,f_{ijk}f_{i\ell m}\,G_{j\mu}G_{k\nu}G^\mu_\ell G^\nu_m,\end{aligned} \tag{8.54}$$

where

$$G^i_{\mu\nu} = \partial_\mu G^i_\nu - \partial_\nu G^i_\mu \,. \tag{8.55}$$

To derive the gluon propagator, we isolate the kinetic term in (54)

$$\mathcal{L}^0_{\mathrm{G}} = -\frac{1}{4} G_{i\mu\nu}(x)\, G^{\mu\nu}_i(x)\,, \tag{8.56}$$

which corresponds to the action

$$\begin{aligned}\int \mathrm{d}^4x\, \mathcal{L}^0_{\mathrm{G}} &= -\frac{1}{2}\int \mathrm{d}^4x\, \partial_\mu G_{i\nu}\left(\partial^\mu G^\nu_i - \partial^\nu G^\mu_i\right)\\ &= -\frac{1}{2}\int \frac{\mathrm{d}^4p}{(2\pi)^4}\, G^\mu_i(-p)\, p^2\left(g_{\mu\nu} - \frac{p_\mu p_\nu}{p^2}\right) G^\nu_i(p)\,. \end{aligned}\tag{8.57}$$

The integrand contains the reciprocal of the propagator. In order to invert it, one would find it convenient to introduce first the transverse and longitudinal projection operators

$$P^{\mathrm{T}}_{\mu\nu} = g_{\mu\nu} - \frac{p_\mu p_\nu}{p^2}\,, \qquad P^{\mathrm{L}}_{\mu\nu} = \frac{p_\mu p_\nu}{p^2}\,, \tag{8.58}$$

which have the properties

$$\left(P^{\mathrm{T}}_{\mu\nu}\right)^2 = P^{\mathrm{T}}_{\mu\nu}\,;\ \left(P^{\mathrm{L}}_{\mu\nu}\right)^2 = P^{\mathrm{L}}_{\mu\nu}\,;\ P^{\mathrm{L}}_{\mu\nu}P^{\mathrm{T}}_{\mu\nu} = 0\,;\ P^{\mathrm{T}}_{\mu\nu} + P^{\mathrm{L}}_{\mu\nu} = g_{\mu\nu}\,. \tag{8.59}$$

Then the inverse propagator from (57) may be rewritten as

$$\left[D_{\mu\nu}(p)\right]^{-1} = -p^2 P^{\mathrm{T}}_{\mu\nu} + 0 P^{\mathrm{L}}_{\mu\nu}\,. \tag{8.60}$$

It is a purely transverse, singular operator, and therefore cannot be inverted. This difficulty stems from the fact that, just as for the photon, not all components of the gluon fields are physical. In order to calculate physical quantities, it is necessary to exclude the unphysical field components and select a definite gauge in which calculations are to be done. In the Lagrangian formalism, the gauge selection may be made from the start by introducing an extra term into the Lagrangian itself. This gauge-fixing term may be chosen as

$$\mathcal{L}_{\mathrm{GF}} = \frac{-1}{2\xi}\, \partial_\mu G^\mu_i\, \partial_\nu G^\nu_i = \frac{1}{2\xi} G^\mu_i \partial_\mu \partial_\nu G^\nu_i + \text{total derivative}\,, \tag{8.61}$$

where $\xi$ is a real parameter corresponding to different gauges (e.g. $\xi = 1$ for Feynman gauge and $\xi = 0$ for Landau gauge) and should not affect physical quantities. The terms quadratic in the fields in the action then yield

$$\int \mathrm{d}^4x\, (\mathcal{L}^0_{\mathrm{G}} + \mathcal{L}_{\mathrm{GF}}) = \frac{1}{2}\int \frac{\mathrm{d}^4p}{(2\pi)^4}\, G^\mu_i(-p)\left[-g_{\mu\nu}p^2 + (1-\xi^{-1})p_\mu p_\nu\right] G^\nu_i(p)\,.$$

The inverse gluon propagator can be immediately read off:

$$\begin{aligned}D^{-1}_{\mu\nu}(p) &= -g_{\mu\nu}p^2 + (1-\xi^{-1})p_\mu p_\nu\\ &= -p^2 P^{\mathrm{T}}_{\mu\nu} - \xi^{-1}p^2 P^{\mathrm{L}}_{\mu\nu}\,. \end{aligned}\tag{8.62}$$

It is now nonsingular as long as $\xi \neq \infty$, and admits as its inverse

$$D_{\mu\nu}(p) = -\frac{1}{p^2 + \mathrm{i}\varepsilon} P^{\mathrm{T}}_{\mu\nu} - \frac{\xi}{p^2 + \mathrm{i}\varepsilon} P^{\mathrm{L}}_{\mu\nu} \,. \tag{8.63}$$

To each gluon internal line, we thus assign the expression

$$\mathrm{i}D_{\mu\nu}(p)\delta_{ij} = \frac{\mathrm{i}}{p^2 + \mathrm{i}\varepsilon} \left[ -g_{\mu\nu} + \frac{(1-\xi)p_\mu p_\nu}{p^2 + \mathrm{i}\varepsilon} \right] \delta_{ij} \,. \tag{8.64}$$

To obtain the Feynman rules for the gluon interaction vertices, we Fourier transform the remaining terms in (54). For the three-gluon coupling we get

$$\begin{aligned} \mathrm{i} \int \mathrm{d}^4x \, \mathcal{L}^1_{\mathrm{G}}(\text{three gluons}) &= \frac{1}{2}\, g_{\mathrm{s}} f_{ijk} \int \frac{\mathrm{d}^4p\, \mathrm{d}^4q\, \mathrm{d}^4r}{(2\pi)^{12}} (2\pi)^4 \delta^4(p+q+r) \\ &\quad \times \left[ g_{\lambda\nu} p_\mu - g_{\lambda\mu} p_\nu \right] G_i^\lambda(p) G_j^\mu(q) G_k^\nu(r) \,. \end{aligned} \tag{8.65}$$

Using the total antisymmetry of $f_{ijk}$ in its three indices and the invariance of the whole expression under simultaneous permutations of its indices, we interchange $i$, $\lambda$, $p$ with $j$, $\mu$, $q$, and $i$, $\lambda$, $p$ with $k$, $\nu$, $r$ to obtain a fully symmetrized expression

$$\begin{aligned} \mathrm{i} \int \mathrm{d}^4x \, \mathcal{L}^1_{\mathrm{G}} &= \frac{1}{6}\, g_{\mathrm{s}} f_{ijk} \int \frac{\mathrm{d}^4p\, \mathrm{d}^4q\, \mathrm{d}^4r}{(2\pi)^{12}} (2\pi)^4 \delta^4(p+q+r)\, G_i^\lambda(p) G_j^\mu(q) G_k^\nu(r) \\ &\quad \times \left[ g_{\lambda\nu} p_\mu - g_{\lambda\mu} p_\nu - g_{\mu\nu} q_\lambda + g_{\lambda\mu} q_\nu - g_{\lambda\nu} r_\mu + g_{\mu\nu} r_\lambda \right] \\ &= \frac{1}{3!}\, g_{\mathrm{s}} f_{ijk} \int \frac{\mathrm{d}^4p\, \mathrm{d}^4q\, \mathrm{d}^4r}{(2\pi)^{8}} \delta^4(p+q+r)\, G_i^\lambda(p) G_j^\mu(q) G_k^\nu(r) \\ &\quad \times \left[ g_{\lambda\nu}(p-r)_\mu + g_{\lambda\mu}(q-p)_\nu + g_{\mu\nu}(r-q)_\lambda \right] . \end{aligned} \tag{8.66}$$

The Feynman rule for three-gluon vertex can then be read off:

$$g_{\mathrm{s}}\, f_{ijk} \left[ g_{\lambda\nu}(p-r)_\mu + g_{\lambda\mu}(q-p)_\nu + g_{\mu\nu}(r-q)_\lambda \right] , \tag{8.67}$$

subject to four-momentum conservation

$$p + q + r = 0 \,. \tag{8.68}$$

The four-gluon coupling can be similarly treated, by first symmetrizing its action

$$\begin{aligned} \mathrm{i} \int \mathrm{d}^4x \, \mathcal{L}^2_{\mathrm{G}}(\text{four gluons}) &= \frac{1}{4} \int \frac{\mathrm{d}^4p\, \mathrm{d}^4q\, \mathrm{d}^4r\, \mathrm{d}^4s}{(2\pi)^{16}} (2\pi)^4 \delta^4(p+q+r+s) \\ &\quad \times G_{i\lambda}(p) G_{j\mu}(q) G_k^\nu(r) G_\ell^\rho(s) \left(-\mathrm{i}g_{\mathrm{s}}^2\right) f_{nij} f_{nk\ell}\, g_{\lambda\nu} g_{\mu\rho} \\ = \frac{1}{4!} \int \frac{\mathrm{d}^4p\, \mathrm{d}^4q\, \mathrm{d}^4r\, \mathrm{d}^4s}{(2\pi)^{12}} &\delta^4(p+q+r+s)\, G_{i\lambda}(p) G_{j\mu}(q) G_k^\nu(r) G_\ell^\rho(s) \\ \times \left(-\mathrm{i}g_{\mathrm{s}}^2\right) \big[ f_{nij} f_{nk\ell} (g_{\lambda\nu} g_{\mu\rho} - g_{\mu\nu} g_{\lambda\rho}) &+ f_{nkj} f_{ni\ell} (g_{\lambda\nu} g_{\mu\rho} - g_{\lambda\mu} g_{\nu\rho}) \\ + f_{nik} f_{nj\ell} (g_{\lambda\mu} g_{\nu\rho} - g_{\mu\nu} g_{\lambda\rho}) \big] \,. & \end{aligned} \tag{8.69}$$

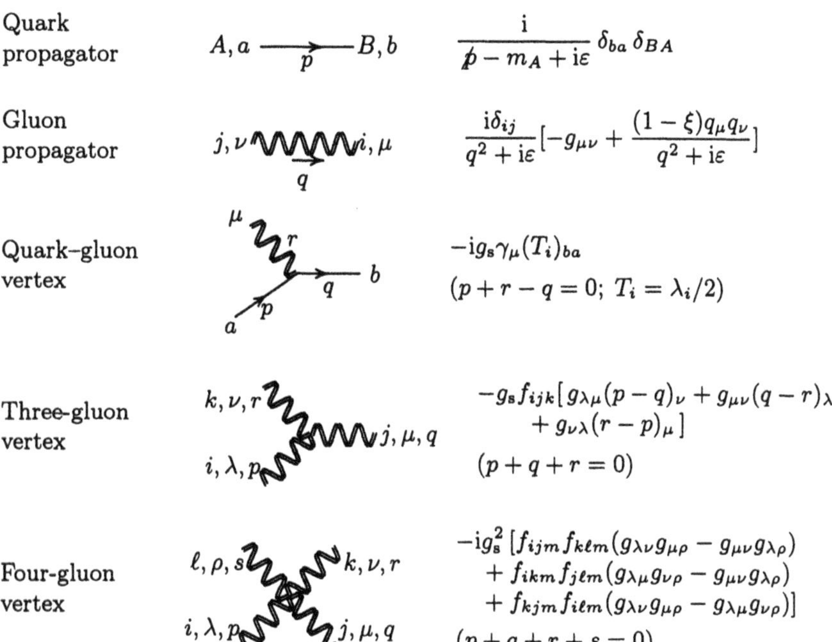


**Fig. 8.2.** Feynman rules for QCD tree diagrams

This yields the Feynman rule for the four-gluon vertex:

$$\begin{aligned}(-\mathrm{i}g_s^2)\big[&f_{nij}f_{nk\ell}(g_{\lambda\nu}g_{\mu\rho} - g_{\mu\nu}g_{\lambda\rho}) + f_{nkj}f_{ni\ell}(g_{\lambda\nu}g_{\mu\rho} - g_{\lambda\mu}g_{\nu\rho})\\ &+ f_{nik}f_{nj\ell}(g_{\lambda\mu}g_{\nu\rho} - g_{\mu\nu}g_{\lambda\rho})\big]\end{aligned} \tag{8.70}$$

with four-momentum conservation at the vertex

$$p + q + r + s = 0\,. \tag{8.71}$$

The Feynman rules thus derived from the Lagrangian (46) are summarized in Fig. 8.2. However, as rules for QCD, they are not complete. In a full quantum formulation of QCD in a covariant gauge like (61), an additional, nonphysical (*ghost*) field has to be introduced, whose main effect is to suppress the nontransverse components of real gluons while preserving gauge invariance (see Chap. 15). A complete list of the Feynman rules for QCD is given in the Appendix.

## 8.5 Spontaneous Breaking of Global Symmetries

Experiment shows that quantum electrodynamics is a gauge theory consistent in all aspects with the principle of gauge invariance applied to the $U_Q(1)$ group. In particular, the photon can be identified with the massless gauge field of the group and interacts just as expected with the conserved fermion current that follows from the gauge symmetry.

On the other hand, in spite of its apparently distinctive properties (e.g. a much shorter force range, a greater diversity in transition modes), the weak interaction gives clear hints to its close parentage with the electromagnetic interaction. In particular, the currents found in many weak processes are electrically charged and have precisely the form implied by a non-Abelian symmetry based on a certain semisimple group. It is thus quite possible that there exists a gauge theory that can describe both weak and electromagnetic interactions. However, as we have seen earlier in this chapter, the gauge fields required by gauge invariance must apparently be massless and must therefore generate long-range forces. In order to construct a gauge theory of this kind for weak interactions, one is then confronted with the problem of reconciling the presence of massive gauge fields needed to generate the short-range weak forces actually observed with the preservation in some sense of gauge invariance essential for a renormalizable theory.

One way of generating masses for vector bosons without destroying the underlying gauge symmetry of the theory is by 'spontaneously' breaking that symmetry. This phrase refers to a process in which, from a set of degenerate minimum energy states that are equivalent by symmetry, one arbitrarily selects a member of the multiplet as *the* physical ground state of the system in apparent violation of the underlying symmetry. But in reality the symmetry is not lost in the process, it is merely hidden and can be recovered through special relations between masses and couplings. That it is possible to pick the ground state in this way simply reflects the fact, fairly widespread in nature, that physical states may exist with a symmetry apparently lower than that of the basic equations of motion.

### 8.5.1 The Basic Idea

To understand the idea of spontaneous breakdown of symmetry, let us mentally consider a large sample of ferromagnetic material at 0° K in the absence of any external field. A ferromagnet is viewed in the Heisenberg model as an infinite regular array of spin-$\frac{1}{2}$ magnetic dipoles with spin–spin interactions between nearest neighbors such that neighboring dipoles tend to align. Although the Hamiltonian describing the system is rotationally invariant, the ground state is not always. At high temperatures, thermal agitations will make the magnetic moments flutter at random in different directions, so that there is no net magnetization, which results in a rotationally symmetric state endowed with the same symmetry as the law of interaction. If the ferromagnet is now sufficiently cooled down (below a certain critical temperature, called the Curie temperature), all the atomic dipole moments will tend to align parallel to each other and to some arbitrary direction, leading to a nonzero magnetization for the sample. This is one of the infinitely many degenerate lowest-energy states that exist for an infinite ferromagnet, and the symmetry resides hidden in the equivalence of these states through rotations. Transitions between these states are not possible, because for an infinite ferromagnet any single transition would require an infinite amount of energy. The particular ground state the system 'spontaneously' falls into as it cools down cannot be foreseen, and certainly is not symmetric since the magnetization points in a definite direction; it corresponds to a magnetization vector $\boldsymbol{M}$ with magnitude $M$ such that the free energy $F$ of the ferromagnet is minimum, as shown in Fig. 8.3.

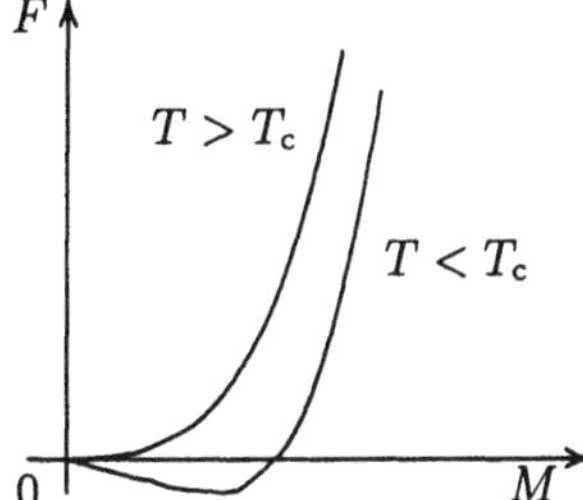


**Fig. 8.3.** Free energy $F$ of a ferromagnet as function of magnetization $M$

We now attempt to transfer this insight to relativistic quantum mechanics, substituting a particle Hamiltonian (or Lagrangian) for the ferromagnet Hamiltonian, the particle vacuum for the ferromagnet ground state and some other symmetry for rotational symmetry. Specifically, assuming nature to possess symmetries that are not manifest to us because we live in an imperfectly symmetric universe, we will take the Lagrangian that defines the particle theory to be invariant in some *internal symmetry*, but the particle vacuum to be lacking this symmetry. This asymmetric state is realized by requiring that the vacuum-to-vacuum expectation value of some field be nonvanishing, much as the ferromagnetic ground state was determined by a nonzero magnetization. The field in question cannot have a nonzero spin because otherwise the vacuum would be characterized by a nonzero angular momentum and rotational invariance would have been broken. Since the vacuum is observed to be rotationally invariant the field must be spinless, and the internal symmetry of the theory must be broken by a scalar field acquiring a nonzero vacuum expectation value. As translation invariance is also an observed symmetry of particle physics, this expectation value must not depend on space-time in the absence of any source. The basic conjecture is that there exist, beside matter and gauge fields, one or more spin-0 fields, called the *Higgs fields*, which would assume uniform nonzero values even in the vacuum and which could couple to each other and to other, massless particles to give them masses.

A quantum-mechanical vacuum is a complex state filled with pairs of virtual particles and antiparticles continuously being created and annihilated. If those virtual particles interact strongly enough among themselves, they might form a permanent state of high density, called a *vacuum condensate.*

There is a thermodynamic transition point separating the vacuum without a condensate from the vacuum with a condensate. For a condensate to form, there must be a strong enough attraction among the particles at low density, but also a strong enough repulsion at high density to prevent a runaway situation to occur. It is believed that such a situation exists in the world of the fundamental particles, where the vacuum is filled with a high density of Higgs fields, the *Higgs condensate.* By interacting with the light particles (bosons and fermions) that populate this vacuum, the condensate drags them sufficiently down to make them massive. Such is in simplified terms the physics of the *Higgs mechanism,* as this mass-generating process is called.

## 8.5.2 Breakdown of Discrete Symmetry

Let us begin with the simplest model having a discrete symmetry and containing a single massless *real* scalar field which plays the role of the Higgs field. The part of the Lagrangian relevant to the present discussion is

$$\mathcal{L}_s = \tfrac{1}{2}\, \partial_\mu \phi\, \partial^\mu \phi - V(\phi)\,, \tag{8.72}$$

with a potential parameterized by two real constants, $\lambda$ and $\mu^2$,

$$V(\phi) = \frac{1}{2}\,\mu^2\,\phi^2 + \frac{\lambda}{4!}\,\phi^4\,. \tag{8.73}$$

The corresponding energy density is given by

$$\mathcal{H}_s = \tfrac{1}{2}\,\left[(\partial_0\phi)^2 + (\nabla\phi)^2\right] + V(\phi)\,. \tag{8.74}$$

The only internal symmetry of the model is its invariance to field reflection

$$\phi(x) \to \phi'(x) \equiv -\phi(x)\,. \tag{8.75}$$

The energy minimum of the system is determined at the classical level by the condition

$$\frac{\partial V}{\partial \phi} = \phi\left(\mu^2 + \frac{\lambda}{6}\phi^2\right) = 0\,. \tag{8.76}$$

When $\lambda < 0$ the potential $V(\phi)$ has no stable minima for finite $\phi$ (see Fig. 8.4a). We will therefore assume $\lambda \geq 0$. Then, for $\mu^2 > 0$ the potential has a unique minimum at $\phi = 0$ (as shown in Fig. 8.4b) and the symmetry of the vacuum is manifest. More interesting is the case $\mu^2 < 0$ when $V$ has minima at the nonzero field values

$$\phi = \pm\sqrt{-6\mu^2/\lambda} \tag{8.77}$$

(shown in Fig. 8.4c). This is precisely the situation in which $V$ is attractive at small values of $\phi$ but becomes strongly repulsive at large values. The field

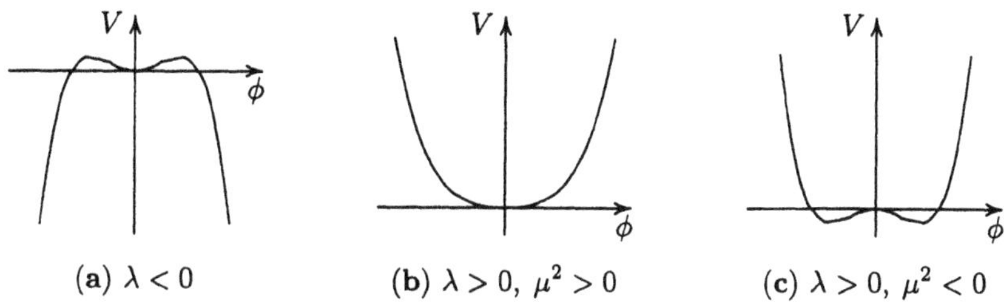


**Fig. 8.4a–c.** Potential for the scalar field with reflection symmetry, for different values of the parameters

values given in (77) are independent of $x$ and correspond to the quantum-mechanical vacuum expectation value of the field operator, denoted by $\langle\phi\rangle$ or $\langle 0\,|\,\phi\,|\,0\rangle$. Because of the reflection symmetry (75) of the model, whichever solution is chosen will lead to the same physics; but once the choice is made, the symmetry of the system is (spontaneously) broken. Let us arbitrarily select the positive value of $\phi$ at minimum $V$ to define the vacuum:

$$\langle\phi\rangle = v = \sqrt{-6\mu^2/\lambda}\,. \tag{8.78}$$

In order to do any calculations beyond the ground state, it is convenient to introduce a new field

$$\chi(x) = \phi(x) - v\,, \tag{8.79}$$

which is designed to have a zero vacuum expectation value. It measures field oscillations about the uniform background $\phi = v$. In terms of $\chi$, the Lagrangian density becomes

$$\mathcal{L}_s = \frac{1}{2}\left[\partial_\mu\chi\,\partial^\mu\chi - (-2\mu^2)\chi^2\right] - \frac{\lambda}{4!}\left(4v\,\chi^3 + \chi^4\right) - \frac{1}{4}\,\mu^2 v^2\,. \tag{8.80}$$

It can now be interpreted in the usual way, with no more concern about the properties of the vacuum, since the dynamic field vanishes in the vacuum, $\langle\chi\rangle = 0$. It simply describes the dynamics of a spin-0 field with real mass $\sqrt{-2\mu^2}$. Even though it is the same Lagrangian as before, the presence of the cubic term $\chi^3$ gives us no reason to suspect that a symmetry actually lies in the background.

### 8.5.3 Breakdown of Abelian Symmetry

We are eventually interested in theories with continuous gauge symmetries. The simplest model that exhibits such a symmetry is the *complex* scalar field theory described by the Lagrangian

$$\begin{aligned} \mathcal{L}_s &= \partial_\mu\varphi\,\partial^\mu\varphi^* - V(\varphi,\varphi^*)\,, \\ V(\varphi,\varphi^*) &= \mu^2\,\varphi\varphi^* + \tfrac{1}{4}\,\lambda\,(\varphi\varphi^*)^2\,, \end{aligned} \tag{8.81}$$

which is evidently invariant under the *global* phase transformation

$$\begin{aligned} \varphi &\to \mathrm{e}^{-\mathrm{i}\alpha}\varphi\,, \\ \varphi^* &\to \mathrm{e}^{\mathrm{i}\alpha}\varphi^*\,, \end{aligned} \tag{8.82}$$

with an arbitrary real constant $\alpha$.

For the same reason as before, we assume $\lambda > 0$. Then, for $\mu^2$ positive, $V$ acquires an absolute minimum at $\varphi = 0$ and the vacuum has manifestly the same symmetry as the Lagrangian. But if $\mu^2$ is negative, the system has the lowest energy for

$$|\varphi|^2 = -2\mu^2/\lambda\,.$$

Thus, there is an infinite number of degenerate minima lying on a circle of radius $\sqrt{-2\mu^2/\lambda}$ (see Fig. 8.5) and differing from one another by a relative phase factor, but all equivalent through the phase transformations (82) and all leading to the same physics. Any particular choice of $\langle\varphi\rangle$ will spontaneously break the symmetry; so we may as well let its phase-angle be zero and select the vacuum such that

$$\langle\varphi\rangle = \frac{v}{\sqrt{2}} \tag{8.83}$$

for $v = \sqrt{-4\mu^2/\lambda}$. It is significant to note that in this choice only the real part of $\varphi$ acquires a nonzero vacuum expectation value, fixing the direction of the symmetry breakdown.

We now define a shifted complex field $\chi$ such that

$$\varphi = \langle\varphi\rangle + \tfrac{1}{\sqrt{2}}\chi = \tfrac{1}{\sqrt{2}}(v + \chi_1 + \mathrm{i}\chi_2)\,. \tag{8.84}$$

Both real fields $\chi_1$ and $\chi_2$ have zero vacuum expectation values. They measure excitations of the fields from the vacuum in the directions radial and tangential to the circle of degenerate minima. In terms of these fields, we have for the potential

$$V = -\mu^2\chi_1^2 + \frac{\lambda}{16}(\chi_1^2 + \chi_2^2)\left[4v\chi_1 + \chi_1^2 + \chi_2^2\right] + \frac{\mu^2 v^2}{4}\,, \tag{8.85}$$

and for the Lagrangian

$$\begin{aligned} \mathcal{L}_\mathrm{s} = {}& \frac{1}{2}\left[\partial_\mu\chi_1\,\partial^\mu\chi_1 - (-2\mu^2)\chi_1^2\right] + \frac{1}{2}\,\partial_\mu\chi_2\,\partial^\mu\chi_2 \\ & - \frac{\lambda}{16}(\chi_1^2 + \chi_2^2)(4v\chi_1 + \chi_1^2 + \chi_2^2) - \frac{1}{4}\mu^2 v^2\,. \end{aligned} \tag{8.86}$$

Evidently, the phase symmetry has been spontaneously broken. The field $\chi_1$, which represents fluctuations in the direction of symmetry breakdown,

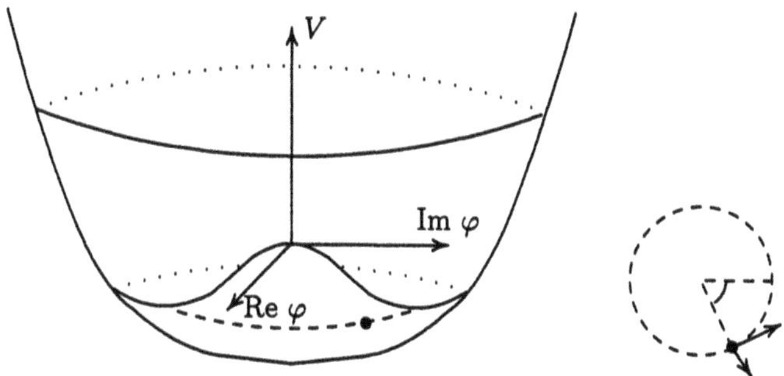


**Fig. 8.5.** Symmetry-breaking ground state in a potential that exhibits invariance under continuous symmetry transformations

acquires a mass, just as in the real scalar model. But the field $\chi_2$, which measures deviations in the direction of symmetry conservation, remains massless – a new feature, absent when it is a discrete symmetry that breaks down. In geometrical terms, as the vacuum is selected at some point on the circle of degenerate minima of $V$, it is an absolute minimum for the potential curve in the radial direction, and excitations from such a point always require energy, which implies massive modes. On the other hand, the selected vacuum has precisely the same potential energy $(-\mu^4/\lambda)$ as any other minimum, and deviations in the tangential direction, in which $V$ is flat and the total energy constant, describe the zero-frequency motion around the minimum circle. Such massless and spinless modes that arise from a spontaneous breaking of a continuous symmetry are called the *Nambu–Goldstone bosons* in particle physics. This property is not particular to the present model but is a general feature of spontaneous breakdown of symmetry.

## 8.5.4 Breakdown of Non-Abelian Symmetry

Let us turn now to a general non-Abelian gauge symmetry G. Since a complex representation can always be replaced by a real one by doubling the basis vectors of the space on which it is defined, we need to consider only real representations. Thus, we take $n$ *real* scalar fields, $\phi_1, \ldots, \phi_n$, to form a column vector $\phi$ which transforms as a (generally reducible) representation of G:

$$\phi \to \phi' = U\phi, \tag{8.87}$$

where $U$ is a real, orthogonal $n \times n$ constant matrix. In the usual parameterization we write it as

$$U = \mathrm{e}^{-\mathrm{i}g\omega_j T_j}, \tag{8.88}$$

where $g$ and $\omega_j$ are real constants and $T_j$ for $j = 1, \ldots, N$ are the $n \times n$ matrices satisfying the Lie algebra associated with the group. These matrices are Hermitian, $T_j^\dagger = T_j$, because $U$ is unitary; as well as imaginary and antisymmetric, $T_j^* = T_j^{\rm T} = -T_j$, because $U$ is also real (an upper index T denotes a transposed matrix).

The Lagrangian for these fields is taken to be

$$\begin{aligned}\mathcal{L}_{\rm s} &= \tfrac{1}{2}\,\partial_\mu \boldsymbol{\phi}\,\partial^\mu \boldsymbol{\phi} - V(\boldsymbol{\phi})\,,\\ V(\boldsymbol{\phi}) &= \tfrac{1}{2}\mu^2\,\boldsymbol{\phi}^{\rm T}\boldsymbol{\phi} + \tfrac{1}{16}\,\lambda\left(\boldsymbol{\phi}^{\rm T}\boldsymbol{\phi}\right)^2,\end{aligned} \tag{8.89}$$

where $\lambda > 0$. As in previous cases, nothing noteworthy happens when $\mu^2$ is positive; but when $\mu^2$ turns negative, $V$ acquires an infinite set of degenerate nonzero minima at

$$|\boldsymbol{\phi}|^2 = -\frac{4\mu^2}{\lambda}\,. \tag{8.90}$$

The group symmetry is spontaneously broken when the vacuum is selected, such that, for example,

$$\langle \boldsymbol{\phi} \rangle = \boldsymbol{v}\,, \tag{8.91}$$

for some real constant $n$-dimensional vector satisfying $|\boldsymbol{v}|^2 = -4\mu^2/\lambda$. Proceeding as before, we define the shifted field

$$\boldsymbol{\chi} = \boldsymbol{\phi} - \boldsymbol{v}\,, \tag{8.92}$$

which has a zero vacuum value, $\langle \boldsymbol{\chi} \rangle = 0$. In terms of $\boldsymbol{\chi}$ the potential becomes

$$V = \frac{1}{4}\,\mu^2 \boldsymbol{v}^2 + \frac{\lambda}{4}\,(\boldsymbol{\chi}^{\rm T}\boldsymbol{v})^2 + \frac{\lambda}{16}\,(\boldsymbol{\chi}^{\rm T}\boldsymbol{\chi})\left[4\,(\boldsymbol{\chi}^{\rm T}\boldsymbol{v}) + (\boldsymbol{\chi}^{\rm T}\boldsymbol{\chi})\right], \tag{8.93}$$

and the Lagrangian assumes the form

$$\begin{aligned}\mathcal{L}_{\rm s} &= \frac{1}{2}\left[\partial_\mu \chi_a \partial^\mu \chi_a - \tfrac{1}{2}\,\lambda\, v_a v_b\, \chi_a \chi_b\right]\\ &\quad - \frac{\lambda}{16}\,(\boldsymbol{\chi}^{\rm T}\boldsymbol{\chi})\left[4\,(\boldsymbol{\chi}^{\rm T}\boldsymbol{v}) + (\boldsymbol{\chi}^{\rm T}\boldsymbol{\chi})\right] - \frac{1}{4}\,\mu^2 \boldsymbol{v}^2\,.\end{aligned} \tag{8.94}$$

The masses of the fields are not apparent from (94) because they reside in the nondiagonalized quadratic terms which give the squared-mass matrix

$$\left(\mathcal{M}_{\rm B}^2\right)_{ab} = \tfrac{1}{2}\,\lambda\, v_a v_b\,, \qquad \text{for} \quad a,\, b = 1, \ldots, n\,. \tag{8.95}$$

To find the allowed eigenvalues of $\mathcal{M}_{\rm B}^2$, let it operate on any vector $T_i\,\boldsymbol{v}$:

$$\begin{aligned}\mathcal{M}_{\rm B}^2\, T_i\, \boldsymbol{v} &= \tfrac{1}{2}\,\lambda\, \boldsymbol{v}\,(\boldsymbol{v}^{\rm T} T_i \boldsymbol{v}) \;= \tfrac{1}{2}\,\lambda\, \boldsymbol{v}\,(\boldsymbol{v}^{\rm T} T_i \boldsymbol{v})^{\rm T}\\ &= \tfrac{1}{2}\,\lambda\, \boldsymbol{v}\,(\boldsymbol{v}^{\rm T} T_i^{\rm T} \boldsymbol{v}) \;= -\tfrac{1}{2}\,\lambda\, \boldsymbol{v}\,(\boldsymbol{v}^{\rm T} T_i \boldsymbol{v})\,,\end{aligned} \tag{8.96}$$

so that

$$\mathcal{M}_{\mathrm{B}}^2 T_i \boldsymbol{v} = 0\,, \qquad \text{for} \quad i = 1, \ldots, N\,. \tag{8.97}$$

On the other hand, since the symmetry is broken by setting $\langle\boldsymbol{\phi}\rangle = \boldsymbol{v}$,

$$\boldsymbol{v} \neq U\boldsymbol{v} \approx \boldsymbol{v} - \mathrm{i}g\omega_j T_j\, \boldsymbol{v}\,,$$

and so there must exist at least one $T_k$ such that

$$T_k\, \boldsymbol{v} \neq 0\,. \tag{8.98}$$

For each such $T_k$, the matrix $\mathcal{M}_{\mathrm{B}}^2$ has a zero-eigenvalue, as required by (97); this zero-eigenvalue corresponds to a Nambu–Goldstone mode.

Let S be the maximum subgroup of G that survives as a symmetry of the vacuum after the breakdown of G; let $M$ ($M \leq N$) be its dimension. We can always choose the generators $T_i$ of G such that the first $M$ generators, $T_j$ for $j = 1, \ldots, M$, generate S. Then, since the vacuum remains invariant under subgroup S,

$$T_j\, \boldsymbol{v} = 0\,, \qquad \text{for} \quad j = 1, \ldots, M\,; \tag{8.99}$$

but for the remaining generators,

$$T_k\, \boldsymbol{v} \neq 0\,, \qquad \text{for} \quad k = M+1, \ldots, N\,, \tag{8.100}$$

and (97) tells us that $\mathcal{M}_{\mathrm{B}}^2$ admits $N-M$ zero-eigenvalues. Since the $N-M$ vectors $T_k\boldsymbol{v}$, for $k = M+1, \ldots, N$, are evidently linearly independent, there must be $N-M$ massless Nambu–Goldstone bosons in the theory, one for each symmetry-breaking generator. The other $(n-N+M)$ bosons in the system have, in general, nonvanishing masses.

**Example 8.1 Orthogonal Group**

The orthogonal group G $=$ O$(n)$ has $N = \frac{1}{2}n(n-1)$ generators. We take $n$ real scalar fields to form the $n$-dimensional vector representation $\boldsymbol{\phi}$, and let their potential $V$ acquire a minimum for $|\boldsymbol{\phi}|^2 = v^2$. Among the infinite number of possible minima, a particular vector $\boldsymbol{v}$ of squared modulus $v^2$ is chosen to define the vacuum. The vacuum symmetry consists of all rotations that leave $\boldsymbol{v}$ invariant. These are the rotations that act on a space with one less dimension, and together form an orthogonal group O$(n-1)$ with $M = \frac{1}{2}(n-1)(n-2)$ independent generators. In particular, if we choose the axes in the representation space such that the vacuum vector $\boldsymbol{v}$ points along the $n$th axis, so that $v_a = v\delta_{an}$, the elements of O$(n-1)$ do not mix the $n$th component of $\boldsymbol{v}$ with the others. If $L_{ij}$ denote the generators of O$(n)$,

$$(L_{ij})_{ab} = -\mathrm{i}(\delta_{ia}\,\delta_{jb} - \delta_{ib}\,\delta_{ja})\,, \qquad \text{for} \quad i, j,\, a, b = 1, \ldots, n\,,$$

the vacuum vector $\boldsymbol{v}$ satisfies the conditions

$$\begin{aligned} (L_{ij}\boldsymbol{v})_a &= 0 \qquad &\text{for} \quad i,j = 1,\ldots,n-1\,; \\ (L_{kn}\boldsymbol{v})_a &= -\mathrm{i} v\,\delta_{ka} \qquad &\text{for} \quad k = 1,\ldots,n-1\,. \end{aligned}$$

It follows that $L_{ij}$ with $i,j = 1,\ldots,n-1$ generate the vacuum symmetry group, while $L_{kn}$ for $k = 1,\ldots,n-1$ lead to nontrivial vectors when applied on $\boldsymbol{v}$. There are, as expected, $N - M = n - 1$ massless Nambu–Goldstone bosons; and since we started out with $n$ fields in all, there remains just one Higgs boson with mass $M_{\mathrm{H}}^2 = \lambda v^2/2$, given by the single element of $\mathcal{M}_{\mathrm{B}}^2$.

Up to now we have parameterized field deviations from the vacuum in the obvious way, that is, as in (92). Another possibility which might come handy can be illustrated by the present example. Let us start with $\boldsymbol{\phi} = \boldsymbol{v} + \boldsymbol{\chi}$ as in (92), with $v_a = v\,\delta_{an}$ for $a = 1,\ldots,n$, and construct the $n \times n$ matrix

$$U(\omega) = \exp\left(-\mathrm{i}\sum_{k=1}^{n-1} \omega_k\, L_{kn}\right).$$

Under this rotation, $\boldsymbol{\phi}$ transforms into

$$\boldsymbol{\phi}' = U\,\boldsymbol{\phi} = U\,(\boldsymbol{v} + \chi)\,.$$

Assuming that both the fluctuations $\chi_a$ and the transformation parameters $\omega_i$ are infinitesimal, we obtain up to linear terms

$$\begin{aligned} \phi'_a &\approx v_a + \chi_a - \mathrm{i}\sum_k \omega_k (L_{kn})_{ab}\, v_b \\ &\approx (v + \chi_n)\,\delta_{an} + (\chi_a - v\,\omega_a)(1 - \delta_{an})\,, \qquad a = 1,\ldots,n. \end{aligned}$$

Thus, if we choose $\omega_a = \chi_a/v$, the transformed field $\boldsymbol{\phi}'$ will align with the $n$th axis, in the same direction as $\boldsymbol{v}$, so that

$$\phi'_a \approx (v + \chi_n)\,\delta_{an}\,.$$

Inversely, a general vector $\boldsymbol{\phi}$ may be obtained from the vector with components $(v + \chi_n)\,\delta_{an}$ by the rotation $U(-\omega)$. To summarize, an alternative to (92) is the parameterization

$$\boldsymbol{\phi} = \exp\left(\frac{\mathrm{i}}{v}\sum_{k=1}^{n-1} \xi_k L_{kn}\right) \boldsymbol{\phi}_{\parallel}\,, \tag{8.101}$$

where $\boldsymbol{\phi}_{\parallel}$ is an $n$-component vector with a single nonvanishing component, $(\boldsymbol{\phi}_{\parallel})_a = (v + \eta)\,\delta_{an}$. The two parameterizations are equivalent to first order, $\eta \approx \chi_n$ and $\xi_k \approx \chi_k$ for $k = 1,\ldots,n-1$. ■

## 8.6 Spontaneous Breaking of Local Symmetries

The Nambu–Goldstone bosons have the amazing property that, when it is a *local* gauge symmetry that is spontaneously broken, they disappear and simultaneously the normally massless gauge fields become massive, giving the associated long-range gauge forces a finite range. This shielding effect is akin to the Meissner effect in superconductivity, which makes an external magnetic field attenuate beyond a surface layer inside a superconductor.

### 8.6.1 Abelian Symmetry

We first study the simple Abelian model of scalar electrodynamics; when spontaneously broken, it is called the *Higgs model.* Even though it does not provide practically useful results, it will illustrate many of the ideas to be found in a more general model. The model is defined by

$$\mathcal{L} = D_\mu\varphi\, D^\mu\varphi^* - \mu^2\varphi\varphi^* - \frac{1}{4}\lambda\,(\varphi\varphi^*)^2 - \frac{1}{4}\,F_{\mu\nu}F^{\mu\nu}\,, \tag{8.102}$$

where $\varphi$ is a complex scalar field, with covariant derivatives

$$\begin{aligned} D_\mu\varphi &= (\partial_\mu + \mathrm{i}qA_\mu)\varphi\,, \\ D_\mu\varphi^* &= (\partial_\mu - \mathrm{i}qA_\mu)\varphi^*\,, \end{aligned} \tag{8.103}$$

and $F_{\mu\nu}$ is the gauge-invariant field strength associated with the gauge field $A_\mu$. This Lagrangian is, of course, the version of (81) made invariant under the $U(1)$ local gauge transformations

$$A_\mu \to A'_\mu = A_\mu + \partial_\mu\omega\,, \tag{8.104}$$

$$\varphi \to \varphi' = \mathrm{e}^{-\mathrm{i}q\omega}\,\varphi\,. \tag{8.105}$$

When $\mu^2$ is positive the Lagrangian (102) just describes a scalar particle of mass $\mu$ and charge $q$ interacting with an electromagnetic field. We are rather interested in the case of negative $\mu^2$ when the potential develops minima at the field values $|\varphi|^2 = -2\mu^2/\lambda$. Then the symmetry may be hidden by selecting the vacuum so that the field acquires the vacuum expectation value

$$\langle\varphi\rangle = \tfrac{1}{\sqrt{2}}\,v\,, \tag{8.106}$$

for the real number $v = \sqrt{-4\mu^2/\lambda}$. Now, define the real fields $\chi_1$ and $\chi_2$ through

$$\varphi(x) = \tfrac{1}{\sqrt{2}}(v + \chi_1 + \mathrm{i}\chi_2)\,. \tag{8.107}$$

Then the covariant derivative of the field becomes

$$D_\mu\varphi = \frac{1}{\sqrt{2}}\left[\partial_\mu\chi_1 + \mathrm{i}qv\,(A_\mu + \frac{1}{qv}\partial_\mu\chi_2) + \mathrm{i}qA_\mu\,(\chi_1 + \mathrm{i}\chi_2)\right]\,, \tag{8.108}$$

leading to the expression for the kinetic term

$$K = D_\mu\varphi\, D^\mu\varphi^* \\ = \frac{1}{2}\left(\partial_\mu\chi_1 - qA_\mu\chi_2\right)^2 + \frac{1}{2}\left(\partial_\mu\chi_2 + qvA_\mu + qA_\mu\chi_1\right)^2 . \tag{8.109}$$

As in the complex scalar model with global symmetry, here $\chi_1$ acquires a mass too, but a clear interpretation of $\chi_2$ and $A_\mu$ is difficult to have because they are coupled together in the second order. What is significant is that this coupling comes as part of the expression

$$\frac{1}{2}(qv)^2\left(A_\mu + \frac{1}{qv}\partial_\mu\chi_2\right)\left(A^\mu + \frac{1}{qv}\partial^\mu\chi_2\right), \tag{8.110}$$

which could be regarded as a mass term for a redefined vector field

$$A'_\mu = A_\mu + \frac{1}{qv}\partial_\mu\chi_2 . \tag{8.111}$$

This field redefinition appears as a gauge transformation (104) of $A_\mu$ with the local transformation parameter $\omega = \chi_2/qv$; it tells us that $\chi_2$ has no real physical significance and might be eliminated by an appropriate gauge transformation. With this in mind, let us rewrite the gauge transformation (105) for the real fields $\chi_i$:

$$\chi_1 \to \chi'_1 = -v + (\cos q\omega)(v+\chi_1) + (\sin q\omega)\chi_2 , \\ \chi_2 \to \chi'_2 = (\cos q\omega)\chi_2 - (\sin q\omega)(v+\chi_1) . \tag{8.112}$$

For an infinitesimal $\omega$ this gives

$$\chi'_1 \approx \chi_1 + q\omega\,\chi_2 , \\ \chi'_2 \approx \chi_2 - q\omega\,\chi_1 - q\omega\, v . \tag{8.113}$$

We see that the field $\chi_2$ transforms with an inhomogeneous term, just like $A_\mu$, so that separately neither can have a direct physical meaning. In fact the gauge invariance of the theory allows us to make a gauge transformation that completely removes $\chi_2$. It suffices to choose as parameter

$$\omega(x) = \frac{1}{q}\tan^{-1}\left(\frac{\chi_2}{v+\chi_1}\right) . \tag{8.114}$$

In this gauge, only two fields survive: $\chi'_1$, which will be renamed $H$, and

$$A'_\mu = A_\mu + \partial_\mu\omega \approx A_\mu + \frac{1}{qv}\partial_\mu\chi_2 + \text{higher-order terms} , \tag{8.115}$$

which will be simply called $A_\mu$. We then have, for the potential,

$$V = -\mu^2 H^2 + \frac{1}{16}\lambda H^2 (4vH + H^2) + \frac{1}{4}\mu^2 v^2, \tag{8.116}$$

and, for the Lagrangian,

$$\begin{aligned}\mathcal{L} =& \frac{1}{2}[\partial_\mu H\, \partial^\mu H + 2\mu^2 H^2] - \frac{1}{4}F_{\mu\nu}F^{\mu\nu} + \frac{1}{2}(qv)^2 A_\mu A^\mu \\ &+ \frac{1}{2}q^2 A_\mu A^\mu H(H+2v) - \frac{1}{16}\lambda H^3(H+4v) - \frac{1}{4}\mu^2 v^2 .\end{aligned} \tag{8.117}$$

This result can now be naturally interpreted as the Lagrangian for a neutral scalar particle of mass $\sqrt{-2\mu^2}$ and a massive vector particle with mass $M_{\mathrm{A}} = qv$, conveniently decoupled from each other in the second order. The would-be Goldstone boson is completely gone; it has been gauged away, absorbed as the newly formed longitudinal polarization state of the vector field, as indicated by (115). Thus, two massless particles have been disposed of: the vector meson has gained mass and the Goldstone boson has been eliminated. Instead of a massless gauge boson with its two transverse modes and a complex scalar field composed of two real components, we have, after the symmetry breaking, a single real spin-0 field $H$ and a massive spin-1 meson with three spin states (two transverse and one longitudinal). The number of degrees of freedom has not changed; it remains four.

In the gauge specified by (114) all fields that survive the symmetry breakdown are all physical fields; fictitious particles, whose Green's functions would have singularities that violate unitarity, are absent. But the Lagrangian (117) contains a massive vector field, whose propagator for large momentum behaves as $1/M_{\mathrm{A}}^2$ rather than as $1/k^2$ characteristic of massless vector fields and, therefore, does not lead to an obviously renormalizable theory. This gauge, manifestly unitary but not manifestly renormalizable, is called the *unitary* (or U) gauge. A surprising result, obtained by G. 't Hooft, is that the renormalizability of the theory, though not manifest in (117), has in fact been preserved in the spontaneous symmetry breaking; it is not apparent simply because of the particular gauge being used. 't Hooft's proof of renormalizability of spontaneously broken gauge theories relies on the discovery that it is useful to adopt a class of more general gauges, called $\mathrm{R}_\xi$, which even though not manifestly unitary, are explicitly renormalizable; that is, the ultraviolet divergences that arise will behave no worse than those occurring in QED.

The $\mathrm{R}_\xi$-gauges may be enforced by adding to the Lagrangian (117) the gauge-fixing term

$$\begin{aligned}\mathcal{L}_{\mathrm{GF}} &= -\frac{1}{2\xi}(\partial_\mu A^\mu - f)^2 \\ &= -\frac{1}{2\xi}(\partial_\mu A^\mu)^2 + \frac{1}{\xi}(\partial_\mu A^\mu)\, f - \frac{1}{2\xi}f^2 ,\end{aligned} \tag{8.118}$$

where $\xi$ is a positive real constant that defines the gauge, while $f$ will be chosen so as to cancel the awkward quadratic coupling of $A_\mu$ and $\chi_2$ found in (109). This is done by requiring

$$\xi^{-1}(\partial_\mu A^\mu)\, f + qv\, A^\mu \partial_\mu \chi_2 = 0\,,$$

which is satisfied up to a total derivative, provided that

$$f = \xi\, qv\, \chi_2\,. \tag{8.119}$$

Together with the gauge-fixing term, the Lagrangian becomes in this gauge

$$\begin{aligned}
\mathcal{L} = &\frac{1}{2}[(\partial_\mu\chi_1)^2 + (\partial_\mu\chi_2)^2 + (qv)^2 A_\mu^2 + 2qv\, A^\mu\partial_\mu\chi_2 \\
&+ 2q\, A^\mu(\chi_1\partial_\mu\chi_2 - \chi_2\partial_\mu\chi_1) + q^2\, A_\mu^2(2v\chi_1 + \chi_1^2 + \chi_2^2)] \\
&- \frac{1}{4}F_{\mu\nu}^2 - \frac{1}{2\xi}(\partial_\mu A^\mu)^2 - \frac{1}{2\xi}(\xi qv)^2\,\chi_2^2 \\
&+ \mu^2\chi^2 - \frac{\lambda}{16}(\chi_1^2 + \chi_2^2)(4v\chi_1 + \chi_1^2 + \chi_2^2) - \frac{\mu^2 v^2}{4} \\
= &\frac{1}{2}\,[(\partial_\mu\chi_1)^2 + 2\mu^2\chi_1^2] + \frac{1}{2}\,[(\partial_\mu\chi_2)^2 - \xi(qv)^2\chi_1^2] \\
&- \frac{1}{4}F_{\mu\nu}^2 - \frac{1}{2\xi}(\partial_\mu A^\mu)^2 + \frac{1}{2}(qv)^2 A_\mu^2 + \text{higher-order terms.}
\end{aligned} \tag{8.120}$$

Thus, in a general R$_\xi$-gauge three fields are involved, decoupled from each other in the second order: the vector field of mass $M_A = qv$, the Higgs boson of mass $M_H = \sqrt{-2\mu^2}$, and the former Goldstone boson now with mass $\sqrt{\xi}\, M_A$. The dependence of the latter on the gauge parameter reveals the inherently nonphysical character of the Goldstone field.

The propagators of these fields can now be found from their respective quadratic terms in (120). Thus, the Goldstone mode has the propagator

$$\Delta(p, \sqrt{\xi} M_\mathrm{A}) = [p^2 - \xi M_\mathrm{A}^2]^{-1}\,, \tag{8.121}$$

and the Higgs boson has the propagator

$$\Delta(p, M_\mathrm{H}) = [p^2 - M_\mathrm{H}^2 + \mathrm{i}\varepsilon]^{-1}\,. \tag{8.122}$$

To find the propagator for the vector field, we consider the quadratic terms in $A_\mu$ in the Lagrangian which are, up to total derivatives,

$$\begin{aligned}
&-\frac{1}{4}\,F_{\mu\nu}^2 - \frac{1}{2\xi}\,(\partial_\mu A^\mu)^2 + \frac{1}{2}\,(qv)^2\,A_\mu^2 \\
&\quad = \frac{1}{2}\,A^\mu\,[g_{\mu\nu}(\partial^2 + q^2v^2) - (1 - \xi^{-1})\,\partial_\mu\partial_\nu\,]\,A^\nu\,,
\end{aligned} \tag{8.123}$$

from which the inverse propagator can be immediately read off:

$$[D_{\mu\nu}(p)]^{-1} = g_{\mu\nu}\,(-p^2 + M_{\rm A}^2) + (1-\xi^{-1})\,p_\mu p_\nu\,, \tag{8.124}$$

or in terms of projection operators,

$$[D_{\mu\nu}(p)]^{-1} = -(p^2 - M_{\rm A}^2)\left(g_{\mu\nu} - \frac{p_\mu p_\nu}{p^2}\right) - \xi^{-1}(p^2 - \xi M_{\rm A}^2)\frac{p_\mu p_\nu}{p^2}\,.$$

The propagator itself is then

$$D_{\mu\nu}(p) = -(p^2 - M_{\rm A}^2 + {\rm i}\varepsilon)^{-1}\left(g_{\mu\nu} - \frac{p_\mu p_\nu}{p^2}\right) - \xi(p^2 - \xi M_{\rm A}^2)^{-1}\frac{p_\mu p_\nu}{p^2}\,,$$

which reduces to

$$D_{\mu\nu}(p) = \frac{-g_{\mu\nu} + (1-\xi)p_\mu p_\nu/(p^2 - \xi\, M_{\rm A}^2)}{p^2 - M_{\rm A}^2 + {\rm i}\varepsilon}\,. \tag{8.125}$$

Note that the poles at $p^2 = \xi M_{\rm A}^2$ in (121) and (125) are unphysical and need not be defined with an i$\varepsilon$ term. They will be canceled out in any physical transition amplitude (as shown by 't Hooft).

When $M_A = 0$, one recovers the propagator for the photon or gluon,

$$D_{\mu\nu}(p) = \frac{-g_{\mu\nu} + (1-\xi)p_\mu p_\nu/p^2}{p^2 + {\rm i}\varepsilon}\,. \tag{8.126}$$

Familiar gauges correspond to special values of $\xi$. With $\xi = 1$, we recover the Feynman gauge often used in QED,

$$D_{\mu\nu}(p) = \frac{-g_{\mu\nu}}{p^2 - M_A^2 + {\rm i}\varepsilon}\,, \tag{8.127}$$

while for $\xi = 0$, it is the Landau gauge, in which the propagator depends only on the transverse projection operator,

$$D_{\mu\nu}(p) = \frac{-g_{\mu\nu} + p_\mu p_\nu/p^2}{p^2 - M_{\rm A}^2 + {\rm i}\varepsilon}\,. \tag{8.128}$$

We can see that for any finite value of $\xi$, the propagator at large momentum behaves as $1/p^2$, just as in the massless case, and the corresponding gauge is manifestly renormalizable. But as $\xi \to \infty$, (125) tends to the ordinary propagator for a massive vector field

$$D_{\mu\nu}(p) = \frac{-g_{\mu\nu} + p_\mu p_\nu/M_A^2}{p^2 - M_A^2 + {\rm i}\varepsilon}\,, \tag{8.129}$$

while the propagator for the erstwhile Nambu–Goldstone mode in (121) tends to zero, suggesting that this field will drop out of the system. The gauge $\xi \to \infty$ coincides with the unitary gauge.

### 8.6.2 Non-Abelian Symmetry

We now generalize the above considerations to local gauge symmetry by introducing the general Yang–Mills fields to make the model of Sect. 8.5.4 invariant under a local gauge group G. We consider a real $n$-dimensional representation of G spanned by $n$-component scalar fields $\boldsymbol{\phi}$ and in which the transformations are generated by $N$ imaginary and antisymmetric $n \times n$ matrices $T_i$ for $i = 1, \dots, N$. To each generator corresponds a vector gauge field, $A_{i\mu}$, so as to satisfy G gauge invariance (by which we mean gauge invariance under the symmetry group $G$). Thus, we consider the Lagrangian

$$\mathcal{L} = \frac{1}{2}\,(\boldsymbol{D}_\mu \boldsymbol{\phi})^{\mathrm{T}} \boldsymbol{D}^\mu \boldsymbol{\phi} - V(\boldsymbol{\phi}) - \frac{1}{4}\, F_{i\mu\nu} F_i^{\mu\nu}\,, \tag{8.130}$$

with the G-covariant derivative

$$\boldsymbol{D}_\mu \boldsymbol{\phi} = (\partial_\mu + \mathrm{i}g\boldsymbol{A}_\mu)\,\boldsymbol{\phi} = (\partial_\mu + \mathrm{i}gA_{j\mu}T_j)\,\boldsymbol{\phi}\,,$$

and the G-covariant field tensors associated with the gauge fields

$$F_i^{\mu\nu} = \partial^\mu A_i^\nu - \partial^\nu A_i^\mu - g\, f_{ijk}\, A_j^\mu A_k^\nu\,.$$

An explicit form of $V$ in terms of $\boldsymbol{\phi}$ is not essential, all we need is that it respects the G symmetry and develops degenerate minima at nonzero constant field values $|\boldsymbol{\phi}|^2 = v^2$. Then, the gauge invariance is spontaneously broken when the system arbitrarily selects for itself a vacuum state such that $\langle \boldsymbol{\phi} \rangle$ is some constant vector $\boldsymbol{v}$, which we call the *vacuum vector*, satisfying the condition that $\boldsymbol{v}^2$ minimizes $V$. Now we suppose that the symmetry breaking leaves the vacuum invariant under a subgroup S of G and that the generators $T_i$ of G are chosen so that $T_j$, for $j = 1, \dots, M$ and $M < N$, generate S . Since by assumption $\boldsymbol{v}$ is invariant under S but noninvariant under its complement in G, we have

$$T_j \boldsymbol{v} = 0\,, \qquad j = 1, \dots, M\,; \tag{8.131}$$

$$T_k \boldsymbol{v} \neq 0\,, \qquad k = M+1, \dots, N\,. \tag{8.132}$$

Just as the algebra g of G is defined by all the linear combinations $\sum_{i=1}^N c_i T_i$, so too is the algebra $\mathrm{g_S}$ of the subgroup S defined by all the combinations $\sum_{j=1}^M c_j T_j$. Subalgebra $\mathrm{g_S}$ has dimension $M$; it annihilates the vacuum. The orthogonal complement to $\mathrm{g_S}$ in g has dimension $N-M$; its elements applied on the vacuum vector yield $N-M$-dimensional vectors, $\sum_{k=M+1}^N c_k T_k \boldsymbol{v}$, which span the space of the Nambu–Goldstone modes.

Taking advantage of the gauge invariance of (130), we may perform a gauge transformation on all the fields without changing the underlying physics. The particular gauge transformation $U$ is so chosen to cancel the quadratic coupling between the gauge fields and the scalar fields, which is equivalent to requiring

$$(U\boldsymbol{\phi})^{\mathrm{T}}(x)\,T_i\boldsymbol{v} = 0 \qquad \text{for all } i \text{ and all } x. \tag{8.133}$$

That such a (unitary) gauge always exists was proved by Weinberg. Given this general result, let us define an $n$-component field $\boldsymbol{H}(x)$ orthogonal to all $T_k\boldsymbol{v}$ for $k = M+1,\ldots,N$. As $T_i$ are antisymmetric matrices (as in Sect. 8.5.4), we also have $\boldsymbol{v}^{\mathrm{T}}\,T_i\boldsymbol{v} = 0$, and therefore,

$$(\boldsymbol{v}+\boldsymbol{H})^{\mathrm{T}}\,T_k\boldsymbol{v} = 0\,, \qquad k = M+1,\ldots,N\,, \tag{8.134}$$

in addition to the identity

$$(\boldsymbol{v}+\boldsymbol{H})^{\mathrm{T}}\,T_j\boldsymbol{v} = 0\,, \qquad j = 1,\ldots,M\,, \tag{8.135}$$

which holds because of the invariance of the vacuum vector under S. Taken together, these relations yield

$$(\boldsymbol{v}+\boldsymbol{H})^{\mathrm{T}}\,T_i\boldsymbol{v} = 0\,, \qquad i = 1,\ldots,N\,. \tag{8.136}$$

As we will now see, it proves useful to adopt a parameterization of fields similar in form to (101):

$$\boldsymbol{\phi} = \exp\left(\frac{\mathrm{i}}{v}\sum_{k=M+1}^{N}\xi_k T_k\right)(\boldsymbol{v}+\boldsymbol{H})\,. \tag{8.137}$$

In this parameterization, the independent fields are the $N-M$ would-be Nambu–Goldstone modes $\xi_k$, with $k = M+1,\ldots,N$, and the $n-N+M$ independent components of $\boldsymbol{H}$ representing the Higgs bosons. It now becomes clear that the transformation $U$ needed to satisfy (133) or (136) is

$$U = \exp\left(-\frac{\mathrm{i}}{v}\sum_{k=M+1}^{N}\xi_k T_k\right)\,. \tag{8.138}$$

Since the Lagrangian (130) is invariant under the local group G, it remains unchanged with the fields written in the new gauge

$$\begin{aligned}
\boldsymbol{\phi}' &= U\boldsymbol{\phi} = \boldsymbol{v}+\boldsymbol{H}\,,\\
\boldsymbol{A}'_\mu &= U\boldsymbol{A}_\mu U^\dagger + \frac{\mathrm{i}}{g}(\partial_\mu U)U^\dagger\,,\\
\boldsymbol{F}'_{\mu\nu} &= U\boldsymbol{F}_{\mu\nu}U^\dagger\,.
\end{aligned} \tag{8.139}$$

Thus, in this U-gauge, the fields $\xi_k$ have completely vanished, and (130) involves only $\boldsymbol{H}$ and $A'_{j\mu}$, which we will now simply write $A_{j\mu}$,

$$\mathcal{L} = \frac{1}{2}\left[\boldsymbol{D}_\mu(\boldsymbol{v}+\boldsymbol{H})\right]^{\mathrm{T}} \boldsymbol{D}^\mu(\boldsymbol{v}+\boldsymbol{H}) - V(\boldsymbol{v}+\boldsymbol{H}) - \frac{1}{2}\partial_\mu A_{i\nu}\left(\partial^\mu A_i^\nu - \partial^\nu A_i^\mu\right). \tag{8.140}$$

The 'kinetic' part can be expanded as

$$\begin{aligned}
&(\boldsymbol{D}_\mu(\boldsymbol{v}+\boldsymbol{H}))^{\mathrm{T}}\, \boldsymbol{D}^\mu(\boldsymbol{v}+\boldsymbol{H}) \\
&= \left(\partial_\mu \boldsymbol{H}^{\mathrm{T}} + \mathrm{i}g(\boldsymbol{v}+\boldsymbol{H})^{\mathrm{T}}\boldsymbol{A}_\mu^{\mathrm{T}}\right)\left(\partial^\mu \boldsymbol{H} + \mathrm{i}g\boldsymbol{A}^\mu(\boldsymbol{v}+\boldsymbol{H})\right) \\
&= \partial_\mu \boldsymbol{H}^{\mathrm{T}}\, \partial^\mu \boldsymbol{H} + 2\mathrm{i}g\, \partial_\mu \boldsymbol{H}^{\mathrm{T}}\, \boldsymbol{A}^\mu(\boldsymbol{v}+\boldsymbol{H}) + g^2(\boldsymbol{v}+\boldsymbol{H})^{\mathrm{T}}\boldsymbol{A}_\mu \boldsymbol{A}^\mu\,(\boldsymbol{v}+\boldsymbol{H}).
\end{aligned} \tag{8.141}$$

We observe that the quadratic mixing term of $\boldsymbol{A}_\mu$ and $\boldsymbol{H}$ can now be disposed of, as expected from (136),

$$2\mathrm{i}g\, \partial_\mu \boldsymbol{H}^{\mathrm{T}}\, \boldsymbol{A}^\mu \boldsymbol{v} = 2\mathrm{i}g\, A_{i\mu}\, \partial^\mu \boldsymbol{H}^{\mathrm{T}} T_i \boldsymbol{v} = 0. \tag{8.142}$$

The Lagrangian then reduces to

$$\begin{aligned}
\mathcal{L} &= \frac{1}{2}\left(\partial_\mu \boldsymbol{H}^{\mathrm{T}}\, \partial^\mu \boldsymbol{H} - \boldsymbol{H}^{\mathrm{T}}\mathcal{M}_{\mathrm{B}}^2 \boldsymbol{H}\right) \\
&\quad - \frac{1}{2}\,\partial_\mu A_{i\nu}\left(\partial^\mu A_i^\nu - \partial^\nu A_i^\mu\right) + \frac{1}{2}g^2 \boldsymbol{v}^{\mathrm{T}}\boldsymbol{A}_\mu \boldsymbol{A}^\mu\, \boldsymbol{v} \\
&\quad + \mathrm{i}\,g\,\partial_\mu \boldsymbol{H}^{\mathrm{T}}\, \boldsymbol{A}^\mu \boldsymbol{H} + g^2 \boldsymbol{v}^{\mathrm{T}}\boldsymbol{A}_\mu \boldsymbol{A}^\mu \boldsymbol{H} + \frac{1}{2}\,g^2 \boldsymbol{H}^{\mathrm{T}}\boldsymbol{A}_\mu \boldsymbol{A}^\mu \boldsymbol{H} + \dots,
\end{aligned} \tag{8.143}$$

where ... indicates the cubic and quartic self-coupling terms in $\boldsymbol{H}$, whose details depend on the assumed potential $V$. The surviving real scalar Higgs fields are massive, with squared masses determined by the matrix

$$\left(\mathcal{M}_{\mathrm{B}}^2\right)_{ab} = \left.\frac{\partial^2 V}{\partial\phi_a\, \partial\phi_b}\right|_{\phi=v}. \tag{8.144}$$

The squared-mass matrix for the vector mesons

$$g^2(\boldsymbol{v}^{\mathrm{T}} T_\ell T_k\, \boldsymbol{v}) \tag{8.145}$$

is real, symmetric, and positive-definite, and is nonvanishing for $\ell, k = \mathrm{M}+1, \dots, N$. Of all the gauge fields, only those associated with the symmetry-breaking part of G, that is, $A_k^\mu$ with $k = M+1, \dots, N$, acquire *masses* and *longitudinal components*, as shown by the presence of the inhomogeneous terms in the variations under the transformation (139):

$$\delta A_{k\mu} = \frac{1}{gv}\partial_\mu \xi_k - \frac{1}{v} f_{km\ell} A_{m\mu} \xi_\ell\,, \tag{8.146}$$

whereas the others, $A_j^\mu$ with $j = 1, \dots, M$, associated with the surviving symmetry, remain *massless, transversely polarized*, and transform homogeneously under (139). The number of independent degrees of freedom remains the same before and after the symmetry breaking. The original $n$ real scalar fields and $N$ massless gauge mesons, producing altogether $n + 2N$ degrees of freedom, are replaced after the symmetry breaking by $n - N + M$ Higgs bosons, $M$ massless vector fields, and $N - M$ massive vector fields for a total of $n - N + M + 2M + 3(N - M) = n + 2N$ degrees of freedom.

As we have discussed above, a unitary gauge leads to a formalism that is simple to interpret, but has the disadvantage of not being manifestly renormalizable. It is more useful for practical calculations to adopt a manifestly renormalizable gauge so that the powerful techniques developed for renormalizable theories can be applied. This can be accomplished by a generalization of the gauge-fixing Lagrangian (118),

$$\mathcal{L}_{\text{GF}} = -\frac{1}{2\xi}(\partial_\mu A_i^\mu - \mathrm{i}g\xi\boldsymbol{\phi}^{\mathrm{T}}T_i\boldsymbol{v})^2 , \tag{8.147}$$

designed so that the quadratic mixing terms between $A_{i\mu}$ and $\partial_\mu\boldsymbol{\phi}$ found here and in the Lagrangian (130) exactly cancel out.

We have limited ourselves in this chapter to a discussion of the mass generation of gauge bosons. When matter fields are introduced, they may not be allowed by gauge invariance to have explicit masses in the basic Lagrangian. It is possible however to induce their masses by coupling matter fields to the Higgs fields in a gauge-invariant way. We will study how this mechanism can make quarks and leptons massive in the context of the standard model of the electroweak interaction.

## Problems

**8.1 Equations of motion.** Show that the equations of motion corresponding to the Lagrangian (42) are

$$\begin{aligned}(\mathrm{i}\gamma^\mu \boldsymbol{D}_\mu - m)\,\psi &= 0\,,\\ \boldsymbol{D}^\mu F^j_{\mu\nu} &= g\,\overline{\psi}\gamma_\nu T_j\psi\,.\end{aligned}$$

**8.2 Group multiplication in gauge groups.** Show that, for any element $h$ of the gauge group, the transformation rule for the gauge field

$$h : \boldsymbol{A}_\mu \to \boldsymbol{A}'_\mu = U\boldsymbol{A}_\mu U^\dagger + \frac{\mathrm{i}}{g}(\partial_\mu U)U^\dagger$$

satisfies the group multiplication law, that is, if $h : \boldsymbol{A}_\mu \to \boldsymbol{A}'_\mu$ and $h' : \boldsymbol{A}'_\mu \to \boldsymbol{A}''_\mu$, then $h'' : \boldsymbol{A}_\mu \to \boldsymbol{A}''_\mu$, where $h'' = h'h$.

**8.3 The linear $\sigma$–$\pi$ model.** Consider a model for a real field $\sigma$, transforming as an isosinglet and three real scalar fields $\phi_i$, forming an isotriplet.

Call the conjugate momenta of fields $\pi^i_\mu = \partial_\mu \phi_i$, $\pi^4_\mu = \partial_\mu \sigma$ and their time components $\pi^i_0 = \pi_i$ and $\pi^4_0 = \pi_4$. The fields are quantum operators satisfying the canonical commutation relations at equal times. The Lagrangian of the model is given by

$$\mathcal{L}_\mathrm{s} = \frac{1}{2}(\partial_\mu \phi_i \partial^\mu \phi^i + \partial_\mu \sigma \partial^\mu \sigma) - V(\boldsymbol{\phi}^2 + \sigma^2)\,. \tag{1}$$

(a) Show that the model is invariant under the following two global transformations of their internal degrees of freedom ($\sigma \to \sigma + \delta_i \sigma\,, \boldsymbol{\phi} \to \boldsymbol{\phi} + \delta_i \boldsymbol{\phi}$):
– isospin rotation:

$$\delta_i \sigma = 0\,; \quad \delta_i \phi_j = \epsilon_{ijk} \omega_i \phi_k\,, \text{ (no sum over } i). \tag{2}$$

– chiral transformation:

$$\delta_i \sigma = \omega_i \phi_i\,; \quad \delta_i \phi_j = -\delta_{ij} \omega_i \sigma\,, \text{ (no sum over } i). \tag{3}$$

(b) Show that the associated conserved isospin and axial currents are $V_{i\mu} = \epsilon_{ijk} \phi_j \pi^k_\mu$, and $A_{i\mu} = \pi^i_\mu \sigma - \pi^4_\mu \phi_i$; and that the corresponding conserved charges, $Q_i$ and $Q^5_i$ satisfy

$$[Q_i,\, Q_j] = \mathrm{i}\epsilon_{ijk}\, Q_k\,,$$
$$[Q_i,\, Q^5_j] = \mathrm{i}\epsilon_{ijk}\, Q^5_k\,,$$
$$[Q^5_i,\, Q^5_j] = \mathrm{i}\epsilon_{ijk}\, Q_k\,.$$

Show that $Q^+_i \equiv \frac{1}{2}(Q_i + Q^5_i)$ and $Q^-_i \equiv \frac{1}{2}(Q_i - Q^5_i)$ form two independent commuting SU(2) algebras, so that the algebra of the model is a semisimple algebra, SU(2) × SU(2).
(c) Assuming that $V = \frac{1}{2}\mu^2(\sigma^2 + \boldsymbol{\phi}^2) + \frac{1}{4}\lambda\,(\sigma^2 + \boldsymbol{\phi}^2)^2$, where $\lambda > 0$ and $\mu^2 < 0$. The potential $V$ has minima for fields satisfying $\sigma^2 + \boldsymbol{\phi}^2 = -\mu^2/\lambda$. Select the vacuum such that $\langle \phi_i \rangle = 0\,,$ and $\langle \sigma \rangle = v = \sqrt{-\mu^2/\lambda}$, thus provoking a symmetry breakdown. Now define $\sigma' = \sigma - v$. Calculate the masses of $\phi_i$ and $\sigma'$. Calculate the commutation relations of $Q_i$ and $Q^5_i$ with $\sigma'$ and $\phi_i$, and show that $Q^5_i$ generate a symmetry that is broken in the vacuum.
(d) To $\mathcal{L}_\mathrm{s}$, add the following Lagrangian for an isodoublet of fermions, interacting with $\sigma$ and $\phi_i$,

$$\mathcal{L}_\mathrm{F} = \overline{\psi}(\mathrm{i}\gamma \cdot \partial - m_0)\psi - g\,\overline{\psi}(\sigma + \mathrm{i}\tau_j \phi_j \gamma_5)\psi\,. \tag{4}$$

Together $\mathcal{L}_\mathrm{s}$ and $\mathcal{L}_\mathrm{F}$ define the Gell-Mann–Levy model. The isospin and chiral transformations (2) and (3) are supplemented by the following for $\psi$:

$$\delta_i \psi = \tfrac{1}{2}\mathrm{i}\omega_i \tau_i \psi\,, \quad \text{(no sum over } i).$$
$$\delta_i \psi = \tfrac{1}{2}\mathrm{i}\omega_i \tau_i \gamma_5 \psi\,, \quad \text{(no sum over } i).$$

Show that $\mathcal{L}_\mathrm{F}$ is invariant to isospin rotations for $m$ arbitrary, but is invariant to chiral transformation only for $m_0 = 0$. Assume now $m_0 = 0$. Consider the full system described by $\mathcal{L}_\mathrm{s} + \mathcal{L}_\mathrm{F}$, and show that when there is spontaneous symmetry breaking by (8), the fermion acquires a mass, $m = gv$. Express the parameters $\mu^2$, $\lambda$, and $v$ in terms of g, $m$, and $m_{\sigma'}$.

# Suggestions for Further Reading

*General references:*
Abers, E. S. and Lee, B. W., Phys. Rep **9C** (1973) 1
Aitchison, I. J. R. and Hey, A. J. G., *Gauge Theories in Particle Physics* (Second edition). Adam Hilger, Bristol 1989; Chap. 9
Bailin, D. and Love, A., *Introduction to Gauge Field Theory.* Adam Hilger, Bristol 1986; Chaps. 9, 13
Cheng, T.-P. and Li, L.-F., *Gauge Theory of Elementary Particle Physics.* Oxford U. Press, New York 1984
Coleman, S., in *Laws of Hadronic Matter.* Academic Press, New York 1975
*On gauge field theories:*
Gell-Mann, M. and Glashow, S. L., Ann. Phys. (NY), **15** (1961) 437
Utiyama, R., Phys. Rev. **101** (1956) 1597
Yang, C. N. and Mills, R. L., Phys. Rev. **96** (1954) 191
*Quantum electrodynamics:*
Kinoshita, T. (ed.), *Quantum Electrodynamics.* World Scientific, Singapore 1990
Schweber, S. S., *QED and the Men Who Made It: Dyson, Feynman, Schwinger, and Tomonaga.* Princeton U. Press, Princeton 1994
Schwinger, Julian (ed.) *Selected Papers on Quantum Electrodynamics.* Dover, New York 1958
*Quantum chromodynamics:*
Fritzsch, H., Gell-Mann, M. and Leutwyler, H., Phys. Lett. **47B** (1973) 365
Gross, D. J. and Wilczek, F., Phys. Rev. **D8** (1973) 3633
Marciano, W. and Pagels, H., Phys. Rep. **36C** (1978) 137
Politzer, H. D., Phys. Rep. **14C** (1974) 129
Weinberg, S., Phys. Rev. Lett. **31** (1973) 494
Wilczek, F., Ann. Rev. Nucl. and Part. Sci. **32** (1982) 177
Ynduráin, F. J., *Quantum Chromodynamics, An Introduction to the Theory of Quarks and Flavors.* Springer, Berlin, Heidelberg 1983
*Lattice gauge theory:*
Creutz, M., *Quarks, Gluons, and Lattices.* Cambridge U. Press, Cambridge 1983
Lattice 96, Nucl. Phys. (Proc. Suppl.) **B53** (1997) February 1997
Rebbi, C., *Lattice Gauge Theories and Monte Carlo Simulations.* World Scientific, Singapore 1983
Wilson, K. G., Phys. Rev. **D10** (1974) 2445; Rev. Mod. Phys. **47** (1975) 773
*On Nambu–Goldstone's theorem:*
Nambu, Y., Phys. Rev. Lett. **4** (1960) 380
Goldstone, J., Nuovo Cimento **19** (1961) 154
Goldstone, J., Salam, A. and Weinberg, S., Phys. Rev. **127** (1962) 965
*The Higgs phenomenon:*
Englert, F. and Brout, R., Phys. Rev. Lett. **13** (1964) 321
Guralnik, G., Hagen, C. and Kibble, T., Phys. Rev. Lett. **13** (1964) 585
Higgs, P., Phys. Lett. **12** (1964) 132; Phys. Rev. **145** (1966) 1156
Kibble, T., Phys. Rev. **155** (1967) 1554
*Renormalization:*
't Hooft, G., Nucl. Phys. **B33** (1971) 173; **B35** (1971) 167

# 9 The Standard Model of the Electroweak Interaction

In this chapter we describe the unified theory of weak and electromagnetic interactions that is often referred to as 'the standard model' (of the electroweak interaction). It is a non-Abelian gauge theory in which the local phase invariance is hidden, or spontaneously broken, so that the weak gauge forces may acquire a finite range as experimentally observed without sacrificing the renormalizability expected of a physically meaningful theory. We shall first review the developments that have led to the formulation of the current theory. We next describe in some detail the model of the electron and its neutrino, including the identification of the gauge symmetry group, its subsequent spontaneous breaking and the attendant mass generation for the electron and the gauge bosons. Introducing one family of quarks into the model poses no particular problem of principle, but when several families of leptons and quarks enter, extra care must be taken to distinguish between the gauge eigenstates and the mass eigenstates in the fermion sector, which gives rise to several novel and powerful predictions by the theory.

## 9.1 The Weak Interaction Before the Gauge Theories

Before the advent of the gauge models in the late 1960s, weak transitions have been described by a local two-current interaction originally due to Fermi (Chap. 5) and generalized to its present form by Feynman and Gell-Mann,

$$\mathcal{H}_{\text{weak}} = \frac{G_{\text{F}}}{\sqrt{2}} J'^{\mu} J'^{\dagger}_{\mu} . \tag{9.1}$$

Here the Lorentz vector $J'_{\mu}$ is a *charged current*, so called because the charge of the particle entering the interaction vertex differs by one unit from that of the particle leaving the vertex. (It equals twice the current $J_{\mu}$ to be introduced later in the chapter.) With a current that incorporates both leptons and hadrons,

$$J'_{\mu} = L_{\mu} + H_{\mu} , \tag{9.2}$$

the interaction Hamiltonian (1) provides a complete description of the weak processes at low energies, the only energy region where it is regarded as applicable. The only notable exception to this general success is the phenomenon of violation of time-reversal invariance discovered in the neutral K meson system by Christenson, Cronin, Fitch, and Turlay in 1964, which appears to require completely new ideas (Chap. 11).

The coupling constant in (1) is not dimensionless, being given by

$$G_{\mathrm{F}} = 1.166\,39 \times 10^{-5}\,\mathrm{GeV}^{-2}\,. \tag{9.3}$$

Numerically, $G_{\mathrm{F}}$ is very small, but having the dimension of the inverse squared mass, it leads to a nonrenormalizable interaction. Corrections beyond the tree-diagram level, which are given by loop graphs with internal particle lines, involve higher powers of $G_{\mathrm{F}}$, or of mass in the denominator, and hence higher powers of momentum in the numerator. This leads to increasingly divergent terms in successive orders of the perturbation theory, which cannot be rearranged so as to be absorbed into a small number of 'bare' parameters and fields to yield a finite theory. The theory is not renormalizable.

But if on dimensional grounds we pose

$$4\sqrt{2}\,G_{\mathrm{F}} = e^2/M_{\mathrm{W}}^2\,, \tag{9.4}$$

the resulting 'mass' $M_{\mathrm{W}} \approx 37$ GeV may be viewed as indicative that the weak interaction might not be inherently feeble after all and its apparent weakness might just come from the presence of a very massive quantum exchanged between interacting particles. Like the electromagnetic current $j_\mu^{\mathrm{em}}$, the weak charged current $J'_\mu$ is a Lorentz four-vector, and we may use the familiar form of the electromagnetic interaction, $-e\, j_\mu^{\mathrm{em}} A^\mu$, as a model to construct the basic weak interaction, coupling $J'_\mu$ to a new *massive charged field* $W_\mu$ of mass $M_{\mathrm{W}}$, in the form

$$\mathcal{L}_{\mathrm{weak}} = -\frac{g}{2\sqrt{2}}(J'^{\mu} W_\mu^\dagger + J'^{\mu\dagger} W_\mu)\,. \tag{9.5}$$

Then to second order, $\mathcal{L}_{\mathrm{weak}}$ will generate (1) as an effective low-energy weak interaction with coupling constant $G_{\mathrm{F}}/\sqrt{2} = g^2/(8M_{\mathrm{W}}^2)$, where $M_{\mathrm{W}}^2$ comes from the $W_\mu$ field propagator in the limit of *small momentum*. Even though the new coupling constant is dimensionless, the theory is still not manifestly renormalizable because, as we have seen in the last chapter, the propagator of a massive vector particle reduces (also) to $M_{\mathrm{W}}^{-2}$ at *large momentum*, leading to divergent integrals in higher-order diagrams. Nevertheless, such theories can be renormalized provided that gauge invariance holds. Thus, gauge invariance is the key. The problem lies in formulating a gauge theory of weak interactions containing massive gauge fields while preserving renormalizability that can be meshed with the electromagnetic interaction theory into a unified theory of the electroweak interaction.

## 9.2 Gauge-Invariant Model of One-Lepton Family

In this section, we construct a gauge model for one family of leptons, the electron and its neutrino. It already contains many of the main properties found in the complete unified theory of the electroweak interaction. We first determine the simplest group to be gauged that would give rise to the key features of both the electromagnetic and weak interactions of the leptons. Next, we describe the gauge-invariant model involving leptons and scalar fields. Finally, we discuss in detail how spontaneous symmetry breaking generates masses for both matter and gauge fields.

We can limit ourselves to one family of leptons because, as we have seen previously, there exist electron-type conservation laws: the number of electrons $e^-$ plus the number of electronic neutrinos $\nu_e$, minus the number of the corresponding antiparticles, $e^+$ and $\overline{\nu}_e$, is conserved; and similarly for the muon- and tau-type leptons. In the following, we denote the field of a particle by its usual symbol, e.g. $\nu_e$ designates the spinor field for the neutrino, and $e$, the spinor field for the electron.

Any Dirac spinor field can be decomposed into left- and right-handed components

$$\chi(x) = \chi_{\mathrm{L}}(x) + \chi_{\mathrm{R}}(x)\,, \tag{9.6}$$

where one defines

$$\chi_{\mathrm{L}}(x) = a_{\mathrm{L}}\,\chi(x)\,, \qquad \chi_{\mathrm{R}}(x) = a_{\mathrm{R}}\,\chi(x)\,, \tag{9.7}$$

in terms of $\chi$ by application of the left and right chiral projection operators

$$\begin{aligned} a_{\mathrm{L}} &\equiv \tfrac{1}{2}(1-\gamma_5)\,,\\ a_{\mathrm{R}} &\equiv \tfrac{1}{2}(1+\gamma_5)\,. \end{aligned} \tag{9.8}$$

Note in particular the expressions of their adjoint conjugates

$$\begin{aligned} \overline{\chi}_{\mathrm{L}} &= \chi_{\mathrm{L}}^{\dagger}\gamma_0 = \chi^{\dagger}\,a_{\mathrm{L}}\gamma_0 = \overline{\chi}\,a_{\mathrm{R}}\,,\\ \overline{\chi}_{\mathrm{R}} &= \chi_{\mathrm{R}}^{\dagger}\gamma_0 = \chi^{\dagger}\,a_{\mathrm{R}}\gamma_0 = \overline{\chi}\,a_{\mathrm{L}}\,. \end{aligned} \tag{9.9}$$

As we have seen in Chap. 3, breakup (6) has no Lorentz-invariant meaning when the field is massive. But if on the contrary the mass of the field is zero, either of the two chiral components, which then coincides with a helicity eigenstate, may provide a complete representation of the Lorentz group.

The key experimental fact is that in the spectra of weak decays, such as $\mathrm{n}\to\mathrm{p}+\mathrm{e}^-+\overline{\nu}_{\mathrm{e}}$, $\mu^-\to\mathrm{e}^-+\overline{\nu}_{\mathrm{e}}+\nu_\mu$, and $\pi^-\to\mu^-+\overline{\nu}_\mu$, only left-handed leptons and right-handed antileptons show up, so that the decay amplitudes can be described in terms of a charged current that involves only the left chiral components of the fields,

$$\begin{aligned} L_\mu(x) &= 2\,\overline{e}_{\mathrm{L}}(x)\gamma_\mu\nu_{\mathrm{eL}}(x) + (\text{other lepton types})\\ &= \overline{e}(x)\gamma_\mu(1-\gamma_5)\nu_{\mathrm{e}}(x) + \ldots\,. \end{aligned} \tag{9.10}$$

This expression resembles the isovector current introduced in Chap. 6, and suggests that $\nu_{eL}$ and $e_L$ should be gathered into a two-component vector which can be associated with an SU(2) group, the simplest group having a complex doublet representation. On the other hand, the right chiral components $\nu_{eR}$ and $e_R$, which do not interact with any other particles, should be left in one-dimensional representations. But while $e_R$ should certainly stay because it has the same nonvanishing charge and mass as $e_L$, the right chiral component of the neutrino $\nu_{eR}$ may be immediately dropped because the neutrino is observed left-handed and electrically neutral, and is assumed to be exactly massless.

Therefore, in this model of the electron family, we have as matter fields a doublet, $\psi_L$, and a singlet, $\psi_R$, of an SU(2) group,

$$\psi_L = \begin{pmatrix} \nu_{eL} \\ e_L \end{pmatrix}, \qquad \psi_R = e_R \,. \tag{9.11}$$

As this group acts nontrivially just on the left chiral fermions, it is sometimes denoted by $SU_L(2)$ and referred to as the *weak-isospin* group. We will be using a simplified notation in this section, for instance $\nu$ for $\nu_e$, when no risks of confusion can arise.

### 9.2.1 Global Symmetry

The free Lagrangian for the (massless) fields in (11) is

$$\begin{aligned}\mathcal{L}_0 &= \overline{\psi}_L \, i\gamma^\lambda \partial_\lambda \, \psi_L + \overline{\psi}_R \, i\gamma^\lambda \partial_\lambda \, \psi_R \\ &= \overline{\nu}_L \, i\gamma^\lambda \partial_\lambda \, \nu_L + \overline{e} \, i\gamma^\lambda \partial_\lambda \, e \,.\end{aligned} \tag{9.12}$$

**Weak Isospin.** By construction, $\mathcal{L}_0$ is invariant to SU(2) transformations

$$U_2(\omega) = e^{-ig\omega_i t_i} \,, \tag{9.13}$$

where $\omega_i$, for $i = 1, 2, 3$, are the transformation constant parameters, and $t_i$ is equal to $t_{iL} = \frac{1}{2}\tau_i$ (the usual Pauli matrices) when it operates on $\psi_L$, and to $t_{iR} = 0$ when it operates on $\psi_R$. For infinitesimal transformations, we write

$$\begin{aligned} &U_2\psi_L = \psi_L + \delta\psi_L \,, \qquad && \delta\psi_L \approx -i\tfrac{1}{2} \, g\omega_i \tau_i \, \psi_L \,; \\ &U_2\psi_R = \psi_R + \delta\psi_R \,, && \delta\psi_R = 0 \,. \end{aligned} \tag{9.14}$$

In general, the conserved currents associated with continuous global symmetries parameterized by real $\alpha_i$ are defined, as in Chap. 2,

$$j_i^\mu = \frac{\partial \mathcal{L}}{\partial(\partial_\mu \varphi_a)} \frac{\delta \varphi_a}{\delta \alpha_i} + \mathcal{L} \frac{\delta x^\mu}{\delta \alpha_i} \,. \tag{9.15}$$

For transformations on internal space, $\delta x^\mu = 0$ and the last term on the right-hand side is absent.

In the present case, we may choose $\alpha_i = g\omega_i$ in (15) so that

$$\frac{\delta\psi_\mathrm{L}}{\delta(g\omega_i)} = -\mathrm{i}\frac{\tau_i}{2}\psi_\mathrm{L}\,, \qquad \frac{\delta\psi_\mathrm{R}}{\delta(g\omega_i)} = 0\,, \tag{9.16}$$

leading to the conserved weak-isospin currents

$$j_i^\mu = \overline{\psi}_\mathrm{L}\gamma^\mu\frac{\tau_i}{2}\psi_\mathrm{L} \quad (i = 1,\, 2,\, 3)\,. \tag{9.17}$$

Of course, they act only on the left chiral component. The corresponding conserved charges are the weak-isopin operators

$$T_i = \int \mathrm{d}^3x\, j_i^0(x) = \int \mathrm{d}^3x\, \psi_\mathrm{L}^\dagger\frac{\tau_i}{2}\psi_\mathrm{L}\,, \tag{9.18}$$

or, more explicitly,

$$T_1 = \frac{1}{2}\int \mathrm{d}^3x\,(\nu_\mathrm{L}^\dagger e_\mathrm{L} + e_\mathrm{L}^\dagger\nu_\mathrm{L})\,, \tag{9.19}$$

$$T_2 = -\frac{\mathrm{i}}{2}\int \mathrm{d}^3x\,(\nu_\mathrm{L}^\dagger e_\mathrm{L} - e_\mathrm{L}^\dagger\nu_\mathrm{L})\,, \tag{9.20}$$

$$T_3 = \frac{1}{2}\int \mathrm{d}^3x\,(\nu_\mathrm{L}^\dagger \nu_\mathrm{L} - e_\mathrm{L}^\dagger e_\mathrm{L})\,. \tag{9.21}$$

From the usual canonical commutation relations of fermion fields at equal times and the familiar commutation relations of the Pauli matrices,

$$[\tau_i,\, \tau_j] = 2\,\mathrm{i}\epsilon_{ijk}\,\tau_k\,, \tag{9.22}$$

it is a simple exercise to prove that

$$[T_i,\, T_j] = \mathrm{i}\epsilon_{ijk}\,T_k\,, \tag{9.23}$$

or, alternatively,

$$[T^{(+)},\, T^{(-)}] = 2\,T_3\,, \qquad [T^{(\pm)},\, T_3] = \mp T^{(\pm)}\,, \tag{9.24}$$

using the definition of the raising and lowering operators $T^{(\pm)} \equiv T_1 \pm \mathrm{i}T_2$.

A mass term of the kind

$$-m\,\overline{e}e = -m(\overline{e}_\mathrm{R}e_\mathrm{L} + \overline{e}_\mathrm{L}e_\mathrm{R}) \tag{9.25}$$

would break $\mathrm{SU_L}(2)$ invariance. So that for $\mathcal{L}_0$ to have weak-isospin symmetry, both fermions, the electron and the neutrino, must be massless.

**Weak Hypercharge.** Evidently, the Lagrangian (12) is also invariant to general phase transformations

$$U(\omega) = \mathrm{e}^{-\mathrm{i}\omega f F},$$

which form a U(1) group for a given quantum number matrix $F$ acting as generator. A constant $f$, to be identified with a coupling constant, has been separated from the parameter $\omega$. For infinitesimal transformations, we have $U \approx 1 - \mathrm{i}\omega f F$ and

$$\begin{aligned} \psi &\rightarrow \psi' = U(\omega)\, \psi \approx \psi + \delta\psi\,, \\ \delta\psi &= -\,\mathrm{i}\omega f\, F\psi\,. \end{aligned} \tag{9.26}$$

The order of the matrix $F$ is given by the dimension of the representation $\psi$. Substituting the derivatives

$$\frac{\delta\psi_{\mathrm{L}}}{\delta(f\omega)} = -\mathrm{i}F\,\psi_{\mathrm{L}}\,, \quad \frac{\delta\psi_{\mathrm{R}}}{\delta(f\omega)} = -\mathrm{i}F\,\psi_{\mathrm{R}} \tag{9.27}$$

into (15) yields the conserved current operator

$$j_\mu^{\mathrm{F}} = \overline{\psi}_{\mathrm{L}}\gamma_\mu F\psi_{\mathrm{L}} + \overline{\psi}_{\mathrm{R}}\gamma_\mu F\psi_{\mathrm{R}}\,, \tag{9.28}$$

which in turn leads to the associated conserved charge operator

$$\boldsymbol{F} = \int \mathrm{d}^3x\, j_0^{\mathrm{F}}(x) = \int \mathrm{d}^3x\, \left(\psi_{\mathrm{L}}^\dagger F\psi_{\mathrm{L}} + \psi_{\mathrm{R}}^\dagger F\psi_{\mathrm{R}}\right). \tag{9.29}$$

Not any U(1) symmetry of $\mathcal{L}_0$ is compatible with $\mathrm{SU_L}(2)$. When $\boldsymbol{F} = Q$, the electrical charge number operator, we can identify (28) and (29) with the electromagnetic current and charge-number operators, with the usual values of the electrical charges for the neutrino and electron,

$$j_\mu^{\mathrm{em}} = Q_{\mathrm{e}}\,(\overline{e}_{\mathrm{L}}\gamma_\mu e_{\mathrm{L}} + \overline{e}_{\mathrm{R}}\gamma_\mu e_{\mathrm{R}}) = -\overline{e}\gamma_\mu e\,; \tag{9.30}$$

$$Q = \int \mathrm{d}^3x\, j_0^{\mathrm{em}}(x) = -\int \mathrm{d}^3x\, e^\dagger e\,. \tag{9.31}$$

In order for the associated group, $U_{\mathrm{Q}}(1)$, to coexist with $\mathrm{SU_L}(2)$, the charge operator must commute with the isospin operators. But since the two components of the doublet $\psi_{\mathrm{L}}$ have different charges, the charge number is clearly not a good quantum number in $\mathrm{SU_L}(2)$. In other words,

$$\begin{aligned} Q &\equiv Q_{\mathrm{L}} + Q_{\mathrm{R}} \\ &= -\int \mathrm{d}^3x\, \psi_{\mathrm{L}}^\dagger \tfrac{1}{2}\,(1 - \tau_3)\psi_{\mathrm{L}} - \int \mathrm{d}^3x\, e_{\mathrm{R}}^\dagger e_{\mathrm{R}} \end{aligned} \tag{9.32}$$

does not commute with all $T_i = \frac{1}{2}\tau_i$, but rather gives

$$[Q, T_i] = [T_3, T_i] = \mathrm{i}\epsilon_{3ij} T_j \,. \tag{9.33}$$

Therefore, $U_{\mathrm{Q}}(1)$ and $\mathrm{SU_L}(2)$ cannot be simultaneous symmetries of $\mathcal{L}_0$. However, (33) tells us that $\alpha(Q - T_3)$ commutes with $T_i$ for all $i$ and arbitrary constant $\alpha$, and hence may be regarded as the generator of a U(1) group commuting with $\mathrm{SU_L}(2)$. We choose $\alpha = 2$ and call $2(Q - T_3)$ the *weak-hypercharge* operator, $Y$, in analogy with the strong hypercharge which was defined for hadrons. The relation

$$Q = T_3 + \tfrac{1}{2} Y \tag{9.34}$$

provides then a connection between electricity and the weak interaction for it gives the electric charge of an electron or associated neutrino (or, as it turns out, of any weakly interacting particle) in terms of its weak-isospin $z$ component and its weak hypercharge. An explicit expression for $Y$ may be obtained from (21) and (32) as follows:

$$\begin{aligned} Y &= -\int \mathrm{d}^3x\, \psi_{\mathrm{L}}^\dagger (1 - \tau_3)\psi_{\mathrm{L}} - 2\int \mathrm{d}^3x\, \psi_{\mathrm{R}}^\dagger \psi_{\mathrm{R}} - \int \mathrm{d}^3x\, \psi_{\mathrm{L}}^\dagger \tau_3 \psi_{\mathrm{L}} \\ &= -\int \mathrm{d}^3x\, \psi_{\mathrm{L}}^\dagger \psi_{\mathrm{L}} - 2\int \mathrm{d}^3x\, \psi_{\mathrm{R}}^\dagger \psi_{\mathrm{R}} \,. \end{aligned} \tag{9.35}$$

Identifying this result with (29), $\boldsymbol{F} = Y$, we have for the electron family

$$Y_{\mathrm{L}} = -1 \quad \text{and} \quad Y_{\mathrm{R}} = -2 \,. \tag{9.36}$$

For a given isomultiplet, $Y = 2\langle Q - T_3\rangle = 2\langle Q\rangle$, i.e. twice the average charge of the multiplet. The corresponding conserved current

$$j_\mu^{\mathrm{Y}} = Y_{\mathrm{L}}\, \overline{\psi}_{\mathrm{L}}\gamma_\mu\psi_{\mathrm{L}} + Y_{\mathrm{R}}\, \overline{\psi}_{\mathrm{R}}\gamma_\mu\psi_{\mathrm{R}} \tag{9.37}$$

is simply related to the electromagnetic current and the isospin current by

$$j_\mu^{\mathrm{em}} = j_\mu^3 + \tfrac{1}{2}\, j_\mu^{\mathrm{Y}} \,. \tag{9.38}$$

To summarize, the free-lepton Lagrangian $\mathcal{L}_0$ is invariant under the direct product group $\mathrm{SU_L}(2) \times \mathrm{U_Y}(1)$ of global transformations

$$\mathrm{SU_L}(2): \qquad U_2(\omega) = \mathrm{e}^{-\mathrm{i}g\omega_i \frac{1}{2}\tau_i} \,, \tag{9.39}$$

$$\mathrm{U_Y}(1): \qquad U_1(\omega) = \mathrm{e}^{-\mathrm{i}\frac{1}{2} g'\omega Y} \,, \tag{9.40}$$

where $g$ and $g'$ are eventually identified with coupling constants. We show in Table 9.1 the classification and the assigned quantum numbers of the electron family in $\mathrm{SU_L}(2) \times \mathrm{U_Y}(1)$. It is the symmetry group to be gauged.

**Table 9.1.** Classification and assigned quantum numbers of the electron family

| | $T$ | $T_3$ | $Y$ | $Q$ |
|---|---|---|---|---|
| $\begin{pmatrix} \nu_{\mathrm{eL}} \\ e_{\mathrm{L}} \end{pmatrix}$ | $\frac{1}{2}$ | $\pm\frac{1}{2}$ | $-1$ | $0$<br>$-1$ |
| $e_{\mathrm{R}}$ | $0$ | $0$ | $-2$ | $-1$ |

### 9.2.2 Gauge Invariance

We now proceed to transform the free-particle Lagrangian $\mathcal{L}_0$ into an interacting particle model by applying the gauge invariance principle (Chap. 8). This means four vector boson fields corresponding to the four generators of SU(2) × U(1) will have to be introduced. One of these fields will remain massless to generate the electromagnetic force, and the remaining three will acquire mass via the Higgs mechanism so as to produce the observed short-range weak forces. To make this mechanism work properly, one needs at least one weak-isospin doublet of complex scalar fields, one of which is electrically neutral. These fields interact with each other via self-coupling so that hiding gauge invariance becomes feasible, and also with the electron field in order to eventually give it a mass.

The free-field Lagrangian $\mathcal{L}_0$ is accordingly replaced by its corresponding gauge-invariant form

$$\mathcal{L}_\ell = \overline{\psi}_{\mathrm{L}}\, \mathrm{i}\gamma^\mu D^{\mathrm{L}}_\mu\, \psi_{\mathrm{L}} + \overline{\psi}_{\mathrm{R}}\, \mathrm{i}\gamma^\mu D^{\mathrm{R}}_\mu\, \psi_{\mathrm{R}}\,, \tag{9.41}$$

where the covariant derivatives of fields are

$$D^{\mathrm{L}}_\mu \psi_{\mathrm{L}} = \left(\partial_\mu + \mathrm{i}gA_{i\mu}\frac{\tau_i}{2} + \mathrm{i}g' B_\mu \frac{Y_{\mathrm{L}}}{2}\right)\psi_{\mathrm{L}}\,, \tag{9.42}$$

$$D^{\mathrm{R}}_\mu \psi_{\mathrm{R}} = \left(\partial_\mu + \mathrm{i}g' B_\mu \frac{Y_{\mathrm{R}}}{2}\right)\psi_{\mathrm{R}}\,. \tag{9.43}$$

Here $A_{i\mu}$ are the three vector gauge fields associated with $\mathrm{SU_L}(2)$, and $B_\mu$ is the $\mathrm{U_Y}(1)$ gauge field. The dynamics of these fields is contained in the Lagrangian

$$\mathcal{L}_{\mathrm{G}} = -\frac{1}{4}\, W^i_{\mu\nu} W_i^{\mu\nu} - \frac{1}{4}\, B_{\mu\nu} B^{\mu\nu}\,, \tag{9.44}$$

where

$$W^i_{\mu\nu} = \partial_\mu A^i_\nu - \partial_\nu A^i_\mu - g\epsilon_{ijk}\, A^j_\mu A^k_\nu\,, \tag{9.45}$$

$$B_{\mu\nu} = \partial_\mu B_\nu - \partial_\nu B_\mu\,. \tag{9.46}$$

The Lagrangian $\mathcal{L}_\ell$ and $\mathcal{L}_{\mathrm{G}}$ are invariant to the $\mathrm{SU_L}(2) \times \mathrm{U_Y}(1)$ group of local transformations $U_2[\omega(x)]$ and $U_1[\omega(x)]$ with space-time coordinate-dependent parameters. The transformed fields are given by

$$\psi'_{\mathrm{L}} = U_2 U_1 \psi_{\mathrm{L}}\,, \qquad \psi'_{\mathrm{R}} = U_1 \psi_{\mathrm{R}}\,; \tag{9.47}$$

$$B'_\mu = B_\mu + \partial_\mu \omega\,; \tag{9.48}$$

$$\boldsymbol{A}'_\mu = U_2 \boldsymbol{A}_\mu U_2^\dagger + \frac{\mathrm{i}}{g}(\partial_\mu U_2) U_2^\dagger \approx \boldsymbol{A}_\mu + \partial_\mu \boldsymbol{\omega} + \mathrm{i}g\,[\boldsymbol{A}_\mu, \boldsymbol{\omega}]\,; \tag{9.49}$$

where $\boldsymbol{A}_\mu = \frac{1}{2}\,\tau_i A^i_\mu$ is the Hermitian gauge field matrix.

In order to eventually hide gauge invariance, two complex scalar fields forming an SU(2) doublet (having weak hypercharge called $Y_{\mathrm{H}}$) are now introduced:

$$\phi = \begin{pmatrix} \varphi^+ \\ \varphi^0 \end{pmatrix} . \tag{9.50}$$

Their dynamics in a self-coupling potential is represented by the gauge-invariant Lagrangian

$$\mathcal{L}_s = (D_\mu \phi)^\dagger (D^\mu \phi) - V(\phi) \equiv (D_\mu \phi)^\dagger (D^\mu \phi) - \mu^2 \phi^\dagger \phi - \lambda (\phi^\dagger \phi)^2. \tag{9.51}$$

Here the covariant derivatives of the scalar fields are given by

$$\begin{aligned} D_\mu \phi &= \left( \partial_\mu + \mathrm{i} g \boldsymbol{A}_\mu + \mathrm{i} g' B_\mu \frac{Y_{\mathrm{H}}}{2} \right) \phi , \\ (D_\mu \phi)^\dagger &= \partial_\mu \phi^\dagger - \mathrm{i} g \phi^\dagger \boldsymbol{A}_\mu - \mathrm{i} g' B_\mu \frac{Y_{\mathrm{H}}}{2} \phi^\dagger . \end{aligned} \tag{9.52}$$

Finally, with a view to generating the electron mass, we introduce a gauge-invariant Yukawa coupling between scalars and fermions,

$$\mathcal{L}_{\ell Y} = -C_{\mathrm{e}} \left[ \overline{\psi}_{\mathrm{R}} (\phi^\dagger \psi_{\mathrm{L}}) + (\overline{\psi}_{\mathrm{L}} \phi) \psi_{\mathrm{R}} \right] , \tag{9.53}$$

where $C_{\mathrm{e}}$, an additional parameter, gives the strength of this coupling. $\mathcal{L}_{\ell Y}$ is evidently invariant under $\mathrm{SU_L}(2)$ – which is precisely why we need a doublet of scalars – while its invariance under $\mathrm{U_Y}(1)$ is guaranteed by requiring that $\phi$ have weak hypercharge $Y_{\mathrm{H}} = Y_{\mathrm{L}} - Y_{\mathrm{R}}$, that is, $Y_{\mathrm{H}} = 1$. From this assignment and $Q = T_3 + Y/2$, it follows that $\varphi^+$ has charge $Q = +1$ and $\varphi^0$ has charge $Q = 0$. The presence of such an electrically neutral member in the doublet makes it possible for $\phi$ to develop a $U_{\mathrm{Q}}(1)$-invariant expectation value and for one gauge boson to remain massless.

### 9.2.3 Spontaneous Symmetry Breaking

As we have already discussed in Sect. 8.6, the potential $V(\phi)$ with positive $\lambda$ and negative $\mu^2$ has minima, $(\partial V / \partial \phi) = 0$, at values of $\phi$ given by

$$|\phi|^2 = -\frac{\mu^2}{2\lambda} \equiv \frac{v^2}{2} , \tag{9.54}$$

so that when the scalar doublet develops a vacuum expectation value

$$\langle 0 | \phi | 0 \rangle = \boldsymbol{v} = \begin{pmatrix} 0 \\ v/\sqrt{2} \end{pmatrix} , \tag{9.55}$$

with real constant $v = \sqrt{-\mu^2/\lambda}$, we have spontaneous symmetry breakdown. It is clear then that neither $T_i$ nor $Y$ cancels $\boldsymbol{v}$. In particular,

$$T_3\,\boldsymbol{v} = -\tfrac{1}{2}\,\boldsymbol{v}\,,$$
$$Y\boldsymbol{v} = Y_{\mathrm{H}}\,\boldsymbol{v} = \boldsymbol{v}\,;$$

but

$$Q\boldsymbol{v} = (T_3 + \tfrac{1}{2}Y)\,\boldsymbol{v} = 0\,. \tag{9.56}$$

Thus, SU(2) and U(1) are completely broken separately, but the product group SU(2) × U(1) is not: after symmetry breaking there remains a residual symmetry generated by $Q$. This pattern of symmetry breakdown is described by the reduction equation

$$\mathrm{SU(2)_L} \times \mathrm{U_Y(1)} \to \mathrm{U_Q(1)}\,. \tag{9.57}$$

It proves convenient to reparameterize $\boldsymbol{\phi}$ in polar field variables

$$\boldsymbol{\phi} = \begin{pmatrix} \varphi^+ \\ \varphi^0 \end{pmatrix} = \exp\left(\frac{\mathrm{i}}{v}\sum \xi_i T_i\right) \begin{pmatrix} 0 \\ \frac{1}{\sqrt{2}}(v+H) \end{pmatrix}, \tag{9.58}$$

so that the original two complex scalar fields $\varphi^+$ and $\varphi^0$ are replaced by four real scalar fields $H$, $\xi_1$, $\xi_2$, and $\xi_3$. All these fields have zero vacuum expectation values:

$$\langle 0 \,|\, \xi_i \,|\, 0 \rangle = \langle 0 \,|\, H \,|\, 0 \rangle = 0\,. \tag{9.59}$$

We will now reformulate the model in the *unitary gauge* where the three would-be Goldstone bosons $\xi_i$ are transformed away, bringing out in a particularly transparent way the spectra and interactions of the remaining physical particles. First, apply the unitary transformation

$$\mathcal{S} = \exp\left(-\frac{\mathrm{i}}{v}\sum \xi_i T_i\right) \tag{9.60}$$

on all fields, resulting in the transformed fields

$$\boldsymbol{\phi}' = \mathcal{S}\boldsymbol{\phi} = \frac{1}{\sqrt{2}}[v + H(x)]\chi\,, \quad \text{with} \quad \chi = \begin{pmatrix} 0 \\ 1 \end{pmatrix};$$
$$\psi'_{\mathrm{L}} = \mathcal{S}\psi_{\mathrm{L}}\,; \qquad \psi'_{\mathrm{R}} = \psi_{\mathrm{R}}\,;$$
$$B'_\mu = B_\mu\,;$$
$$\boldsymbol{A}'_\mu = \mathcal{S}\boldsymbol{A}_\mu\,\mathcal{S}^\dagger + \frac{\mathrm{i}}{g}\,(\partial_\mu \mathcal{S})\,\mathcal{S}^\dagger\,. \tag{9.61}$$

The Lagrangian of the model is of course invariant to these transformations. In terms of the new fields, its different parts become

$$\mathcal{L}_s = (D'_\mu \phi')^\dagger (D'^\mu \phi') - \mu^2 \phi'^\dagger \phi' - \lambda \left(\phi'^\dagger \phi'\right)^2 ; \tag{9.62}$$

$$\mathcal{L}_G = -\frac{1}{4} W'^i_{\mu\nu} W'^{\mu\nu}_i - \frac{1}{4} B'_{\mu\nu} B'^{\mu\nu} ; \tag{9.63}$$

$$\mathcal{L}_\ell = \overline{\psi}'_L \mathrm{i}\gamma^\mu D'^L_\mu \psi'_L + \overline{\psi}'_R \mathrm{i}\gamma^\mu D'^R_\mu \psi'_R ; \tag{9.64}$$

$$\mathcal{L}_{\ell Y} = -C_e \left[ \overline{\psi}'_R (\phi'^\dagger \psi'_L) + (\overline{\psi}'_L \phi') \psi'_R \right] . \tag{9.65}$$

Each of these parts is now examined in turn. For brevity we will drop the prime accents on the field symbols, identifying for example $\psi'$ with $\psi$.

**Scalar Fields.** The main role of the scalar fields is to generate masses for the gauge bosons and the electron. The gauge field matrix may be written explicitly as

$$\begin{aligned} \boldsymbol{A}_\mu = \tfrac{1}{2} \tau_i A_{i\mu} &= \tfrac{1}{2} (\tau_1 A_{1\mu} + \tau_2 A_{2\mu} + \tau_3 A_{3\mu}) \\ &= \tfrac{1}{\sqrt{2}} (\tau_+ W_\mu + \tau_- W^\dagger_\mu) + \tfrac{1}{2} \tau_3 A_{3\mu} , \end{aligned}$$

with the definitions $\tau_\pm = \frac{1}{2}(\tau_1 \pm \mathrm{i}\tau_2)$ and $W_\mu = \frac{1}{\sqrt{2}}(A_{1\mu} - \mathrm{i}A_{2\mu})$. Then the covariant derivative of the scalar is

$$\begin{aligned} D_\mu \phi &= \left( \partial_\mu + \mathrm{i}g \, \boldsymbol{A}_\mu + \mathrm{i}g' \, B_\mu \frac{Y_H}{2} \right) \frac{v + H}{\sqrt{2}} \chi \\ &= \frac{1}{\sqrt{2}} \begin{pmatrix} \frac{1}{\sqrt{2}} \mathrm{i}g \, W_\mu (v + H) \\ \partial_\mu H - \frac{1}{2}\mathrm{i}(g \, A_{3\mu} - g' \, B_\mu)(v + H) \end{pmatrix} . \end{aligned} \tag{9.66}$$

The vector meson masses are found in the 'kinetic term'

$$\begin{aligned} &(D_\mu \phi)^\dagger (D_\mu \phi) \\ &= \tfrac{1}{4} g^2 (v + H)^2 W^\dagger_\mu W^\mu + \tfrac{1}{2} \left[ \partial_\mu H \, \partial^\mu H + \tfrac{1}{4} (v + H)^2 (g \, A_{3\mu} - g' \, B_\mu)^2 \right] \\ &= \tfrac{1}{4} g^2 v^2 \, W^\dagger_\mu W^\mu + \tfrac{1}{8} v^2 (g A_{3\mu} - g' B_\mu)^2 + \tfrac{1}{2} \partial_\mu H \, \partial^\mu H \\ &\quad + \tfrac{1}{4} (2vH + H^2) \left[ g^2 \, W^\dagger_\mu W^\mu + \tfrac{1}{2} (g \, A_{3\mu} - g' \, B_\mu)^2 \right] . \end{aligned} \tag{9.67}$$

In general, the expected mass term for a complex vector field is of the form $M^2_W \, W^\dagger_\mu W^\mu$. So that the mass for the charged vector mesons can be read off from (67):

$$M_W = \frac{1}{2} g v . \tag{9.68}$$

On the other hand, the quadratic terms in the neutral fields,

$$\frac{1}{8}v^2\,(g\,A_{3\mu} - g'\,B_\mu)^2\,, \tag{9.69}$$

contain the nondiagonal matrix

$$\frac{1}{8}\,v^2\begin{pmatrix} g^2 & -gg' \\ -gg' & g'^2 \end{pmatrix}.$$

To diagonalize it, we introduce the orthogonal combinations that give the mass eigenstates for the two neutral fields

$$A_\mu = \sin\theta_\mathrm{W}\;A_{3\mu} + \cos\theta_\mathrm{W}\;B_\mu\,, \tag{9.70}$$

$$Z_\mu = \cos\theta_\mathrm{W}\;A_{3\mu} - \sin\theta_\mathrm{W}\;B_\mu\,; \tag{9.71}$$

or, inversely,

$$A_{3\mu} = \sin\theta_\mathrm{W}\;A_\mu + \cos\theta_\mathrm{W}\;Z_\mu\,, \tag{9.72}$$

$$B_\mu = \cos\theta_\mathrm{W}\;A_\mu - \sin\theta_\mathrm{W}\;Z_\mu\,; \tag{9.73}$$

where $\theta_\mathrm{W}$ is a mixing angle (called the Weinberg angle) yet to be determined. Substituting these expressions into (69) yields

$$\begin{aligned} &\tfrac{1}{8}v^2\,(g\,A_{3\mu} - g'\,B_\mu)^2 \\ &\quad = \tfrac{1}{8}v^2\left[A_\mu^2\,(g\sin\theta_\mathrm{W} - g'\cos\theta_\mathrm{W})^2 + Z_\mu^2\,(g\cos\theta_\mathrm{W} + g'\sin\theta_\mathrm{W})^2\right. \\ &\qquad \left. + \,2A_\mu\,Z^\mu\,(g\sin\theta_\mathrm{W} - g'\cos\theta_\mathrm{W})(g\cos\theta_\mathrm{W} + g'\sin\theta_\mathrm{W})\right]. \end{aligned} \tag{9.74}$$

As we have seen in (57), $\mathrm{U_Q}(1)$ is unbroken and the associated gauge boson (the photon) remains massless. If we let the corresponding field be $A_\mu$, the diagonalization of (74) yields the condition

$$g\sin\theta_\mathrm{W} = g'\cos\theta_\mathrm{W}\,. \tag{9.75}$$

Thus, the mixing angle $\theta_\mathrm{W}$ gives a measure of the relative strength of the SU(2) and U(1) group factors; it may be calculated from the relations

$$\cos\theta_\mathrm{W} = \frac{g}{\sqrt{g^2+g'^2}}\,; \qquad \sin\theta_\mathrm{W} = \frac{g'}{\sqrt{g^2+g'^2}}\,. \tag{9.76}$$

Since these functions will recur again and again in the following, a simplified notation is called for:

$$c_\mathrm{W} \equiv \cos\theta_\mathrm{W}\,, \qquad s_\mathrm{W} \equiv \sin\theta_\mathrm{W}\,. \tag{9.77}$$

The quadratic form (74) must reduce to the expected mass term for a neutral vector field, $\frac{1}{2}\,M_\mathrm{Z}^2 Z_\mu Z^\mu$, so that the mass of the field $Z_\mu$ is

$$M_\mathrm{Z} = \frac{1}{2}v\,(g c_\mathrm{W} + g' s_\mathrm{W}) = \frac{1}{2}v\sqrt{g^2+g'^2} = \frac{gv}{2\,c_\mathrm{W}}\,. \tag{9.78}$$

Note that the masses of the two weak gauge fields satisfy the identity

$$M_{\mathrm{W}} = c_{\mathrm{W}}\, M_{\mathrm{Z}} \,. \tag{9.79}$$

On the other hand, the potential becomes after symmetry breaking

$$\begin{aligned} V(\phi) &= \tfrac{1}{2}\,\mu^2 (v+H)^2\,(\chi^\dagger\chi) + \tfrac{1}{4}\lambda\,(v+H)^4\,(\chi^\dagger\chi)^2 \\ &= \tfrac{1}{4}\mu^2 v^2 - \mu^2 H^2 + \lambda\,(vH^3 + \tfrac{1}{4}H^4)\,, \end{aligned} \tag{9.80}$$

where $v^2 = -\mu^2/\lambda$, from which the mass of the surviving scalar can be identified:

$$M_{\mathrm{H}}^2 = -2\mu^2 \,. \tag{9.81}$$

The Lagrangian $\mathcal{L}_{\mathrm{s}}$ thus becomes in the unitary gauge

$$\begin{aligned} \mathcal{L}_{\mathrm{s}} &= (D_\mu\phi)^\dagger (D_\mu\phi) - V(\phi) \\ &= \frac{1}{2}(\partial_\mu H \partial^\mu H - M_{\mathrm{H}}^2 H^2) - \frac{g M_{\mathrm{H}}^2}{4\,M_{\mathrm{W}}}\,H^3 - \frac{g^2 M_{\mathrm{H}}^2}{32\,M_{\mathrm{W}}^2}\,H^4 \\ &\quad + g M_{\mathrm{W}}\left(H + \frac{g}{4M_{\mathrm{W}}}H^2\right) W_\mu^\dagger W^\mu + M_{\mathrm{W}}^2 W_\mu^\dagger W^\mu \\ &\quad + \frac{1}{2}\,\frac{g M_{\mathrm{Z}}}{c_{\mathrm{W}}}\left(H + \frac{g}{4c_{\mathrm{W}}\,M_{\mathrm{Z}}}H^2\right) Z_\mu Z^\mu + \frac{1}{2}\,M_{\mathrm{Z}}^2\,Z_\mu Z^\mu \,. \end{aligned} \tag{9.82}$$

Here we have replaced the original parameters by the particle masses; in particular

$$v = \frac{2M_{\mathrm{W}}}{g} = \frac{2M_{\mathrm{Z}}}{\sqrt{g^2+g'^2}}\,; \qquad \lambda = -\frac{\mu^2}{v^2} = \frac{M_{\mathrm{H}}^2}{2\,v^2} = \frac{g^2 M_{\mathrm{H}}^2}{8\,M_{\mathrm{W}}^2}\,. \tag{9.83}$$

The field $H$, being electrically neutral, is not coupled to the electromagnetic field, but $\mathcal{L}_{\mathrm{s}}$ is nonetheless $U_{\mathrm{Q}}(1)$-invariant.

**Gauge Fields.** The Lagrangian for the gauge fields $\mathcal{L}_{\mathrm{G}}$ will be split into free-field and interacting-field parts:

$$\mathcal{L}_{\mathrm{G}} = \mathcal{L}_{\mathrm{G}}^0 + \mathcal{L}_{\mathrm{G}}^1 + \mathcal{L}_{\mathrm{G}}^2\,; \tag{9.84}$$

$$\mathcal{L}_{\mathrm{G}}^0 = -\frac{1}{4}\,A_{\mu\nu}^i A_i^{\mu\nu} - \frac{1}{4}\,B_{\mu\nu}B^{\mu\nu}\,, \tag{9.85}$$

$$\mathcal{L}_{\mathrm{G}}^1 = \frac{1}{2}\,g\epsilon_{ijk}\,A_\mu^j A_\nu^k\,A_i^{\mu\nu}\,, \tag{9.86}$$

$$\mathcal{L}_{\mathrm{G}}^2 = -\frac{1}{4}\,g^2\epsilon_{ijk}\epsilon_{i\ell m}\,A_\mu^j A_\nu^k A_\ell^\mu A_m^\nu\,. \tag{9.87}$$

The two interacting-field terms are characteristic of non-Abelian theories, with the structure constants of SU(2) algebra. Here the customary symbols for field strengths have been used:

$$A^i_{\mu\nu} \equiv \partial_\mu A^i_\nu - \partial_\nu A^i_\mu\,, \qquad B_{\mu\nu} \equiv \partial_\mu B_\nu - \partial_\nu B_\mu\,. \tag{9.88}$$

Our main task is to re-express $\mathcal{L}_{\mathrm{G}}$ in terms of the mass eigenstates $A_\mu$, $W_\mu$, and $Z_\mu$. It is convenient for this purpose to introduce their respective field strengths:

$$\begin{aligned} A_{\mu\nu} &= \partial_\mu A_\nu - \partial_\nu A_\mu\,, \\ W_{\mu\nu} &= \partial_\mu W_\nu - \partial_\nu W_\mu\,, \qquad Z_{\mu\nu} = \partial_\mu Z_\nu - \partial_\nu Z_\mu\,. \end{aligned} \tag{9.89}$$

Noting that

$$\begin{aligned} A^1_{\mu\nu}A_1^{\mu\nu} + A^2_{\mu\nu}A_2^{\mu\nu} &= 2\,W^\dagger_{\mu\nu}W^{\mu\nu}\,, \\ A^3_{\mu\nu}A_3^{\mu\nu} + B_{\mu\nu}B^{\mu\nu} &= A_{\mu\nu}A^{\mu\nu} + Z_{\mu\nu}Z^{\mu\nu}\,, \end{aligned}$$

we readily get the *kinetic part*

$$\mathcal{L}^0_{\mathrm{G}} = -\frac{1}{2}\,W^\dagger_{\mu\nu}W^{\mu\nu} - \frac{1}{4}\,Z_{\mu\nu}Z^{\mu\nu} - \frac{1}{4}\,A_{\mu\nu}A^{\mu\nu}\,. \tag{9.90}$$

The *three-field coupling* $\mathcal{L}^1_{\mathrm{G}}$ contains two factors, the first of which, $\epsilon_{ijk}\,A^j_\mu A^k_\nu$, gives the terms

$$\begin{aligned} \epsilon_{1jk}A_{j\mu}A_{k\nu} &= A_{2\mu}A_{3\nu} - A_{2\nu}A_{3\mu} \\ &= -\tfrac{\mathrm{i}}{\sqrt{2}}\,\left[(W^\dagger_\mu - W_\mu)(s_{\mathrm{W}}A_\nu + c_{\mathrm{W}}Z_\nu) - (W^\dagger_\nu - W_\nu)(s_{\mathrm{W}}A_\mu + c_{\mathrm{W}}Z_\mu)\right], \\ \epsilon_{2jk}A_{j\mu}A_{k\nu} &= A_{3\mu}A_{1\nu} - A_{3\nu}A_{1\mu} \\ &= \tfrac{1}{\sqrt{2}}\,\left[-(W^\dagger_\mu + W_\mu)(s_{\mathrm{W}}A_\nu + c_{\mathrm{W}}Z_\nu) + (W^\dagger_\nu + W_\nu)(s_{\mathrm{W}}A_\mu + c_{\mathrm{W}}Z_\mu)\right], \\ \epsilon_{3jk}A_{j\mu}A_{k\nu} &= A_{1\mu}A_{2\nu} - A_{1\nu}A_{2\mu} = \mathrm{i}(W^\dagger_\mu W_\nu - W^\dagger_\nu W_\mu)\,; \end{aligned}$$

and the second, $A_i^{\mu\nu}$, the following:

$$\begin{aligned} A^1_{\mu\nu} &= \tfrac{1}{\sqrt{2}}(W^\dagger_{\mu\nu} + W_{\mu\nu})\,, \\ A^2_{\mu\nu} &= -\tfrac{\mathrm{i}}{\sqrt{2}}(W^\dagger_{\mu\nu} - W_{\mu\nu})\,, \\ A^3_{\mu\nu} &= c_{\mathrm{W}}Z_{\mu\nu} + s_{\mathrm{W}}A_{\mu\nu}\,. \end{aligned}$$

Together, they lead to

$$\begin{aligned} \mathcal{L}^1_{\mathrm{G}} = \frac{1}{2}\,g\,\epsilon_{ijk}\,A_{j\mu}A_{k\nu}A_i^{\mu\nu} &= \mathrm{i}gW^{\mu\dagger}W^\nu(s_{\mathrm{W}}A_{\mu\nu} + c_{\mathrm{W}}Z_{\mu\nu}) \\ &\quad + \mathrm{i}g\,(W^\mu W^\dagger_{\mu\nu} - W^{\mu\dagger}W_{\mu\nu})\,(s_{\mathrm{W}}A^\nu + c_{\mathrm{W}}Z^\nu)\,. \end{aligned} \tag{9.91}$$

As for the *four-field coupling* $\mathcal{L}_{\rm G}^2$, it suffices to note that

$$A_{j\mu}A_{j\nu} = A_{1\mu}A_{1\nu} + A_{2\mu}A_{2\nu} + A_{3\mu}A_{3\nu}$$
$$= W_\mu^\dagger W_\nu + W_\mu W_\nu^\dagger + (s_{\rm W} A_\mu + c_{\rm W} Z_\mu)(s_{\rm W} A_\nu + c_{\rm W} Z_\nu)\,. \quad (9.92)$$

Then $\mathcal{L}_{\rm G}^2$ can be rewritten in the desired form

$$\begin{aligned}\mathcal{L}_{\rm G}^2 &= -\frac{1}{4}g^2\epsilon_{ijk}\epsilon_{i\ell m}A_\mu^j A_\nu^k A_\ell^\mu A_m^\nu \\ &= -\frac{1}{4}g^2\left[(A_{j\mu}A_j^\mu)(A_{k\nu}A_k^\nu) - (A_{j\mu}A_j^\nu)(A_{k\nu}A_k^\mu)\right] \\ &= -\frac{1}{2}g^2\left(W_\mu^\dagger W^\mu\, W_\nu^\dagger W^\nu - W^{\mu\dagger}W_\mu^\dagger\, W^\nu W_\nu\right) \\ &\quad - g^2\, W_\mu^\dagger W^\mu (s_{\rm W}^2 A_\nu A^\nu + c_{\rm W}^2 Z_\nu Z^\nu + 2\, s_{\rm W}\, c_{\rm W} A_\nu Z^\nu) \\ &\quad + g^2 W_\mu^\dagger W_\nu\left[s_{\rm W}^2 A^\mu A^\nu + c_{\rm W}^2 Z^\mu Z^\nu + s_{\rm W}c_{\rm W}(A^\mu Z^\nu + A^\nu Z^\mu)\right]. \end{aligned} \quad (9.93)$$

The terms that depend only on the neutral fields (e.g. $A^2Z^2$ or $Z^4$) have canceled out, so that all the remaining terms in $\mathcal{L}_{\rm G}^2$ involve charged bosons. These must be coupled to $A_\mu$ in a $\mathrm{U_Q}(1)$-gauge-invariant way. To check that this is really so, we sum the terms

| | | |
|---|---|---|
| $-\frac{1}{2}\, W_{\mu\nu}^\dagger W^{\mu\nu}$ | from | $\mathcal{L}_{\rm G}^0$, |
| $\mathrm{i}gs_{\rm W}\, A^\nu (W_{\mu\nu}^\dagger W^\mu - W^{\mu\dagger} W_{\mu\nu})$ | from | $\mathcal{L}_{\rm G}^1$, |
| $-g^2 s_{\rm W}^2\, (W_\mu^\dagger W^\mu A_\nu A^\nu - W_\mu^\dagger W_\nu\, A^\mu A^\nu)$ | from | $\mathcal{L}_{\rm G}^2$ |

into a single expression

$$\begin{aligned}&-\tfrac{1}{2}\left[W_{\mu\nu}^\dagger W^{\mu\nu} - 2\,\mathrm{i}gs_{\rm W}\, A^\nu (W_{\mu\nu}^\dagger W^\mu - W^{\mu\dagger} W_{\mu\nu})\right. \\ &\left.+ 2\,(gs_{\rm W})^2\,(W_\mu^\dagger W^\mu A_\nu A^\nu - W_\mu^\dagger A^\mu W_\nu A^\nu)\right] \\ &\quad = -\tfrac{1}{2}\,(D_\mu W_\nu - D_\nu W_\mu)^\dagger (D^\mu W^\nu - D^\nu W^\mu)\,. \end{aligned} \quad (9.94)$$

Here the covariant derivative of $W_\mu$ is defined as

$$D_\mu W_\nu = (\partial_\mu + \mathrm{i}\, gs_{\rm W}\, A_\mu) W_\nu\,. \quad (9.95)$$

It now becomes clear that $W_\mu$ is a field carrying a positive electrical charge $gs_{\rm W}$ interacting with the electromagnetic field via an expected $\mathrm{U_Q}(1)$-invariant coupling. Not surprisingly, both $A_\mu$ and $Z_\mu$ are neutral.

**The Lepton Sector.** We now consider $\mathcal{L}_\ell$ and $\mathcal{L}_{\ell\rm Y}$. We anticipate that after symmetry breaking the electron becomes massive from its coupling to the scalars. It is indeed the case because with

$$\bar\psi_{\rm L}\phi = (\bar\nu_{\rm L} \quad \bar e_{\rm L})\begin{pmatrix}0\\1\end{pmatrix}\frac{v+H}{\sqrt{2}} = \frac{v+H}{\sqrt{2}}\,\bar e_{\rm L}\,, \quad (9.96)$$

the Yukawa couplings become

$$\begin{aligned}\mathcal{L}_{\ell Y} &= -C_e \left[\overline{\psi}_R (\phi^\dagger \psi_L) + (\overline{\psi}_L \phi)\psi_R\right] \\ &= -\frac{C_e}{\sqrt{2}}(v+H)(\bar{e}_R e_L + \bar{e}_L e_R) = -\frac{C_e}{\sqrt{2}}(v+H)\,\bar{e}e\,. \end{aligned} \tag{9.97}$$

The term quadratic in the electron field should be recognized as the Dirac mass term for the electron, $-m_e\,\bar{e}e$, with mass

$$m_e = \frac{C_e v}{\sqrt{2}}\,, \tag{9.98}$$

which may be inverted to give the Yukawa coupling strength

$$C_e = \frac{\sqrt{2}\,m_e}{v} = \frac{g m_e}{\sqrt{2}\,M_W}\,. \tag{9.99}$$

The scalar–electron coupling now reduces to

$$\mathcal{L}_{\ell Y} = -m_e\,\bar{e}e - \frac{g m_e}{2\,M_W}\,H\bar{e}e\,. \tag{9.100}$$

Finally, we come to the gauge-invariant part of the electron and its neutrino

$$\mathcal{L}_\ell = \overline{\psi}_L\,\mathrm{i}\gamma^\mu D_\mu^L\,\psi_L + \overline{\psi}_R\,\mathrm{i}\gamma^\mu D_\mu^R\,\psi_R\,. \tag{9.101}$$

In more detail, one has for the first term on the right-hand side

$$\begin{aligned}\overline{\psi}_L \mathrm{i}\gamma^\mu D_\mu^L\,\psi_L &= \overline{\psi}_L \mathrm{i}\gamma^\mu(\partial_\mu + \mathrm{i}g\boldsymbol{A}_\mu - \tfrac{\mathrm{i}}{2}g' B_\mu)\psi_L \\ &= \overline{\psi}_L \mathrm{i}\gamma^\mu\left[\partial_\mu + \tfrac{\mathrm{i}}{\sqrt{2}}g(W_\mu\tau_+ + W_\mu^\dagger\tau_-) + \tfrac{\mathrm{i}}{2}(gA_{3\mu}\tau_3 - g'B_\mu)\right]\psi_L\,; \end{aligned}$$

and for the second term

$$\begin{aligned}\overline{\psi}_R\,\mathrm{i}\gamma^\mu D_\mu^R\,\psi_R &= \overline{\psi}_R\,\mathrm{i}\gamma^\mu\left(\partial_\mu + \mathrm{i}g'\frac{Y_R}{2}B_\mu\right)\psi_R \\ &= \bar{e}_R \mathrm{i}\gamma^\mu\partial_\mu e_R + g'\,\bar{e}_R\gamma^\mu e_R\,B_\mu\,. \end{aligned}$$

They contain the expected kinetic terms for the two leptons

$$\overline{\psi}_L \mathrm{i}\gamma^\mu\partial_\mu\psi_L + \overline{\psi}_R \mathrm{i}\gamma^\mu\partial_\mu\psi_R = \bar{\nu}_L\,\mathrm{i}\gamma^\mu\,\partial_\mu\nu_L + \bar{e}\,\mathrm{i}\gamma^\mu\,\partial_\mu e \tag{9.102}$$

as well as their various interactions with the gauge bosons. First, we have the charge-changing couplings

$$\begin{aligned}\mathcal{L}_{cc}^\ell &= -\frac{g}{\sqrt{2}}\left[(\overline{\psi}_L\gamma^\mu\tau_+\psi_L)\,W_\mu + (\overline{\psi}_L\gamma^\mu\tau_-\psi_L)\,W_\mu^\dagger\right] \\ &= -\frac{g}{\sqrt{2}}\,(J^{\mu\dagger}W_\mu + J^\mu W_\mu^\dagger)\,, \end{aligned} \tag{9.103}$$

where the charged currents are defined as

$$J_\mu = j^1_\mu - \mathrm{i} j^2_\mu = \overline{\psi}_\mathrm{L}\gamma_\mu \tau_- \psi_\mathrm{L} = \overline{e}_\mathrm{L}\gamma_\mu \nu_\mathrm{L}\,; \tag{9.104}$$

$$J^\dagger_\mu = j^1_\mu + \mathrm{i} j^2_\mu = \overline{\psi}_\mathrm{L}\gamma_\mu \tau_+ \psi_\mathrm{L} = \overline{\nu}_\mathrm{L}\gamma_\mu e_\mathrm{L}\,. \tag{9.105}$$

Next, we reshape the remaining terms which describe the couplings to the neutral gauge fields in a similar form:

$$\begin{aligned}\mathcal{L}^\ell_\mathrm{nc} &= -g\, A_{3\mu}\left(\overline{\psi}_\mathrm{L}\gamma^\mu \frac{\tau_3}{2}\psi_\mathrm{L}\right) - \frac{1}{2}g'\, B_\mu\left(-\overline{\psi}_\mathrm{L}\gamma^\mu\psi_\mathrm{L} - 2\overline{\psi}_\mathrm{R}\gamma^\mu\psi_\mathrm{R}\right)\\ &= -g\, j^3_\mu A^\mu_3 - \frac{1}{2}g'\, j^\mathrm{Y}_\mu B^\mu\,,\end{aligned} \tag{9.106}$$

where we have used (17) and (37). Re-expressing the gauge group eigenstates $A^3_\mu$ and $B_\mu$ in terms of the mass eigenstates $A_\mu$ and $Z_\mu$, the couplings take the form

$$\mathcal{L}^\ell_\mathrm{nc} = -\left(g s_\mathrm{W}\, j^3_\mu + \frac{1}{2}g' c_\mathrm{W}\, j^\mathrm{Y}_\mu\right)A^\mu - \left(g c_\mathrm{W}\, j^3_\mu - \frac{1}{2}g' s_\mathrm{W}\, j^\mathrm{Y}_\mu\right)Z^\mu\,. \tag{9.107}$$

As the various neutral currents are related through $j^\mathrm{em}_\mu = j^3_\mu + \frac{1}{2}j^\mathrm{Y}_\mu$, the vector current to which $A_\mu$ is coupled may be written as

$$\begin{aligned}g\, s_\mathrm{W}\, j^3_\mu + \tfrac{1}{2}g'\, c_\mathrm{W}\, j^\mathrm{Y}_\mu &= g'\, c_\mathrm{W}\, j^\mathrm{em}_\mu + (g\, s_\mathrm{W} - g'\, c_\mathrm{W})\, j^3_\mu\\ &= g s_\mathrm{W}\, j^\mathrm{em}_\mu + \tfrac{1}{2}(g' c_\mathrm{W} - g s_\mathrm{W})\, j^\mathrm{Y}_\mu\,.\end{aligned}$$

The second terms on the right-hand sides of the last two equations vanish by (75), and the first terms should be recognized as the electromagnetic current, $e j^\mathrm{em}_\mu$. From this follows a relation between the coupling constants associated with the original symmetries and the residual symmetry:

$$e = g s_\mathrm{W} = g' c_\mathrm{W} \qquad \text{or} \qquad \frac{1}{e^2} = \frac{1}{g^2} + \frac{1}{g'^2}\,. \tag{9.108}$$

On the other hand, the vector field coupled to $Z^\mu$ is

$$g c_\mathrm{W}\, j^3_\mu - \tfrac{1}{2}g' s_\mathrm{W}\, j^\mathrm{Y}_\mu = g\, c^{-1}_\mathrm{W}\,(j^3_\mu - s^2_\mathrm{W}\, j^\mathrm{em}_\mu) \equiv g\, c^{-1}_\mathrm{W} j^\mathrm{Z}_\mu\,. \tag{9.109}$$

Here we have introduced the weak neutral current

$$\begin{aligned}j^\mathrm{Z}_\mu &= j^3_\mu - s^2_\mathrm{W}\, j^\mathrm{em}_\mu\\ &= \overline{\psi}_\mathrm{L}\gamma_\mu T_3\psi_\mathrm{L} - s^2_\mathrm{W}\left[Q_\mathrm{e}(\overline{e}_\mathrm{L}\gamma_\mu e_\mathrm{L} + \overline{e}_\mathrm{R}\gamma_\mu e_\mathrm{R})\right]\\ &= \overline{\psi}_\mathrm{L}\gamma_\mu Z_\mathrm{L}\psi_\mathrm{L} + \overline{\psi}_\mathrm{R}\gamma_\mu Z_\mathrm{R}\psi_\mathrm{R}\,,\end{aligned} \tag{9.110}$$

where, in analogy with the electric charges, we have defined the 'weak charges'

$$Z_\mathrm{L} = T_{3\mathrm{L}} - Q s^2_\mathrm{W}\,, \quad Z_\mathrm{R} = -Q s^2_\mathrm{W}\,.$$

The weak neutral current is a novel feature of the unified model. It differs from the charged current in many ways – by having a characteristic charge $T_3 - Qs_{\mathrm{W}}^2$, by being diagonal in flavor, and by containing both left and right chiral components of the electron. It also differs from the electromagnetic current in that it involves both neutral and charged lepton fields.

After these transformations, the complete neutral current couplings reduce to a compact expression

$$\mathcal{L}_{\mathrm{nc}}^{\ell} = -e\, j_\mu^{\mathrm{em}} A^\mu - \frac{g}{c_{\mathrm{W}}}\, j_\mu^{\mathrm{Z}} Z^\mu . \tag{9.111}$$

The e.m. coupling $-e\, j_\mu^{\mathrm{em}} A^\mu$, together with the kinetic terms (102),

$$\overline{\nu}_{\mathrm{L}} \mathrm{i}\gamma^\mu \partial_\mu \nu_{\mathrm{L}} + \overline{e} \mathrm{i}\gamma^\mu \partial_\mu e + e\, \overline{e}\gamma_\mu e\, A^\mu = \overline{\nu}_{\mathrm{L}} \mathrm{i}\gamma^\mu \partial_\mu \nu_{\mathrm{L}} + \overline{e} \mathrm{i}\gamma^\mu (\partial_\mu - \mathrm{i}e\, A_\mu) e ,$$

makes explicit the familiar minimal coupling and the $\mathrm{U_Q}(1)$ gauge invariance of the lepton Lagrangian.

### 9.2.4 Feynman Rules for One-Lepton Family

For convenience we now gather together the results of this section, rewriting them in a slightly more logical arrangement. The $\mathrm{W}^\pm$ and Z mass terms from $\mathcal{L}_{\mathrm{s}}$ are joined with $\mathcal{L}_{\mathrm{G}}$ to give

$$\begin{aligned} \mathcal{L}_{\mathrm{G}}^{\prime 0} = & -\frac{1}{2} W_{\mu\nu}^\dagger W^{\mu\nu} + M_{\mathrm{W}}^2 W_\mu^\dagger W^\mu \\ & -\frac{1}{4} Z_{\mu\nu} Z^{\mu\nu} + \frac{1}{2} M_{\mathrm{Z}}^2 Z_\mu Z^\mu - \frac{1}{4} A_{\mu\nu} A^{\mu\nu}; \end{aligned} \tag{9.112}$$

$$\begin{aligned} \mathcal{L}_{\mathrm{G}}^{1} = & \mathrm{i}g s_{\mathrm{W}} \left( W^{\mu\dagger} W^\nu A_{\mu\nu} + W_{\mu\nu}^\dagger W^\mu A^\nu - W_{\mu\nu} W^{\mu\dagger} A^\nu \right) \\ & + \mathrm{i}g c_{\mathrm{W}} \left( W^{\mu\dagger} W^\nu Z_{\mu\nu} + W_{\mu\nu}^\dagger W^\mu Z^\nu - W_{\mu\nu} W^{\mu\dagger} Z^\nu \right) ; \end{aligned} \tag{9.113}$$

$$\begin{aligned} \mathcal{L}_{\mathrm{G}}^{2} = & -(g s_{\mathrm{W}})^2 \left( W_\mu^\dagger W^\mu A_\nu A^\nu - W_\mu^\dagger W_\nu A^\mu A^\nu \right) \\ & -(g c_{\mathrm{W}})^2 \left( W_\mu^\dagger W^\mu Z_\nu Z^\nu - W_\mu^\dagger W_\nu Z^\mu Z^\nu \right) \\ & -(g s_{\mathrm{W}})(g c_{\mathrm{W}}) \left[ 2 W_\mu^\dagger W^\mu A_\nu Z^\nu - W_\mu^\dagger W_\nu (A^\mu Z^\nu + A^\nu Z^\mu) \right] \\ & + \frac{1}{2} g^2 \left( W_\mu^\dagger W^{\mu\dagger} W_\nu W^\nu - W_\mu^\dagger W^\mu W_\nu^\dagger W^\nu \right) . \end{aligned} \tag{9.114}$$

The original Lagrangian for scalars, minus the $\mathrm{W}^\pm$ and Z mass terms, represents the Higgs field and its interactions with the vector bosons:

$$\begin{aligned} \mathcal{L}_{\mathrm{s}}^{\prime} = & \frac{1}{2} (\partial_\mu H \partial^\mu H - M_{\mathrm{H}}^2 H^2) - \frac{1}{3!} \frac{3\, g M_{\mathrm{H}}^2}{2\, M_{\mathrm{W}}} H^3 - \frac{1}{4!} \frac{3\, g^2 M_{\mathrm{H}}^2}{4\, M_{\mathrm{W}}^2} H^4 \\ & + g M_{\mathrm{W}} \left( H + \frac{g}{2\, M_{\mathrm{W}}} \frac{1}{2} H^2 \right) W_\mu^\dagger W^\mu \\ & + \frac{1}{2} \frac{g M_{\mathrm{Z}}}{c_{\mathrm{W}}} \left( H + \frac{g}{2\, c_{\mathrm{W}} M_{\mathrm{Z}}} \frac{1}{2} H^2 \right) Z_\mu Z^\mu . \end{aligned} \tag{9.115}$$

The Yukawa coupling, minus the electron mass term, reduces to the electron–Higgs coupling

$$\mathcal{L}'_{\ell\mathrm{H}} = -\frac{g m_\mathrm{e}}{2\,M_\mathrm{W}}\,\overline{e}e\,H\,. \tag{9.116}$$

Finally, the lepton sector is described by $\mathcal{L}_\ell$ augmented by the electron mass term from the original $\mathcal{L}_\mathrm{s}$,

$$\mathcal{L}'^{\ell}_{\mathrm{kin}} = \overline{\nu}_\mathrm{L}\mathrm{i}\gamma^\mu\partial_\mu\nu_\mathrm{L} + \overline{e}\,[\mathrm{i}\gamma^\mu(\partial_\mu - \mathrm{i}eA_\mu) - m_\mathrm{e}\,]\,e\,; \tag{9.117}$$

$$\mathcal{L}^{\ell}_{\mathrm{cc}} = -\frac{g}{2\sqrt{2}}\left[\overline{e}\gamma^\mu(1-\gamma_5)\nu\,W^\dagger_\mu + \overline{\nu}\gamma^\mu(1-\gamma_5)e\,W_\mu\right]\,; \tag{9.118}$$

$$\mathcal{L}^{\ell}_{\mathrm{nc}} = -\frac{g}{4c_\mathrm{W}}\,\overline{\nu}\gamma^\mu(1-\gamma_5)\nu\,Z_\mu - \frac{g}{4c_\mathrm{W}}\overline{e}\gamma^\mu\left[(-1+4s^2_\mathrm{W})+\gamma_5\right]e\,Z_\mu\,. \tag{9.119}$$

The sum of (112)–(119) gives the Lagrangian of the gauge model of the electroweak interaction for the electron-type lepton family. This model contains five independent parameters. Before symmetry breaking, they are the SU(2) coupling $g$, the $\mathrm{U_Y}(1)$ coupling $g'$, the scalar potential parameters $\lambda$ and $\mu^2$, and the Yukawa coupling $C_\mathrm{e}$. After symmetry breaking, they may be equivalently replaced by the absolute value of the electron charge $e$, the Weinberg mixing angle $\theta_\mathrm{W}$, the electron mass $m_\mathrm{e}$, the Higgs mass $M_\mathrm{H}$, and the mass of the charged boson $M_\mathrm{W}$. The mass of the neutral boson is not independent, being $M_\mathrm{Z} = M_\mathrm{W}/\cos\theta_\mathrm{W}$. The two sets of parameters are related through

$$\begin{aligned}
\tan\theta_\mathrm{W} &= \frac{g'}{g}\,,\\
e &= g\sin\theta_\mathrm{W}\,,\\
m_\mathrm{e} &= \frac{1}{\sqrt{2}}C_\mathrm{e}v = \frac{1}{\sqrt{2}}C_\mathrm{e}\sqrt{-\mu^2/\lambda}\,,\\
M_\mathrm{W} &= \frac{1}{2}\,gv = \frac{1}{2}\,g\sqrt{-\mu^2/\lambda}\,,\\
M_\mathrm{H} &= \sqrt{-2\mu^2}\,.
\end{aligned} \tag{9.120}$$

The fact that interactions of all gauge fields are determined by the electric charge and one free parameter (the Weinberg angle) is noteworthy. It is proof that the standard model is a unified theory of the weak and electromagnetic interactions, but also that the unification is not complete. A free parameter, $\theta_\mathrm{W}$, appears in addition to $e$ because the symmetry group on which the model is based is a direct product of two simple groups, and would be unnecessary in a larger simple group.

The Feynman rules for this model are obtained from (112)–(119) in the same way as in Chap. 8 for QCD. They are given in Fig. 9.1a–c.

| | |
|---|---|
| Photon propagator | $\dfrac{-\mathrm{i}g_{\mu\nu}}{q^2+\mathrm{i}\varepsilon}$ |
| $W^\pm$, Z propagators | $\dfrac{\mathrm{i}}{q^2-M^2+\mathrm{i}\varepsilon}[-g_{\mu\nu}+\dfrac{q_\mu q_\nu}{M^2}]$ |
| Higgs propagator | $\dfrac{\mathrm{i}}{p^2-M_{\mathrm{H}}^2+\mathrm{i}\varepsilon}$ |
| Neutrino propagator | $\dfrac{1-\gamma_5}{2}\dfrac{\mathrm{i}}{\not p+\mathrm{i}\varepsilon}\dfrac{1+\gamma_5}{2}$ |
| Lepton propagator | $\dfrac{\mathrm{i}}{\not p-m_\ell+\mathrm{i}\varepsilon}$ |

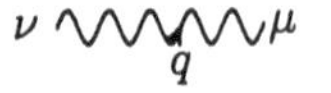

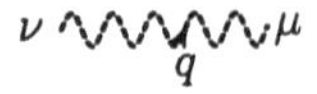

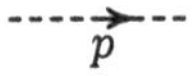

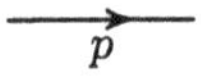

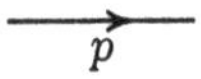

**Fig. 9.1.** (a) Propagators in the gauge-invariant model of one-lepton family

| | |
|---|---|
| WW$\gamma$ vertex | $\mathrm{i}e[(r-q)_\lambda g_{\mu\nu}+(q-p)_\nu g_{\lambda\mu}+(p-r)_\mu g_{\lambda\nu}]$ $(p+r+q=0)$ |
| WWZ vertex | $\mathrm{i}g\cos\theta_{\mathrm{W}}[(r-q)_\lambda g_{\mu\nu}+(q-p)_\nu g_{\lambda\mu}+(p-r)_\mu g_{\lambda\nu}]$ $(p+r+q=0)$ |
| $\mathrm{W}^4$ vertex | $-\mathrm{i}g^2(g_{\lambda\rho}g_{\mu\nu}+g_{\lambda\mu}g_{\rho\nu}-2g_{\lambda\nu}g_{\mu\rho})$ |
| $\mathrm{W}^2\gamma\gamma$ vertex | $\mathrm{i}e^2(g_{\lambda\rho}g_{\mu\nu}+g_{\lambda\nu}g_{\rho\mu}-2g_{\lambda\mu}g_{\rho\nu})$ |
| $\mathrm{W}^2\mathrm{Z}^2$ vertex | $\mathrm{i}(g\cos\theta_{\mathrm{W}})^2(g_{\lambda\rho}g_{\mu\nu}+g_{\lambda\nu}g_{\rho\mu}-2g_{\lambda\mu}g_{\rho\nu})$ |
| $\mathrm{W}^2\mathrm{Z}\gamma$ vertex | $\mathrm{i}eg\cos\theta_{\mathrm{W}}(g_{\lambda\rho}g_{\mu\nu}+g_{\lambda\nu}g_{\rho\mu}-2g_{\lambda\mu}g_{\rho\nu})$ |

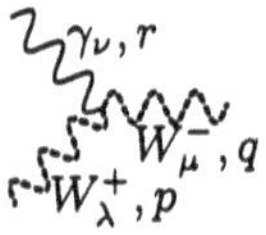

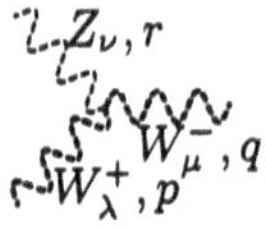

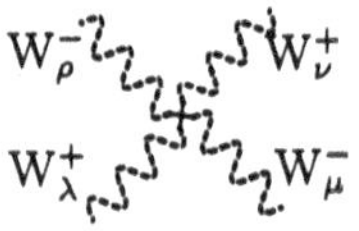

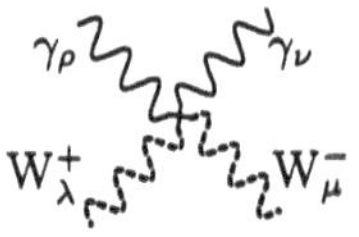

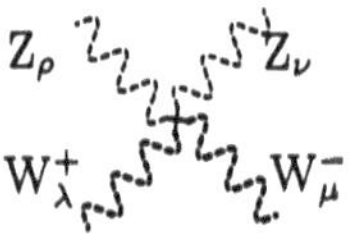

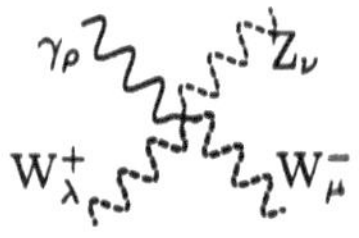

**Fig. 9.1.** (b) Interaction vertices in the gauge-invariant model of one-lepton family: gauge boson self-couplings

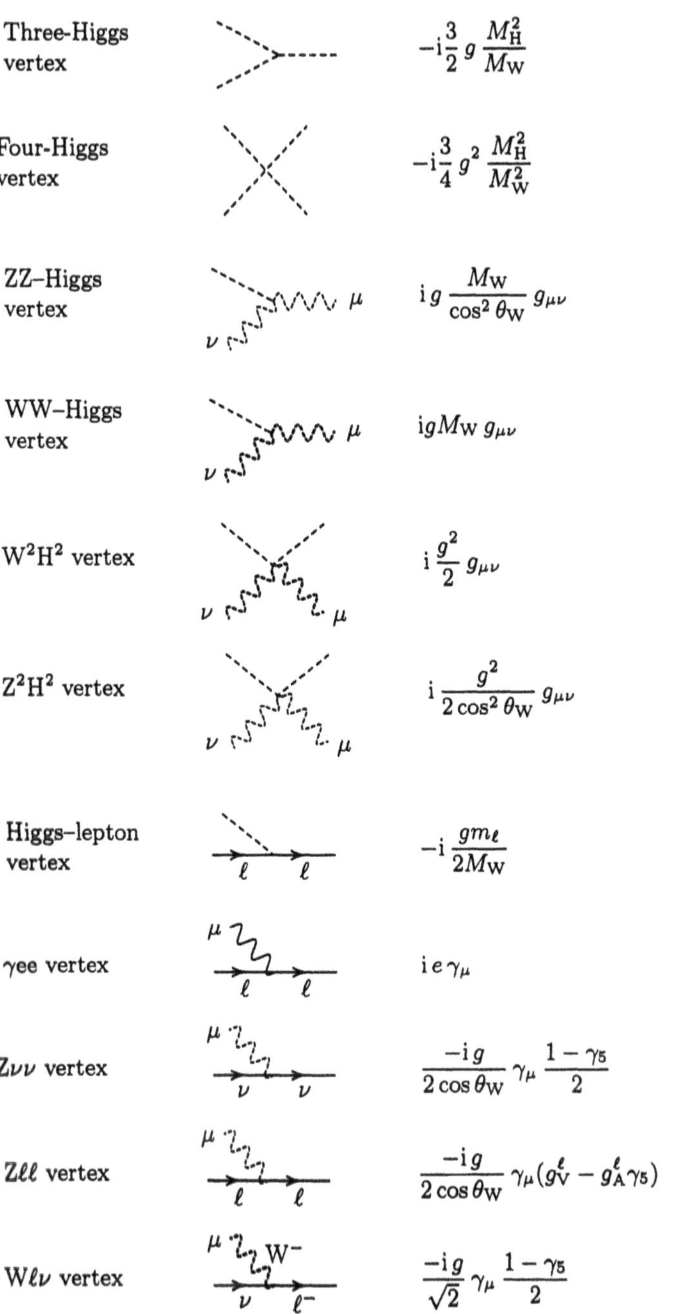


**Fig. 9.1.** (c) Interaction vertices in the gauge-invariant model of one-lepton family: couplings in the Higss boson and lepton sectors

## 9.3 Including u and d Quarks

We now introduce the u and d quarks, which form with the electron and the neutrino $\nu_e$ the first generation of fundamental fermions. The quarks and the leptons enter the model in rather similar ways, in spite of their distinctive characteristics. First, even though quarks are colored and leptons are not, no complications result because the electroweak interactions are insensitive to color. For this reason we will suppress the color label, with the understanding that the implied color indices are summed over where necessary. Second, quarks differ from leptons in their electric charges. However, the fact that $Q_u - Q_d = Q_\nu - Q_e = 1$ and the well-established observation that the weak charged currents of hadrons are left-handed suggest that the left chiral components of the quarks should be grouped, similarly to the leptons, into weak-isospin doublets. Finally, both u and d quarks are massive, whereas the neutrino is (believed to be) massless. This implies that the right chiral components of both quarks should appear in the model, to be compared with the sole $e_R$ in the lepton sector. Therefore the quark sector should include a doublet $\psi_L$ plus two singlets $u_R$ and $d_R$ in the weak-isospin group SU(2):

$$\psi_L = \begin{pmatrix} u_L \\ d_L \end{pmatrix}; \quad u_R, \quad d_R; \tag{9.121}$$

and the model described by the Lagrangians in Sect. 9.2.2 should be amended to include the appropriate gauge-invariant terms for the quarks and the appropriate scalar–quark Yukawa couplings.

The Lagrangian for the free quarks is given by

$$\begin{aligned}\mathcal{L}_q^0 &= \overline{u}\,\mathrm{i}\gamma^\mu\partial_\mu u + \overline{d}\,\mathrm{i}\gamma^\mu\partial_\mu d \\ &= \overline{\psi}_L\,\mathrm{i}\gamma^\mu\partial_\mu\psi_L + \overline{u}_R\,\mathrm{i}\gamma^\mu\partial_\mu u_R + \overline{d}_R\,\mathrm{i}\gamma^\mu\partial_\mu d_R\,.\end{aligned} \tag{9.122}$$

It is clearly invariant under global $SU_L(2) \times U_Y(1)$. To this symmetry correspond the conserved currents

$$j^i_\mu = \overline{\psi}_L\gamma_\mu\frac{\tau_i}{2}\psi_L\,, \qquad (i = 1,2,3)\,, \tag{9.123}$$

$$j^Y_\mu = Y_L\,\overline{\psi}_L\gamma_\mu\psi_L + Y_R^u\,\overline{u}_R\gamma_\mu u_R + Y_R^d\,\overline{d}_R\gamma_\mu d_R\,; \tag{9.124}$$

and the conserved charges $T_3$ and $Y$ are related as usual to the electric charge number $Q$ through $Q = T_3 + \frac{1}{2}Y$. The values assigned to these quantum numbers for the u–d quark multiplets are listed in Table 9.2.

The $SU(2) \times U_Y(1)$ local gauge-invariant form of (122) is

$$\begin{aligned}\mathcal{L}_q &= \overline{\psi}_L\mathrm{i}\gamma^\mu D^L_\mu\psi_L + \overline{u}_R\mathrm{i}\gamma^\mu D^R_\mu u_R + \overline{d}_R\mathrm{i}\gamma^\mu D^R_\mu d_R \\ &= \overline{\psi}_L\mathrm{i}\gamma^\mu(\partial_\mu + \mathrm{i}g\boldsymbol{A}_\mu + \tfrac{\mathrm{i}}{2}g'Y_L B_\mu)\psi_L \\ &\quad + \overline{u}_R\mathrm{i}\gamma^\mu(\partial_\mu + \tfrac{\mathrm{i}}{2}g'Y_R^u B_\mu)u_R + \overline{d}_R\mathrm{i}\gamma^\mu(\partial_\mu + \tfrac{\mathrm{i}}{2}g'Y_R^d B_\mu)d_R\,.\end{aligned} \tag{9.125}$$

**Table 9.2.** Classification of the u–d quark family and assigned quantum numbers

| | $T$ | $T_3$ | $Y$ | $Q$ |
|---|---|---|---|---|
| $\begin{pmatrix} u_\mathrm{L} \\ d_\mathrm{L} \end{pmatrix}$ | $\frac{1}{2}$ | $\pm\frac{1}{2}$ | $\frac{1}{3}$ | $\frac{2}{3}$, $-\frac{1}{3}$ |
| $u_\mathrm{R}$ | 0 | 0 | $\frac{4}{3}$ | $\frac{2}{3}$ |
| $d_\mathrm{R}$ | 0 | 0 | $-\frac{2}{3}$ | $-\frac{1}{3}$ |

The scalar–quark interactions will include the couplings $(\overline{\psi}_\mathrm{L}\phi)d_\mathrm{R}$ and $\overline{d}_\mathrm{R}(\phi^\dagger\psi_\mathrm{L})$ similar to those found in the lepton sector. In order to couple $u_\mathrm{R}$ to scalars in a gauge-invariant way, we also need $\varphi^-$ and $\overline{\varphi}^0$, the charge conjugates to $\varphi^+$ and $\varphi^0$, which form a doublet conjugate to $\phi$, that is,

$$\phi^\mathrm{c} = \mathrm{i}\tau_2\phi^* = \begin{pmatrix} \overline{\varphi}^0 \\ -\varphi^- \end{pmatrix} . \tag{9.126}$$

It has weak hypercharge $Y_{\mathrm{H^c}} = -Y_\mathrm{H} = -1$. The Yukawa quark couplings require two coupling constants, $C_\mathrm{u}$ and $C_\mathrm{d}$, and assume the general form

$$\mathcal{L}_\mathrm{qY} = -C_\mathrm{u}\left[(\overline{\psi}_\mathrm{L}\phi^\mathrm{c})u_\mathrm{R} + \overline{u}_\mathrm{R}(\phi^{\mathrm{c}\dagger}\psi_\mathrm{L})\right] - C_\mathrm{d}\left[(\overline{\psi}_\mathrm{L}\phi)d_\mathrm{R} + \overline{d}_\mathrm{R}(\phi^\dagger\psi_\mathrm{L})\right] . \tag{9.127}$$

Gauge invariance of these couplings under $U_\mathrm{Y}(1)$ is guaranteed by the assigned weak hypercharges of the particles: $Y_\mathrm{L} - Y_\mathrm{R}^\mathrm{u} = Y_{\mathrm{H^c}}$ and $Y_\mathrm{L} - Y_\mathrm{R}^\mathrm{d} = Y_\mathrm{H}$.

After breaking symmetry, one goes to the unitary gauge just as before, so that the scalar doublets become

$$\phi \to \mathcal{S}\phi = \frac{1}{\sqrt{2}}(v+H)\chi , \qquad \chi \equiv \begin{pmatrix} 0 \\ 1 \end{pmatrix} ; \tag{9.128}$$

$$\phi^\mathrm{c} \to \mathcal{S}\phi^\mathrm{c} = \frac{1}{\sqrt{2}}(v+H)\chi^\mathrm{c} , \qquad \chi^\mathrm{c} \equiv \begin{pmatrix} 1 \\ 0 \end{pmatrix} . \tag{9.129}$$

In the unitary gauge, the Yukawa interaction takes the form

$$\mathcal{L}_\mathrm{qY} = -\frac{1}{\sqrt{2}}(v+H)\left(C_\mathrm{u}\,\overline{u}u + C_\mathrm{d}\,\overline{d}d\right) , \tag{9.130}$$

which shows that through the Higgs mechanism the u and d quarks acquire the masses

$$m_\mathrm{u} = \tfrac{1}{\sqrt{2}} C_\mathrm{u} v \quad \text{and} \quad m_\mathrm{d} = \tfrac{1}{\sqrt{2}} C_\mathrm{d} v . \tag{9.131}$$

Inversely, the Yukawa couplings can be expressed in terms of the quark masses

$$C_\mathrm{u} = \frac{\sqrt{2}\, m_\mathrm{u}}{v} = \frac{g m_\mathrm{u}}{\sqrt{2}\, M_\mathrm{W}} \quad \text{and} \quad C_\mathrm{d} = \frac{\sqrt{2}\, m_\mathrm{d}}{v} = \frac{g m_\mathrm{d}}{\sqrt{2}\, M_\mathrm{W}} ,$$

so that the Lagrangian $\mathcal{L}_{\rm qY}$ assumes the form

$$\mathcal{L}_{\rm qY} = -m_{\rm u}\,\overline{u}u - m_{\rm d}\,\overline{d}d - \frac{gm_{\rm u}}{2\,M_{\rm W}}\,\overline{u}u\,H - \frac{gm_{\rm d}}{2\,M_{\rm W}}\,\overline{d}d\,H\,. \tag{9.132}$$

The quark Lagrangian $\mathcal{L}_{\rm q}$ is very similar to the $\mathcal{L}_\ell$ considered in the last section. Written now in the unitary gauge, it includes, besides the usual kinetic terms for $u$ and $d$ fields, the following interaction terms. First, there are the contributions of the quark fields to the charged current interaction:

$$\begin{aligned}\mathcal{L}^{\rm q}_{\rm cc} &= -\tfrac{1}{\sqrt{2}}\,g\,\overline{\psi}_{\rm L}\gamma^\mu\left(\tau_+W_\mu + \tau_-W^\dagger_\mu\right)\psi_{\rm L}\\ &= -\tfrac{1}{\sqrt{2}}\,g\,\left(J^{\mu\dagger}W_\mu + J^\mu W^\dagger_\mu\right)\,;\end{aligned} \tag{9.133}$$

and then their contributions to the neutral current interaction:

$$\begin{aligned}\mathcal{L}^{\rm q}_{\rm nc} &= -\tfrac{1}{2}\,\overline{\psi}_{\rm L}\gamma_\mu\left(gA^\mu_3\tau_3 + g'B^\mu Y_{\rm L}\right)\psi_{\rm L} - \tfrac{1}{2}\,g'Y^i_{\rm R}\,\overline{\psi}^i_{\rm R}\gamma_\mu\psi^i_{\rm R}\,B^\mu\\ &= g\,j^3_\mu\,A^\mu_3 - \tfrac{1}{2}\,g'\,j^{\rm Y}_\mu\,B^\mu\,,\end{aligned} \tag{9.134}$$

which read in terms of the photon and the $\rm Z^0$ fields

$$\begin{aligned}\mathcal{L}^{\rm q}_{\rm nc} &= -(gs_{\rm W}\,j^3_\mu + \tfrac{1}{2}\,g'c_{\rm W}\,j^{\rm Y}_\mu)\,A^\mu - (gc_{\rm W}\,j^3_\mu - \tfrac{1}{2}\,g's_{\rm W}\,j^{\rm Y}_\mu)\,Z^\mu\\ &= -e\,j^{\rm em}_\mu\,A^\mu - \frac{g}{c_{\rm W}}\,j^{\rm Z}_\mu Z^\mu\,.\end{aligned} \tag{9.135}$$

Here the $Z$-current for the u–d quarks may be written as

$$\begin{aligned}j^{\rm Z}_\mu &= j^3_\mu - s^2_{\rm W}\,j^{\rm Y}_\mu = \overline{\psi}\gamma^\mu(T_3 - s^2_{\rm W}\,Q)\,\psi\\ &= \overline{\psi}_{\rm L}\gamma^\mu Z_{\rm L}\psi_{\rm L} + Z^{\rm u}_{\rm R}\,\overline{u}_{\rm R}\gamma^\mu u_{\rm R} + Z^{\rm d}_{\rm R}\,\overline{d}_{\rm R}\gamma^\mu d_{\rm R}\,.\end{aligned}$$

It preserves quark flavors and couples to both chiral components of quarks. Alternatively, it may be written as

$$j^{\rm Z}_\mu = \tfrac{1}{2}\,\overline{u}\gamma_\mu\left(g^{\rm u}_{\rm V} - g^{\rm u}_{\rm A}\gamma_5\right)u + \tfrac{1}{2}\,\overline{d}\gamma_\mu\left(g^{\rm d}_{\rm V} - g^{\rm d}_{\rm A}\gamma_5\right)d\,. \tag{9.136}$$

The weak charges and the weak neutral current coupling constants are defined as before, $Z_{\rm L} = T_3 - s^2_{\rm W}\,Q$ and $Z_{\rm R} = -s^2_{\rm W}\,Q$, or alternatively, $g_{\rm V} = Z_{\rm L} + Z_{\rm R}$ and $g_{\rm A} = Z_{\rm L} - Z_{\rm R}$. They depend on a single parameter, the Weinberg angle $\theta_{\rm W}$. Their values are listed in Table 9.3.

**Table 9.3.** Charges and coupling constants of the weak neutral current for leptons and quarks in the standard model

| f | $Q$ | $Z^{\rm f}_{\rm L}$ | $Z^{\rm f}_{\rm R}$ | $g^{\rm f}_{\rm V}$ | $g^{\rm f}_{\rm A}$ |
|---|---|---|---|---|---|
| $\nu$ | 0 | $\frac{1}{2}$ | 0 | $\frac{1}{2}$ | $\frac{1}{2}$ |
| e | $-1$ | $-\frac{1}{2} + \sin^2\theta_{\rm W}$ | $\sin^2\theta_{\rm W}$ | $-\frac{1}{2} + 2\,\sin^2\theta_{\rm W}$ | $-\frac{1}{2}$ |
| u | $\frac{2}{3}$ | $\frac{1}{2} - \frac{2}{3}\,\sin^2\theta_{\rm W}$ | $-\frac{2}{3}\,\sin^2\theta_{\rm W}$ | $\frac{1}{2} - \frac{4}{3}\sin^2\theta_{\rm W}$ | $\frac{1}{2}$ |
| d | $-\frac{1}{3}$ | $-\frac{1}{2} + \frac{1}{3}\,\sin^2\theta_{\rm W}$ | $\frac{1}{3}\,\sin^2\theta_{\rm W}$ | $-\frac{1}{2} + \frac{2}{3}\sin^2\theta_{\rm W}$ | $-\frac{1}{2}$ |

To close, we summarize the results found in this section. The unified model of the electroweak interaction for the first generation of fermions is described by the Lagrangian that includes (112)–(119) from the e–$\nu_e$ family and the following contributions from the u and d quarks:

$$\mathcal{L}_q + \mathcal{L}_{qY} = \mathcal{L}_q^0 + \mathcal{L}_{qH} + \mathcal{L}_{cc}^q + \mathcal{L}_{nc}^q \,. \tag{9.137}$$

The first term on the right-hand side gives the kinetic part

$$\mathcal{L}_q^0 = \overline{u}\,(i\gamma^\mu\partial_\mu - m_u)\,u + \overline{d}\,(i\gamma^\mu\partial_\mu - m_d)d\,; \tag{9.138}$$

the second represents the couplings of quarks to the Higgs boson

$$\mathcal{L}_{qH} = -\frac{g m_u}{2\,M_W}\,\overline{u}u\,H - \frac{g m_d}{2\,M_W}\,\overline{d}d\,H\,; \tag{9.139}$$

while the remaining terms represent the couplings of the gauge bosons to the charged and neutral currents for quarks:

$$\begin{aligned} \mathcal{L}_{cc}^q &= -\frac{g}{\sqrt{2}}\,\left(J_\mu^\dagger W^\mu + J_\mu W^{\mu\dagger}\right) \\ &= -\frac{g}{\sqrt{2}}\,\left(\overline{u}_L\gamma_\mu d_L\,W^\mu + \overline{d}_L\gamma_\mu u_L\,W^{\mu\dagger}\right)\,; \end{aligned} \tag{9.140}$$

$$\begin{aligned} \mathcal{L}_{nc}^q &= -\,e\,j_\mu^{em}A^\mu - \frac{g}{c_W}\,j_\mu^Z Z^\mu \\ &= -\,e\left(\frac{2}{3}\,\overline{u}\gamma_\mu u - \frac{1}{3}\,\overline{d}\gamma_\mu d\right)A^\mu \\ &\quad - \frac{g}{2\,c_W}\,\left[\overline{u}\gamma_\mu(g_V^u - g_A^u\,\gamma_5)u + \overline{d}\gamma_\mu(g_V^d - g_A^d\,\gamma_5)d\right]Z^\mu\,. \end{aligned} \tag{9.141}$$

A comparison with similar results for leptons in (112)–(119) shows the close parallel between the u, d quarks and the e, $\nu_e$ leptons in their electroweak interactions. Their charged current interactions are identical, and they both break parity in the strongest possible way. Their electromagnetic and neutral current interactions differ only because of their different electric and weak charges, and their couplings to the Higgs boson differ only in strengths because of their different masses .

Including u and d quarks adds two more parameters to the model, the quark masses, $m_u$ and $m_d$. Thus, for one generation of quarks and leptons, the model requires seven independent parameters: in the original gauge-invariant Lagrangian, they are the two gauge couplings $g$ and $g'$, the two scalar self-couplings $\lambda$ and $\mu^2$, and the three Yukawa couplings $C_e$, $C_u$, and $C_d$; after symmetry breaking, they are replaced by $e$, $\theta_W$, $M_W$, $M_H$, $m_e$, $m_u$, and $m_d$. The content of Fig. 9.1 is now complemented by the additional Feynman rules derived from (138)–(141) and listed in Fig. 9.2.

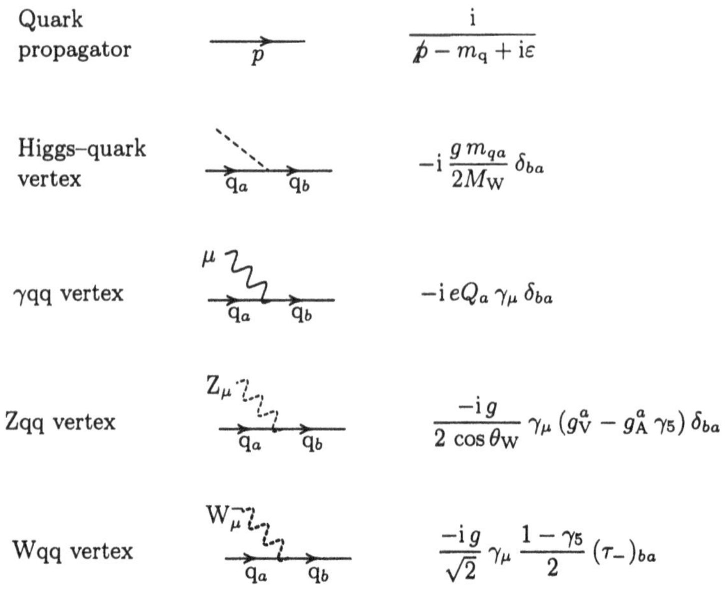

Quark propagator $\qquad \xrightarrow[p]{} \qquad \dfrac{\mathrm{i}}{\not{p} - m_{\mathrm{q}} + \mathrm{i}\varepsilon}$

Higgs–quark vertex $\qquad q_a \quad q_b \qquad -\mathrm{i}\,\dfrac{g\,m_{qa}}{2M_{\mathrm{W}}}\,\delta_{ba}$

$\gamma$qq vertex $\qquad \mu \quad q_a \quad q_b \qquad -\mathrm{i}\,eQ_a\,\gamma_\mu\,\delta_{ba}$

Zqq vertex $\qquad Z_\mu \quad q_a \quad q_b \qquad \dfrac{-\mathrm{i}\,g}{2\cos\theta_{\mathrm{W}}}\,\gamma_\mu\,(g_{\mathrm{V}}^{a} - g_{\mathrm{A}}^{a}\,\gamma_5)\,\delta_{ba}$

Wqq vertex $\qquad W_\mu^{-} \quad q_a \quad q_b \qquad \dfrac{-\mathrm{i}\,g}{\sqrt{2}}\,\gamma_\mu\,\dfrac{1-\gamma_5}{2}\,(\tau_-)_{ba}$

**Fig. 9.2.** Feynman diagrams for the electroweak interaction of quarks

## 9.4 Multigeneration Model

It is now a well-established experimental fact that there exist six leptons – $\mathrm{e}^-$, $\nu_{\mathrm{e}}$, $\mu^-$, $\nu_\mu$, $\tau^-$, and $\nu_\tau$ – and six quarks – u, d, c, s, t, and b – plus their corresponding antiparticles (cf. Table 7.9). Together they constitute the complete fermionic content of the standard model of the electroweak interactions. Although a casual look at the data on leptonic weak decays might indicate otherwise, the incorporation of all known leptons and quarks in the model involves much more than a mere replication of the formulation with a single family, as was presented in the last two sections for $\mathrm{e}^-$, $\nu_{\mathrm{e}}$, u, and d. Over the years, observations of a multitude of weak processes have brought out many novel features; some have helped to shape the emerging theory, while others might yet find in it a possible explanation. These features include a certain mixing of the quark fields and the absence of a similar mixing of the lepton fields, the suppression of flavor-changing neutral currents and the phenomenon of CP violation.

### 9.4.1 The GIM Mechanism

The purely leptonic decay $\mu^- \to \mathrm{e}^-\overline{\nu}_{\mathrm{e}}\nu_\mu$ and the related scattering process $\mathrm{e}^-\nu_\mu \to \mu^-\nu_{\mathrm{e}}$ can be described to a good accuracy by the purely leptonic part, $L^\alpha L_\alpha^\dagger$, of the effective Hamiltonian (1),

$$\mathcal{H}_{\mathrm{weak}}^{\mathrm{l}} = \frac{G_{\mathrm{F}}}{\sqrt{2}}\,[\overline{\nu}_\mu\gamma^\alpha(1-\gamma_5)\mu]\,[\overline{\mathrm{e}}\gamma_\alpha(1-\gamma_5)\nu_{\mathrm{e}}]\,. \tag{9.142}$$

By comparing the calculated $\mu^-$ lifetime, including radiative corrections, with the measured lifetime, $\tau_\mu = 2.179 \times 10^{-5}$ s, one gets the value of the decay strength $G_{\rm F}$, as is given in (3).

On the other hand, the amplitudes of $\beta$-decays may be calculated with the semileptonic coupling terms, $L^\alpha H_\alpha^\dagger + L_\alpha^\dagger H^\alpha$, or

$$\mathcal{H}_{\rm weak}^{\rm sl} = \frac{G^{(0)}}{\sqrt{2}} \left[\overline{p}\gamma^\alpha(1 - c_{\rm A}\gamma_5)n\right] \left[\overline{e}\gamma_\alpha(1-\gamma_5)\nu_{\rm e}\right] + \text{h.c.}\,, \tag{9.143}$$

where, in order to fit data,

$$G^{(0)}/G_{\rm F} \approx 0.975\,, \qquad c_{\rm A} = 1.2573 \pm 0.0028\,, \tag{9.144}$$

both of which are close to but unmistakably different from 1. The apparent similarity between (142) and (143) suggests there should be universality in the structure of interactions at the quark level,

$$\mathcal{H}_{\rm weak} = \frac{G^{(0)}}{\sqrt{2}} \left[\overline{u}\gamma^\alpha(1-\gamma_5)d\right] \left[\overline{e}\gamma_\alpha(1-\gamma_5)\nu_{\rm e}\right] + \text{h.c.}\,. \tag{9.145}$$

While it is expected that the *axial-vector* current coupling gets modified by the QCD effects when hadrons are involved, causing $c_{\rm A}$ to deviate from 1, there is a strong belief that, just as the electromagnetic current is conserved, so too is the weak *vector* current, so that the matrix element of this current should not get renormalized at the hadronic level: conservation of vector current, $\partial^\mu(\overline{u}\gamma_\mu d) = 0$, implies no renormalization by strong interactions, $\langle p\,|\,\overline{u}\gamma_\mu d\,|\,n\rangle = \overline{p}\gamma_\mu n$, at zero invariant momentum transfer. Therefore, the deviation of $G^{(0)}$ from $G_{\rm F}$ must be a genuine effect, persisting even after radiative corrections are taken into account, and thus must have a deeper physical origin: these strangeness-conserving charged quark currents alone cannot generate an SU(2) group in the way the lepton currents do.

The general structure of the weak interaction is found to be of the V–A type, as in (142) and (143), but the strength of the strangeness-changing decay, $G^{(1)}$, is consistently smaller than that of the $\beta$-decay: $G^{(1)} \approx 0.22\,G_{\rm F}$ (from comparing, for example, the rate of $\Lambda \to {\rm pe}^-\overline{\nu}$ with that of ${\rm n} \to {\rm pe}^-\overline{\nu}$). To reflect this fact, one introduces a parameter $\theta_{\rm C}$ (called the Cabibbo mixing angle), such that $\cos\theta_{\rm C} = G^{(0)}/G_{\rm F}$ and $\sin\theta_{\rm C} = G^{(1)}/G_{\rm F}$ may give a measure of the relative strength of the strangeness-changing charged current. The amplitudes of strangeness-conserving and strangeness-changing decays can be obtained from the *charged current*

$$H_\mu^\dagger = \overline{u}\gamma_\mu(1-\gamma_5)d_{\rm C}\,, \tag{9.146}$$

which differs from the corresponding current in (145) by replacing $d$ with

$$d_{\rm C} = d\,\cos\theta_{\rm C} + s\,\sin\theta_{\rm C}\,, \tag{9.147}$$

where $s$ is the strange quark field. However, a similar substitution $d \to d_{\mathrm{C}}$ in the *neutral current* $j^{\mathrm{Z}}_{\mu}$ of (136) would give

$$\bar{d}_{\mathrm{C}}\, \gamma_\mu \,(g^{\mathrm{d}}_{\mathrm{V}} - g^{\mathrm{d}}_{\mathrm{A}}\gamma_5)\, d_{\mathrm{C}}\,, \tag{9.148}$$

and would lead to terms like

$$\cos\theta_{\mathrm{C}}\, \sin\theta_{\mathrm{C}}\, \left[\bar{d}\,\gamma_\mu\,(g^{\mathrm{d}}_{\mathrm{V}} - g^{\mathrm{d}}_{\mathrm{A}}\gamma_5)\, s + \bar{s}\,\gamma_\mu\,(g^{\mathrm{d}}_{\mathrm{V}} - g^{\mathrm{d}}_{\mathrm{A}}\gamma_5)\, d\right]\,. \tag{9.149}$$

The neutral flavor-changing transition $\mathrm{s} \to \mathrm{d}$ (as in $\mathrm{K}^- \to \pi^-\mathrm{e}^+\mathrm{e}^-$) would then be possible at a strength roughly comparable to that of $\mathrm{s} \to \mathrm{u}$ processes (as in $\mathrm{K}^- \to \pi^0\mathrm{e}^-\bar{\nu}$), in sharp disagreement with data. Again, the neutral current in this form does not seem to be complete.

Glashow, Iliopoulos, and Maiani (GIM) suggested in 1970 that an additional flavor, the *charm* c, should exist and should form with $s_{\mathrm{C}}$, the orthogonal complement to $d_{\mathrm{C}}$, a second quark doublet. The lower components of the two doublets $(u,\, d_{\mathrm{C}})$ and $(c,\, s_{\mathrm{C}})$ are related to the physical quarks d and s by an orthogonal transformation,

$$\begin{pmatrix} d_{\mathrm{C}} \\ s_{\mathrm{C}} \end{pmatrix} = \boldsymbol{V}_{\mathrm{C}} \begin{pmatrix} d \\ s \end{pmatrix} \equiv \begin{pmatrix} \cos\theta_{\mathrm{C}} & \sin\theta_{\mathrm{C}} \\ -\sin\theta_{\mathrm{C}} & \cos\theta_{\mathrm{C}} \end{pmatrix} \begin{pmatrix} d \\ s \end{pmatrix}\,. \tag{9.150}$$

Then, as long as the coupling strengths of the two doublets to $Z_\mu$ are equal, the flavor-changing neutral processes, as in $\mathrm{s} \to \mathrm{d}$, should be suppressed to all orders of perturbation because of the orthogonality of the Cabibbo rotation, $\boldsymbol{V}^{\mathrm{T}}_{\mathrm{C}}\boldsymbol{V}_{\mathrm{C}} = 1$, so that

$$\bar{d}_{\mathrm{C}} d_{\mathrm{C}} + \bar{s}_{\mathrm{C}} s_{\mathrm{C}} = \bar{d}d + \bar{s}s\,. \tag{9.151}$$

To sum up, the contributions of quarks to the charged and neutral currents that are needed to reproduce the $\Delta S = 0$ and $|\Delta S| = 1$ weak transitions at low energies should have the forms

$$J_\mu(\text{quarks}) = \tfrac{1}{2}\,\bar{d}_{\mathrm{C}}\,\gamma_\mu\,(1-\gamma_5)u + \tfrac{1}{2}\,\bar{s}_{\mathrm{C}}\,\gamma_\mu(1-\gamma_5)c\,, \tag{9.152}$$

and

$$\begin{aligned} j^{\mathrm{Z}}_\mu(\text{quarks}) &= \tfrac{1}{2}\,\bar{u}\,\gamma_\mu\,(g^{\mathrm{u}}_{\mathrm{V}} - g^{\mathrm{u}}_{\mathrm{A}}\gamma_5)\,u + \tfrac{1}{2}\,\bar{d}\,\gamma_\mu(g^{\mathrm{d}}_{\mathrm{V}} - g^{\mathrm{d}}_{\mathrm{A}}\gamma_5)\,d \\ &\quad + \tfrac{1}{2}\,\bar{c}\gamma_\mu\,(g^{\mathrm{u}}_{\mathrm{V}} - g^{\mathrm{u}}_{\mathrm{A}}\gamma_5)\,c + \tfrac{1}{2}\,\bar{s}\,\gamma_\mu(g^{\mathrm{d}}_{\mathrm{V}} - g^{\mathrm{d}}_{\mathrm{A}}\gamma_5)\,s\,. \end{aligned} \tag{9.153}$$

Note that $u$ and $c$ have been assigned equal weak charges; similarly, $d$ and $s$. The forms of these currents indicate that the left chiral components of fields should be considered weak-isospin doublets, $(u_{\mathrm{L}},\, d_{\mathrm{CL}})$ and $(c_{\mathrm{L}},\, s_{\mathrm{CL}})$, while all right chiral components, $u_{\mathrm{R}}$, $d_{\mathrm{R}}$, $c_{\mathrm{R}}$, and $s_{\mathrm{R}}$, weak-isospin singlets. The presence of a mixing angle in the left-handed quark sectors means that a clear distinction must be made between *gauge symmetry eigenstates* (also referred to as *weak interaction eigenstates*), $d_{\mathrm{C}}$ and $s_{\mathrm{C}}$, endowed with definite gauge transformation properties, and *mass eigenstates*, $d$ and $s$, having definite masses acquired through spontaneous symmetry breaking. As $u_{\mathrm{R}}$ and $d_{\mathrm{R}}$ do not couple to $c_{\mathrm{R}}$ or $s_{\mathrm{R}}$, no mixing occurs among the right-handed quarks.

### 9.4.2 Classification Scheme for Fermions

When all known leptons and quarks are introduced into the model, we must similarly distinguish between the gauge symmetry basis and the physical (mass) basis. It is the gauge symmetry eigenstates (which will be marked by a prime accent, as in $f'$) that describe the fermionic content of the gauge-invariant model. Just as in the one-family model, so too in the general model the left chiral components of fields transform as isodoublets and the right chiral components transform as isosinglets. They are shown in Table 9.4 together with their quantum numbers in $SU(2)_L \times U_Y(1)$. We denote by $\boldsymbol{l}$ and $\boldsymbol{q}$ the vectors in generation space, with components $\ell_{AL}$ and $q_{AL}$, for $A = 1, 2, 3$, designating the three lepton and quark doublets. We will also need vectors $\boldsymbol{\nu}'$, $\boldsymbol{e}'$, $\boldsymbol{u}'$, and $\boldsymbol{d}'$, which have as components lepton or quark fields of equal charges:

$$\begin{aligned}
\boldsymbol{\nu}' &= (\nu'_e, \nu'_\mu, \nu'_\tau);\\
\boldsymbol{e}' &= (e', \mu', \tau');\\
\boldsymbol{u}' &= (u', c', t');\\
\boldsymbol{d}' &= (d', s', b').
\end{aligned} \tag{9.154}$$

**Table 9.4.** SU(2)×U(1) classification and quantum numbers of the fundamental fermions in the standard model

| | 1 | 2 | 3 | $T$ | $T_3$ | $Y$ | $Q$ |
|---|---|---|---|---|---|---|---|
| $\ell_{AL}$ | $\begin{pmatrix}\nu'_{eL}\\ e'_L\end{pmatrix}$ | $\begin{pmatrix}\nu'_{\mu L}\\ \mu'_L\end{pmatrix}$ | $\begin{pmatrix}\nu'_{\tau L}\\ \tau'_L\end{pmatrix}$ | $\frac{1}{2}$ | $\pm\frac{1}{2}$ | $-1$ | $0$<br>$-1$ |
| $e'_{AR}$ | $e'_R$ | $\mu'_R$ | $\tau'_R$ | 0 | 0 | $-2$ | $-1$ |
| $q_{AL}$ | $\begin{pmatrix}u'_L\\ d'_L\end{pmatrix}$ | $\begin{pmatrix}c'_L\\ s'_L\end{pmatrix}$ | $\begin{pmatrix}t'_L\\ b'_L\end{pmatrix}$ | $\frac{1}{2}$ | $\pm\frac{1}{2}$ | $\frac{1}{3}$ | $\frac{2}{3}$<br>$-\frac{1}{3}$ |
| $u'_{AR}$ | $u'_R$ | $c'_R$ | $t'_R$ | 0 | 0 | $\frac{4}{3}$ | $\frac{2}{3}$ |
| $d'_{AR}$ | $d'_R$ | $s'_R$ | $b'_R$ | 0 | 0 | $-\frac{2}{3}$ | $-\frac{1}{3}$ |

### 9.4.3 Fermion Families and the CKM Matrix

The incorporation of additional fermions in the model leaves the gauge and scalar sectors unchanged. It only affects the fermion–gauge and fermion–scalar couplings. The gauge-invariant fermion Lagrangian now reads:

$$\begin{aligned}
\mathcal{L}_F &= \overline{\psi}_L i\gamma^\mu D^L_\mu \psi_L + \overline{\psi}_R i\gamma^\mu D^R_\mu \psi_R\\
&= \overline{\ell}_{AL} i\gamma^\mu D^L_\mu \ell_{AL} + \overline{e}'_{AR} i\gamma^\mu D^R_\mu e'_{AR}\\
&\quad + \overline{q}_{AL} i\gamma^\mu D^L_\mu q_{AL} + \overline{u}'_{AR} i\gamma^\mu D^R_\mu u'_{AR} + \overline{d}'_{AR} i\gamma^\mu D^R_\mu d'_{AR},
\end{aligned} \tag{9.155}$$

where the covariant derivatives of fields are

$$\begin{aligned}
D^{\mathrm{L}}_\mu \ell_{A\mathrm{L}} &= (\partial_\mu + \mathrm{i}\tfrac{1}{2} g \tau_i A^i_\mu - \mathrm{i}\tfrac{1}{2} g' B_\mu)\, \ell_{A\mathrm{L}}\,, \\
D^{\mathrm{R}}_\mu e'_{A\mathrm{R}} &= (\partial_\mu - \mathrm{i} g' B_\mu)\, e'_{A\mathrm{R}}\,, \\
D^{\mathrm{L}}_\mu q_{A\mathrm{L}} &= (\partial_\mu + \mathrm{i}\tfrac{1}{2} g \tau_i A^i_\mu + \mathrm{i}\tfrac{1}{6} g' B_\mu)\, q_{A\mathrm{L}}\,, \\
D^{\mathrm{R}}_\mu u'_{A\mathrm{R}} &= (\partial_\mu + \mathrm{i}\tfrac{2}{3} g' B_\mu)\, u'_{A\mathrm{R}}\,, \\
D^{\mathrm{R}}_\mu d'_{A\mathrm{R}} &= (\partial_\mu - \mathrm{i}\tfrac{1}{3} g' B_\mu)\, d'_{A\mathrm{R}}\,.
\end{aligned} \tag{9.156}$$

With all the neutrinos assumed to be exactly massless, the Yukawa couplings for the remaining fermions are

$$\begin{aligned}
\mathcal{L}_{\mathrm{Y}} = -\ &\big[C^{\mathrm{e}}_{AB}\,(\bar{\ell}_{A\mathrm{L}}\phi)\, e'_{B\mathrm{R}} \\
&+ C^{\mathrm{u}}_{AB}\,(\bar{q}_{A\mathrm{L}}\phi^{\mathrm{c}})\, u'_{B\mathrm{R}} + C^{\mathrm{d}}_{AB}\,(\bar{q}_{A\mathrm{L}}\phi)\, d'_{B\mathrm{R}} + \text{h.c.}\big]\,.
\end{aligned} \tag{9.157}$$

The coupling strengths are given by three $3\times 3$ matrices $\boldsymbol{C}^f$, with $f=$ e, u, d, one matrix for each set of equally charged fermions.

**Fermion Mass Matrix.** Upon spontaneous symmetry breaking and going to the usual unitary gauge, which implies in particular

$$\begin{aligned}
\phi &\to \tfrac{1}{\sqrt{2}}\,(v+H)\chi\,, \\
\phi^{\mathrm{c}} &\to \tfrac{1}{\sqrt{2}}\,(v+H)\chi^{\mathrm{c}}\,,
\end{aligned} \tag{9.158}$$

the various Yukawa couplings become

$$\begin{aligned}
(\bar{\ell}_{A\mathrm{L}}\phi) &= \tfrac{1}{\sqrt{2}}\,(v+H)\,\bar{e}'_{A\mathrm{L}}\,, \\
(\bar{q}_{A\mathrm{L}}\phi) &= \tfrac{1}{\sqrt{2}}\,(v+H)\,\bar{d}'_{A\mathrm{L}}\,, \\
(\bar{q}_{A\mathrm{L}}\phi^{\mathrm{c}}) &= \tfrac{1}{\sqrt{2}}\,(v+H)\,\bar{u}'_{A\mathrm{L}}\,.
\end{aligned}$$

The corresponding terms contribute to the Yukawa Lagrangian

$$\mathcal{L}_{\mathrm{Y}} = -\left(1+\frac{H}{v}\right)\left(\overline{\boldsymbol{e}}'_{\mathrm{L}}\mathcal{M}'_{\mathrm{e}}\boldsymbol{e}'_{\mathrm{R}} + \overline{\boldsymbol{u}}'_{\mathrm{L}}\mathcal{M}'_{\mathrm{u}}\boldsymbol{u}'_{\mathrm{R}} + \overline{\boldsymbol{d}}'_{\mathrm{L}}\mathcal{M}'_{\mathrm{d}}\boldsymbol{d}'_{\mathrm{R}} + \text{h.c.}\right), \tag{9.159}$$

which is expressed in terms of the fermionic mass matrices in the gauge eigenstate basis

$$\mathcal{M}'_f = \frac{v}{\sqrt{2}}\boldsymbol{C}^f\,, \quad \text{for} \quad f = \mathrm{e},\ \mathrm{u},\ \mathrm{d}\,. \tag{9.160}$$

These matrices are generally neither symmetric nor Hermitian, but they can still be diagonalized by biunitary transformations. For each fundamental massive fermion $f$, one can write $\mathcal{M}'_f$ as the product of a Hermitian matrix $\boldsymbol{H}$ and a unitary matrix $\boldsymbol{T}$ as follows:

$$\mathcal{M}'_f = \boldsymbol{H}_f\,\boldsymbol{T}_f\,;\quad \boldsymbol{H}_f = \sqrt{\mathcal{M}'_f\mathcal{M}'^{\dagger}_f}\,.$$

Given that by construction $\boldsymbol{H}_f$ is Hermitian and positive-definite, the matrix $\boldsymbol{T}_f$ can be shown to be unitary, and $\boldsymbol{H}_f$ can be diagonalized by another unitary matrix, $\boldsymbol{S}_f$,

$$\boldsymbol{S}_f \boldsymbol{H}_f \boldsymbol{S}_f^\dagger = \mathcal{M}_f \,, \tag{9.161}$$

so that

$$\mathcal{M}'_f = \boldsymbol{S}_f^\dagger \mathcal{M}_f \boldsymbol{S}_f \boldsymbol{T}_f \,. \tag{9.162}$$

This means that for each term in (159) one may transform the mass matrix written there in the gauge eigenstate basis, $\mathcal{M}'_f$, into a diagonal matrix in the mass eigenstate basis, $\mathcal{M}_f$,

$$\overline{\boldsymbol{\psi}}'_{f\mathrm{L}} \mathcal{M}'_f \boldsymbol{\psi}'_{f\mathrm{R}} = \overline{\boldsymbol{\psi}}_{f\mathrm{L}} \mathcal{M}_f \boldsymbol{\psi}_{f\mathrm{R}} \,, \quad \text{for} \quad f = \mathrm{e,\ u,\ d} \,. \tag{9.163}$$

Here the *mass eigenstates* $\boldsymbol{\psi}_i$ are related to the *gauge eigenstates* $\boldsymbol{\psi}'_i$ by linear transformations:

$$\begin{aligned} \boldsymbol{\psi}_{f\mathrm{L}} &\equiv \boldsymbol{B}_{f\mathrm{L}} \boldsymbol{\psi}'_{f\mathrm{L}} = \boldsymbol{S}_f \boldsymbol{\psi}'_{f\mathrm{L}} \,, \\ \boldsymbol{\psi}_{f\mathrm{R}} &\equiv \boldsymbol{B}_{f\mathrm{R}} \boldsymbol{\psi}'_{f\mathrm{R}} = \boldsymbol{S}_f \boldsymbol{T}_f \boldsymbol{\psi}'_{f\mathrm{R}} \,, \end{aligned} \tag{9.164}$$

and the matrix $\mathcal{M}_f$ is diagonal,

$$(\mathcal{M}_f)_{AB} = m_A^f \, \delta_{AB} \,, \tag{9.165}$$

with the diagonal elements identified with the masses of the nine massive fermions emerging from the model:

$$\begin{aligned} \mathcal{M}_\mathrm{e} &= \mathrm{diagonal}\,(m_1^\mathrm{e},\, m_2^\mathrm{e},\, m_3^\mathrm{e}) = \mathrm{diagonal}\,(m_\mathrm{e},\, m_\mu,\, m_\tau) \,, \\ \mathcal{M}_\mathrm{u} &= \mathrm{diagonal}\,(m_1^\mathrm{u},\, m_2^\mathrm{u},\, m_3^\mathrm{u}) = \mathrm{diagonal}\,(m_\mathrm{u},\, m_\mathrm{c},\, m_\mathrm{t}) \,, \\ \mathcal{M}_\mathrm{d} &= \mathrm{diagonal}\left(m_1^\mathrm{d},\, m_2^\mathrm{d},\, m_3^\mathrm{d}\right) = \mathrm{diagonal}\,(m_\mathrm{d},\, m_\mathrm{s},\, m_\mathrm{b}) \,. \end{aligned} \tag{9.166}$$

The Yukawa Lagrangian now assumes a simpler form

$$\mathcal{L}_\mathrm{Y} = -\left(1 + \frac{H}{v}\right) \left[m_A^\mathrm{e}\,(\bar{e}_A e_A) + m_A^\mathrm{u}\,(\bar{u}_A u_A) + m_A^\mathrm{d}\,(\bar{d}_A d_A)\right] \,, \tag{9.167}$$

where, as before, $v = 2M_\mathrm{W}/g$. The strengths of the coupling of fermions to the Higgs depend linearly on $m_A^f/M_\mathrm{W}$, a factor which ranges from $6 \times 10^{-6}$ for the electron to 2 for the top quark.

**Fermion Currents.** Let us turn now to the fermion Lagrangian (155). Exactly as in the situation with one generation, here too the interaction terms describe the couplings of the gauge fields to the neutral and charged currents, which now, however, involve all quarks and leptons. The *electromagnetic current* is

$$j_\mu^{\mathrm{em}} = \overline{\boldsymbol{e}}'\gamma_\mu Q\boldsymbol{e}' + \overline{\boldsymbol{u}}'\gamma_\mu Q\boldsymbol{u}' + \overline{\boldsymbol{d}}'\gamma_\mu Q\boldsymbol{d}'$$
$$= \overline{\boldsymbol{e}}\gamma_\mu \boldsymbol{B}_{\mathrm{e}} Q\boldsymbol{B}_{\mathrm{e}}^\dagger \boldsymbol{e} + \overline{\boldsymbol{u}}\gamma_\mu \boldsymbol{B}_{\mathrm{u}} Q\boldsymbol{B}_{\mathrm{u}}^\dagger \boldsymbol{u} + \overline{\boldsymbol{d}}\gamma_\mu \boldsymbol{B}_{\mathrm{d}} Q\boldsymbol{B}_{\mathrm{d}}^\dagger \boldsymbol{d}\,. \tag{9.168}$$

Since fermions with the same charge and the same helicity have the same transformation properties under the gauge group $\mathrm{SU}(2)\times\mathrm{U}(1)$, the matrices $\boldsymbol{B}_{f\mathrm{L}}$ and $\boldsymbol{B}_{f\mathrm{R}}$ commute with the charge operator $Q$. And since both matrices are unitary, one immediately has

$$\boldsymbol{B}_{f\mathrm{L}} Q\boldsymbol{B}_{f\mathrm{L}}^\dagger = Q\,, \qquad \boldsymbol{B}_{f\mathrm{R}} Q\boldsymbol{B}_{f\mathrm{R}}^\dagger = Q\,. \tag{9.169}$$

In other words, in each case, $Q = T_3 + \frac{1}{2}Y$ is proportional to the identity matrix, with the same proportionality coefficients in both bases. Therefore,

$$j_\mu^{\mathrm{em}} = \overline{\boldsymbol{e}}\gamma_\mu Q\boldsymbol{e} + \overline{\boldsymbol{u}}\gamma_\mu Q\boldsymbol{u} + \overline{\boldsymbol{d}}\gamma_\mu Q\boldsymbol{d}\,. \tag{9.170}$$

By the same token, the weak *neutral current*

$$j_\mu^{\mathrm{Z}} = \overline{\boldsymbol{\nu}}_{\mathrm{L}}'\gamma_\mu Z\boldsymbol{\nu}_{\mathrm{L}}' + \overline{\boldsymbol{e}}'\gamma_\mu Z\boldsymbol{e}' + \overline{\boldsymbol{u}}'\gamma_\mu Z\boldsymbol{u}' + \overline{\boldsymbol{d}}'\gamma_\mu Z\boldsymbol{d}' \tag{9.171}$$

is unchanged in form when written in the mass eigenstate basis:

$$j_\mu^{\mathrm{Z}} = \overline{\boldsymbol{\nu}}_{\mathrm{L}}\gamma_\mu Z\boldsymbol{\nu}_{\mathrm{L}} + \overline{\boldsymbol{e}}\gamma_\mu Z\boldsymbol{e} + \overline{\boldsymbol{u}}\gamma_\mu Z\boldsymbol{u} + \overline{\boldsymbol{d}}\gamma_\mu Z\boldsymbol{d}\,. \tag{9.172}$$

This is again because $\boldsymbol{B}_{f\mathrm{L}}$ and $\boldsymbol{B}_{f\mathrm{R}}$ are unitary and commute with the weak charge operator $Z$:

$$\boldsymbol{B}_{f\mathrm{L}}(T_3 - Qs_{\mathrm{W}}^2)\boldsymbol{B}_{f\mathrm{L}}^\dagger = T_3 - Qs_{\mathrm{W}}^2\,, \quad \boldsymbol{B}_{f\mathrm{R}}(Qs_{\mathrm{W}}^2)\boldsymbol{B}_{f\mathrm{R}}^\dagger = Qs_{\mathrm{W}}^2\,. \tag{9.173}$$

Thus, the crucial property that the neutral currents are flavor-diagonal survives intact the transformation of basis; each chiral component of fermion goes to itself after emitting or absorbing a $Z_\mu$. Note that in the above we have defined $\boldsymbol{\nu}_{\mathrm{L}} \equiv \boldsymbol{S}_\nu \boldsymbol{\nu}_{\mathrm{L}}'$ for any arbitrary unitary matrix, arbitrary because it is not constrained by any mass matrix since the neutrinos are assumed to be mass-degenerate, i.e. massless.

Let us now consider the *charged current*

$$J_\mu^\dagger = j_\mu^1 + \mathrm{i}j_\mu^2 = \overline{\boldsymbol{l}}_{\mathrm{L}}\gamma_\mu\tau_+\boldsymbol{l}_{\mathrm{L}} + \overline{\boldsymbol{q}}_{\mathrm{L}}\gamma_\mu\tau_+\boldsymbol{q}_{\mathrm{L}}$$
$$= \overline{\boldsymbol{\nu}}_{\mathrm{L}}'\gamma_\mu \boldsymbol{e}_{\mathrm{L}}' + \overline{\boldsymbol{u}}_{\mathrm{L}}'\gamma_\mu \boldsymbol{d}_{\mathrm{L}}'\,.$$

In the lepton sector,

$$\overline{\boldsymbol{\nu}'}_{\mathrm{L}}\gamma_\mu \boldsymbol{e}_{\mathrm{L}}' = \overline{\boldsymbol{\nu}}_{\mathrm{L}}\,\gamma_\mu\,\boldsymbol{S}_\nu \boldsymbol{S}_{\mathrm{e}}^\dagger\,\boldsymbol{e}_{\mathrm{L}}\,. \tag{9.174}$$

Since $\boldsymbol{S}_\nu$ is an arbitrary unitary matrix, as we have already noted, it may be chosen so that $\boldsymbol{V}_\ell \equiv \boldsymbol{S}_\nu \boldsymbol{S}_\mathrm{e}^\dagger = 1$. This convention is allowed as long as the neutrinos remain exactly massless; but if it turns out that some or all of them acquire a nonnegligible mass, then $\boldsymbol{V}_\ell \neq 1$ necessarily (see Chap. 12). With $\boldsymbol{V}_\ell = 1$ so chosen, the lepton term reduces to

$$\overline{\boldsymbol{\nu}'}_\mathrm{L}\gamma_\mu \boldsymbol{e}'_\mathrm{L} = \overline{\boldsymbol{\nu}}_\mathrm{L}\gamma_\mu \boldsymbol{e}_\mathrm{L} = \overline{\nu}_{AL}\gamma_\mu e_{AL} \,.$$

In other words, it remains the same in form in the mass eigenstate basis and is diagonal in the generation labels.

As for the quarks, they contribute to the charged current

$$\overline{\boldsymbol{u}}'_\mathrm{L}\gamma_\mu \boldsymbol{d}'_\mathrm{L} = \overline{\boldsymbol{u}}_\mathrm{L}\gamma_\mu \,\boldsymbol{S}_\mathrm{u}\boldsymbol{S}_\mathrm{d}^\dagger\, \boldsymbol{d}_\mathrm{L} \,. \tag{9.175}$$

Since $\boldsymbol{S}_\mathrm{u}\boldsymbol{S}_\mathrm{d}^\dagger \neq 1$ generally, different generations are all mixed up in the quark mass eigenstates. The mixing may be entirely limited to either the u-type quarks or the d-type quarks. But it is customary to leave the three quarks of charge $Q = 2/3$ unmixed and let all the mixing be confined to the $Q = -1/3$ charge sector. Accordingly, with the shorthand notation

$$\boldsymbol{d}'' \equiv \boldsymbol{S}_\mathrm{u}\boldsymbol{S}_\mathrm{d}^\dagger\, \boldsymbol{d} \,,$$

the complete charged current for the model becomes

$$J_\mu^\dagger = \bar{\boldsymbol{\nu}}_\mathrm{L}\gamma_\mu \boldsymbol{e}_\mathrm{L} + \bar{\boldsymbol{u}}_\mathrm{L}\gamma_\mu \boldsymbol{d}''_\mathrm{L} \,. \tag{9.176}$$

Through this current, a neutrino converts itself into its corresponding charged lepton, conserving the lepton type, whereas a u-type quark can couple to any flavor of the d-type quarks, which results in a far greater variety of hadronic weak processes.

The unitary matrix $\boldsymbol{V} \equiv \boldsymbol{S}_\mathrm{u}\boldsymbol{S}_\mathrm{d}^\dagger$ is the generalization to three quark families of the Cabibbo rotation matrix. It was first introduced by Kobayashi and Maskawa (1973) and for this reason is referred to as the Cabibbo–Kobayashi–Maskawa (CKM) matrix. Explicitly,

$$\begin{pmatrix} d'' \\ s'' \\ b'' \end{pmatrix} = \begin{pmatrix} V_\mathrm{ud} & V_\mathrm{us} & V_\mathrm{ub} \\ V_\mathrm{cd} & V_\mathrm{cs} & V_\mathrm{cb} \\ V_\mathrm{td} & V_\mathrm{ts} & V_\mathrm{tb} \end{pmatrix} \begin{pmatrix} d \\ s \\ b \end{pmatrix} . \tag{9.177}$$

In general, a unitary $N \times N$ matrix can be parameterized by $N^2$ independent real quantities ($2N^2$ real parameters minus $N^2$ unitarity relations). Of these, $N(N-1)/2$ may be taken as the Euler angles associated with rotations in $N$-dimensional space. The remaining $N(N+1)/2$ are called phases, not all of which have physical meaning as some may be removed by redefining the quark fields that form the basis of the matrix representation. Of these $2N$ field phases ($N$ from the up-type quarks and another $N$ from the down-type quarks), $2N-1$ are not measurable. Thus, the number of measurable phases in the matrix is $\frac{1}{2}N(N+1)-(2N-1)=\frac{1}{2}(N-1)(N-2)$. In the standard model, with $N=3$ quark families, $\boldsymbol{V}$ contains 3 angles and 1 phase. A complex mixing matrix of this kind provides a mechanism for CP violation. Since no such a phase can appear in the presence of only two families of quarks, the existence of a third family could have been inferred from the observed CP violation in the neutral K mesons before the actual discoveries of the b and t quarks. Various equivalent parameterizations of $\boldsymbol{V}$ are possible; a popular one is

$$\boldsymbol{V}=\begin{pmatrix} c_{12}\,c_{13} & s_{12}\,c_{13} & s_{13}\,\mathrm{e}^{-\mathrm{i}\delta_{13}} \\ -s_{12}\,c_{23}-c_{12}\,s_{23}\,s_{13}\,\mathrm{e}^{\mathrm{i}\delta_{13}} & c_{12}\,c_{23}-s_{12}\,s_{23}\,s_{13}\,\mathrm{e}^{\mathrm{i}\delta_{13}} & s_{23}\,c_{13} \\ s_{12}\,s_{23}-c_{12}\,c_{23}\,s_{13}\,\mathrm{e}^{\mathrm{i}\delta_{13}} & -c_{12}\,s_{23}-s_{12}\,c_{23}\,s_{13}\,\mathrm{e}^{\mathrm{i}\delta_{13}} & c_{23}\,c_{13} \end{pmatrix}. \quad (9.178)$$

Here $c_{AB}=\cos\theta_{AB}$ and $s_{AB}=\sin\theta_{AB}$, with $A,B=1,2,3$ being generation labels. The real angles $\theta_{12}$, $\theta_{13}$, and $\theta_{23}$ can be made to lie in the first quadrant by properly choosing the quark field phases; then $c_{AB}\geq 0$, $s_{AB}\geq 0$ and $0\leq\delta_{13}\leq 2\pi$. In the limit $\theta_{13}=\theta_{23}=0$, the third generation decouples, and the situation reduces to the Cabibbo mixing of the first two generations, with $\theta_{12}$ identified with $\theta_{\mathrm{C}}$, the Cabibbo angle.

### 9.4.4 Summary and Extensions

In the original gauge-invariant Lagrangian, all fields behave as eigenstates of the gauge group. In particular, the fermion fields display a repetitive pattern and may be grouped into generations, each generation composed of a doublet of left-handed leptons, a doublet of left-handed quarks, a right-handed charged lepton, and two right-handed quarks. It is assumed that all neutrinos are massless. For three generations, $N_{\mathrm{G}}=3$, we have 21 chiral fields in all. General gauge-invariant Yukawa interactions couple generations together and lead to nondiagonal mass matrices, one matrix for each set of equal charge fermions. When the quark mass matrices are diagonalized, the gauge eigenstates of the d-type quarks appear as linear combinations of the mass eigenstates via unitary transformations. As for the leptons, the mass eigenstates are unmixed because of the assumed mass degeneracy (that is, complete absence of mass) of the neutrinos.

The final form of the Lagrangian of the standard model is written in the mass eigenstates. The gauge field and scalar field sectors are given in (112) and (115). The fermion sector, studied in this section, appears grouped into three families, differing from one another by their masses and flavor quantum numbers (see Table 9.5) but having essentially identical electroweak interaction properties. Their dynamics is described by the following terms.

Free-quark fields:

$$\begin{aligned}\mathcal{L}_{\mathrm{F}}^{0} &= \overline{\nu}_A\,\mathrm{i}\gamma^{\mu}\partial_{\mu}\,a_{\mathrm{L}}\nu_A+\overline{e}_A\,(\mathrm{i}\gamma^{\mu}\partial_{\mu}-m_A^{\mathrm{e}})\,e_A \\ &\quad +\overline{u}_A\,(\mathrm{i}\gamma^{\mu}\partial_{\mu}-m_A^{\mathrm{u}})\,u_A+\overline{d}_A\,(\mathrm{i}\gamma^{\mu}\partial_{\mu}-m_A^{\mathrm{d}})\,d_A\,. \end{aligned} \quad (9.179)$$

**Table 9.5.** Family pattern of the fundamental fermions in the standard model after spontaneous symmetry breaking, with $d''$, $s''$, and $b''$ denoting orthogonal combinations of $d$, $s$, and $b$ defined by the CKM matrix, $d''_A = V_{AB} d_B$

| 1 | 2 | 3 | $Q$ |
|---|---|---|---|
| $\begin{pmatrix} \nu_{\rm eL} \\ e_{\rm L} \end{pmatrix}$ | $\begin{pmatrix} \nu_{\mu\rm L} \\ \mu_{\rm L} \end{pmatrix}$ | $\begin{pmatrix} \nu_{\tau\rm L} \\ \tau_{\rm L} \end{pmatrix}$ | $\begin{matrix} 0 \\ -1 \end{matrix}$ |
| $e_{\rm R}$ | $\mu_{\rm R}$ | $\tau_{\rm R}$ | $-1$ |
| $\begin{pmatrix} u_{\rm L} \\ d''_{\rm L} \end{pmatrix}$ | $\begin{pmatrix} c_{\rm L} \\ s''_{\rm L} \end{pmatrix}$ | $\begin{pmatrix} t_{\rm L} \\ b''_{\rm L} \end{pmatrix}$ | $\begin{matrix} \frac{2}{3} \\ -\frac{1}{3} \end{matrix}$ |
| $u_{\rm R}$ | $c_{\rm R}$ | $t_{\rm R}$ | $\frac{2}{3}$ |
| $d_{\rm R}$ | $s_{\rm R}$ | $b_{\rm R}$ | $-\frac{1}{3}$ |

Higgs couplings:

$$\mathcal{L}_{\rm FH} = -\frac{g}{2\,M_{\rm W}} \left( m^{\rm e}_A\, \overline{e}_A e_A + m^{\rm u}_A\, \overline{u}_A u_A + m^{\rm d}_A\, \overline{d}_A d_A \right) H\,. \tag{9.180}$$

Electromagnetic coupling:

$$\mathcal{L}_{\rm em} = -\, e\big( Q_{\rm e}\, \overline{e}_A \gamma_\mu e_A + Q_{\rm u}\, \overline{u}_A \gamma_\mu u_A + Q_{\rm d}\, \overline{d}_A \gamma_\mu d_A \big)\, A^\mu\,. \tag{9.181}$$

Neutral current coupling:

$$\begin{aligned}\mathcal{L}_{\rm nc} = -\,\frac{g}{2\,c_{\rm W}} \big[ &\overline{\nu}_A \gamma_\mu (g^\nu_{\rm V} - g^\nu_{\rm A}\gamma_5) \nu_A + \overline{e}_A \gamma_\mu (g^{\rm e}_{\rm V} - g^{\rm e}_{\rm A}\gamma_5)\, e_A \\ &+ \overline{u}_A \gamma_\mu (g^{\rm u}_{\rm V} - g^{\rm u}_{\rm A}\gamma_5)\, u_A + \overline{d}_A \gamma_\mu (g^{\rm d}_{\rm V} - g^{\rm d}_{\rm A}\gamma_5)\, d_A \big]\, Z^\mu\,. \end{aligned} \tag{9.182}$$

Charged current coupling:

$$\mathcal{L}_{\rm cc} = -\,\frac{g}{\sqrt{2}} \left[ \big( \overline{\nu}_A\, \gamma_\mu\, a_{\rm L}\, e_A + \overline{u}_A\, \gamma_\mu\, a_{\rm L}\, V_{AB}\, d_B \big)\, W^\mu + \text{h.c.} \right], \tag{9.183}$$

where $a_{\rm L} = \frac{1}{2}(1-\gamma_5)$.

The weak charged currents display the required V–A structure, a well-established fact of low-energy physics. The W bosons couple with the same strength, $g/2\sqrt{2}$, to all charged fermionic currents (up to CKM mixing factors in the quark sector). The weak neutral currents, a new feature introduced by the unified model, conserve flavor and display universality in interaction: they couple to the Z field with the same coupling strengths in all generations, the values of $g^{\rm f}_{\rm V}$ and $g^{\rm f}_{\rm A}$ given in Table 9.3 for the first generation being also valid in general for every lepton or quark having the indicated charge.

In the gauge and scalar sectors, the model contains four parameters, $g$, $g'$, $\lambda$, and $\mu^2$, or alternatively, $e$, $\theta_{\rm W}$, $M_{\rm W}$, and $M_{\rm H}$. In the fermion sector, with $N_{\rm G} = 3$, thirteen parameters are needed: three charged lepton masses and six quark masses plus three quark mixing angles and one phase, all originating from the unknown Yukawa couplings $C^f$.

The model can be readily extended to include the *strong force* treated as the gauge interaction based on the color SU(3) group. As leptons are insensitive to this force, they are regarded as singlets under $\mathrm{SU_c(3)}$ (so also are the Higgs fields), but quarks belong to the fundamental triplet representations. Evidently, as the generators of the color group commute with the weak isospin and the weak hypercharge, the group of symmetry to be gauged is the direct product group

$$\mathrm{SU_c(3) \times SU_L(2) \times U_Y(1)}\,. \tag{9.184}$$

It is assumed that $\mathrm{SU_c(3)}$ remains unbroken, whereas $\mathrm{SU_L(2) \times U_Y(1)}$ spontaneously breaks down to $\mathrm{U_Q}$. This symmetry breaking is represented schematically by

$$\mathrm{SU_c(3) \times SU_L(2) \times U_Y(1) \to SU_c(3) \times U_Q(1)}\,. \tag{9.185}$$

Since (184) is a direct product group, no complications arise in the formulation, but neither can any relationships between the strong and the electroweak forces emerge from this juxtaposition of two gauge theories. The resulting Lagrangian for the gauge group $\mathrm{SU_c(3) \times SU_L(2) \times U_Y(1)}$ is essentially just the sum of (8.46), (112)–(115), and (179)–(183). The number of parameters has now increased by one, adding the strong coupling constant $g_s$, for a total of 18 in the case of three complete generations of fermions.

Thus, we now have the basic elements of a theory that proves to be consistent with the present state of our knowledge of particle physics and that in some cases (e.g. in weak neutral current processes) can pass stringent experimental tests at a very high degree of precision. In the following chapters, we shall re-examine the underlying assumptions of the theory, study a number of predictions, and introduce further theoretical concepts essential for a more complete theory.

Among the ingredients of the standard model, none is more important than the 'elementarity' of quarks. The evidence for this key property, as found in deep inelastic electron–nucleon and neutrino–nucleon scattering, will be considered in Chaps. 10 and 12. The latter chapter also dwells on the assumption of massless neutrinos (and hence that of the conservation of the lepton numbers), while Chap. 13 presents further proof of the universality of the left-handed structure of the weak charged current.

As mentioned above, the gauge sector of the electroweak theory contains three independent parameters, a convenient choice of which is

$$\begin{cases} \dfrac{4\pi}{e^2} = \alpha^{-1} = 137.0359895 \pm 0.0000061\,, \\ G_\mathrm{F} = (1.16639 \pm 0.00002) \times 10^{-5}\,\mathrm{GeV}^{-2}\,, \\ M_\mathrm{Z} = (91.1888 \pm 0.0044)\,\mathrm{GeV}\,. \end{cases} \tag{9.186}$$

The fine structure constant $\alpha$ can be determined from the quantum Hall effect, the Fermi constant $G_\mathrm{F}$ from the muon lifetime formula, and the $\mathrm{Z}^0$

gauge boson mass $M_Z$ from $e^+e^-$ and $p\bar{p}$ collider experiments. The model makes several very definite predictions, the simplest being on the charged boson mass and the weak mixing angle:

$$M_W^2 s_W^2 = \pi\alpha/(\sqrt{2}\,G_F)\,, \qquad s_W^2 = 1 - M_W^2/M_Z^2\,. \tag{9.187}$$

These relations, which follow from $e = g s_W$, $G_F/\sqrt{2} = g^2/(8M_W^2)$, and $M_W^2 = M_Z^2 c_W^2$, yield $M_W \approx 80.94$ GeV and $s_W^2 \approx 0.212$. The small differences with the measured values ($M_W = 80.33\,\text{GeV}$ and $s_W^2 = 0.2315$) are attributable to higher-order quantum corrections. Similar corrections, which turn out to be more substantial in several important observables, will be the subject of discussion in Chaps. 14 and 15. (The latter contains also a detailed study of essential properties of QCD.)

Whereas the weak *neutral* current is well known, as it depends only on $\alpha$ and $\theta_W$, this is not the case for the weak *charged* current for quarks since its coupling constant depends also on the CKM parameters. In principle, the magnitudes of the matrix elements $V_{AB}$ can be evaluated directly from the rates of the quark transitions $q_A \to q_B \ell^- \bar{\nu}_\ell$. However, since quarks are confined, the relevant processes are the corresponding leptonic weak decays of hadrons $H \to H'\ell^-\bar{\nu}_\ell$; their amplitudes always involve hadronic matrix elements of the weak charged currents – a nonperturbative QCD problem. For those matrix elements relating b and c quarks, the heavy-flavor symmetry allows a clean calculation of the hadronic form factor, but for those relating to the top quark, they can be accessed only indirectly, e.g. through the top's participation in the $B^0$–$\overline{B}^0$ mixing. Considerations of this kind are found in Chap. 16. The Kobayashi–Maskawa phase is even harder to come by; the best that can be done is to subject its value to constraints derived from the CP violation parameters $\varepsilon$ and $\varepsilon'$ of the neutral K mesons and from the $K^0$–$\overline{K}^0$ and $B^0$–$\overline{B}^0$ mixings (Chaps. 11 and 16).

Finally, the mass of the physical neutral Higgs scalar is not predicted by the model. It remains the most poorly known parameter in the model, and the existence and real nature of the Higgs boson is the object of active research (Chap. 17).

## Problems

**9.1 Necessity of conserved current.** Consider the coupling of a vector field of mass $M$ to a current of the form $j_\mu(x)A^\mu(x)$. Assume that in momentum space the vector field satisfies the Lorentz condition $k_\mu A^\mu = 0$, for any particle momentum $k_\mu$, to give three independent physical components. For a nonvanishing mass $M$, decompose $A^\mu$ into a transverse part $A^\mu_\perp$ (defined by $k_\mu A^\mu_\perp = 0$ and $\boldsymbol{k}\cdot\boldsymbol{A} = 0$) and its orthogonal longitudinal complement, $A^\mu_\parallel$. Let the first-order transition matrix be $\mathcal{M} = T_\mu A^\mu$, where

$T_\mu = \langle \mathrm{f} \,|\, j_\mu \,|\, \mathrm{i} \rangle$. Show that the longitudinal vector $A_\parallel^\mu$ increases with energy and causes ultraviolet divergences in $\mathcal{M}$ unless the current is conserved.

**9.2 Massive neutrino.** In the one-lepton family, make the appropriate modifications when the neutrino is massive, and justify the neglect of the right-handed component of the neutrino field in the limit of zero mass.

**9.3 Numerical estimates of parameters of the model.** It is convenient to take as inputs to the model the three parameters in (186) plus $M_\mathrm{H}$ and the fermion masses. Define also $A = \pi\alpha/\sqrt{2}G_\mathrm{F} = M_\mathrm{W}^2 s_\mathrm{W}^2$. From these data, calculate $M_\mathrm{W}$, $\theta_\mathrm{W}$, and $v$. In addition, with $m_\mathrm{e} = 0.511$ MeV, calculate the Yukawa coupling constant $C_\mathrm{e}$. Note that since $M_\mathrm{H}^2 = 2\lambda v^2$ and since there is no simple way of obtaining $\lambda$, it is not possible to predict its value.

**9.4 Decay width of $\mathrm{W}^\pm$.** Assuming the electron is massless and the W–lepton coupling given in the form $-(g/2\sqrt{2})\bar{e}\gamma^\mu(1-\gamma_5)\nu\, W_\mu^\dagger + \mathrm{h.c.}$, calculate the decay width $\Gamma(\mathrm{W}^+ \to \mathrm{e}^+\nu)$. Assuming the quarks are also massless, calculate to lowest order the decay widths of $\mathrm{W}^+$ to various allowed quark–antiquark channels, and give an estimate of the total decay width of W.

**9.5 Decay width of $\mathrm{Z}^0$.** The coupling of the Z boson to fermions is given by $(-g/c_W) j_\mu^Z\, Z^\mu$, where $j_\mu^Z$ is the weak neutral current for fermions. Calculate to lowest order the decay width $\Gamma(\mathrm{Z}^0 \to \nu\bar{\nu})$ and compare it with $\Gamma(\mathrm{W}^+ \to \mathrm{e}^+\nu)$. Also calculate to lowest order the rates of decay of Z to various allowed lepton–antilepton and quark–antiquark channels. Give an estimate of the total decay width of Z.

**9.6 Front–back asymmetry for $\mathrm{e}^+ + \mathrm{e}^- \to \mathrm{f}^+ + \mathrm{f}^-$.** In the reaction $\mathrm{e}^-(p) + \mathrm{e}^+(p') \to \mathrm{f}^-(k) + \mathrm{f}^+(k')$, where f is a charged fermion, there is a relative difference between the probabilities of observing $\mathrm{f}^-$ traveling in the forward ($\sigma_\mathrm{F}$) and backward ($\sigma_\mathrm{B}$) directions due to an interference between the contributions of the photon $\gamma$ and the weak boson $\mathrm{Z}^0$ exchanged in the s-channel. Compute the asymmetry of the total cross-sections, $A_\mathrm{FB} = (\sigma_\mathrm{F} - \sigma_\mathrm{B})/(\sigma_\mathrm{F} + \sigma_\mathrm{B})$. A measure of this quantity would give the Weinberg angle.

## Suggestions for Further Reading

*The classic papers:*

Glashow, S. L., Nucl. Phys. **22** (1961) 579

Salam, A., in *Elementary Particle Theory* (ed. by N. Svartholm). Almquist and Wiksells, Stockholm 1968, p. 367

Weinberg, S., Phys. Rev. Lett. **19** (1967) 1264

*Introduction of the charmed quark:*

Glashow, S. L., Iliopoulos, J. and Maiani, L., Phys. Rev. **D2** (1970) 1285

*Quark mixing matrix:*

Cabibbo, N., Phys. Rev. Lett. **10** (1963) 531

Kobayashi, M. and Maskawa, T., Progr. Theor. Phys. **49** (1973) 652

# 10 Electron–Nucleon Scattering

This chapter begins with an introduction to the notion of form factors and structure functions which play a central role in all electromagnetic and weak processes involving hadrons. As functions of the Lorentz-invariant momentum transfer $q^2$, form factors parameterize the interactive effects of the constituents of the hadrons. First, we give an intuitive physical interpretation of the electromagnetic form factor as the charge distribution of the hadron and associate its slope with the hadron size. Next, we look at the form factors of the weak interaction. Their normalization and dependence on $q^2$ are also discussed. A brief survey is made of their analytic property through dispersion relations and pole dominance. The nucleon form factors can be measured by elastic lepton–nucleon scattering, and the physical meaning of each term in the Rosenbluth formula for the cross-section is explained in detail.

We recall that a particle (of four-momentum $q_\mu$) is virtual or off-mass-shell if its invariant mass squared $q^2 = q_0^2 - |\boldsymbol{q}|^2$ is not necessarily equal to its true mass squared $m^2$, e.g. a virtual photon has $q^2 \neq 0$. The invariant mass of the virtual photon exchanged in electron–nucleon scattering can be varied by changing the energy and/or the angle of the scattered electron. The possibility of varying $q^2$ in deep inelastic scattering provides a powerful probe of the detailed structure of the nucleon, showing that quarks are pointlike constituents of matter. We introduce the Bjorken scaling law of the nucleon structure functions and its interpretation by Feynman with the quark–parton picture, which describes so well experimental data. Evidence for gluons as hadronic constituents insensitive to electroweak interactions is also given.

## 10.1 Electromagnetic and Weak Form Factors

In our present understanding, based on direct and indirect experimental data, there is every reason to believe that leptons and quarks – the fundamental constituents of matter – are structureless down to a distance scale of $10^{-16}$ cm, independently of their other properties. From the very light or even massless neutrinos to the top quark as heavy as the Au nucleus, all of these twelve fermionic constituents are assumed to be pointlike in spite of the huge differences in their masses.

On the other hand, mesons and baryons (hadrons) have structure. Their static and dynamic properties deduced from their production and decay modes, together with their spectra, all indicate that hadrons are effectively bound states of quarks (see Chap. 7). Like any composite object, the hadrons naturally carry complicated spatial structures and behave differently from pointlike leptons in their electromagnetic and weak interactions. For instance, the lepton–hadron cross-sections decrease rapidly with $q^2$, in sharp constrast with the lepton–lepton cross-sections.

This can be understood intuitively as follows: in lepton–lepton scattering, if a pointlike lepton is hit by a photon emitted from the other lepton, the only effect is that its momentum will be changed in a way consistent with energy-momentum conservation; the strength of the interaction and therefore the cross-section is insensitive to the momentum transfer $q^2$. On the other hand, in lepton–hadron scattering, because of the interaction between the constituents of the hadron and the photon emitted from the lepton, the strength of the interaction will depend on $q^2$; the more $q^2$ increases, the more the inner structure of the composite target can be probed and the nature of the interaction between the constituents can be revealed.

To describe the hadron structure, the standard approach is to introduce a *form factor*, which is the Fourier transform in momentum space of the spatial structure of the hadron. Its physical meaning is illustrated by the following example.

Let us consider the scattering of an electron by the static Coulomb field of a heavy nucleus (Rutherford experiment at the beginning of the century). In our contemporary language, the electron–nucleus interaction is governed by the exchange of a virtual spacelike photon between the projectile (electron) and the target (nucleus). When the three-momentum $\boldsymbol{q} = \boldsymbol{p}_f - \boldsymbol{p}_i$ transferred to the photon by the incoming electron $\boldsymbol{p}_i$ and the outgoing electron $\boldsymbol{p}_f$ is small, say $|\boldsymbol{q}| \sim 20\,\text{KeV} \sim 10^9/\text{cm} = 1/(10^{-9}\text{cm})$, the electromagnetic probe cannot penetrate the interior of the nucleus, which has a much smaller size $\sim 10^{-12}\,\text{cm}$, as if the latter were simply a pointlike positive charge $Ze$, $e > 0$. As the transferred momentum increases, say up to $20\,\text{MeV} = 1/(10^{-12}\,\text{cm})$ or higher, the complexity of the nucleus becomes more and more transparent and the photon starts to *see* the protons with their electric charges distributed inside the nucleus. The Coulomb potential of a *pointlike nucleus* should be replaced by that of an extended object

$$\frac{Ze}{4\pi\, r} \equiv \frac{Ze}{4\pi\, |\boldsymbol{x}|} \longrightarrow \int \mathrm{d}^3 y \frac{\rho(\boldsymbol{y})}{4\pi\, |\boldsymbol{x} - \boldsymbol{y}|} \equiv \frac{1}{4\pi} V(r)\,, \tag{10.1}$$

where $\rho(\boldsymbol{y})$ is the charge density of protons inside the nucleus, normalized to $Ze$: $\int \rho(\boldsymbol{y}) \mathrm{d}^3 y = Ze$. An unrealistic, structureless nucleus may be considered as a special case where all the protonic charges are concentrated at a single point: $\rho(\boldsymbol{y}) = Ze\, \delta^3(\boldsymbol{y})$. The Fourier transforms in momentum space of the potentials $Ze/4\pi\, r$ and $V(r)/4\pi$ are denoted by $V_{\text{pt}}(\boldsymbol{q})$ and $V(\boldsymbol{q})$ respectively,

$$V_{\text{pt}}(\boldsymbol{q}) \equiv \frac{Ze}{4\pi} \int \mathrm{d}^3 x\, \mathrm{e}^{-\mathrm{i}\boldsymbol{q}\cdot\boldsymbol{x}} \frac{1}{r} \quad , \quad V(\boldsymbol{q}) \equiv \frac{1}{4\pi} \int \mathrm{d}^3 x\, \mathrm{e}^{-\mathrm{i}\boldsymbol{q}\cdot\boldsymbol{x}} V(r)\,. \tag{10.2}$$

Our purpose is to show that $V(\boldsymbol{q})$ and $V_{\rm pt}(\boldsymbol{q})$ are related by a measurable nuclear form factor $F_{\rm N}(q^2)$ defined in (8) and (9). In (2), the three-dimension Fourier transformation proceeds only with a three-vector $\boldsymbol{q}$ adapted to the nonrelativistic case of a heavy nucleus of mass $M$ considered here. The transferred energy $q_0 \equiv \sqrt{M^2+|\boldsymbol{q}|^2} - M$ is practically zero, only $\boldsymbol{q}$ enters, and the fourth-component Fourier transform $\int {\rm d}t\ {\rm e}^{{\rm i}q_0 t} = 2\pi\,\delta(q_0)$ simply refers to this fact. Let us first consider $V_{\rm pt}(\boldsymbol{q})$,

$$\begin{aligned}\int {\rm d}^3x\ {\rm e}^{-{\rm i}\boldsymbol{q}\cdot\boldsymbol{x}}\frac{1}{r} &= \lim_{\mu\to 0}\int {\rm d}^3x\ {\rm e}^{-{\rm i}\boldsymbol{q}\cdot\boldsymbol{x}}\frac{{\rm e}^{-\mu r}}{r} \\ &= \lim_{\mu\to 0}\ 2\pi\int_0^\infty \frac{{\rm e}^{-\mu r}}{r} r^2 {\rm d}r \int_{-1}^{1} {\rm e}^{-{\rm i}|\boldsymbol{q}|\cdot r\cos\theta}{\rm d}\cos\theta \\ &= \lim_{\mu\to 0}\frac{4\pi}{(|\boldsymbol{q}|^2+\mu^2)} = \frac{4\pi}{|\boldsymbol{q}|^2}\ . \end{aligned} \tag{10.3}$$

The parameter $\mu$ (which has the dimension of mass or inverse of length, since $\mu r$ is dimensionless) is introduced to make the integral easier to handle, the final result is however independent of it. We have

$$V_{\rm pt}(\boldsymbol{q}) = \frac{Ze}{|\boldsymbol{q}|^2}\ . \tag{10.4}$$

Instead of (4), where everything is expressed in terms of the three-vector $\boldsymbol{q}$, it would be nice to have a covariant form with the Lorentz-invariant $q^2 = q_0^2 - |\boldsymbol{q}|^2$ of the four-vector $q_\mu$. The Fourier transform of the static Coulomb potential corresponds, as we have seen, to zero energy transfer for which the invariant $q^2$ takes the $-|\boldsymbol{q}|^2$ value. One then naturally arrives at the prescription: $|\boldsymbol{q}|^2$ is to be replaced by $-q^2$. Note that the invariant $q^2$ can be timelike ($q^2 > 0$) or spacelike ($q^2 < 0$), however in all scattering processes, $q^2$ is necessarily spacelike and the substitution $q^2 \to -|\boldsymbol{q}|^2 < 0$ is natural. We rewrite (4) in the form

$$V_{\rm pt}(q^2) = \frac{-Ze}{q^2}\ . \tag{10.5}$$

**Yukawa potential.** We can make a remark about (3): from (2) and (4), the nonrelativistic limit $1/|\boldsymbol{q}|^2$ of the propagator $-1/q^2$ of a massless boson mediated between the electron and the nucleus gives rise to a potential proportional to $1/r$.

By the same trick, when going back from the bottom to the top right-hand side of (3), we realize that the nonrelativistic limit $1/(|\boldsymbol{q}|^2+\mu^2)$ of $-1/(q^2-\mu^2)$ (propagator of an exchanged spinless particle of mass $\mu$) can generate the potential ${\rm e}^{-\mu r}/r$. The range of the force is $1/\mu$ (since for a distance beyond this range, $\mu r > 1$ , ${\rm e}^{-\mu r}$ becomes exponentially negligible).

It results that the exchange of a meson through its nonrelativistic propagator is the source of the interacting potential between two particles

$$\int \frac{d^3q\, e^{i\boldsymbol{q}\cdot\boldsymbol{x}}}{(2\pi)^3} \frac{1}{|\boldsymbol{q}|^2} = \frac{1}{4\pi\, r} \quad ; \quad \int \frac{d^3q\, e^{i\boldsymbol{q}\cdot\boldsymbol{x}}}{(2\pi)^3} \frac{1}{|\boldsymbol{q}|^2+\mu^2} = \frac{e^{-\mu r}}{4\pi\, r} \ .$$

This fundamental concept that brings together the two-body potential $e^{-\mu r}/r$ and the mass $\mu$ of their mediated meson was discovered by Yukawa in 1935. Its physical meaning is already mentioned in Chap. 1 (Fig. 1.1). Knowing the nuclear force range of about 1 or 2 fm [1 fm $= 10^{-13}$cm $\approx (200\,\text{MeV})^{-1}$], Yukawa then predicted the existence and the mass between 100–200 MeV of a spinless particle, which later turns out to be the $\pi$ meson exchanged between nucleons. When $\mu \to 0$, we recover the Coulomb potential; the infinite range of the electromagnetic force is a direct consequence of massless photons. Similarly, if the Coulomb potential between two charges $e_1$ and $e_2$ is $e_1e_2/4\pi r$, the nuclear potential produced by the $\pi$ meson exchanged between the two nucleons would be $g^2_{\pi\text{NN}}e^{-r\,m_\pi}/4\pi r$, where $g_{\pi\text{NN}}$ is the pion–nucleon coupling constant ($g^2_{\pi\text{NN}}/4\pi \approx 13.5$). The spin effect of the nucleon can also be incorporated and yields the one-pion-exchange (OPE) nucleon–nucleon force (Problem 10.1). It is important not to confuse the notion of form factors considered here with the Yukawa mechanism that provides the interacting potential $e^{-\mu r}/r$ from an exchanged boson of mass $\mu$.

Let us go back to the potential $V(\boldsymbol{q})$ in (1) and (2):

$$V(\boldsymbol{q}) = \frac{1}{4\pi}\int d^3x\, e^{-i\boldsymbol{q}\cdot\boldsymbol{x}} \int d^3y \frac{\rho(\boldsymbol{y})}{|\boldsymbol{x}-\boldsymbol{y}|} \ . \tag{10.6}$$

Putting $\boldsymbol{x}-\boldsymbol{y}=\boldsymbol{z}$ and using (3), we obtain

$$V(\boldsymbol{q}) = \frac{1}{4\pi}\int d^3z \frac{e^{-i\boldsymbol{q}\cdot\boldsymbol{z}}}{|\boldsymbol{z}|} \int d^3y\, e^{-i\boldsymbol{q}\cdot\boldsymbol{y}}\rho(\boldsymbol{y}) = \frac{F_\text{N}(q^2)}{|\boldsymbol{q}|^2} \ . \tag{10.7}$$

$V(\boldsymbol{q})$ is the nonrelativistic version of $V(q^2)$ and $F_\text{N}(q^2)$ defined by

$$F_\text{N}(q^2=-|\boldsymbol{q}|^2) \equiv \int d^3y\, e^{-i\boldsymbol{q}\cdot\boldsymbol{y}}\rho(\boldsymbol{y}) \tag{10.8}$$

is the Fourier transform of $\rho(\boldsymbol{y})$, the proton charge distribution in the nucleus. $F_\text{N}(q^2)$ is called the electromagnetic form factor of the nucleus, with the normalization $F_\text{N}(0) = Ze$, as can be seen by putting $|\boldsymbol{q}| = 0$ in (8) and remembering that $\int d^3y\, \rho(\boldsymbol{y}) = Ze$. From (4) to (7) we get

$$V(q^2) = V_\text{pt}(q^2)\frac{F_\text{N}(q^2)}{F_\text{N}(0)} = \frac{-F_\text{N}(q^2)}{q^2} \ . \tag{10.9}$$

The meaning of form factors can be seen by comparing (5) with (9).

Now in (8) we expand $\mathrm{e}^{-\mathrm{i}\boldsymbol{q}\cdot\boldsymbol{y}} = 1 - \mathrm{i}\,\boldsymbol{q}\cdot\boldsymbol{y} - \frac{1}{2}|\boldsymbol{q}|^2 r^2 \cos^2\theta + \cdots$, then after the integration over $\mathrm{d}^3y$, we obtain $F_\mathrm{N}(q^2) = F_\mathrm{N}(0)[1 + \frac{1}{6}\langle r^2\rangle q^2 + \cdots]$ (remember the $|\boldsymbol{q}|^2 \to -q^2$ substitution). The integration of the linear term $\boldsymbol{q}\cdot\boldsymbol{y}$ vanishes by the spatial symmetry, and the coefficient $1/6 = (1/2)\ (1/3)$ comes from averaging $\cos^2\theta$. The quantity

$$\langle r^2\rangle \equiv \frac{6}{F_\mathrm{N}(0)} \left| \frac{\mathrm{d}F_\mathrm{N}(q^2)}{\mathrm{d}q^2} \right|_{q^2=0}$$

represents the squared radius of the nucleus.

Equations (8) and (9) show that the notion of form factor is appropriate for describing the particle structure, the slope of the form factor at $q^2 = 0$ gives the hadron size; *the greater the object is, the faster its form factor decreases.* The electromagnetic and weak form factors of a pointlike object are $q^2$-independent. Form factors are directly measurable physical quantities. We will see later in (37) that by performing electron scattering on nuclei, we can measure $F_\mathrm{N}(q^2)$ via the Rutherford or Mott cross-section, and once $F_\mathrm{N}(q^2)$ is obtained, by the inverse Fourier transformation of (8), we get $\rho(\boldsymbol{y})$, the distribution of protons inside the nucleus:

$$\rho(\boldsymbol{y}) = \frac{1}{(2\pi)^3} \int \mathrm{d}^3q\, \mathrm{e}^{+\mathrm{i}\boldsymbol{q}\cdot\boldsymbol{y}} F_\mathrm{N}(q^2 = -|\boldsymbol{q}|^2)\ .$$

The notion of the nucleus form factor $F_\mathrm{N}(q^2)$, taken as an illustrative example, can be generalized to hadrons. The electromagnetic or weak properties of the latter are influenced by strong interactions of their constituents (quarks and gluons) which in turn provide them with form factors, in the same way as the protons – constituents of nuclei – induce the nucleus form factor $F_\mathrm{N}(q^2)$.

For example, the $\pi^\pm$ and $\mathrm{K}^\pm$ mesons are not pointlike, their radii can be measured by $\mathrm{e}^+ + \mathrm{e}^- \to \pi^+ + \pi^-$ or $\mathrm{K}^+ + \mathrm{K}^-$ (Fig. 10.1).

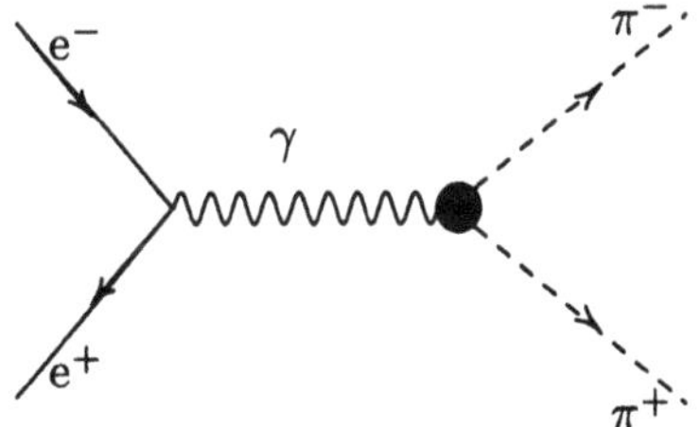


**Fig. 10.1.** Measurement of the pion form factor $F_\pi(q^2)$ and its radius $\langle r_\pi\rangle$ by $\mathrm{e}^+ + \mathrm{e}^- \to \pi^+ + \pi^-$

The *anomalous* magnetic moment of the proton is another manifestation of the proton structure. In the following Gordon decomposition of a pointlike fermionic current $\overline{u}(P')\gamma_\mu u(P)$ (see the Appendix),

$$e\overline{u}(P')\gamma_\mu u(P) A^\mu = \frac{e}{2M}\overline{u}(P')\left[(P'+P)_\mu + \mathrm{i}\sigma_{\mu\nu}(P'-P)^\nu\right] u(P) A^\mu\ , \quad (10.10)$$

the second term $(e/2M)\overline{u}(P')\left[\mathrm{i}\sigma_{\mu\nu}(P'-P)^\nu\right]u(P)A^\mu$ – which can be written in the nonrelativistic limit as $(e/2M)\overline{u}(P')\boldsymbol{\sigma}\cdot\boldsymbol{B}u(P)$ – clearly indicates that the magnetic moment of a pointlike fermion of charge $e$ and mass $M$ is equal to one Bohr magneton $\mu_\mathrm{B}\equiv(e/2M)$. Here $\boldsymbol{B}$ is the external magnetic field, and $\frac{1}{2}\boldsymbol{\sigma}$ is the proton spin.

Experimentally, the magnetic moment of the proton turns out to be 2.79 $\mu_\mathrm{B}$. The difference of 1.79 $\mu_\mathrm{B}$, which obviously reflects the complex structure of the proton, is called the *anomalous* magnetic moment. Compared with the anomalous term of the pointlike electron which is very close to $\alpha_\mathrm{em}/2\pi=0.001161$ [due to QED radiative corrections first computed by Schwinger, see (14.15)], the proton anomalous magnetic moment is very large indeed. Moreover, experiments show that the neutron also has a magnetic moment equal to $-1.91\,\mu_\mathrm{B}$ and not zero as naively expected for a neutral pointlike particle. As discussed in Sect. 7.4.3, these anomalous magnetic moments can be viewed as the effects of the quark constituents of the nucleon.

In brief, probed by electromagnetic (weak) currents, the electromagnetic (weak interaction) properties of hadrons can be described by form factors which encapsulate strong interaction effects of the hadronic constituents. Form factors are usually parameterized as $F(q^2)=F(0)/(1-q^2/\Lambda^2)^n$; the powers $n=1,2$ correspond to monopole and diplole respectively, and $\Lambda$ is the pole mass. Since electroweak cross-sections and decay rates of hadrons are functions of form factors, the importance of $F(0)$ and of the $q^2$ behavior is evident. In principle, the strong interaction dynamics of the quark and gluon constituents should determine the $q^2$ dependence and the value $F(0)$ of the hadronic form factors; however their determination is far from being achieved at present although there has been significant progress. The main reason is that we are in the low-energy QCD regime (whimsically called infrared slavery to contrast with ultraviolet asymptotic freedom corresponding to high energies) which deals with bound state problems for which the strong coupling constant is not small, and a perturbative treatment is inadequate. Nonperturbative methods, such as quark models, QCD sum rules, or lattice gauge theory, not discussed in this book, are still inconclusive at present.

However, even without any guidance from the dynamics of the hadronic constituents, the form factors can be obtained kinematically by using only a few principles: Lorentz invariance, conservation of currents, and *heavy flavor symmetry* (Chap. 16). In some cases, the normalizations are also fixed by these principles. We give in the following three typical examples: the pion and the nucleon electromagnetic form factors, and the weak form factors involved in the semileptonic decay $\overline{\mathrm{B}}\to\mathrm{D}+\ell^-+\overline{\nu}_\ell$ of the bottom B meson.

**Example 10.1 Electromagnetic Pion Form Factor**
The most general amplitude for the interaction of the charged pions $\pi^\pm$ (of initial four-momentum $k$ and final four-momentum $k'$) with the photon $\varepsilon_\mu(q)$ can be written as $\pm e\,\varepsilon_\mu(q)\,T^\mu(k',k)$, where

$$T^\mu(k',k)\equiv\langle\pi(k')\,|\,J^\mu_\mathrm{em}(0)\,|\,\pi(k)\rangle\,.$$

The quantity $T^\mu$ must obey $q_\mu T^\mu = 0$, $q_\mu = (k' - k)_\mu$ since the electromagnetic current $J^\mu_{\rm em}$ is conserved, i.e. $\partial_\mu J^\mu_{\rm em} = 0$. With spinless particles, we have at our disposal only their momenta $k$ and $k'$ as degrees of freedom, hence the most general Lorentz four-vector $T^\mu$ should have the form $a(k'+k)^\mu + bq^\mu$, which is reduced to $b = 0$ by $q_\mu T^\mu = bq^2 = 0$. Consequently, $(k'+k)^\mu F_\pi(q^2)$ is the only possible form of $T^\mu(k', k)$, and $F_\pi(q^2)$ is called the charged pion form factor, normalized by the condition $F_\pi(0) = 1$ (Problem 10.2),

$$\langle \pi(k') \,|\, J^\mu_{\rm em}(0) \,|\, \pi(k) \rangle = (k'+k)^\mu F_\pi(q^2) , \qquad F_\pi(0) = 1 . \tag{10.11}$$

For simplicity, we omit the standard one-particle state normalization factor $[1/\left[(2\pi)^3\sqrt{4E_k E'_k}\right]$ in (11). The electromagnetic interaction of a pointlike spinless particle $\pm e\,\varepsilon_\mu(q)(k'+k)^\mu$ becomes $\pm e\,\varepsilon_\mu(q)(k'+k)^\mu F_\pi(q^2)$ for the $\pi^\pm$ meson

$$\begin{array}{lcl} \pm\, e\,(k'+k)^\mu & \longrightarrow & \pm e\,(k'+k)^\mu F_\pi(q^2) , \\ \text{pointlike pion} & \longrightarrow & \text{physical pion} . \end{array}$$

The above substitution is reminiscent of (9) for a spinless nucleus with its form factor $F_{\rm N}(q^2)$. The proton distribution inside the nucleus is responsible of $F_{\rm N}(q^2)$, similarly the strong interaction effects due to the quark and gluon constituents of the pion produce the form factor $F_\pi(q^2)$ (Fig. 10.2a). This form factor is already measured in $\mathrm{e}^- + \mathrm{e}^+ \to \pi^- + \pi^+$, where $q^2 \geq 4m_\pi^2$ is timelike (Problem 10.3). It may be also measured by the pion scattering on atomic electrons $\mathrm{e}^- + \pi^\pm \to \mathrm{e}^- + \pi^\pm$ or by pion electroproduction, in which $q^2 \leq 0$ is spacelike [see (36) below]. The size of the pion, $\sqrt{\langle r_\pi^2 \rangle}$, is given by

$$\langle r_\pi^2 \rangle = 6 \left| \frac{\mathrm{d}F_\pi(q^2)}{\mathrm{d}q^2} \right|_{q^2=0} .$$

Of course, all other charged spinless flavored mesons, such as $\mathrm{K}^\pm$, $\mathrm{D}^\pm$, and $\mathrm{B}^\pm$ have similar electromagnetic form factors $F_{\rm K}(q^2), F_{\rm D}(q^2)$, and $F_{\rm B}(q^2)$, normalized by $F_{\rm K,D,B}(0) = 1$ as in the pion case. We emphasize that these normalizations are *model-independent results*, due to the conserved electromagnetic current. These values at (and only at) $q^2 = 0$ are not modified by strong interaction effects of the hadronic constituents (quarks and gluons). This important result is known as the nonrenormalized form factors.

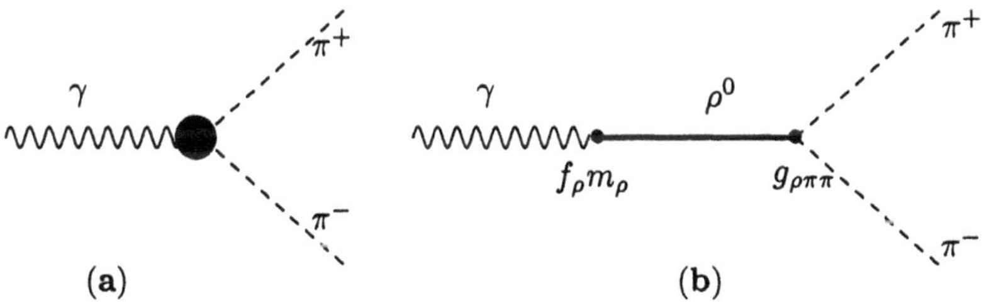


**Fig. 10.2.** (a) Pion form factor $F_\pi(q^2)$; (b) $\rho^0$ dominance of $F_\pi(q^2)$

**Example 10.2 Electromagnetic Nucleon Form Factors**
To describe the electromagnetic current $J^\mu_{\rm em}$ of the nucleon on the most general grounds, we note that there are in all four matrices of the vectorial type to be inserted between the nucleon spinors $\overline{u}(P')$ and $u(P)$. They are $\gamma_\mu$ , $\mathrm{i}\sigma_{\mu\nu}q^\nu$ , $q_\mu \equiv (P'-P)_\mu$, and $(P'+P)_\mu$. However, from the Gordon decomposition (10), the term $(P'+P)_\mu$ can be written as a combination of $\gamma_\mu$ and $\mathrm{i}\sigma_{\mu\nu}q^\nu$; in addition, the conservation of the electromagnetic current implies that the term proportional to $q_\mu$ must be zero, using both $q_\mu\overline{u}(P')\gamma^\mu u(P)=0$ and the antisymmetry of $\sigma_{\mu\nu}$. Therefore, the most general form of the nucleon electromagnetic current is expressed in terms of only two dimensionless form factors $F_1(q^2)$ and $F_2(q^2)$:

$$
\begin{aligned}
e\,\overline{u}(P')\gamma_\mu u(P) \quad &\longrightarrow \quad e\,\overline{u}(P')\left[\gamma_\mu F_1(q^2)+\mathrm{i}\frac{\sigma_{\mu\nu}q^\nu}{2M}F_2(q^2)\right]u(P)\,, \\
\text{pointlike nucleon} \quad &\longrightarrow \quad \text{physical nucleon}\;, \qquad (10.12)
\end{aligned}
$$

with $F_1^{\rm p}(0)=1$ (proton electric charge), $F_1^{\rm n}(0)=0$ (neutron electric charge), $F_2^{\rm p}(0)=1.79$ (proton anomalous magnetic moment), and $F_2^{\rm n}(0)=-1.91$ (neutron anomalous magnetic moment).

As in the case of meson form factors $F_{\rm M}(0)=1$ (M stands for mesons), the normalizations $F_1^{\rm p}(0)=1$ , $F_1^{\rm n}(0)=0$ are exact results due to the conserved electric current. On the other hand, the anomalous magnetic terms $F_2^{\rm p,n}(0)$ cannot be computed from first principle.

We may consider the proton and neutron as the $I_3=+1/2$ and $I_3=-1/2$ components of an isospin doublet $I=1/2$, and define the isoscalar $F_i^0(q^2)$ and isovector $F_i^1(q^2)$ form factors by $(i=1,2)$:

$$
F_i^0(q^2)=F_i^{\rm p}(q^2)+F_i^{\rm n}(q^2)\,; \qquad F_i^1(q^2)=F_i^{\rm p}(q^2)-F_i^{\rm n}(q^2)\,. \qquad (10.13)
$$

With $\frac{1}{2}(1+\tau_3)u=u_{\rm p}$ , $\frac{1}{2}(1-\tau_3)u=u_{\rm n}$, (12) may be rewritten as

$$
\frac{e}{2}\overline{u}(P')\left\{\left[\gamma_\mu F_1^0(q^2)+\mathrm{i}\frac{\sigma_{\mu\nu}q^\nu}{2M}F_2^0(q^2)\right]+\left[\gamma_\mu F_1^1(q^2)+\mathrm{i}\frac{\sigma_{\mu\nu}q^\nu}{2M}F_2^1(q^2)\right]\tau_3\right\}u(P).
$$

These form factors can be measured by the elastic scattering $\mathrm{e}^-+\mathrm{N}\to\mathrm{e}^-+\mathrm{N}$ ($q^2\leq 0$, see Fig. 4.9 and Sect. 10.3 below) or by the annihilation $\mathrm{e}^++\mathrm{e}^-\to\mathrm{N}+\overline{\mathrm{N}}$ ($q^2\geq 4M^2$). The isovector form factors $F_1^1(q^2)$ , $F_2^1(q^2)$ are useful when we study weak currents in nucleon $\beta$-decay together with the conserved vector current (CVC) property (Chap. 12). The same CVC relates the electromagnetic pion form factor $F_\pi(q^2)$ to the weak one, the latter appears for example in pion $\beta$-decay $\pi^-\to\pi^0+\mathrm{e}^-+\overline{\nu}_{\rm e}$ (Problem 10.4).

**Example 10.3 $\overline{\mathrm{B}}\to\mathrm{D}$ Weak Decay Form Factors**
The amplitude $\mathcal{A}$ of the semileptonic decay $\overline{\mathrm{B}}(p)\to\mathrm{D}(p')+\ell^-(k_1)+\overline{\nu}_\ell(k_2)$ (the four-momenta $p,p',k_1,k_2$ of these particles are indicated in parentheses)

is a product of two matrix elements, that of the V − A left-handed hadronic weak current $H^\mu$ taken between these meson states and that of the V − A leptonic current $L_\mu = \bar{\ell}\gamma_\mu(1-\gamma_5)\nu_\ell$ taken between the vacuum and the lepton pair $\ell$-$\bar{\nu}_\ell$, where $\langle \ell^-(k_1), \bar{\nu}_\ell(k_2) \,|\, L_\mu \,|\, 0\rangle = \bar{u}(k_1)\gamma_\mu(1-\gamma_5)v(k_2)$:

$$\mathcal{A} = \frac{G_\mathrm{F}}{\sqrt{2}} V_\mathrm{cb} \langle \ell^-(k_1), \bar{\nu}_\ell(k_2) \,|\, L_\mu \,|\, 0\rangle \times \langle \mathrm{D}(p') \,|\, H^\mu \,|\, \overline{\mathrm{B}}(p)\rangle \ . \tag{10.14}$$

The hadronic current $H^\mu = \bar{c}\gamma^\mu(1-\gamma_5)b \equiv V^\mu - A^\mu$ is written in terms of the relevant charm and bottom quark fields, $V_\mathrm{cb}$ is the corresponding Cabibbo–Kobayashi–Maskawa (CKM) flavor mixing, and $G_\mathrm{F}$ is the Fermi coupling constant (Chap. 9). The second-quantized $b(x)$ field represents the annihilation operator of the b quark, while the $\bar{c}(x)$ field refers to the creation of the c quark, corresponding to the decay $\mathrm{b} \to \mathrm{c} + \ell^- + \bar{\nu}_\ell$ we are considering.

Since both $\overline{\mathrm{B}}(\mathrm{b}\bar{\mathrm{q}})$ and $\mathrm{D}(\mathrm{c}\bar{\mathrm{q}})$ are pseudoscalar ($J^P = 0^-$) mesons, one has $\langle \mathrm{D}(p') \,|\, A^\mu \,|\, \overline{\mathrm{B}}(p)\rangle = 0$ from general considerations of Lorentz covariance and parity. Indeed, with only two momenta $p_\alpha$ and $p'_\beta$ as degrees of freedom at our disposal, there is no way to build up a matrix element of an axial-vector $A^\mu$ sandwiched between two spinless mesons having the *same intrinsic parity*. The matrix element of $A^\mu$ in this case must have the structure $\varepsilon^{\mu\alpha\beta\gamma}p_\alpha p'_\beta P_\gamma$ suited to its $J^P = 1^+$ property. But an independent third vector $P_\gamma$ is lacking to construct such a term. So, $\langle 0^\pm \,|\, A^\mu \,|\, 0^\pm\rangle = 0$ for all $J^P = 0^\pm$ mesons.

There remains the vector part $\langle 0^\pm \,|\, V^\mu \,|\, 0^\pm\rangle \neq 0$. If one particle is pseudoscalar, the other is scalar (or the vacuum), then the roles of $V^\mu$ and $A^\mu$ are interchanged, i.e. $\langle 0^\pm \,|\, V^\mu \,|\, 0^\mp\rangle = 0$ while $\langle 0^\pm \,|\, A^\mu \,|\, 0^\mp\rangle \neq 0$. The best example is $\langle 0 \,|\, A^\mu \,|\, \pi(k)\rangle = \mathrm{i} f_\pi k^\mu$. From Lorentz covariance, the most general matrix element of $V^\mu$ sandwiched between the two $J^P = 0^-$ mesons is

$$\langle \mathrm{D}(p') \,|\, V^\mu \,|\, \overline{\mathrm{B}}(p)\rangle = f_+(q^2)(p+p')^\mu + f_-(q^2)(p-p')^\mu \ , \tag{10.15}$$

where $q \equiv p - p' = k_1 + k_2$ is the four-momentum transfer and $f_+(q^2)$ and $f_-(q^2)$ are the dimensionless weak transition $\overline{\mathrm{B}} \to \mathrm{D}$ form factors which are functions of the invariant $q^2$. Contrary to the electromagnetic current $J^\mu_\mathrm{em}$ of charged pions considered in (11), the weak vector current $V^\mu = \bar{c}\gamma^\mu b$ is not conserved $[q_\mu V^\mu \propto (m_b - m_c) \neq 0]$, hence we have two form factors $f_\pm(q^2)$ instead of a single $F_\pi(q^2)$ as in the pion case. Here the timelike $q^2 = (k_1+k_2)^2$ represents also the squared invariant mass of the lepton pair, it varies within the range $m_\ell^2 \leq q^2 \leq (M_\mathrm{B} - M_\mathrm{D})^2 \equiv q^2_\mathrm{max}$.

Unlike the case of electromagnetic interactions with the exact results (11), the normalizations of the weak form factors are in general unknown. However, in the limit of infinitely heavy quark masses, $\Lambda_\mathrm{QCD} \ll M_\mathrm{B}, M_\mathrm{D} \to \infty$, a new *heavy flavor symmetry* (Chap. 16) appears in the effective Lagrangian of the standard model. This symmetry provides *model-independent* normalization of the weak form factors $f_\pm(q^2_\mathrm{max})$ at $q^2_\mathrm{max}$, using a method similar to the derivation of $F_\pi(0) = 1$. The results are

$$f_+(q^2_\mathrm{max}) = \frac{M_\mathrm{B} + M_\mathrm{D}}{2\sqrt{M_\mathrm{B}M_\mathrm{D}}} \quad , \quad f_-(q^2_\mathrm{max}) = -\frac{M_\mathrm{B} - M_\mathrm{D}}{2\sqrt{M_\mathrm{B}M_\mathrm{D}}} \ . \tag{10.16}$$

The normalizations (16) are of great importance for the determination of $V_{bc}$. Its proof based on the heavy flavor symmetry will be given in (16.75).

## 10.2 Analyticity and Dispersion Relation

We remark that form factors are analytic functions in the complex $q^2$ plane except for singularities on the real timelike $q^2 \geq 0$ axis. This property is illustrated by a calculation of the magnetic form factor in Chap. 14; we also show, in another explicit example of the vacuum polarization (Chap. 15), that Feynman loop amplitudes are analytic. The singularities (poles or cuts) exist whenever the variable $q^2$ has values for which it is possible for all the particles in an intermediate state to be on the mass shell, i.e. to be physical [see Fig. 10.2b and (15.32)]. Let us take the simplest example of $F_\pi(q^2)$. For $q^2 \geq 4m_\pi^2$, the virtual photon can produce two on-mass-shell pions, $F_\pi(q^2)$ becomes a complex function with a cut starting at $q^2 \geq 4m_\pi^2$.

The discontinuity of $F_\pi(q^2)$ above and below the cut gives the imaginary part of $F_\pi(q^2)$: $F_\pi(q^2+\mathrm{i}\varepsilon) - F_\pi(q^2-\mathrm{i}\varepsilon) = 2\mathrm{i}\,\mathrm{Im}F_\pi(q^2)$. An isolated single intermediate state gives rise to a pole.

On general grounds, the analyticity of the scattering amplitudes (or form factors) was first derived by Gell-Mann, Goldberger, and Thirring from the condition of macroscopic causality, which states that commutators of field operators vanish when the points at which the operators are evaluated are separated by a spacelike interval. Then using the Cauchy theorem, we can write dispersion relations relating them to their imaginary parts. The once-subtracted dispersion relation for $F_\pi(q^2)$ is

$$F_\pi(q^2) = F_\pi(0) + \frac{q^2}{\pi}\int_{4m_\pi^2}^{\infty} \frac{\mathrm{Im}F_\pi(s)}{s(s-q^2-\mathrm{i}\epsilon)}\mathrm{d}s\,. \tag{10.17}$$

Of course the unsubtracted dispersion relation is equally valid, provided that the function $F_\pi(z)$ decreases rapidly $[F_\pi(z) \to 0$ as $|z| \to \infty]$ to allow the integral to converge. In this case the imaginary part obeys the sum rule

$$F_\pi(q^2) = \frac{1}{\pi}\int_{4m_\pi^2}^{\infty} \frac{\mathrm{Im}F_\pi(s)}{s-q^2-\mathrm{i}\epsilon}\mathrm{d}s\,, \quad F_\pi(0) = 1 = \frac{1}{\pi}\int_{4m_\pi^2}^{\infty} \frac{\mathrm{Im}F_\pi(s)}{s}\mathrm{d}s\,. \tag{10.18}$$

The computation of the form factor then reduces to evaluating its imaginary part. $\mathrm{Im}[F_\pi(s)]$ can be obtained from $\mathrm{e}^+ + \mathrm{e}^- \to \gamma^* \to \pi^+ + \pi^-$ data for which the propagator of the $\rho(770)$ meson dominates in the $s \sim 0.6$ GeV$^2$ region (see Fig. 10.2b). In the zero-width approximation of this $\rho$ meson, we have from its propagator [note that $\mathrm{Im}(x \pm \mathrm{i}\epsilon)^{-1} = \mp\pi\delta(x)$],

$$\mathrm{Im}F_\pi(s) = \pi g_{\rho\pi\pi} f_\rho m_\rho \delta(s - m_\rho^2) + \cdots\,, \tag{10.19}$$

where the residue at the $\rho$ pole is $g_{\rho\pi\pi} f_\rho m_\rho$ as shown by Fig. 10.2b. The dots denote other contributions beyond the $\rho^0$. The two parameters $g_{\rho\pi\pi}$ and

$f_\rho$ in (19) are defined as follows: The dimensionless coupling constant $g_{\rho\pi\pi}$ can be determined from the $\rho(k) \to \pi(p) + \pi(p')$ strong decay for which the effective Lagrangian $\mathcal{L}_{\text{eff}}$ and the decay amplitude may be written as

$$\mathcal{L}_{\text{eff}} = g_{\rho\pi\pi}\rho^\mu(x) \left[\pi(x) \overleftrightarrow{\partial_\mu} \pi(x)\right] \ , \quad \mathcal{A}(\rho \to \pi\pi) = g_{\rho\pi\pi}\varepsilon^\mu(k)(p-p')_\mu \ .$$

$$\text{This gives :} \quad \Gamma_\rho \equiv \Gamma(\rho \to \pi\pi) = \frac{m_\rho g^2_{\rho\pi\pi}}{48\pi}\left(1 - \frac{4m_\pi^2}{m_\rho^2}\right)^{3/2} \ . \tag{10.20}$$

The electromagnetic decay constant of the $\rho^0$, called $f_\rho$, is defined similarly to the weak decay constant of the pion $f_\pi \approx 131\,\text{MeV}$. This $f_\pi$, which governs the weak decay $\pi^+ \to \text{W}^+ \to \text{e}^+ + \nu$, is defined by $\langle 0 \,|\, \text{A}^\mu \,|\, \pi^+(k)\rangle = \text{i} f_\pi k^\mu$ [Fig. 10.3a and (12.55), (13.32)]. The decay constant $f_\rho$ determines the amplitude $\mathcal{A}(\rho^0 \to \gamma^* \to \text{e}^+ + \text{e}^-)$ (Fig. 10.3b):

$$\langle 0 \,|\, J^\mu_{\text{em}} \,|\, \rho(k)\rangle = f_\rho m_\rho \varepsilon^\mu \ , \quad \left\langle \text{e}^+(p'), \text{e}^-(p) \,\middle|\, J^{\text{em}}_\mu \,\middle|\, 0\right\rangle = \overline{u}(p)\gamma_\mu v(p') \ ,$$
$$\mathcal{A}(\rho^0 \to \text{e}^+ + \text{e}^-) = \text{i}\,(-\text{i}e)^2 \left\langle \text{e}^+(p'), \text{e}^-(p) \,\middle|\, J^{\text{em}}_\mu \,\middle|\, 0\right\rangle \frac{-\text{i}}{k^2} \langle 0 \,|\, J^\mu_{\text{em}} \,|\, \rho(k)\rangle \ ,$$

$$\text{from which :} \quad \Gamma(\rho^0 \to \text{e}^+ + \text{e}^-) = \frac{4\pi\alpha^2 f_\rho^2}{3m_\rho} \ . \tag{10.21}$$

The decay constant $f_\rho$ has the dimension of mass, as does $f_\pi$.

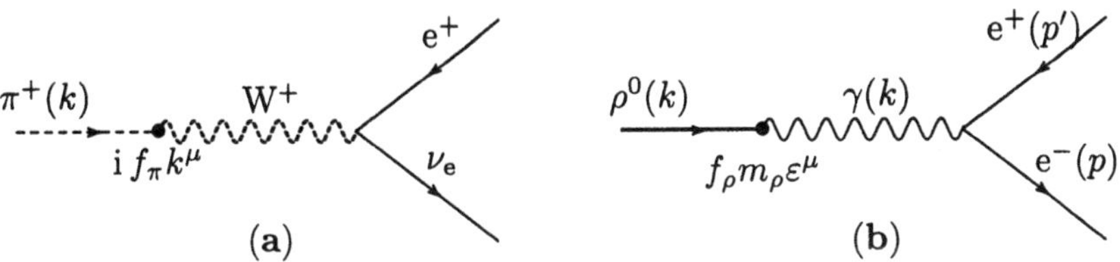


**Fig. 10.3.** (a) $\pi^+ \to \text{e}^+ + \nu_\text{e}$ weak decay; (b) $\rho^0 \to \text{e}^+ + \text{e}^-$ electromagnetic decay

Putting (19) into (17), we get

$$F_\pi(q^2) = 1 + \frac{g_{\rho\pi\pi} f_\rho}{m_\rho} \frac{q^2}{m_\rho^2 - q^2} + \cdots \ . \tag{10.22}$$

It turns out from (20) and (21) that data give $g^2_{\rho\pi\pi}/4\pi \approx 2.88$ , $f_\rho \approx$ 150 MeV. From these numbers, we get $g_{\rho\pi\pi} f_\rho / m_\rho \approx 1.17$, close to 1, which is the value corresponding to the universal $\rho$ dominance hypothesis, according to which the $\rho$ meson completely determines the electromagnetic form factors of low-lying hadrons (pion, nucleon). In the dispersion relation approach, this hypothesis consists in neglecting all contributions from non-resonant background ($\gamma^* \to$ continuum background $\to \pi^+ + \pi^-$), as well as from other resonances $f_0(1300)$, $\rho(1450)$, etc. This value 1 obviously satisfies the sum rule (18). Then, in the zero-width approximation of the $\rho$ meson:

$$F_\pi(q^2) = \frac{1}{1-\frac{q^2}{m_\rho^2}} \longrightarrow \text{ pion radius } \sqrt{\langle r_\pi^2 \rangle} = \frac{\sqrt{6}}{m_\rho} \approx 0.64\,\text{fm}\,. \tag{10.23}$$

A more sophisticated expression of $F_\pi(q^2)$ with a Breit–Wigner form and a $q^2$-dependent of the $\rho$-width, is given in (13.40) and (13.41).

We turn now to the weak form factors of $\overline{\text{B}} \to \text{D}$ transition. Instead of $f_\pm(q^2)$ in (15), we introduce another parameterization that singles out the spin character of the current,

$$\begin{aligned}\langle \text{D}(p') \,|\, \text{V}^\mu \,|\, \overline{\text{B}}(p)\rangle = G_1(q^2) &\left[(p+p')^\mu - \frac{M_B^2 - M_D^2}{q^2} q^\mu\right] \\ &+ G_0(q^2) \frac{M_B^2 - M_D^2}{q^2} q^\mu\,. \end{aligned}\tag{10.24}$$

Indeed let $q_\mu = (p-p')_\mu$ act on both the left and right members of the above equation, then we realize that $G_0(q^2)$ represents the matrix element of the operator $q_\mu \text{V}^\mu$ which behaves like a $J^P = 0^+$ scalar object. For a conserved current, this spin-0 form factor vanishes. The $G_1(q^2)$ form factor corresponds to the spin-1 part of the current since its associated operator is orthogonal to $q_\mu$, i.e. $q_\mu\left[(p+p')^\mu - (M_B^2 - M_D^2) q^\mu/q^2\right] = 0$. $G_1(q^2)$ and $G_0(q^2)$ are subject to $G_1(0) = G_0(0)$, which eliminates the spurious pole at $q^2 = 0$.

One advantage of considering $G_1(q^2)$ and $G_0(q^2)$ lies in the fact that their $q^2$ dependences are easy to guess, since the imaginary parts of these form factors are associated respectively with the vector $\text{B}_c^*$ and scalar $\text{B}_c$ resonances bearing the bottom–charm quantum numbers. To be more explicit, let us mimic the $F_\pi(q^2)$ case with the $\rho$ dominance, then the $G_1(q^2)$ and $G_0(q^2)$ form factors as visualized by Fig. 10.4 are respectively dominated by the vector $\text{B}_c^*$ and scalar ($J^P = 0^+, \text{B}_c$) bottom–charm $\bar{\text{c}}\text{b}$ resonances of masses around 6.5 GeV, for which the $q^2$ dependences are monopoles, as in (23):

$$G_1(q^2) = \frac{G_1(0)}{1 - q^2/M_{B_c^*}^2}\,, \qquad G_0(q^2) = \frac{G_0(0)}{1 - q^2/M_{B_c}^2}\,. \tag{10.25}$$

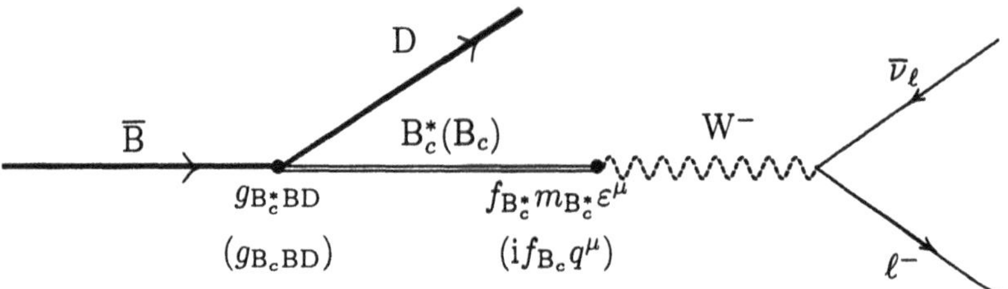


**Fig. 10.4.** $\text{B}_c^*\,(\text{B}_c)$ pole dominance of the $G_1(q^2)\,(G_0(q^2))$ form factors in weak decay $\overline{\text{B}} \to \text{D} + \ell^- + \overline{\nu}_\ell$

It must be emphasized that these $q^2$ behaviors, although plausible, are model dependent, they are derived from the assumption of the nearest resonance dominance. It could be a good approximation in the largest-$q^2$ region near the poles. Assuming this monopole $q^2$ dependence (25), and using (16), the weak form factors $G_{1,0}(q^2)$ are theoretically determined, including their normalizations $G_{1,0}(0)$. B meson decays will be discussed in detail in Chap. 16; as shown by (16.86) the $q^2$ dependence of $G_1(q^2) = f_+(q^2)$ can be experimentally measured by looking at the $q^2$ distribution of $\mathrm{d}\Gamma(\mathrm{B} \to \overline{\mathrm{D}} + e^+ + \nu_e)/\mathrm{d}q^2$, therefore the $G_1(q^2)$ extracted from data can be confronted with theoretical models of form factor. As for the scalar form factor $G_0(q^2) = f_+(q^2) + [q^2/(M_{\mathrm{B}}^2 - M_{\mathrm{D}}^2)]\, f_-(q^2)$, its contribution to $\Gamma(\mathrm{B} \to \overline{\mathrm{D}} + \ell^+ + \nu_\ell)$ via $f_-(q^2)$ is proportional to the squared lepton mass [see (16.87)], hence only the $\tau$ in the decay products is sensitive to $G_0(q^2)$.

This section ends with a discussion of the Watson theorem according to which, below the inelastic threshold, the phases of electromagnetic (weak) amplitudes are equal to the strong interaction elastic phase-shifts of the hadrons involved. The demonstration is based on the unitarity of the $S$ matrix together with the time-reversal invariance of the amplitudes. Since $S^\dagger S = 1$ , $S = 1 - \mathrm{i}\mathcal{T}$,

$$\mathrm{i}(\mathcal{T}_{\mathrm{f\,i}} - \mathcal{T}^*_{\mathrm{i\,f}}) = \sum_n \mathcal{T}^*_{n\,\mathrm{f}} \mathcal{T}_{n\,\mathrm{i}} \;, \; \mathcal{T}_{\mathrm{f\,i}} \equiv \langle \mathrm{f} \,|\, \mathcal{T} \,|\, \mathrm{i} \rangle \;.$$

Consider the electromagnetic transition $A \to B + C$ followed by an elastic final-state strong interaction: $A \xrightarrow{\text{em}} B + C \xrightarrow{\text{strong}} B + C$, i.e. we have i$= A$ and $n$=f$=B + C$. Time-reversal invariance implies $\mathcal{T}_{f\,i} = \mathcal{T}_{\bar{i}\,\bar{f}}$ (where $|\bar{i}\rangle = \mathcal{T}\,|i\rangle$), such that the left-hand side of the above unitarity relation reduces to $-2\mathrm{Im}\,\mathcal{T}_{f\,i}$ and both sides are real valued. Denoting the electromagnetic amplitude by $|T_{\mathrm{em}}|\exp(\mathrm{i}\,\varphi)$, and the strong amplitude by $|T_{\mathrm{sg}}|\exp(\mathrm{i}\,\delta)$, then $\varphi = \delta$. For instance, consider the electromagnetic form factor $\gamma^* \xrightarrow{\text{em}} \pi + \pi$ followed by a final-state strong interaction between the pions $\pi + \pi \xrightarrow{\text{strong}} \pi + \pi$, then $F_\pi(q^2)$ is given by $|F_\pi(q^2)|\exp[\mathrm{i}\delta_{\pi\pi}(q^2)]$. Since the strong phase-shifts can in principle be extracted from experimental data, in particular from $\pi + \pi$ scattering, the form factor $F_\pi(q^2)$ can be obtained from the p-wave phase-shift $\delta_{\pi\pi}(q^2)$ via dispersion relation (Problem 10.7). The same theorem applies to the weak form factor $G_1(q^2)$ [or $G_0(q^2)$], its phase is the p-wave (or s-wave) phase-shift of the B+D $\to$ B+D strong scattering.

## 10.3 Exclusive Reaction: Elastic Scattering

According to the Feynman rules, the one-photon exchange elastic amplitude $\mathrm{e}^-(p) + \mathrm{N}(P) \to \mathrm{e}^-(p') + \mathrm{N}(P')$ (Figs. 4.9 and 10.6a) may be written as

$$\mathcal{M} = \mathrm{i}\,(+\mathrm{i}e)\overline{u}(p')\gamma^\mu u(p)\left(\frac{-\mathrm{i}}{q^2}\right)(-\mathrm{i}e)H_\mu\,,$$

$$H_\mu \equiv \overline{u}(P')\left[\gamma_\mu F_1(q^2) + \mathrm{i}\frac{\sigma_{\mu\nu}q^\nu}{2M}F_2(q^2)\right]u(P)\,, \tag{10.26}$$

where $q = P' - P$. It can be deduced directly from the $e^- + \mu^+ \to e^- + \mu^+$ amplitude by a simple substitution $\gamma_\mu \Rightarrow \gamma_\mu F_1(q^2) + \mathrm{i}[\sigma_{\mu\nu} q^\nu / 2M] F_2(q^2)$ at the muon vertex.

We put $K = P + P'$ , $s = (P+p)^2$ , $-t = -q^2 = Q^2 \geq 0$. The nucleon and electron masses are respectively denoted by $M$ and $m$. The cross-section is given by formulas (4.57) and (4.62):

$$\frac{\mathrm{d}\sigma}{\mathrm{d}Q^2} = \frac{1}{16\pi\lambda(s, M^2, m^2)} \left( \frac{1}{4} \sum_{\text{spins}} |\mathcal{M}|^2 \right) . \tag{10.27}$$

The factor $1/4 = (1/2)(1/2)$ in (27) represents the spin averaging of the incoming electron and proton; for undetected spins in the final states, their summation is understood. With the Gordon decomposition, let us rewrite the nucleon current $H_\mu$ as $\overline{u}(P') \left[ (F_1 + F_2)\gamma_\mu - (K_\mu / 2M) F_2 \right] u(P)$, then

$$\begin{aligned}
\frac{1}{4} \sum_{\text{spins}} |\mathcal{M}|^2 &= \left( \frac{e^2}{q^2} \right)^2 l^{\mu\nu} H_{\mu\nu} \, , \\
H_{\mu\nu} &= \frac{1}{2} \Big\{ (F_1 + F_2)^2 \mathrm{Tr} \left[ \not{P}' \gamma_\mu \not{P} \gamma_\nu + M^2 \gamma_\mu \gamma_\nu \right] \\
&\quad + F_2^2 \frac{K_\mu K_\nu}{4M^2} \mathrm{Tr} \left[ \not{P}' \not{P} + M^2 \right] - 4 F_2 (F_1 + F_2) K_\mu K_\nu \Big\} \, , \\
l^{\mu\nu} &= \frac{1}{2} \mathrm{Tr} \left[ \not{p}' \gamma^\mu \not{p} \gamma^\nu + m^2 \gamma^\mu \gamma^\nu \right] = 2 \left( p^\mu p'^\nu + p^\nu p'^\mu + \frac{q^2 g^{\mu\nu}}{2} \right) \, ,
\end{aligned} \tag{10.28}$$

so that

$$\begin{aligned}
\frac{1}{4} \sum_{\text{spins}} |\mathcal{M}|^2 &= 4 \left( \frac{e^2}{q^2} \right)^2 \Big\{ (F_1 + F_2)^2 \left[ (s - M^2 - m^2)^2 + q^2 (s + \frac{q^2}{2}) \right] \\
&\quad - \left[ 2F_1 F_2 + \left( 1 + \frac{q^2}{4M^2} \right) F_2^2 \right] \left[ (s - M^2 - m^2)^2 + q^2 (s - m^2) \right] \Big\} \\
&= 4 \left( \frac{e^2}{q^2} \right)^2 \Big\{ \left( F_1^2 - \frac{q^2}{4M^2} F_2^2 \right) \left[ (s - M^2 - m^2)^2 + q^2 (s - m^2) \right] \\
&\quad + (F_1 + F_2)^2 q^2 \left( m^2 + \frac{q^2}{2} \right) \Big\} \, .
\end{aligned} \tag{10.29}$$

Following Sachs, let us define the electric $G_{\mathrm{E}}(q^2)$ and magnetic $G_{\mathrm{M}}(q^2)$ form factors as a combination of $F_1(q^2)$ and $F_2(q^2)$:

$$\begin{aligned}
&G_{\mathrm{E}}(q^2) = F_1(q^2) + \frac{q^2}{4M^2} F_2(q^2) \quad , \quad G_{\mathrm{M}}(q^2) = F_1(q^2) + F_2(q^2) \, , \\
&G_E^{\mathrm{p}}(0) = 1 \ , \ G_E^{\mathrm{n}}(0) = 0 \ , \ G_M^{\mathrm{p}}(0) = 2.79 \ , \ G_M^{\mathrm{n}}(0) = -1.91 \, , \\
&F_1^2 - \frac{q^2}{4M^2} F_2^2 = \frac{G_E^2 - (q^2/4M^2) G_M^2}{1 - (q^2/4M^2)} \, .
\end{aligned} \tag{10.30}$$

In the *laboratory* system $P = (M, \mathbf{0})$ , $p = (E, \mathbf{p})$ , $p' = (E', \mathbf{p}')$. Using $P'^2 = M^2 = (P+q)^2$, we deduce

$$q^2 + 2M(E - E') = 0 \quad , \quad q^2 = -2M\nu \ , \ \nu \equiv E - E' \ . \tag{10.31}$$

In the limit $s \gg m^2$, the incoming and outgoing electrons are extremely relativistic, the formula (4.69) for the cross-section is adapted to this case. We also have

$$\mathbf{p} \cdot \mathbf{p}' \approx EE' \cos\theta \quad ; \quad q^2 \approx -2EE'(1 - \cos\theta) = -4EE' \sin^2\frac{\theta}{2} \, . \tag{10.32}$$

From (31) and (32): $q^2 = -2M(E - E') = -2EE'(1 - \cos\theta)$, we remark that for fixed $E$ the two quantities $E'$ and $\cos\theta$ are not independent because

$$\frac{E'}{E} = \frac{1}{[1 + \frac{E}{M}(1 - \cos\theta)]} = \frac{1}{1 + \frac{2E}{M}\sin^2\frac{\theta}{2}} \, . \tag{10.33}$$

Coming back to the first term of the right-hand side of (29), we find $(s - M^2 - m^2)^2 + q^2(s - m^2) \approx 4M^2EE'\cos^2\frac{\theta}{2}$. When we rewrite the $q^2(m^2 + q^2/2) \approx q^2(q^2/2)$ of the second term as $(-q^2/2M^2)\tan^2\frac{\theta}{2}$ times the common factor $4M^2EE'\cos^2\frac{\theta}{2}$, we obtain the following results, using (4.64), (4.69) and the relation $\mathrm{d}\Omega_{\mathrm{lab}} = \pi\, \mathrm{d}Q^2/E'^2$, $(e^2/4\pi = \alpha \approx 1/137)$:

$$\frac{\mathrm{d}\sigma}{\mathrm{d}Q^2} = \frac{\pi\,\sigma_{\mathrm{Mott}}}{EE'}\left[\frac{G_{\mathrm{E}}^2(q^2) - \frac{q^2}{4M^2}G_{\mathrm{M}}^2(q^2)}{1 - \frac{q^2}{4M^2}} - \frac{q^2}{2M^2}G_{\mathrm{M}}^2(q^2)\tan^2\frac{\theta}{2}\right] ,$$

$$\frac{\mathrm{d}\sigma}{\mathrm{d}\Omega_{\mathrm{lab}}} = \left(\frac{\mathrm{d}\sigma}{\mathrm{d}\Omega_{\mathrm{lab}}}\right)_{\mathrm{NS}}\left[\frac{G_{\mathrm{E}}^2(q^2) - \frac{q^2}{4M^2}G_{\mathrm{M}}^2(q^2)}{1 - \frac{q^2}{4M^2}} - \frac{q^2}{2M^2}G_{\mathrm{M}}^2(q^2)\tan^2\frac{\theta}{2}\right] ,$$

$$\left(\frac{\mathrm{d}\sigma}{\mathrm{d}\Omega_{\mathrm{lab}}}\right)_{\mathrm{NS}} \equiv \frac{\sigma_{\mathrm{Mott}}}{1 + \frac{2E}{M}\sin^2\frac{\theta}{2}} \ , \ \text{and} \ \sigma_{\mathrm{Mott}} \equiv \left(\frac{\alpha\,\cos\frac{\theta}{2}}{2E\sin^2\frac{\theta}{2}}\right)^2 \, . \tag{10.34}$$

**Remarks.** From the Rosenbluth formula (34), three remarks can be made: (a) For a *structureless* nucleon: $F_1^{\mathrm{p}}(q^2) = 1$ , $F_1^{\mathrm{n}}(q^2) = F_2^{\mathrm{p}}(q^2) = F_2^{\mathrm{n}}(q^2) = 0$; $G_{\mathrm{E}}^{\mathrm{p}}(q^2) = G_{\mathrm{M}}^{\mathrm{p}}(q^2) = 1$ , $G_{\mathrm{E}}^{\mathrm{n}}(q^2) = G_{\mathrm{M}}^{\mathrm{n}}(q^2) = 0$. The elastic cross-section $\mathrm{e}^- +$ *pointlike* neutron is identically vanishing. We also recover the formula (4.161) of a *pointlike* proton. This formula (4.161) or (34) also gives the elastic $\mathrm{e}^- + \mu^+$ cross-section

$$\frac{\mathrm{d}\sigma(\mathrm{e}^- + \mu^+ \to \mathrm{e}^- + \mu^+)}{\mathrm{d}Q^2} = \frac{\pi\sigma_{\mathrm{Mott}}}{EE'}\left(1 - \frac{q^2}{2M^2}\tan^2\frac{\theta}{2}\right) ,$$

$$\frac{\mathrm{d}\sigma(\mathrm{e}^- + \mu^+ \to \mathrm{e}^- + \mu^+)}{\mathrm{d}\Omega_{\mathrm{lab}}} = \left(\frac{\mathrm{d}\sigma}{\mathrm{d}\Omega_{\mathrm{lab}}}\right)_{NS}\left(1 - \frac{q^2}{2M^2}\tan^2\frac{\theta}{2}\right) . \tag{10.35}$$

The subscript NS in $(\mathrm{d}\sigma/\mathrm{d}\Omega_{\mathrm{lab}})_{\mathrm{NS}}$ denotes *no-structure* cross-section. The first term $\sigma_{\mathrm{Mott}} \equiv \left[(\alpha \cos\frac{\theta}{2})/2E\sin^2\frac{\theta}{2}\right]^2$ in (34) [already met in (4.159)] corresponds to the scattering of a spin-$\frac{1}{2}$ pointlike charged particle by a spinless pointlike target of charge $\pm e$. The second term $1/(1+2E/M\sin^2\frac{\theta}{2})$ represents the target recoil where the generic $M$ denotes the target mass. Without the $\cos^2\frac{\theta}{2}$ term, the Mott cross-section is identical to the Rutherford cross-section $\sigma_{\mathrm{R}} \equiv [\alpha/2E\sin^2\frac{\theta}{2}]^2$ which corresponds to the scattering of a scalar projectile by an ultra heavy scalar target, both are pointlike with charges $\pm e$. The angular distribution $\cos^2\frac{\theta}{2}$ reflects the spin-$\frac{1}{2}$ of the *projectile.*
(b) The term $\tan^2\frac{\theta}{2}$ of (34), (35) comes from $G_{\mathrm{M}}^2 \equiv (F_1+F_2)^2$ presented on the last line of the right-hand side of (29). Through the magnetic moment, it represents the target spin. We recall that a Dirac particle, even pointlike, has a magnetic moment which is equal to one Bohr magneton. The $\tan^2\frac{\theta}{2}$ corresponds to the spin-$\frac{1}{2}$ effect of the *target*: if the latter is spinless, as in electron scattering by pion, this term is absent, and the Mott cross-section $\sigma_{\mathrm{Mott}}$ is recovered. Note the coefficient $q^2/2M^2$ of the magnetic term which becomes important for large $q^2$. In brief, the $\cos^2\frac{\theta}{2}$ refers to the spin-$\frac{1}{2}$ of the projectile, the $\tan^2\frac{\theta}{2}$ carries the spin-$\frac{1}{2}$ character of the target. The electron–pion scattering therefore can be immediately deduced:

$$\frac{\mathrm{d}\sigma(\mathrm{e}^- + \pi^\pm \to \mathrm{e}^- + \pi^\pm)}{\mathrm{d}\Omega_{\mathrm{lab}}} = \left(\frac{\mathrm{d}\sigma}{\mathrm{d}\Omega_{\mathrm{lab}}}\right)_{\mathrm{NS}} |F_\pi(q^2)|^2 \,. \tag{10.36}$$

Similarly, we get the cross-section of electron scattering by a spinless nucleus $\mathcal{N}$ of charge $Ze$ considered as an example of form factors in Sect. 10.1. Thus,

$$\frac{\mathrm{d}\sigma(\mathrm{e}^- + \mathcal{N} \to \mathrm{e}^- + \mathcal{N})}{\mathrm{d}\Omega_{\mathrm{lab}}} = \left(\frac{\mathrm{d}\sigma}{\mathrm{d}\Omega_{\mathrm{lab}}}\right)_{\mathrm{NS}} |F_{\mathrm{N}}(q^2)/e|^2 \,, \tag{10.37}$$

which tends to $Z^2\,(\mathrm{d}\sigma/\mathrm{d}\Omega_{\mathrm{lab}})_{\mathrm{NS}}$ for a pointlike spinless nucleus.

We will see that in deep inelastic lepton–nucleon scattering, quarks are revealed as constituents of the nucleon through the $\tan^2\frac{\theta}{2}$ term. The physical meaning of the $\cos^2\frac{\theta}{2}$ and $\tan^2\frac{\theta}{2}$ terms (or $\sin^2\frac{\theta}{2}$ for the latter if we do not factorize the common $\cos^2\frac{\theta}{2}$) can be understood as follows.

The electromagnetic interactions of the projectile electron may be split into two parts: the first, composed of $\overline{u}_{\mathrm{L}}\gamma^\mu u_{\mathrm{L}}$ and $\overline{u}_{\mathrm{R}}\gamma^\mu u_{\mathrm{R}}$, corresponds to its electric interaction with another charged target for which the outgoing and incoming electrons conserve their helicities. The second, $\overline{u}_{\mathrm{L}}[i\sigma^{\mu\nu}q_\nu]u_{\mathrm{R}}$ and $\overline{u}_{\mathrm{R}}[i\sigma^{\mu\nu}q_\nu]u_{\mathrm{L}}$, corresponds to its magnetic interaction with the target spin for which the outgoing and incoming electron have opposite helicities. Angular momentum conservation implies that the forward (backward) scattering is allowed (forbidden) by the electric interaction and forbidden (allowed) by the magnetic interaction. The helicity conserving amplitude is proportional to the familiar rotation matrix $d^{1/2}_{1/2,1/2}(\theta) = d^{1/2}_{-1/2,-1/2}(\theta) = \cos\frac{\theta}{2}$, it characterizes the electron scattered by any charged target (spinless or not); while the helicity reversing amplitude is related to $d^{1/2}_{1/2,-1/2}(\theta) = -d^{1/2}_{-1/2,1/2}(\theta) = \sin\frac{\theta}{2}$; it can only occur with an electron scattered by a spin-1/2 target having a magnetic moment [remember $\sigma_{\mu\nu} \sim \boldsymbol{\sigma}$ with (10)]. The diagrams in Fig. 10.5 clearly illustrate the meaning of the $\cos^2\frac{\theta}{2}$ and $\sin^2\frac{\theta}{2}$ angular distributions.

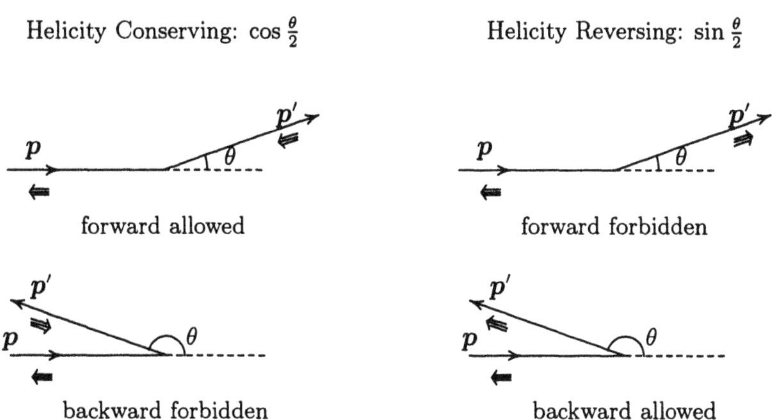


**Fig. 10.5.** Angular distributions from helicity arguments

(c) The angular distribution $[\sin\frac{\theta}{2}]^{-4}$ is intimately connected with the discovery of nuclei. Until today, from atoms to nuclei, from nuclei to nucleons, from nucleons to quarks, the successive layers of matter are revealed in striking similarity with the Rutherford discovery of nuclei. Before Rutherford, atoms were considered to be the most fundamental objects. Since an atom is electrically neutral and contains light particles of negative charges (electrons), the question was how its very heavy (compared to the electron) and positive charged particles were distributed. If the positive charges were uniformly spread over the whole atomic volume as was generally thought at the time (in particular by Thomson, the electron's discoverer), then by projecting energetic He ions ($\alpha$ particles) on thin metal foils, Geiger and Marsden expected that the $\alpha$ beam would be deflected by a small angle. To their surprise, they observed that a non-negligible number of projectiles were bounced back by an angle $\theta \geq 90^0$. It took Rutherford two years to find the explanation.

If the Thomson model of charge distribution were correct, such large deflections could only arise, according to Rutherford, from a multitude of scatterings by a huge number of atoms of matter. However a simple calculation based on the theory of probability shows that the odds for finding an event with $\theta \geq 90^0$ are vanishingly small, something like $10^{-3500}$ and not 1/20 000 as found by Geiger and Marsden. The simple power law $[\sin\frac{\theta}{2}]^{-4}$ computed by Rutherford and experimentally verified by his group can only be explained if the projectile strikes a hard obstacle, its charge must be concentrated in an extremely small region inside the atom, it cannot be spread over the large atomic volume. The physical meaning of the angular distribution $[\sin\frac{\theta}{2}]^{-4}$ is clear. This term proportional to $(1/q^2)^2$ comes from the propagator of a photon probing a pointlike charged target found at the center of a much larger atom. If this were not the case, i.e. if the charges were distributed at random over the whole atomic volume and hit by the photon, then the atomic form factor, which decreases very rapidly for large $q^2$ (large $\theta$) because of its size, would make vanishingly small the probability for observing an event with large $\theta$. Thus was born the notion of a pointlike nucleus of the atom, in which the atom contains a hard constituent: a small, massive and positively charged nucleus. After nearly one hundred years, this remarkable feature is still relevant today. In all scattering processes, a large deflection of the projectile (or large transverse momentum $P_T$ in our contemporary language) is synonymous with a hard constituent in the target, whether the latter is an atom or a hadron. Events in which a projectile is bounced back by $\approx 180^0$ are really spectacular and may reveal new things. ■

After these remarks, we come back to the Rosenbluth formula (34). For each fixed $q^2$, if we plot experimental data of $\mathrm{d}\sigma/\mathrm{d}\Omega_{\mathrm{lab}}$ as a function of $\tan^2\frac{\theta}{2}$, its linear form $A(q^2)+B(q^2)\tan^2\frac{\theta}{2}$ allows us to separate $A(q^2)$ and $B(q^2)$ and to extract the form factors $G_{\mathrm{E}}(q^2)$ and $G_{\mathrm{M}}(q^2)$. From studies of the elastic scattering $\mathrm{e}+\mathrm{N}\to\mathrm{e}+\mathrm{N}$, $G_{\mathrm{E}}(q^2)$ and $G_{\mathrm{M}}(q^2)$ are found to decrease as a *dipole mode*,

$$\frac{G_E^{\mathrm{p}}(q^2)}{G_E^{\mathrm{p}}(0)}=\frac{G_M^{\mathrm{p}}(q^2)}{G_M^{\mathrm{p}}(0)}=\frac{G_M^{\mathrm{n}}(q^2)}{G_M^{\mathrm{n}}(0)}=\frac{1}{\left(1-\frac{q^2}{\Lambda^2}\right)^2}\quad;\quad G_E^{\mathrm{n}}(q^2)=0\,, \tag{10.38}$$

with $\Lambda=0.84\,\mathrm{GeV}$. By doing inverse Fourier transform of the nonrelativistic limit of $[1-(q^2/\Lambda^2)]^{-2}$, we obtain $\rho(r=|\boldsymbol{x}|)$, the electric charge distribution in the proton, together with its squared radius $\langle r_{\mathrm{p}}^2\rangle$ :

$$\rho(r)=\int\frac{\mathrm{d}^3q}{(2\pi)^3}\frac{\mathrm{e}^{\mathrm{i}\boldsymbol{q}\cdot\boldsymbol{x}}}{\left(1+\frac{|\boldsymbol{q}|^2}{\Lambda^2}\right)^2}=\frac{\Lambda^3}{8\pi}\mathrm{e}^{-\Lambda r}\ ,\quad\int\mathrm{d}^3x\rho(r)=1\,,$$

$$\sqrt{\langle r_{\mathrm{p}}^2\rangle}=\frac{\sqrt{12}}{\Lambda}\approx 0.8\ \mathrm{fm}\,. \tag{10.39}$$

The above distribution $\rho(r)=\Lambda^3\mathrm{e}^{-\Lambda r}/(8\pi)$ has the property that $\rho(r)\to$ constant as $r\to 0$, indicating that there is no hard core in the nucleon, it is not like a plum with a stone in the middle. The hard core corresponds for example to a monopole decreasing form factor that gives a spatial distribution $\rho(r)=\Lambda^2\mathrm{e}^{-\Lambda r}/(4\pi r)$ tending to infinity as $r\to 0$ (Problem 10.6). The question is: Is the proton like jelly or like a pomegranate? Intuitively we may imagine the former configuration as an interpretation of the bootstrap nuclear democracy concept, for which hadrons are composite of themselves and no one is more elementary than the others. For example the proton may be a bound state of $\mathrm{n}+\pi^+$ or $\Lambda$+K$^+$ or $\Sigma^+$+K$^0$. As we will see, deep inelastic scattering provides a definite answer in favor of the second interpretation according to which the proton contains pointlike constituents. As for the neutron, it is *a priori* not at all guaranteed that the electromagnetic e–n cross-section turns out to be also nonzero, as the neutron is electrically neutral. Since this is the case, the constituents of the latter must be charged and must be the sources of the neutron anomalous magnetic moment $G_M^{\mathrm{n}}(0) = -1.91\mu_B$. Finally, to connect the electromagnetic form factors to the weak form factors via the conserved vector current (CVC) property of the $\Delta S = 0$ weak vector current (Chap. 12), it is useful to write down the isovector and isoscalar form factors previously defined in (13) and (30), using experimental data on $G_{\mathrm{E}}(q^2)$ and $G_{\mathrm{M}}(q^2)$ in (38). We have

$$F_1^1(q^2) = \frac{1-4.70\,u(q^2)}{(1-u(q^2))[1-v(q^2)]^2}\ ;\ F_1^0(q^2) = \frac{1-0.88\,u(q^2)}{(1-u(q^2))[1-v(q^2)]^2}\ ;$$

$$F_2^1(q^2) = \frac{3.70}{(1-u(q^2))[1-v(q^2)]^2}\ ;\ F_2^0(q^2) = \frac{-0.12}{(1-u(q^2))[1-v(q^2)]^2}\ ,$$

$$\text{where}\quad u(q^2) \equiv \frac{q^2}{4M^2}\ ,\quad v(q^2) \equiv \frac{q^2}{\Lambda^2}\ . \tag{10.40}$$

By CVC, the isovector form factors $F_1^1(q^2)$ and $F_2^1(q^2)$ are the same form factors of the vectorial $V$ part of the V – A weak charged current involved in nucleon $\beta$-decay and in neutrino–nucleon elastic scattering (Chap. 12).

To close the section, let us mention that the experimental range of the momentum transfer $q^2$ reached today at the electron–proton (positron–proton) HERA collider in Hamburg is $Q^2 \geq 2000\ \mathrm{GeV}^2$ for which the form factors $G_{\mathrm{E,M}}(q^2)$ squared, as given by (38), decreases from 1 to $10^{-14}$ or less, in sharp contrast with the nearly constant behavior of the structure functions that we are going to discuss now. The reason is that the nucleon contains pointlike constituents, as we will see.

## 10.4 Inclusive Reaction: Deep Inelastic Scattering

An exclusive reaction is a process in which the final state contains a limited number of particles effectively observed. Elastic or quasielastic lepton scattering from nucleon $\mathrm{e}+\mathrm{N} \to \mathrm{e}+\mathrm{N}$ , $\mathrm{e}+\mathrm{N} \to \mathrm{e}+\mathrm{N}^*$ , $\nu_\mu+\mathrm{N} \to \mu^- +\mathrm{N}+\pi$ are some examples.

On the other hand, in an inclusive reaction $\mathrm{e}+\mathrm{N} \to \mathrm{e}+X$ (more generally $\ell+\mathrm{N} \to \ell'+X$) only the final lepton $\ell'$ is detected, no attempt is made to select a particular hadronic channel. These unobserved hadrons are symbolically designated by $X$. Since each exclusive cross-section decreases sharply as the squared of its corresponding form factors, the inclusive one, which is nothing but the sum of all possible exclusive modes, would at first sight be negligible for large $q^2$. When the inclusive reaction $\mathrm{e}+\mathrm{N} \to \mathrm{e}+X$ was getting ready to be measured at SLAC (Stanford) in the 1960s, the general impression was rather pessimistic about the number of events that could be collected. That was without allowing for the remarkable intuition of Bjorken who among the first showed that deep inelastic scattering (inclusive reaction at large $q^2$) was an ideal tool to probe the nucleon constituents. He also predicted the *scaling law* for the nucleon structure functions, later confirmed by experiments. This law tells us that, once the term $1/Q^4$ is subtracted, the deep inelastic cross-section would be constant [see (10.66) below], rather than rapidly decreasing as the squared form factors of an exclusive reaction. This $1/Q^4$ term (from the one-photon exchange) is common to both inclusive and exclusive cross-sections [see (29) and (34)].

This surprising discovery is at the origin of the *parton* model, the name was proposed by Feynman to denote the free and pointlike constituents of hadrons in his explanation of the Bjorken scaling law. At high energy, it is intuitively conceivable that deep inelastic reaction represents the sum of scatterings by partons. Since the latter are pointlike, there are no decreasing form factor, the cross-section would behave differently from the exclusive one.

### 10.4.1 Structure Functions

Before considering the details of the dynamics, let us first examine the kinematics of exclusive (elastic) and inclusive electron–nucleon reactions shown in Figs. 10.6a and 10.6b respectively. For a two-body → two-body scattering, for example the $\mathrm{e}(p)+\mathrm{N}(P) \to \mathrm{e}(p')+\mathrm{N}(P')$ or $\mathrm{e}(p)+\mathrm{N}(P) \to \mathrm{e}(p')+\mathrm{N}^*(P')$, the cross-section depends kinematically on two independent variables which can be chosen as the Mandelstam invariants $s \equiv (P+p)^2 = (P'+p')^2$ and $t \equiv q^2 = (p-p')^2 \equiv -Q^2$. In the laboratory system $P = (M, \mathbf{0})$ , $p = (E, \boldsymbol{p})$ , $p' = (E', \boldsymbol{p}')$, another kinematic variable $\nu = P.q/M = E - E'$ which represents the energy loss is frequently used. The two independent variables can be taken as $E$ and $\nu$, instead of $s$ and $t$.

On the other hand, in an inclusive reaction $\mathrm{e}(p) + \mathrm{N}(P) \to \mathrm{e}(p') + X$, since no specific hadron is selected, the squared invariant mass of unobserved hadronic states $X$ is free to take all continuous values $\geq M^2$. Unlike (31), now $p_X^2 \equiv (P+q)^2 = M^2 + q^2 + 2M\nu$ is no more constrained to be equal to the squared mass of any specific hadron observed in a two-body reaction, i.e. $q^2$ and $\nu$ are independent. The inclusive cross-section depends on three kinematic variables that may be taken as $s$, $q^2$, and $\nu$. In the one-photon exchange, the elastic cross-section $\mathrm{d}\sigma_{\mathrm{el}}$ : $\mathrm{e}(p) + \mathrm{N}(P) \to \mathrm{e}(p') + \mathrm{N}(P')$ and the inclusive one $\mathrm{d}\sigma_{\mathrm{in}}$ : $\mathrm{e}(p) + \mathrm{N}(P) \to \mathrm{e}(p') + X$ are given by (Chap. 4)

$$\mathrm{d}\sigma_{\mathrm{el}} = \frac{(2\pi)^4\delta^4(p'+P'-p-P)}{2\sqrt{\lambda(s,M^2,m^2)}}\left(\frac{e^2}{q^2}\right)^2 l^{\mu\nu}H_{\mu\nu}\frac{\mathrm{d}^3p'}{(2\pi)^3 2E_{p'}}\frac{\mathrm{d}^3P'}{(2\pi)^3 2E_{P'}} ,$$

$$\mathrm{d}\sigma_{\mathrm{in}} = \frac{1}{2\sqrt{\lambda(s,M^2,m^2)}}\left(\frac{e^2}{q^2}\right)^2 l^{\mu\nu}W_{\mu\nu}\frac{\mathrm{d}^3p'}{(2\pi)^3 2E_{p'}} , \tag{10.41}$$

in which the leptonic and the nucleonic tensors $l^{\mu\nu}$ , $H_{\mu\nu}$ are defined in (28).

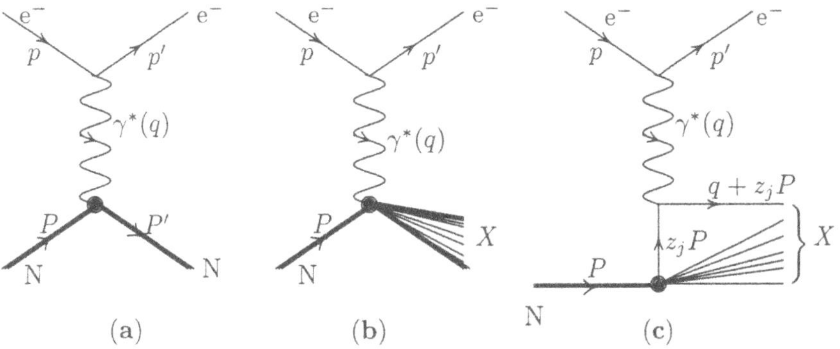


**Fig. 10.6.** (**a**) Elastic electron–nucleon (e–N) scattering; (**b**) deep inelastic e–N scattering; (**c**) electron–parton scattering

Using the relation $2s\,\mathrm{d}Q^2 = \lambda(s, M^2, m^2)\,\mathrm{d}\cos\theta$ and the Appendix formulas, the integration over the solid angle $\mathrm{d}\Omega = 2\pi\,\mathrm{d}\cos\theta$ in $\mathrm{d}\sigma_{\mathrm{el}}$ gives

$$\int \frac{\mathrm{d}^3p'}{(2\pi)^3 2E_{p'}} \frac{\mathrm{d}^3P'}{(2\pi)^3 2E_{P'}} (2\pi)^4\delta^4(p' + P' - p - P) = \int \frac{\mathrm{d}Q^2}{8\pi\sqrt{\lambda(s, M^2, m^2)}} ,$$

and (27) is recovered. In (41) the sum over the unobserved hadronic states denoted as $W_{\mu\nu}(P, q)$ can be written similarly to $H_{\mu\nu}(P, q)$:

$$\begin{aligned} H_{\mu\nu}(P,q) &= \frac{1}{2}\langle \mathrm{N}(P)\,|\,J_\mu^{\mathrm{em}}\,|\,\mathrm{N}(P')\rangle\,\langle \mathrm{N}(P')\,|\,J_\nu^{\mathrm{em}}\,|\,\mathrm{N}(P)\rangle\ , \\ W_{\mu\nu}(P,q) &= \frac{1}{2}\sum_X \{\langle \mathrm{N}(P)\,|\,J_\mu^{\mathrm{em}}\,|\,X, p_X\rangle\,\langle X, p_X\,|\,J_\nu^{\mathrm{em}}\,|\,\mathrm{N}(P)\rangle \\ &\quad \times (2\pi)^4\delta^4(P + q - p_X)\}\ . \end{aligned} \tag{10.42}$$

In (42) the sum over final spin states is understood, and the average of the initial nucleon spin is explicit with the factor $\frac{1}{2}$. If we count the dimensions of different terms in the left-hand and the right-hand sides of (41) [note that $\delta^4(K)$ has the $(\mathrm{mass})^{-4}$ dimension], we deduce that $H_{\mu\nu}(P, q)$ must have a $(\mathrm{mass})^2$ dimension which is confirmed by (28). On the other hand, $W_{\mu\nu}(P, q)$ must be dimensionless, in agreement with the limit $\sigma_{\mathrm{in}} \to \sigma_{\mathrm{el}}$ where the sum over the $X$ states, represented by the symbol $\Sigma_X$ in (42), is reduced to a simple nucleon $\mathrm{N}(P') : \Sigma_X \ \rightarrow\ \int \mathrm{d}^3P'/[(2\pi)^3 2E_{P'}]$. The latter, which has the $(\mathrm{mass})^2$ dimension, makes $W_{\mu\nu}(P, q)$ dimensionless.

By analogy with $H_{\mu\nu}(P, q)$ in (28), the most general form of $W_{\mu\nu}(P, q)$ depends on $g_{\mu\nu}$ and three other tensors made up of $P, q$, and is symmetric in the interchange $\mu \leftrightarrow \nu$ since $l^{\mu\nu}$ is. They are $(P_\mu q_\nu + q_\mu P_\nu)$, $P_\mu P_\nu$, and $q_\mu q_\nu$. Moreover, from the conserved electromagnetic current, $W_{\mu\nu}(P, q)$ must satisfy the two conditions $q^\mu W_{\mu\nu}(P, q) = q^\nu W_{\mu\nu}(P, q) = 0$. The four terms $g_{\mu\nu}$ , $P_\mu P_\nu$ , $q_\mu q_\nu$, and $(P_\mu q_\nu + P_\nu q_\mu)$ are then reduced to two that can be chosen as the dimensionless tensors $T^1_{\mu\nu}$ and $T^2_{\mu\nu}$, which are separately conserved ($q^\mu T^1_{\mu\nu} = q^\nu T^1_{\mu\nu} = 0$ , $q^\mu T^2_{\mu\nu} = q^\nu T^2_{\mu\nu} = 0$),

$$T^1_{\mu\nu} = -g_{\mu\nu} + \frac{q_\mu q_\nu}{q^2} \quad , \quad T^2_{\mu\nu} = \frac{1}{M^2}\left(P_\mu - \frac{P.q}{q^2}q_\mu\right)\left(P_\nu - \frac{P.q}{q^2}q_\nu\right) .$$

Conventionally, we define

$$W_{\mu\nu}(P,q) = 4\pi\left[T^1_{\mu\nu}W_1(q^2,\nu) + T^2_{\mu\nu}W_2(q^2,\nu)\right] . \tag{10.43}$$

$W_1(q^2,\nu)$ and $W_2(q^2,\nu)$ are called the nucleon *structure functions*. Like $W_{\mu\nu}$, they are dimensionless and depend on two variables usually taken as $q^2$ and $\nu$. In (43), we make explicit the factor $4\pi$ coming from the angular integration of the one-particle state $\mathrm{d}^3p_X$ in (42). Putting (43) into (41), we find in the nucleon rest frame

$$l^{\mu\nu}T^1_{\mu\nu} = 8EE'\sin^2\frac{\theta}{2} \quad , \quad l^{\mu\nu}T^2_{\mu\nu} = 4EE'\cos^2\frac{\theta}{2} \quad , \quad \frac{\mathrm{d}^3p'}{2E_{p'}} = \frac{\pi}{2E}\mathrm{d}Q^2\mathrm{d}\nu .$$

Then (41) becomes

$$\frac{\mathrm{d}\sigma_{\mathrm{in}}(\mathrm{e}+\mathrm{N}\to\mathrm{e}+X)}{\mathrm{d}Q^2\mathrm{d}\nu} = \frac{\pi\,\sigma_{\mathrm{Mott}}}{MEE'}\left[W_2(q^2,\nu) + 2\,W_1(q^2,\nu)\tan^2\frac{\theta}{2}\right] . \tag{10.44}$$

This equation is to be compared with (35), the elastic cross-section of electron scattered by a pointlike fermion $\mathrm{N_{pt}}$. In this case, using the relation (31), i.e. $q^2 = -2M\nu$ or $\int \mathrm{d}\nu\,\delta(\nu + \frac{q^2}{2M}) = 1$, (35) can be rewritten as

$$\frac{\mathrm{d}\sigma_{\mathrm{el}}(\mathrm{e}+\mathrm{N_{pt}}\to\mathrm{e}+\mathrm{N_{pt}})}{\mathrm{d}Q^2\mathrm{d}\nu} = \frac{\pi\,\sigma_{\mathrm{Mott}}}{EE'}\left[1 + 2\left(\frac{-q^2}{4M^2}\right)\tan^2\frac{\theta}{2}\right]\delta\left(\nu + \frac{q^2}{2M}\right) . \tag{10.45}$$

Comparing (44) with (45), we deduce that $W_{1,2}(q^2,\nu)$ tend to the following simplest expressions of the elastic scattering by a pointlike nucleon:

$$\frac{W_2(q^2,\nu)}{M} \longrightarrow \delta\left(\nu + \frac{q^2}{2M}\right) \quad , \quad \frac{W_1(q^2,\nu)}{M} \longrightarrow \left(\frac{-q^2}{4M^2}\right)\delta\left(\nu + \frac{q^2}{2M}\right) . \tag{10.46}$$

It is possible to extract $W_1(q^2,\nu)$ and $W_2(q^2,\nu)$ by plotting data of $\mathrm{d}\sigma_{\mathrm{in}}$ given by (44) as a function of $\tan^2\frac{\theta}{2}$ for fixed $q^2$, exactly as in the case of $\mathrm{d}\sigma_{\mathrm{el}}$ with $G_{\mathrm{E}}(q^2)$ and $G_{\mathrm{M}}(q^2)$ discussed previously in (34).

On the other hand, we can always write $W_{1,2}(q^2,\nu) \equiv W_{1,2}(q^2,x)$ as functions of the Bjorken variable $x \equiv -q^2/2M\nu$ and $q^2$. The invariant mass $M_X$ of the final hadronic state is given by $M_X^2 \equiv (p+q)^2 = M^2 + Q^2(1-x)/x$, for fixed momentum transfer $Q^2$, each value of $x$ may be associated with a specific hadronic final state: $x \approx 1$ corresponds to the few-body quasielastic scatterings, and $x \approx 0$ (large $M_X$) to multi-particles. Also, at $x \approx 0$, the nucleon–sea is probed.

It was a great surprise in 1968 when, for the first time, the SLAC–MIT experiments showed that at large $q^2$, the deep inelastic cross-section appeared *much larger than expected.* In other words, the structure functions $W_{1,2}(x,q^2)$ extracted from data are found to behave very differently from the form factors squared $G^2_{E,M}(q^2)$. Indeed, data show that for fixed values of $x$, when $Q^2$ varies from 1 to 25 GeV$^2$, the maximal energies reached at that time, $W_1(x,q^2)$ and $(\nu/M)W_2(x,q^2)$ are practically constant whereas in the same $Q^2$ range, the $G^2_{E,M}(q^2)$ drop from 1 to $10^{-6}$. Nowadays for $Q^2$ as large as a few thousands of GeV$^2$ (Fig. 10.7) at HERA, the structure functions remain astonishingly constant or slightly increasing at small $x$, while the form factors squared decrease from 1 to $10^{-14}$.

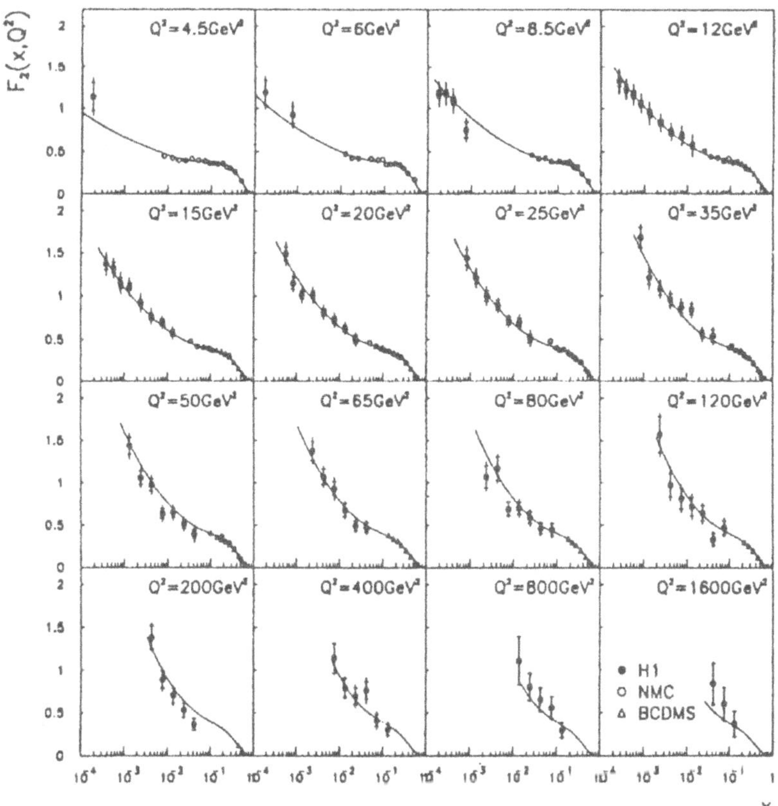


**Fig. 10.7.** The structure function $F_2(x,q^2)$ from Dainton, J. in *Proc. Workshop on Deep Inelastic Scattering and QCD* (eds. Laporte, JF. and Sirois, Y.). Editions de l'Ecole Polytechnique, Paris 1995

### 10.4.2 Bjorken Scaling and the Feynman Quark Parton

This surprising behavior was in fact already anticipated by Bjorken in 1966. From considerations based on the Gell-Mann quark model current algebra commutators, he discovered the scaling law according to which in the limit

$$-q^2 \equiv Q^2 \to \infty \quad , \quad \nu \to \infty \quad , \quad \text{with } \frac{Q^2}{2M\nu} \equiv x \text{ fixed} , \tag{10.47}$$

the structure functions depend only on $x$ :

$$W_1(q^2,\nu) \underset{q^2,\nu\to\infty}{\longrightarrow} F_1(x) \quad , \quad \frac{\nu}{M} W_2(q^2,\nu) \underset{q^2,\nu\to\infty}{\longrightarrow} F_2(x) . \tag{10.48}$$

The physical content of the Bjorken scaling law (48) lies essentially in the *finite limit* of the structure functions $F_1(x)$ and $F_2(x)$, since one can always write $W_1(q^2,\nu) \equiv F_1(x,q^2)$ and $(\nu/M)W_2(q^2,\nu) \equiv F_2(x,q^2)$. For each fixed value of $x$, when $-q^2 \to \infty$, the limits of $F_1(x,q^2)$ and $F_2(x,q^2)$ can depend only on $x$. In principle, they may tend to infinity or zero, the latter possibility is naively expected when we notice that, in a sense, the structure functions represent just an incoherent sum of squared form factors, each tending quickly to zero for large $q^2$. Bjorken assured us that $F_1(x)$ and $F_2(x)$ are *finite*.

How did Feynman interpret deep inelastic scattering data ? Since the experiments on elastic electron–proton scattering by Hofstadter and his group, who in the 1960s found that the $G_{\rm E}(q^2)$ and $G_{\rm M}(q^2)$ form factors decrease as dipole distributions according to (38), it is known that the nucleon has a structure and must be a bound state. But what are its constituents called partons by Feynman, and what is the nature of their interactions ? Two experimental facts obtained from deep inelastic scatterings are crucial: (i) the structure functions are almost independent of $q^2$, and remarkably do not tend to zero as $q^2 \to \infty$; (ii) the $\tan^2\frac{\theta}{2}$ term is present in (44), i.e. $W_1(q^2,x) \to F_1(x) \neq 0$. We recognize that (i) hints at a loosely bound pointlike parton probed by the virtual photon whereas (ii) suggests that this pointlike constituent is a fermion [remember the second remark (b) after the Rosenbluth formula]. So pointlike quarks naturally emerge from these observations as fundamental constituents of matter. As will be discussed later, the loosely bound parton is a consequence of the QCD asymptotic freedom.

On the other hand, if the proton is a bound state of $\rm n\pi^+$ or $\rm \Lambda K^+$ for instance, then $W_j(q^2,\nu)$, with $j = 1,2$, would be strongly dependent on $q^2$ (since the constituents n and $\Lambda$, unlike the partons, are themselves composite objects like p), and $F_1(x)$ would vanish (since the photon would probe the spinless 'constituent' meson $\pi^+$ or $\rm K^+$ of the proton). Also, the quark *color* degrees of freedom get their dramatic confirmation in the total cross-section $\sigma(\rm e^+ + e^- \to \text{hadrons})$, and in the decay rate $\Gamma(\tau \to \nu_\tau + \text{hadrons})$, to mention only two examples (Sect. 7.5 and Chaps. 13–14).

Let us assume that the constituents of nucleons are *valence* quarks $\rm u_v$ , $\rm d_v$, linked by gluons through non-Abelian QCD interaction and surrounded by pairs of quark–antiquark referred to as the nucleon *sea*. Besides the light quark pairs $\overline{\mathrm{u}}_{\mathrm{s}}, \mathrm{u}_{\mathrm{s}}$ and $\overline{\mathrm{d}}_{\mathrm{s}}, \mathrm{d}_{\mathrm{s}}$, the sea may also contain strange $\overline{\mathrm{s}}, \mathrm{s}$ and charm $\overline{\mathrm{c}}, \mathrm{c}$ pairs. We distinguish the valence $\mathrm{u}_{\mathrm{v}}, \mathrm{d}_{\mathrm{v}}$ from the sea $\mathrm{u}_{\mathrm{s}}, \mathrm{d}_{\mathrm{s}}$ and denote their sums as u and d : $\mathrm{u} = \mathrm{u}_{\mathrm{v}} + \mathrm{u}_{\mathrm{s}}$, $\mathrm{d} = \mathrm{d}_{\mathrm{v}} + \mathrm{d}_{\mathrm{s}}$. When this hypothesis is confronted with experiments, all the data are excellently described: the partons turn out to be quarks, antiquarks, and gluons.

Indeed for large $q^2$, the photon penetrates more and more deeply into the nucleon and strikes the nucleon partons. It interacts on the one hand with quarks u, d, antiquarks $\overline{\mathrm{u}}, \overline{\mathrm{d}}$ and with other flavored pairs $\overline{\mathrm{q}}\mathrm{q}$ of the sea; on the other hand, like the other gauge bosons W and Z, the photon is insensitive to gluons. The interaction between partons becomes weaker and weaker, at large $q^2$, the struck quarks (antiquarks) behave as if they were loosely bound, i.e. almost noninteracting or free. They interact softly with the remaining partons, so that when hit by photons, the outgoing partons materialize as jets of hadrons collinear with the directions of the struck partons.

On the other hand, if quarks are not probed by high $q^2, \nu$ photons, they are strongly interacting and firmly bound in the hadron. This strange behavior of quarks – their mutual interactions are stronger at low energy than at high energy – can only be understood by the asymptotic freedom of non-Abelian QCD. This property will be discussed in Chap. 15.

Of course we do not observe partons in the final state, only hadrons. Somehow the scattered and unscattered partons must recombine to form hadrons. The basic assumption is that the collision occurs in two steps. First, a parton is hit during the collision time interval $t_1$ defined by the energy transfer i.e. $t_1 \sim \hbar/\nu$. At a much later time $t_2$, the partons recombine to form hadrons of invariant mass $M_X$, i.e. $t_2 \geq \hbar/M_X$, or in the laboratory frame $t_2 \geq \hbar\nu/M_X^2$. Since $M_X^2 \sim 2M\nu$, we have $t_2 \geq 1/2M$, and $t_2 \gg t_1$ is equivalent to $\nu \gg M$ which is the Bjorken limit. Scaling implies that during such a rapid scattering, interactions among the partons are negligible, they are nearly 'free'. The cross-section depends foremost on the dynamics of the first step and very weakly on the complexities of recombination into hadrons in the second step. High-energy deep inelastic experimental data are described as an *incoherent* sum of elastic electron–quarks (or electron–antiquarks) scatterings. Only incoherent additions take place, because the struck quarks (antiquarks) are noninteracting and independent of each other. As we will see later, the structure functions $W_{1,2}$ are essentially total cross-sections, so in a constituent model, it is natural that cross-sections should be independently additive without interference.

To go further, let us denote by $z_j$ the nucleon fractional momentum carried away by a parton $j$, i.e. the parton momentum $k_j^\mu$ is equal to $z_j P^\mu$ where $P^\mu$ is the nucleon momentum (Fig. 10.6c). For on-shell partons, $(q+z_j P)^2 = m_j^2$, and we have

$$q^2 + 2M\nu\, z_j = 0 \quad \Longrightarrow z_j = x \equiv \frac{Q^2}{2M\nu} .$$

The relation $z_j = x$ allows us to interpret the Bjorken variable $x$ as the fraction of the nucleon momentum carried away by partons. Since the invariant mass of the unobserved hadrons is larger than the nucleon mass, $p_X^2 \equiv (P+q)^2 \geq M^2 \implies q^2 + 2M\nu \geq 0$, we get $0 \leq x \leq 1$. The parton mass $m_j$ is equal to $Mz_j$, since $m_j^2 \equiv k_j^2 = (z_j P)^2 = M^2 z_j^2$. The cross-section $\mathrm{d}\sigma_j$ of electron scattered by a quark (or antiquark) $j$ of charge $e_j$ (in units of $e > 0$) can be obtained from (35) in which the mass $M$ is to be replaced by $m_j = Mz_j$ together with an overall common factor $e_j^2$. From (45), we get

$$\frac{\mathrm{d}\sigma_j}{\mathrm{d}Q^2 \mathrm{d}\nu} = \frac{\pi\, \sigma_{\mathrm{Mott}}}{EE'} \left[ e_j^2 + 2\, e_j^2 \left( \frac{Q^2}{4m_j^2} \right) \tan^2 \frac{\theta}{2} \right] \delta \left( \nu - \frac{Q^2}{2m_j} \right) . \tag{10.49}$$

The contribution of each parton $j$ to the structure functions is immediately recognized by comparing (44) and (49). We call them $w_1(j)$ and $w_2(j)$:

$$\begin{aligned} w_1(j) &= Me_j^2 \frac{Q^2}{4m_j^2} \delta \left( \nu - \frac{Q^2}{2m_j} \right) = e_j^2 \frac{Q^2}{4Mz_j^2} \delta \left( \nu - \frac{Q^2}{2Mz_j} \right) , \\ w_2(j) &= Me_j^2 \delta \left( \nu - \frac{Q^2}{2Mz_j} \right) . \end{aligned} \tag{10.50}$$

Since each parton contributes incoherently to the cross-section, to obtain the structure functions $W_{1,2}(q^2, \nu)$, we simply add up the $w_{1,2}(j)$, each being weighted by the probability $\mathcal{F}_j(z_j)$ for the parton $j$ to have a four-momentum $z_j P^\mu$, and finally we integrate over the whole range of $z_j$. Note that

$$\delta \left( \nu - \frac{Q^2}{2Mz_j} \right) = \delta \left[ \frac{\nu}{z_j} (z_j - \frac{Q^2}{2M\nu}) \right] = \frac{z_j}{\nu} \delta(z_j - x) . \tag{10.51}$$

We identify partons as quarks or antiquarks in the construction of $W_{1,2}(q^2, \nu)$. On the other hand, the partonic gluons which are insensitive to electroweak interactions do not contribute to the structure functions. The probabilities $\mathcal{F}_j(z_j = x)$ for finding the partons $j$ are nothing but the distributions in the nucleon of $u(x), d(x), s(x), \overline{u}(x), \overline{d}(x), \overline{s}(x)$ [may be $c(x), \overline{c}(x)$]. For the moment, we keep the generic $\mathcal{F}_j(z_j = x)$ and get from (50) and (51):

$$\begin{aligned} W_1(q^2, \nu) &= \sum_j \int_0^1 \mathrm{d}z_j \mathcal{F}_j(z_j) w_1(j) = \sum_j \int_0^1 \mathrm{d}z_j \mathcal{F}_j(z_j) \frac{e_j^2 Q^2}{4Mz_j^2} \frac{z_j}{\nu} \delta(z_j - x) \\ &= \sum_j \int_0^1 \mathrm{d}z_j \mathcal{F}_j(z_j) e_j^2 \frac{x}{2z_j} \delta(z_j - x) = \sum_j \frac{e_j^2 \mathcal{F}_j(x)}{2} \equiv F_1(x) , \\ W_2(q^2, \nu) &= \sum_j \int_0^1 \mathrm{d}z_j \mathcal{F}_j(z_j) w_2(j) = \sum_j \int_0^1 \mathrm{d}z_j \mathcal{F}_j(z_j) e_j^2 M \frac{z_j}{\nu} \delta(z_j - x) \\ &= \frac{M}{\nu} \sum_j e_j^2\, x\, \mathcal{F}_j(x) \implies \frac{\nu}{M} W_2(q^2, \nu) = \sum_j e_j^2\, x\, \mathcal{F}_j(x) \equiv F_2(x). \end{aligned} \tag{10.52}$$

From the above equation, we deduce directly the Callan–Gross relation

$$2xF_1(x) = F_2(x) \tag{10.53}$$

which expresses the simple fact that the parton hit by the photon is a spin-1/2 object; a spinless parton would give $F_1(x) = 0$. Since by definition, partons are pointlike and devoid of form factors, we easily understand why the structure functions are $q^2$-independent, at least when QCD effects are ignored.

Let us mention that QCD corrections induce a smooth logarithmic $q^2$ dependence for the structure functions in a definite way. As $q^2$ increases, $F_{1,2}(x, q^2)$ slightly increase for small $x$. For large $x$ the tendency is reversed, the structure functions become smaller. Only non-Abelian gauge theory with its asymptotic freedom property can offer such a behavior. In other words, the Bjorken scaling law is predicted to be smoothly violated; such a violation is indeed experimentally confirmed (see Fig. 10.7). The message can hardly be clearer: first, one needs spin-1/2 pointlike constituents to describe the Bjorken scaling; second, the scaling violation is predicted in a clear-cut way. The observation of the QCD effects on electromagnetic and weak processes (Chaps. 14 and 16) is one of the great triumphs of the standard model.

The Callan–Gross relation may also be interpreted as follows. If we look at the diagram Fig. 10.6a for $\mathrm{e} + \mathrm{N} \to \mathrm{e} + X$ as a two-step reaction $\mathrm{e} \to \mathrm{e} + \gamma^*$ , $\gamma^* + \mathrm{N} \to X$, where $\gamma^*$ is a virtual photon, then we realize that the structure functions $W_{1,2}(q^2, \nu)$ represent in fact $\sigma_{\mathrm{tot}}(\gamma^* + \mathrm{N})$, the total photoabsorption cross-sections by nucleon. In the nucleon rest frame, $\nu$ is just the energy of the virtual photon, and $q^2$ is the square of its invariant mass. With a nonzero mass, the virtual photon not only has two transverse polarizations $\varepsilon^\mu_\pm(q)$, but also a longitudinal polarization $\varepsilon^\mu_{\mathrm{L}}(q)$, and the corresponding total cross-sections are given by

$$\begin{aligned}\sigma_\pm(q^2, \nu) &= K\ \varepsilon^\mu_\pm(q)\varepsilon^{*\rho}_\pm(q) W_{\mu\rho}(q^2, \nu)\ , \\ \sigma_{\mathrm{L}}(q^2, \nu) &= K\ \varepsilon^\mu_{\mathrm{L}}(q)\varepsilon^{*\rho}_{\mathrm{L}}(q) W_{\mu\rho}(q^2, \nu)\ .\end{aligned} \tag{10.54}$$

Following conventions appropriate for real photons, the kinematic flux factor denoted by $K$ is given by $K = 8\pi^2\alpha/(q^2 + 2M\nu)$. With

$$p_\mu = (M, 0, 0, 0)\ ;\ q_\mu = (\nu, 0, 0, q_3 = \sqrt{Q^2 + \nu^2})\ ;\ \varepsilon^\mu_\pm = \frac{-1}{\sqrt{2}}(0, 1, \pm i, 0)\ ;$$

$$\varepsilon^\mu_{\mathrm{L}} = \frac{1}{\sqrt{Q^2}}(\sqrt{Q^2 + \nu^2}, 0, 0, \nu)\ ;\ \varepsilon^\mu_\pm \varepsilon^{*\rho}_\pm g_{\mu\rho} = -1\ ;\ \varepsilon^\mu_{\mathrm{L}} \varepsilon^{*\rho}_{\mathrm{L}} g_{\mu\rho} = +1\ ,$$

hence

$$\varepsilon^\mu_\pm \varepsilon^{*\rho}_\pm T^1_{\mu\rho} = 1\ ,\ \varepsilon^\mu_\pm \varepsilon^{*\rho}_\pm T^2_{\mu\rho} = 0\ ,\ \varepsilon^\mu_{\mathrm{L}} \varepsilon^{*\rho}_{\mathrm{L}} T^1_{\mu\rho} = -1\ ,\ \varepsilon^\mu_{\mathrm{L}} \varepsilon^{*\rho}_{\mathrm{L}} T^2_{\mu\rho} = \frac{Q^2 + \nu^2}{Q^2}\ .$$

Then (54) becomes

$$\sigma_{\rm T}(q^2,\nu) \equiv \frac{1}{2}[\sigma_+(q^2,\nu)+\sigma_-(q^2,\nu)] = K\,W_1(q^2,\nu)\,,$$
$$\sigma_{\rm L}(q^2,\nu) = K\left[-W_1(q^2,\nu)+\left(1+\frac{\nu^2}{Q^2}\right)W_2(q^2,\nu)\right]\,. \tag{10.55}$$

In the Bjorken limit where $Q^2$ , $\nu \to \infty$,

$$\sigma_{\rm T} \to K\,F_1(x) \quad , \quad \sigma_{\rm L} \to K\,\frac{F_2(x)-2xF_1(x)}{2x}\,,$$
$$R \equiv \frac{\sigma_{\rm L}(q^2,\nu)}{\sigma_{\rm T}(q^2,\nu)} \to \frac{F_2(x)-2xF_1(x)}{2xF_1(x)}\,. \tag{10.56}$$

The Callan–Gross relation which is equivalent to $R \equiv \sigma_{\rm L}/\sigma_{\rm T} \to 0$ agrees with data. A spinless parton would give $F_1(x) = 0$ or $R \to \infty$. Such a possibility is certainly ruled out by experiments. The reason why $R$ is not identically equal to zero is that the photon also hits the nucleon sea $\overline{\rm q}{\rm q}$ which is presumably spinless. The fact that $R \ll 1$ implies that the sea contribution is globally much smaller than the quark contributions.

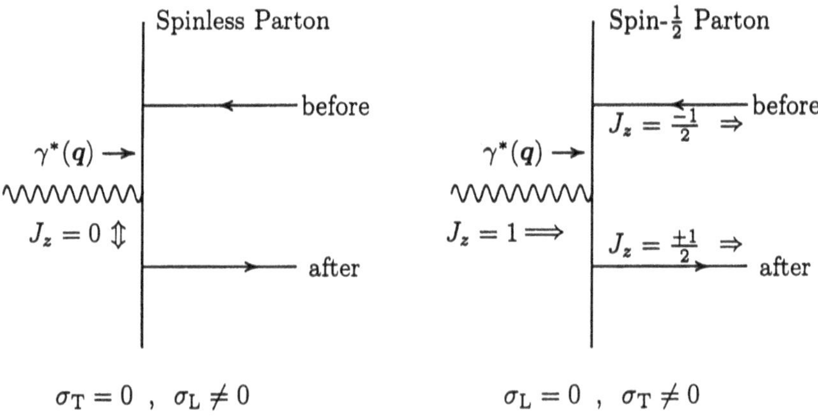


**Fig. 10.8.** Absorption of a virtual photon by a parton in the Breit frame

We may understand the relation $\sigma_{\rm L} \to 0$ by considering the helicities of photon and parton in the Breit frame (Fig. 10.8). Indeed in this frame, the three-momenta of the photon and the incoming parton are collinear and opposite, moreover the parton momenta before and after the collision are reversed and exactly equal. By angular momentum conservation, a spinless parton cannot absorb the transverse components of a photon (with its spin $J_z = \pm 1$ aligned along its three-momentum vector $\boldsymbol{q}$) but may absorb its longitudinal component (corresponding to $J_z = 0$); in other words, with spinless parton, one gets $\sigma_{\rm T} = 0$ and $\sigma_{\rm L} \neq 0$. On the other hand, along the $\boldsymbol{q}$ axis, an incoming spin-$1/2$ parton with its helicities $\mp 1/2$ can catch a $\pm 1$ photon's spin and go away with $\pm 1/2$ helicities. In this Breit frame, the parton helicity conservation implies a change by $\pm 1$ unit along the $\boldsymbol{q}$ axis before and after the collision that only a transverse photon can satisfy, hence $\sigma_L = 0$ and $\sigma_T \neq 0$.

We rewrite the probabilities $\mathcal{F}_j(z_j = x)$ in (52) as the distributions of quarks and antiquarks in the nucleon:

$$F_2^{\mathrm{p}}(x) = x\left\{\frac{4}{9}[u(x)+\overline{u}(x)] + \frac{1}{9}[d(x)+\overline{d}(x)+s(x)+\overline{s}(x)] + \cdots\right\}, \tag{10.57}$$

where the dots stand for possible charm quarks $c(x), \overline{c}(x)$ in the nucleon sea. Invariance of the nucleon properties to isospin rotations implies that the same $u(x), d(x)$ quark distributions in the proton can be used for the neutron. The isospin symmetry by up $\leftrightarrow$ down quarks interchange in proton $\leftrightarrow$ neutron makes the $u^{\mathrm{p}}(x)$ [respectively $d^{\mathrm{p}}(x)$] distribution in the proton equal to the $d^{\mathrm{n}}(x)$ [respectively $u^{\mathrm{n}}(x)$] distribution in the neutron:

$$u^{\mathrm{p}}(x) = d^{\mathrm{n}}(x) \equiv u(x) \ , \quad d^{\mathrm{p}}(x) = u^{\mathrm{n}}(x) \equiv d(x) \, . \tag{10.58}$$

From (58) we have

$$F_2^{\mathrm{n}}(x) = x\left\{\frac{4}{9}[d(x)+\overline{d}(x)] + \frac{1}{9}[u(x)+\overline{u}(x)+s(x)+\overline{s}(x)] + \cdots\right\}. \tag{10.59}$$

Since the nucleon has neither strangeness nor charm, one has

$$\int_0^1 \mathrm{d}x[s(x)-\overline{s}(x)] = \int_0^1 \mathrm{d}x[c(x)-\overline{c}(x)] = 0 \, . \tag{10.60}$$

Similarly, since the proton and neutron charges are respectively 1 and 0,

$$\begin{aligned} &\int_0^1 \mathrm{d}x\left\{\frac{2}{3}[u(x)-\overline{u}(x)] - \frac{1}{3}[d(x)-\overline{d}(x)]\right\} = 1 \, , \\ &\int_0^1 \mathrm{d}x\left\{\frac{2}{3}[d(x)-\overline{d}(x)] - \frac{1}{3}[u(x)-\overline{u}(x)]\right\} = 0 \, , \end{aligned} \tag{10.61}$$

from which

$$\int_0^1 \mathrm{d}x[u(x)-\overline{u}(x)] = 2 \quad \text{and} \quad \int_0^1 \mathrm{d}x[d(x)-\overline{d}(x)] = 1 \, . \tag{10.62}$$

We now remark that the quark and antiquark distributions also appear in deep inelastic neutrino scattering by nucleon (Chap. 12). Electromagnetic and weak inclusive reactions are related, thus the quark parton model can also be tested by neutrinos as projectiles. From experimental data on $F_2^{\mathrm{p,n}}(x)$

(by electromagnetic scattering) and on $F^{\nu}_{2,3}(x)$ (by neutrino scattering), we can extract the distributions $u(x), d(x), s(x)$, $\overline{u}(x)$, $\overline{d}(x)$, and $\overline{s}(x)$ which are constrained by the sum rules (60)–(62).

Indirect evidence for gluons emerges from the observation that the nucleon total momentum must be shared by all of its contituents, gluons included. Since each parton has a momentum $z_j P^\mu$, one must have the sum rule

$$\sum_j \int_0^1 (z_j P^\mu)\mathcal{F}_j(z_j)\mathrm{d}z_j = P^\mu \implies \sum_j \int_0^1 x\mathcal{F}_j(x)\mathrm{d}x = 1 \,. \qquad (10.63)$$

To estimate the quark and antiquark contributions to the above sum rule, we now replace in (63) $\mathcal{F}_j(x)$ by the $\mathrm{q}(x), \overline{\mathrm{q}}(x)$ distributions. With (57) and (59), their contributions can be written as

$$F_2^{\mathrm{p}}(x) + F_2^{\mathrm{n}}(x) = \frac{5}{9}x\left[u(x) + d(x) + \overline{u}(x) + \overline{d}(x)\right] + \frac{2}{9}x\left[s(x) + \overline{s}(x)\right] \,. \quad (10.64)$$

The sum $F_2^{\mathrm{p}}(x) + F_2^{\mathrm{n}}(x) \equiv F_2^{I=0}(x)$ is obtained from data of electron scattering by an isoscalar target (deuteron). We rewrite (64) under the integrated form

$$\frac{9}{5}\int_0^1 \mathrm{d}x F_2^{I=0}(x) = \int_0^1 \mathrm{d}x\, x\left\{u(x) + d(x) + \overline{u}(x) + \overline{d}(x) + \frac{2}{5}\left[s(x) + \overline{s}(x)\right]\right\} .$$

The integration of the left-hand side of (64) is found to be $0.245 \pm 0.02$. Provided that $\frac{3}{5}x\left[s(x) + \overline{s}(x)\right] \ll 1$, which is plausible for the sea, the above equation may be written as

$$\sum_j \int_0^1 \mathrm{d}x\, x\left[q_j(x) + \overline{q}_j(x)\right] \approx \frac{9}{5}\,(0.245 \pm 0.02) = 0.44 \pm 0.02 \ .$$

Comparing the above equation with (63), we find that the remaining $(56\mp 2)\%$ of the nucleon momentum is carried away not by quarks and antiquarks but by objects insensitive to photons, and gluons are natural candidates.

Finally, in terms of the scaling structure function $F_2(x)$ and the ratio $R \equiv \sigma_{\mathrm{L}}/\sigma_{\mathrm{T}} = -1 + F_2(x)/2xF_1(x)$, the deep inelastic cross-section (44) can be rewritten with a new variable $y \equiv \nu/E = (E - E')/E$ which evidently satisfies $0 \leq y \leq 1$. This energy loss variable $y$ is also useful in the description of deep inelastic neutrino scattering (Chap. 12). Since $\mathrm{d}Q^2\mathrm{d}\nu = 2ME\,\nu\,\mathrm{d}x\,\mathrm{d}y$, we have

$$\begin{aligned}\frac{\mathrm{d}\sigma(\mathrm{e} + \mathrm{N} \to \mathrm{e} + X)}{\mathrm{d}x\mathrm{d}y} &= \frac{(2ME\,\nu)\mathrm{d}\sigma}{\mathrm{d}Q^2\mathrm{d}\nu} = \frac{4\pi\alpha^2\, s}{Q^4}\left\{F_2(x)(1-y) + xF_1(x)y^2\right\} \\ &= \frac{4\pi\alpha^2\, s}{Q^4}F_2(x)\left\{(1-y) + \frac{y^2}{2(1+R)}\right\} . \qquad (10.65)\end{aligned}$$

The above result is obtained by using the identity $2E(1-y)\sin^2(\theta/2) = Mxy$ and neglecting $M^2 \ll s \approx 2ME$. Since $Q^2 = s\,xy$, (65) may be written as

$$\frac{\mathrm{d}\sigma(\mathrm{e}+\mathrm{N}\to\mathrm{e}+X)}{\mathrm{d}x\mathrm{d}Q^2} = \frac{4\pi\alpha^2}{Q^4}\frac{F_2(x)}{x}\left\{(1-y)+\frac{y^2}{2(1+R)}\right\}. \tag{10.66}$$

For each fixed value of $x$, the $\mathrm{d}\sigma/\mathrm{d}y$ (or $\mathrm{d}\sigma/\mathrm{d}Q^2$) distribution allows us to extract the structure function $F_2(x)$. When the violation of the Bjorken scaling law is incorporated, (65) is usually written with $F_2(x,q^2)$, its $q^2$ dependence is smoothly logarithmic as predicted by QCD and experimentally confirmed.

In summary, the photon is a powerful probe of the ultimate structure of matter. The main reason for the photon usefulness is that in the one-photon exchange of deep inelastic scattering, via $\ell^- \to \ell^- + \gamma^*$, the invariant mass squared $q^2$ and the energy $\nu$ of the virtual photon $\gamma^*$ can be varied when the latter strikes the hadronic constituents to explore their nature. We have seen the crucial role of high $q^2, \nu$ in elastic and deep inelastic scatterings. In pion–nucleon collision taken as an example of the more general hadron–hadron scattering, one cannot take advantage of the well-known photon exchanged mechanism of the electromagnetic interaction, i.e. first we make two unknown hadrons collide on each other, second we can only vary the pion energy but we cannot change its mass which must be $m_\pi$. In this context, hadron–hadron collisions are not the right way to discover the hadronic constituents. On the other hand, in electromagnetic e–N scattering, only the strong interaction of the struck hadron N is unknown. The more familiar photon, with both its invariant mass and its energy increasing, is capable of revealing the hadronic layers. Once the quarks and gluons are fully recognized as the fundamental constituents of matter, hadron–hadron collisions can be interpreted as quark–quark, quark–gluon, gluon–gluon reactions in a subsequent step. Further evidence for quarks is provided by the neutrino deep inelastic collision (Chap. 12) and by the $\mathrm{e}^+ + \mathrm{e}^-$ annihilation into hadrons. The total cross-section of the latter process fits so well with the $\mathrm{e}^+ + \mathrm{e}^- \to \mathrm{q} + \overline{\mathrm{q}}$ (see Sect. 7.5 and Chap. 14 where the QCD radiative corrections are included) that the reality of color quarks can hardly be denied.

## Problems

**10.1 OPE Yukawa nucleon potential.** Show that the matrix element of the one pion exchange (OPE) between the two nucleons 1,2 can be written as the difference of the direct and the exchange terms:

$$\mathcal{M} = g^2_{\pi\mathrm{NN}}\frac{[\overline{u}(p_1',s_1')\,\gamma_5\,\tau_j\,u(p_1,s_1)]\,[\overline{u}(p_2',s_2')\,\gamma_5\,\tau_j\,u(p_2,s_2)]}{(p_1-p_1')^2-m_\pi^2}$$

$$-\ \text{exchange term}\ [(p_1',s_1')\Leftrightarrow(p_2',s_2')].$$

Why the minus sign? Show that $\overline{u}(p_1',s_1')\,\gamma_5\,u(p_1,s_1) = \chi_1'^{\dagger}(\boldsymbol{\sigma}\cdot\boldsymbol{q})\chi_1 \stackrel{\text{def}}{=} \boldsymbol{\sigma}_1\cdot\boldsymbol{q}$ where $\boldsymbol{q} = \boldsymbol{p}_1 - \boldsymbol{p}_1'$ in the center-of-mass system. $\chi$ is the standard two-component Pauli spinor describing the $\pm 1/2$ spin states of the nucleon (Chap. 3). Then the nonrelativistic limit of the direct term is

$$\mathcal{M}_{\rm dir}(\boldsymbol{q}) = \left[\frac{-g^2_{\pi\rm NN}\,(\boldsymbol{\tau}_1\cdot\boldsymbol{\tau}_2)(\boldsymbol{\sigma}_1\cdot\boldsymbol{q})(\boldsymbol{\sigma}_2\cdot\boldsymbol{q})}{|\boldsymbol{q}|^2+m_\pi^2}\right] .$$

The nucleon–nucleon potential $V_{\rm dir}(\boldsymbol{x})$ is obtained from the Fourier transform of $V_{\rm dir}(\boldsymbol{q})$, the latter is related to the invariant nucleon–nucleon scattering matrix element $\mathcal{M}$ by $V_{\rm dir}(\boldsymbol{q}) = \mathcal{M}_{\rm dir}(\boldsymbol{q})/4M_{\rm N}^2$ in the nonrelativistic limit

$$V_{\rm dir}(\boldsymbol{x}) = \frac{1}{(2\pi)^3}\int {\rm d}^3q\, {\rm e}^{{\rm i}\boldsymbol{q}\cdot\boldsymbol{x}}\frac{\mathcal{M}_{\rm dir}(\boldsymbol{q})}{4M_{\rm N}^2} .$$

Explain the origin of the denominator $4M_{\rm N}^2$. Show that ($r = |\boldsymbol{x}|$)

$$V_{\rm dir}(\boldsymbol{x}) = \frac{g^2_{\pi NN}}{4\pi}\frac{m_\pi^2}{12M_{\rm N}^2}(\boldsymbol{\tau}_1\cdot\boldsymbol{\tau}_2)\Big\{\boldsymbol{\sigma}_1\cdot\boldsymbol{\sigma}_2\frac{{\rm e}^{-m_\pi r}}{r} + \left[\frac{3\sigma_1\cdot\boldsymbol{x}\sigma_2\cdot\boldsymbol{x}}{r^2} - \boldsymbol{\sigma}_1\cdot\boldsymbol{\sigma}_2\right]\frac{{\rm e}^{-m_\pi r}}{r}\left(1+\frac{3}{m_\pi r}+\frac{3}{(m_\pi r)^2}\right)\Big\} .$$

**10.2 Normalization of $F_\pi(0)$.** Show that at $q^2 = 0$, the pion form factor satisfies $F_\pi(0) = 1$. Generalize it to the nucleon case and get $F_1^{\rm p}(0) = 1$, $F_1^{\rm n}(0) = F_2^{\rm n}(0) = 0$.

**10.3 Meson form factors.** In terms of the electromagnetic form factor $F_\pi(q^2)$, write the amplitude of the reaction ${\rm e}^+(p')+{\rm e}^-(p)\to\pi^+(k')+\pi^-(k)$. The timelike photon exchange (Fig. 10.1) is in the channel of the Mandelstam variable $s\equiv q^2 = (k'+k)^2 > 0$. In the center-of-mass (cm) system of ${\rm e}^+{\rm e}^-$, show that the cross-section is given by

$$\frac{{\rm d}\sigma}{{\rm d}\Omega_{\rm cm}} = \frac{\alpha^2|F_\pi(s)|^2}{8\,s}\left(1-\frac{4m_\pi^2}{s}\right)^{3/2}(1-\cos^2\theta_{\rm cm}) , \quad \boldsymbol{p}\cdot\boldsymbol{k} = |\boldsymbol{p}|\cdot|\boldsymbol{k}|\cos\theta_{\rm cm} .$$

So $\sigma$ has a broad peak at $s\approx m^2_{\rho^0}$. Compare this result with the formula (36) of the reaction ${\rm e}^-(p)+\pi^\pm(k)\to{\rm e}^-(p')+\pi^\pm(k')$, the spacelike photon exchange is now in the $t\equiv(k'-k)^2\le 0$ channel. What do we expect of the cross-sections ${\rm e}^+ + {\rm e}^-\to{\rm K}^+ + {\rm K}^-$ , ${\rm D}^+ + {\rm D}^-$ , ${\rm B}^+ + {\rm B}^-$?

**10.4 $\beta$-decay of $\pi^\pm$.** The conserved vector current (CVC) relates the (isospin 1) electromagnetic current $J^\mu_{\rm em}$ to $V^\mu$, the *flavor-conserving vector* part of the $V^\mu - A^\mu$ charged current in the d $\to$ u transition occurred for example in neutron $\beta$ decay. Show that

$$\langle\pi^0(p')\,|\,V^\mu\,|\,\pi^\pm(p)\rangle = \sqrt{2}(p'+p)^\mu\,F_\pi(q^2) .$$

Compute the rate $\pi^+\to\pi^0+{\rm e}^+ +\nu_{\rm e}$. Generalize to $\overline{\rm K}^0\to{\rm K}^- + {\rm e}^+ + \nu_{\rm e}$. Why is the factor $\sqrt{2}$ absent in the K case ?

**10.5 $\rho^0$ decay.** By dimensional argument, show that the couplings between $\pi^\pm - W^\pm$ and $\gamma - \rho^0$ have the dimension of [mass]$^2$. Write the amplitudes $\pi^+ \to e^+ + \nu_e$ and $\rho^0 \to e^+ + e^-$ in terms of the decay constants $f_\pi$ and $f_{\rho^0}$; both have the dimension of [mass] (Fig. 10.3a, b). The former amplitude is proportional to the electron mass, whereas the latter is not. Therefore $\Gamma(\pi^+ \to e^+ + \nu_e) \ll \Gamma(\pi^+ \to \mu^+ + \nu_\mu)$. Using CVC, compute the $\rho^+ \to e^+ + \nu_e$ rate. Check the formulas (20), (21) of $\rho^0 \to \pi^+ + \pi^-$ and $\rho^0 \to e^+ + e^-$.

**10.6 Inverse Fourier transform of form factors.** The distribution $\rho(r)$ of charge in hadrons can be obtained from the inverse Fourier transform of the corresponding form factors in the nonrelativistic limit ($q^2 \to -|\boldsymbol{q}|^2$). Show that

$$\int \frac{d^3q}{(2\pi)^3} \frac{e^{i\boldsymbol{q}\cdot\boldsymbol{x}}}{1 - \frac{q^2}{\Lambda^2}} = \frac{\Lambda^2}{4\pi} \frac{e^{-\Lambda r}}{r}, \quad \int \frac{d^3q}{(2\pi)^3} e^{i\boldsymbol{q}\cdot\boldsymbol{x}} e^{\frac{-q^2}{\Lambda^2}} = \frac{\Lambda^3 e^{-\frac{\Lambda^2 r^2}{4}}}{8\pi\sqrt{\pi}}.$$

**10.7 Omnès–Muskhelishvili form factor representation.** According to the Watson theorem, we have $F_\pi(s) = |F_\pi(s)|e^{i\delta(s)}$, where $\delta(s)$ is the p-wave phase-shift of $\pi$–$\pi$ strong interaction. Write the dispersion relation for $G(s) \equiv \log F_\pi(s)$. Show that

$$F_\pi(s) = \exp \frac{s}{\pi} \int_{4m_\pi^2}^{\infty} \frac{\delta(s')}{s'(s'-s)} ds.$$

If the form factor $F_\pi(s)$ has zeros in the complex $s$ plane, to avoid difficulties when taking its logarithm, one may factorize $F_\pi(s) = P(s)\overline{F}_\pi(s)$ where the polynome $P(s)$ contains all the zeros of $F_\pi(s)$. Repeat the analysis with $\overline{F}_\pi(s)$. Consider now the function

$$H(s) \equiv \frac{\log \overline{F}_\pi(s)}{\sqrt{s - 4m_\pi^2}}.$$

First find the imaginary of $H(s)$ in terms of $|F_\pi(s)|$, the latter is measured directly in $e^+ + e^- \to \pi^+ + \pi^-$ (Problem 10.3); then write the dispersion relation for $H(s)$.

## Suggestions for Further Reading

*General reading, form factors, elastic* e–N *scattering:*


Bjorken, J. D. and Drell, S., *Relativistic Quantum Mechanics*. McGraw-Hill, New York 1964

De Wit, B. and Smith, J., *Field Theory in Particle Physics* (Vol. I). North-Holland, Amsterdam 1986

Gourdin, M., Phys. Rep. **11C** (1974) 29

Hofstadter, R. (ed.), *Electron Scattering and Nuclear and Nucleon Structure*. Benjamin, New York 1964

Perl, M. L., *High Energy Hadron Physics.* Wiley-Interscience, New York 1974

*Deep inelastic* e–N *scattering, Bjorken scaling, parton model:*

Bjorken, J. D., Phys. Rev. **179** (1969) 1547

Drell, S., Peccei, R. and Taylor, R., in *Lepton–Hadron Scattering*, Proc. 19th SLAC Summer Institute on Particle Physics (ed. Hawthorne, J.). SLAC Report-398, Stanford 1991

Feynman, R. P., *Photon–Hadron Interactions.* Benjamin, Reading, MA 1972

Halzen, F. and Martin, A., *Quarks and Leptons: An Introductory Course in Modern Particle Physics.* Wiley, New York 1984

Nachtmann, O., *Elementary Particle Physics.* Springer, Berlin, Heidelberg 1990

Quigg, C., *Gauge Theories of Strong, Weak and Electromagnetic Interactions.* Addison-Wesley, Reading, MA 1983

# 11 Neutral K Mesons and CP Violation

The neutral K mesons, with their medium-sized masses and their capacity of interacting both weakly and strongly, seem to be specially selected by nature to demonstrate through a few typical phenomena the reality of quantum effects. Even if they did not exist, as L. B. Okun once said, we would have invented them in order to illustrate the fundamental principles of quantum physics.

Four of these phenomena will be studied in this chapter. The first to be considered arises from the existence, in the presence of strong interactions only, of two degenerate states of opposite strangeness quantum numbers, called $K^0$ and $\overline{K}^0$. These states are mixed by the weak interaction, which does not conserve strangeness, to produce two states quite similar in their masses (which differ only by $\Delta m = 3.49 \times 10^{-6}\,\text{eV} = 5.30 \times 10^{9}\,\text{s}^{-1}$) but very dissimilar in their distinctive decay modes and their lifetimes. One short-lived, $K_S$, with lifetime $\tau_S = 8.92 \times 10^{-11}$ s, and the other long-lived, $K_L$, with lifetime $\tau_L = 5.17 \times 10^{-8}$ s. The existence of these states results in a second property called *strangeness oscillations*: a pure strangeness eigenstate, say $K^0$, produced at a given time becomes at a later time a mixture of $K^0$ and $\overline{K}^0$. The amplitude of such a meson beam oscillates in time with a period of $T = 2\pi/\Delta m = 1.18 \times 10^{-9}$ s.

The third property to be studied concerns the *regeneration* of $K_S$. Consider a beam of $K^0$ produced at time $t = 0$ by some strong interaction process. After a lapse of time of the order of $\tau_S$, every K meson decays into two pions through its $K_S$ components; but this process ceases to occur after $\tau_S \ll t \ll \tau_L$ when all the $K_S$ have gone, leaving only the $K_L$ in the beam. If a block of matter is now placed on the beam path, the $K_L$ will interact with matter and partially transforms itself into $K_S$.

Last but not least, we will discuss the CP violation by weak interactions, a phenomenon first reported in 1964 and to this day observed only in the neutral K meson system. It is one of the most fundamental but least understood properties of particle physics. To explore the physics of heavy flavors, several B meson factories are being constructed in the U. S. A. and Japan. Among the first projects to be carried out at these laboratories will certainly see the investigation of the nature, the origin, and the mechanism of the CP violation. The importance of this study may even have a larger impact if the enormous disproportion of matter (baryons) and antimatter (antibaryons) existing in the universe is regarded as a direct consequence of a CP violation that occurred just after the Big Bang.

## 11.1 The Two Neutral K Mesons

From Dirac's work (Chap. 3), we know that each particle corresponds to an antiparticle, both having equal masses, spins, and lifetimes. But their charges of all types (electric, leptonic, baryonic, flavor, or color) are equal in magnitudes but opposite in signs. Among the electrically neutral particles, the neutron is distinct from the antineutron, but certain particles such as the photon, the mesons $\pi^0$ and $\eta$ are identical to their respective antiparticles. In contrast, the neutral K mesons have peculiarly mixed identities.

Recall first that the pseudoscalar mesons $K^+$, $K^0$ and their conjugates $K^-$ and $\overline{K}^0$ are bound states, composed mainly of quarks u, d, and s. In Table 11.1, their quantum numbers (strangeness, isospin) and quark contents are given.

**Table 11.1.** Strange pseudoscalar mesons

| K Mesons | Quark contents | $I_3$ | Strangeness $S$ |
|---|---|---|---|
| $K^+$ | $\bar{s}u$ | $+\frac{1}{2}$ | +1 |
| $K^0$ | $\bar{s}d$ | $-\frac{1}{2}$ | +1 |
| $\overline{K}^0$ | $\bar{d}s$ | $+\frac{1}{2}$ | −1 |
| $K^-$ | $\bar{u}s$ | $-\frac{1}{2}$ | −1 |

The two mesons $\overline{K}^0$ and $K^0$ are quite distinct in the presence of strong interactions which conserve strangeness. Consider for example the strong production process $\pi^- + p \to K^0 + \Lambda$. The initial state has zero strangeness; therefore, since $\Lambda \equiv$ sdu has strangeness $S = -1$, it is a $K^0$ with $S = +1$ that is produced in the final state, and not a $\overline{K}^0$. If strange particles are produced in a strong interaction (normally from nonstrange particles), they are always produced in pairs of particles of opposite strangeness quantum numbers so as to conserve total strangeness (a phenomenon called *associated production*). On the other hand, since the baryonic number is equally conserved by the strong interaction and since the initial state (the proton) has baryonic number $N_B = 1$ in the present example, the final state cannot be $\overline{K}^0 + \overline{\Lambda}$ in spite of its correct strangeness. Thus, from the point of view of the strong interaction, $K^0$ is as distinct from $\overline{K}^0$ as the neutron is from the antineutron.

The difference between these two pairs $K^0$–$\overline{K}^0$ and n–$\bar{n}$ appears in the presence of the weak interaction. As far as we know, the baryonic number is conserved in all situations (the proton lifetime is greater than $10^{39}$ s), but strangeness is not, being broken in weak processes. The conservation of $N_B$ forbids transitions between the neutron and the antineutron because there exist no common intermediate states connecting those two states. On the other hand, both $\overline{K}^0$ and $K^0$ can decay into pions via strangeness-violating weak transitions. Thus, the transmutation of $K^0$ into $\overline{K}^0$, or inversely of $\overline{K}^0$ into $K^0$, can proceed through common intermediates states of pions, as in

$$K^0 \longrightarrow (2\pi,\ 3\pi) \longrightarrow \overline{K}^0 . \tag{11.1}$$

On the quark level, the transitions $K^0 \leftrightarrow \overline{K}^0$ are represented by the Feynman box diagrams in Fig. 11.1. Via the weak interaction, the $\bar{s}$ and d quarks of $K^0$ annihilate into a pair $W^+W^-$ (Fig. 11.1a) or a pair of quark–antiquark $Q_i\overline{Q}_j$ in all possible combinations of the three u, c, t quarks (Fig. 11.1b). These pairs $W^+W^-$ and $Q_i\overline{Q}_j$ then transform by the same weak interaction into s and $\bar{d}$, giving a $\overline{K}^0$ in the final state. Since these transitions change strangeness by two units ($|\Delta S| = 2$), they must proceed through the weak interaction.

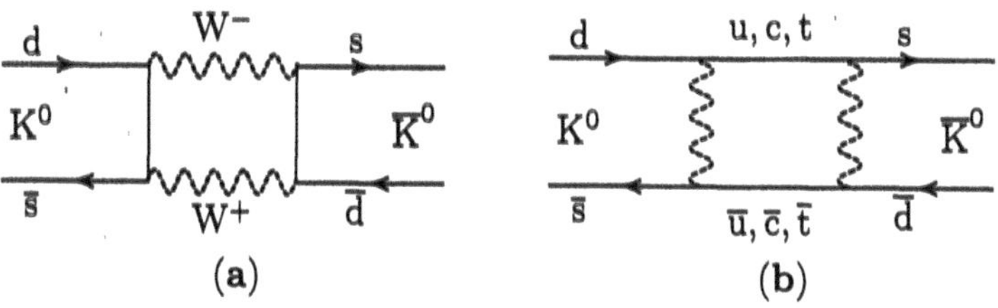


**Fig. 11.1a,b.** $K^0 \leftrightarrow \overline{K}^0$ transition through $\bar{s}\,d \to \bar{d}\,s$

As implicitly understood in the above discussion, $K^0$ and $\overline{K}^0$ are respectively strangeness eigenstates with eigenvalues $S = 1$ and $S = -1$, a meaningful statement only in the absence of weak interactions. Since $S$ is not conserved by weak interactions, another quantum number has to be used to classify the neutral K mesons wherever the weak interaction operates. It turns out that the combined charge conjugation and parity operators, $\mathcal{CP}$, provide an almost perfect candidate for this role, 'almost' because the symmetry they represent is in fact only slightly violated (by a factor of $10^{-3}$), while parity $\mathcal{P}$ and charge conjugation $\mathcal{C}$ are each separately violated at the highest level by weak interactions.

Leaving temporarily aside the question of its violation, we assume in this and the next three sections that CP is a good symmetry even in the presence of the weak interaction. As the K mesons are pseudoscalar, $\mathcal{P}\left|\overline{K}^0\right\rangle = -\left|\overline{K}^0\right\rangle$ and $\mathcal{P}\left|K^0\right\rangle = -\left|K^0\right\rangle$, and as $\overline{K}^0$ and $K^0$ are charge conjugates to each other, $\mathcal{C}\left|K^0\right\rangle = \left|\overline{K}^0\right\rangle$ by an appropriate choice of phase, one gets

$$\mathcal{CP}\left|K^0\right\rangle = -\left|\overline{K}^0\right\rangle \quad , \quad \mathcal{CP}\left|\overline{K}^0\right\rangle = -\left|K^0\right\rangle . \tag{11.2}$$

From this it follows that the orthogonal linear combinations

$$|K_1^0\rangle \equiv \frac{1}{\sqrt{2}}\left(|K^0\rangle - |\overline{K}^0\rangle\right) \quad , \quad |K_2^0\rangle \equiv \frac{1}{\sqrt{2}}\left(|K^0\rangle + |\overline{K}^0\rangle\right) , \tag{11.3}$$

are eigenstates of CP :

$$\mathcal{CP}\,|K_1^0\rangle = +\,|K_1^0\rangle \quad , \quad \mathcal{CP}\,|K_2^0\rangle = -\,|K_2^0\rangle \,. \tag{11.4}$$

The state $K_1^0$ is even and $K_2^0$ is odd under CP.

To determine their decay modes into pions, it suffices to find the corresponding CP-parities of the multipion states. It turns out (Problem 11.1) that neutral two-pion states ($\pi^0 + \pi^0$ and $\pi^+ + \pi^-$) are CP-even and neutral three-pion states ($\pi^0 + \pi^0 + \pi^0$ and $\pi^0 + \pi^+ + \pi^-$) are CP-odd. Since CP is assumed to be conserved, the only allowed decay modes are

$$K_1^0 \to 2\pi \;, \qquad K_2^0 \to 3\pi \,. \tag{11.5}$$

It is noteworthy that these are the only hadronic decay channels open to the neutral K mesons, a fact attributed to the particular value of their mass, just slightly over three times the mass of the pion. This fact also implies that kinematics favors the two-pion mode over the three-pion mode because the kinetic energy available to the reaction products is larger, and the phase space is correspondingly larger in the former than in the latter. Therefore, $K_1^0$ has a proportionally larger decay width or equivalently a shorter lifetime than $K_2^0$.

This was a result predicted by Gell-Mann and Pais and later confirmed by experiment. What was observed was that the neutral K mesons decayed in two different hadronic channels at two different time scales. The first type goes through two-pion channels, with lifetime $\tau_S = 8.92\times10^{-11}$ s, and is called $K_S$. The second type, which can decay into three pions with characteristic time $\tau_L = 5.17 \times 10^{-8}$ s, is called $K_L$. Assuming CP conservation, one may identify $K_S$ with the CP-even state $K_1^0$, and $K_L$ with the CP-odd state $K_2^0$.

This double property, mass degeneracy and distinct lifetimes, is unique to the neutral K mesons. The situation is completely different with the neutral D mesons and the neutral B mesons, which yet parallel the K mesons in their quark compositions only with heavier flavors.

## 11.2 Strangeness Oscillations

We now consider the evolution in time of the amplitudes of states $K_S$ and $K_L$. They are of the general form

$$a(t) = a(0)\exp[-\mathrm{i}(E - \tfrac{\mathrm{i}}{2}\Gamma)t] \,, \tag{11.6}$$

here $E$ is the energy of the state and $\Gamma$ its total decay width. The term $\mathrm{i}\Gamma/2$ is what is needed to yield the familiar exponential decay law

$$I(t) = a(t)a(t)^* = |a(0)|^2 \mathrm{e}^{-\Gamma t} \,, \tag{11.7}$$

which says that the particle decays at the rate given by $\Gamma$. For a stable noninteracting particle, $\Gamma = 0$ and the amplitude $a(t)$ is given by the usual $\exp(-ip \cdot x)$. In the particle rest frame, the particle energy equals its mass $E = m$, and its lifetime is given by $\tau = \hbar/\Gamma$.

Thus, the amplitudes that describe the time evolution of $\mathrm{K_S}$ and $\mathrm{K_L}$ are respectively

$$a_{\mathrm{S}}(t) = a_{\mathrm{S}}(0)\mathrm{e}^{-(\frac{\Gamma_{\mathrm{S}}}{2} + \mathrm{i}m_{\mathrm{S}})t} \quad , \quad a_{\mathrm{L}}(t) = a_{\mathrm{L}}(0)\mathrm{e}^{-(\frac{\Gamma_{\mathrm{L}}}{2} + \mathrm{i}m_{\mathrm{L}})t} . \tag{11.8}$$

There is no reason to expect here that the corresponding masses $m_{\mathrm{S}}$ and $m_{\mathrm{L}}$ be equal even though by CPT invariance the masses of $\overline{\mathrm{K}}^0$ and $\mathrm{K}^0$ must be identical (Chap. 5). However, as the decay modes and the lifetimes of $\mathrm{K_S}$ and $\mathrm{K_L}$ are bound to differ from effects of the $|\Delta S| = 2$ effective interactions mentioned above, it is expected that $\Delta m \equiv m_{\mathrm{L}} - m_{\mathrm{S}} \neq 0$. We will now discuss how $\Delta m$ is calculated and measured, and how it gives rise to the oscillation phenomenon observed in neutral K meson beams.

Suppose at $t = 0$ a $\mathrm{K}^0$ beam is produced, e.g. by the strong production process $\pi^- + \mathrm{p} \to \mathrm{K}^0 + \Lambda$. Its amplitude written in terms of $\mathrm{K_S} = \mathrm{K}_1^0$ and $\mathrm{K_L} = \mathrm{K}_2^0$ is

$$\mathrm{K}^0 = \frac{1}{\sqrt{2}}(\mathrm{K_S} + \mathrm{K_L}) \quad , \quad a_{\mathrm{K}^0}(t) = \frac{1}{\sqrt{2}}\left[a_{\mathrm{S}}(t) + a_{\mathrm{L}}(t)\right] .$$

The beam intensity for $\mathrm{K}^0$, call it $I_0(t)$, is given by

$$I_0(t) = \frac{1}{2}\left[a_{\mathrm{S}}(t) + a_{\mathrm{L}}(t)\right]\left[a_{\mathrm{S}}(t) + a_{\mathrm{L}}(t)\right]^* , \tag{11.9}$$

or equivalently, making use of (8), the normalized density $N(t) = I(t)/I(0)$ is given by

$$N_0(t) = \frac{1}{4}\left[\mathrm{e}^{-\Gamma_{\mathrm{S}}t} + e^{-\Gamma_{\mathrm{L}}t} + 2\,\mathrm{e}^{-(\Gamma_{\mathrm{S}}+\Gamma_{\mathrm{L}})\frac{t}{2}}\cos(\Delta m\, t)\right] . \tag{11.10}$$

Similarly, we may write the normalized density for the $\overline{\mathrm{K}}^0$ beam (produced at $t = 0$ for example by the reaction $\pi^+ + \mathrm{p} \to \overline{\mathrm{K}}^0 + \mathrm{K}^+ + \mathrm{p}$) as

$$N_{\bar{0}}(t) = \frac{1}{4}\left[\mathrm{e}^{-\Gamma_{\mathrm{S}}t} + e^{-\Gamma_{\mathrm{L}}t} - 2\,\mathrm{e}^{-(\Gamma_{\mathrm{S}}+\Gamma_{\mathrm{L}})\frac{t}{2}}\cos(\Delta m\, t)\right] . \tag{11.11}$$

Thus, the two beams oscillate with frequency $\Delta m/2\pi$ (Fig. 11.2). Since $\Delta m\,\tau_{\mathrm{S}} = 0.47$, the oscillation waves will be clearly visible at $t$ of the order of a few $\tau_{\mathrm{S}}$, before all the $\mathrm{K_S}$ have died out, leaving only the $\mathrm{K_L}$ in the beam. As $c\tau_{\mathrm{S}} = 2.67$ cm and $c\tau_{\mathrm{L}} = 1551$ cm, the $\mathrm{K_L}$ will survive long after all the $\mathrm{K_S}$ have gone. In a beam made up entirely of $\mathrm{K}^0$ at time $t = 0$, mesons $\overline{\mathrm{K}}^0$ will appear far from the production source through their presence in $\mathrm{K_L}$ with

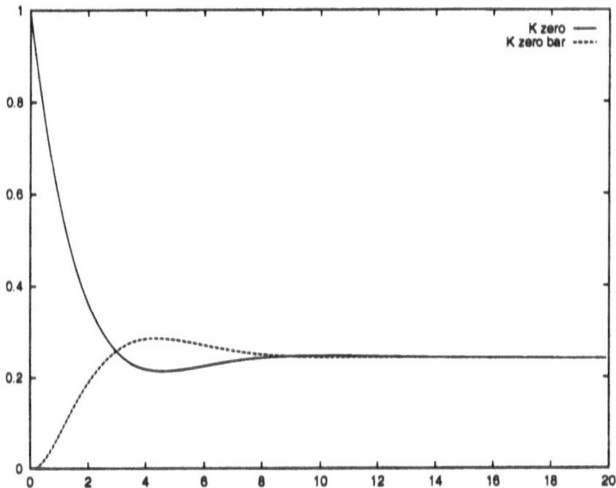


**Fig. 11.2.** Plots of $N_0(t)$ and $N_{\bar{0}}(t)$, $t$ in units of $10^{-10}$ s

equal probability as $K^0$. Similarly, an initially pure $\overline{K}^0$ will progressively become contaminated with $K^0$.

The oscillations can be detected from observing the $K_{\ell 3}$ decay modes. In fact, from the quark contents of $K^0(\bar{s}d)$ and $\overline{K}^0(s\bar{d})$, it follows that $\bar{s} \to \bar{u} + \ell^+ + \nu_\ell$ ($s \to u + \ell^- + \overline{\nu}_\ell$) and therefore the $\overline{K}^0$ and $K^0$ decays are governed by the $\Delta S = \Delta Q$ rule:

$$K^0 \to \pi^- + \ell^+ + \nu_l \quad , \quad \overline{K}^0 \to \pi^+ + \ell^- + \bar{\nu}_l \,. \tag{11.12}$$

Indeed from the initial $K^0$ to the final $\pi^-$ hadronic states, there is a change in strangeness, $\Delta S = (+1) - (0) = +1$, and in electric charge $\Delta Q = (0) - (-1) = +1$, so $\Delta S = \Delta Q$. The same change $\Delta S = \Delta Q = -1$ occurs in $\overline{K}^0 \to \pi^+$. Then from this rule, we can *identify* $K^0$ by its decay product $e^+$, and the $\overline{K}^0$ by its $e^-$. By recording the number of the emitted positrons ($N^+$) and electrons ($N^-$), one can measure the charge asymmetry

$$\delta(t) \equiv \frac{N^+ - N^-}{N^+ + N^-} = \frac{N_0(t) - N_{\bar{0}}(t)}{N_0(t) + N_{\bar{0}}(t)} \,, \tag{11.13}$$

whose time evolution as given by (10) and (11) is

$$\delta(t) \sim 2\, e^{-(\Gamma_S + \Gamma_L)\frac{t}{2}} \cos(\Delta m\, t) \,. \tag{11.14}$$

This asymmetry displays a sinusoidal time oscillation, from which one may infer the magnitude of the mass difference (Fig. 11.3). Experiment gives the mass difference $|\Delta m| \simeq 3.49 \times 10^{-6}\,\text{eV} \simeq 5.30 \times 10^9\,\text{s}^{-1}$ (the sign of $\Delta m$ will be considered in the following section).

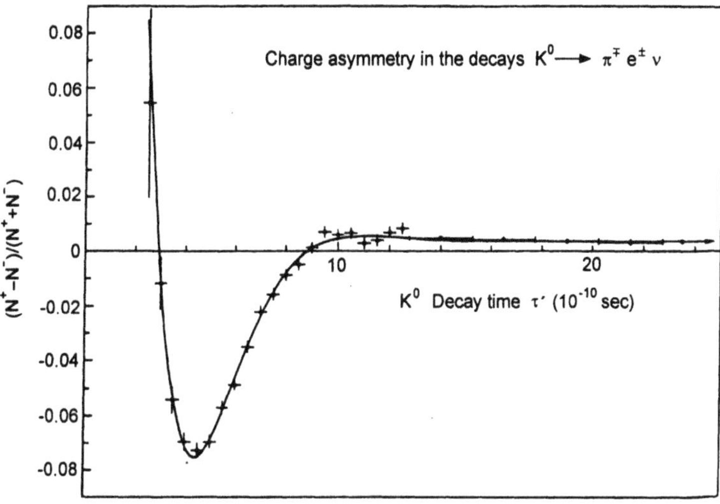


**Fig. 11.3.** The charge asymmetry $\delta(t)$ in the decays $K^0 \to \pi^{\mp}e^{\pm} + \nu$, as measured by Gjesdal, S. et al., Phys. Lett. **52B** (1974) 113. Reprinted by permission of Elsevier Science

## 11.3 Regeneration of $K_S^0$

Let us consider again a $K^0$ beam produced at $t = 0$, and suppose that, long after the vanishing of all the $K_S$ from the beam so that only the $K_L$ component remains, we place on the path of the beam a block of matter which may be regarded for all practical purposes as composed of protons and neutrons. Pais and Piccioni suggested in 1955 that, as the $K^0$ and $\overline{K}^0$ components present in $K_L$ interact very differently with matter, the $K_S$ would eventually reappear in the beam. That this is indeed the case can be seen by examining the hadronic interactions of $K^0$ and $\overline{K}^0$ with the neutron and the proton. While there are no differences in their elastic scatterings, only $\overline{K}^0$ can be absorbed by matter through $\overline{K}^0 + p \to \Lambda + \pi^+$ or $\overline{K}^0 + n \to \Lambda + \pi^0$. Similar processes for $K^0$ are forbidden by the strangeness conservation: thus $K^0 + p \not\to \Lambda + \pi^+$ and $K^0 + n \not\to \Lambda + \pi^0$. This difference in the absorption properties of K mesons by matter is essential to understanding the regeneration of $K_S$ as the beam is traveling through matter. If we call $f$ and $\bar{f}$ the amplitudes of scattering of $K^0$ and $\overline{K}^0$ on the atomic nuclei in matter, then, as we have just seen, $f \neq \bar{f}$. Now, the wave function that enters the block is that of $K_L$:

$$\psi_i = \psi_{K_L} = \frac{K^0 + \overline{K}^0}{\sqrt{2}} . \qquad (11.15)$$

Interactions of the mesons with matter will cause the wave function to change, so that at exit it is given by

$$\psi_f = \frac{1}{\sqrt{2}}(f\,\mathrm{K}^0 + \bar{f}\,\overline{\mathrm{K}}^0) = \frac{1}{2}[(f+\bar{f})\,\mathrm{K_L} + (f-\bar{f})\mathrm{K_S}]\,.$$

In other terms, the $\mathrm{K_L}$ amplitude will become $\mathrm{K_L} + r\mathrm{K_S}$, where

$$r = \frac{f-\bar{f}}{f+\bar{f}} \tag{11.16}$$

parameterizes the regeneration process, a factor which depends on the properties of the sample of matter that the beam has gone through. Since $r \neq 0$, there must be regeneration, and the phenomenon has been observed.

The $\mathrm{K_S}$ regeneration can be exploited to measure the mass difference $\Delta m$, a quantity of key importance in the studies of neutral K mesons. For this purpose, let us place on the path of the beam two blocks of matter separated by a distance $d$ that may be varied at will. When $\mathrm{K_L}$ and the regenerated $\mathrm{K_S}$ go through the second block, their oscillations interfere in a manner different than in the vacuum and that depends on $d$. By letting $d$ vary, one may determine the sign of $\Delta m$ from the observed interference effects between the $\mathrm{K_L}$ and $\mathrm{K_S}$. It turns out that $\Delta m > 0$, that is, $m_\mathrm{L} > m_\mathrm{S}$. The $\mathrm{K_S}$ regeneration phenomenon, which arises from an interplay between strangeness hadronic eigenstates and weakly interacting CP eigenstates, displays many similarities with the oscillations of neutrinos in matter, which we will discuss in the next chapter.

The two pairs $(\mathrm{K}^0, \overline{\mathrm{K}}^0)$ and $(\mathrm{K_L}, \mathrm{K_S})$, which represent respectively the eigenstates of strangeness S and discrete symmetry CP, are *dual* from the quantum physics viewpoint. The neutral K system which decays into $\pi^-\mathrm{e}^+\nu_\mathrm{e}$ and $\pi^+\mathrm{e}^-\bar{\nu}_\mathrm{e}$ are respectively $\mathrm{K}^0$ and $\overline{\mathrm{K}}^0$. These modes involve the strangeness $\mathcal{S}$ operator. The same neutral K system which decays into two-pion and three-pion modes are respectively $\mathrm{K_S}$ and $\mathrm{K_L}$. These decays concern the $\mathcal{CP}$ operator. The semileptonic and hadronic decay modes are used to select the eigenstates of $\mathcal{S}$ and $\mathcal{CP}$ respectively.

In some sense, the pairs $(\mathrm{K}^0, \overline{\mathrm{K}}^0)$ and $(\mathrm{K_L}, \mathrm{K_S})$ are similar to the spin states $\sigma_x$ and $\sigma_y$ of the electron. The operators $\mathcal{S}$ and $\mathcal{CP}$ play the role of two orthogonal magnetic fields $H_x$ and $H_y$ which project out the spins $\sigma_x$ and $\sigma_y$. According to the Heisenberg uncertainty principle and demonstrated by the Stern–Gerlach experiment, it is impossible to quantize simultaneously the components $\sigma_x$ and $\sigma_y$ of the electron spin since they do not commute. Similarly, the two pairs $(\mathrm{K}^0, \overline{\mathrm{K}}^0)$ and $(\mathrm{K_L}, \mathrm{K_S})$ cannot be simultaneously determined, the operators $\mathcal{S}$ and $\mathcal{CP}$ also do not commute in weak interaction. We have a simple illustration of the two familiar concepts in quantum physics, state superposition and quantization.

## 11.4 Calculation of $\Delta m$

All three phenomena that we have just described depend on the mass difference between $K_L$ and $K_S$. It would therefore be interesting to see how $\Delta m$ is viewed in the standard model. Let us write $m_L$ and $m_S$ as the real part of the expectation values of a certain Hamiltonian operator $H = H^{(0)} + H^{(2)}$, the imaginary part is related to their decay widths (the subscripts 0 and 2 correspond to the $\Delta S = 0$ and $|\Delta S| = 2$ transitions):

$$
\begin{aligned}
m_L &\equiv \mathrm{Re}\left[\langle K_L \,|\, H \,|\, K_L\rangle\right] = \mathrm{Re}\left[\frac{1}{\sqrt{2}}\left\langle K^0 + \overline{K}^0 \,|\, H \,|\, K^0 + \overline{K}^0\right\rangle\right] , \\
m_S &\equiv \mathrm{Re}\left[\langle K_S \,|\, H \,|\, K_S\rangle\right] = \mathrm{Re}\left[\frac{1}{\sqrt{2}}\left\langle K^0 - \overline{K}^0 \,|\, H \,|\, K^0 - \overline{K}^0\right\rangle\right] ,
\end{aligned} \tag{11.17}
$$

from which is deduced

$$
\Delta m \equiv m_L - m_S = \mathrm{Re}\left[\left\langle K^0 \middle| H^{(2)} \middle| \overline{K}^0\right\rangle + \left\langle \overline{K}^0 \middle| H^{(2)} \middle| K^0\right\rangle\right] . \tag{11.18}
$$

This relation tells us that $\Delta m$ stems from the effective $|\Delta S| = 2$ interaction $H^{(2)}$ that causes the transitions between $K^0$ and $\overline{K}^0$, represented by the diagrams in Fig. 11.1. To illustrate the calculation of the matrix elements involved, at first we keep only the contributions from the u quark in the intermediate states and assume that the four-momenta of the external quarks may be neglected. Applying the Feynman rules for the standard model (Chap. 9), we obtain for the process represented in Fig. 11.1a the transition amplitude

$$
\begin{aligned}
\mathcal{M}_a = \mathrm{i}\left[\frac{-\mathrm{i}g}{2\sqrt{2}}V_{\mathrm{ud}}^*\right]^2\left[\frac{-\mathrm{i}g}{2\sqrt{2}}V_{\mathrm{us}}\right]^2 &\int \frac{\mathrm{d}^4k}{(2\pi)^4}\bar{u}(\mathrm{s})\gamma_\lambda(1-\gamma_5)\frac{\mathrm{i}(\not{k}+m_\mathrm{u})}{k^2-m_\mathrm{u}^2}\gamma_\rho(1-\gamma_5)v(\mathrm{d}) \\
\times\, \bar{v}(\mathrm{s})\gamma_\alpha(1-\gamma_5)\frac{\mathrm{i}(\not{k}+m_\mathrm{u})}{k^2-m_\mathrm{u}^2}\gamma_\sigma(1-\gamma_5)u(\mathrm{d})\,&\frac{-\mathrm{i}g^{\lambda\sigma}}{k^2-M_\mathrm{W}^2}\,\frac{-\mathrm{i}g^{\alpha\rho}}{k^2-M_\mathrm{W}^2} .
\end{aligned} \tag{11.19}
$$

The Feynman–'t Hooft ($\xi = 1$) gauge is used for the W-boson propagator. Of course, the end result [(see (36) below] should be independent of the particular gauge used in the calculation, and (19) can be written as

$$
\mathcal{M}_a = \frac{\mathrm{i}g^4(V_{\mathrm{ud}}^* V_{\mathrm{us}})^2}{16} I_{\mu\nu} T^{\mu\nu} , \quad \text{with} \tag{11.20}
$$

$$
\begin{aligned}
I_{\mu\nu} &= \int \frac{\mathrm{d}^4k}{(2\pi)^4}\frac{k_\mu k_\nu}{(k^2-M_\mathrm{W}^2)^2(k^2-m_\mathrm{u}^2)^2} , \quad \text{and} \\
T^{\mu\nu} &= \left[\bar{u}(\mathrm{s})\gamma_\lambda\gamma^\mu\gamma_\rho(1-\gamma_5)v(\mathrm{d})\right]\left[\bar{v}(\mathrm{s})\gamma^\rho\gamma^\nu\gamma^\lambda(1-\gamma_5)u(\mathrm{d})\right] .
\end{aligned} \tag{11.21}
$$

Neglecting $m_\mathrm{u}^2/M_\mathrm{W}^2$ and using the integral formulas in the Appendix, the integral $I_{\mu\nu}$ can be easily performed, leading to the result

$$
I_{\mu\nu} = -\mathrm{i}\,\frac{g_{\mu\nu}}{64\pi^2 M_\mathrm{W}^2} . \tag{11.22}
$$

With the help of the identity

$$\gamma_\lambda\gamma_\mu\gamma_\rho = g_{\lambda\mu}\gamma_\rho + g_{\mu\rho}\gamma_\lambda - g_{\lambda\rho}\gamma_\mu - \mathrm{i}\epsilon_{\lambda\mu\rho\alpha}\gamma_5\gamma^\alpha \,, \tag{11.23}$$

one gets

$$\begin{aligned} g_{\mu\nu}T^{\mu\nu} &= 4\ [\bar{u}(\mathrm{s})\gamma_\lambda(1-\gamma_5)v(\mathrm{d})]\ [\bar{v}(\mathrm{s})\gamma^\lambda(1-\gamma_5)u(\mathrm{d})] \\ &\Longrightarrow 4\ \Theta^{|\Delta S|=2} \,, \end{aligned} \tag{11.24}$$

where $\Theta^{|\Delta S|=2} \equiv [\bar{s}\gamma_\lambda(1-\gamma_5)d]\ [\bar{s}\gamma^\lambda(1-\gamma_5)d]$.

The spinors $u$ and $v$ have been replaced by the corresponding quark fields, so that an effective operator on the Hilbert space, $H_a^{|\Delta S|=2}$, may be obtained from the transition amplitude. Similarly, one may calculate the effective operator corresponding to the process represented in Fig. 11.1b. It turns out to be exactly equal to $H_a^{|\Delta S|=2}$. Putting together (20), (22), and (24), the full interaction operator with only u quark in the intermediate states is simply

$$H^{(2)} = 2H_a^{|\Delta S|=2} = \frac{G_\mathrm{F}^2}{4\pi^2}(V_\mathrm{ud}^* V_\mathrm{us})^2 M_\mathrm{W}^2\ \Theta^{|\Delta S|=2} \,, \tag{11.25}$$

where $g^2/8M_\mathrm{W}^2 = G_\mathrm{F}/\sqrt{2}$ is used.

The calculation of $\Delta m$ then boils down to the calculation of the matrix element of $\Theta^{|\Delta S|=2}$ between states $\mathrm{K}^0$ and $\overline{\mathrm{K}}^0$. We will calculate it in the *vacuum insertion approximation*, which implies keeping the vacuum as the only intermediate state. First, using Fierz's rearrangement (Appendix), the operator $\Theta^{|\Delta S|=2}$ can also be written as

$$\bar{s}_a\gamma_\mu(1-\gamma_5)d_a\ \bar{s}_b\gamma^\mu(1-\gamma_5)d_b = \bar{s}_a\gamma_\mu(1-\gamma_5)d_b\ \bar{s}_b\gamma^\mu(1-\gamma_5)d_a \,, \tag{11.26}$$

where $a, b = 1, 2, 3$ label the quark colors. The minus sign in the Fierz's transformation combined with the anticommutation rules for the fermionic field operators yields a positive sign in (26). Now we insert $|0\rangle\langle 0|$ between the two bilinear spinor products in all possible ways and define

$$\langle 0\,|\,A^\mu_{ab}\,|\,\mathrm{K}^0\rangle \equiv \langle 0\,|\,\bar{s}_b\gamma^\mu\gamma_5 d_a\,|\,\mathrm{K}^0\rangle = \langle 0\,|\,\bar{s}_b\gamma^\mu\gamma_5 u_a\,|\,\mathrm{K}^+\rangle = \frac{\mathrm{i}f_\mathrm{K}\,q^\mu}{\sqrt{2m_\mathrm{K}}}\frac{\delta_{ab}}{3} \,, \tag{11.27}$$

where the decay constant of the K meson, $f_\mathrm{K} \approx 160$ MeV, is extracted from the $\mathrm{K}^+ \to \mu^+ + \nu_\mu$ rate, like $f_\pi \approx 131$ MeV from $\pi^+ \to \mu^+ + \nu_\mu$ (Chap. 10). The denominator $\sqrt{2m_\mathrm{K}}$ comes from the K meson one-particle state normalization. We obtain in the vacuum insertion approximation

$$\left\langle \mathrm{K}^0 \middle| \Theta^{|\Delta S|=2} \middle| \overline{\mathrm{K}}^0 \right\rangle = \left(\frac{2}{3}\right)\frac{f_\mathrm{K}^2 m_\mathrm{K}^2}{2m_\mathrm{K}} \,. \tag{11.28}$$

In (28), the factor $\frac{2}{3} = \frac{1}{2}\left(1+\frac{1}{3}\right)$ comes from the rearrangement of the color indices by using (26) and (27).

If the vacuum insertion approximation is relaxed, the result must be modified, but the necessary changes may be simply parameterized by a multiplicative $B$ factor, so that (28) may be replaced by

$$\left\langle \mathrm{K}^0 \left| \Theta^{|\Delta S|=2} \right| \overline{\mathrm{K}}^0 \right\rangle = \left(\frac{2}{3}\right) \frac{f_{\mathrm{K}}^2 m_{\mathrm{K}}^2}{2m_{\mathrm{K}}} B\,, \tag{11.29}$$

$B = 1$ corresponds to the vacuum insertion (28). In a wide variety of models and in lattice gauge calculations, estimates for $B$ differ from 1, within a factor of two. From (18), (25), and (29) we obtain the expression for $\Delta m$:

$$\Delta m = \mathrm{Re}\left[\frac{G_F^2}{6\pi^2}(V_{\mathrm{ud}}^* V_{\mathrm{us}})^2 f_{\mathrm{K}}^2 m_{\mathrm{K}} M_{\mathrm{W}}^2\right] B\,. \tag{11.30}$$

The resulting value for $\Delta m$ is larger than the measured value by a factor of $3 \times 10^3$. Clearly, something is fundamentally wrong, not so much with letting $B \approx 1$ as with neglecting the presence of other quarks, and especially of the c quark in the intermediate states. This oversight can be easily amended.

From the unitarity of the CKM matrix (Chap. 9), the matrix elements that are relevant to our calculation satisfy the relation

$$V_{\mathrm{ud}}^* V_{\mathrm{us}} + V_{\mathrm{cd}}^* V_{\mathrm{cs}} + V_{\mathrm{td}}^* V_{\mathrm{ts}} = 0\,. \tag{11.31}$$

Neglecting the last term, which is actually very small, being of the order of $10^{-4}$, this relation yields

$$V_{\mathrm{ud}}^* V_{\mathrm{us}} = -V_{\mathrm{cd}}^* V_{\mathrm{cs}}\,. \tag{11.32}$$

This equation indicates there is a destructive interference between the u and c quarks in the transitions under consideration, an effect that will drastically reduce $\Delta m$ to the level actually observed. It is precisely this interference effect between the u and c quarks in their actions in weak processes that underlies the GIM mechanism (Chap. 9). With (32), keeping both u and c quarks in the intermediate states has the effect of replacing in (19) the factor

$$\left(\frac{1}{k^2 - m_{\mathrm{u}}^2}\right)^2 \text{ by } \left(\frac{1}{k^2 - m_{\mathrm{u}}^2} - \frac{1}{k^2 - m_{\mathrm{c}}^2}\right)^2.$$

Therefore, the old expression (22) for the integral $I_{\mu\nu}$ is now replaced with

$$I_{\mu\nu}^{(\mathrm{u,c})} \equiv \int \frac{d^4k}{(2\pi)^4} \frac{k_\mu k_\nu (m_{\mathrm{c}}^2 - m_{\mathrm{u}}^2)^2}{(k^2 - m_{\mathrm{u}}^2)^2 (k^2 - m_{\mathrm{c}}^2)^2 (k^2 - M_{\mathrm{W}}^2)^2}\,. \tag{11.33}$$

With $m_{\mathrm{u}} \ll m_{\mathrm{c}} \ll M_{\mathrm{W}}$,
$$I_{\mu\nu}^{(\mathrm{u,c})} = \frac{-\mathrm{i}\, g_{\mu\nu}}{64\pi^2} \frac{m_{\mathrm{c}}^2 - m_{\mathrm{u}}^2}{M_{\mathrm{W}}^4}\,, \tag{11.34}$$

from which we finally get

$$\Delta m = \mathrm{Re}\left[\frac{G_F^2}{6\pi^2}(V_{cd}^* V_{cs})^2 f_K^2 m_K \left(m_c^2 - m_u^2\right)\right] B\,. \tag{11.35}$$

It is the GIM mechanism that allows replacing $M_W^2$ in (30) with $m_c^2 - m_u^2$ in (35), making possible a correct prediction for $\Delta m$, with $m_c \approx 1.5$ GeV. Historically, Gaillard and Lee exploited this mechanism to predict the mass of the c quark before the discovery of the charmomium $J/\psi$ in 1974. Let us also note that if u and c quarks are mass-degenerate, $m_u = m_c$, the suppression effect of the GIM mechanism would be complete, and $\Delta m$ would vanish and there would be no transitions between $K^0$ and $\overline{K}^0$ at all. This suppression can be easily understood in the framework of the standard model of the electroweak interaction. We saw in Chap. 9 that if the $Q = 2/3$ quarks (u, c, and t), or alternatively the $Q = -1/3$ quarks (d, s, and b), were *mass-degenerate*, there would be no need for a CKM mixing matrix, and there would be no flavor mixing between quarks having the same charge. More precisely, in that case each up-type quark would couple only to its own down-type weak-isospin partner by the charged currents, so that cross-family couplings would be completely absent, i.e. $V_{us} = V_{cd} = 0$.

As already mentioned, the GIM mechanism, which was invented primarily in order to suppress the flavor-changing neutral currents at the tree level, is still potent in higher-order loop diagrams. Processes depicted by the box diagrams in Fig. 11.1 are examples of its power.

Finally, let us examine the contributions of the top quark to $\Delta m$. A rough estimate using (35), with parameters for t replacing those for c, shows that a top quark mass of about 180 GeV, large as it is, cannot compensate for the smallness of the CKM matrix elements $V_{td}^* V_{ts}$ to yield a significant contribution to $\Delta m$. The exact unitarity of $V_{CKM}$ and a more careful calculation in the $R_\xi$ gauge, keeping $x_c \equiv m_c^2/M_W^2$, $x_t \equiv m_t^2/M_W^2$, lead to

$$\Delta m = \mathrm{Re}\left\{\frac{G_F^2}{6\pi^2} f_K^2 m_K m_c^2 \Big[(V_{cd}^* V_{cs})^2 g(x_c) + (V_{td}^* V_{ts})^2 \frac{x_t}{x_c} g(x_t) + 2\left(V_{td}^* V_{ts} V_{cd}^* V_{cs}\right) h(x_c, x_t)\Big]\right\} B\,, \tag{11.36}$$

where the functions $g(x)$ and $h(x,y)$ are given by[1]

$$\begin{aligned} g(x) &= \frac{1}{4} + \frac{9}{4(1-x)} - \frac{3}{2(1-x)^2} - \frac{3}{2}\frac{x^2}{(1-x)^3}\log x \underset{x\to 0}{\longrightarrow} 1\,, \\ h(x,y) &= y\left\{\frac{\log x}{x-y}\left[\frac{1}{4} + \frac{3}{2(1-x)} - \frac{3}{4(1-x)^2}\right] + (x \leftrightarrow y) \right. \\ &\quad \left. - \frac{3}{4(1-x)(1-y)}\right\} \underset{x,y\to 0}{\longrightarrow} 0\,. \end{aligned} \tag{11.37}$$

[1] Inami, T. and Lim, C. S., Prog. Theor. Phys. **65** (1981) 297; *ibid.* **65** (1981) 1772 (E)

We note that in the $R_\xi$ gauge, besides the contributions of the $W^\pm$ weak boson (Fig. 11.1), in the box diagrams there are also contributions from the would-be Goldstone unphysical scalar bosons $w^\pm$ (those absorbed by $W^\pm$ to become massive). Since the coupling of these internal $w^\pm$ to the fermions $Q_i$ = u, c, t is proportional to the $Q_i$ mass, the role of the top is dominant in these $w^\pm$ contributions. Nevertheless, the top overall part amounts to only 8% of that of the c quark. Terms proportional to $V^*_{\rm ud}V_{\rm us}$, which one expects to appear in (36), has been eliminated by the unitarity relation (31).

## 11.5 CP Violation

Discrete symmetries C, P, and T in particle physics have been discussed in Chap. 5. As explained there, the CPT theorem implies that T invariance in strong and electromagnetic interactions is equivalent to CP conservation in these interactions. The simplest test of the CPT theorem is the equality of the masses of a particle and its antiparticle, and the best test comes from the mass difference between $K^0$ and $\overline{K}^0$ which may be related to $\Delta m$ :

$$\frac{m_{K^0} - m_{\overline{K}^0}}{m_{K^0}} \approx \frac{\Delta m}{m_{K^0}} \approx 9 \times 10^{-19} .$$

Any such difference contributes to the CP-violating parameter $\epsilon$ which we will introduce later. After the discovery in 1956 of the maximum P and C violation in weak interactions, physicists still believed that T (or CP) were nevertheless invariant in weak interactions, as in the strong and electromagnetic interactions. Imagine the great surprise when CP violations in weak decays of the neutral K system was discovered in 1964.

The standard electroweak Lagrangian with *three quark families* provides a natural framework for CP violation through the Kobayashi–Maskawa (KM) *nonzero phase* in the CKM quark mixing matrix. As remarkably shown by KM, the two left-handed doublets $(\mathrm{u}, \mathrm{d}'')$ and $(\mathrm{c}, \mathrm{s}'')$ that GIM used to cancel strangeness-changing neutral current are not enough to incorporate a complex phase in the quark flavor mixing matrix in order to have CP violation. With $N$ quark families, the number of measurable nonzero phases is $\frac{1}{2}(N-1) \times (N-2)$ according to the discussion after (9.177). Thus $N$ must be larger than 2 to yield a nonzero phase, i.e. at least *a third doublet* $(\mathrm{t}, \mathrm{b}'')$ is needed. It is noteworthy that this observation was made by KM in 1973 even before charm was discovered. We will call this KM theory the *standard* CP-violating mechanism.

### 11.5.1 General Formalism

As we have seen in the previous sections, $K^0$ and $\overline{K}^0$ are mixed by the weak interaction. Thus, in the presence of this type of interaction, the two states are inseparable; they form a basis for a two-dimensional subspace. Similarly, $K_S$ and $K_L$ form another, equivalent basis of the same space.

We may then write, for example,

$$|\mathrm{K}^0\rangle = \begin{pmatrix} 1 \\ 0 \end{pmatrix}, \qquad |\overline{\mathrm{K}}^0\rangle = \begin{pmatrix} 0 \\ 1 \end{pmatrix}. \tag{11.38}$$

Let $\psi(t)$ be an arbitrary state of such space,

$$|\psi(t)\rangle = A(t)\,|\mathrm{K}^0\rangle + B(t)\,|\overline{\mathrm{K}}^0\rangle = \begin{pmatrix} A(t) \\ B(t) \end{pmatrix}. \tag{11.39}$$

The Hamiltonian in (17) may be split into two parts; one, $H^{(0)}$, conserves strangeness (strong and electromagnetic interactions), and the other, $H^{(2)}$, violates strangeness by $|\Delta S| = 2$. Their sum $H = H^{(0)} + H^{(2)}$ should be non-Hermitian and therefore may be decomposed into a Hermitian and an anti-Hermitian part, responsible respectively for the mass and the width of unstable particles

$$\langle j|H|k\rangle = M_{jk} - \frac{\mathrm{i}}{2}\Gamma_{jk}. \tag{11.40}$$

In the two-dimensional space under consideration, $H$ has the matrix representation

$$H \equiv \begin{pmatrix} M_{11} - \frac{\mathrm{i}}{2}\Gamma_{11} & M_{12} - \frac{\mathrm{i}}{2}\Gamma_{12} \\ M_{21} - \frac{\mathrm{i}}{2}\Gamma_{21} & M_{22} - \frac{\mathrm{i}}{2}\Gamma_{22} \end{pmatrix}$$
$$= \begin{pmatrix} \langle \mathrm{K}^0|H^{(0)}|\mathrm{K}^0\rangle & \langle \mathrm{K}^0|H^{(2)}|\overline{\mathrm{K}}^0\rangle \\ \langle \overline{\mathrm{K}}^0|H^{(2)}|\mathrm{K}^0\rangle & \langle \overline{\mathrm{K}}^0|H^{(0)}|\overline{\mathrm{K}}^0\rangle \end{pmatrix}. \tag{11.41}$$

CPT invariance of $H$ implies that $M_{11} = M_{22} \equiv M_0$, with $M_0 = m_{\mathrm{K}^0} = m_{\overline{\mathrm{K}}^0}$, and $\Gamma_{11} = \Gamma_{22} \equiv \Gamma_0$, with $\Gamma_0$ identified with the (common) total decay width of $\mathrm{K}^0$ or $\overline{\mathrm{K}}^0$. Since by construction, both $M$ and $\Gamma$ are Hermitian, $M_0$ and $\Gamma_0$ are both real numbers, and $M_{21} = M_{12}^*$ and $\Gamma_{21} = \Gamma_{12}^*$. Hence

$$H \equiv \begin{pmatrix} M_0 - \frac{\mathrm{i}}{2}\Gamma_0 & M_{12} - \frac{\mathrm{i}}{2}\Gamma_{12} \\ M_{12}^* - \frac{\mathrm{i}}{2}\Gamma_{12}^* & M_0 - \frac{\mathrm{i}}{2}\Gamma_0 \end{pmatrix}. \tag{11.42}$$

If $H^{(2)}$ is invariant under time reversal T, or equivalently under CP, one has $M_{21} = M_{12}$ and $\Gamma_{21} = \Gamma_{12}$, from which $M_{12}^* = M_{12}$ and $\Gamma_{12}^* = \Gamma_{12}$, i.e. these matrix elements are real. On the other hand, if T (or equivalently CP) is not a symmetry of $H^{(2)}$, one may either have $M_{21} \neq M_{12}$ or $\Gamma_{21} \neq \Gamma_{12}$, or both possibilities. This implies either $M_{12}$ or $\Gamma_{12}$, or both, may be complex.

We proceed now to determine the physical eigenstates $\mathrm{K_L}$ and $\mathrm{K_S}$ having respective masses $m_\mathrm{L}$ and $m_\mathrm{S}$ and full widths $\Gamma_\mathrm{L}$ and $\Gamma_\mathrm{S}$. Diagonalization of (42) immediately gives

$$\mathrm{K_L} = \frac{1}{\sqrt{1+|\bar\epsilon|^2}}\left(\frac{\mathrm{K}^0 + \overline{\mathrm{K}}^0}{\sqrt{2}} + \bar\epsilon\,\frac{\mathrm{K}^0 - \overline{\mathrm{K}}^0}{\sqrt{2}}\right) \equiv \frac{\mathrm{K}_2^0 + \bar\epsilon\,\mathrm{K}_1^0}{\sqrt{1+|\bar\epsilon|^2}},$$
$$\mathrm{K_S} = \frac{1}{\sqrt{1+|\bar\epsilon|^2}}\left(\frac{\mathrm{K}^0 - \overline{\mathrm{K}}^0}{\sqrt{2}} + \bar\epsilon\,\frac{\mathrm{K}^0 + \overline{\mathrm{K}}^0}{\sqrt{2}}\right) \equiv \frac{\mathrm{K}_1^0 + \bar\epsilon\,\mathrm{K}_2^0}{\sqrt{1+|\bar\epsilon|^2}}. \tag{11.43}$$

We have re-expressed $K_L$ and $K_S$ in terms of $K_1^0$ and $K_2^0$, the even- and odd-CP states already defined before. The parameter $\bar{\epsilon}$ is given by

$$\bar{\epsilon} = \frac{\sqrt{M_{12} - \frac{i}{2}\Gamma_{12}} - \sqrt{M_{12}^* - \frac{i}{2}\Gamma_{12}^*}}{\sqrt{M_{12} - \frac{i}{2}\Gamma_{12}} + \sqrt{M_{12}^* - \frac{i}{2}\Gamma_{12}^*}} \equiv \frac{p-q}{p+q} . \tag{11.44}$$

If CP is conserved, the quantities $M_{12}$ and $\Gamma_{12}$ are real, as mentioned above, and therefore $\bar{\epsilon}$ vanishes. Then, $K_S = K_1^0$ and $K_L = K_2^0$, and there is no mixing between $K_1^0$ and $K_2^0$. Such a mixing can occur only if CP is a broken symmetry. In general, the eigenvalues of $M - i\Gamma/2$ corresponding to (43) are

$$m_L - \frac{i}{2}\Gamma_L = M_0 - \frac{i}{2}\Gamma_0 + \sqrt{\left(M_{12} - \frac{i}{2}\Gamma_{12}\right)\left(M_{12}^* - \frac{i}{2}\Gamma_{12}^*\right)} ,$$

$$m_S - \frac{i}{2}\Gamma_S = M_0 - \frac{i}{2}\Gamma_0 - \sqrt{\left(M_{12} - \frac{i}{2}\Gamma_{12}\right)\left(M_{12}^* - \frac{i}{2}\Gamma_{12}^*\right)} , \tag{11.45}$$

from which

$$(m_L - m_S) + \frac{i}{2}(\Gamma_S - \Gamma_L) = 2\sqrt{\left(M_{12} - \frac{i}{2}\Gamma_{12}\right)\left(M_{12}^* - \frac{i}{2}\Gamma_{12}^*\right)} \equiv 2pq . \tag{11.46}$$

If CP is conserved, this result becomes

$$\Delta m \equiv (m_L - m_S) = 2M_{12} \ , \quad \Delta\gamma \equiv (\Gamma_S - \Gamma_L) = -2\Gamma_{12} . \tag{11.47}$$

Let us now turn to the CP violation in neutral K mesons. In 1964, Christensen, Cronin, Fitch, and Turlay observed that $K_L$ decays not only via the three-pion mode, $K_L \to 3\pi$, which was natural given its CP parity, but also via the two-pion mode, $K_L \to 2\pi$, which was truly unexpected. To make sure that it was the $K_L$ and not the regenerated $K_S$ that provoked the observed events, they surrounded the $K_L$ beam with a great quantity of helium to eliminate any possible regeneration of $K_S$.

Since the same particle, $K_L$, can decay through two channels of opposite CP parities, it is clear that CP symmetry is violated in the pionic decay modes of $K_L$. This experiment is comparable in nature and significance to another result obtained eight years earlier, when it was realized that another K meson, the charged $K^+$, could also decay through two different channels, $K^+ \to \pi^+\pi^0$ and $K^+ \to \pi^+\pi^+\pi^-$ (or $\pi^+\pi^0\pi^0$). Now, a two-pion state in the s-wave is parity-even, while a three-pion state with zero angular momentum is parity-odd. Therefore, parity (space inversion) symmetry must be broken in these decay modes of $K^+$. Of course, the discovery of parity violation in weak decays was one of the most important events in modern physics (Chap. 5).

But while parity suffers maximal breakdown in weak processes (for example, in $K^+$ to $2\pi$ and $3\pi$, the branching ratios are comparable in magnitudes after phase-space corrections), CP symmetry is only slightly broken (the $K_L$ to $3\pi$ mode clearly dominates the $2\pi$ mode). In any case, violation of P, C, and CP symmetries by the weak interaction is a well-established fact.

CP violation manifests itself through the presence of a nonvanishing complex quantity $\bar{\epsilon}$ which may be estimated from the branching ratios

$$\rho \equiv \frac{\Gamma(K_L \to \pi^+\pi^-)}{\Gamma(K_S \to \pi^+\pi^-)} = \frac{Br(K_L \to \pi^+\pi^-)}{Br(K_S \to \pi^+\pi^-)} \left(\frac{\tau_S}{\tau_L}\right) \approx 5.1 \times 10^{-6}\,, \tag{11.48}$$

which gives $|\bar{\epsilon}| \approx \sqrt{\rho} = 2.25 \times 10^{-3}$. To obtain the complex phase of $\bar{\epsilon}$, we may proceed through (44) which takes the form

$$\bar{\epsilon} = \frac{p-q}{p+q} = \frac{p^2 - p^2}{4pq + (p-q)^2} \approx \frac{p^2 - q^2}{4pq}\,, \tag{11.49}$$

where the magnitude $|\bar{\epsilon}| \approx 10^{-3}$ justifies dropping the quadratic term $(p-q)^2$, which yields the approximation

$$\bar{\epsilon} \approx \mathrm{i}\, \frac{\mathrm{Im}(M_{12}) - \frac{\mathrm{i}}{2}\mathrm{Im}(\Gamma_{12})}{\Delta m + \frac{\mathrm{i}}{2}\Delta\gamma}\,. \tag{11.50}$$

Again, with $|\bar{\epsilon}| \approx 10^{-3}$, one may deduce

$$\frac{\mathrm{Im}(M_{12})}{\mathrm{Re}(M_{12})} \ll 1\,, \quad \frac{\mathrm{Im}(\Gamma_{12})}{\mathrm{Re}(\Gamma_{12})} \ll 1\,. \tag{11.51}$$

Hence, to a very good approximation, (46) yields

$$\Delta m = 2\,\mathrm{Re}(M_{12}), \quad \Delta\gamma = -2\,\mathrm{Re}(\Gamma_{12})\,. \tag{11.52}$$

On the other hand, as the ratio $\Delta m/\Delta\gamma = 0.477$ is experimentally known, we may write

$$\frac{\mathrm{i}}{\Delta m + \frac{\mathrm{i}}{2}\Delta\gamma} = \frac{\mathrm{e}^{\mathrm{i}(43.37)^\circ}}{\sqrt{2.098}\,\Delta m} \approx \frac{\mathrm{e}^{\mathrm{i}\pi/4}}{\sqrt{2}\,\Delta m}\,, \tag{11.53}$$

and the approximate expression for $\bar{\epsilon}$, (50) becomes

$$\bar{\epsilon} \approx \frac{\mathrm{e}^{\mathrm{i}\pi/4}}{\sqrt{2}\,\Delta m}\left[\mathrm{Im}(M_{12}) - \frac{\mathrm{i}}{2}\mathrm{Im}(\Gamma_{12})\right]. \tag{11.54}$$

In order to determine the phase of $\bar{\epsilon}$, or equivalently its real part, we consider again the charge asymmetry $\delta(t)$ as defined in (14). This quantity was previously obtained assuming exact CP conservation, i.e. with $K_L = K_2^0$ and

$K_S = K_1^0$. Now, in the presence of CP violation, one should replace $K_L$ and $K_S$ by their exact expressions (43). This replacement transforms (14) into

$$\delta(t) = 2\left[\mathrm{Re}(\bar{\epsilon}) + \mathrm{e}^{-(\Gamma_S+\Gamma_L)\frac{t}{2}}\cos(\Delta m\, t)\right], \tag{11.55}$$

which includes the CP violation effect. This result tells us that at $t \approx \tau_S$ the time oscillation yields a measure of $\Delta m$, while at $t \gg \tau_S$ the amplitude of $\delta(t)$ directly yields $\mathrm{Re}(\bar{\epsilon})$. Figure 11.3 clearly shows the nonvanishing value of $\mathrm{Re}(\bar{\epsilon})$. We may understand the above result as follows: at large $t$ only $K_L$ survives; because of CP violation as shown in (43), $K^0$ and $\overline{K}^0$ exist in unequal proportions in $K_L$, and their difference is given by $2\bar{\epsilon}$.

### 11.5.2 Model-Independent Analysis of $K_L \to 2\pi$

Let us now examine in some detail the CP-violating modes $K_L \to \pi^+ + \pi^-$ and $K_L \to \pi^0 + \pi^0$. The eigenstates (43) indicate that there are two possible not mutually exclusive scenarios in which these decays may proceed.

First, it is the $K_2^0$ component of $K_L$ itself that decays into two pions. As $K_2^0$ and $\pi\pi$ states have opposite CP parities, this decay mode can occur only if the transition violates CP symmetry. As this is a direct violation by the amplitude itself, the effective coupling strength for this interaction is of the order of $10^{-3}$ of the strength of the CP-conserving interaction. It is a $|\Delta S| = 1$ transition and is referred to as a *milliweak* transition. In $K_L \to \pi + \pi$, its effects are parameterized by a quantity called $\epsilon'$ defined later.

In the second scenario, the decay $K_L \to \pi + \pi$ is viewed as being due to the $\bar{\epsilon} K_1^0$ component, with $\bar{\epsilon} \neq 0$. In contrast to the first scenario, in this case the decay amplitude, which is that of $K_1^0 \to \pi\pi$ multiplied by $\bar{\epsilon}$, exactly conserves CP. The CP violation actually observed in the $K_L$ decay stems from the $K^0$–$\overline{K}^0$ mixing through the mass matrix in (41). It is this mass matrix that violates CP. As the violation occurs through $|\Delta S| = 2$ transitions, which are $\sim G_F^2$, Wolfenstein dubbed it *superweak*.

Of course, it is quite possible that the CP violation in $K_L \to \pi + \pi$ proceeds via both scenarios. Actually, it is what happens in the *standard* CP-violating mechanism defined earlier. The two scenarios are represented by the diagrams in Fig. 11.4.

*milliweak* : $\Delta S = 1$ — Direct CP violation in amplitude

*superweak* : $\Delta S = 2$ — CP violation in mass matrix

**Fig. 11.4.** Two scenarios of CP violation in $K_L \to 2\pi$

In order to show how experiment may bring out these two scenarios, let us introduce the following complex quantities which give the ratios of the two-pion decay amplitudes of $K_L$ and $K_S$ :

$$\begin{aligned} \eta^{+-} \equiv |\eta^{+-}|e^{i\phi^{+-}} &\equiv \frac{A(K_L \to \pi^+\pi^-)}{A(K_S \to \pi^+\pi^-)} , \\ \eta^{00} \equiv |\eta^{00}|e^{i\phi^{00}} &\equiv \frac{A(K_L \to \pi^0\pi^0)}{A(K_S \to \pi^0\pi^0)} . \end{aligned} \tag{11.56}$$

We will discuss how their complex phases and their magnitudes can be determined. In particular, a very precise measure of their magnitudes will teach us a great deal about the nature and the mechanism of the CP violation.

A method used in such experiments is based on the interference phenomenon, analogous to the one considered in Sect. 11.2. At time $t = 0$, a beam of pure $K^0$ mesons is produced. Using (43), its amplitude of transition into a $2\pi$ final state is given by

$$A(K^0 \to 2\pi) = \sqrt{\frac{1+|\bar{\epsilon}|^2}{2(1+\bar{\epsilon})^2}} \left[A(K_S \to 2\pi) + A(K_L \to 2\pi)\right] . \tag{11.57}$$

The time evolution of $K_L$ and $K_S$ are

$$K_{S,L}(t) = K_{S,L}(0) \exp\left[\frac{-t\,\Gamma_{S,L}}{2}\right] e^{-i(m_{S,L})t} . \tag{11.58}$$

From (56), the $t$-dependence of the probability for the $\pi^+\pi^-$ or $2\pi^0$ mode is

$$I_{\pi\pi}(t) = I_{\pi\pi}(0)\left[e^{-\Gamma_S t} + |\eta|^2 e^{-\Gamma_L t} + 2|\eta| e^{-(\Gamma_S+\Gamma_L)\frac{t}{2}} \cos(\Delta m\, t - \phi)\right] . \tag{11.59}$$

As $\Delta m$ is already known from independent measurements (Sect. 11.2), observation of the oscillations of $I_{\pi\pi}(t)$ for $t$ of the order of a few $\tau_S$ yields the phase $\phi$, while for $t \gg \tau_S$, when the first term in (59) has died down, one can make a measure of $|\eta|$. Data from the two most recent experiments give

$$\begin{array}{rll} \phi^{+-} & = (46.9 \pm 2.2)^\circ & \text{CERN} , \\ \phi^{+-} & = (43.53 \pm 0.97)^\circ & \text{FNAL} , \\ \phi^{00} & = (47.1 \pm 2.8)^\circ & \text{CERN} , \\ \Delta\phi \equiv \phi^{00} - \phi^{+-} & = (0.62 \pm 1.03)^\circ & \text{FNAL} , \\ |\eta^{00}/\eta^{+-}| & = (0.9931 \pm 0.002 & \text{CERN} , \\ |\eta^{00}/\eta^{+-}| & = (0.9904 \pm 0.0084 \pm 0.0036) & \text{FNAL} . \end{array}$$

In order to show how these results serve to distinguish the two CP-violating scenarios, we have to make a detailed analysis of the decay amplitudes. The final products of $K^0$ decays, $\pi^+\pi^-$ and $2\pi^0$, have only zero orbital angular momentum (s-wave). These two pions cannot have isospin 1 (forbidden by Bose statistics) but may have isospin 0 or 2. The isospin decomposition of the decay amplitudes may be written down with the help of the Clebsch–Gordan coefficients as follows:

$$A(\mathrm{K}^0 \to \pi^+ + \pi^-) = \frac{1}{\sqrt{3}}[A_2 + \sqrt{2}A_0] \, ,$$
$$A(\mathrm{K}^0 \to \pi^0 + \pi^0) = \frac{1}{\sqrt{3}}[\sqrt{2}A_2 - A_0] \, , \tag{11.60}$$

where $A_0$ and $A_2$ are respectively weak decay amplitudes into isospin $I = 0$ and isospin $I = 2$ states of the two-pion system:

$$A_0 \equiv \langle \pi\pi, I = 0 | H_{\mathrm{w}} | \mathrm{K}^0 \rangle \quad , \quad A_2 \equiv \langle \pi\pi, I = 2 | H_{\mathrm{w}} | \mathrm{K}^0 \rangle \, . \tag{11.61}$$

As $\mathrm{K}^0$ has isospin $1/2$, the matrix element $A_2$ is nonvanishing only if $H_{\mathrm{w}}$ behaves as an isospin $I = 3/2$ or $I = 5/2$ operator. Similarly, for a nontrivial $A_0$, $H_{\mathrm{w}}$ must transform as an isospin $I = 1/2$ operator. Now, from the empirical rule of isospin weak transitions $\Delta I = 1/2$ (Chap. 6), one should expect $a_{1/2} = \left\langle \beta \middle| H_{\mathrm{w}}^{I=1/2} \middle| \alpha \right\rangle$ to be substantially larger than $a_{3/2} = \left\langle \beta \middle| H_{\mathrm{w}}^{I=3/2} \middle| \alpha \right\rangle$. In fact, it is found in various hadronic decays of strange particles $\alpha$ into non-strange particles $\beta$ that their ratio, $|a_{1/2}/a_{3/2}|$, varies between 15 and 30. It follows that $|A_0| \gg |A_2|$. From $\mathrm{K_S} \to \pi^+\pi^-$ and $\mathrm{K_S} \to \pi^0\pi^0$ experimental data, one gets $|A_0/A_2| \approx 22$.

From our previous discussion, it is expected that in the case of exact CP symmetry, $A_0$ and $A_2$ may be taken to be real, or more exactly their relative phase may be chosen to be zero. But if CP symmetry is broken and if this symmetry breakdown resides in the decay amplitude, as the case of the first scenario, the isospin amplitudes $A_I$ are in general complex. This implies necessarily complex matrix elements $\Gamma_{12}$, i.e. $\mathrm{Im}\,\Gamma_{12} \neq 0$.

Now for the $\overline{\mathrm{K}}^0$ case, the corresponding two-pion $\overline{\mathrm{K}}^0$ decay amplitudes can be obtained from (60) by assuming exact CPT symmetry. Let us define

$$\bar{A}_0 \equiv \langle \pi\pi, I = 0 | H_{\mathrm{w}} | \overline{\mathrm{K}}^0 \rangle \quad , \quad \bar{A}_2 \equiv \langle \pi\pi, I = 2 | H_{\mathrm{w}} | \overline{\mathrm{K}}^0 \rangle \, . \tag{11.62}$$

Upon CPT transformations,

$$|\mathrm{K}^0\rangle \longrightarrow -\langle \overline{\mathrm{K}}^0 |$$
$$\langle \pi\pi, I = 0 | \longrightarrow +|\pi\pi, I = 0\rangle \quad , \quad \langle \pi\pi, I = 2 | \longrightarrow +|\pi\pi, I = 2\rangle \, ,$$

and so the assumed CPT symmetry of $H_{\mathrm{w}}$ yields

$$\bar{A}_0 = -A_0^* \quad , \quad \bar{A}_2 = -A_2^* \; ; \tag{11.63}$$

then from (60), the $\overline{\mathrm{K}}^0$ decay amplitudes are

$$A(\overline{\mathrm{K}}^0 \to \pi^+\pi^-) = -\frac{1}{\sqrt{3}}[A_2^* + \sqrt{2}A_0^*] ,$$
$$A(\overline{\mathrm{K}}^0 \to \pi^0\pi^0) = \frac{1}{\sqrt{3}}[-\sqrt{2}A_2^* + A_0^*] . \tag{11.64}$$

It remains to include the effects of the final-state *strong interaction* that acts between the $\pi$ mesons in the final states, i.e. the decay products of the $\mathrm{K_L}$. These effects are parameterized, as usual, by the pion–pion scattering phase-shifts $\delta_0(m_K)$ and $\delta_2(m_K)$ for isospin $I = 0$ and $I = 2$ respectively. These pion–pion phase-shifts are measured and known. With these strong interaction effects included, the four relevant amplitudes are

$$\sqrt{3}\,A(\mathrm{K}^0 \to \pi^+\pi^-) = \left[e^{i\delta_2} A_2 + \sqrt{2}\, e^{i\delta_0} A_0\right] ,$$
$$\sqrt{3}\,A(\mathrm{K}^0 \to \pi^0\pi^0) = \left[\sqrt{2}\, e^{i\delta_2} A_2 - e^{i\delta_0} A_0\right] ,$$
$$\sqrt{3}\,A(\overline{\mathrm{K}}^0 \to \pi^+\pi^-) = -\left[e^{i\delta_2} A_2^* + \sqrt{2}\, e^{i\delta_0} A_0^*\right] ,$$
$$\sqrt{3}\,A(\overline{\mathrm{K}}^0 \to \pi^0\pi^0) = \left[-\sqrt{2}\, e^{i\delta_2} A_2^* + e^{i\delta_0} A_0^*\right] . \tag{11.65}$$

Re-expressing $\mathrm{K}^0$ and $\overline{\mathrm{K}}^0$ in terms of $\mathrm{K_L}$ and $\mathrm{K_S}$, one gets

$$\eta^{+-} = \epsilon + \epsilon' \ , \ \eta^{00} = \epsilon - 2\epsilon' , \ \text{with} \tag{11.66}$$

$$\epsilon = \bar{\epsilon} + i\,\frac{\mathrm{Im}(A_0)}{\mathrm{Re}(A_0)} , \ \text{and} \tag{11.67}$$

$$\epsilon' = \frac{i\,w}{\sqrt{2}} e^{i(\delta_2-\delta_0)} \left[\frac{\mathrm{Im}(A_2)}{\mathrm{Re}(A_2)} - \frac{\mathrm{Im}(A_0)}{\mathrm{Re}(A_0)}\right] , \tag{11.68}$$

where $w = \mathrm{Re}\, A_2/\mathrm{Re}\, A_0 \approx |A_2/A_0| \approx 1/22$.

From the above results we learn the following. First, observation of CP violation effects generally involves two interfering transition amplitudes; for $\epsilon'$, it is the interference between $A_0$ and $A_2$, while for $\epsilon$, it is the interference between the amplitudes $\mathrm{K}^0 \to 2\pi$ and $\overline{\mathrm{K}}^0 \to 2\pi$ through $\bar{\epsilon}$.

Second, the parameter $\epsilon'$ stems exclusively from the first scenario through nonzero values of $\mathrm{Im}\, A_0$ or $\mathrm{Im}\, A_2$, or of both of them. With (68), the phase of $\epsilon'$ as taken from ($i\ e^{i(\delta_2-\delta_0)}$) is $\pi/2 + \delta_2 - \delta_0$. Since both $\delta_2$ and $\delta_0$ are experimentally known, the phase of $\epsilon'$ is $48° \pm 4°$. In fact, the direct CP violation through the first scenario is completely determined from the knowledge of $\eta^{+-} - \eta^{00} = 3\epsilon'$ and $|\eta^{00}/\eta^{+-}|^2 = 1 - 6\,\mathrm{Re}(\epsilon'/\epsilon)$. Due to an unfortunate circumstance, $\epsilon'$ is strongly suppressed by $w$, reduced by the $\Delta I = 1/2$ rule (Chap. 6), and also by a possibly small difference $(\mathrm{Im}A_2/\mathrm{Re}A_2) - (\mathrm{Im}A_0/\mathrm{Re}A_0)$. Note that while $A_2 \ll A_0$, the ratios of their imaginary/real parts may be comparable, since $\mathrm{Re}A_2$ is in the denominator.

From available data on $\eta^{+-}$ and $\eta^{00}$,

$$|\epsilon'/\epsilon| = \begin{array}{ll} 0.00074 \pm 0.00059 & \text{FNAL, E731}\ , \\ 0.0023 \pm 0.00065 & \text{CERN, NA31}\ . \end{array} \tag{11.69}$$

The question of whether $\epsilon'$ vanishes or not is still inconclusively answered.

Finally, from the $\Delta I = 1/2$ rule, the isospin scalar amplitude predominates in $\mathrm{K} \to 2\pi$, hence the CP-violating component in $\Gamma_{12}$, i.e. $\mathrm{Im}\,\Gamma_{12}$ must come mainly from $\mathrm{Im}\,A_0$, so that

$$\mathrm{Im}(\Gamma_{12}) \approx \Gamma_{\mathrm{S}} \frac{\mathrm{Im}(A_0)}{\mathrm{Re}(A_0)} \approx 2\Delta m \frac{\mathrm{Im}(A_0)}{\mathrm{Re}(A_0)}\ . \tag{11.70}$$

Substituting this result for $\mathrm{Im}\,\Gamma_{12}$ in the expression (54) for $\bar{\epsilon}$, and using (67), one gets

$$\begin{aligned} \epsilon = \bar{\epsilon} + \mathrm{i}\,\frac{\mathrm{Im}(A_0)}{\mathrm{Re}(A_0)} &\approx \frac{\mathrm{e}^{\mathrm{i}\pi/4}}{\sqrt{2}}\left[\frac{\mathrm{Im}(M_{12})}{\Delta m} - \mathrm{i}\,\frac{\mathrm{Im}(A_0)}{\mathrm{Re}(A_0)}\right] + \mathrm{i}\frac{\mathrm{Im}(A_0)}{\mathrm{Re}(A_0)} \\ &\approx \frac{\mathrm{e}^{\mathrm{i}\pi/4}}{\sqrt{2}}\left[\frac{\mathrm{Im}(M_{12})}{2\,\mathrm{Re}(M_{12})} + \frac{\mathrm{Im}(A_0)}{\mathrm{Re}(A_0)}\right]\ . \end{aligned} \tag{11.71}$$

The expressions for $\epsilon'$ and $\epsilon$ given in (68) and (71) have the following important property. As a general rule in physics, the absolute phase of an isolated amplitude has no physical meaning; it is only when it is measured relatively to the phase of another amplitude that it takes on a meaning. Therefore, to see whether there are actually CP violation effects or not, it is necessary to determine the relative phase of two interfering amplitudes in some appropriate weak transition. Also the absolute phase of a state function has no physical meaning, thus one expects that making the changes

$$\mathrm{K}^0 \to \mathrm{e}^{\mathrm{i}\alpha}\mathrm{K}^0, \qquad \overline{\mathrm{K}}^0 \to \mathrm{e}^{-\mathrm{i}\alpha}\overline{\mathrm{K}}^0 \qquad \text{or equivalently,}$$

$$\frac{\mathrm{Im}(A_{0,2})}{\mathrm{Re}(A_{0,2})} \to \frac{\mathrm{Im}(A_{0,2})}{\mathrm{Re}(A_{0,2})} + \alpha \quad , \quad \frac{\mathrm{Im}(M_{12})}{\mathrm{Re}(M_{12})} \to \frac{\mathrm{Im}(M_{12})}{\mathrm{Re}(M_{12})} - 2\alpha\ , \tag{11.72}$$

would leave $\epsilon$ and $\epsilon'$ unchanged. This is indeed the case, which shows that these parameters are physically meaningful and truly represent the CP violation effects in K decays.

The phenomenological model-independent analysis given above is served to confront experimental data with different CP-violating mechanisms provided by theoretical models. These theories must give predictions for $\mathrm{Im}(M_{12})$ and $\mathrm{Im}(\Gamma_{12})$ (or Im $(A_{0,2})$) from which we obtain $\epsilon$ and $\epsilon'$.

In the standard model we will show that the gluonic penguin diagram implies $\mathrm{Im}\,A_0 \neq 0$, i.e. direct CP violation in the decay amplitude (first scenario). Thus a measure of $\epsilon'$ through that of $\eta^{+-}/\eta^{00}$ is crucial for the validity of the standard KM mechanism of CP violation, which predicts a small but nonvanishing value of $\epsilon'$. An exactly vanishing value of $\epsilon'$ would be a fatal blow to this *standard* KM mechanism.

In the second scenario (superweak) discussed below, CP violation in $K_L \to 2\pi$ is only due to an unknown mechanism which has, by assumption, an effective $\Delta S = 2$ *complex* mass matrix element $M_{12}$. This complex $M_{12}$, put in by hand to fit $K_L \to 2\pi$ data, is translated into an effective $\bar{\epsilon} \neq 0$. In superweak model, except in the neutral K meson system, the discrete symmetry CP is nowhere violated. This scenario is completely different from the standard KM nonzero phase of the electroweak theory which predicts that, outside the K meson system with nonzero $\epsilon'$, large CP violations are expected in many decay channels of B mesons.

### 11.5.3 The Superweak Scenario

As already mentioned, the decay amplitudes in the second scenario are CP conserving. It is only because of $\bar{\epsilon} \neq 0$ that $K_L \to 2\pi$. In this case, the amplitudes $A_0$ and $A_2$ can be taken as real, so $\epsilon'$ is *identically zero*. The CP violation observed in $K_L$ decay would arise from the $K_1^0$–$K_2^0$ mixing through a complex mass matrix element $M_{12}$. Note that $\Gamma_{12}$ is, in contrast, a real number.

The $K_1^0$–$K_2^0$ mixing may be calculated as

$$\begin{aligned}\langle K_1^0|H_w|K_2^0\rangle &= \left(\frac{1}{\sqrt{2}}\right)^2 \langle K^0 - \overline{K}^0|H_w|K^0 + \overline{K}^0\rangle \\ &= \tfrac{1}{2}\left[\langle K^0|H_w|\overline{K}^0\rangle - \langle \overline{K}^0|H_w|K^0\rangle\right] \\ &= \tfrac{1}{2}\left(M_{12} - M_{12}^*\right) = \mathrm{i}\,\mathrm{Im}(M_{12})\,. \end{aligned} \qquad (11.73)$$

Since $\mathrm{Im}\,A_0 = 0$ and $\mathrm{Im}\,\Gamma_{12} = 0$, this yields, through (54) and (67),

$$|\epsilon| = |\bar{\epsilon}| = \frac{\mathrm{Im}(M_{12})}{\sqrt{2}\Delta m}\ , \quad \epsilon' = 0\,. \qquad (11.74)$$

Since $\Delta m$ is of the order of $G_F^2$, the superweak mixing matrix Im $(M_{12})$ is further reduced below this level by $\epsilon$. The effective coupling strength of the superweak, which is proportional to Im $(M_{12})$, has roughly the magnitude of $|\epsilon| G_F m_c^2/6\pi^2 \approx 10^{-10}$ compared to weak interaction strength $\sim G_F$. Nevertheless, the superweak interaction may still manifest itself through the mixing factor $\bar{\epsilon}$ because in (74), the denominator $\Delta m$ is also very small.

In this second scenario, all CP violation effects depend on the single parameter $\bar{\epsilon}$. It predicts

$$\eta^{+-} = \frac{\langle \pi^+\pi^-|K_L\rangle}{\langle \pi^+\pi^-|K_S\rangle} = \frac{\langle \pi^+\pi^-|\bar{\epsilon}K_1^0\rangle}{\langle \pi^+\pi^-|K_1^0\rangle} = \bar{\epsilon}\,, \qquad (11.75)$$

$$\eta^{00} = \frac{\langle \pi^0\pi^0|K_L\rangle}{\langle \pi^0\pi^0|K_S\rangle} = \frac{\langle \pi^0\pi^0|\bar{\epsilon}K_1^0\rangle}{\langle \pi^0\pi^0|K_1^0\rangle} = \bar{\epsilon}\,, \qquad (11.76)$$

so $\eta^{+-} = \eta^{00}$ and $\phi^{+-} = \phi^{00} = \phi^{\bar{\epsilon}} = \arctan\dfrac{2\Delta m}{\Delta\gamma} = (43.37 \pm 0.2)^\circ$. (11.77)

To date, all data on $\eta^{+-}|$, $|\eta^{00}|$, $\phi^{+-}$, and $\phi^{00}$ are consistent with the superweak scenario. This mechanism also makes the prediction that $K_L \to 2\pi$ and $K_S \to 3\pi$ give rise to quantitatively equal CP violation effects:

$$\eta^{+-0} \equiv \frac{\langle \pi^+\pi^-\pi^0 | K_S \rangle}{\langle \pi^+\pi^-\pi^0 | K_L \rangle} = \bar{\epsilon} = \eta^{000} \equiv \frac{\langle \pi^0\pi^0\pi^0 | K_S \rangle}{\langle \pi^0\pi^0\pi^0 | K_L \rangle} . \tag{11.78}$$

So $\eta^{+-} = \eta^{00} = \eta^{+-0} = \eta^{000}$. Unfortunately, the three-pion $K_S$ decay modes are extremely difficult to detect because they are suppressed by the smallness of the CP symmetry breakdown ($10^{-3}$) and by the reduction of the available phase space in the final state.

As shown in the next subsection, the standard model also gives results for $\epsilon$ close to the data, although, as already noted, CP violation operates through both scenarios and $\epsilon' \neq 0$ is predicted.

### 11.5.4 Calculations of $\epsilon$ and $\epsilon'$ in the Standard Model

In order to obtain $\epsilon$ and $\epsilon'$, defined in (68) and (71), we compute the quantities $M_{12}$, $A_0$ and $A_2$. Let us first introduce the Wolfenstein parameterization of the CKM matrix ($V_{CKM}$) which is particularly suitable for CP violation analyses.

As in any version, $V_{CKM}$ possesses four parameters: three Euler angles and one phase. Wolfenstein's version expands the matrix elements in power of $\lambda = \sin\theta_C = 0.2205 \pm 0.0018$ ($\theta_C$ is the Cabibbo angle). The other three parameters are $A$, $\rho$, and $\eta$. To order $\lambda^3$ for the real parts and order $\lambda^4$ for the imaginary parts, $V_{CKM}$ reads

$$\begin{pmatrix} 1-\frac{\lambda^2}{2} & \lambda & A\lambda^3[\rho - i\eta(1-\frac{\lambda^2}{2})] \\ -\lambda & 1-\frac{\lambda^2}{2} - i\eta A^2\lambda^4 & A\lambda^2(1+i\eta\lambda^2) \\ A\lambda^3(1-\rho-i\eta) & -A\lambda^2 & 1 \end{pmatrix} . \tag{11.79}$$

The parameter $\eta$ represents the complex phase responsible for CP violation. $A$, $\rho$, and $\eta$ can be extracted from data on B meson decays (see Chap. 16), with the results: $A = 0.794 \pm 0.054$, $\sqrt{\rho^2+\eta^2} = 0.363 \pm 0.073$.

**Calculation of $\epsilon$.** Contributions to the matrix element $M_{12}$ in the standard model are represented by diagrams in Fig. 11.1. As expressed by (52), its real part, $\Delta m/2$, has been calculated in (36), with the final result obtained by taking the real part of a certain complex expression. Up to a factor of 2, due to the uncertainty of the parameter $B$ in (35), the imaginary part is

$$\begin{aligned} \mathrm{Im}(M_{12}) = & \frac{G_F^2}{12\pi^2} f_K^2 m_K m_c^2 \Big\{ g(x_c)\,\mathrm{Im}(V_{cd}^* V_{cs})^2 + \frac{x_t}{x_c}\, g(x_t)\,\mathrm{Im}(V_{td}^* V_{ts})^2 \\ & + 2h(x_c, x_t)\,\mathrm{Im}(V_{td}^* V_{ts} V_{cd}^* V_{cs}) \Big\} B , \end{aligned} \tag{11.80}$$

where $g(x)$ and $h(x,y)$ are known from (37). Therefore, the calculation reduces to that of the product of four complex matrix elements $V_{qd}^* V_{qs} V_{q'd}^* V_{q's}$.

Let us start by studying some properties of the CKM matrix related to CP violation. Define the product

$$\Delta_{\gamma k} \equiv V_{\alpha i} V_{\beta j} V^*_{\alpha j} V^*_{\beta i} \, , \tag{11.81}$$

where $(\alpha, \beta, \gamma) = (1,2,3) \equiv$ (u,c,t) or any other cyclic permutation, and similarly, $(i, j, k) = (1,2,3) \equiv$ (d, s, b). These nine complex numbers $\Delta_{\gamma k}$, though very different in magnitudes and in their real parts, have exactly equal imaginary parts. Their common imaginary parts will be denoted by $J$, from Cecilia Jarlskog who was the first to point out this remarkable property. As will be shown further on,

$$\mathrm{Im}(\Delta_{\gamma k}) \equiv \mathrm{Im}(V_{\alpha i} V_{\beta j} V^*_{\alpha j} V^*_{\beta i}) = J \sum_{\gamma,k} \epsilon_{\alpha\beta\gamma} \epsilon_{ijk} \, . \tag{11.82}$$

The term $J$ is a universal number in the sense that it does not depend on how the CKM matrix is parameterized. It shares this property with $|V_{\alpha i}|$. In addition, $J$ as well as $|V_{\alpha i}|$ are also invariant to the phase redefinition of the quark fields that define the matrix representation, e.g. $q \to q\mathrm{e}^{\mathrm{i}\theta}$ implies $V_{\gamma k} \to V_{\gamma k} \mathrm{e}^{\mathrm{i}(\theta_k - \theta_\gamma)}$. $J$ is both invariant to phase redefinition of the quark fields (called rephasing-invariant) and independent on the parameterization of the CKM matrix. Since any physical quantity that violates CP symmetry is proportional to $J$, that quantity must equally possess these properties.

The reason all nine $\Delta_{\gamma k}$ have the same imaginary part can be seen as follows. First multiply both sides of the unitarity relation

$$V_{\alpha i} V^*_{\alpha j} = - \sum_{\delta \neq \alpha} V_{\delta i} V^*_{\delta j} \, , \qquad \text{with } i \neq j \, , \tag{11.83}$$

by $V_{\beta j} V^*_{\beta i}$, and using the cyclic character of the indices, the $\Delta_{\gamma k}$ in (81) is transformed into

$$\Delta_{\gamma k} = -\Delta^*_{\alpha k} - |V_{\beta i}|^2 |V_{\beta j}|^2 \, . \tag{11.84}$$

From (84) it is evident that the nine $\Delta_{\gamma k}$ have equal imaginary parts, and that there exists just one independent $\Delta_{\gamma k}$, the other eight being expressible in terms of it and of the magnitudes of the CKM matrix elements. The common imaginary part is given by

$$J = A^2 \lambda^6 \eta \, , \quad \text{or} \quad J = |(c_{13})^2 \, c_{23} \, c_{12} \, s_{12} \, s_{13} \, s_{23} \, \sin\delta_{13}| \tag{11.85}$$

in the Wolfenstein's version or in the version (9.178) of $V_{\mathrm{CKM}}$.

It can also be shown that $J$ is given by twice the area of any one of the six triangles defined by the following six unitarity relations: three obtained by fixing any two columns $i$ and $j$, and three others obtained by fixing any two rows $\beta$ and $\gamma$ :

$$\sum_{\substack{\alpha=1 \\ i\neq j}}^{3} V_{\alpha i} V^*_{\alpha j} = 0 \quad , \quad \sum_{\substack{k=1 \\ \beta\neq\gamma}}^{3} V_{\beta k} V^*_{\gamma k} = 0.$$

These six relations may be represented by six triangles in the complex plane. For example, the three complex numbers considered as vectors $\mathbf{A}_1 = V_{11}V_{13}^*$, $\mathbf{A}_2 = V_{21}V_{23}^*$, and $\mathbf{A}_3 = V_{31}V_{33}^*$, which sum up to zero, define one such triangle. They are called *unitarity triangles*, their significance to heavy flavor physics will be discussed later in Chap. 16.

Although very dissimilar in their shapes, all those triangles have equal areas, given by $\frac{1}{2}|\mathbf{A}_1||\mathbf{A}_2|\sin(\mathbf{A}_1 \cdot \mathbf{A}_2)$, which is $\frac{1}{2}\,\mathrm{Im}[\mathbf{A}_1 \cdot \mathbf{A}_2^*] = \frac{1}{2}\,\mathrm{Im}\,\Delta_{32} = \frac{1}{2}\,J$. Just as the area of a triangle is given by the lengths of its sides, so $J$ is also given by the various $|V_{ij}|$. In particular $J$ vanishes if one of the nine $V_{ij}$ does. Hence the necessary condition for CP violation (i.e. $J \neq 0$) is that none of the nine matrix elements $V_{ij}$ is zero. Returning now to (80), we express the three factors found in it as

$$\begin{aligned}\mathrm{Im}(V_{\mathrm{cd}}^* V_{\mathrm{cs}})^2 &= -2J\,,\\ \mathrm{Im}(V_{\mathrm{td}}^* V_{\mathrm{ts}})^2 &= 2A^2\lambda^4(1-\rho)J\,,\\ \mathrm{Im}(V_{\mathrm{td}}^* V_{\mathrm{ts}} V_{\mathrm{cd}}^* V_{\mathrm{cs}}) &= +J\,,\end{aligned} \tag{11.86}$$

where $J = A^2\lambda^6\eta$. From (54), (71), (80) and the above equation, one gets

$$|\epsilon| = \frac{G_\mathrm{F}^2 f_\mathrm{K}^2 m_\mathrm{c}^2 m_\mathrm{K}}{6\sqrt{2}\pi^2\Delta m}\, J \left[-g(x_\mathrm{c}) + A^2\,\lambda^4\,(1-\rho)\,\frac{x_\mathrm{t}}{x_\mathrm{c}} g(x_\mathrm{t}) + h(x_\mathrm{c}, x_\mathrm{t})\right] B\,. \tag{11.87}$$

We have approximated $|\epsilon| \approx |\bar{\epsilon}|$ by neglecting $\mathrm{Im}(A_0)$, i.e. by neglecting $\epsilon'$ in $\epsilon$, [note that $|\epsilon'/\epsilon| = \mathcal{O}(10^{-4})$]. Since $J$ is small, it is not surprising that $\epsilon$ is in agreement with experiment. With the $\epsilon$ measurement alone, the standard KM mechanism cannot be differentiated from the superweak scenario, hence the crucial role of $\epsilon'$ for testing different CP-violating mechanisms.

**How do we compute $\epsilon'$ ?** Whereas $\epsilon$ is related to the $\Delta S = 2$ mixing matrix $M_{12}$ as depicted in Fig. 11.1, the parameter $\epsilon'$ describes direct CP violation with the $\Delta S = 1$ transition s→ d shown in Fig. 11.5a. As will be shown later, for large $W$ boson mass, this diagram is equivalent to the 'penguin' diagram of Fig. 11.5b.

The calculation of $\epsilon'$ is more subtle because it involves the amplitudes $A_0$ and $A_2$, the first of which being dominant, according to the $\Delta I = 1/2$ empirical rule. It will be seen later (Chap. 16) that QCD does indeed amplify the $I = 1/2$ transitions at the expense of the $I = 3/2$ transitions, giving a qualitative explanation for the $\Delta I = 1/2$ rule. There are several reasons for the penguin diagram to be the key to the calculation of $A_0$, and hence to direct CP violation in the amplitude (first scenario).

First, the transition s→d changes strangeness by one unit, exactly as required by the first scenario.

Second, we will see later in (94) that the gluonic penguin operator has the form $\alpha_s[\bar{d}\gamma_\mu(1-\gamma_5)\lambda^j s]\,[\bar{q}\gamma^\mu\lambda^j q]$. This is an $I = 1/2$ operator, because it is the product of an $I = 1/2$ operator, $\bar{d}\gamma_\mu(1-\gamma_5)\lambda^j s$, with an $I = 0$

gluonic current $\bar{q}\gamma^\mu\lambda^j q$. Thus the matrix element $A_0$ of an $I = 1/2$ transition is obtained. Let us mention in passing that $A_2$ may be obtained from a diagram similar to that in Fig. 11.5, with the gluon replaced by a photon or a $Z^0$ and is called electroweak penguin. Now there is an electromagnetic or a weak neutral current of isospins 0 and 1 coupled to the $I = 1/2$ operator to give, among others, isospin-$3/2$ transition and hence $A_2$.

Finally, the penguin diagram involves the matrix elements $V^*_{\mathrm{Qd}}V_{\mathrm{Qs}}$ for Q= u, c, t; hence the $A_0$ depends in particular on $V^*_{\mathrm{cd}}V_{\mathrm{cs}}$ and $V^*_{\mathrm{td}}V_{\mathrm{ts}}$, and so must be complex. These complex matrix elements give rise to direct CP violation in the amplitude.

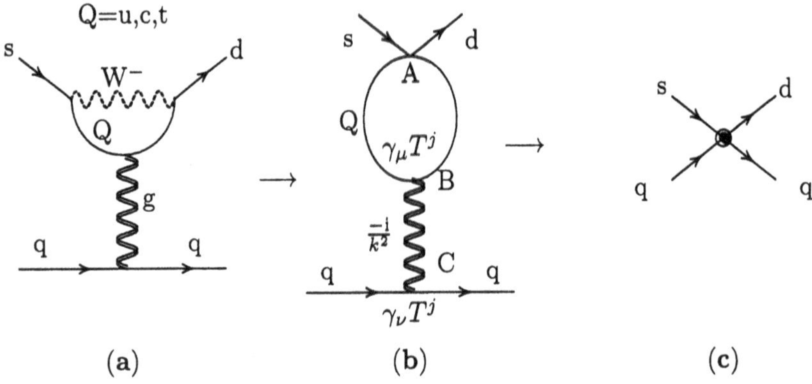


**Fig. 11.5.** (**a**) $\mathrm{s + q \to d + q}$ transition; (**b**) penguin: same as (**a**) but the W propagator is squeezed into the point A; (**c**) $H^{\mathrm{glu}}_{\mathrm{pen}}$ [see (93)] is a local operator because the gluon propagator $1/k^2$ is compensated by $k^2$ of the loop integral

### 11.5.5 The Gluonic Penguin and $|\epsilon'/\epsilon|$

To lowest order of the Fermi coupling $G_{\mathrm{F}}$, the flavor-changing neutral current $\mathrm{s \to d}$ is forbidden by the GIM cancellation mechanism (Chap. 9). Induced by QCD, the $\mathrm{s \to d}$ transition with one gluon emitted, as shown by Fig. 11.5a, gives rise to an effective interaction $H^{\mathrm{glu}}_{\mathrm{pen}}$ which yields CP violation in the decay amplitude $\mathrm{K_L} \to 2\pi$ with a strength $\sim G_{\mathrm{F}}\alpha_{\mathrm{s}}/\pi$.

Our purpose is to compute this effective interaction $H^{\mathrm{glu}}_{\mathrm{pen}}$. As can be shown below, the diagram 11.5a gives the same result as the diagram 11.5b, once the sum over Q=u, c, t is performed to overcome the divergences in the loop integrals of the diagrams 11.5a, b.

We first compute the Q quark loop represented by Fig. 11.5b. At the gluon vertex (shown by B in Fig. 11.5b), there is a QCD quark current $\bar{Q}\gamma_\mu T^j Q$ which contains the SU(3) color matrix $T^j \equiv \frac{1}{2}\lambda^j$. On the other hand, the weak interaction vertex (shown by A in Fig. 11.5b) is a four-point vertex of the type $[\bar{d}\gamma_\mu(1-\gamma_5)Q]\,[\bar{Q}\gamma^\mu(1-\gamma_5)s]$ which of course is a color singlet. To calculate the effective interaction operator which is under consideration, a trace over the color labels is to be taken in the quark loop integral. Since the expression of the transition amplitude contains an explicit color matrix $T^j$ from the gluon vertex but none from the weak interaction four-point vertex, it is convenient to make a decomposition of the color content of the latter. This can be done with the help of the identity (Problem 11.5)

$$\delta_{eh}\delta_{gf} = \frac{1}{3}\delta_{ef}\delta_{gh} + \frac{1}{2}\sum_{j=1}^{8}(\lambda^j)_{ef}(\lambda^j)_{gh}\,, \tag{11.88}$$

where $e, f, g, h$ are color indices running from 1 to 3. Actually, this relation is very useful in QCD corrections to weak decays, and we will exploit it again in Chap. 16. Using (88) together with the Fierz rearrangement, we have

$$\begin{aligned}[]
[\bar{d}\gamma_\mu(1-\gamma_5)\,Q][\bar{Q}\gamma^\mu(1-\gamma_5)s] &= \frac{1}{3}[\bar{d}\gamma_\mu(1-\gamma_5)s][\bar{Q}\gamma^\mu(1-\gamma_5)Q] \\
&+ \frac{1}{2}[\bar{d}\gamma_\mu(1-\gamma_5)\lambda^j s][\bar{Q}\gamma^\mu(1-\gamma_5)\lambda^j\,Q]\,.
\end{aligned} \tag{11.89}$$

The current $\bar{Q}\gamma^\mu(1-\gamma_5)\lambda^j Q$ of the last term in (89) will couple to the QCD current $\bar{Q}\gamma_\mu T^j\,Q$ and leads to a nonvanishing trace in color space.

For each internal quark line Q, one has the following contribution from the loop where the external momenta of s and d are neglected, as in (19),

$$\begin{aligned}
\Gamma^\nu_Q(k^2) &= \mathrm{i}\left(\frac{-\mathrm{i}g}{2\sqrt{2}}\right)^2\left(\frac{\mathrm{i}}{M_\mathrm{W}^2}\right)(-\mathrm{i}g_s)\,\frac{1}{2}\,[\bar{d}\,\gamma_\mu(1-\gamma_5)\lambda^j\,s\,]\,V^*_{Qd}V_{Qs} \\
&\times(-)\int\frac{\mathrm{d}^4p}{(2\pi)^4}\mathrm{Tr}\left[\gamma^\mu(1-\gamma_5)\lambda^j\frac{\mathrm{i}}{\not{p}-m_Q}\gamma^\nu T^l\frac{\mathrm{i}}{\not{p}-\not{k}-m_Q}\right].
\end{aligned}$$

Here $(-\mathrm{i}g_s)$ and $(-\mathrm{i}g/2\sqrt{2})$ are the strong and weak coupling constants; $\mathrm{i}/M_\mathrm{W}^2$ is the approximation of the W-boson propagator $-\mathrm{i}/(p^2-M_\mathrm{W}^2)$ when we go from Fig. 11.5a to Fig. 11.5b. Its justification will be given later. There is a minus sign due to the anticommutation rule of fermions in the loop, and $k$ is the gluon momentum. Setting $g^2/8M_\mathrm{W}^2 = G_\mathrm{F}/\sqrt{2}$ and using $\mathrm{Tr}\,(\lambda^j\lambda^l) = 2\delta^{jl}$, the above expression becomes

$$\Gamma^\nu_\mathrm{Q}(k^2) = \frac{G_\mathrm{F}}{\sqrt{2}}(-\mathrm{i}g_s)\,[\bar{\mathrm{d}}\,\gamma_\mu(1-\gamma_5)\,T^j\,\mathrm{s}]\;V^*_\mathrm{Qd}V_\mathrm{Qs}\;I^{\mu\nu}_\mathrm{Q}(k^2)\,, \tag{11.90}$$

where

$$I^{\mu\nu}_\mathrm{Q}(k^2) = (-1)\int\frac{\mathrm{d}^4p}{(2\pi)^4}\,\mathrm{Tr}\left[\gamma^\mu(1-\gamma_5)\frac{\mathrm{i}}{\not{p}-m_\mathrm{Q}}\gamma^\nu\frac{\mathrm{i}}{\not{p}-\not{k}-m_\mathrm{Q}}\right]. \tag{11.91}$$

Besides the factor $(1-\gamma_5)$ which is irelevant because of the trace, the above integral is familiar and will be discussed in Chap. 15, in relation with the *vacuum polarization* and the *running coupling*. The divergent part $I^{\mu\nu}_\mathrm{div}$ of (91), coming from $p \gg M_\mathrm{W}$ and $p \gg m_\mathrm{Q}$, turns out to be independent of Q.

It does not contribute to the transition amplitude because of the unitarity of the CKM matrix, or the GIM cancelation mechanism:

$$\sum_{\mathrm{Q}} V^*_{\mathrm{Qd}} V_{\mathrm{Qs}} I^{\mu\nu}_{\mathrm{div}} = I^{\mu\nu}_{\mathrm{div}} \sum_{\mathrm{Q}} V^*_{\mathrm{Qd}} V_{\mathrm{Qs}} = 0 \ .$$

Since the divergences in both diagrams of Fig. 11.5a and Fig. 11.5b vanish by the GIM mechanism, the upper limit of their $p$-integrals can be taken at any value lower than the W mass, and the substitution $-\mathrm{i}/(p^2 - M_{\mathrm{W}}^2)$ by $\mathrm{i}/M_{\mathrm{W}}^2$ is justified. The W mass plays the role of the momentum cutoff, this in turn explains why the diagram in Fig. 11.5a gives the same result as that in Fig. 11.5b.

Only the finite part of the integral (91) remains to be evaluated. The calculation will be done in Chap. 15 and given in (15.6) and (15.30). The dominant finite term can be directly taken from (15.6) which gives

$$I^{\mu\nu}_{\mathrm{Q}}(k^2) = -\mathrm{i}(k^2 g^{\mu\nu} - k^\mu k^\nu)\frac{1}{12\pi^2}\log\frac{m_{\mathrm{Q}}^2}{\mu^2} \ . \tag{11.92}$$

In the course of the computation, a mass scale $\mu$ is introduced in (15.6) or (92) for dimensional reason. However $\mu^2$ will disappear in the final result, as can be seen below in (94). Once the loop integral is known, we attach (92) to the external quark current $\bar{q}\gamma_\nu T^j q$ (indicated by C in Fig. 11.5b) via the gluon propagator. Note that the first factor, $k^2 g^{\mu\nu}$, when multiplied by the gluon propagator $1/k^2$, yields a local operator, i.e. a $k^2$-independent finite term. The contribution of the second factor, $k^\mu k^\nu$, vanishes when it operates on the conserved current $\bar{\mathrm{q}}\gamma_\nu T^j \mathrm{q}$.

Finally, summation over internal quarks Q = u, c, t and application of the unitarity of $V_{\mathrm{CKM}}$ lead to the penguin operator

$$\begin{aligned} H^{\mathrm{glu}}_{\mathrm{pen}} =& \frac{G_{\mathrm{F}}}{\sqrt{2}}(-\mathrm{i}g_{\mathrm{s}})\ [\bar{d}\,\gamma_\mu(1-\gamma_5)\,T^j\,s] \sum_{\mathrm{Q=u,c,t}} V^*_{\mathrm{Qd}} V_{\mathrm{Qs}} \frac{1}{12\pi^2}\log\frac{m_{\mathrm{Q}}^2}{\mu^2} \\ & \times (-\mathrm{i}k^2 g^{\mu\nu})\left(\frac{-\mathrm{i}}{k^2}\right)(-\mathrm{i}g_{\mathrm{s}})\ [\bar{q}\,\gamma_\nu\,T^j\,q] \ , \end{aligned} \tag{11.93}$$

which can also be written as

$$\begin{aligned} H^{\mathrm{glu}}_{\mathrm{pen}} &= \frac{G_{\mathrm{F}}}{\sqrt{2}}\frac{\alpha_{\mathrm{s}}}{12\pi}\left\{V^*_{\mathrm{td}} V_{\mathrm{ts}}\log\frac{m_{\mathrm{t}}^2}{m_{\mathrm{c}}^2} - V^*_{\mathrm{ud}} V_{\mathrm{us}}\log\frac{m_{\mathrm{c}}^2}{m_{\mathrm{u}}^2}\right\}\mathcal{O}_{\mathrm{pen}} \ , \\ \mathcal{O}_{\mathrm{pen}} &= [\bar{d}\,\gamma_\mu(1-\gamma_5)\lambda^j\,s]\ [\bar{q}\,\gamma^\mu\lambda^j\,q] \ . \end{aligned} \tag{11.94}$$

Using $V^*_{\mathrm{cd}} V_{\mathrm{cs}} = -\{V^*_{\mathrm{td}} V_{\mathrm{ts}} + V^*_{\mathrm{ud}} V_{\mathrm{us}}\}$, the term inside the curly brackets {} of (94) is derived from the identity

$$\sum_{\mathrm{Q=u,c,t}} V^*_{\mathrm{Qd}} V_{\mathrm{Qs}}\log\frac{m_{\mathrm{Q}}^2}{\mu^2} = V^*_{\mathrm{td}} V_{\mathrm{ts}}\log\frac{m_{\mathrm{t}}^2}{m_{\mathrm{c}}^2} - V^*_{\mathrm{ud}} V_{\mathrm{us}}\log\frac{m_{\mathrm{c}}^2}{m_{\mathrm{u}}^2} \ .$$

As explicitly shown by (94), this effective s $\to$ d strangeness-changing neutral current is a four-fermion operator in the bilinear form with a total isospin $I = 1/2$, as explained earlier.
The calculation of the imaginary part of $A_0 \equiv \langle \pi(k)\pi(p)\,|\,H^{\text{glu}}_{\text{pen}}\,|\,\mathrm{K}^0(P)\rangle$ involves the consideration of the matrix element

$$\langle \pi(k)\pi(p)\,|\,[\bar{d}\gamma_\mu(1-\gamma_5)\lambda^j\; s]\;[\bar{q}\gamma^\mu\lambda^j q]\,|\,\mathrm{K}^0(P)\rangle\;. \tag{11.95}$$

From (88), we write $\lambda^j\lambda^j$ in the above equation as a product of the color singlet currents, then applying a Fierz's transformation on the latter, and finally, using the factorization approximation in the calculation of the matrix element [see Sect. 16.4, and (16.95) as an example], we estimate (95) to be

$$\begin{aligned}\langle \pi(k)\,|\,\bar{d}\gamma_\mu\gamma_5 q\,|\,0\rangle\,\langle \pi(p)\,|\,\bar{q}\gamma^\mu s\,|\,\mathrm{K}^0(P)\rangle \qquad&\\ = f_\pi k_\mu\left[(P+p)^\mu f_+(k^2) + k^\mu f_-(k^2)\right]&\\ = f_\pi\left[(m_{\mathrm{K}}^2 - m_\pi^2)\,f_+(m_\pi^2) + m_\pi^2\,f_-(m_\pi^2)\right]&\\ \approx f_\pi(m_{\mathrm{K}}^2 - m_\pi^2)\,,&\end{aligned} \tag{11.96}$$

the last line comes from $f_+(0) \approx 1$. All of these calculations lead to the result

$$\begin{aligned}\mathrm{Im}(A_0) &\approx \frac{G_F}{\sqrt{2}}\frac{\alpha_{\mathrm{s}}}{12\pi} f_\pi(m_K^2 - m_\pi^2)\mathrm{Im}(V^*_{\mathrm{td}}V_{\mathrm{ts}})\log\frac{m_{\mathrm{t}}^2}{m_{\mathrm{c}}^2}\\ &\approx \frac{G_F}{\sqrt{2}}\frac{\alpha_{\mathrm{s}}}{12\pi} f_\pi(m_K^2 - m_\pi^2)A^2\lambda^5\eta\,\log\frac{m_{\mathrm{t}}^2}{m_{\mathrm{c}}^2}\,.\end{aligned} \tag{11.97}$$

The term $V^*_{\mathrm{ud}}V_{\mathrm{us}}$ is real and does not contribute to Im $(A_0)$. With this result for $\mathrm{Im}(A_0)$ and the experimental value for $\mathrm{Re}\,A_0 = 3.3 \times 10^{-4}$ MeV taken from the decay rate of $\mathrm{K_S} \to 2\pi$, one gets $\epsilon'$ from (68) using $\omega = 1/22$ and $\mathrm{Im}(A_2) = 0$. For another estimate of $A_0$, see Problem 16.1.

Finally, from the measured value of $|\epsilon| = 2.258 \times 10^{-3}$, one obtains the ratio $|\epsilon'/\epsilon| \approx 10^{-3}A^2\eta$. Thus the standard model predicts a small but definitely nonzero ratio $|\epsilon'/\epsilon| \approx 10^{-4}$ with a large uncertainty by a factor of 3, the uncertainty essentially comes from the difficult evaluation of the matrix element in (95) because of its nonperturbative character. Nevertheless, this prediction provides an important test of the CP violation in the KM fashion. A ratio that is either vanishingly small or greater than about $10^{-3}$ would indicate that this mechanism is inadequate and that an explanation beyond the standard model is called for. Experimental measurements of $\epsilon'$ are being planned at CERN and FNAL. The ultimate accuracy of these experiments would be better by an order of magnitude than (69), and may resolve this important issue.

## Problems

**11.1 CP-even and -odd eigenvalues of pions in $K^0$ decay.** We consider the neutral K meson decays into pions. Show that the two-pion system $\pi^0\pi^0$ and $\pi^+\pi^-$ has even eigenvalue $\mathcal{CP}\,|\pi\pi\rangle = +\,|\pi\pi\rangle$. Also show that the three-pion system $\pi^0\pi^0\pi^0$ has odd eigenvalue $\mathcal{CP}\,|\pi^0\pi^0\pi^0\rangle = -\,|\pi^0\pi^0\pi^0\rangle$. How about the CP eigenvalue of $\pi^+\pi^-\pi^0$ ?

**11.2 $\Delta I = 1/2$ rule in the decays of strange particles.** Show that $\Gamma(\mathrm{K}^0 \to \pi^+ + \pi^-) = 2\Gamma(\mathrm{K}^0 \to \pi^0 + \pi^0)$, if $\Delta I = 1/2$ strictly holds. The deviation is used to measure the ratio $\omega = |A_2/A_0| = 1/22$ of the amplitudes $A_0$ and $A_2$ mentioned in the text. With the $\Delta I = 1/2$ rule, show that the amplitudes $a_+ \equiv \mathcal{A}(\Sigma^+ \to \mathrm{n} + \pi^+)$, $a_- \equiv \mathcal{A}(\Sigma^- \to \mathrm{n} + \pi^-)$, and $a_0 \equiv \mathcal{A}(\Sigma^+ \to \mathrm{p} + \pi^0)$ satisfy $a_+ + \sqrt{2}\,a_0 = a_-$. This relation, represented by a rectangular triangle, can be translated into $\Gamma(\Sigma^+ \to \mathrm{n}+\pi^+) = \Gamma(\Sigma^- \to \mathrm{n} + \pi^-) = \Gamma(\Sigma^+ \to \mathrm{p} + \pi^0)$. Compare this prediction with the data.

**11.3 Long and short neutral D and B mesons.** Explain why for the flavored neutral meson systems : $\mathrm{D}^0 = (\bar{\mathrm{u}}\mathrm{c})$, $\mathrm{B}^0_\mathrm{d}(\bar{\mathrm{b}}\mathrm{d})$, and $\mathrm{B}^0_\mathrm{s}(\bar{\mathrm{b}}\mathrm{s})$, the CP-even and -odd eigenstates cannot appear as the long and short components to be easily identified, contrary to the neutral K system.

**11.4 Mass difference $\Delta m_\mathrm{B}$.** For the two eigenstates coming from the $\mathrm{B}^0_\mathrm{d}$–$\overline{\mathrm{B}}^0_\mathrm{d}$ mixing, while one cannot make a distinction between their lifetimes, their mass difference $\Delta m_\mathrm{B}$ can however be measured (Chap. 16). It turns out that $\Delta m_\mathrm{B} = (3 \pm 0.12) \times 10^{-4}$ eV $\approx 10^2 \times \Delta m_\mathrm{K}$. With such value of $\Delta m_\mathrm{B}$, show that one can predict a lower bound of the top quark mass, before its discovery in 1994. Explain why the mass difference $\Delta m^\mathrm{s}_\mathrm{B}$ of the two $\mathrm{B}^0_\mathrm{s}$ eigenstates is again much larger than $\Delta m_\mathrm{B}$. Estimate $\Delta m^\mathrm{s}_\mathrm{B}$.

**11.5 The relation (11.88).** Derive this useful relation.

## Suggestions for Further Reading

$\mathrm{K_L}$, $\mathrm{K_S}$, *strangeness oscillations, regeneration:*

Commins, E. D. and Bucksbaum, P. H., *Weak Interactions of Leptons and Quarks.* Cambridge U. Press, Cambridge 1983

Okun, L. B., *Leptons and Quarks.* North-Holland, Amsterdam 1982

Perkins, D. H., *Introduction to High Energy Physics* (Third edition). Addison-Wesley, Menlo Park, CA 1987

*Penguin mechanism:*

Shifman, M., Vainshtein, A. and Zakharov, V., Nucl. Phys. **B120** (1977) 316

CP *violation:*

Frère, J. M., in *Ecole d'Été de Physique des Particules* GIF 91. IN2P3, Paris 1991

Jarlskog, C., in CP *violation* (ed. Jarlskog, C.). World Scientific, Singapore 1989

Kleinknecht, K., in CP *Violation* (op. cit.)

# 12 The Neutrinos

"The neutrino is the smallest bit of material reality ever conceived of by man; the largest is the universe. To attempt to understand something of one in terms of the other is to attempt to span the dimension in which lie all manifestations of natural law." These comments were made in 1956 by Cowan and Reines in their report on the definite evidence of the neutrino,[1] the elementary particle that Pauli postulated 26 years earlier in his attempt to explain the continuous energy spectrum of the electrons emitted by $\beta$-decays of nuclei: $\mathcal{N}_1(Z) \to \mathcal{N}_2(Z+1) + \mathrm{e}^- + \overline{\nu}_\mathrm{e}$.

Many years later, the acute insight of the neutrino discoverers remains astonishingly topical. Because of their abundance in nature, if the neutrinos have a tiny but nonzero mass, they would play a crucial role in the evolution of the universe and fulfill their mission of bridging the gap separating the two extreme scales of physics. So the first three sections are devoted to the question of their masses, through the fascinating possibility for neutrino species to transmute into each other (a process called neutrino oscillations) and a related problem known as the solar $\nu_\mathrm{e}$ deficit. Next, the crucial role of neutrinos in the discovery of weak neutral currents is emphasized, in relation to the neutrino scattering by the electron. The evidence for neutral currents in turn leads to the confirmation of the standard model and the prediction of the gauge boson $\mathrm{W}^\pm$ and $\mathrm{Z}^0$ masses, long before their observations. Finally, deep inelastic neutrino–nucleon collision is shown to be a powerful probe of the quark and gluon constituents of matter. Neutrinos and electrons play complementary roles in their respective weak and electromagnetic reactions which may be exploited to determine the quark fractional charges. All of these topics constitute the core of the standard electroweak theory and its possible extensions for which an active research on the neutrino masses is crucial.

## 12.1 On the Neutrino Masses

Three neutrino species are known to exist: the electron neutrino $\nu_\mathrm{e}$, the neutrino $\nu_\mu$ associated with the muon and the neutrino $\nu_\tau$ associated with

---

[1] *Nature* **178** (1956) 446. In fact, it was the antineutrino $\overline{\nu}_\mathrm{e}$ emitted in nuclear $\beta$-decay (Savannah River reactor).

the $\tau$ lepton. Until now evidence for the existence of $\nu_\tau$ is only indirect from the $\tau$ decay modes, in contrast with the first two $\nu_e$ and $\nu_\mu$ which are directly observed. Altogether, there are now six leptons in nature: three neutral ($\nu_e$ , $\nu_\mu$ , $\nu_\tau$) and three charged ($e^-$ , $\mu^-$ , $\tau^-$), as well as the six corresponding antileptons. One of the most remarkable experiments performed on the LEP collider at CERN is the establishment of the number of neutrino species that have exactly the *same properties* as the $\nu_e$ (identical V – A coupling, massless or almost massless). There must exist only three neutrino families, otherwise the $Z^0$ width would exceed its current value by at least 167 MeV (see Problem 9.5).

In distinction with all other fermions, the neutrinos are sensible only to weak interactions. The following example may illustrate the distinctive character of these unique particles: of the sixty billions or so of neutrinos that come out of the sun and that pass through each cm$^2$ of the earth surface per second, very few will interact with matter, the cross-section of neutrino interacting with matter being so vanishingly small.

### 12.1.1 General Properties

In the Glashow–Salam–Weinberg (GSW) standard model, the following assumptions on the neutrinos are explicitly made:

(i) their masses are identically zero;

(ii) only their left-handed components $\psi_L \equiv \frac{1}{2}(1-\gamma_5)\psi$ are operative in physical processes.

The right-handed components of neutrinos $\psi_R \equiv \frac{1}{2}(1+\gamma_5)\psi$, even if they exist, do not interact with other particles and are thus absent from the Lagrangian. We also recall that for a massless fermion, $\boldsymbol{\sigma}\cdot\widehat{\boldsymbol{p}}\,\psi_L = -\psi_L$, i.e. the left-handed neutrino is also the eigenstate of the helicity operator $\boldsymbol{\sigma}\cdot\widehat{\boldsymbol{p}}$ with eigenvalue $-1$, its spin $\boldsymbol{\sigma}$ is antiparallel to its three-momentum $\boldsymbol{p}$. If the neutrino is left-handed, the antineutrino is right-handed (its spin is then parallel to its momentum). These properties are explicit in the Weyl representation, suitable for two-component massless neutrinos (Chap. 3).

The second assumption (ii) is based on, among others, the experimental observation of the electron asymmetry from a polarized nucleus in its $\beta$-decay (Sect. 5.1), on the energy and asymmetry distributions of the electron in $\mu$ and $\tau$ decays (Chap. 13), and on the direct determination of the neutrino helicity in a key experiment by Goldhaber et al. (Further Reading). All of these data definitely establish the V – A character of the charged currents.

Because of these assumptions, there is a distinction between the leptons and the quarks in their weak interactions with the gauge bosons $W^\pm$, $Z^0$. To describe these interactions (see Table 9.5), the left-handed fermions are put in SU(2) doublets and the right-handed fermions in U(1) singlets. Only left-handed doublets are coupled to $W^\pm$, while both left and right components couple to $Z^0$. In the leptonic sector, we remark the absence of right-handed neutrinos $\nu_R$ and of mixing between the lepton families, to be contrasted with the Cabibbo–Kobayashi–Maskawa (CKM) mixing among the quark families.

### 12.1.2 Dirac or Majorana Neutrino?

A neutral fermion may exist either as a Dirac particle (fermion $\neq$ antifermion) or as a Majorana particle (fermion $\equiv$ antifermion). For a Dirac fermion (neutral or charged), the mass term is $-m\overline{\psi}\psi = -m(\overline{\psi}_R + \overline{\psi}_L)(\psi_R + \psi_L) = -m(\overline{\psi}_R\psi_L + \overline{\psi}_L\psi_R)$ since $\overline{\psi}_R\psi_R$ and $\overline{\psi}_L\psi_L$ vanish using (9.7)–(9.9). The mass term always connects the opposite chiral components of the same field. The absence of either, $\psi_R$ or $\psi_L$, automatically leads to $m = 0$.

If the neutrinos are of the Majorana type, even in the absence of right-handed components, we can build a mass term by using the antiparticle which is identical to its conjugate, only with opposite chirality. Indeed, contrary to charged fermions, the neutrino and the antineutrino, being chargeless, can be self-conjugated $\nu_M \equiv \nu_M^c$. They are called the Majorana neutrino $\nu_M$.

To each fermionic field $\psi$ there corresponds the field of its antiparticle, denoted by $\psi^c$, obtained with the help of the charge conjugation operator $C = \mathrm{i}\gamma^2\gamma^0$ (Chap. 5). We have $\psi^c \equiv C\psi C^{-1} = \mathrm{i}\gamma^2\gamma^0\overline{\psi}^T = \mathrm{i}\gamma^2\psi^*$. The field of a fermion $F$ is $\psi$ and the field of its antifermion $\overline{F}$ is $\psi^c$.

While for a charged fermion $m\overline{\psi}\psi$ is the only possible mass term, for a neutral fermion there are other possibilities. In addition to the standard term $\overline{\psi}\psi$, the terms $\overline{\psi^c}\psi^c$ , $\overline{\psi}^c\psi$, and $\overline{\psi}\psi^c$ are equally valid. The first $\overline{\psi}^c\psi^c$ is equivalent to $\overline{\psi}\psi$, but the last two, $\overline{\psi}^c\psi$ and $\overline{\psi}\psi^c$, may be written respectively as $\overline{\psi}_L^c\psi_L + \overline{\psi}_R^c\psi_R$ and $\overline{\psi}_L\psi_L^c + \overline{\psi}_R\psi_R^c$. Indeed

$$\psi_L^c \equiv (\psi_L)^c = C\psi_L C^{-1} = \mathrm{i}\gamma^2\psi_L^* = \tfrac{1}{2}(1+\gamma_5)\psi^c\,,$$

$$\psi_R^c \equiv (\psi_R)^c = C\psi_R C^{-1} = \mathrm{i}\gamma^2\psi_R^* = \tfrac{1}{2}(1-\gamma_5)\psi^c\,,$$

$$\overline{\psi}_L^c = \overline{\psi}^c\,\tfrac{1}{2}(1-\gamma_5), \qquad \overline{\psi}_R^c = \overline{\psi}^c\,\tfrac{1}{2}(1+\gamma_5)\,.$$

If the neutrino is a Majorana fermion, we can always construct a mass term $\overline{\psi}_L^c\psi_L + \overline{\psi}_L\psi_L^c$ without the right-handed component $\psi_R$ *precisely because* $\psi_L^c$ *is right-handed with positive helicity.*

The existence of Majorana neutrinos implies that their interactions violate the leptonic number $L_\ell$. Since $\nu_M$ is $(\psi + \psi^c)/\sqrt{2}$, the weak charged current connecting the electron to the Majorana neutrino contains both $L_e = \pm 1$ terms. The most spectacular manifestation of $\nu_M$ would be the neutrinoless double $\beta$-decay of nuclei $\mathcal{N}_1(Z) \to \mathcal{N}_2(Z+2) + \mathrm{e}^- + \mathrm{e}^-$ (Fig. 12.1a), denoted by $(\beta\beta)_{0\nu}$. The initial state has zero leptonic quantum number ($L_e = 0$), while the final state with two electrons has $L_e = 2$. In $(\beta\beta)_{0\nu}$, the Majorana neutrino $\nu_M$ emitted by $\mathrm{n} \to \mathrm{p} + \mathrm{e}^- + \nu_M$ can be absorbed by the second neutron $\mathrm{n}'$ to become $\mathrm{p}' + \mathrm{e}^-$. This is because $\nu_M$ does not have a well-defined lepton number; when emitted by n, it has $L_e = -1$ and when reabsorbed by $\mathrm{n}'$, it has $L_e = +1$.

On the other hand, with the Dirac neutrino for which the leptonic number is conserved, double $\beta$-decay $\mathcal{N}_1(Z) \to \mathcal{N}_2(Z+2) + \mathrm{e}^- + \mathrm{e}^- + \overline{\nu}_e + \overline{\nu}_e$ (Fig. 12.1b), referred to as $(\beta\beta)_{2\nu}$, can only occur with two antineutrinos $\overline{\nu}_e$

emitted together with two electrons. Unlike the $\nu_M$, the Dirac $\overline{\nu}_e$ emitted in $n \to p + e^- + \overline{\nu}_e$ cannot be absorbed by n′ to become $p' + e^-$.

By energy-momentum conservation, the energy spectrum of the two-electron system in $(\beta\beta)_{2\nu}$ decay with Dirac neutrinos is continuous. In $(\beta\beta)_{0\nu}$ by Majorana neutrinos, the same two-electron energy spectrum has a sharp peak (ideally a delta function) which is the distinctive signature of this decay mode. The amplitudes of both $(\beta\beta)_{2\nu}$ and $(\beta\beta)_{0\nu}$ are of the second order in the Fermi constant $G_F$, therefore their rates are very low; nevertheless positive results of the standard decay mode $(\beta\beta)_{2\nu}$ have been reported[2] for nine different isotopes, with half-lives in the range of $10^{19} - 10^{24}$ years. Experiments have been carried out to observe neutrinoless $(\beta\beta)_{0\nu}$ decays of the $^{136}$Xe, $^{76}$Ge , $^{48}$Ca isotopes, but the results were not conclusive.[2] If the electron–$\nu_M$ mixing ($V_{lep}$ below) is small, and/or if the $\nu_M$ mass is too small (through the $\nu_M$ propagator effect), then $(\beta\beta)_{0\nu}$ may still escape observation.

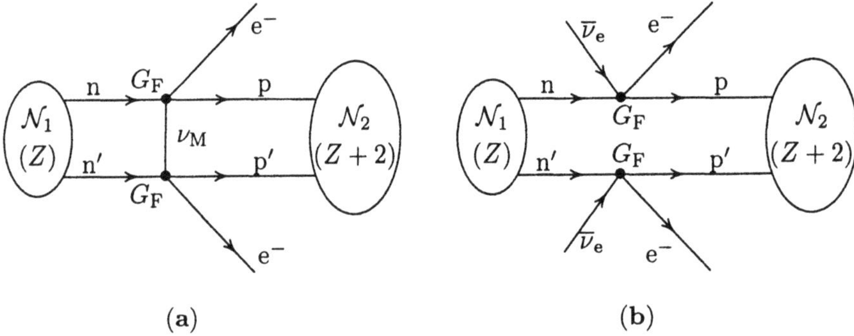


**Fig. 12.1.** (**a**) Double neutrinoless $(\beta\beta)_{0\nu}$ decay by Majorana neutrino; (**b**) double $(\beta\beta)_{2\nu}$ decay by Dirac neutrino

*Remarks.* (i) Unlike the electromagnetic U(1) *local* symmetry, the leptonic number symmetry does not govern the dynamics; rather it is a consequence of the dynamics and the field contents of the standard model. In other words, there is nothing sacred about the leptonic number conservation. If this quantum number is broken, the left-handed neutrino $\nu_L$ and the right-handed antineutrino $\nu_R^c$ will constitute the left- and right-handed components of the same field (the Majorana neutrino) and a mass term with only $\nu_L$ can be constructed. This self-conjugacy is the reason why a Majorana field can be described only by two-component complex spinors, while the Dirac field needs four-component complex spinors. The former has only half as many degrees of freedom as the latter. The situation is analogous to the neutral $\pi$ and K mesons: the $\pi^0$, which is its own antiparticle, can be represented by a real scalar field, while it is necessary to have a complex scalar field to distinguish $K^0$ from $\overline{K}^0$.

[2] M.K.Moe, *Neutrino* 94, Nucl. Phys. (Proc. Suppl.) **B38** (1995)

(ii) For different reasons, both the Majorana and the Weyl fields are two-component spinors. For the Majorana particle, because it is self-conjugate; for the Weyl particle, which is distinct from its antiparticle, because it is massless.

### 12.1.3 Lepton Mixing

In any case, whether of the Dirac type or of the Majorana type, massless neutrinos of different families do not mix up, contrary to quarks. If the neutrinos are massless, i.e. degenerate in mass, the leptonic flavors are not mixed. All states with degenerate masses are physically equivalent and are the eigenstates of their common mass operator. This implies the absence of nondiagonal charged currents like $\overline{\nu}_e\gamma_\lambda(1-\gamma_5)\mu$ (symbolically written as $\overline{\nu}_e\mu$). The six nondiagonal charged currents $\overline{\nu}_e\mu$, $\overline{\nu}_e\tau$, $\overline{\nu}_\mu e$, $\overline{\nu}_\mu\tau$, $\overline{\nu}_\tau e$, and $\overline{\nu}_\tau\mu$ do not exist, there remain only three diagonal currents $\overline{\nu}_e e$, $\overline{\nu}_\mu\mu$, and $\overline{\nu}_\tau\tau$ that separately conserve their respective leptonic numbers $L_e, L_\mu, L_\tau$. Consequently, all leptonic flavor-changing reactions like $\nu_\mu + n \to e^- + p$, $\mu^\pm \to e^\pm + \gamma$, etc. (Problem 12.2) are forbidden, whereas hadronic flavor-changing reactions, like $D \to \overline{K} + e^+ + \nu_e$, $B \to \overline{D}^* + \rho$, and $K^\pm \to \pi^\pm + \pi^0 + \gamma$ coming respectively from $c \to s$, $\overline{b} \to \overline{c}$, and $s \to (u,c) \to d$ are allowed and observed. The latter mode, although rare because of higher-order effects (penguin diagrams, as in Chap. 11), nevertheless exists.

In the standard model, neutrinos are assumed massless simply because a firm proof of nonzero lower bounds of their masses is still lacking, the averages of terrestrial (noncosmic) direct measurements give only their upper bounds $m(\nu_e) < 15\,\text{eV}$, $m(\nu_\mu) < 170\,\text{KeV}$, $m(\nu_\tau) < 19.3\,\text{MeV}$.

Nevertheless, the two hypotheses of the GSW standard model mentioned above demand close scrutiny for many reasons: first, the neutrino helicity is measured with large errors (at 10% of accuracy at best); second, it seems impossible to demonstrate experimentally that the neutrino mass is identically zero. Moreover, the masslessness of fermions has no deep theoretical foundation, in contrast to the massless photon demanded by local gauge invariance. If the neutrinos turn out to be massive, then like the three quark families, the three lepton families could get mixed up, and the presumably small neutrino masses could be indirectly revealed by the oscillation phenomenon analogous to the neutral K-meson oscillations considered in the previous chapter. The mixing of massive neutrinos may follow one of two different scenarios. The first, identical to that for quarks, involves Dirac neutrinos which acquire masses through the usual Higgs mechanism. The second involves Majorana neutrinos whose masses are generated only in models beyond the standard model.

The existence and the size of the neutrino masses are of essential importance in particle physics and astrophysics. In particular, given the enormous abundance of the neutrinos in the universe, solutions to the problems of dark matter, the missing mass, and the expansion rate of the universe will depend crucially on whether the neutrinos are massive or not.

Through their oscillations, massive neutrinos could explain the solar neutrino deficit observed continuously for the last thirty years in the Homestake mines (USA) and actively investigated in different underground experiments: GALLEX (Italy), Kamiokande (Japan), SNO (Canada), and SAGE (Russia). The solar neutrino deficit may be briefly described as follows. The $\nu_e$ flux – produced inside the sun by thermonuclear reactions and measured in these experiments – is lower than predicted by sophisticated calculations within the standard solar model. The $\nu_e$ loss, if it is true, could be attributed to its conversion into $\nu_\mu$ (and/or $\nu_\tau$) through oscillations due to their nonzero masses.

## 12.2 Oscillations in the Vacuum

The quantum oscillation phenomenon occurs when a particle produced by a reaction is not identically the same as the particle that subsequently propagates and decays. The best-known example is the neutral K mesons considered previously. The system $K^0, \overline{K}^0$ produced by strong interactions are distinct from the set $K_L$, $K_S$ which are governed by weak interactions. The $K^0$ and $\overline{K}^0$ are distinguished by their associated production (Chap. 11); whereas the $K_L$, $K_S$, each with a distinctive mass, are characterized by their decay modes. In this context, let us call the former the eigenstates of the strong interaction and the latter, the eigenstates of the weak interaction. The neutral K system oscillates, as we know, since there exists a transition connecting the strong interaction eigenstates $K^0, \overline{K}^0$ to the weak eigenstates $K_L$, $K_S$.

Following the example of $K_L$ and $K_S$ defined as a combination of $K^0$ and $\overline{K}^0$ through (11.1), let us introduce two mass eigenstates $\nu_1$ and $\nu_2$ (of masses $m_1$ and $m_2$) such that we can imagine the physical weakly interacting eigenstates $\nu_e$ and $\nu_\mu$ as linear combinations of $\nu_1$ and $\nu_2$:

$$\begin{pmatrix} \nu_e \\ \nu_\mu \end{pmatrix} \equiv U(\theta) \begin{pmatrix} \nu_1 \\ \nu_2 \end{pmatrix} \equiv \begin{pmatrix} \cos\theta & \sin\theta \\ -\sin\theta & \cos\theta \end{pmatrix} \begin{pmatrix} \nu_1 \\ \nu_2 \end{pmatrix} . \tag{12.1}$$

We can pursue the analogy with the quark sector. We recall that the three left-handed quark doublets in Table 9.5 [thoses defined by (9.176) and (9.177)] can also be written as:

$$\begin{pmatrix} u'' \\ d \end{pmatrix}_L , \begin{pmatrix} c'' \\ s \end{pmatrix}_L , \begin{pmatrix} t'' \\ b \end{pmatrix}_L , \text{ where } \begin{pmatrix} u'' \\ c'' \\ t'' \end{pmatrix} = \begin{pmatrix} V_{ud} & V_{us} & V_{ub} \\ V_{cd} & V_{cs} & V_{cb} \\ V_{td} & V_{ts} & V_{tb} \end{pmatrix}^\dagger \begin{pmatrix} u \\ c \\ t \end{pmatrix}$$

In the same way that the weak interaction eigenstates $u'', c'', t''$ are linear combinations of the mass eigenstates u, c, t of masses $m_u$, $m_c$, $m_t$ via the $V^\dagger_{CKM}$ mixing matrix, the weakly interacting neutrino eigenstates $\nu_e$, $\nu_\mu$, $\nu_\tau$ – analogous to $u'', c'', t''$ – are linear combinations of $\nu_1$, $\nu_2$, $\nu_3$, the neutrino mass eigenstates of masses $m_1, m_2, m_3$.

The lepton mixing is realized by a $3 \times 3$ unitary matrix $\mathrm{V_{lep}}$:

$$\begin{pmatrix} \nu_e \\ e^- \end{pmatrix}_L , \begin{pmatrix} \nu_\mu \\ \mu^- \end{pmatrix}_L , \begin{pmatrix} \nu_\tau \\ \tau^- \end{pmatrix}_L \quad \text{where} \quad \begin{pmatrix} \nu_e \\ \nu_\mu \\ \nu_\tau \end{pmatrix} = \mathrm{V_{lep}} \begin{pmatrix} \nu_1 \\ \nu_2 \\ \nu_3 \end{pmatrix} .$$

The idea of neutrino oscillations was put forth for the first time by Pontecorvo, and the mixing (1) was suggested by Maki, Nakagawa, and Sakata even before its analog in the hadronic sector was proposed by Cabibbo. Like the $V_{CKM}$ quark-mixing matrix, the $\mathrm{V_{lep}}$ can only be determined by experiment. In the present state of our knowledge, the standard model does not pretend to predict either the masses of the fermions nor their mixings. The determination of these parameters is one of the most fascinating problems and is actively investigated in particle physics. Observations of neutrino oscillations seem to be the best (maybe unique) method to measure their eventual nonzero tiny masses and $\mathrm{V_{lep}}$. To illustrate the phenomenon, let us only consider the two families $\nu_e$ and $\nu_\mu$ using the submatrix $U(\theta)$ of $\mathrm{V_{lep}}$. This simplification avoids complications of a $3 \times 3$ matrix without losing any physical understanding.

First, we show that when the muon-neutrino $\nu_\mu$ propagates, it oscillates between $\nu_\mu$ and $\nu_e$ because of the mass difference, $m_1 \neq m_2$. Then $\nu_\mu$ is partially converted into $\nu_e$, just as $\mathrm{K}^0$ becomes partially $\overline{\mathrm{K}}^0$. Indeed, the evolution of the mass eigenstates $\nu_1(t), \nu_2(t)$ at the time $t > 0$ is given by

$$\nu_1(t) = \nu_1(0)e^{-iE_1 t} \quad , \quad \nu_2(t) = \nu_2(0)e^{-iE_2 t} , \tag{12.2}$$

where $E_j^2 = |\boldsymbol{p}_j|^2 + m_j^2$. For relativistic neutrinos, which is always the case since $m_j \ll |\boldsymbol{p}_j|$, we have $|\boldsymbol{p}_1| = |\boldsymbol{p}_2| = |\boldsymbol{p}| \approx E$ and $E_j \approx E + m_j^2/2E$. Putting (2) into (1) we get

$$\begin{aligned} \nu_\mu(t) &= [e^{-iE_1 t}\sin^2\theta + e^{-iE_2 t}\cos^2\theta]\nu_\mu(0) + \cos\theta\sin\theta[e^{-iE_2 t} - e^{-iE_1 t}]\nu_e(0) , \\ \nu_e(t) &= [e^{-iE_1 t}\cos^2\theta + e^{-iE_2 t}\sin^2\theta]\nu_e(0) + \cos\theta\sin\theta[e^{-iE_2 t} - e^{-iE_1 t}]\nu_\mu(0) . \end{aligned} \tag{12.3}$$

The probability for a muon-neutrino $\nu_\mu$ produced at $t = 0$ remains the same particle $\nu_\mu$ at $t > 0$ is then given by

$$\begin{aligned} P(\nu_\mu \to \nu_\mu, t) &\equiv |\langle \nu_\mu(t) | \nu_\mu(0) \rangle|^2 \\ &= 1 - \frac{1}{2}\sin^2 2\theta + \frac{1}{2}\sin^2 2\theta \ \cos\left(\frac{\Delta m_{21}^2}{2E} t\right) \\ &= 1 - \sin^2 2\theta \ \sin^2\left(\frac{\Delta m_{21}^2}{4E} t\right) = P(\nu_e \to \nu_e, t) , \end{aligned} \tag{12.4}$$

where $\Delta m_{21}^2 \equiv m_2^2 - m_1^2$. The probability for $\nu_\mu$ to be converted into $\nu_e$ at $t > 0$ is then

$$P(\nu_\mu \to \nu_e, t) = \sin^2 2\theta \ \sin^2\left(\frac{\Delta m_{21}^2}{4E} t\right) = P(\nu_e \to \nu_\mu, t) . \tag{12.5}$$

Because of their presumed tiny masses, neutrinos are ultra-relativistic. The distance they travel from their production source to a detector is $L = t$ ($c = 1$ in natural units), such that if $L$ is much larger than $2E/|\Delta m^2_{21}|$, the rapidly fluctuating cosine term in (4) vanishes on the average, and the transitions $\nu_\mu \to \nu_\mu, \nu_\mu \to \nu_e$ become constant in space $L$ and time $t$. The oscillations average to zero.

The conditions for oscillations to appear are: both $\theta$ and $\Delta m^2_{21}$ have nonzero values, and the traveling distance $L$ of the neutrino must not differ too much from the oscillation length $L_{osc}$ defined by

$$L_{osc} \equiv \frac{4\pi E}{|\Delta m^2_{21}|} = 2.48 \times \frac{E/(\mathrm{MeV})}{|\Delta m^2_{21}|/(\mathrm{eV})^2}\,\mathrm{m}\,. \tag{12.6}$$

If $L \gg L_{osc}$, $\cos(\Delta m^2_{21}t/2E) = \cos(2\pi L/L_{osc})$ is zero on the average and oscillations cannot be observed. In this case, the sine term in (4) and (5) i.e. $\sin^2(\Delta m^2_{21}t/4E) = \sin^2(\pi L/L_{osc})$ can be effectively replaced with $1/2$.

When $\Delta m^2_{21}$ is expressed in $(\mathrm{eV})^2$, $E$ in MeV, and $L$ in meters, (4) and (5) are written numerically as

$$P(\nu_\mu \to \nu_\mu, t) = 1 - \sin^2 2\theta \ \sin^2\left(\frac{1.27\Delta m^2_{21}L}{E}\right),$$

$$P(\nu_\mu \to \nu_e, t) = \sin^2 2\theta \ \sin^2\left(\frac{1.27\Delta m^2_{21}L}{E}\right). \tag{12.7}$$

These equations tell us that oscillations could be observed in many different experiments, provided that $|\Delta m^2_{21}|$ belongs to the ranges given in Table 12.1. In turn this Table shows that explorations of several neutrino sources are necessary to cover the completely unknown domain of $|\Delta m^2_{21}|$.

**Table 12.1.** Typical ranges of parameters in neutrino oscillations

| Source | Energy $E$ (MeV) | Distance $L$ (m) | $\lvert\Delta m^2_{21}\rvert$ $(\mathrm{eV})^2$ |
|---|---|---|---|
| Reactor | $1 - 10$ | $10 - 100$ | $1 - 10^{-2}$ |
| Accelerator | $10^3 - 10^5$ | $10^2 - 10^3$ | $10^3 - 1$ |
| Atmosphere | $10^2 - 10^3$ | $10^4 - 10^7$ | $10^{-1} - 10^{-5}$ |
| Solar core | $10^{-1} - 10$ | $10^{11}$ | $10^{-10} - 10^{-12}$ |

The neutrino transmutations are usually plotted in the plane $x = \sin^2 2\theta$, $y = |\Delta m^2_{21}|$ (in units of $\mathrm{eV}^2$) where the allowed (forbidden) regions are exhibited. The data are illustrated in Fig. 12.2 in which the point ($x = 0$, $y = 0$) is not definitely excluded for the time being. The very difficult experiments to observe neutrino oscillations are actively pursued in America, Asia, and Europe.

The question about the neutrino mass – either by direct study of the end point of the electron energy spectrum measured in $^3H \rightarrow ^3He + e^- + \overline{\nu}_e$ (tritium $\beta$-decay) or by observation of neutrino oscillations in terrestrial experiments – still has no definite answer at present.

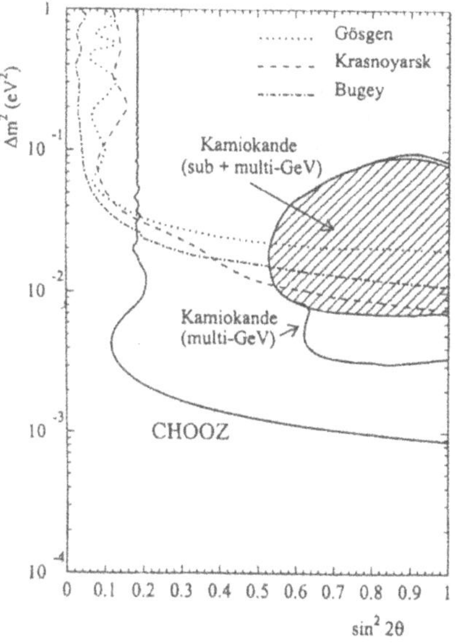


**Fig. 12.2.** Limits on $\nu_e \leftrightarrow \nu_\mu$ in the $\sin^2 2\theta$, $|\Delta m^2_{21}|/(eV^2)$ plane from Chooz reactor neutrinos, compared with experiments from Bugey, Gösgen, Krasnoyarsk where the excluded regions are above and to the right of the contours. The allowed area from Kamiokande with atmospheric neutrinos suggests that oscillations might involve tau neutrinos. Courtesy CERN Courier, February 1998

## 12.3 Oscillations in Matter

In most realistic situations, the neutrinos move not in the vacuum but in matter. For instance, the solar neutrino is produced in the central part of the sun and moves to its surface through the solar material medium. We must therefore consider the effects of the surroundings on the particle oscillations. When a neutrino propagates in a medium filled with other particles, its interaction with matter modifies its oscillations. The reason is that the interaction of the neutrino with matter would change its *effective mass*, just as the well-known example of electromagnetic waves. Massless in vacuum, the photon passes through a medium with a velocity smaller than $c$, as it gets an effective mass by interacting with matter. In conventional optics, the phenomenon is described by an index of refraction $n \neq 1$ of the medium.

Similarly, as a neutrino goes through matter, its effective mass is modified by its interaction with other particles in the medium. As we will see, since the $\nu_e$ interacts with the solar matter differently than $\nu_\mu$ or $\nu_\tau$, the $\nu_e$ oscillations in the sun are different from those of the other $\nu_\mu$ or $\nu_\tau$. This difference causes significant changes in the masses and mixing angles of neutrinos and could give rise to dramatic *resonance oscillations*, known as the Mikheyev–Smirnov–Wolfenstein (MSW) effect.

### 12.3.1 Index of Refraction, Effective Mass

When a neutrino propagates in matter, its interaction with other particles results in either coherent or incoherent transitions. In a coherent process, the medium remains unchanged, allowing scattered and unscattered neutrino wave functions to interfere. The initial and final states in a scattering in the medium must remain exactly the same, requiring *forward elastic scattering* of neutrinos by particles in the medium. As in conventional optics, these coherent elastic forward scatterings are responsible for optical phenomena and provide effective masses to the neutrinos, as first pointed out by Wolfenstein. On the other hand, any change in the states would produce incoherent waves which cannot give rise to optical phenomena. The index of refraction $n$ in *neutrino optics* may be derived from an effective potential $\mathcal{V}$ that the neutrino 'feels' when it travels in and interacts with the medium. The $\mathcal{V}$ gives masses to the neutrinos and changes their mixing angles. Our purpose is to compute $\mathcal{V}$ and show how the masses of the neutrinos and their mixings in the vacuum are modified by $\mathcal{V}$.

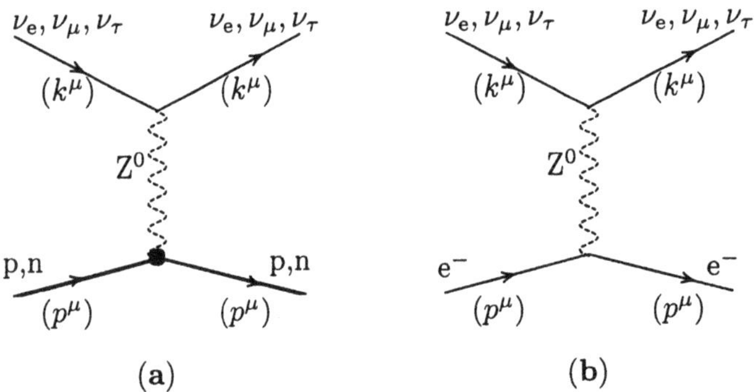


**Fig. 12.3.** (**a**) Neutrino–proton and neutrino–neutron forward elastic scatterings by $Z^0$ exchange; (**b**) neutrino–electron forward elastic scatterings by $Z^0$ exchange

As a specific example, let us consider a neutrino propagating inside the sun, its medium is composed of protons, neutrons, and electrons. We are only interested in elastic scattering of the neutrinos with p, n, and $e^-$ for the reason mentioned above. The $Z^0$-exchange elastic scatterings of $\nu_e, \nu_\mu, \nu_\tau$ are identical since there is no difference between the three neutrino families in their interactions with matter by neutral currents (Fig. 12.3). On the other hand, only the $\nu_e$ can have an elastic scattering with $e^-$ by the W-exchange charged current $\nu_e + e^- \to e^- + \nu_e$ (Fig. 12.4). In normal matter like the sun, there are neither muons nor $\tau$ leptons, therefore the W-exchange elastic reactions $\nu_\mu + \mu^- \to \mu^- + \nu_\mu$ , $\nu_\tau + \tau^- \to \tau^- + \nu_\tau$ do not arise for lack of targets $\mu^-, \tau^-$. Inelastic scattering by charged currents $\nu_{\mu,\tau} + \mathrm{e}^- \to (\mu^-, \tau^-) + \nu_\mathrm{e}$ can occur; however these incoherent reactions cannot give rise to optical phenomena responsible for oscillations in matter.

The effective potential is then the sum of $\mathcal{V}_\mathrm{N}$ (from $\mathrm{Z}^0$ exchange) and $\mathcal{V}_\mathrm{C}$ (from W exchange). We first compute $\mathcal{V}_\mathrm{C}$; its important role will become clear later when we discuss the MSW effect. $\mathcal{V}_\mathrm{C}$ is derived from an effective Hamiltonian $H_\mathrm{C}(x)$ built up by charged currents acting in the medium:

$$\mathcal{V}_\mathrm{C} \equiv \left\langle \nu_\mathrm{e}(k) \middle| \int \mathrm{d}^3x H_\mathrm{C}(x) \middle| \nu_\mathrm{e}(k) \right\rangle ,$$

$$H_\mathrm{C}(x) = -\mathcal{L}_\mathrm{C}(x) = -\frac{G_\mathrm{F}}{\sqrt{2}} \int \mathrm{d}^3p\, f(E,T) \left\langle e(p) \middle| J^\lambda(x) J_\lambda^\dagger(x) \middle| e(p) \right\rangle ,$$

$$J^\lambda(x) = \overline{\psi}_\mathrm{e}(x)\gamma^\lambda(1-\gamma_5)\psi_{\nu_\mathrm{e}}(x) \equiv \overline{\psi}_\mathrm{e}(x)\mathcal{O}^\lambda\psi_{\nu_\mathrm{e}}(x) . \tag{12.8}$$

The function $f(E,T)$ in (8) is the energy distribution of electrons in the medium at temperature $T$, normalized to 1, i.e. $\int \mathrm{d}^3p\ f(E,T) = 1$. By Fierz's rearrangement (Appendix):

$$\overline{\psi}_\mathrm{e}(x)\mathcal{O}^\lambda\psi_{\nu_\mathrm{e}}(x)\,\overline{\psi}_{\nu_\mathrm{e}}(x)\mathcal{O}_\lambda\psi_\mathrm{e}(x) = -\overline{\psi}_{\nu_\mathrm{e}}(x)\mathcal{O}^\lambda\psi_{\nu_\mathrm{e}}(x)\,\overline{\psi}_\mathrm{e}(x)\mathcal{O}_\lambda\psi_\mathrm{e}(x) ,$$

$$H_\mathrm{C}(x) = \frac{G_\mathrm{F}}{\sqrt{2}}\overline{\psi}_{\nu_\mathrm{e}}(x)\mathcal{O}^\lambda\psi_{\nu_\mathrm{e}}(x) \int \mathrm{d}^3p\, f(E,T) \left\langle e(p) \,|\, \overline{\psi}_\mathrm{e}(x)\mathcal{O}_\lambda\psi_\mathrm{e}(x) \,|\, e(p) \right\rangle .$$

With the standard normalization of the electron state

$$\psi_\mathrm{e}(x) = \sqrt{\frac{1}{L^3}}\sqrt{\frac{1}{2E}}\, u(p)\mathrm{e}^{-\mathrm{i}px} , \quad \int_{L^3} \mathrm{d}^3x\, \psi_\mathrm{e}^\dagger(x)\psi_\mathrm{e}(x) = 1 ,$$

where $L^3$ is the volume of the box in which the electron state is normalized in the medium, the continuum limit ($L \to \infty$) is obtained by the replacement $1/L^3 \to \mathrm{d}^3p/(2\pi)^3$. We have

$$\begin{aligned}\left\langle \mathrm{e}^-(p) \,|\, \overline{\psi}_\mathrm{e}(x)\mathcal{O}_\lambda\psi_\mathrm{e}(x) \,|\, \mathrm{e}^-(p) \right\rangle &= \frac{1}{2E\langle L^3\rangle}\overline{u}(p)\mathcal{O}_\lambda u(p) \\ &= \frac{1}{2}\frac{\mathrm{Tr}\{(m+\not{p})\mathcal{O}_\lambda\}}{2E\langle L^3\rangle} = N_\mathrm{e}\frac{p_\lambda}{E} .\end{aligned} \tag{12.9}$$

The factor $\frac{1}{2}$ in the above equation takes care of the spin average of the initial target electron, and $N_\mathrm{e} \equiv 1/\langle L^3\rangle$ is the number of electrons per unit volume (electron number density), which has the dimension of $(\mathrm{mass})^3$. Since

$$\gamma^\lambda \int \mathrm{d}^3p\, f(E,T)\frac{p_\lambda}{E} = \int \mathrm{d}^3p\, f(E,T)\left[\gamma^0 - \frac{\boldsymbol{\gamma}\cdot\boldsymbol{p}}{E}\right] = \gamma^0 ,$$

one has for left-handed neutrino $\psi_\mathrm{L}(x)$:

$$H_\mathrm{C} = \frac{G_\mathrm{F}N_\mathrm{e}}{\sqrt{2}}\overline{\psi}_{\nu_\mathrm{e}}(x)\gamma^0(1-\gamma_5)\psi_{\nu_\mathrm{e}}(x) = \sqrt{2}\,G_\mathrm{F}\,N_\mathrm{e}\,\psi_\mathrm{L}^\dagger(x)\psi_\mathrm{L}(x) .$$

With $\int \mathrm{d}^3x \left\langle \nu_\mathrm{e} \middle| \psi_\mathrm{L}^\dagger(x)\psi_\mathrm{L}(x) \middle| \nu_\mathrm{e} \right\rangle = 1$, then using (8), we get

$$\mathcal{V}_\mathrm{C} = \sqrt{2}\, G_\mathrm{F} N_\mathrm{e} \,. \tag{12.10}$$

The potential $\mathcal{V}$ has the dimension of mass, as it should. We will see later in (18), $\mathcal{V}_\mathrm{C}$ is proportional to the amplitude $\mathcal{M}_\mathrm{C}(E_\nu, q^2 = 0)$ of the forward elastic scattering by charged currents $\nu_\mathrm{e}(k) + \mathrm{e}^-(p) \to \nu_\mathrm{e}(k) + \mathrm{e}^-(p)$ (Fig. 12.4) for which the momentum transfer $q^2$ is zero.

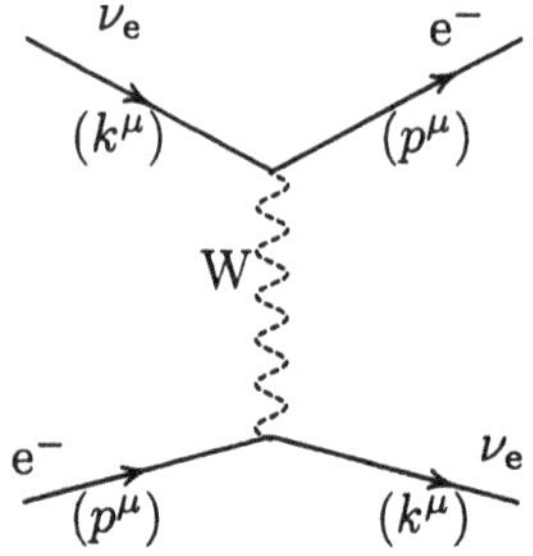


**Fig. 12.4.** $\nu_\mathrm{e}$–$\mathrm{e}^-$ forward elastic scatterings by W boson exchange

By the same method, it is a straightforward task to compute $\mathcal{V}_\mathrm{N}$, the potential due to $Z^0$-exchange contributions. The forward elastic amplitude by neutral currents of Fig. 12.3b is denoted as $\mathcal{M}_\mathrm{N}(E_\nu, q^2 = 0)$. The corresponding effective Hamiltonian $H_\mathrm{N}(x)$ is

$$H_\mathrm{N}(x) = \frac{-G_\mathrm{F}}{\sqrt{2}}\, \overline{\psi}_{\nu_\mathrm{e}}(x) O^\lambda \psi_{\nu_\mathrm{e}}(x) \int \mathrm{d}^3p\, f(E,T)\, \langle \mathrm{e}(p) \,|\, \overline{\psi}_\mathrm{e}(x)\Gamma_\lambda \psi_\mathrm{e}(x) \,|\, \mathrm{e}(p) \rangle \,,$$
$$\Gamma_\lambda = \gamma_\lambda(g_\mathrm{V}^\mathrm{e} - \gamma_5 g_\mathrm{A}^\mathrm{e}) \quad , \quad g_\mathrm{V}^\mathrm{e} = -\tfrac{1}{2} + 2\sin^2\theta_\mathrm{W} \,, \; g_\mathrm{A}^\mathrm{e} = -\tfrac{1}{2} \,.$$

The relative sign between the charged current and neutral current amplitudes, $\mathcal{M}_\mathrm{C}$ and $\mathcal{M}_\mathrm{N}$, must be negative, because of the anticommutation relations of the fermionic creation and destruction operators (Sect. 12.4). Since $\mathcal{V}_\mathrm{C}$ and $\mathcal{V}_\mathrm{N}$ are proportional to $\mathcal{M}_\mathrm{C}$ and $\mathcal{M}_\mathrm{N}$ respectively, this relative sign is reflected in $\mathcal{V}_\mathrm{C}$ and $\mathcal{V}_\mathrm{N}$. Taking this minus sign into account, the contribution of the target electron to $\mathcal{V}_\mathrm{N}$ is found to be

$$\mathcal{V}_\mathrm{N}^\mathrm{e} = \frac{-G_\mathrm{F}}{\sqrt{2}}\,(1 - 4\sin^2\theta_\mathrm{W}) N_\mathrm{e} \,. \tag{12.11}$$

For a more general case, we get

$$\mathcal{V}_\mathrm{N} = \sqrt{2}\, G_\mathrm{F} \sum_f \left\{ (T_3^f)_\mathrm{L} - 2\sin^2\theta_\mathrm{W} Q^f \right\} N_f \,.$$

In the case of the sun, the sum over $f$ corresponds to the three targets: p, n, and $\mathrm{e}^-$. With $(T_3^\mathrm{p})_\mathrm{L} = -(T_3^\mathrm{e})_\mathrm{L} = +1/2$ , $Q^\mathrm{p} = -Q^\mathrm{e} = +1$, and $N_\mathrm{p} = N_\mathrm{e}$

for an electrically neutral medium, the contributions to $\mathcal{V}_{\rm N}$ from protons and electrons exactly cancel out each other, only those of neutrons remain. With $(T_3^{\rm n})_{\rm L} = -1/2$ , $Q^{\rm n} = 0$ and $N_{\rm n}$ is the neutron number density, we get

$$\mathcal{V}_{\rm N} = -(G_{\rm F}/\sqrt{2})N_{\rm n} \,. \tag{12.12}$$

It is instructive to have some numerical values of the number density $N$ for typical media. Since $5.98 \times 10^{23}$ protons weigh one gram, a ground rock with a density of about $4{\rm g/cm}^3$ has $N_{\rm e} = N_{\rm p} \approx N_{\rm n} \approx (\frac{4}{2}) \times 6 \times 10^{23}/{\rm cm}^3$. The solar core with a density of about $100{\rm g/cm}^3$ has $N_{\rm e} = N_{\rm p} \approx 3N_{\rm n} \approx 75 \times 6 \times 10^{23}/{\rm cm}^3$ (we have neglected the electron mass). Supernova density is about $10^{14}{\rm g/cm}^3$. Also $G_{\rm F}/{\rm cm}^3 \approx 8.96 \times 10^{-38}$ eV $= 4.54 \times 10^{-33}/{\rm cm}$.

One consequence of the potential $\mathcal{V}$ felt by the neutrino traveling in matter is the modification of the relation $E_\nu^2 = |\boldsymbol{p}|^2 + m_\nu^2$ in the vacuum. In matter it reads $E_\nu^2 = |\boldsymbol{p}|^2 + m_\nu^2 + 2|\boldsymbol{p}|\mathcal{V}$ for $|\mathcal{V}| \ll |\boldsymbol{p}|$ . We may interpret this modification as an effective mass acquired by the neutrino

$$m_\nu^2 \longrightarrow m_\nu^2 + 2|\boldsymbol{p}|\mathcal{V} \,. \tag{12.13}$$

The index of refraction in the vacuum is

$$n = \frac{|\boldsymbol{p}|}{E_\nu} \approx 1 - \frac{m_\nu^2}{2E_\nu^2} + \cdots \,. \tag{12.14}$$

The propagation of a neutrino in the vacuum has a phase

$$\exp[{\rm i}(n-1)E_\nu t] = \exp\left(-{\rm i}m_\nu^2 t/2E_\nu\right) \,,$$

which is responsible for oscillations in the vacuum, and (4) is recovered. In a material medium composed of particles – collectively denoted by P – which interact with neutrinos, similar to the conventional photon-optics, the index of refraction is given by

$$n \approx 1 - \frac{m_\nu^2}{2E_\nu^2} + \frac{N_{\rm P}}{4mE_\nu^2}\,\mathcal{M}(E_\nu, q^2 = 0) \,, \tag{12.15}$$

where $\mathcal{M}(E_\nu, q^2 = 0)$ is the dimensionless forward elastic amplitude of the neutrino scattered by the particle P, $E_\nu$ is the neutrino energy in the rest frame of P (of mass $m$), and $N_{\rm P}$ is the number density of P in the medium.

The well-known optical theorem relates the imaginary part of the forward elastic amplitude $\mathcal{M}(E_\nu, q^2 = 0)$ to the total cross-section of the neutrino scattered by P, $\sigma_{\rm tot}(E_\nu)$. According to the theorem [see (15.94)], we have Im $\mathcal{M}(E_\nu, q^2 = 0) = 2m|\boldsymbol{p}|\,\sigma_{\rm tot}(E_\nu) \approx 2mE_\nu\,\sigma_{\rm tot}(E_\nu)$. Note that our definition of amplitudes $\mathcal{M}$ coincides with the amplitudes that enter the general formulas of differential cross-sections given in (4.59) and (4.64). The real and imaginary parts of the index of refraction are

$$\begin{aligned} {\rm Re}\,(n) &\approx 1 - \frac{m_\nu^2}{2E_\nu^2} + \frac{N_{\rm P}}{4mE_\nu^2}\,{\rm Re}\mathcal{M}(E_\nu, q^2 = 0) \,, \\ {\rm Im}\,(n) &\approx \frac{N_{\rm P}}{4mE_\nu^2}\,{\rm Im}\mathcal{M}(E_\nu, q^2 = 0) = \frac{N_{\rm P}}{2E_\nu}\sigma_{\rm tot} = \frac{1}{2\,l\,E_\nu} \,, \end{aligned} \tag{12.16}$$

where $l$ is the mean free path of the neutrino in the medium. On the other hand, from (13), we also have

$$\mathrm{Re}\,(n) \approx 1 - \frac{m_\nu^2 + 2|\boldsymbol{p}|\mathcal{V}}{2E_\nu^2} = 1 - \frac{m_\nu^2}{2E_\nu^2} - \frac{\mathcal{V}}{E_\nu}\,. \tag{12.17}$$

Matching with (16), we get

$$\mathrm{Re}\;\mathcal{M}(E_\nu, q^2 = 0) = \frac{-4mE_\nu}{N_\mathrm{P}}\mathcal{V}\,. \tag{12.18}$$

If the target P is an electron, $\mathcal{V} = \mathcal{V}_\mathrm{C} + \mathcal{V}_\mathrm{N}^\mathrm{e}$, where $\mathcal{V}_\mathrm{C}$ and $\mathcal{V}_\mathrm{N}^\mathrm{e}$ are given by (10) and (11). Putting $m = m_\mathrm{e}$ in the above equation, we obtain the real part of the forward elastic scattering amplitude of $\nu_\mathrm{e}$ by an electron. It is easy to check, by a direct calculation [see (48) below], that Re $\mathcal{M}_\mathrm{e}(E_\nu, q^2 = 0)$ as given by the $\mathrm{Z}^0$ exchange of Fig. 12.3b and the W exchange of Fig. 12.4 is

$$\mathrm{Re}\;\mathcal{M}_\mathrm{e}(E_\nu, q^2 = 0) = -2\sqrt{2}\,G_\mathrm{F}\,m_\mathrm{e}\,E_\nu\,(1 + 4\sin^2\theta_\mathrm{W}) \tag{12.19}$$

and we recover (18).

For the antineutrino, the signs of the forward scattering amplitudes are reversed. Hence in the same medium, the potentials $\mathcal{V}_\mathrm{C}$ and $\mathcal{V}_\mathrm{N}$ felt by the antineutrinos have the opposite sign to the potentials (10) and (12) felt by neutrinos. Once $\mathcal{V}$ is computed, the neutrino effective masses in various media characterized by $N_\mathrm{P}$ can be estimated. It is important to note that the oscillation length $\widetilde{L}_\mathrm{osc}$ in matter – related to $\mathcal{V}$ by $\widetilde{L}_\mathrm{osc} \sim 2\pi/\mathcal{V}$ using (6) and (13) – is proportional to $G_\mathrm{F}^{-1}$, whereas the mean free path $l$ depends on $\sigma_\mathrm{tot}^{-1} \sim G_\mathrm{F}^{-2}$. Therefore $\widetilde{L}_\mathrm{osc}$ is many orders of magnitude smaller than $l$, so that oscillations in matter are in principle accessible.

### 12.3.2 The MSW Effect

For a quantitative treatment, let us again stick to the simplest case with two flavors $\nu_\mathrm{e}$ and $\nu_\mu$ and let us first recapitulate the $2 \times 2$ matrix formalism of oscillations in the vacuum suitable for generalization to material medium. The evolution equation of the mass eigenstates $\nu_1$ and $\nu_2$ propagating in the vacuum can be written as

$$\mathrm{i}\frac{\mathrm{d}}{\mathrm{d}t}\begin{pmatrix}\nu_1(t)\\ \nu_2(t)\end{pmatrix} = H\begin{pmatrix}\nu_1(t)\\ \nu_2(t)\end{pmatrix}, \tag{12.20}$$

where $H$ is diagonal in this basis:

$$H = \begin{pmatrix}E_1 & 0\\ 0 & E_2\end{pmatrix} = E_\nu + \begin{pmatrix}m_1^2/2E_\nu & 0\\ 0 & m_2^2/2E_\nu\end{pmatrix}. \tag{12.21}$$

Using (1) for the mixing matrix $U(\theta)$ which connects the mass eigenstates $\nu_1, \nu_2$ to the flavor physical states $\nu_e, \nu_\mu$, we rewrite the above equation as

$$\mathrm{i}\frac{\mathrm{d}}{\mathrm{d}t}\begin{pmatrix}\nu_e(t)\\ \nu_\mu(t)\end{pmatrix} = H'\begin{pmatrix}\nu_e(t)\\ \nu_\mu(t)\end{pmatrix}, \quad \text{where} \quad \Delta \equiv \Delta m_{21}^2 = m_2^2 - m_1^2 ,$$
$$H' = U(\theta)HU^\dagger(\theta) = E_\nu + \frac{m_1^2 + m_2^2}{4E_\nu} + \frac{\Delta}{4E_\nu}\begin{pmatrix}-\cos 2\theta & \sin 2\theta\\ \sin 2\theta & \cos 2\theta\end{pmatrix} . \quad (12.22)$$

In terms of the matrix elements $H'_{ij}$, we write $\theta$, the mixing angle in the vacuum, in the following form, which will be useful later:

$$\tan 2\theta = \frac{2\, H'_{21}}{H'_{22} - H'_{11}} . \quad (12.23)$$

We consider now the problem of neutrinos traveling through a material medium. The evolution equation for $\nu_e, \nu_\mu$ still keeps the same matrix structure (91), with $H'$ replaced by $\widetilde{H}$,

$$\widetilde{H} = E_\nu + \frac{m_1^2 + m_2^2}{4E_\nu} + \mathcal{V}_\mathrm{N} + \begin{pmatrix}-\frac{\Delta}{4E_\nu}\cos 2\theta + \mathcal{V}_\mathrm{C} & \frac{\Delta}{4E_\nu}\sin 2\theta\\ \frac{\Delta}{4E_\nu}\sin 2\theta & \frac{\Delta}{4E_\nu}\cos 2\theta\end{pmatrix} . \quad (12.24)$$

The potential $\mathcal{V}_\mathrm{N}$ acts on both flavor neutrinos $\nu_e$ and $\nu_\mu$. It contributes equally to the common effective mass of $\nu_e$ and $\nu_\mu$ through (13). On the other hand, the potential $\mathcal{V}_\mathrm{C}$ acts only on $\nu_e$ and not on $\nu_\mu$. Therefore, in the matrix $\widetilde{H}$, $\mathcal{V}_\mathrm{N}$ is diagonal whereas $\mathcal{V}_\mathrm{C}$ appears only in $\widetilde{H}_{11}$.

The mixing angle $\theta$ in the vacuum now becomes $\Phi$, the mixing angle in matter, using the general formula (23):

$$\tan 2\Phi \equiv \frac{2\widetilde{H}_{21}}{\widetilde{H}_{22} - \widetilde{H}_{11}} = \frac{\Delta \sin 2\theta}{A_\mathrm{R} - A_\mathrm{C}} , \quad \text{or}$$
$$\sin^2 2\Phi = \frac{\tan^2 2\Phi}{1 + \tan^2 2\Phi} = \frac{\Delta^2 \sin^2 2\theta}{(A_\mathrm{C} - A_\mathrm{R})^2 + \Delta^2 \sin^2 2\theta} , \quad (12.25)$$

$$\text{where } A_\mathrm{R} = \Delta \cos 2\theta , \qquad A_\mathrm{C} = 2E_\nu \mathcal{V}_\mathrm{C} = 2\sqrt{2} G_\mathrm{F} N_e E_\nu . \quad (12.26)$$

When $N_e = 0$, both $\mathcal{V}_\mathrm{C}$ and $A_\mathrm{C}$ vanish and $\Phi$ is equal to $\theta$ as it should be.

The crucial point of (25), first noticed by Mikheyev and Smirnov, is the *resonance behavior* which dramatically changes the mixing angle $\Phi$ and reveals unexpected features of the oscillation phenomenon. The angle $\Phi$ shown by Fig. 12.5 as a function of $A_\mathrm{C}$ is clearly a resonance, peaking at the pole $A_\mathrm{C} = A_\mathrm{R}$ with a width $\Delta^2 \sin^2 2\theta$.

The sign of $A_\mathrm{C}$ or $\mathcal{V}_\mathrm{C}$ is essential for a resonance behavior to appear. With neutrinos, $A_\mathrm{R}$ and $A_\mathrm{C}$ have the same signs, and the denominator of (25) could vanish. However, with antineutrinos in the same electron-rich medium, the

$+A_C$ becomes $-A_C$ and the resonance behavior is absent. If $A_C \approx A_R$, then (25) shows that even starting with a very small $\theta \approx 0$ in the vacuum, the mixing angle $\Phi$ could become $\approx \pi/4$ in matter. In other words, as soon as it is produced, the neutrino $\nu_e$, by its interaction with electrons in matter, becomes equally split into $\nu_e$ and $\nu_\mu$ in the resonance region characterized by $A_C \approx A_R$. We notice the important role of $A_C$ in the MSW effect.

To go further, let us write down the eigenvalues $E_{1,2}$ of $\widetilde{H}$ to understand the exact meaning of this maximum mixing. They are

$$E_{1,2} = |\boldsymbol{p}| + \frac{(\mu_{1,2})^2}{2|\boldsymbol{p}|} + \dots ,$$

$$(\mu_1)^2 = \frac{1}{2}\left[m_1^2 + m_2^2 + A_C + 2A_N - \sqrt{(A_R - A_C)^2 + (\Delta \sin 2\theta)^2}\right] ,$$

$$(\mu_2)^2 = \frac{1}{2}\left[m_1^2 + m_2^2 + A_C + 2A_N + \sqrt{(A_R - A_C)^2 + (\Delta \sin 2\theta)^2}\right], \quad (12.27)$$

where $A_N = 2E_\nu \mathcal{V}_N = -\sqrt{2}G_F N_n E_\nu$. The above formula clearly indicates that even if the neutrinos are massless in the vacuum, i.e. $m_{1,2} = 0$, they can acquire effective masses $\mu_{1,2} \neq 0$ in matter. In the vacuum the relevant quantities are $\Delta$ and $\theta$, in matter they become $\widetilde{\Delta}$ and $\Phi$, where $\widetilde{\Delta} = (\mu_2)^2 - (\mu_1)^2 = \sqrt{(A_R - A_C)^2 + (\Delta \sin 2\theta)^2}$, and $\Phi$ is given by (25). To illustrate the surprising behavior of neutrino oscillations in matter, let us assume that $\theta$ is extremely small, such that in the vacuum, by (1) the light mass eigenstate $\nu_1$ is nearly $\nu_e$, and the heavy mass eigenstate $\nu_2$ is almost $\nu_\mu$.

In the other extreme condition of matter, we assume that $A_C \gg A_R$, from (25) the mixing angle $\Phi \approx \pi/2$. Now $\theta$ is replaced by $\Phi$, and $\nu_{1,2}$ by $\widetilde{\nu}_{1,2}$ (the mass eigenstates in matter). Since $\nu_e = \widetilde{\nu}_1 \cos\Phi + \widetilde{\nu}_2 \sin\Phi \approx \widetilde{\nu}_2$, we note that $\nu_e$, produced in the region where $A_C \gg A_R$, is nearly a $\widetilde{\nu}_2$ which has

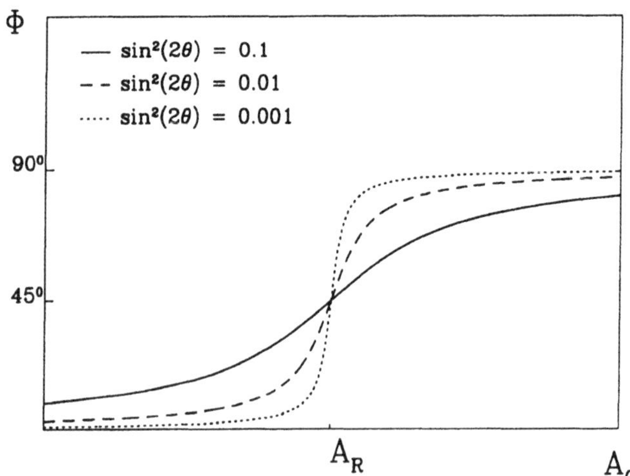


**Fig. 12.5.** The mixing angle $\Phi$ in matter as a function of $A_C$

an effective mass $\mu_2$ greater than the effective mass $\mu_1$ of $\nu_\mu$. The neutrino $\nu_e$ which is lighter than the neutrino $\nu_\mu$ in the vacuum becomes heavier than $\nu_\mu$ in an electron-rich medium.

In matter where $A_C \gg A_R$, $\nu_e$ starts to be a $\widetilde{\nu}_2$, it propagates along the path of the latter (if the adiabatic condition discussed in the following is satisfied). There is not much transition (since the corresponding oscillation length in this region $\widetilde{L}_{osc} \sim 2\pi/\mathcal{V}_C$ is very short) until it arrives in the resonance region ($A_C \sim A_R$) for which $\Phi \approx \pi/4$, the oscillations are enhanced, and $\widetilde{\nu}_2$ is composed of $\nu_e$ and $\nu_\mu$ in equal parts. At the solar surface (i.e. the vacuum), $A_C$ gradually decreases, $\Phi$ tends towards $\theta$, $\widetilde{\nu}_2$ gradually becomes $\nu_\mu \cos\theta + \nu_e \sin\theta$, and finally comes out nearly as a $\nu_\mu$ in the vacuum. The evolution of neutrinos in matter, expressed by (27) and illustrated by Fig. 12.6, is called a *level crossing* : Produced as a $\nu_e$ in an electron-dense solar core, the traveling neutrino becomes almost a $\nu_\mu$ when it reaches the solar surface. The depletion is spectacular indeed.

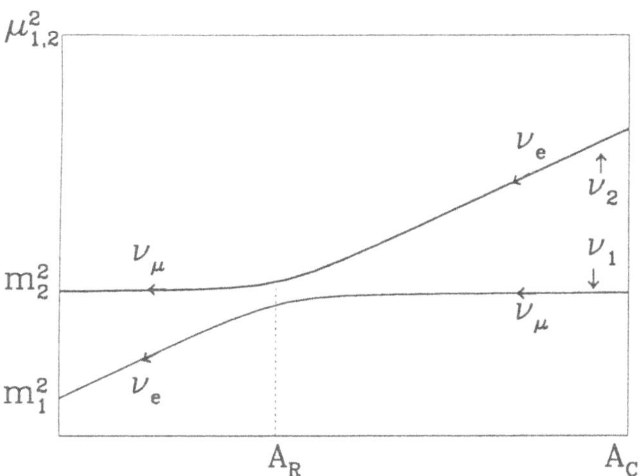


**Fig. 12.6.** Adiabatic MSW effect: Following the $\widetilde{\nu}_2$ path, a $\nu_e$ produced in the solar core becomes a $\nu_\mu$ at the solar surface

## 12.3.3 Adiabaticity

So far we have assumed that the solar density $N$ is homogeneous everywhere. It is constant throughout the region covered by the traveling neutrinos. This is actually not the case of the sun, and we must accordingly take into account the variations of $N(r) = N(t)$. We have taken $r = ct = t$ appropriate for relativistic neutrinos, $r$ being the distance from the center of the sun. The mixing angle $\Phi$ and the effective masses $\mu_{1,2}$ – given respectively by (25) and (27) – are no longer constant in $r$, hence in $t$. The mixing angle in matter, always expressed through its analytic form (25), is now a function of $t$ since $A_C(t)$ depends on it via $N(t)$. First we write $\widetilde{\nu}_1(t)$ and $\widetilde{\nu}_2(t)$, the mass eigenstates in matter, as a mixing of $\nu_e$ and $\nu_\mu$ with the angle $\Phi(t)$,

$$\begin{pmatrix} \widetilde{\nu}_1(t) \\ \widetilde{\nu}_2(t) \end{pmatrix} = \begin{pmatrix} \cos\Phi(t) & -\sin\Phi(t) \\ \sin\Phi(t) & \cos\Phi(t) \end{pmatrix} \begin{pmatrix} \nu_e \\ \nu_\mu \end{pmatrix} .$$

In the evolution equation (18) of $\nu_e$ and $\nu_\mu$, we keep the $t$ dependence of $\mathcal{V}_C(t)$. After rewriting $\nu_e$ and $\nu_\mu$ in terms of $\widetilde{\nu}_1$ and $\widetilde{\nu}_2$, we get

$$\mathrm{i}\frac{\mathrm{d}}{\mathrm{d}t}\begin{pmatrix} \widetilde{\nu}_1 \\ \widetilde{\nu}_2 \end{pmatrix} = \frac{1}{2E_\nu}\begin{pmatrix} \mu_1^2(t) & -\mathrm{i}\delta\mu^2(t) \\ +\mathrm{i}\delta\mu^2(t) & \mu_2^2(t) \end{pmatrix} \begin{pmatrix} \widetilde{\nu}_1 \\ \widetilde{\nu}_2 \end{pmatrix} , \tag{12.28}$$

where

$$\delta\mu^2(t) = 2E_\nu \frac{\mathrm{d}\Phi(t)}{\mathrm{d}t} = \frac{E_\nu \Delta \sin 2\theta}{[\mu_2^2(t) - \mu_1^2(t)]^2} \left| \frac{1}{N_e}\frac{\mathrm{d}N_e}{\mathrm{d}t} \right| A_C(t) . \tag{12.29}$$

The oscillations depend now on an additional parameter denoted by $h(t)$:

$$h(t) \equiv \left| \frac{1}{N_e}\frac{\mathrm{d}N_e}{\mathrm{d}t} \right| .$$

In general, (28) is solved by numerical methods. We remark that if $N_e(t)$ is constant, $\delta\mu^2(t)$ vanishes, and $\widetilde{\nu}_{1,2}$ are stationary states. For a varying density $N(r)$, we can only define the stationary states at a given point $r$. Nevertheless, if (28) is *almost diagonal*, i.e. if $\delta\mu^2(t) \ll [\mu_2^2(t) - \mu_1^2(t)]$ (a relation referred to as the adiabatic condition), then as long as this condition is satisfied, the matter eigenstates $\widetilde{\nu}_{1,2}$ move in the medium without undergoing transitions between themselves, with the relative admixture of $\nu_e, \nu_\mu$ determined according to the value of $N_e(r)$ at a given point $r$. The adiabatic condition can also be rewritten as

$$\frac{\Delta \sin 2\theta}{[\mu_2^2(t) - \mu_1^2(t)]} E_\nu\, A_C(t)\, h(t) \ll \mu_2^2(t) - \mu_1^2(t) . \tag{12.30}$$

In the resonance region where $A_C = A_R \equiv \Delta \cos 2\theta$, we note from (27) that the right-hand side of the above equation, i.e. $[\mu_2^2 - \mu_1^2]$ reaches its minimum value which is equal to $\Delta \sin 2\theta$, whereas the left-hand side is maximum (because $[\mu_2^2 - \mu_1^2]$ is now in the denominator). Provided that $h(t)$ is monotonously changing (this is the case of the sun, see below), if (30) is satisfied at the resonance, it is satisfied everywhere. At the resonance, the adiabatic condition (30) becomes

$$\frac{\Delta \sin^2 2\theta}{E_\nu \cos 2\theta} \gg h_{\mathrm{Res}} \ , \quad h_{\mathrm{Res}} = \left| \frac{1}{N_e}\frac{\mathrm{d}N_e}{\mathrm{d}t} \right|_{\mathrm{Res}} , \tag{12.31}$$

where $h_{\text{Res}}$ is the value of $h(t)$ at the resonance. Physically, the adiabatic condition corresponds to the case of many oscillations that take place in the resonance region. This region is characterized by the resonance oscillation length $\widetilde{L}_{\text{Res}} = L_{\text{osc}}/\sin 2\theta$, where $L_{\text{osc}}$ is given by (6). For the standard solar density, $N_{\text{e}}(r) = N_{\text{e}}(0)\text{e}^{-ar/R_\odot}$, where $a \simeq 10.54$ and $R_\odot \simeq 7 \times 10^8$m is the radius of the sun, we get $h_{\text{Res}} \simeq 3 \times 10^{-15}\,\text{eV} = 1.52 \times 10^{-10}/\text{cm}$. When $\Delta$ is expressed in $(\text{eV})^2$ and $E_\nu$ in MeV, the adiabatic condition (31) is

$$\frac{\sin^2 2\theta(\Delta/\text{eV}^2)}{\cos 2\theta(E_\nu/\text{MeV})} \gg 3 \times 10^{-9} \,.$$

If the mixing angle in the vacuum $\theta$ satisfies the above inequality, i.e. if (28) is almost diagonal, then for a given $E_\nu$ and $\Delta$, the $r$ dependence of $N(r)$ is harmless and the level crossing can be fully achieved. Let us explain in detail the MSW effect in matter satisfying the adiabatic condition. Similar to the vacuum case (3), the amplitude $\widetilde{\mathcal{A}}(\nu_{\text{e}} \to \nu_{\text{e}}; t)$ in matter is written as

$$\widetilde{\mathcal{A}}(\nu_{\text{e}} \to \nu_{\text{e}}; t) = \sum_{a,b} \langle \nu_{\text{e}}(t)\,|\widetilde{\nu}_b(t)\rangle \, \langle \widetilde{\nu}_b(t)\,|\widetilde{\nu}_b(t_R)\rangle \\ \times \langle \widetilde{\nu}_b(t_R)\,|\widetilde{\nu}_a(t_R)\rangle \, \langle \widetilde{\nu}_a(t_R)\,|\widetilde{\nu}_a(t_{\text{O}})\rangle \, \langle \widetilde{\nu}_a(t_{\text{O}})\,|\nu_{\text{e}}(t_{\text{O}})\rangle \,, \quad (12.32)$$

where $t_{\text{O}}$ and $t_R$ are the traveling time (or distance) from the solar center to the $\nu_{\text{e}}$ production region and to the resonance localization respectively, and reading from right to left, the first term is

$$\langle \widetilde{\nu}_a(t_{\text{O}})\,|\nu_{\text{e}}(t_{\text{O}})\rangle = U^*_{\text{e}a}(\Phi) \,,$$

where $U$ is the mixing matrix in matter with the angle $\Phi$, similar to (1) in the vacuum. To simplify, we consider only the two-family case $(a, b = 1, 2)$. Under the adiabatic condition, the stationary mass-eigenstates $\widetilde{\nu}_1$ , $\widetilde{\nu}_2$ propagate from the core to the surface without mixing, i.e. $\widetilde{\nu}_1$ remains $\widetilde{\nu}_1$, and $\widetilde{\nu}_2$ remains $\widetilde{\nu}_2$ in the whole distance covered. Then the three factors in the middle of (32) are simply

$$\langle \widetilde{\nu}_b(t)\,|\widetilde{\nu}_b(t_R)\rangle \, \langle \widetilde{\nu}_b(t_R)\,|\widetilde{\nu}_a(t_R)\rangle \, \langle \widetilde{\nu}_a(t_R)\,|\widetilde{\nu}_a(t_{\text{O}})\rangle \\ = \delta_{ab} \exp[\text{i} \int_{t_{\text{O}}}^{t} E_a(t')\text{d}t'] \equiv \delta_{ab} \exp[\text{i}\, E_a(t)] \,.$$

Note that $E_a(t)$ is a function of time (or distance) because the effective mass $\mu_a$ given by (27) changes as it propagates in matter (Fig. 12.6). The factor $\langle \nu_{\text{e}}(t)\,|\widetilde{\nu}_b(t)\rangle$ in the extreme left of the right-hand side of (32) which projects out $\nu_{\text{e}}$ from $\widetilde{\nu}_b$ with the mixing angle $\theta$ in the vacuum is

$$\langle \nu_{\text{e}}(t)\,|\widetilde{\nu}_b(t)\rangle = U_{\text{e}b}(\theta) \,.$$

The transition probability becomes

$$\widetilde{P}(\nu_e \to \nu_e, t) = \left| \sum_{a=1,2} U_{ea}(\theta) U^*_{ea}(\Phi) \exp\left[\frac{-iE_a(t)}{2E}\right] \right|^2$$

$$= \cos^2\theta \cos^2\Phi + \sin^2\theta \sin^2\Phi + \tfrac{1}{2}\sin 2\theta \sin 2\Phi \cos\frac{\widetilde{\delta}(t)}{2E} ,$$

$$\widetilde{\delta}(t) = \int_{t_0}^{t} dt' [\mu_2^2(t') - \mu_1^2(t')] = \int_{t_0}^{t} dt' \sqrt{[\Delta \cos 2\theta - A_C(t')]^2 + (\Delta \sin 2\theta)^2} .$$

In practice, the oscillating term that depends on $t$ can be neglected, and the time average of $\widetilde{P}(\nu_e \to \nu_e, t)$ is

$$\widetilde{P}(\nu_e \to \nu_e) = \cos^2\theta \cos^2\Phi + \sin^2\theta \sin^2\Phi .$$

Since $\Phi$ depends on $A_C$, $\widetilde{P}(\nu_e \to \nu_e)$ is a function of the localization where neutrinos are produced. When they are produced in the region $A_C \gg A_R$, $\Phi \approx 90^0$, we get $\widetilde{P}(\nu_e \to \nu_e) = \sin^2\theta$, so that depending on the value of $\theta$ in the vacuum, we can have any amount of depletion. Figure 12.6 illustrates the situation. This is in sharp contrast to the vacuum depletion given by (4), where $P(\nu_e \to \nu_e) = \cos^4\theta + \sin^4\theta = 1 - \frac{1}{2}\sin^2 2\theta$ is larger than $\frac{1}{2}$ for all $\theta$.

**Summary.** The nonzero mass of neutrinos is of great importance not only in particle physics but also in astrophysics and cosmology. If the three neutrino families have nondegenerate masses, they could mix together like the quark families and oscillations would occur. The answer to the question on the existence of neutrino masses depends mainly on possible observations of oscillations either in the vacuum or in a material medium. This may be the only experimental method to measure their vanishingly small masses. To cover the large spectrum of $\Delta m^2$ between $10^{-12}$ to $10^3\,\mathrm{eV}^2$ (see Table 12.1), several sources of neutrino production should be exploited. From the sun to the particle accelerators and nuclear reactors, each source – with its specific energy and distance to the detectors – brings an answer appropriate to each range of values of $\Delta m^2$. Finally, the solar neutrino deficit may find its explanation in the MSW mechanism.

## 12.4 Neutral Currents by Neutrino Scattering

We recall that weak interactions were historically discovered by processes involving charged currents, their first manifestation at the beginning of the century was the $\beta$-radioactivity of nuclei for which the neutron disintegration $n \to p + e^- + \overline{\nu}_e$ represents the simplest mode. The amplitude of this decay is obtained from the product of two charged currents: the hadronic one $V_{ud}\overline{u}\gamma_\mu(1-\gamma_5)d$ which may be written as a $d \to u$ transition between the quark u , d fields connected by the CKM matrix element $V_{ud}$, and the leptonic one $\overline{e}\gamma^\mu(1-\gamma_5)\nu_e$ constructed from the $e^-$ and $\nu_e$ fields. All charged currents share the universal V − A property symbolized by $\gamma_\mu(1-\gamma_5)$. There are in all $9 = 3 \times 3$ hadronic charged currents, only in this specific d → u transition is flavor conserved.

### 12.4.1 Neutral Currents, Why Not?

From the beginning of the $\beta$-radioactivity period to the formulation of the standard model (SM) in the 1970s, physicists had always wondered why only charged currents are involved in weak interactions and not neutral currents, since *a priori* there is no deep reason to suppress the latter. Moreover, in every non-Abelian gauge theory that may underlie weak interactions – the SM is a prototype – the neutral currents (NC) naturally emerge on an equal footing with the charged currents (CC). The problem is to demonstrate experimentally the existence of the neutral currents.

We illustrate the situation by an example. The hadronic charged currents $V_{\rm ud}\overline{u}\gamma_\mu(1-\gamma_5)d$ and $V_{\rm us}\overline{u}\gamma_\mu(1-\gamma_5)s$ are respectively responsible for the decays $\pi^+ \to {\rm e}^+ + \nu_{\rm e}$ and ${\rm K}^+ \to \mu^+ + \nu_\mu$ (Fig. 12.7). If the hadronic neutral and charged currents have comparable couplings, as they do in the case of non-Abelian gauge theories, we would expect that $\Gamma(\pi^0 \to {\rm e}^+ + {\rm e}^-) \approx \Gamma(\pi^+ \to {\rm e}^+ + \nu_{\rm e})$ and $\Gamma({\rm K}^0 \to \mu^+ + \mu^-) \approx \Gamma({\rm K}^+ \to \mu^+ + \nu_\mu)$. But nothing of the kind happens for the latter case, the rate $\Gamma({\rm K}^0_{\rm L} \to \mu^+ + \mu^-)$ is very suppressed, being $\approx 2.72 \times 10^{-9}\ \Gamma({\rm K}^+ \to \mu^+ + \nu_\mu)$. Another example of the strongly suppressed strangeness-changing neutral current is the rate of ${\rm K}^+ \to \pi^+ + {\rm e}^+ + {\rm e}^-$, which is much weaker than the rate of strangeness-changing charged current involved in ${\rm K}^+ \to \pi^0 + {\rm e}^+ + \nu_{\rm e}$ (Sect. 7.6). Obviously, there must exist a cancelation mechanism that *forbids strangeness-changing neutral current*, and at the same time *allows strangeness-changing charged current*. As explained in Chap. 9, these two constraints are realized by the Glashow–Iliopoulos–Maiani (GIM) mechanism, via the unitarity of the Cabibbo–Kobayashi–Maskawa (CKM) matrix. At the lowest order $G_{\rm F}$ tree-diagram level, flavors (strangeness, charm, bottom, top) are systematically conserved in neutral currents but generally not in charged currents.

The absence of ${\rm K}^0_{\rm L} \to \mu^+ + \mu^-$ at the tree diagram level is illustrated by Fig. 12.8b. The amplitudes of all flavor-changing neutral currents (FCNC) can only come from loop diagrams similar to the penguin loop considered in Chap. 11 where the gluon is replaced by the photon or the ${\rm Z}^0$. Compared to the charged current tree amplitude of order $G_{\rm F}$, these FCNC loop amplitudes are of the order of $G_{\rm F}\alpha_{\rm em}/\pi$, its computation is similar to (11.94).

But how about $\pi^0 \to {\rm e}^+ + {\rm e}^-$, the flavor-conserving neutral current process (Fig. 12.8a) which can occur at the tree level ? Unsuppressed by GIM, its weak decay rate could be similar to the usual $\pi^+ \to {\rm e}^+ + \nu_{\rm e}$. The reason why the existence of neutral currents was not suspected and the $\pi^0 \to {\rm e}^+ + {\rm e}^-$ mode – a typical manifestation of neutral current – was not actively searched for, is simply that electromagnetic interactions also govern this decay through the chain $\pi^0 \to \gamma + \gamma \to {\rm e}^+ + {\rm e}^-$. This electromagnetic transition dominates the weak decay $\pi^0 \to Z^0 \to e^+ + e^-$ by many orders of magnitude (Problem 12.4). Therefore $\pi^0 \to e^+ + e^-$, contaminated by electromagnetic interactions, is not a clean process for proving the existence of neutral currents.

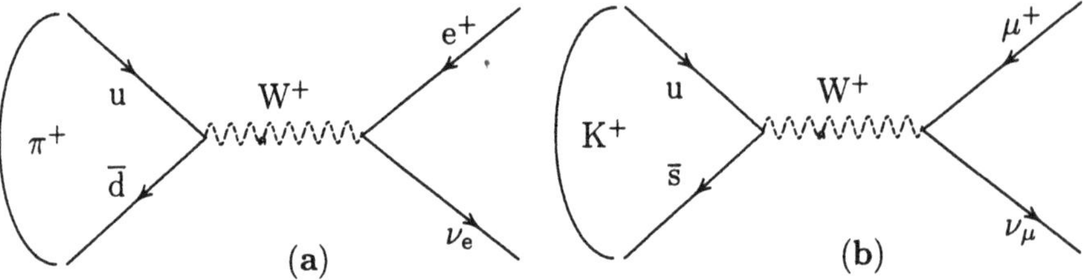


**Fig. 12.7.** Decays by charged currents: (**a**) $\pi^+ \to e^+ + \nu_e$ ; (**b**) $K^+ \to \mu^+ + \nu_\mu$

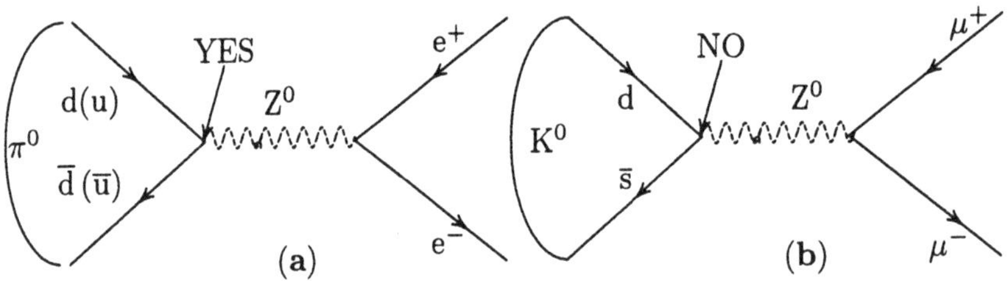


**Fig. 12.8.** Decays by neutral currents: (**a**) $\pi^0 \to e^+ + e^-$ flavor-conserving neutral current is allowed; (**b**) $K^0 \to \mu^+ + \mu^-$ flavor-changing neutral current is forbidden

From these considerations, we learn that at low energies, electromagnetic processes always dominate weak neutral current ones. For the latter to show up, one should consider reactions in which electromagnetic interactions are absent. We come to the crucial role of neutrinos in the discovery of weak neutral currents which consecrates the standard model. Since neutrinos are insensitive to electromagnetic forces, it suffices to observe the absence of charged leptons in neutrino scatterings to prove the existence of neutral currents. For example, $\nu_e + n \to e^- + p$ is due to charged currents but $\nu_e + p \to \nu_e + p$ can only come from neutral currents. More generally, in the scattering of neutrinos by a target T, if the cross-section $\sigma(\nu_\ell + T \to$ *without* $\ell^- + \cdots)$ is comparable to $\sigma(\nu_\ell + T \to$ *with* $\ell^- + \cdots)$ , then the existence of neutral currents is irrefutable. It was precisely how the latter were discovered at CERN by the Gargamelle collaboration in 1973, ten years before the neutral current carrier $Z^0$ was found at CERN and SLAC.

### 12.4.2 Neutrino–Electron Scattering

The scattering of neutrino by electron, a purely leptonic reaction, is difficult to observe since the cross-section, being proportional to the electron mass, is small ($\sigma \simeq 10^{-42}\text{cm}^2$). This experimental difficulty is compensated by clean theoretical treatments, since with pointlike leptons, theoretical treatments do not suffer from uncertainties due to weak form factors inherent to hadrons. From the neutrino–electron scattering, one can deduce the neutral-current properties, the Weinberg angle $\theta_W$, and the $W^{\pm}$ and $Z^0$ masses.

We consider the following reactions:

$$\nu_\mu + e^- \to \nu_\mu + e^- \quad , \quad \overline{\nu}_\mu + e^- \to \overline{\nu}_\mu + e^- \ , \tag{I}$$

$$\nu_e + e^- \to \nu_e + e^- \quad , \quad \overline{\nu}_e + e^- \to \overline{\nu}_e + e^- \ , \tag{II}$$

$$\nu_\mu + e^- \to \mu^- + \nu_e \ . \tag{III}$$

The reactions (I) are governed only by neutral currents (NC), in (III) are involved charged currents (CC), while both NC and CC participate in (II). The sources of $\overline{\nu}_e$ are mainly from nuclear reactors, their energies are in the MeV range, while the $\nu_\mu, \overline{\nu}_\mu$ mainly come from the decays of $\pi$ and K mesons produced by accelerators. Their energies can reach a few hundred GeV.

We start with the pure NC reaction $\nu_\mu(k_1) + e^-(p_1) \to \nu_\mu(k_2) + e^-(p_2)$; the corresponding Feynman diagram is similar to Fig. 12.3b. For non-forward scattering considered here, $k_1 \neq k_2$ and $p_1 \neq p_2$. The kinematics of two-body $\to$ two-body reactions is conveniently described by the Mandelstam variables

$$\begin{aligned} s &\equiv (k_1 + p_1)^2 = (k_2 + p_2)^2 \ , \\ t &\equiv (k_1 - k_2)^2 = (p_2 - p_1)^2 \ , \\ u &\equiv (k_1 - p_2)^2 = (k_2 - p_1)^2 \ , \end{aligned} \tag{12.33}$$

only two of which are independent since $s + t + u = \Sigma_j m_j^2 = 2m_e^2 + 2m_\nu^2$. In the following, we take $m_\nu = 0$ and put $m_e = m$.

In the center-of-mass system $\boldsymbol{k}_1 + \boldsymbol{p}_1 = \boldsymbol{0} = \boldsymbol{k}_2 + \boldsymbol{p}_2$, and $\boldsymbol{k}_1 \cdot \boldsymbol{k}_2 = |\boldsymbol{k}_1||\boldsymbol{k}_2| \cos\theta_{cm}$ , $\sqrt{s}$ is the total energy of the ingoing (or outgoing) particles, and $|\boldsymbol{k}_1| = |\boldsymbol{k}_2| = (s - m^2)/2\sqrt{s}$. The momentum transfer is denoted by $q_\mu = (k_1 - k_2)_\mu$, such that $t = q^2 = -|\boldsymbol{k}_1 - \boldsymbol{k}_2|^2$. We also write $Q^2 = -q^2 \geq 0$.

In the laboratory system for which the target electron is at rest, $\boldsymbol{p}_1 = \boldsymbol{0}$, we have $s = m^2 + 2mE_\nu$ where $E_\nu$ is the incoming neutrino energy, and $t = -2m(E_e - m) = -2mT_e$. $E_e$ is the outgoing electron energy. Another laboratory variable frequently used is $y \equiv T_e/E_\nu = -t/(s - m^2)$. The following relations may be useful :

$$\begin{aligned} &k_1 \cdot p_2 = k_2 \cdot p_1 = m(m + E_\nu - E_e) = mE_\nu(1 - y) = (s + t - m^2)/2 \ , \\ &k_1 \cdot p_1 = k_2 \cdot p_2 = mE_\nu - (s - m^2)/2 \ , \quad k_1 \cdot k_2 = -t/2 \ , \\ &Q^2 = 2mE_\nu y \quad ; \quad 0 \leq Q^2 \leq (s - m^2)^2/s \ ; \ 0 \leq y \leq 1 - m^2/s \ . \end{aligned} \tag{12.34}$$

The two-body $\to$ two-body cross-section always depends on two independent variables that can be chosen as $s$ and $t$, or $E_\nu$ and $y$ in the laboratory frame, or $\sqrt{s}$ and $\theta_{cm}$ in the center-of-mass system. Using the general formulas

(4.59) and (4.62), we have

$$\frac{\mathrm{d}\sigma}{\mathrm{d}\cos\theta_{\mathrm{cm}}} = \frac{1}{32\pi\, s}\left(\tfrac{1}{2}\sum_{\mathrm{spin}}|\mathcal{M}_{\mathrm{Z}}|^2\right),$$

$$\frac{\mathrm{d}\sigma}{\mathrm{d}Q^2} = \frac{1}{16\pi\,(s-m^2)^2}\left(\tfrac{1}{2}\sum_{\mathrm{spin}}|\mathcal{M}_{\mathrm{Z}}|^2\right). \tag{12.35}$$

The symbol $\frac{1}{2}\sum_{\mathrm{spin}}$ refers to the averaging over the incoming electron spins, since it is unnecessary to do spin averaging for the incoming left-handed neutrino which has only one helicity. The amplitude of $\nu_\mu(k_1) + \mathrm{e}^-(p_1) \to \nu_\mu(k_2) + \mathrm{e}^-(p_2)$ obtained from the Feynman rules is

$$\mathcal{M}_{\mathrm{Z}} = \mathrm{i}\left(\frac{-\mathrm{i}g}{2\sqrt{2}\cos\theta_{\mathrm{W}}}\right)^2 \overline{u}(k_2)\gamma^\mu(1-\gamma_5)u(k_1)$$
$$\times\ \frac{-\mathrm{i}(g_{\mu\nu} - q_\mu q_\nu/M_{\mathrm{Z}}^2)}{q^2 - M_{\mathrm{Z}}^2}\ \overline{u}(p_2)\gamma^\nu(g_{\mathrm{V}}^{\mathrm{e}} - g_{\mathrm{A}}^{\mathrm{e}}\gamma_5)u(p_1)$$
$$g_{\mathrm{V}}^{\mathrm{e}} = -\tfrac{1}{2} + 2\sin^2\theta_{\mathrm{W}}\,; \qquad g_{\mathrm{A}}^{\mathrm{e}} = -\tfrac{1}{2}\,. \tag{12.36}$$

The product $q_\mu\, \overline{u}(k_2)\gamma^\mu(1-\gamma_5)u(k_1)$ vanishes with massless neutrinos, leaving only the $g_{\mu\nu}$ to the $\mathrm{Z}^0$ propagator. With $G_{\mathrm{F}}/\sqrt{2} = g^2/8M_{\mathrm{Z}}^2\cos^2\theta_{\mathrm{W}}$,

$$\mathcal{M}_{\mathrm{Z}} = \frac{G_{\mathrm{F}}}{\sqrt{2}}\frac{\overline{u}(k_2)\gamma^\mu(1-\gamma_5)u(k_1)\ \overline{u}(p_2)\gamma_\mu(g_{\mathrm{V}}^{\mathrm{e}} - g_{\mathrm{A}}^{\mathrm{e}}\gamma_5)u(p_1)}{(1+Q^2/M_{\mathrm{Z}}^2)}, \tag{12.37}$$

so that

$$\tfrac{1}{2}\sum_{\mathrm{spin}}|\mathcal{M}_{\mathrm{Z}}|^2 = \frac{G_{\mathrm{F}}^2}{2(1+Q^2/M_{\mathrm{Z}}^2)^2}\ \left(\tfrac{1}{2}A_{\mu\rho}\,B^{\mu\rho}\right),$$

$$A_{\mu\rho} = \sum_{\mathrm{spin}}\overline{u}(p_2)\gamma_\mu(g_{\mathrm{V}}^{\mathrm{e}} - g_{\mathrm{A}}^{\mathrm{e}}\gamma_5)u(p_1)\ \overline{u}(p_1)\gamma_\rho(g_{\mathrm{V}}^{\mathrm{e}} - g_{\mathrm{A}}^{\mathrm{e}}\gamma_5)u(p_2)$$
$$= \mathrm{Tr}\left[\not{p}_2\gamma_\mu\not{p}_1\gamma_\rho[(g_{\mathrm{V}}^{\mathrm{e}})^2 + (g_{\mathrm{A}}^{\mathrm{e}})^2 - 2g_{\mathrm{V}}^{\mathrm{e}}g_{\mathrm{A}}^{\mathrm{e}}\gamma_5] + m^2[(g_{\mathrm{V}}^{\mathrm{e}})^2 - (g_{\mathrm{A}}^{\mathrm{e}})^2]\gamma_\mu\gamma_\rho\right],$$
$$B^{\mu\rho} = \sum_{\mathrm{spin}}\overline{u}(k_2)\gamma^\mu(1-\gamma_5)u(k_1)\ \overline{u}(k_1)\gamma^\rho(1-\gamma_5)u(k_2)$$
$$= 2\,\mathrm{Tr}\left[\not{k}_2\gamma^\mu\not{k}_1\gamma^\rho(1-\gamma_5)\right]. \tag{12.38}$$

Using the relation

$$\mathrm{Tr}[\gamma^\lambda\gamma^\mu\gamma^\sigma\gamma^\rho(a-b\gamma_5)] \times \mathrm{Tr}[\gamma_\alpha\gamma_\mu\gamma_\beta\gamma_\rho(c-d\gamma_5)]$$
$$= 32\left[ac(\delta_\alpha^\lambda\delta_\beta^\sigma + \delta_\alpha^\sigma\delta_\beta^\lambda) + bd(\delta_\alpha^\lambda\delta_\beta^\sigma - \delta_\alpha^\sigma\delta_\beta^\lambda)\right], \tag{12.39}$$

we obtain

$$A_{\mu\rho}\,B^{\mu\rho} = 64\left[(g_{\mathrm{V}}^{\mathrm{e}} + g_{\mathrm{A}}^{\mathrm{e}})^2(k_1\cdot p_1)(k_2\cdot p_2)\right.$$
$$\left.+(g_{\mathrm{V}}^{\mathrm{e}} - g_{\mathrm{A}}^{\mathrm{e}})^2(k_1\cdot p_2)(k_2\cdot p_1) + [(g_{\mathrm{A}}^{\mathrm{e}})^2 - (g_{\mathrm{V}}^{\mathrm{e}})^2]m^2(k_1\cdot k_2)\right]. \tag{12.40}$$

Putting (40) into (34), (35), and (38), we have

$$\frac{\mathrm{d}\sigma(\nu_\mu + \mathrm{e}^- \to \nu_\mu + \mathrm{e}^-)}{\mathrm{d}Q^2} = \frac{G_\mathrm{F}^2}{4\pi(s-m^2)^2} \frac{1}{(1+Q^2/M_\mathrm{Z}^2)^2} \times \tag{12.41}$$
$$\left[(g_\mathrm{V}^\mathrm{e}+g_\mathrm{A}^\mathrm{e})^2(s-m^2)^2 + (g_\mathrm{V}^\mathrm{e}-g_\mathrm{A}^\mathrm{e})^2(s-m^2-Q^2)^2 + 2[(g_\mathrm{A}^\mathrm{e})^2-(g_\mathrm{V}^\mathrm{e})^2]m^2Q^2\right] .$$

Neglecting $m^2 \ll s, Q^2$, the integrated cross-section becomes

$$\sigma \equiv \int_0^s \frac{\mathrm{d}\sigma}{\mathrm{d}Q^2}dQ^2 = \frac{G_\mathrm{F}^2\, s}{4\pi}\left\{\frac{(g_\mathrm{V}^\mathrm{e}+g_\mathrm{A}^\mathrm{e})^2}{(1+s/M_\mathrm{Z}^2)} + \right. \tag{12.42}$$
$$\left. + (g_\mathrm{V}^\mathrm{e}-g_\mathrm{A}^\mathrm{e})^2\frac{M_\mathrm{Z}^2}{s}\left[1+\frac{2M_\mathrm{Z}^2}{s}-\frac{2M_\mathrm{Z}^2}{s}\left(1+\frac{M_\mathrm{Z}^2}{s}\right)\log\left(1+\frac{s}{M_\mathrm{Z}^2}\right)\right]\right\} .$$

For $s \ll M_\mathrm{Z}^2$, we develop the logarithm term of (42) in powers of $s/M_\mathrm{Z}^2$, then in the first approximation, the cross-section depends linearly on $s$:

$$\sigma(\nu_\mu + \mathrm{e}^- \to \nu_\mu + \mathrm{e}^-) \approx \frac{G_\mathrm{F}^2\, s}{4\pi}\left[(g_\mathrm{V}^\mathrm{e}+g_\mathrm{A}^\mathrm{e})^2 + \frac{1}{3}(g_\mathrm{V}^\mathrm{e}-g_\mathrm{A}^\mathrm{e})^2\right] . \tag{12.43}$$

The $\mathrm{Z}^0$ propagator effect through $(1+Q^2/M_\mathrm{Z}^2)^{-2}$ in (41) is very important at high energies, since for $s \gg M_\mathrm{Z}^2$, the cross-section (42) ceases to increase with $s$, it bends down and tends asymptotically towards a constant

$$\lim_{s\to\infty} \sigma(\nu_\mu + \mathrm{e}^- \to \nu_\mu + \mathrm{e}^-) = \frac{G_\mathrm{F}^2 M_\mathrm{Z}^2}{2\pi}[(g_\mathrm{V}^\mathrm{e})^2 + (g_\mathrm{A}^\mathrm{e})^2] . \tag{12.44}$$

The physical significance of (43) and (44) is worth emphasizing. A cross-section cannot increase forever as a linear function of energy without violating the unitarity of the $S$-matrix. Based on the most general properties of the latter, Froissart and Martin show that a total cross-section – hence *a fortiori* an elastic cross-section considered here – cannot grow asymptotically faster than $(\log s)^2$. At low energies, the linear dependence of (43) on $s$ is only approximate; actually, at high energies the cross-section (44) tends to a constant in accordance with the asymptotic theorem (Froissart bound).

In the laboratory frame, at low energy $mE_\nu \ll M_\mathrm{Z}^2$, we neglect $Q^2/M_\mathrm{Z}^2$ and use (34), then (41) and (43) can be written as

$$\frac{\mathrm{d}\sigma(\nu_\mu + \mathrm{e}^- \to \nu_\mu + \mathrm{e}^-)}{\mathrm{d}y} = \frac{G_\mathrm{F}^2 m E_\nu}{2\pi}\left[(g_\mathrm{V}^\mathrm{e}+g_\mathrm{A}^\mathrm{e})^2 + (g_\mathrm{V}^\mathrm{e}-g_\mathrm{A}^\mathrm{e})^2(1-y)^2\right] ,$$

$$\sigma(\nu_\mu + \mathrm{e}^- \to \nu_\mu + \mathrm{e}^-) = \frac{G_\mathrm{F}^2 m E_\nu}{2\pi}\left[(g_\mathrm{V}^\mathrm{e}+g_\mathrm{A}^\mathrm{e})^2 + \tfrac{1}{3}(g_\mathrm{V}^\mathrm{e}-g_\mathrm{A}^\mathrm{e})^2\right] . \tag{12.45}$$

The $y$ distribution as well as the integrated cross-section enable us to extract $g_\mathrm{V}^\mathrm{e}, g_\mathrm{A}^\mathrm{e}$, i.e. $\sin^2\theta_\mathrm{W}$.

For the antineutrino $\overline{\nu}_\mu$ scattering $\overline{\nu}_\mu(k_1) + \mathrm{e}^-(p_1) \to \overline{\nu}_\mu(k_2) + \mathrm{e}^-(p_2)$, its cross-section can be deduced from $\nu_\mu(k_1) + \mathrm{e}^-(p_1) \to \nu_\mu(k_2) + \mathrm{e}^-(p_2)$ by a simple substitution $g_\mathrm{R}^2 \equiv (g_\mathrm{V}^\mathrm{e} + g_\mathrm{A}^\mathrm{e})^2 \leftrightarrow g_\mathrm{L}^2 \equiv (g_\mathrm{V}^\mathrm{e} - g_\mathrm{A}^\mathrm{e})^2$. This rule can be traced back to (37) for which the current $\overline{u}(k_2)\gamma^\lambda(1-\gamma_5)u(k_1)$ is replaced by $\overline{v}(k_1)\gamma^\lambda(1-\gamma_5)v(k_2)$, i.e. $k_1 \leftrightarrow k_2$, and the substitution $g_R^2 \leftrightarrow g_L^2$ comes from (40) in which the last term proportional to $m^2$ is neglected. Thus

$$\frac{\mathrm{d}\sigma(\overline{\nu}_\mu + \mathrm{e}^- \to \overline{\nu}_\mu + \mathrm{e}^-)}{\mathrm{d}y} = \frac{G_\mathrm{F}^2 m E_{\overline{\nu}}}{2\pi}\left[(g_\mathrm{V}^\mathrm{e} - g_\mathrm{A}^\mathrm{e})^2 + (g_\mathrm{V}^\mathrm{e} + g_\mathrm{A}^\mathrm{e})^2(1-y)^2\right] ,$$

$$\sigma(\overline{\nu}_\mu + \mathrm{e}^- \to \overline{\nu}_\mu + \mathrm{e}^-) = \frac{G_\mathrm{F}^2 m E_{\overline{\nu}}}{2\pi}\left[(g_\mathrm{V}^\mathrm{e} - g_\mathrm{A}^\mathrm{e})^2 + \tfrac{1}{3}(g_\mathrm{V}^\mathrm{e} + g_\mathrm{A}^\mathrm{e})^2\right] . \quad (12.46)$$

Numerically, with $G_\mathrm{F}^2 m_\mathrm{e} E_\nu = 27.05 \times 10^{-42}\mathrm{cm}^2(E_\nu/\ \mathrm{GeV})$, we get

$$\sigma(\nu_\mu + \mathrm{e}^- \to \nu_\mu + \mathrm{e}^-) = 4.3\,\frac{E_\nu}{\mathrm{GeV}}\left[(2\sin^2\theta_\mathrm{W} - 1)^2 + \tfrac{4}{3}\sin^4\theta_\mathrm{W}\right] 10^{-42}\mathrm{cm}^2,$$

$$\sigma(\overline{\nu}_\mu + \mathrm{e}^- \to \overline{\nu}_\mu + \mathrm{e}^-) = 4.3\,\frac{E_{\overline{\nu}}}{\mathrm{GeV}}\left[4\sin^4\theta_\mathrm{W} + \tfrac{1}{3}(2\sin^2\theta_\mathrm{W} - 1)^2\right] 10^{-42}\mathrm{cm}^2.$$

The ratio of the neutrino/antineutrino cross-sections, which is given by

$$R_\mathrm{N} \equiv \frac{\sigma(\nu_\mu + \mathrm{e}^- \to \nu_\mu + \mathrm{e}^-)}{\sigma(\overline{\nu}_\mu + \mathrm{e}^- \to \overline{\nu}_\mu + \mathrm{e}^-)} = 3\,\frac{1 - 4\sin^2\theta_\mathrm{W} + \frac{16}{3}\sin^4\theta_\mathrm{W}}{1 - 4\sin^2\theta_\mathrm{W} + 16\sin^4\theta_\mathrm{W}} ,$$

enables us to extract $\sin^2\theta_\mathrm{W}$. By this method, the electron detection efficiencies cancel in the ratio, and an absolute neutrino flux is not needed. Systematic errors are significantly reduced, resulting in an improvement of the determination of $\sin^2\theta_\mathrm{W} = 0.211 \pm 0.036 \pm 0.011$. From (44) and the rule $g_\mathrm{L}^2 \leftrightarrow g_\mathrm{R}^2$ for $\nu \leftrightarrow \overline{\nu}$, the ratio $R_\mathrm{N}$ tends to 1 as $s \to \infty$. Note that the equality holds independently of energy if $\sin^2\theta_\mathrm{W} = 0.25$, i.e. if $g_\mathrm{V}^\mathrm{e} = 0$.

Both neutral and charged currents contribute to reactions (II):

$$\nu_\mathrm{e}(k_1) + \mathrm{e}^-(p_1) \to \nu_\mathrm{e}(k_2) + \mathrm{e}^-(p_2) , \quad (\mathrm{II}.1)$$

$$\overline{\nu}_\mathrm{e}(k_1) + \mathrm{e}^-(p_1) \to \overline{\nu}_\mathrm{e}(k_2) + \mathrm{e}^-(p_2) . \quad (\mathrm{II}.2)$$

For (II.1), the diagrams are Fig. 12.3b and Fig. 12.4, associated respectively with the $\mathrm{Z}^0$ and W exchange in the $t$ and $u$ channels of the $t$ and $u$ variables defined in (33), i.e. their propagators are $(t - M_\mathrm{Z}^2)^{-1}$ and $(u - M_\mathrm{W}^2)^{-1}$ respectively. The amplitudes are referred to as $\mathcal{M}_\mathrm{Z}$ and $\mathcal{M}_\mathrm{W}$. Since both contribute to the reaction (II.1), their relative sign is important and turns out to be negative. To see how it arises, it may be convenient to go back to the second quantization of the fields that enter the composition of $\mathcal{M}_\mathrm{Z}$ and $\mathcal{M}_\mathrm{W}$. The latter are obtained from the products of the fermionic creation and destruction operators which yield the initial and final states when applied to the vacuum state $|0\rangle$. Since these operators anticommute, their relative order is important. At $g^2$, they come from the time-ordered product $T[H(x)H(y)]$. To determine their relative sign, we consider the combinations

$$\text{for } \mathcal{M}_Z : b_e^\dagger(p_2)b_e(p_1)b_\nu^\dagger(k_2)b_\nu(k_1) \text{ coming from } \overline{\psi}_e(x)\psi_e(x)\overline{\psi}_\nu(y)\psi_\nu(y)\ ,$$

$$\text{for } \mathcal{M}_W : b_e^\dagger(p_2)b_\nu(k_1)b_\nu^\dagger(k_2)b_e(p_1) \text{ coming from } \overline{\psi}_e(x)\psi_\nu(x)\overline{\psi}_\nu(y)\psi_e(y)\ ,$$

where $b^\dagger$ $(b)$ is the creation (destruction) fermionic operator (Chap. 3). In writing these amplitudes, we keep only in $T[H(x)H(y)]$ the order of the fermion fields. In $\mathcal{M}_Z$, using the anticommutation relations of $b_j^\dagger, b_j$, we have

$$b_e^\dagger b_e b_\nu^\dagger b_\nu = +b_e^\dagger b_\nu^\dagger b_\nu b_e = -b_e^\dagger b_\nu b_\nu^\dagger b_e\ .$$

The extreme left (right) member of the above equation is related to $\mathcal{M}_Z$ ($\mathcal{M}_W$), so that the relative sign between $\mathcal{M}_Z$ and $\mathcal{M}_W$ is definitely $-1$. The expression of $\mathcal{M}_Z(\nu_e + e^- \to \nu_e + e^-)$ is identical to that of $\mathcal{M}_Z(\nu_\mu + e^- \to \nu_\mu + e^-)$ given above in (37). According to Feynman rules, the amplitude $\mathcal{M}_W[\nu_e(k_1) + e^-(p_1) \to \nu_e(k_2) + e^-(p_2)]$ is

$$\begin{aligned}\mathcal{M}_W &= \frac{G_F}{\sqrt{2}}\frac{\overline{u}(k_2)\gamma^\mu(1-\gamma_5)u(p_1)\,\overline{u}(p_2)\gamma_\mu(1-\gamma_5)u(k_1)}{1-u/M_W^2}\\ &= \frac{-G_F}{\sqrt{2}}\frac{\overline{u}(k_2)\gamma^\mu(1-\gamma_5)u(k_1)\,\overline{u}(p_2)\gamma_\mu(1-\gamma_5)u(p_1)}{1-u/M_W^2}\ ,\end{aligned}\tag{12.47}$$

after a Fierz rearrangement (Appendix). For low neutrino energy, we may neglect $-t/M_Z^2$ and $-u/M_W^2$ in (37) and (47). The relative minus sign can be conventionally put into $\mathcal{M}_Z$, so the total amplitude of the reaction (II.1) is $\mathcal{M} = \mathcal{M}_W - \mathcal{M}_Z$. Combining (37) and the second line of (47), we get

$$\begin{aligned}&\mathcal{M} = \frac{-G_F}{\sqrt{2}}\overline{u}(k_2)\gamma^\mu(1-\gamma_5)u(k_1)\,\overline{u}(p_2)\gamma_\mu(g'_V - g'_A\gamma_5)u(p_1)\ ,\\ &g'_V = 1 + g_V^e = +\tfrac{1}{2} + 2\sin^2\theta_W\ ;\ \ g'_A = 1 + g_A^e = +\tfrac{1}{2}\ .\end{aligned}\tag{12.48}$$

The forward amplitude $\mathcal{M}(E_\nu, q^2 = 0)$ of (II.1) can be readily obtained from (48) by putting $k_1 = k_2$, $p_1 = p_2$ and we recover (19) after summing and averaging over the electron spin states. The cross-section is now readily obtained using (45) and (46) as examples. We have

$$\begin{aligned}\frac{d\sigma(\nu_e + e^- \to \nu_e + e^-)}{dy} &= \frac{G_F^2 m E_\nu}{2\pi}\left[(g'_V + g'_A)^2 + (g'_V - g'_A)^2(1-y)^2\right]\\ &= \frac{G_F^2 m E_\nu}{2\pi}\left[(g_V^e + g_A^e + 2)^2 + (g_V^e - g_A^e)^2(1-y)^2\right]\ ,\end{aligned}$$

$$\sigma(\nu_e + e^- \to \nu_e + e^-) = \frac{G_F^2 m E_\nu}{2\pi}\left[(g_V^e + g_A^e + 2)^2 + \frac{1}{3}(g_V^e - g_A^e)^2\right]\ .\tag{12.49}$$

The amplitude of the reaction (II.1) in (48) is to be compared with (37) of the reactions (I). The cross-section of (II.2) is deduced from (49) by the substitution $g'_A \leftrightarrow -g'_A$ or $1 \leftrightarrow (1-y)^2$:

$$\frac{d\sigma(\overline{\nu}_e + e^- \to \overline{\nu}_e + e^-)}{dy} = \frac{G_F^2 m E_\nu}{2\pi}\left[(g_V^e - g_A^e)^2 + (g_V^e + g_A^e + 2)^2(1-y)^2\right] ,$$

$$\sigma(\overline{\nu}_e + e^- \to \overline{\nu}_e + e^-) = \frac{G_F^2 m E_\nu}{2\pi}\left[(g_V^e - g_A^e)^2 + \frac{1}{3}(g_V^e + g_A^e + 2)^2\right] . \quad (12.50)$$

Finally, the scattering amplitude of the pure charged currents reaction (III), $\nu_\mu(k_1) + e^-(p_1) \to \mu^-(p_2) + \nu_e(k_2)$ is similar to the $\mu^- \to e^- + \nu_\mu + \overline{\nu}_e$ decay. It is given by

$$\mathcal{M}(\nu_\mu + e^- \to \mu^- + \nu_e) = \frac{G_F}{\sqrt{2}}\overline{u}(p_2)\gamma^\mu(1-\gamma_5)u(k_1)\,\overline{u}(k_2)\gamma_\mu(1-\gamma_5)u(p_1) .$$

Using (45) with $g_V^e = g_A^e = 1$, the corresponding cross-section is

$$\sigma(\nu_\mu + e^- \to \nu_e + \mu^-) = \frac{2G_F^2 m E_\nu}{\pi}\left[1 - \frac{m_\mu^2}{2mE_\nu}\right]^2 . \quad (12.51)$$

The last factor $(1 - m_\mu^2/2mE_\nu)^2$ is purely kinematic. Comparing the above equation with (45), the ratio of NC over CC cross-sections is

$$R_{\mathrm{NC/CC}} \equiv \frac{\sigma(\nu_\mu + e^- \to \nu_\mu + e^-)}{\sigma(\nu_\mu + e^- \to \nu_e + \mu^-)} = \frac{\frac{1}{4} - \sin^2\theta_W + \frac{4}{3}\sin^4\theta_W}{[1 - (m_\mu^2/2mE_\nu)]^2} .$$

The integrated cross-sections of the reactions (I), (II) as given by (45), (46), (49), (50) can be represented in the $(g_V^e, g_A^e)$ plane by four ellipses. Their intersections give two solutions for $g_V^e, g_A^e$ since the equations are symmetric by $(g_V^e, g_A^e) \longleftrightarrow -(g_V^e, g_A^e)$. Precise measurements of the purely leptonic cross-sections come from the CHARM II collaboration at CERN, which gives $\sin^2\theta_W = 0.2324 \pm 0.012$.

With this value of $\sin^2\theta_W$, the gauge boson masses $W^\pm$ and $Z^0$ could be estimated long before their discoveries. Using the formulas in Chap. 9, we get

$$M_W^2 = \frac{\pi\alpha_{em}}{\sqrt{2}G_F\sin^2\theta_W} \longrightarrow M_W \approx 77.34 \text{ GeV} \ , \quad M_Z \approx 88.12 \text{ GeV} .$$

Electroweak corrections at one-loop level will increase these tree-level masses by about 0.038%. The corrected masses agree very well with the experimental data, $M_W = 80.33 \pm 0.15$ GeV, and $M_Z = 91.187 \pm 0.007$ GeV.

From these studies of neutrino–electron scatterings, we draw another important conclusion: The *linear rise* of the cross-section with $E_\nu$ is characteristic of the neutrino scattered by a pointlike fermion. If the target has structure, its cross-section cannot increase at large $E_\nu$.

## 12.5 Neutrino–Nucleon Elastic Scattering

As another example of the effects of the target structure, we consider the neutrino–nucleon scattering

$$\nu_\mu + \mathrm{N} \to \mu^- + \mathrm{N}' \quad , \quad \overline{\nu}_\mu + \mathrm{N} \to \mu^+ + \mathrm{N}' .$$

The study of these reactions enables us to familiarize ourself with the two important properties of the flavor-conserving V – A charged weak current $V_{\mathrm{ud}}\, \overline{u}\, \gamma_\mu (1-\gamma_5)\, d$, to wit, the conserved vector current (CVC) and the partial conservation of the axial current (PCAC) which are natural consequences of the standard electroweak model.

The amplitude $\nu_\mu(k_1) + \mathrm{n}(p_1) \to \mu^-(k_2) + \mathrm{p}(p_2)$ can be obtained from that of $\nu_\mu(k_1) + \mathrm{e}^-(p_1) \to \mu^-(k_2) + \nu_\mathrm{e}(p_2)$ by replacing the pointlike e–$\nu_\mathrm{e}$ current $\overline{u}(p_2)\gamma_\mu(1-\gamma_5)u(p_1)$ by the nucleon n–p current:

$$V_{\mathrm{ud}} \langle \mathrm{p}(p_2) \,|\, V_\mu - A_\mu \,|\, \mathrm{n}(p_1) \rangle = V_{\mathrm{ud}}\, \overline{u}(p_2) \Big\{ \gamma_\mu f_1(q^2) + \frac{\mathrm{i}\sigma_{\mu\nu} q^\nu}{2M} f_2(q^2) \\ - g_1(q^2)\gamma_\mu\gamma_5 - g_3(q^2) \frac{q_\mu}{M} \gamma_5 \Big\} u(p_1) \, , \quad (12.52)$$

where $M$ is the nucleon mass and $q_\mu = (p_2 - p_1)_\mu$.

The nucleon form factors of $V_\mu$ are denoted by $f_{1,2}(q^2)$, those of the $A_\mu$ by $g_{1,3}(q^2)$. They are real by time-reversal invariance. From considerations of Lorentz covariance alone, the most general matrix element of $V_\mu - A_\mu$ has six form factors, three for $V_\mu$ and three for $A_\mu$. Since form factors are induced by strong interactions which conserve G-parity (Chap. 6), only the terms even by G-parity transformations are kept. The four form factors in (52) satisfying this condition are said to be of the first class, following Weinberg. On the other hand, the two other terms odd under G-parity $f_3(q^2)$ and $g_2(q^2)$ respectively proportional to $q_\mu$ and $\mathrm{i}\sigma_{\mu\nu}q^\nu\gamma_5$ (second class current), are discarded. In any case, the $q_\mu$ term does not contribute if the current $V_\mu$ is conserved, i.e. if $q^\mu V_\mu = 0$ [see also (10.12)].

According to the CVC hypothesis postulated by Feynman and Gell-Mann, the vector part $V_\mu$ of the weak charged current, its Hermitian conjugate $V_\mu^\dagger$, and the isospin $I = 1$ component of the electromagnetic current, form an isotriplet.

Now CVC is a direct consequence of the isospin structure of the weak charged current $\overline{u}\gamma_\mu d$ in the standard model. Indeed with the doublet $q$ for the $u, d$ quark fields, the three currents: $V_\mu = \overline{q}\gamma_\mu\tau^+ q = \overline{u}\gamma_\mu d$, $V_\mu^\dagger = \overline{q}\gamma_\mu\tau^- q = \overline{d}\gamma_\mu u$, and $J_\mu^{\mathrm{em}}(I=1) = \frac{1}{2}\overline{q}\gamma_\mu\tau^3 q = \frac{1}{2}(\overline{u}\gamma_\mu u - \overline{d}\gamma_\mu d)$ are the three components of an isovector (Chap. 9). CVC implies that the weak form factors $f_{1,2}(q^2)$ are equal to the electromagnetic form factors $F^1_{1,2}(q^2)$ which are given by (10.13) and (10.40) from electron–nucleon elastic scattering.

The contribution of $g_3(q^2)$ is proportional to the muon mass and can be neglected. We take $m_\mu = 0$ in the following. The contributions of

$f_1(q^2)$, $g_1(q^2)$ to the differential cross-section can be obtained from (41) with the replacement $g_V^e \longrightarrow f_1$, $g_A^e \longrightarrow g_1$. We only have to compute the contributions of $f_2(q^2)$ and get ($Q^2 = -q^2 > 0$)

$$\frac{d\sigma}{dQ^2} = \frac{G_F^2|V_{ud}|^2}{4\pi}\left[(f_1+g_1)^2 + (f_1-g_1)^2\left(1-\frac{Q^2}{2ME_\nu}\right)^2 + (g_1^2 - f_1^2)\frac{Q^2}{2E_\nu^2}\right.$$
$$\left. + f_2^2\frac{Q^2}{2M^2}\left(1-\frac{Q^2}{2ME_\nu}+\frac{Q^2}{4E_\nu^2}\right) + f_1 f_2\frac{Q^4}{2M^2E_\nu^2} + g_1 f_2\left(\frac{2Q^2}{ME_\nu} - \frac{Q^4}{2M^2E_\nu^2}\right)\right].$$

To obtain the antineutrino–nucleon cross-section $\sigma(\overline{\nu}_\mu + p \to \mu^+ + n)$, we follow the discussions preceding (46) and simply replace $g_1(q^2)$ by $-g_1(q^2)$ in the above equation. The value of the differential cross-section at $q^2 = 0$ is *independent* of the incident neutrino energy and takes a simple form

$$\left.\frac{d\sigma(\nu_\mu + n \to \mu^- + p)}{dQ^2}\right|_{q^2=0} = \frac{G_F^2|V_{ud}|^2}{2\pi}\left[f_1^2(0) + g_1^2(0)\right] .$$

As stated, the form factors $f_{1,2}(q^2)$ are equal to $F_{1,2}^1(q^2)$ [cf. (10.40)]:

$$f_1(q^2) = F_1^1(q^2)\ ;\ f_2(q^2) = F_2^1(q^2)\ ;\text{hence } f_1(0) = 1\ ;\ f_2(0) = 3.7 .$$

Therefore measurements of the neutrino–nucleon differential cross-section $d\sigma/dq^2$ enable us to determine the remaining $g_1(q^2)$, in particular $g_1(0)$. The value $g_1(0) \approx 1.25$ can also be determined from neutron $\beta$-decay in which the same flavor-conserving charged current is involved (Problem 13.6). Experiments show that the $q^2$ dependence of $g_1(q^2)$ is of the dipole type, $g_1(q^2) = 1.25 \times \left(1 - q^2/M_A^2\right)^{-2}$, with a pole mass $M_A \approx 0.95$ GeV.

**PCAC and the Goldberger–Treiman Relation.** The special case of zero momentum transfer $q^\mu = 0$ is illuminating. At $q^2 = 0$, the matrix element of the nucleon in (52) looks like the pointlike V – A quark current $V_{ud}\,\overline{u}\gamma_\mu(1-\gamma_5)d$, the only change is in the form factor $g_1(0)$, which shifts to 1.25 from 1. Indeed, from (52) with $q^\mu = 0$, we have

$$V_{ud}\,\langle p(k)\,|\,V_\mu - A_\mu\,|\,n(k)\rangle = V_{ud}\,\overline{u}(k)\,\gamma_\mu\,[1 - g_1(0)\gamma_5]u(k) .$$

We know from CVC that $f_1(0)$ must be equal to 1, i.e. at $q^2 = 0$ the vector form factor $f_1(q^2)$ is not renormalized by the strong interaction. The pointlike vector coupling of quarks is exactly reflected on the hadronic level at $q^2 = 0$, because the hadronic vector current is conserved, i.e. $q^\mu V_\mu = 0$.

We would like to show that the value of $g_1(0) \approx 1.25$ has something to do with the partial conservation of the axial current (PCAC) which is a natural consequence of the small u, d quark mass $m$, $q^\mu A_\mu = 2m\ \overline{u}\gamma_5 d$. Let us examine the consequences of the massless quark ($m = 0$) or equivalently of the conserved axial current. For that, we multiply the left- and the right-hand sides of (52) with $q^\mu$. Thus,

$$q^\mu \langle \mathrm{p}(k_2)\,|\, A_\mu \,|\, \mathrm{n}(p_1)\rangle = \overline{u}(k_2)\left[2M\, g_1(q^2) + \frac{q^2}{M} g_3(q^2)\right]\gamma_5\, u(p_1)\,.$$

If the axial current is conserved, i.e. $q^\mu A_\mu = 0$, then

$$g_1(0) = \lim_{q^2\to 0} \frac{-q^2}{2M^2}\, g_3(q^2)\,. \tag{12.53}$$

Since $g_1(0) \neq 0$, the form factor $g_3(q^2)$ must have a pole at $q^2 = 0$ to cancel the numerator $q^2$. Such a pole implies the presence of a physical massless particle. There is one available, the nearly massless pion considered as a Goldstone–Nambu boson in the context of massless up and down quarks. The fact that the form factor $g_3(q^2)$ has a pion pole is clearly indicated in Fig. 12.9a, from which we derive

$$\overline{u}(k_2)\, g_3(q^2)\frac{q^\mu\gamma_5}{M}\, u(p_1)\, W_\mu = \left\{\sqrt{2} g_{\pi\mathrm{NN}}\overline{u}(k_2)\,\gamma_5\, u(p_1)\right\}\frac{\mathrm{i}}{q^2 - m_\pi^2}\,[\mathrm{i} f_\pi q^\mu]\, W_\mu.$$

In the above equation, $\sqrt{2}\, g_{\pi\mathrm{NN}}$ is the charged pion–proton–neutron coupling constant in the effective pion–nucleon interaction $g_{\pi\mathrm{NN}}\overline{N}\,\gamma_5\,\tau^k\, N\phi_k$, where $\phi_k(x)$ is the pion field with the isospin index $k = 1,2,3$ [see (6.57–58)]. We then deduce

$$g_3(q^2) = \lim_{m_\pi^2\to 0} \frac{-\sqrt{2} f_\pi M g_{\pi\mathrm{NN}}}{q^2 - m_\pi^2}\,.$$

Together with (53), one gets the Goldberger–Treiman (GT) relation which gives $g_1(0)$ in terms of $g_{\pi\mathrm{NN}}$ and the pion decay constant $f_\pi \approx 131$ MeV:

$$g_1(0) = \frac{f_\pi g_{\pi\mathrm{NN}}}{\sqrt{2} M}\,, \quad \text{GT relation}\,. \tag{12.54}$$

With $g_{\pi\mathrm{NN}}^2/(4\pi) \approx 13.5$, the GT relation is satisfied to 5% accuracy. PCAC is also written in a form which says that we may use $\partial^\mu A_\mu^k(x)$ to interpolate the pion field $\phi^k(x)$:

$$\partial^\mu A_\mu^k(x) - \frac{f_\pi m_\pi^2}{\sqrt{2}}\,\phi^k(x)\,, \quad \text{from } \langle 0\,|\, A_\mu^k \,|\, \pi^j(q)\rangle = \frac{\mathrm{i} f_\pi}{\sqrt{2}}\, q_\mu \delta^{kj}\,. \tag{12.55}$$

The above equation also tells us that the axial current is conserved in the limit $m_\pi \to 0$ of the Goldstone–Nambu massless pion.

We cannot leave PCAC without emphasizing that the conservation of the axial current with massless quarks is only valid at the tree level. Due to quantum effects (illustrated by the triangle loop in Fig. 12.9b,c), even with massless quarks, for the isospin component $k = 3$ associated with $\pi^0$, the $\partial^\mu A_\mu^{k=3}$ no longer vanishes. In the presence of electromagnetic interactions, the conservation of both vector and axial currents is incompatible by loop corrections. To maintain gauge invariance (conservation of the vector current), we are led to $\partial^\mu A_\mu^3 = (e^2/16\pi^2)\varepsilon^{\alpha\beta\rho\sigma} F_{\alpha\beta} F_{\rho\sigma}$ where $F_{\alpha\beta}$ is the electromagnetic field tensor defined in (2.132). This is called the Adler–Bell–Jackiw (ABJ) anomaly, which has a number of remarkable consequences, the most famous being the decay $\pi^0 \to 2\gamma$ for which the three colors of quarks exhibit their glaring evidence (see Further Reading).

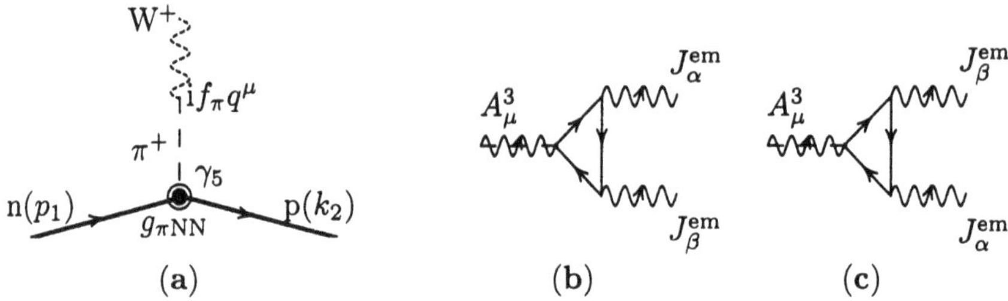


**Fig. 12.9.** (**a**) Pion pole dominance of the axial current ; (**b-c**) the ABJ anomaly of the axial–vector–vector currents, related to the $\pi^0 \to 2\gamma$ decay

Another case of interest in the neutrino–nucleon cross-section is the high energy limit $2ME_\nu \gg Q^2$. The differential cross-section decreases quickly with $Q^2$ as the square of dipole form factors:

$$\frac{\mathrm{d}\sigma(\nu_\mu + \text{n} \to \mu^- + \text{p})}{\mathrm{d}Q^2} = \frac{G_\text{F}^2 |V_\text{ud}|^2}{2\pi} \left[ f_1^2(q^2) + g_1^2(q^2) + \frac{Q^2}{4M^2} f_2^2(q^2) \right] ,$$

and the integrated cross-section $\sigma$ is a constant $\sim G_\text{F}^2 \Lambda^2 / 3\pi$, where $\Lambda \approx 1$ GeV is the pole mass of the form factors. With pointlike targets, $\sigma$ is linearly rising with $E_\nu$; the contrast is striking. Numerically, this exclusive cross-section $\sigma(\nu_\mu + \text{n} \to \mu^- + \text{p}) \lesssim 10^{-38}$ cm$^2$ constitutes a tiny portion of the inclusive $\sigma(\nu_\mu + \text{N} \to \mu^- + X)$ which we consider in the following section.

## 12.6 Neutrino–Nucleon Deep Inelastic Collision

One of the most dramatic evidences for quarks as fundamental constituents of hadrons is provided by data on deep inelastic neutrino–nucleon, its cross-section shows up as a *linearly rising* function of the incident neutrino energy. From experiments at CERN, FermiLab, and Serpukhov, with $E_\nu$ ranging over two orders of magnitude, from 2 to 260 GeV, the neutrino and antineutrino cross-sections continue to increase impressively (Fig. 12.10). This behavior is what we expect from neutrino scattering by a pointlike particle, illustrated by the previous study with the target electron. We remark that for $E_\nu \sim 300$ GeV, $s \sim 2ME_\nu \ll M_\text{W}^2$, so that the linear approximation of the cross-section is still valid and the propagator effect of the W boson can be neglected.

### 12.6.1 Deep Inelastic Cross-Section

Deep inelastic neutrino–nucleon cross-section can be easily transcribed from that of deep inelastic electron–nucleon scattering; the photon exchange of the latter is replaced by the weak boson $W^{\pm}$ or $Z^0$ of the former, depending on whether $\nu_\ell + N \to \ell^- + X$ or $\nu_\ell + N \to \nu_\ell + X$ process is observed. For definiteness, let us concentrate on charged current deep inelastic scattering (Fig. 12.11). The cross-section $\nu_\mu(p) + N(P) \to \mu^-(p') + X$ can be obtained from $e(p) + N(P) \to e(p') + X$ by (10.41) by the replacements

$$\text{couplings and propagators : } \frac{e^2}{q^2} \longrightarrow \left(\frac{g}{2\sqrt{2}}\right)^2 \frac{1}{q^2 - M_W^2} = \frac{-G_F}{\sqrt{2}(1+\frac{Q^2}{M_W^2})} ,$$

$$\text{lepton vertex : } \bar{\ell}\gamma^\mu \ell \longrightarrow \bar{\ell}\gamma^\mu(1-\gamma_5)\nu_\ell ,$$

$$\text{hadron vertex : } J_\mu^{\text{em}} \equiv \bar{q}\gamma_\mu q \longrightarrow V_{Qq}(V_\mu - A_\mu) \equiv V_{Qq}\overline{Q}\gamma_\mu(1-\gamma_5)q . \quad (12.56)$$

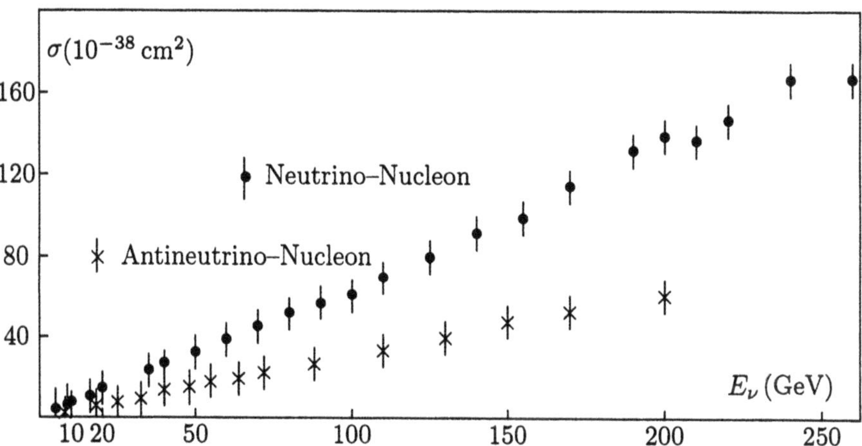


**Fig. 12.10.** $\sigma(\nu_\mu + N \to \mu^- + X)$ , $\sigma(\overline{\nu}_\mu + N \to \mu^+ + X)$

Taking the square of the matrix element, the leptonic tensor $l^{\mu\nu}$ in (10.28) becomes now $\widetilde{l}^{\mu\nu}$. We have

$$\begin{aligned} \widetilde{l}^{\mu\nu}(p,p') &= 2\,\mathrm{Tr}[\not{p}'\gamma^\mu \not{p}\gamma^\nu(1-\gamma_5)] \\ &= 8\big(p^\mu p'^\nu + p^\nu p'^\mu - g^{\mu\nu} p\cdot p' - \mathrm{i}\,\varepsilon^{\mu\nu\alpha\beta} p_\alpha p'_\beta\big) . \end{aligned} \quad (12.57)$$

Compared with $l^{\mu\nu}$ in (10.28), we note the absence of the factor $(\frac{1}{2})$ of the spin average before the trace of $\widetilde{l}^{\mu\nu}$, since the incoming neutrino has only one helicity, contrary to the electron in $l^{\mu\nu}$. We have a factor of 2 because of $(1-\gamma_5)^2 = 2(1-\gamma_5)$ in $\widetilde{l}^{\mu\nu}$. For antineutrino $\overline{\nu}_\mu(p) + N(P) \to \mu^+(p') + X$,

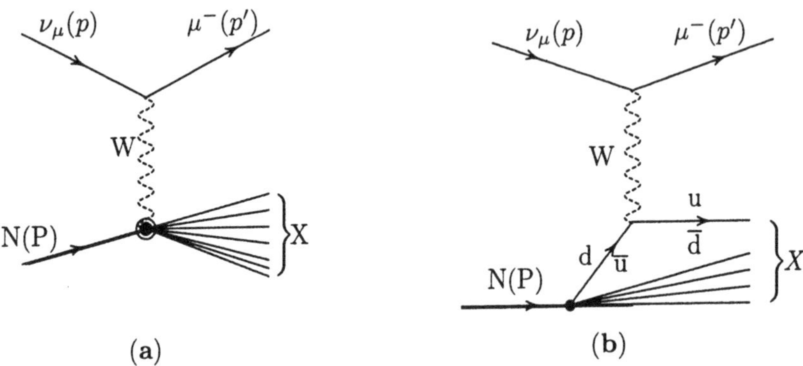


**Fig. 12.11.** (**a**) Deep inelastic neutrino–nucleon scattering by charged current; (**b**) $\nu_\mu + \mathrm{d}(\text{or } \overline{\mathrm{u}}) \to \mu^- + \mathrm{u}(\text{or } \overline{\mathrm{d}})$ at the parton level

the modification in $\widetilde{l}^{\mu\nu}(p,p')$ is the interchange $p \leftrightarrow p'$. Hence we have $+\mathrm{i}\,\varepsilon^{\mu\nu\alpha\beta}p_\alpha p'_\beta$ in the corresponding $\widetilde{l}^{\mu\nu}$ of antineutrinos. As for the dimensionless hadronic tensor $\widetilde{W}_{\mu\nu}(P,q)$ – defined analogously to $W_{\mu\nu}(P,q)$ in (10.42) with the replacement (56) – it has both symmetric and antisymmetric parts due to the parity violating V×A of the weak currents. Contrary to electromagnetic currents which are conserved, weak currents are not: $q^\mu \widetilde{W}_{\mu\nu}(P,q) \neq 0$, therefore $\widetilde{W}_{\mu\nu}(P,q)$ has the maximum number of Lorentz-invariant terms. Following (10.43), we write

$$\widetilde{W}_{\mu\nu}(P,q) = 4\pi\Big\{-g_{\mu\nu}\widetilde{W}_1(q^2,\nu) + \frac{P_\mu P_\nu}{M^2}\widetilde{W}_2(q^2,\nu) - \frac{\mathrm{i}\,\varepsilon_{\mu\nu\alpha\beta}}{2M^2}P^\alpha q^\beta \widetilde{W}_3(q^2,\nu)$$
$$+ \frac{q_\mu q_\nu}{M^2}\widetilde{W}_4(q^2,\nu) + \frac{P_\mu q_\nu + P_\nu q_\mu}{2M^2}\widetilde{W}_5(q^2,\nu) + \frac{\mathrm{i}\,(P_\mu q_\nu - P_\nu q_\mu)}{2M^2}\widetilde{W}_6(q^2,\nu)\Big\}\,.$$

When $\widetilde{W}_{\mu\nu}(P,q)$ is multiplied by the leptonic tensor $\widetilde{l}^{\mu\nu}(p,p')$, the antisymmetric $(P_\mu q_\nu - P_\nu q_\mu)$ factor of $\widetilde{W}_6$ vanishes when contracted with $\varepsilon^{\mu\nu\alpha\beta}$ (because only three of the vectors $p, p', q, P$ are independent). The other factors involved in $\widetilde{W}_{4,5}$ are proportional to the squared mass of the muon, and can be neglected. In the following, we take $m_\mu = 0$. Only three structure functions $\widetilde{W}_{1,2,3}(q^2,\nu)$ remain in the $\nu$–N cross-section, instead of two $W_{1,2}(q^2,\nu)$ in the electron–nucleon deep inelastic cross-section. Using the general formula (10.41) together with the replacements (56), the neutrino deep inelastic cross-section $\mathrm{d}\sigma_{\text{in}}$ is given by

$$\mathrm{d}\sigma_{\text{in}}(\nu_\mu + \mathrm{N} \to \mu^- + X) = \frac{1}{2(s-M^2)}\,\frac{G_\mathrm{F}^2}{2}\,\frac{1}{(1+\frac{Q^2}{M_W^2})^2}\,\widetilde{l}^{\mu\nu}\widetilde{W}_{\mu\nu}\frac{\mathrm{d}^3p'}{(2\pi)^3\,2E_{p'}}\,.$$

In the laboratory system $P = (M, \mathbf{0})$ , $p = (E = |\boldsymbol{p}|, \boldsymbol{p})$ , $p' = (E' = |\boldsymbol{p}'|, \boldsymbol{p}')$, $q^2 = (p-p')^2 = -2EE'\sin^2\frac{\theta}{2}$, $\nu \equiv P\cdot q/M = (E - E')$, we find

$$\widetilde{l}^{\mu\nu}\widetilde{W}_{\mu\nu} = 64\pi EE'\left\{2\widetilde{W}_1\sin^2\frac{\theta}{2} + \widetilde{W}_2\cos^2\frac{\theta}{2} + \widetilde{W}_3\frac{E+E'}{M}\sin^2\frac{\theta}{2}\right\}.$$

For $Q^2 \ll M_W^2$, we can neglect the W boson propagator effects and obtain

$$\frac{d\sigma^{\nu,\overline{\nu}}}{dQ^2 d\nu} = \frac{G_F^2}{2\pi M}\frac{E'}{E}\left[2\widetilde{W}_1(q^2,\nu)\sin^2\frac{\theta}{2} + \widetilde{W}_2(q^2,\nu)\cos^2\frac{\theta}{2} \right.$$
$$\left. \pm \widetilde{W}_3(q^2,\nu)\frac{E+E'}{M}\sin^2\frac{\theta}{2}\right] . \quad (12.58)$$

From the $p \leftrightarrow p'$ interchange mentioned above, the $+(-)$ sign corresponds to $\nu_\mu(\overline{\nu}_\mu)$ . Like the electron scattering in (10.65), the neutrino cross-section may be conveniently recast in terms of the Bjorken variable $x$ and the energy loss variable $y = \nu/E$. With $dQ^2 d\nu = 2ME\,\nu\, dx\, dy$, (58) becomes

$$\frac{d\sigma^{\nu,\overline{\nu}}}{dx dy} = \frac{G_F^2 ME}{\pi}\left[\widetilde{W}_1(x,q^2)x\,y^2 + \frac{\nu}{M}\widetilde{W}_2(x,q^2)\,(1-y) \right.$$
$$\left. \pm \frac{\nu}{M}\widetilde{W}_3(x,q^2)\,x\,y\left(1-\frac{y}{2}\right)\right] . \quad (12.59)$$

### 12.6.2 Quarks as Partons

We immediately see in (59) that, when $q^2$ becomes very large and the structure functions $\widetilde{W}(x,q^2)$ do not vanish, the deep inelastic $\nu$–N cross-section rises linearly with the neutrino energy $E$, exactly as if the neutrino were hitting a pointlike object. This feature is dramatically confirmed by experiments (Fig. 12.10) and is similarly found in the e–N deep inelastic scattering studied in Chap. 10. Both electromagnetic and weak currents are probing the same pointlike spin-1/2 constituents of the nucleon. In analogy with e–N deep inelastic scattering, we identify the partons as quarks and antiquarks. The parton picture discussed in the electromagnetic case can be extended to deep inelastic neutrino scattering. Let us then write the charged current cross-section of $\nu_\mu$ scattered by a pointlike spin-1/2 object [quark $\mathrm{Q_j}$ or antiquark $\overline{\mathrm{Q}}_k$ of mass $m_{j,k}$ and four-momentum $k^\mu_{j,k} = z_{j,k}P^\mu$]. Using (45) and (46), with $g_V = g_A = 1$, in addition to the appropriate CKM mixing, and similar to (10.51) with the trick $\int dx\,\delta(z-x) = 1$, we have

$$\frac{d\sigma(\nu_\mu + \mathrm{Q_j} \to \mu^- + \mathrm{q_1})}{dx dy} = |V_{\mathrm{Q_j q_1}}|^2\,\frac{2G_F^2 m_j E}{\pi}\,\delta(z_j - x) ,$$
$$\frac{d\sigma(\nu_\mu + \overline{\mathrm{Q}}_k \to \mu^- + \overline{\mathrm{q}}_2)}{dx dy} = |V_{\mathrm{Q_k q_2}}|^2\,\frac{2G_F^2 m_k E}{\pi}(1-y)^2\,\delta(z_k - x) . \quad (12.60)$$

We remark that if a quark is hit by a neutrino, there is no $y$ dependence; but when an antiquark is probed, the dependence is $(1-y)^2$. Similarly, the antineutrino–antiquark cross-section is $y$ independent, while the antineutrino–quark cross-section [see (46)] varies as $(1-y)^2$. These distributions correspond to the V − A charged currents of the standard model. In models beyond the

standard model with a V + A coupling for hypothetic new quarks, we simply interchange 1 with $(1-y)^2$ or $Q(x)$ with $\overline{Q}(x)$ in (60).
Deep inelastic cross-section is then the sum of parton cross-sections, each contribution is weighted by the distribution $Q_j(z_j)$, $\overline{Q}_k(z_k)$ in the nucleon.

As in (10.50)–(10.52), the contributions of quarks and antiquarks to the cross-section can be obtained from (60) (remember $m_{j,k} = M z_{j,k}$, where $M$ is the nucleon mass):

$$\sum_j \int \mathrm{d}z_j \frac{2G_\mathrm{F}^2 m_j E}{\pi} Q_j(z_j)\delta(z_j - x) = \sum_j \frac{2G_\mathrm{F}^2 ME}{\pi} x Q_j(x) ,$$

$$\sum_k \int \mathrm{d}z_k \frac{2G_\mathrm{F}^2 m_k E}{\pi} \overline{Q}_k(z_k)\delta(z_k - x)(1-y)^2 = \sum_k \frac{2G_\mathrm{F}^2 ME}{\pi} x \overline{Q}_k(x)(1-y)^2 .$$

We have

$$\frac{\mathrm{d}\sigma^\nu}{\mathrm{d}x\mathrm{d}y} = \frac{2G_\mathrm{F}^2 ME}{\pi} x \sum_{j,k} \left[|V_{\mathrm{Q_j q_1}}|^2 Q_j(x) + |V_{\mathrm{Q_k q_2}}|^2 \overline{Q}_k(x)(1-y)^2\right] ,$$

$$\frac{\mathrm{d}\sigma^{\overline{\nu}}}{\mathrm{d}x\mathrm{d}y} = \frac{2G_\mathrm{F}^2 ME}{\pi} x \sum_{j,k} \left[|V_{\mathrm{Q_k q_2}}|^2 \overline{Q}_k(x) + |V_{\mathrm{Q_j q_1}}|^2 Q_j(x)(1-y)^2\right] . \quad (12.61)$$

Let us rewrite (59) as a power series in $(1-y)$:

$$\frac{\mathrm{d}\sigma^{\nu,\overline{\nu}}}{\mathrm{d}x\mathrm{d}y} = \frac{G_\mathrm{F}^2 ME}{\pi} \left\{ x\left[\widetilde{W}_1 \pm \frac{\nu}{2M}\widetilde{W}_3\right] + \left[\frac{\nu}{M}\widetilde{W}_2 - 2x\widetilde{W}_1\right](1-y) \right.$$
$$\left. + x\left[\widetilde{W}_1 \mp \frac{\nu}{2M}\widetilde{W}_3\right](1-y)^2 \right\} . \quad (12.62)$$

In the parton picture, (61) is identified with (62). By comparing the coefficients of $(1-y)^n$ for $n = 0, 1, 2$ in the expressions in (61) and (62), we get

$$\widetilde{W}_1(x,q^2) \to \widetilde{F}_1(x) \ ; \ \frac{\nu}{M}\widetilde{W}_2(x,q^2) \to \widetilde{F}_2(x) \ ; \ \frac{\nu}{M}\widetilde{W}_3(x,q^2) \to \widetilde{F}_3(x) ;$$

$$\widetilde{F}_1(x) = \sum_{j,k} \left[|V_{\mathrm{Q_j q_1}}|^2 Q_j(x) + |V_{\mathrm{Q_k q_2}}|^2 \overline{Q}_k(x)\right] \quad ; \quad \widetilde{F}_2(x) = 2x\widetilde{F}_1(x) ;$$

$$\widetilde{F}_3(x) = 2\sum_{j,k} \left[|V_{\mathrm{Q_j q_1}}|^2 Q_j(x) - |V_{\mathrm{Q_k q_2}}|^2 \overline{Q}_k(x)\right] . \quad (12.63)$$

The structure functions $\widetilde{F}_2(x)$ and $\widetilde{F}_3(x)$ can be separated by writing the sum and difference of (62) for neutrino and antineutrino:

$$\frac{\mathrm{d}\sigma^{\nu\mathrm{N}} + \mathrm{d}\sigma^{\overline{\nu}\mathrm{N}}}{\mathrm{d}x\mathrm{d}y} = \frac{G_\mathrm{F}^2 ME}{\pi} \widetilde{F}_2(x)\left[1 + (1-y)^2\right] ,$$

$$\frac{\mathrm{d}\sigma^{\nu\mathrm{N}} - \mathrm{d}\sigma^{\overline{\nu}\mathrm{N}}}{\mathrm{d}x\mathrm{d}y} = \frac{G_\mathrm{F}^2 ME}{\pi} x\widetilde{F}_3(x)\left[1 - (1-y)^2\right] . \quad (12.64)$$

In (64), N stands either for the proton or the neutron. Let us specify what $Q_j(x)$ and $\overline{Q}_k(x)$ are. At the parton level (Fig. 12.11b), below the charmed hadron threshold, we have for reactions involving neutrinos

$$\nu_\mu + \mathrm{d} \to \mu^- + \mathrm{u} \quad , \quad \nu_\mu + \mathrm{s} \to \mu^- + \mathrm{u} \, ,$$
$$\nu_\mu + \overline{\mathrm{u}} \to \mu^- + \overline{\mathrm{d}} \quad , \quad \nu_\mu + \overline{\mathrm{u}} \to \mu^- + \overline{\mathrm{s}} \, .$$

We then deduce (taking for simplicity $|V_{\mathrm{ud}}|^2 + |V_{\mathrm{us}}|^2 = 1$)

$$\begin{aligned} \widetilde{F}_2^{\nu,\mathrm{p}}(x) &= 2x\Big[|V_{\mathrm{ud}}|^2 d(x) + |V_{\mathrm{us}}|^2 s(x) + \overline{u}(x)\Big] = 2x\widetilde{F}_1^{\nu,\mathrm{p}}(x) \, , \\ \widetilde{F}_3^{\nu,\mathrm{p}}(x) &= 2\Big[|V_{\mathrm{ud}}|^2 d(x) + |V_{\mathrm{us}}|^2 s(x) - \overline{u}(x)\Big] \, , \end{aligned} \tag{12.65}$$

where $u(x), d(x)$, and $s(x)$ are the up, down, and strange quark distributions inside the proton.

For antineutrino reactions, the corresponding structure functions are obtained from the above equation with the replacements of $q_j(x)$ with $\overline{q}_j(x)$ in $\widetilde{F}_{1,2}$ and $q_j(x)$ with $-\overline{q}_j(x)$ in $\widetilde{F}_3$, i.e.

$$\begin{aligned} \widetilde{F}_2^{\overline{\nu},\mathrm{p}}(x) &= 2x\Big[u(x) + |V_{\mathrm{ud}}|^2 \overline{d}(x) + |V_{\mathrm{us}}|^2 \overline{s}(x)\Big] = 2x\widetilde{F}_1^{\overline{\nu},\mathrm{p}}(x) \, , \\ \widetilde{F}_3^{\overline{\nu},\mathrm{p}}(x) &= 2\Big[u(x) - |V_{\mathrm{ud}}|^2 \overline{d}(x) - |V_{\mathrm{us}}|^2 \overline{s}(x)\Big] \, . \end{aligned} \tag{12.66}$$

As already discussed in (10.58), the up quark distribution in the neutron is $d(x)$ and the down quark distribution in the neutron is $u(x)$, by isospin invariance. Then

$$\begin{aligned} \widetilde{F}_2^{\nu,\mathrm{n}}(x) &= 2x\Big[|V_{\mathrm{ud}}|^2 u(x) + |V_{\mathrm{us}}|^2 s(x) + \overline{d}(x)\Big] \, , \\ \widetilde{F}_3^{\nu,\mathrm{n}}(x) &= 2\Big[|V_{\mathrm{ud}}|^2 u(x) + |V_{\mathrm{us}}|^2 s(x) - \overline{d}(x)\Big] \, . \end{aligned} \tag{12.67}$$

The structure function of an isoscalar target (sum of proton and neutron) probed by the neutrino is obtained using (65) and (67) (in which we put $|V_{\mathrm{ud}}|^2 \approx .95 = 1$, $|V_{\mathrm{us}}|^2 \approx 0.048 = 0$ for simplicity):

$$\widetilde{F}_2^{I=0} \equiv [\widetilde{F}_2^{\nu,\mathrm{p}}(x) + \widetilde{F}_2^{\nu,\mathrm{n}}(x)] = 2x\Big[u(x) + d(x) + \overline{u}(x) + \overline{d}(x)\Big] \, . \tag{12.68}$$

Comparing the above equation with (10.64) which gives the electromagnetic structure function $F_2^{I=0}$ of the same isoscalar target (deuteron), we get

$$\frac{\widetilde{F}_2^{I=0}}{F_2^{I=0}} \le \frac{18}{5} \, . \tag{12.69}$$

The equality holds if in (10.64) we neglect the contribution of sea quarks $s(x), \overline{s}(x)$ to the electromagnetic structure function $F_2^{I=0}(x)$, which amounts to at most 13.5%. If quarks had integral (rather than fractional) charges, this ratio would be $\leq 2$.

The structure functions $\frac{18}{5}F_2^{I=0}(x)$ and $\widetilde{F}_2^{I=0}(x)$, which are the main quantities measured in deep inelastic scattering, are plotted in Fig. 12.12. The agreement of data with (69) is remarkable and provides another strong confirmation of the fractional charges of quarks. With the photon $\gamma$ and weak-boson $W^{\pm}$ probes, the electron and the neutrino see the same constituents of the nucleon, and the quark fractional charges can be revealed.

Finally, from (61), the ratio of the integrated cross-sections $\sigma^{\nu,\mathrm{N}}/\sigma^{\overline{\nu},\mathrm{N}}$ gives the antiquark (sea) content of the nucleon:

$$R = \frac{\sigma^{\nu,\mathrm{N}}}{\sigma^{\overline{\nu},\mathrm{N}}} = \frac{3+\alpha}{1+3\alpha}\,, \qquad \alpha \equiv \frac{\int_0^1 \mathrm{d}x\, x\overline{Q}(x)}{\int_0^1 \mathrm{d}x\, xQ(x)} < 1\,. \tag{12.70}$$

If the sea is absent, the ratio would be 3. From data plotted in Fig. 12.10, the ratio $R$ turns out to be $\approx 0.67E_\nu/0.3E_\nu \approx 2.23$, which corresponds to $\alpha \approx 0.135$.

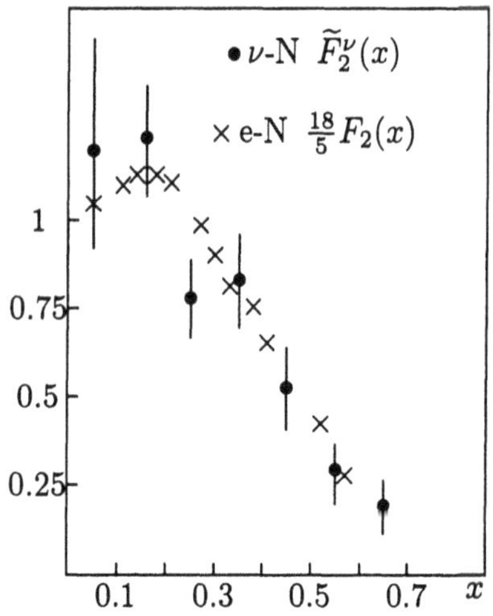


**Fig. 12.12.** Comparison of the isoscalar structure functions $\frac{18}{5}F_2(x)$ and $\widetilde{F}_2^\nu(x)$ as measured in electron and neutrino deep inelastic scatterings on nucleons. Data are taken from Fig. 8.12a of Perkins, D. H., *An Introduction to High Energy Physics* Addition-Wesley 1987. Adapted with permission of Addition-Wesley Longman Inc.

## Problems

**12.1 Leptons mixing.** If the neutrinos are massive, the three doublets of leptons are mixed by $V_{\text{lep}}$, analogously to the $V_{\text{CKM}}$ of the quark sector. Then the leptonic flavors (or numbers) are no longer conserved, the decay $\mu^{\pm} \to e^{\pm} + \gamma$ can occur. Draw the corresponding Feynman diagrams. Show qualitatively that the rate is proportional to $V_{\text{lep}}$ and to the neutrino masses.

**12.2 Solar neutrino flux.** The solar heat flux received on earth is $\approx 1.95\,\text{cal/cm}^2/\text{min}$, and the distance sun-earth is $\approx 1.5 \times 10^8$ km. Deduce that the solar luminosity is $\approx 3.86 \times 10^{33}$ erg /sec. According to the standard solar model, the sun shines by converting protons into helium ($\alpha$):

$$4\text{p} \to (\text{He})^4_2 + 2\text{e}^+ + 2\nu_\text{e} + \gamma\,.$$

For every four protons consumed, this fusion produces about 26 MeV = 4.16 $\times 10^{-5}$ erg of thermal energy. How many fusions take place in the sun every second ? Show that there are $\approx 1.8 \times 10^{38}$ neutrinos produced by the sun per second. Deduce that the solar neutrino flux at the surface of the earth is $\approx 6.4 \times 10^{10}/\text{cm}^2/\text{sec}$.

**12.3 Effective neutrino mass in matter.** Compute the effective mass of a neutrino of energy 10 MeV traveling in a supernova core (density $10^{14}\,\text{g/cm}^3$) and in the solar core (density $100\,\text{g/cm}^3$). Estimate the mean free path $l$ of the neutrino.

**12.4 Electromagnetic and weak decays of the $\pi^0$.** Compute the weak decay rate $\pi^0 \to Z^0 \to e^+e^-$, using $\pi^+ \to W^+ \to e^+\nu_e$ as a guide. The electromagnetic decay width $\Gamma_{\text{em}}(\pi^0 \to \gamma + \gamma \to e^+ + e^-)$ can be estimated to be $\sim (2\frac{\alpha m_e}{M_\pi} \log \frac{m_e}{M_\pi})^2 \Gamma(\pi^0 \to \gamma + \gamma)$. Draw the Feynman diagram of this cascade decay, and explain the origin of the coefficient $(\alpha m_e/M_\pi)^2$. The logarithm term comes from a loop integral. From experimental data, the width $\Gamma(\pi^0 \to \gamma + \gamma)$ is $2.5 \times 10^{+12}$ larger than $\Gamma_W(\pi^+ \to e^+ + \nu_e)$; deduce that $\Gamma_W(\pi^0 \to e^+e^-) \ll \Gamma_{\text{em}}(\pi^0 \to e^+ + e^-)$.

**12.5 Neutrino sum rule.** Just as for (10.62), show that

$$\int_0^1 \mathrm{d}x \widetilde{F}_3^{\nu,\text{N}}(x) \equiv \frac{1}{2} \int_0^1 \mathrm{d}x \left[\widetilde{F}_3^{\nu,\text{p}}(x) + \widetilde{F}_3^{\nu,\text{n}}(x)\right] = 3\,,$$

which tells us that the number of valence quarks in the nucleon is three. Again this sum rule agrees remarkably well with experiments.

**12.6 Neutral current deep inelastic scattering.** Write the differential cross-sections in terms of the parton distributions $u(x), d(x)$,etc. ,

$$\frac{\mathrm{d}\sigma}{\mathrm{d}x\mathrm{d}y}(\nu + \text{N} \to \nu + X) \quad \text{and} \quad \frac{\mathrm{d}\sigma}{\mathrm{d}x\mathrm{d}y}(\overline{\nu} + \text{N} \to \overline{\nu} + X)\,.$$

Derive the Paschos–Wolfenstein relation

$$\frac{\mathrm{d}\sigma^{\nu}_{\mathrm{NC}} - \mathrm{d}\sigma^{\overline{\nu}}_{\mathrm{NC}}}{\mathrm{d}\sigma^{\nu}_{\mathrm{CC}} - \mathrm{d}\sigma^{\overline{\nu}}_{\mathrm{CC}}} = |u_{\mathrm{L}}|^2 + |d_{\mathrm{L}}|^2 - |u_{\mathrm{R}}|^2 - |d_{\mathrm{R}}|^2 = \frac{1}{2} - \sin^2\theta_{\mathrm{W}} ,$$

which may be used to determine the Weinberg angle.

## Suggestions for Further Reading

*Neutrino mass, oscillations, MSW mechanism, solar neutrino deficit:*

Bahcall, J., Calaprice, F., McDonald, A. and Toksuka, Y., *Solar Neutrino Experiments: The Next Generation.* Physics Today, July 1996

Bilenky, S. M. and Petcov, S. T., Rev. Mod. Phys. **59** (1987) 671

Kim, C. W. and Pevsner, A., *Neutrinos in Physics and Astrophysics.* Harwood Academic Publishers, Chur 1993

*Neutral currents:*

Binétruy, P., Maiani, L., Pessard, H., Rozanov, A., Smirnov, A. and Veltman, M., in *Neutral Currents Twenty Years Later*, Proc. Int. Conf. Paris 1993. (ed. Nguyen-Khac, Ung and Lutz, A. M.). World Scientific, Singapore 1994

*Neutrino helicity (Goldhaber experiment); deep inelastic $\nu$–N scattering:*

Commins, E. D. and Bucksbaum, P. H., *Weak Interactions of Leptons and Quarks.* Cambridge U. Press, Cambridge 1983

Perkins, D. H., *Introduction to High Energy Physics* (Third edition). Addison-Wesley, Menlo Park, CA 1987

*Anomalies*:

Adler, S. L., in *Lectures on Elementary Particles Physics and Quantum Field Theory*, Proc. 1970 Brandeis Summer Institute (ed. Deser, S., Grisaru, M. and Pendleton, H.). MIT-Press, Cambridge 1970

Jackiw, R., in *Current Algebra and Anomalies* (ed. Treiman, S., Jackiw, R., Zumino, B. and Witten, E.). World Scientific, Singapore 1985

CVC, PCAC, *etc.* :

Scheck, F., *Electroweak and Strong Interactions.* Springer, Berlin, Heidelberg 1996

# 13 Muon and Tau Lepton Decays

In this chapter we study the $\tau^\pm$ (tau lepton) and $\mu^\pm$ (muon) decays governed by weak interactions only. Since the leptonic numbers are assumed to be strictly conserved in the standard model, the electromagnetic decay modes $\tau^\pm \to \mu^\pm(\mathrm{e}^\pm)+\gamma$ and $\mu^\pm \to \mathrm{e}^\pm+\gamma$ cannot occur. See Problem 12.1, however.

The importance of the subject is two-fold. On the one hand, the leptonic decay of $\tau$ (or $\mu$) is the simplest and cleanest process which unambiguously determines the left-handed V − A structure of the weak charged currents. This left-handed structure is universal in the sense that it describes weak reactions of all particles, whether they are leptons, mesons, or baryons. The violation of discrete symmetries P, C, and CP is well illustrated in the $\tau$ leptonic modes. On the other hand, $\tau$ is the only lepton massive enough to disintegrate into hadrons. Its semileptonic modes in both exclusive and inclusive channels are ideal for studying the strong interaction in the best possible conditions. The $\tau$ semileptonic decays offer an extremely favorable testing ground for both perturbative QCD radiative corrections and nonperturbative QCD topics such as decay constants, form factors, and the conserved vector current (CVC).

To have a global viewpoint and to put the subject in context, we first recall the general framework of weak decays before going into the specific $\tau$ and $\mu$ cases.

## 13.1 Weak Decays: Classification and Generalities

As the lightest particles in their categories, the proton and the electron are the only stable charged fermions in nature, a consequence of the conservation of the baryon and lepton numbers (at least to a very high degree of accuracy). On the other hand, when a hadron or charged lepton is produced, it decays more or less quickly into other particles. Hadrons can be grouped into two categories: resonances and low-lying metastable particles. A glance at the some two hundred and fifty existing hadrons in the Review of Particle Physics[1] shows that most of the hadrons are resonances. Examples of resonances are mesons $\rho(770)$, $\mathrm{K}^*(892)$, the charmonia, the bottomonia, as well as the baryons $\Delta(1620)$, $\Sigma_\mathrm{C}(2455)$, etc. Resonances decay quickly by

[1] Phys. Rev. **D54** (1996) 1

strong interactions, for instance $\rho \to 2\pi$, $K^* \to K+\pi$, $\Delta \to N+\pi+\pi$; their lifetimes are very short $\sim 10^{-23}$ s. Other neutral particles like the $\pi^0$, $\eta$, $\Sigma^0$ have smaller total widths (or longer lifetimes $\sim 10^{-19}$ s) because they only decay by electromagnetic interactions which violate isospin or G-parity (Chap. 6): $\pi^0 \to \gamma+\gamma$, $\Sigma^0 \to \Lambda+\gamma$, $\eta \to 3\pi$ (the G-parity of $\eta$ is +1, whereas an odd number of pions have G-parity $-1$).

Generally, strong decays conserve quantum numbers such as isospin, flavors, and discrete C, P, T symmetries. Electromagnetic decays only violate isospin, whereas in weak interactions, isospin, flavors and, discrete symmetries are violated.

The weak interaction governs the decay of low-lying metastable hadrons: $\pi^\pm$, charged and neutral flavored mesons K, D, $D_s$, B, $B_s^0$ as well as baryons n, $\Lambda$, $\Sigma^\pm$, $\Xi$, $\Omega^-$, $\Lambda_c^+$, $\Lambda_b^0$. Their widths are much smaller than those of the resonances, their lifetimes range from $10^3$ s for the free neutron to $10^{-13}$ s for charmed or bottom mesons. Weak decays of leptons, mesons, and baryons appear at first glance to have little in common. Lepton decays involve either only leptons, for example $\mu^- \to \nu_\mu + e^- + \overline{\nu}_e$, or a neutrino and hadrons, e.g. $\tau^- \to \nu_\tau + \pi^- + \pi^0$ (semileptonic modes). Mesons can decay into only lepton pairs, for instance $K^+ \to \mu^+ + \nu_\mu$, or into hadrons and a pair of leptons (semileptonic) such as $D^0 \to \rho^- + e^+ + \nu_e$, or into pure hadrons, e.g. $B \to J/\psi + K^*$. Baryons can have both semileptonic and purely hadronic channels. Isospin, parity, and charge conjugation are violated in weak decays. In most cases, hadronic flavors (strangeness, charm, bottomness) change. Because of the conservation rules mentioned above, strong and electromagnetic interactions are forbidden in the decays of low-lying metastable hadrons, otherwise weak interactions are swamped by them.

To lowest order of $G_F$, the neutral gauge boson Z does not participate in flavor-changing weak decays because of the GIM mechanism. Flavor-conserving weak decays by neutral currents are many orders of magnitude smaller than electromagnetic decays (Chap. 12). Therefore the Z-mediated weak decays will not be considered, only charged current processes are studied. It is remarkable that weak decays of particles, in spite of the huge differences in the various channels, partial rates, and lifetimes, all share a common feature. They can be quantitatively described by an *effective Lagrangian* which is the product of two left-handed V − A charged currents mediated by the gauge bosons W, as mentioned in (9.1)–(9.5):

$$\begin{aligned}\mathcal{L}_{\text{eff}} &= \mathrm{i} \lim_{q^2 \ll M_W^2} \left(\frac{-ig}{2\sqrt{2}}\right)^2 \left(L^\lambda + H^\lambda\right) \frac{-ig_{\lambda\rho}}{q^2 - M_W^2} \left(L^\rho + H^\rho\right)^\dagger \\ &= \frac{G_F}{\sqrt{2}} \left(L^\lambda + H^\lambda\right)\left(L_\lambda^\dagger + H_\lambda^\dagger\right) , \end{aligned} \tag{13.1}$$

where $L^\lambda$ and $H^\lambda$ are the leptonic and hadronic currents, with the Fermi coupling constant $G_F$ defined by $G_F/\sqrt{2} = g^2/8M_W^2$. Weak decays of all particles, in particular the modes in Table 6.6, are described by (1):

- leptonic decays of leptons by $L^\lambda L^\dagger_\lambda$ ,
- leptonic modes of mesons and semileptonic decays by $L^\lambda H^\dagger_\lambda + H^\lambda L^\dagger_\lambda$,
- nonleptonic or hadronic decays of hadrons by $H^\lambda H^\dagger_\lambda$ .

These V − A charged currents are expressed in terms of lepton and quark fields, they may be written as

$$L_\lambda = \sum_{\ell=\mathrm{e},\mu,\tau} \overline{\nu}_\ell(x)\gamma_\lambda(1-\gamma_5)\ell(x) \ ,$$

$$H_\lambda = \sum_{\mathrm{Q},\mathrm{q}} V_{\mathrm{Qq}} H^{Qq}_\lambda \ , \quad \text{where } H^{Qq}_\lambda = \overline{Q}(x)\gamma_\lambda(1-\gamma_5)q(x) \ ,$$

$$Q = u\,,\ c \ \text{ and } \ q = d\,,\ s\,,\ b \text{ fields.} \tag{13.2}$$

In (2), $V_{\mathrm{Qq}}$ is a CKM matrix element, $Q$ stands for the up and charmed quark fields, while $q$ represents the down, strange, and bottom quark fields. The reason for the absence of the top quark is that top is heavier than the gauge boson W and can decay directly into t → W + b with the coupling $g \sim M_{\mathrm{W}}\sqrt{G_{\mathrm{F}}}$ without passing through the virtual W propagator as in (1).

The universal weak effective Lagrangian (1) has a long history, starting with the neutron and muon decays together with the crucial discovery of parity violation in 1956. This Lagrangian is now the core of the standard electroweak theory. The $\tau$ decay offers a powerful test of (1) in both leptonic and semileptonic channels. Moreover in semileptonic modes, the interplay between QCD and weak interactions can be fully exploited.

Feynman diagrams for the $\mu$ and $\tau$ decays are drawn in Fig. 13.1. At the quark level, only the doublet (u, d″) enters, where d″ = $V_{\mathrm{ud}}$ d + $V_{\mathrm{us}}$ s. Since the $\tau$ is lighter than the charm, the other doublet (c, s″) where s″ = $V_{\mathrm{cs}}$ s + $V_{\mathrm{cd}}$ d does not intervene. The $\tau$ decay products are light mesons formed by u, d, and s quarks, such as the unflavored $\pi, \rho$, and $\mathrm{a_1}(1260)$ associated with $|V_{\mathrm{ud}}|^2 \approx (0.97)^2$ for the Cabibbo-favored modes, and the strange K, K*(892) with $|V_{\mathrm{us}}|^2 \approx (0.22)^2$ for the Cabibbo-suppressed modes.

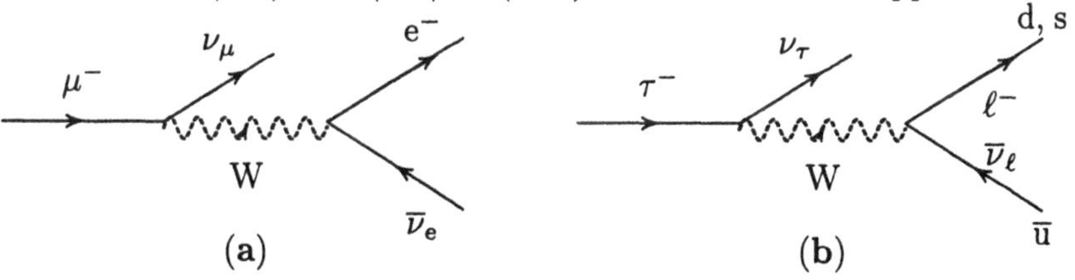


**Fig. 13.1.** (a) The only possible decay mode of the light $\mu^-$ ; (b) leptonic and semileptonic decays of the heavy $\tau^-$

Let us close the section with one remark. In any reaction, the branching ratio of a parent particle $P$ decaying into any particular channel $F$ is the first quantity to be measured:

$$\mathrm{B}_F \equiv \frac{\Gamma(P \to F)}{\Gamma_{\mathrm{total}}(P)} = \tau_P \times \Gamma(P \to F) \ , \quad (\tau_P \text{ is the lifetime of } P) \ .$$

Important weak decay dynamics can be revealed by the branching ratios of processes into different final states.

## 13.2 Leptonic Modes

Let us start with the important leptonic decay mode of $\tau$:

$$\tau^-(P) \to \nu_\tau(k_1) + \ell^-(p) + \overline{\nu}_\ell(k_2) \,, \tag{13.3}$$

where $\ell^-$ stands for the electron or the muon; the four-momentum of these particles are specified in parentheses. The $\tau^-$ and the $\ell^-$ masses are denoted by $M$ and $m$ respectively. The neutrinos are assumed massless. This mode is decisive for the discovery of $\tau$. Produced in $\mathrm{e}^+ + \mathrm{e}^- \to \tau^+ + \tau^-$, the $\tau^+$ and $\tau^-$ leptons subsequently decay with a distinctive signature rarely found in other particle decays. Indeed, from $\tau^- \to \nu_\tau + \mathrm{e}^- + \overline{\nu}_\mathrm{e}$ and $\tau^+ \to \overline{\nu}_\tau + \mu^+ + \nu_\mu$, one observes in the decay product a pair $\mathrm{e}^-\mu^+$ + invisible neutrinos whose presence is revealed by the apparent missing energy. It was through this special signature $\mathrm{e}^\mp\mu^\pm$ that the $\tau^\pm$ leptons were discovered.

### 13.2.1 Leptonic Branching Ratio

We first note that a very naive estimate of the leptonic branching ratio $\mathrm{Br}(\tau^- \to \nu_\tau + \mathrm{e}^- + \overline{\nu}_\mathrm{e}) \equiv \mathrm{B_e}$ can be made by a simple counting rule. Indeed the $\tau$ has two pure leptonic modes leading to emission of electrons and muons. It also has the semileptonic modes, its *inclusive decay* defined as $\tau \to \nu_\tau +$ any hadron is symbolically written as $\tau \to \nu_\tau + X$, where $X$ stands for the sum of all kinematically allowed mesons. This inclusive semileptonic process may be described by $\tau$ decays into its own neutrino $\nu_\tau$ and a quark+antiquark pair. In the parton model spirit, this approach is justified by a large energy released by the $\tau$. From the closure argument, these quark pairs saturate the sum of all the hadronic modes, since once quarks are produced by a weak decay, they can only form hadrons. The inclusive semileptonic rate is given by

$$\Gamma(\tau^- \to \nu_\tau + X) = \Gamma(\tau^- \to \nu_\tau + \mathrm{d}_j + \overline{\mathrm{u}}_j) + \Gamma(\tau^- \to \nu_\tau + \mathrm{s}_j + \overline{\mathrm{u}}_j) \,,$$

where $j$ is the color index. The first (second) term on the right-hand side of the above equation is the decay rate into kinematically allowed nonstrange (strange) mesons. They are respectively associated with $|V_\mathrm{ud}|^2 \approx (0.97)^2$ and $|V_\mathrm{us}|^2 \approx (0.22)^2$. As we will see later in (22) and (62), the rate depends on the fermionic masses in the final state. However, in the first approximation we may neglect their masses, so that for each color $j$ of quarks, we have

$$\Gamma(\tau^- \to \nu_\tau + \mathrm{e}^- + \overline{\nu}_\mathrm{e}) \approx \Gamma(\tau^- \to \nu_\tau + \mathrm{q}_j + \overline{\mathrm{u}}_j) \,, \quad \text{where } \mathrm{q} = \mathrm{d}, \mathrm{s} \,. \tag{13.4}$$

Since quarks have $N_\mathrm{c} = 3$ colors and $|V_\mathrm{ud}|^2 + |V_\mathrm{us}|^2 \approx 1$, the inclusive semileptonic width $\Gamma(\tau^- \to \nu_\tau + X)$ is three times $\Gamma(\tau^- \to \nu_\tau + \mathrm{e}^- + \overline{\nu}_\mathrm{e})$, and the total width is five times the latter (do not forget the muon), so the leptonic branching ratio $B_e$ is $\frac{1}{5}$ by this counting rule to be compared with the experimental data of $(17.83 \pm 0.06)\%$. The difference can be explained by the mass and QCD correction effects at the quark level which amount to about 10%. This simple counting rule supports the quark–parton picture which in turn indicates that the energy released by the $\tau$ is large enough to make legitimate the use of the parton model. This color-counting argument holds also in high energy $e^+ + e^-$ annihilation into hadrons. The similarity of $\Gamma(\tau^- \to \nu_\tau + \text{hadrons})$ and $\sigma(e^+ + e^- \to \text{hadrons})$ may be seen as follows:

$$\tau^- \to \nu_\tau + X = \tau^- \to \nu_\tau + W^* \ , \ \text{followed by } W^* \to q_j\,\overline{u}_j \Rightarrow \text{hadrons} \ ,$$
$$e^+ + e^- \to \gamma^* \ , \ \text{followed by } \gamma^* \to Q_j\,\overline{Q}_j \ \Rightarrow \text{hadrons} \ , \qquad (13.5)$$

where $W^*$ and $\gamma^*$ are the virtual W boson and photon.

### 13.2.2 Parity Violation. Energy Spectrum

We go further by computing the $\ell^-$ angular distribution and its asymmetry with respect to the $\tau$ polarization axis, the $\ell^-$ energy spectrum, and finally the integrated leptonic width $\Gamma(\tau^- \to \nu_\tau + \ell^- + \overline{\nu}_\ell)$. All of these physical quantities are of great importance in the determination of the $\tau$ properties, in particular its weak coupling strength and the structure of its charged current. Their measurements can give a definite answer to the question: is the $\tau$ lepton a replica of the electron and the muon, or is there any deviation from the standard model? To allow for possible deviations from the pure left-handed V − A current of the $\tau$–$\nu_\tau$ system, let us write the (current × current) decay amplitude of (3) in a more general form:

$$\mathcal{M} = \frac{G_F}{\sqrt{2}} \ \{\overline{u}(k_1)\gamma_\lambda(a - b\gamma_5)u(P)\} \ \{\overline{u}(p)\gamma^\lambda(1-\gamma_5)v(k_2)\} \ . \qquad (13.6)$$

A V − A current of the $\tau$–$\nu_\tau$ system corresponds to $a = b$, and in the standard electroweak model $a = b = 1$ (universality of the three lepton families). This property is well established for the e–$\nu_e$ and $\mu$–$\nu_\mu$ systems, as explicitly shown by the second factor $\{\overline{u}(p)\gamma^\lambda(1-\gamma_5)v(k_2)\}$ in (6). A V + A structure of the $\tau$–$\nu_\tau$ current corresponds to $a = -b$. Arbitrary $a$ and $b$ correspond to a mixture of left-handed and right-handed currents. We first evaluate $|\mathcal{M}|^2 = \frac{1}{2}\,G_F^2\,(T_1)_{\lambda\rho}(T_2)^{\lambda\rho}$, where

$$(T_1)_{\lambda\rho} = \mathrm{Tr}\big[u(k_1)\overline{u}(k_1)\gamma_\lambda(a-b\gamma_5)u(P)\overline{u}(P)\gamma_\rho(a-b\gamma_5)\big] \ ,$$
$$(T_2)^{\lambda\rho} = \mathrm{Tr}\big[u(p)\overline{u}(p)\gamma^\lambda(1-\gamma_5)v(k_2)\overline{v}(k_2)\gamma^\rho(1-\gamma_5)\big] \ . \qquad (13.7)$$

In order to study the angular distribution of $\ell^-$ with respect to the $\tau$ spin, we sum only the spins of the final state, but still keep untouched $S^\mu_\tau$, the spin polarization of the initial state $\tau$. In the $\tau$ rest frame $P^\mu = (M, \mathbf{0})$, its spin vector $S^\mu_\tau$ takes the form $S^\mu_\tau = (0, \widehat{\boldsymbol{S}})$ , with $|\widehat{\boldsymbol{S}}| = 1$, and we recall that

$$u(P)\overline{u}(P) = \tfrac{1}{2}(\not{P} + M)[1 + \gamma_5 \not{S}_\tau] \ .$$

We find

$$\begin{aligned}(T_1)_{\lambda\rho} &= \left(\frac{a+b}{2}\right)^2 \mathrm{Tr}\Big[\not{k}_1\gamma_\lambda(\not{P} - M\not{S}_\tau)\gamma_\rho(1-\gamma_5)\Big] \\ &+ \left(\frac{a-b}{2}\right)^2 \mathrm{Tr}\Big[\not{k}_1\gamma_\lambda(\not{P} + M\not{S}_\tau)\gamma_\rho(1+\gamma_5)\Big]\,, \\ (T_2)^{\lambda\rho} &= 2\,\mathrm{Tr}\Big[\not{p}\gamma^\lambda \not{k}_2\gamma^\rho(1-\gamma_5)\Big]\,. \end{aligned} \tag{13.8}$$

The relation (12.39) is useful to distinguish the effect of (V ∓ A) × (V ∓ A) product of currents from the (V ± A) × (V ∓ A) one. We get

$$\begin{aligned}\sum_{\text{spins}} |\mathcal{M}|^2 &= 64\, G_F^2 \Big\{ \left(\frac{a+b}{2}\right)^2 [p\cdot k_1]\, [(P - MS_\tau)\cdot k_2] \\ &+ \left(\frac{a-b}{2}\right)^2 [k_1\cdot k_2]\, [(P + MS_\tau)\cdot p] \Big\}\,. \end{aligned} \tag{13.9}$$

The general formula for the computation of decay widths is given by (4.70). In our case, with three particles in the final state, we have

$$\begin{aligned}\mathrm{d}\Gamma &= \frac{1}{2M}\frac{\mathrm{d}^3p}{2E}\int\!\!\int \frac{\mathrm{d}^3k_1}{2E_1}\,\frac{\mathrm{d}^3k_2}{2E_2}\,\frac{\delta^4(k_1+k_2-q)}{(2\pi)^5}\sum_{\text{spins}}|\mathcal{M}|^2, \quad q = P - p \\ &= \frac{64\, G_F^2}{2M}\frac{\mathrm{d}^3p}{2E}\Big\{\left(\frac{a+b}{2}\right)^2 p_\mu\,(P - MS_\tau)_\nu + \left(\frac{a-b}{2}\right)^2 (P + MS_\tau)\cdot p\; g_{\mu\nu}\Big\} \\ &\times \frac{1}{(2\pi)^5}\int\!\!\int \frac{\mathrm{d}^3k_1}{2E_1}\,\frac{\mathrm{d}^3k_2}{2E_2}\,\delta^4(k_1+k_2-q)\, k_1^\mu\, k_2^\nu\,, \end{aligned} \tag{13.10}$$

there is no factor $\frac{1}{2}$ in $\sum_{\text{spins}}|\mathcal{M}|^2$ on the right-hand side of (10) because the $\tau$ spins are not averaged. Since the neutrinos are unobserved, we first integrate over their three-momenta $\boldsymbol{k}_1$ and $\boldsymbol{k}_2$. On the other hand, to study the energy and angular distributions of $\ell^-$, we keep its momentum $\boldsymbol{p}$ untouched at the beginning. The last term on the right-hand side of (10) can be computed using formulas in the Appendix. With massless neutrinos, the integration is simple (Problem 5.2) and we get

$$I^{\mu\nu} \equiv \int \frac{\mathrm{d}^3k_1}{2E_1}\,\frac{\mathrm{d}k_2^3}{2E_2}\,\delta^4(k_1+k_2-q)\, k_1^\mu\, k_2^\nu = \frac{\pi}{24}\left(q^2 g^{\mu\nu} + 2q^\mu q^\nu\right)\,. \tag{13.11}$$

The product of $I^{\mu\nu}$ with the quantity in the curly brackets of (10) is

$$\begin{aligned}&\frac{\pi}{24}\Big\{\left(\frac{a+b}{2}\right)^2\Big[q^2[(P - MS_\tau)\cdot p] + 2\,p\cdot q\,[(P - MS_\tau)\cdot q]\Big] \\ &+ \left(\frac{a-b}{2}\right)^2 6\,q^2\,[(P + MS_\tau)\cdot p]\Big\}\,. \end{aligned} \tag{13.12}$$

To distinguish the effects of the V − A current from those of a possible V + A current, it is convenient to arrange (12) into two parts. The first part is a sum of left and right chiral current contributions *with equal weights*, and the second part belongs to their unequal mixture. This separation enables us to introduce later the Michel parameters $\rho$, $\xi$, and $\delta$, which are important measurable quantities to test whether or not the heavy lepton $\tau$ is a replica of the muon and the electron. This decomposition turns out to be a powerful method of investigating the decay dynamics, as we will see. With the coefficient $\pi/24$ implicitly understood, let us rewrite (12) in a form in which the mentioned separation is explicit [note that $S_\tau \cdot q \equiv S_\tau \cdot (P-p) = -S_\tau \cdot p$]. The quantity inside the curly brackets of (12) is

$$
\begin{aligned}
&6\,q^2\,P\cdot p\Big[\Big(\frac{a+b}{2}\Big)^2+\Big(\frac{a-b}{2}\Big)^2\Big]+\Big[2p\cdot q\,P\cdot q-5\,q^2\,P\cdot p\Big]\Big(\frac{a+b}{2}\Big)^2\\
&+MS_\tau\cdot p\Big\{2\,q^2\Big[\Big(\frac{a+b}{2}\Big)^2+\Big(\frac{a-b}{2}\Big)^2\Big]+(2\,p\cdot q-3\,q^2)\Big(\frac{a+b}{2}\Big)^2\\
&+4\,q^2\Big(\frac{a-b}{2}\Big)^2\Big\}\,.
\end{aligned}
\tag{13.13}
$$

In (13), this separation applies to both the spin-dependent $MS_\tau \cdot p$ and the spin-independent $q^2\,P\cdot p$ terms. In the $\tau$ rest frame, $\theta$ denotes the angle between the three-momentum $\boldsymbol{p}$ of the $\ell^-$ and the spin $\widehat{\boldsymbol{S}}$ of the $\tau^-$, thus

$$
\begin{aligned}
&p^\mu=(E,\boldsymbol{p})\ ,\ S_\tau\cdot p=-|\boldsymbol{p}|\cos\theta\ ,\ \mathrm{d}^3p=2\pi\,\mathrm{d}(\cos\theta)\,|\boldsymbol{p}|\,E\,\mathrm{d}E\ ,\\
&q^2=M^2-2ME+m^2\ .\ \text{Since } q^2\geq 0\Rightarrow m\leq E\leq\frac{M^2+m^2}{2M}\equiv E_{\max}\ ,\\
&P\cdot p=ME\ \ ,\ \ P\cdot q=M(M-E)\ \ ,\ \ p\cdot q=ME-m^2\ .
\end{aligned}
\tag{13.14}
$$

Putting (10), (13), and (14) together, we obtain

$$
\frac{\mathrm{d}\Gamma}{\mathrm{d}\cos\theta\,\mathrm{d}E}=\frac{G_\mathrm{F}^2\,|\boldsymbol{p}|\,E}{3(2\pi)^3}\,\frac{a^2+b^2}{2}\{X\}\ ,\quad\text{where}
\tag{13.15}
$$

$$
\begin{aligned}
\{X\}=\Big\{&6\,q^2+\frac{(a+b)^2}{2(a^2+b^2)}\,[8ME-3M^2-m^2[3+(2M/E)]]\\
&-\frac{|\boldsymbol{p}|}{E}\cos\theta\Big[2\,q^2\Big(1+\frac{(a-b)^2}{a^2+b^2}\Big)+\frac{(a+b)^2}{2(a^2+b^2)}\,(8ME-3M^2-5m^2)\Big]\Big\}.
\end{aligned}
$$

In the above expression of $\{X\}$, the $\cos\theta$-independent term (isotropic part) gives the energy spectrum of the emitted $\ell^-$. The $\cos\theta$ term (anisotropic part) on the last line represents the angular correlation between the spin $\boldsymbol{S}$ of the decaying particle and the three-momentum $\boldsymbol{p}$ of the outgoing $\ell^-$. This correlation is a crucial quantity to reveal the parity violation of weak interaction. A short recall of the discussions given in Chap. 5 might be useful.

**Parity Violation.** The parity violation phenomenon can only appear as a *pseudoscalar term* constructed from experimentally measurable quantities, for which the anisotropic part of (15) is the simplest example. This apparently trivial fact had never been noticed before 1956, when Lee and Yang pointed out that if one did not look for a pseudoscalar measurable quantity, the nonconservation of parity could never be experimentally discovered, even if the interaction violates space inversion (or parity) symmetry P.

In the current × current amplitude, the interference V × A is a pseudoscalar quantity. The electron energy spectrum and the integrated rate are two examples of scalar quantities coming from the V × V and A × A products of two curents. Their measurements cannot tell whether the P symmetry is broken by weak interactions or not.

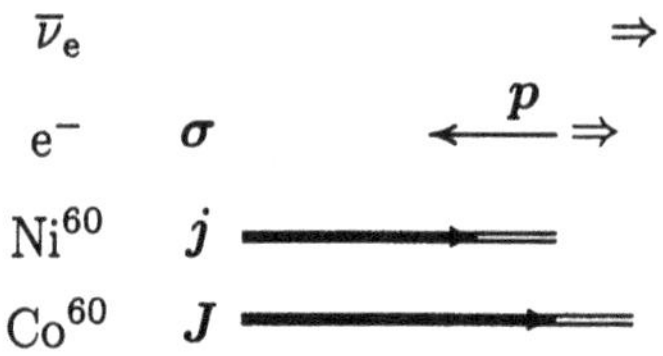

$\overline{\nu}_e$ $\Rightarrow$

$e^-$ $\sigma$ $\xleftarrow{\;\boldsymbol{p}\;}\Rightarrow$

$Ni^{60}$ $j$

$Co^{60}$ $J$

**Fig. 13.2.** $\boldsymbol{J}\cdot\boldsymbol{p}$ correlation in $Co^{60} \to Ni^{60} + e^- + \overline{\nu}_e$

As explained in Chap. 5, under space inversion P, $\boldsymbol{x} \to -\boldsymbol{x}$, $\boldsymbol{S} \to +\boldsymbol{S}$ while $\boldsymbol{p} \to -\boldsymbol{p}$, the nonconservation of parity manifests itself by a nonzero value of the coefficient of the pseudoscalar term $\boldsymbol{S}\cdot\boldsymbol{p} = |\boldsymbol{p}|\cos\theta$ which occurs in the matrix element of the operator product V × A. A forward–backward asymmetry in the emission of $l$ with respect to the spin $\boldsymbol{S}$ constitutes an unequivocal proof of parity violation. Very similar to the $\boldsymbol{S}\cdot\boldsymbol{p}$ correlation considered here is the electron asymmetry with respect to the polarization axis of the cobalt nucleus in $Co^{60} \to Ni^{60} + e^- + \overline{\nu}_e$ observed by C. S. Wu, who gave the first experimental demonstration of parity violation. A sample of $Co^{60}$ was kept at a very low temperature, its spin $\boldsymbol{J}$ ($J = 5$) is aligned and the final $Ni^{60}$ has spin $j = 4$. The electron is distributed according to

$$\mathcal{F}(\theta) = 1 + \alpha\,\frac{\boldsymbol{J}\cdot\boldsymbol{p}}{|\boldsymbol{J}|\,E}\ ,$$

$\boldsymbol{p}$ and $E$ are the momentum and energy of the electron. If the coefficient $\alpha$ is found to be definitely nonzero, a parity violation is proven. This was indeed the case, and the electron was found to be emitted preferentially antiparallel to $\boldsymbol{J}$, i.e. $\alpha = -1$. The difference by one unit of spin between the initial and final nuclei on the one hand, and the conservation of the $z$ component of the angular momentum along $\boldsymbol{J}$ on the other hand, imply that the electron spin $\boldsymbol{\sigma}$ must point in the direction $\boldsymbol{J}$. It shows that the electron emitted in nuclear $\beta$-decay is antiparallel to its spin $\boldsymbol{\sigma}$, i.e. the electron is left-handed. As illustrated in Fig. 13.2, this is the first experimental indication of the V − A character of the charged current.

On the other hand, scalar quantities, such as the isotropic energy spectrum in (15), have no bearing on the question of parity violation and cannot be used to test it.

**Energy Spectrum.** Integrating (15) over $\theta$, we obtain the energy spectrum of $\ell^-$. The anisotropic term, which is linear in $\cos\theta$, vanishes, while the isotropic term is doubled. We have

$$\frac{\mathrm{d}\Gamma}{\mathrm{d}E} = \frac{a^2+b^2}{2}\frac{G_{\mathrm{F}}^2\,|\boldsymbol{p}|\,E}{3(2\pi)^3}\Big\{12\,(M^2-2ME+m^2) + \frac{8}{3}\rho\Big[8ME-3M^2-m^2[3+(2M/E)]\Big]\Big\}\,, \tag{13.16}$$

where we define the Michel parameter $\rho$ by

$$\rho = \frac{3}{8}\,\frac{(a+b)^2}{a^2+b^2}\,. \tag{13.17}$$

For historic reasons (see below), conventionally, we write $\rho$ with the coefficients 8/3 in (16) or 3/8 in (17), since $\rho$ turns out to be 3/4 in the four-fermion interaction of the type (V − A) × (V − A) product of currents. We now see that the $\ell^-$ energy spectrum $\mathrm{d}\Gamma/\mathrm{d}E$ in $\tau$ decay (Fig. 13.3) is very useful because it distinguishes the V ± A property of the $\tau$-$\nu_\tau$ weak current. Although the shape $\mathrm{d}\Gamma/\mathrm{d}E$ for V − A is distinct from the V + A one, their integrated rates $\Gamma = \int \frac{\mathrm{d}\Gamma}{\mathrm{d}E}\mathrm{d}E$ are identical. Hence measurement of $\Gamma$ alone cannot distinguish the chirality of weak currents.

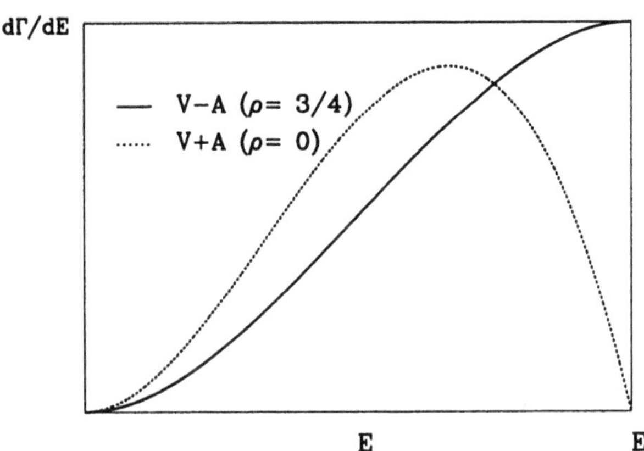


**Fig. 13.3.** The electron energy spectrum $\mathrm{d}\Gamma/\mathrm{d}E$ in $\tau^- \to \nu_\tau + \mathrm{e}^- + \overline{\nu}_\mathrm{e}$

In the 1950s, the theory of weak interaction was still in an embryonic state, and it was not known whether the four-fermion $\beta$-decay of nuclei or muon was of the current × current form, e.g. V × V or $[\overline{\psi}_1\gamma_\mu\psi_2]\,[\overline{\psi}_3\gamma^\mu\psi_4]$, as postulated by Fermi in analogy with the electromagnetic interaction. At that time, without data on the parity violation, the weak interaction could be *a priori* any combination of the scalars obtained from the covariant products S × S, V × V, T × T, A × A, P × P (Chap. 5). Michel's idea is to introduce in the electron energy spectrum a parameter $\rho$ to separate terms which are *common* to the structures S, V, T, A, P from other terms which are *sensitive* to some of these structures. For example, a pure product S $\times$ S would give $\rho = 0$, while a pure T $\times$ T would give $\rho = 1$.

Following Michel, we are led to rearrange (16) into two terms separated by a parameter $\rho$. The first term $12\,(M^2 - 2ME + m^2)$, which is independent of $a$ and $b$, cannot distinguish the V $-$ A from the V + A structure of the $\tau$-$\nu_\tau$ current. The second term [last line of (16)], which depends on $a$ and $b$, is sensitive to V $\mp$ A and can be used to determine this V $\mp$ A structure. We emphasize that in the decay $\tau^- \to \nu_\tau + \ell^- + \overline{\nu}_\ell$, once the current of the final state $\ell$–$\nu_\ell$ is known to have the V $-$ A structure, then the V $\pm$ A current of the initial state $\tau$–$\nu_\tau$ can be determined by measuring the parameter $\rho$. We extract $\rho$ by fitting the electron energy distribution (16) with data and obtain important information on the dynamics.

From (17), $\rho$ is always $\leq 3/4$. The V $-$ A ($a = b$) of the $\tau$–$\nu_\tau$ current in (6) corresponds to $\rho = 3/4$, a pure V ($b = 0$) or a pure A ($a = 0$) would result in $\rho = 3/8$, while a V + A type ($a = -b$) would imply $\rho = 0$.

Recent data[2] give $\rho_\tau = 0.742 \pm 0.027$, which is in excellent agreement with the V $-$ A charged current involved in $\tau$ decays. For the muon, the $\rho_\mu$ parameter measured in muon decay $\mu^- \to \nu_\mu + \mathrm{e}^- + \overline{\nu}_\mathrm{e}$ is $0.7518 \pm 0.0026$.

### 13.2.3 Angular Distribution. Decay Rate

The confirmation of $a = b$ is again found in the $\boldsymbol{p} \cdot \boldsymbol{S}$ correlation. We rewrite the anisotropic part of (15) in terms of two additional Michel parameters usually denoted as $\xi$ and $\delta$:

$$\{X\} = 6\,(M^2 - 2ME + m^2) + \frac{4}{3}\rho\left[8ME - 3M^2 - m^2\left(3 + \frac{2M}{E}\right)\right]$$

$$-\,\xi\,\frac{|\boldsymbol{p}|}{E}\cos\theta\Big[2\,(M^2 - 2ME + m^2) + \frac{4}{3}\,\delta\,(8ME - 3M^2 - 5m^2)\Big]\,, \quad (13.18)$$

$$\xi = 1 + \frac{(a-b)^2}{a^2 + b^2}\quad , \quad \delta = \frac{3}{8}\,\frac{(a+b)^2}{a^2 + b^2 + (a-b)^2}\,. \quad (13.19)$$

For fixed $E$, the parameters $\xi$ and $\delta$ can be obtained by fitting the $\theta$ distribution with experiments. Data[2] give $\xi = 1.03 \pm 0.12$, and $\xi\,\delta = 0.76 \pm 0.11$; they are in excellent agreement with $a = b$. For a V + A current, we would get $\xi = 3$, $\delta = 0$, in sharp contrast with $\xi = 1$, $\delta = 3/4$ for a V $-$ A.

Equivalently, for fixed $E$, the angular distribution of the $\ell^-$ with respect to the spin direction of the $\tau^-$ in (15) can be written as $D(\theta)$. With an overall normalization factor not explicitly shown and neglecting $m^2$ in (15), $D(\theta)$ can be written as

$$D(\theta) = 1 - \alpha\cos\theta\ , \quad \text{where } \alpha = \frac{4E - M}{3M - 4E} \quad \text{for V} - \text{A}$$

$$\alpha = +1 \qquad \text{for V + A}\,. \quad (13.20)$$

---

[2] Phys. Rev. **D54** (1996) 1

The forward–backward asymmetry is also an important physical measurable quantity that can fix the relative sign between $a$ and $b$; its measurement is therefore useful. For a V – A current, the asymmetry parameter $\alpha$ increases from $-1/3$ at $E = 0$ to $+1$ when $E$ reaches $E_{\max} = M/2$, whereas $\alpha$ is constant in the V + A case.

From now on, we put $a = b$ but still let the overall normalization factor $a$ be arbitrary. We now see that data fix $a$ to 1, i.e. the *universality* of e, $\mu$, $\tau$ is experimentally confirmed.

Putting $\rho = 3/4$ and $a = b$, we now integrate (16) to obtain the full leptonic rate; the integration range for $E$ is given in (14):

$$\Gamma = \frac{2\,a^2\,G_{\mathrm{F}}^2}{3\,(2\pi)^3} \int_m^{E_{\max}} \sqrt{E^2 - m^2}\;\left[ME\,(3M - 4E) - m^2\,(2M - 3E)\right]\,\mathrm{d}E\,.$$

The result of the above integration is

$$\Gamma(\tau^- \to \nu_\tau + \ell^- + \overline{\nu}_\ell) = a^2 f(m^2/M^2)\,\Gamma_0\ ,\quad \Gamma_0 \equiv \frac{G_{\mathrm{F}}^2 M^5}{192\,\pi^3}\,, \tag{13.21}$$

$$f(x) = 1 - 8x + 8x^3 - x^4 - 12\,x^2\,\log x\,;\quad f(m_\mu^2/M^2) = 0.9728\,. \tag{13.22}$$

The formula $\Gamma_0 \equiv G_{\mathrm{F}}^2 M^5/192\,\pi^3$ – which gives the decay rate of a fermion of mass $M$ into three massless fermions (Problem 5.2) – will be repeatedly used. The phase space correction $f(m^2/M^2)$ takes into account one massive fermion among the three in the final state, the other two are massless. The numerical value of $\Gamma(\tau^- \to \nu_\tau + \mathrm{e}^- + \overline{\nu}_\mathrm{e})$ can be obtained using data of both the $\tau$ lifetime and the electronic branching ratio

$$\Gamma(\tau^- \to \nu_\tau + \mathrm{e}^- + \overline{\nu}_\mathrm{e}) = \frac{\mathrm{B_e}}{\tau_\tau} = \frac{0.1783 \pm 0.0006}{2.91 \pm 0.015} \times 10^{13}\ \mathrm{s}^{-1}\,. \tag{13.23}$$

To determine the coefficient $a$, we compare (23) with the theoretical rate (21), in which the phase space correction due to the electron mass $m_\mathrm{e}$ is completely negligible, i.e. $f(m_\mathrm{e}^2/M^2)$ is taken as 1. However the radiative correction 0.996 given below in (28) is included, so that

$$\frac{G_{\mathrm{F}}^2 M^5}{192\,\pi^3} \times 0.996 = 4.033 \times 10^{-13}\ \mathrm{GeV} = 0.06127 \times 10^{13}\,\mathrm{s}^{-1} \Rightarrow a = 1 \pm 0.006.$$

Having shown that $a = b = 1$, we rewrite the previous formula:

$$\frac{\mathrm{d}\Gamma(\tau^\mp \to \ell^\mp + \nu + \overline{\nu})}{\mathrm{d}\cos\theta\,\mathrm{d}E} = \frac{G_{\mathrm{F}}^2\,|\boldsymbol{p}|}{24\,\pi^3}\Big\{ME\,(3M - 4E) - m^2\,(2M - 3E) \mp |\boldsymbol{p}|\cos\theta\,[M\,(4E - M) - 3m^2]\Big\}\,. \tag{13.24}$$

The angular distribution for $\tau^+$ can be obtained from the angular distribution of the $\tau^-$ by $u(P) \leftrightarrow \overline{v}(P)$, implying $MS_\tau \leftrightarrow -MS_\tau$ in (9), from which (24) follows.

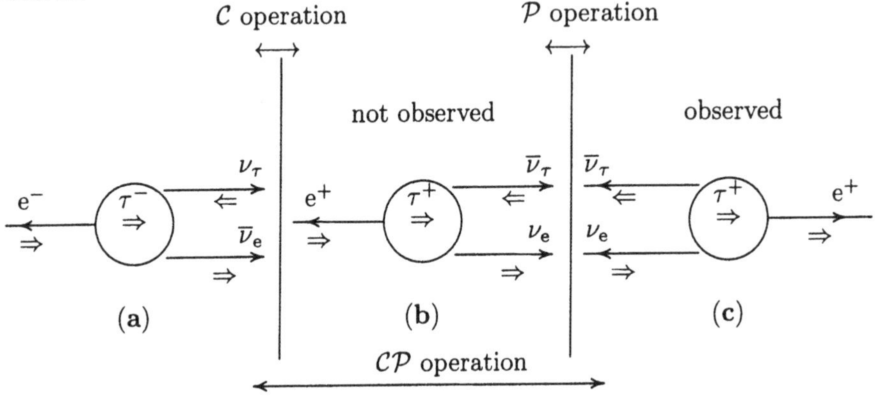


**Fig. 13.4.** (**a**) At $E = E_{\max}$, the angular distribution of the $e^-$ with respect to the $\tau^-$ polarization; (**b**) the charge-conjugation states of (**a**); (**c**) $\mathcal{CP}$ operated on (**a**) gives the angular distribution of the $e^+$ with respect to the $\tau^+$ polarization

The V − A character of the $\tau$–$\nu_\tau$ and e–$\nu_e$ currents is shown in Fig. 13.4a. In the $\tau$ rest frame, when the electron energy is maximum ($E \approx E_{\max}$), kinematics implies that the two neutrinos $\nu_\tau$ and $\overline{\nu}_e$ are emitted in one direction (for instance the $+x$ axis), whereas the electron is emitted in the opposite direction ($-x$ axis). Since the two neutrinos have opposite helicities, the total angular momentum conservation in the $x$ axis forces the spin of $e^-$ to be parallel to the spin of $\tau^-$. Since the electron has negative helicity at high momentum, it must be emitted antiparallel to the $\tau^-$ spin.

This correlation between spin and three-momentum is also described by the asymmetry parameter $\alpha$ in (20). When $E = E_{\max} = M/2$, $\alpha = +1$, and the electron is likely emitted antiparallel to the $\tau^-$ spin direction, whereas the positron prefers to be emitted parallel to the $\tau^+$ spin. This $E_{\max}$ configuration is illustrated in Fig. 13.4.

Near the lower end $E = 0$, exactly the opposite configuration appears, since $\alpha = -1/3$.

Moreover, Fig. 13.4 shows why the spin and momentum correlation in the $\tau^+$ decay can be derived from that of the $\tau^-$, assuming CP invariance of the weak interaction involving leptons. Starting with the $\tau^- \to \nu_\tau + e^- + \overline{\nu}_e$ decay in Fig. 13.4a, let us consider its charge conjugate states in Fig. 13.4b. The configuration of the latter cannot be observed because all of the $e^+$, $\nu_e$ and $\overline{\nu}_\tau$ have the wrong helicities. The charge conjugation symmetry C is manifestly violated by weak interactions. We then go further by letting the space reversal $\mathcal{P}$ operate on Fig. 13.4b, which becomes Fig. 13.4c. The combination of $\mathcal{C}$ and $\mathcal{P}$, i.e. $\mathcal{CP}$, transforms Fig. 13.4a into Fig. 13.4c. While the violations of P and C are strongest possible, the product CP is to a good approximation conserved by weak interactions. If CP is invariant, Fig. 13.4c must be physically observable. The angular distribution of the $e^+$, with respect to the $\tau^+$ polarization, can be obtained from that of the $e^-$ in $\tau^-$ decay, by a simple change of sign $\cos\theta \leftrightarrow -\cos\theta$ in (24). This substitution rule $\cos\theta \leftrightarrow -\cos\theta$ in $\tau^- \leftrightarrow \tau^+$ may be used as a test of CP violation.

Finally, the energy spectrum and the width are given by

$$\frac{\mathrm{d}\Gamma}{\mathrm{d}E} = \frac{G_F^2\,|\boldsymbol{p}|}{12\,\pi^3}\Big[ME(3M-4E) - m^2(2M-3E)\Big]\,, \tag{13.25}$$

$$\Gamma = \frac{G_F^2 M^5}{192\,\pi^3} f\left(\frac{m^2}{M^2}\right) = \Gamma_0\, f\left(\frac{m^2}{M^2}\right)\,. \tag{13.26}$$

The integrated width $\Gamma$ depends on *the fifth power of the energy released by the decaying particle* (which is $M$, in the case of massless decay products). This power law is easy to understand since the dimension of $G_F^2$ is $(\text{mass})^{-4}$ and that of the width is $(\text{mass})^{+1}$. For a fermion of mass $M$ decaying into three massless fermions, the only mass involved is $M$, so $G_F^2\,M^5$ naturally appears. The huge difference by a factor of $10^{16}$ in the lifetimes of weakly decaying particles (for instance between the charm D meson and the neutron) essentially comes from this fifth power of the energy released. For neutron, the energy liberated $\approx m_n - m_p - m_e$ is only 0.78 MeV, whereas for charmed or bottom-flavored mesons, it can reach a few GeVs. A more accurate estimate calls for more sophisticated computations, however this fifth power can explain the huge differences in the lifetimes of weakly decaying particles.

Beside these tree diagram results, we should add the electromagnetic radiative corrections to leptonic weak interactions. These corrections are due to virtual photons in loops involving the charged $\tau$ and $\ell$, as well as to real photons emitted by them (bremsstrahlung). There are in all five diagrams similar to the five drawn in Figs. 14.2–3 with photons replacing gluons. The calculation could be done similarly to that in Chap. 14. These corrections[3] yield

$$1 - \frac{\alpha_{\text{em}}}{2\pi}\left(\pi^2 - \frac{25}{4}\right)\,. \tag{13.27}$$

Another type of corrections involves the W propagator effect if we do not neglect $q^2 \ll M_W^2$ in (1). This gives $1 + 3/5\, M^2/M_W^2 - 2\,m^2/M_W^2$ . These two types of corrections multiplied by (21) give

$$\Gamma = a^2\Gamma_0\, f\left(\frac{m^2}{M^2}\right)\left[1 - \frac{\alpha_{\text{em}}}{2\pi}\left(\pi^2 - \frac{25}{4}\right)\right]\left[1 + \frac{3}{5}\frac{M^2}{M_W^2} - 2\,\frac{m^2}{M_W^2}\right]\,. \tag{13.28}$$

[3] Berman, S., Phys. Rev. **112** (1958) 267; Kinoshita, T. and Sirlin, A., Phys. Rev. **113** (1959) 1652

These last two corrections are numerically small $\approx 4 \times 10^{-3}$, i.e. the product of the last two factors of the above equation is 0.996. We note, however, that one-loop QCD radiative corrections – relevant to the inclusive semileptonic $\tau$ decay described by $\tau \to \nu_\tau +$ a quark pair $(\mathrm{q_i} + \overline{\mathrm{q}}_j)$, in which gluons replace photons – are much more important, simply because $(-\alpha_{\mathrm{em}}/2\pi) \times (\pi^2 - 25/4)$ is replaced by $+\alpha_s/\pi$, where the running coupling $\alpha_s(M)$ turns out to be $\approx 0.37$ at the appropriate scale $M$ of the $\tau$ decays. As we will see in Chap. 14, these one-loop QCD corrections, which enhance the $\Gamma(\tau \to \nu_\tau + \mathrm{q}_i + \overline{\mathrm{q}}_j)$ rate, will consequently pull down the leptonic branching ratio from its naive 0.2 value (Sect. 13.2) to 0.186, closer to the observed $\mathrm{B_e} = 0.1783 \pm 0.0006$.

Finally, we note that (24)–(26) and (28) also apply to $\mu^- \to \nu_\mu + \mathrm{e}^- + \overline{\nu}_\mathrm{e}$, for which $M$ and $m$ are the muon and electron masses respectively.

## 13.3 Semileptonic Decays

The $\tau$ is the only lepton massive enough to decay into hadrons. Its semileptonic channels in both exclusive and inclusive modes are ideal for studying strong interaction in clean conditions.

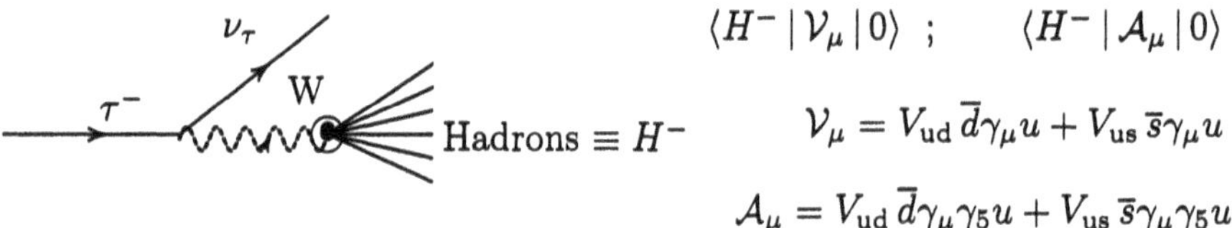


**Fig. 13.5.** Semileptonic decays of $\tau^-$

These decays probe the matrix element of the V and A parts of the charged current between the vacuum and the final hadronic state $H$ (Fig. 13.5). Since these matrix elements are just the decay constants (if $H$ is a single particle) or form factors (if $H$ represents several particles), the importance of semileptonic decays cannot be underrated. The two-pion modes [including the $\rho(770)$] and more gererally the $2n$-pion modes are the cleanest hadronic channels in which the CVC property of the charged current $\overline{d}\gamma^\mu u$ (a consequence of its isospin structure) can be unambiguously tested at relatively high momentum transfer $q^2$ released by $\tau$.

### 13.3.1 The One-Pion Mode: $\tau^- \to \nu_\tau + \pi^-$

The relevant hadronic current sandwiched between a pseudoscalar pion and the vacuum is $V_{\mathrm{ud}}\, A_\mu = V_{\mathrm{ud}}\, \overline{d}\gamma_\mu\gamma_5 u$ and $\langle \pi^-(p) \,|\, A_\mu \,|\, 0\rangle \equiv \mathrm{i}\, f_\pi p_\mu$. We note that the pion decay constant $f_\pi$ is the same parameter that enters the $\pi_{\ell_2}$ mode $\pi \to \ell + \overline{\nu}_\ell$ of Fig. 10.3a, and $f_\pi$ is one of the most fundamental constants frequently met in different circumstances in particle physics.

The amplitude $\tau^-(P) \to \nu_\tau(k) + \pi^-(p)$ can be written as

$$\begin{aligned}\mathcal{M} &= \left(\frac{-ig}{2\sqrt{2}}\right)^2 \overline{u}(k)\gamma_\nu(1-\gamma_5)u(P)\frac{-ig^{\mu\nu}+\frac{p^\mu p^\nu}{M_W^2}}{p^2-M_W^2}\, V_{\rm ud}\, \langle \pi^-(p)\,|\,A_\mu\,|\,0\rangle \\ &= \frac{G_{\rm F}}{\sqrt{2}} V_{\rm ud}\, M\, f_\pi\, \overline{u}(k)(1+\gamma_5)u(P)\,. \end{aligned} \tag{13.29}$$

Averaging the intial state $\tau$ spin, we get

$$\tfrac{1}{2}\sum_{\rm spin} |\mathcal{M}|^2 = \frac{G_{\rm F}^2}{2}|V_{\rm ud}|^2\, M^2\, f_\pi^2\, {\rm Tr}\Big[\not k \not P(1-\gamma_5)\Big] = G_{\rm F}^2 |V_{\rm ud}|^2 f_\pi^2 M^4 \left(1-\frac{m_\pi^2}{M^2}\right).$$

Applying (4.73) for the decay with two particles in the final state, we have

$$\Gamma(\tau \to \nu_\tau + \pi) = \frac{1}{2\,M}\int \frac{{\rm d}^3k}{2E_k}\frac{{\rm d}^3p}{2E_p}\frac{\delta^4(k+p-P)}{(2\pi)^2}\ \tfrac{1}{2}\sum_{\rm spin}|\mathcal{M}|^2\,.$$

Using the two-body phase space integral formula in the Appendix,

$$\int \frac{{\rm d}^3k}{2E_k}\frac{{\rm d}^3p}{2E_p}\delta^4(k+p-P) = \frac{\pi}{2}\frac{\sqrt{\lambda(P^2,0,p^2)}}{P^2} = \frac{\pi}{2}\left(1-\frac{m_\pi^2}{M^2}\right)\,, \tag{13.30}$$

we obtain

$$\Gamma = \frac{G_{\rm F}^2|V_{\rm ud}|^2}{16\pi}\, f_\pi^2\, M^3 \left(1-\frac{m_\pi^2}{M^2}\right)^2 = 12\,\pi^2\,|V_{\rm ud}|^2\,\frac{f_\pi^2}{M^2}\left(1-\frac{m_\pi^2}{M^2}\right)^2\Gamma_0\,. \tag{13.31}$$

This formula is to be compared with $\pi^- \to \mu^- + \overline{\nu}_\mu$ (Problem 5.3):

$$\Gamma(\pi^- \to \ell^- + \overline{\nu}_\ell) = \frac{G_{\rm F}^2\,|V_{\rm ud}|^2}{8\pi} f_\pi^2 m_\ell^2\, m_\pi \left(1-\frac{m_\ell^2}{m_\pi^2}\right)^2\,. \tag{13.32}$$

From $\Gamma(\pi^- \to \mu^- + \overline{\nu}_\mu) = 0.384 \times 10^8\ {\rm s}^{-1} = 2.53 \times 10^{-14}$ MeV, one gets $f_\pi \approx 131$ MeV. Comparing (26) with (31), the ratio of the branching fractions

$$\frac{{\rm B}_\pi}{{\rm B}_{\rm e}} \equiv \frac{\Gamma(\tau^- \to \nu_\tau + \pi^-)}{\Gamma(\tau^- \to \nu_\tau + {\rm e}^- + \overline{\nu}_{\rm e})} = 12\pi^2\,|V_{\rm ud}|^2 \frac{f_\pi^2}{M^2}\left(1-\frac{m_\pi^2}{M^2}\right)^2 = 0.60$$

is in good agreement with data. A straightforward generalization can be made for the Cabibbo-suppressed mode $\tau^- \to \nu_\tau + {\rm K}^-$, with the interchange $V_{\rm ud} \leftrightarrow V_{\rm us}$ and $f_\pi, m_\pi \leftrightarrow f_{\rm K}, m_{\rm K}$ in (31). The decay constant $f_{\rm K} \approx 160$ MeV is obtained from the ${\rm K}^- \to \mu^- + \overline{\nu}_\mu$ rate, similar to (32) for $f_\pi$. The ratio

$$\frac{{\rm B}_{\rm K}}{{\rm B}_\pi} \equiv \frac{\Gamma(\tau^- \to \nu_\tau + {\rm K}^-)}{\Gamma(\tau^- \to \nu_\tau + \pi^-)} = \frac{|V_{\rm us}|^2}{|V_{\rm ud}|^2}\frac{f_{\rm K}^2}{f_\pi^2}\left(\frac{M^2-m_{\rm K}^2}{M^2-m_\pi^2}\right)^2 = 0.066$$

is in agreement with data.

### 13.3.2 The $2n$-Pion Mode and CVC

Before the discovery of $\tau$, CVC was only tested at low momentum transfer $q^2$ in nuclear physics and neutrino–nucleon scattering (Chap. 12). For the first time CVC can be tested at high momentum $q^2$ released by $\tau$. The $\tau^-(P) \to \nu_\tau(k) + \pi^-(p_1) + \pi^0(p_2)$ amplitude can be obtained from the weak vector current $V_\mu = \overline{d}\gamma_\mu u$:

$$\mathcal{M} = \frac{G_F V_{ud}}{\sqrt{2}} \overline{u}(k)\gamma^\mu(1-\gamma_5)u(P) \left\langle \pi^-(p_1)\pi^0(p_2) \,|\, V_\mu \,|\, 0 \right\rangle . \tag{13.33}$$

Using CVC, we can relate the two-pion matrix element of $V_\mu$ to that of the electromagnetic current $J_\mu^{\mathrm{em}}$:

$$\begin{aligned} \left\langle \pi^-(p_1)\pi^0(p_2) \,|\, V_\mu \,|\, 0 \right\rangle &= \sqrt{2} \left\langle \pi^-(p_1)\pi^+(p_2) \,\middle|\, J_\mu^{\mathrm{em}} \,\middle|\, 0 \right\rangle \\ &= \sqrt{2}\,(p_1 - p_2)_\mu F_\pi(q^2) , \end{aligned} \tag{13.34}$$

where $q_\mu = (p_1 + p_2)_\mu$. $F_\pi(q^2)$ is the pion electromagnetic form factor, already introduced in (10.11). The momentum transfer $q^2$ is $\geq 4m_\pi^2$. We calculate

$$\begin{aligned} \tfrac{1}{2}\sum_{\mathrm{spin}} |\mathcal{M}|^2 &= 4\,G_F^2 |V_{ud}|^2 |F_\pi(q^2)|^2 \left\{Y\right\} , \\ \left\{Y\right\} &= 2\left[k\cdot(p_1 - p_2)\right]\left[P\cdot(p_1 - p_2)\right] - (P\cdot k)(p_1 - p_2)^2 . \end{aligned} \tag{13.35}$$

Using $P = k + q$ and $P\cdot(p_1 - p_2) = k\cdot(p_1 - p_2)$, we rewrite $\left\{Y\right\}$ in the following form which is convenient for the phase space integration:

$$\left\{Y\right\} = 6\,(k\cdot p_1)^2 + 2\,(k\cdot p_2)^2 - 2\,(M^2 - q^2)\,(k\cdot p_1) + \frac{(M^2 - q^2)\,(q^2 - 4m_\pi^2)}{2} .$$

With (35), the decay rate is given by

$$\begin{aligned} \Gamma &= 4\,G_F^2 |V_{ud}|^2 \frac{1}{2M} \frac{1}{(2\pi)^5} \int_{\mathcal{PS}3} \left\{Y\right\} |F_\pi(q^2)|^2 , \quad \text{where} \\ \int_{\mathcal{PS}3} &\equiv \int \frac{\mathrm{d}^3 p_1}{2E_1} \frac{\mathrm{d}^3 p_2}{2E_2} \frac{\mathrm{d}^3 k}{2E} \delta^4(p_1 + p_2 + k - P) . \end{aligned} \tag{13.36}$$

Since (35) is symmetric under $p_1 \leftrightarrow p_2$, the phase space integration $\int_{\mathcal{PS}3}$ of $(k\cdot p_2)^2$ in $\left\{Y\right\}$ is equal to that of $(k\cdot p_1)^2$. With the help of formulas in the Appendix, we get for different terms in $\left\{Y\right\}$:

$$\begin{aligned} \int_{\mathcal{PS}3} (k\cdot p_1)^2 &= \frac{\pi^2}{48} \int_{4m_\pi^2}^{M^2} \mathrm{d}q^2 \left(1 - \frac{q^2}{M^2}\right) \sqrt{1 - \frac{4m_\pi^2}{q^2}}\,(1 - \frac{m_\pi^2}{q^2})\,\left(M^2 - q^2\right)^2 , \\ \int_{\mathcal{PS}3} (k\cdot p_1) &= \frac{\pi^2}{16} \int_{4m_\pi^2}^{M^2} \mathrm{d}q^2 \left(1 - \frac{q^2}{M^2}\right) \sqrt{1 - \frac{4m_\pi^2}{q^2}}\,(M^2 - q^2) , \\ \int_{\mathcal{PS}3} &= \frac{\pi^2}{4} \int_{4m_\pi^2}^{M^2} \mathrm{d}q^2 \left(1 - \frac{q^2}{M^2}\right) \sqrt{1 - \frac{4m_\pi^2}{q^2}} . \end{aligned} \tag{13.37}$$

Putting together (35) and (37) into (36), we finally obtain

$$\Gamma = \frac{G_F^2\,|V_{ud}|^2\,M^3}{384\,\pi^3}\int_{4m_\pi^2}^{M^2} \mathrm{d}\,q^2\left(1-\frac{4m_\pi^2}{q^2}\right)^{3/2}\left(1-\frac{q^2}{M^2}\right)^2\left(1+\frac{2\,q^2}{M^2}\right)|F_\pi(q^2)|^2$$
$$= \frac{\Gamma_0\,|V_{ud}|^2}{2}\int_{4m_\pi^2}^{M^2}\frac{\mathrm{d}\,q^2}{M^2}\left(1-\frac{4m_\pi^2}{q^2}\right)^{3/2}\left(1-\frac{q^2}{M^2}\right)^2\left(1+\frac{2\,q^2}{M^2}\right)|F_\pi(q^2)|^2\,.$$

The pion form factor $|F_\pi(q^2)|$ can be directly measured from experiments $e^+ + e^- \to \pi^+ + \pi^-$, its cross-section is given by (Problem 10.3)

$$\sigma_{e^+ + e^- \to \pi^+ + \pi^-}(q^2) = \frac{\pi\,\alpha^2}{3\,q^2}\left(1-\frac{4m_\pi^2}{q^2}\right)^{3/2}|F_\pi(q^2)|^2 \equiv \sigma(q^2)\,, \qquad (13.38)$$

$$\text{then } \Gamma = \Gamma_0\,\frac{3\,|V_{ud}|^2}{2\pi\,\alpha^2}\int_{4m_\pi^2}^{M^2}\frac{\mathrm{d}\,q^2}{M^2}\left(1-\frac{q^2}{M^2}\right)^2\left(1+\frac{2\,q^2}{M^2}\right)q^2\,\sigma(q^2)\,. \qquad (13.39)$$

Putting the data[4] of $\sigma(q^2)$ into (39) and performing numerical integration, the resulting branching ratio $B_{2\pi} = (23.54 \pm 1.2)\%$ is in agreement with experiment $(25.24 \pm 0.16)\%$. Using only the $e^+ + e^- \to \pi^+ + \pi^-$ data as input, this result constitutes a powerful test of CVC.

Also, for an even number of pions, the rate $\Gamma(\tau \to \nu_\tau + 2n$ pions) can be directly obtained by CVC from the cross-sections $\sigma(e^+ + e^- \to 2n$ pions) using (39). In particular the branching ratios for $\tau^- \to \nu_\tau + \pi^- + \pi^+ + \pi^- + \pi^0$ and $\tau^- \to \nu_\tau + \pi^- + \pi^0 + \pi^0 + \pi^0$ are computed to be respectively $(4.9 \pm 2)\%$ and $(0.98 \pm 0.4)\%$, using $\sigma(e^+ + e^- \to 4$ pions) data. They are again in good agreement with experiments.

We note that the two pions must be in an isospin $I = 1$ state since it is created from the vacuum by the $I = 1$ vector current $\bar{d}\gamma^\mu u$. Bose statistics implies that the dipion is in p-wave. It turns out that in the energy range $\sim 1$ GeV of $\tau$ decay, the p-wave two-pion state resonates to form the $\rho(770)$ meson: $\tau^- \to \nu_\tau + \pi^- + \pi^0 \approx \tau^- \to \nu_\tau + \rho^-$. This $\rho(770)$ dominance of the two-pion state enhances the pion form factor $F_\pi(q^2)$ in the region $q^2 \approx m_\rho^2$. As discussed in Chap. 10 (Fig. 10.2), the form factor $F_\pi(q^2)$ may be parameterized by the Breit–Wigner resonance form

$$F_\pi(q^2) = \frac{m_\rho f_\rho g_{\rho\pi\pi}}{m_\rho^2 - q^2 - \mathrm{i}\,m_\rho\Gamma_\rho} \longrightarrow \frac{m_\rho f_\rho g_{\rho\pi\pi}}{m_\rho^2 - q^2 - \mathrm{i}\sqrt{q^2}\Gamma_\rho(q^2)}\,. \qquad (13.40)$$

In (40), instead of keeping the constant $\mathrm{i}\,m_\rho\Gamma_\rho$, it may be more appropriate to take into account the $q^2$ dependence of the $\rho$ width, i.e.

$$m_\rho\Gamma_\rho \longrightarrow \sqrt{q^2}\Gamma_\rho(q^2)\,, \text{ where } \Gamma_\rho(q^2) = \Gamma_\rho\frac{m_\rho^2}{q^2}\left(\frac{q^2 - 4m_\pi^2}{m_\rho^2 - 4m_\pi^2}\right)^{3/2}\,. \qquad (13.41)$$

[4] L.M. Barkov et al., Nucl. Phys. **B256** (1985) 365

Putting (40) into (38), after doing the $q^2$ integration, the rate obtained is again in excellent agreement with data. In the narrow width approximation of the $\rho$, the Breit–Wigner factor becomes a delta function. Using

$$\delta(x) = \lim_{\varepsilon\to 0} \frac{\varepsilon}{\pi}\frac{1}{x^2+\varepsilon^2} \,,$$

we write

$$|F_\pi(q^2)|^2 = \frac{\left(m_\rho f_\rho g_{\rho\pi\pi}\right)^2}{(q^2-m_\rho^2)^2+\Gamma_\rho^2 m_\rho^2} \longrightarrow \frac{\pi\,\delta(q^2-m_\rho^2)}{\Gamma_\rho m_\rho}\left(m_\rho f_\rho g_{\rho\pi\pi}\right)^2 . \quad (13.42)$$

We recall that the strong coupling $g_{\rho\pi\pi}$ is related to the $\rho$ width $\Gamma_\rho$ by (10.20)

$$\Gamma_\rho = \frac{g_{\rho\pi\pi}^2 m_\rho}{48\pi}\left(1-\frac{4m_\pi^2}{m_\rho^2}\right)^{3/2} .$$

Combining the above relation with (38) and (42), we get

$$\sigma(\mathrm{e}^+ + \mathrm{e}^- \to \rho^0) \longrightarrow 16\alpha^2\pi^3\frac{f_\rho^2}{m_\rho^2}\delta(q^2-m_\rho^2) \,. \quad (13.43)$$

Putting (43) in (39), we obtain

$$\Gamma(\tau^- \to \nu_\tau + \rho^-) = 24\,\pi^2\,|V_{\mathrm{ud}}|^2\frac{f_\rho^2}{M^2}\left(1+\frac{2m_\rho^2}{M^2}\right)\left(1-\frac{m_\rho^2}{M^2}\right)^2 \Gamma_0 \,. \quad (13.44)$$

According to (10.21) and Fig. 10.3, the $\rho^0$ decay constant is $f_\rho \approx 150 \pm 10$ MeV coming from the $\rho^0 \to \mathrm{e}^+ + \mathrm{e}^-$ width. We get

$$\begin{aligned} \frac{\mathrm{B}_\rho}{\mathrm{B}_\mathrm{e}} &\equiv \frac{\Gamma(\tau^- \to \nu_\tau + \rho^-)}{\Gamma(\tau^- \to \nu_\tau + \mathrm{e}^- + \overline{\nu}_\mathrm{e})} = 24\pi^2|V_{\mathrm{ud}}|^2\frac{f_\rho^2}{M^2}\left(1+\frac{2m_\rho^2}{M^2}\right)\left(1-\frac{m_\rho^2}{M^2}\right)^2 \\ &= 1.44 \pm 0.2 \quad \Longrightarrow \mathrm{B}_\rho = 0.256 \pm 0.035 \,. \end{aligned} \quad (13.45)$$

A straightforward generalization can also be made for the Cabibbo-suppressed mode $\tau^- \to \nu_\tau + \mathrm{K}^{*-}(892)$ by the substitution $V_{\mathrm{ud}} \to V_{\mathrm{us}}$, $m_\rho \to m_{\mathrm{K}^*}$, $f_\rho \to f_{\mathrm{K}^*}$ in (44). However, the decay constant $f_{\mathrm{K}^*}$, unlike the $f_\rho$, is difficult to determine by experiment. It may be estimated by the SU(3) flavor symmetry, which gives $m_{\mathrm{K}^*} f_{\mathrm{K}^*} = m_\rho f_\rho$. We get $\mathrm{B}_{\mathrm{K}^*} = (1.1 \pm 0.1)\%$ for the branching ratio of $\tau \to \nu_\tau + \mathrm{K}^*$ to be compared with data $(1.43 \pm 0.31)\%$.

The method for obtaining $\tau^- \to \nu_\tau + \pi^- + \pi^0$ can be generalized to the $\tau^- \to \nu_\tau + \mathrm{K}^- + \mathrm{K}^0$ decay, which is also a Cabibbo-favored mode. The $\mathrm{K}^- + \mathrm{K}^0$ pair ($S = 0$) can be created by the same conserved vector current $\overline{d}\gamma^\mu u$ from the sea $\overline{\mathrm{s}}\mathrm{s}$ in the vacuum. The only replacements in (38) and (39) are $F_\pi(q^2) \leftrightarrow F_\mathrm{K}(q^2)$ and $m_\pi \leftrightarrow m_\mathrm{K}$. However, compared with the two-pion case, the $\mathrm{K}^-+\mathrm{K}^0$ mode is suppressed by both kinematic and dynamic reasons. Kinematic suppression is due to the factor $(1/2)\,[1-(4m_\mathrm{K}^2/q^2)]^{3/2}$, and the $q^2$ phase space integration range is smaller. Dynamic suppression occurs because an $I = 1, J^P = 1^-$ resonance for $\mathrm{K} + \overline{\mathrm{K}}$ is lacking, the form factor $F_\mathrm{K}(q^2)$ is not enhanced and is smaller than $F_\pi(q^2)$. The sea $\overline{\mathrm{s}}\mathrm{s}$ contribution is also negligible.

## 13.4 The Method of Spectral Functions

Instead of considering semileptonic exclusive channels with one or two particles in the final state as we have just done, we introduce now the notion of spectral functions which represent a more systematic way of dealing with multiparticle system. The aim is to derive a general formulation, valid for any hadronic channels, and all the formulas developed in the previous section could be recovered. In addition, the formalism is useful later for the computation of the inclusive rate.

The semileptonic decay width $\tau^-(P) \to \nu_\tau(k) + H^-$, where $H^-$ is any hadronic system of one or several particles, can be written as

$$\Gamma(\tau^- \to \nu_\tau + H^-) = \frac{1}{2M}\frac{G_F^2}{2}\int \frac{d^3k}{2E_k(2\pi)^3}\,\frac{1}{2}\{2\,\mathrm{Tr}[\not{k}\gamma_\mu \not{P}\gamma_\nu(1-\gamma_5)]\}\mathcal{H}^{\mu\nu},$$

$$\mathcal{H}^{\mu\nu} \equiv \sum_{PS_H} \langle 0\,|\,J^\mu\,|\,H^-(p_H)\rangle\,\langle H^-(p_H)\,|\,J^{\dagger\nu}\,|\,0\rangle\;(2\pi)^4\delta^4(p_H + k - P). \quad (13.46)$$

In (46), the coefficient 1/2 represents the averaging of the $\tau$ spin, while the coefficient 2 before the trace comes from $(1-\gamma_5)^2 = 2(1-\gamma_5)$ of the leptonic tensor. As in (10.42), the hadronic tensor $\mathcal{H}^{\mu\nu}$ is the product of the matrix element of the current $J^\mu$ with that of the current $J^{\nu\dagger}$, where the V – A current $J^\mu$ is the sum of two parts $V_{ud}\,\overline{d}\gamma^\mu(1-\gamma_5)u$ and $V_{us}\,\overline{s}\gamma^\mu(1-\gamma_5)u$. The first term $V_{ud}\,\overline{d}\gamma^\mu(1-\gamma_5)u$ carries isospin $I=1$ and strangeness $S=0$, while the second term $V_{us}\overline{s}\gamma^\mu(1-\gamma_5)u$ has $I=1/2$ and $S=-1$. The symbol $\sum_{PS_H}$ of $\mathcal{H}^{\mu\nu}$ in (46) denotes the phase space integration of the final state in $H^-$ (including the spin summation):

$$\sum_{PS_H} \equiv \int \prod_{j\in H}\left[\frac{d^3p_j}{2E_j(2\pi)^3}\right], \text{ with } p_H \equiv \sum_j p_j = P - k \equiv q\,.$$

No matter how complicated $\mathcal{H}^{\mu\nu}$ is, it has two general features:

(i) $\mathcal{H}^{\mu\nu}$ has the dimension of (mass)$^2$, we can check this point by inspecting the dimension of the different terms in (46).

(ii) After doing the phase space integration of all the particles in $H^-$ and summing over their spins, the four-momentum transfer $q$ remains as the only dynamical variable in $\mathcal{H}^{\mu\nu}(q)$. Therefore the latter must be a function of the Lorentz-invariant $q^2$ and can only depend on the tensors $q^\mu q^\nu$ and $g^{\mu\nu}$. They can be conveniently put into two independent sets $(-q^2g^{\mu\nu}+q^\mu q^\nu)$ and $q^\mu q^\nu$ which are respectively orthogonal and parallel to the four-momentum $q$, i.e. the former satisfies $q_\mu(-q^2g^{\mu\nu}+q^\mu q^\nu)=0$, $q_\nu(-q^2g^{\mu\nu}+q^\mu q^\nu)=0$. It is important to note that the hadronic state $H^-$ (whether a single or multiparticle system) has a total angular momentum $j=1$ or 0, since $H^-$ is created from the vacuum by the virtual W boson which carries at most $j=1$.

These two features enable us to write the most general form for $\mathcal{H}^{\mu\nu}$ in terms of the dimensionless spectral functions

$$\sum_{PS_{\rm H}} \langle 0 \,|\, J^\mu \,|\, H^-(p_{\rm H})\rangle \langle H^-(p_{\rm H}) \,|\, J^{\nu\dagger} \,|\, 0\rangle \; (2\pi)^4\delta^4(p_{\rm H}-q)$$
$$= (-q^2 g^{\mu\nu} + q^\mu q^\nu)\,[v_1(q^2) + a_1(q^2] + q^\mu q^\nu\,[v_0(q^2) + a_0(q^2)] \,. \quad (13.47)$$

The $v_{1,0}(q^2)$ come from the vector currents, the $a_{1,0}(q^2)$ from the axial currents. The subscript 1 in $v_1(q^2)$ and $a_1(q^2)$ corresponds to the total angular momentum $j = 1$ of $H^-$, whereas $v_0(q^2)$ and $a_0(q^2)$ are associated with $j = 0$. Note that the vector current $V^\mu = \bar{d}\gamma^\mu u$ is conserved $(q_\mu \bar{d}\gamma^\mu u = 0)$, the product of the matrix element of $V^\mu$ with that of $V^{\nu\dagger}$ gives rise, after the phase space $\sum_{PS_H}$ operation, to the tensor $(-q^2 g^{\mu\nu} + q^\mu q^\nu)$. This implies that only $v_1(q^2)$ exists, while $v_0(q^2) = 0$.

On the other hand, the axial current $A^\mu = \bar{d}\gamma^\mu\gamma_5 u$ is not conserved, the product of the matrix element of $A^\mu$ with that of $A^{\nu\dagger}$ contains both $(-q^2 g^{\mu\nu} + q^\mu q^\nu) a_1(q^2)$ and $q^\mu q^\nu a_0(q^2)$. Because of the G-parity, the current $V^\mu$ $(A^\mu)$ produces an even (odd) numbers of pions in the final state. Table 13.1 summarizes the point.

**Table 13.1.** Spectral functions of currents and the associated final states

| Currents | Spectral functions | $J^P$ | Final states |
|---|---|---|---|
| $V^\mu = \bar{d}\gamma^\mu u$ | $v_1(q^2)\,,\; v_0(q^2) = 0$ | $1^-$ | $\rho^-\,,\; 2\pi\,,\; 4\pi\,,\; \mathrm{K}^- + \mathrm{K}^0$ |
| $A^\mu = \bar{d}\gamma^\mu\gamma_5 u$ | $a_1(q^2)\,,\; a_0(q^2)$ | $0^-\,,\; 1^+$ | $\pi^-\,,\; 3\pi\,,\; \mathrm{a}_1^-(1260)$ |
| $V^\mu_{\rm S} = \bar{s}\gamma^\mu u$ | $v_1^{\rm S}(q^2)\,,\; v_0^{\rm S}(q^2)$ | $0^+\,,\; 1^-$ | $\mathrm{K}_0^{*-}(1430)\,,\; \mathrm{K}^{*-}(892)$ |
| $A^\mu_{\rm S} = \bar{s}\gamma^\mu\gamma_5 u$ | $a_1^{\rm S}(q^2)\,,\; a_0^{\rm S}(q^2)$ | $0^-\,,\; 1^+$ | $\mathrm{K}^-\,,\; \mathrm{K}_1^-(1270)$ |

Since (47) is symmetric in $\mu \leftrightarrow \nu$, when we put $\mathcal{H}^{\mu\nu}$ into (46), the role of $\gamma_5$ in the leptonic part $\mathrm{Tr}[\not{k}\gamma_\mu \not{P}\gamma_\nu(1-\gamma_5)]$ is automatically superfluous, its antisymmetric tensor $\mathrm{i}\,\varepsilon_{\mu\nu\alpha\beta}\, k^\alpha P^\beta$ does not contribute. Note that

$$\int \frac{\mathrm{d}^3 k}{2E_k} = \frac{\pi}{2M^2}\int \sqrt{\lambda(M^2,k^2,q^2)}\,\mathrm{d}q^2 \underset{k^2=0}{\longrightarrow} \frac{\pi}{2M^2}\int (M^2-q^2)\,\mathrm{d}q^2 \,. \quad (13.48)$$

With massless neutrino ($k^2 = 0$), using (46), (47), and (48), we obtain for the Cabibbo-favored modes associated with $V_{\rm ud}$ and the current $\bar{d}\gamma^\mu(1-\gamma_5)u$

$$\Gamma^{S=0} = \frac{G_{\rm F}^2|V_{\rm ud}|^2}{32\pi^2 M^3}\int_{m_\pi^2}^{M^2} \mathrm{d}q^2\,(M^2-q^2)^2$$
$$\times \Big\{(M^2+2q^2)[v_1(q^2)+a_1(q^2)] + M^2 a_0(q^2)\Big\}. \quad (13.49)$$

Associated with $V_{us}$, the vector part $\bar{s}\gamma^\mu u$ of the $I = 1/2, S = -1$ current $\bar{s}\gamma^\mu(1-\gamma_5)u$ is not conserved, so the corresponding spectral function $v_0^S(q^2)$ is not vanishing. This Cabibbo-suppressed width is

$$\Gamma^{S=-1} = \frac{G_F^2|V_{us}|^2}{32\pi^2 M^3} \int_{m_K^2}^{M^2} dq^2\,(M^2-q^2)^2 \times \left\{(M^2+2q^2)[v_1^S(q^2)+a_1^S(q^2)] + M^2[v_0^S(q^2)+a_0^S(q^2)]\right\}. \quad (13.50)$$

Formulas (49) and (50) show that the calculation of the rate is reduced to a computation of the spectral functions.

We can check the general structure (47) of $\mathcal{H}^{\mu\nu}(q)$ with one-particle state by putting $H^- = \pi^-$, $\rho^-$ in (47) and find

$$a_0^\pi(q^2) = 2\pi f_\pi^2\,\delta(q^2-m_\pi^2) \quad ; \quad v_1^\rho(q^2) = 2\pi(\sqrt{2}f_\rho)^2\,\delta(q^2-m_\rho^2)\,. \quad (13.51)$$

The easiest way to obtain $a_0^\pi(q^2)$ in (51) is to use

$$\langle 0\,|\,A^\mu\,|\,\pi(p_H)\rangle = \mathrm{i}\,f_\pi\,p_H^\mu \quad \text{and} \quad \frac{d^3p_H}{(2\pi)^3\,2E_H} = \frac{d^4p_H}{(2\pi)^3}\,\delta(p_H^2-m_\pi^2)\theta(p_H^0)\,.$$

The tensor $(-q^2g^{\mu\nu}+q^\mu q^\nu)$ associated with $v_1^\rho(q^2)$ comes from

$$\sum_{\rho\ \mathrm{spin}} \varepsilon^\mu(q)\,\varepsilon^\nu(q) = -g^{\mu\nu} + q^\mu\,q^\nu/m_\rho^2\,.$$

Putting (51) into (49), we recover (31) and (44).

For the two-pion state $\pi^-+\pi^0$, the corresponding spectral function $v_1^{\pi\pi}(q^2)$ can be obtained by putting (34) into (47) and using the formulas in the Appendix for the two-particle phase space integration. We get

$$v_1^{\pi\pi}(q^2) = \left(1-\frac{4m_\pi^2}{q^2}\right)^{3/2}\frac{|F_\pi(q^2)|^2}{12\pi}\,. \quad (13.52)$$

Putting (52) into (49), we again recover (38) and (39).

### 13.4.1 The Three-Pion Mode

As a first illustration of the spectral function method, let us consider the $\tau^- \to \nu_\tau + 3\pi$ decay which has a substantial branching ratio of about 18% for the sum of two modes $\pi^-+\pi^++\pi^-$ and $\pi^-+\pi^0+\pi^0$. By the G-parity conservation, the vector current $V^\mu = \bar{d}\gamma^\mu u$, which has G=+1, does not contribute. The relevant axial current $A^\mu = \bar{d}\gamma^\mu\gamma_5 u$, responsible for $\tau \to \nu_\tau + 3\,\pi$, is the same current that intervenes in the one-pion mode discussed previously. However, in contrast to the one-pion or the two-pion decays for which everything in the amplitude (29) or (34) is known, the matrix element

$\langle 3\pi \,|\, A^\mu \,|\, 0\rangle$ for the three-pion mode is more complicated and poorly known. Its most general expression depends on three form factors $H_1$, $H_2$, $H_3$, each of which is a function of three kinematical variables $q^2$, $s_1$, $s_2$ defined below, and consequently the decay rate calculation is rather model dependent. The three-pion decay amplitude is given by

$$\mathcal{M} = \frac{G_\mathrm{F}}{\sqrt{2}} V_\mathrm{ud}\overline{u}(k)\gamma_\mu(1-\gamma_5)u(P)\ \langle \pi_1(p_1)\,,\ \pi_2(p_2)\,,\ \pi_3(p_3)\,|\,A^\mu\,|\,0\rangle\ ,$$
$$\langle \pi_1(p_1)\,,\ \pi_2(p_2)\,,\ \pi_3(p_3)\,|\,A^\mu\,|\,0\rangle = \sum_{i=1,2,3} H_i(q^2, s_1, s_2)\frac{p_i^\mu}{m_\pi}\ ,$$
$$q^2 = (p_1+p_2+p_3)^2\ ,\ s_1 = (p_2+p_3)^2\ ,\ s_2 = (p_1+p_3)^2\ . \tag{13.53}$$

The existence of three form factors is easy to understand, since the only degrees of freedom for spinless pions are their momenta. The most general covariant structure of $\langle 3\pi \,|\, A^\mu \,|\, 0\rangle$ can be expressed in terms of these three independent momenta $p_i^\mu$ taken as a basis, their coefficients are the form factors $H_i$. Kinematically, each of these form factors is considered as a *four-point function* which connects the off-mass-shell W gauge boson to the three on-shell pions. The four-point function depends on $q^2$ (the virtual mass squared of the W) as well as on two independent Mandelstam variables that can be chosen as $s_1$ and $s_2$. These form factors $H_i(q^2, s_1, s_2)$ are not well determined, unlike the more familiar $F_\pi(q^2)$ in the two-pion case.

To compute the three-pion rate, the first approximation consists of considering the $3\pi$ as a quasi-two-body state $\rho(770)+\pi$ followed by $\rho \to 2\pi$. Then the matrix element may be written as

$$\langle \rho(p), \pi(p')\,|\,A^\mu\,|\,0\rangle = m_\rho\,\varepsilon^\mu K_1(q^2) + \frac{\varepsilon\cdot q}{m_\rho}\left[(p-p')^\mu K_2(q^2) + q^\mu K_3(q^2)\right],$$

where $\varepsilon^\mu$ is the four-vector polarization of the $\rho$ and the three dimensionless form factors $K_i(q^2)$ depend only on $q^2 = (p+p')^2 = (p_1+p_2+p_3)^2$. These form factors $K_i(q^2)$ may be considered as linear combinations of the $H_i(q^2, s_1, s_2)$ in the limit where two pions form a $\rho$ resonance. Then from the $\rho$ propagator, the dependence of $H_i(q^2, s_1, s_2)$ on the variables $s_{j=1,2}$ would take the Breit–Wigner shape similar to (40), i.e. the $s_j$ are concentrated near $m_\rho^2$ as $\left[m_\rho^2 - s_j - \mathrm{i}\sqrt{s_j}\Gamma_\rho(s_j)\right]^{-1}$ (Fig. 13.6).

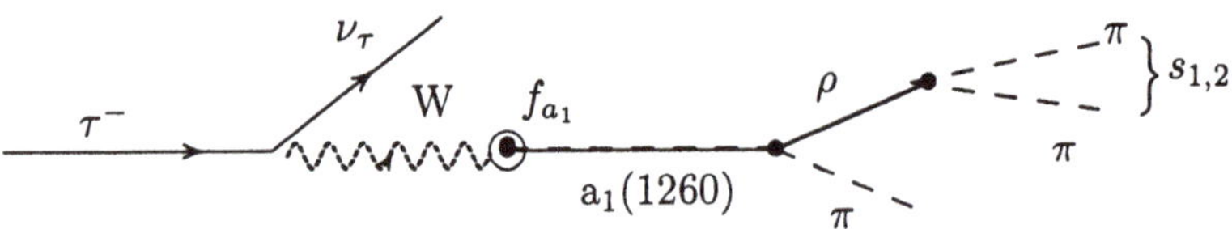


**Fig. 13.6.** $\tau^- \to \nu_\tau + \rho + \pi$ followed by $\rho \to 2\pi$

The next approximation is to assume the dominance of two resonances, the axial $\mathrm{a}_1(1260)$ meson ($J^P = 1^+$) and the pseudoscalar $\pi(1300) \equiv \pi'$ meson. In analogy with the decay $\tau \to \nu_\tau + \rho$ followed by $\rho \to 2\pi$ considered previously, the three-pion mode may be approximated by $\tau \to \nu_\tau + \mathrm{a}_1(1260)$ and $\tau \to \nu_\tau + \pi'$ followed by $\mathrm{a}_1(1260) \to 3\pi$ and $\pi' \to 3\pi$. The spectral functions

$$a_1^{3\pi}(q^2) = 2\pi f_{\mathrm{a}_1}^2 \, \delta(q^2 - m_{\mathrm{a}_1}^2) \ , \quad a_0^{3\pi}(q^2) = 2\pi f_{\pi'}^2 \, \delta(q^2 - m_{\pi'}^2) \ , \tag{13.54}$$

correspond to the zero-width approximation of both $\mathrm{a}_1(1260)$ and $\pi'$, where the decay constants $f_{\mathrm{a}_1}$ and $f_{\pi'}$ of $\mathrm{a}_1(1260)$ and $\pi'$ are defined by

$$\langle \mathrm{a}_1^-(1260) \,|\, A^\mu \,|\, 0 \rangle = m_{\mathrm{a}_1} f_{\mathrm{a}_1} \varepsilon^\mu \quad , \quad \langle \pi'(p) \,|\, A^\mu \,|\, 0 \rangle = \mathrm{i}\, f_{\pi'} p^\mu \ .$$

Unlike $f_\pi$ and $f_\rho$, the $f_{\pi'}$ and $f_{\mathrm{a}_1}$ are not well determined experimentally. Weinberg sum rules[5] may be useful for an estimate of $f_{\mathrm{a}_1}$. In our notation, the first Weinberg sum rule is written as

$$f_{\rho^\pm}^2 - f_{\mathrm{a}_1}^2 = f_\pi^2 \ , \text{ where } \ f_{\rho^\pm} = \sqrt{2} f_\rho \ . \tag{13.55}$$

The crudest estimation consists in keeping only $a_1^{3\pi}(q^2)$ and neglecting $a_0^{3\pi}(q^2)$. Using (49), the three-pion rate is given by

$$\begin{aligned} \frac{\Gamma(\tau^- \to \nu_\tau + 3\pi)}{\Gamma(\tau^- \to \nu_\tau + 2\pi)} &= \frac{\Gamma(\tau^- \to \nu_\tau + \mathrm{a}_1^-)}{\Gamma(\tau^- \to \nu_\tau + \rho^-)} \\ &= \frac{f_{\mathrm{a}_1}^2}{f_{\rho^\pm}^2} \left( \frac{M^2 - m_{\mathrm{a}_1}^2}{M^2 - m_\rho^2} \right)^2 \frac{M^2 + 2m_{\mathrm{a}_1}^2}{M^2 + 2m_\rho^2} \ . \end{aligned} \tag{13.56}$$

Combining (56) with (55), the branching ratio $\mathrm{Br}(\tau^- \to \nu_\tau + 3\pi)$ is found to be $\approx 9\%$, lower than the data by a factor of 2. Since both the $\rho$ and the $\mathrm{a}_1(1260)$ are broad resonances, this zero-width approximation is rather poor. One may improve the estimation by replacing $\delta(q^2 - m_{\mathrm{a}_1}^2)$ with a Breit–Wigner factor, in analogy with the $\rho$ case discussed in (41) where the $q^2$ dependence of the width is taken into account. We write

$$\begin{aligned} \delta(q^2 - m_{\mathrm{a}_1}^2) &\longrightarrow \frac{\Gamma_{a_1}(q^2)\sqrt{q^2}}{\pi} \, \frac{1}{(q^2 - m_{\mathrm{a}_1}^2)^2 + q^2\, \Gamma_{a_1}^2(q^2)} \ , \\ \Gamma_{a_1}(q^2) &= \Gamma_{a_1} \frac{m_{\mathrm{a}_1}^2}{q^2} \left( \frac{q^2 - 9\, m_\pi^2}{m_{\mathrm{a}_1}^2 - 9\, m_\pi^2} \right)^{3/2} \ , \text{ where } \Gamma_{a_1} \sim 400 \text{ MeV} \ , \\ a_1^{3\pi}(q^2) &= \frac{2\, f_{\mathrm{a}_1}^2 \, \Gamma_{a_1}(q^2)\sqrt{q^2}}{(q^2 - m_{\mathrm{a}_1}^2)^2 + q^2\, \Gamma_{a_1}^2(q^2)} \ . \end{aligned}$$

One may use this formula for $a_1^{3\pi}(q^2)$ and plug it into (49) to extract the decay constant $f_{\mathrm{a}_1}$ from the experimental branching ratio. The result is encouraging and gives $f_{\mathrm{a}_1} \approx 250$ MeV.

[5] S. Weinberg, Phys. Rev. Lett. **18** (1967) 507

### 13.4.2 Spectral Functions of Quark Pairs

We calculated in the previous sections many exclusive modes $\tau^- \to \nu_\tau + H^-$, where $H^-$ are $\pi^-$, $\mathrm{K}^-$, $\pi^- + \pi^0$, (including $\rho^-$, $\mathrm{K}^{*-}$), $\mathrm{K}^- + \overline{\mathrm{K}}^0$, $3\pi$. The sum of these dominant decays almost saturates the observed semileptonic width, but still leaves out many other channels. We would like however to sum over all the exclusive semileptonic rates, which is, by definition, the inclusive decay rate $\Gamma(\tau^- \to \nu_\tau +$ hadrons). Doing such a sum, i.e. computing the whole spectral functions $v_{1,0}(q^2)$ and $a_{1,0}(q^2)$ for several particles, is not only cumbersome but also involves large uncertainties due to our lack of knowledge of the involved form factors. We must look for another approach.

Fortunately, the energy released by the heavy $\tau$ is large enough (on the QCD scale) that the quark-parton model is presumably valid. Then the inclusive rate may be saturated by $\tau^- \to \nu_\tau + \mathrm{d} + \overline{\mathrm{u}}$ and $\tau^- \to \nu_\tau + \mathrm{s} + \overline{\mathrm{u}}$. This saturation has the name of quark-hadron duality. For the calculation of these two rates, we need the spectral functions of quark pairs which replace those of all the exclusive hadronic states.

Let us denote the pair carrying the same color $j$ (since it is created from the colorless $\mathrm{W}^\pm$) by $\mathrm{q}_2 + \overline{\mathrm{q}}_3$ with momentum $p_2$, $p_3$ and mass $m_2$, $m_3$ respectively. Defined in (46), the hadronic tensor $\mathcal{H}^{\mu\nu}_{\mathrm{q}_2\mathrm{q}_3}$ associated with the quark pair $\mathrm{q}_2+\overline{\mathrm{q}}_3$ in $\tau^-(P) \to \nu_\tau(k) + \mathrm{q}_2(p_2) + \overline{\mathrm{q}}_3(p_3)$ is

$$\mathcal{H}^{\mu\nu}_{\mathrm{q}_2\mathrm{q}_3} = |V_{\mathrm{q}_2\mathrm{q}_3}|^2 \int \frac{\mathrm{d}^3p_2}{2E_2}\frac{\mathrm{d}^3p_3}{2E_3}\frac{\delta^4(p_2+p_3-q)}{(2\pi)^2}\, 8\left[p_2^\mu p_3^\nu + p_2^\nu p_3^\mu - g^{\mu\nu}(p_2\cdot p_3)\right];$$

the last factor $8\left[p_2^\mu p_3^\nu + p_2^\nu p_3^\mu - g^{\mu\nu}(p_2\cdot p_3)\right]$ comes from the trace of $|\overline{u}(p_2)\gamma^\mu(1-\gamma_5)v(p_3)|^2$. In fact, the latter is a sum of a vectorial and an axial parts, which are given respectively by $4\left[p_2^\mu p_3^\nu + p_2^\nu p_3^\mu - g^{\mu\nu}(p_2\cdot p_3 \pm m_2 m_3)\right]$. Their interference $8\,\mathrm{i}\,\varepsilon^{\mu\nu\alpha\beta}(p_2)_\alpha(p_3)_\beta$ does not contribute, since the integration is symmetric in $p_2$, $p_3$. We calculate the integration of the vectorial part first. With the implicit factor $|V_{\mathrm{q}_2\mathrm{q}_3}|^2/\pi^2$ , the following quantity is to be evaluated:

$$\int \frac{\mathrm{d}^3p_2}{2E_2}\frac{\mathrm{d}^3p_3}{2E_3}\,\delta^4(p_2+p_3-q)\left[p_2^\mu p_3^\nu + p_2^\nu p_3^\mu - g^{\mu\nu}(p_2\cdot p_3 + m_2 m_3)\right].\tag{13.57}$$

This equation has the general form $-A\,q^2\,g^{\mu\nu} + B\,q^\mu q^\nu$ which can be rewritten as $(-q^2\,g^{\mu\nu} + q^\mu q^\nu)\,A + q^\mu q^\nu\,(B-A)$. Multiplying respectively (57) by $g_{\mu\nu}$ and $q_\mu q_\nu$, we get two equations for two unknowns $A$ and $B$:

$$q^2\,(B-4A) = \left[-q^2 + m_2^2 + m_3^2 - 4m_2m_3\right]\int \frac{\mathrm{d}^3p_2}{2E_2}\frac{\mathrm{d}^3p_3}{2E_3}\,\delta^4(p_2+p_3-q)\,,$$

$$q^4(B-A) = \frac{(m_2-m_3)^2\left[q^2-(m_2+m_3)^2\right]}{2}\int \frac{\mathrm{d}^3p_2}{2E_2}\frac{\mathrm{d}^3p_3}{2E_3}\,\delta^4(p_2+p_3-q)\,.$$

Using (30) for the double integral of the above equation, the analytic expressions of $A$ and $B$ are obtained. Comparing with (47), we can identify the term $A$ as $v_1(q^2)$ and $(B-A)$ as $v_0(q^2)$ (modulo the factor $|V_{q_2q_3}|^2/\pi^2$ ). For each color, the spectral functions of the quark pair $q_2+\overline{q}_3$ are found to be

$$v_1(q^2) = \frac{|V_{q_2q_3}|^2}{12\,\pi}\frac{\sqrt{\lambda(q^2,m_2^2,m_3^2)}}{q^2}\left[2-\frac{m_2^2+m_3^2-6\,m_2\,m_3}{q^2}-\frac{(m_2^2-m_3^2)^2}{q^4}\right],$$

$$v_0(q^2) = \frac{|V_{q_2q_3}|^2}{4\,\pi}\frac{\sqrt{\lambda(q^2,m_2^2,m_3^2)}}{q^2}\left[\frac{(m_2-m_3)^2}{q^2}-\frac{(m_2^2-m_3^2)^2}{q^4}\right]. \quad (13.58)$$

If $m_2 = m_3$, the vector current $\overline{q}_2\gamma^\mu q_3$ is conserved, and as expected, we get $v_0(q^2)=0$. The multiplicative color factor $N_c=3$ must be included on the right-hand side of (58), since we sum over the quark colors. The $a_1(q^2)$ $[a_0(q^2)]$ can be deduced respectively from $v_1(q^2)$ $[v_0(q^2)]$ by changing only $m_3\to -m_3$ in (58). We get

$$\rho_1(q^2)\equiv v_1(q^2)+a_1(q^2) = \frac{C(q^2)}{6\,\pi}\left[2-\frac{m_2^2+m_3^2}{q^2}-\frac{(m_2^2-m_3^2)^2}{q^4}\right],$$

$$\rho_0(q^2)\equiv v_0(q^2)+a_0(q^2) = \frac{C(q^2)}{2\pi}\left[\frac{m_2^2+m_3^2}{q^2}-\frac{(m_2^2-m_3^2)^2}{q^4}\right], \quad (13.59)$$

where $C(q^2)\equiv N_c\,|V_{q_2q_3}|^2\,\sqrt{\lambda(q^2,m_2^2,m_3^2)}/q^2$ . Putting (59) into (49) or (50) accordingly, the width $\Gamma(\tau^-\to\nu_\tau+q_2+\overline{q}_3)$ is given by

$$\Gamma = N_c\frac{G_F^2\,|V_{q_2q_3}|^2}{192\pi^3M^3}\int_{(m_2+m_3)^2}^{M^2}dq^2(M^2-q^2)^2\frac{\sqrt{\lambda(q^2,m_2^2,m_3^2)}}{q^2}$$
$$\times\left[(M^2+2q^2)\left(2-\frac{\sigma}{q^2}-\frac{\delta^2}{q^4}\right)+3M^2\left(\frac{\sigma}{q^2}-\frac{\delta^2}{q^4}\right)\right],$$
$$\sigma\equiv m_2^2+m_3^2\,,\ \delta\equiv m_2^2-m_3^2\,,\qquad \lim_{m_2,m_2\to 0}\Gamma = N_c\,|V_{q_2q_3}|^2\,\Gamma_0\,. \quad (13.60)$$

Using this formula, we compute the Cabibbo-favored decay width $\Gamma^{S=0}\equiv\Gamma(\tau^-\to\nu_\tau+d+\overline{u})$ and the Cabibbo-suppressed $\Gamma^{S=-1}\equiv\Gamma(\tau^-\to\nu_\tau+s+\overline{u})$. Their sum saturates the inclusive semileptonic width $\Gamma(\tau^-\to\nu_\tau+\text{hadrons})$.

For $\Gamma^{S=0}$, we take $m_d=m_u=m$ and define $t=q^2/M^2$ , $\eta=m^2/M^2$. The result is

$$\Gamma^{S=0} = 2\,N_c\,|V_{ud}|^2\,\Gamma_0\int_{4\eta}^1 dt\,(1-t)^2\sqrt{1-4\frac{\eta}{t}}\left\{(1+2t)\left(1-\frac{\eta}{t}\right)+3\frac{\eta}{t}\right\},$$
$$= N_c\,|V_{ud}|^2\,\Gamma_0\,G\left(\frac{m^2}{M^2},\frac{m^2}{M^2}\right),$$
$$G(x,x) = \sqrt{1-4x}\,\left[1-14x-2x^2-12x^3\right]$$
$$+\,24\,x^2\,(1-x^2)\log\frac{1+\sqrt{1-4x}}{1-\sqrt{1-4x}}\,. \quad (13.61)$$

The $\Gamma^{S=-1}$ can be obtained similarly. Here the two masses $m_2 = m_s$ and $m_3 = m_u$ are unequal, the phase space integration in (60) is more involved,

$$\Gamma^{S=-1} = N_c\,|V_{us}|^2\,\Gamma_0\,G\left(\frac{m_2^2}{M^2},\frac{m_3^2}{M^2}\right)\;,\;\text{where}$$

$$G(x,y) = \sqrt{\lambda(1,x,y)}\,\left[1-(x+y)(7+6xy)-(x^2+y^2)+(x-y)^2(x+y-6)\right] + 12\left[x^2(1-y^2)\log\frac{1+x-y+\sqrt{\lambda(1,x,y)}}{1+x-y-\sqrt{\lambda(1,x,y)}}+(x\leftrightarrow y)\right]\;. \quad (13.62)$$

Note that $G(x,y)=G(y,x)$. When $x$ or $y$ vanishes, (22) is recovered:

$$G(x,0)=G(0,x)=f(x)=1-8x+8x^3-x^4-12x^2\log x\;.$$

The phase space factor (62) multiplied by $\Gamma_0$ is the width of a fermion F of mass $M$ decaying into three other fermions $\mathrm{F}\to \mathrm{f}_1+\mathrm{f}_2+\overline{\mathrm{f}}_3$, when one of the three final fermions is massless. If the three fermions are massive with masses $m_k \neq 0$ , $k=1,2,3$, the width is obtained by combining (46), (48), and (59). Thus $\Gamma(\mathrm{F}\to \mathrm{f}_1+\mathrm{f}_2+\overline{f}_3)=\Gamma_0\,I(\eta_1,\eta_2,\eta_3)$ , with $\eta_k = m_k^2/M^2$:

$$I(\eta_1,\eta_2,\eta_3) = \frac{1}{M^8}\int_{(m_2+m_3)^2}^{(M-m_1)^2}\frac{\mathrm{d}q^2}{q^2}\sqrt{\lambda(M^2,m_1^2,q^2)}\,\sqrt{\lambda(q^2,m_2^2,m_3^2)} \times\left[\left(2-\frac{\sigma}{q^2}-\frac{\delta^2}{q^4}\right)\,\Phi_1(q^2)+3\left(\frac{\sigma}{q^2}-\frac{\delta^2}{q^4}\right)\,\Phi_0(q^2)\right],$$

$$\Phi_1(q^2) = (M^2-q^2)(M^2+2q^2)-m_1^2\,(2M^2-q^2-m_1^2)\;,$$
$$\Phi_0(q^2) = (M^2-q^2)M^2-m_1^2\,(2M^2+q^2-m_1^2)\;. \quad (13.63)$$

This phase space factor $I(\eta_1,\eta_2,\eta_3)$ is *totally symmetric* in the permutation of the three arguments (Problem 13.8). Of course, $I(0,x,y)=G(x,y)$, $I(0,x,x)=G(x,x)$, $I(0,0,x)=f(x)$, and $I(0,0,0)=1$.

The formula for $I(\eta_1,\eta_2,\eta_3)$ is particularly important when we study the heavy–flavored D and B meson decays (Chap. 16). In the limit $m_k=0$ for all $k$, we recover $\Gamma^{S=0}=N_c\,|V_{ud}|^2\,\Gamma_0$, and $\Gamma^{S=-1}=N_c\,|V_{us}|^2\,\Gamma_0$, thus

$$\frac{\Gamma(\tau^-\to\nu_\tau+\text{hadrons})}{\Gamma(\tau^-\to\nu_\tau+\mathrm{e}^-+\overline{\nu}_\mathrm{e})} = N_c\,(|V_{ud}|^2+|V_{us}|^2)\approx N_c\;. \quad (13.64)$$

The relation (64), like the ratio $R$ defined in (7.132), may also be derived by the color-counting argument (5). We have (see also Sect. 7.5),

$$R=\frac{\sigma(\mathrm{e}^++\mathrm{e}^-\to\text{hadrons})}{\sigma(\mathrm{e}^++\mathrm{e}^-\to\mu^++\mu^-)} = N_c\sum_k Q_k^2\;, \quad (13.65)$$

where $Q_k$ is the charge (in units of $e>0$) of the quark $k$.

We close this section with a remark. First, we consider many exclusive hadronic channels of the $\tau$ decay, their rates are given by (31),(39), (56) in which enter parameters taken from elsewhere, like $f_\pi$, $f_\rho$, $f_{a_1}$ and $\sigma(\mathrm{e}^+ + \mathrm{e}^- \to \pi^+ + \pi^-)$. These parameters are not at all directly related to the $\tau$ properties. When all of these exclusive decay rates are summed up, it is quite possible *a priori* that the sum exceeds the inclusive rate $N_\mathrm{c}\Gamma_0\, G(x,y)$ described by the quark picture, which would be disastrous. Remarkably, the sum approaches $N_\mathrm{c}\Gamma_0\, G(x,y)$ from below and nearly saturates $N_\mathrm{c}\Gamma_0\, G(x,y)$. These quantitatively correct results support the quark-hadron duality.

## Problems

**13.1 The forbidden mode $\pi^- + \eta$.** Explain why $\Gamma(\tau^- \to \nu_\tau + \pi^- + \eta) \ll \Gamma(\tau^- \to \nu_\tau + \mathrm{K}^- + \eta) \ll \Gamma(\tau^- \to \nu_\tau + \pi^- + \pi^0)$. Also why $\Gamma(\eta \to 2\pi) \ll \Gamma(\eta \to 3\pi)$.

**13.2 Angular distribution in $\tau^- \to \nu_\tau + \pi^-$.** For a $\tau^-$ at rest, its neutrino and the $\pi^-$ come out back-to-back. Since $\nu_\tau$ has helicity $-1$, show that the $\pi^-$ prefers to be emitted parallel to the spin direction $\boldsymbol{S}$ of the $\tau^-$. This property is confirmed by the following angular distribution. Show that

$$\frac{\mathrm{d}\Gamma(\tau^\mp \to \nu + \pi^\mp)}{\mathrm{d}\cos\theta} = \frac{G_\mathrm{F}^2\, |V_\mathrm{ud}|^2}{32\pi} f_\pi^2 M^3 \left(1 - \frac{m_\pi^2}{M^2}\right)^2 (1 \pm \cos\theta)\,,$$

where $\theta$ is the angle between $\boldsymbol{S}$ and the three-momentum of the $\pi$. Notice the $\pm\cos\theta$ of the above equation is opposite to $\mp\cos\theta$ in (24). Draw diagrams similar to Fig. 13.4 to explain this change of sign.

**13.3 Michel parameter $\rho$ and Fierz rearrangement.** Consider the weak decay of a fermion F into three massless fermions, $\mathrm{F} \to \mathrm{f}_1 + \mathrm{f}_2 + \overline{\mathrm{f}}_3$. In the standard model, its matrix element has the structure $(\mathrm{V}-\mathrm{A}) \times (\mathrm{V}-\mathrm{A})$. Let us write it as $K_{12}[(\mathrm{V}-\mathrm{A}) \times (\mathrm{V}-\mathrm{A})] \equiv \overline{f}_1\gamma^\mu(1-\gamma_5)F\,\overline{f}_2\gamma_\mu(1-\gamma_5)f_3$, the Fermi coupling $G_\mathrm{F}/\sqrt{2}$ is omitted. Show that the amplitudes

$$K_{12}[(\mathrm{V} \mp \mathrm{A}) \times (\mathrm{V} \mp \mathrm{A})]\,,\ K_{12}[\mathrm{V} \times \mathrm{V}]\,,\text{and}\ \ K_{12}[\mathrm{A} \times \mathrm{A}]$$

give rise to $\rho = 3/4$, whereas $K_{12}[\mathrm{V} \times (\mathrm{V} \pm \mathrm{A})]$ and $K_{12}[\mathrm{A} \times (\mathrm{V} \pm \mathrm{A})]$ yield $\rho = 3/8$. For $K_{12}[(\mathrm{V} \mp \mathrm{A}) \times (\mathrm{V} \pm \mathrm{A})]$, one gets $\rho = 0$. Using the Fierz rearrangement formulas given in the Appendix, $K_{12}[(\mathrm{V} \mp \mathrm{A}) \times (\mathrm{V} \pm \mathrm{A})] \leftrightarrow K_{21}[\mathrm{S} \times \mathrm{P}]$, we then again understand why $\rho = 0$.

**13.4 Experimental determination of $f_{\mathrm{K}^*}$, $f_{\pi'}$, and $f_{a_1}$.** Why are these decay constants difficult to be measured, contrary to $f_\pi$ and $f_\rho$ ?

**13.5 Strong evidence against the S, P structure from $\pi$ decay.** Write the matrix element of $\pi \to \ell + \nu_\ell$ if the weak interaction is of the S, P types. Compute the rate and show that

$$\frac{\pi^- \to \mathrm{e}^- + \nu_\mathrm{e}}{\pi^- \to \mu^- + \nu_\mu} \sim 1\,,\quad \text{which is in strong disagreement with experiments}\,.$$

**13.6 Neutron decay** $\mathrm{n} \to \mathrm{p} + \mathrm{e}^- + \overline{\nu}_\mathrm{e}$**.** We recall from (12.52) that the matrix element of the weak current $V_{\mathrm{ud}}\overline{u}\gamma_\mu(1-\gamma_5)d$, sandwiched between the neutron and the proton, has four form factors. Explain why only two of them dominate the neutron $\beta$-decay matrix element. The other two could be neglected, which ones? Compute the decay rate, using $m_\mathrm{n} = 939.5656$ MeV, $m_\mathrm{p} = 938.2723$ MeV, $m_\mathrm{e} = 0.5109$ MeV, $|V_{\mathrm{ud}}| = 0.9736$. Compare it with the neutron lifetime 887 $\pm 2$ s. Deduce $g_1(0)$.

**13.7 W propagator effect.** Derive the last factor of (28)

$$1 + \frac{3}{5}\frac{M^2}{M_\mathrm{W}^2} - 2\frac{m^2}{M_\mathrm{W}^2} .$$

**13.8 Decay rate of a fermion into three massive fermions.** Let us write the matrix element of $\mathrm{F}(P) \to \mathrm{f}_1(p_1) + \mathrm{f}_2(p_2) + \overline{\mathrm{f}}_3(p_3)$ as

$$\mathcal{M} = \frac{G_\mathrm{F}}{\sqrt{2}}\overline{u}(p_1)\gamma_\mu(1-\gamma_5)u(P)\,\overline{u}(p_2)\gamma^\mu(1-\gamma_5)v(p_3) .$$

First compute $Y \equiv 1/2\sum_{\mathrm{spin}}|\mathcal{M}|^2$, keeping all final-state masses $m_{1,2,3} \neq 0$. Express $Y$ in terms of $s = (p_1+p_2)^2$ and $M^2, m_k^2$. Show that $\Gamma = \Gamma_0 J(\eta_1,\eta_2,\eta_3)$, $\eta_k = m_k^2/M^2$, where the phase space factor $J(\eta_1,\eta_2,\eta_3)$ is

$$\begin{aligned} J(\eta_1,\eta_2,\eta_3) =& \frac{12}{M^8}\int_{(m_1+m_2)^2}^{(M-m_3)^2}\frac{\mathrm{d}s}{s}\sqrt{\lambda(M^2,m_3^2,s)}\,\sqrt{\lambda(s,m_1^2,m_2^2)} \\ & \times (s-m_1^2-m_2^2)\,(M^2+m_3^2-s) , \end{aligned} \tag{13.66}$$

implying that $J(\eta_1,\eta_2,\eta_3)$ is symmetric in the permutation of $m_1$ and $m_2$. Compare $J(\eta_1,\eta_2,\eta_3)$ with $I(\eta_1,\eta_2,\eta_3)$ in (63). Show that $J(\eta_1,\eta_2,\eta_3)$ must be equal to $I(\eta_1,\eta_2,\eta_3)$. Consequently, $J(x,y,z) = I(x,y,z)$ is totally symmetric by permutation of their three arguments. What happens if the interaction is of the $(\mathrm{V}+\mathrm{A})\times(\mathrm{V}-\mathrm{A})$ type ? Show that the integrated width is identical to that of the $(\mathrm{V}-\mathrm{A})\times(\mathrm{V}-\mathrm{A})$ type.

## Suggestions for Further Reading

*Parity violation in weak interactions and spin-momentum correlation:*

Lee, T. D. and Yang, C. N., Phys. Rev. **104** (1956) 254

Wu, C. S. et al., Phys. Rev. **105** (1957) 1413

*Michel parameters as probe of weak interaction structures:*

Bouchiat, C. and Michel, L., Phys. Rev. **106** (1957) 170

Jarlskog, C., Nucl. Phys. **75** (1966) 659

Marshak, R. E., Riazuddin and Ryan, C. P., *Theory of Weak Interactions in Particle Physics*. Wiley-Interscience, New York 1969

*Systematic analyses of $\tau$ decay:*

Alemany, R., Davier, M. and Höcker, A., Eur. Phys. J. **C2** (1998) 123

Braaten, E., Narison, S. and Pich, A., Nucl. Phys. **B373** (1992) 581

Davier, M., in *Proc. Second Workshop on Tau Lepton Physics* (ed. Gan, K. K.). World Scientific, Singapore 1992

Perl, M. L., Reports Progress Phys. **55** (1992) 653

Tsai, Yung-Su, Phys. Rev. **D4** (1971) 2821

# 14 One-Loop QCD Corrections

Until now our calculations have been limited for the most part to the lowest perturbative order or 'tree' graphs. But all processes receive higher-order contributions – usually called radiative corrections – from diagrams that contain 'loops', even those (such as flavor changing neutral reactions) for which the lowest-order tree amplitudes are absent. Then new effects can only arise from loops, some typical examples are the $\mathrm{K}^0$–$\overline{\mathrm{K}}^0$ mixing and the effective $\Delta S = 1$ neutral current (penguin) considered in Chap. 11. Weak decays offer an excellent opportunity for the study of radiative corrections due to either QCD or electroweak interactions. An $n$-loop diagram has an implicit factor $(\hbar)^n$; a tree (zero-loop) diagram has $(\hbar)^0$. The $(\hbar)^n$ factor shows that the tree graphs are equivalent to classical (Born) approximation, whereas loops are synonymous with quantum effects. This chapter introduces some important concepts and basic calculational methods of quantum corrections.

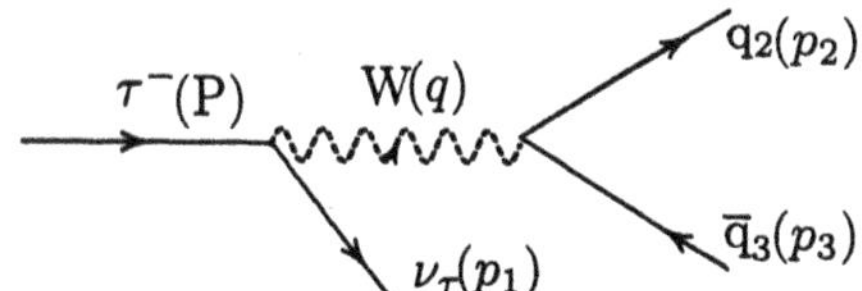


**Fig. 14.1.** Tree diagram of weak decay $\tau^- \to \nu_\tau + \mathrm{q}_2 + \overline{\mathrm{q}}_3$

As a first illustration, let us consider the one-loop QCD corrections to the inclusive semileptonic decay of the lepton $\tau$ as described by $\tau \to \nu_\tau +$ quark pairs. This mode has been studied in Sect. 13.4 at the tree diagram level (Fig. 14.1). The QCD corrections to order $g_s^2 = 4\pi\alpha_s$ are represented by the five diagrams in Fig. 14.2 and Fig. 14.3.

Although the diagrams of Fig. 14.2 are plagued with *ultraviolet* divergences due to high momenta of virtual particles in the loop integrals, these divergences will be removed by the *renormalization* procedure, which ultimately yields a finite (renormalized) amplitude.

To order $\alpha_s$, the radiatively corrected rate of $\tau \to \nu_\tau + \mathrm{q}_2 + \overline{\mathrm{q}}_3$ has two parts. One, $\Gamma_{\mathrm{Vi}}$ with virtual gluons, comes from the interference between the tree amplitude (Fig. 14.1) and the renormalized amplitude (Fig. 14.2). The other, $\Gamma_{\mathrm{Re}}$ with real gluons, comes from the square of the bremsstrahlung amplitude (Fig. 14.3). Both $\Gamma_{\mathrm{Vi}}$ and $\Gamma_{\mathrm{Re}}$ are still ill defined because of another kind of divergence, viz. the *infrared* divergence, which originates from low momenta of virtual or real massless gluons. What is well defined is the sum $\Gamma_{\text{Vi}} + \Gamma_{\text{Re}}$, in which the infrared divergences cancel each other. This chapter shows how to deal with both ultraviolet (UV) and infrared (IR) divergences encountered in quantum corrections and how to calculate the finite parts. We remark that five diagrams similar to those considered here also represent the $\alpha_s$-order QCD corrections to the $e^+ + e^- \to$ hadrons cross-section, discussed in (13.5) and (13.65). The only difference is that a photon (connecting the $e^+e^-$ with the quark pair) replaces the W boson.

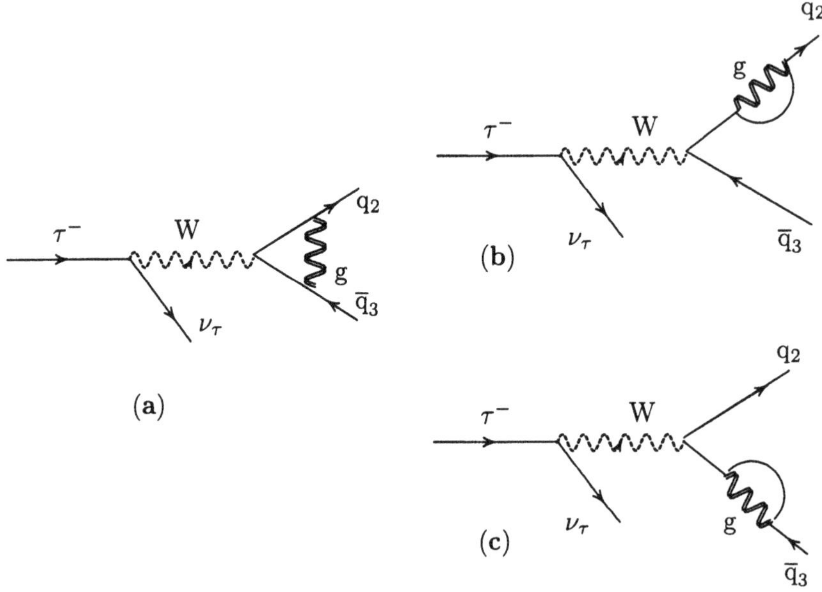


**Fig. 14.2a–c.** Virtual gluon corrections to weak decay $\tau^- \to \nu_\tau + q_2 + \overline{q}_3$

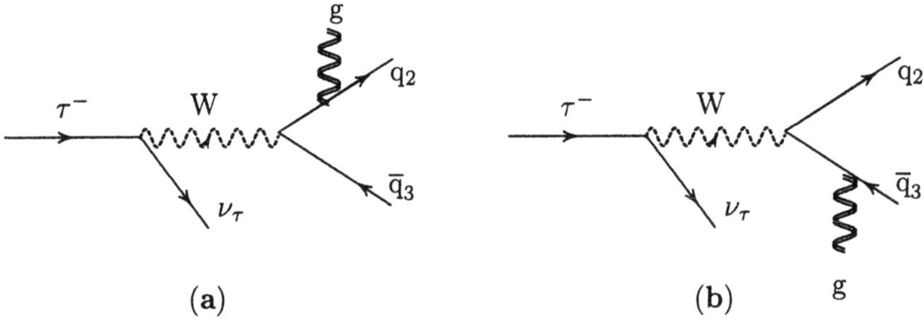


**Fig. 14.3a, b.** Real gluon emission in weak decay $\tau^- \to \nu_\tau + q_2 + \overline{q}_3 + g$

Before computing these QCD radiative corrections, we first evaluate the vertex function $\Gamma^\mu(p_2, p_3)$ and the quark self-energy $\Sigma(p)$ which represent the essential parts of the diagrams in Fig. 14.2.

## 14.1 Vertex Function

To zero order of the QCD coupling constant $g_s$ (Fig. 14.4a), the weak interaction vertex $W(q) \to q_2(p_2) + \overline{q}_3(p_3)$ can be written as $\gamma^\mu(1-\gamma_5)$ to be inserted between the spinors $\overline{u}(p_2)$ and $v(p_3)$. The weak coupling constant $g_W = -\mathrm{i}\, g\, V_{q_2q_3}/2\sqrt{2}$ is implicit. When effects of virtual gluons to order $g_s^2$ (Fig. 14.4b) are considered, the tree-level weak vertex $\gamma^\mu(1-\gamma_5)$ becomes the QCD-corrected weak vertex function $\Gamma^\mu(p_2,p_3)$. Thus

$$\overline{u}(p_2)\,\gamma^\mu(1-\gamma_5)v(p_3) \longrightarrow \overline{u}(p_2)\left[\gamma^\mu(1-\gamma_5) + \Gamma^\mu(p_2,p_3)\right]v(p_3)\,, \qquad (14.1)$$

where $\Gamma^\mu(p_2,p_3)$ has the following expression from Feynman rules:

$$\Gamma^\mu(p_2,p_3) = \int \frac{\mathrm{d}^4k}{(2\pi)^4}(-\mathrm{i}\, g_s\gamma^\rho \tfrac{\lambda_j}{2})\frac{\mathrm{i}}{\not k + \not p_2 - m + \mathrm{i}\eta}\gamma^\mu(1-\gamma_5)\frac{\mathrm{i}}{\not k - \not p_3 - m + \mathrm{i}\eta}$$
$$\times\, (-\mathrm{i}\, g_s\gamma^\sigma \tfrac{\lambda_j}{2})\frac{\mathrm{i}\left[-g_{\rho\sigma} + (1-\xi)k_\rho k_\sigma/k^2\right]}{k^2 + \mathrm{i}\eta}\,. \qquad (14.2)$$

For simplicity, the quark masses are taken to be equal ($m_2 = m_3 = m$).

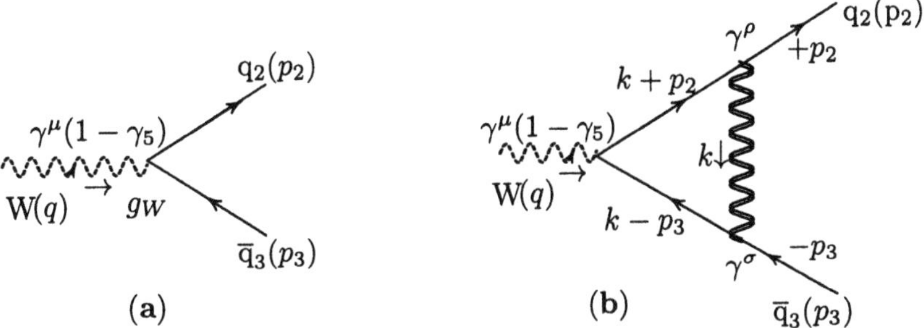


**Fig. 14.4.** (a) Bare vertex $\gamma^\mu(1-\gamma_5)$; (b) dressed vertex function $\Gamma^\mu(p_2,p_3)$

The dependence of $\Gamma^\mu(p_2,p_3)$ on the gauge-fixing term $\xi$ of the gluon propagator in fact will be canceled by the dependence of $\delta_q$ on $\xi$ [this quantity $\delta_q$ which comes from the self-energy term $\Sigma(p)$ of Fig. 14.2b, c, will be introduced later in (40)]. Therefore, when we sum over the three diagrams of Fig. 14.2 to obtain the *ultraviolet-convergent renormalized vertex function*, the $\xi$-dependent contributions are canceled so that we can use the Feynman–'t Hooft gauge ($\xi = 1$) from the outset. For SU($N_c$) color group, using (7.45) and (7.46), the sum over $j$ of $(\frac{\lambda_j}{2})(\frac{\lambda_j}{2})$ in (2) yields

$$\sum_j \frac{\lambda_j}{2}\frac{\lambda_j}{2} = C_2(N_c)\mathbf{I} = \frac{N_c^2-1}{2N_c}\mathbf{I}\,, \qquad (14.3)$$

which is equal to $\frac{4}{3}\mathbf{I}$ for $N_c = 3$. All of these operations lead to

$$\Gamma^\mu(p_2,p_3) = \frac{-4\mathrm{i}g_s^2}{3}\int\frac{\mathrm{d}^4k}{(2\pi)^4}\frac{\gamma^\rho(\not k + \not p_2 + m)\gamma^\mu(1-\gamma_5)(\not k - \not p_3 + m)\gamma_\rho}{(k^2+2k\cdot p_2+\mathrm{i}\eta)(k^2-2k\cdot p_3+\mathrm{i}\eta)(k^2+\mathrm{i}\eta)}\,. \qquad (14.4)$$

When $\Gamma^\mu(p_2,p_3)$ is inserted between $\overline{u}(p_2)$ and $v(p_3)$ and the Dirac equation $\overline{u}(p_2)(\not p_2 - m) = 0 = (\not p_3 + m)v(p_3)$ together with $\not a \not b + \not b \not a = 2a\cdot b$ are used, the numerator $\gamma^\rho(\not k + \not p_2 + m)\gamma^\mu(1-\gamma_5)(\not k - \not p_3 + m)\gamma_\rho$ of (4) can be rewritten as $\mathcal{N}^\mu(k,p_2,p_3)$ defined by

$$\mathcal{N}^\mu(k,p_2,p_3) \equiv \Big\{[-4p_2\cdot p_3 + 4k\cdot(p_2-p_3)]\gamma^\mu + 4mk^\mu - 4\not k\,(p_2-p_3)^\mu + \gamma^\rho \not k\gamma^\mu \not k\gamma_\rho\Big\}(1-\gamma_5) + 4m\not k\gamma^\mu\gamma_5 \ . \tag{14.5}$$

As we will see, most of the integrals in loop diagrams diverge, we must cut off or regularize the momentum variable to handle the infinities encountered in the integrations. There are at least two methods: the Pauli–Villars covariant momentum cutoff (Problem 14.5), and the *dimensional regularization.* The latter, invented by 't Hooft and Veltman, also independently by Bollini and Giambiagi and by Ashmore, is used here. No matter which regularization method is adopted, any choice is as good as any other, provided that the symmetries of the theory (e.g. Lorentz and gauge invariances) are preserved throughout the calculations.

The idea of dimensional regularization is simple, and is based on the observation that the degree of divergences of loop integrals may be decreased by lowering the space-time dimension. Dimensional regularization amounts to computing divergent integrals in an arbitrary space-time dimension $n$ smaller than 4, resulting in a finite result, as long as the parameter $\varepsilon \equiv 4-n$ is nonvanishing. Then the result is analytically continued to $n \geq 4$. The original infinities in the physical $n=4$ world then arise as poles at $\varepsilon$, in the form of $\Gamma(2-\frac{n}{2}) \sim 2/\varepsilon$, where $\Gamma(x)$ is the Euler function. Appropriate formulas are given in the Appendix. The advantage of this scheme is that Feynman rules and the gauge symmetry of the theory do not depend on $n$.

**Remark.** While there is no problem with the extension of $\gamma^\mu = g^{\mu\nu}\gamma_\nu$ to arbitrary $n$ dimensions (the metric tensor $g^{\mu\nu}$ associated with $\{\gamma^\mu,\gamma^\nu\}$ can be defined for $\mu,\nu = 0,\cdots n-1$), the generalization of $\gamma_5 = \gamma^5$ to $n \neq 4$

$$\gamma_5 = \mathrm{i}\gamma^0\gamma^1\gamma^2\gamma^3 = -\mathrm{i}\gamma_0\gamma_1\gamma_2\gamma_3 = \frac{\mathrm{i}}{4!}\varepsilon_{\mu\nu\rho\sigma}\gamma^\mu\gamma^\nu\gamma^\rho\gamma^\sigma = \frac{\mathrm{i}}{4!}\varepsilon^{\mu\nu\rho\sigma}\gamma_\mu\gamma_\nu\gamma_\rho\gamma_\sigma,$$

does present a problem, since the antisymmetric tensor $\varepsilon_{\mu\nu\rho\sigma}$ is only defined in 4 dimensions. One may still define the equivalent of $\gamma_5$ in $n$ dimensions as

$$\widetilde{\gamma} = \frac{\mathrm{i}}{n!}\varepsilon_{\mu_1\cdots\mu_n}\gamma^{\mu_1}\cdots\gamma^{\mu_n} \ . \tag{14.6}$$

However, when taking the trace of products of $\gamma_\mu$ matrices and $\widetilde{\gamma}$, one will get into trouble. For instance, the trace $\mathrm{Tr}[\gamma_\mu\gamma_\nu\gamma_\rho\gamma_\sigma\widetilde{\gamma}]$ which gives $4\mathrm{i}\,\varepsilon_{\mu\nu\rho\sigma} \neq 0$ for $n=4$, would vanish if we use (6) for $n \neq 4$. One might worry about the presence of $\gamma_5$ in our calculation. For the problem considered here, $\gamma_5$

is innocuous. The trace of $\gamma_\mu\gamma_\nu\cdots\gamma_5$ never occurs in our integral loops, it arises only in the computation of the rate. However, for the latter, it is even unnecessary to perform the trace with $\gamma_5$ because we first integrate over the symmetric phase space $\int d\boldsymbol{p}_2\, d\boldsymbol{p}_3$, such that the contributions of $\gamma_5$ vanish (they are antisymmetric in $\boldsymbol{p}_2 \leftrightarrow \boldsymbol{p}_3$). We will explicitly show that $\gamma_5$ is harmless in the course of the calculation.

Furthermore, the $n$-dimensional Dirac matrices obey $\mathrm{Tr}(\gamma^\mu\gamma^\nu) = 2^{n/2}g^{\mu\nu}$ ($n$ even) or $2^{n-2}g^{\mu\nu}$ ($n$ odd), however, the $n$ dependence of $2^{n/2}$ or $2^{n-2}$ does not affect the momentum integrals, these factors can be safely replaced by 4, their value at $n \to 4$, when we compute physical quantities. As a result,[1] the traces of Dirac matrices *without* the $\gamma_5$ depend only on their anticommutation relations, and can be safely taken from the corresponding four-dimensional formulas, for instance $\mathrm{Tr}(\not{a}\,\not{b}) = 4\,a\cdot b$. However, $g^{\mu\nu}g_{\mu\nu} = \delta^\mu_\mu = n$. ∎

With (5), the $\Gamma^\mu(p_2,p_3)$ becomes in $n$ space-time dimensions

$$\Gamma^\mu(p_2,p_3) = \frac{-4\mathrm{i}g_s^2}{3}\int\frac{d^nk}{(2\pi)^n}\frac{\mathcal{N}^\mu(k,p_2,p_3)}{(k^2+2k\cdot p_2)(k^2-2k\cdot p_3)(k^2)}\,. \tag{14.7}$$

In (7) and what follows, the small $+\mathrm{i}\eta$ in the denominator of (4) is omitted, although we keep in mind that the Feynman prescription $+\mathrm{i}\eta$ in the propagators is important for the evaluation of the $d^nk$ integrals, using the Wick rotation (Appendix). On the second line of (5), the first term $\gamma^\rho\,\not{k}\gamma^\mu\,\not{k}\gamma_\rho$, which is quadratic in $k$ makes the integral divergent as $k\to\infty$. Therefore, we keep $\gamma^\rho\,\not{k}\gamma^\mu\,\not{k}\gamma_\rho = (2-n)\,\not{k}\gamma^\mu\,\not{k}$ in $n$ dimensions (Appendix), instead of $-2\,\not{k}\gamma^\mu\,\not{k}$, when $n=4$. The difference between the $(2-n)$ and $-2$ of the $\not{k}\gamma^\mu\,\not{k}$ term, when multiplied by the pole $1/\varepsilon$, yields finite terms as $\varepsilon\to 0$.

From now on, we adopt a convenient method due to Feynman which consists in writing every product in the denominator of a product of propagators as an integral over auxiliary variables (Appendix). Accordingly, the denominator of (7) can be written as

$$\frac{1}{(k^2+2k\cdot p_2)(k^2-2k\cdot p_3)\,k^2} = 2\int_0^1 dx\int_0^{1-x}\frac{dy}{[k^2+2k\cdot(-p_3x+p_2y)]^3}\,.$$

Putting it back into (7), we have

$$\Gamma^\mu(p_2,p_3) = -2\mathrm{i}\left(\frac{4g_s^2}{3}\right)\int_0^1 dx\int_0^{1-x}dy\int\frac{d^nk}{(2\pi)^n}\frac{\mathcal{N}^\mu(k,p_2,p_3)}{[k^2+2k\cdot(-p_3x+p_2y)]^3}\,.$$

The $\int d^nk$ of the six terms of $\mathcal{N}^\mu(k,p_2,p_3)/[k^2+2k\cdot(-p_3x+p_2y)]^3$ – where $\mathcal{N}^\mu(k,p_2,p_3)$ is given in (5) – can be read off directly from the Appendix. For instance, we get for the first term of $\mathcal{N}^\mu(k,p_2,p_3)$

$$\int d^nk\frac{-4p_2\cdot p_3\gamma^\mu(1-\gamma_5)}{[k^2+2k\cdot(-p_3x+p_2y)]^3} = C\times\left[2(2m^2-q^2)\gamma^\mu(1-\gamma_5)\right]\,,$$

[1] Marciano, W., Nucl. Phys. **B84** (1975) 132

where $C$ is an overall factor, with $q^2 = (p_2 + p_3)^2 \geq 4m^2$ :

$$C \equiv \frac{i\pi^{\frac{n}{2}}\,\Gamma(3-\frac{n}{2})}{2\,[\mathcal{D}(x,y)]^{3-\frac{n}{2}}}\,, \quad \mathcal{D}(x,y) = -m^2(x+y)^2 + q^2xy.$$

The result of the integration is now written as follows, with the symbol $\Rightarrow$

$$-4p_2\cdot p_3\gamma^\mu(1-\gamma_5) \;\Rightarrow\; C\times\left[2(2m^2-q^2)\gamma^\mu(1-\gamma_5)\right] .$$

In this way, the $\int \mathrm{d}^n k$ integration of the six terms in $\mathcal{N}^\mu(k,p_2,p_3)$ are

$$\begin{aligned}
-4p_2\cdot p_3\gamma^\mu(1-\gamma_5) &\Rightarrow C\times\left[2(2m^2-q^2)\gamma^\mu(1-\gamma_5)\right] ,\\
+4k\cdot(p_2-p_3)\gamma^\mu(1-\gamma_5) &\Rightarrow C\times\left[2(q^2-4m^2)(x+y)\gamma^\mu(1-\gamma_5)\right] ,\\
+4mk^\mu(1-\gamma_5) &\Rightarrow C\times\left[-2m(x+y)(p_2-p_3)^\mu(1-\gamma_5)\right] ,\\
-4\,\not{k}(p_2-p_3)^\mu(1-\gamma_5) &\Rightarrow C\times\left[+4m(x+y)(p_2-p_3)^\mu\right] ,\\
+4m\,\not{k}\gamma^\mu\gamma_5 &\Rightarrow C\times\left[-4m^2(x+y)\gamma^\mu\gamma_5 + 8m\,x\,p_3^\mu\gamma_5\right] ,\\
(2-n)\,\not{k}\gamma^\mu\,\not{k}(1-\gamma_5) &\Rightarrow \Big\{(2-n)\left[\mathcal{D}(x,y)\gamma^\mu(1-\gamma_5) + (x+y)^2m(p_2-p_3)^\mu\right]\\
&\quad + \frac{(2-n)^2}{2}\frac{\Gamma(2-\frac{n}{2})}{\Gamma(3-\frac{n}{2})}\mathcal{D}(x,y)\gamma^\mu(1-\gamma_5)\Big\} \times C. \qquad (14.8)
\end{aligned}$$

The final result is put into the form

$$\Gamma^\mu(p_2,p_3) = \frac{\frac{4}{3}g_s^2}{(4\pi)^{\frac{n}{2}}}\int_0^1 \mathrm{d}x \int_0^{1-x} \mathrm{d}y\, \frac{\Gamma(3-\frac{n}{2})\,\mathcal{G}^\mu(x,y,p_2,p_3)}{[\mathcal{D}(x,y)]^{3-\frac{n}{2}}}\,, \qquad (14.9)$$

where $\mathcal{G}^\mu(x,y,p_2,p_3)$ is the sum of the six terms on the right-hand side of (8), without the overall factor $C$. The $x \leftrightarrow y$ symmetry of both the denominator $[\mathcal{D}(x,y)]^{3-n/2}$ and the integration domain in (9) implies that the antisymmetric $(x-y)$ terms in (8) vanish after the $x, y$ integrations. These antisymmetric $(x-y)$ terms are therefore not shown in (8). We regroup $\mathcal{G}^\mu(x,y,p_2,p_3)$ into the combination

$$\mathcal{G}^\mu(x,y,p_2,p_3) = a\gamma^\mu + b\frac{(p_2-p_3)^\mu}{2m} - \left[c\gamma^\mu + d\frac{(p_2+p_3)^\mu}{2m}\right]\gamma_5\,, \qquad (14.10)$$

with $a = 2(2m^2-q^2) + 2(q^2-4m^2)(x+y) + (2-n)\mathcal{D}(x,y)$
$$+\frac{(2-n)^2}{2}\frac{\Gamma(2-\frac{n}{2})}{\Gamma(3-\frac{n}{2})}\mathcal{D}(x,y)\,,$$
$$b = 2m^2(x+y)[2+(2-n)(x+y)]\,,\quad d = -4m^2(x+y)\,,\quad c = a-d\,.$$

Using the Gordon decomposition, we replace $(p_2-p_3)^\mu$ with $2m\gamma^\mu - \mathrm{i}\sigma^{\mu\nu}q_\nu$ and group all terms on the right-hand side of (10) into two bases, the vectorial part $\gamma^\mu$, $\mathrm{i}\sigma^{\mu\nu}q_\nu$ and the axial one $\gamma^\mu\gamma_5$, $q^\mu\gamma_5$. These sets belong to the first-class currents with respect to the G-parity classification, like (12.52). Thus

$$\Gamma^\mu(p_2,p_3) = \gamma^\mu F_1(q^2) + \frac{\mathrm{i}\sigma^{\mu\nu}q_\nu}{2m}F_2(q^2) - \left[\gamma^\mu G_1(q^2) + \frac{q^\mu}{2m}G_3(q^2)\right]\gamma_5 \,. \quad (14.11)$$

At the tree level, the weak vertex is just a constant $\gamma^\mu(1-\gamma_5)$. When dressed by the gluon to order $g_s^2$, the one-loop diagram of Fig. 14.4b gives rise to the vector form factors $F_1(q^2)$ and $F_2(q^2)$ as well as to the axial form factors $G_1(q^2)$ and $G_3(q^2)$. The emergence of form factors is intuitively understandable, since surrounded by clouds of gluons and quark pairs, the pointlike quark continuously emits and absorbs virtual particles. It behaves physically as a composite system of virtual particles and hence develops a structure with its own form factors. We have:

$$\begin{aligned}
F_1(q^2) &= \tfrac{4}{3}\frac{g_s^2}{(4\pi)^{n/2}}\frac{(n-2)^2}{2}\int_0^1 \mathrm{d}x\int_0^{1-x}\mathrm{d}y\,\frac{\Gamma(2-\frac{n}{2})}{[-m^2(x+y)^2+q^2xy]^{2-\frac{n}{2}}} \\
&+ \tfrac{4}{3}\frac{g_s^2}{8\pi^2}\int_0^1 \mathrm{d}x\int_0^{1-x}\mathrm{d}y\,\frac{m^2[-2+2(x+y)+(x+y)^2]+q^2(1-x)(1-y)}{m^2(x+y)^2-q^2xy}\,, \\
F_2(q^2) &= \tfrac{4}{3}\frac{g_s^2}{4\pi^2}\int_0^1 \mathrm{d}x\int_0^{1-x}\mathrm{d}y\,\frac{m^2(x+y)(1-x-y)}{m^2(x+y)^2-q^2xy}\,, \\
G_1(q^2) &= F_1(q^2) - 2F_2(q^2) \quad ; \quad G_3(q^2) = 2F_2(q^2)\,. \qquad (14.12)
\end{aligned}$$

In (10) and (12), $\varepsilon = 4-n \neq 0$ is kept only for terms coming along with the singular function $\Gamma(2-\frac{n}{2}) = \Gamma(\varepsilon/2)$. For the regular terms associated with $\Gamma(3-\frac{n}{2})$, the limit $\varepsilon\to 0$ can be safely taken right away. From (3), the typical QCD factor $\frac{4}{3}$ becomes 1 in QED when the gluon in Fig. 14.4b is replaced by a photon. Making the change $u = x+y$, $uv = x-y$, we obtain

$$\begin{aligned}
F_1(q^2) &= \frac{4}{3}\frac{g_s^2}{(4\pi)^{n/2}}\frac{(n-2)^2}{2}\int_0^1\frac{\mathrm{d}u}{u^{3-n}}\int_{-1}^1\left(\frac{\mathrm{d}v}{2}\right)\frac{\Gamma(2-\frac{n}{2})}{\left[-m^2+\frac14 q^2(1-v^2)\right]^{2-\frac{n}{2}}} \\
&+ \frac{4}{3}\frac{g_s^2}{8\pi^2}\int_0^1\frac{\mathrm{d}u}{u}\int_{-1}^1\left(\frac{\mathrm{d}v}{2}\right)\frac{m^2[-2+2u+u^2]+q^2[1-u+\frac14 u^2(1-v^2)]}{m^2-\frac14 q^2(1-v^2)}\,, \\
F_2(q^2) &= \frac{4}{3}\frac{g_s^2}{4\pi^2}\int_0^1 \mathrm{d}u\int_{-1}^1\left(\frac{\mathrm{d}v}{2}\right)\frac{m^2(1-u)}{m^2-\frac14 q^2(1-v^2)}\,. \qquad (14.13)
\end{aligned}$$

For $F_2(q^2)$, the $u, v$ integrations are straightforward and give

$$\begin{aligned}
F_2(q^2) &= \frac{4}{3}\frac{g_s^2}{16\pi^2}\frac{1}{\sqrt{\eta(\eta-1)}}\left\{\log\frac{\sqrt{\eta}-\sqrt{\eta-1}}{\sqrt{\eta}+\sqrt{\eta-1}}+\mathrm{i}\pi\right\} \quad \text{for } \eta\equiv\frac{q^2}{4m^2}>1\,, \\
&= \frac{4}{3}\frac{g_s^2}{16\pi^2}\frac{1}{\sqrt{\eta(\eta-1)}}\log\frac{\sqrt{1-\eta}+\sqrt{-\eta}}{\sqrt{1-\eta}-\sqrt{-\eta}} \quad \text{for } \eta<0 \quad \xrightarrow[q^2\to 0]{} \frac{4}{3}\frac{g_s^2}{8\pi^2}\,, \\
&= \frac{4}{3}\frac{g_s^2}{8\pi^2}\frac{1}{\sqrt{\eta(1-\eta)}}\tan^{-1}\sqrt{\frac{\eta}{1-\eta}} \quad \text{for } 0\le\eta\le 1\,. \qquad (14.14)
\end{aligned}$$

This formula is an example showing that form factors are analytic functions in the complex $q^2$ plane, with cuts on the timelike axis, as mentioned in Chap. 10. By a simple replacement $\frac{4}{3}g_s^2 \to e^2$, we recover the QED corrections to the electron magnetic moment $F_2^{\rm QED}(0)$, first derived by Schwinger,

$$F_2^{\rm QED}(0) = \frac{e^2}{8\pi^2} = \frac{\alpha_{\rm em}}{2\pi} \; . \tag{14.15}$$

The anomalous magnetic moment of the electron, i.e. the deviation from its Bohr magneton value $\mu_e = -e/2m_e$, is then $\mu_e(\alpha_{\rm em}/2\pi)$. A pointlike electron with a unit electric form factor and a zero anomalous magnetic moment at the tree level develops its electromagnetic form factors from QED radiative corrections, changing 1 into $1 + F_1^{\rm QED}(q^2)$ and 0 into $F_2^{\rm QED}(q^2)$.

Similarly, we find that QCD generates new weak form factors for quarks. As can be seen from (1) and (11), the modification is

$$1 \to \mathcal{F}_1^{\rm cor}(q^2) = 1 + F_1(q^2) \; , \; 0 \to \left[F_2(q^2) \, , \, G_3(q^2)\right] \; , \; 1 \to 1 + G_1(q^2) \, . \tag{14.16}$$

While quantum corrections make unambiguous finite predictions to $F_2(q^2)$ and $G_3(q^2)$, all the complexities of loop integrals are present in $F_1(q^2)$ [and hence in $G_1(q^2) = F_1(q^2) - 2F_2(q^2)$]. Since the variable of integration $k^2$ spans the whole range from 0 to $\infty$, a glance at (4) or (13) reveals two major problems of radiative corrections, both present only in $F_1(q^2)$:

(i) As $k^2 \to \infty$, the integral diverges like $\mathrm{d}^4k/k^4$, and this divergence is transformed into a pole $2/\varepsilon$ via the $\Gamma(2 - n/2)$ term contained in the first line of (13). The ultraviolet divergence (UV) has its origin in the *locality* of field theories ($x \to 0$ or equivalently $k \to \infty$); its removal is treated by the *renormalization* program.

(ii) At the lower limit $k^2 \to 0$ of the integration domain, the term $k^2$ in the denominator of (4) causes a different kind of infinity called infrared divergence (IR). It is identified with the factor $\mathrm{d}u/u$ of $F_1(q^2)$ as $u \to 0$ in (13). Originating from the *massless gluon* via its propagator $1/k^2$, this IR divergence will be canceled by the same IR divergence of the soft gluon bremsstrahlung (Fig. 14.3) that accompanies all radiative processes (Sect. 14.5).

Thus, the study of $F_1(q^2)$ provides an excellent exercise for the understanding of both UV and IR divergences. Looking at the numerator of (7) or at the six terms in (8), we realize that IR and UV divergences come respectively from the contant term $-4p_2 \cdot p_3 \gamma^\mu$ and the quadratic term $(2-n) \not{k} \gamma^\mu \not{k}$ in the integration variable $k$. The four linear terms in $k$ yield finite results. Let us explicitly compute $F_1(q^2)$ in order to single out these two types of divergences.

Consider the first part of $F_1(q^2)$ in the first line of (13), i.e. the part that contains the singular $\Gamma(2 - n/2)$ factor, we call it $F_{1,\rm uv}(q^2)$. Besides $\Gamma(2 - n/2) = 2/\varepsilon - \gamma_E + O(\varepsilon)$, where $\gamma_E \approx 0.5772$ is the Euler constant, we also expand all other terms of $F_{1,\rm uv}(q^2)$ up to order $\varepsilon$ such that they can cancel the singularity $2/\varepsilon$ when multiplied by the latter. The result is

$$F_{1,\text{uv}}(q^2) = \frac{4}{3}\frac{g_s^2}{16\pi^2}\left\{-\int_0^1 \mathrm{d}v \log\left[1 - \frac{q^2}{4m^2}(1-v^2)\right] + \frac{2}{\varepsilon} - \gamma_E + \log(4\pi) - 1 - \log\frac{m^2}{\mu^2}\right\}$$

$$\underset{q^2\to 0}{\longrightarrow} \frac{4}{3}\frac{g_s^2}{16\pi^2}\left[\frac{2}{\varepsilon} - \gamma_E + \log(4\pi) - 1 - \log\frac{m^2}{\mu^2}\right] . \tag{14.17}$$

On dimensional grounds, a mass scale factor $\mu$ must enter the denominator $\Delta^{2-\frac{n}{2}} \equiv [-m^2 + q^2(1-v^2)/4]^{2-\frac{n}{2}}$ of (13) in order to make the latter dimensionally correct for $n \neq 4$. Indeed, for small $\varepsilon \neq 0$, one can always write the term $\Delta^{\frac{\varepsilon}{2}}$ as $1 + \frac{\varepsilon}{2}\log(\Delta/\mu^2) + \mathcal{O}(\varepsilon^2)$. This arbitrary factor $\mu^2$ serves as a reminder that the decomposition of a divergent integral into an infinite term and a finite term always involves an ambiguity. Only a coherent procedure that removes the $\varepsilon$ pole and gives finite values *independent of* $\mu^2$ to physical quantities is meaningful. This is the essence of the renormalization.

The second part of $F_1(q^2)$ on the second line of (13) – the part without the singular $\Gamma(2-\frac{n}{2})$ – is either finite or infrared divergent. The factor $\mathrm{d}u/u$ is the source of IR divergence when $u \to 0$. To neutralize $1/u$, we group in the numerator of the integrand all the terms which are linear and quadratic in the variable $u$. They are $m^2(2u+u^2) + q^2(-u + u^2(1-v^2)/4)$. These terms cancel $1/u$ and yield a finite result denoted by $F_{1,\text{fi}}(q^2)$ after the $u$, $v$ integrations:

$$F_{1,\text{fi}}(q^2) = \frac{4}{3}\frac{g_s^2}{16\pi^2}\left\{-1 + \frac{3-4\eta}{\sqrt{\eta(\eta-1)}}\left[\log\frac{\sqrt{\eta}-\sqrt{\eta-1}}{\sqrt{\eta}+\sqrt{\eta-1}} + \mathrm{i}\pi\right]\right\} ,$$

$$\underset{q^2\to 0}{\longrightarrow} \frac{4}{3}\frac{5g_s^2}{16\pi^2} . \tag{14.18}$$

The remainder $(-2m^2 + q^2)$, which cannot cancel $1/u$, will give rise to an infrared divergent term, $F_{1,\text{ir}}(q^2)$. We have

$$F_{1,\text{ir}}(q^2) = \frac{4}{3}\frac{g_s^2}{8\pi^2}\int_0^1 \mathrm{d}v \frac{-2m^2+q^2}{m^2 - \frac{1}{4}q^2(1-v^2)}\int_0^1 \frac{\mathrm{d}u}{u} ,$$

$$\underset{q^2\to 0}{\longrightarrow} -\frac{4}{3}\frac{g_s^2}{4\pi^2}\int_0^1 \frac{\mathrm{d}u}{u} . \tag{14.19}$$

The sum of the three contributions (17), (18), (19) at the limit $q^2 = 0$ gives

$$F_1(0) = \frac{4}{3}\frac{g_s^2}{16\pi^2}\left\{\frac{2}{\varepsilon} - \gamma_E + \log(4\pi) - \log\left(\frac{m^2}{\mu^2}\right) + 4 - 4\int_0^1 \frac{\mathrm{d}u}{u}\right\} . \tag{14.20}$$

The UV and IR divergences are represented by $2/\varepsilon$ and $\int_0^1 \mathrm{d}u/u$ respectively. The reason for writing the explicit expression of $F_1(0)$ will be clear later. We have seen that $\Gamma^\mu(p_2, p_3)$, hence $F_1(q^2)$, has two types of divergence. Let us concentrate first on UV, leaving the treatment of IR to Sect. 14.4.

## 14.2 Quark Self-Energy

To order $g_s^2$ corrections to the tree vertex $\gamma^\mu(1-\gamma_5)$, besides the diagram of Fig. 14.2a that we have just considered, there are two more diagrams shown in Fig. 14.2b, c which also contribute. They are related to the quark self-energy $\Sigma(p)$, in which the generic $p$ stands for the external momenta.
The expression for the self-energy of Fig. 14.5 can be written as

$$-\mathrm{i}\Sigma(p) \stackrel{\text{def}}{=} \int \frac{\mathrm{d}^4k}{(2\pi)^4}\left(-\mathrm{i}g_s\gamma^\rho\frac{\lambda_j}{2}\right)\frac{\mathrm{i}}{\not p - \not k - m + \mathrm{i}\eta}\left(-\mathrm{i}g_s\gamma^\sigma\frac{\lambda_j}{2}\right)\frac{-\mathrm{i}g_{\rho\sigma}}{k^2+\mathrm{i}\eta}. \quad (14.21)$$

Like the vertex function $\Gamma^\mu(p_2,p_3)$ discussed in (4), we can at the outset use the Feynman–'t Hooft gauge $\xi = 1$ for the gluon propagator in (21). In the definition of $-\mathrm{i}\Sigma(p)$, we note that an additional factor $(-\mathrm{i})$ is included, its convenience is clearly seen in (22). Apart from a common factor, the three amplitudes of Fig. 14.2 can be written respectively from left to right as

$$\Gamma^\mu(p_2,p_3)\ ,\quad \frac{\mathrm{i}\left[-\mathrm{i}\Sigma(p_2)\right]}{\not p_2 - m}\gamma^\mu = \frac{\Sigma(p_2)}{\not p_2 - m}\gamma^\mu\ , \quad \text{and} \quad \gamma^\mu\frac{\Sigma(p_3)}{\not p_3 - m}\ . \quad (14.22)$$

A glance at the integral in (21) shows that both UV and IR divergences are present in $\Sigma(p)$. After writing the denominator of (21) as an integral of the Feynman auxiliary variable $x$ and integrating over $k$, we obtain

$$\begin{aligned}\Sigma(p) &= -\left(\frac{4}{3}\right)\mathrm{i}g_s^2\int_0^1\mathrm{d}x\int\frac{\mathrm{d}^nk}{(2\pi)^n}\frac{nm-(n-2)(\not p - \not k)}{[k^2-2p.kx+(p^2-m^2)x]^2}\\ &\equiv mA(p^2) + \not p B(p^2)\ ,\end{aligned}$$

$$\begin{aligned}A(p^2) &= \frac{4}{3}\frac{g_s^2}{(4\pi)^{n/2}}\int_0^1\mathrm{d}x\,\frac{n\Gamma(2-\frac{n}{2})}{[p^2x(1-x)-m^2x]^{2-\frac{n}{2}}} = \frac{4}{3}\frac{g_s^2}{4\pi^2}\left(\frac{2}{\varepsilon}+\cdots\right),\\ B(p^2) &= \frac{4}{3}\frac{g_s^2}{(4\pi)^{n/2}}\int_0^1\mathrm{d}x\,\frac{(2-n)\Gamma(2-\frac{n}{2})(1-x)}{[p^2x(1-x)-m^2x]^{2-\frac{n}{2}}} = \frac{4}{3}\frac{-g_s^2}{16\pi^2}\left(\frac{2}{\varepsilon}+\cdots\right). \quad (14.23)\end{aligned}$$

Besides their UV divergences with the $\varepsilon$ pole, the dimensionless quantities $A(p^2)$, $B(p^2)$ also have their IR divergences coming from $x \to 0$.

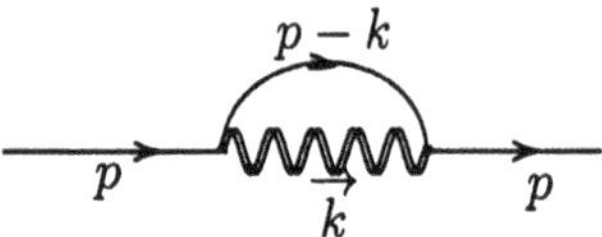


**Fig. 14.5.** Fermionic self-energy $\Sigma(p)$

The removal of the ultraviolet divergences in $\Gamma^\mu(p_2,p_3)$ and $\Sigma(p)$ is at the heart of the renormalization program which can be stated as follows: The ultraviolet infinities encountered in $F_1(q^2)$ and $\Sigma(p)$ can be consistently removed and the final results become finite if they are expressed in terms of the *renormalized quark mass* together with the *renormalized quark field.* We now go further by introducing the notion of *bare* and *renormalized* quantities.

## 14.3 Mass and Field Renormalization

Let us recall the relation between the propagator and the mass of a particle. The latter is associated with the pole of the former, so the mass is the solution of the equation obtained by setting to zero the inverse of its propagator. How can the mass of a fermion be determined if viewed as a cloud of virtual particles that are continuously being created and destroyed? Starting from a bare mass $m_0$, the fermion develops a change in $m_0$ by interacting with gluons through loop diagrams, the simplest example $\Sigma(p)$ is shown in Fig. 14.5. $\Sigma(p)$ is called self-energy, i.e. the extra amount of rest mass energy *generated by quantum corrections.* For the moment this change is infinite. If the mass is changed, so is the propagator. The bare fermion propagator is then replaced by its *full* or *dressed* propagator which includes all possible loop diagrams having two external fermion lines, also called the two-point Green's function.

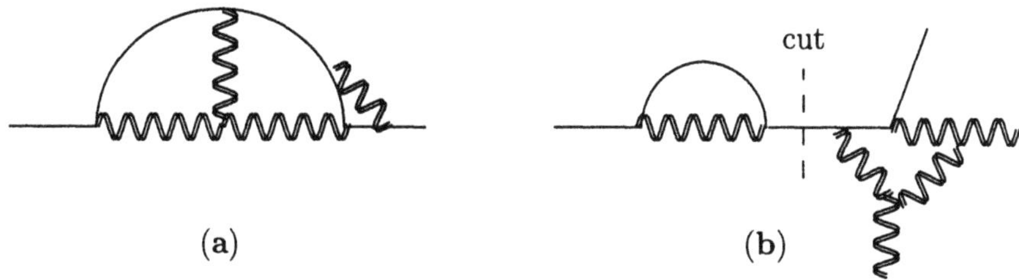


**Fig. 14.6.** (**a**) Irreducible $1PI$ diagram; (**b**) reducible $1PR$ diagram

To compute the dressed propagator, let us define a *one-particle irreducible* $1PI$ diagram to be any loop diagram that cannot be split into two disconnected ones by cutting only a single internal line. Every diagram is either irreducible ($1PI$) or reducible ($1PR$). An example of $1PI$ is Fig. 14.6a while Fig. 14.6b illustrates a $1PR$ diagram. The reason to select $1PI$ is that any $1PR$ can be decomposed into a set of $1PI$ without further loops, such that, if we know how to handle the UV divergences in $1PI$, then these UV divergences are also under control in $1PR$.

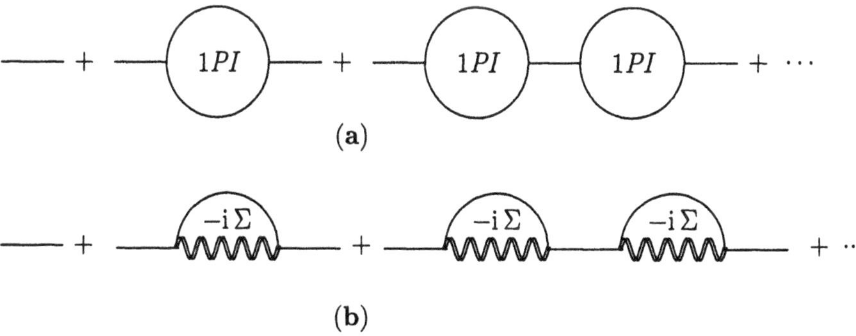


**Fig. 14.7.** (**a**) Geometric sum of $1PI$ diagrams; (**b**) dressed propagator $S_D(p)$

**Dressed Propagator.** A dressed propagator is an infinite geometric sum of $1PI$ diagrams with two external lines (Fig. 14.7a). Each $1PI$ by itself is an infinite sum of diagrams of orders $g_s^2, g_s^4 \cdots$, the lowest $1PI$ starting at $g_s^2$ is the diagram of Fig. 14.5 denoted by $-\mathrm{i}\Sigma(p)$ in (21).

To this order $g_s^2$ shown in Fig. 14.7b, the dressed quark propagator $S_\mathrm{D}(p)$ is obtained by summing up the geometric series of $-\mathrm{i}\Sigma(p)$. We have

$$\begin{aligned} S_\mathrm{D}(p) &= \frac{\mathrm{i}}{\not p - m_0} + \frac{\mathrm{i}}{\not p - m_0}[-\mathrm{i}\Sigma(p)]\frac{\mathrm{i}}{\not p - m_0} + \\ &+ \frac{\mathrm{i}}{\not p - m_0}[-\mathrm{i}\Sigma(p)]\frac{\mathrm{i}}{\not p - m_0}[-\mathrm{i}\Sigma(p)]\frac{\mathrm{i}}{\not p - m_0} + \cdots \\ &= \frac{\mathrm{i}}{\not p - m_0}\left[1 + \frac{\Sigma(p)}{\not p - m_0} + \left(\frac{\Sigma(p)}{\not p - m_0}\right)^2 + \cdots\right] \\ &= \frac{\mathrm{i}}{\not p - m_0}\left(1 - \frac{\Sigma(p)}{\not p - m_0}\right)^{-1} = \frac{\mathrm{i}}{\not p - m_0 - \Sigma(p)} \,. \end{aligned} \tag{14.24}$$

With quantum corrections, the physical mass $m$ is now associated with the pole of the dressed propagator $S_\mathrm{D}(p)$ in (24). The pole is no longer the bare mass $m_0$ but is shifted to $m$, solution to the equation

$$\begin{aligned} &\not p - m_0 - \Sigma(\not p = m) = 0\,, \qquad \text{or} \\ &m = m_0 + \Sigma(\not p = m) = m_0 + m_0 A(m^2) + m B(m^2)\,. \end{aligned} \tag{14.25}$$

It is important to remark that in the above equation, $\Sigma(p)$ can be considered as a function of $\not p$, for $p^2 = (\not p)^2$. For instance $\Sigma(m)$ is understood as $\Sigma(\not p = m)$. In the expressions of $A(p^2)$ and $B(p^2)$ given by (23), $m$ must be understood as $m_0$, since $\Sigma(p)$ was computed with the bare mass $m_0$. For example the $B(m^2)$ in (25) must be read as

$$B(m^2) = \frac{4}{3}\frac{(2-n)\,g_s^2}{(4\pi)^{n/2}}\Gamma\left(2 - \tfrac{n}{2}\right)\int_0^1 \mathrm{d}x \, \frac{(1-x)}{[m^2 x(1-x) - m_0^2 x]^{2-\frac{n}{2}}} \,. \tag{14.26}$$

Close to the physical pole $\not p \approx m$, we write the Taylor expansion

$$\Sigma(p) = \Sigma(m) + (\not p - m)\left.\frac{\mathrm{d}\Sigma(p)}{\mathrm{d}\not p}\right|_{\not p = m} + \mathcal{O}(p^2 - m^2)\,. \tag{14.27}$$

With $\Sigma(m) = m - m_0$, the denominator of $S_\mathrm{D}(p)$ in (24) has the following form for $\not p \approx m$:

$$\begin{aligned} \not p - m_0 - \Sigma(p) &\approx \not p - m_0 - \Sigma(m) - (\not p - m)\left.\frac{\mathrm{d}\Sigma(p)}{\mathrm{d}\not p}\right|_{\not p = m}\,, \\ &\approx (\not p - m)\left(1 - \left.\frac{\mathrm{d}\Sigma(p)}{\mathrm{d}\not p}\right|_{\not p = m}\right)\,. \end{aligned} \tag{14.28}$$

The shift of the mass $m_0 \to m = m_0 + \Sigma(m)$ not only brings a new pole at $\not{p} = m$ to the dressed propagator $S_{\rm D}(p)$ but also changes its residue at the physical pole $m$:

$$\begin{array}{ccc} \text{Bare propagator} & & \text{Dressed propagator} \\ \dfrac{\mathrm{i}}{\not{p} - m_0} & \longrightarrow & S_{\rm D}(p) = \dfrac{\mathrm{i}\, Z_{\rm q}}{\not{p} - m}\,, \end{array} \tag{14.29}$$

where $Z_{\rm q}$ is given by

$$\frac{1}{Z_{\rm q}} = 1 - \left.\frac{\mathrm{d}\Sigma(p)}{\mathrm{d}\,\not{p}}\right|_{\not{p}=m} . \tag{14.30}$$

The above expression for $Z_{\rm q}$ is derived by comparing (28) with the inverse of the dressed propagator given in (24). Indeed

$$S_{\rm D}^{-1}(p) = \frac{1}{\mathrm{i}}(\not{p} - m_0 - \Sigma(p)) = \frac{1}{\mathrm{i}}(\not{p} - m)\left(1 - \left.\frac{\mathrm{d}\Sigma(p)}{\mathrm{d}\,\not{p}}\right|_{\not{p}=m}\right) \equiv \frac{\not{p} - m}{\mathrm{i}\, Z_{\rm q}} .$$

Equation (30), which relates the residue $Z_{\rm q}$ to the derivative of the self-energy $\Sigma(p)$ at the physical mass $\not{p} = m$, is in principle formally valid when $\Sigma(p)$ and its derivative are finite. However, if they are infinite, (30) might be misleading since it would give $Z_{\rm q} \to 0$ for $\mathrm{d}\Sigma(p)/\mathrm{d}\,\not{p}|_{\not{p}=m} \to \infty$ which is absurd. To understand this subtlety, let us remark that without interactions, i.e. at zero order of the coupling constant $g_{\rm s}$, $Z_{\rm q} \equiv 1$, $\Sigma(p) \equiv 0$, which is of course satisfied by (30). The perturbative quantum corrections start at order $g_{\rm s}^2 : \Sigma(p) = \mathcal{O}(g_{\rm s}^2)$, $Z_{\rm q} = 1 + \mathcal{O}(g_{\rm s}^2)$. Within the framework of perturbative calculations in which $Z_{\rm q}$ is derived, (30) should be correctly written as

$$\left.\frac{\mathrm{d}\Sigma(p)}{\mathrm{d}\,\not{p}}\right|_{\not{p}=m} = 1 - \frac{1}{Z_{\rm q}} = \frac{Z_{\rm q} - 1}{Z_{\rm q}} = Z_{\rm q} - 1 + \mathcal{O}(g_{\rm s}^4) . \tag{14.31}$$

The above relation satisfies the $\mathcal{O}(g_{\rm s}^2)$ expansion. In the spirit of perturbative calculations, one has

$$Z_{\rm q} = 1 + \left.\frac{\mathrm{d}\Sigma(p)}{\mathrm{d}\,\not{p}}\right|_{\not{p}=m} , \quad \sqrt{Z_{\rm q}} = 1 + \tfrac{1}{2}\left.\frac{\mathrm{d}\Sigma(p)}{\mathrm{d}\,\not{p}}\right|_{\not{p}=m} . \tag{14.32}$$

Let us recall that the propagator is the Fourier transform of a two-point function. The bare propagator is associated with $\langle 0 | T[\psi_0(x)\overline{\psi}_0(y)] | 0\rangle$ of the bare field $\psi_0(x)$, the dressed propagator $S_D(p)$ is computed from the bare field $\psi_0(x)$ too [by summation over the geometric series in (24)].

This two-point function definition of the propagator suggests that if we scale the bare field $\psi_0$ by $1/\sqrt{Z_{\rm q}}$ and define the *renormalized* field $\psi$ by $\psi_0(x) = \sqrt{Z_{\rm q}}\psi(x)$, the infinite residue $Z_{\rm q}$ of $S_{\rm D}(p)$ can be absorbed by $\psi_0$, and $S_{\rm D}(p)$ is promoted to the *renormalized propagator* $\widetilde{S}_{\rm ren}(p)$ with *pole* at $\not p = m$ and with *residue* $= 1$. Thus,

$$S_{\rm D}(p) = \int {\rm d}x\, {\rm e}^{-{\rm i}p\cdot x}\, \langle 0 \,|\, T(\psi_0(x)\overline{\psi}_0(0))\,|\, 0\rangle = \frac{{\rm i}\, Z_{\rm q}}{\not p - m}\,.$$

Putting $\psi_0(x) = \sqrt{Z_{\rm q}}\psi(x)$ , one has

$$\widetilde{S}_{\rm ren}(p) = \int {\rm d}x\, {\rm e}^{-{\rm i}p\cdot x}\, \langle 0 \,|\, T(\psi(x)\overline{\psi}(0))\,|\, 0\rangle = \frac{{\rm i}}{\not p - m}\,. \qquad (14.33)$$

Let us clarify again the meaning of the shift of the fields proposed in (33). We have two quantities $m$ and $Z_{\rm q}$ governed by two equations (25) and (31). If they can be solved, we would obtain physical quantities $m$ and $Z_{\rm q}$ and could stop here. It must be emphasized that the renormalization of masses and fields has nothing directly to do with infinities encountered in the computation of the self-energy $\Sigma(p)$. It will still be necessary even in a theory in which loop integrals are convergent. However, since $\Sigma(p)$ is infinite, $m$ cannot be computed but is only fixed by the physical mass. The bare mass parameter $m_0$ (which is infinite) is adjusted to $\Sigma(\not p = m)$ in (25) to cancel its UV divergences and giving the physical finite mass $m$. Similarly, the infinite $Z_{\rm q}$ is adjusted to bare field $\psi_0$ (which is also infinite) to define the renormalized field $\psi = \psi_0/\sqrt{Z_{\rm q}}$, such that the propagator of a renormalized field has a pole at the physical mass and has a residue equal to unit, thus

$$\begin{array}{ccccccc} \text{Bare} & & \text{Interaction} & & \text{Dressed} & & \text{Renormalized} \\ \dfrac{{\rm i}}{\not p - m_0} & \longrightarrow & \Sigma(p) & \longrightarrow & \dfrac{{\rm i}\, Z_{\rm q}}{\not p - m} & \longrightarrow & \dfrac{{\rm i}}{\not p - m}\,. \end{array} \qquad (14.34)$$

The procedure by which the UV divergence in $Z_{\rm q}$ is removed from the theory is referred to as the *field strength renormalization* or the *wave function renormalization*, likewise, the removal of the UV divergence in $\Sigma(p)$ is referred to as the *mass renormalization.* The role of $Z_{\rm q}$ (and other $Z_j$ introduced later) is essential here and in the next chapters.

The concept of removal of the infinities is not restricted to quantum field theory. Even in classical electrodynamics, the self-energy of a pointlike electron which interacts with its own electric field is also divergent. The idea of subtraction of infinities – the core of the renormalization concept – was first suggested by Kramers. He observed that although the self-energy of a pointlike electron is infinite, actually the meaningful quantity is the *difference* between the self-energy of the free electron and that of the electron bound in an atom. Both of these self-energies diverge, but their difference is finite.

**Counterterms.** A counterterm is formally an infinite parameter introduced in the Lagrangian to remove (or absorb) the UV divergences by adjusting the bare parameters. After a subtraction of the infinities, the finite parts (also called the renormalized quantities) are constrained to obey *renormalization conditions* (also called normalization or subtraction conditions or prescriptions). In the case discussed here, the renormalization conditions dictate that the residue of the renormalized propagator is 1, and $\Sigma(m)+m_0 = m$. The number of counterterms must be limited, independently of the perturbative orders, otherwise the theory is nonrenormalizable.

For this renormalization program to work, it is essential that the original Lagrangian includes all interactions generated by the UV parts of Feynman amplitudes. This is the case of the standard electroweak theory and QCD. If for some reason, the loop integrals produce a UV term which has a covariant structure that the original Lagrangian does not possess, this UV divergence cannot be removed. For instance, if the magnetic moment $F_2(q^2)$ found in (14) were infinite, there is no way to absorb its divergence into the original QCD or QED Lagrangian since the latter does not possess such a Pauli magnetic interaction $(g_s/m)\,[\partial_\mu A^k_\nu(x) - \partial_\nu A^k_\mu(x)]\,\overline{\psi}(x)\lambda_k\sigma^{\mu\nu}\psi(x)$ from the start. The fact that $F_2(q^2)$ is finite is not an accident, it reveals the renormalizability of QCD. The renormalizability of QCD and QED forbids the presence of the Pauli magnetic interaction although Lorentz and gauge invariances allow such a term. The magnetic interaction will provoke an avalanche of infinities in loop integrals since the dimensional coupling constant $g_s/m$ implies a momentum dependence of the vertex. In turn it makes the loop integrals more and more divergent with increasing perturbative orders and an infinite numbers of counterterms are needed. This interaction is nonrenormalizable.

### 14.3.1 Renormalized Form Factor $\widetilde{F}_1^{\rm ren}(q^2)$

Having outlined the concept of renormalization, we now apply the method to our problem of removing the ultraviolet divergences in $F_1(q^2)$ and $\Sigma(p)$. This procedure promotes $F_1(q^2)$ to the renormalized form factor $\widetilde{F}_1^{\rm ren}(q^2)$, and $\Sigma(p)$ to the renormalized self-energy $\widetilde{\Sigma}_{\rm ren}(p)$; both of them are free of UV divergences. Let us see how it works.

In the Lagrangian $\mathcal{L}$, the bare quark fields $\psi^0(x)$ coupled with the boson field $W_\mu(x)$ can be written as (the weak coupling constant $g_{\rm W}$ is implicit)

$$\mathcal{L} = \overline{\psi}^0_{\rm q_2}(x)\gamma^\mu(1-\gamma_5)\psi^0_{\rm q_3}(x)W_\mu(x) + \text{ kinetic term}\,, \tag{14.35}$$

where the kinetic term $\overline{\psi^0}(x)(\mathrm{i}\,\partial\!\!\!/ - m_0)\psi^0(x)$ is applied to both quark fields $\psi^0_{\rm q_2}(x)$ and $\psi^0_{\rm q_3}(x)$. We define the renormalized fields $\psi_{\rm q_2}(x)$ and $\psi_{\rm q_3}(x)$ by introducing the counterterms $\sqrt{Z_{\rm q_2}}$ and $\sqrt{Z_{\rm q_3}}$. The W boson field $W_\mu(x)$, not affected by gluons, is untouched. We write

$$\psi^0_{\rm q_j}(x) = \sqrt{Z_{\rm q_j}}\,\psi_{\rm q_j}(x)\;,\;\; \mathrm{j} = 2,3\,. \tag{14.36}$$

Putting $\psi^0_{\rm q_j}(x) = \sqrt{Z_{\rm q_j}}\,\psi_{\rm q_j}(x)$ back into (35), the original $\mathcal{L}$ is now split into two parts:

$$\begin{aligned}
&\mathcal{L} = \mathcal{L}_{\rm ren} + \mathcal{L}_{\rm ct}\,;\\
&\mathcal{L}_{\rm ren} = \overline{\psi}_{\rm q_2}(x)\gamma^\mu(1-\gamma_5)\psi_{\rm q_3}(x)W_\mu(x)\,,\\
&\mathcal{L}_{\rm ct} = \delta_{\rm q}\,\overline{\psi}_{\rm q_2}(x)\gamma^\mu(1-\gamma_5)\psi_{\rm q_3}(x)W_\mu(x)\;;\;\; \delta_{\rm q} = \sqrt{Z_{\rm q_2}}\sqrt{Z_{\rm q_3}} - 1\,,
\end{aligned} \tag{14.37}$$

the kinetic term of (35), which should appear in the above equation, will be given later in (44). The role of $\mathcal{L}_{\rm ct}$ is to cancel the UV divergences of $\mathcal{L}_{\rm ren}$. Physical quantities, which are obtained by summing the contributions from both $\mathcal{L}_{\rm ren}$ and $\mathcal{L}_{\rm ct}$, are free of UV divergences.

To first order of $\delta_{\rm q}$, $\mathcal{L}_{\rm ct}$ gives a contribution $\delta_{\rm q}\,\gamma^\mu(1-\gamma_5)$ to the vertex $\gamma^\mu(1-\gamma_5)$. Note that $\delta_{\rm q}$ is $\mathcal{O}(g_{\rm s}^2)$. To the same $g_{\rm s}^2$ order, $\mathcal{L}_{\rm ren}$ gives $\Gamma^\mu(p_2,p_3)$ in (12) for which $m$ is now understood as the renormalized mass. The sum of $\delta_{\rm q}\,\gamma^\mu(1-\gamma_5)$ with $\Gamma^\mu(p_2,p_3)$ is the ultraviolet-convergent renormalized vertex function $\widetilde{\Gamma}^\mu_{\rm ren}(p_2,p_3)$ which replaces $\Gamma^\mu(p_2,p_3)$:

$$\Gamma^\mu(p_2,p_3) \longrightarrow \widetilde{\Gamma}^\mu_{\rm ren}(p_2,p_3) = \Gamma^\mu(p_2,p_3) + \delta_{\rm q}\,\gamma^\mu(1-\gamma_5)\,. \tag{14.38}$$

Comparing (38) with (11), we get the renormalized form factor $\widetilde{F}_1^{\rm ren}(q^2)$ which replaces $F_1(q^2)$:

$$F_1(q^2) \longrightarrow \widetilde{F}_1^{\rm ren}(q^2) = F_1(q^2) + \delta_{\rm q}\,. \tag{14.39}$$

Our next task is to compute $\delta_{\rm q}$. From (32) and (37), we have

$$\delta_{\rm q} = \frac{1}{2}\left(\frac{{\rm d}\Sigma(p_2)}{{\rm d}\,\not p_2}\bigg|_{\not p_2=m_2} + \frac{{\rm d}\Sigma(p_3)}{{\rm d}\,\not p_3}\bigg|_{\not p_3=m_3}\right) = \frac{{\rm d}\Sigma(p)}{{\rm d}\,\not p}\bigg|_{\not p=m}\,, \tag{14.40}$$

since we take $m_2=m_3=m$. The derivative of $\Sigma(p)$ can be readily performed using $\Sigma(p)$ given by (23). We obtain

$$\begin{aligned}\delta_{\rm q} = \frac{{\rm d}\Sigma(p)}{{\rm d}\,\not p}\bigg|_{\not p=m} &= B(m^2) + 2m^2\,\frac{{\rm d}A(p^2)}{{\rm d}p^2}\bigg|_{p^2=m^2} + 2m^2\,\frac{{\rm d}B(p^2)}{{\rm d}p^2}\bigg|_{p^2=m^2}\\ &= \frac{4}{3}\frac{g_{\rm s}^2}{(4\pi)^{\frac{n}{2}}}\Gamma\left(2-\frac{n}{2}\right)\int_0^1 \frac{{\rm d}x(1-x)}{[-m^2x^2]^{2-\frac{n}{2}}}\left(-2+3\varepsilon+\frac{2\varepsilon}{x}\right).\end{aligned}$$

Developing all the terms of the above equation in a series of $\varepsilon$, we get

$$\delta_{\rm q} = \frac{4}{3}\frac{g_{\rm s}^2}{16\pi^2}\left[\frac{-2}{\varepsilon} + \gamma_E - \log(4\pi) + \log\left(\frac{m^2}{\mu^2}\right) - 4 + 4\int_0^1\frac{{\rm d}x}{x}\right]. \tag{14.41}$$

Comparing (41) with $F_1(0)$ as given by (20), we get a remarkable result; to wit, the quantity $\delta_{\rm q}$ *exactly cancels* $F_1(0)$:

$$\delta_{\rm q} + F_1(0) = 0\ ,\quad \text{or}\quad \frac{{\rm d}\Sigma(p)}{{\rm d}\,\not p}\bigg|_{\not p=m} + F_1(0) = 0\,. \tag{14.42}$$

This important result, due to the current conservation, in fact mimics the Ward identity in QED (Problem 14.3) according to which the apparently unrelated quantities $F_1(0)$ and ${\rm d}\Sigma(p)/{\rm d}\not p|_{\not p=m}$ turn out to satisfy (42). In the case considered here, since we take $m_2 = m_3 = m$, the weak vector current is conserved, hence we also obtain (42). With (39), we have

$$\widetilde{F}_1^{\text{ren}}(q^2) = F_1(q^2) - F_1(0)\ . \tag{14.43}$$

The expression (43) is pleasantly simple. Both $F_1(q^2)$ and $F_1(0)$ are separately divergent, but their difference $\widetilde{F}_1^{\text{ren}}(q^2)$ is finite (free of UV divergences more precisely). The $2/\varepsilon$ pole of $F_{1,\text{uv}}(q^2)$ in (17) cancels the same $2/\varepsilon$ pole of $F_1(0)$. The counterterm used to absorb the ultraviolet divergence in $F_1(q^2)$ is precisely the value of $F_1(q^2)$ itself at $q^2 = 0$.

From now on, $F_1(q^2)$ is replaced with $\widetilde{F}_1^{\text{ren}}(q^2) = F_1(q^2) - F_1(0)$. Notice that the equation $\widetilde{F}_1^{\text{ren}}(0) = 0$ represents the *renormalization condition.* When we look back at $\mathcal{F}_1^{\text{cor}}(q^2)|_{\text{ren}} = 1 + \widetilde{F}_1^{\text{ren}}(q^2)$ in (16), we realize that at $q^2 = 0$, QCD does not bring any modification to $\mathcal{F}_1^{\text{cor}}(0)|_{\text{ren}}$.

### 14.3.2 Important Consequence of Mass Renormalization

To renormalize the quark mass, we need a counterterm $\Delta m_0$ to be added to the bare mass $m_0$ to get the physical mass $m$, $\Delta m_0 + m_0 = m$. Together with $Z_{\text{q}}$, this $\Delta m_0$ brings in $\mathcal{L}_{\text{ct}}$ a new term to be added to $\Sigma(p)$. Their sum is the renormalized self-energy $\widetilde{\Sigma}_{\text{ren}}(p)$, which is found to be

$$\widetilde{\Sigma}_{\text{ren}}(p) = \Sigma(p) - (Z_{\text{q}} - 1)(\not{p} - m) - Z_{\text{q}}\,\Delta m_0\ . \tag{14.44}$$

The generic $p$ stands for $p_2$ or $p_3$ and $m$ stands for $m_2$ or $m_3$. Like $\delta_{\text{q}}$ in (37), the last two terms of (44) arise when we put $\psi^0 = \sqrt{Z_{\text{q}}}\,\psi$ in the kinetic expression $\overline{\psi^0}(\mathrm{i}\,\not{\partial} - m_0)\psi^0$ of $\mathcal{L}$ in (35). Like (38) where $\Gamma^\mu(p_2,p_3)$ is replaced with $\widetilde{\Gamma}^\mu_{\text{ren}}(p_2,p_3)$, now $\Sigma(p)$ is replaced with $\widetilde{\Sigma}_{\text{ren}}(p)$. Thus, the amplitudes of Fig. 14.2b, c are now expressed in terms of $\widetilde{\Sigma}_{\text{ren}}(p)$. From (22), they are given by

$$\frac{\widetilde{\Sigma}_{\text{ren}}(p_2)}{\not{p}_2 - m_2}\gamma^\mu \quad \text{and} \quad \gamma^\mu \frac{\widetilde{\Sigma}_{\text{ren}}(p_3)}{\not{p}_3 - m_3}\ . \tag{14.45}$$

Now comes an important result which eliminates the contributions of both Fig. 14.2b, c. The renormalization condition, which sets the pole of the renormalized propagator $\widetilde{S}_{\text{ren}}(p)$ to be $m$, implies that $\widetilde{\Sigma}_{\text{ren}}(\not{p} = m) = 0$. This equation is obtained from (25) and (44). Indeed, keeping in mind that $Z_{\text{q}} = 1 + \mathcal{O}(g_{\text{s}}^2)$ and $\Delta m_0$ is also $\mathcal{O}(g_{\text{s}}^2)$, one has

$$\widetilde{\Sigma}_{\text{ren}}(\not{p} = m) = \Sigma(\not{p} = m) - Z_{\text{q}}\Delta m_0 = \Sigma(\not{p} = m) - \Delta m_0 = 0\ . \tag{14.46}$$

Not only is $\widetilde{\Sigma}_{\text{ren}}(\not{p} = m)$ equal to zero, its derivative at $\not{p} = m$ also vanishes, using (44) and (40) and remembering that $\delta_{\text{q}} = Z_{\text{q}} - 1$. We therefore have

$$\widetilde{\Sigma}_{\text{ren}}(\not{p} = m) = 0 \quad \text{and} \quad \left.\frac{\mathrm{d}\widetilde{\Sigma}_{\text{ren}}(\not{p})}{\mathrm{d}\,\not{p}}\right|_{\not{p}=m} = 0\ . \tag{14.47}$$

Equation (47) has an important consequence which forces (45) to vanish. The demonstration is straightforward using the Taylor expansion similar to (27). Indeed, when

$$\widetilde{\Sigma}_{\rm ren}(\not p = m) = 0 \quad \text{and} \quad \frac{{\rm d}\,\widetilde{\Sigma}_{\rm ren}(\not p)}{{\rm d}\not p}\Big|_{\not p = m} = 0 \ , \text{ then } \frac{\widetilde{\Sigma}_{\rm ren}(m)}{\not p - m} = 0 \ . \quad (14.48)$$

From (45) and (48), we note that after the quark gets its renormalized mass, the self-energy amplitude of each of the two diagrams (Fig. 14.2b, c) vanishes. This important property, established for fermionic fields, can be generalized to boson fields too. We arrive at the essential point: *for external particles (fermion or boson) on the mass-shell, it is not necessary to include self-energy corrections.* Via the self-energy $\Sigma(p)$, the two diagrams in Fig. 14.2b, c have no other role other than providing indirectly the counterterm $\delta_{\rm q}$ to the vertex $\gamma^\mu(1-\gamma_5)$ in (38), then cease to contribute.

The detailed studies in the three previous sections can be summarized as follows. The ultraviolet divergences of the three diagrams in Fig. 14.2 are absorbed by the renormalization procedure into the bare parameters (fields and masses) of the Lagrangian. It results in replacing $\Gamma^\mu(p_2,p_3)$ with the renormalized $\widetilde{\Gamma}^\mu_{\rm ren}(p_2,p_3)$, leading to the key result $F_1(q^2) \longrightarrow \widetilde{F}_1^{\rm ren}(q^2) = F_1(q^2) - F_1(0)$ which is ultraviolet convergent.

## 14.4 Virtual Gluon Contributions

We have determined $\widetilde{F}_1^{\rm ren}(q^2)$, the most important quantity in the calculation of virtual gluon corrections to the $\tau(P) \to \nu_\tau(p_1) + q_2(p_2) + \overline{q}_3(p_3)$ decay rate. As noted before, to order $g_{\rm s}^2$ correction to the rate, we need the sum – denoted by $\mathcal{M}_{\rm Vi}$ – of the tree amplitude of Fig. 14.1 with the renormalized loop amplitude of Fig. 14.2. Their interference in the square $|\mathcal{M}_{\rm Vi}|^2$ is $\mathcal{O}(g_{\rm s}^2)$:

$$\mathcal{M}_{\rm Vi} = \frac{G_{\rm F}V_{{\rm q}_2{\rm q}_3}}{\sqrt{2}}\overline{u}(p_1)\gamma_\mu(1-\gamma_5)u(P)\Big\{\overline{u}(p_2)[\gamma^\mu(1-\gamma_5) + \widetilde{\Gamma}^\mu_{\rm ren}(p_2,p_3)]v(p_3)\Big\} \ .$$

To lighten the computations of $\Gamma^\mu(p_2,p_3)$, hence of $\widetilde{\Gamma}^\mu_{\rm ren}(p_2,p_3)$ or $\mathcal{M}_{\rm Vi}$, we assume from now on $m=0$. With $m=0$, the expression for $\Gamma^\mu(p_2,p_3)$ is considerably simpler. The vector current is conserved, and of course the Ward identity holds in (42). Going back to (12), we see that only $F_1(q^2) = G_1(q^2)$ survives, while $F_2(q^2) = G_3(q^2) = 0$. Then

$$\widetilde{\Gamma}^\mu_{\rm ren}(p_2,p_3) = \gamma^\mu(1-\gamma_5)\widetilde{F}_1^{\rm ren}(q^2) \ .$$

Until now, we have postponed the problem of infrared divergence (IR) which occurs in $F_1(q^2)$, hence in $\widetilde{F}_1^{\rm ren}(q^2)$. Now is the time to treat it together with the real gluon emission process, also called bremsstrahlung.

The second part of $F_1(q^2)$ in the second line of (12) – that without the singular $\Gamma(2-\frac{n}{2})$ – is free of UV divergence. It has only IR divergence. To handle the latter, since with $\int_0^1 du/u$ in (19) we can go no further, we may try to parameterize the IR of $F_1(q^2)$ again as poles in $\varepsilon$, on equal footing with UV.

This can be done by going back to our original (9) and keeping everywhere the $\Gamma(3-\frac{n}{2})$ which is equal to 1 for $n=4$.

The denominator $-\mathcal{D}(x,y) = m^2(x+y)^2 - q^2xy$ on the second line of (12) is always written under the original form $[-\mathcal{D}(x,y)]^{3-\frac{n}{2}}$ as in (9), i.e. we keep $\varepsilon \neq 0$ even if the UV divergences are absent because we would like to parameterize the IR divergences by the $\varepsilon$ pole too. As for the first part of $F_1(q^2)$ appearing on the first line of (12) with $\Gamma(2-\frac{n}{2})$, we rewrite it in terms of $\Gamma(3-\frac{n}{2})$ using $\Gamma(2-\frac{n}{2}) = 2\,\Gamma(3-\frac{n}{2})/(4-n)$. With the relations

$$\int_0^1 \mathrm{d}x\, x^{a-1}(1-x)^{b-1} = \frac{\Gamma(a)\Gamma(b)}{\Gamma(a+b)}\ , \quad \text{and} \quad z\,\Gamma(z) = \Gamma(1+z)\ , \tag{14.49}$$

we get altogether from (12), with $m=0$,

$$\begin{aligned} F_1(q^2) &= \frac{4}{3}\frac{g_\mathrm{s}^2}{2^n\pi^{n/2}}\frac{\Gamma(3-\frac{n}{2})[\Gamma(\frac{n}{2}-1)]^2}{\Gamma(n-2)}\left[\frac{-8}{(4-n)^2}+\frac{2}{4-n}-2\right](-q^2)^{\frac{n}{2}-2}\ ,\\ &\equiv g_\mathrm{s}^2\,\mathcal{A}(n)\ (+q^2)^{\frac{n}{2}-2}\exp\left(\mathrm{i}\,\frac{n\,\pi}{2}\right)\ , \end{aligned} \tag{14.50}$$

where

$$\mathcal{A}(n) = \frac{4}{3}\frac{\Gamma(3-\frac{n}{2})[\Gamma(\frac{n}{2}-1)]^2}{\Gamma(n-2)\,2^n\pi^{n/2}}\left[\frac{-8}{(4-n)^2}+\frac{2}{4-n}-2\right]\ .$$

Due to the masslessness of both the gluon and the fermion, the double pole $1/\varepsilon^2$ in (50) has a pure IR origin. It comes from

$$\int_0^1 \mathrm{d}x \int_0^{1-x} \mathrm{d}y\,\frac{1}{(x\,y)^{3-\frac{n}{2}}} = \frac{4}{(4-n)^2}\,\frac{[\Gamma(\frac{n}{2}-1)]^2}{\Gamma(n-3)}\ .$$

Since we are free to continue analytically any result to $n \geq 4$, from (50) we find that $F_1(0)=0$, and (38) becomes $\widetilde{\Gamma}^\mu_\mathrm{ren}(p_2,p_3) = \gamma^\mu(1-\gamma_5)F_1(q^2)$. Then

$$\begin{aligned} \mathcal{M}_\mathrm{Vi} = &\frac{G_\mathrm{F}V_{\mathrm{q_2q_3}}}{\sqrt{2}}\overline{u}(p_1)\gamma_\mu(1-\gamma_5)u(P)\\ &\times\left\{\overline{u}(p_2)\gamma^\mu(1-\gamma_5)\left[1+F_1(q^2)\right]\,v(p_3)\right\}\ . \end{aligned} \tag{14.51}$$

*Remark.* As $F_1(0) = 0 = \delta_\mathrm{q}$, one may suspect that the UV divergence in $F_1(q^2)$ could not be removed since $\widetilde{F}_1^\mathrm{ren}(q^2) = F_1(q^2)$. In fact, $F_1(0)$ vanishes only in the approximation $m=0$. However, as explicitly written in (20),

$F_1(0)$ contains both UV and IR divergences, its vanishing is equivalent to the cancellation of a UV by a IR. Since these two kinds of divergence are parameterized by the same $4-n$ poles, the simple pole $2/(4-n)$ in (50) is understood as coming from an IR divergence which replaces a UV, the latter is implicitly removed by the interchange of the UV and IR poles which is imposed by $F_1(0)=0$. In other words, all the poles in (50) represent IR. ∎

For timelike $q^2 > 4m^2$, the form factor $F_1(q^2)$ develops its imaginary part, represented by $\exp(\mathrm{i}\,\frac{n\pi}{2})$. Putting the expression of $F_1(q^2)$ in (50) into $\mathcal{M}_{\mathrm{Vi}}$ in (51), after averaging the spin of the initial $\tau$ lepton state, as well as summing the spins and colors of the final states, we obtain *to order* $g_s^2$

$$\frac{1}{2}\sum_{\text{spin,color}} |\mathcal{M}_{\mathrm{Vi}}|^2 = g_s^2\, \mathcal{B}(q^2)\, \mathrm{Tr}[\not{p}_1\gamma_\mu \not{P}\gamma_\nu(1-\gamma_5)]\, \mathrm{Tr}[\not{p}_2\gamma^\mu\, \not{p}_3\gamma^\nu(1-\gamma_5)]\,,$$

$$\mathcal{B}(q^2) \equiv N_c\, G_F^2 |V_{q_2q_3}|^2 \left\{2\mathcal{A}(n)\,\cos\frac{n\pi}{2}\right\}(q^2)^{\frac{n}{2}-2}\,. \tag{14.52}$$

This $g_s^2$ order of $|\mathcal{M}_{\mathrm{Vi}}|^2$ is the interference between the tree amplitude of Fig. 14.1 and the renormalized loop amplitude of Fig. 14.2. Having obtained $|\mathcal{M}_{\mathrm{Vi}}|^2$ in (52), we go on to compute the decay rate $\Gamma_{\mathrm{Vi}}$ by using the formulas of Chap. 4 as in 4 dimensions, with only the replacement of 4 by $n$ for the phase space integral. Thus,

$$\Gamma_{\mathrm{Vi}} = \frac{1}{2M}\, g_s^2 \int \frac{\mathrm{d}^{n-1}p_1}{2E_1\,(2\pi)^{n-1}}\, \mathcal{B}(q^2)\, \mathrm{Tr}[\not{p}_1\gamma_\mu \not{P}\gamma_\nu(1-\gamma_5)]\, \frac{(2\pi)^n}{(2\pi)^{2n-2}}$$
$$\times \int\!\!\int \frac{\mathrm{d}^{n-1}p_2}{2E_2}\frac{\mathrm{d}^{n-1}p_3}{2E_3}\, \mathrm{Tr}[\not{p}_2\gamma^\mu\, \not{p}_3\gamma^\nu(1-\gamma_5)]\delta^{(n)}(p_2+p_3-q)\,. \tag{14.53}$$

The double integration $\int\!\int \mathrm{d}^{n-1}p_2\ \mathrm{d}^{n-1}p_3$ of (53) is symmetric by interchanging $p_2$ and $p_3$. Therefore the contribution of the $\gamma_5$ in the integrand $\mathrm{Tr}[\not{p}_2\gamma^\mu\, \not{p}_3\gamma^\nu(1-\gamma_5)]$ vanishes because of its $p_2$–$p_3$ antisymmetric character. Only terms symmetric in the $\mu$–$\nu$ permutation remain. In turn, it renders the $\gamma_5$ of $\mathrm{Tr}[\not{p}_1\gamma_\mu \not{P}\gamma_\nu(1-\gamma_5)]$ on the first line of (53) superfluous.

We compute now the double integration of (53) denoted by $J^{\mu\nu}(q^2)$:

$$J^{\mu\nu}(q^2) \equiv \int\!\!\int \frac{\mathrm{d}^{n-1}p_2}{2E_2}\frac{\mathrm{d}^{n-1}p_3}{2E_3}\, \mathrm{Tr}[\not{p}_2\gamma^\mu\, \not{p}_3\gamma^\nu]\, \delta^{(n)}(p_2+p_3-q)\,. \tag{14.54}$$

First we notice that $q_\mu J^{\mu\nu}(q^2)=0$ because of $m^2=p_2^2=p_3^2=0$, implying that $J^{\mu\nu}(q^2)$ must have the structure $J^{\mu\nu}(q^2)=(-g^{\mu\nu}q^2+q^\mu q^\nu)L(q^2)$.

Since $L(q^2)$ is a function of $q^2$, it is convenient to use the center-of-mass frame where $\boldsymbol{q}=\boldsymbol{p}_2+\boldsymbol{p}_3=\boldsymbol{0}$. The variable $q^\mu$ is $(\sqrt{q^2}, \boldsymbol{q}=\boldsymbol{0})$ and $2E_2 = 2E_3 = \sqrt{q^2}$. We multiply the left and the right sides of (54) by $g_{\mu\nu}$, and integrate over $\mathrm{d}^{n-1}p_3$ to get rid of $\delta^{n-1}(p_2+p_3-q)$. The result is

$$(1-n)L(q^2) = 2(2-n)\int \frac{\mathrm{d}^{n-1}p_2}{2E_2}\frac{1}{2E_3}\,\delta(E_2+E_3-\sqrt{q^2})\,. \tag{14.55}$$

As remarked before, $\mathrm{Tr}[\not{p}_2\gamma^\mu \not{p}_3\gamma^\nu] = 4\,[p_2^\mu p_3^\nu + p_2^\nu p_3^\mu - (p_2\cdot p_3)g^{\mu\nu}]$ for $n \neq 4$ in (54). The $n$-dimensional solid angle is defined by $n-1$ angles: $\theta, \theta_1, \cdots, \theta_{n-2}$,

$$\Omega^n = \int \mathrm{d}\Omega^n = \int_0^\pi \mathrm{d}\theta(\sin\theta)^{n-2} \int_0^\pi \mathrm{d}\theta_1(\sin\theta_1)^{n-3} \cdots \int_0^{2\pi} \mathrm{d}\theta_{n-2}\,.$$

Using (49), we now demonstrate the following useful formulas

$$I_k \equiv \int_0^\pi \mathrm{d}\theta(\sin\theta)^k = \int_{-1}^{+1} \mathrm{d}\cos\theta(1-\cos^2\theta)^{\frac{k-1}{2}} = \int_0^1 \mathrm{d}x\, x^{-\frac{1}{2}}\,(1-x)^{\frac{k-1}{2}}$$

$$= \frac{\Gamma(\frac{1}{2})\Gamma[\frac{1}{2}(k+1)]}{\Gamma[\frac{1}{2}(k+2)]}\,, \quad I_0 = \pi \longrightarrow \Gamma(\tfrac{1}{2}) = \sqrt{\pi}\,,$$

$$\Omega^n = \left\{\frac{\sqrt{\pi}\,\Gamma[\frac{1}{2}(n-1)]}{\Gamma(\frac{n}{2})}\right\} \cdots \left\{\frac{\sqrt{\pi}\Gamma(\frac{1}{2}[n-(n-2)])}{\Gamma(\frac{1}{2}[n-(n-3)])}\right\} \times (2\pi) = \frac{2\,\pi^{n/2}}{\Gamma(\frac{n}{2})}\,. \tag{14.56}$$

With (56), the quantity $\mathrm{d}^{n-1}p_2/2E_2$ in (55) can be written as

$$\frac{\mathrm{d}^{n-1}p_2}{2E_2} = \frac{E_2^{n-3}\,\mathrm{d}E_2}{2}\,\Omega^{n-1} = E_2^{n-3}\,\frac{\pi^{\frac{n-1}{2}}}{\Gamma(\frac{n-1}{2})}\,\mathrm{d}E_2\,. \tag{14.57}$$

Putting (57) back into (55), we integrate over $E_2$ to eliminate the last function $\delta(2E_2 - \sqrt{q^2})$. Using $(n-1)\Gamma[\frac{1}{2}(n-1)] = 2\,\Gamma[\frac{1}{2}(n+1)]$, we obtain

$$J^{\mu\nu}(q^2) = \frac{(n-2)\,\pi^{\frac{n-1}{2}}}{2^{n-2}\,\Gamma(\frac{n+1}{2})}\,(q^2)^{\frac{n}{2}-2}\,\left[-q^2\,g^{\mu\nu} + q^\mu q^\nu\right]\,. \tag{14.58}$$

We now insert the above expression of $J^{\mu\nu}(q^2)$ into (53); the product of $\mathrm{Tr}[\not{p}_1\gamma_\mu \not{P}\gamma_\nu]$ with the tensor $[-q^2\,g^{\mu\nu} + q^\mu q^\nu]$ is found to be

$$[-q^2\,g^{\mu\nu} + q^\mu q^\nu]\,[\mathrm{Tr}[\not{p}_1\gamma_\mu \not{P}\gamma_\nu] = 2M^4(1-\xi)[1+(n-2)\xi]\,, \tag{14.59}$$

where $\xi = q^2/M^2$. The remaining integral $\mathrm{d}^{n-1}p_1$ in (53) is simple in the rest frame of the decaying $\tau$ lepton. Using $(P-p_1)^2 = (M^2 - 2ME_1) = q^2$,

$$\int \frac{\mathrm{d}^{n-1}p_1}{2E_1\,(2\pi)^{n-1}} = \Omega^{n-1}\int \frac{E_1^{n-2}\,\mathrm{d}E_1}{2E_1\,(2\pi)^{n-1}} = \int \frac{E_1^{n-3}\,\mathrm{d}E_1}{\Gamma(\frac{n-1}{2})\,(4\pi)^{\frac{n-1}{2}}}$$

$$= \frac{M^{n-2}}{2^{2n-3}\,\pi^{\frac{n-1}{2}}\,\Gamma(\frac{n-1}{2})}\int_0^1 \mathrm{d}\xi\,(1-\xi)^{n-3}\,. \tag{14.60}$$

Putting together (52)–(60), we get

$$\frac{\Gamma_{\mathrm{Vi}}}{\Gamma_0} = \frac{[N_c|V_{q_2q_3}|^2\,\alpha_s]\;2^{5\varepsilon}\,\pi^{\frac{3}{2}\varepsilon}\,\Gamma(2-\frac{1}{2}\varepsilon)}{M^{3\varepsilon}\Gamma[\frac{1}{2}(3-\varepsilon)]\,\Gamma[\frac{1}{2}(5-\varepsilon)]}\int_0^1 \mathrm{d}\xi\,\xi^{-\varepsilon}\,(1-\xi)^{2-\varepsilon}[1+(2-\varepsilon)\,\xi]$$

$$\times\left\{\frac{\Gamma(1+\frac{1}{2}\varepsilon)\Gamma(1-\frac{1}{2}\varepsilon)}{\Gamma(2-\varepsilon)}\cos(\tfrac{1}{2}\pi\varepsilon)\left[-\frac{4}{\varepsilon^2}+\frac{1}{\varepsilon}-1\right]\right\}\,, \tag{14.61}$$

where $\Gamma_0 = G_F^2 M^5/192\pi^3$ already given in (13.20) is the QCD uncorrected decay rate $\tau^- \to \nu_\tau + q_2 + \overline{q}_3$ of Fig. 14.1, in which colors (of massless quarks) are not yet summed, i.e. the factor $N_c$ is not yet included.

As we will see, the IR divergent *singular* $\varepsilon$ pole terms in (61) are exactly canceled by those of the bremsstrahlung rate that we are going now to compute from the two diagrams in Fig. 14.3, such that the sum of (61) and (81) is *infrared convergent.* We remark further that both (61) and (81) share a common regular term represented by the first factor, i.e. the first line on the right hand side of (61). Therefore, we keep untouched this factor and we only expand up to $\mathcal{O}(\varepsilon^2)$ the regular terms $\Gamma(1+\frac{1}{2}\varepsilon)\Gamma(1-\frac{1}{2}\varepsilon)/\Gamma(2-\varepsilon)] \times \cos(\frac{1}{2}\pi\varepsilon)$ in the second factor (second line) represented by the curly brackets $\{\}$ of (61). When these $\varepsilon^2$ expansions are multiplied by the singular poles $-4/\varepsilon^2 + 1/\varepsilon - 1$ in $\{\}$, some finite terms emerge. This expansion of the regular terms up to second order in $\varepsilon$ is therefore mandatory. We obtain for the curly brackets [second line of (61)] the following result

$$\{\} = -\frac{4}{\varepsilon^2} + \frac{4\gamma_E - 3}{\varepsilon} - 2\gamma_E^2 + 3\gamma_E - 4 + \frac{2\pi^2}{3} + \mathcal{O}(\varepsilon)\,, \tag{14.62}$$

using (49) together with the expansion up to $z^2$ of $\Gamma(1+z)$ for $z \ll 1$,

$$\Gamma(1+z) \underset{z\to 0}{\longrightarrow} 1 - \gamma_E\, z + \frac{6\gamma_E^2 + \pi^2}{12}\, z^2 + \mathcal{O}(z^3)\,. \tag{14.63}$$

## 14.5 Real Gluon Contributions

The amplitude of the two diagrams in Fig. 14.3 can be written as

$$\begin{aligned}\mathcal{M}_{\rm Re} =& \frac{G_F\, V_{q_2q_3}\, g_s}{\sqrt{2}}\ \overline{u}(p_1)\gamma_\mu(1-\gamma_5)u(P) \qquad (14.64)\\ &\times \overline{u}(p_2)\left[\not{\varepsilon}_{k'}\ \frac{\lambda_i}{2}\ \frac{\not{p}_2 + \not{k}'}{(p_2+k')^2}\gamma^\mu - \gamma^\mu\ \frac{\not{p}_3 + \not{k}'}{(p_3+k')^2}\ \not{\varepsilon}_{k'}\ \frac{\lambda_i}{2}\right](1-\gamma_5)\ v(p_3).\end{aligned}$$

In (64), $\varepsilon^\alpha(k', i)$ is the gluon polarization vector associated with the eight SU(3) color matrices $\lambda_i$. To simplify the trace calculation of many $\gamma$ matrices, we write $\not{\varepsilon}_{k'}\ \not{p}_2 = -\ \not{p}_2\ \not{\varepsilon}_{k'} + 2p_2\cdot\varepsilon_{k'}$, and $\not{p}_3\ \not{\varepsilon}_{k'} = -\ \not{\varepsilon}_{k'}\ \not{p}_3 + 2p_3\cdot\varepsilon_{k'}$, and apply the Dirac equation to the spinors $\overline{u}(p_2)$, $v(p_3)$ in (64). The transition probability $|\mathcal{M}_{\rm Re}|^2$, summed and averaged over spins and colors in the usual manner, can best be expressed in terms of a tensor $\mathcal{T}^{\mu\nu}(p_2, p_3, k')$ defined below. We find

$$\frac{1}{2}\sum_{\rm color,spins} |\mathcal{M}_{\rm Re}|^2 = g_s^2\ \frac{N_c}{3} G_F^2\ |V_{q_2q_3}|^2\ {\rm Tr}[\not{p}_1\gamma_\mu\ \not{P}\gamma_\nu(1-\gamma_5)]\ \mathcal{T}^{\mu\nu}(p_2,p_3,k'),$$

$$\begin{aligned}\mathcal{T}^{\mu\nu}(p_2,p_3,k') =& {\rm Tr}\Bigg\{\not{p}_2\left[\frac{2p_2^\alpha + \gamma^\alpha\ \not{k}'}{p_2\cdot k'}\gamma^\mu - \gamma^\mu\frac{2p_3^\alpha + \not{k}'\gamma^\alpha}{p_3\cdot k'}\right]\\ &\times \not{p}_3\left[\gamma^\nu\frac{2p_2^\beta + \not{k}'\gamma^\beta}{p_2\cdot k'} - \frac{2p_3^\beta + \gamma^\beta\ \not{k}'}{p_3\cdot k'}\gamma^\nu\right](1-\gamma_5)\Bigg\}(-g_{\alpha\beta}),\end{aligned} \tag{14.65}$$

the factor $-g_{\alpha\beta}$ comes from the summation over the gluon polarizations. Using the fact that the bremsstrahlung rate is obtained by integration over the symmetric phase space in $p_2, p_3, k'$, it is easy to show that the $\gamma_5$ can be dropped from $\mathcal{T}^{\mu\nu}(p_2,p_3,k')$. Then the integration of the latter will result in a term symmetric in $\mu$ and $\nu$. This in turn renders irrelevant the $\gamma_5$ in $\mathrm{Tr}[\not{p}_1\gamma_\mu \not{P}\gamma_\nu(1-\gamma_5)]$ of (65). Thus both $\gamma_5$ can be eliminated.

### 14.5.1 Infrared Divergence

As explicitly shown in $\mathcal{T}^{\mu\nu}(p_2,p_3,k')$, the denominators $p_2\cdot k'$ and $p_3\cdot k'$ indicate that the real gluon emission rate is divergent in the limit where the energy-momentum $k'$ of the gluon tends to zero. For massless quarks, these denominators vanish also when both the gluon and the quark are emitted in parallel, regardless of the gluon energy. In these limits, the processes with radiated gluons cannot be distinguished from those without gluons. The bremsstrahlung is thus an essential part of radiative corrections in this $\tau$ decay as well as in all other QCD (QED) reactions with real gluons (photons) emitted. The tensor $\mathcal{T}^{\mu\nu}(p_2,p_3,k')$ is found to be

$$\begin{aligned}
\frac{1}{4}\mathcal{T}^{\mu\nu}(p_2,p_3,k') = &-g^{\mu\nu}\left[\frac{8s^2}{ut}+\frac{8s}{u}+\frac{8s}{t}+2(n-2)\left(\frac{u}{t}+\frac{t}{u}\right)+4(n-4)\right] \\
&- p_2^\mu p_2^\nu\left[\frac{16}{t}\right] - p_3^\mu p_3^\nu\left[\frac{16}{u}\right] - k'^\mu k'^\nu\left[\frac{8(n-4)s}{ut}\right] \\
&+ (p_2^\mu p_3^\nu + p_2^\nu p_3^\mu)\left(\frac{16s}{ut}+\frac{8}{u}+\frac{8}{t}\right) \\
&+ (p_2^\mu k'^\nu + p_2^\nu k'^\mu)\left[\frac{8s}{ut}+\frac{4(n-4)}{t}+\frac{4(n-2)}{u}\right] \\
&+ (p_3^\mu k'^\nu + p_3^\nu k'^\mu)\left[\frac{8s}{ut}+\frac{4(n-4)}{u}+\frac{4(n-2)}{t}\right], \qquad (14.66)
\end{aligned}$$

where the three invariants $s,t,u$ are defined as follows, only two of them are independent on account of the momentum conservation

$$s = 2p_2\cdot p_3\ ,\quad t = 2p_2\cdot k'\ ,\quad u = 2p_3\cdot k'\ ,\quad s+t+u = q^2\ .$$

For $m_2 = m_3 = m$, and *a fortiori* for $m = 0$, the conservation of the vector current at the vertex $W\mathrm{q}_2\overline{\mathrm{q}}_3$ in Fig. 14.3 implies that $q_\mu\mathcal{T}^{\mu\nu}(p_2,p_3,k') = q_\nu\mathcal{T}^{\mu\nu}(p_2,p_3,k') = 0$, where $q = p_2+p_3+k'$. We can verify this property by multiplying the right-hand side of (66) by $(p_2+p_3+k')_\mu$, and check that the product effectively is equal to zero.

Using (65), the decay width $\Gamma_{\mathrm{Re}} \equiv \Gamma(\tau\to\nu_\tau+\mathrm{q}_2+\overline{\mathrm{q}}_3+\mathrm{g})$ is

$$\begin{aligned}
\Gamma_{\mathrm{Re}} = &\frac{1}{2M}\frac{N_c}{3}G_F^2\,|V_{\mathrm{q_2q_3}}|^2\,g_s^2\int\frac{\mathrm{d}^{n-1}p_1}{(2\pi)^{n-1}2E_1}\mathrm{Tr}[\not{p}_1\gamma_\mu\not{P}\gamma_\nu] \\
&\times\frac{(2\pi)^n}{(2\pi)^{3n-3}}\int_{PS3}\mathcal{T}^{\mu\nu}(p_2,p_3,k')\ , \qquad (14.67)
\end{aligned}$$

where the three-body phase space integral is denoted by $\int_{\mathcal{P}S3}$,

$$\int_{\mathcal{P}S3} \equiv \int \frac{\mathrm{d}^{n-1}p_2}{2E_2} \frac{\mathrm{d}^{n-1}p_3}{2E_3} \frac{\mathrm{d}^{n-1}k'}{2E_{k'}} \delta^n(p_2 + p_3 + k' - q) \,. \tag{14.68}$$

No matter how complicated the integration $\int_{\mathcal{P}S3} \mathcal{T}^{\mu\nu}(p_2, p_3, k')$ is, it results only in a function of the momentum transfer $q^2$. Moreover, this phase space integration $\int_{\mathcal{P}S3} \mathcal{T}^{\mu\nu}(p_2, p_3, k')$ must have the structure $-q^2 g^{\mu\nu} + q^\mu q^\nu$, due to $q_\mu \mathcal{T}^{\mu\nu}(p_2, p_3, k') = q_\nu \mathcal{T}^{\mu\nu}(p_2, p_3, k') = 0$, so that

$$\int_{\mathcal{P}S3} \mathcal{T}^{\mu\nu}(p_2, p_3, k') = (-q^2 g^{\mu\nu} + q^\mu q^\nu) H(q^2) \,. \tag{14.69}$$

To compute $H(q^2)$, we multiply the left- and the right-hand sides of (69) by $g_{\mu\nu}$. Using the expression of $\mathcal{T}^{\mu\nu}(p_2, p_3, k')$ in (66), we get

$$q^2 H(q^2) = \frac{8(2-n)}{1-n} \int_{\mathcal{P}S3} \left\{ 4\left[\frac{s^2}{ut} + \frac{s}{u} + \frac{s}{t}\right] + (n-2)\left[\frac{u}{t} + \frac{t}{u}\right] + 2(n-4) \right\}.$$

Since $\int_{\mathcal{P}S3}$ is completely symmetric in the three integration variables $p_2$, $p_3$, and $k'$ and all of these three particles are massless, the integration of the three variables $s, t$, and $u$ are completely equivalent because of the possible interchange among $p_2$, $p_3$, and $k'$. More precisely, we have

$$\int_{\mathcal{P}S3} \left(\frac{u}{t}\right) = \int_{\mathcal{P}S3} \left(\frac{t}{u}\right) \,.$$

Then using $(s + t + u) = q^2$, we obtain a simple form for $H(q^2)$:

$$H(q^2) = \frac{16}{q^2} \frac{n-2}{n-1} \int_{\mathcal{P}S3} \left[ \frac{2\, sq^2}{ut} + (n-2)\frac{u}{t} + (n-4) \right] \,. \tag{14.70}$$

Our next task is to evaluate the three-body phase space integral $\int_{\mathcal{P}S3}$.

### 14.5.2 Three-Particle Phase Space

It is instructive to see how $\int_{\mathcal{P}S3}$ can be decomposed and computed, first in $n = 4$ dimensions. By energy-momentum conservation, there are in all $3 \times 3 - 4 = 5$ independent variables that describe the three-body phase space. Since $\int_{\mathcal{P}S3}$ is a function of $q^2$, it may be convenient to choose the rest frame of $q^\mu$, i.e. $\boldsymbol{q} = \boldsymbol{p}_2 + \boldsymbol{p}_3 + \boldsymbol{k}' = \boldsymbol{0}$: $q^\mu = (\sqrt{q^2}, \boldsymbol{0})$. The three momenta $\boldsymbol{p}_2, \boldsymbol{p}_3, \boldsymbol{k}'$ define a plane $\mathcal{P}$ in this frame. The vector $\boldsymbol{k}'$ is fixed by $E_2, E_3$ and the angle $\theta$ between $\boldsymbol{p}_2$ and $\boldsymbol{p}_3$. Energy conservation restricts $\theta$ in terms of $E_2, E_3$. So only $E_2$ and $E_3$ are independent. Thus for the three massless particles:

$$E_{k'} = \sqrt{E_2^2 + E_3^2 + 2E_2\, E_3 \cos\theta} \;,$$

$$2E_2\, E_3 \cos\theta = q^2 + 2E_2\, E_3 - 2\sqrt{q^2}(E_2 + E_3) \,. \tag{14.71}$$

The three remaining independent variables can be chosen as the angular orientation of $\boldsymbol{p}_2, \boldsymbol{p}_3$, and $\boldsymbol{k}'$. For instance, two angles, denoted by $\Omega$ to determine the vector $\boldsymbol{p}_2$, and one angle $\Phi$ to fix the plane $\mathcal{P}$ around $\boldsymbol{p}_2$. Using $\int \mathrm{d}^3k' \, \delta^3(p_2+p_3+k'-q) = 1$, we have for $m_2 = m_3 = 0$,

$$\int_{\mathcal{PS}3} (n=4) = \int \mathrm{d}\Omega \int_0^{2\pi} \mathrm{d}\Phi \int \frac{E_2 \mathrm{d}E_2}{2} \int \frac{E_3 \mathrm{d}E_3}{2} \times \int_{-1}^{+1} \frac{\mathrm{d}\cos\theta}{2E_{k'}} \delta\left(E_2 + E_3 + E_{k'} - \sqrt{q^2}\right) . \tag{14.72}$$

We remove the last $\delta$-function by performing the $\cos\theta$ integration, using

$$\delta\{f(x)\} = \frac{\delta(x-x_0)}{|f'(x)|_{x=x_0}} \to \delta\left(E_2 + E_3 + E_{k'} - \sqrt{q^2}\right) = \frac{\delta(\cos\theta - z)}{\frac{\mathrm{d}E_{k'}}{\mathrm{d}\cos\theta}} .$$

From (71),

$$\frac{\mathrm{d}E_{k'}}{\mathrm{d}\cos\theta} = \frac{E_2 E_3}{E_{k'}} \Longrightarrow \int \frac{\mathrm{d}\cos\theta}{2E_{k'}} \delta\left(E_2 + E_3 + E_{k'} - \sqrt{q^2}\right) = \frac{1}{2E_2E_3} , \tag{14.73}$$

and we obtain

$$\int_{\mathcal{PS}3} (n=4) = \frac{1}{2^3} \int \mathrm{d}\Omega \int_0^{2\pi} \mathrm{d}\Phi \int_\triangle \mathrm{d}E_2 \, \mathrm{d}E_3 = \pi^2 \int_\triangle \mathrm{d}E_2 \, \mathrm{d}E_3 . \tag{14.74}$$

Due to (71), the $E_2$ and $E_3$ integration domain $\triangle$ is restricted by

$$|q^2 + 2E_2 E_3 - 2\sqrt{q^2}(E_2+E_3)| \le 2E_2E_3 , \tag{14.75}$$

which is translated into a rectangular isosceles triangle limited by three lines in the $(E_2, E_3)$ plane: $E_2 = \frac{1}{2}\sqrt{q^2}$, $E_3 = \frac{1}{2}\sqrt{q^2}$, and $E_2 + E_3 = \frac{1}{2}\sqrt{q^2}$.

**A digression.** In the general case of a particle $A$ decaying into 3 others, say $a_1 + a_2 + a_3$, the squared decay amplitude multiplied by the phase space volume (72) will give the double distribution $\mathrm{d}\Gamma/(\mathrm{d}E_j \, \mathrm{d}E_\ell)$ of $a_j, a_\ell$ energies. The kinematic result in (74) indicates that a plot of $\mathrm{d}\Gamma/(\mathrm{d}E_j \, \mathrm{d}E_\ell)$ is a powerful tool for investigating the decay dynamics, as suggested by Dalitz.

Indeed, if the amplitude is constant, the events will be uniformly distributed in the $\triangle$ domain according to (74). On the other hand, any dynamical specific structure of the amplitude will be immediately revealed by a characteristic density of events in this plot. A concentration of events which cluster along a line in the $\triangle$ domain of $E_j, E_\ell$ corresponds to the presence of a resonance formed by $a_j$ and $a_\ell$: $A \to a_k + B$ follows by $B \to a_j + a_\ell$. Many hadronic resonances are found by this method. The angular $\Omega$ and $\Phi$ distributions $\mathrm{d}\Gamma/(\mathrm{d}\Omega\mathrm{d}\Phi)$, on the other hand, provide useful information on the spin and intrinsic parity of the decaying particle $A$, once the decay amplitude squared is incorporated in (72). ∎

### 14.5.3 Bremsstrahlung Rate

Going back to our case of (68) and (70), we first integrate over the gluon momentum $\mathrm{d}^{n-1}k'$ to get rid of the $\delta^{n-1}(p_2+p_3+k'-q)$, similar to (72). Together with (56) and (57), we write

$$\frac{\mathrm{d}^{n-1}p_3}{2E_3} = \frac{E_3^{n-3}\mathrm{d}E_3}{2}\,\Omega^{n-2}\,\mathrm{d}\theta(\sin\theta)^{n-3}\;,\;\text{with}\;\Omega^{n-2} = \frac{2\pi^{\frac{n-2}{2}}}{\Gamma(\frac{n-2}{2})}\;,$$
$$\mathrm{d}\theta(\sin\theta)^{n-3} = -\mathrm{d}\cos\theta\,(1-\cos\theta)^{\frac{n}{2}-2}\,(1+\cos\theta)^{\frac{n}{2}-2}\;. \qquad (14.76)$$

From the isosceles triangle in (75), one has $\sqrt{q^2} \le 2(E_2+E_3) \le 2\sqrt{q^2}$, it proves convenient to introduce two variables $x, y$ related to $E_2, E_3$ by

$$E_2 = \frac{\sqrt{q^2}\,x}{2}\;,\quad E_3 = \frac{\sqrt{q^2}(1-xy)}{2}\;,\quad \mathrm{d}E_2\,\mathrm{d}E_3 = \frac{q^2}{4}x\,\mathrm{d}x\,\mathrm{d}y\;, \qquad (14.77)$$

then from (71), we get

$$1-\cos\theta = \frac{2(1-y)}{1-xy}\;,\quad 1+\cos\theta = \frac{2y(1-x)}{1-xy}\;. \qquad (14.78)$$

Using (73), the remaining $\delta(E_2+E_3+E_{k'}-\sqrt{q^2})$ is replaced by $E_{k'}/(E_2E_3)$ after the $\cos\theta$ integration. With (57), (76), (77), and (78), the $n$-dimension three-body phase space $\int_{\mathcal{P}S3}$ defined in (68) can now be written as

$$\int_{\mathcal{P}S3} = \frac{\pi^{n-\frac{3}{2}}}{2^{n-1}}\frac{[q^2]^{n-3}}{\Gamma[\frac{1}{2}(n-1)]\,\Gamma[(\frac{1}{2}(n-2)]}$$
$$\times\int_0^1 \mathrm{d}x\,x^{n-3}\,(1-x)^{\frac{n}{2}-2}\int_0^1 \mathrm{d}y\,[y\,(1-y)]^{\frac{n}{2}-2}\;. \qquad (14.79)$$

Using (77) and (78), we have

$$s = q^2\,x\,(1-y)\;,\quad t = q^2 x\,y\;,\quad u = q^2\,(1-x)\;. \qquad (14.80)$$

With (79) and (80) plugged into (70) and using (49), we finally get

$$H(q^2) = \frac{[q^2]^{n-4}\pi^{n-\frac{3}{2}}}{2^{n-5}}\,\frac{\Gamma(\frac{n}{2})\Gamma(\frac{n}{2}-1)}{\Gamma(\frac{n+1}{2})\Gamma(\frac{3n}{2}-3)}\left[\frac{2n(n-2)}{(n-4)^2}+\frac{n^2-4}{(n-4)}+(n-4)\right]\;.$$

The $\varepsilon = (n-4)$ double and simple poles in the above equation come from the integration of $2sq^2/ut$ and $(n-2)u/t$ of (70) respectively.

Putting the quantity $H(q^2)$ back into (69) and (67) and using again (59) and (60), we obtain the bremsstrahlung rate

$$\frac{\Gamma_{\mathrm{Re}}}{\Gamma_0} = \frac{[\mathrm{N_c}|V_{q_2q_3}|^2\,\alpha_{\mathrm{s}}]\;2^{5\varepsilon}\,\pi^{\frac{3}{2}\varepsilon}\,\Gamma(2-\frac{1}{2}\varepsilon)}{M^{3\varepsilon}\Gamma[\frac{1}{2}(3-\varepsilon)]\,\Gamma[\frac{1}{2}(5-\varepsilon)]}\int_0^1 \mathrm{d}\xi\,\xi^{-\varepsilon}\,(1-\xi)^{2-\varepsilon}[1+(2-\varepsilon)\,\xi]$$
$$\times\frac{\Gamma(1-\frac{1}{2}\varepsilon)}{\Gamma(3-\frac{3}{2}\varepsilon)}\left(\frac{8}{\varepsilon^2}-\frac{12}{\varepsilon}+5\right)\;. \qquad (14.81)$$

The first factor on the first line of (81) is exactly the same as the first factor of (61) for the virtual gluon correction rate, so we only need to expand the regular terms in the second line of (81) up to $\varepsilon^2$. Using (63), we find

$$\frac{\Gamma(1-\frac{1}{2}\varepsilon)}{\Gamma(3-\frac{3}{2}\varepsilon)}\left[\frac{8}{\varepsilon^2}-\frac{12}{\varepsilon}+5\right]=\frac{4}{\varepsilon^2}-\frac{4\gamma_E-3}{\varepsilon}+2\gamma_E^2-3\gamma_E+\frac{19}{4}-\frac{2\pi^2}{3}\,. \tag{14.82}$$

## 14.6 Final Result

As explicitly shown, the $\varepsilon$ poles of (82) exactly cancel those of (62), i.e. the sum of the right-hand sides of (62) and (82) is finite and equal to $-4+\frac{19}{4}$. The sum of virtual and real gluon corrections to the rate is now free of IR divergences, so we put $\varepsilon = 0$ in the first line of $\Gamma_{\mathrm{Vi}}$ and $\Gamma_{\mathrm{Re}}$ in (61) and (81).

Let us summarize. The ultraviolet divergences of loop diagrams in Fig. 14.2 are removed by replacing $F_1(q^2)$ with the renormalized form factor $F_1^{\mathrm{ren}}(q^2) = F_1(q^2) - F_1(0)$. Both $F_1(q^2)$ and $F_1(0)$ are UV divergent, but their difference is free of UV divergences. The next step deals with the infrared divergences in both $\Gamma_{\mathrm{Vi}}$ and $\Gamma_{\mathrm{Re}}$. They are found to cancel exactly each other to yield a finite result free of both UV and IR divergences. The final result for the radiative corrections is the sum of (61) with (81)

$$\begin{aligned}\Gamma_{\mathrm{rad.}} &= \Gamma_{\mathrm{Vi}}+\Gamma_{\mathrm{Re}} = \Gamma_0\frac{N_{\mathrm{c}}\,|V_{\mathrm{q_2q_3}}|^2\,\alpha_{\mathrm{s}}}{\Gamma(\frac{3}{2})\Gamma(\frac{5}{2})}\int_0^1 \mathrm{d}\xi\,(1-\xi)^2\,(1+2\xi)\left(-4+\frac{19}{4}\right)\\ &= N_{\mathrm{c}}\Gamma_0\,|V_{\mathrm{q_2q_3}}|^2\,\frac{\alpha_{\mathrm{s}}}{\pi} = N_{\mathrm{c}}\frac{G_{\mathrm{F}}^2\,M^5}{192\pi^3}\,|V_{\mathrm{q_2q_3}}|^2\,\frac{\alpha_{\mathrm{s}}}{\pi}\,.\end{aligned} \tag{14.83}$$

In the limit of massless quarks when the tree diagram rate $\Gamma_0\,N_{\mathrm{c}}\,|V_{\mathrm{q_2q_3}}|^2$ as given by (13.60) is added to the one-loop QCD corrections (83), then together with $|V_{\mathrm{ud}}|^2+|V_{\mathrm{us}}|^2 \approx 1$, the inclusive semileptonic decay width of the $\tau$ lepton is given by

$$\Gamma(\tau \to \nu_\tau + \text{ hadrons }) = N_{\mathrm{c}}\frac{G_{\mathrm{F}}^2\,M^5}{192\pi^3}\left[1+\frac{\alpha_{\mathrm{s}}}{\pi}\right]. \tag{14.84}$$

As already mentioned at the beginning, this QCD correction is identical to the correction of the ratio $R$ defined in (13.65) for $\mathrm{e}^+ + \mathrm{e}^-$ annihilation into hadrons. One has the same five diagrams, except that the photon replaces the W weak boson and the vertex $\gamma^\mu$ replaces $\gamma^\mu(1-\gamma_5)$. Thus

$$R = \frac{\sigma(\mathrm{e}^+ + \mathrm{e}^- \to \mathrm{hadrons})}{\sigma(\mathrm{e}^+ + \mathrm{e}^- \to \mu^+ + \mu^-)} = N_{\mathrm{c}}\sum_j Q_j^2\left[1+\frac{\alpha_{\mathrm{s}}}{\pi}\right]. \tag{14.85}$$

We notice that the factor $\pi^2$ in (62) and (82) comes from the second derivative of the $\Gamma(x)$ function [see (63)]. In the sum $\Gamma_{\mathrm{Vi}}+\Gamma_{\mathrm{Re}}$, it happens that the $\pi^2$ terms cancel out. However, there are circumstances in which the $\pi^2$ term remains. Examples of such cases are the one-loop QED correction to muon or tau lepton decays, as given by (13.27), and the one-loop QCD correction to quark decays $\mathrm{Q} \to \mathrm{q}_1 + \mathrm{q}_2 + \overline{\mathrm{q}}_3$ illustrated by (16.6).

This chapter ends with a remark. When the fermionic masses $m_2 \neq m_3 \neq 0$ are taken into account, calculations of radiative corrections are exceedingly complicated. We simply report the result obtained by replacing in (83) $\Gamma_0$ by $\Gamma_0\, G(x_2, x_3)$, and $(\alpha_s/\pi)$ by $(\alpha_s/\pi) K(x_2, x_3)$ where $x_k = m_k^2/M^2$. Of course $G(0,0) = K(0,0) = 1$. The results in (83) and (84) become

$$\Gamma_{\text{rad.}} = N_c\, |V_{\mathrm{q_2 q_3}}|^2 \left\{ \frac{G_F^2\, M^5}{192\pi^3} G\left(\frac{m_2^2}{M^2}, \frac{m_3^2}{M^2}\right) \right\} \left[ \frac{\alpha_s}{\pi} K\left(\frac{m_2^2}{M^2}, \frac{m_3^2}{M^2}\right) \right] , \quad (14.86)$$

$$\Gamma(\tau \to \nu_\tau + \text{ hadrons }) = N_c \frac{G_F^2\, M^5}{192\pi^3}\, G\left(\frac{m_2^2}{M^2}, \frac{m_3^2}{M^2}\right) \left[ 1 + \frac{\alpha_s}{\pi} K\left(\frac{m_2^2}{M^2}, \frac{m_3^2}{M^2}\right) \right] .$$

The analytic expression of $G(x, y)$ – corresponding to the tree diagram of Fig. 14.1, uncorrected by QCD – is already given by (13.62). Some numerical values of $K(x, x)$ and $K(x, 0) = K(0, x)$ together with $G(x, x)$ and $G(x, 0) = G(0, x)$ are given in Table 14.1. The decrease of $G(x, y)$ is expected from kinematic phase space effect. What is surprising in the radiative corrections is the spectacular increase of $K(x, y)$ with growing $x$ and $y$. The mass effect in $K(x, y)$ finds its full application in heavy flavor physics. Its relevance to the decay $\mathrm{b} \to \mathrm{c} + \mathrm{s} + \overline{\mathrm{c}}$ is an example and will be discussed in Chap. 16.

**Table 14.1.** $G(x,0)$, $G(x,x)$ and $K(x,0)$, $K(x,x)$

| $\sqrt{x}$ | 0 | 0.1 | 0.2 | 0.3 | 0.4 |
|---|---|---|---|---|---|
| $G(x,0)$ | 1 | 0.93 | 0.74 | 0.52 | 0.32 |
| $G(x,x)$ | 1 | 0.85 | 0.52 | 0.20 | 0.026 |
| $K(x,0)$ | 1 | 1.62 | 2.8 | 4.47 | 7 |
| $K(x,x)$ | 1 | 2.26 | 4.63 | 8.15 | 16.27 |

# Problems

**14.1 Noninterference between tree and bremsstrahlung diagrams.** To order $\mathcal{O}(g_s^2)$, one can draw two diagrams similar to Fig. 14.3 with two gluons (instead of one) emitted. We call them Fig. 14.3bis. While there is an interference between the diagram of Fig. 14.1 and the loop diagrams of Fig. 14.2 to obtain the $\mathcal{O}(g_s^2)$ corrections to the rate, there is no interference between Fig. 14.1 and Fig. 14.3bis for the same order $g_s^2$ corrections to the rate. Explain why.

**14.2 Analytic expressions of $F_{1,\text{uv}}(q^2)$ and $F_{1,\text{ir}}(q^2)$.** For $F_{1,\text{uv}}(q^2)$ compute the integral in the second line of (17). For $F_{1,\text{ir}}(q^2)$ in (19), the IR divergence is symbolically written as $\int_0^1 \mathrm{d}u/u$. There are two different ways to parameterize this IR, either by (i) putting a fictitious small mass $\zeta$ to the gluon, i.e. by replacing its propagator $1/k^2$ in (4) with $1/(k^2-\zeta^2)$, or by (ii) using dimensional regularization as in Sect. 14.4 so that the IR divergence is also represented by a pole $1/\varepsilon$ as the UV divergence. Derive an analytic expression of $F_{1,\text{ir}}(q^2)$ in both cases (i) and (ii).

**14.3 $F_1(0)+\delta_\text{q}=0$ from Ward identity.** We have seen the trivial role of $\gamma_5$ in our QCD corrections to the rate, so let us forget it in the expression of $\Gamma^\mu(p_2,p_3)$ as given by (4). We are considering only the vector current, i.e. QED where photons replace gluons. Multiply $q_\mu=(p_2+p_3)_\mu$ by $\Gamma^\mu(p_2,p_3)$ (without $\gamma_5$), and using the expression of $\Sigma(p)$ in (21), show that

$$(p_2+p_3)_\mu\Gamma^\mu(p_2,p_3)=\Sigma(-p_3)-\Sigma(p_2)\,. \tag{14.87}$$

This QED equation, known as the generalized Ward identity, was derived by Takahashi. The original one, pioneered by Ward, can be obtained by letting $p_2$ tend to $-p_3$. In this limit, we have

$$\Gamma^\mu(p,p)=-\frac{\partial}{\partial p_\mu}\Sigma(p)\,. \tag{14.88}$$

Notice that in (4), if $m_2\neq m_3$, the vector current is not conserved, show that (87) cannot hold. From (12), the left-hand side of (88) is $\Gamma^\mu(p,p)=\gamma^\mu F_1(0)$. Its right-hand side is $-\gamma^\mu\delta_\text{q}$, since $\Sigma(p)=\Sigma(m)+\delta_\text{q}(\not p-m)$ from (27) and (31). Then $F_1(0)+\delta_\text{q}=0$. In QED, the counterterm $Z_\text{q}$ is usually denoted by $Z_2=1+\delta_2$, so the Ward identity (88) is written as $F_1(0)=1-Z_2$.
On the other hand, the QED counterterm of the vertex denoted by $Z_1$, which is used to cancel the UV divergence of the form factor $F_1(q^2)$, is given by $F_1(0)=(1/Z_1)-1=1-Z_1+\mathcal{O}(e^4)$, i.e. $F_1(0)=1-Z_1$ (see 15.26). Then together with (88), one has $Z_1=Z_2$.
In brief, the vertex function counterterm $Z_1=1-F_1(0)$ in QED is equal to the fermion field counterterm $Z_2=1+\mathrm{d}\Sigma(p)/\mathrm{d}\not p|_{\not p=m}$ . As we will see in the next chapter, the relation $Z_1=Z_2$ does not hold in QCD.

**14.4 $F_2^\text{em}(0)$ from Higgs boson contribution.** If we replace in Fig. 14.4b, the vertex $\gamma^\mu(1-\gamma_5)$ by $\gamma^\mu$, and the internal gluon by an internal photon, then we have one-loop QED corrections. We are interested in the finite form factor $F_2^\text{em}(0)$ as given by $e^2/(8\pi^2)=\alpha_\text{em}/2\pi$ [see (14)]. The magnetic moment of the electron is usually written as $\frac{1}{2}g\mu_\text{e}$, i.e. $g=2$ corresponds to its pointlike value of $\mu_\text{e}=-e/2m_\text{e}$. Its anomalous magnetic moment, i.e. the deviation from its pointlike value, is $\frac{1}{2}(g-2)=\alpha_\text{em}/2\pi$.
Consider the field $\phi(x)$ of the Higgs boson H which has an interaction with a charged lepton field $\psi(x)$ of mass $m_\ell$: $f_\ell\phi(x)\overline{\psi}(x)\psi(x)$. In fact, the coupling

constant is $f_\ell = m_\ell / v$ where $v = (\sqrt{2}G_F)^{-1/2}$. Similar to Fig. 14.4b, instead of QED corrections, the internal photon field is now replaced by H. Compute the anomalous magnetic moment of the electron $F_2(0)$ due to this virtual H. Experimentally, the deviation $\frac{1}{2}(g-2)$ is known to be 0.0011597 for the electron. What limits on $M_H$ can we deduce from this number?

**14.5 Fermion mass generated by the gap equation.** The Pauli–Villars regularization procedure introduces a large cutoff $\Lambda$, i.e. the gluon propagator $1/(k^2-\zeta^2)$ is replaced by

$$\frac{1}{k^2-\zeta^2} \longrightarrow \frac{1}{k^2-\zeta^2} - \frac{1}{k^2-\Lambda^2}\,, \tag{14.89}$$

where $\zeta$ is a small gluon mass introduced to regulate the infrared divergence. Compute $\Sigma(p)$ and $B(m^2)$ in terms of $\Lambda$. In (25), if the bare mass $m_0$ is assumed to be zero, then the renormalized mass $m$ in (25) obeys the *gap equation* $B(m^2)=1$. Find $m$ in terms of $\Lambda$. This mechanism of mass generation is known as the Nambu–Jona-Lasinio model.

**14.6 Mass effect in two-body and three-body phase space.** Using (56) in $n$ dimensions, first compute $J^{\mu\nu}(q^2)$ as defined by (54) where $m_2$ and $m_3$ are not neglected. Compute the three-body phase space

$$\int \frac{d^3p_1}{2E_1}\frac{d^3p_2}{2E_2}\frac{d^3p_3}{2E_3}\delta^4(p_1+p_2+p_3-q) \tag{14.90}$$

as an integral over $E_j, E_\ell$ where all the masses are $\neq 0$. Give an equivalent of (75) with the three nonzero masses. Show that the $E_j, E_\ell$ domain of the Dalitz plot is no longer an isosceles triangle, but resembles an ellipse.

## Suggestions for Further Reading

*Dimensional regularization:*

Veltman, M., *Diagrammatica.* Cambridge U. Press, Cambridge 1994

*Radiative corrections; mass and field renormalization; infrared effects:*

Aoki, K., Hioki, Z., Kawabe, R., Konuma, M. and Muta, T., *Electroweak Theory.* Supp. Prog. Theor. Phys. **73** (1982) 1

Weinberg, S., *The Quantum Theory of Fields* (Vol. I). Cambridge U. Press, Cambridge 1995

*Phase space integrals in four-dimensions:*

Pietschmann, H., *Formulae and Results in Weak Interactions.* Springer, Wien, New York 1974

*Infrared-safe radiative corrections, using dimensional regularization:*

Field, R., *Applications of Perturbative* QCD. Addison-Wesley, Redwood, CA 1989

Guberina, B., Peccei, R. D. and Ruckl, R., Nucl. Phys. **B171** (1980) 333

Marciano, W. J., Phys. Rev. **D12** (1975) 3861

*QCD corrections to weak decays, taking full account of all fermionic masses:*

Czarnecki, A., Jezabek, M. and Kühn, J. H., Phys. Lett. **B346** (1995) 335

Ho-Kim, Q. and Pham, Xuan-Yem, Ann. of Phys. (N.Y.) **155** (1984) 202

# 15 Asymptotic Freedom in QCD

From the Bjorken scaling of the nucleon structure functions, we learn that the hadronic constituents probed at small distance or at high energy behave as if they were almost noninteracting or 'free'. We are confronted with an apparent paradox, since in quantum field theory virtual particles exchanged between partons can have arbitrarily high momenta, quantum fluctuations associated with them naturally occur at short distances. Why do these fluctuations turn themselves off, and the partons behave as if they were free at high energy, whereas at low energy they are strongly bound? How can a model of noninteracting quarks be reconciled with a force that is extremely strong in other circumstances?

The resolution to this paradox lies in the *asymptotic freedom* property of QCD. We first show that in field theory, quantum effects cause the coupling constants to vary with energy; they are *no more constant* but *running*. In QED we will find that the charge, i.e. the coupling becomes stronger at large momenta and weaker at small momenta. The opposite behavior happens in QCD, its interaction is strong at low energy and vanishes asymptotically at infinitely high energy, hence the name of asymptotic freedom. 't Hooft, Politzer, and Gross and Wilczek discovered that non-Abelian gauge theories are asymptotically free. Later, Coleman and Gross show that only Yang–Mills fields possess this unique feature in four-dimensional space-time. The success of QCD as the fundamental theory of strong interactions is due to this discovery. Here the Bjorken scaling finds its natural explanation: at infinite energy, the strong coupling vanishes and the quarks are noninteracting. However, the slowly logarithmic variations of the strong coupling $\alpha_s(q^2)$ should induce smooth violations of scaling since, as large as $q^2$ may be, $\alpha_s(q^2)$ is small but still nonzero, and perturbative calculations with the strong coupling are fully justifiable. This violation, expressed by a smooth $q^2$-evolution of the structure functions, is governed by the Gribov-Lipatov-Altarelli-Parisi (GLAP) equation. As discussed in Chaps. 14 and 16, the use of perturbative QCD in heavy particle decays or in the $e^+ + e^-$ annihilation into hadrons at high energy is legitimate, since $\alpha_s(q^2)$ is small in these reactions.

On the other hand, at low energy and for sufficiently large $\alpha_s(q^2)$, QCD exhibits *confinement of color*. The only finite-energy asymptotic states of QCD are color-singlets. If one attempts to separate a color-singlet state into its color constituents, for instance by breaking a meson into a quark and an antiquark, a tube of gluon would form between these two color sources. With sufficiently strong coupling, this tube would have a fixed radius, so the energy cost of separating color sources would grow proportionally with the separation distance. This picture is similar to that of a magnetic bar in which the north and south poles can never be isolated.

## 15.1 Running Coupling Constant

We show how a coupling constant can be running by an example of the Abelian QED interaction between the electron and the photon. The variation with energy of the coupling constant is entirely due to quantum effects provided by loop diagrams. To lowest order of the coupling constant $e > 0$, the photon–electron vertex is given by

$$-\mathrm{i}(-e)\overline{u}(p')\gamma^\mu u(p) = +\mathrm{i}\, e\,\overline{u}(p')\gamma^\mu u(p)\,. \tag{15.1}$$

Up to order $e^2$, four diagrams contribute to the corrections of the charge in (1). Three of them (Fig. 15.1a–c) are similar to Fig. 14.2a–c respectively. The new one Fig. 15.2 is related to the photon self-energy (also called the vacuum polarization) $\Pi^{\mu\nu}(q)$.

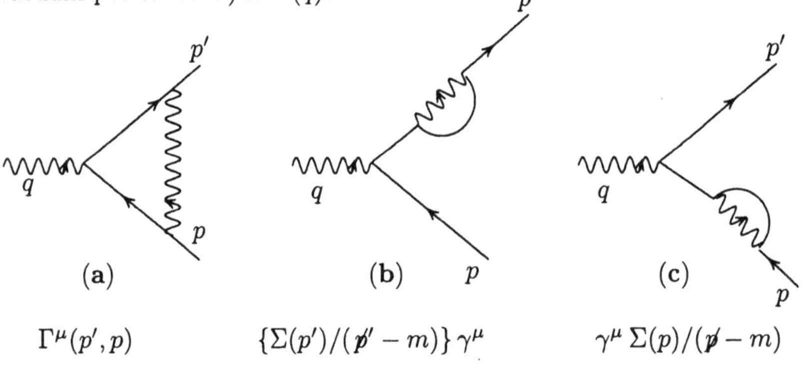


**Fig. 15.1.** (**a**) Correction to the charge by the vertex function; (**b, c**) correction to the charge by the fermion self-energy

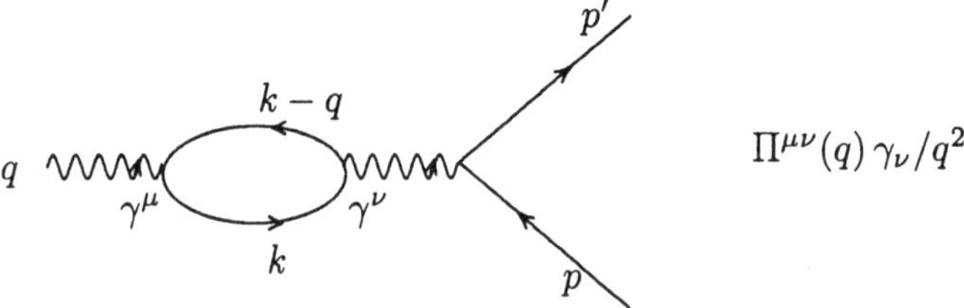


**Fig. 15.2.** Correction to the charge by the photon self-energy

The first thing to do is to calculate these four diagrams. In the first three sections of Chap. 14, we have studied in detail the diagrams of Fig. 14.4b and Fig. 14.5 which are the same diagrams of Fig. 14.2a–c without the $\tau$–$\nu_\tau$ vertex. The result can be directly used for the three diagrams of Fig. 15.1a–c with the replacement $\frac{4}{3}g_s^2 \leftrightarrow e^2$, $p_2 \leftrightarrow p'$, and $p_3 \leftrightarrow -p$. Some formulas for $F_1(q^2)$ and $Z_q$, such as (14.12), (14.14), (14.23), (14.31) and the Ward identity (14.42), are relevant to this chapter.

The quark masses $m_2, m_3$ taken in Chap. 14 to be $m$ are understood here as the electron mass. There, the $\Gamma^\mu(p_2,p_3)$ is the weak current corrected by QCD. Here, $\Gamma^\mu(p',p)$ is the electromagnetic current corrected by QED. Besides the $\gamma_5$ terms in (14.11) which are absent here, formally these $\Gamma^\mu$ are identical after the replacement $\frac{4}{3}g_s^2 \leftrightarrow e^2$. Also in Chap. 14, $\Sigma(p)$ is the quark self-energy induced by QCD, here it represents the QED corrected electron self-energy. The $Z_q$ is now denoted by $Z_2$, a notation commonly used for the field-strength renormalization of fermions. $Z_2$ is given by the derivative of $\Sigma(p)$ taken at $\not{p} = m$ according to (14.31). Also, $\delta_2 = Z_2 - 1$.

### 15.1.1 Vacuum Polarization

Our next step is to compute the photon self-energy $\Pi^{\mu\nu}(q)$ associated with the diagram of Fig. 15.2. For this loop integral, Feynman rules give

$$+\mathrm{i}\,\Pi^{\mu\nu}(q) \stackrel{\mathrm{def}}{=} (-1)\int\frac{\mathrm{d}^4k}{(2\pi)^4}\mathrm{Tr}\left[(+\mathrm{i}e\gamma^\mu)\frac{\mathrm{i}}{\not{k}-m}(+\mathrm{i}e\gamma^\nu)\frac{\mathrm{i}}{\not{k}-\not{q}-m}\right], \quad (15.2)$$

the factor $(-1)$ comes from the anticommuting property of the fermionic loop. Using again dimensional regularization and Feynman auxiliary variable $x$ for the product of two propagators in (2), the latter can be re-expressed as

$$\begin{aligned}\Pi^{\mu\nu}(q) &= 4\mathrm{i}\,e^2\int_0^1\mathrm{d}x\int\frac{\mathrm{d}^nk}{(2\pi)^n}\frac{2k^\mu k^\nu - k^\mu q^\nu - k^\nu q^\mu - g^{\mu\nu}(k^2 - k\cdot q - m^2)}{[k^2 - 2k\cdot qx + q^2x - m^2]^2}\\ &= \frac{-4e^2}{(4\pi)^{n/2}}\Gamma(2-\frac{n}{2})\int_0^1\mathrm{d}x\frac{N^{\mu\nu}(q,x)}{[m^2 - q^2x(1-x)]^{2-\frac{n}{2}}}\,. \end{aligned} \quad (15.3)$$

The $\mathrm{d}^nk$ integration gives $C \equiv \mathrm{i}\,\pi^{n/2}\,\Gamma(2-\frac{n}{2})\,[m^2 - q^2x(1-x)]^{\frac{n}{2}-2}$ times $N^{\mu\nu}(q,x)$, where $N^{\mu\nu}(q,x)$ is the sum of all the terms inside the curly brackets on the right-hand sides of the following equation:

$$\begin{aligned}\int\mathrm{d}^nk\,\frac{2k^\mu k^\nu}{D(k)} &= C\left\{2q^\mu q^\nu x^2 - g^{\mu\nu}[m^2 - q^2x(1-x)]\frac{\Gamma(1-\frac{n}{2})}{\Gamma(2-\frac{n}{2})}\right\},\\ \int\mathrm{d}^nk\,\frac{-k^\mu q^\nu - k^\nu q^\mu}{D(k)} &= -C\left\{2q^\mu q^\nu x\right\},\\ \int\mathrm{d}^nk\,\frac{-g^{\mu\nu}k^2}{D(k)} &= -C\left\{g^{\mu\nu}\left[q^2x^2 - \frac{n}{2}[m^2 - q^2x(1-x)]\frac{\Gamma(1-\frac{n}{2})}{\Gamma(2-\frac{n}{2})}\right]\right\},\\ \int\mathrm{d}^nk\,\frac{g^{\mu\nu}(k.q + m^2)}{D(k)} &= C\left\{g^{\mu\nu}(q^2x + m^2)\right\},\end{aligned}$$

with $D(k) \equiv [k^2 - 2k.qx + q^2x - m^2]^2$. (15.4)

Grouping together all the terms of the right-hand sides of (4), we obtain

$$\Pi^{\mu\nu}(q) = \frac{-4e^2}{(4\pi)^{n/2}}\Gamma\left(2-\frac{n}{2}\right)\int_0^1 \mathrm{d}x \left\{\frac{2x(1-x)[q^2g^{\mu\nu}-q^\mu q^\nu]}{[m^2-q^2x(1-x)]^{2-\frac{n}{2}}}\right.$$
$$\left. + \frac{g^{\mu\nu}\left[1-(1-\frac{n}{2})\frac{\Gamma(1-\frac{n}{2})}{\Gamma(2-\frac{n}{2})}\right]}{[m^2-q^2x(1-x)]^{1-\frac{n}{2}}}\right\} . \qquad (15.5)$$

The coefficient $1-(1-\frac{n}{2})\frac{\Gamma(1-\frac{n}{2})}{\Gamma(2-\frac{n}{2})}$ of $g^{\mu\nu}$ turns out to be identically zero since $a\,\Gamma(a) = \Gamma(a+1)$, so the final expression of $\Pi^{\mu\nu}(q)$ is proportional to the gauge-invariant tensor $(q^2g^{\mu\nu}-q^\mu q^\nu)$, in accordance with the conservation of the electromagnetic current which implies $q_\mu\Pi^{\mu\nu}(q) = q_\nu\Pi^{\mu\nu}(q) = 0$. Thus,

$$\Pi^{\mu\nu}(q) = (q^2g^{\mu\nu}-q^\mu q^\nu)\Pi(q^2) ,$$
$$\Pi(q^2) \equiv \frac{-4e^2}{(4\pi)^{n/2}}\Gamma\left(2-\frac{n}{2}\right)\int_0^1 \mathrm{d}x\, \frac{2x(1-x)}{[m^2-q^2x(1-x)]^{2-\frac{n}{2}}}$$
$$= \frac{-e^2}{12\pi^2}\left\{\frac{2}{\varepsilon} - \gamma_E + \log(4\pi) - \log\frac{m^2}{\mu^2}\right.$$
$$\left. -6\int_0^1 \mathrm{d}x\, x(1-x)\log\left[1-\frac{q^2}{m^2}x(1-x)\right] + \mathcal{O}(\varepsilon)\right\} . \qquad (15.6)$$

Note that

$$\Pi(0) = \frac{-e^2}{12\pi^2}\left(\frac{4\pi}{m^2}\right)^{\frac{\varepsilon}{2}}\Gamma\left(\frac{\varepsilon}{2}\right) \quad , \quad \varepsilon = 4-n .$$

To arrive at the two last lines in (6), we have used

$$\left[m^2-q^2x(1-x)\right]^{\frac{-\varepsilon}{2}} = 1-\frac{\varepsilon}{2}\log\frac{m^2-q^2x(1-x)}{\mu^2} + \mathcal{O}(\varepsilon^2) . \qquad (15.7)$$

Like (14.17), we notice that beside the pole $\varepsilon$, an arbitrary mass scale $\mu$ must enter in (6) to make the regular part dimensionally correct.

A nice feature of the dimensional regularization is that it automatically satisfies the gauge invariance of the theory. Other regularizations (such as the Pauli–Villars procedure with a covariant cutoff of the integral $\mathrm{d}^4k$ by a large mass $\Lambda$) give the same result for $\Pi^{\mu\nu}(q)$ but only after a long manipulation. Naively the integral in (2) $\sim \mathrm{d}^4k/k^2 \sim \Lambda^2/\mu^2 \sim \Gamma(1-\frac{n}{2})$ is quadratically UV divergent, however it disappears because it is contained in the gauge noninvariant $g^{\mu\nu}$ term, the coefficient of which is identically zero, the remaining divergence is only logarithmic $[\mathrm{d}^4k/k^4 \sim \log(\Lambda^2/\mu^2) \sim \Gamma(2-\frac{n}{2})]$. Gauge invariance constraint usually alleviates UV divergences. Due to nonvanishing fermion mass in the loop ($m \neq 0$), infrared divergences are absent in the vacuum polarization $\Pi^{\mu\nu}(q)$. In (6), when $x \to 0$ or 1, the integral does not blow up. Note that the pole $\Gamma(2-\frac{n}{2})$ is identified with $\log(\Lambda^2/\mu^2)$ in the Pauli–Villars cutoff (Problem 15.1).

If we write the tree vertex of (1) as $\gamma^\mu$ to be inserted between $+\mathrm{i}\, e\bar{u}(p')$ and $u(p)$, then the corrections to $\gamma^\mu$, as represented by the four diagrams of Fig. 15.1a–c and Fig. 15.2, are respectively

$$\Gamma^\mu(p',p) \quad , \quad \frac{\Sigma(p')}{\not{p}'-m}\gamma^\mu \quad , \quad \gamma^\mu\frac{\Sigma(p)}{\not{p}-m} \quad , \quad \frac{\Pi^{\mu\nu}(q)\gamma_\nu}{q^2} \,. \tag{15.8}$$

These four operators are also inserted between $+\mathrm{i}\, e\bar{u}(p')$ and $u(p)$; all of them are $\mathcal{O}(e^2)$. Here $q = p' - p$.

### 15.1.2 Dressed and Renormalized Photon Propagator

In Fig. 14.7, the dressed electron propagator $S_D(p)$ was obtained by summing the geometric series of $\Sigma(p)$. For photon, the corresponding lowest $1PI$ is the vacuum polarization $\Pi^{\mu\nu}(q)$, and the *dressed photon propagator* $D_{\mu\nu}(q)$ can be similarly generated by summing the geometric series of $\Pi^{\mu\nu}(q)$. Thus

$$\begin{aligned} D_{\mu\nu}(q) = &\left(\frac{-\mathrm{i}\, g_{\mu\nu}}{q^2}\right) + \left(\frac{-\mathrm{i}\, g_{\mu\rho}}{q^2}\right)[+\mathrm{i}\,\Pi^{\rho\sigma}(q)]\left(\frac{-\mathrm{i}\, g_{\sigma\nu}}{q^2}\right) + \\ &+ \left(\frac{-\mathrm{i}\, g_{\mu\rho}}{q^2}\right)[+\mathrm{i}\,\Pi^{\rho\sigma}(q)]\left(\frac{-\mathrm{i}\, g_{\sigma\lambda}}{q^2}\right)[+\mathrm{i}\,\Pi^{\lambda\alpha}(q)]\left(\frac{-\mathrm{i}\, g_{\alpha\nu}}{q^2}\right) + \cdots \end{aligned} \tag{15.9}$$

This summation is visualized in Fig. 15.3:

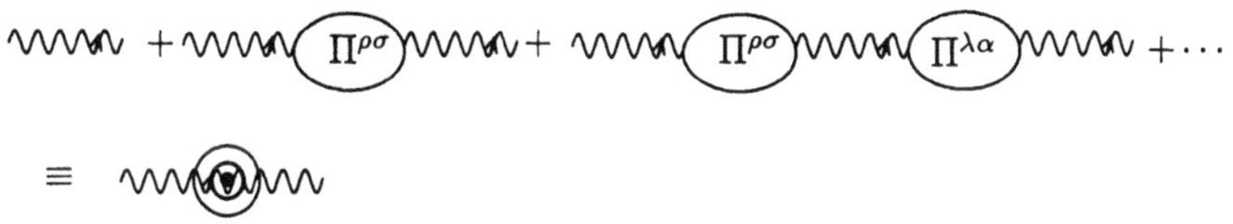

**Fig. 15.3.** Dressed photon propagator $D_{\mu\nu}(q)$

The second term on the right-hand side of (9) can be written as

$$\frac{-\mathrm{i}\, g_{\mu\rho}}{q^2}(g^{\rho\sigma} - \frac{q^\rho q^\sigma}{q^2})g_{\sigma\nu}\Pi(q^2) = \frac{-\mathrm{i}}{q^2}\left(g_{\mu\nu} - \frac{q_\mu q_\nu}{q^2}\right)\Pi(q^2)\,.$$

The third term is found to be $\frac{-\mathrm{i}}{q^2}\left(g_{\mu\nu} - \frac{q_\mu q_\nu}{q^2}\right)\Pi^2(q^2)$, then

$$\begin{aligned} D_{\mu\nu}(q) &= \frac{-\mathrm{i}\,(g_{\mu\nu} - q_\mu q_\nu/q^2)}{q^2}[1 + \Pi(q^2) + \Pi^2(q^2) + \cdots] - \mathrm{i}\,\frac{q_\mu q_\nu}{q^4} \\ &= \frac{-\mathrm{i}\,(g_{\mu\nu} - q_\mu q_\nu/q^2)}{q^2[1-\Pi(q^2)]} - \mathrm{i}\,\frac{q_\mu q_\nu}{q^4}\,. \end{aligned} \tag{15.10}$$

In any computation of physical amplitudes, the photon propagator is always attached to at least one fermion line, so in (10), the $q_\mu$ and $q_\nu$ terms vanish when they are contracted with a fermionic conserved current. Therefore we can omit them for our purposes of calculating physical amplitudes, i.e. we only keep $-\mathrm{i}\, g_{\mu\nu}$. The bare and dressed propagators can be abbreviated as

$$\frac{-\mathrm{i}\, g_{\mu\nu}}{q^2} \longrightarrow D_{\mu\nu}(q) = \frac{-\mathrm{i}\, g_{\mu\nu}}{q^2[1-\Pi(q^2)]} \,. \tag{15.11}$$

The above equation indicates that as long as the dimensionless quantity $\Pi(q^2)$ is *regular at* $q^2 = 0$, i.e. $\Pi(q^2) \neq \frac{\mu^2}{q^2}$, the dressed propagator $D_{\mu\nu}(q)$ *always has a pole at* $q^2 = 0$, implying that the photon remains massless. The only source for $\Pi(q^2)$ to be singular at $q^2 = 0$ would be the existence of a single massless particle coupled to the photon in the intermediate state. This does not happen in any $1PI$ diagram, at least in four-dimensional space-time. The photon cannot acquire mass in spite of quantum corrections in contrast to the fermion case.

Close to the pole $q^2 = 0$ in (11), the residue of the dressed propagator no longer equals 1 but becomes $Z_3$ defined below. Indeed, the analog of (14.29) for the photon case is

Bare propagator $\longrightarrow$ Dressed propagator

$$\frac{-\mathrm{i}\, g_{\mu\nu}}{q^2} \longrightarrow D_{\mu\nu}(q) \underset{q^2\to 0}{=} \frac{-\mathrm{i}\, g_{\mu\nu}}{q^2[1-\Pi(0)]} \equiv \frac{-\mathrm{i}\, g_{\mu\nu}}{q^2} Z_3 \,, \tag{15.12}$$

where $Z_3$ is given by $(Z_3)^{-1} = 1 - \Pi(0)$. For the same reason already explained in (14.31), the perturbative expression of $Z_3$ is

$$\Pi(0) = 1 - \frac{1}{Z_3} = \frac{Z_3 - 1}{Z_3} = Z_3 - 1 + \mathcal{O}(e^4) \,. \tag{15.13}$$

From (6) and (13), we get

$$\begin{aligned} Z_3 &= 1 - \frac{e^2}{12\pi^2}\left(\frac{4\pi}{m^2}\right)^{\frac{1}{2}\varepsilon}\Gamma(\tfrac{1}{2}\varepsilon) = 1 - \frac{e^2}{12\pi^2}\left(\frac{2}{\varepsilon} - \gamma_E + \log\frac{4\pi\mu^2}{m^2}\right) + \mathcal{O}(e^4)\,, \\ \sqrt{Z_3} &= 1 - \left(\frac{1}{2}\right)\frac{e^2}{12\pi^2}\left(\frac{2}{\varepsilon} - \gamma_E + \log\frac{4\pi\mu^2}{m^2}\right) + \mathcal{O}(e^4)\,. \end{aligned} \tag{15.14}$$

Let us relate the pole in $\Gamma(\frac{1}{2}\varepsilon)$ to the divergence $\log(\Lambda^2/\mu^2)$ found in the Pauli–Villars procedure

$$\left(\frac{4\pi}{m^2}\right)^{\frac{1}{2}\varepsilon}\Gamma(\tfrac{1}{2}\varepsilon) \longleftrightarrow \log\frac{\Lambda^2}{\mu^2}\,, \quad \text{so } Z_3 = 1 - \frac{e^2}{12\pi^2}\log\frac{\Lambda^2}{\mu^2} + \mathcal{O}(e^4)\,. \tag{15.15}$$

Now from (14.34) we recall that the bare fermion field $\psi_0(x)$ is absorbed by $\sqrt{Z_2}$ to become the renormalized field $\psi(x)$, from which the renormalized fermion propagator with residue equal to 1 is built. Similarly, $\sqrt{Z_3}$ is absorbed by the bare photon field $A^0_\mu(x)$ which becomes the renormalized $A_\mu(x)$. With $A^0_\mu(x) = \sqrt{Z_3}A_\mu(x)$, the renormalized photon propagator has residue equal to 1 at the pole $q^2 = 0$. We have

$$\text{Dressed propagator } D_{\mu\nu}(q) = \frac{-\mathrm{i}\, Z_3\, g_{\mu\nu}}{q^2} \longleftrightarrow \langle 0\,|\,T(A^0_\mu(x)\; A^0_\nu(y))\,|\,0\rangle \;,$$

$$\text{Renormalized propagator } \frac{-\mathrm{i}\, g_{\mu\nu}}{q^2} \longleftrightarrow \langle 0\,|\,T(A_\mu(x)\; A_\nu(y))\,|\,0\rangle \;. \quad (15.16)$$

Now the renormalized charge $e$ appears accompanied by $Z_3$. As already noted, since the fermionic charges are always attached to each end of the photon propagator in all amplitudes, the infinite $Z_3$ in (14) can be absorbed by the infinite bare charge $e_0$ to become the finite renormalized charge $e$: $e_0 \to e = e_0\sqrt{Z_3}$. Indeed, as shown in Fig. 15.4, when a fermion line is attached to each end of the propagator in (12), the amplitude has an additional factor $e_0^2$ or $e^2$ times the corresponding propagator:

$$\begin{array}{ccccccc} \text{Bare} & & \text{Interaction} & & \text{Dressed} & & \text{Renormalized} \\ \dfrac{-\mathrm{i}\, g_{\mu\nu}}{q^2} e_0^2 & \longrightarrow & \Pi(q^2) & \longrightarrow & \dfrac{-\mathrm{i}\, g_{\mu\nu} Z_3}{q^2} e_0^2 & \longrightarrow & \dfrac{-\mathrm{i}\, g_{\mu\nu}}{q^2} e^2 \,. \end{array} \quad (15.17)$$

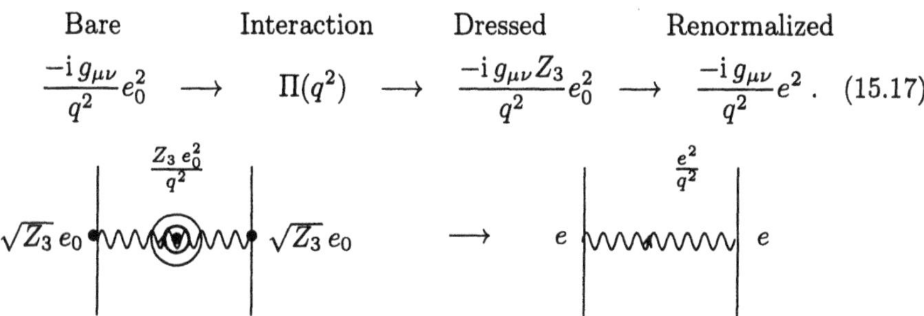


**Fig. 15.4.** Dressed and renormalized photon propagators attached to the bare and renormalized fermion charges respectively

Note that $e_0\,\psi_0(x) = e\,\psi(x)$. In spite of its apparent simplicity, the relation $e = e_0\sqrt{Z_3}$ has a far-reaching consequence, since it implies that the renormalized charge $e$ depends *only on the universal photon field-strength renormalization* $Z_3$ and not at all on other properties of the charged fermions to which the photon is coupled. In particular, in spite of the huge mass difference between the heavy tau lepton and the electron, their renormalized charges are identical, because they depend only on the photon $Z_3$. If the renormalized charge $e$ depended also on the fermion self-energy $\Sigma(p)$ and its electric form factor $F_1(q^2)$, the universality of the renormalized charge $e$ would be lost. In fact, $e$ does depend on $\Sigma(p)$ and $F_1(q^2)$ via the three diagrams of Fig. 15.1. To recover universality, there must exist a cancelation among the quantities related to $\Sigma(p)$ and $F_1(q^2)$, so that the charge renormalization is only due to $Z_3$. We now show that the cancelation indeed takes place, due to the Ward identity.

### 15.1.3 Vertex Renormalization

Mass and field renormalizations have been considered in Chap. 14. The last but not the least important quantity to undergo renormalization is the coupling constant. Let us proceed in three steps.

A – Similar to (14.35), we start with the bare quantities $e_0$, $m_0$, $\psi_0(x)$, and $A^0_\mu(x)$ of the QED Lagrangian

$$\mathcal{L} = -\frac{1}{4}(\partial_\mu A^0_\nu - \partial_\nu A^0_\mu)^2 + \overline{\psi}_0(\mathrm{i}\,\partial\!\!\!/ - m_0)\psi_0 + e_0\overline{\psi}_0\gamma^\mu\psi_0 A^0_\mu \,. \tag{15.18}$$

Then we define the *renormalized* quantities $e$, $m$, $\psi(x)$, and $A_\mu(x)$ by introducing at first three counterterms $Z_2, Z_3$, and $\Delta m_0$ with which we are already acquainted in Chap. 14 and (14). $Z_2$ and $Z_3$ are used to cancel the UV divergences of the fermion and photon self-energies respectively, while $\Delta m_0$ renormalizes the fermion mass by absorbing the bare mass $m_0$. We put

$$\psi_0(x) = \sqrt{Z_2}\psi(x) \quad , \quad A^0_\mu(x) = \sqrt{Z_3}A_\mu(x) \quad , \quad m_0 = m - \Delta m_0 \,. \tag{15.19}$$

Because the $\psi_0(x)$ and $A^0_\mu(x)$ fields are scaled by $\sqrt{Z_2}$ and $\sqrt{Z_3}$ respectively, when we substitute (19) in the interaction $e_0\overline{\psi}_0\gamma^\mu\psi_0 A^0_\mu$ of (18), we have a factor $e_0 Z_2\sqrt{Z_3}$ in front of $\overline{\psi}\gamma^\mu\psi A_\mu$. Let us multiply this factor by 1:

$$1 = \frac{Z_1}{Z_1} = \frac{1}{Z_1} + \frac{\delta_1}{Z_1} \;, \quad \delta_1 = Z_1 - 1 \;, \quad \text{then}$$

$$e_0 Z_2\sqrt{Z_3} = \frac{e_0\, Z_2\sqrt{Z_3}}{Z_1}(1+\delta_1) = e(1+\delta_1)\;, \text{where } e \equiv \frac{e_0\, Z_2\sqrt{Z_3}}{Z_1} \,. \tag{15.20}$$

We will see [the last line of (24)] that in addition to the counterterms $Z_2$ and $Z_3$, one more counterterm $Z_1$ is needed to subtract the UV divergence of the QED vertex function, i.e. the UV divergence of the form factor $F_1(q^2)$. The latter has already been computed in (14.17), we only need to replace $\frac{4}{3}g_s^2$ of (14.17) by $e^2$ in our case here.

When we put (19) into (18) and use (20), the original $\mathcal{L}$ in (18) can be rewritten as a sum of two parts having exactly the same forms, differing only in their coefficients. They are $\mathcal{L}_{\mathrm{ren}}$ (the renormalized Lagrangian) and $\mathcal{L}_{\mathrm{ct}}$ (the Lagrangian of counterterms):

$$\begin{aligned}
\mathcal{L} &= \mathcal{L}_{\mathrm{ren}} + \mathcal{L}_{\mathrm{ct}} \;; \\
\mathcal{L}_{\mathrm{ren}} &= -\frac{1}{4}(\partial_\mu A_\nu - \partial_\nu A_\mu)^2 + \overline{\psi}(\mathrm{i}\,\partial\!\!\!/ - m)\psi + e\overline{\psi}\gamma^\mu\psi A_\mu \;, \\
\mathcal{L}_{\mathrm{ct}} &= -\frac{1}{4}\delta_3(\partial_\mu A_\nu - \partial_\nu A_\mu)^2 + \overline{\psi}(\mathrm{i}\,\delta_2\,\partial\!\!\!/ - \delta_m)\psi + e\delta_1\,\overline{\psi}\gamma^\mu\psi A_\mu \;, \\
\delta_1 &= Z_1 - 1 \;, \; \delta_2 \equiv Z_2 - 1 \;, \; \delta_3 \equiv Z_3 - 1 \;, \; \delta_m \equiv m\delta_2 - Z_2\Delta m_0 \,.
\end{aligned} \tag{15.21}$$

$\mathcal{L}_{\mathrm{ren}}$ becomes formally identical to the bare one in (18) by interchanging $m\,,\, e\,,\, \psi\,,\, A_\mu$ with $m_0\,,\, e_0\,,\, \psi_0\,,\, A^0_\mu$. Therefore the quantities $\Gamma^\mu(p',p)$, $\Sigma(p)$, and $\Pi(q^2)$ that we have computed with the bare Lagrangian can be identified with the ones issued from $\mathcal{L}_{\rm ren}$ by the interchange $(m_0\,,\,e_0) \leftrightarrow (m\,,\,e)$.

As for the counterterm Lagrangian $\mathcal{L}_{\rm ct}$ in (21), all are of second- or higher-order terms in $e$.

B – The Feynman rules with these renormalized fields are: in addition to the familiar propagators and vertex from $\mathcal{L}_{\rm ren}$, all are expressed in terms of the physical mass $m$ and charge $e$, there are three contributions coming from $\mathcal{L}_{\rm ct}$. To first order in $\delta_1, \delta_2\,,\,\delta_3\,,\,\delta_m$, they can be easily obtained from the three terms of $\mathcal{L}_{\rm ct}$ and visualized in Fig. 15.5:

$$\mathrm{i}\,(-q^2 g^{\mu\nu} + q^\mu q^\nu)\delta_3\,, \qquad \mathrm{i}\,(\not p\delta_2 - \delta_m)\,, \qquad \mathrm{i}\,e\gamma^\mu\delta_1\,. \tag{15.22}$$

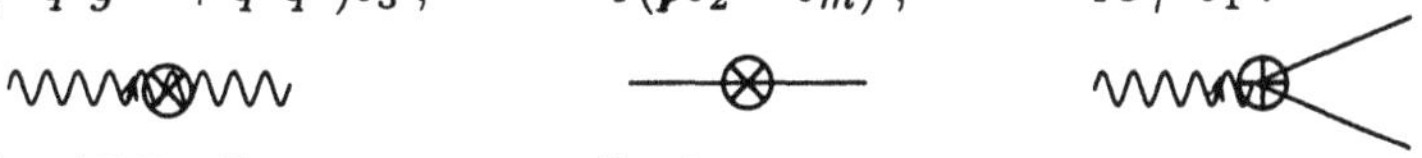

**Fig. 15.5.** Counterterm contributions

The counterterms of the photon self-energy, of the electron self-energy, and of the electron–photon vertex are shown from left to right in (22). For instance, the term $(-q^2 g^{\mu\nu} + q^\mu q^\nu)\,\delta_3$ is obtained when we integrate by parts the kinetic term $-(\frac{1}{4})(\partial_\mu A_\nu - \partial_\nu A_\mu)^2\,\delta_3$ of (21) to get $(\frac{1}{2})A_\mu(q)(-q^2 g^{\mu\nu} + q^\mu q^\nu)A_\nu(q)\,\delta_3$ in momentum space. Lastly, we multiply by two because of the symmetry $\mu,\nu$ of the two-photon line. The two other terms in (22) are straightforward from (21). Their unique role is to cancel the UV infinities in loop integrals computed from $\mathcal{L}_{\rm ren}$.

In parallel with the three quantities from left to right in (22), we have also three corresponding terms coming from $\mathcal{L}_{\rm ren}$, which we have already computed. They are given by (6), (14.23) and (14.11):

$$\mathrm{i}\,(q^2 g^{\mu\nu} - q^\mu q^\nu)\Pi(q^2)\,, \quad -\mathrm{i}\,\Sigma(p)\,, \quad \mathrm{i}\,e\Gamma^\mu(p',p)\,. \tag{15.23}$$

C – The sum of the two contributions from $\mathcal{L}_{\rm ren}$ and $\mathcal{L}_{\rm ct}$ are the renormalized quantities $\widetilde{\Pi}_{\rm ren}(q^2)\,,\,\widetilde{\Sigma}_{\rm ren}(p)\,,$ and $\widetilde{\Gamma^\mu_{\rm ren}}(p',p)$, which are free of UV divergences. Taken from (22) and (23), they are

$$\begin{aligned}
&\widetilde{\Pi}_{\rm ren}(q^2) = \Pi(q^2) - \delta_3\,,\\
&\widetilde{\Sigma}_{\rm ren}(p) = \Sigma(p) - (\not p\delta_2 - \delta_m) = \Sigma(p) - \delta_2(\not p - m) - Z_2\Delta m_0\,,\\
&\widetilde{\Gamma^\mu_{\rm ren}}(p',p) = \Gamma^\mu(p',p) + \delta_1\gamma^\mu \Longrightarrow \widetilde{F_1}^{\rm ren}(q^2) = F_1(q^2) + \delta_1\,.
\end{aligned} \tag{15.24}$$

As a general rule stated in Chap. 14, these renormalized quantities obey renormalization conditions. For $\widetilde{\Pi}_{\rm ren}(q^2)$ and $\widetilde{\Sigma}_{\rm ren}(p)$, the conditions are

$$\widetilde{\Pi}_{\rm ren}(q^2 = 0) = 0\,, \quad \left.\frac{\mathrm{d}\widetilde{\Sigma}_{\rm ren}(\not p)}{\mathrm{d}\,\not p}\right|_{\not p = m} = 0\,, \quad \widetilde{\Sigma}_{\rm ren}(\not p = m) = 0\,, \text{ equivalently}$$

$$\Pi(q^2 = 0) = \delta_3\,, \quad \left.\frac{\mathrm{d}\Sigma(p)}{\mathrm{d}\,\not p}\right|_{\not p = m} = \delta_2\,, \quad \Sigma(\not p = m) = Z_2\Delta m_0\,. \tag{15.25}$$

From left to right in (25), the first two terms fix the residues of the renormalized photon and electron propagators to be 1, and we recover (13) and (14.31) respectively. The third term sets the pole of the renormalized electron propagator to be $m$ , and as expected we recognize (14.46), remember that $Z_2 = 1 + \mathcal{O}(e^2)$ and $\Delta m_0 = m - m_0 = \mathcal{O}(e^2)$. See also (14.47).

Concerning the vertex part, the third line of (24) clearly indicates the role of $\delta_1$. It is used to cancel the UV divergence of $\Gamma^\mu(p', p)$, that of $F_1(q^2)$ precisely. The renormalization condition for this vertex part is

$$\widetilde{\Gamma^{\mu}_{\text{ren}}}(p,p) = \Gamma^\mu(p,p) + \delta_1\gamma^\mu = 0 \implies \widetilde{F_1}^{\text{ren}}(0) = F_1(0) + \delta_1 = 0 \,. \tag{15.26}$$

This condition defines the renormalized charge as the electron–photon coupling at vanishing four-momentum transfer $[q^\mu = (p' - p)^\mu = 0]$, i.e. $\widetilde{F_1}^{\text{ren}}(q^2)$ *must vanish at* $q^2 = 0$. For this specific value $q^2 = 0$, the charge does not receive quantum corrections, it is *nonrenormalized.* See also Problem 10.2.

On the other hand, direct computations – done in Chap. 14 through (14.20), (14.41) resulting in (14.42) – give the Ward identity (see Problem 14.3):

$$F_1(0) + \left.\frac{\mathrm{d}\Sigma(p)}{\mathrm{d}\,\not{p}}\right|_{\not{p}=m} \equiv F_1(0) + \delta_2 = 0 \,,$$

so that with (26), we have $\delta_1 = \delta_2 \implies Z_1 = Z_2$, then using (20), we have

$$e = e_0\sqrt{Z_3} \,. \tag{15.27}$$

These detailed studies show that $Z_1 = Z_2$ for any type of charged fermions. The three diagrams of Fig. 15.1a–c do not contribute to the charge renormalization, the latter depends only on $Z_3$. The Ward identity actually is a nonperturbative result, valid for all perturbative orders.

This completes the renormalization of the coupling constant, the sum of the four terms in (8) results in the QED renormalized electric form factor $\widetilde{F_1}^{\text{ren}}(q^2)$ which is free of UV divergences, and satisfies the renormalization condition $\widetilde{F_1}^{\text{ren}}(0) = 0$. We have from (8)

$$\widetilde{F_1}^{\text{ren}}(q^2) = F_1(q^2) - F_1(0) + \widetilde{\Pi}_{\text{ren}}(q^2) \,, \tag{15.28}$$

the last term $\widetilde{\Pi}_{\text{ren}}(q^2)$ is given below in (29). Compared with the QCD corrected weak form factor (14.43), the last term $\widetilde{\Pi}_{\text{ren}}(q^2)$ is missing. This is expected since the equivalent of Fig. 15.2 is absent from Fig. 14.2.

### 15.1.4 Renormalized Vacuum Polarization $\widetilde{\Pi}_{\rm ren}(q^2)$

The total $\mathcal{L}_{\rm ren}+\mathcal{L}_{\rm ct}$ gives us $\widetilde{\Pi}_{\rm ren}(q^2)=\Pi(q^2)-\delta_3$ written in (24). Together with the renormalization condition $\widetilde{\Pi}_{\rm ren}(0)=0$ in (25), we get

$$\widetilde{\Pi}_{\rm ren}(q^2)=\Pi(q^2)-\Pi(0)\ . \qquad (15.29)$$

While both $\Pi(q^2)$ and $\Pi(0)$ are UV divergent, their *difference is unambiguously finite.* Indeed, from (6), $\widetilde{\Pi}_{\rm ren}(q^2)$ is given by

$$\widetilde{\Pi}_{\rm ren}(q^2)=\frac{+e^2}{2\pi^2}\int_0^1 {\rm d}x\, x(1-x)\,\log\left[1-\frac{q^2}{m^2}x(1-x)\right]\ . \qquad (15.30)$$

The renormalization condition fixes the subtraction point at $q^2=0$, i.e. we subtract the divergence of $\Pi(q^2)$ by the divergence of $\Pi(0)$. This choice ensures that at the $q^2=0$ pole, the residue of the renormalized photon propagator is 1. In practice, the renormalization can be seen as a simple replacement of the divergent $\Pi(q^2)$ in (6) by the finite $\widetilde{\Pi}_{\rm ren}(q^2)\equiv\Pi(q^2)-\Pi(0)$ in (30) symbolized by the following picture:

Dressed Propagator Renormalized Propagator

$$\frac{-{\rm i}\,g_{\mu\nu}}{q^2[1-\Pi(q^2)]}\underset{q^2\to 0}{\longrightarrow}\frac{-{\rm i}\,g_{\mu\nu}}{q^2[1-\Pi(0)]}\ ,\quad \frac{-{\rm i}\,g_{\mu\nu}}{q^2[1-\widetilde{\Pi}_{\rm ren}(q^2)]}\underset{q^2\to 0}{\longrightarrow}\frac{-{\rm i}\,g_{\mu\nu}}{q^2}\ . \qquad (15.31)$$

**Analyticity of $\widetilde{\Pi}_{\rm ren}(q^2)$.** The analytic property of $\widetilde{\Pi}_{\rm ren}(q^2)$ in (30) is illuminating. For spacelike photon $q^2<0$, the argument $1-q^2x(1-x)/m^2$ of the logarithm in (30) is positive, and $\widetilde{\Pi}_{\rm ren}(q^2)$ is real. For timelike photon, the argument is negative for $q^2\geq 4m^2$ [since $0\leq x(1-x)\leq 1/4$], hence $\widetilde{\Pi}_{\rm ren}(q^2)$ becomes complex with a branch cut starting at $q^2=4m^2$ which is the threshold for the creation of an on-shell electron–positron pair. The imaginary part of $\widetilde{\Pi}_{\rm ren}(q^2)$ can be directly obtained from that of the logarithm, using the relation ${\rm Im}\,[\log(-(X^2\pm{\rm i}\varepsilon))]=\mp\pi$. We have

$$\begin{aligned}{\rm Im}\,[\widetilde{\Pi}_{\rm ren}(q^2\pm{\rm i}\epsilon)] &= \frac{e^2(\mp\pi)}{2\pi^2}\int_{X^-}^{X^+}{\rm d}x\, x(1-x)\\ &=\mp\frac{e^2}{12\pi}\sqrt{1-\frac{4m^2}{q^2}}(1+\frac{2m^2}{q^2})\ ,\end{aligned} \qquad (15.32)$$

where $X^{\pm}=\left(1\pm\sqrt{1-\frac{4m^2}{q^2}}\right)$. This is an example of singularities or cuts in the complex $q^2$ plane when the particles in any intermediate state are on-shell. It also illustrates the analytic property of physical amplitudes from which dispersion relations can be derived (Chap. 10).

From (30), the real part of $\widetilde{\Pi}_{\text{ren}}(q^2)$ is (with $\rho = \frac{q^2}{4m^2} > 1$)

$$\text{Re}\,[\widetilde{\Pi}_{\text{ren}}(q^2)] = \frac{-e^2}{2\pi^2}\left[\frac{3+5\rho}{18\rho} + \frac{1+\rho-2\rho^2}{12\rho\sqrt{\rho(\rho-1)}}\log\frac{\sqrt{\rho}+\sqrt{\rho-1}}{\sqrt{\rho}-\sqrt{\rho-1}}\right] . \tag{15.33}$$

For $0 < q^2 \leq 4m^2$, the explicit expression of $\widetilde{\Pi}_{\text{ren}}(q^2)$ is found to be

$$\widetilde{\Pi}_{\text{ren}}(q^2) = \frac{-e^2}{2\pi^2}\left[\frac{3+5\eta}{18\eta} + \frac{2\eta^2-\eta-1}{6\eta\sqrt{\eta(1-\eta)}}\tan^{-1}\sqrt{\frac{\eta}{1-\eta}}\right] , \; \eta \equiv \frac{q^2}{4m^2} \leq 1 .$$

For spacelike $q^2 \equiv -Q^2 \leq 0$, we get (with $\xi = \frac{Q^2}{4m^2} \geq 0$)

$$\widetilde{\Pi}_{\text{ren}}(Q^2) = \frac{-e^2}{2\pi^2}\left[\frac{-3+5\xi}{18\xi} + \frac{1-\xi-2\xi^2}{12\xi\sqrt{\xi(1+\xi)}}\log\frac{\sqrt{1+\xi}+\sqrt{\xi}}{\sqrt{1+\xi}-\sqrt{\xi}}\right] . \tag{15.34}$$

The two extreme limits $Q^2 \ll m^2$ and $Q^2 \gg m^2$ of $\widetilde{\Pi}_{\text{ren}}(q^2)$ can also be easily obtained from (30):

$$\widetilde{\Pi}_{\text{ren}}(q^2) \underset{q^2\to 0}{\longrightarrow} \frac{e^2}{2\pi^2}\left(\frac{-q^2}{m^2}\right)\int_0^1 \mathrm{d}x\, x^2(1-x)^2 = \frac{-\alpha}{15\pi}\frac{q^2}{m^2} , \tag{15.35}$$

$$\begin{aligned}\widetilde{\Pi}_{\text{ren}}(q^2) &\underset{q^2\to -\infty}{\longrightarrow} \frac{e^2}{2\pi^2}\int_0^1 \mathrm{d}x\, x(1-x)\left\{\log\left(\frac{-q^2}{m^2}\right) + \log\left[x(1-x)\right]\right\} \\ &= \frac{e^2}{12\pi^2}\left[\log\left(\frac{Q^2}{m^2}\right) - \frac{5}{3}\right] = \frac{+\alpha}{3\pi}\log\left(\frac{Q^2}{C\,m^2}\right) ,\end{aligned} \tag{15.36}$$

where $C = \exp(5/3)$. The (+) sign in (36) for the asymptotic limit of the vacuum polarization is an important property of QED. Aside the common factor $e^2/2\pi^2 = 2\alpha/\pi$, the real and imaginary parts of the function $\widetilde{\Pi}_{\text{ren}}(q^2)$ are plotted in Fig. 15.6 for both spacelike and timelike $q^2$.

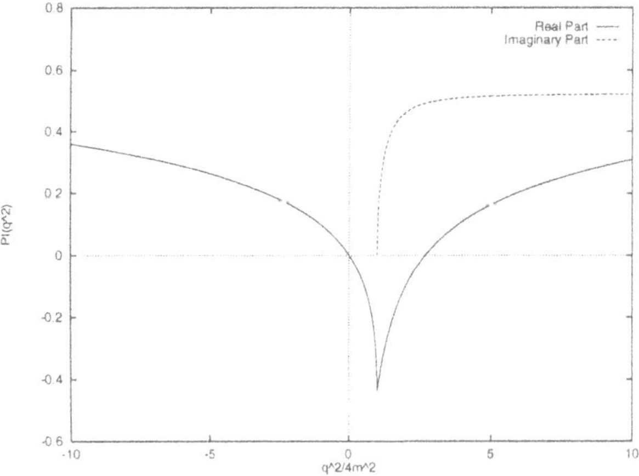


**Fig. 15.6.** Real and imaginary parts of $\widetilde{\Pi}_{\text{ren}}(q^2)$ (without the overall factor $2\alpha/\pi$) are plotted as functions of $q^2/4m^2$

### 15.1.5 Physical Effects of $\widetilde{\Pi}_{\rm ren}(q^2)$

We recall that a charged fermion line is always attached to each end of a photon propagator. To lowest order of the coupling constant, the electron scattering amplitude is given by $e^2$ times the photon propagator. How does $\widetilde{\Pi}_{\rm ren}(q^2)$ modify this amplitude? The effect of replacing the free photon propagator by the fully interacting photon propagator is equivalent to the substitution of $e^2$ by $e^2(q^2)$. The latter is defined following (31):

$$\frac{-\mathrm{i}e^2 g_{\mu\nu}}{q^2} \longrightarrow \frac{-\mathrm{i}e^2\, g_{\mu\nu}}{q^2[1-\widetilde{\Pi}_{\rm ren}(q^2)]} \equiv \frac{-\mathrm{i}e^2(q^2)\, g_{\mu\nu}}{q^2}\,,$$

$$\text{where } e^2(q^2) = \frac{e^2}{1-\widetilde{\Pi}_{\rm ren}(q^2)} \quad \text{or } \alpha \longrightarrow \alpha_{\rm eff}(q^2) \equiv \frac{\alpha}{1-\widetilde{\Pi}_{\rm ren}(q^2)}\,. \tag{15.37}$$

At small distance or high $-q^2 \equiv Q^2 > 0$, using (36) we get

$$\alpha_{\rm eff}(Q^2) = \frac{\alpha}{1-\frac{\alpha}{3\pi}\log(Q^2/C\,m^2)}\,, \quad C = \exp\left(\tfrac{5}{3}\right)\,. \tag{15.38}$$

We can also write $\alpha_{\rm eff}(r)$ as a function of the distance $r$ by performing a Fourier transformation of $\alpha_{\rm eff}(q^2)$. Because of the *minus sign* in the denominator of (38), the effective electromagnetic coupling constant $\alpha_{\rm eff}(q^2)$ becomes larger at higher energies or at smaller distances, as we penetrate the screening cloud of virtual fermion pairs considered as a dielectrically polarized medium filled up with effective dipoles of length $\sim 1/m$. An intuitive interpretation is represented in Fig. 15.7.

The quantum vacuum is *not empty* but filled with virtual charged fermion pairs. The introduction of a bare electron polarizes the vacuum in much the

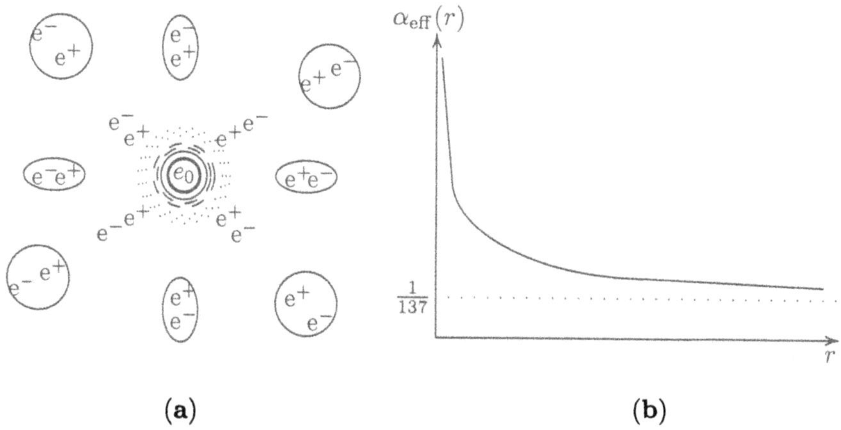


(a) (b)

**Fig. 15.7.** (**a**) Screening of the bare charge $e_0$ by virtual pairs $e^+e^-$; (**b**) qualitative dependence of the QED $\alpha_{\rm eff}(r)$ on the distance $r$

same way as a classical charge polarizes a dielectric medium. Attracted by $e_0$, the virtual positrons spend more time near the negative bare charge before annihilation, while the virtual electrons are repulsed and spend more time away. The cloud of virtual positrons that surrounds $e_0$ screens the bare charge and acts like a dielectric medium with a dielectric constant $\epsilon > 1$. This makes the charge stronger at shorter distances.

The fact that $e_0$ is larger than the effective renormalized charge $e$ can be seen indirectly through $e = e_0\sqrt{Z_3} < e_0$. In (15), we notice the *sign* of the *coefficient of* $\log(\Lambda/\mu)$ in $Z_3$ which implies $\sqrt{Z_3} = 1 - \mathcal{O}(e^2)\log(\Lambda/\mu) < 1$. Its justification will be given later by the renormalization group equation. The *minus sign of the coefficient of* $\log(\Lambda/\mu)$ is crucial to the asymptotic behavior of the QED coupling. This minus sign is intimately related to the anticommutation property of the fermionic loop in (2) which gives $Z_3$.

The $q^2$-dependence effect of the QED coupling $\alpha_{\text{eff}}(q^2)$ is experimentally observed in high energy Bhabha scattering.[1] Between $q^2 = 0$ and $-q^2 = 30\,(\text{GeV})^2$, the effective coupling increases by about 5%. Also at the weak boson $Z^0$ mass, $\alpha_{\text{eff}}(M_Z^2)$ is $\approx 1/128.896 > 1/137.036$.

On the opposite side, for small $q^2$ ($mr \gg 1$), the effective charge $\alpha_{\text{eff}}(r)$ decreases like $\exp(-2mr)/(mr)^{3/2}$. It modifies the Coulomb potential and contributes to lowering the energy levels of atomic states. This is called the Uehling effect (Problem 15.3).

The concept of a running coupling, which is no longer constant but energy-dependent, emerges from these remarkable quantum effects and constitutes a major theme of QCD. As we will see, due to the gluonic non-Abelian interaction, the color dielectric constant $\epsilon_{\text{color}} < 1$ has an antiscreening effect and makes the running QCD coupling weaker at shorter distances.

## 15.2 The Renormalization Group

We have illustrated the notion of the running coupling constant by the effects of quantum corrections. The problem can be better understood in the most generality using the remarkable renormalization group concept.

From (6) taken as an example, we notice a crucial point already mentioned after (14.17): there always exists *an arbitrariness* (via an arbitrary mass scale $\mu$) in the *finite parts* of $F_1(q^2)$ and $\Pi(q^2)$, after their UV divergences are removed. Indeed, since loops always have $q^2$ and $m^2$ terms in the integrals, they must be scaled by a certain parameter $\mu^2$ to make the answers dimensionally correct for $\varepsilon \neq 0$. An explicit example is shown in (7). This arbitrariness is unavoidable in any regularization method, as can be seen merely on dimensional grounds. For example, in the Pauli–Villars regularization by a large mass $\Lambda$ cutoff of the divergent integrals $\mathrm{d}^4k$, a scale $\mu$ must enter via the factors $(\Lambda/\mu)^l$ and $\log(\Lambda/\mu)$ in order to make dimensionally correct the quantities we compute. When $\Lambda$ is removed by renormalization

[1] HRS collaboration, Derrick, M. et al., Phys. Rev. **D34** (1986) 3286.

(as do the $\varepsilon$ poles in dimensional regularization), some functions of $q^2/\mu^2$, $m^2/\mu^2$, etc. in the finite parts always remain. Of course we can take a specific choice $\mu^2 = m^2$, but why $m^2$ and not $8\,m^2$? And what happens in the massless case? This case is not merely academic, because we are considering the asymptotic behavior of physical quantities for which $m^2 \ll q^2$.

The arbitrariness associated with the mass scale $\mu$ is reflected in the renormalization conditions which fix the finite part of the renormalized physical quantity at some kinematic point chosen arbitrarily (Sect. 8.4). The renormalization conditions for coupling constants are usually made by convention, sometimes guided by experiments. For instance, the renormalization condition (26) for the electric form factor $\widetilde{F}_1^{\text{ren}}(0) = 0$ is dictated by the definition of the charge $e$ as the coupling of an electron to a photon at zero momentum transfer. This coupling is given by the Thomson formula (4.223) of the e–$\gamma$ cross-section measured at $q^2 = 0$.

But we can equally define the charge – denoted now by $e'$ – as extracted for instance from the cross-section $e^+ + e^- \to \mu^+ + \mu^-$ measured at $q^2 \neq 0$, say at $30\,(\text{GeV})^2$, for which *a posteriori* we know that $e'^2/4\pi = \alpha[30\,(\text{GeV})^2] > \alpha(0) = 1/137.0359895$. The cross-sections, calculated by two physicists using two different definitions of the coupling constant, appear to be different by an overall constant. But it is immaterial, since the coupling constant is extracted from experiments by the value of the cross-section measured at some energy scale. In the present state of our knowledge, we cannot calculate the coupling constant, we only define it at some kinematic point, and then use this definition to compute other physical quantities like decay rates.

How about QCD? To obtain the coupling constant $\alpha_s$, it does not make sense to measure the quark–gluon cross-section at the Thomson limit. Obviously, as any dimensionless coupling constant, $\alpha_s$ needs a scale.

Let us briefly summarize the concept before considering its realization through the renormalization group equation. We start with a bare Lagrangian, then we compute quantum effects. To handle UV divergences in loop integrals, we choose some regularization scheme. After the subtraction of UV infinities, we impose renormalization conditions on the finite parts of the computed quantities at some kinematic point. Clearly, physics should not depend on this arbitrary choice of either the regularization procedures or the subtraction points. Any choice is as good as any other. This apparent arbitrariness in fact is subtle and turns out to imply powerful constraints on the asymptotic behavior of the theory. There must exist physical quantities invariant under the transformations which merely change the renormalization conditions as well as the regularizations.

This is the gist of the *renormalization group* independently formulated by Callan and Symanzik in a concrete equation. The original work went back to Stueckelberg and Petermann, and was exploited by Gell-Mann and Low to get information on the asymptotic behavior of the photon propagator, and developed by Bogoliubov and Shirkov.

### 15.2.1 The Callan–Symanzik Equation

We start by considering a bare Lagrangian with $\Phi_0$ and $g_0$ as bare field and coupling constant. $\Phi_0$ can be a self-interacting real scalar field with one coupling (for instance $\lambda\phi^4/4!$), or $\Phi_0$ can designate collectively many fields, for example the fields of a charged fermion $\psi$ and a photon $A^\mu$ in QED ($g_0 = e_0$), or the fields of color quarks and color gluons in QCD. Also, $g_0$ may generically designate one or many dimensionless couplings.

Exactly as in (14.36) and (19), we renormalize the interacting field $\Phi_0$ by a multiplicative counterterm $\sqrt{Z}$ and get the renormalized field $\Phi$, i.e. $\Phi_0 = \Phi\sqrt{Z}$. If there are many different fields $\phi^j$, there are as many $Z_j$. By the explicit examples of $Z_2$ and $Z_3$ for fermion and photon fields respectively in (14.41) and (14), we are familiar with the dependence of $Z$ on an arbitrary scale $\mu$. $Z$ is also a function of $g_0$, and we write $Z(\mu, g_0)$.

The bare coupling constant $g_0$ accordingly becomes the renormalized coupling constant $g$ which is $g_0$ times a counterterm denoted by $Z_{\rm cpl}$: $g \equiv g_0 Z_{\rm cpl}$. The crucial role of $Z_{\rm cpl}$ will be clear later in the renormalization group equation. The important point is the general formula (39) given below, which tells us how to build $Z_{\rm cpl}$. This formula is a generalization of (20) and is illustrated in Fig. 15.8. For QED, $e = e_0\sqrt{Z_2}\sqrt{Z_2}\sqrt{Z_3}/Z_1$, so $Z_{\rm cpl} = Z_2\sqrt{Z_3}/Z_1 = \sqrt{Z_3}$ (remember $Z_1 = Z_2$).

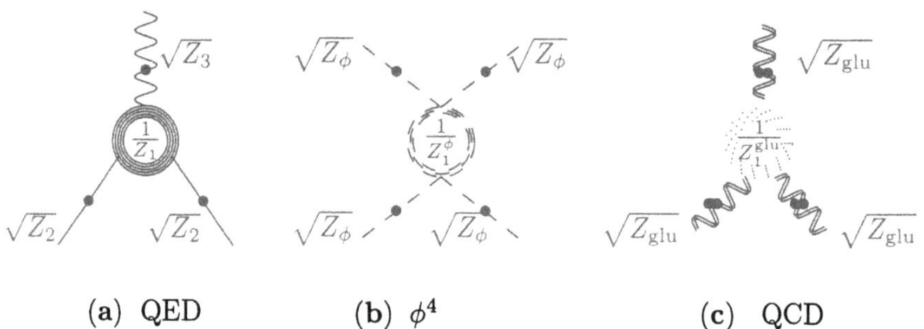


**Fig. 15.8.** Coupling constant counterterm $Z_{\rm cpl}$ in (**a**) QED; (**b**) $\phi^4$; and (**c**) QCD

In general, $Z_{\rm cpl}$ is the *product* $\prod_{\rm ext}(\sqrt{Z_{\rm ext}})$ of all the external lines which are connected to the vertex loop then this product is *divided* by the vertex counterterm generically denoted by $\widetilde{Z}_1$. The vertex loop and its counterterm $\widetilde{Z}_1$ naturally contribute to the renormalized coupling constant $g$ because the zero-loop vertex is by definition the bare coupling constant $g_0$.

The presence of $\widetilde{Z}_1$ in the denominator of $Z_{\rm cpl}$ comes about exactly as in (20) by the trick $1 = \widetilde{Z}_1/\widetilde{Z}_1 = (1/\widetilde{Z}_1) + (\widetilde{\delta}_1/\widetilde{Z}_1)$. The first term $(1/\widetilde{Z}_1)$ enters $Z_{\rm cpl}$ to define the renormalized coupling constant $g$; the second $(\widetilde{\delta}_1/\widetilde{Z}_1)$ is a counterterm used to subtract the vertex loop divergence, similar to (26).

Thus, the most general form of the renormalized coupling constant is

$$g = g(\mu) = g_0\, Z_{\text{cpl}}(\mu) = \frac{g_0}{\widetilde{Z}_1} \prod_{\text{ext}} \sqrt{Z_{\text{ext}}}\ . \tag{15.39}$$

Two more examples are shown in Fig. 15.8b, c respectively for $\phi^4$ and QCD. For the $\phi^4$ self-interacting scalar field, we have $Z_{\text{cpl}} = (\sqrt{Z_\phi})^4/Z_1^\phi$. In non-Abelian Lagrangian with only bosons, e.g. QCD without quarks, via the three-boson self-coupling, we have $Z_{\text{cpl}} = (\sqrt{Z_{\text{glu}}})^3/Z_1^{\text{glu}}$. Each Lagrangian has its own $Z_{\text{ext}}$ and $\widetilde{Z}_1$ to be calculated. We are not without experience of their tedious computations in (14.41) and (14).

We go now to the next step by introducing the Green's functions which are convenient for the discussion of quantum loop effects. A $k$-point function or Green's function is defined by

$$G^k(x_1, \cdots, x_{\text{k}}) = \langle 0\,|\, T[\Phi(x_1) \cdots \Phi(x_{\text{k}})]\,|\,0\rangle\ ,$$

and its Fourier transform in momentum space is denoted by $G^k(p_1, \cdots, p_{\text{k}})$. $\Phi(x)$ may designate a single field or many different fields.

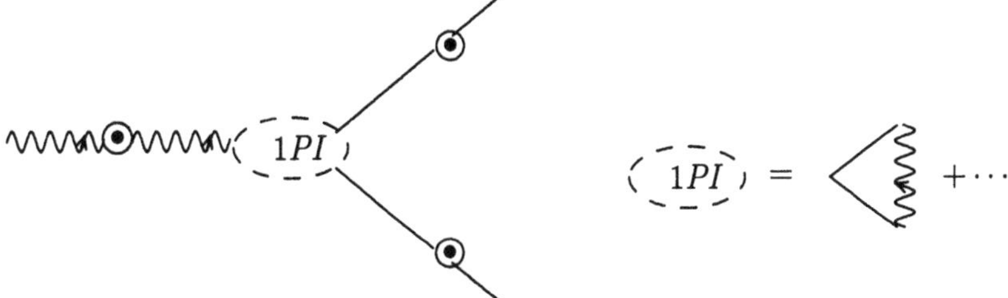


**Fig. 15.9.** QED three-point function

A typical fermion two-point function is its full (dressed) propagator, i.e. the geometric sum of $1PI$ fermionic self-energy with two external fermion lines, similar to Fig. 14.7a. Note that a $1PI$ self-energy is already an infinite sum of graphs in perturbative calculations. A QED three-point function (with two charged fermion fields and one photon field) is the full vertex function which includes both a three-leg tree diagram and a $1PI$ closed loop with three external lines, the latter are full propagators of these fields (schematically drawn in Fig. 15.9 with • ). When these external propagators are removed, we have an amputated three-point function.

The $k$-point Green's function is defined similarly with $k$ external propagators connected to closed loops. The $S$-matrix element with which we compute cross-section or decay rate is an amputated Green's function.

As it stands, it is a herculean task to compute Green's functions in their most general form. In perturbative orders, the simplest Green's functions are the tree amplitude, the one-loop amplitude, and so on. Renormalization group equation provides, as we will see, a powerful method for 'partial summation' over all perturbative orders; by partial summation we mean the leading logarithm terms of Green functions.

Let us now consider a bare $k$-point function $G_0^k(p_i, g_0)$ derived from a bare Lagrangian. For simplicity, we take $\Phi$ as a single field, for instance the scalar field $\phi$ in the $g\,\phi^4/4!$ Lagrangian. The changes $\Phi_0 = \Phi\sqrt{Z}$ and $g_0 = g\,Z_{\rm cpl}^{-1}$ induce the change of $G_0^k(p_i, g_0)$ into the renormalized Green's function $G_{\rm ren}^k(p_i, \mu, g)$. After the rescaling by $k$ powers of $\sqrt{Z}$ for $\Phi_0$, and the elimination of $g_0$ in favor of $g$ , we obtain $G_{\rm ren}^k(p_i, \mu, g)$ which is numerically equal to the bare $G_0^k(p_i, g_0)$. The former depends explicitly and also implicitly through $g(\mu)$ on $\mu$. All of these operations are equivalent to a reparameterization. Thus,

$$G_{\rm ren}^k(p_i, \mu, g) = [Z(\mu, g_0)]^{-\frac{k}{2}}\, G_0^k(p_i, g_0)\ . \tag{15.40}$$

Equation (40) is the generalization of (16) from a two-point function to a $k$-point function. We first remark that the bare $G_0^k(p_i, g_0)$ makes no reference to the scale $\mu$, it depends only on $g_0$. The independence with respect to the variable $\mu$ of $G_0^k(p_i, g_0)$ implies that $[Z(\mu, g(\mu))]^{k/2}\, G_{\rm ren}^k(p_i, \mu, g)$ is independent of $\mu$ too. So we have

$$\frac{\mathrm{d}}{\mathrm{d}\mu}\left(Z^{\frac{k}{2}} G_{\rm ren}^k\right) = Z^{\frac{k}{2}}\left(\frac{\mathrm{d}}{\mathrm{d}\mu} + \frac{k}{2}\frac{1}{Z}\frac{\mathrm{d}Z}{\mathrm{d}\mu}\right) G_{\rm ren}^k = 0\ . \tag{15.41}$$

Applying the differentiation

$$\frac{\mathrm{d}}{\mathrm{d}\mu} = \frac{\partial}{\partial\mu} + \frac{\partial}{\partial g}\frac{\partial g(\mu)}{\partial\mu}$$

on (41) and multiplying it by $\mu$ to make operators dimensionless yields the Callan–Symanzik (CS) equation

$$\left(\mu\frac{\partial}{\partial\mu} + \beta\frac{\partial}{\partial g} + \frac{k}{2}\gamma\right) G_{\rm ren}^{\rm k}(p_i, \mu, g) = 0\ , \quad \text{where} \tag{15.42}$$

$$\beta = \mu\left.\frac{\partial g(\mu)}{\partial\mu}\right|_{g_0\,\text{fixed}}\ , \tag{15.43}$$

$$\gamma = \mu\left.\frac{\partial \log Z(\mu, g_0)}{\partial\mu}\right|_{g_0\,\text{fixed}}\ . \tag{15.44}$$

Equation (42) expresses the fact that any change in the subtraction point $\mu$ amounts to a change in the coupling constant and a change in the field-strength, i.e. the physical content of the theory is not affected by a mere change of this parameterization. For each type of Lagrangian, there exists two corresponding functions $\beta$ and $\gamma$, called Callan–Symanzik functions; the former governs the evolution of the coupling constant and the latter that of the field-strength. The method can be easily generalized to other theories with many fields and dimensionless couplings. For instance in QED, (42) becomes

$$\left\{\mu\frac{\partial}{\partial\mu}+\beta(e)\frac{\partial}{\partial e}+\tfrac{1}{2}\left[N_{\rm e}\,\gamma_2(e)+N_{\rm ph}\,\gamma_3(e)\right]\right\}G^k_{\rm ren}(p_i,\mu,e)=0\;, \qquad (15.45)$$

where $N_{\rm e}$ and $N_{\rm ph}$ are respectively the number of external electron and photon fields in the $k$-point Green's function $G^k_{\rm ren}(p_i,\mu,e)$, with $k=N_{\rm e}+N_{\rm ph}$. As in (44), $\gamma_2$ and $\gamma_3$ are respectively the derivatives of the field-strength counterterms $Z_2$ and $Z_3$ of the electron and photon fields.

According to (43), the $\beta$-function represents the rate of change of $g$ with respect to $\mu$. A positive sign for the $\beta$-function indicates that $g$ increases at large momenta and decreases at small momenta, and inversely for a negative sign of $\beta$. Therefore the $\beta$-function, in particular its sign, is of great importance for the evolutionary behavior of the coupling. Before looking for the solution of (42), let us see how we can compute the $\beta$- and $\gamma$-functions.

### 15.2.2 Calculation of the $\beta$- and $\gamma$-Functions

In free field theory without interactions ($g=0, Z=1$), the $\beta$ and $\gamma$ functions obviously vanish. If the interacting Lagrangian represents a finite theory, i.e. if all loop integrals were finite, then the renormalization group equation (42) is trivial and empty. Since $Z$ is finite, we do not need regulators and the presence of $\mu$ is superfluous. For a finite theory, $Z(\mu,g_0)$ and $g(\mu)$ do not depend on $\mu$, the CS functions are also identically zero.

This observation clarifies the role of counterterms, since nonzero $\beta$- and $\gamma$-functions appear when the theory, although not primitively finite, is renormalizable by the absorption of the counterterms. The infinities encountered in loop integrals turn out to be providentially useful. The counterterm $Z$ and consequently the dimensionless $g$ are now functions of $\mu$ and of the regulator $n-4=\varepsilon$; the latter can be replaced by a large mass scale $\Lambda$ [the equivalence between the pole $1/\varepsilon$ and $\log(\Lambda/\mu)$ is explicitly written in (15)]. In fact, since the function $Z$ is dimensionless, its dependence on $\mu$ must be through the ratio $\Lambda/\mu$, more precisely through $\log(\Lambda/\mu)$.

The $\beta$-function is obtained from (39) and (43) by taking the derivative of $g=g_0\,Z_{\rm cpl}(\mu)$ with respect to $\mu$. We immediately recognize that via the ratio $\Lambda/\mu$, the dependence of $Z_{\rm cpl}$ on $\Lambda$ is *essential.* To compute the $\beta$-function, it suffices to keep only the $\Lambda$ dependence of the $Z$s , i.e. *only their UV divergent parts*; their finite terms are irrelevant. Note that whereas $Z\sim\log(\Lambda/\mu)$ is divergent, the derivative of $Z$ is finite, so too is the $\beta$-function. Let us write

$$\left(\mu\,\frac{\partial}{\partial\mu}+\Lambda\,\frac{\partial}{\partial\Lambda}\right)\log\frac{\Lambda}{\mu}=0\;, \qquad (15.46)$$

using $\mu\,\dfrac{\partial}{\partial\mu}\log\dfrac{\Lambda}{\mu}=-\Lambda\,\dfrac{\partial}{\partial\Lambda}\log\dfrac{\Lambda}{\mu}=-1\;.$

We have from (39)

$$\beta(g) = g_0\,\mu\frac{\partial}{\partial\mu}\,Z_{\rm cpl}\left(\frac{\Lambda}{\mu},g_0\right)\bigg|_{g_0,\Lambda\ \rm fixed} = -g_0\Lambda\frac{\partial}{\partial\Lambda}\,Z_{\rm cpl}\left(\frac{\Lambda}{\mu},g_0\right)\bigg|_{g_0,\mu\ \rm fixed}$$
$$= -g\Lambda\frac{\partial}{\partial\Lambda}\log\left[Z_{\rm cpl}\left(\frac{\Lambda}{\mu},g_0\right)\right]\bigg|_{g_0,\mu\ \rm fixed}, \tag{15.47}$$

using $g\,\partial\log Z_{\rm cpl} = g_0\,\partial Z_{\rm cpl}$. With $Z\equiv 1+\delta$ and $\log Z\approx\delta$, i.e. to lowest orders in $g$ for the $\beta$-function, we get from (39) and (47)

$$\beta(g) = -g\,\Lambda\frac{\partial}{\partial\Lambda}\left[-\log\widetilde{Z}_1 + \frac{1}{2}\sum_{\rm ext}\log Z_{\rm ext}\right]$$
$$= g\,\mu\frac{\partial}{\partial\mu}\left[-\widetilde{\delta}_1(\mu,g) + \frac{1}{2}\sum_{\rm ext}\delta_{\rm ext}(\mu,g)\right]. \tag{15.48}$$

The existence of $\beta$ and its dependence on the renormalized coupling $g$ are due to the renormalizability of the theory. When the cutoff $\Lambda$ goes to infinity, Green's functions remain finite when expressed in terms of the renormalized parameters, in particular of the renormalized coupling $g$. Since the CS functions $\beta$ and $\gamma$ are finite, they cannot depend on the cutoff $\Lambda$, hence they cannot depend *explicitly* on $\mu$ via the ratio $\Lambda/\mu$; they depend only on the dimensionless renormalized coupling $g$, and *implicitly* on $\mu$ via $g(\mu)$.

The CS function $\gamma_j(g)$ defined in (44) for each field $\Phi_j$, can be similarly obtained by

$$\gamma_j(g) = \mu\frac{\partial}{\partial\mu}\,\delta_j(\mu,g)\,. \tag{15.49}$$

In brief, the computation of the $\beta$- and $\gamma$-functions reduces to the calculation of the vertex counterterm $\widetilde{Z}_1$ and the field-strength counterterms $Z_{\rm ext}$.

**One-Loop $\beta$- and $\gamma$-Functions in QED.** As the first illustration, let us compute $\beta(e)$ and $\gamma(e)$ in QED. From (48) and $e = e_0\,Z_2\sqrt{Z_3}/Z_1$,

$$\beta(e) = e\,\mu\frac{\partial}{\partial\mu}\left[-\delta_1 + \frac{1}{2}(2\delta_2+\delta_3)\right]. \tag{15.50}$$

We have already computed the $\delta_2$ of QCD in Chap. 14 and given it in (14.41). We now get the $\delta_2$ of QED by a simple substitution:

$$\frac{4}{3}g_s^2 \leftrightarrow e^2 \quad\text{and}\quad \Gamma\left(2-\frac{n}{2}\right)\sim\frac{2}{\varepsilon}\leftrightarrow\log\frac{\Lambda^2}{\mu^2}\,. \tag{15.51}$$

As for $\delta_1$ and $\delta_3$, they are given by (26) and (15) respectively,

$$\delta_1=\delta_2=\frac{-e^2}{16\,\pi^2}\log\frac{\Lambda^2}{\mu^2}+(\text{ ift })\ ,\quad \delta_3=\frac{-e^2}{12\pi^2}\log\frac{\Lambda^2}{\mu^2}+(\text{ ift })\,,$$

where ( ift ) means irrelevant finite terms. We get from (50)

$$\beta(e) = e\,\mu\,\frac{\partial(\frac{1}{2}\delta_3)}{\partial\mu} = e\,\mu\,\frac{\partial}{\partial\mu}\left[\frac{-e^2}{24\pi^2}\log\frac{\Lambda^2}{\mu^2}\right] = \frac{+e^3}{12\pi^2}\,. \tag{15.52}$$

To check that the pole $\Gamma(2-\frac{n}{2})$ can be identified with $\log(\Lambda^2/\mu^2)$ in (51), we directly compute $\beta(g)$ from the $\delta$'s which are written in terms of $\Gamma(2-\frac{n}{2})$ without passing by the cutoff $\Lambda$. With (14) for QED,

$$\delta_3(\mu) = \frac{-e^2}{12\pi^2}\frac{\Gamma(2-\frac{n}{2})}{[\mu^2]^{2-\frac{n}{2}}} + (\text{ ift })\,, \tag{15.53}$$

$$\beta(e) = e\,\mu\,\frac{\partial}{\partial\mu}\left[\frac{-e^2}{24\pi^2}\frac{\Gamma(2-\frac{n}{2})}{(\mu^2)^{2-\frac{n}{2}}}\right] = \frac{-e^3}{24\pi^2}\left[(-2)\,\mu^{n-4}\,\Gamma\left(3-\frac{n}{2}\right)\right] \to \frac{+e^3}{12\pi^2}\,.$$

$$\gamma_2(e) = \mu\frac{\partial}{\partial\mu}\left[\frac{-e^2}{16\pi^2}\frac{\Gamma(2-\frac{n}{2})}{(\mu)^{4-n}}\right] = \frac{e^2}{8\pi^2}\quad,\quad \gamma_3(e) = \frac{e^2}{6\pi^2}\,. \tag{15.54}$$

As expected, the *positive* sign of $\beta(e)$ results in an increase of the running charge with energy.

### 15.2.3 Running Coupling from the Renormalization Group

Now that we know how to compute the $\beta$- and $\gamma$-functions, we go further by looking for solutions of (42). In fact, we obtain a relation connecting the Green's function evaluated at two different scales $\mu_1$ and $\mu_2$, valid for all perturbative orders (but limited only to the leading logarithmic terms).

The CS equation is an exact consequence of the renormalizability of the theory. An amplitude computed perturbatively to a given order does not in general satisfy (42), therefore the renormalization group equation may complement perturbative calculations, and can be used to 'improve' them. We first give the solution of (42), the demonstration will follow in the next subsection.

We start with an amplitude $\mathcal{A}(p_i,g)$ evaluated at the momenta $p_i$. Then at a different scale $\eta\,p_i$ of the momenta ($\eta$ is an arbitrary number), the amplitude $\mathcal{A}(\eta\,p_i,g)$ is related to the initial $\mathcal{A}(p_i,g)$ by replacing the constant $g$ with the running $\overline{g}(t)$:

$$\mathcal{A}(\eta\,p_i,g) = \mathcal{A}(p_i,\overline{g}(t))\exp\left(\int_0^t \widetilde{\gamma}[\overline{g}(t')]\,\mathrm{d}t'\right)\,,\ \text{where}\ t=\log\eta\,, \tag{15.55}$$

where the running $\overline{g}(t)$ is a solution to the equation

$$\frac{\mathrm{d}\,\overline{g}(t)}{\mathrm{d}t} = \beta\left(\overline{g}\right)\,,\ \text{with the initial condition}\ \overline{g}(0)=g\,. \tag{15.56}$$

In this form, (55) is easy to interpret. The effect of rescaling the momenta $p_i$ in the amplitude is equivalent to replacing the coupling $g$ by the running coupling $\overline{g}$, apart from an overall multiplicative factor represented by the exponential of $\widetilde{\gamma}(g)$. The latter will be defined later. So the most important quantity is the running coupling $\overline{g}(t)$ which can be computed from (56), once the $\beta$-function is known.

We now discuss the physical implication of the solution (55). If $\eta \sim \mathcal{O}(1)$, $t \sim 0$, (55) would essentially be the ordinary perturbative evaluation of the amplitude. On the other hand, if $\eta \gg 1$, the replacement of $g$ by $\overline{g}(t)$ in (55) is a powerful method which tells us how to sum large logarithm terms of the amplitude. Indeed, in many circumstances, the Feynman perturbation series depend not only on $g$ but also on the product $g^n\,[\log(p^2/\mu^2)]^l$. Even when $g$ is small, but $\log(p^2/\mu^2)$ large, the perturbative result is not reliable. The result (55) of the renormalization group equation solves this problem by reorganizing the dependence of the amplitude into a function of the running $\overline{g}$ and an exponential scale factor. The separate criteria $g^2 \ll 1$ and $g^2 \log(p^2/\mu^2) \ll 1$ – which could fail in conventional perturbation theory when large logarithm factors appear at each order – are saved if $\overline{g}^2(p^2) \ll 1$. This running coupling $\overline{g}(p^2)$ enables us to sum the leading logarithm terms of Green's functions to all perturbative orders, as illustrated in Fig. 15.10.

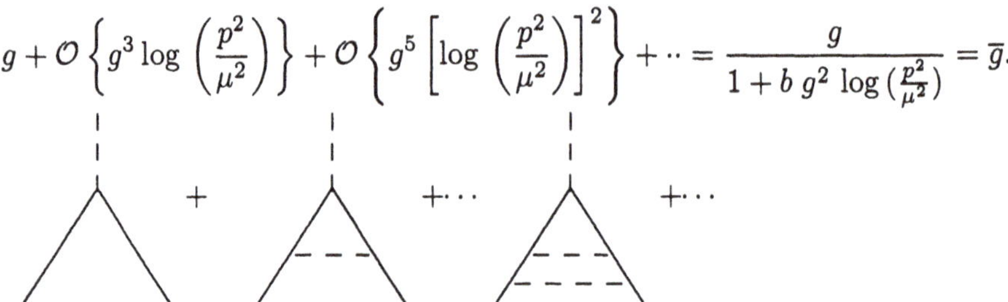


**Fig. 15.10.** The meaning of $\overline{g}$ as $\sum_n g^{2n+1}\,[\log(\frac{p^2}{\mu^2})]^n$ through the $\beta$-function

The fact that the above equation is a geometric series in $g^2 \log(p^2/\mu^2)$, reminiscent of (38), would be impossible to verify by a direct calculation of Feynman diagrams, while it is an elementary consequence of the renormalization group equation, as we will see later.

The exponential factor of $\widetilde{\gamma}(\overline{g})$ in (55) also has a simple interpretation. It represents the accumulation of the field-strength rescaling from $\mu$ to the actual momentum $p$ at which the amplitude is evaluated.

### 15.2.4 Solution of the Renormalization Group Equation

We first remark that the Green's function, which is built up from coupling constants, masses, external fields, etc., has the dimension $(\text{mass})^d$, where $d$ is the canonical or 'naive' dimension. For instance the naive dimension of the strong coupling $g_s$ in QCD is $d = 0$, a fermion field has $d = 3/2$, a boson field has $d = 1$. A two-point Green's function of the photon field $\Pi^{\mu\nu}(q)$ in (2) has $d = 2$. Its amputated two-point Green's function, or the photon propagator, has $d = -2$. A QED three-point function has $d = -4$; when the three external propagators are removed, it has $d = 0$.

Next, to cast (42) in its most useful form, instead of varying $\mu$ and fixing $p_i$, we change $p_i$ into $\eta p_i$ (where $\eta$ is an arbitrary large number) and keep $\mu$ fixed in the Green's function $G(p_i, \mu, g)$. One may always write

$$G(p_i, \mu, g) = \mu^d \, F\left(\frac{p_i}{\mu}, g\right) ,$$

where $F(p_i/\mu, g)$ is an arbitrary dimensionless function, and the naive dimension $\mu^d$ of $G(p_i, \mu, g)$ is explicit. For instance, the $d = -2$ propagator $\mathrm{i}/p^2$ can be written as

$$\frac{\mathrm{i}}{p^2} = \mu^{-2} \frac{\mathrm{i}}{(p/\mu)^2} ,$$

and so on for other Green's functions. Next, we have

$$G(\eta\, p_i, \mu, g) = \mu^d F\left(\eta \frac{p_i}{\mu}, g\right) . \tag{15.57}$$

Since the dimensionless $F$ depends on $\eta$ and $\mu$ only through the ratio $\eta/\mu$, we can check from dimensional analysis that the following equation holds:

$$\left(\mu \frac{\partial}{\partial \mu} + \eta \frac{\partial}{\partial \eta}\right) F\left(\eta \frac{p_i}{\mu}, g\right) = 0 . \tag{15.58}$$

This familiar equation is already seen in (46). Combining (57) with (58) and putting $t = \log \eta$, one finds

$$\left(\mu \frac{\partial}{\partial \mu} + \frac{\partial}{\partial t} - d\right) G(\eta\, p_i, \mu, g) = 0 ,$$

the substitution of the above equation into (42) gives

$$\left[\frac{\partial}{\partial t} - \beta(g) \frac{\partial}{\partial g} - \tilde{\gamma}(g)\right] G(\eta\, p_i, \mu, g) = 0 , \tag{15.59}$$

where $\tilde{\gamma}(g) = d + \frac{k}{2}\gamma(g)$, the $\tilde{\gamma}(g)$ is called the *anomalous dimension* since the amplitude has no longer the canonical naive dimension $d$. To find the solution to (59) with two independent variables $t = \log \eta$ and $g$, let us consider a similar differential equation with two variables $t$ and $x$ of a function $D(t, x)$, where the initial condition $D(0, x)$ is given. Thus

$$\left[\frac{\partial}{\partial t} - \beta(x) \frac{\partial}{\partial x}\right] D(t, x) = R(x) D(t, x) . \tag{15.60}$$

The solution to the inhomogeneous differential equation (60) will immediately follow from the solution to the corresponding homogeneous equation. To find it, we first introduce a new variable $\overline{g}(t,x)$ obeying

$$\frac{\mathrm{d}\,\overline{g}(t,x)}{\mathrm{d}\,t} = \beta(\overline{g}) \ , \quad \text{with the initial condition } \ \overline{g}(0,x) = x \ . \tag{15.61}$$

Then we integrate (61) with respect to $t$ and get

$$t = \int_x^{\overline{g}(t,x)} \frac{\mathrm{d}\,y}{\beta(y)} \ .$$

Next, we differentiate both sides of the above equation with respect to $x$,

$$\frac{\mathrm{d}t}{\mathrm{d}x} = 0 = \frac{1}{\beta(\overline{g})}\frac{\mathrm{d}\overline{g}}{\mathrm{d}x} - \frac{1}{\beta(x)} \Longrightarrow \beta(x)\frac{\mathrm{d}\overline{g}}{\mathrm{d}x} = \beta(\overline{g}) = \frac{\mathrm{d}\overline{g}}{\mathrm{d}t} \ ,$$

which shows that, in addition to its own equation (61), $\overline{g}(t,x)$ equally obeys the homogeneous equation

$$\left[\frac{\partial}{\partial t} - \beta(x)\frac{\partial}{\partial x}\right]\overline{g}(t,x) = 0 \ . \tag{15.62}$$

Once $\overline{g}(t,x)$ – which is the solution to the homogeneous equation (62) – is known, the solution to the inhomogeneous equation (60) can be easily derived by replacing in the initial function $D(0,x)$ the argument $x$ by $\overline{g}$. After this replacement, the result is multiplied by an exponential of $R[\overline{g}(t')]$ integrated from 0 to $t$. We have

$$D(t,x) = D[0,\overline{g}(t,x)] \times \exp\left(\int_0^t R[\overline{g}(t',x)]\mathrm{d}t'\right) \ . \tag{15.63}$$

By the substitution $g \leftrightarrow x$, $\widetilde{\gamma}(g) \leftrightarrow R(x)$ in (63), the solution (55) of the renormalization group equation (42) is then obtained.

**The QED Running Charge $\overline{e}(t)$.** Once $\beta(e)$ is known, we determine the running coupling $\overline{e}(t)$ of QED by solving (56) with (52):

$$\frac{\mathrm{d}\overline{e}(t)}{\mathrm{d}t} = \beta\left(\overline{e}\right) \ , \quad \text{where } \beta(e) = \frac{e^3}{12\pi^2} \ .$$

The solution is

$$\frac{\mathrm{d}\overline{e}(t)}{\overline{e}^3(t)} = \frac{\mathrm{d}\,t}{12\pi^2} \longrightarrow \frac{1}{\overline{e}^2(t)} - \frac{1}{\overline{e}^2(0)} = \frac{-t}{6\pi^2} \longrightarrow \overline{e}^2(t) = \frac{\overline{e}^2(0)}{1-\overline{e}^2(0)\,t/6\pi^2} \ ,$$
$$\alpha_{\mathrm{em}}(q^2) = \frac{\alpha_{\mathrm{em}}}{1-\frac{\alpha_{\mathrm{em}}}{3\pi}\log(\frac{q^2}{\mu^2})} \quad \text{for } q^2 \gg \mu^2 \ , \quad t = \log\eta = \log\frac{q}{\mu} \ , \tag{15.64}$$

and we recover our previous result (38) if we identify $\mu^2$ with $Cm^2$.

## 15.3 One-Loop Computation of the QCD $\beta$-Function

In (50), we have seen that the $\beta$-function of QED is derived from the divergent counterterms $Z_1$ (the electron–photon vertex), $Z_2$ (the electron self-energy), and $Z_3$ (the photon self-energy). Since the first two terms cancel each other by the Ward identity, only $Z_3$ contributes to the QED $\beta$-function. In the non-Abelian QCD case, with quark and gluon replacing electron and photon respectively, all of these three terms contribute. Thus from (48)

$$\beta_{\mathrm{QCD}}(g_{\mathrm{s}}) = g_{\mathrm{s}}\, \mu \frac{\partial}{\partial \mu} \left[ -\delta_1(\mu) + \delta_{\mathrm{q}}(\mu) + \frac{1}{2}\delta_{\mathrm{glu}}(\mu) \right] . \tag{15.65}$$

When we compute these $\delta(\mu)$ functions separately, they depend on the gauge parameter $\xi$ of the gluon propagator, however their $\xi$ dependences cancel each other in their combination (65), so that $\beta(g_{\mathrm{s}})$ is gauge-independent as it should be. As in Chap. 14, we first use the $\xi = 1$ Feynman–'t Hooft gauge, then step by step show that the final result for $\beta_{\mathrm{QCD}}(g_{\mathrm{s}})$ is independent of $\xi$.

### 15.3.1 Quark Self-Energy Counterterm $Z_{\mathrm{q}}$

In (14.41), we have already computed the quark self-energy $\Sigma(p)$, and its derivative at $\not p = m$ gives $\delta_{\mathrm{q}} = Z_{\mathrm{q}} - 1$. Let us rewrite it, using (14.23) and (14.41),

$$\delta_{\mathrm{q}}(\mu) = \frac{-g_{\mathrm{s}}^2}{16\pi^2} C_2(N_{\mathrm{c}}) \,\frac{\Gamma(2-\frac{n}{2})}{\mu^{4-n}} = \frac{-g_{\mathrm{s}}^2}{16\pi^2} \left[\frac{N_{\mathrm{c}}^2 - 1}{2\,N_{\mathrm{c}}}\right] \frac{\Gamma(2-\frac{n}{2})}{\mu^{4-n}} . \tag{15.66}$$

For a general $\xi$ gauge, in fact, we get

$$\delta_{\mathrm{q}}(\mu) = \frac{-g_{\mathrm{s}}^2}{16\pi^2} \left[\frac{N_{\mathrm{c}}^2 - 1}{2N_{\mathrm{c}}}\,\xi\right] \frac{\Gamma(2-\frac{n}{2})}{\mu^{4-n}} , \tag{15.67}$$

which tells us that in the Landau gauge ($\xi = 0$), $\delta_{\mathrm{q}}$ vanishes at one-loop level.

### 15.3.2 Quark–Gluon Vertex Counterterm $Z_1$

For the quark–gluon vertex $Z_1$, there are two diagrams, and not only one as in the Abelian QED case. In addition to Fig. 15.11a – for which the external as well as the internal photons of Fig. 15.1a are replaced by gluons – there is one more graph shown in Fig. 15.11b.

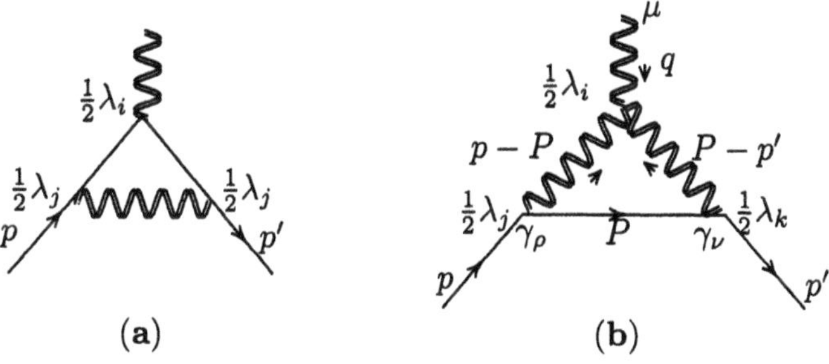


**Fig. 15.11a, b.** $\delta_1$ from quark–gluon vertex function

The expression of the vertex function (Fig. 15.11a) can be directly obtained from (14.2), with a single modification: the $\gamma^\mu(1-\gamma_5)$ in (14.2) is now replaced by $-\mathrm{i}\, g_\mathrm{s}\, \gamma^\mu T_i$, where $T_i \equiv \frac{1}{2}\lambda_i$. Because of this additional color matrix $T_i$, the Casimir operator $T_jT_j$ in (14.4) is now replaced by

$$T_jT_j \longrightarrow T_jT_iT_j = T_jT_jT_i + T_j[T_i,T_j] = C_2(N_\mathrm{c})\,T_i + \mathrm{i}\,T_j\, f_{ijk}T_k$$
$$= C_2(N_\mathrm{c})T_i + \mathrm{i}\, f_{ijk}\left[\frac{1}{2}\,\mathrm{i} f_{jkl}T_l\right] = \left(\frac{N_\mathrm{c}^2-1}{2N_\mathrm{c}} - \frac{1}{2}N_\mathrm{c}\right)T_i\,,$$

such that, after including the factor $-\mathrm{i}\, g_\mathrm{s}$, the overall substitution is

$$\frac{4}{3} = \frac{N_\mathrm{c}^2-1}{2N_\mathrm{c}} \longrightarrow -\mathrm{i}\, g_\mathrm{s}\left(\frac{N_\mathrm{c}^2-1}{2N_\mathrm{c}} - \frac{1}{2}N_\mathrm{c}\right)\, T_i\,. \tag{15.68}$$

This substitution tells us that the $\Gamma^\mu(p_2,p_3)$ obtained previously in (14.11) and (14.12), in particular its divergent part associated with $\Gamma(2-\frac{n}{2})$ of $F_1(q^2)$, can be immediately adapted to $\Gamma^\mu_\mathrm{A}(p',p)$ of Fig.15.11a by the substitution (68). We denote it by $\Gamma^\mu_\mathrm{A}(p',p)|_\mathrm{div}$, where

$$\Gamma^\mu_\mathrm{A}(p',p)\big|_\mathrm{div} = -\mathrm{i}g_\mathrm{s}\gamma^\mu T_i\left\{\frac{g_\mathrm{s}^2}{(4\pi)^{\frac{n}{2}}}\left[\frac{N_\mathrm{c}^2-1}{2N_\mathrm{c}} - \tfrac{1}{2}N_\mathrm{c}\right]\frac{(n-2)^2}{4}\frac{\Gamma(2-\frac{n}{2})}{\mu^{4-n}}\right\}$$
$$= -\mathrm{i}g_\mathrm{s}\gamma^\mu T_i\left\{\frac{g_\mathrm{s}^2}{16\pi^2}\left[\frac{N_\mathrm{c}^2-1}{2N_\mathrm{c}} - \tfrac{1}{2}N_\mathrm{c}\right]\frac{\Gamma(2-\frac{n}{2})}{\mu^{4-n}}\right\}\,. \tag{15.69}$$

Like (53), the denominator $[-m^2(x+y)^2+q^2xy]^{2-\frac{n}{2}}$ in (14.12) is simply taken as $(\mu^2)^{2-\frac{n}{2}} = \mu^{4-n}$, since only the divergent part is needed. For later use, we keep $n \neq 4$ in the factor $\Gamma(2-\frac{n}{2})/\mu^{4-n}$, because we will take its derivative with respect to $\mu$ to get the $\beta$-function.

The diagram of Fig. 15.11b is given, always in the Feynman–'t Hooft gauge $\xi = 1$, by

$$\Gamma^\mu_\mathrm{B}(p',p) = \int\frac{\mathrm{d}^4P}{(2\pi)^4}(-\mathrm{i}\, g_\mathrm{s}\,\gamma_\nu T_k)\frac{\mathrm{i}(\not P+m)}{P^2-m^2}(-\mathrm{i}\, g_\mathrm{s}\,\gamma_\rho T_j)\frac{-\mathrm{i}}{(p-P)^2}\frac{-\mathrm{i}}{(P-p')^2}$$
$$\times\,(-g_\mathrm{s}\, f_{ijk})[g^{\mu\rho}(q-p+P)^\nu + g^{\rho\nu}\,(p+p'-2P)^\mu + g^{\nu\mu}\,(P-p'-q)^\rho]\,.$$

With $f_{ijk}T_k\,T_j = -\mathrm{i}\,\frac{1}{2}N_\mathrm{c}T_i$, we have

$$\Gamma^\mu_\mathrm{B}(p',p) = \frac{-g_\mathrm{s}^3}{2}N_\mathrm{c}T_i\int\frac{\mathrm{d}^nP}{(2\pi)^n}\frac{\gamma_\nu(\not P+m)\gamma_\rho T^{\mu\nu\rho}(p',p,P)}{(P^2-m^2)(p-P)^2(P-p')^2}, \tag{15.70}$$
$$T^{\mu\nu\rho}(p',p,P) \equiv [g^{\mu\rho}(q-p+P)^\nu + g^{\rho\nu}\,(p+p'-2P)^\mu + g^{\nu\mu}\,(P-p'-q)^\rho]\,.$$

To compute the divergent part of $\Gamma^\mu_\mathrm{B}(p',p)$, i.e. the coefficient of $\Gamma(2-\frac{n}{2})$, we only need the quadratic power of the integration variable $P$ in the numerator of (70). This is found to be

$$(\gamma_\nu\,\not P\,\gamma_\rho)\,[g^{\mu\rho}\,P^\nu - 2g^{\rho\nu}\,P^\mu + g^{\nu\mu}\,P^\rho] = 2P^2\,\gamma^\mu + 2(n-2)\,\not P\,P^\mu\,. \tag{15.71}$$

After writing the product of the three terms in the denominator of (70) as an integration over the auxiliary Feynman variables $x$ and $y$ and using (71) in the numerator, we finally get for the divergent part of $\int \mathrm{d}^n P$,

$$\frac{\mathrm{i}}{(4\pi)^{\frac{n}{2}}} \int_0^1 \mathrm{d}x \int_0^{1-x} \mathrm{d}y \frac{2(n-1)\Gamma(2-\frac{n}{2})\,\gamma^\mu}{[m^2(1-x-y)^2 - q^2 xy]^{2-\frac{n}{2}}} \longrightarrow \frac{3\mathrm{i}}{16\pi^2} \frac{\Gamma(2-\frac{n}{2})}{\mu^{4-n}} \gamma^\mu .$$

Like (53) or (69), the denominator $[m^2(1-x-y)^2 - q^2xy]^{2-\frac{n}{2}}$ is taken as $\mu^{4-n}$ for the divergent part. With (70), we obtain

$$\Gamma^\mu_{\mathrm{B}}(p',p)\big|_{\mathrm{div}} = -\mathrm{i}g_{\mathrm{s}}\gamma^\mu T_i \left[\frac{g_{\mathrm{s}}^2}{16\pi^2}\left(\frac{3N_{\mathrm{c}}}{2}\right)\frac{\Gamma(2-\frac{n}{2})}{\mu^{4-n}}\right] . \tag{15.72}$$

The QCD vertex function $\Gamma^\mu_{\mathrm{QCD}}(p',p)$ is the sum $\Gamma^\mu_{\mathrm{A}}(p',p) + \Gamma^\mu_{\mathrm{B}}(p',p)$. By definition, the QCD counterterm $(-\mathrm{i}g_{\mathrm{s}}\gamma^\mu T_i)\,\delta_1$ is used to cancel the divergence of the vertex function $\Gamma^\mu_{\mathrm{QCD}}(p',p)$, just as the QED counterterm $\mathrm{i}e\gamma^\mu\delta_1$ of (22) was used to cancel the divergence of $\mathrm{i}e\Gamma^\mu(p',p)$ in (23). Note that the factor $(-\mathrm{i}g_{\mathrm{s}}\gamma^\mu T_i)$ already appears in (69) and (72). Thus

$$-\mathrm{i}g_{\mathrm{s}}\gamma^\mu T_i\,\delta_1 + \Gamma^\mu_{\mathrm{A}}(p',p)\big|_{\mathrm{div}} + \Gamma^\mu_{\mathrm{B}}(p',p)\big|_{\mathrm{div}} = 0 ,$$

from which we get

$$\delta_1(\mu) = \frac{-g_{\mathrm{s}}^2}{16\pi^2}\left(\frac{N_{\mathrm{c}}^2-1}{2N_{\mathrm{c}}} + N_{\mathrm{c}}\right)\frac{\Gamma(2-\frac{n}{2})}{\mu^{4-n}} . \tag{15.73}$$

In the most general gauge parameter $\xi$, one finds

$$\delta_1(\mu) = \frac{-g_{\mathrm{s}}^2}{16\pi^2}\left[\frac{N_{\mathrm{c}}^2-1}{2N_{\mathrm{c}}}\,\xi + \left(1-\frac{1-\xi}{4}\right)N_{\mathrm{c}}\right]\frac{\Gamma(2-\frac{n}{2})}{\mu^{4-n}} . \tag{15.74}$$

From (67) and (74), we notice that $\delta_1$ is not equal to $\delta_{\mathrm{q}}$ as it is in QED.

### 15.3.3 Gluon Self-Energy Counterterm $Z_{\mathrm{glu}}$

The self-energy of the gluon field, $\delta_{\mathrm{glu}}$, remains to be evaluated. Its calculation is more involved and the corresponding diagrams are shown in Fig. 15.12. The three diagrams Fig. 15.12a–c are evaluated from the Feynman rules given in Chap. 8 for QCD. These rules which only involve the 'tree' diagrams do not apply to Fig. 15.12d.

We are already familiar with Fig. 15.12a. Except for the additional SU(3) color matrices $T_i$ and $T_j$ at each point where a gluon is attached to the fermionic loop, this diagram is identical to the photon self-energy computed in (2). We have

$$\mathrm{i}\,\Pi^{\mu\nu}_{\mathrm{A}}(q) \stackrel{\mathrm{def}}{=} (-1)\int\frac{\mathrm{d}^4P}{(2\pi)^4}\mathrm{Tr}\left[(-\mathrm{i}g_{\mathrm{s}}\gamma^\mu T_i)\frac{\mathrm{i}}{\not{P}-m}(-\mathrm{i}g_{\mathrm{s}}\gamma^\nu T_j)\frac{\mathrm{i}}{\not{P}-\not{q}-m}\right] . \tag{15.75}$$

So $\Pi_{\mathrm{A}}^{\mu\nu}(q)$ and the corresponding QCD counterterm of Fig. 15.12a, called $\delta_{\mathrm{glu}}^{\mathrm{A}}(\mu)$, can be directly taken from (6) and (14) by including the trace factor $\mathrm{Tr}[T_i T_j] = \frac{1}{2}\delta_{ij}$. Since there are $N_{\mathrm{f}}$ flavors of quarks in the loop integral – all in the fundamental $\mathrm{SU_c}(3)$ representation – their contributions must be summed up and it results in a multiplicative factor $N_{\mathrm{f}}$. Below the threshold of heavy quarks, $N_{\mathrm{f}} = 3$, if we only count the u, d, s light quarks. For higher energy, $N_{\mathrm{f}} = 6$, which includes c, b, t quarks. From (6), the divergent part of $\Pi_{\mathrm{A}}^{\mu\nu}(q)$ is

$$\Pi_{\mathrm{A}}^{\mu\nu}(q)\big|_{\mathrm{div}} = N_{\mathrm{f}}\,\frac{\delta_{ij}}{2}\left[\frac{-g_{\mathrm{s}}^2}{12\pi^2}\,\frac{\Gamma(2-\frac{n}{2})}{(\mu)^{4-n}}\right](q^2 g^{\mu\nu} - q^\mu q^\nu)\,. \tag{15.76}$$

From (14), we deduce

$$\delta_{\mathrm{glu}}^{\mathrm{A}}(\mu) = \frac{N_{\mathrm{f}}}{2}\left[\frac{-g_{\mathrm{s}}^2}{12\pi^2}\,\frac{\Gamma(2-\frac{n}{2})}{(\mu)^{4-n}}\right].$$

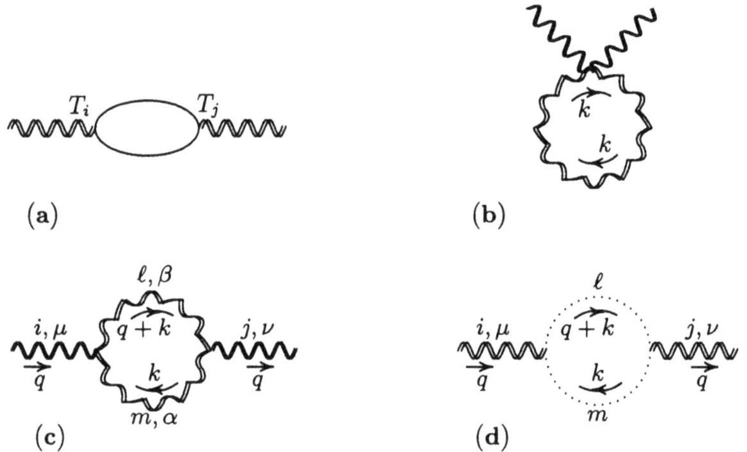


**Fig. 15.12a–d.** $\delta_{\mathrm{glu}}$ from gluon self-energy

The next three diagrams Fig. 15.12b–d are typically non-Abelian and absent in QED. The diagram of Fig. 15.12b does not contribute to $\delta_{\mathrm{glu}}$; this can be seen by considering the degree of divergence of the corresponding integral which is quadratic and not logarithmic. Indeed, besides the four-gauge-boson coupling constant, the relevant loop integral of Fig. 15.12b is

$$\int\frac{\mathrm{d}^n k}{k^2} = \lim_{v\to 0}\int\frac{\mathrm{d}^n k}{k^2+v^2} \sim \lim_{v\to 0} v^{n-2}\,\frac{\Gamma(2-\frac{n}{2})}{1-\frac{n}{2}} \longrightarrow 0 \text{ for } n>2\,. \tag{15.77}$$

To compute $\delta_{\mathrm{glu}}^{\mathrm{B}}(\mu)$, we need the coefficient of $\Gamma(2-\frac{n}{2})$. It is $v^{n-2}$ that vanishes as $n\to 4$, then $\delta_{\mathrm{glu}}^{\mathrm{B}}(\mu) = 0$.

With the same convention as in (75) for $+\mathrm{i}\,\Pi_{\mathrm{A}}^{\mu\nu}(q)$, the gluon self-energy $+\mathrm{i}\Pi_{\mathrm{C}}^{\mu\nu}(q)$ of Fig. 15.12c is obtained using Feynman rules:

$$+\mathrm{i}\Pi_{\mathrm{C}}^{\mu\nu}(q) \stackrel{\mathrm{def}}{=} \frac{1}{2}\int \frac{\mathrm{d}^n k}{(2\pi)^n}\frac{-\mathrm{i}}{k^2}\frac{-\mathrm{i}}{(q+k)^2}(-g_{\mathrm{s}}f_{iml})\,(-g_{\mathrm{s}}f_{jml})\,N^{\mu\nu}(k,q)\;,$$

$$\text{where } N^{\mu\nu}(k,q) = \left[g^{\mu\alpha}\,(q-k)^{\beta} + g^{\alpha\beta}\,(2k+q)^{\mu} + g^{\beta\mu}\,(-k-2q)^{\alpha}\right] \times \left[\delta_{\alpha}^{\nu}\,(k-q)_{\beta} + g_{\alpha\beta}\,(-2k-q)^{\nu} + \delta_{\beta}^{\nu}\,(k+2q)_{\alpha}\right]\;, \quad (15.78)$$

the factor $\frac{1}{2}$ in front of the right-hand side of the above equation is due to two identical internal bosons in the loop. We find

$$N^{\mu\nu}(k,q) = -\,g^{\mu\nu}(2k^2+2k\cdot q+5q^2) + (6-4n)k^{\mu}k^{\nu} + (3-2n)(k^{\mu}q^{\nu}+k^{\nu}q^{\mu}) + (6-n)q^{\mu}q^{\nu}\;.$$

After introducing the auxiliary variable $x$ for the product $k^{-2}\,(q+k)^{-2}$ in the denominator of (78) following Feynman, the $\int \mathrm{d}^n k$ integration can be done, using formulas in the Appendix. Together with $f_{iml}\,f_{jml} = N_{\mathrm{c}}\,\delta_{ij}$, we get

$$\Pi_{\mathrm{C}}^{\mu\nu}(q) = \frac{g_{\mathrm{s}}^2\,N_{\mathrm{c}}\,\delta_{ij}}{2(4\pi)^{\frac{n}{2}}}\int_0^1 \mathrm{d}x\,\frac{\Gamma(2-\frac{n}{2})}{[q^2x(1-x)]^{2-\frac{n}{2}}}\Bigg\{3\,q^2g^{\mu\nu}\,(n-1)x(1-x)\frac{\Gamma(1-\frac{n}{2})}{\Gamma(2-\frac{n}{2})} + q^2g^{\mu\nu}[5-2x(1-x)] - q^{\mu}q^{\nu}\left[6(1-x+x^2)-n(1-2x)^2\right]\Bigg\}\;.$$

Using $(1-\frac{n}{2})\Gamma(1-\frac{n}{2}) = \Gamma(2-\frac{n}{2})$, we finally obtain

$$\Pi_{\mathrm{C}}^{\mu\nu}(q) = \frac{g_{\mathrm{s}}^2\,N_{\mathrm{c}}\,\delta_{ij}}{(4\pi)^{\frac{n}{2}}}\int_0^1 \mathrm{d}x\,\frac{\Gamma(2-\frac{n}{2})}{[q^2x(1-x)]^{2-\frac{n}{2}}}\Bigg\{q^2g^{\mu\nu}\left[\frac{5}{2}-\frac{(4n-5)\,x(1-x)}{n-2}\right] + q^{\mu}q^{\nu}\left[\frac{n}{2}(1-2x)^2-3(1-x+x^2)\right]\Bigg\},$$

$$\Pi_{\mathrm{C}}^{\mu\nu}(q)\big|_{\mathrm{div}} = \frac{g_{\mathrm{s}}^2\,N_{\mathrm{c}}\,\delta_{ij}}{16\pi^2}\left(\frac{19}{12}q^2g^{\mu\nu}-\frac{11}{6}q^{\mu}q^{\nu}\right)\frac{\Gamma(2-\frac{n}{2})}{\mu^{4-n}}\;. \quad (15.79)$$

For the most general $\xi$ gauge, we have for $\Pi_{\mathrm{C}}^{\mu\nu}(q)|_{\mathrm{div}}$,

$$\frac{g_{\mathrm{s}}^2\,N_{\mathrm{c}}\,\delta_{ij}}{16\pi^2}\left\{\left[\frac{19}{12}q^2g^{\mu\nu}-\frac{11}{6}q^{\mu}q^{\nu}\right]+\frac{1-\xi}{2}(q^2g^{\mu\nu}-q^{\mu}q^{\nu})\right\}\frac{\Gamma(2-\frac{n}{2})}{\mu^{4-n}}\;. \quad (15.80)$$

Gauge invariance of QCD implies that the gluon self-energy, like the photon self-energy, must satisfy $q_{\mu}\Pi^{\mu\nu}(q) = q_{\nu}\Pi^{\mu\nu}(q) = 0$, i.e. it must have the structure $\Pi^{\mu\nu}(q) = (q^2g^{\mu\nu}-q^{\mu}q^{\nu})\Pi(q^2)$. However, in contrast to (76), this structure is not found in (80) for $\Pi_{\mathrm{C}}^{\mu\nu}(q)$. There must exist something new in the Feynman rules for non-Abelian theories when *pure gluons in loops are connected to external gluons*, as in Fig. 15.12c.

A systematic solution of this problem is referred to as the Faddeev–Popov (FP) prescription, according to which contributions of *fictitious scalar particles obeying Fermi statistics*, called FP 'ghosts', must be included. They only *appear in closed loops with a minus sign.* These color-octet ghosts *couple only to gauge bosons* and not to fermions. Their Feynman rules (propagator and coupling) are indicated in Fig. 15.13. The two-ghost–two-gluon four-point vertex does not exist. We will come back to ghosts later.

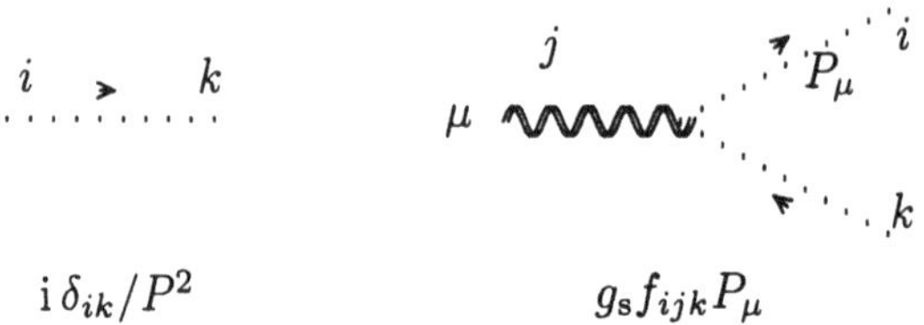


$\mathrm{i}\,\delta_{ik}/P^2$ $\qquad$ $g_{\mathrm{s}} f_{ijk} P_\mu$

**Fig. 15.13.** Feynman rules for FP ghosts (*dotted line*) coupled to gauge boson

According to this rule, the FP ghost contribution (Fig. 15.12d) to the gluon self-energy is

$$+\mathrm{i}\Pi_{\mathrm{D}}^{\mu\nu}(q) \stackrel{\mathrm{def}}{=} (-1)\int \frac{\mathrm{d}^n k}{(2\pi)^n}\frac{\mathrm{i}}{k^2}\frac{\mathrm{i}}{(k+q)^2}(g_{\mathrm{s}} f_{lim})\,(g_{\mathrm{s}} f_{mjl})\,(q+k)^\mu k^\nu \,.$$

With $f_{lim}\, f_{mjl} = -f_{iml}\, f_{jml} = -N_{\mathrm{c}}\,\delta_{ij}$, and using formulas from the Appendix,

$$\Pi_{\mathrm{D}}^{\mu\nu}(q) = \frac{-g_{\mathrm{s}}^2 N_{\mathrm{c}}\,\delta_{ij}}{(4\pi)^{\frac{n}{2}}}\int_0^1 \mathrm{d}x\frac{x(1-x)\,\Gamma(2-\frac{n}{2})}{[q^2x(1-x)]^{2-\frac{n}{2}}}\left[\tfrac{1}{2}q^2 g^{\mu\nu}\frac{\Gamma(1-\frac{n}{2})}{\Gamma(2-\frac{n}{2})} - q^\mu q^\nu\right],$$

$$\Pi_{\mathrm{D}}^{\mu\nu}(q)\big|_{\mathrm{div}} = \frac{g_{\mathrm{s}}^2 N_{\mathrm{c}}\,\delta_{ij}}{16\pi^2}\left(\frac{q^2 g^{\mu\nu}}{12}+\frac{q^\mu q^\nu}{6}\right)\frac{\Gamma(2-\frac{n}{2})}{\mu^{4-n}}\,. \tag{15.81}$$

Adding (80) to (81), gauge invariance is restored in the sum $\Pi_{\mathrm{C}}^{\mu\nu}(q)+\Pi_{\mathrm{D}}^{\mu\nu}(q)$:

$$\Pi_{\mathrm{C}}^{\mu\nu}(q)\big|_{\mathrm{div}} + \Pi_{\mathrm{D}}^{\mu\nu}(q)\big|_{\mathrm{div}} = \frac{g_{\mathrm{s}}^2 N_{\mathrm{c}}\,\delta_{ij}}{16\pi^2}\left(\frac{5}{3}+\frac{1-\xi}{2}\right)\frac{\Gamma(2-\frac{n}{2})}{\mu^{4-n}}\,(q^2 g^{\mu\nu} - q^\mu q^\nu)\,.$$

When the above equation is added to $\Pi_{\mathrm{A}}^{\mu\nu}(q)$ in (76), we obtain the divergent part of the gluon self-energy described by the four diagrams of Fig. 15.12 in its totality. Thus,

$$\Pi_{\mathrm{glu}}^{\mu\nu}(q)\big|_{\mathrm{div}} = \frac{g_{\mathrm{s}}^2\,\delta_{ij}}{16\pi^2}\left[\left(\frac{5}{3}+\frac{1-\xi}{2}\right)N_{\mathrm{c}} - \frac{2}{3}N_{\mathrm{f}}\right]\frac{\Gamma(2-\frac{n}{2})}{\mu^{4-n}}\,(q^2 g^{\mu\nu} - q^\mu q^\nu)\,,$$

$$\delta_{\mathrm{glu}}(\mu) = \frac{g_{\mathrm{s}}^2}{16\pi^2}\frac{\Gamma(2-\frac{n}{2})}{\mu^{4-n}}\left[\left(\frac{5}{3}+\frac{1-\xi}{2}\right)N_{\mathrm{c}} - \frac{2}{3}N_{\mathrm{f}}\right]. \tag{15.82}$$

### 15.3.4 The Running QCD Coupling

Plugging (67), (74), and (82) into (65), we obtain

$$\beta(g_s) = \frac{-g_s^3}{16\pi^2}\left(\frac{11}{3}N_c - \frac{2\,N_f}{3}\right) . \tag{15.83}$$

As announced, the gauge-dependence $\xi$ disappears in $\beta(g_s)$ when the various contributions of (67), (74) and (82) are added. Furthermore, for $N_f < (11/2)\,N_c$, i.e. for the number of quark species less than 16, the Callan–Symanzik $\beta$-function is *negative*, then according to (43) the running coupling decreases for increasing energy, and so QCD is asymptotically free. The counterterms $Z_j$ that we have computed are summarized in Table 15.1.

**Table 15.1.** Coefficients of the counterterms at the one-loop level

| *Coefficients* | QCD : $\frac{g_s^2}{16\pi^2}\log\frac{\Lambda^2}{\mu^2}$ | QED : $\frac{e^2}{16\pi^2}\log\frac{\Lambda^2}{\mu^2}$ |
|---|---|---|
| $\delta_1$ (vertex) | $-\frac{N_c^2-1}{2N_c}\xi - \left(1-\frac{1-\xi}{4}\right)N_c$ | $-\xi$ |
| $\delta_2$ (fermion) | $-\frac{N_c^2-1}{2N_c}\xi$ | $-\xi$ |
| $\delta_3$ (gauge boson) | $\left(\frac{5}{3}+\frac{1-\xi}{2}\right)N_c - \frac{2}{3}N_f$ | $-\frac{2}{3}\,N_f(1+\frac{5}{9}N_c)^\dagger$ |

$^\dagger$1 from e$^-$ , $\frac{5}{9} = (\frac{2}{3})^2 + (\frac{-1}{3})^2$ from u, d quarks

From (48) or (65), we infer a general rule: a negative $\delta_1$ would render the theory asymptotically free, just as a positive $\delta_{ext}$ ($\delta_2$ and $\delta_3$) would. Their roles can be seen in Table 15.1. A glance at the various coefficients $\delta_j$ reveals that the negative sign $(-1)N_c$ in $\delta_1$ and the positive sign $(+5/3)N_c$ in $\delta_3$ are mainly responsible of the asymptotic freedom. This is due to the gluonic non-Abelian character of Fig. 15.11b and Fig. 15.12c. On the other hand, due to the fermionic loop in Fig. 15.12a, the minus sign in $-(2/3)N_f$ of $\delta_3$ is the main obstacle for QCD or the Abelian QED to be asymptotically free.

Finally, we remark that for the pure Yang-Mills gauge fields, e.g. QCD without quarks, the corresponding $\beta(g_s)$ can be obtained from Fig. 15.8c. This has no equivalent in QED, and is computed according to

$$\beta(g_s) = g_s\,\mu\frac{\partial}{\partial\,\mu}\left[-\Delta_1^{\rm glu}(\mu) + \frac{3}{2}\delta^*_{\rm glu}(\mu)\right] , \tag{15.84}$$

where $Z_1^{\rm glu} = 1+\Delta_1^{\rm glu}$ is the three-gluon loop vertex counterterm derived from the diagrams of Fig. 15.14. Only pure gluon fields, and not quarks, come in $Z_1^{\rm glu}$. The $\delta^*_{\rm glu}(\mu)$ is taken from (82) but without the $N_f$ term, i.e. without Fig. 15.12a .

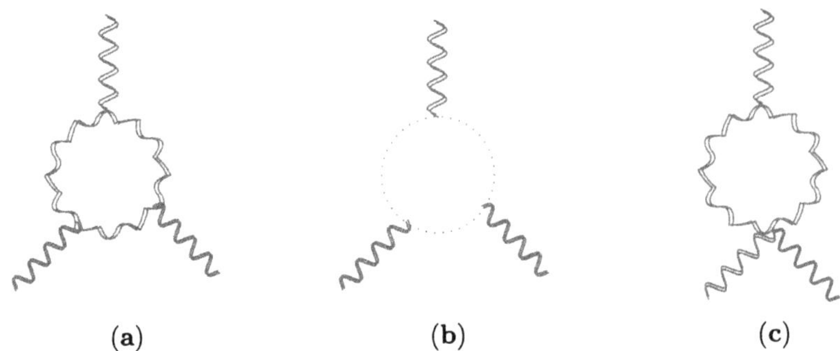


**Fig. 15.14a–c.** Loop diagrams of the three-gluon vertex $\Delta_1^{\mathrm{glu}}$

The easiest way to understand asymptotic freedom is by considering the *magnetic screening properties* of the vacuum.[2] In a relativistic theory (with natural units $c = 1$), the dielectric constant $\varepsilon$ is equal to $1/\mu$ where $\mu$ is the magnetic permeability. In QED, all media are diamagnetic, i.e. $\mu < 1$. The origin of the diamagnetism may be understood by the Lenz law: In response to an external magnetic field, electrically charged particles will set up a current which itself induces a magnetic field that acts to decrease the external magnetic field. Then $\varepsilon > 1$, which implies the screening effect.

However in non-Abelian QCD, paramagnetism is possible. Gluons are *color charged* particles of spin one, they behave as color magnetic dipoles that align themselves parallel to the external chromomagnetic field, thus increasing its magnitude and producing $\mu_{\mathrm{color}} > 1$, hence the color charge is antiscreened with $\varepsilon_{\mathrm{color}} < 1$. We can therefore regard the QCD vacuum as a color paramagnetic media, and asymptotic freedom arises if the antiscreening due to gluons overcomes the screening due to quarks.

On the other hand, as one goes toward the infrared low energy regime, $\varepsilon_{\mathrm{color}} \to 0$, the effective coupling strength would diverge. This would cause an effect similar to the QED Meissner phenomenon in which the magnetic field is repelled by the superconducting material. In QCD, it would be the chromoelectric field that exhibits the Meissner effect. This in turn would squeeze the chromoelectric lines of force between a quark and an antiquark into a narrow tube, and, furthermore, the energy of the tube, i.e. the quark potential $V(r)$, would increase proportionally to the distance $r$ separating q and $\overline{\mathrm{q}}$. This suggests confinement of quarks and gluons.

Now the QCD running coupling can be obtained by putting (83) into (56). Exactly as in (64), the solution for $g_{\mathrm{s}}(q^2)$ and $\alpha_{\mathrm{s}}(q^2) = g_{\mathrm{s}}^2/4\pi$ is

$$\alpha_{\mathrm{s}}(q^2) = \frac{\alpha_{\mathrm{s}}(\mu^2)}{1 + b_0 \frac{\alpha_{\mathrm{s}}(\mu^2)}{4\pi} \log \frac{q^2}{\mu^2}} , \quad \text{with } b_0 = \frac{11}{3} N_{\mathrm{c}} - \frac{2}{3} N_{\mathrm{f}} . \tag{15.85}$$

[2] Nielsen, N. K., Am. J. Phys. **49** (1981) 1171, Hughes, R. J., Nucl. Phys. **B186** (1981) 376

Since $b_0$ is positive with $N_c = 3$ and $N_f = 6$, the strong coupling $\alpha_s(q^2)$ *decreases with growing* $q^2$, in opposite direction to the electromagnetic coupling $\alpha_{em}(q^2)$ as given in (64). Also in contrast to QED where $\alpha_{em}(0)$ can be directly measured, for QCD the corresponding Thomson limit is not accessible: at $q^2 = 0$, the strong coupling $g_s$ is *very strong*. The solution (85) only gives us the evolution of the running coupling as a function of $q^2$, but the absolute value of $\alpha_s$ at a certain fixed value of $q^2$, i.e. the constant of integration of the differential equation (56), must be determined by experiment.

A conventional choice is to introduce a mass parameter $\Lambda_{\overline{MS}}$ which provides a parameterization of $\alpha_s(q^2)$. By defining $\Lambda_{\overline{MS}}$ as follows, one can rearrange (85) in a simple form:

$$1 = g_s^2 \frac{b_0}{8\pi^2} \log \frac{\mu}{\Lambda_{\overline{MS}}} \Longrightarrow \alpha_s(q) = \frac{2\pi}{b_0 \log(q/\Lambda_{\overline{MS}})} \quad \text{for } q \gg \Lambda_{\overline{MS}} \,. \qquad (15.86)$$

Experiments (Fig. 15.15) yield a value of $\Lambda_{\overline{MS}}$ ranging from 150 to 250 MeV, and perturbative QCD theory is valid when $q^2$ is much larger than $\Lambda^2_{\overline{MS}}$. Depending on $q^2$, the number of the quark flavor $N_f$ in $b_0$ is taken between 3 and 6. For instance, at the $\tau$ lepton mass ($N_f = 3$), then $\alpha_s(M_\tau) \approx 0.37$, and at the $Z^0$ weak boson mass ($N_f = 5$), $\alpha_s(M_{Z^0}) \approx 0.12$. On the other hand, the strong interaction becomes strong at distance larger than $1/\Lambda_{\overline{MS}}$ which is the size of the light hadrons.

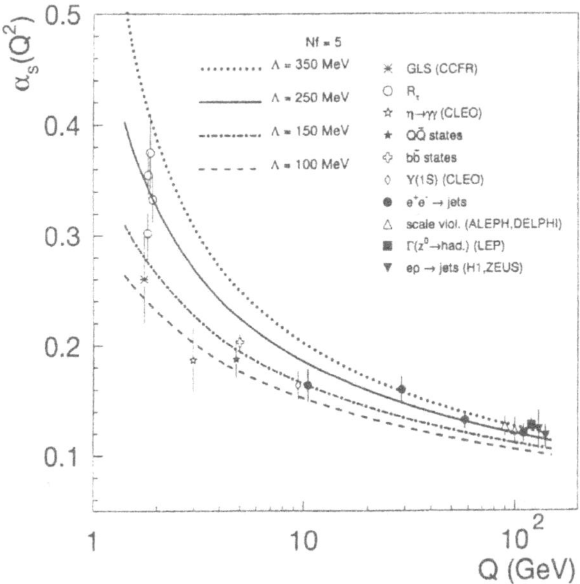


**Fig. 15.15.** Various measurements of $\alpha_s(Q)$ from Yamauchi M., *Proc. 17th Int. Symp. on Lepton–Photon Interactions*, 1995 Beijing (ed. Zheng Zhi-Peng and Chen He-Sheng). World Scientific, Singapore 1996 (Courtesy Masanori Yamauchi)

As an immediate application of (86), let us illustrate the implication of the CS equation for the QCD corrected decay rate $\Gamma(\tau^- \to \nu_\tau + \text{hadrons})$. In (14.84), we have computed this quantity which depends on $\alpha_s$, the latter must be defined at some scale $\mu$. This is in contrast to $\alpha_{em}$ naturally defined by on-mass-shell electron and photon, which is impossible for quarks and gluons. CS equation allows us to avoid this problem.

Since $\Gamma(M_\tau, \mu, \alpha_s) \equiv \Gamma(\tau^- \to \nu_\tau + \text{hadrons})$ is a physical measurable quantity, it is independent of any scale $\mu$ and therefore obeys

$$\left[\mu \frac{\partial}{\partial \mu} + \beta(g_s) \frac{\partial}{\partial g_s}\right] \Gamma(M_\tau, \mu, \alpha_s) = 0 \,. \tag{15.87}$$

Such physical quantities change only because of their dependence on the coupling constant and the renormalization point, and therefore they satisfy (87). The corresponding CS functions $\gamma$ are not present for physical quantities whose normalizations $Z$ follow from their definitions.

On dimensional grounds, we can write $\Gamma(M_\tau, \mu, \alpha_s)$ as $\Gamma_0 F(M_\tau/\mu, \alpha_s)$. The above equation tells us that the dimensionless $F$ depends on its arguments only through the running coupling constant $\alpha_s(\mu)$ evaluated at $\mu = M_\tau$. This allows us to replace in (14.84) the fixed $\alpha_s$ with the running coupling constant $\alpha_s(M_\tau)$: $\Gamma_0 N_c \left[1 + \frac{1}{\pi}\alpha_s(M_\tau)\right]$.

## 15.4 Ghosts

The appearance of ghosts in Fig. 15.12d and the corresponding Feynman rules in Fig. 15.13 at first sight seem to be entirely ad hoc. It is not so. At the origin, ghosts represent a prescription proposed by Faddeev and Popov to cure some difficulties encountered in the *quantization of massless gauge fields* (see Chaps. 2 and 8), whether the gauge symmetry is spontaneously broken (electroweak) or not (QCD). The problem has to do with the redundant degrees of freedom of the gauge field $A_\mu$, its polarization vector $\varepsilon_\mu$ has only two transverse physical components (see Sect. 2.5), whereas all of its four components are treated *covariantly in the quantization.* This redundancy is common to both massless photon and gluon fields; it must be removed.

### 15.4.1 The Faddeev–Popov Gauge-Fixing Method

To quantize these fields and to single out their essential difference, one could adopt the functional path integral formulation which is not used in this book, but some rudimentary notions would be useful. In this formalism, the key ingredients for the quantization are the functional integrals

$$\int \mathcal{D}A_\mu \exp\left(\mathrm{i}\, S[A_\mu]\right) \;\text{ for QED}\;, \quad \int \mathcal{D}A^j_\mu \exp\left(\mathrm{i}\, S[A^j_\mu]\right) \;\text{ for QCD}\,, \tag{15.88}$$

where $S[A_\mu]$ and $(S[A^j_\mu])$ are the actions of the photon and gluon fields

$$S[A_\mu] = \int \mathrm{d}^4x \, \left[-\tfrac{1}{4}(F_{\mu\nu})^2\right] \;, \quad S[A^j_\mu] = \int \mathrm{d}^4x \, \left[-\tfrac{1}{4}(F^j_{\mu\nu})^2\right] \,.$$

$A^j_\mu, j = 1 \cdots 8$ are the gluon fields in the adjoint representation of the $\mathrm{SU_c}(3)$ color group. As explained in Chap. 8, the transformations, parameterized by an arbitrary function $\omega(x)$ $[\omega^i(x)]$ applied to the field $A_\mu(x)$ $[A^i_\mu(x)]$:

$$\begin{aligned} &A_\mu(x) \longrightarrow A_\mu(x) + \partial_\mu \omega(x) \equiv A[\omega](x) , \\ \text{or } &A^i_\mu(x) \longrightarrow A^i_\mu(x) + \partial_\mu \omega^i(x) - g_s f^{ijk} A^j_\mu(x) \omega^k(x) \equiv A^i[\omega](x) , \end{aligned} \tag{15.89}$$

leave $F_{\mu\nu}$ $(F^i_{\mu\nu})$ and the actions unchanged. We can use the freedom of adding an arbitrary $\omega(x)$ $[\omega^i(x)]$ to $A_\mu$ $[A^i_\mu]$ to make the latter satisfy some relations. Choosing $A_\mu$ $(A^i_\mu)$ to satisfy some conditions is called fixing the gauge. In the functional integrals (88), we redundantly integrate over an infinite number of physically equivalent gauge field configurations $A_\mu$ $(A^i_\mu)$. Without fixing the gauge, we overquantize unphysical variables. We must remove the redundancy by some gauge-fixing conditions in the course of quantization. To count each configuration only once, Faddeev and Popov use a constraint $G(A[\omega]) = 0$ as a gauge-fixing condition. For instance $G(A[\omega]) = \partial^\mu A_\mu - h(x)$, where $h(x)$ is an arbitrary function and $h(x) = 0$ is the familiar covariant Lorentz gauge condition [see (2.140) for QED]. A similar condition applies to the non-Abelian $A^i[\omega]$ field, cf (8.61), (8.118). We could force the integral to cover only the configurations satisfying $G(A[\omega]) = 0$ by inserting a delta functional $\delta\{G(A[\omega])\}$ in (88). This functional change implies that a Jacobian factor (which is the determinant $\det \partial G(A[\omega])/\partial\omega$) enters into the integrand. We omit the Lorentz index $\mu$ and the color index $i$ to abbreviate the notations. Since

$$1 = \int \mathcal{D}\omega(x)\, \delta\{G(A[\omega])\} \det \frac{\partial G(A[\omega])}{\partial \omega} , \tag{15.90}$$

equation (88) can be rewritten as

$$\int \mathcal{D}\omega \int \mathcal{D}A \, \exp(\mathrm{i}\, S[A])\, \delta\{G(A[\omega])\} \det \frac{\partial G(A[\omega])}{\partial \omega} . \tag{15.91}$$

Now in the above functional integral, the only difference between QED and QCD is that in the former, the $\partial G(A[\omega])/\partial\omega$ is *independent* of the field $A$, whereas the corresponding $\partial G(A^i[\omega])/\partial\omega$ of the non-Abelian case depends on the field $A^i$ because of the third term $f^{ijk} A^j \omega^k$ in (89).

In the Abelian case, we can factor the determinant of $\partial G(A[\omega])/\partial\omega$ out of the functional integral $\int \mathcal{D}A$ and use a Gaussian function (weighted by an arbitrary finite number $\xi$) to integrate $\exp(\mathrm{i}\, S[A])\, \delta\{G(A[\omega])\}$, the result of the integration yields a term $(1/2\xi)[\partial^\mu A_\mu]^2$ to be added to the original Lagrangian $-\frac{1}{4}(F_{\mu\nu})^2$. This $(1/2\xi)[\partial^\mu A_\mu]^2$ term coming from $\delta\{G(A[\omega])\}$ gives rise to the propagator of massless gauge fields obtained in Chap. 8. This propagator, which depends on an arbitrary gauge parameter $\xi$, is common to the photon and gluon:

$$\frac{-\mathrm{i}}{q^2 + \mathrm{i}\epsilon} \left[ g^{\mu\nu} - \frac{(1-\xi)\, q^\mu q^\nu}{q^2} \right] .$$

Now comes the crucial difference between the Abelian and the non-Abelian cases. For the latter, because the determinant depends on $A^i_\mu$, it cannot be put outside $\int \mathcal{D}A$, therefore, after the integration has been done, we obtain a new term in addition to $(1/2\xi)\,[\partial^\mu A^i_\mu]^2$. This new term is

$$\overline{s}^i(-\partial^2\,\delta^{ik} + g_s f^{ijk}\partial^\mu A^j_\mu)s^k\,, \tag{15.92}$$

where $s^i$ is a scalar field obeying the anticommutation rule, like a fermionic field, hence the name of ghost. This object is a representation of

$$\begin{aligned}\det\frac{\partial\, G(A^i[\omega])}{\partial\omega} &= \det\,(\partial^\mu\, D_\mu) \;\text{ where }\; D_\mu\omega^i = \partial_\mu\omega^i - g_s f^{ijk} A^j_\mu\omega^k \\ &= \int \mathcal{D}s\,\mathcal{D}\overline{s}\,\exp\left[\mathrm{i}\int \mathrm{d}^4x\,\overline{s}\,(-\partial^\mu\, D_\mu\,)s\right]\,,\end{aligned}$$

using the general Gaussian integral formula involving a Hermitian matrix X with eigenvalues $x_j$:

$$\begin{aligned}\det \mathrm{X} = \prod_i x_i &= \left(\prod_i \int \mathrm{d}\theta_i^*\mathrm{d}\theta_i\right)\exp\left[-\Sigma_i\theta_i^*\,x_i\,\theta_i\right] \\ &= \left(\prod_i \int \mathrm{d}\theta_i^*\mathrm{d}\theta_i\right)\exp\left[-\theta_i^*\,X_{ij}\,\theta_j\right]\,.\end{aligned}$$

The functional integration variable $\theta$ must be a Grassmann (anticommuting) function; if we use an ordinary commuting function, we would get the inverse of the determinant, $(\det\mathrm{X})^{-1}$, and not $\det\mathrm{X}$. This may be seen by observing the generalization of the one variable integral $\int \mathrm{d}x\exp[-Ax^2] = \sqrt{\pi/A}$:

$$(\det\mathrm{X})^{-N} = C\int \mathcal{D}\Phi\mathcal{D}\Phi^*\exp[-\Phi^*\mathrm{X}\Phi]\,,$$

where $\Phi$ is a complex scalar field with $N$ components. Feynman rules are derived for closed loops of $\Phi$ fields, each loop goes with a factor $N$. Since we want $N$ to be $-1$, our closed loops will go with a factor $-1$, just like the rules for fermions.

The kinetic term $\partial^2\,\delta^{ik}\overline{s}^i s^k$ in (92) gives rise to the ghost propagator, whereas the second term $g_s f^{ijk}\overline{s}^i\,\partial^\mu A^j_\mu\,s^k$ represents the gluon–ghost–ghost vertex; both are displayed through the Feynman rules of Fig. 15.13. Ghosts, derived from non-Abelian gauge fields, couple only to the latter and *appear only in loops*; they do not exist as external free particles. Because of the self-interacting three-boson coupling which is typically non-Abelian, the ghost term (92) occurs not only in QCD but also in electroweak interactions.

In the above discussion we considered only massless gauge fields. When these fields become massive through the Higgs mechanism, their quantization is similar to the massless case, and ghosts are also present.

Absent at the tree level, the elastic photon–photon scattering and the decay $Z^0 \to 3$ photons are pure loop effects (Fig. 15.16). Because of the non-Abelian $\gamma W^+W^-$ and $Z^0W^+W^-$ couplings, ghosts appear in the loops of $\gamma+\gamma\to\gamma+\gamma$ and $Z^0\to 3\gamma$ processes. They are massive and charged.

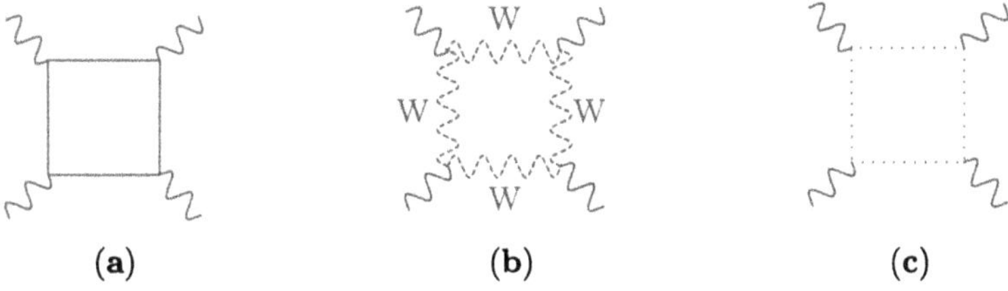


**Fig. 15.16.** Contributions to $\gamma\gamma \to \gamma\gamma$ scattering: (**a**) fermion box; (**b**) W-boson box; and (**c**) ghost box

## 15.4.2 Ghosts and Unitarity

We have seen that ghosts restore the gauge-invariant structure $-q^2 g^{\mu\nu} + q^\mu q^\nu$ of the gluon self-energy $\Pi^{\mu\nu}_{\text{glu}}(q)$. Next, ghosts arise from the Faddeev–Popov gauge-fixing quantization by path integrals. Now we show that ghosts are even more indispensable. Without them, unitarity would not be satisfied, which would be as disastrous as a nonconservation of probability.

The amplitude of any physical process computed to some perturbative order must satisfy unitarity to that order. The requirement of unitarity for the $S$-matrix operator $S\,S^\dagger = S^\dagger\,S = 1$ or $\sum_n S_{an}\,S^*_{bn} = \delta_{ab}$, implies that the matrix element of the operator $T$, defined by $S = 1 + \mathrm{i}\,T$, satisfies

$$2\,\mathrm{Im}\,\langle b\,|\,T\,|\,a\rangle = \sum_j \int \mathrm{d}\rho_j\,\langle b\,|\,T\,|\,j\rangle\,\langle j\,|\,T^\dagger\,|\,a\rangle\ ,$$

where $\mathrm{d}\rho_j$ is the phase space of the intermediate state $|j\rangle$. The imaginary part of the scattering amplitude $\langle a\,|\,T\,|\,b\rangle$ is directly related to the sum over the products of physical amplitudes connecting any intermediate state $|j\rangle$ to the initial state $|a\rangle$ and the final state $|b\rangle$. For a two-body $\to$ two-body amplitude, the unitarity constraint is expressed as

$$2\,\mathrm{Im}\,\mathcal{A}(k+k' \to p+p') = \sum_n \int \prod_{j=2}^{n} \frac{\mathrm{d}^3[q_j]}{(2\pi)^3\,2E_j}\,(2\pi)^4\delta^4([q_j]-k-k') \times \mathcal{A}(k+k' \to [q_j])\,\mathcal{A}^*(p+p' \to [q_j]) \quad (15.93)$$

where $[q_j]$ collectively denotes momenta of the intermediate state $|j\rangle$. In the particular case of two-body forward elastic scattering $k = p$ and $k' = p'$, the right-hand side of (93) contains a squared amplitude $|\mathcal{A}|^2$. To calculate a cross-section from $|\mathcal{A}|^2$, we must include the kinematic factor $S/4F$ according to (4.54). When we multiply this kinematic factor by both sides of (93), we obtain the optical theorem relating the imaginary part of the two-body *forward elastic* scattering amplitude to the total cross-section $\sigma_{\text{tot}}$ for the production of all states. Thus,

$$\mathrm{Im}\,\mathcal{A}_{\text{elas.}}(k+k' \to k+k') = [2E_{\text{cm}}K_{\text{cm}}]\,\sigma_{\text{tot}}(k+k' \to \text{anything})\ ,$$
$$\mathrm{Im}\,\mathcal{A}_{\text{elas.}}(s,t=0) = \sqrt{\lambda(s,m_1^2,m_2^2)}\,\sigma_{\text{tot}}(s)\ , \quad (15.94)$$

where $E_{\rm cm}$ is the total energy of the two-body system and $K_{\rm cm}$ is the momentum of either particle in the center-of-mass frame. The variables $E_{\rm cm}$ and $K_{\rm cm}$ are written in terms of the Mandelstam variables $s$ and $t$ in (94).

As an application of (93), we consider the quark–antiquark scattering where the intermediate state are two gluons, as described by Fig. 15.17. Unitarity relates the imaginary part of the $\mathrm{q}(p_1)+\overline{\mathrm{q}}(p_2)\to\mathrm{q}(p_3)+\overline{\mathrm{q}}(p_4)$ amplitude at order $g_{\rm s}^4$ to the $\mathrm{q}(p_1)+\overline{\mathrm{q}}(p_2)\to\mathrm{g}(k_1)+\mathrm{g}(k_2)$ squared amplitude:

$$2\,\mathrm{Im}\mathcal{A}(\mathrm{q}+\overline{\mathrm{q}}\to\mathrm{q}+\overline{\mathrm{q}})=\tfrac{1}{2}\int\mathrm{d}\rho_2\,|\mathcal{A}(\mathrm{q}+\overline{\mathrm{q}}\to 2\text{ gluons})|^2\,,\quad\text{where}$$

$$\int\mathrm{d}\rho_2\equiv\int\frac{\mathrm{d}^3k_1}{(2\pi)^3\,2E_1}\frac{\mathrm{d}^3k_2}{(2\pi)^3\,2E_2}(2\pi)^4\delta^4(k_1+k_2-p_1-p_2)\qquad(15.95)$$

is the two-gluon phase space integral, and the symmetry factor 1/2 on the right-hand side of (95) is due to two identical bosons in the final state.

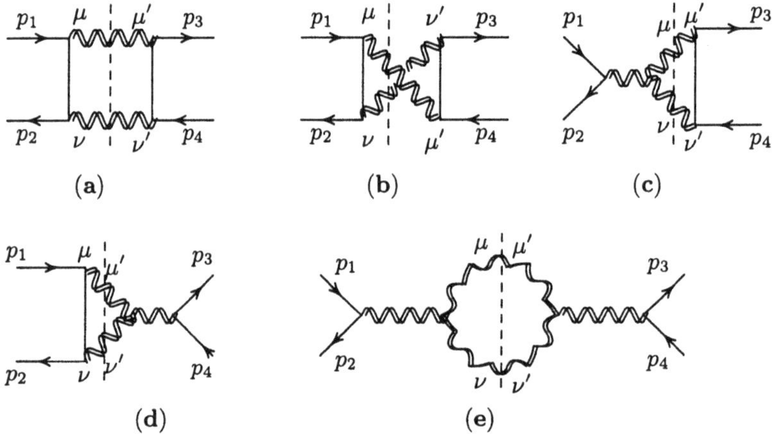


**Fig. 15.17a–e.** The cuts (*dashed lines*) give the imaginary part of the amplitude (to order of $g_s^4$) for $\mathrm{q}+\bar{\mathrm{q}}\to\mathrm{q}+\bar{\mathrm{q}}$ via two-gluon exchange

The crucial point is the following: The two *real* gluons in the final state (Fig. 15.18) have only physical *transverse polarizations*; while the *virtual* gluons in Fig. 15.17, via their propagators with the tensor $g^{\mu\nu}$, can have unphysical *nontransverse polarization states* [see (99) below where $g^{\mu\nu}$ contains nontransverse polarizations $G_{\rm L}^{\mu\nu}(k)$]. How can we be sure *a priori* that in (95), the unphysical polarizations of the left-hand side would disappear on the right-hand side? If they are present, something must exist to cancel them. Historically, by inspecting (95), Feynman recognized that unitarity hints at a missing piece in the non-Abelian theory at one-loop level.

To verify explicitly the unitarity relation (95), let us proceed in three steps. First, the polarizations of massless gauge particles (see also Sect. 2.5) are discussed, next we go to the $\overline{\mathrm{q}}+\mathrm{q}\to\mathrm{g}+\mathrm{g}$ amplitude, i.e. the right-hand side of (95), then its left-hand side is computed lastly.

**Massless Gauge-Particle Polarizations.** A massive spin-1 particle has three polarization states, hence its polarization four-vector $e^\mu(k,\sigma)$, $\sigma = 1,2,3$ subject to the constraint $k_\mu e^\mu(k,\sigma) = 0$ is well defined. On the other hand, a massless gauge particle has only two physical polarization states $e^\mu(k,\lambda)$, $\lambda = 1,2$. They are transverse $\boldsymbol{k}\cdot\boldsymbol{e}(k,\lambda) = 0$. These two spacelike transverse polarization vectors satisfy

$$e^\mu(k,\lambda)e^*_\mu(k,\lambda') = -\delta_{\lambda\lambda'} \quad , \quad k_\mu e^\mu(k,\lambda) = 0 \,. \tag{15.96}$$

The first one is simply the relativistic version of $\boldsymbol{e}(k,\lambda)\cdot\boldsymbol{e}^*(k,\lambda') = \delta_{\lambda\lambda'}$. The three independent vectors $k^\mu$, $e^\mu(k,1)$, and $e^\mu(k,2)$ cannot completely span the four-dimensional space-time, i.e. the conditions (96) do not uniquely specify the transverse polarization states. Indeed, any $e^\mu(k,\lambda)+\omega k^\mu$ (where $\omega$ is an arbitrary parameter) also satisfies the conditions (96). To uniquely define the two transverse states of a massless gauge particle, one must impose one more condition to determine $\omega$, i.e. one must fix the gauge.

To achieve an orthogonal basis, we introduce an independent lightlike four-vector $\eta^\mu$ *orthogonal* to the transverse polarization vectors $e^\mu(k,\lambda)$. Together with $k^\mu$, $e^\mu(k,1)$, and $e^\mu(k,2)$, the $\eta^\mu$ closes the basis. Since $k^2 = 0$ and $\eta^\mu$ is independent of $k^\mu$, we must have $\eta_\mu k^\mu \neq 0$. So $\eta^\mu$ satisfies

$$\eta^2 = 0 \text{ with } \eta_\mu e^\mu(k,\lambda) = 0 \ , \ \eta_\mu k^\mu \neq 0 \,. \tag{15.97}$$

The two unphysical nontransverse polarization states can now be built up from $k^\mu$ and $\eta^\mu$. When we compute the right-hand side of (95), we have to sum over the transverse polarization states. Thus, we define

$$g_{\mathrm{T}}^{\mu\nu}(k) \equiv \sum_{\lambda=1,2} e^\mu(k,\lambda)\, e^{\nu*}(k,\lambda) \,. \tag{15.98}$$

The tensor $g_{\mathrm{T}}^{\mu\nu}(k)$ which must have the form $ag^{\mu\nu} + b\eta^\mu\eta^\nu + ck^\mu k^\nu + dk^\mu\eta^\nu + e\eta^\mu k^\nu$ is reduced to

$$g_{\mathrm{T}}^{\mu\nu}(k) = -g^{\mu\nu} + G_{\mathrm{L}}^{\mu\nu}(k) \,, \qquad \text{where } G_{\mathrm{L}}^{\mu\nu}(k) = \frac{k^\mu\eta^\nu + \eta^\mu k^\nu}{k\cdot\eta} \tag{15.99}$$

using (96) and (97). The extra term $G_{\mathrm{L}}^{\mu\nu}(k)$ subtracts out of $g^{\mu\nu}$ the longitudinal and time components of $e^\mu$, i.e. the two unphysical nontransverse polarizations, called $e_{\mathrm{L}}^\mu(k)$, with $k_\mu e_{\mathrm{L}}^\mu(k) \neq 0$. The important thing is

$$k_\mu g_{\mathrm{T}}^{\mu\nu}(k) = k_\nu g_{\mathrm{T}}^{\mu\nu}(k) = 0 \text{ while } k_\mu G_{\mathrm{L}}^{\mu\nu}(k) = k^\nu \ , \ k_\nu G_{\mathrm{L}}^{\mu\nu}(k) = k^\mu \,. \tag{15.100}$$

For the problem on hand, with two gluons, a convenient choice of the nontransverse polarizations of the first (second) gluon $k_1(k_2)$ is the one in which the vector $\eta^\mu$ could be taken as the momentum $k_2^\mu(k_1^\mu)$ of the other gluon in their *center-of-mass*.

This frame is required, since according to (97), $\eta_\mu$ must satisfy $\eta_\mu e^\mu(k,\lambda) = 0$. In this frame $k_1 = (E_0 = |\boldsymbol{k}|, \boldsymbol{k})$ , $k_2 = (E_0, -\boldsymbol{k})$, such that $k_2^\mu e_\mu(k_1,\lambda) = 0 = k_1^\mu e_\mu(k_2,\lambda)$ for transverse $e^\mu(k,\lambda)$ having zero time component. With this choice, we have

$$G_{\mathrm{L}}^{\mu\nu}(k_1) = \frac{k_1^\mu k_2^\nu + k_1^\nu k_2^\mu}{2E_0^2} = \frac{k_1^\mu k_2^\nu + k_1^\nu k_2^\mu}{k_1 \cdot k_2} = G_{\mathrm{L}}^{\mu\nu}(k_2) \,. \tag{15.101}$$

This explicit form of $G_{\mathrm{L}}^{\mu\nu}$ is convenient for subsequent calculations.

**The Amplitude for** $\mathrm{q}(p_1)+\overline{\mathrm{q}}(p_2) \to \mathrm{g}(k_1)+\mathrm{g}(k_2)$ **.** To order $g_{\mathrm{s}}^2$, there are three diagrams drawn in Fig. 15.18 that contribute to the amplitude. Let us write it as $\mathcal{A} = e^{*\mu}(k_1)e^{*\nu}(k_2)\overline{v}(p_2)\mathcal{M}_{\mu\nu}u(p_1)$, where Feynman rules give

$$\begin{aligned}\mathcal{M}_{\mu\nu} &= (-\mathrm{i}g_{\mathrm{s}})^2 \left\{\gamma_\nu T^j \frac{\mathrm{i}}{\not p_1 - \not k_1 - m}\gamma_\mu T^i + \gamma_\mu T^i \frac{\mathrm{i}}{\not k_1 - \not p_2 - m}\gamma_\nu T^j\right\} \\ &+ \frac{g_{\mathrm{s}}^2 f^{ijk}\gamma^\rho T^k}{(k_1+k_2)^2}\Big[g_{\mu\nu}(k_2-k_1)_\rho - g_{\nu\rho}(2k_2+k_1)_\mu + g_{\rho\mu}(2k_1+k_2)_\nu\Big] . \end{aligned} \tag{15.102}$$

Note that $\mathcal{M}_{\mu\nu}$ is symmetric under the interchange of the two final state gluons $(k_1,\mu,T^i) \leftrightarrow (k_2,\nu,T^j)$. The two terms inside the curly brackets { } of (102), corresponding to the two diagrams of Fig. 15.18a–b, are similar to the $\mathrm{e}^+ + \mathrm{e}^- \to \gamma + \gamma$ amplitude in QED (except for the color matrices $T^i$ and the coupling $g_{\mathrm{s}}$). We call this amplitude $\mathcal{M}^1_{\mu\nu}$. When we multiply $\mathcal{M}^1_{\mu\nu}$ by $k_1^\mu$ and use $\overline{v}(p_2)[\not p_2 + m] = 0 = [\not p_1 - m]u(p_1)$, we obtain

$$\overline{v}(p_2)k_1^\mu \mathcal{M}^1_{\mu\nu}u(p_1) = -\mathrm{i}g_{\mathrm{s}}^2\overline{v}(p_2)\gamma_\nu[T^i,T^j]u(p_1) = +g_{\mathrm{s}}^2 f^{ijk}T^k\overline{v}(p_2)\gamma_\nu u(p_1) \,,$$

which *vanishes in Abelian* QED.

The last term associated with the brackets [ ] of (102) corresponds to Fig. 15.18c. Absent in QED, this amplitude is denoted by $\mathcal{M}^2_{\mu\nu}$. When it is multiplied by $k_1^\mu$, using $\overline{v}(p_2)[\not k_1 + \not k_2]u(p_1) = \overline{v}(p_2)[\not p_2 + \not p_1]u(p_1) = 0$, we get

$$\overline{v}(p_2)k_1^\mu \mathcal{M}^2_{\mu\nu}u(p_1) = +g_{\mathrm{s}}^2 f^{ijk}T^k\overline{v}(p_2)\left[\frac{k_{2\nu}\not k_1}{(k_1+k_2)^2} - \gamma_\nu\right]u(p_1) \,.$$

The sum $k_1^\mu[\mathcal{M}^1_{\mu\nu} + \mathcal{M}^2_{\mu\nu}] = k_1^\mu \mathcal{M}_{\mu\nu}$ is

$$\overline{v}(p_2)k_1^\mu \mathcal{M}_{\mu\nu}u(p_1) = g_{\mathrm{s}}^2 f^{ijk}T^k\overline{v}(p_2)\left[\frac{k_{2\nu}\not k_1}{(k_1+k_2)^2}\right]u(p_1) \equiv k_{2\nu}\mathcal{P} \,, \tag{15.103}$$

where $\mathcal{P}$ is defined by (103), i.e.

$$\begin{aligned}\mathcal{P} &= g_{\mathrm{s}}^2 f^{ijk}T^k\overline{v}(p_2)\frac{\not k_1}{(k_1+k_2)^2}u(p_1) \\ &= g_{\mathrm{s}}^2 f^{ijk}T^k\overline{v}(p_2)\frac{-\not k_2}{(k_1+k_2)^2}u(p_1) \,. \end{aligned} \tag{15.104}$$

Note that $\mathcal{P}$ is symmetric under the interchange of the two gluons, therefore

$$\overline{v}(p_2)k_2^\nu \mathcal{M}_{\mu\nu}u(p_1) = k_{1\mu}\mathcal{P}. \tag{15.105}$$

From now on, to simplify the notations, we will write $k_1^\mu \mathcal{M}_{\mu\nu}$ and $k_2^\nu \mathcal{M}_{\mu\nu}$ without $\overline{v}(p_2)$ and $u(p_1)$ although the insertion is implicitly understood.

We learn from this analysis that a two-photon (or two-gluon) amplitude can always be written as $e^\mu(k_1,\lambda)\, e^\nu(k_2,\lambda)\mathcal{M}_{\mu\nu}$. When we replace $e^\mu(k_1,\lambda)$ by $k_1^\mu$ or $e^\nu(k_2,\lambda)$ by $k_2^\nu$, we get

$$\text{QED}\;:\; k_1^\mu \mathcal{M}_{\mu\nu} = 0 = k_2^\nu \mathcal{M}_{\mu\nu}\;,\quad \text{QCD}\;:\; k_1^\mu \mathcal{M}_{\mu\nu} \neq 0\;,\; k_2^\nu \mathcal{M}_{\mu\nu} \neq 0\,. \tag{15.106}$$

However, for QCD, from (96), (103), and (105), we have a weaker constraint

$$k_1^\mu k_2^\nu \mathcal{M}_{\mu\nu} = k_1^\mu e^\nu(k_2,\lambda)\mathcal{M}_{\mu\nu} = k_2^\nu e^\mu(k_1,\lambda)\mathcal{M}_{\mu\nu} = 0\,. \tag{15.107}$$

Using (103) and (105), most important is the following nonzero term for QCD

$$k_1^\mu k_1^\nu \mathcal{M}_{\mu\nu} = k_2^\mu k_2^\nu \mathcal{M}_{\mu\nu} = k_1 \cdot k_2 \mathcal{P}\,. \tag{15.108}$$

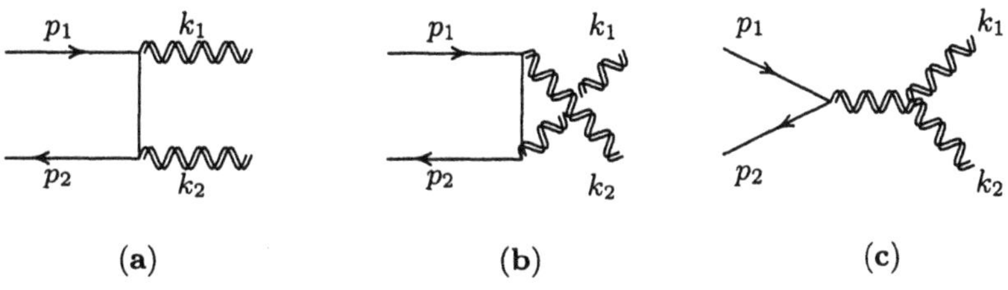


**Fig. 15.18a–c.** Amplitude (to order $g_s^2$) of $\mathrm{q} + \overline{\mathrm{q}} \to \mathrm{g} + \mathrm{g}$

**The Imaginary Part of $\mathcal{A}(\mathrm{q} + \overline{\mathrm{q}} \to \mathrm{q} + \overline{\mathrm{q}})$.** The amplitude of Fig. 15.17 – which is constructed from the two vertices $\mathcal{M}_{\mu\nu}$ and $\mathcal{M}^*_{\mu'\nu'}$ connected by two gluons exchanged in the loop – can be written as [with $P = \frac{1}{2}(p_1 + p_2)$]

$$\mathcal{A} = \tfrac{1}{2}\int \frac{\mathrm{d}^4 K}{(2\pi)^4}\mathcal{M}_{\mu\nu}\frac{-\mathrm{i}\, g^{\mu\mu'}}{(K+P)^2 + \mathrm{i}\epsilon}\;\frac{-\mathrm{i}\, g^{\nu\nu'}}{(P-K)^2 + \mathrm{i}\epsilon}\mathcal{M}^*_{\mu'\nu'}\,, \tag{15.109}$$

the factor 1/2 in front takes into account the two identical gluons in the loop.

The imaginary part of the amplitude is obtained – according to the Cutkosky rules – by putting the two virtual gluons on the mass-shell, i.e. by replacing their propagators with the delta functions

$$\frac{1}{p^2 + \mathrm{i}\epsilon} \longrightarrow (-2\pi\,\mathrm{i})\,\delta(p^2)\,.$$

Using $\int \frac{d^4K}{(2\pi)^4} = \int \frac{d^4k_1}{(2\pi)^4} \int \frac{d^4k_2}{(2\pi)^4} (2\pi)^4 \delta^4(k_1 + k_2 - 2P)$, and

$\int \frac{d^3k_j}{2E_j} = \int d^4k_j \delta(k_j^2)\theta(E_j)$ , $j = 1, 2$,

it can be shown that when the four-dimensional integral $\int d^4K/(2\pi)^4$ in (109) is multiplied by the two delta functions, it becomes the dimensionless two-gluon phase space integral $\int d\rho_2$ of (95). We find that the imaginary part of $\mathcal{A}(q + \bar{q} \to q + \bar{q})$, i.e. the left-hand side of (95) is

$$2\,\mathrm{Im}\,\mathcal{A}(q + \bar{q} \to q + \bar{q}) = \frac{1}{2} \int d\rho_2\, \mathcal{M}_{\mu\nu} \left[ g^{\mu\mu'} g^{\nu\nu'} \right] \mathcal{M}^*_{\mu'\nu'}$$

$$= \frac{1}{2} \int d\rho_2\, \mathcal{M}_{\mu\nu} \left[ \left( -g_{\mathrm{T}}^{\mu\mu'}(k_1) + G_{\mathrm{L}}^{\mu\mu'}(k_1) \right) \left( -g_{\mathrm{T}}^{\nu\nu'}(k_2) + G_{\mathrm{L}}^{\nu\nu'}(k_2) \right) \right] \mathcal{M}^*_{\mu'\nu'} \,. \tag{15.110}$$

Equation (110) clearly shows that unphysical polarizations appear in $\mathrm{Im}\,\mathcal{A}$.

On the other hand, the right-hand side of (95) with real gluons, i.e. the quantity $\frac{1}{2} \int d\rho_2 \left| \mathcal{A}(q + \bar{q} \to 2 \text{ gluons}) \right|^2$, is equal to

$$\frac{1}{2} \int d\rho_2\, \mathcal{M}_{\mu\nu} \left[ g_{\mathrm{T}}^{\mu\mu'}(k_1) g_{\mathrm{T}}^{\nu\nu'}(k_2) \right] \mathcal{M}^*_{\mu'\nu'} \,. \tag{15.111}$$

As stated by (95), if unitarity is satisfied, (110) and (111) must be identical. Actually this is the case of $e^+ + e^- \to \gamma + \gamma$ in Abelian QED. Indeed, because of (106), the contraction of $G_{\mathrm{L}}^{\mu\mu'}(k_1)$ with $\mathcal{M}_{\mu\nu}\mathcal{M}^*_{\mu'\nu'}$ yields vanishing result, and the same thing happens with $G_{\mathrm{L}}^{\nu\nu'}(k_2)$. More generally, *in Abelian processes*, we can safely take $g_{\mathrm{T}}^{\mu\nu}(k) = -g^{\mu\nu}$.

For QCD in (110), the cross terms $g_{\mathrm{T}}^{\mu\mu'}(k_1)G_{\mathrm{L}}^{\nu\nu'}(k_2) + g_{\mathrm{T}}^{\nu\nu'}(k_2)G_{\mathrm{L}}^{\mu\mu'}(k_1)$ also vanish when they are contracted with $\mathcal{M}_{\mu\nu}$ and $\mathcal{M}_{\mu'\nu'}$ due to (107). The difference between (110) and (111) which comes from $G_{\mathrm{L}}^{\mu\mu'}(k_1)G_{\mathrm{L}}^{\nu\nu'}(k_2)$ would violate unitarity if it did not vanish. Using (101) together with (107) and (108), this difference actually is nonzero and equals to

$$\frac{1}{2} \int d\rho_2 \frac{\mathcal{M}_{\mu\nu}}{(k_1 \cdot k_2)^2} \left[ k_1^\mu k_1^\nu k_2^{\mu'} k_2^{\nu'} + k_2^\mu k_2^\nu k_1^{\mu'} k_1^{\nu'} \right] \mathcal{M}^*{}_{\mu'\nu'} = \int d\rho_2 \mathcal{P}\mathcal{P}^* . \tag{15.112}$$

To cancel the right-hand side of (112), the ghost contribution coming from the imaginary part of the amplitude $\mathcal{A}_{\mathrm{ghost}}(q + \bar{q} \to q + \bar{q})$ in Fig. 15.19 is added to (110).

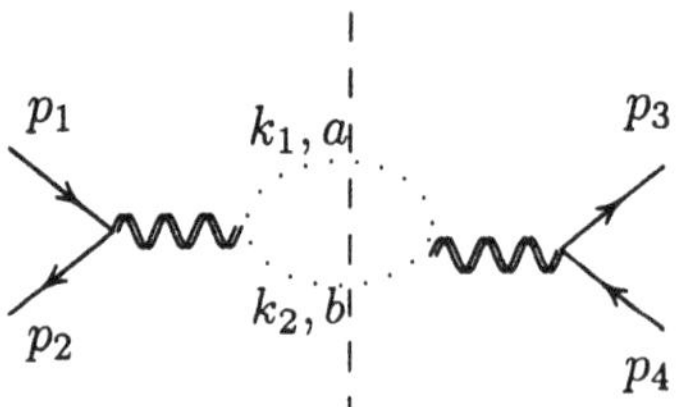


**Fig. 15.19.** $\mathcal{A}_{\mathrm{ghost}}$ from ghost loop. The dashed line cut gives the imaginary part of $\mathcal{A}_{\mathrm{ghost}}$

The imaginary part of $\mathcal{A}_{\text{ghost}}(\mathrm{q}+\overline{\mathrm{q}} \to \mathrm{q}+\overline{\mathrm{q}})$ is calculated similarly to (109) and (110) by the following operations:

(i) like the case of gluons, we replace the ghost propagators with the delta functions. After this replacement, the ghost loop integral becomes the phase space $\mathrm{d}\rho_2$ of (95).

(ii) substitute the $\mathrm{q}+\overline{\mathrm{q}} \to \mathrm{g}+\mathrm{g}$ amplitude by the $\mathrm{q}+\overline{\mathrm{q}} \to s_i+\overline{s}_j$ amplitude. The latter is obtained from Feynman rules,

$$(-\mathrm{i}g_s)\overline{v}(p_2)\gamma_\mu T^k u(p_1)\frac{-\mathrm{i}}{(k_1+k_2)^2}[g_s f^{ikj} k_1^\mu] = \mathcal{P}\,. \tag{15.113}$$

Because of the fermionic character of ghosts, there is a minus sign in the loop integral, but there is no bosonic symmetry factor of $\frac{1}{2}$. From all of these operations, we get

$$2\,\mathrm{Im}\,\mathcal{A}_{\text{ghost}} = -\int \mathrm{d}\rho_2 \mathcal{P}\mathcal{P}^*\,. \tag{15.114}$$

So (112) and (114) cancel each other. Without the ghost contribution as given by (114), the difference between (110) and (111), i.e. (112), would not vanish, and unitarity would be violated.

This example illustrates a general interpretation of ghosts as agents that neutralize the unphysical polarizations of the gauge bosons. Their presence was anticipated by Feynman well before the Faddeev–Popov quantization method was proposed.

## Problems

**15.1 Computation of $\Sigma(p)$ by the Pauli–Villars regularization.** Use the Pauli–Villars procedure (14.89) to compute the fermionic self-energy $\Sigma(p)$ defined in (14.21) and (14.23). As $\Lambda \to \infty$, one recovers the original theory. Compute $Z_2$ and check that the pole $\Gamma(2-\frac{n}{2})$ can be identified with $\log(\Lambda^2/\mu^2)$.

**15.2 Massive photon in space-time two dimensions.** In one-space and one-time dimension (two-dimensional QED), derive the dimensions of the photon field, the fermion field, and the coupling $e$. Using (5) with $n=2$, show that the photon has a mass $e/\sqrt{\pi}$ when it couples to massless fermions. In two dimensions, the $\Pi(q^2)$ can have a pole at $q^2=0$ which arises from the massless fermion-antifermion intermediate state. As discussed after (11), this is the reason why the photon can be massive (Schwinger's mechanism).

**15.3 Correction to the Coulomb potential. The Uehling effect.** From (37), the running charge $e$ is $q^2$ dependent. Derive the Fourier transform of the Coulomb potential $e^2(q^2)/q^2$, first in the nonrelativistic case $q^2 \ll m^2$, then in the $q^2 \gg m^2$ case. The modified Coulomb potential shifts the atomic electron energy levels (Uehling effect).

**15.4 The $\beta$- and $\gamma$-functions in $\lambda\phi^4$ at one-loop.** Draw to order $\mathcal{O}(\lambda^2)$ the vertex of four interacting scalar fields. There are three loop diagrams, corresponding to the three Mandelstam variables $s, t, u$. Show that $\beta(\lambda) = 3\lambda^2/(16\pi^2)$. The one-loop $\mathcal{O}(\lambda)$ correction to the two-point function gives the field-strength counterterm $Z_\phi$ from which is derived the CS-function $\gamma(\lambda)$. Show that $\gamma(\lambda)$ vanishes.

**15.5 The pure Yang–Mills $\beta$-function by (84).** Consider QCD without quarks but only gluons with their three-gluon and four-gluon interactions. Compute the $Z_1^{\text{glu}} = 1 + \Delta_1^{\text{glu}}$ from Fig. 15.14. Then show that the corresponding $\beta$-function is $\beta(g_s) = (-g_s^3/16\pi^2)\frac{11}{3}N_c$.

## Suggestions for Further Reading


*Field Theory, Renormalization:*

Collins, J. C., *Renormalization.* Cambridge U. Press, Cambridge 1984

Hatfield, B., *Quantum Field Theory of Point Particles and Strings.* Addison-Wesley, Redwood, CA 1992

Itzykson, C. and Zuber, J. B., *Quantum Field Theory.* McGraw-Hill, New York 1980

Kaku, M., *Quantum Field Theory.* Oxford U. Press, New York 1993

Ramond, P., *Field theory: A Modern Primer* (Second edition). Addison-Wesley, Redwood, CA 1989

*Quantization of Yang–Mills fields:*

Faddeev, L. D. and Slavnov, A. A., *Gauge Fields: Introduction to Quantum theory.* Benjamin, Reading, MA 1980

't Hooft, G., *Under the Spell of the Gauge Principle.* World Scientific, Singapore 1994

*Vacuum polarization $\Pi^{\mu\nu}(q)$, vertex function $\Gamma^\mu(p',p)$, fermionic self-energy $\Sigma(p)$:*

De Wit, B. and Smith, J., *Field Theory in Particle Physics* (Vol. I). North-Holland, Amsterdam 1986

Peskin, M. E. and Schroeder, D. V., *An Introduction to Quantum Field Theory.* Addison-Wesley, Reading, MA 1995

*Renormalization group method and $\beta$-functions:*

Cheng, Ta-Pei and Li, Ling-Fong, *Gauge Theory of Elementary Particle Physics.* Oxford U. Press, New York 1984

Peskin, M. E. and Schroeder, D. V., (op. cit.)

Weinberg, S., *The Quantum Theory of Fields* (Vol. II.) Cambridge U. Press, Cambridge 1996

*Ghosts and unitarity, Cancelation of gauge-dependent $\xi$ terms:*

Aitchison, I. J. R. and Hey, A. J. G., *Gauge Theories in Particle Physics* (Second edition). Adam Hilger, Bristol 1989

Feng, Y. J. and Lam, C. S., Phys. Rev. **D53** (1996) 2115

Gross, F., *Relativistic Quantum Mechanics and Field Theory.* Wiley-Interscience, New York 1993

*Perturbative* QCD, GLAP *equations:*

Field, R. D., *Applications of Perturbative* QCD. Addison-Wesley, Redwood, CA 1989

# 16 Heavy Flavors

As shown in Fig. 15.15, the energy point $\simeq 1$ GeV separates the QCD scale into regions of large and small running coupling $\alpha_s(\mu)$, or equivalently, of low and high energy, or light and heavy particles. The world of heavy flavors begins with the charm quark and the $\tau$ lepton and extends to the bottom and top quarks in the fermionic sector. The list of heavy particles in the standard model also includes the weak bosons $W^{\pm}$ and $Z^0$, and the neutral Higgs boson. Of these, only the Higgs boson is still not experimentally observed at present.

Once a particle is produced, its decay provides the traditional way to determine its intrinsic properties and its characteristic interactions with other particles. Let us recall a few examples: P and CP violations were revealed by $K^+$ and $K^0$ decays; the extremely narrow width of the heavy $J/\psi$ signaled its presence; charm was discovered by its typically dominant decay into strangeness; the $\tau^{\pm}$ lepton pair was recognized by the distinctive signature $e^{\pm}\mu^{\mp}$ left by its leptonic decays.

For the B meson studied in this chapter, its decays are particularly interesting for the following reasons:

(i) due to the QCD asymptotic freedom and the large masses and momenta released by heavy flavors, electroweak and strong interactions are closely correlated. Their interplay in perturbative calculations can be further improved by the renormalization group methods. The first section is devoted to the QCD renormalization of weak interactions which provides a basis for nonleptonic decays.

(ii) a new symmetry – called heavy flavor symmetry (HFS) – appears in an effective Lagrangian derived from QCD in the limit $M \to \infty$ ($M$ being the heavy quark mass). This symmetry allows the determination of the form factors involved in the exclusive decay modes. Some of these predictions play a crucial role in determining the CKM matrix elements. The $1/M$ expansion provides a solid theoretical framework for the spectator model in which only the heavy quark undergoes decay while the light constituents are spectators. Semileptonic decays are the best way to understand many properties of the b-flavored hadrons and to measure $V_{cb}$ and $V_{ub}$.

(iii) the physics of heavy particles has their impact on various quantities through their quantum effects in loops. For instance, from the observed $B^0$–$\overline{B}^0$ mixing, a lower bound of the top mass is predicted before its discovery.

(iv) the physics of heavy flavors plays an essential role in CP violation too, and may open windows on the mechanism of the gauge symmetry breaking, i.e. the Higgs sector.

## 16.1 QCD Renormalization of Weak Interactions

In (13.1) and (13.2), the effective Lagrangian for nonleptonic (or hadronic) decays of hadrons is given in the most general form by

$$\frac{G_F}{\sqrt{2}} H^\mu H_\mu^\dagger \ , \quad \text{with } H_\mu = \sum_{Q,q} V_{\mathrm{Qq}} H_\mu^{\mathrm{Qq}} \ , \quad H_\mu^{\mathrm{Qq}} = \overline{Q}\gamma_\mu(1-\gamma_5)\, q \,. \tag{16.1}$$

This universal effective Lagrangian governs the hadronic decays of all flavored hadrons, from strange to charm and bottom. Unflavored hadrons (like the neutron and the pion) weakly decay only in semileptonic modes by the lack of a sufficiently large phase volume, while the top quark directly decays into the real W boson and the b quark, t→ W + b, without passing by (1). As for the semileptonic decays of hadrons, they are governed by

$$\frac{G_F}{\sqrt{2}} \left[ H^\mu L_\mu^\dagger + L^\mu H_\mu^\dagger \right] , \tag{16.2}$$

where $L_\mu$ is the leptonic current given by (13.2).

In weak decays of hadrons, we note the prominent role of the quark current $H^\mu$. Since quarks interact through QCD, the effect of the hard (energetic) gluons on $H^\mu$ alone, as well as on the product $H^\mu H_\mu^\dagger$, must be taken into account. This is called the QCD renormalization of weak interactions to which this section is devoted. The reason for considering hard gluons is that they can be treated within the QCD perturbative framework, due to the asymptotic freedom. Nonperturbative soft gluons effects related to form factors will be studied within the framework of HFS in Sect. 3.

Let us start by considering as an example the hadronic decays of the $\overline{\mathrm{B}}$ mesons ($\mathrm{b}\overline{\mathrm{q}}$ bound states) into charmed and unflavored hadrons. These modes are described by the decay $\mathrm{b} \to \mathrm{c} + \mathrm{d} + \overline{\mathrm{u}}$ (Fig. 16.1). This is the spectator model where the b constituent of the $\overline{\mathrm{B}}$ undergoes decay while the other light constituent $\overline{\mathrm{q}}$ is a spectator.

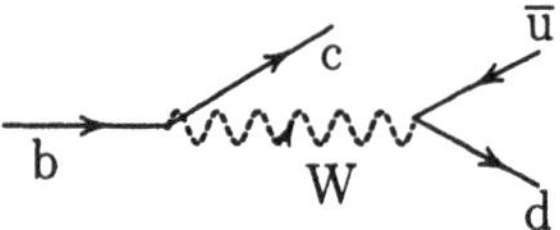


**Fig. 16.1.** $\mathrm{b} \to \mathrm{c} + \mathrm{d} + \overline{\mathrm{u}}$ at the electroweak tree level

The corresponding effective Lagrangian is the product of the two Cabibbo-favored currents $V_{\mathrm{cb}} H_\mu^{\mathrm{bc}}$ and $V_{\mathrm{ud}}^* H_{\mathrm{ud}}^{\mu\dagger}$

$$\mathrm{i}\left(\frac{-\mathrm{i}g}{2\sqrt{2}}\right)^2 V_{\mathrm{cb}} V_{\mathrm{ud}}^* \frac{-\mathrm{i}}{k^2 - M_{\mathrm{W}}^2} (H_\mu^{\mathrm{bc}})\,(H_{\mathrm{ud}}^{\mu\dagger}) \underset{k^2 \ll M_{\mathrm{W}}^2}{\longrightarrow} \frac{G_F}{\sqrt{2}} V_{\mathrm{cb}} V_{\mathrm{ud}}^* \,\mathcal{O}_{\mathrm{A}},$$

$$\text{where} \quad \mathcal{O}_{\mathrm{A}} = (H_\mu^{\mathrm{bc}})\,(H_{\mathrm{ud}}^{\mu\dagger}) = [\overline{c}\,\gamma_\mu(1-\gamma_5)\,b]\,[\overline{d}\,\gamma^\mu(1-\gamma_5)\,u] \,. \tag{16.3}$$

### 16.1.1 Corrections to Single Currents

Before considering the gluonic corrections to the product $H^\mu H^\dagger_\mu$ which governs nonleptonic decays, let us discuss the QCD effect on the single currents $H^\mu$; these corrections concern also the semileptonic modes. In fact, we are already familiar with the latter in Chap. 14 which treats the correction to $H^\mu_{\rm ud}$ (we call it from now on the 'right vertex' [ud]). The corrections apply to $H^{\rm bc}_\mu$, the 'left vertex' [bc] too.

For each of these two vertices ([ud] or [bc]) taken separately, there are in all five Feynman diagrams shown in Fig. 14.2 (virtual gluons) and Fig. 14.3 (real gluons) that contribute to the corrections. They are symbolically represented by a • in Fig. 16.2 for the right vertex and in Fig. 16.3 for the left vertex.

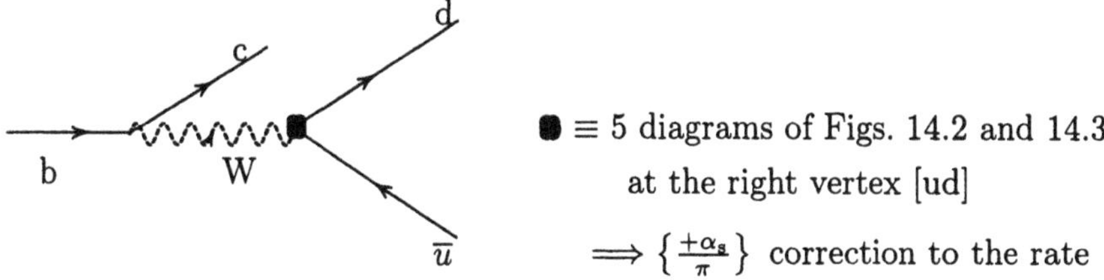


**Fig. 16.2.** QCD corrections to the right vertex [ud] in $\rm b \to c + d + \bar{u}$

For the right vertex [ud] part, the one-loop QCD correction to the *rate* of $\rm b \to c + d + \bar{u}$ is equal to the one found in $\tau^- \to \nu_\tau + \rm d + \bar{u}$, i.e.

$$N_{\rm c}\frac{G_{\rm F}^2 M^5}{192\pi^3}|V_{\rm cb}V^*_{\rm ud}|^2\left\{\frac{\alpha_{\rm s}}{\pi}\right\} \equiv N_{\rm c}\,\Gamma_0|V_{\rm cb}V^*_{\rm ud}|^2\left\{\frac{\alpha_{\rm s}}{\pi}\right\}\,, \tag{16.4}$$

where $\Gamma_0 = G_{\rm F}^2 M^5/(192\pi^3)$ as given by (13.21) is the width of the fermion b of mass $M$ decaying into three massless fermions c, d, and $\rm\bar{u}$. For massive c, d and $\rm\bar{u}$ quarks, the rate $\Gamma_0$ is to be multiplied by the phase space suppression factor $I(x,y,z)$ as given by (13.63). The rate $\Gamma_0$ and the coefficient $|V_{\rm cb}V^*_{\rm ud}|^2$ are implicitly understood in the following. The factor $\alpha_{\rm s}/\pi$ represents the one-loop QCD correction to the right vertex [ud] (see 14.83).

To convert the $\rm\bar{u}d$ pair into hadrons in the inclusive rate $\tau^- \to \nu_\tau +$ hadrons (or $\rm\overline{B} \to$ hadrons), the color factor $N_{\rm c} = 3$ must be included in (4) [see also (14.83) and (69) below].

For massive d and $\rm\bar{u}$ quarks associated with this right [ud] vertex, the correction becomes larger than $\alpha_{\rm s}/\pi$ [see the function $K(x,y)$ in Table 14.1]. In $\rm b \to c + d + \bar{u}$, for any fixed value of the c quark mass of the left part, the right vertex correction grows with the increasing masses of its associated u, d quarks and is always positive. This dynamical enhancement is completely different from the purely kinematic phase space suppression effect due to the c quark mass. For instance, in $\rm b \to c + s + \bar{c}$, the s $\rm\bar{c}$ pair replaces the d $\rm\bar{u}$ pair, the correction to the right vertex [cs] is considerably larger than $\alpha_{\rm s}/\pi$ because the s and $\rm\bar{c}$ masses largely exceed the d and $\rm\bar{u}$ masses.

The QCD corrections to the left vertex [bc] shown in Fig. 16.3 can be computed with the five similar diagrams as in Chap. 14. The result

$$-\frac{2}{3}\frac{\alpha_s}{\pi}\left(\pi^2-\frac{25}{4}\right)=-2.41\frac{\alpha_s}{\pi} \tag{16.5}$$

is taken from the electromagnetic corrections to $\mu^- \to \nu_\mu + e^- + \overline{\nu}_e$ already given by (13.27), with the substitution $e^2 \to \frac{4}{3}g_s^2$.

As noted in Chap. 14, the factors $\pi^2$ – coming from the second derivative of the $\Gamma(x)$ function in both real and virtual gluon diagrams – do not exactly cancel each other in this left vertex [bc] (while they do in the right vertex), so we have the $\pi^2$ term in (5).

The formula (5) is valid only for massless c, u, and d quarks in the final state. When quarks are massive, the result while remaining negative is reduced in magnitude. For instance, with $m_c = 0.3\,M$ and $m_u = m_d = 0$, the QCD correction to the rate is $-0.87\ (\alpha_s/\pi)$ instead of $-2.41\ (\alpha_s/\pi)$, always with the multiplicative factor $\Gamma_0|V_{cb}V^*_{ud}|^2$.

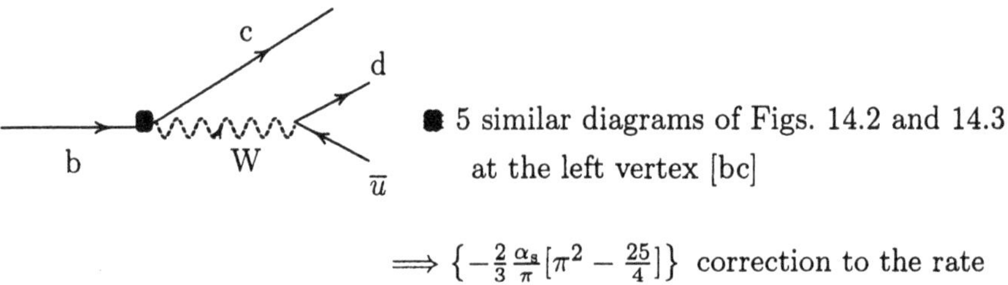


**Fig. 16.3.** QCD corrections to the left vertex [bc] in $b \to c + d + \overline{u}$

The overall corrections to nonleptonic decays $b \to c + d + \overline{u}$ coming from the left and right vertices are (the common factor $N_c\,\Gamma_0|V_{cb}V^*_{ud}|^2$ in (4) and (5) is implicit):

$$\frac{-2\alpha_s}{3\pi}\left(\pi^2-\frac{25}{4}\right)+\frac{\alpha_s}{\pi}=\frac{-2\alpha_s}{3\pi}\left(\pi^2-\frac{31}{4}\right)=-1.41\frac{\alpha_s}{\pi}\,. \tag{16.6}$$

This formula is valid for the three massless quarks in the final state. For $m_c = 0.3\,M$ and $m_u = m_d = 0$, the overall corrections are no more (6) but change into $[-0.87+0.52](\alpha_s/\pi) = -0.35\ (\alpha_s/\pi)$. The $-0.87(\alpha_s/\pi)$ factor comes from corrections to the left vertex [bc] as mentioned above, while the $+0.52\,(\alpha_s/\pi)$ comes from corrections to the right vertex [ud]. The purely kinematic phase space suppression effect of the c quark mass on the correction to the right vertex [ud] is given by $+0.52\,(\alpha_s/\pi)$, instead of $(\alpha_s/\pi)$ when c, u, and d quarks are massless.

In the decay $b \to c + s + \overline{c}$ (Fig. 16.4) responsible for the mode $B \to J/\psi + K$ (Fig. 16.9), we note that the quark mass effect in QCD correction is much more important.[1] Instead of the negative value $-1.41\,(\alpha_s/\pi)$ for the three massless quarks in the final state as given by (6) , the mass effect reverses the sign resulting in an overall positive sign $+3.02\,(\alpha_s/\pi)$. This is because of the massive s and $\bar{c}$ quarks in the right vertex [cs] which replace the d and $\bar{u}$ [again see the function $K(x,y)$]. This B→ $J/\psi$ + K mode is particularly important for the study of CP violation in B mesons (Sect. 5).

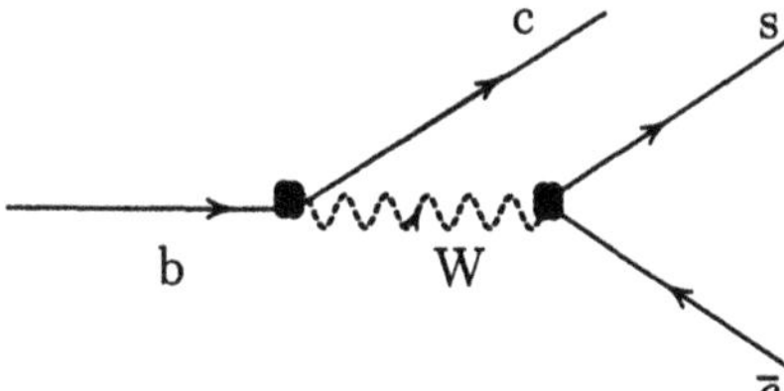


**Fig. 16.4.** QCD corrections to the left and right vertices in b → c + s + $\bar{c}$

These two types of corrections involving the single currents $H_\mu$ apply also to the semileptonic modes. For instance, the right vertex [$q_2q_3$] correction which occurs in $\tau \to \nu_\tau + q_2 + \bar{q}_3$ is already studied in Chap. 14. The left vertex [bc] correction, which occurs in the $b \to c + e^- + \bar{\nu}_e$ decay and describing the inclusive semileptonic decays of B mesons, is given below by (56). They are QCD corrections that have no large logarithmic factors, in contrast to the large logarithmic corrections to the product $H^\mu H^\dagger_\mu$ that we are considering now.

### 16.1.2 Corrections to Product of Currents

A new type of QCD correction applies to the product $H^{\rm bc}_\mu H^{\mu\dagger}_{\rm ud}$ in which one gluon is exchanged between the two vertices [bc] and [ud] in all possible ways. This correction concerning only hadronic decays has no equivalent in semileptonic modes since gluons are insensitive to leptons. Shown in Fig. 16.5a, the gluon connects the c to the d quark, and in Fig. 16.5b the c to the $\bar{u}$ quark. There are four diagrams in all, the other two are similar (with a gluon exchanged between b and $\bar{u}$ and between b and d) and are not shown.

Our first task is to compute these diagrams. The loop integral with one gluon exchanged between the c and d quarks in Fig. 16.5a yields the term denoted by $I_{\rm c\leftrightarrow d}$, which is given by

$$\mathrm{i}\left(\frac{-\mathrm{i}g}{2\sqrt{2}}\right)^2(-\mathrm{i}g_s)^2\int\frac{\mathrm{d}^4k}{(2\pi)^4}\,\frac{-\mathrm{i}}{k^2}\,\frac{-\mathrm{i}}{k^2-M_W^2}\left[\bar{d}\,T^j\,\gamma^\lambda\,\frac{\mathrm{i}(-\not{k}+m_\mathrm{d})}{k^2-m_\mathrm{d}^2}\gamma^\mu(1-\gamma_5)\,u\right]$$
$$\times\left[\bar{c}\,T^j\,\gamma_\lambda\,\frac{\mathrm{i}(\not{k}+m_\mathrm{c})}{k^2-m_\mathrm{c}^2}\gamma_\mu(1-\gamma_5)\,b\right], \qquad (16.7)$$

[1] Ho-Kim, Q. and Pham, Xuan-Yem, Ann. of Phys. (N.Y.) **155** (1984) 202; Bagan, E., Ball, P., Braun, V. M. and Gosdzinsky, P., Phys. Lett. **342B** (1995) 362; Neubert, M. and Sachrajda, C. T., Nucl. Phys. **B483** (1997) 339.

the external momenta are taken to be zero in the first approximation, however, we can no longer neglect the integration variable $k^2$ as in (3). The summation over the index $j = 1, \cdots, 8$ of the SU(3) color matrix $T^j = \frac{1}{2}\lambda^j$ is understood. Neglecting $m_\mathrm{d}$, the integral (7) is rewritten as

$$I_{\mathrm{c}\leftrightarrow\mathrm{d}} = \frac{-\mathrm{i}\, G_\mathrm{F}\, g_\mathrm{s}^2}{\sqrt{2}} \int \frac{\mathrm{d}^4 k}{(2\pi)^4} \frac{1}{k^2} \frac{M_\mathrm{W}^2}{k^2 - M_\mathrm{W}^2} \frac{1}{k^2 - m_\mathrm{c}^2}$$

$$\times \sum_{j=1}^{8} [\bar{d}\,(T^j)_{ef}\, \gamma^\lambda \gamma^\rho \gamma^\mu (1-\gamma_5)\, u]\; [\bar{c}\,(T^j)_{gh}\, \gamma_\lambda \gamma_\rho \gamma_\mu (1-\gamma_5)\, b] \;, \quad (16.8)$$

where $e, f, g, h$ are the color indices, running from 1 to 3. Writing the product of the three propagators in (8) as an integral over the Feynman auxiliary variables $x, y$, and applying formulas in the Appendix, we obtain

$$\int \frac{\mathrm{d}^4 k}{(2\pi)^4} \frac{1}{k^2} \frac{M_\mathrm{W}^2}{k^2 - M_\mathrm{W}^2} \frac{1}{k^2 - m_\mathrm{c}^2} = \frac{-\mathrm{i}}{16\pi^2} \log \frac{M_\mathrm{W}^2}{m_\mathrm{c}^2} \quad (16.9)$$

after integration first over $\int \mathrm{d}^4 k$, then over $y$ and $x$. This result uses (7) in which the external momenta are neglected. If these momenta are not neglected, the correct lower limit of the logarithm in (9) actually must be set equal to the external momenta which is dominated by the decaying b quark, i.e. $\log(M_\mathrm{W}^2/m_\mathrm{c}^2)$ should be replaced by $\log(M_\mathrm{W}^2/M^2)$.

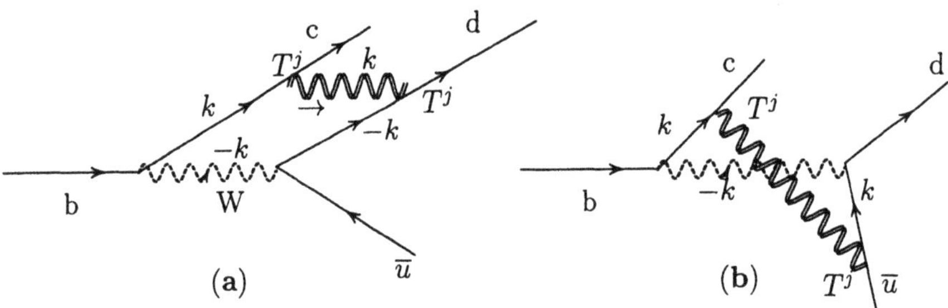


**Fig. 16.5a, b.** QCD corrections to nonleptonic decays through the product $H^\mu H_\mu^\dagger$. There are two more graphs with a gluon between b and ū, and between b and d.

This lower limit fixed by the mass $M$ of the decaying particle is easy to understand from the diagram with one gluon exchanged between the b and d (or ū) quarks, i.e. in (8), we replace $m_\mathrm{c}$ by $M$ in their respective propagators.

It remains to treat the second line of (8), which is

$$\{XY\} \equiv \sum_{j=1}^{8} [\bar{d}\,(T^j)_{ef}\, \gamma^\lambda \gamma^\rho \gamma^\mu (1-\gamma_5)\, u]\; [\bar{c}\,(T^j)_{gh}\, \gamma_\lambda \gamma_\rho \gamma_\mu (1-\gamma_5)\, b] \,.$$

We have already met this product without the color matrix $T^j$ in (11.21) and (11.24). Due to (11.23), we have

$$[\bar{d}\gamma^\lambda \gamma^\rho \gamma^\mu (1-\gamma_5)\, u]\; [\bar{c}\gamma_\lambda \gamma_\rho \gamma_\mu (1-\gamma_5)\, b] = 4\,\mathcal{O}_\mathrm{A} \;, \quad (16.10)$$

where $\mathcal{O}_{\mathrm{A}}$ is given in (3). However, QCD, via the color matrix $T^j$, drastically changes the dynamics. This can be seen when we use again (11.88). To abbreviate the notations, let us denote $[\overline{d}\,(T^j)_{ef}\;\gamma^\lambda\,\gamma^\rho\,\gamma^\mu(1-\gamma_5)\,u]$ as $[\overline{d}\,(T^j)_{ef}\,u]$ without the Dirac matrices. Similarly the second factor in (8) is written as $[\overline{c}\,(T^j)_{gh}\,b]$. Then from (11.88) we have

$$\sum_{j=1}^{8}\left[\overline{d}\,(T^j)_{ef}\,u\right]\,\left[\overline{c}\,(T^j)_{gh}\,b\right]=\frac{1}{2}\,\left[\overline{d}_e\,u_f\right]\,\left[\overline{c}_f\,b_e\right]-\frac{1}{6}\,\left[\overline{d}_e\,u_e\right]\,\left[\overline{c}_g\,b_g\right]\,.\tag{16.11}$$

Now we reintroduce the Dirac matrices $\gamma^\lambda\,\gamma^\rho\,\gamma^\mu(1-\gamma_5)$ and $\gamma_\lambda\,\gamma_\rho\,\gamma_\mu(1-\gamma_5)$ into the right-hand side of (11), and using again (11.23), we gain a factor of 4 like the right-hand side of (10). The final result is

$$\begin{aligned}\frac{1}{4}\,\{XY\}=&\,\frac{1}{2}[\overline{d}_e\,\gamma^\mu(1-\gamma_5)\,u_f]\,[\overline{c}_f\,\gamma_\mu(1-\gamma_5)\,b_e]\\&-\frac{1}{6}\,[\overline{d}_e\,\gamma^\mu(1-\gamma_5)\,u_e]\,[\overline{c}_g\,\gamma_\mu(1-\gamma_5)\,b_g]\,.\end{aligned}\tag{16.12}$$

The second line of the above equation is the product of two color-singlet quark currents, and is nothing but the operator $-\frac{1}{6}\mathcal{O}_{\mathrm{A}}$ in (3). The first line is the product of two color-non-singlet quark currents having the familiar structure $(\mathrm{V}-\mathrm{A})\times(\mathrm{V}-\mathrm{A})$. This naturally suggests the use of the Fierz transformation which let the structure $(\mathrm{V}-\mathrm{A})\times(\mathrm{V}-\mathrm{A})$ stay intact (see Appendix). The minus sign from the Fierz rearrangement, when combined with the minus sign from interchanging the order of the anticommuting quark fields, yields an overall positive sign in

$$[\overline{d}_e\,\gamma^\mu(1-\gamma_5)\,u_f]\,[\overline{c}_f\,\gamma_\mu(1-\gamma_5)\,b_e]=[\overline{d}_e\,\gamma^\mu(1-\gamma_5)\,b_e]\,[\overline{c}_f\,\gamma_\mu(1-\gamma_5)\,u_f]\,.$$

The right-hand side of the above equation is a product of two color-singlet currents. It is symbolically written as $(\overline{d}\,b)\,(\overline{c}\,u)$ and replaces now the first line of (12). Consequently, the quantity $\{XY\}$ can also be expressed in terms of two operators; each one is the product of *bilinear color-singlet* quark currents: the original one $\mathcal{O}_{\mathrm{A}}\sim(\overline{d}\,u)\,(\overline{c}\,b)$ in (3) and a new operator $\mathcal{O}_{\mathrm{B}}\sim(\overline{d}\,b)\,(\overline{c}\,u)$ emerging from QCD corrections. We have

$$\begin{aligned}&\{XY\}=\,2\,\mathcal{O}_{\mathrm{B}}-\frac{2}{3}\mathcal{O}_{\mathrm{A}}\,,\quad\text{where}\\&\mathcal{O}_{\mathrm{B}}=[\overline{d}\,\gamma^\mu(1-\gamma_5)\,b]\,[\overline{c}\,\gamma_\mu(1-\gamma_5)\,u]\,.\end{aligned}\tag{16.13}$$

The two color-singlet currents $(\overline{d}\,b)$ and $(\overline{c}\,u)$ in $\mathcal{O}_{\mathrm{B}}$ should not be confused with the flavor-changing neutral currents (FCNC) $d\leftrightarrow b$ and $c\leftrightarrow u$ which never exist because of the GIM mechanism. These apparent FCNC forms arise simply from the Fierz rearrangement. Putting together (8), (9), and (13), we get the contribution of the diagram in Fig. 16.5a:

$$I_{\mathrm{c}\leftrightarrow\mathrm{d}}=\frac{G_{\mathrm{F}}}{\sqrt{2}}\left(\frac{\alpha_{\mathrm{s}}}{4\pi}\log\frac{M_{\mathrm{W}}^2}{M^2}\right)\left(\frac{2}{3}\mathcal{O}_{\mathrm{A}}-2\mathcal{O}_{\mathrm{B}}\right)\,.\tag{16.14}$$

The second contribution $I_{c\leftrightarrow\bar{u}}$ coming from the diagram of Fig. 16.5b (with the gluon exchanged between the c and $\bar{u}$ quarks) is calculated similarly. Compared with $I_{c\leftrightarrow d}$ in (7), there are two modifications which can be seen explicitly below:

$$\mathrm{i}\left(\frac{-\mathrm{i}g}{2\sqrt{2}}\right)^2(-\mathrm{i}g_s)^2\int\frac{\mathrm{d}^4k}{(2\pi)^4}\,\frac{-\mathrm{i}}{k^2}\,\frac{-\mathrm{i}}{k^2-M_W^2}\left[\bar{d}\,\gamma^\mu(1-\gamma_5)\,\frac{\mathrm{i}(+\not{k}+m_u)}{k^2-m_u^2}\gamma^\lambda\,T^j\,u\right]$$
$$\times\left[\bar{c}\,T^j\,\gamma_\lambda\,\frac{\mathrm{i}(\not{k}+m_c)}{k^2-m_c^2}\gamma_\mu(1-\gamma_5)\,b\right]. \qquad (16.15)$$

With the outgoing $\bar{u}$ replacing the outgoing d quark, from (7) to (15) we note an interchange $(+\not{k}\leftrightarrow-\not{k})$ in their propagators. Also, we cannot use (11.23) which provides the factor of 4 as in (10). Instead, we simply have

$$[\bar{d}\,\gamma^\mu\,(1-\gamma_5)\,\gamma^\rho\gamma^\lambda\,u]\;[\bar{c}\,\gamma_\lambda\,\gamma_\rho\,\gamma_\mu(1-\gamma_5)\,b]=\mathcal{O}_A\,.$$

The result is $I_{c\leftrightarrow\bar{u}}=-\frac{1}{4}\,I_{c\leftrightarrow d}$, thus

$$I_{c\leftrightarrow d}+I_{c\leftrightarrow\bar{u}}=\frac{3}{4}I_{c\leftrightarrow d}=\frac{G_F}{\sqrt{2}}\left(\frac{\alpha_s}{4\pi}\log\frac{M_W^2}{M^2}\right)\left(\frac{1}{2}\mathcal{O}_A-\frac{3}{2}\mathcal{O}_B\right). \qquad (16.16)$$

The two remaining contributions $I_{b\leftrightarrow\bar{u}}$ and $I_{b\leftrightarrow d}$ are computed exactly as in (7) and (15). They are respectively equal to $I_{c\leftrightarrow d}$ and $I_{c\leftrightarrow\bar{u}}$, such that the total contributions (from gluon exchanged in all possible ways between the left [bc] and right [ud] vertices) are doubled. Thus

Bare Electroweak    Renormalization by QCD

$$\frac{G_F}{\sqrt{2}}\mathcal{O}_A\quad\Longrightarrow\quad\frac{G_F}{\sqrt{2}}\left(\frac{\alpha_s}{4\pi}\log\frac{M_W^2}{M^2}\right)[\mathcal{O}_A-3\mathcal{O}_B]\,. \qquad (16.17)$$

Compared with the QCD uncorrected (3), the effective operators for nonleptonic decays are modified as follows:

$$\mathcal{O}_A\;\Longrightarrow\;c_A\mathcal{O}_A=\left(1+\frac{\alpha_s}{4\pi}\log\frac{M_W^2}{M^2}\right)\mathcal{O}_A\,,$$
$$0\;\Longrightarrow\;c_B\mathcal{O}_B=-3\left(\frac{\alpha_s}{4\pi}\log\frac{M_W^2}{M^2}\right)\mathcal{O}_B\,. \qquad (16.18)$$

The detailed calculation is illustrative. Not only does QCD renormalize the original operator $\mathcal{O}_A$, it also brings in a new operator $\mathcal{O}_B$. The operator $\mathcal{O}_B$, which is absent when QCD is neglected, now emerges when gluons enter. Starting from zero, we get $-3\,[\alpha_s/4\pi](\log M_W^2/M^2)\mathcal{O}_B$.

Furthermore, there is a *logarithmic enhancement* $\log(M_W^2/M^2)$ in the corrections to the product $H^\mu H_\mu^\dagger$ that we do not find in the corrections (4) and (5) to the single current $H^\mu$. Hence the name of leading logarithm by contrast with the nonleading logarithm. This leading logarithmic enhancement concerns only the product $H^\mu H^\dagger_\mu$, i.e. only nonleptonic weak decays. The reason why we get enhancement when the gluon in Fig. 16.5 is parallel to or crosses the weak gauge boson W is simple. The logarithmic ultraviolet divergence in the vertex corrections to the single current $H^\mu$ is canceled by the same type of divergence as in the quark self-energy (Chap. 14). Whereas for the product $H^\mu H^\dagger_\mu$, the loop integrals of the diagrams in Fig. 16.5 are convergent, we do not need counterterms here, and so the logarithmic enhancement is not removed.

The method just obtained for the b→ c+ d + $\overline{\mathrm{u}}$ can be immediately extended to the nonleptonic decays of all other flavored hadrons described by $\mathrm{Q} \to \mathrm{q}_1 + \mathrm{q}_2 + \overline{\mathrm{q}}_3$, for instance s→ u+ d + $\overline{\mathrm{u}}$ and c → s+ u + $\overline{\mathrm{d}}$. The only change is in the lower limit (collectively denoted now by $\mu$) of the logarithm $\log(M^2_{\mathrm{W}}/\mu^2)$ in (17). Since the upper limit is always the W mass, the mass scale $\mu$ for the lower limit is naturally set equal to the mass of the decaying particles, such that the scale $\mu$ is $m_{\mathrm{s}}$ ($m_{\mathrm{c}}$) for strange (charm) decaying quarks. The logarithmic enhancement is more pronounced in strange than in charm (and *a fortiori* in bottom) nonleptonic decays.

Once we get the effective hadronic weak Lagrangian $H^\mu H^\dagger_\mu$ renormalized by perturbative QCD, we can examine the correction to the decay rate. In Chap. 14, we remark that because of the interference – between the weak-interaction tree amplitude of Fig. 16.1 and the QCD correction to the current $H^\mu$ (Figs. 16.2–3) – we get the $\mathcal{O}(\alpha_{\mathrm{s}})$ correction to the rate as given by (4) and (5). This interference no longer applies to the product $H^\mu H^\dagger_\mu$. Since the color matrix $T^j$ in the diagram of Fig. 16.5 is traceless, the interference between the latter diagram and the tree diagram in Fig. 16.1 vanishes, so we have to square the amplitude of (17) and get $\mathcal{O}(\alpha^2_{\mathrm{s}})$ corrections to the rate. For the product $H^\mu H^\dagger_\mu$ we gain a logarithmic enhancement but we have a smaller $\alpha^2_{\mathrm{s}}$ factor in the corrected rate.

### 16.1.3 Renormalization Group Improvement

In (17), the logarithmic enhancement $\log(M^2_{\mathrm{W}}/m^2_{\mathrm{Q}})$ for the decaying Q quark is about 6 for bottom (Q=b) and 10 for strange (Q=s) particles. The perturbative correction to the *decay amplitude* is $\sim \alpha_{\mathrm{s}} \log(M^2_{\mathrm{W}}/m^2_{\mathrm{Q}})$ which is of order 1, so higher-order corrections $[\alpha_{\mathrm{s}} \log(M^2_{\mathrm{W}}/m^2_{\mathrm{Q}})]^n$ cannot be ignored. These $\alpha^n_{\mathrm{s}}$ corrections correspond to multiple hard gluons exchanged between the two vertices. The renormalization group method discussed in Chap. 15 and illustrated by Fig. 15.10 is particularly adapted to the situation. While it is an impossible task to compute and to sum over all individual Feynman diagrams, the renormalization group equation can sum over the leading logarithmic terms to all orders of $\alpha_{\mathrm{s}}$. Let us rewrite the QCD effects on $\mathcal{O}_{\mathrm{A}}$ taken from (17):

$$\mathcal{O}_{\mathrm{A}} \Longrightarrow \mathcal{O}_{\mathrm{A}} + \left[\frac{\alpha_{\mathrm{s}}}{2\pi} \log\left(\frac{M_{\mathrm{W}}}{\mu}\right)\right] \left[\mathcal{O}_{\mathrm{A}} - 3\,\mathcal{O}_{\mathrm{B}}\right] , \tag{16.19}$$

and note that there is always a mixing between the two operators $\mathcal{O}_{\rm A}$ and $\mathcal{O}_{\rm B}$. The gluonic corrections to $\mathcal{O}_{\rm B}$ yield a similar result:

$$\mathcal{O}_{\rm B} \Longrightarrow \mathcal{O}_{\rm B} + \left[\frac{\alpha_{\rm s}}{2\pi}\log\left(\frac{M_{\rm W}}{\mu}\right)\right]\left[\mathcal{O}_{\rm B} - 3\,\mathcal{O}_{\rm A}\right] . \tag{16.20}$$

Local operators appear often in quantum loop calculations (see Chap. 11 for box and penguin diagrams), and if they have the same dimension (as in the case of $\mathcal{O}_{\rm A}$ and $\mathcal{O}_{\rm B}$ with the dimension six), they are usually mixed. We would like to find combinations of operators that are unmixed by QCD. From (19) and (20), we remark that the operators

$$\mathcal{O}_{\pm} = \tfrac{1}{2}\left[\mathcal{O}_{\rm A} \pm \mathcal{O}_{\rm B}\right] \tag{16.21}$$

satisfy this requirememt. They are form-invariant, i.e. unmixed. Using (19) and (20), we note that the bare $\mathcal{O}_{\pm}$ are *multiplicatively renormalized* by QCD with the coefficient $c_{\pm}$ :

$$\mathcal{O}_{\pm} \Longrightarrow c_{\pm}\,\mathcal{O}_{\pm}\ , \qquad \text{where} \quad c_{\pm} = 1 + d_{\pm}\left[\frac{\alpha_{\rm s}}{\pi}\log\left(\frac{M_{\rm W}}{\mu}\right)\right] , \tag{16.22}$$

with $d_{+} = -1$ and $d_{-} = +2$ .

Then the right hand side of (19) becomes $\mathcal{O}_{\rm NL} = c_{+}\mathcal{O}_{+} + c_{-}\mathcal{O}_{-}$.

We go further by summing over all $\sum_n[\alpha_{\rm s}\log(M_{\rm W}/\mu)]^n$ terms with the renormalization group equation. Like the bare and renormalized Green's functions, the operators in field theories can be defined as bare and renormalized operators through their matrix elements. Green's function associated with the operator $\mathcal{O}_{\pm}$ can be constructed from the four quark fields participating in the decay of $\mathrm{Q} \to \mathrm{q}_1 + \mathrm{q}_2 + \overline{\mathrm{q}}_3$. It is defined as

$$\langle 0\,|\,\mathcal{O}_{\pm}\, q_1\, q_2\, q_3\, Q\,|\,0\rangle\ , \tag{16.23}$$

where the creation and destruction operators of the quark fields applied to the vacuum $\langle 0|$ and $|0\rangle$ yield the matrix element of $\mathcal{O}_{\pm}$. Therefore, through their associated Green's functions, the renormalized operators also obey the Callan–Symanzik (CS) equation, which states that a change in the scale $\mu$ must be compensated by their anomalous dimensions through the running coupling constant $\alpha_{\rm s}(\mu)$, leaving the bare operators independent of $\mu$. We may define the corresponding rescaling factor $Z_{\mathcal{O}}(\mu)$ of any operator $\mathcal{O}$ similarly to the field-strength renormalization functions $Z(\mu)$ which we are familiar with in Chap. 15. Let us define

$$\mathcal{O}^{\rm bare} = Z_{\mathcal{O}}(\mu)\ \mathcal{O}^{\rm ren}\ .$$

For the problem on hand, the calculation of $Z_{\mathcal{O}_{\pm}}(\mu)$ is already performed.

To see how $Z_{\mathcal{O}_\pm}(\mu)$ is calculated, let us return again to the logarithm in (19) and note that at the scale $\mu = M_W$, QCD does not bring in any correction to the weak-interaction operators from which we start. The cutoff $\Lambda$ [or $\Gamma(2-\frac{n}{2})$] in the regularization of loop integrals is naturally replaced with $M_W$ in the case considered here. It can also be seen by putting $M_W \to \infty$ in (9), which becomes

$$-\int \frac{d^4k}{(2\pi)^4} \frac{1}{k^2} \frac{1}{k^2 - m_c^2} = \frac{-i}{16\pi^2} \frac{\Gamma(2-\frac{n}{2})}{\mu^{4-n}} ,$$

using the familiar dimensional regularization method. Comparing the above equation with (9), we note that $\log(M_W^2/\mu^2)$ is replaced with $\Gamma(2-\frac{n}{2})/\mu^{4-n}$.

From the high mass cutoff $M_W$ going down to the scale $\mu = m_Q$ of the decay process $Q \to q_1 + q_2 + \overline{q}_3$, the bare operators $\mathcal{O}_\pm$ are multiplicatively renormalized by $c_\pm$ according to (22). So the corresponding $Z_{\mathcal{O}_\pm}(\mu)$ are nothing but $c_\pm(\mu)$. According to (15.44), the anomalous dimensions $\gamma_\pm(g_s)$ which govern the evolution of the operators $\mathcal{O}_\pm(\mu)$ can be obtained by taking the derivative of the function $\log Z_{\mathcal{O}_\pm}(\mu)$. We have, from (22),

$$\gamma_\pm(g_s) \equiv \mu \frac{\partial \log Z_{\mathcal{O}_\pm}(\mu)}{\partial \mu} = \mu \frac{\partial \log c_\pm(\mu)}{\partial \mu} = - \left[ d_\pm \frac{\alpha_s(\mu)}{\pi} \right] . \tag{16.24}$$

The fact that the product $c_\pm(\mu)\, \mathcal{O}_\pm^{\text{ren}}(\mu) = \mathcal{O}_\pm^{\text{bare}}$ is independent of $\mu$ implies

$$\left( \mu \frac{d}{d\mu} + \gamma_\pm \right) \mathcal{O}_\pm^{\text{ren}}(\mu) = 0 , \quad \left( \mu \frac{d}{d\mu} - \gamma_\pm \right) c_\pm(\mu) = 0 .$$

The latter may be rewritten as

$$\frac{d}{d \log(\mu/M_W)} c_\pm(\mu) = \gamma_\pm(g_s) c_\pm(\mu) , \tag{16.25}$$

with the initial condition $c_\pm(M_W) = 1$.
This differential equation is easy to solve. When we put into (25) the explicit expression (24) of $\gamma_\pm(g_s)$ written in terms of the QCD running coupling $\alpha_s(\mu)$ as given by (15.86), i.e.

$$\alpha_s(\mu) = \frac{2\pi}{b_0 \log(\mu/\Lambda_{\overline{\text{MS}}})} , \quad \text{where } b_0 = \frac{11}{3} N_c - \frac{2}{3} N_f ,$$

then we find that the solutions of (25) are

$$\begin{aligned} c_\pm(\mu) &= \left[ \frac{\log(M_W/\Lambda_{\overline{\text{MS}}})}{\log(\mu/\Lambda_{\overline{\text{MS}}})} \right]^{2d_\pm/b_0} = \left[ \frac{\alpha_s(\mu)}{\alpha_s(M_W)} \right]^{2d_\pm/b_0} \\ &= \left[ 1 + b_0 \frac{\alpha_s(\mu)}{4\pi} \log \frac{M_W^2}{\mu^2} \right]^{2d_\pm/b_0} . \end{aligned} \tag{16.26}$$

This expression of $c_\pm(\mu)$ manifestly represents the summation over the series $[\alpha_s \log(M_W/\mu)]^n$. It is gratifying to check that for $\alpha_s \ll 1$, we recover (22):

$$\left[1 + b_0 \frac{\alpha_s(\mu)}{4\pi} \log \frac{M_W^2}{\mu^2}\right]^{2d_\pm/b_0} \underset{\alpha_s \ll 1}{\longrightarrow} 1 + d_\pm \left\{\frac{\alpha_s(\mu)}{\pi} \log \frac{M_W}{\mu}\right\} .$$

The renormalization group analyses replace the $c_\pm(\mu)$ in (22) with the new expressions in (26):

$$\begin{aligned} c_+ &= 1 - \frac{\alpha_s}{\pi} \log\left(\frac{M_W}{\mu}\right) \Longrightarrow \left[\frac{\alpha_s(\mu)}{\alpha_s(M_W)}\right]^{-2/b_0} , \\ c_- &= 1 + 2\, \frac{\alpha_s}{\pi} \log\left(\frac{M_W}{\mu}\right) \Longrightarrow \left[\frac{\alpha_s(\mu)}{\alpha_s(M_W)}\right]^{+4/b_0} . \end{aligned} \tag{16.27}$$

Renormalized by QCD according to (19), the original $\mathcal{O}_A$ in (3) now becomes the new $\mathcal{O}_{NL} = c_+\mathcal{O}_+ + c_-\mathcal{O}_-$ which constitutes the effective Lagrangian for nonleptonic decays:

$$\begin{aligned} &\mathcal{O}_A \Longrightarrow \mathcal{O}_{NL} = c_+\, \mathcal{O}_+ + c_-\, \mathcal{O}_- = c_A\mathcal{O}_A + c_B\mathcal{O}_B , \\ &\mathcal{O}_\pm = \tfrac{1}{2}\left\{[\bar{c}\gamma^\mu(1-\gamma_5)b]\,[\bar{d}\gamma_\mu(1-\gamma_5)u] \pm [\bar{c}\gamma^\mu(1-\gamma_5)u]\,[\bar{d}\gamma_\mu(1-\gamma_5)b]\right\} , \\ &c_A = \tfrac{1}{2}(c_+ + c_-) \ , \quad c_B = \tfrac{1}{2}(c_+ - c_-) . \end{aligned} \tag{16.28}$$

We note that $(c_+)^2 = 1/c_-$, and $c_+ < 1 < c_-$ for all scales $\mu$. The operator $\mathcal{O}_-$ receives an enhancement by $c_-$, while the $\mathcal{O}_+$ is suppressed by $c_+$. From (3) to (28), the important result

$$H^\mu H^\dagger_\mu \Longrightarrow \sum_n c_n \mathcal{O}_n$$

which illustrates the Wilson operator product expansion (OPE) method, is the starting point of all phenomenological analyses of inclusive as well as exclusive hadronic decays, and will be extensively used in the next sections.

### 16.1.4 The $\Delta I = \frac{1}{2}$ in Strangeness Hadronic Decays

As the first application, the general result (28) is now used to study nonleptonic decays of strange particles described by s$\rightarrow$ u + d + $\bar{\text{u}}$. Historically, the works on QCD renormalization[2] of weak interaction were motivated by this $\Delta I = \frac{1}{2}$ empirical rule (Chap. 6). We start from

$$\frac{G_F}{\sqrt{2}} V_{us}V^*_{ud}[\bar{d}\gamma^\mu(1-\gamma_5)u]\ [\bar{u}\gamma_\mu(1-\gamma_5)s] \equiv \frac{G_F}{\sqrt{2}} V_{us}V^*_{ud}\ \mathcal{O}_A^{S=1} , \tag{16.29}$$

where $\mathcal{O}_A^{S=1}$ is an equal mixture of the $I = 1/2$ and $I = 3/2$ isospin components.

[2] Gaillard, M. K. and Lee, B. W., Phys. Rev. Lett. **33** (1974) 108; Altarelli, G. and Maiani, L., Phys. Lett. **52B** (1974) 351

Indeed, the $s$ field is an isoscalar ($I = 0$) object (as all other flavored quarks), only the unflavored $u, d$ fields form an isospin doublet. Therefore the $\overline{d}\gamma^\mu(1-\gamma_5)u$ current behaves as an $I = 1$ operator (since $I_3 = 1$), whereas the other current $\overline{u}\gamma_\mu(1-\gamma_5)s$ has $I = 1/2$. The product $1 \otimes 1/2$ of the two isospin currents is a mixture of isospin $1/2 \oplus 3/2$. A priori, the nonleptonic operator $\mathcal{O}_{\mathrm{A}}^{S=1}$ described by the product of these two currents can have the component $I = 3/2$ as important as the $I = 1/2$ component; they are on the same footing.

This isospin analysis strongly disagrees with experiments. As explained in Chap. 6, the isospin $I = 1/2$ part of all nonleptonic decay amplitudes largely dominates the $I = 3/2$ component. Nonleptonic weak decays of strange particles (the mesons K as well as the hyperons $\Lambda$, $\Sigma$, $\Xi$, $\Omega^-$) regularly obey the $\Delta I = 1/2$ rule, the ratio $(A_{1/2})/(A_{3/2})$ of the decay amplitudes ranges between 15–30, i.e. the corresponding $\Delta I = 1/2$ rates are a few hundred times faster than the rates having only the $\Delta I = 3/2$ like the $\mathrm{K}^+ \to \pi^+ + \pi^0$ mode.

Let us indicate how QCD partially solves this difficult problem which has been with us since the 1950s and is still not completely understood at present. From (28) the 'tree' operator $\mathcal{O}_{\mathrm{A}}^{S=1}$ in (29) gets renormalized by QCD and becomes

$$\mathcal{O}_{\mathrm{A}}^{S=1} \Longrightarrow c_- \mathcal{O}^{(1/2)} + c_+ \mathcal{O}^{(3/2)} \ .$$

We identify $\mathcal{O}_-^S$ with $\mathcal{O}^{(1/2)}$ and $\mathcal{O}_+^S$ with $\mathcal{O}^{(3/2)}$. The subscripts 1/2 and 3/2 refer to the isospin content of these operators. Let us first show that $\mathcal{O}_-^S$ is a pure $I = 1/2$ operator. Indeed

$$\mathcal{O}_-^S = \tfrac{1}{2}\Big\{ [\overline{d}\gamma^\mu(1-\gamma_5)u]\,[\overline{u}\gamma_\mu(1-\gamma_5)s] - [\overline{u}\gamma^\mu(1-\gamma_5)u]\,[\overline{d}\gamma_\mu(1-\gamma_5)s] \Big\}$$

is antisymmetric under the interchange of $\overline{u} \leftrightarrow \overline{d}$. The antisymmetric state of these $\overline{u}$ and $\overline{d}$ fields has total isospin $I = 0$, so that the whole $\mathcal{O}_-^S$ is purely an $I = 1/2$ object. The $I = 1/2$ structure of $\mathcal{O}_-^S$ can also be recognized by using the raising $I_+$ and lowering $I_-$ isospin operators defined by

$$I_+ d = u \quad , \quad I_+ \overline{u} = -\overline{d} \quad , \quad I_- u = d \quad , \quad I_- \overline{d} = -\overline{u}\ .$$

When applying $I_+$ on $\mathcal{O}_-^S$, we find $I_+\mathcal{O}_-^S = 0$. Since $\mathcal{O}_-^S$ has $I_3 = +1/2$, this implies that the total isospin $I$ of $\mathcal{O}_-^S$ must be $1/2$, otherwise we would not get a vanishing result with the raising $I_+$ operator. On the other hand, $\mathcal{O}_+^S$ is a mixture of $I = 1/2$ and $I = 3/2$ since by applying the lowering operator $I_-$ on $\mathcal{O}_+^S$, one gets $I_-\mathcal{O}_+^S \neq 0$.

QCD renormalization of the weak operator $\mathcal{O}_A^{S=1}$ in (29) enhances the $I = 1/2$ part and suppresses the $I = 3/2$ part. At the scale $\mu = m_{\mathrm{K}}$ of K decays and using (27), the associated coefficients are $c_-(m_{\mathrm{K}}) = 2.1$ and

$c_+(m_K) = 0.7$. The $\Delta I = 1/2$ enhancement (by the coefficient $c_-$) over the $\Delta I = 3/2$ suppression (by the coefficient $c_+$) is about 3. This is encouraging but not large enough. For instance, in $K \to \pi + \pi$, the ratio $(A_{1/2})/(A_{3/2})$ is found to be $\approx 22$ (see Chap. 11).

To obtain the decay amplitude $A_I$, in addition to the coefficients $c_-$ and $c_+$, we also need the matrix elements of the associated operators $\mathcal{O}_-^S$ and $\mathcal{O}_+^S$ inserted between the hadronic states K and $\pi + \pi$. This part is determined by nonperturbative QCD dynamics of the light K and $\pi$ mesons which is likely governed by the chiral symmetry. This topic is not covered in this book. Quantitatively, it is not clear how $c_- \langle \pi\pi \,|\, \mathcal{O}_- \,|\, \mathrm{K} \rangle$ could dominate $c_+ \langle \pi\pi \,|\, \mathcal{O}_+ \,|\, \mathrm{K} \rangle$ by a factor of 22.

Finally, we mention that the penguin operator (11.93) arising from QCD corrections has also the dimension six as $\mathcal{O}_A^{S=1}$ and $\mathcal{O}_B^{S=1}$, so it can mix with them too. The penguin operator is not of a $(\mathrm{V-A}) \times (\mathrm{V-A})$ type as are $\mathcal{O}_A^{S=1}$ and $\mathcal{O}_B^{S=1}$, rather it has the structure $\mathrm{V} \times (\mathrm{V-A})$. The mixing is not as simple as the $\mathcal{O}_\pm^{S=1}$ in (21) since we cannot use the Fierz rearrangement together with (11.88). The final result[3] is that there are four additional terms related to gluonic penguin and four to electroweak penguin (diagrams with photon and $\mathrm{Z}^0$ replacing the gluon), so in total there are ten operators instead of two, $\mathcal{O}_A^{S=1}$ and $\mathcal{O}_B^{S=1}$. As emphasized in Chap. 11, the gluonic penguin is also a pure $I = 1/2$ operator. Due to the Fierz transformation $\mathrm{V} \times (\mathrm{V-A}) \sim \mathrm{S} + \mathrm{P}$ (see the Appendix), its matrix element taken between the K and $2\pi$ states may be arranged into $\langle 0 \,|\, P \,|\, K \rangle$, $\langle \pi \,|\, P \,|\, 0 \rangle$, and $\langle \pi \,|\, S \,|\, K \rangle$. These quantities are proportional to $1/m_s$ and $1/m_d$ so they can be large for light quark masses $m_s$ and $m_d$ (Problem 16.1). However, the coefficients associated with these penguin matrix elements are small, and the overall contributions may not be sufficiently large.

In brief, the $\Delta I = 1/2$ rule is still an open question and only semiquantitatively understood. Presumably because the s quark is not heavy, we are in the low-energy regime of strangeness decay, for which nonperturbative QCD effect is important and should be included. On the other hand, for heavy particles, perturbative QCD is more reliable due to asymptotic freedom.

## 16.2 Heavy Flavor Symmetry

Mesons as bound states of quarks and antiquarks can be grouped into three categories $q\bar{q}$, $Q\bar{Q}$ and $Q\bar{q}$ where q stands for light u, d, or s quarks while Q for heavy c or b (the top is too heavy to form hadronic bound states before it decays).

These bound states are characterized by a large separation of mass scales: $M_Q \approx$ few GeV and $\Lambda_{\mathrm{QCD}} \equiv \Lambda_{\overline{MS}} \approx 0.2$ GeV, or equivalently, of length scales

---

[3] Buras, A. J., Jamin, M., Lautenbacher, M. E. and Weisz, P. H., Nucl. Phys. **B370** (1992) 69; *ibid* **B375** (1992) 501. Adel, K. and Yao, Y. P., Phys. Rev. **D49** (1994) 4945.

$\lambda_Q \simeq 1/M_Q$ and $R_{\text{had}} \simeq 1/\Lambda_{\text{QCD}}$. The sizes of $q\overline{q}$ or $Q\overline{q}$ hadrons are more or less determined by $R_{\text{had}} \approx 1$ fm which sets the nonperturbative confining regime of quarks and gluons.

On the other hand, the QCD coupling $\alpha_s(M_Q)$ of heavy quarks is small due to the asymptotic freedom, implying that on length scales comparable to $\lambda_Q \ll R_{\text{had}}$, the strong interaction is similar to the electromagnetic interactions. Presumably, for this reason the quarkonium system $Q\overline{Q}$, whose size is of the order of $\lambda_Q/\alpha_s(M_Q) \ll R_{\text{had}}$, behaves like the positronium (Chap. 7).

### 16.2.1 Basic Physical Pictures

At first sight a heavy hadron, say a meson $Q\overline{q}$, seems to be more difficult to treat because its size is still determined by the long-distance confining regime $\sim R_{\text{had}}$ on the one hand, while the Compton wavelength $\lambda_Q$ of the heavy quark is much smaller on the other hand.

However, the crucial point is that the typical momenta exchanged between the light constituents ($\overline{q}$, gluons) and the heavy quark Q are only of the order $\Lambda_{\text{QCD}}$ in the confining state $Q\overline{q}$. An energetic hard probe would be required to resolve the properties of the heavy quark Q (for instance its flavors and its spin orientations). The soft momenta exchanged between the light constituents and the heavy Q can only probe distances much larger than $\lambda_Q$. Therefore, to the extent that charm and bottom are considered heavy, the light constituents of the mesons $Q\overline{q}$ are *insensitive* to the flavor (charm or bottom), mass ($M_c$ or $M_b$), and spin orientations of the heavy quark Q.

The heavy Q acts as a static source of color electric field localized at the origin, relativistic effects such as color magnetism vanish as $M_Q \to \infty$. Therefore, the spin of Q decouples, and the light quark and gluon cannot recognize the spin orientations of Q. This is completely different from the situation found in $q\overline{q}$ and $Q\overline{Q}$ systems. The irrelevant effect of $M_Q$ on the properties of $Q\overline{q}$ can be seen as follows. In the rest frame of the heavy hadron, the heavy quark Q is at rest too, and it is almost on its mass-shell. The wave function of the light constituents follows from a solution of the field equations of QCD subject to the boundary condition of the *static color source* Q. This boundary condition is independent of $M_Q$, and so is the solution of the light constituents. This is the physical picture of the heavy flavor symmetry (HFS) concerning both flavors and spins of the heavy quarks.

The situation bears some similarity with atoms where the nucleus plays the role of Q and the surrounding electrons that of $\overline{q}$. Various isotopes of a given atomic element have, to a good approximation, the same chemical properties. Since the electron is governed only by the total electric charge of the nucleus, adding some neutrons to or removing some from the nucleus may not change its chemistry. The equivalent of the isotopes are the mesons B, D, $B^*$, $D^*$, and the baryons $\Lambda_b$, $\Lambda_c$. Their light constituents cannot distinguish their heavy partners Q in the limit $M_Q \to \infty$. The spin symmetry is analogous to the degenerate hyperfine levels in atoms.

Although, for the moment, this observation still does not allow any concrete calculation, and since $M_{\rm Q}$ is not infinite, corrections to large but finite $M_{\rm Q}$ must be somehow treated, the idea nevertheless provides some interesting relations between the properties of such particles and can be experimentally checked. This is illustrated by the following examples.

**Spectroscopic Consequences of HFS.** In the limit $M_{\rm Q} \to \infty$, the spin of the heavy quark and the total angular momentum $j$ of the light degrees of freedom inside a hadron are decoupled. The mass $M_{\rm Q}$ is irrelevant, the dynamics is independent of the spin and mass of the heavy quark. Heavy flavored hadronic states can thus be characterized by the light flavor, spin, parity of the light constituents. The heavy flavor symmetry relates the properties of different bottom and charm particles, while the spin symmetry predicts that, for a fixed $j \neq 0$, there is a doublet of degenerate states with total spin $J = j \pm \frac{1}{2}$.

In general, the mass of a hadron $\rm H_Q$ containing a heavy quark Q can be written in the form

$$M_{\rm H} = M_{\rm Q} + \overline{\Lambda}_{\rm q} + \frac{\Delta m^2}{2M_{\rm Q}} + \mathcal{O}(1/M_{\rm Q}^2) \ , \tag{16.30}$$

where the parameter $\overline{\Lambda}_{\rm q}$ represents contributions arising from all terms in the effective Lagrangian that are independent of $M_{\rm Q}$, while $\Delta m^2$ originates from all terms of order $1/M_{\rm Q}$. For the moment, the details of these terms are not important, they will be determined later by the Heavy Quark Effective Theory (HQET). For the ground states $J^P = 0^-$ and $J^P = 1^-$, one can parameterize the $\Delta m^2$ in terms of the two quantities $\lambda_1$ and $\lambda_2$,

$$\Delta m^2 = \lambda_1 + 2[J(J+1) - \tfrac{3}{2}]\lambda_2 \ , \tag{16.31}$$

where $J$ is the total spin of the meson $\rm H_Q$. All parameters $\overline{\Lambda}_{\rm q}, \lambda_1, \lambda_2$ are independent of $M_{\rm Q}$, they are functions of the light constituents. So

$$M_{\rm B_s} - M_{\rm B_d} = \overline{\Lambda}_{\rm s} - \overline{\Lambda}_{\rm d} + \mathcal{O}(1/M_{\rm b}) \ , \ \ M_{\rm D_s} - M_{\rm D_d} = \overline{\Lambda}_{\rm s} - \overline{\Lambda}_{\rm d} + \mathcal{O}(1/M_{\rm c}) \ ,$$

where the value of $\overline{\Lambda}_{\rm q}$ depends on the light constituents collectively denoted by q in (30). Since both charm and bottom are considered as heavy, the mass splitting should be equal. It is confirmed by experiments:

$$M_{\rm B_s} - M_{\rm B_d} = 90 \pm 3 \ \text{MeV} \quad \text{and} \quad M_{\rm D_s} - M_{\rm D_d} = 99 \pm 1 \ \text{MeV}.$$

For the pseudoscalar and vector meson mass splitting, from (31), we get

$$M_{\rm B^*}^2 - M_{\rm B}^2 = 4\lambda_2 + \mathcal{O}(1/M_{\rm b}) \ \ \text{and} \ \ M_{\rm D^*}^2 - M_{\rm D}^2 = 4\lambda_2 + \mathcal{O}(1/M_{\rm c}) \ . \tag{16.32}$$

Again this prediction is compatible with data, which give

$$M_{\rm B^*}^2 - M_{\rm B}^2 \simeq 0.49 \ \text{GeV}^2 \ \ \text{and} \ \ M_{\rm D^*}^2 - M_{\rm D}^2 \simeq 0.55 \ \text{GeV}^2 \ . \tag{16.33}$$

### 16.2.2 Elements of Heavy Quark Effective Theory (HQET)

The Fermi theory of weak interactions represented by the right-hand side of (3) is a typical example of an effective Lagrangian. The W weak boson effect is 'integrated out' and replaced by the dimensional coupling $G_F$ and calculations can be done at low energies far below the W mass. In this effective Fermi theory, we only use $G_F$. Neither the fundamental weak coupling $g$ nor the W mass are considered in the computation. Physical quantities computed from the effective theory are the same as if they are calculated with the fundamental Lagrangian in a certain kinematic region. It is only at much higher energies that the $W^{\pm}$ and $Z^0$ effects can be felt, and the difference appears. We have met one example: the neutrino cross-section (12.43) approximates at low energies the exact results (12.42) and (12.44) derived from the fundamental Lagrangian with the full weak boson effects.

A similar calculational method may be formulated within HQET as a systematic expansion in terms of local operators with powers of $\Lambda_{QCD}/M$. The long-distance physics of several observables is described by a few parameters which can be defined and calculable in terms of the matrix elements of these operators.

The starting point is the *velocity* $v^{\mu}$ of the heavy hadron H defined by $M_H v^{\mu} = P^{\mu}$, where $M_H$ and $P^{\mu}$ are the mass and four-momentum of H containing the heavy quark Q. We note that $v^2 = v^{\mu} v_{\mu} = 1$.

To construct an effective theory in which the heavy quark mass $M$ (we drop the index Q) becomes irrelevant, we let $M$ tend to infinity while keeping fixed the four-velocity as in classical mechanics. The momentum $P^{\mu}_{Q}$ of the heavy quark may be written as

$$P^{\mu}_{Q} = M v^{\mu} + k^{\mu} \ , \tag{16.34}$$

where $k^{\mu}$ is much smaller than $M v^{\mu}$. The heavy quark is almost on-mass-shell, i.e. $P^2_Q \approx M^2$. Consider the transition of the hadron H into a new state of the same heavy quark (with velocity $v'$) and $k^{\mu}$ is the momentum transfer. The transition changes the residual momentum $\delta k \sim \Lambda_{QCD}$, but the changes in the heavy quark velocity $\delta v = v - v'$ vanish as $\Lambda_{QCD}/M \to 0$, i.e. the velocity $v^{\mu}$ is a conserved quantity.

The next step is to introduce the large component $H_v(x)$ and small component $h_v(x)$ of the quark field $Q(x)$ by

$$Q(x) = \mathrm{e}^{-\mathrm{i}Mv\cdot x} \left[ H_v(x) + h_v(x) \right] \ , \tag{16.35}$$

$$H_v(x) = \mathrm{e}^{\mathrm{i}Mv\cdot x} \mathcal{P}_+ Q(x) \ , \ \ h_v(x) = \mathrm{e}^{\mathrm{i}Mv\cdot x} \mathcal{P}_- Q(x) \ , \tag{16.36}$$

where $\mathcal{P}_{\pm}$ are the projection operators defined as

$$\mathcal{P}_{\pm} = \frac{1 \pm \not{v}}{2} \quad \text{with} \quad \not{v} = \gamma_{\mu} v^{\mu} \ . \tag{16.37}$$

The main dependence on $M$ has been factored out and contained in the exponential. These new fields satisfy $\not{v}H_v = H_v$ and $\not{v}h_v = -h_v$.

The heavy quark moving with a fixed velocity has its wave function essentially contained in $H_v$. Compared to $H_v$, the component $h_v$ is suppressed by $\Lambda_{\text{QCD}}/M$ and is absent for an on-shell heavy quark Q. Recall that in the Dirac equation (Chap. 3), there are the two-component upper and lower parts of the full four-component spinor [see (3.45)]. The large and small components are respectively the upper and lower parts of the full four-component spinor of Q in the rest frame $v^\mu = (1,0,0,0)$. The field $H_v$ destroys a heavy quark, while $h_v$ creates a heavy antiquark with the same velocity $v$.

The QCD Lagrangian quark–gluon interaction $\mathcal{L}_Q = \overline{Q}\,(\mathrm{i}\,\not{D} - M)\,Q$ is now rewritten in terms of the new large and small components $H_v$ and $h_v$:

$$\begin{aligned}\mathcal{L}_Q = \overline{H}_v(\,\mathrm{i}v\cdot D)\,H_v \;\; &- \overline{h}_v\,(\mathrm{i}v\cdot D + 2M)h_v \\ &+ \left[\overline{H}_v\,(\mathrm{i}\,\not{D}_\perp)\,h_v + \overline{h}_v\,(\mathrm{i}\,\not{D}_\perp)\,H_v\right]\;,\end{aligned} \tag{16.38}$$

where $D^\mu$ is the QCD covariant derivative, and $D^\mu_\perp = D^\mu - (v\cdot D)v^\mu$ is orthogonal to the velocity, i.e. $v\cdot D_\perp = 0$.

Clearly, $H_v$ describes massless degrees of freedom, while $h_v$ corresponds to fluctuations with mass twice the mass $M$. The $h_v$, representing the heavy degrees of freedom, will be eliminated in the construction of HQET. Using the equation of motion $(\mathrm{i}\,\not{D} - M)Q = 0$, we can re-express $h_v$ in terms of $H_v$. Indeed,

$$h_v = \left[\frac{\mathrm{i}\,\not{D}_\perp}{2M + \mathrm{i}\,v\cdot D}\right]H_v\;,$$

which shows that the small component $h_v$ is of order $1/M$. The above equation is now put back into (38), and we get a nonlocal effective Lagrangian

$$\mathcal{L}_{\text{eff}} = \overline{H}_v(\,\mathrm{i}v\cdot D)\,H_v + \overline{H}_v(\,\mathrm{i}\,\not{D}_\perp)\,\frac{1}{2M + \mathrm{i}\,v\cdot D}\,(\,\mathrm{i}\,\not{D}_\perp)\,H_v\;.$$

Because of the phase factor in (36), the $x$ dependence of the effective field $H_v$ is weak. In momentum space, derivatives acting on $H_v$ represent powers of the residual momentum $k^\mu$ which are smaller than $M$. Therefore, the nonlocal Lagrangian can be expanded in powers of $\not{D}/M$:

$$\mathcal{L}_{\text{eff}} = \overline{H}_v(\,\mathrm{i}v\cdot D)\,H_v + \frac{1}{2M}\sum_{n=0}^{\infty}\overline{H}_v(\,\mathrm{i}\,\not{D}_\perp)\left[-\frac{\mathrm{i}\,v\cdot D}{2M}\right]^n(\,\mathrm{i}\,\not{D}_\perp)\,H_v\;.$$

Since $\mathcal{P}_+(\mathrm{i}\,\not{D}_\perp)(\mathrm{i}\,\not{D}_\perp)\,\mathcal{P}_+ = \mathcal{P}_+\left\{(\mathrm{i}\,\not{D}_\perp)^2 + \frac{g_\mathrm{s}}{2}\sigma_{\mu\nu}G^{\mu\nu}\right\}\mathcal{P}_+\;,$

with $[\mathrm{i}D^\mu, \mathrm{i}D^\nu] = \mathrm{i}g_\mathrm{s}G^{\mu\nu}$ is the gluon field-strength tensor, one finds

$$\mathcal{L}_{\text{eff}} = \overline{H}_v(\,\mathrm{i}v\cdot D)\,H_v + \frac{1}{2M}\overline{H}_v(\,\mathrm{i}\,\not{D}_\perp)^2\,H_v + \frac{g_\mathrm{s}}{4M}\,\overline{H}_v\,(\sigma_{\mu\nu}G^{\mu\nu})\,H_v + \cdots\;, \tag{16.39}$$

where the dots represents $\mathcal{O}(1/M^2)$. The first term

$$\overline{H}_v(\,\mathrm{i}v\cdot D)\,H_v \tag{16.40}$$

clearly indicates that the strong interaction of a heavy quark is independent of its mass and spin. Since the Dirac matrices are absent, interactions of the heavy quark with gluons leave its spin unchanged. If there are $N_{\mathrm{H}}$ heavy quarks moving with the same velocity $v$, then (40) is generalized to

$$\sum_{j=1}^{N_{\mathrm{H}}} \overline{H}^j_v(\mathrm{i}v\cdot D)\,H^j_v\;.$$

Its invariance under the rotation in flavor space is manifest. When combined with the spin symmetry, we get the SU($2N_{\mathrm{H}}$) group which represents the spin and heavy flavor symmetries. Corrections to the spectator model are provided by the two other terms in the $1/M$ expansion (39).

The second operator

$$\mathcal{O}_{\mathrm{kin}} = \frac{1}{2M}\overline{H}_v(\,\mathrm{i}\,\not{D}_\perp)^2\,H_v \longrightarrow -\frac{1}{2M}\overline{H}_v(\mathrm{i}\mathbf{D})^2 H_v$$

is the gauge-covariant generalization of the kinetic energy arising from the residual motion of the nearly on-mass-shell heavy quark Q. In the rest frame of the heavy particle H, $(\mathrm{i}\mathbf{D})^2 = (\mathrm{i}v_\mu D^\mu)^2 - (\mathrm{i}D_\mu)(\mathrm{i}D^\mu)$ is the square of the operator representing the spatial momentum of the heavy quark.

Analogous to the Pauli magnetic interaction, the third operator describes the chromomagnetic coupling of the heavy quark spin to the gluon field

$$\mathcal{O}_{\mathrm{mag}} = \frac{g_{\mathrm{s}}}{4M}\,\overline{H}_v(\sigma_{\mu\nu}G^{\mu\nu})\,H_v \longrightarrow -\frac{g_{\mathrm{s}}}{M}\,\overline{H}_v\,(\mathbf{S}\cdot\mathbf{B}_{gl})\,H_v\;,$$

where $\mathbf{S} = \frac{1}{2}\boldsymbol{\sigma}$ is the spin operator ($\sigma^i$ are the three Pauli matrices), and the space components of the chromomagnetic field $\mathbf{B}_{\mathrm{gl}}$ are $\mathrm{B}^i_{\mathrm{gl}} = -\frac{1}{2}\epsilon^{ijk}G_{jk}$. The three terms in (39) constitute the basis for the computation of decay rates, in particular of the inclusive widths of heavy flavored hadrons. In the limit $M\to\infty$, only (40) remains and represents HQET.

## 16.3 Inclusive Decays

As the first application of HQET, let us calculate the two typical inclusive decays of a heavy particle, say of the $\overline{\mathrm{B}}$ meson taken as an example: the semileptonic $\Gamma_{\mathrm{SL}}$ and the nonleptonic $\Gamma_{\mathrm{NL}}$ widths

$$\Gamma_{\mathrm{SL}} \equiv \Gamma(\overline{\mathrm{B}}\to\ell^- + \overline{\nu}_\ell + \mathrm{hadrons}) \quad \text{and} \quad \Gamma_{\mathrm{NL}} \equiv \Gamma(\overline{\mathrm{B}}\to\mathrm{hadrons})\;.$$

By definition, hadrons are not identified in these inclusive modes. Neglecting some rare decay modes like $\overline{\mathrm{B}}\to\gamma+\overline{\mathrm{K}}^*$ (coming from higher-order penguin loop diagram $\mathrm{b} \to \mathrm{s} + \gamma$), the sum of these inclusive decays saturates the total width or the inverse of the $\overline{\mathrm{B}}$ meson lifetime. The latter by definition describes the most inclusive decay one can imagine, since none of the particles in the final state is observed. Inclusive semileptonic decays are relatively easy to measure, only the charged lepton $\ell^-$ is identified in the decay product. The inclusive nonleptonic rate may be obtained by subtraction from the total width the sum over $\ell^- = \mathrm{e}^-, \mu^-$, and $\tau^-$ of the inclusive $\Gamma_{\mathrm{SL}}$.

From the theoretical viewpoint, inclusive decays of heavy hadrons have two unique features. First, bound-state effects related to the decaying $\overline{\mathrm{B}}$ such as the motion of the heavy b quark inside the initial state can be systematically accounted for by using the $1/M$ expansion. The leading term describes the free b quark decay, i.e. the spectator parton model. Secondly, since the final states contain so many hadrons in all possible channels, the bound-state effects of individual decay products may be eliminated (more exactly averaged out). This situation of course does not hold for light hadrons which have a very limited number of decay products. The second property is based on a concept called quark–hadron duality, a typical example of which is the inclusive semileptonic decays of the heavy lepton $\tau$ discussed in Chap. 13.

According to the quark–hadron duality, the inclusive decay rate is calculable by QCD, i.e. at the quark and gluon level, after a smearing (or averaging) procedure has been applied. In semileptonic decays $\overline{\mathrm{B}} \to \ell^- + \overline{\nu}_\ell + X$, the integration over the lepton pair phase space provides a smearing over the invariant mass squared $(P_{\mathrm{B}} - q)^2$ of the hadrons $X$, where $q = p_\ell + p_\nu$. In $\tau \to \nu_\tau + X$, the integration over the $\nu_\tau$ momentum provides the smearing.

### 16.3.1 General Formalism

From the optical theorem, the inclusive decay rate of the meson $\overline{\mathrm{B}}$ may be obtained from the imaginary part of the forward transition amplitude $\overline{\mathrm{B}} \to \overline{\mathrm{B}}$ to second-order $G_{\mathrm{F}}^2$ as shown by Fig. 16.6. The inclusive width can be written as

$$\Gamma(\overline{\mathrm{B}} \to X) = \frac{1}{2M_{\mathrm{B}}} \left\{ 2\,\mathrm{Im}\, \langle \overline{\mathrm{B}} \,|\, \mathbf{T} \,|\, \overline{\mathrm{B}} \rangle \right\} . \tag{16.41}$$

The operator $\mathbf{T}$ is given by

$$\mathbf{T} = \mathrm{i} \int \mathrm{d}^4x \, T\left\{ \mathcal{L}_{\mathrm{W}}(x) \mathcal{L}_{\mathrm{W}}(0) \right\} . \tag{16.42}$$

Inserting a complete set of states $X$ in the time-ordered $T$ product, we recover the standard formula for the decay rate:

$$\Gamma(\overline{\mathrm{B}} \to X) = \frac{1}{2M_{\mathrm{B}}} \sum_X (2\pi)^4 \delta^4(P_{\mathrm{B}} - P_X) |\langle X \,|\, \mathcal{L}_{\mathrm{W}} \,|\, \overline{\mathrm{B}} \rangle|^2 . \tag{16.43}$$

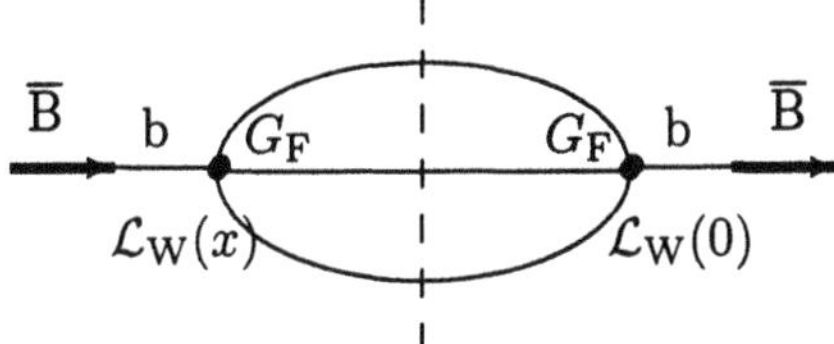


**Fig. 16.6.** Imaginary part of the forward scattering amplitude $\overline{\mathrm{B}} \to \overline{\mathrm{B}} \Longleftrightarrow$ Decay rate of $\mathrm{b} \to X \to \mathrm{b}$, to order $G_\mathrm{F}^2$. Dashed line cut gives the imaginary part

In (42), $\mathcal{L}_\mathrm{W}(x)$ is the effective weak Lagrangian built up from the quark fields and corrected by QCD as outlined in Sect. 16.1. Most importantly, the $b(x)$ quark field in $\mathcal{L}_\mathrm{W}(x)$ is governed by the HQET Lagrangian.

Because of the large b quark mass, the momenta flowing through the internal lines in the diagram of Fig. 16.6 are large, thus, following the Wilson operator product expansion method, the nonlocal operator $\mathbf{T}$ may be represented as a sum of local operators $\sum_n \mathcal{O}_n(b)$ containing the $b(x)$. The decay rate may therefore be written as

$$\Gamma(\overline{\mathrm{B}} \to X) = \frac{G_\mathrm{F}^2\, M^5}{192\pi^3} \sum_n a_n(\mathrm{b} \to X)\, \frac{1}{2M_\mathrm{B}} \langle \overline{\mathrm{B}} \,|\, \mathcal{O}_n(b) \,|\, \overline{\mathrm{B}} \rangle \ . \qquad (16.44)$$

On the right-hand side of the above equation, the first common factor $\Gamma_0 = G_\mathrm{F}^2 M^5/192\pi^3$ represents the free heavy quark b decay width as given by the spectator model symbolically depicted in Fig. 16.1. The coefficient $a_n(\mathrm{b} \to X)$ appropriately specifies the mode in which the b quark decays into a definite final state $X$, e.g. the semileptonic or the nonleptonic mode. The last factor $\frac{1}{2M_\mathrm{B}} \langle \overline{\mathrm{B}} \,|\, \mathcal{O}_n(b) \,|\, \overline{\mathrm{B}} \rangle$ parameterizes the long-distance bound-state effect of the b quark inside the $\overline{\mathrm{B}}$ meson. The $1/2M_\mathrm{B}$ coefficient is conveniently introduced for dimensional reason, as can be seen later in (53). HQET allows us to determine the last factor by an expansion in powers of $1/M$, the leading term is model independent and represented by (53).

By a simple dimensional argument, operators $\mathcal{O}_n(b)$ with higher dimension are suppressed by higher powers of $1/M$. Constructed from the b quark and gluon fields, and since $\mathcal{O}_n(b)$ is a scalar object by definition, the operator having the lowest dimension is $\bar{b}\,b$ with dimension 3. The next operators of dimension 4 are absent. The reason is simple, since the only gauge-invariant operator of dimension $d = 4$ is $\bar{b}(\mathrm{i}\, \not{D})\, b$. However, when inserted between physical quark states, this operator can be reduced to $M\,\bar{b}\,b$ using the equation of motion of the $b$ field. The next-to-leading operator of dimension $d = 5$ is the one with the gluon field $g_\mathrm{s}\bar{b}(\sigma_{\mu\nu}G^{\mu\nu})\, b$. So we have

$$\sum_n \mathcal{O}_n(b) = \bar{b}\,b + \frac{1}{M^2} g_\mathrm{s}\bar{b}(\sigma_{\mu\nu}G^{\mu\nu})\, b + \mathcal{O}\left(\frac{1}{M^3}\right) \ . \qquad (16.45)$$

We proceed in two steps to calculate the inclusive rate $\Gamma(\overline{\mathrm{B}} \to X)$ from (44). The first step concerns the coefficients $a_n(\mathrm{b} \to X)$; they are considered at the quark level. For instance, the inclusive semileptonic decays are described by $\mathrm{b} \to \mathrm{q}_1 + \ell^- + \overline{\nu}_\ell$, where $\mathrm{q}_1$ stands for the charm and the up quarks. As for the inclusive nonleptonic decays, we compute $\mathrm{b} \to \mathrm{q}_1 + \mathrm{q}_2 + \overline{\mathrm{q}}_3$, where the $\mathrm{q}_2$–$\overline{\mathrm{q}}_3$ pair stands for d–$\overline{\mathrm{u}}$ and s–$\overline{\mathrm{c}}$ (Cabibbo-favored modes due to $|V_{\mathrm{ud}}| \approx |V_{\mathrm{cs}}| \approx 1$) or s–$\overline{\mathrm{u}}$ and d–$\overline{\mathrm{c}}$ (Cabibbo-suppressed reactions, $|V_{\mathrm{us}}| \approx |V_{\mathrm{cd}}| \approx 0.22$).

The second step concerns the matrix elements of the two operators $\bar{b}\, b$ and $\bar{b}\sigma_{\mu\nu}G^{\mu\nu}\, b$ in (45) inserted between the $\overline{\mathrm{B}}$ states. The operator $\bar{b}b$ dominates over the second $\bar{b}\sigma_{\mu\nu}G^{\mu\nu}\, b$ by $1/M^2$, so let us first concentrate on the former. Using (39), we develop the $\bar{b}\ \not{v} b$ as another series in inverse powers of $M$:

$$\bar{b}\ \not{v} b = \bar{b}\, b + \frac{1}{2M^2}\bar{b}\left[(\mathrm{i}v\cdot D)^2 - (\mathrm{i}D)^2\right] b - \frac{1}{2M^2}\bar{b}\left[\tfrac{1}{2}g_{\mathrm{s}}\sigma_{\mu\nu}G^{\mu\nu}\right] b + \cdots\ , \tag{16.46}$$

where the dots represent the $\mathcal{O}(1/M^3)$ terms. The important thing is that the matrix element of the left-hand side of (46) inserted between the physical $\overline{\mathrm{B}}$ meson states is definitely known. Actually, the $\bar{b}\gamma^\mu b$ is the Noether current for the global bottom quantum number, its expectation value is thus determined by the bottom flavor content of the $\overline{\mathrm{B}}$ meson and must be 1. This can be seen as follows.

The starting point is the elastic transition of a $\overline{\mathrm{B}}$ meson of velocity $v$ into another $\overline{\mathrm{B}}$ meson of velocity $v'$ induced by the vector current $\bar{b}\gamma^\mu b$ of the HQET $b(x)$ field. The most general form of this transition is

$$\frac{1}{M_{\mathrm{B}}}\langle\overline{\mathrm{B}}(v')\,|\,\bar{b}\gamma^\mu b\,|\,\overline{\mathrm{B}}(v)\rangle = \xi(v\cdot v')\ (v+v')^\mu\ , \tag{16.47}$$

where the dimensionless quantity $\xi(v\cdot v')$ is called the Isgur–Wise function. The factor $1/M_{\mathrm{B}}$ on the left-hand side compensates for the dimensional dependence on $(\text{mass})^{-1}$ of the state $|\overline{\mathrm{B}}\rangle$ in the conventional one-particle state normalization

$$\langle\overline{\mathrm{B}}(p')\,|\overline{\mathrm{B}}(p)\,\rangle = 2M_{\mathrm{B}}\ v^0\ (2\pi)^3\ \delta^3(\boldsymbol{p}-\boldsymbol{p}')\ . \tag{16.48}$$

There is no term proportional to $(v-v')^\mu$ on the right-hand side of (47), as can be seen by contracting its left-hand side with $(v-v')^\mu$ and noting that $\not{v} b_v = b_v$ and $\bar{b}_{v'}\ \not{v}' = \bar{b}_{v'}$.

The form (47) is familiar, we have met it in (10.11) with the electromagnetic form factor $F_\pi(q^2)$ of the $\pi$ meson which has an identical Lorentz structure. Like (10.11), the matrix element of the vector current $\bar{b}\gamma^\mu b$ – inserted between the $\overline{\mathrm{B}}$ meson states – defines the $\overline{\mathrm{B}}$ meson electromagnetic form factor $F_{\mathrm{B}}(q^2)$:

$$\langle\overline{\mathrm{B}}(v')\,|\,\bar{b}\gamma^\mu b\,|\,\overline{\mathrm{B}}(v)\rangle = F_{\mathrm{B}}(q^2)(p+p')^\mu\ , \tag{16.49}$$

where $p^\mu = M_{\mathrm{B}}v^\mu$, $p'^\mu = M_{\mathrm{B}}v'^\mu$. Comparing (49) with (47), one gets

$$F_{\mathrm{B}}(q^2) = \xi(v\cdot v')\,,\ \text{where}\ q^2 = (p-p')^2 = 2m_{\mathrm{B}}^2(1-v\cdot v')\ . \tag{16.50}$$

As emphasized in (10.11), the form factor is normalized to unity at $q^2 = 0$ since the current is conserved. This normalization at, and only at $q^2 = 0$, is not modified by strong or electroweak interaction. This is the physical reason for the charge of the electron to be the same as that of the charged $\overline{\mathrm{B}}$ meson (or the antiproton) for instance. Although they are governed by completely different interactions, their electric charges are identical, i.e. their electromagnetic form factors at $q^2 = 0$ are all equal to unity:

$$F_{\mathrm{B}}(0) = F_{\mathrm{D}}(0) = F_{\mathrm{K}}(0) = F_{\pi}(0) = F_1^{\mathrm{p}}(0) = 1 \,. \tag{16.51}$$

From (50) and (51), we get

$$\xi(\omega = 1) = 1 \,, \quad \text{where } \omega \equiv v \cdot v' \,. \tag{16.52}$$

The model-independent normalization $\xi(1) = 1$ turns out to be very useful for the determination of many weak-interaction form factors involved in heavy flavor physics, as we will see in the next section.

Going back to (46), we multiply (47) by $v^\mu$, then using (52), we get for the last factor of (44),

$$\frac{1}{2M_{\mathrm{B}}} \langle \overline{\mathrm{B}} | \bar{b} b | \overline{\mathrm{B}} \rangle = 1 - \frac{1}{2M^2}(\lambda_1 - 3\lambda_2) + \mathcal{O}(\frac{1}{M^3}) \,, \tag{16.53}$$

where

$$\begin{aligned} \lambda_1 &= \frac{1}{2M_{\mathrm{B}}} \langle \overline{\mathrm{B}} | \bar{b} \, (\mathrm{i}\mathbf{D})^2 \, b | \overline{\mathrm{B}} \rangle \;, \quad (\mathrm{i}\mathbf{D})^2 = (\mathrm{i}v \cdot D)^2 - (\mathrm{i}D)^2 \,, \\ 3\,\lambda_2 &= \frac{1}{2M_{\mathrm{B}}} \left\langle \overline{\mathrm{B}} \middle| \bar{b} \, \frac{g_s}{2} \sigma_{\mu\nu} G^{\mu\nu} \, b \middle| \overline{\mathrm{B}} \right\rangle \,. \end{aligned} \tag{16.54}$$

Equation (53) is remarkable in that the normalization (the term 1 on the right-hand side) is not only model-independent and unambiguously determined, but the $1/M$ correction is also absent because there is no dimensional $d = 4$ operators in (45).

Due to HQET, the spectator model receives a solid justification.[4] According to (44), (45), and (53), the decay width $\Gamma(\overline{\mathrm{B}} \to X)$ is given by that of the free b quark decay $\Gamma(\mathrm{b} \to X)$. Nonperturbative corrections due to bound-state effects start only at the $1/M^2$ order associated with the parameters $\lambda_1$ and $\lambda_2$. Their physical meaning may be seen as follows. In the rest frame, the expectation value of the operator $(\mathrm{i}\mathbf{D})^2$ is $< \boldsymbol{k}^2 >$, i.e. $\lambda_1$ represents the spatial momentum of the b quark inside the $\overline{\mathrm{B}}$ . Its contribution is the field theory analog of the Lorentz contraction factor $\sqrt{1 - \boldsymbol{v}_b^2} \approx 1 - (\boldsymbol{k}^2/2M^2)$ in accordance with the increase of the lifetime of a moving particle due to time dilatation. It cannot be computed by first principle, but only by some phenomenological analyses. We may estimate $\lambda_1$ to be $\simeq 0.3\,\mathrm{GeV}^2$ using the mass formulas (30) and (31). As for $\lambda_2$, from (32) we may take

$$\lambda_2 = \frac{1}{4}(M_{\mathrm{B}^*}^2 - M_{\mathrm{B}}^2) \approx 0.12\,\mathrm{GeV}^2 \,.$$

[4] Bigi, I., Blok, B., Shifman, M., Uraltsev, N. and Vainshtein, A., in *B Decays* (ed. Stone, S.). World Scientific, Singapore 1994

### 16.3.2 Inclusive Semileptonic Decay: $\overline{\mathrm{B}} \to \mathrm{e}^- + \overline{\nu}_\mathrm{e} + X_\mathrm{c}$

Experiments indicate that the process $\overline{\mathrm{B}} \to$ charmed hadrons dominates $\overline{\mathrm{B}} \to$ unflavored hadrons, in particular the inclusive semileptonic decay $\overline{\mathrm{B}} \to$ charm $+ \mathrm{e}^- + \overline{\nu}_\mathrm{e}$ largely exceeds $\overline{\mathrm{B}} \to$ charmless $+ \mathrm{e}^- + \overline{\nu}_\mathrm{e}$. This implies $V_\mathrm{cb} \gg V_\mathrm{ub}$ for the CKM matrix.

As the first application of the general formalism just studied, we consider the dominant mode $\overline{\mathrm{B}} \to \mathrm{e}^- + \overline{\nu}_\mathrm{e} + X_\mathrm{c}$, where $X_\mathrm{c}$ represents the sum over all charmed hadrons. According to (44) and (53), the process is reliably described by $\mathrm{b} \to \mathrm{c} + \mathrm{e}^- + \overline{\nu}_\mathrm{e}$, and improved by corrections starting at $1/M^2$. This decay is important, it enables an accurate determination of $V_\mathrm{cb}$ and probes the dynamics of heavy flavors in the cleanest possible conditions, providing a testing ground for QCD to interplay quantitatively with weak interaction.

First, at the electroweak level uncorrected by QCD, $\Gamma(\mathrm{b} \to \mathrm{c} + \ell^- + \overline{\nu}_\ell)$ can be directly obtained from Chap. 13 where the rate of a fermion decaying into three fermions is given by (13.21), (13.62), or (13.63) accordingly. Thus,

$$\Gamma(\mathrm{b} \to \mathrm{c}+\mathrm{e}^- +\overline{\nu}_\mathrm{e}) = \frac{G_\mathrm{F}^2 M^5}{192\pi^3}|V_\mathrm{cb}|^2 f\left(\frac{m_\mathrm{c}^2}{M^2}\right) \equiv \Gamma_0 |V_\mathrm{cb}|^2 f\left(\frac{m_\mathrm{c}^2}{M^2}\right) , \quad (16.55)$$

with $f(x) = 1 - 8x + 8x^3 - x^4 - 12x^2 \log x$ taken from (13.22). We have neglected the electron and the neutrino masses. For the decay $\mathrm{b} \to \mathrm{c}+\tau^- +\overline{\nu}_\tau$ representing $\overline{\mathrm{B}} \to \tau^- + \overline{\nu}_\tau + X_\mathrm{c}$, the $\tau$ lepton mass cannot be neglected, the function $f(x)$ is replaced with the function $G(x,y)$ given by (13.62) where $x = m_\mathrm{c}^2/M^2, y = m_\tau^2/M^2$.

Next we include QCD corrections to $\mathrm{b} \to \mathrm{c} + \mathrm{e}^- + \overline{\nu}_\mathrm{e}$ through the left vertex [bc] as shown in Fig. 16.3 and (5), so that (55) can be improved:

$$\Gamma(\mathrm{b} \to \mathrm{c} + \mathrm{e}^- + \overline{\nu}_\mathrm{e}) = \Gamma_0|V_\mathrm{cb}|^2 \left[f\left(\frac{m_\mathrm{c}^2}{M^2}\right) - \frac{\alpha_\mathrm{s}}{\pi} g\left(\frac{m_\mathrm{c}^2}{M^2}\right)\right] , \quad (16.56)$$

where $g(0) = \frac{2}{3}\left[\pi^2 - \frac{25}{4}\right] \approx 2.41$ is taken from (5).
The analytic expression of the function $g(x)$ is also known,[5] it decreases with increasing $x$, i.e. $g(x) < g(0)$. Nonperturbative bound-state effects (parameterized by $\lambda_1$ and $\lambda_2$) add new contributions to (56) and yield the rate $\Gamma(\overline{\mathrm{B}} \to \mathrm{e}^- + \overline{\nu}_\mathrm{e} + X_\mathrm{c})$ following (44) and (53). Thus

$$\begin{aligned}\Gamma(\overline{\mathrm{B}} \to \mathrm{e}^-\overline{\nu}_\mathrm{e} X_\mathrm{c}) =& \Gamma_0|V_\mathrm{cb}|^2 \left\{\left(1 - \frac{\lambda_1 - 3\lambda_2}{2M^2}\right)\left[f\left(\frac{m_\mathrm{c}^2}{M^2}\right) - \frac{\alpha_\mathrm{s}}{\pi} g\left(\frac{m_\mathrm{c}^2}{M^2}\right)\right]\right.\\ &\left. - \frac{6\lambda_2}{M^2}\left(1 - \frac{m_\mathrm{c}^2}{M^2}\right)^4\right\} . \end{aligned} \quad (16.57)$$

---

[5] Nir, Y., Phys. Lett. **221B** (1989) 184; $g(x)$ is obtained by integrating over $\xi$ the functions $R_\mathrm{u}^\mathrm{v}(\xi) + R_\mathrm{u}^\mathrm{b}(\xi)$ with ($\rho_2 = \rho_3 = 0$) of Ho-Kim, Q. and Pham, Xuan-Yem, Phys. Lett. **122B** (1983) 297

The first nonperturbative term in (57) is the matrix element of the operator $\bar{b}b$ as given by (53), while the last term in (57) comes from the matrix element of the nonleading $1/M^2$ operator $g_s\bar{b}\sigma_{\mu\nu}G^{\mu\nu}\, b$ in (45). It is proportional to $\lambda_2$ defined in (54), accompanied by the phase space factor $(1-x)^4 = f(x) - \frac{1}{2}x\,\mathrm{d}f(x)/\mathrm{d}x$ due to the nonzero c quark mass.

Equation (57) is important, since the width can be directly obtained using the semileptonic branching ratio $\mathrm{B}(\overline{\mathrm{B}} \to \mathrm{e}^- + \overline{\nu}_\mathrm{e} + X_\mathrm{c}) \approx (10.9 \pm 0.46)\%$ together with the lifetime $\tau_\mathrm{B} = (1.549 \pm 0.02) \times 10^{-12}$ s of the B meson: $\Gamma(\overline{\mathrm{B}} \to \mathrm{e}^- + \overline{\nu}_\mathrm{e} + X_\mathrm{c}) = \mathrm{B}(\overline{\mathrm{B}} \to \mathrm{e}^- + \overline{\nu}_\mathrm{e} + X_\mathrm{c})/\tau_\mathrm{B}$. From these experimental data, the numerical value of the left-hand side of (57) is known. As for the right-hand side, using

$$M = (4.8 \pm 0.2)\ \mathrm{GeV}\ ,\ M - m_\mathrm{c} = (3.4 \pm 0.06)\ \mathrm{GeV}\ , \tag{16.58}$$

one can extract the CKM matrix element $|V_\mathrm{cb}|$ and find

$$|V_\mathrm{cb}| = 0.04 \pm 0.004\ . \tag{16.59}$$

The main uncertainty in (59) comes from the b quark mass $M$.

The magnitude of $V_\mathrm{ub}$ involving in b→ u transition may be estimated by the electron energy spectrum in the decay b→ $\mathrm{q}_1 + \mathrm{e}^- + \overline{\nu}_\mathrm{e}$ ($\mathrm{q}_1$ = c or u). Since the electron is emitted with either c or u quark, the electron spectrum is sensitive to the relative contributions of b→ u $+\mathrm{e}^- + \overline{\nu}_\mathrm{e}$ and b→ c $+\mathrm{e}^- + \overline{\nu}_\mathrm{e}$ and hence to the ratio $V_\mathrm{ub}/V_\mathrm{cb}$. Because the c is much heavier than the u quark, at high momentum transfer $q^2 > (M - m_\mathrm{c})^2$, or equivalently at high electron energy, only the b→ u $+\mathrm{e}^- + \overline{\nu}_\mathrm{e}$ contributes. Hence $V_\mathrm{ub}$ can be extracted at the so-called endpoint electron energy spectrum $E_\mathrm{e} > M/2 = 2.4$ GeV, this region is populated only by leptons emitted in b→ u $+\ell^-\overline{\nu}_\ell$. One gets $|V_\mathrm{ub}/V_\mathrm{cb}| \approx 0.08 \pm 0.02$. In the Wolfenstein version (11.79) of the CKM matrix, the value $|V_\mathrm{ub}/V_\mathrm{cb}| = \lambda\sqrt{\rho^2 + \eta^2} \approx 0.08 \pm 0.02$ in turn implies $\sqrt{\rho^2 + \eta^2} = 0.363 \pm 0.073$, using $\lambda = 0.2205 \pm 0.0018$. Also from (11.79), $|V_\mathrm{cb}| = A\,\lambda^2$ and using (59), the parameter $A$ can also be extracted. Thus

$$A = 0.794 \pm 0.054\ ,\ \sqrt{\rho^2 + \eta^2} = 0.363 \pm 0.073\ . \tag{16.60}$$

### 16.3.3 Inclusive Nonleptonic Decay: $\overline{\mathrm{B}} \to$ Hadrons

This dominant mode (about 75% for the branching ratio) can be described by b→ $\mathrm{q}_1 + \mathrm{q}_2 + \overline{\mathrm{q}}_3$ following the discussion after (45). There are eight combinations for the three quarks $\mathrm{q}_1$, $\mathrm{q}_2$, and $\overline{\mathrm{q}}_3$ in the final state. They all contribute to the nonleptonic width and must be added up.

The decays b→ c + d + $\overline{\mathrm{u}}$ and b→ c + s + $\overline{\mathrm{c}}$ are dominant since favored by the CKM matrix elements $V_\mathrm{cb}V^*_\mathrm{ud}$ and $V_\mathrm{cb}V^*_\mathrm{cs}$ respectively, while the two others b→ u + s $+\overline{\mathrm{u}}$ and b→ u + d $+\overline{\mathrm{c}}$ are doubly suppressed by $V_\mathrm{ub}V^*_\mathrm{us}$ and

$V_{\rm ub}V^*_{\rm cd}$. The remaining modes b→ c + s + ū, b→ c + d + c̄, b→ u + d +ū and b→ u + s +c̄ are moderately suppressed at the CKM level.

For definiteness, we consider the b→ c + d + ū decay. The corresponding nonleptonic effective Lagrangian renormalized by QCD is given by (28):

$$\mathcal{L}_{\rm eff} = \frac{G_{\rm F}}{\sqrt{2}} V_{\rm cb}V^*_{\rm ud}\left[c_{\rm A}\mathcal{O}_{\rm A} + c_{\rm B}\mathcal{O}_{\rm B}\right] , \tag{16.61}$$

where $\mathcal{O}_{\rm A}$ is given by (3) and $\mathcal{O}_{\rm B}$ by (13) respectively, while

$$c_{\rm A} = \tfrac{1}{2}[c_+(M) + c_-(M)] , \quad c_{\rm B} = \tfrac{1}{2}[c_+(M) - c_-(M)] . \tag{16.62}$$

The coefficients $c_\pm(\mu)$ are given by (26). At the scale $\mu = M$, numerically we have $c_{\rm A} \approx 1.12$ and $c_{\rm B} \approx -0.28$.

The width $\Gamma(\rm b \to c + d + \bar{u}) \sim |\langle c, d, \bar{u}\,|\, c_{\rm A}\mathcal{O}_{\rm A} + c_{\rm B}\mathcal{O}_{\rm B}\,|\,b\rangle|^2$ computed from the first term $c_{\rm A}\mathcal{O}_{\rm A}$ is familiar. We have met many times in $\mathcal{O}_{\rm A}$ the $(\rm V - A) \times (V - A)$ currents involved in the decay of a fermion into three other fermions. The same remark applies to the second term $c_{\rm B}\mathcal{O}_{\rm B}$. The width obtained from these currents is given by $\Gamma_0\, I(x,y,z)$ where $I(x,y,z)$ is taken[6] from (13.63) or (13.66):

$$I(x,y,z) = 12 \int_{(x+y)^2}^{(1-z)^2} \frac{{\rm d}s}{s}\,(s - x^2 - y^2)(1 + z^2 - s)\left[\lambda(s,x^2,y^2)\lambda(1,z^2,s)\right]^{1/2} .$$

So the width – denoted by $\Gamma_{A^2\oplus B^2}$ – coming from the direct contributions of $\mathcal{O}_{\rm A}$ and $\mathcal{O}_{\rm B}$ before their interference in $|\langle c, d, \bar{u}\,|\, c_{\rm A}\mathcal{O}_{\rm A} + c_{\rm B}\mathcal{O}_{\rm B}\,|\,b\rangle|^2$ is

$$\Gamma_{A^2\oplus B^2} = \frac{G_{\rm F}^2 M^5}{192\pi^3}|V_{\rm cb}V^*_{\rm ud}|^2\, I\left(\frac{m_{\rm c}}{M}, \frac{m_{\rm d}}{M}, \frac{m_{\rm u}}{M}\right)\left[c_{\rm A}^2 + c_{\rm B}^2\right] . \tag{16.63}$$

It remains to calculate the contribution to the width coming from the interference between the two operators $\mathcal{O}_{\rm A}$ and $\mathcal{O}_{\rm B}$. Since the currents involved in these operators are different, it may be convenient to recast them in a form involving the same currents so that the interference can take place. For this purpose, let us rewrite $\mathcal{O}_{\rm B}$ as $\mathcal{O}_{\rm B} \equiv (\bar{d}_e b_e)(\bar{c}_f u_f)$ where for simplification the $\gamma^\mu(1-\gamma_5)$ is omitted in these two color-singlet currents $(\bar{d}b)$ and $(\bar{c}u)$. The color indices $e, f = 1,2,3$ are on the other hand explicitly written out, thus by Fierz rearrangement

$$\mathcal{O}_{\rm B} = (\bar{d}_e b_e)(\bar{c}_f u_f) = (\bar{d}_e u_f)(\bar{c}_f b_e) .$$

The last term on the right-hand side of the above equation can be cast into a form close to $\mathcal{O}_{\rm A}$ in order to interfere with it. Using again (11.88) or (11)

$$(\bar{d}_e u_f)(\bar{c}_f b_e) = \frac{1}{3}(\bar{d}_e u_e)(\bar{c}_g b_g) + 2\sum_j [\bar{d}\,(T^j)_{ef}\,u]\,[\bar{c}\,(T^j)_{gh}\,b] ,$$

[6] Cortes, J. L., Pham, X. Y. and Tounsi, A. Phys. Rev. **D25** (1982) 188

which immediately gives

$$\mathcal{O}_{\rm B} = \frac{1}{3}\mathcal{O}_{\rm A} + 2\sum_{j=1}^{8}[\overline{d}\,T^j\,u]\,[\overline{c}\,T^j\,b]\,. \tag{16.64}$$

Similarly, we also have

$$\mathcal{O}_{\rm A} = \frac{1}{3}\mathcal{O}_{\rm B} + 2\sum_{j=1}^{8}[\overline{d}\,T^j\,b]\,[\overline{c}\,T^j\,u]\,. \tag{16.65}$$

From (64), we see that in $\mathcal{O}_{\rm B}$ only the operator $\frac{1}{3}\mathcal{O}_{\rm A}$ can be used to interfere with $\mathcal{O}_{\rm A}$. The color-octet currents $\overline{d}\,T^j\,u$ and $\overline{c}\,T^j\,b$ cannot interfere with the color-singlet currents in $\mathcal{O}_{\rm A}$. The contribution to the $\rm b \to c + d + \overline{u}$ rate due to the interference between the two operators $c_{\rm A}\mathcal{O}_{\rm A}$ and $c_{\rm B}\mathcal{O}_{\rm B}$ in $|c_{\rm A}\mathcal{O}_{\rm A} + c_{\rm B}\mathcal{O}_{\rm B}|^2$ must come only from $\mathcal{O}_{\rm A}$ or $\mathcal{O}_{\rm B}$. Thus,

$$|(c_{\rm A} + \tfrac{1}{3}c_{\rm B})\mathcal{O}_{\rm A}|^2,\ \text{or}\ |(\tfrac{1}{3}c_{\rm A} + c_{\rm B})\mathcal{O}_{\rm B}|^2 \underset{\text{interference}}{\Longrightarrow} \tfrac{2}{3}c_{\rm A}\,c_{\rm B}|\mathcal{O}_{\rm A}|^2,\ \text{or}\ \tfrac{2}{3}c_{\rm A}\,c_{\rm B}|\mathcal{O}_{\rm B}|^2,$$

$$\Gamma_{A\otimes B} = \frac{G_{\rm F}^2\,M^5}{192\pi^3}|V_{\rm cb}V_{\rm ud}^*|^2 I\left(\frac{m_{\rm c}}{M}, \frac{m_{\rm nd}}{M}, \frac{m_{\rm u}}{M}\right)\left(\frac{2}{3}c_{\rm A}\,c_{\rm B}\right)\,. \tag{16.66}$$

Adding (63) to (66), with $c_{\rm A}^2 + c_{\rm B}^2 + \frac{2}{3}c_{\rm A}c_{\rm B} = \frac{1}{3}(2c_+^2 + c_-^2)$, the $\rm b \to c + d + \overline{u}$ rate derived from the effective Lagrangian (61) is

$$\Gamma({\rm b \to c + d + \overline{u}}) = \frac{G_{\rm F}^2\,M^5}{192\pi^3}|V_{\rm cb}V_{\rm ud}^*|^2\, I\left(\frac{m_{\rm c}}{M}, \frac{m_{\rm d}}{M}, \frac{m_{\rm u}}{M}\right)\left(\frac{2\,c_+^2 + c_-^2}{3}\right)\,. \tag{16.67}$$

A direct calculation of the width from (61) without passing by (64) is tedious, and of course one should recover (67). This general formula can be used for all $\Gamma(\rm b \to q_1 + q_2 + \overline{q}_3)$ with the appropriate changes in $\Gamma_0|V_{\rm bq_1}V_{\rm q_2q_3}^*|^2$ $I\,(m_{\rm q_1}/M, m_{\rm q_2}/M, m_{\rm q_3}/M)$ symbolically denoted by $\widetilde{\Gamma}_0$.

The factor $(2c_+^2 + c_-^2)$, which is the *leading logarithmic* correction to the decay rate, represents the summation $|\sum_n(\alpha_{\rm s}\log M/\mu)^n|^2$ by the renormalization group equation. The nonleading $\alpha_{\rm s}$ QCD corrections to the rate as given by (6) could also be added for completeness. For the decay $\rm b \to c{+}d{+}\overline{u}$, this nonleading correction to be added to (67) is $[\,-1.41\,\alpha_{\rm s}/\pi] \times \widetilde{\Gamma}_0$ if we neglect the three final quark masses. As discussed after (6), for $m_{\rm c} = 0.3\,M$ and $m_{\rm u} = m_{\rm d} = 0$, the correction is in fact $[\,-0.35\,\alpha_{\rm s}/\pi] \times \widetilde{\Gamma}_0$ which is small. However, for the decay $\rm b \to c + s + \overline{c}$, this $\alpha_{\rm s}$ correction is especially important because of the massive $\rm s{+}\overline{c}$ quarks at the right vertex as discussed previously (Fig. 16.4). Instead of $[-0.35\,\alpha_{\rm s}/\pi] \times \widetilde{\Gamma}_0$, one has $[+3.02\,\alpha_{\rm s}/\pi] \times \widetilde{\Gamma}_0$ which enhances the $\rm b \to\ c\ + s\ + \overline{c}$ rate.

To get the inclusive nonleptonic decay rate of $\overline{\mathrm{B}}$ into charmed hadrons, denoted by $\Gamma(\overline{\mathrm{B}} \to X_{\mathrm{c}})$, from $\Gamma(\mathrm{b} \to \mathrm{c} + \mathrm{d} + \overline{\mathrm{u}})$ as given by (67), the color factor $N_{\mathrm{c}} = 3$ must be included by summing over the colors of the pair d+$\overline{\mathrm{u}}$. This is to be compared with the inclusive semileptonic rate $\Gamma(\mathrm{b} \to \mathrm{c} + \ell^- + \overline{\nu}_\ell)$ in (55) where the quark pair d+$\overline{\mathrm{u}}$ replaces the lepton pair $\ell^- + \overline{\nu}_\ell$. Neglecting the u and d massses and the nonleading $\alpha_{\mathrm{s}}$ corrections, we have from (67) and $f(x) = I(x, 0, 0)$

$$\Gamma(\overline{\mathrm{B}} \to X_{\mathrm{c}}) = \left[2\,c_+^2 + c_-^2\right]\, |V_{\mathrm{cb}} V_{\mathrm{ud}}^*|^2 \, f\left(\frac{m_{\mathrm{c}}^2}{M^2}\right) \Gamma_0\,. \tag{16.68}$$

Since $2c_+^2 + c_-^2 > 3$, the leading logarithm QCD renormalization effect *always enhances* the hadronic rate. It is gratifying to recover the old formula when the pure electroweak Lagrangian is not renormalized by QCD, i.e. with $\alpha_{\mathrm{s}} = 0$, $c_{\mathrm{A}} = 1$, and $c_{\mathrm{B}} = 0$ ($c_+ = c_- = 1$), thus

$$\left[2\,c_+^2 + c_-^2\right] \longrightarrow 3 \text{ and } \Gamma(\overline{\mathrm{B}} \to X_{\mathrm{c}}) \longrightarrow 3\,|V_{\mathrm{cb}} V_{\mathrm{ud}}^*|^2 \, f\left(\frac{m_{\mathrm{c}}^2}{M^2}\right) \Gamma_0\,, \tag{16.69}$$

which is three times the inclusive semileptonic rate $\Gamma(\overline{\mathrm{B}} \to X_{\mathrm{c}} + \mathrm{e}^- + \overline{\nu}_{\mathrm{e}})$ naively expected from counting the color number $N_{\mathrm{c}} = 3$.

Finally, the nonperturbative b quark effect inside the B meson can be appropriately improved by the factor $\left[1 - (\lambda_1 - 3\lambda_2)/(2M^2)\right]$ corresponding to the leading term of the $\bar{b}b$ operator, as in (57). We have

$$\Gamma(\overline{\mathrm{B}} \to X_{\mathrm{c}}) = \left[2\,c_+^2 + c_-^2\right]\, |V_{\mathrm{cb}} V_{\mathrm{ud}}^*|^2 \, f\left(\frac{m_{\mathrm{c}}^2}{M^2}\right) \left(1 - \frac{\lambda_1 - 3\lambda_2}{2M^2}\right) \Gamma_0\,. \tag{16.70}$$

The inclusive nonleptonic rate is an important quantity contributing to the total lifetime, from which various branching ratios (in particular the semileptonic branching ratio) can be derived.

## 16.4 Exclusive Decays

Being complementary to the inclusive decays, the exclusive modes are extensively investigated on both experimental and theoretical sides. Exclusive decays are particularly important for testing the dynamics of heavy flavors. The form factors involved in the bottom decaying into charm can be determined within the framework of HFS, since both b and c are heavy. Actually the most precise determination of the CKM matrix element $V_{\mathrm{cb}}$ comes from the exclusive mode $\overline{\mathrm{B}} \to \mathrm{D}^* + \mathrm{e}^- + \overline{\nu}_{\mathrm{e}}$. We first study the semileptonic decays, then discuss the two-body hadronic modes using the factorization method, which turns out to be a good approximation, as we will see.

### 16.4.1 Form Factors in $B_{\ell 3}$ Decays

There are several reasons for the semileptonic decays to play a prominent role in heavy flavor B physics. These decays are the simplest to understand theoretically through the spectator diagram shown in Fig. 16.7. Only the form factors are unknown, but they can be reliably determined from HFS. Secondly, the charge of the detected lepton identifies the flavor of the B hadron according to the $\Delta B = \Delta Q$ rule similar to the $\Delta S = \Delta Q$, i.e. an emitted negative lepton charge shows that a $\overline{\mathrm{B}}$ meson is involved, and conversely a positive lepton charge represents a B meson. Thirdly, semileptonic branching ratios are large in B decays, allowing for extensive experimental investigations. These modes are used to measure the CKM matrix elements $V_{\mathrm{cb}}$ and $V_{\mathrm{ub}}$, and the size of the $\mathrm{B}^0$–$\overline{\mathrm{B}}^0$ mixing.

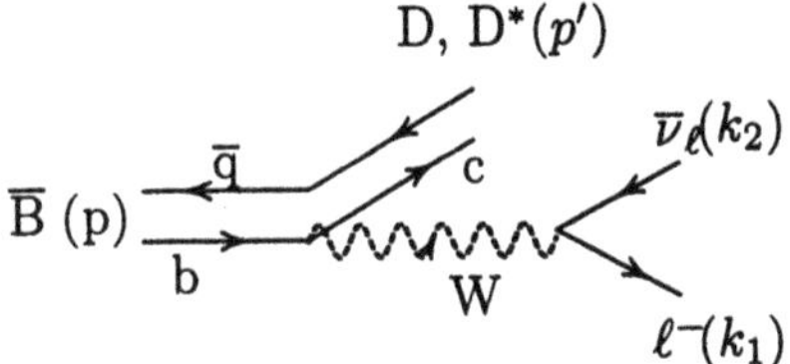


**Fig. 16.7.** $\overline{\mathrm{B}} \to \mathrm{D}\ (\mathrm{D}^*) + \ell^- + \overline{\nu}_\ell$

Like the $\mathrm{K}_{\ell 3}$, the simplest $\mathrm{B}_{\ell 3}$ decay is $\overline{\mathrm{B}}\ (\mathrm{p}) \to \mathrm{D}(p') + \ell^-(k_1) + \overline{\nu}_\ell(k_2)$. Its amplitude can be written as a product of the $\overline{\mathrm{B}} \to \mathrm{D}$ transition induced by the hadronic vector current $V^\mu = \overline{c}\gamma^\mu b$ times the leptonic current

$$\mathcal{A} = \frac{G_\mathrm{F} V_\mathrm{cb}}{\sqrt{2}} \langle \mathrm{D}(p') \,|\, V^\mu \,|\, \overline{\mathrm{B}}(p) \rangle\, \overline{u}(k_1)\gamma_\mu(1-\gamma_5)\, v(k_2)\ . \tag{16.71}$$

As explained in Sect. 10.1, the matrix element of any axial current (in particular $A^\mu = \overline{c}\gamma^\mu\gamma_5\, b$) between the two pseudoscalar mesons must vanish, hence only $V^\mu$ contributes. As discussed in (10.15), on general grounds, the $\langle \mathrm{D}(p') \,|\, V^\mu \,|\, \overline{\mathrm{B}}(p) \rangle$ matrix element is expressed in terms of the two form factors $f_\pm(q^2)$, with $q^\mu = (p-p')^\mu$ :

$$\langle \mathrm{D}(p') \,|\, V^\mu \,|\, \overline{\mathrm{B}}(p) \rangle = f_+(q^2)\,(p+p')^\mu + f_-(q^2)\,(p-p')^\mu\ . \tag{16.72}$$

The heavy flavor symmetry (between the b and c quarks) enables us to determine the normalizations of the form factors at $q^2_\mathrm{max} = (M_\mathrm{B} - M_\mathrm{D})^2$. Starting from (47) we make use of the flavor symmetry to replace the b quark in the final state meson by the c quark, thereby turning B meson into D meson. Then (47) becomes

$$\frac{1}{\sqrt{M_\mathrm{B}\, M_\mathrm{D}}} \langle \mathrm{D}(v') \,|\, \overline{c}\gamma^\mu b \,|\, \overline{\mathrm{B}}(v) \rangle = \xi(\omega)\,(v+v')^\mu\,, \tag{16.73}$$

where the same Isgur–Wise function $\xi(\omega)$ is involved. Comparing (72) with (73), we obtain

$$f_{\pm}(q^2) = \frac{M_{\rm D} \pm M_{\rm B}}{2\sqrt{M_{\rm B}\, M_{\rm D}}}\, \xi(v \cdot v') \,. \tag{16.74}$$

Thus HFS relates two independent form factors $f_{\pm}(q^2)$ to the same function $\xi(v \cdot v')$. Most importantly, with (52), the normalization $\xi(v \cdot v' = 1) = 1$ implies a nontrivial normalization of the form factors $f_{\pm}(q^2)$ at the maximum momentum transfer $q^2_{\rm max}$. Since $q^2 = M^2_{\rm B} + M^2_{\rm D} - 2\, M_{\rm B}\, M_{\rm D}\, v \cdot v'$, the zero-recoil limit $\omega = 1$ is equivalent to $q^2_{\rm max} = (M_{\rm B} - M_{\rm D})^2$, thus

$$f_{\pm}(q^2_{\rm max}) = \frac{M_{\rm D} \pm M_{\rm B}}{2\sqrt{M_{\rm B}\, M_{\rm D}}} \tag{16.75}$$

as already announced in (10.16). This model-independent result is valid in the limit of heavy b and c masses much larger than $\Lambda_{\rm QCD} \simeq 0.2$GeV.

The transition between the pseudoscalar $\overline{\rm B}(p)$ and the vector $\rm D^*(p', \epsilon)$ mesons depends on four independent form factors,[7] one with $V^{\mu}_{\rm cb} = \bar{c}\gamma^{\mu}\, b$ and three with $A^{\mu}_{\rm cb} = \bar{c}\gamma^{\mu}\gamma_5\, b$. They are

$$\begin{aligned}
\langle {\rm D}^*(p', \epsilon)\,|\,\bar{c}\gamma^{\mu}\, b\,|\,\overline{\rm B}(p)\rangle &= 2\,{\rm i}\,\epsilon^{\mu\nu\alpha\beta}\frac{\epsilon_{\nu}\, p'_{\alpha}\, p_{\beta}}{M_{\rm B} + M_{\rm D^*}}\, V(q^2)\,, \\
\langle {\rm D}^*(p', \epsilon)\,|\,\bar{c}\gamma^{\mu}\,\gamma_5\, b\,|\,\overline{\rm B}(p)\rangle &= (M_{\rm B} + M_{\rm D^*})\left[\epsilon^{\mu} - \frac{\epsilon \cdot q\, q^{\mu}}{q^2}\right] A_1(q^2) \\
&\quad - \epsilon \cdot q\left[\frac{(p+p')^{\mu}}{M_{\rm B} + M_{\rm D^*}} - \frac{(M_{\rm B} - M_{\rm D^*})\, q^{\mu}}{q^2}\right] A_2(q^2) \\
&\quad + 2M_{\rm D^*}\, \frac{\epsilon \cdot q\, q^{\mu}}{q^2}\, A_0(q^2)\,.
\end{aligned} \tag{16.76}$$

In the above equations, the three tensors associated with $V(q^2)$, $A_1(q^2)$, and $A_2(q^2)$ are constructed to be orthogonal to $q_{\mu} = (p - p')_{\mu}$ such that they vanish when multiplied by $q_{\mu} = (p - p')_{\mu}$. The linear combination

$$A_3(q^2) \equiv \frac{M_{\rm B} + M_{\rm D^*}}{2M_{\rm D^*}} A_1(q^2) - \frac{M_{\rm B} - M_{\rm D^*}}{2M_{\rm D^*}} A_2(q^2) \tag{16.77}$$

is subject to the constraint $A_3(0) = A_0(0)$ so that no pole occurs at $q^2 = 0$.

The spin symmetry leads to additional relations among these four form factors. Using this symmetry, the vector meson $\rm D^*$ with longitudinal polarization $\epsilon_3$ is related to the D meson by

$$|{\rm D}^*(v', \epsilon_3)\rangle = 2\boldsymbol{S}_3\ |{\rm D}(v')\rangle\ , \tag{16.78}$$

[7] Wirbel, M., Stech, B. and Bauer, M., Z. Phys. **C29** (1985) 637

where $S_3$ is a Hermitian operator acting on the c quark, its matrix representation is denoted by $S_3$. In a general frame, one can define a set of three orthonormal vectors $\epsilon_i$ orthogonal to $v^i$ from which the generators of the spin symmetry may be taken as

$$S_i = \frac{1}{2}\gamma_5 \not{v} \not{\epsilon}_i .$$

From (78), it follows that

$$\langle \mathrm{D}^*(v', \epsilon_3) \,|\, \gamma^\mu(1-\gamma_5) \,|\, \overline{\mathrm{B}}(v)\rangle = \langle \mathrm{D}(v') \,|\, 2\left[S_3, \gamma^\mu(1-\gamma_5)\right] \,|\, \overline{\mathrm{B}}(v)\rangle \ . \quad (16.79)$$

For the evaluation of the above commutators, it is convenient to use the rest frame of the final state D* meson

$$v'^\mu = (1,0,0,0) \ , \ \epsilon_3^\mu = (0,0,0,1) \ , \ S_3 = \frac{1}{2}\gamma_5\gamma^0\gamma_3 \ ,$$

and get

$$2\left[S_3, \gamma^0(1-\gamma_5)\right] = -\gamma^3(1-\gamma_5) \ , \ 2\left[S_3, \gamma^3(1-\gamma_5)\right] = -\gamma^0(1-\gamma_5) \ ,$$
$$2\left[S_3, \gamma^1(1-\gamma_5)\right] = -\mathrm{i}\gamma^2(1-\gamma_5) \ , \ 2\left[S_3, \gamma^2(1-\gamma_5)\right] = +\mathrm{i}\gamma^1(1-\gamma_5) \ .$$

From (79) and the above equation, one can obtain the matrix element of the $\overline{\mathrm{B}}\,(v) \to \mathrm{D}^*(v')$ transition from the $\overline{\mathrm{B}}\,(v) \to \mathrm{D}(v')$ transition, both are related to the universal function $\xi(v \cdot v')$:

$$\frac{1}{\sqrt{M_\mathrm{B} M_{\mathrm{D}^*}}} \langle \mathrm{D}^*(v', \epsilon_3) \,|\, \bar{c}\gamma^\mu\gamma_5 \, b \,|\, \overline{\mathrm{B}}(v)\rangle = \left[(1+\omega)\,\epsilon^\mu - (\epsilon \cdot v)\, v'^\mu\right] \, \xi(v \cdot v') \ ,$$
$$\frac{1}{\sqrt{M_\mathrm{B} M_{\mathrm{D}^*}}} \langle \mathrm{D}^*(v', \epsilon_3) \,|\, \bar{c}\gamma^\mu b \,|\, \overline{\mathrm{B}}(v)\rangle = \mathrm{i}\,\epsilon^{\mu\nu\alpha\beta} \epsilon_\nu \, v'_\alpha \, v_\beta \, \xi(v \cdot v') \ . \quad (16.80)$$

Comparing (80) with (76), the four form factors are now related to $\xi(\omega)$ by

$$\frac{M_\mathrm{B} + M_{\mathrm{D}^*}}{2\sqrt{M_\mathrm{B} M_{\mathrm{D}^*}}} \, \xi(\omega) = V(q^2) = A_2(q^2) = A_0(q^2)$$
$$= \left[1 - \frac{q^2}{(M_\mathrm{B} + M_{\mathrm{D}^*})^2}\right]^{-1} A_1(q^2) \ . \quad (16.81)$$

At $q^2_{\max} = (M_\mathrm{B} - M_{\mathrm{D}^*})^2$ for which $\omega = 1$, their normalizations are

$$\frac{(M_\mathrm{B} + M_{\mathrm{D}^*})^2}{4\,M_\mathrm{B}\,M_{\mathrm{D}^*}} A_1(q^2_{\max}) = V(q^2_{\max}) = A_2(q^2_{\max}) = A_0(q^2_{\max}) = \frac{M_\mathrm{B} + M_{\mathrm{D}^*}}{2\sqrt{M_\mathrm{B} M_{\mathrm{D}^*}}} \ .$$

This remarkable relation is a model-independent result, like (75). The six independent form factors in (72) and (76) are expressed in terms of the universal $\xi(\omega)$ function, normalized by $\xi(1) = 1$. On the other hand, the $q^2$ behavior of the form factors as well as the $\omega$ dependence of $\xi(\omega)$ are not determined by the heavy flavor symmetry.

### 16.4.2 Semileptonic Decay Rates

Equipped with these form factors we are now ready to compute the $\mathrm{B}_{\ell 3}$ decay rates. From the amplitude of $\overline{\mathrm{B}}\,(p) \to \mathrm{D}(p') + \ell^-(k_1) + \overline{\nu}_\ell(k_2)$ in (71), the rate is given according to the general formulas (4.55) and (4.70) by

$$\mathrm{d}\Gamma = \frac{1}{2M_\mathrm{B}\,(2\pi)^5}\frac{\mathrm{d}^3p'}{2E'}\frac{\mathrm{d}^3k_1}{2E_1}\frac{\mathrm{d}^3k_2}{2E_2}\sum_{\text{spins}}|\mathcal{A}|^2\,\delta^4(q-k_1-k_2)\,, \tag{16.82}$$

where $q = p - p'$. Using (74), we have

$$\mathcal{A} = \frac{G_\mathrm{F}\,V_\mathrm{cb}}{\sqrt{2}}\frac{M_\mathrm{B}+M_\mathrm{D}}{\sqrt{M_\mathrm{B}M_\mathrm{D}}}\,\xi(\omega)\Big[\overline{u}(k_1)\;\not{p}\,(1-\gamma_5)v(k_2) - \frac{m_\ell M_\mathrm{B}}{M_\mathrm{B}+M_\mathrm{D}}\overline{u}(k_1)(1-\gamma_5)v(k_2)\Big]\,; \tag{16.83}$$

the last term proportional to the lepton mass $m_\ell$ may be neglected for electron or muon, in which case

$$\sum_{\text{spins}}|\mathcal{A}|^2 = \Big[16\,G_\mathrm{F}^2|V_\mathrm{cb}|^2|f_+(q^2)|^2\Big]\Big[2(p\cdot k_1)\,(p\cdot k_2) - M_\mathrm{B}^2(k_1\cdot k_2)\Big]\,.$$

In the rest frame of the decaying $\overline{\mathrm{B}}$ , $q^2 = M_\mathrm{B}^2 + M_\mathrm{D}^2 - 2M_\mathrm{B}E'$, we have

$$\frac{\mathrm{d}^3p'}{2E'} = 2\pi|\mathbf{p}'|\mathrm{d}E'\;,\quad |\mathbf{p}'| = \frac{\sqrt{\lambda(M_\mathrm{B}^2, M_\mathrm{D}^2, q^2)}}{2M_\mathrm{B}} = M_\mathrm{D}\sqrt{\omega^2-1}\;,\quad E' = \omega\,M_\mathrm{D}\,.$$

Thus

$$\mathrm{d}\Gamma = \frac{2\pi|\mathbf{p}'|\mathrm{d}E'}{2M_\mathrm{B}\,(2\pi)^5}\Big[16\,G_\mathrm{F}^2|V_\mathrm{cb}|^2|f_+(q^2)|^2\Big]\Big(2p_\mu p_\nu - g_{\mu\nu}M_\mathrm{B}^2\Big)\,I^{\mu\nu}\,, \tag{16.84}$$

where $I^{\mu\nu}$ is the phase space integration of the lepton pair $\sim \int \mathrm{d}^3k_1\mathrm{d}^3k_2$ already given by (13.11). Since the neutrino is unobserved, we integrate first (82) over its momentum $\int \mathrm{d}^3k_2$ to obtain the double distributions of the charged lepton energy $E_\ell$ and the D meson energy $E'$, the $E'$ distribution is equivalent to the $q^2$ distribution, since $\mathrm{d}q^2 = 2M_\mathrm{B}\,\mathrm{d}E'$:

$$\frac{\mathrm{d}\Gamma}{\mathrm{d}q^2\,\mathrm{d}E_\ell} = \frac{G_\mathrm{F}^2|V_\mathrm{cb}|^2}{16\,\pi^3}\,|f_+(q^2)|^2\Big[2(M_\mathrm{B}^2 - M_\mathrm{D}^2 + q^2)\frac{E_\ell}{M_\mathrm{B}} - 4E_\ell^2 - q^2\Big]\,.$$

To obtain the $q^2$ distribution, instead of integrating the above equation over $E_\ell$, we take directly (84) and use

$$\Big(2p_\mu p_\nu - g_{\mu\nu}M_\mathrm{B}^2\Big)\,I^{\mu\nu} = \frac{\pi}{24}\lambda(M_\mathrm{B}^2, M_\mathrm{D}^2, q^2) \equiv \frac{\pi}{6}M_\mathrm{B}^2\,|\mathbf{p}'|^2\,. \tag{16.85}$$

Putting (85) into (84), the result is

$$\frac{\mathrm{d}\Gamma(\overline{\mathrm{B}} \to \mathrm{D} + \mathrm{e}^- + \overline{\nu}_\mathrm{e})}{\mathrm{d}q^2} = \frac{G_\mathrm{F}^2 |V_\mathrm{cb}|^2}{24\,\pi^3} |\boldsymbol{p}'|^3 \, |f_+(q^2)|^2 \,,$$

$$\text{or} \quad \frac{\mathrm{d}\Gamma}{\mathrm{d}\omega} = \frac{G_\mathrm{F}^2 |V_\mathrm{cb}|^2}{48\,\pi^3} (M_\mathrm{B} + M_\mathrm{D})^2 \, M_\mathrm{D}^3 \left(\sqrt{\omega^2 - 1}\right)^3 \xi^2(\omega) \,,$$

$$m_\ell^2 \leq q^2 \leq (M_\mathrm{B} - M_\mathrm{D})^2 \;, \;\; 1 \leq \omega \leq \frac{M_\mathrm{B}^2 + M_\mathrm{D}^2 - m_l^2}{2\,M_\mathrm{B}\,M_\mathrm{D}} \,. \qquad (16.86)$$

For $\overline{\mathrm{B}} \to +\mathrm{D} + \tau^- + \overline{\nu}_\tau$, the tau lepton mass $m_\ell$ cannot be neglected, then

$$\frac{\mathrm{d}\Gamma}{\mathrm{d}q^2} = \frac{G_\mathrm{F}^2 |V_\mathrm{cb}|^2}{24\,\pi^3} |\boldsymbol{p}'|^3 \, (1 - 2\rho)^2 \Big\{ (1 + \rho) |f_+(q^2)|^2 + 3\,\rho\, |f_0(q^2)|^2 \Big\} \,, \qquad (16.87)$$

where

$$\rho = \frac{m_\ell^2}{2\,q^2} \quad \text{and} \quad f_0(q^2) = \frac{(M_\mathrm{B}^2 - M_\mathrm{D}^2) f_+(q^2) + q^2 f_-(q^2)}{2 M_\mathrm{B}\, |\boldsymbol{p}'|} \,. \qquad (16.88)$$

The distribution (86) determines the $q^2$ dependence of the form factor $f_+(q^2)$, specially the $\omega$ dependence of the universal function $\xi(\omega)$. Furthermore, taking advantage of the normalization $\xi(1) = 1$, the only unknown in (86) is $V_\mathrm{cb}$, therefore once the kinematic term $\left(\sqrt{\omega^2 - 1}\right)^3$ is subtracted from the differential rate $\mathrm{d}\Gamma/\mathrm{d}\omega$ , its value at $\omega \to 1$ determines $|V_\mathrm{cb}|$.

In fact, this method is used in the decay $\overline{\mathrm{B}} \to \mathrm{D}^* + \mathrm{e}^- + \overline{\nu}_\mathrm{e}$ which has the largest branching ratio. Furthermore, with the vector meson $\mathrm{D}^*$, measurements of its helicity amplitudes by the angular distributions provide a detailed study of the form factors $V(q^2), A_1(q^2)$, and $A_2(q^2)$, and the relation (81) can be checked. The longitudinal $H_0$ and transverse $H_\pm$ helicity amplitudes of the $\mathrm{D}^*$ are related to the form factors by

$$H_0(q^2) = \frac{M_\mathrm{B} + M_{\mathrm{D}^*}}{2 M_\mathrm{D}^* \sqrt{q^2}} \left[ (M_\mathrm{B}^2 - M_{\mathrm{D}^*}^2 - q^2) A_1(q^2) - \frac{4 M_\mathrm{B}^2 |\boldsymbol{p}'|^2}{(M_\mathrm{B} + M_{\mathrm{D}^*})^2} A_2(q^2) \right],$$

$$H_\pm(q^2) = (M_\mathrm{B} + M_{\mathrm{D}^*}) \left[ A_1(q^2) \mp \frac{2 M_\mathrm{B} |\boldsymbol{p}'|}{(M_\mathrm{B} + M_{\mathrm{D}^*})^2} V(q^2) \right] \,. \qquad (16.89)$$

There are three angular distributions that could be used to separate the helicity amplitudes: the angle of the $\pi^+$ in the rest frame of the $\mathrm{D}^*$ which decays into $\mathrm{D} + \pi$, the $\mathrm{e}^-$ angle in the lepton pair rest frame and the correlation between these two decay planes. The $q^2$ distribution is found to be

$$\frac{\mathrm{d}\Gamma(\overline{\mathrm{B}} \to \mathrm{D}^* + \mathrm{e}^- + \overline{\nu}_\mathrm{e})}{\mathrm{d}q^2} = \frac{G_\mathrm{F}^2 |V_\mathrm{cb}|^2}{96\,\pi^3} \frac{q^2\, |\boldsymbol{p}'|}{M_\mathrm{B}^2} \sum_{i=0,\pm} |H_i(q^2)|^2 \,. \qquad (16.90)$$

Using (81), (89), and (90), one gets

$$\frac{\mathrm{d}\Gamma(\overline{\mathrm{B}} \to \mathrm{D}^* + \mathrm{e}^- + \overline{\nu}_\mathrm{e})}{\mathrm{d}\omega} = \frac{G_\mathrm{F}^2 |V_\mathrm{cb}|^2}{48\,\pi^3} (M_\mathrm{B} - M_{\mathrm{D}^*})^2 \, M_{\mathrm{D}^*}^3 \sqrt{\omega^2 - 1}\,(\omega + 1)^2 \times \left[1 + \frac{4\omega}{\omega + 1} \frac{q^2(\omega)}{(M_\mathrm{B} - M_{\mathrm{D}^*})^2}\right] \xi^2(\omega) \,, \tag{16.91}$$

where $q^2(\omega) = M_\mathrm{B}^2 + M_{\mathrm{D}^*}^2 - 2\,\omega\, M_{\mathrm{D}^*}\, M_\mathrm{B}$ .

In principle (91) should be corrected for the fact that the heavy quarks are not infinitely large. Fortunately, there exists a theorem[8] which states that at $\omega = 1$, the leading $1/M$ correction vanishes, such that only corrections of order $1/M^2$ are needed for the modes with a vector meson in the final state, e.g. the $\mathrm{D}^*$ considered here.

Once the kinematic terms in (91) are subtracted from the $\mathrm{d}\Gamma/\mathrm{d}\omega$ distribution, the value of the latter at $\omega = 1$ allows the determination of $V_\mathrm{cb}$ by taking advantage of $\xi(1) = 1 + \mathcal{O}(1/M^2)$. One gets[9]

$$|V_\mathrm{cb}| = 0.039 \pm 0.003 \,, \tag{16.92}$$

which is in remarkable agreement with the inclusive semileptonic decay (59).

### 16.4.3 Two-Body Hadronic Decays

Nonleptonic decays of B mesons have been extensively observed and hundreds of channels have been identified, most of them being the two-body modes like $\overline{\mathrm{B}} \to \mathrm{D} + \pi$, or quasi-two-body like $\overline{\mathrm{B}} \to \overline{\mathrm{K}}^* + \psi$. Above all, the studies of nonleptonic decays must deal with the *matrix element* of the operator product $H^\mu\, H_\mu^\dagger$ inserted between the decaying parent hadron $\mathcal{P}$ and the hadrons in the final state collectively denoted by $\mathcal{F}$. This is in contrast to the semileptonic decay modes $\mathcal{P} \to \mathcal{F} + \overline{\ell} + \nu_\ell$ for which only the single operator $H^\mu$ enters and the decay amplitude is a product of the well-determined matrix element of the leptonic current $\langle \overline{\ell} + \nu_\ell \,|\, L_\mu^\dagger \,|\, 0 \rangle = \overline{u}_\nu \gamma_\mu (1 - \gamma_5) v_\ell$ times the matrix element $\langle \mathcal{F} \,|\, H^\mu \,|\, \mathcal{P} \rangle$. The latter is expressed in terms of form factors which are theoretically more or less tractable, and experimentally measurable. This property for an amplitude to be equal to the product of the matrix elements of separate currents is called the *factorization* property.

**Factorization.** The factorization ansatz for nonleptonic decays is inspired by the semileptonic decay amplitudes which are always factorized into a product of two matrix elements of the quark current $H^\mu$ and the lepton current $L_\mu$. Quarks and leptons are separated by the W boson propagator, and gluons cannot connect them; a typical example is (71).

[8] Luke, M. E., Phys. Lett. **252B** (1990) 447; analog of the Ademollo–Gatto theorem in Phys. Rev. Lett. **13** (1964) 264

[9] Stone, S., in B Decays, World Scientific, Singapore 1994; Neubert, M., Int. J. Mod. Phys. **A11** (1996) 4173

This is not the case of nonleptonic decays, generically described by Q$\rightarrow$ $q_1 + q_2 + \overline{q}_3$, where soft gluons exchanged among quarks of the two currents $H^{\mu}_{Qq_1} = \overline{q}_1\gamma^{\mu}(1-\gamma_5)Q$ and $H^{\mu}_{q_2q_3} = \overline{q}_2\gamma^{\mu}(1-\gamma_5)q_3$ make factorization *a priori* inapplicable. For the hadronic decays $\mathcal{P} \rightarrow \mathcal{F}$, the matrix element $\langle \mathcal{F} \,|\, H^{\mu} H^{\dagger}_{\mu} \,|\, \mathcal{P} \rangle$ is involved; if we insert hadronic intermediate states between the product $H^{\mu}\, H^{\dagger}_{\mu}$, we would not know how many states to include.

However, for the two-body decays of a heavy particle, there is a physical reason supporting factorization; it may be undestood through the intuitive[10] 'color transparency argument'. Because of the large mass of the decaying particle in energetic two-body transitions, hadronization by soft gluons exchanged among the decay products does not occur until they have traveled some distance away from each other. The reason is that once the two color-singlet quarks are formed in $H^{\mu}$ and $H^{\dagger}_{\mu}$ separately, soft gluons are ineffective in rearranging them. Created in a pointlike interaction b$\rightarrow$ c + d + $\overline{\mathrm{u}}$, a fast-moving colorless $\overline{\mathrm{u}}$d quark pair leaves the interaction region in the same direction with a velocity close to the speed of light and will hadronize only after a time given by its $\gamma = [\sqrt{1-(v^2/c^2)}]^{-1}$ factor multiplied by a typical hadronization time $\tau_{\mathrm{had}} \sim 1$ fm/c. In this example, this means that hadronization occurs about 20 fm away from the remaining c +$\overline{\mathrm{q}}$ quarks ($\overline{\mathrm{q}}$ being the spectator light constituent of the $\overline{\mathrm{B}}$ meson). Inside the weak interaction region, the $\overline{\mathrm{u}}$d colorless pair behaves like a hadron and does not interact with the remaining colorless c$\overline{\mathrm{q}}$ pair which forms the second hadron.

The factorization is a working hypothesis which turns out to be a good approximation and can be tested, at least for a class of two-body reactions. Using factorization, nonleptonic two-body decays not only are the cleanest channels to study but also provide an instrument for exploring the most interesting aspect of nonperturbative QCD. Indeed, since the decaying heavy particle can be isolated and the weak transition has a well-determined structure, a detailed analysis of the decay products with different spin and flavor provides precious information about the long-range forces that govern the internal structure of hadrons. Especially interesting are the form factors and the decay constants, which describe the quark–antiquark attraction inside a hadron. As many of them are not experimentally accessible in leptonic or electromagnetic processes, nonleptonic decays provide a way to extract them.

To illustrate the factorization method, we consider the two-body modes $\overline{\mathrm{B}} \rightarrow \mathrm{D} + \pi$, $\overline{\mathrm{B}} \rightarrow \mathrm{D}^* + \pi$, and $\overline{\mathrm{B}} \rightarrow \mathrm{D} + \rho$ for both charged and neutral decaying $\overline{\mathrm{B}}$ meson. The corresponding Lagrangian which governs these decays are given by (61) and (62). The mesons in the final state can be directly generated by the quark current carrying the appropriate flavor quantum numbers. The three quarks c, d, and $\overline{\mathrm{u}}$ in the final state then combine with the light $\overline{\mathrm{q}}$ constituent of the $\overline{\mathrm{B}}$ meson to form the charmed and unflavored hadrons in the decay products, for instance $\overline{\mathrm{B}} \rightarrow \mathrm{D} + \pi$. There are essentially two ways for the spectator $\overline{\mathrm{q}}$ to combine with the other three quarks c, d, and $\overline{\mathrm{u}}$ to

[10] Bjorken, J. D., Nucl. Phys. (Proc. Suppl.) **B11** (1989) 325

generate D and $\pi$. Either D comes from the combination c+$\overline{\mathrm{q}}$ and $\pi$ from d+$\overline{\mathrm{u}}$ (Fig. 16.8a), or D comes from c+$\overline{\mathrm{u}}$, and $\pi$ from the other combination d+$\overline{\mathrm{q}}$ (Fig. 16.8b). The latter is sometimes called the color-suppressed mode, for a reason that will be clear soon.

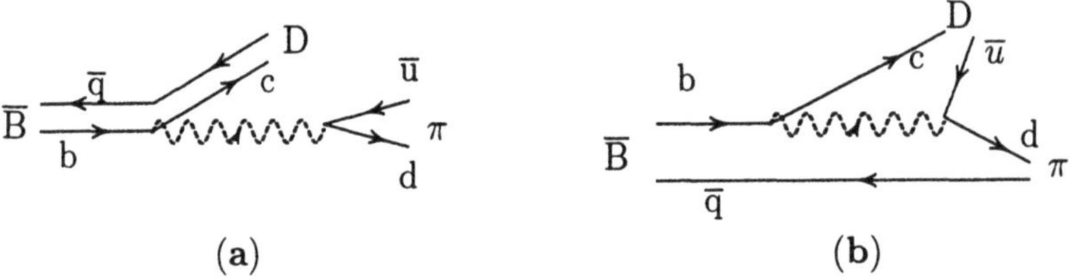


**Fig. 16.8.** Spectator picture for (**a**) $\overline{\mathrm{B}}^0 \to \mathrm{D}^+ + \pi^-$, and (**b**) $\overline{\mathrm{B}}^0 \to \mathrm{D}^0 + \pi^0$, the light constituent $\overline{\mathrm{q}} = \overline{\mathrm{d}}$ of $\overline{\mathrm{B}}^0$ is spectator

Thus, like the semileptonic formula (71) and Fig. 16.7, the amplitude factorizes into the product of the matrix elements of two color-singlet quark currents. For instance, the amplitude of $\overline{\mathrm{B}}^0(p) \to \mathrm{D}^+(p')+\pi^-(q)$ in Fig. 16.8a may be written as

$$\mathcal{M}_{\pi^-\mathrm{D}^+} = a_1 \frac{G_\mathrm{F}}{\sqrt{2}} V_\mathrm{cb} V_\mathrm{ud}^* \,\langle \pi^-(q) \,|\, \overline{d}\gamma_\mu \gamma_5 \, u \,|\, 0 \rangle \left\langle \mathrm{D}^+(p') \,|\, \overline{c}\gamma^\mu b \,|\, \overline{\mathrm{B}}^0(p) \right\rangle , \quad (16.93)$$

where the coefficient $a_1$ is defined as follows: we go back to (61), (62), and especially using (64), to write the nonleptonic effective Lagrangian as

$$\begin{aligned} \mathcal{L}_\mathrm{eff} &= \frac{G_\mathrm{F}}{\sqrt{2}} V_\mathrm{cb} V_\mathrm{ud}^* \left\{ \left( c_\mathrm{A} + \frac{1}{N_\mathrm{c}} c_\mathrm{B} \right) \mathcal{O}_\mathrm{A} + 2 \sum_j [\overline{d}\, T^j \, u][\overline{c}\, T^j \, b] \right\} \\ &= \frac{G_\mathrm{F}}{\sqrt{2}} V_\mathrm{cb} V_\mathrm{ud}^* \left[ a_1 \, \mathcal{O}_\mathrm{A} + \ldots \right] \ , \text{where } \ a_1 = c_\mathrm{A} + \frac{1}{N_\mathrm{c}} c_\mathrm{B} \ , \end{aligned} \quad (16.94)$$

the dots represent the second term with color-non-singlets $T^j$ currents.

Unlike (93), the matrix element of the second operator vanishes when we insert colorless hadrons between $T^j \cdot T^j$, e.g. $\langle \pi | \, \overline{d}\, T^j \, u \, |0\rangle \times \langle \mathrm{D} | \, \overline{c}\, T^j \, b \; |\mathrm{B}\rangle = 0$. Whereas for $\mathcal{O}_\mathrm{A}$, which is the product $[\overline{c}\gamma_\mu(1-\gamma_5)\, b] \times [\overline{d}\gamma^\mu(1-\gamma_5)\, u]$ of two color-singlet currents, the insertion of hadronic states (or vacuum) between these currents yields nonzero result as explicitly shown by (93). The $\langle \mathrm{D}\pi | \; \overline{d}\, T^j \, u \, \overline{c}\, T^j \, b \; |\mathrm{B}\rangle$ term which constitutes the nonfactorizable contribution is discarded.

Since everything in (93) is known,

$$\begin{aligned} \langle \pi^-(q) \,|\, \overline{d}\gamma_\mu \gamma_5 \, u \,|\, 0 \rangle &= \mathrm{i} f_\pi q_\mu = \mathrm{i} f_\pi (p-p')_\mu \ , \\ \left\langle \mathrm{D}^+(p') \,|\, \overline{c}\gamma^\mu b \,|\, \overline{\mathrm{B}}^0(p) \right\rangle &= (p+p')^\mu f_+(q^2) + q^\mu f_-(q^2) \ , \end{aligned}$$

we obtain

$$\mathcal{M}_{\pi^- \mathrm{D}^+} = \mathrm{i}\, a_1 \frac{G_\mathrm{F}}{\sqrt{2}} V_{\mathrm{cb}} V_{\mathrm{ud}}^* \, f_\pi \left[ (M_\mathrm{B}^2 - M_\mathrm{D}^2) f_+(m_\pi^2) + m_\pi^2 f_-(m_\pi^2) \right] \,, \quad (16.95)$$

where $f_\pm(q^2)$ are given in (74); the last term $\sim m_\pi^2$ may be neglected.
The process $\overline{\mathrm{B}}^0 \to \mathrm{D}^+ + \pi^-$ associated with $a_1\, \mathcal{O}_\mathrm{A}$ of (94) is shown in Fig. 16.8a. This $\overline{\mathrm{B}}^0 \to \mathrm{D}^+ + \pi^-$ decay is similar to the semileptonic mode $\overline{\mathrm{B}}^0 \to \mathrm{D}^+ + \ell^- + \overline{\nu}_\ell$ shown in Fig. 16.7 with the $\pi^-$ replacing the $\ell^- + \overline{\nu}_\ell$ pair. Let us call class I the modes governed by the $a_1\, \mathcal{O}_\mathrm{A}$ operator. All decay modes $\overline{\mathrm{B}}^0 \to H_\mathrm{c}^+ + h^-$, where $H_\mathrm{c}^+$ are charmed mesons ($\mathrm{D}^+$ , $\mathrm{D}^{*+}$ etc. ) and $h^-$ unflavored mesons ($\pi^-, \rho^-, \mathrm{a}_1^-$ (1260), etc. ) belong to class I.

The case of Fig. 16.8b associated with $\mathcal{O}_\mathrm{B}$ in $\mathcal{L}_\mathrm{eff}$ is new and has no direct relation with semileptonic decays; it belongs to class II. From (61) and (65), we have

$$\begin{aligned} \mathcal{L}_\mathrm{eff} &= \frac{G_\mathrm{F}}{\sqrt{2}} V_{\mathrm{cb}} V_{\mathrm{ud}}^* \left\{ \left( c_\mathrm{B} + \frac{1}{N_\mathrm{c}} c_\mathrm{A} \right) \mathcal{O}_\mathrm{B} + 2 \sum_j [\overline{d}\, T^j\, b][\overline{c}\, T^j\, u] \right\} \\ &= \frac{G_\mathrm{F}}{\sqrt{2}} V_{\mathrm{cb}} V_{\mathrm{ud}}^* \left[ a_2\, \mathcal{O}_\mathrm{B} + \cdots \right] \,, \text{where } a_2 = c_\mathrm{B} + \frac{1}{N_\mathrm{c}} c_\mathrm{A} \,. \end{aligned} \quad (16.96)$$

The class II is also called color-suppressed decay, since $c_\mathrm{B} < c_\mathrm{A}$ and in $a_2$ the dominant coefficient $c_\mathrm{A}$ is suppressed by the factor $1/N_\mathrm{c}$. All decay modes $\overline{\mathrm{B}}^0 \to H_\mathrm{c}^0 + h^0$, where $H_\mathrm{c}^0$ denotes neutral charmed mesons ($\mathrm{D}^0$ , $\mathrm{D}^{*0}$, etc. ) and $h^0$ neutral unflavored mesons ($\pi^0, \rho^0, \mathrm{a}_1^0$ (1260), etc. ), belong to class II. Also the mode $\mathrm{B} \to \mathrm{J}/\psi + \mathrm{K}$ is color-suppressed (Fig. 16.9).

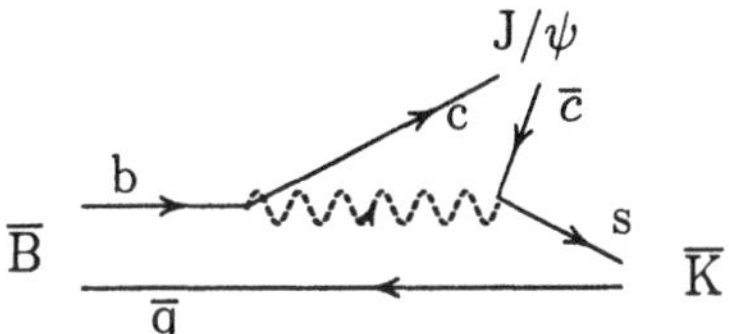


**Fig. 16.9.** Color-suppressed decay $\overline{\mathrm{B}} \to \mathrm{J}/\psi + \overline{\mathrm{K}}$

Finally, both $a_1\, \mathcal{O}_\mathrm{A}$ and $a_2\, \mathcal{O}_\mathrm{B}$ contribute to the decays $\mathrm{B}^- \to H_\mathrm{c}^0 + h^-$ which belong to classe III and illustrated by Fig. 16.10.

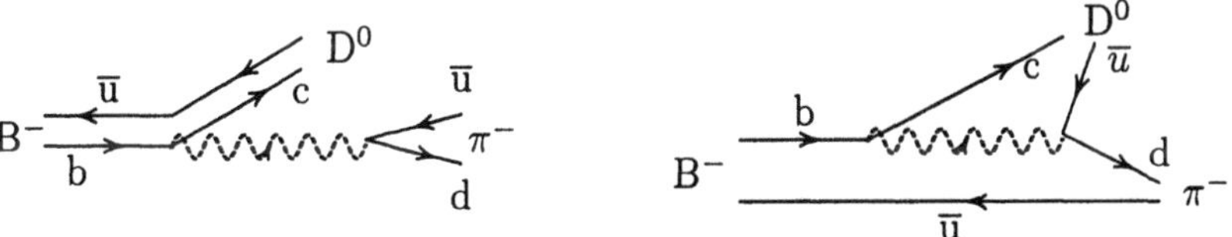


**Fig. 16.10.** Class III decay: $\mathrm{B}^- \to \mathrm{D}^0 + \pi^-$

The decay width $\Gamma(\overline{\mathrm{B}}^0 \to \mathrm{D}^+ + \pi^-)$ is directly obtained using the two-body phase space integral given in the Appendix together with (95):

$$\Gamma(\overline{\mathrm{B}}^0 \to \mathrm{D}^+ + \pi^-) = \frac{1}{2M_\mathrm{B}} \frac{1}{(2\pi)^2} \int \frac{\mathrm{d}^3 p'}{2E'} \frac{\mathrm{d}^3 q}{2E_\pi} |\mathcal{M}_{\pi^- \mathrm{D}^+}|^2 \delta^4(p - p' - q)$$
$$= \frac{a_1^2\, G_\mathrm{F}^2\, |V_\mathrm{cb} V_\mathrm{ud}^*|^2}{32\pi} M_\mathrm{B}^3 f_\pi^2 \left(1 - \frac{M_\mathrm{D}^2}{M_\mathrm{B}^2}\right)^3 |f_+(0)|^2 . \quad (16.97)$$

In (97), the pion mass is neglected. Like (93), the amplitude for $\overline{\mathrm{B}}^0(p) \to \mathrm{D}^+(p') + \rho^-(q)$ is obtained straightforwardly. Using (13.45) for the $\rho^-$ decay constant $f_{\rho^-} = \sqrt{2}\, f_{\rho^0} \approx \sqrt{2} \times 150$ MeV, one has

$$\mathcal{M}_{\rho^- \mathrm{D}^+} = a_1 \frac{G_\mathrm{F}}{\sqrt{2}} V_\mathrm{cb} V_\mathrm{ud}^*\, m_\rho\, f_{\rho^-}\ [(p + p') \cdot \epsilon_\rho]\ f_+(m_\rho^2) ,$$

$$\Gamma(\overline{\mathrm{B}}^0 \to \mathrm{D}^+ + \rho^-) = \frac{a_1^2\, G_\mathrm{F}^2\, |V_\mathrm{cb} V_\mathrm{ud}^*|^2 f_{\rho^-}^2}{32\pi M_\mathrm{B}^3} [\lambda(M_\mathrm{B}^2, M_\mathrm{D}^2, m_\rho^2)]^{\frac{3}{2}}\, |f_+^2(m_\rho^2)| . \quad (16.98)$$

For $\overline{\mathrm{B}}^0 \to \mathrm{D}^{*+} + \pi^-$, we need the $\overline{\mathrm{B}}^0 \to \mathrm{D}^{*+}$ transition form factors as defined in (76), and only the $A_0(q^2)$ contributes when (76) is multiplied by the pion four-momentum $q_\mu$ in $\langle \pi(q) | \bar{d} \gamma_\mu \gamma_5 u | 0 \rangle = \mathrm{i}\, f_\pi q_\mu$:

$$\mathcal{M}_{\pi^- \mathrm{D}^{*+}} = \mathrm{i}\, a_1 \frac{G_\mathrm{F}}{\sqrt{2}} V_\mathrm{cb} V_\mathrm{ud}^*\, f_\pi\ [2M_{\mathrm{D}^*}]\, [q \cdot \epsilon_{\mathrm{D}^*}]\, A_0(m_\pi^2) ,$$

$$\Gamma(\overline{\mathrm{B}}^0 \to \mathrm{D}^{*+} + \pi^-) = \frac{a_1^2\, G_\mathrm{F}^2\, |V_\mathrm{cb} V_\mathrm{ud}^*|^2}{32\pi} M_\mathrm{B}^3 f_\pi^2 \left(1 - \frac{M_{\mathrm{D}^*}^2}{M_\mathrm{B}^2}\right)^3 |A_0(0)|^2 , \quad (16.99)$$

where again the pion mass is neglected. In the class I decays governed by $a_1 \mathcal{O}_\mathrm{A}$, the form factors $f_+(q^2)$, $A_0(q^2)$ correspond to the 'heavy-to-heavy' $b \to c$ transitions. With the heavy flavor symmetry (HFS), they are described by the universal function $\xi(\omega)$ and given in (74) and (81).

For the class II decays with $a_2 \mathcal{O}_\mathrm{B}$, the involved form factors correspond to the 'heavy-to-light' $b \to d$ transitions. For instance, the amplitude of $\overline{\mathrm{B}}^0(p) \to \mathrm{D}^0(p') + \pi^0(q)$ is

$$\mathcal{M}_{\pi^0 \mathrm{D}^0} = a_2 \frac{G_\mathrm{F}}{\sqrt{2}} V_\mathrm{cb} V_\mathrm{ud}^*\ \langle \mathrm{D}^0(p') \,|\, \bar{c} \gamma_\mu \gamma_5\, u \,|\, 0 \rangle \left\langle \pi^0(q) \,|\, \bar{d} \gamma^\mu\, b \,|\, \overline{\mathrm{B}}^0(p) \right\rangle . \quad (16.100)$$

The first factor $\langle \mathrm{D}^0(p') \,|\, \bar{c} \gamma_\mu \gamma_5\, u \,|\, 0 \rangle$ can be related to the decay constant $f_\mathrm{D}$ defined by $\langle 0 \,|\, \bar{c} \gamma_\mu \gamma_5\, d \,|\, \mathrm{D}^+(p') \rangle = \mathrm{i} f_\mathrm{D} p'_\mu$. In principle, the latter can be extracted from $\mathrm{D}^+ \to \mu^+ + \nu_\mu$, similar to the $\pi^+ \to \mu^+ + \nu_\mu$ which gives $f_\pi$. As for the second factor $\left\langle \pi^0 \,|\, \bar{d} \gamma^\mu\, b \,|\, \overline{\mathrm{B}}^0 \right\rangle$, it can also be related to $\left\langle \pi^+ \,|\, \bar{u} \gamma^\mu\, b \,|\, \overline{\mathrm{B}}^0 \right\rangle$ by isospin rotation:

$$\left\langle \pi^0(q) \,|\, \bar{d} \gamma^\mu\, b \,|\, \overline{\mathrm{B}}^0(p) \right\rangle = \frac{1}{\sqrt{2}} \left\langle \pi^+(q) \,|\, \bar{u} \gamma^\mu\, b \,|\, \overline{\mathrm{B}}^0(p) \right\rangle$$
$$= \frac{1}{\sqrt{2}} \left\{ (p + q)^\mu \widetilde{f}_+(p'^2) + (p - q)^\mu \widetilde{f}_-(p'^2) \right\} .$$

Thus

$$\mathcal{M}_{\pi^0 D^0} = i\, a_2\, \frac{G_F}{2} V_{cb} V^*_{ud}\, f_D \left[ (M_B^2 - m_\pi^2)\widetilde{f}_+(M_D^2) + M_D^2 \widetilde{f}_-(M_D^2) \right] ,$$

from which the rate $\Gamma(\overline{B}^0 \to D^0 + \pi^0)$ is easily obtained. The heavy-to-light form factors $\widetilde{f}_\pm$ are those involved in the decay $\overline{B}^0 \to \pi^+ + e^- + \overline{\nu}_e$. This mode, represented at the quark level by $b \to u + e^- + \overline{\nu}_e$, is suppressed by $|V_{ub}|^2 \ll |V_{cb}|^2$. Unlike the 'heavy-to-heavy' case, these 'heavy-to-light' form factors cannot take advantage of HFS, its theoretical determination may be uncertain and model dependent. Nevertheless, the normalizations $\widetilde{f}_\pm(0)$ and the $q^2$ dependences of $\widetilde{f}_\pm(q^2)$ can be determined by $\overline{B}^0 \to \pi^+ + \ell^- + \overline{\nu}_l$ data, although it is a difficult experimental task.

**Tests of Factorization.** A glance at Fig. 16.7 and Fig. 16.8a together with the corresponding formulas (86) and (97) or (98) immediately gives us the following relation which constitutes a test of factorization:

$$R_h \equiv \frac{\Gamma(\overline{B}^0 \to {H_c}^+ + h^-)}{d\Gamma(\overline{B}^0 \to {H_c}^+ + e^- + \overline{\nu}_e)/dq^2\, |_{q^2 = m_h^2}} = 6\,\pi^2 |a_1|^2 |V_{ud}|^2 f_h^2 X_h, \quad (16.101)$$

where $h^-$ is an unflavored meson like $\pi^-$, $\rho^-$, $a_1(1260)^-$ with their respective decay constants $f_h$ and their corresponding phase space ratios $X_h = M_B^6 \,/\, [\lambda(M_B^2, M_{H_c}^2, m_h^2)]^{3/2} \simeq 1$. From the $q^2$ distribution of the semileptonic decay $d\Gamma(\overline{B}^0 \to {H_c}^+ + e^- + \overline{\nu}_e)/dq^2$ at different values of $q^2 = m_h^2$, the relation (101) can be tested. For many modes, the ratios $R_h$ are found to be in good agreement with data[11], thus supporting the factorization.

The method is easily extended to other class I decays, in particular the $\overline{B}^0 \to D^+ + D_s^-$ and $\overline{B}^0 \to D^+ + D_s^{*-}$ modes, which are also favored by the CKM matrix elements $V_{cb} V^*_{cs}$. Using factorization, one can extract the decay constants $f_{D_s}$ and $f_{D_s^*}$. The $f_{D_s} \approx 240$MeV obtained from factorization is in agreement with data coming from $D_s^+ \to \mu^+ + \nu_\mu$.

The decay constant $f_{D_s^*}$ (like $f_{K^*}$), which cannot be determined by the leptonic mode (since the weak decays $D_s^{*+} \to \mu^+ + \nu_\mu$, $K^* \to \mu^+ + \nu_\mu$ are swamped by the strong decays $D_s^* \to D_s + \pi$, $K^* \to K + \pi$), nevertheless can be extracted from $\overline{B}^0 \to D^+ + D_s^{*-}$. Also, the decay constant of the $a_1(1260)$ meson defined in (13.56) can be determined using

$$\frac{\Gamma(\overline{B^0} \to D^+ + {a_1}^-)}{\Gamma(\overline{B^0} \to D^+ + \rho^-)} = \frac{f_{a_1}^2}{f_\rho^2} \left[ \frac{\lambda(M_B^2, M_D^2, m_{a_1}^2)}{\lambda(M_B^2, M_D^2, m_\rho^2)} \right]^{3/2} \frac{|f_+(m_{a_1}^2)|^2}{|f_+(m_\rho^2)|^2} .$$

One obtains $f_{a_1} \approx 250$ MeV in agreement with the value derived in Chap. 13 from $\tau \to \nu_\tau + 3\pi$.

---

[11] Browder, T. E. and Honscheid, K., Prog. Nucl. Part. Phys. **35** (1995) 81; Neubert, M. and Stech, B., CERN preprint –TH 97-99 (1997)

## 16.5 CP Violation in B Mesons

Since 1964, with the discovery of CP violation in $K_L \to 2\pi$, no new CP-violating phenomenon has been observed although the standard KM mechanism offers a theoretical framework for this effect and opens to CP violation prospects outside the K system, in particular in the bottom sector. B meson decays address three fundamental questions:

(i) Nature and sources of CP violation. Is the KM phase the only mechanism for CP violation ? Is it true that only charged currents violate CP whereas neutral currents are CP-conserving as the standard model tells us? As explained in Chap. 11, while the experimental value of the parameter $\epsilon$ in the neutral K meson can be accommodated by the standard model, it does not by itself provide a test for it. On the other hand, the $\epsilon'$ parameter is subject to theoretical uncertainties because we are in the low-energy regime of nonperturbative QCD, and experimental data for $\epsilon'$ are still inconclusive as to whether or not there exists a direct CP violation as predicted by KM.

(ii) CP asymmetries in $B^0$ decays provide an alternative method to measure the CKM parameters. This is important since the traditional semileptonic modes used for the determination of $V_{ub}$ cannot take advantage of HFS (the u quark is light) and therefore is subject to theoretical uncertainties. Also, the precise values of the CKM matrix elements, in particular those related to the top quark $V_{td}, V_{ts}$ which cannot be directly measured, can be determined. The unitarity of the 3×3 CKM matrix can then be tested, so that the prospect of new fermionic families arises if unitarity does not hold.

(iii) CP asymmetries in the B sector are sensitive to the possible existence of 'new physics' beyond the standard model induced by the quantum loop effects of virtual new particles.

### 16.5.1 $B^0$–$\overline{B}^0$ Mixing

In Chap. 11, the general formalism for the neutral K mesons mixing has been outlined and the formulas (11.39)–(11.49) can be directly applied also to the systems $B^0_d \equiv \bar{b}d$ and $B^0_s \equiv \bar{b}s$. We discuss the case of $B^0_d$ and omit the subscript d for convenience, so that $B^0_d = B^0$.

Since $B^0$ can mix with $\overline{B}^0$ by weak interactions (box diagrams of Fig. 16.11 similar to Fig. 11.1), a neutral B meson weak-interaction eigenstate can be

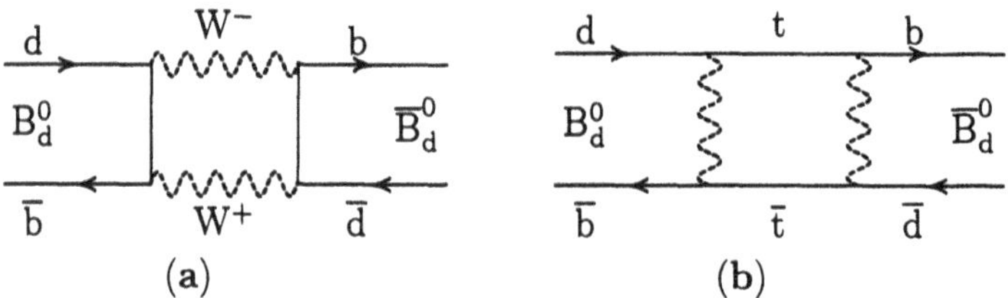


**Fig. 16.11a, b.** $B^0_d \leftrightarrow \overline{B}^0_d$ transition through $\bar{b}\,d \to \bar{d}\,b$

written as a superposition of $B^0$ and $\overline{B}^0$ using a $2\times2$ mixing matrix $\widetilde{M}_{ij}-\frac{i}{2}\widetilde{\Gamma}_{ij}$ as in (11.41). The off-diagonal terms are responsible for the $B^0$–$\overline{B}^0$ mixing, $\widetilde{M}_{12}$ corresponds to virtual transition between the $B^0$ and $\overline{B}^0$, while $\widetilde{\Gamma}_{12}$ describes real transition due to decay modes which are common to both $B^0$ and $\overline{B}^0$. Similar to (11.43), we introduce a heavy component $B_h$ and a 'less heavy' $B_l$ component. They are mixtures of the CP eigenstates, $B_1 = (B^0 - \overline{B^0})/\sqrt{2}$ and $B_2 = (B^0 + \overline{B^0})/\sqrt{2}$ of eigenvalues $\pm1$ respectively:

$$\begin{aligned} B_l &= p\, B^0 + q\, \overline{B}^0 = \frac{1}{\sqrt{1+|\widetilde{\epsilon}|^2}}\,(B_2 + \widetilde{\epsilon} B_1)\,, \\ B_h &= p\, B^0 - q\, \overline{B}^0 = \frac{1}{\sqrt{1+|\widetilde{\epsilon}|^2}}\,(B_1 + \widetilde{\epsilon} B_2)\,, \\ \frac{q}{p} &= \frac{1-\widetilde{\epsilon}}{1+\widetilde{\epsilon}} = \frac{\sqrt{\widetilde{M}^*_{12} - \frac{i}{2}\widetilde{\Gamma}^*_{12}}}{\sqrt{\widetilde{M}_{12} - \frac{i}{2}\widetilde{\Gamma}_{12}}} = \frac{M^*_{12} - \frac{i}{2}\Gamma^*_{12}}{\sqrt{\left[\widetilde{M}^*_{12} - \frac{i}{2}\widetilde{\Gamma}^*_{12}\right]\left[\widetilde{M}_{12} - \frac{i}{2}\widetilde{\Gamma}_{12}\right]}}\,. \end{aligned} \tag{16.102}$$

Indirect CP violation in $B^0$–$\overline{B}^0$ mixing arises if

$$\left|\frac{q}{p}\right| \neq 1 \longrightarrow \widetilde{\epsilon} \neq 0 \tag{16.103}$$

which results from the fact that the physical flavor eigenstates $B_h$ and $B_l$ are different from the CP eigenstates $B_1$ and $B_2$.
Let us define $\Delta m_B = M_{B_h} - M_{B_l}$ and $\Delta\Gamma_B = \Gamma_{B_h} - \Gamma_{B_l}$. Similarly to (11.45) and (11.46), the diagonalization of the $2\times2$ matrix $\widetilde{M}_{12} - \frac{i}{2}\widetilde{\Gamma}_{12}$ gives

$$\begin{aligned} M_{B_l} - \frac{i}{2}\Gamma_{B_l} &= M_0 - \frac{i}{2}\Gamma_0 + \sqrt{\left[\widetilde{M}^*_{12} - \frac{i}{2}\widetilde{\Gamma}^*_{12}\right]\left[\widetilde{M}_{12} - \frac{i}{2}\widetilde{\Gamma}_{12}\right]}\,, \\ M_{B_h} - \frac{i}{2}\Gamma_{B_h} &= M_0 - \frac{i}{2}\Gamma_0 - \sqrt{\left[\widetilde{M}^*_{12} - \frac{i}{2}\widetilde{\Gamma}^*_{12}\right]\left[\widetilde{M}_{12} - \frac{i}{2}\widetilde{\Gamma}_{12}\right]}\,, \\ \frac{q}{p} &= \frac{-2\,(M^*_{12} - \frac{i}{2}\Gamma^*_{12})}{\Delta m_B - \frac{i}{2}\Delta\Gamma_B} = \frac{\Delta m_B - \frac{i}{2}\Delta\Gamma_B}{-2\,(M_{12} - \frac{i}{2}\Gamma_{12})}\,, \end{aligned} \tag{16.104}$$

so that

$$\begin{aligned} (\Delta m_B)^2 - \frac{1}{4}(\Delta\Gamma_B)^2 &= 4\,|\widetilde{M}_{12}|^2 - |\widetilde{\Gamma}_{12}|^2\,, \\ (\Delta m_B)\,(\Delta\Gamma_B) &= 4\,\mathrm{Re}(\widetilde{M}_{12}\widetilde{\Gamma}^*_{12})\,. \end{aligned} \tag{16.105}$$

The mass difference $\Delta m_B$ has been measured to be $\approx (0.467\pm0.017)\ (\mathrm{ps})^{-1} = (3.07\pm0.11)\times10^{-4}$ eV by the $B^0$ and $\overline{B}^0$ oscillations as well as by the observation of the same sign dilepton events (see below). From (105) we deduce that once $\Delta m_{\mathrm{B}}$ is known, a tiny $\widetilde{\Gamma}_{12} \ll \widetilde{M}_{12}$ implies $|\Delta\Gamma_{\mathrm{B}}| \ll \Delta m_{\mathrm{B}}$ and $\Delta m_{\mathrm{B}} \approx 2\,|\widetilde{M}_{12}|$. We first show that $\widetilde{\Gamma}_{12} \ll \widetilde{M}_{12}$.

For the neutral $\mathrm{B}^0$ and $\overline{\mathrm{B}}^0$ system, the off-diagonal term $\widetilde{\Gamma}_{12}$ is extremely small since the overlap in the decay products of $\mathrm{B}^0$ and $\overline{\mathrm{B}}^0$ is rare. Indeed, the $\mathrm{B}^0$ decays mostly into anticharmed and unflavored particles described by $\overline{\mathrm{b}} \to \overline{\mathrm{c}}+\overline{\mathrm{d}}+\mathrm{u}$, while the $\overline{\mathrm{B}}^0$ decays into charm from $\mathrm{b} \to \mathrm{c} + \mathrm{d} + \overline{\mathrm{u}}$. These final decay products are completely distinct. There exist only a few common channels into which both $\mathrm{B}^0$ and $\overline{\mathrm{B}}^0$ decay. They are $\mathrm{B}^0 \to \mathrm{D}^+ + \mathrm{D}^-$ ( or $\pi^+ + \pi^-$) $\leftarrow \overline{\mathrm{B}}^0$ coming from $\mathrm{b} \to \mathrm{c} + \mathrm{d} + \overline{\mathrm{c}}$ (or $\mathrm{b} \to \mathrm{u} + \mathrm{d} + \overline{\mathrm{u}}$). However, their rates are suppressed by $|V_{\mathrm{cb}}V^*_{\mathrm{cd}}|^2$ (or $|V_{\mathrm{ub}}V^*_{\mathrm{ud}}|^2$). Experimentally, only an upper bound $< 10^{-3}$ is known for the branching ratio into the common decays $\mathrm{B}^0 \to X_{\mathrm{com}}$ and $\overline{\mathrm{B}}^0 \to X_{\mathrm{com}}$.

Therefore, the lifetime difference $\Delta\Gamma_{\mathrm{B}}$ between $\mathrm{B}_h$ and $\mathrm{B}_l$ is tiny and almost impossible to measure. Neither the long nor the short neutral B mesons are experimentally accessible; however, a mass difference $\Delta m_{\mathrm{B}}$ exists between the two neutral weak-interaction eigenstates $\mathrm{B}_h$ and $\mathrm{B}_l$, which have almost equal lifetimes $\tau_{\mathrm{B}^0} = 1/\Gamma_{\mathrm{B}} \approx (1.549 \pm 0.020) \times 10^{-12}$ s.

This is in sharp contrast to the neutral K meson system where the off-diagonal $\Gamma_{12}$ is large since both $\mathrm{K}^0$ and $\overline{\mathrm{K}}^0$ can decay into the common $2\pi$ and $3\pi$ channels. With a large $\Gamma_{12}$, the $\mathrm{K_L}$ and $\mathrm{K_S}$ have a large lifetime difference, and therefore they are called K-long and K-short. This unique situation is due to the particular mass scale of the K mesons which have only two hadronic decay modes into $2\pi$ and $3\pi$, as discussed in Chap. 11. Both $\Delta m_{\mathrm{K}} = m_{\mathrm{L}} - m_{\mathrm{S}}$ and $\Delta\gamma = \Gamma_{\mathrm{S}} - \Gamma_{\mathrm{L}}$ are comparable, as given in (11.52).

**The Ratio $(q/p)_{\mathrm{B}}$ in the B System.** For the K system, we know that $2\,\mathrm{Re}\,(\overline{\epsilon}) \approx (3.27 \pm 0.12) \times 10^{-3}$ as measured by the electron–positron asymmetry $\delta_{\mathrm{K}}(t)$ in (11.55). Therefore, the ratio $(q/p)_{\mathrm{K}}$ defined in (11.44) for the K system is

$$\left|\frac{q}{p}\right|_{\mathrm{K}} \approx 1 - 2\,\mathrm{Re}\,(\overline{\epsilon}) \approx 1 - \mathcal{O}(10^{-3})\ ,$$

a small deviation from 1 of the ratio $|(q/p)_{\mathrm{K}}|$ for the K system is due to CP violation in the $\Delta S = 2$ transition through the $\mathrm{K}^0$–$\overline{\mathrm{K}}^0$ mixing.

For the B system, on the other hand, independently of the question of CP violation, we can already anticipate from the fact $\widetilde{\Gamma}_{12} \ll \widetilde{M}_{12}$, i.e. $\widetilde{\epsilon} \ll 1$ that the ratio $(q/p)_{\mathrm{B}}$ must be very close to 1. Indeed,

$$\left(\frac{q}{p}\right)_{\mathrm{B}} = \frac{\sqrt{\widetilde{M}^*_{12} - \frac{\mathrm{i}}{2}\widetilde{\Gamma}^*_{12}}}{\sqrt{\widetilde{M}_{12} - \frac{\mathrm{i}}{2}\widetilde{\Gamma}_{12}}} \approx \frac{-\widetilde{M}^*_{12}}{|\widetilde{M}_{12}|}\left(1 - \tfrac{1}{2}\mathrm{Im}\left[\frac{\widetilde{\Gamma}_{12}}{\widetilde{M}_{12}}\right]\right)\ ,$$

$$\left|\frac{q}{p}\right|_{\mathrm{B}} \approx 1 - \mathcal{O}(10^{-3})\ . \tag{16.106}$$

Therefore, indirect CP violation in the $\Delta B = 2$ transition through $B^0$–$\overline{B}^0$ mixing must be a very small effect, as in the K system. The $\widetilde{\epsilon}$ of the B mesons is presumably even smaller than the $\widetilde{\epsilon}$ of the K mesons. However, as we will see, direct CP violation in the $\Delta B = 1$ transitions, i.e. in B decays, is expected to be large according to the standard model. This is again in sharp contrast with the K meson in which the $\Delta S = 1$ direct CP violation is vanishingly small [recall the parameter $\epsilon' \ll \epsilon$ (Chap. 11)].

**The Mass Difference $\Delta m_B$.** Once the neutral B mesons are produced in pairs, their semileptonic decays (inclusive or exclusive) provide an excellent method to measure the $B^0$–$\overline{B}^0$ mixing. From their respective quark contents, $B^0$ decays into a positive charged lepton $\ell^+$ while $\overline{B}^0$ goes into a negative $\ell^-$. If $B^0$ and $\overline{B}^0$ do not mix, the produced pair $B^0$+$\overline{B}^0$ would have a distinctive signature of a dilepton with opposite signs $\ell^+ + \ell^-$. Therefore, a fully reconstructed $\mu^+ + \mu^+$ same-sign event would unambiguously demonstrate the conversion of a $\overline{B}^0$ into a $B^0$, i.e. the pair $B^0$–$\overline{B}^0$ becomes two $B^0$ which subsequently decay into $\mu^+ + \mu^+$. This event indeed was found[12] and shows that mixing must exist. Since then, the $B^0$–$\overline{B}^0$ mixing has a much better statistics.[13]

The mass difference $\Delta m_B$ is a measure of the frequency of the change from a $B^0$ into a $\overline{B}^0$ or vice versa. This change is reflected in either the time-dependent oscillations (similar to Fig. 11.2 for the K mesons) or in the time-integrated rates corresponding to the dilepton events having the same sign (see (112) below). Similar to the $K_L$ and $K_S$ system in (11.8), let us write

$$
\begin{aligned}
|B_h(t)\rangle &= \left[e^{-t\Gamma_B/2}\right] \left(e^{-it M_B}\right) e^{-it\Delta m_B/2} \; |B_h(0)\rangle \; , \\
|B_l(t)\rangle &= \left[e^{-t\Gamma_B/2}\right] \left(e^{-it M_B}\right) e^{+it\Delta m_B/2} \; |B_l(0)\rangle \; .
\end{aligned}
\tag{16.107}
$$

The evolution (107) of the mass eigenstates $B_h(t)$ and $B_l(t)$ when combined with (102) gives the time evolution of $B^0(t)$ and $\overline{B}^0(t)$:

$$
\begin{aligned}
|B^0(t)\rangle &= h_+(t)\,|B^0(0)\rangle + \frac{q}{p}\,h_-(t)\,\left|\overline{B}^0(0)\right\rangle \; , \\
\left|\overline{B}^0(t)\right\rangle &= \frac{p}{q}\,h_-(t)\,|B^0(0)\rangle + h_+(t)\,\left|\overline{B}^0(0)\right\rangle \; ,
\end{aligned}
\tag{16.108}
$$

$$
\begin{aligned}
h_+(t) &= e^{-t\Gamma_B/2}\,e^{-it M_B}\,\cos(t\,\Delta m_B/2) \; , \\
h_-(t) &= i\,\left[e^{-t\Gamma_B/2}\,e^{-it M_B}\,\sin(t\,\Delta m_B/2)\right] \; .
\end{aligned}
\tag{16.109}
$$

[12] Albrecht, H. et al., Phys. Lett. **192B** (1987) 245

[13] Wu Sau Lan, in *Proc. 17th Int. Symp. on Lepton–Photon Interactions*, (1995) Beijing (ed. Zheng Zhi-Peng and Chen He-Sheng). World Scientific, Singapore 1996

As in (11.10) and (11.11), starting at $t = 0$ with an initially pure $B^0$, the probability for finding a $B^0$ ($\overline{B}^0$) at time $t \neq 0$ is given by $|h_+(t)|^2$ ($|h_-(t)|^2$). Taking $|q/p| = 1$, one gets

$$|h_\pm(t)|^2 = \frac{1}{2} e^{-t\Gamma_B} \left[1 \pm \cos(t\,\Delta m_B)\right] . \tag{16.110}$$

Conversely, from an initially pure $\overline{B}^0$ at $t = 0$, the probability for finding a $\overline{B}^0$ ($B^0$) at time $t \neq 0$ is also given by $|h_+(t)|^2$ ($|h_-(t)|^2$). The oscillations of $B^0$ or $\overline{B}^0$ as shown by (110) give $\Delta m_B$ directly. Integrating $|h_\pm(t)|^2$ from $t = 0$ to $t = \infty$, we get

$$\int_0^\infty \mathrm{d}t |h_\pm(t)|^2 = \frac{1}{2} \left[ \frac{1}{\Gamma_B} \pm \frac{\Gamma_B}{\Gamma_B^2 + (\Delta m_B)^2} \right] . \tag{16.111}$$

The ratio

$$r = \frac{B^0 \leftrightarrow \overline{B}^0}{B^0 \leftrightarrow B^0} = \frac{\int_0^\infty \mathrm{d}t |h_-(t)|^2}{\int_0^\infty \mathrm{d}t |h_+(t)|^2} = \frac{x^2}{2 + x^2} , \quad \text{where } x \equiv \frac{\Delta m_B}{\Gamma_B} , \tag{16.112}$$

reflects the change of a pure $B^0$ into a $\overline{B}^0$, or vice versa. This change is manifested by the same-sign dilepton events compared to the opposite sign dilepton and yields

$$x = \frac{\Delta m_B}{\Gamma_B} = 0.71 \pm 0.04 . \tag{16.113}$$

This result, when combined with the oscillation measurement, gives the average $\Delta m_B = (3.07 \pm 0.12) \times 10^{-4}$ eV, which is a hundred times larger than the corresponding $\Delta m_K$ of the K meson system.

**The CKM Matrix Element $V_{td}$ from $\Delta m_B$.** Similar to the $\Delta m_K$ in Chap. 11, the mass difference $\Delta m_B$ is calculated from the box diagrams of Fig. 16.11 which give $\widetilde{M}_{12}$. Contrary to the K meson case where both the charm and the top quark contributions are important, we find that when we apply the formulas (11.36) and (11.37) to the B meson case, the top quark largely dominates $\widetilde{M}_{12}$. Since $\widetilde{\Gamma}_{12} \ll \widetilde{M}_{12}$ from (105), one has $\Delta m_B \approx 2|\widetilde{M}_{12}|$:

$$\Delta m_B \approx 2|\widetilde{M}_{12}| = \frac{G_F^2\, m_t^2\, M_B\, f_B^2}{6\,\pi^2}\, g(x_t)\, \eta_t\, |V_{td}^*\, V_{tb}|^2\, B , \tag{16.114}$$

where $\eta_t \approx 0.55$ is the gluonic correction[14] to the box diagrams, $g(x_t)$ is already given in (11.37), $f_B$ is the B meson decay constant involved in $B^+ \to \tau^+ + \nu_\tau$ similar to the decay constants $f_\pi, f_K, f_D$. The last factor $B$, like the one defined in (11.29), represents the correction to the vacuum insertion used in the evaluation of $\Delta m_B$. The decay constant $f_B$, not yet determined by experiments, is taken to be $180 \pm 50$ MeV and the parameter $B$ is taken as

[14] Buras, A., Jamin, M. and Weizs, P. H., Nucl. Phys. **B347** (1990) 491

$1 \pm 0.2$, both values obtained from lattice calculations.[15] Then, (114) gives

$$|V_{\rm td}| = (9.2 \pm 3) \times 10^{-3} = A\,\lambda^3\,\sqrt{(1-\rho)^2 + \eta^2}\,, \tag{16.115}$$

which represents one more constraint on the parameters $\rho$ and $\eta$ of the CKM matrix, in addition to (60):

$$|1 - \rho - {\rm i}\eta| = 1.01 \pm 0.22\,.$$

From (112) and (114), we realize that the $\mathrm{B}^0 - \overline{\mathrm{B}}^0$ mixing can be observed because of the large $\Delta m_{\rm B}$ or, equivalently, because of the large top quark mass which compensates for the small $|V^*_{\rm td}|^2$.

Using (106) and (114), one finds the following expression frequently used:

$$\left(\frac{q}{p}\right)_{\rm B} = \frac{\sqrt{\widetilde{M}^*_{12} - \frac{\rm i}{2}\widetilde{\Gamma}^*_{12}}}{\sqrt{\widetilde{M}_{12} - \frac{\rm i}{2}\widetilde{\Gamma}_{12}}} \approx \frac{\sqrt{\widetilde{M}^*_{12}}}{\sqrt{\widetilde{M}_{12}}} = \frac{\sqrt{(V_{\rm td}V^*_{\rm tb})^2}}{\sqrt{(V^*_{\rm td}V_{\rm tb})^2}} = \frac{V_{\rm td}V^*_{\rm tb}}{V^*_{\rm td}V_{\rm tb}} \equiv {\rm e}^{-2\,{\rm i}\,\beta}, \tag{16.116}$$

which tells us that to a very good approximation the ratio $q/p$ is a pure phase. The angle $\beta$ which characterizes the standard CP violating mechanism is an important quantity in the unitarity triangle we are considering now.

**Unitarity Triangles.** As discussed in (11.85), the unitarity of the CKM matrix is visualized by six triangles, all of which have the same area $\frac{1}{2}J$, and the standard CP violating mechanism is reflected by $J \neq 0$. In a phase reparameterization of the quark fields that build the CKM matrix, the triangles change their orientation in the plane, but their shape remains unaffected. Among the six triangles, only two have a regular shape due to the fact that their three sides are not dissimilar and proportional to $\lambda^3$ where $\lambda = \sin\theta_{\rm C} \approx 0.22$. The remaining four triangles always have one side much smaller than the other two sides. These two regular triangles are those which connect the CKM matrix elements that belong either to the first and the third columns, or to the first and the third rows. The triangle illustrated in Fig. 16.12 comes from

$$V_{\rm ud}\,V^*_{\rm ub} + V_{\rm cd}\,V^*_{\rm cb} + V_{\rm td}\,V^*_{\rm tb} = 0\,. \tag{16.117}$$

Figure 16.12b is taken from Fig. 16.12a by dividing all sides by $V_{\rm cd}\,V^*_{\rm cb}$. This quantity by convention is taken to be real. The rescaled triangle has the coordinates $(0,0), (1,0)$ and $(\overline{\rho}, \overline{\eta})$ with $\overline{\rho} = \rho\,(1 - \frac{1}{2}\lambda^2)$, $\overline{\eta} = \eta\,(1 - \frac{1}{2}\lambda^2)$. Physical quantities measuring CP violation can be expressed in terms of $J$ or, equivalently, of the angles $\alpha, \beta, \gamma$. Now we show that these angles can be obtained by measuring the differences between the B and $\overline{\rm B}$ decay rates into various channels due to CP violation.

---

[15] Michael, C., in *Proc. 17th Int. Symp. on Lepton–Photon Interactions*, (1995) Beijing (ed. Zheng Zhi-Peng and Chen He-Sheng). World Scientific, Singapore 1996

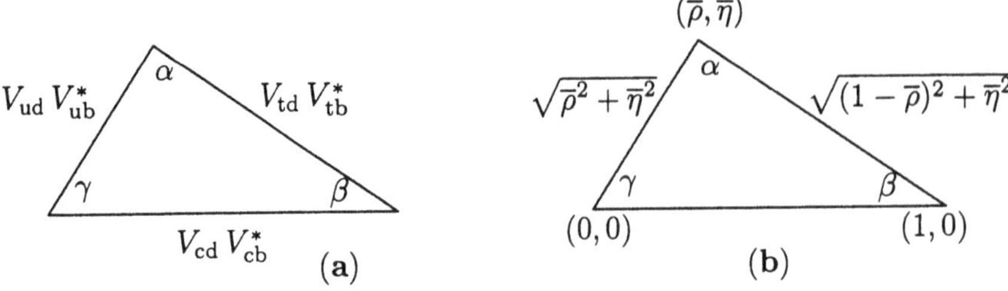


**Fig. 16.12a,b.** The angles $\alpha, \beta, \gamma$ of the unitarity triangle

### 16.5.2 CP Asymmetries in Neutral B Meson Decays

The most promising method of measuring CP violation is to look for an asymmetry between the $\Gamma(\mathrm{B}^0 \to f_{\mathrm{cp}})$ and $\Gamma(\overline{\mathrm{B}}^0 \to \overline{f}_{\mathrm{cp}})$, where $f_{\mathrm{cp}}$ is a hadronic state having a well-defined CP eigenvalue $\pm 1$. These states of definite CP parities are called CP eigenstates. We have $\overline{f}_{\mathrm{cp}} = \pm f_{\mathrm{cp}}$ depending on the CP parity of $f_{\mathrm{cp}}$. Some examples of CP eigenstates are the two-particle systems: $\psi + \mathrm{K_S}$ (CP parity = -1), $\pi^+ + \pi^-$ (CP parity =+1), and $\rho^0 + \mathrm{K_S}$ (CP parity = -1). Next, we define the amplitudes $A$, $\overline{A}$ and the parameter $\xi$ as

$$A \equiv \langle f_{\mathrm{cp}} | H_W | \mathrm{B}^0 \rangle \ , \ \overline{A} \equiv \left\langle \overline{f}_{\mathrm{cp}} | H_W | \overline{\mathrm{B}}^0 \right\rangle \ , \ \xi \equiv \frac{q}{p} \frac{\overline{A}}{A} \ . \tag{16.118}$$

The time evolution of the decay amplitudes can be written as

$$\begin{aligned} \langle f_{\mathrm{cp}} | H_W | \mathrm{B}^0(t) \rangle &= A \ [h_+(t) + \xi \, h_-(t)] \ , \\ \left\langle \overline{f}_{\mathrm{cp}} | H_W | \overline{\mathrm{B}}^0(t) \right\rangle &= \frac{p}{q} \, A \ [h_-(t) + \xi \, h_+(t)] \ . \end{aligned} \tag{16.119}$$

The rates for an initial pure $\mathrm{B}^0$ (or $\overline{\mathrm{B}}^0$) to decay into a CP eigenstate $f_{\mathrm{cp}}$ (or $\overline{f}_{\mathrm{cp}}$) at time $t$ are given by

$$\Gamma\left(\mathrm{B}^0(t) \to f_{\mathrm{cp}}\right) = C \left[ \frac{1 + |\xi|^2}{2} + \frac{1 - |\xi|^2}{2} \cos(\Delta m_{\mathrm{B}} \, t) - \mathrm{Im}(\xi) \, \sin(\Delta m_{\mathrm{B}} \, t) \right] ,$$

$$\Gamma(\overline{\mathrm{B}}^0(t) \to \overline{f}_{\mathrm{cp}}) = C \left[ \frac{1 + |\xi|^2}{2} - \frac{1 - |\xi|^2}{2} \cos(\Delta m_{\mathrm{B}} \, t) + \mathrm{Im}(\xi) \, \sin(\Delta m_{\mathrm{B}} \, t) \right] ,$$

where $C = |A|^2 \mathrm{e}^{-\Gamma_{\mathrm{B}} t}$. The time-dependent CP asymmetry is defined as

$$\begin{aligned} a(t) &= \frac{\Gamma(\mathrm{B}^0(t) \to f_{\mathrm{cp}}) - \Gamma(\overline{\mathrm{B}}^0(t) \to \overline{f}_{\mathrm{cp}})}{\Gamma(\mathrm{B}^0(t) \to f_{\mathrm{cp}}) + \Gamma(\overline{\mathrm{B}}^0(t) \to \overline{f}_{\mathrm{cp}})} \\ &= \frac{(1 - |\xi|^2) \cos(\Delta m_{\mathrm{B}} \, t) - 2 \, \mathrm{Im}(\xi) \, \sin(\Delta m_{\mathrm{B}} \, t)}{1 + |\xi|^2} \ . \end{aligned} \tag{16.120}$$

We have seen in (116) that $|q/p| = 1$. Furthermore, if $|\overline{A}/A| = 1$ so that $|\xi| = 1$, the asymmetry $a(t)$ simplifies considerably:

$$a(t) = -\mathrm{Im}(\xi) \, \sin(\Delta m_{\mathrm{B}} \, t) \ . \tag{16.121}$$

The total rate asymmetry is obtained by integrating the numerator and denominator of (120). For $|\xi| = 1$, we have

$$\overline{a} = \frac{\int_0^\infty \mathrm{d}t \left\{\Gamma(\mathrm{B}^0(t) \to f_{\mathrm{cp}}) - \Gamma(\overline{\mathrm{B}}^0(t) \to \overline{f}_{\mathrm{cp}})\right\}}{\int_0^\infty \mathrm{d}t \left\{\Gamma(\mathrm{B}^0(t) \to f_{\mathrm{cp}}) + \Gamma(\overline{\mathrm{B}}^0(t) \to \overline{f}_{\mathrm{cp}})\right\}} = \frac{-x}{1+x^2}\,\mathrm{Im}(\xi)\,, \quad (16.122)$$

where $x$ is given in (113). We note that the integrated asymmetry $\overline{a}$ is suppressed when $x \ll 1$ (case of charmed $\mathrm{D}^0$ mesons, since $\Gamma_{\mathrm{D}} \gg \Delta m_{\mathrm{D}}$) or when $x \gg 1$ (case of $\mathrm{B}^0_\mathrm{s} \equiv \overline{\mathrm{b}}\mathrm{s}$). In these cases, the time-dependent asymmetry $a(t)$ is appropriately useful.

When $|\xi| = 1$, the quantity $\xi$, which may be written as $\xi = \mathrm{e}^{\mathrm{i}\,\theta}$, is very interesting since its imaginary part, i.e. $\sin\theta$, is directly related to the CKM matrix elements, such that measurements of the asymmetries directly determine the CKM angles. So our next task is to look for the conditions that guarantee $|\overline{A}/A| = 1$.

As already illustrated by (11.64) and (11.65) for $\mathrm{K}^0 \to 2\pi$ and $\overline{\mathrm{K}}^0 \to 2\pi$ processes, in general the amplitudes of $\mathrm{B}^0$ and $\overline{\mathrm{B}}^0$ decaying into an arbitrary state can be written as the sum of various contributions,

$$A = \sum_k A_k \mathrm{e}^{\mathrm{i}\,\delta_k}\, \mathrm{e}^{\mathrm{i}\,\phi_k} \quad , \quad \overline{A} = \sum_k A_k \mathrm{e}^{\mathrm{i}\,\delta_k}\, \mathrm{e}^{-\mathrm{i}\,\phi_k} \,, \quad (16.123)$$

where $\phi_k$ is the weak interaction CKM phase that represents CP violation while $\delta_k$ is the strong-interaction phase-shift due to rescattering effects among the hadrons in the final state. The $\delta_k$ enters $A$ and $\overline{A}$ without changing sign since the strong interactions conserve CP. Thus $|A| = |\overline{A}|$ if the various contributions $A_k$ have the same CKM phase, or in particular, if there is only one dominant contribution. In these special cases, the hadronic uncertainties essentially due to the unknown $\delta_k$ are eliminated in the ratio $\overline{A}/A = 1$.

Generally, $\overline{A}/A \neq 1$, since nonleptonic decays in (123) receive contributions from both the 'tree' and 'penguin' amplitudes. The former – if favored by the CKM matrix elements (say $V_{\mathrm{cb}}V^*_{\mathrm{ud}}$ or $V_{\mathrm{cb}}V^*_{\mathrm{cs}}$) – would dominate the latter which are suppressed by QCD perturbative higher orders. However, it is not difficult to find some counterexamples where 'tree' amplitudes are either forbidden or suppressed by $V_{\mathrm{ub}}V^*_{\mathrm{us}} \sim 10^{-3}\, V_{\mathrm{cb}}V^*_{\mathrm{ud}}$. Furthermore, tree and penguin amplitudes have different CKM phases in general.

Fortunately, there exist a few cases where $\overline{A}/A = 1$. In particular, we are interested in $\mathrm{b} \to \mathrm{s} + \mathrm{c} + \overline{\mathrm{c}}$ (responsible for the decay mode $\mathrm{B} \to \mathrm{J}/\psi + \mathrm{K}$) which in turn allows the determination of the angle $\beta$. This mode receives contributions from both tree and penguin diagrams. However, we will see that $|\overline{A}/A| = 1$ still holds.

Generically, the penguin diagram is always of the type $\mathrm{b} \to \mathrm{s}\ (\mathrm{d}) + \mathrm{q} + \overline{\mathrm{q}}$ and shown in Fig. 16.13a. We note that the internal loop is dominated by the top quark because of its mass and also because $V_{\mathrm{tb}} \approx 1$ is largest. The dominant penguin amplitude corresponds to the process b→ s +q + $\overline{\text{q}}$ (for instance b→ s +c + $\overline{\text{c}}$ or b→ s +u + $\overline{\text{u}}$). The amplitude can be directly taken from $H^{\text{glu}}_{\text{pen}}$ given in (11.94). Let us write it as

$$\frac{\alpha_s}{12\pi} V_{\text{tb}} V^*_{\text{ts}} \log \frac{m_t^2}{m_b^2} , \tag{16.124}$$

where the factor $G_F/\sqrt{2}$, the quark fields, and the Dirac matrices explicitly written in (11.94) are omitted here for simplicity. The nondominant penguin diagram b→ d +q+$\overline{\text{q}}$ can be taken from (124) by replacing $V^*_{\text{ts}}$ with $V^*_{\text{td}} < V^*_{\text{ts}}$.

The same b→ s +c + $\overline{\text{c}}$ amplitude due to the tree diagram (Fig. 16.13b) is proportional to $V_{\text{cb}} V^*_{\text{cs}}$, which is larger than (124).

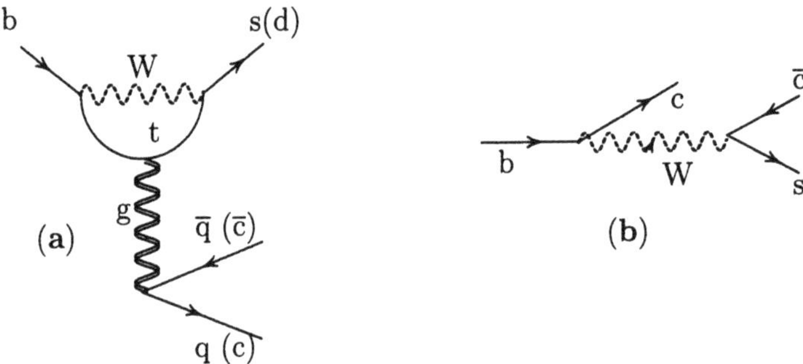


**Fig. 16.13.** The transition b → s + c + $\overline{\text{c}}$ by (**a**) penguin (**b**) tree diagram

Within the standard KM mechanism of CP violation, the decay mode $\overline{\text{B}} \to \text{J}/\psi + \text{K}_S$ is very interesting since, not only the tree amplitude dominates the penguin amplitude, but both amplitudes have a common CKM phase such that independently of their relative magnitude, the condition $|A| = |\overline{A}|$ is fulfilled. Adding the penguin part to the tree diagram amplitude in (123) does not affect $|\xi| = 1$, so that the asymmetry in (121) or (122) is theoretically very clean. For this reason, the asymmetry of this 'gold-plated' mode would be the first to be measured in the future B meson factories. This decay has already been observed with a branching ratio $\sim 10^{-3}$. So the next experimental task is to measure the asymmetry, which needs a much larger statistics.

The decay $\overline{\text{B}} \to \text{J}/\psi + \text{K}_S$ issued from $\overline{\text{B}} \to \text{J}/\psi + \overline{\text{K}}$ is governed at the quark level by b→ c +s + $\overline{\text{c}}$ for which the amplitude of the tree diagram is proportional to $V_{\text{cb}} V^*_{\text{cs}}$ as mentioned previously. The other details of the amplitude are irrelevant. The transition of $\overline{\text{K}}$ into $\text{K}_S$ will be discussed later. The penguin amplitude for the same reaction $\overline{\text{B}} \to \text{J}/\psi + \overline{\text{K}}$, as shown by (124), has the $V_{\text{tb}} V^*_{\text{ts}}$ factor. This factor has the same phase as the tree amplitude $V_{\text{cb}} V^*_{\text{cs}}$, so even with the sum of the tree and penguin amplitudes in (123), we still have $|\overline{A}/A| = 1$ since

$$\frac{\overline{A}}{A} = \frac{V_{\text{cb}} V^*_{\text{cs}}}{V^*_{\text{cb}} V_{\text{cs}}} . \tag{16.125}$$

Let us emphasize that, independently of the details of both the tree (Fig. 16.9) and penguin (Fig. 16.14) decay amplitudes, the ratio $\overline{A}/A$ is model independent and always given by (125). This ratio is sufficient for a prediction of the asymmetry.

There remains the factor $q/p$ for $\xi$ in (118). In fact, there are two $q/p$. One, associated with the B system $(q/p)_{\mathrm{B}}$, is already known in (116) and the other, $(q/p)_{\mathrm{K}}$, is still to be determined since, with a $\mathrm{K_S}$ in the final state, one has to take into account the mixing of $\mathrm{K}^0$ and $\overline{\mathrm{K}}^0$ in the $\mathrm{K_S}$ too. For this purpose, we must look for the ratio

$$\left(\frac{q}{p}\right)_{\mathrm{K}} \equiv \frac{\langle \mathrm{K_S} | \overline{\mathrm{K}} \rangle}{\langle \mathrm{K_S} | \mathrm{K} \rangle} \,.$$

We write $\mathrm{K_S} = p_{\mathrm{K}}\, \mathrm{K}^0 + q_{\mathrm{K}}\, \overline{\mathrm{K}}^0$, then $\langle \mathrm{K_S} | \overline{\mathrm{K}} \rangle = q_{\mathrm{K}}^*$, $\langle \mathrm{K_S} | \mathrm{K} \rangle = p_{\mathrm{K}}^*$, so

$$\left(\frac{q}{p}\right)_{\mathrm{K}} = \frac{q_{\mathrm{K}}^*}{p_{\mathrm{K}}^*} \,.$$

The ratio $q_{\mathrm{K}}/p_{\mathrm{K}}$ comes from $M_{12}$ in (11.80) of the box diagram in Fig. 11.1. This ratio $q_{\mathrm{K}}/p_{\mathrm{K}}$ can be directly taken from $q_{\mathrm{B}}/p_{\mathrm{B}}$ with the substitution b↔s and t↔c which is suggested by comparing the box diagrams in Fig. 11.1 and Fig. 16.11. Thus

$$\left(\frac{q}{p}\right)_{\mathrm{K}} = \left(\frac{V_{\mathrm{cs}}^* V_{\mathrm{cd}}}{V_{\mathrm{cs}} V_{\mathrm{cd}}^*}\right)^* = \frac{V_{\mathrm{cs}} V_{\mathrm{cd}}^*}{V_{\mathrm{cs}}^* V_{\mathrm{cd}}} \,. \tag{16.126}$$

Using (116), (125), and (126), the parameter $\xi$ in (118) for $\mathrm{B} \to \mathrm{J}/\psi + \mathrm{K_S}$ is

$$\begin{aligned}
\xi_{\psi \mathrm{K_S}} &= \left(\frac{q}{p}\right)_{\mathrm{B}} \left(\frac{q}{p}\right)_{\mathrm{K}} \frac{V_{\mathrm{cb}} V_{\mathrm{cs}}^*}{V_{\mathrm{cb}}^* V_{\mathrm{cs}}} = \left(\frac{V_{\mathrm{td}} V_{\mathrm{tb}}^*}{V_{\mathrm{td}}^* V_{\mathrm{tb}}}\right) \left(\frac{V_{\mathrm{cs}} V_{\mathrm{cd}}^*}{V_{\mathrm{cs}}^* V_{\mathrm{cd}}}\right) \left(\frac{V_{\mathrm{cb}} V_{\mathrm{cs}}^*}{V_{\mathrm{cb}}^* V_{\mathrm{cs}}}\right) \\
&= \left(\frac{V_{\mathrm{td}} V_{\mathrm{tb}}^*}{V_{\mathrm{td}}^* V_{\mathrm{tb}}}\right) \left(\frac{V_{\mathrm{cd}}^* V_{\mathrm{cb}}}{V_{\mathrm{cd}} V_{\mathrm{cb}}^*}\right) = \left(\frac{V_{\mathrm{td}} V_{\mathrm{tb}}^*}{V_{\mathrm{td}}^* V_{\mathrm{tb}}}\right) = \mathrm{e}^{-2\mathrm{i}\beta}
\end{aligned}$$

since $V_{\mathrm{cd}}\, V_{\mathrm{cb}}^*$ is real (see Fig. 16.12). Note that the CP eigenstate $\mathrm{J}/\psi + \mathrm{K_S}$ has its intrinsic CP parity $-1$; the ratio $\overline{A}/A$ has an extra minus sign so that actually $\xi_{\psi \mathrm{K_S}} = -\exp(-2\mathrm{i}\beta)$, and consequently the asymmetries $a(t)$ and $\overline{a}$ defined in (121) and (122) are proportional to $-\sin 2\beta$.

Therefore, a measure of the asymmetry in $\mathrm{B} \to \mathrm{J}/\psi + \mathrm{K_S}$ directly gives the $\beta$ angle. From our present knowledge of the CKM parameters $\rho$ and $\eta$ as given by (60), (115) and depicted in Fig. 16.12, one can take $0.3 \leq \sin 2\beta \leq 0.8$, such that the asymmetry $\overline{a}$ is expected to be in the range $-0.4 \leq \overline{a} \leq -0.15$ with a *negative sign.*

We finally mention that asymmetries are expected in many other modes, for instance in charged $\mathrm{B}^{\pm}$ meson decays. The condition

$$\left|\frac{\overline{\mathcal{A}}}{\mathcal{A}}\right| \neq 1 \tag{16.127}$$

implies direct CP violation, where $\mathcal{A} = \langle \mathcal{F} \,|\, H_{\mathrm{W}} \,|\, \mathrm{B}^{+} \rangle$ and $\overline{\mathcal{A}} = \langle \overline{\mathcal{F}} \,|\, H_{\mathrm{W}} \,|\, \mathrm{B}^{-} \rangle$. Note that contrary to the case where $\mathcal{F} = f_{\mathrm{cp}}$ which gives $|\overline{A}/A| = 1$ in (118), since the final state $\mathcal{F}$ is not a CP eigenstate, the condition (127) requires at least two partial amplitudes $A_k$ in (123) that differ in both their weak and strong interaction $\phi_k$ and $\delta_k$ phases.

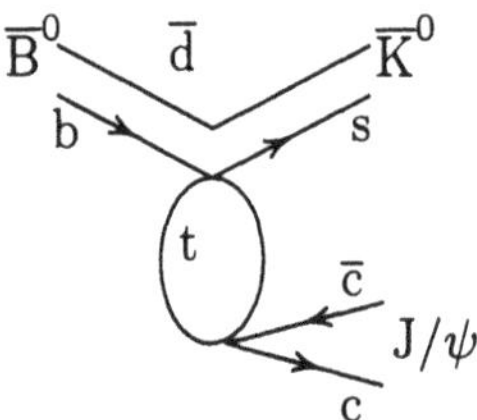


**Fig. 16.14.** $\overline{\mathrm{B}} \to \overline{\mathrm{K}} + \mathrm{J}/\psi$ by penguin transition

The standard CP-violating KM mechanism is very predictive since all CP violation phenomena are described by a single parameter which can be taken as $J$. The interesting asymmetry in $\mathrm{B} \to \mathrm{J}/\psi + \mathrm{K_S}$, which is free of hadronic uncertainties in the evaluation of the decay amplitude, is the best example; it can be used as a consistency check of the KM mechanism. Therefore, once the various CP asymmetries in B mesons are measured, deviations from these predictions, if they arise, would provide clues about the new physics beyond the standard model.

## Problems

**16.1 The $|\Delta I| = \frac{1}{2}$ rule and the penguin operator.** Show that the matrix element of the penguin operator $\mathcal{O}_{\mathrm{pen}} = [\bar{d}\, \gamma_\mu (1 - \gamma_5) \lambda^b\, s]\, [\bar{q}\, \gamma^\mu \lambda^b\, q]$ in (11.94) inserted between the K and $\pi\pi$ states can be enhanced by terms proportional to $1/m_{\mathrm{q}}$ where q are light u, d, or s quarks. So they contribute to the $|\Delta I| = \frac{1}{2}$ rule in the right direction.

**16.2 The decay constant $f_{\mathrm{B}}$.** To extract the decay constant $f_{\mathrm{B}}$, what are the relevant decay modes of the B meson to look for? Estimate the corresponding branching ratios.

**16.3 The rare decay $\mathrm{B} \to \gamma + \mathrm{K}^*$.** At the quark level, show that this mode is governed by $\mathrm{b} \to \gamma + \mathrm{s}$. Draw the corresponding Feynman diagram and estimate the amplitude using the calculation outlined in Chap. 11 for gluonic penguin. Explain why the mode $\mathrm{B} \to \gamma + \mathrm{K}$ is forbidden (more generally, any transition of the pseudoscalar mesons $0^- \to 0^- + \gamma$ vanishes). The rare modes are interesting due to possible effects of new particles in the loops.

**16.4 Relations among $\mathrm{B} \to \mathrm{K}(\mathrm{K}^*)$ and $\mathrm{D} \to \mathrm{K}(\mathrm{K}^*)$ form factors.** These form factors are of the type 'heavy-to-light' transitions, hence heavy flavor symmetry cannot be applied. However, using HFS, one may derive

$$\frac{\langle \mathrm{K}(p') \,|\, V_\mu \,|\, \mathrm{B}(v)\rangle}{\sqrt{M_\mathrm{B}}} = \frac{\langle \mathrm{K}(p') \,|\, V_\mu \,|\, \mathrm{D}(v)\rangle}{\sqrt{M_\mathrm{D}}}$$

up to corrections $\mathcal{O}(1/M)$. Since the form factors are evaluated with the *velocity* transfer $(v - v')^2$, show that

$$f_\pm^{\mathrm{B}\to\mathrm{K}}(t_\mathrm{B}) = \left[\frac{M_\mathrm{B} + M_\mathrm{D}}{2\sqrt{M_\mathrm{B} M_\mathrm{D}}} f_\pm^{\mathrm{D}\to\mathrm{K}}(t_\mathrm{D}) - \frac{M_\mathrm{B} - M_\mathrm{D}}{2\sqrt{M_\mathrm{B} M_\mathrm{D}}} f_\mp^{\mathrm{D}\to\mathrm{K}}(t_\mathrm{D})\right] , \quad (16.128)$$

where $t_\mathrm{B}$ and $t_\mathrm{D}$ are related by

$$v_\mathrm{B} \cdot v_\mathrm{K} = v_\mathrm{D} \cdot v_\mathrm{K} \Longrightarrow M_\mathrm{D} t_\mathrm{B} - M_\mathrm{B} t_\mathrm{D} = [M_\mathrm{B} - M_\mathrm{D}][M_\mathrm{B} M_\mathrm{D} - M_\mathrm{K}^2] . \quad (16.129)$$

The interest of these relations lies in the fact that the D→ K (K*) form factors can be determined by experiments on semileptonic D decays; whereas the B→ K (K*) form factors are not accessible by semileptonic modes. By factorization, the B→ K (K*) form factors are involved in nonleptonic color-suppressed decays. Compute the B→ $\mathrm{J}/\psi + \mathrm{K}$ rate in terms of the D→ K form factors.

**16.5 Inclusive B meson semileptonic branching ratio.** Compute the branching ratio $\mathrm{B_{sl}}$, using the formulas (57), (67), and (68)

$$\mathrm{B_{sl}} \equiv \frac{\Gamma(\mathrm{b} \to \mathrm{c} + \mathrm{e}^- + \overline{\nu}_\mathrm{e})}{\Gamma_\mathrm{tot}} .$$

**16.6 The CKM angles $\alpha$ and $\gamma$.** Show that the CP asymmetry between $\mathrm{B}^0 \to \pi^+ + \pi^-$ and $\overline{\mathrm{B}}^0 \to \pi^+ + \pi^-$ gives the $\alpha$ angle. The angle $\gamma$ comes from the asymmetry between the $\mathrm{B}_s^0 \to \rho^0 + \mathrm{K_S}$ and $\overline{\mathrm{B}}_s^0 \to \rho^0 + \mathrm{K_S}$. Show that the tree and penguin amplitudes of these modes do not have the same phase, in contradistinction with the best mode $\mathrm{J}/\psi + \mathrm{K_S}$.

## Suggestions for Further Reading

QCD *corrections to weak interaction, operator product expansion:*

Peskin, M. E. and Schroeder, D. V., *An Introduction to Quantum Field Theory.* Addison-Wesley, Reading, MA 1995

Wilson, K., Phys. Rev **179** (1969) 1499

*Heavy flavor symmetry:*

Isgur, N. and Wise, M. B., in B *Decays* (ed. Stone, S.). World Scientific, Singapore 1994

CP *violation:*

Bigi, I. I., Khoze, V. A., Uraltsev, N. G. and Sanda, A. I., in CP *violation* (ed. Jarlskog, C.). World Scientific, Singapore 1989

Buras, A. J. and Harlander, M. K., in *Heavy Flavors* (ed. Buras, A. J. and Lindner, M.). World Scientific, Singapore 1992

Nir, Y. and Quinn, H., in *B Decays* (op. cit.)

Rosner, J. L., *Present and Future Aspect of* CP *Violation*, Proc. of the VIII J. A. Swieca Summer School on Particles and Fields (ed. Barcelos-Neto, J., Novaes, S. F. and Rivelles, V. O.). World Scientific, Singapore 1996

# 17 Status and Perspectives of the Standard Model

As described in the preceding chapters, the $\mathrm{SU(3)_c \times SU(2)_L \times U(1)_Y}$ local gauge theory of strong and electroweak interactions accommodates all known elementary particles and incorporates the proven symmetries and successes of quantum electrodynamics, the Fermi–Feynman–Gell-Mann $V - A$ theory of weak interaction and the quark model. Despite intense experimental scrutiny, the theory has not yet displayed any signs of disagreement or inconsistency. The list of the predictions that are successfully tested is impressive; particularly noteworthy are the demonstrations of the existence and the properties of the neutral currents, of the $\mathrm{W}^{\pm}$ and $\mathrm{Z}^0$ gauge bosons, and of the $\tau$ lepton, the charm, bottom and top quarks.

The recent discovery of the top quark, in particular, is quite remarkable in this respect. Although the top quark is anticipated to exist as the weak-isospin partner of the bottom quark, the standard model gives no clear indications as to its mass. However, various ingenious theoretical arguments have succeeded in placing more and more stringent constraints on this property. Thus, at the beginning, a lower bound of $40\,\mathrm{GeV} < m_\mathrm{t}$ was inferred from the observation of the $\mathrm{B}^0$–$\overline{\mathrm{B}}^0$ mixing. Later, sophisticated calculations at the quantum level of the basic properties of the gauge bosons gave results that could be compared with data at the $10^{-3}$ level of accuracy, which led to improved predictions of the top quark mass. In particular, extremely precise data on the mass and width of $\mathrm{Z}^0$, and on the forward–backward asymmetry of fermions produced in the decay $\mathrm{Z}^0 \to \mathrm{f} + \bar{\mathrm{f}}$ were available which in turn determine $\sin^2\theta_\mathrm{W}$ precisely.
The most important contributions beyond the tree level to these processes come from the $\mathrm{Z}^0$ and W self-energies (Fig. 17.1) for which the top quark

**Fig. 17.1.** Electroweak corrections to the W and Z gauge boson propagators by the top quark

plays an essential role in the precise value of $\sin^2\theta_W$, making the contribution

$$\delta_{\sin^2\theta_W} = \frac{-3\alpha_{em}}{16\pi\sin^2\theta_W}\frac{m_t^2}{M_Z^2} . \tag{17.1}$$

From this result, it is inferred[1] that $m_t = 169 \pm 24$ GeV, to be compared with the experimental value of $180 \pm 12$ GeV.

Given its successes, the standard model should be accepted as a proven paradigm against which future experimental findings and alternative theories must be confronted. Yet, in spite of its many achievements, the model leaves unresolved several fundamental issues, most of which revolve around the breaking of the electroweak gauge symmetry. As we know, such a symmetry breaking is a necessary condition for the bosons and fermions in the theory to acquire masses by absorption of the Goldstone bosons associated with a spontaneous breaking of symmetry. But we do not have any idea what breaks it. In the standard model, this spontaneous symmetry breaking is realized via an ingenious although artificial process called the Higgs mechanism. The mechanism leaves behind a scalar boson as its distinctive signature, a signature as yet undetected. Therefore, the experimental observation of the Higgs boson is crucial for the confirmation of the standard model. We will examine in the following section some salient properties of the standard Higgs boson through its productions and decays. We next indicate some possible extensions of the standard electroweak model.

## 17.1 Production and Decay of the Higgs Boson

As explained in Chap. 9, a doublet of complex scalar fields $\phi(x)$ is postulated in the standard model. The dynamics of this scalar field is governed by the potential $V(\phi) = \lambda(\phi^\dagger\phi)^2 + \mu^2\phi^\dagger\phi$, on which the condition $\mu^2 < 0$ is imposed in order to spontaneously break the gauge symmetry and give masses to the gauge bosons, the quarks, and the charged leptons. These masses can be expressed in terms of the vacuum expectation value $\sqrt{2}\langle\phi\rangle = v = \sqrt{-\mu^2/\lambda}$ and nine arbitrary Yukawa couplings $C_f = \sqrt{2}m_f/v$, where f stands for one of the six quarks or one of the three charged leptons. Since their right-hand components are absent, the neutrinos are decoupled from the Higgs field and remain massless. Note that $v = (\sqrt{2}\,G_F)^{-1/2} \approx 246$ GeV.

The complex scalar field doublet $\phi$ has by definition four real components which may be written as

$$\phi = \begin{pmatrix}\varphi^+\\ \varphi^0\end{pmatrix} = \frac{1}{\sqrt{2}}\begin{pmatrix}\varphi^1 + i\varphi^2\\ \varphi^3 + i\varphi^4\end{pmatrix} . \tag{17.2}$$

The three massless Goldstone bosons $w^\pm = (\varphi^1 \mp i\varphi^2)/\sqrt{2}$ and $z = \varphi^4$ interact with the initially massless gauge fields $A^1_\mu$, $A^2_\mu$, $A^3_\mu$, and $B_\mu$ to become the

[1] Hollick, W. and Marciano, W. in *Precision Tests of the Standard Electroweak Model* (ed. Langacker, P.). World Scientific, Singapore 1995

longitudinal components of the massive physical gauge bosons $W^\pm$ and $Z^0$, while the component $\varphi^3 - v \equiv H$ emerges as the physical neutral Higgs scalar boson with mass $M_H = |\mu|\sqrt{2} = v\sqrt{2\lambda}$. This mass is a free parameter in the model. In a sense, the discovery of $W^\pm$ and $Z^0$ is equivalent to that of $w^\pm$ and $z$. The detection of the Higgs scalar boson stands out as a major goal of high-energy physics today and in the near future.

How could H be discovered ? Since the production of H and its subsequent decay depend on the Higgs boson mass, we first discuss the constraints on this parameter. The large electron–positron collider (LEP) experiments at CERN give 71 GeV as a lower bound for the Higgs boson mass. This lower bound is derived from the negative result to find H in the reaction $e^+ + e^- \to Z^0 + H$, which is shown by the diagram in Fig. 17.2. The Higgs boson could also be produced in proton–proton and proton–antiproton high-energy collisions; here the basic processes are the quark-pair annihilation $q\bar{q} \to V^* \to V + H$ and the gluon fusion gg→H (Fig. 17.3), with the latter being the dominant production mechanism.

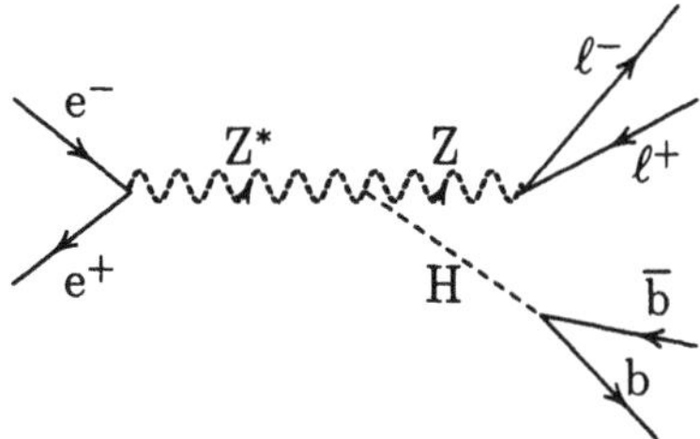


**Fig. 17.2.** Production of H by $Z^* \to Z + H$ and subsequent decay of Z and H

In general, the presence of a Higgs boson could be uncovered through its decay products, which may proceed through one of the following modes, with partial widths predicted by the simplest calculations in the standard model:
(i) $H \to V + V$, $H \to V + V^*$ (V is a real vector boson W or Z, and $V^*$ a virtual vector boson):

$$\Gamma(H \to Z^0 + Z^0) = \frac{1}{2}\frac{\alpha_{em}}{16\sin^2\theta_W}\frac{M_H^3}{M_W^2}\sqrt{1-x_z}\,[1 - x_z + \tfrac{3}{4}x_z^2]\,;$$

$$\Gamma(H \to W^+ + W^-) = \frac{\alpha_{em}}{16\sin^2\theta_W}\frac{M_H^3}{M_W^2}\sqrt{1-x_w}\,[1 - x_w + \tfrac{3}{4}x_w^2]\,; \qquad (17.3)$$

with $x_v = 4M_V^2/M_H^2$, the factor $\frac{1}{2}$ comes from identical $Z^0 + Z^0$ bosons.
(ii) $H \to f + \bar{f}$ (f is a lepton or a quark):

$$\Gamma(H \to f + \bar{f}) = \frac{\alpha_{em}\,N_c}{8\sin^2\theta_W}\frac{m_f^2}{M_W^2}\left(1 - \frac{4m_f^2}{M_H^2}\right)^{3/2} M_H\,, \qquad (17.4)$$

where $N_c = 3$ for quarks and 1 for leptons.

(iii) $\mathrm{H} \to \mathrm{g} + \mathrm{g}$ (g is a gluon):

$$\Gamma(\mathrm{H} \to \mathrm{g}+\mathrm{g}) = \frac{\alpha_{\mathrm{em}}}{\sin^2\theta_{\mathrm{W}}} \frac{\alpha_s^2}{32\pi^2} \frac{M_{\mathrm{H}}^3}{M_{\mathrm{W}}^2} \left| \sum_{\mathrm{q}} \tau_{\mathrm{q}} [1 + (1-\tau_{\mathrm{q}})\, h(\tau_{\mathrm{q}})] \right|^2 ,$$

$$h(\tau) = \left(\arcsin(\sqrt{1/\tau})\right)^2 \text{ or } h(\tau) = -\frac{1}{4}\left(\log \frac{1+\sqrt{1-\tau}}{1-\sqrt{1-\tau}} - \mathrm{i}\pi\right)^2 , \quad (17.5)$$

depending on $\tau \geq 1$ or $< 1$, where $\tau_{\mathrm{q}} = 4\, m_{\mathrm{q}}^2/M_{\mathrm{H}}^2$, and q stands for quarks.
(iv) $\mathrm{H} \to \gamma + \gamma$:

$$\Gamma(\mathrm{H} \to \gamma+\gamma) = \frac{\alpha_{\mathrm{em}}^3}{256\pi^2 \sin^2\theta_{\mathrm{W}}} \frac{M_{\mathrm{H}}^3}{M_{\mathrm{W}}^2} \left| \sum_j N_{c,j}(e_j^2)\, F_j(\tau_j) \right|^2 ,$$

$$F_1(\tau_1) = 2 + 3\tau_1[1 + (2-\tau_1)\, h(\tau_1)] , \qquad F_0(\tau_0) = \tau_0\, [1 - \tau_0\, h(\tau_0)] ,$$
$$F_{1/2}(\tau_{1/2}) = -2\tau_{1/2}\, [1 + (1-\tau_{1/2})\, h(\tau_{1/2})] . \quad (17.6)$$

Contributions to the amplitude for $\mathrm{H} \to \gamma\gamma$ in the standard model are represented by five loop diagrams, one with fermions, two with the gauge bosons W, and two with the scalar Goldstone bosons $w$ (those which become the longitudinal components of W) circulating in loops. In (6) the sum over $j$ stands for the contributions of the spin 0, 1/2, and 1 of these internal virtual particles. The electric charge $e_j$ is in units of $e$, and $N_{c,j}$ is the color multiplicity, i.e. 1 for W, $w$ and leptons, and 3 for quarks. The function $h(\tau)$ is given in (5), and $\tau_j = 4m_j^2/M_{\mathrm{H}}^2$.

Since the decay amplitudes (5) and (6) only come at $\mathcal{O}(gg_s^2)$ and $\mathcal{O}(g^3)$ perturbative orders, they are also sensitive to the physics beyond the standard model due to possibly unknown particles circulating in loop diagrams.

Note that compared with $\mathrm{H} \to \mathrm{VV}$ or $\mathrm{H} \to \mathrm{VV}^*$, the rates for $\mathrm{H} \to \mathrm{f\bar{f}}$, $\mathrm{H} \to \mathrm{gg}$, and $\mathrm{H} \to \gamma\gamma$ are reduced respectively by $\sim m_{\mathrm{f}}^2/M_{\mathrm{H}}^2$, $\alpha_s^2/2\pi^2$, and $\alpha_{\mathrm{em}}/16\pi^2$. Which of those modes would actually occur depends on the energy available. Thus, it is convenient from the observational point of view to divide the possible mass range of H into three distinct regions.

(a) In the mass range $M_{\mathrm{V}} < M_{\mathrm{H}} < 2M_{\mathrm{V}}$ (where V stands for W or Z), all modes, *except* the production of real vector bosons and top quarks, could occur, but the dominant decay mode, $\mathrm{H} \to \mathrm{V} + \mathrm{V}^*$, where $\mathrm{V}^*$ is an off-shell virtual V, is similar to the production mechanism $\mathrm{Z}^* \to \mathrm{Z} + \mathrm{H}$ of Fig. 17.2. The LEP and the $\mathrm{p\bar{p}}$ collider at Fermilab can cover the range $90 < M_{\mathrm{H}} < 130$ GeV. In this mass range, the productions and decays are dominated by the mechanism $\mathrm{V}^* \to \mathrm{V} + \mathrm{H}$.

For the relatively low Higgs mass in this range, the modes $\mathrm{H} \to \mathrm{Z} + \gamma$ and $\mathrm{H} \to \gamma + \gamma$ are interesting because of their distinctive signatures. The partial width of $\mathrm{H} \to \gamma + \gamma$ is reduced by a factor proportional to $\alpha_{\mathrm{em}}/16\pi^2$ to less than 2% of $M_{\mathrm{H}}$, even for $M_{\mathrm{H}}$ as large as $2M_{\mathrm{Z}}$.

(b) If $2M_V < M_H < 2m_t$, the Higgs boson may decay into *real* $W^+ + W^-$ or $Z^0 + Z^0$, mostly into their longitudinal components ($w^\pm$ or $z$) since these states arise from the Higgs mechanism. An explicit calculation gives

$$\frac{\Gamma(H \to V_T + V_T)}{\Gamma(H \to V_L + V_L)} = \frac{x_v^2}{2(1 - \frac{1}{2}x_v)^2} \quad \text{with} \quad x_v = \frac{4M_V^2}{M_H^2},$$

where $V_T$ ($V_L$) is a transversely (longitudinally) polarized gauge boson V. The third power dependence of the width on $M_H$ in (3) makes the width of H very large for a heavy Higgs boson. Numerically, the sum of the decay widths of H into $Z^0\,Z^0$ and $W^+\,W^-$ is $\approx \frac{1}{2}(M_H/\text{TeV})^3 \times$ TeV. However for $M_H \approx 300$ GeV, it is important to note that the Higgs boson width is still narrow; with $M_H \approx 300$ GeV, the width of H decay into the two dominant channels $W^+W^-$ and $Z^0\,Z^0$ is less than 10 GeV. These modes, having the signatures of four highly energetic charged leptons in the final state (by the subsequent decays of VV), would be spectacular. The next decay mode is the $b\bar{b}$-quark pair.

(c) A very heavy boson, $M_H > 2m_t$, could be seen at the future proton–proton large hadron collider (LHC) at CERN. The fusion of two gluons as depicted by the diagram in Fig. 17.3 is likely to be the dominant production mechanism with an effective Higgs boson–gluon–gluon coupling. This coupling may be obtained by computing the triangle loop diagram (Fig. 17.3) in which the top quark contribution dominates (Problem 17.3).

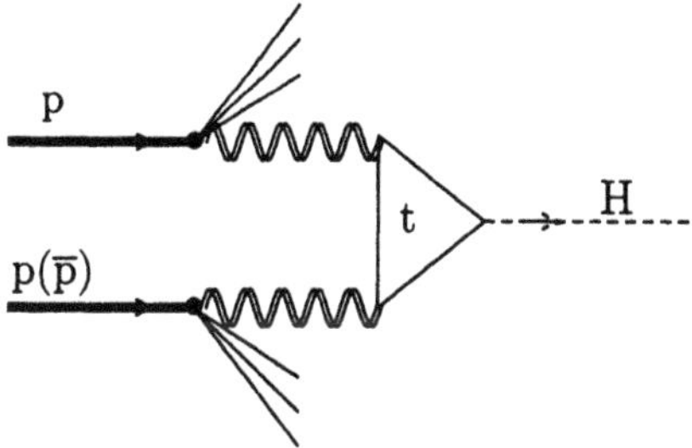


**Fig. 17.3.** Production of the Higgs boson by fusion of two gluons

## 17.2 Why Go Beyond the Standard Model?

If the standard model works so well, why must there be any new physics at all? Actually, there are many reasons why the electroweak theory cannot have the final word. For one, the standard model is largely silent on the issue of the origin of the Higgs boson, and this is because the spontaneous breaking of the electroweak SU(2) × U(1) symmetry is postulated in the model by the device of introducing a potential of scalar fields, $V(\phi)$, constructed just so that it can lead to such a breaking. And the condition for this to happen is $\mu^2 < 0$. This raises the question, what drives $\mu^2$ negative? Clearly the answer, if any, lies beyond the realm of the model, perhaps in some more complex, even fundamental mechanism. The spontaneous breaking of the electroweak $SU(2) \times U(1)$ symmetry may possibly be realized by an unknown mechanism where the scalar fields would not be needed.

Let us recall that the Higgs mechanism of the standard model, though inspired by Anderson's approach[2] of the screening current in solid state physics, lacks a dynamical basis, in contrast to superconductivity. As pointed out by Anderson, there are several phenomena in solid state physics which could be interpreted in terms of an effective, massive electromagnetic field, a typical example being the Meissner effect. If a normal metal is cooled down below its superconducting transition temperature, the flux of an external applied magnetic field will be abruptly expelled. The expulsion of the magnetic field from a conductor can be interpreted as if the photon were massive, with a mass $M$, such that the magnetic field $\boldsymbol{B}(x)$ derived from the solution of equation (8) behaves like $\boldsymbol{B}(x) = \boldsymbol{B}_0\,\mathrm{e}^{-Mx}$, and the field can only penetrate the conductor within a distance $x$ of order $M^{-1}$. The range $M^{-1}$ is called the screening length in condensed matter physics.

For a photon described by an electromagnetic field $A^\rho(t, \boldsymbol{x})$ or a static field $\boldsymbol{A}(\boldsymbol{x})$ that obeys the Maxwell equation

$$\left(\frac{\partial^2}{\partial t^2} - \nabla^2\right) A^\rho = J^\rho \implies \nabla^2 \boldsymbol{A} = -\boldsymbol{J} \tag{17.7}$$

to acquire a mass $M$, it must place itself in a situation where it satisfies the equation

$$\left(\frac{\partial^2}{\partial t^2} - \nabla^2 + M^2\right) A^\rho = 0 \implies \nabla^2 \boldsymbol{A} = M^2 \boldsymbol{A}\,. \tag{17.8}$$

In other words, the driving effective current $\boldsymbol{J}$ must be proportional to the field $\boldsymbol{A}$, i.e. $\boldsymbol{J} = -M^2 \boldsymbol{A}$ (or its covariant generalization $J^\rho$ is proportional to $A^\rho$). This relation which connects the current to the gauge field constitutes the Anderson mechanism that generates an effective mass for the photon, from which the Higgs model finds inspiration.

A specific example of magnetic screening is provided by a single nonrelativistic particle of mass $m$ and charge $e$. The associated current $\boldsymbol{J}$ is given by the usual quantum mechanics rules

$$\boldsymbol{J}_{\text{free}} = \frac{e}{2m}\left[\psi^*(-\mathrm{i}\nabla\psi) + \text{h.c.}\right]\,. \tag{17.9}$$

In the presence of an external electromagnetic field, $\nabla$ is replaced by the covariant derivative $\mathbf{D} = \nabla - \mathrm{i}e\boldsymbol{A}$, so that $\boldsymbol{J}$ becomes

$$\boldsymbol{J} = \boldsymbol{J}_{\text{free}} - \frac{e^2}{m}|\psi|^2 \boldsymbol{A}\,. \tag{17.10}$$

[2] Anderson, P. W., Phys. Rev. **130** (1963) 439

The term proportional to $\boldsymbol{A}$ will therefore yield an effective mass $M^2 = e^2|\psi|^2/m$ for the photon, and induces a diamagnetic screening of atomic electrons, characterized by a screening length $\sqrt{m}/|e|\,|\psi|$.

If this idea is applied to a superconductor, a physical interpretation must be found for the wave function $\psi$. Cooper showed that, in spite of the repulsive Coulomb force between them, two electrons with opposite spins can bind under certain circumstances, no matter how weak their mutual attraction is (the attraction force comes from the vibrations of lattice ions in the superconductor). Such a stable pair is called a Cooper pair which behaves as a boson of charge $-2e$. And the wave function $\psi$ that causes screening represents just the macroscopic coherent state of the plasma of Cooper electron pairs which undergo the Bose condensation.

Applied to particle physics, the vacuum expectation value $\langle\phi\rangle = v/\sqrt{2}$ plays the role of the condensated Cooper pairs, $\langle\overline{\psi}\psi\rangle$. The current of the scalar Higgs field $J^\rho = \phi^*(\partial^\rho\phi)+$ h.c. combined with the covariant derivative $D^\rho = \partial^\rho + \mathrm{i}\,g\mathrm{W}^\rho$ contains a term $(\frac{1}{2}g\,v)^2\mathrm{W}^\rho$. This term of $J^\rho$ being proportional to the gauge field $\mathrm{W}^\rho$, is responsible for its mass. However, contrary to superconductivity where the electron pairing can be derived from the principles of quantum mechanics, the Higgs mechanism of particle physics is described by an ad hoc assumption, $\mu^2 < 0$, that yields $v \neq 0$. The origin and the nature of this spontaneous symmetry-breaking must be found. Following the discovery of the top quark, it has been speculated that these very heavy quarks would somehow condensate and the resulting pairs would be the Cooper pairs of the electroweak particle physics. If so, the force that binds the top quark would be novel and should be found outside the standard model. The standard model would be the effective form of a certain more fundamental theory, valid only below a certain critical energy which defines its domain of validity. The new physics lies beyond this domain.

## 17.3 The Standard Model as an Effective Theory

It would be instructive to have another look at the 'old' weak interaction as formulated by Fermi, Feynman and Gell-Mann, Sudarshan and Marshak. As is well known, this theory is an effective low-energy limit of the Glashow–Salam–Weinberg electroweak theory. Starting from a four-fermion interaction

$$\frac{G_\mathrm{F}}{\sqrt{2}}[\overline{\psi}\gamma^\mu(1-\gamma_5)\psi]\,[\overline{\psi}\gamma_\mu(1-\gamma_5)\psi]\,, \tag{17.11}$$

one may calculate, for example, the fermion–fermion scattering amplitude f+f $\to$ f+f in the lowest order of $G_\mathrm{F}$. Since the amplitude is dimensionless, a simple dimensional argument shows that the leading term must be proportional to $G_\mathrm{F}E^2$, up to a numerical factor, where $E$ is the center-of-mass energy. At high energy, where the fermion masses may be neglected, explicit calculation gives $G_\mathrm{F}E^2(1+\cos\theta)$, i.e. at most only s- and p-partial waves contribute. There is a quick way to estimate the critical energy $E_\mathrm{cr}$ at which the Fermi theory becomes inconsistent and useless. It is a consequence of *unitarity* or probability conservation that any partial wave of the two-body scattering amplitude (parameterized by $e^{i\delta_l}\sin\delta_l$) must be bounded by 1, the real part of the partial wave amplitude is bounded by 1/2. This yields

$$E_{\rm cr} \approx \left(\frac{\sqrt{2}\,\pi}{G_{\rm F}}\right)^{1/2} \approx 600\ \text{GeV}\,. \tag{17.12}$$

This value tells us that the Fermi theory as described by (11) is at most a phenomenological model valid only at energies below $E_{\rm cr}$. Above this limit, the theory violates unitarity and becomes self-contradictory.

For the Fermi theory, the new physics is represented by the appearance of the W and Z bosons (Fig. 17.4), which must exist to regulate the growth of the four-fermion amplitude at high energy.

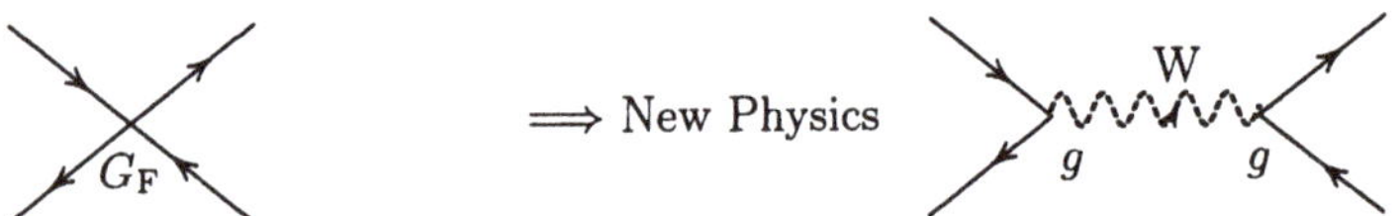


**Fig. 17.4.** For the Fermi theory, "new physics" is represented by the W boson

The presence of these intermediate bosons radically changes the energy dependence of the amplitude $G_{\rm F}E^2$ of the Fermi theory, so that

$$G_{\rm F}\,E^2 \Longrightarrow \frac{g^2E^2}{E^2 - M_{\rm W}^2} \tag{17.13}$$

and tends to a constant $g^2$ as $E \to \infty$. The unitarity constraint is well illustrated by the neutrino–electron cross-section given by (12.42)–(12.44). Without the W-boson propagator, this cross-section would increase linearly with energy and would violate Froissart–Martin bound.

The expansion parameter is a small coupling $g$ with $g^2/4\pi < 1$, so that the theory is predictive at all energies. The fact that new physics ($\sim$ 100 GeV) enters long before the critical energy 600 GeV can be related to a small coupling strength $g$.

### 17.3.1 Problems with the Standard Model

For the Fermi theory, the new physics which resolves the disastrous growth of the partial wave amplitudes is contained in the standard model. However, the electroweak model is not without its own problems.

To see this, we now consider the scattering W+W→ W+W of the gauge bosons as shown in Fig. 17.5. At high energy, the longitudinal component $\varepsilon_{\rm L}^{\mu}$ of the W dominates the amplitude, as can be seen directly from $\varepsilon_{\rm L}^{\mu} \sim p^{\mu}/M_{\rm W}$.

With only the gauge bosons $Z^0$ and $\gamma$ exchanged in the s- and t-channels (Fig. 17.5a), the W–W scattering amplitude $\mathcal{M}$ grows like $g^2E^2/M_W^2$, similar to $G_F E^2$ of the f+f → f+f scattering amplitude. The reason for $\mathcal{M} \sim E^2$ is that the three-vertex of the gauge bosons depends on momenta. Including the Higgs boson exchange contribution (Fig. 17.5b) eliminates the rising $\sim E^2$-behavior of $\mathcal{M}$, and the resulting amplitude W+W→ W+W becomes proportional to the Higgs boson mass, i.e. behaves like $g^2M_H^2/M_W^2$ (Problem 17.2). The similarity with (13) is striking:

$$\frac{g^2}{32\pi}\frac{E^2}{M_W^2} \Longrightarrow \frac{g^2}{32\pi}\frac{M_H^2}{M_W^2} = \frac{\lambda}{4\pi}\,, \tag{17.14}$$

which in turn implies that the amplitude is proportional to the self-coupling constant $\lambda$ of the Higgs field. As long as $\lambda$ is small enough, i.e. as long as the Higgs boson mass $M_H < 700$ GeV (see below), everything may be acceptable from the point of view of the perturbation theory, except that we still do not understand why the parameter $\mu^2$ is negative. For $\lambda \gg 1$ or $M_H \geq 700$ GeV, the standard model becomes unpredictive and we do not even know whether nature provides a Higgs boson to regulate the growth in energy of the W–W scattering amplitude.

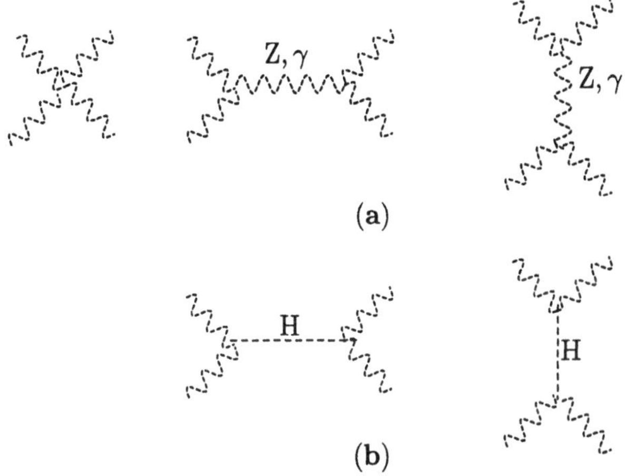


**Fig. 17.5.** W+W → W+W scattering (**a**) without, and (**b**) with the Higgs boson exchanged

Let us summarize the important points that can be drawn from this brief analysis of the W–W scattering.

(i) The Higgs boson is necessary to regulate the $E^2$-growth of the amplitude, otherwise unitarity would be violated.

(ii) If a Higgs boson can be found in future experiments with a relatively small mass, say $M_H < v = 246$ GeV, then we are in the weakly coupled regime of $\lambda$ and the theory remains consistent. But the question why $\mu^2 < 0$ will remain unanswered.

(iii) However, if the Higgs boson is too heavy, i.e. if the coupling constant $\lambda \gg 1$, terms of higher orders in $\lambda$ become increasingly more important and will get out of control, and the perturbative approach loses its usefulness.

All of these considerations strongly hint at the possibility that the electroweak standard model would somehow be embedded in a more fundamental theory. Which one, that is the most compelling question of today's particle physics. The upper bound of $\approx$ 700 GeV considered as the critical Higgs boson mass which separates the weakly coupled regime from the strongly coupled regime can be estimated from the renormalization group equation.

### 17.3.2 Renormalization Group Equation Analysis

In (2), the complex scalar doublet, conveniently considered as a set of four real fields, is governed by the self-interaction $\lambda(\phi^\dagger\phi)^2 = \frac{\lambda}{4}(\sum_i \varphi_i^2)^2, i = 1, \ldots, 4$. In the $\lambda\phi^4/4!$ theory for one real scalar field, the Callan–Symanzik $\beta$ function is $3\lambda^2/16\pi^2$ (Problem 15.4). In $\lambda(\phi^\dagger\phi)^2$ considered here, an additional combinatorial factor $8 = 4 \times 2$ enters and the corresponding $\beta(\lambda)$ function for the Higgs boson field is calculated to be $\beta(\lambda) = 8 \times (3\lambda^2/16\pi^2) = 3\lambda^2/2\pi^2$. The renormalization group equation for $\lambda$ is

$$\frac{\mathrm{d}\lambda}{\mathrm{d}\log(Q/v)} = \frac{3}{2\pi^2}\lambda^2 . \tag{17.15}$$

In this $\beta$-function, we have neglected contributions from fermions and gauge bosons coupled to the Higgs boson, since we are interested only in the limit of large $\lambda$ (large $M_\mathrm{H}$) for which the $\lambda^2$ term in the $\beta$-function dominates. The solution to (15) can be rewritten as

$$\frac{1}{\lambda(v)} - \frac{1}{\lambda(Q)} = \frac{3}{2\pi^2}\log(Q/v) \tag{17.16}$$

$$\text{or } \lambda(Q) = \frac{\lambda}{1 - \dfrac{3\lambda}{2\pi^2}\log(Q/v)} , \quad \lambda(v) \equiv \lambda . \tag{17.17}$$

This result shows that, regardless of how small $\lambda$ is, the coupling strength $\lambda(Q)$ grows with increasing energy. Using $M_\mathrm{H}^2 = 2\lambda v^2$, $\lambda(Q)$ becomes infinite at the scale $Q = E_\mathrm{cr}$ where the denominator of (17) vanishes, i.e.

$$\frac{E_\mathrm{cr}}{v} = \exp\left(\frac{4\pi^2 v^2}{3M_\mathrm{H}^2}\right) . \tag{17.18}$$

The formula (18) is remarkable because $M_\mathrm{H}$ is in the denominator of an exponential which makes the correlation between $E_\mathrm{cr}$ and $M_\mathrm{H}$ particularly interesting. Table 17.1 shows $M_\mathrm{H}$ for some selected values of $E_\mathrm{cr}$. For small $M_\mathrm{H} < 150$ GeV, the critical energy scale $E_\mathrm{cr}$ is very high $\sim 10^{18}$ GeV, and the Higgs model is valid at this high-energy scale. However, for large $M_\mathrm{H} \approx 700$

GeV, the critical energy $E_{cr}$ decreases exponentially so quickly that it nearly reaches $M_H$. In this case, the running coupling constant $\lambda$ blows up for $M_H$ not far from 700 GeV. Larger values of the Higgs boson mass are self-contradictory, since the cutoff $E_{cr}$ by definition cannot be smaller than the effective upper limit of the mass spectrum of the theory.

**Table 17.1.** $E_{cr}$ versus $M_H$

| $M_H$ in GeV | $E_{cr}$ in GeV |
|---|---|
| 150 | $6 \times 10^{17}$ |
| 200 | $1 \times 10^{11}$ |
| 300 | $2 \times 10^{6}$ |
| 500 | $6 \times 10^{3}$ |
| 700 | $1 \times 10^{3}$ |

The illustrative result in (18) may be interpreted as follows. Either the Higgs model is an effective Lagrangian of some unknown strong interaction at the scale $E_{cr}$, or at energies below $E_{cr}$, the standard model is embedded in a more fundamental theory where $E_{cr}$ acts as a cutoff.

Whatever the mechanism of the electroweak symmetry breaking, it would have very little impact on the precision electroweak data. Veltman has shown that the Higgs boson contribution to radiative corrections is screened by a slowly-varying logarithm function. For instance, the radiative correction to $\sin^2\theta_W$ by the virtual Higgs boson in loops is

$$\delta_{\sin^2\theta_W} = \frac{+5\alpha_{em}}{24\pi}\log\frac{M_H}{M_Z},$$

and should be compared with the quadratic dependence on $m_t$ of $\delta_{\sin^2\theta_W}$ given in (1). This explains why low-energy observables are relatively insensitive to the Higgs boson mass and illustrates the difficulty in devising experiments that can probe the Higgs sector by virtual quantum loop effects. This also provides an important constraint on any model of symmetry breaking beyond the standard model.

### 17.3.3 Supersymmetry and Technicolor

The dichotomy of weakly coupled regime (small $\lambda$) and the strongly coupled regime (large $\lambda$) provides a framework for examining new physics beyond the standard model. Let us just mention two possible dynamical mechanisms of breaking the electroweak symmetry: Supersymmetry and technicolor which are respectively associated with the relatively light and heavy Higgs mass, i.e. the weak- and strong-coupling regimes of $\lambda$.

**Supersymmetry.** From general considerations, supersymmetry (SUSY) is the only nontrivial extension of the Lorentz group. The simplest extension, called $N = 1$ SUSY, requires the introduction of a single anticommuting degree of freedom to space-time. This implies, for example that a spin-1/2 field $\psi$ is necessarily associated with a scalar spin-0 field $\phi$. Thus, supersymmetry is a symmetry that links bosons and fermions. Why is SUSY relevant to the electroweak symmetry breaking? A quick answer is that SUSY may offer a framework for the scalar Higgs field to participate naturally in the weak interaction on the same footing as leptons and quarks. SUSY is particularly well suited to the weakly coupled regime ($\lambda < 1$). It may give rise to a mechanism of electroweak symmetry breaking associated with the top quark using the renormalization group evolution. It is beyond the scope of this book to explain this point. Let us only mention that in $N = 1$ SUSY, there are two complex Higgs field doublets. One of these doublets, related to the heaviness of the top quark, has an interacting potential unstable by the renormalization group evolution, such that $\mu^2$ could be driven naturally to a negative value.[3]

**Technicolor.** Technicolor is directly inspired by the following fact. In the standard QCD and electroweak interaction, the three initially massless gauge bosons $A^i$, associated with the generators of $\mathrm{SU(2)_L}$ and introduced in (9.42), already acquire a tiny mass equal to $gf_\pi/(2\sqrt{2}) \approx 31$ MeV via the pion considered as a Goldstone boson. This can be seen as follows.

QCD with two massless u and d quarks has a global $\mathrm{SU(2)_L \times SU(2)_R}$ symmetry represented by the doublets $\mathrm{q_L = (u_L, d_L)}$ and $\mathrm{q_R = (u_R, d_R)}$ which can be independently rotated in their respective SU(2) spaces. These two $\mathrm{SU(2)_L}$ and $\mathrm{SU(2)_R}$ groups are linked by the pairing of q and $\overline{\mathrm{q}}$ in the vacuum so that the operator $\overline{\mathrm{q}}\mathrm{q}$ acquires a nonvanishing vacuum expectation value. The overall symmetry $\mathrm{SU(2)_V}$ corresponding to $\mathrm{L \oplus R}$ is unbroken and gives the isospin symmetry of QCD. The other $\mathrm{SU(2)_A}$ (from $\mathrm{L \ominus R}$) associated with the axial current $a^i_\mu = \overline{q}\gamma_\mu\gamma_5\tau^i q$ ($\tau^i$ are the three Pauli matrices) is spontaneously broken, resulting in three Goldstone bosons, or pions. The matrix element of the current $a^i_\mu$ between the pion and the vacuum is well known; it is

$$\langle 0 \,|\, a^i_\mu \,|\, \pi^j(k)\rangle = \mathrm{i}\frac{f_\pi}{\sqrt{2}}\, k_\mu \delta^{ij} \quad , \quad f_\pi \approx 131 \text{ MeV} .$$

Even without the Higgs mechanism, the massless boson $A^i(x)$, when coupled to the current $J^i_\mu = 1/2(v^i_\mu - a^i_\mu)$ built up by the u and d quark fields, allows the creation of a pion with amplitude $\mathrm{i}g(-\frac{1}{2})(\mathrm{i}f_\pi/\sqrt{2}\, k_\mu)$, in which the factor $-1/2$ comes from the coefficient of $a^i_\mu$ in $J^i_\mu = 1/2(v^i_\mu - a^i_\mu)$.

The contribution of the pion to the vacuum polarization $\Pi_{\mu\nu}(k)$ of the $A^i$ boson as depicted in Fig. 17.6 has a singularity $1/k^2$ near $k^2 = 0$. The residue

[3] Ibañez, L. E. and Ross, G. G., in *Perspective on Higgs Physics* (ed. Kane, G.). World Scientific, Singapore 1992

at the $k^2 = 0$ pole is $\rho = [g f_\pi / 2\sqrt{2}]^2$. Together with the conservation of the current $J^i_\mu$, the vacuum polarization must satisfy $k^\mu \Pi_{\mu\nu}(k) = 0$, so that near the $k^2 = 0$ pole, the vacuum polarization $\Pi_{\mu\nu}(k)$ has the form

$$\Pi^{ij}_{\mu\nu}(k) = \left(g_{\mu\nu} - \frac{k_\mu k_\nu}{k^2}\right)\left(\frac{g f_\pi}{2\sqrt{2}}\right)^2 \delta^{ij} = \left(g_{\mu\nu} - \frac{k_\mu k_\nu}{k^2}\right)\rho\, \delta^{ij} .$$

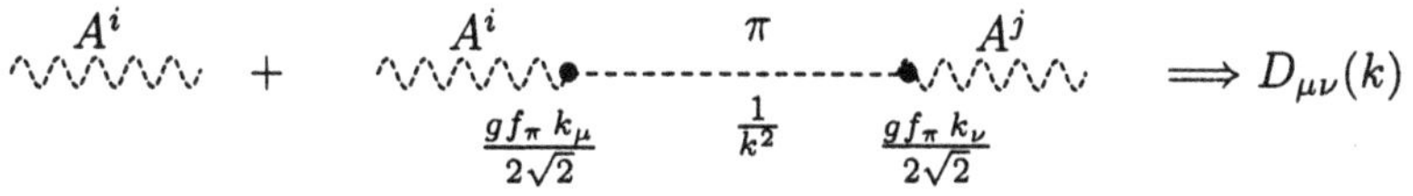

$$A^i \quad + \quad A^i \xrightarrow{\frac{g f_\pi k_\mu}{2\sqrt{2}}} \pi \left(\frac{1}{k^2}\right) \xrightarrow{\frac{g f_\pi k_\nu}{2\sqrt{2}}} A^j \quad \Longrightarrow D_{\mu\nu}(k)$$

**Fig. 17.6.** The Goldstone pion of QCD gives mass to the gauge boson $A^i$

As in (15.9)–(15.11) and following the discussion below (15.11), the propagator of the $A^i$ boson, dressed by the pion, may be written as

$$D_{\mu\nu}(k) = \frac{-i g_{\mu\nu}}{k^2(1 - \rho/k^2)} .$$

It has a pole at $k^2 = \rho$, i.e. the massless $A^i$ boson gets a mass $g f_\pi / 2\sqrt{2}$ by absorbing the Goldstone pion coming from the spontaneous $\mathrm{SU(2)_A}$ symmetry breaking of QCD. So QCD already can give a mass $g f_\pi / 2\sqrt{2} \approx 31$ MeV to the $A^i$ boson, which may eventually emerge as the W boson.

Since the true W boson mass is $gv/2$, Susskind and Weinberg proposed technicolor as a copy of QCD scaled up by the factor $v\sqrt{2}/f_\pi \approx 2600$, with a techni-pion having a decay constant $\widetilde{F}_\pi = v$. This techni-pion is the Goldstone boson built up from $\widetilde{\mathrm{U}}$ and $\widetilde{\mathrm{D}}$ techni-quarks and would be responsible for the weak bosons masses.

Both SUSY and technicolor have rich spectra of new particles. Masses of techni-hadrons are expected in the TeV region, whereas some particles in SUSY may have masses in the range of a few hundred GeVs. The reader is referred to the very abundant literature on the subject (Further Reading).

**Perspectives.** Consistency of the standard model requires that the new physics responsible for mass generation may occur at an energy scale of about 1 TeV or less. The future high-energy colliders, in particular the LHC at CERN, are intended to explore this energy region. However, discoveries may also come from the lower-energy, high-precision, high-intensity physics in which heavy flavors, in particular the B meson, are important. Advances in particle physics may well lie in the least expected directions, but a study of the standard model suggests that we should address the following questions:

(i) Top quark physics: Why are all other fermions so much lighter? Does the top quark have something to do with the gauge symmetry breaking?

(ii) Neutrino masses and mixing: If neutrinos are truly neutral Majorana particles, there is at least one more possibility for the neutrinos to mix than for the quarks (which are strictly Dirac particles). The Majorana neutrino masses can only be generated outside the standard model.

(iii) Nonstandard CP violation: Is there any other mechanism of CP violation than the KM one, where only charged currents are involved ?

## Problems

**17.1 Higgs boson in** $e^+ + e^- \to W^+ + W^-$. How many tree diagrams are there for the above reaction? Show that, without the Higgs boson exchanged, the amplitude blows up as $s^{1/2}$ for the production of longitudinally polarized W + W, where $s^{1/2}$ is the total energy in the center-of-mass system.

**17.2 Amplitude** $W_L + W_L \to W_L + W_L$ **at high energy.** From the five diagrams of Fig. 17.5, write down the amplitude of the longitudinally polarized W–W scattering in the limit $s \gg M_H^2, M_W^2, M_Z^2$, and check (3).

**17.3 H→ g+ g triangle loop.** The amplitude of the Higgs field interacting with two gluons (Fig. 17.3) of momenta $k_1, k_2$ and polarizations $\varepsilon_\mu(k_1)$, $\varepsilon_\nu(k_2)$ has the following form

$$A(k_1,k_2)\,[k_2^\mu k_1^\nu - k_1\cdot k_2 g^{\mu\nu}]\,\varepsilon_\mu(k_1)\,\varepsilon_\nu(k_2) \equiv I^{\mu\nu}\,\varepsilon_\mu(k_1)\,\varepsilon_\nu(k_2)$$

which satisfies the gauge invariance condition $(k_1)_\mu I^{\mu\nu} = (k_2)_\nu I^{\mu\nu} = 0$. The effective Higgs boson–gluon–gluon coupling is described by the coefficient $A(k_1,k_2)$. Compute $A(k_1,k_2)$ from the triangle diagram of Fig. 17.3 with only the internal top quark. One should recover the function $F_{1/2}$ in (6).

## Suggestions for Further Reading

*Production and decay of the Higgs boson:*

Gunion, J. F., Haber, H. E., Kane, G. and Dawson, S., *The Higgs Hunter's Guide.* Addison-Wesley, Menlo Park, CA 1990

*Beyond the standard model:*

Bardeen, W. A. in *Proc. 17th Int. Symp. on Lepton–Photon Interactions, 1995* Beijing (ed. Zheng Zhi-Peng and Chen He-Sheng). World Scientific, Singapore 1996

Fayet, P. in *History of Original Ideas and Basic Discoveries in Particle Physics* (ed. Newman, H. B. and Ypsilantis, T.). Plenum, New York 1995

Kane, G. (ed.), *Perspectives on Higgs Physics.* World Scientific, Singapore 1993

Nilles, H. P., Phys. Rep. **110** (1984) 1

Parsa, Z. (ed.), *Future High Energy Colliders.* AIP Conference Proceedings 397, Woodbury, New York 1997

Peskin, M. E., in *Proc. 1996 European School of High Energy Physics* (ed. Ellis, N. and Neubert, M.). CERN 97 -03

Wilczek, F., in *Critical Problems in Physics* (ed. Fitch, V., Marlow, D. and Dementi, M.). Princeton Series in Physics, Princeton 1997

Witten, E., *Duality, Spacetime and Quantum Mechanics.* Physics Today, May 1997

# Selected Solutions

**1.4** (a) The Bohr's radius is the radius of the lowest stable orbit of the atomic electron defined by the stability condition $dE/dr = 0$, where $E \approx p^2/(2m_e) - \alpha/r \approx 1/(2m_e r^2) - \alpha/r$. This gives $r = 1/(\alpha m_e)$, or $r \approx 5 \times 10^{-9}$ cm.

(b) The energy of the system being $1/(m_N r) - G m_N^2/r$, one gets in the same way $r = 2/G m_N^3$ (note the presence of the reduced mass), which leads to $r = 6 \times 10^{24}$ cm or $6 \times 10^6$ light-years. Alternatively, starting from (i) the equilibrium condition $G m_N^2/r^2 = m_N \omega^2 (r/2)$; and (ii) the quantization $2 m_N \omega (r/2)^2 = n\hbar$, obtain $r = 2n^2\hbar^2/G m_N^3$, which gives the same result for $n = 1$.

**2.1** (a) In their CM frame, the two photons are defined by their momenta $\boldsymbol{k}$, $-\boldsymbol{k}$, and polarizations $\boldsymbol{\epsilon}_1$ and $\boldsymbol{\epsilon}_2$, which satisfy $\boldsymbol{k} \cdot \boldsymbol{\epsilon}_1 = \boldsymbol{k} \cdot \boldsymbol{\epsilon}_2 = 0$. From these vectors, we want to construct a vector $\boldsymbol{A}$ that is a homogeneous linear function of $\boldsymbol{\epsilon}_1$ and $\boldsymbol{\epsilon}_2$, and symmetric in the simultaneous permutations $\boldsymbol{k} \leftrightarrow -\boldsymbol{k}$, $\boldsymbol{\epsilon}_1 \leftrightarrow \boldsymbol{\epsilon}_2$. There are only two combinations compatible with these conditions: $(\boldsymbol{\epsilon}_1 \times \boldsymbol{\epsilon}_2) \times \boldsymbol{k}$ and $(\boldsymbol{\epsilon}_1 \times \boldsymbol{k})(\boldsymbol{\epsilon}_2 \cdot \boldsymbol{k}) + (\boldsymbol{\epsilon}_2 \times \boldsymbol{k})(\boldsymbol{\epsilon}_1 \cdot \boldsymbol{k})$. Both vanish because of the transversality condition for real photons.

(b) Assume that the spin of $\pi^0$ is 1. Consider $\pi \to 2\gamma$. In the $\pi$ rest frame, the initial angular momentum is $J_i = 1$. By angular momentum conservation, the final angular momentum is also $J_f = 1$. Since it is impossible to have a state of two real photons of angular momentum 1 [as in (a)], the assumption $J_\pi = 1$ does not hold.

(c) In $\pi^0$ rest frame, both momentum and angular momentum vanish, hence $J_i = J_{\pi^0} = 0$. By angular momentum conservation, the total angular momentum of the photons is $J_f = 0$ and their individual spins are opposite. It follows that the photons have the same polarizations: $\boldsymbol{J} \cdot \boldsymbol{k} = (-\boldsymbol{J}) \cdot (-\boldsymbol{k})$, where $\boldsymbol{k}$ is the momentum of one photon and $-\boldsymbol{k}$ that of the other. To be definite, let $\boldsymbol{k} = k\hat{z}$. Possible states of polarization are $\phi_{RR} = \epsilon_1(\hat{z},+)\epsilon_2(-\hat{z},+)$ and $\phi_{LL} = \epsilon_1(\hat{z},-)\epsilon_2(-\hat{z},-)$, or their combinations $\phi_{RR} + \phi_{LL} = -\boldsymbol{\epsilon}_1(\hat{z}) \cdot \boldsymbol{\epsilon}_2(-\hat{z})$ and $\phi_{RR} - \phi_{LL} = i\boldsymbol{\epsilon}_1(\hat{z}) \times \boldsymbol{\epsilon}_2(-\hat{z}) \cdot \boldsymbol{k}$.

**2.2** We know that $\pi^0$ has spin $s = 0$. In the decay mode $K^0 \to 2\pi^0$, the angular momentum in the final state is $J_f = \ell$, where $\ell$ is the relative orbital angular momentum of the two mesons. By angular momentum conservation, $J_i = J_f = \ell$. If $K^0$ spin is odd, $\ell$ is also odd. Then, the angular wave function of the final state would change sign in a permutation of the two mesons, $Y_\ell \to (-)^\ell Y_\ell$, in violation of Bose statistics. So $K^0$ cannot have an odd-integral spin.

**2.3** We consider the scale transformation on coordinates and fields defined by $x'_\mu = \lambda^{-1} x_\mu$ and $\phi'(x) = \mathrm{e}^{D \ln \lambda} \phi(\lambda x)$. For infinitesimal $\ln\lambda = \delta\varepsilon$, $\delta x_\mu = -\delta\varepsilon x_\mu$, and $\delta\phi = \delta\varepsilon\,(x^\nu \partial_\nu + D)\,\phi(x)$. Suppose $\mathcal{L} = \frac{1}{2}(\partial\phi)^2 - \frac{1}{2} m^2\phi^2 - \frac{1}{4!} g\phi^4$ is the Lagrangian of the model. Its variation under a scale transformation is

$$(\partial\mathcal{L}/\partial\varepsilon) = x^\nu \partial_\nu \mathcal{L} + (D+1)(\partial\phi)^2 - \tfrac{1}{6} g D \phi^4 - D m^2 \phi^2 .$$

For $D = 1$ this becomes

$$(\partial\mathcal{L}/\partial\varepsilon) = (x^\nu \partial_\nu + 4)\,\mathcal{L} + m^2\phi^2 = \partial_\nu(x^\nu \mathcal{L}) + m^2\phi^2 .$$

The variation of the action includes the variation of $\mathcal{L}$ and the variation of the volume $\delta(\mathrm{d}^4x) = -\delta\varepsilon(\partial_\mu x^\mu)\mathrm{d}^4x = -\delta\varepsilon\,\mathrm{d}^4x$. For a field satisfying the equation of motion, $\mathcal{L} = 0$, and the variation of the volume makes no contribution. Hence the action varies as $\delta S = \delta\varepsilon\, m^2 \int \mathrm{d}^4x\, \phi^2$, which vanishes for $m = 0$. In this case, the model is scale-invariant.

**2.4** (a) The Noether current associated with translation is

$$\mathcal{T}^\mu{}_\nu = \frac{\partial\mathcal{L}}{\partial(\partial_\mu A_\lambda)} \partial_\nu A_\lambda - \delta^\mu{}_\nu \mathcal{L} = -F^{\mu\lambda}\partial_\nu A_\lambda - \delta^\mu{}_\nu \mathcal{L} .$$

For a Lorentz transformation, the Noether current is

$$\mathcal{M}^\mu_{\rho\sigma} = \frac{\partial\mathcal{L}}{\partial(\partial_\mu A_\lambda)}(-\mathrm{i} J_{\rho\sigma} A_\lambda) + \mathcal{L}(\delta^\mu{}_\rho x_\sigma - \delta^\mu{}_\sigma x_\rho) .$$

With $J_{\rho\sigma} = L_{\rho\sigma} + \Sigma_{\rho\sigma}$, $L_{\rho\sigma} = \mathrm{i}(x_\rho\partial_\sigma - x_\sigma\partial_\rho)$, and $(\Sigma_{\rho\sigma})^\alpha{}_\beta = \mathrm{i}(\delta^\alpha_\rho g_{\sigma\beta} - \delta^\alpha_\sigma g_{\rho\beta})$, one finds the result as given. In a gauge where $A^0 = 0$, one has

$$\mathcal{M}^0_{ij} = \sum_{k=1}^{3} \dot{A}^k (x_i\partial_j - x_j\partial_i) A^k \;-\; (\dot{A}_i A_j - \dot{A}_j A_i).$$

(b) In particular, the intrinsic spin part is

$$\begin{aligned} S_{ij} &= -\int \mathrm{d}^3x (\dot{A}_i A_j - \dot{A}_j A_i) \\ &= -\mathrm{i} \sum_k \sum_{\lambda\lambda'} \epsilon_i(k,\lambda')\epsilon_j(k,\lambda)[a^\dagger(\boldsymbol{k},\lambda')a(\boldsymbol{k},\lambda) - a^\dagger(\boldsymbol{k},\lambda)a(\boldsymbol{k},\lambda')]. \end{aligned}$$

With $\boldsymbol{k}$ along $\hat{z}$, we get $\epsilon^i(k,\lambda) = \delta_{i\lambda}$, for $i = \lambda = 1, 2$, and hence the result.

**3.1** For a particle of mass $m$ boosted from rest, Example 2 in Chap. 3 gives the transformation matrix for spinors

$$S_{\mathrm{L}} = \cosh\frac{\omega}{2}\left[1 + \tanh\frac{\omega}{2}\begin{pmatrix} 0 & \boldsymbol{\sigma}\cdot\hat{\boldsymbol{p}} \\ \boldsymbol{\sigma}\cdot\hat{\boldsymbol{p}} & 0 \end{pmatrix}\right] .$$

Here, $E = m\cosh\omega$ and $p = E\tanh\omega$, so that

$$\cosh\frac{\omega}{2} = \sqrt{\frac{E+m}{2m}}\,, \quad \sinh\frac{\omega}{2} = \frac{p}{\sqrt{2m(E+m)}}\,.$$

A particle at rest is described by the wave function $\psi_{0,s} = f(t)u(\mathbf{0},s)$, where $f(t) = \mathrm{e}^{-\mathrm{i}mt}/\sqrt{2m(2\pi)^3}$. Boosting results in

$$\psi_{p,s} = f(t)\cosh\frac{\omega}{2}\begin{pmatrix} 1 \\ \tanh\frac{\omega}{2}\,\boldsymbol{\sigma}\cdot\hat{\boldsymbol{p}}\end{pmatrix}\chi_s\,.$$

As for the space-time part, note that in the same transformation, the new coordinates are given by $t' = t\cosh\omega + x\sinh\omega\,,\quad x' = t\sinh\omega + x\cosh\omega$, so that $mt = Et' - p_x x' = p\cdot x'$. Therefore,

$$S_{\mathrm{L}}\psi_{0,s}(t,x) = \sqrt{\frac{E}{m}}\frac{\mathrm{e}^{-\mathrm{i}p\cdot x'}}{\sqrt{2E(2\pi)^3}}u(\boldsymbol{p},s) = \sqrt{\frac{E}{m}}\,\psi_{p,s}(x')\,.$$

**3.2** See Good, Rev. Mod. Phys. **27** (1955) 187.

**3.3** (a) $(\overline{\psi}\Gamma\psi)^* = \psi^\dagger\Gamma^\dagger\gamma_0\psi = \overline{\psi}\tilde{\Gamma}\psi$. If $\Gamma$ are chosen such that $\tilde{\Gamma} \equiv \gamma_0\Gamma^\dagger\gamma_0 = \Gamma$, then hermiticity is proven.

(b) To obtain the transformation properties of the bilinear covariants, one uses the basic properties of $\gamma_\mu$ and of $\psi(x)$:

$$\overline{\psi}'(x')\Gamma\psi'(x') = \overline{\psi}(x)S^{-1}\Gamma S\psi(x)\,; \qquad S^{-1}\gamma^\mu S = a^\mu{}_\nu\gamma^\nu\,.$$

**3.4** Majorana: $\gamma^\mu_{\mathrm{M}} = S_{\mathrm{M}}\gamma^\mu_{\mathrm{D}}S_{\mathrm{M}}^{-1}$, $\qquad S_{\mathrm{M}} = \frac{1}{\sqrt{2}}\begin{pmatrix} 1 & \sigma_y \\ \sigma_y & -1\end{pmatrix}$;

Majorana spinor: $u_{\mathrm{M}} = S_{\mathrm{M}}u_{\mathrm{D}}(p,s)$.

Weyl: $\gamma^\mu_{\mathrm{W}} = S_{\mathrm{W}}\gamma^\mu_{\mathrm{D}}S_{\mathrm{W}}^{-1}$, $\qquad S_{\mathrm{W}} = \frac{1}{\sqrt{2}}\begin{pmatrix} 1 & -1 \\ 1 & 1\end{pmatrix}$;

Weyl spinor: $u_{\mathrm{W}} = S_{\mathrm{W}}u_{\mathrm{D}}(p,s)$.

**3.7** Start with the Dirac equations $(\mathrm{i}\partial\!\!\!/ - m_1)\psi_1 = 0$ and $\overline{\psi}_2(\mathrm{i}\gamma^\nu\overleftarrow{\partial}_\nu + m_2) = 0$. Multiply the first on the left by $\overline{\psi}_2\gamma_\mu$ or $\overline{\psi}_2\gamma_\mu\gamma_5$, and the second on the right by $\gamma_\mu\psi_1$ or $\gamma_\mu\gamma_5\psi_1$, and take the differences of the resulting expressions to obtain, with the help of the identity $\gamma_\mu\gamma_\nu = g_{\mu\nu} - \mathrm{i}\sigma_{\mu\nu}$,

$$\begin{aligned}
\text{(a)}\quad (m_1+m_2)\overline{\psi}_2\gamma_\mu\psi_1 &= \mathrm{i}\overline{\psi}_2\left(-\overleftarrow{\partial}^\nu\gamma_\nu\gamma_\mu + \gamma_\mu\gamma_\nu\overrightarrow{\partial}^\nu\right)\psi_1 \\
&= \overline{\psi}_2\left(-\mathrm{i}\overleftarrow{\partial}_\mu + \mathrm{i}\overrightarrow{\partial}_\mu\right)\psi_1 + \partial^\nu(\overline{\psi}_2\sigma_{\mu\nu}\psi_1)\,,
\end{aligned}$$

$$\begin{aligned}
\text{(b)}\quad (m_1+m_2)\overline{\psi}_2\gamma_\mu\gamma_5\psi_1 &= \mathrm{i}\overline{\psi}_2\left(-\overleftarrow{\partial}^\nu\gamma_\nu\gamma_\mu\gamma_5 + \gamma_\mu\gamma_5\gamma_\nu\overrightarrow{\partial}^\nu\right)\psi_1 \\
&= -\mathrm{i}\partial_\mu(\overline{\psi}_2\gamma_5\psi_1) + \overline{\psi}_2\left(\overleftarrow{\partial}^\nu - \overrightarrow{\partial}^\nu\right)\sigma_{\mu\nu}\gamma_5\psi_1\,.
\end{aligned}$$

To write these relations in momentum space, take $\psi(x) = e^{-ip\cdot x}u(p)$ or $\psi(x) = e^{+ip\cdot x}v(p)$, depending on whether it is a particle or antiparticle state.

**3.8** Write the basic expansion relation in spinor components $(a, b = 1, \ldots, 4)$

$$(\bar{u}_{1a}\Gamma^i_{ab}u_{2b})(\bar{u}_{3c}\Gamma^j_{cd}u_{4d}) = C^{ij}_{k\ell}(\bar{u}_{1a}\Gamma^k_{ad}u_{4d})(\bar{u}_{3c}\Gamma^\ell_{cb}u_{2b}).$$

We will assume that $u_a$ are all c-valued, so that $\Gamma^i_{ab}\Gamma^j_{cd} = C^{ij}_{k\ell}\Gamma^k_{ad}\Gamma^\ell_{cb}$. (If $u_a$ are anticommuting operators, there is an overalll relative minus sign.) Multiply both sides of this equation by $\Gamma^m_{bc}\Gamma^n_{da}$ and sum over all spinor indices to get

$$\mathrm{Tr}(\Gamma^i\Gamma^m\Gamma^j\Gamma^n) = C^{ij}_{nm}\,N_n N_m\,, \qquad \text{where} \quad \mathrm{Tr}(\Gamma^i\Gamma^j) = N_i\delta_{ij}.$$

**3.10** (a) Consider first the production reaction $\pi^- + p \to K^0 + \Lambda^0$, and choose the center-of-mass frame. The total angular momentum component of the initial state is given by $J^{\mathrm{i}}_z = \ell_z + S^{(\pi)}_z + S^{(p)}_z$. Since $S^{(\pi)} = 0$ and since $\ell_z = 0$ (because the beam direction is chosen along the $z$ axis), one has $J^{\mathrm{i}}_z = S^{(p)}{}_z = \pm 1/2$. In the final state, we observe particles $K$, $\Lambda$ produced in the incident beam direction, and so $L_z = xp_y - yp_x = 0$, so that again $J^{\mathrm{f}}_z = S^\Lambda_z$. By angular momentum conservation, $J^{\mathrm{i}}_z = J^{\mathrm{f}}_z$, or $S^\Lambda_z = \pm 1/2$, which means that $S^\Lambda \geq 1/2$.

(b) Now consider the decay $\Lambda \to \pi + p$ in the $\Lambda$ rest frame. The initial angular momentum is $\boldsymbol{J} = \boldsymbol{S}^\Lambda$, and by angular momentum conservation it is equal to $\vec{\ell} + 1/2$, where $\ell$ is the relative orbital angular momentum of $\pi$–$p$, so that $|S^\Lambda - 1/2| \leq \ell \leq S^\Lambda + 1/2$. The decay amplitude is given by the decomposition of the final wave function in terms of the proton spin states, $\alpha = |+1/2\rangle$ and $\beta = |-1/2\rangle$, and the orbital states $Y_{\ell m}(\theta, \varphi)$. The coefficients of the decomposition are the Clebsch–Gordan coefficients.

- $S^{(\Lambda)} = 1/2$ ; the possible values of $\ell$ are $0, 1$.
  $\ell = 0$: $\left|J = \frac{1}{2}, J_z = \frac{1}{2}\right\rangle = \left\langle \frac{1}{2}\,\frac{1}{2}\,\middle|\ell = 0, 0; s_p = \frac{1}{2}, \frac{1}{2}\right\rangle Y_{00}\,\alpha = Y_{00}\,\alpha.$
  angular distribution: isotropic.
  $\ell = 1$: $\left|J = \frac{1}{2}, J_z = \frac{1}{2}\right\rangle = -\sqrt{1/3}\,Y_{10}\,\alpha + \sqrt{2/3}\,Y_{11}\beta,$
  angular distribution: $\frac{1}{3}|Y_{10}|^2 + \frac{2}{3}|Y_{11}|^2 = \text{const.}$

- $S^{(\Lambda)} = 3/2$ ; the possible values of $\ell$ are $1, 2$.
  $\ell = 1$: $\left|J = 3/2, J_z = \frac{1}{2}\right\rangle = \sqrt{2/3}\,Y_{10}\alpha + \sqrt{1/3}\,Y_{11}\beta,$
  angular distribution: $\frac{2}{3}|Y_{10}|^2 + \frac{1}{3}|Y_{11}|^2 = \frac{1}{6}(1 + 3\cos^2\theta).$
  $\ell = 2$ : same result.

- $S^{(\Lambda)} = 5/2$ ; the possible values of $\ell$ are $2, 3$.
  $\ell = 2$: $\left|J = 5/2, J_z = \frac{1}{2}\right\rangle = \sqrt{3/5}\,Y_{20}\alpha + \sqrt{2/5}\,Y_{21}\beta,$
  angular distribution: $\frac{3}{5}|Y_{20}|^2 + \frac{2}{5}|Y_{21}|^2 = \frac{3}{4}(1 - 2\cos^2\theta + 5\cos^4\theta)$.

**3.11** Using the expansion series (3.91) for $\psi$ and $\overline{\psi}$, one gets the following anticommutation relations for arbitrary $x$ and $y$:

$$\begin{aligned}\{\psi_i(x), \psi^\dagger_j(y)\} &= \sum \left\{b(p)\psi^{(+)}_{ip}(x) + d^\dagger(p)\psi^{(-)}_{i,-p}(x),\, b^\dagger(q)\psi^{(+)*}_{jq}(y) + d(q)\psi^{(-)*}_{j,-q}(y)\right\} \\ &= \sum C^2_p\left[u_i(p,s)u^\dagger_j(p,s)\,e^{-ip\cdot(x-y)} + v_i(p,s)v^\dagger_j(p,s)\,e^{ip\cdot(x-y)}\right].\end{aligned}$$

(a) Assume now $x_0 = y_0$. Change $\boldsymbol{p} \to -\boldsymbol{p}$ in the integral of the second term on the RHS. The RHS then becomes

$$\begin{aligned}\mathrm{RHS} &= \sum_{p} C_p^2 \mathrm{e}^{\mathrm{i}\boldsymbol{p}\cdot(\boldsymbol{x}-\boldsymbol{y})} \sum_{s} [u_i(ps)u_j^\dagger(ps) + v_i(-\boldsymbol{p},s)v_j^\dagger(-\boldsymbol{p},s)] \\ &= \int \mathrm{d}^3p\, C_p^2 \mathrm{e}^{\mathrm{i}\boldsymbol{p}\cdot(\boldsymbol{x}-\boldsymbol{y})}\, \delta_{ij} 2E = \delta_{ij}\, \delta(\boldsymbol{x}-\boldsymbol{y}),\end{aligned}$$

where use has been made of the closure of the spinors and of the normalization $C_p^2 = 1/[2E(2\pi)^3]$. Hence the result

$$\{\psi_i(x),\, \psi_j^\dagger(y)\}\Big|_{x^0=y^0} = \delta_{ij}\, \delta(\boldsymbol{x}-\boldsymbol{y}).$$

(b) If $x_0 \neq y_0$ one reduces the RHS by using the projection operators:

$$\begin{aligned}\mathrm{RHS} &= \int \frac{\mathrm{d}^3p}{2E(2\pi)^3} \left[(\not{p}+m)_{ij}\mathrm{e}^{-\mathrm{i}p\cdot(x-y)} - (-\not{p}+m)_{ij}\mathrm{e}^{\mathrm{i}p\cdot(x-y)}\right] \\ &= \left(\mathrm{i}\gamma\cdot\frac{\partial}{\partial x} + m\right)_{ij} \int \frac{\mathrm{d}^3p}{2E(2\pi)^3} \left[\mathrm{e}^{-\mathrm{i}p\cdot(x-y)} - \mathrm{e}^{\mathrm{i}p\cdot(x-y)}\right] \\ &= \left(\mathrm{i}\gamma\cdot\frac{\partial}{\partial x} + m\right)_{ij} \mathrm{i}\Delta(x-y,\, m) \equiv -\mathrm{i}S_{ij}(x-y,\, m).\end{aligned}$$

Thus, the result $\{\psi_i(x),\, \psi_j^\dagger(y)\} = -\mathrm{i}S_{ij}(x-y,\, m)$.

**4.1** (a) The relation $\delta((f(x)) = [\delta(x-x_0)/|f'(x_0)|]$ holds for any function $f(x)$ with one zero, $x = x_0$, in the interval of interest. It follows that

$$\int \frac{\mathrm{d}^3p}{2E} = \int \mathrm{d}^4p\, \delta(p^2 - m^2)\theta(p_0).$$

(b) Two-particle phase integral ($s = P^2$):

$$R_2(s, m_1^2, m_2^2) = \int \left[\prod_{i=1}^{2} \mathrm{d}^4p_i \delta(p_i^2 - m_i^2)\theta(p_{i0})\right] \delta^{(4)}\left(P - \sum p_i\right).$$

Integration over $p_2$ yields

$$R_2(s, m_1^2, m_2^2) = \int \mathrm{d}^4p_1\, \delta(p_1^2 - m_1^2)\theta(p_{10})\delta[(P-p_1)^2 - m_2^2].$$

In the CM frame, where $\boldsymbol{P} = \boldsymbol{p}_1 + \boldsymbol{p}_2 = 0$, one gets

$$R_2 = \int \frac{\mathrm{d}^3p_1}{2E_1} \delta(s - 2E_1\sqrt{s} + m_1^2 - m_2^2) = \frac{p_1}{4\sqrt{s}} \int \mathrm{d}\Omega_{\mathrm{cm}},$$

where $p_1$ is determined by the zero of the argument of the $\delta$-function in the above equation,

$$p_1 = \frac{1}{2\sqrt{s}}\lambda^{1/2}(s, m_1^2, m_2^2), \qquad \lambda(x,y,z) \equiv (x^2 + y^2 + z^2 - 2xy - 2yz - 2zx).$$

If angular dependence is not relevant, integration over $\Omega$ gives

$$R_2(s, m_1^2, m_2^2) = \frac{\pi}{2s}\lambda^{1/2}(s, m_1^2, m_2^2).$$

(c) Three-particle phase integral ($s = P^2$):

$$R_3(s, m_i^2) = \int d^4p_3\, \delta(p_3^2 - m_3^2)\theta(p_{30})R_2((P-p_3)^2, m_1^2, m_2^2).$$

Making use of the previous result twice, we will get

$$\begin{aligned} R_3 &= \frac{\pi}{2}\int_{a_1}^{\infty}\frac{da}{a}\lambda^{1/2}(a, m_1^2, m_2^2)\int d^4p_3\delta(p_3^2 - m_3^2)\theta(p_{30})\delta((P-p_3)^2 - a) \\ &= \frac{\pi}{2}\int_{a_1}^{\infty}\frac{da}{a}\lambda^{1/2}(a, m_1^2, m_2^2)R_2(s, m_3^2, a) \\ &= \frac{\pi^2}{4s}\int_{a_1}^{a_2}\frac{da}{a}\lambda^{1/2}(a, m_1^2, m_2^2)\lambda^{1/2}(a, s, m_3^2). \end{aligned}$$

In writing down the last line we have used the symmetry of $\lambda(x,y,z)$ in all of its arguments. To find the integration limits, one may use the condition that both $\lambda^{1/2}$ are non-negative. This means all four inequalities $a \geq (m_1+m_2)^2$, $a \geq (m_1-m_2)^2$, $a \leq (\sqrt{s} - m_3)^2$ and $a \leq (\sqrt{s} + m_3)^2$ must be simultaneously satisfied. It suffices to require $a \geq a_1 = (m_1 + m_2)^2$ and $a \leq a_2 = (\sqrt{s} - m_3)^2$.

**4.3** $A_\nu$ can be split into $A_\nu = A_\nu^{\mathrm{T}} + A_\nu^{\mathrm{L}} = A_\nu^{\mathrm{T}} - \dfrac{\lambda}{\mu^2}\partial_\nu(\partial \cdot A)$. The field $A^{\mathrm{T}}$ is transverse: $\partial^\nu A_\nu^{\mathrm{T}} = \partial^\nu A_\nu + \dfrac{\lambda}{\mu^2}\Box(\partial \cdot A) = 0$, and therefore contains only three degrees of freedom. From the equations of motion, the fields can be expanded as

$$\begin{aligned} A_\nu^{\mathrm{T}} &= \sum \left[\phi_k^{(+)}(x)a(k,i)e_\nu(k,i) + \phi_{-k}^{(-)}(x)a^\dagger(k,i)e_\nu^*(k,i)\right], \\ A_\nu^{\mathrm{L}} &= (k_\nu/\mu)\sum\left[\varphi_k^{(+)}(x)a(k,0) + \varphi_{-k}^{(-)}(x)a^\dagger(k,0)\right], \end{aligned}$$

where

$$k^\nu e_\nu(k,i) = 0 \qquad (i = 1,2,3), \qquad \sum_{i=1}^{3} e_\nu(k,i)e_\rho^*(k,i) = -\left(g_{\nu\rho} - \frac{k_\nu k_\rho}{\mu^2}\right),$$

and the operators satisfy the commutation relations ($i,j = 1,2,3$):

$$[a(k,i), a^\dagger(k,j)] = \delta_{ij}\delta(\boldsymbol{k} - \boldsymbol{k}'), \qquad [a(k,0), a^\dagger(k,0)] = -\delta(\boldsymbol{k} - \boldsymbol{k}'),$$

with all other commutation relations vanishing. The symbols $\phi^{(+)}$ and $\varphi^{(+)}$ stand for the positive-energy solutions for masses $\mu$ and $m$ respectively. Similarly for $\phi^{(-)}$ and $\varphi^{(-)}$. With these relations, we can prove that

$$\begin{aligned}&\langle 0 \,|\, \mathrm{T} A_\rho(x) A_\nu(y) \,|\, 0\rangle \\ &\quad = \Big[\theta(x_0 - y_0) \sum_k \phi_k^{(+)}(x) \phi_k^{(+)*}(y) \\ &\qquad + \theta(y_0 - x_0) \sum_k \phi_k^{(+)}(y) \phi_k^{(+)*}(x)\Big] \left(-g_{\rho\nu} + \frac{k_\rho k_\nu}{\mu^2}\right) \\ &\qquad + \Big[\theta(x_0 - y_0) \sum_k \varphi_k^{(+)}(x) \varphi_k^{(+)*}(y) \\ &\qquad + \theta(y_0 - x_0) \sum_k \varphi_k^{(+)}(y) \varphi_k^{(+)*}(x)\Big] \left(-\frac{k_\rho k_\nu}{\mu^2}\right) .\end{aligned}$$

Since the extra factors outside the square brackets are even function of $k$, they will not be affected by flipping the sign of $\boldsymbol{k}$. The remaining calculations are as for the scalar boson case, leading to

$$\begin{aligned}\langle 0 \,|\, \mathrm{T} A_\rho(x) A_\nu(y) \,|\, 0\rangle = &- \mathrm{i} \int \frac{\mathrm{d}^4 k}{(2\pi)^4} \mathrm{e}^{-\mathrm{i}k\cdot(x-y)} \\ &\times \left[\Delta(k, \mu^2) \left(g_{\rho\nu} - \frac{k_\rho k_\nu}{\mu^2}\right) + \Delta(k, m^2) \frac{k_\rho k_\nu}{\mu^2}\right] .\end{aligned}$$

The propagator in momentum space is

$$\Delta_{\rho\nu}(k) = -\frac{g_{\rho\nu} - k_\rho k_\nu/\mu^2}{k^2 - \mu^2 + \mathrm{i}\varepsilon} - \frac{k_\rho k_\nu/\mu^2}{k^2 - (\mu^2/\lambda) + \mathrm{i}\varepsilon} .$$

**4.4** The equation of QED with an external source is

$$\partial_\alpha F^{\alpha\beta} = j^\beta , \tag{1}$$

which also contains the conservation condition $\partial_\alpha j^\alpha = 0$. Since $\partial_0 A_0$ does not occur in $\mathcal{L}$, the conjugate momentum $\pi^0$ is zero and one is free to remove $A_0$. This can be done by taking $\beta = 0$ in (1):

$$\partial_\alpha F^{\alpha 0} = j^0 \;\Rightarrow\; \nabla^2 A^0 + \partial_0 \nabla \cdot \boldsymbol{A} = -j^0 = -\rho . \tag{2}$$

In the Coulomb gauge, $\nabla \cdot \boldsymbol{A} = 0$, this equation can be solved to give the 'instantaneous' (not retarded) potential

$$A^0(t, \boldsymbol{x}) = \frac{1}{4\pi} \int \mathrm{d}^3 x' \, \frac{\rho(t, \boldsymbol{x}')}{|\boldsymbol{x}' - \boldsymbol{x}|} . \tag{3}$$

The space components of (1) are (in general)

$$\Box \boldsymbol{A} + \nabla(\nabla \cdot \boldsymbol{A}) = \boldsymbol{j} - \nabla \partial_0 A^0 . \tag{4}$$

Taking the divergence of (4) and using the current conservation condition and relation (2) in the Coulomb gauge yield $\partial_0^2 \nabla \cdot \boldsymbol{A} = 0$. If $\nabla \cdot \boldsymbol{A} = 0$ and $\partial_0 \nabla \cdot \boldsymbol{A} = 0$ hold at some time $t$, the Coulomb gauge condition holds at all $t$.

Let us separate $\boldsymbol{E} = -\nabla A^0 - \partial_0 \boldsymbol{A}$ into two parts $\boldsymbol{E} = \boldsymbol{E}_\parallel + \boldsymbol{E}_\perp$ such that

$$\begin{aligned} \boldsymbol{E}_\parallel &= -\nabla A_0, \qquad \nabla \times \boldsymbol{E}_\parallel = 0\,, \\ \boldsymbol{E}_\perp &= -\partial_0 \boldsymbol{A}, \qquad \nabla \cdot \boldsymbol{E}_\perp = 0 \ \text{(Coulomb gauge)}. \end{aligned} \tag{5}$$

In the Coulomb gauge, we have

$$\int \mathrm{d}^3x\, \boldsymbol{E}^2 = \int \mathrm{d}^3x\, (\boldsymbol{E}_\perp^2 + \boldsymbol{E}_\parallel^2) = \int \mathrm{d}^3x\, (\boldsymbol{E}_\perp^2 + \rho A_0)\,, \tag{6}$$

where we have used Gauss's theorem twice.

Now, it is generally true that

$$\mathcal{L}_{\mathrm{em}} = -\tfrac{1}{4} F^2 = \tfrac{1}{2}(\boldsymbol{E}^2 - \boldsymbol{B}^2)\,. \tag{7}$$

In the Coulomb gauge, (7) becomes

$$\mathcal{L}_{\mathrm{em}} = \tfrac{1}{2}(\boldsymbol{E}_\perp^2 - \boldsymbol{B}^2 + \rho A_0)\,. \tag{8}$$

Note in particular the presence of the last term. Add $-j_\mu A^\mu$ to (8) and we get the Lagrangian in the Coulomb gauge

$$\mathcal{L} = \tfrac{1}{2}(\boldsymbol{E}_\perp^2 - \boldsymbol{B}^2) - \tfrac{1}{2}\rho A^0 + \boldsymbol{j} \cdot \boldsymbol{A}\,, \tag{9}$$

where $A^0$ is given by (3).

**5.1** (a) In $\pi$–p scattering, there are three variables in CM frame: the relative momenta in initial and final states, $\boldsymbol{p}_i$ and $\boldsymbol{p}_f$, and the nucleon spin variable $\boldsymbol{\sigma}$. Define $\boldsymbol{n} = (\boldsymbol{p}_i \times \boldsymbol{p}_f)/|\boldsymbol{p}_i \times \boldsymbol{p}_f|$. We can construct the general rotationally invariant amplitude

$$M = a + b\boldsymbol{\sigma}\cdot\boldsymbol{n} + c\boldsymbol{\sigma}\cdot\boldsymbol{p}_i + d\boldsymbol{\sigma}\cdot\boldsymbol{p}_f\,,$$

where $a, b, \ldots$ are invariant functions of $\boldsymbol{p}_i$ and $\boldsymbol{p}_f$. Under space inversion, $\mathcal{P}$ : $\boldsymbol{\sigma} \to \boldsymbol{\sigma}$, $\boldsymbol{p}_i \to -\boldsymbol{p}_i$, $\boldsymbol{p}_f \to -\boldsymbol{p}_f$, $\boldsymbol{n} \to \boldsymbol{n}$. Therefore, $P$-invariance requires $c = d = 0$. Under time inversion, $\mathcal{T}$ : $\boldsymbol{\sigma} \to -\boldsymbol{\sigma}$, $\boldsymbol{p}_i \to -\boldsymbol{p}_f$, $\boldsymbol{n} \to -\boldsymbol{n}$. Therefore, T-invariance allows all 4 terms in $M$.

(b) Consider, for example, proton–proton scattering. The basic variables are the momenta $\boldsymbol{p}_i$, $\boldsymbol{p}_f$, and the spins $\boldsymbol{\sigma}_1$, $\boldsymbol{\sigma}_2$. Define $\boldsymbol{n} = \boldsymbol{p}_i \times \boldsymbol{p}_f$, $\boldsymbol{P} = \boldsymbol{p}_i + \boldsymbol{p}_f$ and $\boldsymbol{K} = \boldsymbol{p}_i - \boldsymbol{p}_f$. The most general rotationally invariant amplitude, symmetric in the interchange of the two particles, is

$$\begin{aligned} M = {} & a + b(\boldsymbol{\sigma}_1 + \boldsymbol{\sigma}_2)\cdot\boldsymbol{n} + c(\boldsymbol{\sigma}_1\cdot\boldsymbol{n})(\boldsymbol{\sigma}_2\cdot\boldsymbol{n}) + d(\boldsymbol{\sigma}_1\cdot\boldsymbol{P})(\boldsymbol{\sigma}_2\cdot\boldsymbol{P}) \\ & + e(\boldsymbol{\sigma}_1\cdot\boldsymbol{K})(\boldsymbol{\sigma}_2\cdot\boldsymbol{K}) + f[(\boldsymbol{\sigma}_1\cdot\boldsymbol{P})(\boldsymbol{\sigma}_2\cdot\boldsymbol{K}) + (\boldsymbol{\sigma}_1\cdot\boldsymbol{K})(\boldsymbol{\sigma}_2\cdot\boldsymbol{K})]\,. \end{aligned}$$

It can be checked that P-invariance imposes no restrictions, while T-invariance requires $f = 0$.

**5.2** Consider $\mu(p) \to \mathrm{e}(k) + \bar{\nu}_\mathrm{e}(k') + \nu_\mu(p')$. From Feynman rules, the decay amplitude is

$$\mathrm{i}\mathcal{M} = \frac{-\mathrm{i}G_\mathrm{F}}{\sqrt{2}} [\bar{u}(p')\Gamma_\lambda u(p)] \, [\bar{u}(k)\Gamma^\lambda v(k')] \,,$$

where $\Gamma_\lambda = \gamma_\lambda(1-\gamma_5)$. Summing over all spins in initial and final states gives

$$\begin{aligned}\sum_{\text{spins}} |\mathcal{M}|^2 &= \frac{G_\mathrm{F}^2}{2} \sum \left|\bar{u}(p')\Gamma_\lambda u(p)\, \bar{u}(k)\Gamma^\lambda v(k')\right|^2 \\ &= \tfrac{1}{2} G_\mathrm{F}^2 \,\mathrm{Tr}\,[(\not{k} + m_\mathrm{e})\Gamma^\alpha \not{k}'\Gamma^\beta]\, \mathrm{Tr}\,[\Gamma_\alpha(\not{p} + m_\mu)\Gamma_\beta \not{p}'] \\ &= 128\, G_\mathrm{F}^2 (k{\cdot}p')(p{\cdot}k') \,.\end{aligned}$$

The electron energy spectrum is obtained by integrating over all electron directions and all variables of the unobserved neutrinos:

$$\mathrm{d}\Gamma = \frac{1}{2m_\mu(2\pi)^5} \int \frac{\mathrm{d}^3p'\mathrm{d}^3k'\mathrm{d}^3k}{8E_{p'}E_{k'}E_k} \delta^4(k + k' + p' - p) \frac{1}{2} \sum |\mathcal{M}|^2 \,,$$

(1/2 comes from averaging over the muon spin). We need the Lorentz tensor

$$I^{\alpha\beta} = \int \frac{\mathrm{d}^3p'\mathrm{d}^3k'}{4E_{p'}E_{k'}} p'^\alpha k'^\beta \delta^4(k + k' + p' - p) \,.$$

It can be easily calculated in the CM frame of the two neutrinos ($\boldsymbol{p}' + \boldsymbol{k}' = 0$), and should depend only on $Q_\alpha = p_\alpha - k_\alpha$ [see (A.41)]. Since $I^{\alpha\beta}$ is a Lorentz tensor, the result is valid in any frame:

$$I^{\alpha\beta} = \frac{\pi}{24}(Q^2 g^{\alpha\beta} + 2Q^\alpha Q^\beta).$$

Hence we calculate

$$\mathrm{d}\Gamma = \frac{1}{2m_\mu(2\pi)^5} \int \frac{\mathrm{d}^3k}{2E_k} 64\, G_\mathrm{F}^2\, k_\alpha p_\beta I^{\alpha\beta} .$$

In the $\mu$-rest frame, the electron energy spectrum is

$$\frac{\mathrm{d}\Gamma}{\mathrm{d}E_k} = \frac{G^2}{12\pi^3} m_\mu^2 E_k^2 \left(3 - 4\frac{E_k}{m_\mu}\right), \qquad 0 \le E_k \le m_\mu/2$$

in the limit where $m_e \ll m_\mu$. Integration over $E_k$ gives $\dfrac{1}{t_\mu} = \Gamma_\mu = \dfrac{G_\mathrm{F}^2 m_\mu^5}{192\pi^3}$.

**5.3** The amplitude for $\pi^-(p) \to \ell^-(p_1) + \bar{\nu}(p_2)$ is given by

$$\mathrm{i}\mathcal{M} = \frac{-\mathrm{i}G_\mathrm{F}}{\sqrt{2}} J_\lambda \,[\bar{u}(p_1)\gamma^\lambda(1-\gamma_5)v(p_2)],$$

where $J_\lambda(p) = \langle 0 \,|\, A_\lambda \,|\, \pi(p)\rangle = \mathrm{i} f_\pi p_\lambda$, and $p = p_1 + p_2$. Using the Dirac equations for the two leptons,

$$\mathrm{i}\mathcal{M} = \frac{1}{\sqrt{2}} G_\mathrm{F} f_\pi \, m_\ell \, \bar{u}(p_1)(1-\gamma_5)v(p_2)\,.$$

Summing over the lepton spins yields

$$\begin{aligned}\sum |\mathcal{M}|^2 &= \tfrac{1}{2}(G_\mathrm{F} f_\pi m_\ell)^2 \operatorname{Tr}[(\not{p}_1 + m_\ell)(1-\gamma_5)\not{p}_2(1+\gamma_5)] \\ &= (G_\mathrm{F} f_\pi m_\ell)^2 \operatorname{Tr}(\not{p}_1 \not{p}_2) \\ &= 2(G_\mathrm{F} f_\pi m_\ell m_\pi)^2 (1 - x_\ell); \qquad x_\ell \equiv m_\ell^2/m_\pi^2\,.\end{aligned}$$

Therefore the decay rate is

$$\Gamma(\pi \to \ell\bar{\nu}) = \frac{|\boldsymbol{p}_1|}{8\pi m_\pi^2} \sum |\mathcal{M}|^2 = \frac{1}{8\pi}(G_\mathrm{F} f_\pi)^2 m_\pi^3 x_\ell (1-x_\ell)^2\,,$$

from which follows the ratio

$$\frac{\Gamma(\pi \to \mu\nu)}{\Gamma(\pi \to \mathrm{e}\nu)} = \frac{x_\mu(1-x_\mu)^2}{x_\mathrm{e}(1-x_\mathrm{e})^2}\,.$$

**5.6** See J.J. Amato et al., Phys. Rev. Letters **21** (1968) 1709.

**5.8** We know that $\mathcal{P}\psi(x)\mathcal{P}^{-1} = \eta\gamma_0\psi(x')$ and $\mathcal{C}\psi(x)\mathcal{C}^{-1} = \xi C\bar{\psi}^\mathrm{T}$.

*Parity*:

$$\begin{aligned}\psi^\mathrm{c} &= \xi C\bar{\psi}^\mathrm{T} \xrightarrow{P} \xi C[(\eta\gamma_0\psi)^\dagger\gamma_0]^\mathrm{T} = \xi\eta^* C\gamma_0^\mathrm{T} C^{-1} C\bar{\psi}^\mathrm{T} \\ &= \eta^*(-\gamma_0)\psi^\mathrm{c} = -\eta^*\gamma_0\psi^\mathrm{c}\,.\end{aligned}$$

Hence $\mathcal{P}\,\psi^\mathrm{c}\mathcal{P}^{-1} = -\eta^*\gamma_0\psi^\mathrm{c}$, or $\eta^\mathrm{c} = -\eta^*$.

*Chirality*:

$$\gamma_5\psi^\mathrm{c} = \gamma_5\xi C\bar{\psi}^\mathrm{T} = \xi CC^{-1}\gamma_5 C\bar{\psi}^\mathrm{T} = \xi C\gamma_5^\mathrm{T}\bar{\psi}^\mathrm{T} = -\lambda^*\psi^\mathrm{c}\,.$$

Thus $\lambda^\mathrm{c} = -\lambda^*$.

*Helicity*:

$$\boldsymbol{\Sigma}\cdot\hat{\boldsymbol{p}}\psi^\mathrm{c} = \xi\boldsymbol{\Sigma}\cdot\hat{\boldsymbol{p}}\,C\bar{\psi}^\mathrm{T} = \xi CC^{-1}\boldsymbol{\Sigma}\cdot\hat{\boldsymbol{p}}\,C\bar{\psi}^T = \xi C\boldsymbol{\Sigma}^\mathrm{T}\cdot\hat{\boldsymbol{p}}\,\bar{\psi}^\mathrm{T} = h^*\xi C\bar{\psi}^\mathrm{T}\,,$$

where we have used $\gamma^\dagger = \gamma_0$ and $\boldsymbol{\Sigma}^\dagger = \boldsymbol{\Sigma}$. Conclusion: $h^\mathrm{c} = h^* = h$.

**5.9** The following table gives the relevant transformation rules obtained from Chap. 5 (coordinates of fields are suppressed for simplicity).

| Variables | $\mathcal{P}$ | $\mathcal{C}$ | $\mathcal{T}$ |
|---|---|---|---|
| $A_0,\ A_i$ | $A_0,\ -A_i$ | $-A_0,\ -A_i$ | $A_0, -A_i$ |
| $\gamma^0,\ \gamma^i$ | $\gamma^0,\ -\gamma^i$ | $-\gamma^0,\ -\gamma^i$ | $\gamma^0,\ -\gamma^i$ |
| $A_\mu\gamma^\mu$ | $A_\mu\gamma^\mu$ | $A_\mu\gamma^\mu$ | $A_\mu\gamma^\mu$ |
| $F_{0i},\ F_{ij}$ | $-F_{0i},\ +F_{ij}$ | $-F_{0i},\ -F_{ij}$ | $+F_{0i},\ -F_{ij}$ |
| $\sigma^{0i},\ \sigma^{ij}$ | $-\sigma^{0i},\ +\sigma^{ij}$ | $-\sigma^{0i},\ -\sigma^{ij}$ | $+\sigma^{0i},\ -\sigma^{ij}$ |
| $\mathrm{i}\gamma_5(\sigma^{0i},\ \sigma^{ij})$ | $\mathrm{i}\gamma_5(\sigma^{0i},\ -\sigma^{ij})$ | $\mathrm{i}\gamma_5(-\sigma^{0i},\ -\sigma^{ij})$ | $\mathrm{i}\gamma_5(-\sigma^{0i},\ \sigma^{ij})$ |
| $F_{\mu\nu}\sigma^{\mu\nu}$ | $+F_{\mu\nu}\sigma^{\mu\nu}$ | $+F_{\mu\nu}\sigma^{\mu\nu}$ | $+F_{\mu\nu}\sigma^{\mu\nu}$ |
| $F_{\mu\nu}\gamma_5\sigma^{\mu\nu}$ | $-F_{\mu\nu}\gamma_5\sigma^{\mu\nu}$ | $+F_{\mu\nu}\gamma_5\sigma^{\mu\nu}$ | $-F_{\mu\nu}\gamma_5\sigma^{\mu\nu}$ |

In conclusion, $\bar{\psi}\gamma_\mu\psi A^\mu$ and $F_{\mu\nu}\bar{\psi}\sigma^{\mu\nu}\psi$ are invariant to P, C, T; $F_{\mu\nu}\bar{\psi}\gamma_5\sigma^{\mu\nu}\psi$ is invariant only to C, and changes sign under P and T. All three are invariant under the combined operation PCT.

**6.1** (a) Let $\phi_{1,-1} = nn$. Then from (6.24), $I_+\phi_{1,-1} = \sqrt{2}\phi_{10}$. On the other hand, $\boldsymbol{I} = \boldsymbol{I}^{(1)} + \boldsymbol{I}^{(2)}$, the sum of individual isospins; so that $I_+(nn) = (pn + np)$. Together, the two relations give $\phi_{10} = (pn + np)/\sqrt{2}$. Similarly, $I_+\phi_{10} = \sqrt{2}\phi_{11}$, and $I_+(pn + np)/\sqrt{2} = \sqrt{2}pp$, from which one gets $\phi_{11} = pp$. The combination orthogonal to $(pn + np)/\sqrt{2}$ is $(pn - np)/\sqrt{2}$, which corresponds to $\phi_{00}$.

(b) Starting from $\Phi_{3/2,-3/2} = \pi^- n$, we proceed as above, using both (6.24) and $\boldsymbol{I} = \boldsymbol{I}^{(N)} + \boldsymbol{I}^{(\pi)}$, and the proper sign convention for $\pi^\pm$. Thus, we get

$$I_+\Phi_{3/2,-3/2} = \sqrt{3}\Phi_{3/2,-1/2}, \quad I_+\pi^- n = \pi^- p + \sqrt{2}\pi^0 n$$
$$\Rightarrow \quad \Phi_{3/2,-1/2} = \sqrt{\frac{1}{3}}(\pi^- p + \sqrt{2}\pi^0 n);$$

$$I_+\Phi_{3/2,-1/2} = 2\Phi_{3/2,1/2}, \quad I_+\sqrt{\frac{1}{3}}(\pi^- p + \sqrt{2}\pi^0 n) = \frac{2}{\sqrt{3}}(\sqrt{2}\pi^0 p - \pi^+ n)$$
$$\Rightarrow \quad \Phi_{3/2,1/2} = \sqrt{\frac{1}{3}}(\sqrt{2}\pi^0 p - \pi^+ n);$$

$$I_+\Phi_{3/2,1/2} = \sqrt{3}\Phi_{3/2,3/2}, \quad I_+\frac{1}{\sqrt{3}}(\sqrt{2}\pi^0 p - \pi^+ n) = -\sqrt{3}\pi^+ p$$
$$\Rightarrow \quad \Phi_{3/2,3/2} = -\pi^+ p.$$

The combination orthogonal to $\Phi_{3/2,-1/2}$ is $\Phi_{1/2,-1/2} = \frac{1}{\sqrt{3}}(-\sqrt{2}\pi^- p + \pi^0 n)$. Then applying $\boldsymbol{I}_+$ on this relation, one gets $\Phi_{1/2,1/2} = -\frac{1}{\sqrt{3}}(\pi^0 p + \sqrt{2}\pi^+ n)$.

**6.2** From Table 6.3, one has

$$-\pi^+ p = \left|\tfrac{3}{2}, \tfrac{3}{2}\right\rangle,$$
$$\pi^+ n = \tfrac{1}{\sqrt{3}}\left(\left|\tfrac{3}{2}, \tfrac{1}{2}\right\rangle + \sqrt{2}\left|\tfrac{1}{2}, \tfrac{1}{2}\right\rangle\right),$$
$$\pi^0 p = \tfrac{1}{\sqrt{3}}\left(\sqrt{2}\left|\tfrac{3}{2}, \tfrac{1}{2}\right\rangle - \left|\tfrac{1}{2}, \tfrac{1}{2}\right\rangle\right).$$

Since $\pi^+ p$ is in a pure isospin eigenstate, $R_+ = R_{3/2}$. On the other hand, after scattering from the prepared state $\pi^+ n$, the scattered wave is

$$\begin{aligned} R\left|\pi^+ n\right\rangle &= \sqrt{\frac{1}{3}}R_{\frac{3}{2}}\left|\tfrac{3}{2}, \tfrac{1}{2}\right\rangle + \sqrt{\frac{2}{3}}R_{\frac{1}{2}}\left|\tfrac{1}{2}, \tfrac{1}{2}\right\rangle \\ &= \sqrt{\frac{1}{3}}R_{\frac{3}{2}}\left(\sqrt{\frac{2}{3}}\pi^0 p + \sqrt{\frac{1}{3}}\pi^+ n\right) + \sqrt{\frac{2}{3}}R_{\frac{1}{2}}\left(-\sqrt{\frac{1}{3}}\pi^0 p + \sqrt{\frac{2}{3}}\pi^+ n\right) \\ &= \left(\frac{1}{3}R_{\frac{3}{2}} + \frac{2}{3}R_{\frac{1}{2}}\right)\left|\pi^+ n\right\rangle + \frac{\sqrt{2}}{3}\left(R_{\frac{3}{2}} - R_{\frac{1}{2}}\right)\left|\pi^0 p\right\rangle, \end{aligned}$$

so that

$$R_- = \langle \pi^+ n \,|\, R \,|\, \pi^+ n \rangle = \frac{1}{3}\left( R_{\frac{3}{2}} + 2R_{\frac{1}{2}} \right),$$
$$R_0 = \langle \pi^0 p \,|\, R \,|\, \pi^+ n \rangle = \frac{\sqrt{2}}{3}\left( R_{\frac{3}{2}} - R_{\frac{1}{2}} \right).$$

By isospin invariance, these amplitudes are independent of $I_3$ and, therefore,

$$R_+ = \langle \pi^- n \,|\, R \,|\, \pi^- n \rangle, \quad R_- = \langle \pi^- p \,|\, R \,|\, \pi^- p \rangle, \quad R_0 = \langle \pi^0 n \,|\, R \,|\, \pi^- p \rangle .$$

Since there are just two isospin amplitudes, the physical amplitudes are not independent, they are related by $R_+ - R_- = \sqrt{2}\, R_0$. This relation is a consequence of charge independence, and so can be used to check invariance. When the $I = 3/2$ channel predominates, the ratios of the amplitudes are $R_+ : R_0 : R_- \approx 3 : \sqrt{2} : 1$, corresponding to the ratios of the cross-sections $[\sigma = (\text{kinematics})|R|^2]$: $\sigma_+ : \sigma_0 : \sigma_- \approx 9 : 2 : 1$.

**6.4** (a) The field variations are $\delta_0\psi = -\mathrm{i}\epsilon\psi$, $\delta_0\phi = 0$. Under these transformations, the Lagrangian is invariant and the associated Noether current is

$$J^\mu = \frac{\partial \mathcal{L}}{\partial(\partial_\mu \psi)} \frac{\delta_0 \psi}{\delta \epsilon} = \bar{\psi}\gamma^\mu \psi.$$

Note that since the interaction Lagrangian contains no derivative couplings, one can effectively replace $\mathcal{L}$ by the free-field Lagrangian $\mathcal{L}_0$. Thus the conserved current and charge are given by

$$J^\mu = \bar{\psi}\gamma^\mu\psi = \bar{\psi}_\mathrm{p}\gamma^\mu\psi_\mathrm{p} + \bar{\psi}_\mathrm{n}\gamma^\mu\psi_\mathrm{n},$$
$$N_\mathrm{B} = \int d^3x J^0 = \int d^3x (\psi_\mathrm{p}^\dagger\psi_\mathrm{p} + \psi_\mathrm{n}^\dagger\psi_\mathrm{n}).$$

(b) It is again checked that the Lagrangian is invariant. The associated conserved current is

$$J^\mu = \frac{\partial \mathcal{L}}{\partial(\partial_\mu \psi)} \frac{\delta_0 \psi}{\delta \epsilon} + \frac{\partial \mathcal{L}}{\partial(\partial_\mu \phi)} \frac{\delta_0 \phi}{\delta \epsilon}$$
$$= \bar{\psi}\mathrm{i}\gamma^\mu \left( -\mathrm{i}\frac{1+\tau_3}{2}\psi \right) + \partial^\mu \boldsymbol{\phi} \cdot [-(\boldsymbol{\phi} \times \hat{n}_3)] = \bar{\psi}\gamma^\mu \frac{1+\tau_3}{2}\psi + (\boldsymbol{\phi} \times \partial^\mu \boldsymbol{\phi})_3 .$$

The corresponding conserved charge is

$$Q = \int d^3x J^0 = \int d^3x \psi^\dagger \frac{1+\tau_3}{2}\psi + (\boldsymbol{\phi} \times \dot{\boldsymbol{\phi}})_3 .$$

(c) In checking the Lagrangian invariance, recall that $\tau^\dagger = \tau$ and $\boldsymbol{t}^\mathrm{T} = -\boldsymbol{t}$. To the first order in $\epsilon$, one has (ignoring the $\gamma^\mu$, unimportant for the present purpose)

$$\bar{\psi}'\psi' = \bar{\psi}\psi + \frac{\mathrm{i}}{2}\epsilon\bar{\psi}(\boldsymbol{\tau}^\dagger \cdot \boldsymbol{n} - \boldsymbol{\tau} \cdot \boldsymbol{n})\psi = \bar{\psi}\psi,$$
$$\boldsymbol{\phi}'^T\boldsymbol{\phi}' = \boldsymbol{\phi}^T\boldsymbol{\phi} - \mathrm{i}\epsilon\boldsymbol{\phi}^T(\boldsymbol{t}^\mathrm{T} \cdot \boldsymbol{n} + \boldsymbol{t} \cdot \boldsymbol{n})\boldsymbol{\phi} = \boldsymbol{\phi}^T\boldsymbol{\phi},$$
$$\bar{\psi}'\boldsymbol{\tau}\psi' \cdot \boldsymbol{\phi}' = \bar{\psi}\boldsymbol{\tau}\psi \cdot \boldsymbol{\phi} + \mathrm{i}\epsilon n_a \bar{\psi}[\tfrac{1}{2}(\tau_a\tau_b - \tau_b\tau_a)\phi_b - \tau_b(t_a)_{bc}\phi_c]\psi$$
$$= \bar{\psi}\boldsymbol{\tau}\psi \cdot \boldsymbol{\phi} + \mathrm{i}\epsilon n_a \bar{\psi}[\mathrm{i}\epsilon_{abc}\tau_c\phi_b + \mathrm{i}\epsilon_{abc}\tau_b\phi_c]\psi$$
$$= \bar{\psi}\boldsymbol{\tau}\psi \cdot \boldsymbol{\phi}.$$

The conserved currents are ($a = 1, 2, 3$)

$$J_a^\mu = \frac{\partial\mathcal{L}}{\partial(\partial_\mu\psi)}\frac{\delta_0\psi}{\delta\epsilon n_a} + \frac{\partial\mathcal{L}}{\partial(\partial_\mu\phi)}\frac{\delta_0\phi}{\delta\epsilon n_a}$$
$$= \bar{\psi}\mathrm{i}\gamma^\mu(-\tfrac{\mathrm{i}}{2}\tau_a\psi) + \partial^\mu\phi[-\mathrm{i}t_a\phi] = \tfrac{1}{2}\bar{\psi}\gamma^\mu\tau_a\psi - \mathrm{i}\partial^\mu\phi t_a\phi\,;$$

which yield the corresponding conserved charges

$$Q_a = \int d^3x J_a^0 = \int d^3x(\tfrac{1}{2}\psi^\dagger\tau_a\psi - \mathrm{i}\dot{\phi}t_a\phi)\,.$$

**6.5** In comparison with the hyperon decays, there are at least two advantages in studying the semileptonic decays of the kaons: first, the kaons have zero spin, which simplifies the structure of the interaction, and, second, there is generally a larger energy available to the decay products. In the three-body decays:

$$\mathrm{K}^\pm \to \pi^0 + \ell^\pm + \nu_\ell \text{ ( or } \bar{\nu}_\ell), \quad (\text{called } \mathrm{K}^\pm{}_{\ell 3})\ ,$$
$$\mathrm{K}^0 \to \pi^\mp + \ell^\pm + \nu_\ell \text{ ( or } \bar{\nu}_\ell),$$

($\ell = e, \mu$), the isospin changes are $|\Delta\boldsymbol{I}| = |\boldsymbol{I}_\mathrm{K} - \boldsymbol{I}_\pi| = 1/2$ or $3/2$. Assuming the empirical rule $\Delta I = 1/2$ to be valid, the neutral K decay modes

$$\mathrm{K}^0 \to \pi^+ + \ell^- + \bar{\nu}_\ell \quad \text{and} \quad \bar{\mathrm{K}}^0 \to \pi^- + \ell^+ + \nu_\ell$$

are *forbidden.* Then for the allowed modes one can apply the usual angular-momentum couplings

$$|\tfrac{1}{2}, \tfrac{1}{2}\rangle = -\sqrt{\tfrac{1}{3}}\,\pi^0\mathrm{K}^+ - \sqrt{\tfrac{2}{3}}\,\pi^+\mathrm{K}^0\,,$$
$$|\tfrac{1}{2}, -\tfrac{1}{2}\rangle = -\sqrt{\tfrac{2}{3}}\,\pi^-\bar{\mathrm{K}}^0 + \sqrt{\tfrac{1}{3}}\,\pi^0(-\mathrm{K}^-)$$

to obtain the theoretical ratios

$$\frac{\Gamma(\mathrm{K}^0 \to \pi^-\ell^+\nu_\ell)}{\Gamma(\mathrm{K}^+ \to \pi^0\ell^+\nu_\ell)} = 2 \qquad \text{and} \qquad \frac{\Gamma(\bar{\mathrm{K}}^0 \to \pi^+\ell^-\bar{\nu}_\ell)}{\Gamma(\mathrm{K}^- \to \pi^0\ell^-\bar{\nu}_\ell)} = 2\,.$$

The measured rate is that of the combination $\mathrm{K}_\mathrm{L}^0 = (\mathrm{K}^0 - \bar{\mathrm{K}}^0)/\sqrt{2}$,

$$\mathrm{K}_\mathrm{L}^0 \to (\pi^-\ell^+\nu_\ell) + (\pi^+\ell^-\bar{\nu}_\ell), \qquad (\text{called } \mathrm{K}^0{}_{\ell 3})\,.$$

Hence the prediction for the combination: $[\Gamma(\mathrm{K}_{\ell 3}^0)/\Gamma(\mathrm{K}_{\ell 3}^+)] = [\Gamma(\mathrm{K}_{\ell 3}^0)/\Gamma(\mathrm{K}_{\ell 3}^-)] = 2$. Experimental data for the neutral kaon are

$$\Gamma(\mathrm{K}^0{}_{\ell 3}) = \begin{cases} 0.0474 \times 10^8/\mathrm{s} & \text{for } \ell = \mu^\mp\,; \\ 0.0678 \times 10^8/\mathrm{s} & \text{for } \ell = \mathrm{e}^\mp\,. \end{cases}$$

and for the charged kaons:

$$\Gamma(\mathrm{K}^+{}_{\ell 3}) = \begin{cases} 0.0258 \times 10^8/\mathrm{s} & \text{for } \ell = \mu^+\,; \\ 0.0389 \times 10^8/\mathrm{s} & \text{for } \ell = \mathrm{e}^+\,. \end{cases}$$

The agreement with theoretical values is good but not perfect: there is slight violation of the $\Delta I = \frac{1}{2}$ rule in the $K_{\ell 3}$ decays.

**7.2** Let $S(\boldsymbol{\alpha})$ and $S^*(\boldsymbol{\alpha})$ be representations of 3 and $3^*$ of SU(3):

$$S = 1 + \tfrac{i}{2}\alpha_i\lambda_i + \dots, \qquad S^* = 1 - \tfrac{i}{2}\alpha_i\lambda_i^* + \dots,$$

where $\lambda_i$ $(i = 1, \dots, 8)$ are the Gell-Mann matrices. Note that $\lambda_i^* = \varepsilon_i\lambda_i$ (no sum) with $\varepsilon_i = +1$ for $i = 1, 3, 4, 6, 8$ and $\varepsilon_i = -1$ for $i = 2, 5, 7$. Saying that $S$ and $S^*$ are equivalent means that there is some $\alpha_1^0, \dots, \alpha_n^0$ corresponding to $S_0$ such that $S^* = S_0 S S_0^\dagger$, which may be expressed as

$$S_0\lambda_i S_0^\dagger = -\lambda_i^* = -\varepsilon_i\lambda_i \quad \text{(no sum)}. \tag{1}$$

The Gell-Mann matrices satisfy $\lambda_\ell\lambda_m + \lambda_m\lambda_\ell = \frac{4}{3}\delta_{\ell m} + 2d_{\ell mn}\lambda_n$. Sandwiching both sides of this relation with $S_0$ and $S_0^\dagger$ and using (1), one gets

$$\tfrac{4}{3}\delta_{\ell m} + 2\varepsilon_\ell\varepsilon_m\, d_{\ell mn}\lambda_n = \tfrac{4}{3}\delta_{\ell m} - 2\varepsilon_n\, d_{\ell mn}\lambda_n, \quad \text{(no summation over } \ell, m).$$

Since all eight $\lambda_n$ are independent, one gets $d_{\ell mn} = -\varepsilon_\ell\varepsilon_m\varepsilon_n d_{\ell mn}$ (no summation over $\ell$, $m$). With known $\varepsilon_i$ and $d_{\ell mn}$, one can check that this relation does not hold in general. It must be concluded that 3 and $3^*$ of SU(3) are not equivalent to each other.

**7.5**

$$\begin{aligned} D_8 = d_{8ij}F_iF_j &= \tfrac{-1}{2\sqrt{3}}F_iF_i + \tfrac{\sqrt{3}}{2}(F_1^2 + F_2^2 + F_3^2) - \tfrac{1}{2\sqrt{3}}F_8^2 \\ &= \tfrac{\sqrt{3}}{2}\left(\boldsymbol{I}^2 - \tfrac{1}{4}Y^2 - \tfrac{1}{3}\boldsymbol{F}^2\right). \end{aligned}$$

The Hamiltonian is

$$H = H_0 + H_8 = H_0 - \frac{b}{2\sqrt{3}}\boldsymbol{F}^2 + \frac{\sqrt{3}a}{2}Y + \frac{b\sqrt{3}}{2}\left(\boldsymbol{I}^2 - \frac{1}{4}Y^2\right),$$

which leads to the mass formula for baryons

$$M_B(I, Y) = m_0 + m_1 Y + m_2\left[I(I+1) - \tfrac{1}{4}Y^2\right].$$

**7.6** (a) An irreducible tensor is traceless and symmetric in its indices of the same kind. From two irreducible tensors $M^a{}_b$ and $N^c{}_d$ representing two octets of SU(3), one can construct the following irreducible tensors:

**1** (0,0): $S = M_b^a N_a^b$,
**8** (1,1): $F_b^a = M_c^a N_b^c - N_c^a M_b^c$,
**8** (1,1): $D_b^a = M_c^a N_b^c + N_c^a M_b^c - \frac{2}{3}\delta_b^a S$,
**10** (3,0): $T^{abc} = M_i^a N_j^b \epsilon^{ijc} +$ all permutations of a, b, c,
**10*** (3,0): $T_{abc} = M_a^i N_b^j \epsilon_{ijc} +$ all permutations of a, b, c.

These multiplets contribute 37 components. There remain 27 which can be accounted for by a (2,2) irreducible tensor. Consider the mixed tensor separately symmetric in its upper indices and its lower indices:

$$\tilde{R}_{cd}^{ab} = M_c^a N_d^b + M_c^b N_d^a + M_d^a N_c^b + M_d^b N_c^a.$$

Its trace is $\tilde{R}^{ab}_{cb} = D^a_c + \frac{2}{3}\delta^a_c\, S$. The corresponding irreducible tensor

$$R^{ab}_{cd} = \tilde{R}^{ab}_{cd} + \alpha(\delta^a_c D^b_d + \delta^b_d D^a_c + \delta^a_d D^b_c + \delta^b_c D^a_d) + \beta(\delta^a_c\delta^b_d + \delta^a_d\delta^b_c)\, S$$

is traceless, $R^{ab}_{cb} = 0$, provided that $\alpha = -1/5$ and $\beta = -1/6$. Thus, to complete the decomposition of $\mathbf{8} \times \mathbf{8}$, we have

$\mathbf{27}(2,2):\quad R^{ab}_{cd} = \tilde{R}^{ab}_{cd} - \frac{1}{5}(\delta^a_c D^b_d + \delta^b_d D^a_c + \delta^a_d D^b_c + \delta^b_c D^a_d) - \frac{1}{6}(\delta^a_c\delta^b_d + \delta^a_d\delta^b_c)\, S\,.$

(b) An irreducible tensor of SU(3) is denoted by its rank $(n, m)$. The decomposition of a product of 2 irreducible tensors can be done with Coleman's prescription, which consists of two formulas

$$(n,m)\otimes(n',m') = \sum_{i=0}^{\min n,m'}\ \sum_{j=0}^{\min n',m} (n-i, m-j;\ n'-j, m'-i)\,,$$

$$(n,n';\ m,m') = (n+n', m+m') \oplus \sum_{i=1}^{\min n,n'} (n+n'-2i, m+m'+i)$$

$$\oplus \sum_{i=1}^{\min m,m'} (n+n'+j, m+m'-2j)\,.$$

In our case we have

$$\begin{aligned}(3,0)\times(0,3) &= \sum_{i=0}^{3} (3-i, 0;\ 0, 3-i)\\ &= (3,0;0,3)\oplus(2,0;\ 0,2)\oplus(1,0;\ 0,1)\oplus(0,0;\ 0,0)\\ &= (3,3)\oplus(2,2)\oplus(1,1)\oplus(0,0) = \mathbf{64}\oplus\mathbf{27}\oplus\mathbf{8}\oplus\mathbf{1}\,.\end{aligned}$$

**7.7** Totally antisymmetric: $q^A q^B q^C \pm$ permutations ($A, B, C$ are all different with values $1,\ldots,6$). The possible indices are: $123,\ldots,146,156$ (for 10 components); $234,\ldots,256$ (6); $345, 346, 356$ (3); $456$ (1); giving 20 components in all.

Totally symmetric: $q^A q^B q^C$ + permutations. The possibilities are: all different indices $123,\ldots,146,156$ (10); all identical indices $111, 222,\ldots,666$ (6); two identical indices $112, 113,\ldots,664,665$ (30); giving 56 components.

Mixed symmetry accounts for the remaining $216 - 20 - 56 = 140$ components. There will be two irreducible tensors of dimensions 70.

**7.8** (a) Summation over polarizations gives

$$\sum_{\text{pol}} \varepsilon^*_\mu(k,\lambda)\varepsilon_\nu(k,\lambda) = -g_{\mu\nu} + (k_\mu k_\nu/M^2_{\mathrm{W}})\,.$$

Summation over spins ($\Gamma_\mu = g_{\mathrm{V}}\gamma_\mu - g_{\mathrm{A}}\gamma_\mu\gamma_5$):

$$\begin{aligned}\sum_{ss'} \bar{u}(p)\Gamma_\mu u(P)\,\bar{u}(P)\Gamma_\nu u(p) &= \mathrm{Tr}\,[(\not{p} + m_{\mathrm{q}})\Gamma_\mu(\not{P} + m_{\mathrm{Q}})\Gamma_\nu]\\ &= 4[(g^2_{\mathrm{V}} + g^2_{\mathrm{A}})(p_\mu P_\nu + p_\nu P_\mu - g_{\mu\nu}P\cdot p)\\ &\quad + (g^2_{\mathrm{V}} - g^2_{\mathrm{A}})m_{\mathrm{q}}m_{\mathrm{Q}}g_{\mu\nu} - 2ig_{\mathrm{V}}g_{\mathrm{A}}\epsilon_{\mu\nu\rho\sigma}p^\rho P^\sigma]\,.\end{aligned}$$

Putting the two results together leads to

$$\sum_{\text{spins}}\sum_{\text{pol}}|\mathcal{M}|^2 = 4f^2(g_V^2+g_A^2)\left[p{\cdot}P+\frac{2}{M_W^2}(k{\cdot}p)(k{\cdot}P)\right]-12f^2(g_V^2-g_A^2)m_q m_Q\,,$$

(b) Decay rate (with $\alpha=1$):

$$\Gamma(\mathrm{Q}\to\mathrm{q}+\mathrm{W})=\frac{|\boldsymbol{p}|}{8\pi m_Q^2}\frac{1}{2}\sum|\mathcal{M}|^2=\frac{|\boldsymbol{p}|}{8\pi m_Q^2}\,4\,[p{\cdot}P+\frac{2}{M_W^2}(k{\cdot}p)(k{\cdot}P)]\,.$$

Kinematics gives

$$k{\cdot}p=\tfrac{1}{2}\,(m_Q^2-m_q^2-M_W^2),\quad k{\cdot}P=\tfrac{1}{2}\,(m_Q^2-m_q^2+M_W^2),$$
$$p{\cdot}P=\tfrac{1}{2}\,(m_Q^2+m_q^2-M_W^2)\,,\ |\boldsymbol{p}|^2=\frac{1}{4m_Q^2}\lambda(m_Q^2,m_q^2,M_W^2)\,.$$

**8.2** Call $U=U(h)$ the transformation matrix corresponding to group element $h$; its dependence on $x$ is suppressed. Similarly, $U'=U(h')$ and $U''=U(h'')$ are the matrix representations of $h'$ and $h''$. We have the successive transformations

$$h:\boldsymbol{A}_\mu\to\boldsymbol{A}'_\mu=U\boldsymbol{A}_\mu U^\dagger+\frac{\mathrm{i}}{g}(\partial_\mu U)U^\dagger\,,$$
$$h':\boldsymbol{A}'_\mu\to\boldsymbol{A}''_\mu=U'\boldsymbol{A}'_\mu U'^\dagger+\frac{\mathrm{i}}{g}(\partial_\mu U')U'^\dagger\,.$$

Therefore, $\boldsymbol{A}''_\mu$ results from $\boldsymbol{A}'_\mu$ by application of $h'h$:

$$\begin{aligned}h'h:\boldsymbol{A}_\mu\to\boldsymbol{A}''_\mu&=U'\boldsymbol{A}'_\mu U'^\dagger+\frac{\mathrm{i}}{g}(\partial_\mu U')U'^\dagger\\&=U'U\boldsymbol{A}_\mu U^\dagger U'^\dagger+\frac{\mathrm{i}}{g}U'(\partial_\mu U)U^\dagger U'^\dagger+\frac{\mathrm{i}}{g}(\partial_\mu U')U'^\dagger\\&=U'U\left\{\boldsymbol{A}_\mu+\frac{\mathrm{i}}{g}[\partial_\mu(U'U)]\right\}U^\dagger U'^\dagger\,.\end{aligned}$$

Since for $h''=h'h$, $U(h'')=U(h')U(h)$ or $U''=U'U$, the above relation is equivalent to $g'':\boldsymbol{A}_\mu\to\boldsymbol{A}''_\mu=U''\boldsymbol{A}_\mu U''^\dagger+\frac{\mathrm{i}}{g}(\partial_\mu U'')U''^\dagger$, which verifies the group composition law.

**8.3** (a) Assuming $\phi_i$ and $\sigma$ are classical commuting fields, one checks by direct calculation that $\delta_a\mathcal{L}_s=0$ for $a=1,2,3$ in both isospin rotation and chiral transformation.

(b) Isopspin current: $V_{i\mu}=-{\omega_i}^{-1}\pi^j_\mu\delta_i\phi_j=-\epsilon_{ijk}(\partial_\mu\phi_j)\phi_k$. This is a conserved current because $\partial^\mu V_{i\mu}=-\epsilon_{ijk}(\partial^\mu\partial_\mu)\phi_k$. From the equation of motion for $\phi_i$,

$$\partial^\mu\partial_\mu\phi_i=-2\phi_i\frac{\partial V}{\partial\phi^2}\,,$$

it follows that $\epsilon_{ijk}(\partial^\mu\partial_\mu\phi_j)\phi_k = -2\epsilon_{ijk}\frac{\partial V}{\partial\phi^2}\phi_j\phi_k = 0$. Thus, $\partial^\mu V_{i\mu} = 0$. The conserved isospin charge is

$$Q_i = \int \mathrm{d}^3x\, V_{i0} = -\epsilon_{ijk}\int \mathrm{d}^3x\, \pi_j\phi_k \,.$$

Chiral current: $A_{i\mu} = -\omega_i{}^{-1}(\pi^j_\mu\delta_i\phi_j + \pi^4_\mu\delta_i\sigma) = \pi^i_\mu\sigma - \pi^4_\mu\phi_i$. This is again a conserved current, $\partial^\mu A_{i\mu} = 0$, for fields $\phi_i$ and $\sigma$ satisfying their respective equations of motion. The corresponding chiral charge is

$$Q_i^5 = \int \mathrm{d}^3x\, A_{i0} = \int \mathrm{d}^3x\,(\pi_i\sigma - \pi_4\phi_i)\,.$$

The quantized fields satisfy the commutation relations at equal times:

$$[\pi_i(t,\boldsymbol{x}),\,\phi_j(t,\boldsymbol{x}')] = -\mathrm{i}\delta_{ij}\delta^3(\boldsymbol{x}-\boldsymbol{x}')\,, \qquad [\pi_4(t,\boldsymbol{x}),\,\sigma_j(t,\boldsymbol{x}')] = -\mathrm{i}\delta^3(\boldsymbol{x}-\boldsymbol{x}')\,,$$

while other fields commute. From these rules, the following relations can be derived:

$$\begin{aligned}
&[Q_i,\, V_{j0}] = \mathrm{i}\epsilon_{ijk}\,V_{k0}, && [Q_i,\, Q_j] = \mathrm{i}\epsilon_{ijk}\,Q_k\,;\\
&[Q_i,\, A_{j0}] = \mathrm{i}\epsilon_{ijk}\,A_{k0}, && [Q_i,\, Q_j^5] = \mathrm{i}\epsilon_{ijk}\,Q_k^5\,;\\
&[Q_i^5,\, V_{j0}] = \mathrm{i}\epsilon_{ijk}\,A_{k0}, && [Q_i^5,\, Q_j] = \mathrm{i}\epsilon_{ijk}\,Q_k^5\,;\\
&[Q_i^5,\, V_{j0}] = \mathrm{i}\epsilon_{ijk}\,V_{k0}, && [Q_i^5,\, Q_j^5] = \mathrm{i}\epsilon_{ijk}\,Q_k\,.
\end{aligned}$$

The operators $Q_i^+ = \frac{1}{2}(Q_i + Q_i^5)$ and $Q_i^- = \frac{1}{2}(Q_i - Q_i^5)$ satisfy

$$[Q_i^+,\, Q_j^+] = \mathrm{i}\epsilon_{ijk}\,Q_k^+\,; \quad [Q_i^-,\, Q_j^-] = \mathrm{i}\epsilon_{ijk}\,Q_k^-\,; \quad [Q_i^+,\, Q_j^-] = 0\,,$$

and therefore generate two independent commuting SU(2) algebras, so that the symmetry of the model is that of a semisimple algebra, SU(2) × SU(2).

(c) The vacuum is defined by $\langle\phi_i\rangle = 0$ and $\langle\sigma\rangle = v = \sqrt{-\mu^2/\lambda}$. Upon substitution of $\sigma' = \sigma - v$ into $\mathcal{L}_\mathrm{s}$, one gets

$$\mathcal{L}_\mathrm{s} = \tfrac{1}{2}\,[(\partial_\mu\sigma')^2 + (\partial_\mu\phi)^2 + 2\mu^2\sigma'^2] + \mathcal{L}_\mathrm{int}\,,$$

where $\mathcal{L}_\mathrm{int}$ contains three- and four-field couplings. We thus see that the degeneracy in mass has been removed: $\phi_i$ remain massless but $\sigma'$ acquires a mass of $\sqrt{-2\mu^2}$.

The symmetry algebra of the fields is determined by their commutation with the generators. Specially noteworthy is the relation $[Q_i^5,\, \phi_j] = -\mathrm{i}\delta_{ij}(\sigma' + v)$. Its VEV for any $i = j$, $\langle 0 \,|\, [Q_i^5,\, \phi_i] \,|\, 0\rangle = -\mathrm{i}v \neq 0$ (no sum over $i$), cannot be satisfied unless $Q_i^5\,|0\rangle \neq 0$. This means that $Q_i^5$ generates a symmetry that is broken in the vacuum.

(d) Under isospin rotation the fermion fields change according to $\delta_i\psi = \frac{1}{2}\mathrm{i}\omega_i\tau_i\psi$ and $\delta_i\bar\psi = -\frac{1}{2}\mathrm{i}\omega_i\bar\psi\tau_i$. It can be checked by direct calculation that $\delta_i\mathcal{L}_\mathrm{F} = 0$. While under chiral transformation, $\delta_i\psi = \frac{1}{2}\,\mathrm{i}\,\omega_i\,\tau_i\gamma_5\psi$, $\delta_i\bar\psi = \frac{1}{2}\,\mathrm{i}\,\omega_i\,\bar\psi\tau_i\gamma_5$, so that the kinetic part of the fermion Lagrangian changes as

$$\begin{aligned}
\frac{1}{\omega_i}\delta\mathcal{L}_\mathrm{kin} &= \frac{\mathrm{i}}{2}\bar\psi(\mathrm{i}\gamma\cdot\partial - m_0)\tau_i\gamma_5\psi + \frac{\mathrm{i}}{2}\bar\psi\gamma_5\tau_i(\mathrm{i}\gamma\cdot\partial - m_0)\psi\\
&= -\tfrac{1}{2}\,\bar\psi(\gamma^\mu\gamma_5 + \gamma_5\gamma^\mu)\partial_\mu\psi - \mathrm{i}m_0\bar\psi\tau_i\gamma_5\psi = -\mathrm{i}m_0\bar\psi\tau_i\gamma_5\psi\,.
\end{aligned}$$

The coupling term is invariant. The variation of the total Lagrangian under a chiral transformation is $\delta(\mathcal{L}_s + \mathcal{L}_F) = -im_0\bar{\psi}\tau_i\gamma_5\psi$.

Assume now $m_0 = 0$, so that $\mathcal{L}_s + \mathcal{L}_F$ is invariant under both isospin rotation and chiral transformation. For $\mu^2 < 0$, the symmetry is hidden when $\langle\sigma\rangle = v = \sqrt{-\mu^2/\lambda}$. By defining excitations of $\sigma$ above the constant background, $\sigma' = \sigma - v$, the fermion coupling becomes $\mathcal{L}_{\text{int}} = -gv\bar{\psi}\psi - g\bar{\psi}(\sigma' + i\tau\cdot\phi\gamma_5)\psi$, and the fermion acquires a mass, $m = gv = gm_{\sigma'}/\sqrt{2\lambda}$, simultaneously with $\sigma'$ which now has a mass of $m_{\sigma'} = \sqrt{-2\mu^2}$. The older parameters $\mu^2$, $\lambda$, and $v$ are related to the new parameters $m$, $m_{\sigma'}$ and $g$ by $-\mu^2 = \frac{1}{2}m_{\sigma'}^2$, $\lambda = \frac{1}{2}g^2m_{\sigma'}^2/m^2$, $v = m/g$.

**9.1** The condition $k_\mu A^\mu(k) = 0$ implies that in the rest frame, where $k^\mu = (M;0)$, $A_0 = 0$. In general, normalize $A^\mu$ such that $A_\mu A^\mu = -1$. In the frame of a particle moving in the $z$ direction, where $k^\mu = (\omega; 0, 0, |\boldsymbol{k}|)$ with $\omega = \sqrt{\boldsymbol{k}^2 + M^2}$, the two conditions $k_\mu A^\mu = 0$ and $A_\mu A^\mu = -1$ imply that

$$A^\mu = A^\mu_\parallel \cos\theta + A^\mu_\perp \sin\theta\,,$$

$$A^\mu_\parallel = \left(\frac{|\boldsymbol{k}|}{M}; 0, 0, \frac{\omega}{M}\right), \qquad A^\mu_\perp = (0; A_x, A_y, 0)\,.$$

In the particle rest frame, $\theta$ is the angle between $\boldsymbol{A}$ and the $z$ axis. Note that one may rewrite

$$A^\mu_\parallel = \frac{k^\mu}{M} + a^\mu, \qquad a^\mu \equiv \frac{M}{\omega + |\boldsymbol{k}|}(-1; 0, 0, 1)\,.$$

Since the longitudinal components $A^\mu_\parallel$ increase with energy, so too will the transition amplitudes involving $A^\mu$, leading to nonrenormalizability. Hence the necessity of suppressing these components.

For a coupling of the type $j_\mu A^\mu$, the first-order amplitude is $\mathcal{M} = T_\mu A^\mu$, where $T_\mu = \langle f\,|\,j_\mu\,|\,i\rangle$. We have $\mathcal{M} = \left(\frac{k^\mu}{M} + a^\mu\right) T_\mu \cos\theta + T_\mu A^\mu_\perp \sin\theta$. Since the growth in energy of $A^\mu$ is confined to $k^\mu/M$, it is necessary to suppress this term in $\mathcal{M}$, simply by requiring $k^\mu T_\mu = 0$. This is equivalent to $\partial_\mu j^\mu(x) = 0$, i.e. a conserved current.

**9.3** First, $A = \pi\alpha/\sqrt{2}G_F = (37.3\text{GeV})^2$. The boson masses are given by $M_W = 37.3/s_W$ GeV and $M_Z = M_W/c_W = 74.6/\sin 2\theta_W$ GeV. With $M_Z = 91$ GeV, one gets $\sin^2\theta_W = 0.21$ and $M_W = 81$ GeV. In addition, from $M_W = \frac{1}{2}gv$, it follows that $v = (\sqrt{2}G_F)^{-1/2} = 246$ GeV, and from $C_e = \sqrt{2}m_e/v$, one gets $C_e = 3\times 10^{-6}$.

**9.4** With the coupling by $(-1/2\sqrt{2})\, g\bar{e}\gamma^\mu(1-\gamma_5)\nu\, W^\dagger_\mu + \text{h.c.}$, the amplitude is given by

$$\mathcal{M} = \frac{-ig}{2\sqrt{2}}\bar{u}_e(P)\gamma^\mu(1-\gamma_5)v(p)\,\varepsilon_\mu(k)\,.$$

This is essentially the same amplitude as found in Problem 7.8. Neglecting the masses of the fermions, $P^2 = p^2 = 0$, one gets

$$\sum_{\text{pol}}\sum_{ss'} |\mathcal{M}|^2 = g^2\left[p{\cdot}P + \frac{2}{M_W^2}(k{\cdot}p)(k{\cdot}P)\right] = g^2 M_W^2$$

with the final momentum $|\boldsymbol{P}| = (2M_\mathrm{W})^{-1}\lambda^{1/2}(M_\mathrm{W}^2, 0, 0) = \frac{1}{2}M_\mathrm{W}$. Thus, the decay width for $\mathrm{W} \to \mathrm{e}\bar{\nu}$ is

$$\Gamma(\mathrm{W} \to \mathrm{e}\nu) = \frac{|\boldsymbol{P}|}{8\pi M_\mathrm{W}^2}\frac{1}{3}g^2 M_\mathrm{W}^2 = \frac{g^2 M_\mathrm{W}}{48\pi} = \frac{G_F M_\mathrm{W}^3}{6\sqrt{2}\pi}\,.$$

The factor $\frac{1}{3}$ comes from averaging over the vector boson polarizations. The widths for the other processes are obtained in the standard model Lagrangian,

$$\Gamma(\mathrm{W} \to \mathrm{e}\nu) = \Gamma(\mathrm{W} \to \mu\nu) = \Gamma(\mathrm{W} \to \tau\nu)\,,$$
$$\Gamma(\mathrm{W} \to \mathrm{u}\bar{\mathrm{d}}') = \Gamma(\mathrm{W} \to \mathrm{c}\bar{\mathrm{s}}') = 3\Gamma(\mathrm{W} \to \mathrm{e}\nu)\,.$$

Here, the factor 3 accounts for colors; we have also used $\sum |V_{uj}|^2 = 1$. The total decay width of W is $\Gamma_\mathrm{W}^\mathrm{tot} = 9\Gamma(\mathrm{W} \to \mathrm{e}\bar{\nu}) \approx 9 \times 0.225 = 2.05$ GeV.

**9.5** The coupling is $(-g/c_\mathrm{W})j_\mu^Z Z^\mu$ where the neutral current is

$$j_\mu^Z = \tfrac{1}{2}\,\bar{\nu}\gamma_\mu(g_\mathrm{V}^\nu - g_\mathrm{A}^\nu\gamma_5)\nu + \tfrac{1}{2}\,\bar{e}\gamma_\mu(g_\mathrm{V}^\mathrm{e} - g_\mathrm{A}^\mathrm{e}\gamma_5)e$$
$$+ \tfrac{1}{2}\,\bar{u}\gamma_\mu(g_\mathrm{V}^\mathrm{u} - g_\mathrm{A}^\mathrm{u}\gamma_5)u + \tfrac{1}{2}\,\bar{d}\gamma_\mu(g_\mathrm{V}^\mathrm{d} - g_\mathrm{A}^\mathrm{d}\gamma_5)d\,.$$

The amplitude for decay $\mathrm{Z}\to \mathrm{f}\bar{\mathrm{f}}$ is $\mathcal{M} = \dfrac{-ig}{2c_\mathrm{W}}\bar{u}(P)\gamma_\mu(g_\mathrm{V}^\mathrm{f} - g_\mathrm{A}^\mathrm{f}\gamma_5)v(p)\,\varepsilon_\mu(k)$. In a similar way as in Problem 9.4, one gets $\Gamma(\mathrm{Z} \to \mathrm{f}\bar{\mathrm{f}}) = [(g_\mathrm{V}^\mathrm{f})^2 + (g_\mathrm{A}^\mathrm{f})^2]\,\Gamma_0$, where $\Gamma_0 = G_F M_\mathrm{Z}^3/(6\sqrt{2}\pi) \approx 0.328$ GeV. The neutral weak charges are obtained from Table 9.3. Assuming $s_\mathrm{W}^2 = \sin^2\theta_\mathrm{W} = 0.21$, one gets the following values of $(g_\mathrm{V}^\mathrm{f})^2 + (g_\mathrm{A}^\mathrm{f})^2$:

$$\begin{array}{lll}
\text{for } \nu_e, \nu_\mu, \nu_\tau: & \frac{1}{2} & = 0.5\,, \\
\text{for e}, \mu, \tau: & \frac{1}{2} - 2s_\mathrm{W}^2 + 4s_\mathrm{W}^4 & = 0.25\,, \\
\text{for u, c}: & \frac{1}{2} - \frac{4}{3}s_\mathrm{W}^2 + \frac{16}{9}s_\mathrm{W}^4 & = 0.3\,, \\
\text{for d, s, b}: & \frac{1}{2} - \frac{2}{3}s_\mathrm{W}^2 + \frac{4}{9}s_\mathrm{W}^4 & = 0.38\,.
\end{array}$$

The process $\mathrm{Z}\to\mathrm{t}\bar{\mathrm{t}}$ is not kinematically accessible. Taking account of the color factor, the total decay width is $\Gamma_\mathrm{Z}^\mathrm{tot} = 3 \times (0.5 + 0.25 + 0.6 + 1.14)\Gamma_0 \approx 2.45$ GeV.

**9.6** The amplitudes for $\gamma$ and Z-boson exchanges are:

$$\mathcal{M}_\gamma = \mathrm{i}(-\mathrm{i}e)^2\,\bar{u}(k)\gamma^\mu v(k')\,\bar{v}(p')\gamma^\nu u(p)\left(\frac{-\mathrm{i}g_{\mu\nu}}{s}\right)\,,$$

$$\mathcal{M}_\mathrm{Z} = \mathrm{i}\left(\frac{-\mathrm{i}g}{2\cos\theta_\mathrm{W}}\right)^2 [\bar{u}(k)\gamma^\mu\,(g_\mathrm{V}^\mathrm{f} - \gamma_5 g_\mathrm{A}^\mathrm{f})\,v(k')]\,[\bar{v}(p')\gamma^\nu\,(g_\mathrm{V}^\mathrm{e} - \gamma_5 g_\mathrm{A}^\mathrm{e})\,u(p)]$$
$$\times\left(\frac{-\mathrm{i}[g_{\mu\nu} - k_\mu k_\nu/M_\mathrm{Z}^2]}{s - M_\mathrm{Z}^2 + \mathrm{i}\Gamma_\mathrm{Z} M_\mathrm{Z}}\right)\,; \qquad s = (p + p')^2\,.$$

As an example, consider $\mathrm{e}^+ + \mathrm{e}^- \to \mu^+ + \mu^-$. Neglecting $m_\mathrm{e}, m_\mu$ masses and using (4.181), one finds

$$\frac{\mathrm{d}\sigma}{\mathrm{d}\Omega} = \frac{\alpha^2}{4s}\left[A_0(1 + \cos^2\theta) + A_1\cos\theta\right]\,,$$

where

$$A_0 = 1 + 2\,\mathrm{Re}(z) g_V^2 + |z|^2 (g_V^2 + g_A^2)^2\,, \qquad A_1 = 4\,\mathrm{Re}(z) g_A^2 + 8|z|^2\, g_V^2 g_A^2\,;$$

$$z = \frac{s}{s - M_Z^2 + \mathrm{i}\Gamma_Z M_Z}\frac{1}{(2\sin 2\theta_W)^2}\,, \quad g_A = -\tfrac{1}{2}\,, \quad g_V = -\tfrac{1}{2} + 2\sin^2\theta_W \approx 0\,.$$

With $\sigma_F = \int_0^1 [\mathrm{d}\sigma/\mathrm{d}\Omega]\,\mathrm{d}\Omega$ and $\sigma_B = \int_{-1}^0 [\mathrm{d}\sigma/\mathrm{d}\Omega]\,\mathrm{d}\Omega$, one finds

$$A_{FB} = \frac{3A_1}{8A_0} \approx \frac{3}{2}\mathrm{Re}(z) g_A^2\,.$$

**10.1** The minus sign comes from the anticommuting property of the two nucleons. The potential $V_{\mathrm{dir}}(\boldsymbol{x})$ has the dimension of (mass), so its Fourier transform $V_{\mathrm{dir}}(\boldsymbol{q})$ has the dimension $(\mathrm{mass})^{-2}$,

$$V_{\mathrm{dir}}(\boldsymbol{q}) = \int \mathrm{d}^3 r\, \mathrm{e}^{-\mathrm{i}\boldsymbol{q}\cdot\boldsymbol{x}}\, V_{\mathrm{dir}}(\boldsymbol{x})\,, \quad V_{\mathrm{dir}}(\boldsymbol{x}) = \frac{1}{(2\pi)^3}\int \mathrm{d}^3 q \mathrm{e}^{\mathrm{i}\boldsymbol{q}\cdot\boldsymbol{x}}\, V_{\mathrm{dir}}(\boldsymbol{q})\,.$$

The one-particle state normalization in (4.40) is the origin of the factor $(2M_N)^{-2}$ for the two-nucleon system. The $\boldsymbol{\sigma}_1\cdot\boldsymbol{q}\,\boldsymbol{\sigma}_2\cdot\boldsymbol{q}$ term yields the operator $\boldsymbol{\sigma}_1\cdot\nabla\,\boldsymbol{\sigma}_2\cdot\nabla$; this operator applies to the potential $\mathrm{e}^{-m_\pi r}/r$, which gives $V_{\mathrm{dir}}(\boldsymbol{x})$.

**10.2** First integrate over $\int \mathrm{d}^3 x$ the time ($\mu = 0$)-component of the equation

$$\langle \Phi(\boldsymbol{p}', s')\,|\, J^\mu_{\mathrm{em}}(x)\,|\,\Phi(\boldsymbol{p}, s)\rangle = \mathrm{e}^{\mathrm{i}(p-p')\cdot x}\,\langle \Phi(\boldsymbol{p}', s')\,|\, J^\mu_{\mathrm{em}}(0)\,|\,\Phi(\boldsymbol{p}, s)\rangle\,.$$

From the conserved current $J^\mu_{\mathrm{em}}(x)$, $Q \equiv \int \mathrm{d}^3 x\, J^0_{\mathrm{em}}(x)$ is time independent, then with $Q\,|\Phi(\boldsymbol{p}, s)\rangle = e_Q\,|\Phi(\boldsymbol{p}, s)\rangle$, one gets

$$\langle \Phi(\boldsymbol{p}, s')\,\big|\, J^0_{\mathrm{em}}(0)\,\big|\,\Phi(\boldsymbol{p}, s)\rangle = \frac{e_Q \delta_{s's}}{(2\pi)^3}\,.$$

We apply the above equation to spinless pion $\langle \pi^\pm(\boldsymbol{p})\,|\, J^\mu_{\mathrm{em}}(0)\,|\,\pi^\pm(\boldsymbol{p})\rangle$. With the standard one-particle state normalization $1/\sqrt{2E(2\pi)^3}$, one gets $F_\pi(0) = 1$. By the same method, we have $F_1^{\mathrm{p}}(0) = 1, F_1^{\mathrm{n}}(0) = F_2^{\mathrm{n}}(0) = 0$.

**10.3** The amplitude for this reaction is

$$\mathcal{M} = (-\mathrm{i}e)^2\left(\frac{-\mathrm{i}}{q^2}\right)\overline{v}(p')\gamma_\mu u(p)\;(k - k')^\mu\, F_\pi(q^2)\,.$$

Using (4.59), the cross-section follows. Whereas the pion form factor $F_\pi(q^2)$ is dynamically enhanced by the $\rho^0$ meson which has an isospin $I = 1$, the form factors of the charged K, D, and B mesons would not be enhanced by the lack of the corresponding $I = 1$ resonances decaying into the K, D, B pairs, similar to the $\rho^0$ meson decaying into a $\pi$-pair.

**10.4** According to CVC, the $\Delta S = 0$ vector currents $V_\mu^{1\pm \mathrm{i}2}$ of the weak interaction and the isovector component of the electromagnetic current $V_\mu^3$ form an isospin triplet. The corresponding Clebsch–Gordan coefficient $\sqrt{2}$ comes from $I_-(1,0,0) = \sqrt{2}(0,1,0)$, using (6.53) and (6 .54). The amplitude is

$$\mathcal{M} = 2G_F V_{ud}\, \bar{u}(k')\gamma^0(1-\gamma_5)v(k)\, M_\pi \,.$$

Using the three-particle phase space integration formula in the Appendix,

$$\Gamma = \frac{G_F^2|V_{ud}|^2}{\pi^3}\int_0^{|\boldsymbol{k}|_{\max}} \mathrm{d}|\boldsymbol{k}|\,|\boldsymbol{k}|^2(\Delta - E)^2\,, \qquad E = \sqrt{|\boldsymbol{k}|^2 + m_e^2}\,,$$

where $\boldsymbol{k}$ is the three-momentum of the electron and $\Delta = M_{\pi^-} - M_{\pi^0}$. Thus

$$\Gamma = \frac{G_F^2|V_{ud}|^2\Delta^5}{30\pi^3}\left(1 - \frac{3}{2}\frac{\Delta}{M_{\pi^-}} + \frac{5m_e^2}{\Delta^2}\cdots\right) = 0.393\ \mathrm{s}^{-1}\,,$$

$$\left[\Gamma(\pi^+ \to \pi^0 + \mathrm{e}^+ + \nu_e)/\Gamma(\pi^+ \to \mu^+ + \nu_\mu)\right] = 1\times 10^{-8}\,,$$

which is in excellent agreement with the data.

**10.5** For the $\pi^+ \to \mathrm{e}^+ + \nu_e$ decay amplitude, the solution is given in solution 5.3. The $\rho^0(P) \to \mathrm{e}^+(k') + \mathrm{e}^-(k)$ amplitude is

$$(-\mathrm{i}e)^2\, m_\rho\, f_\rho\, \varepsilon^\mu(P)\left(\frac{-\mathrm{i}}{P^2}\right)\bar{v}(k')\gamma_\mu u(k)\,,$$

with CVC, the $\rho^+(P) \to \mathrm{e}^+(k') + \nu_e(k)$ amplitude is

$$\frac{G_F V_{ud}}{\sqrt{2}}\sqrt{2}\, m_\rho\, f_\rho\, \varepsilon^\mu(P)\, \bar{v}(k')\gamma_\mu(1-\gamma_5)u(k)\,.$$

Using the two-particle phase space integration formula in the Appendix,

$$\Gamma(\rho^+ \to \mathrm{e}^+ + \nu_e) = \frac{G_F^2|V_{ud}|^2}{12\pi} m_\rho^3 f_\rho^2\left(1 - \frac{m_e^2}{m_\rho^2}\right)^2\,.$$

**10.6** Using the trick $\mathrm{e}^{-\mu r} \to 1$ for $\mu \to 0$ as in (10.3), and the Table of Fourier transforms in Erdelyi et. al, McGraw-Hill 1954.

**10.7** Using Im $G(s') = \delta(s')$, write the once-subtracted dispersion relation for $G(s)$, note that $G(0) = 0$. Since the function $H(z)$ is also analytic in the complex $z$ plane, we apply the Cauchy theorem on a closed contour limited by a circle of radius $|z| = \infty$ and a cut above and below the real axis starting at $z = 4m_\pi^2 \pm \mathrm{i}\epsilon$, and get

$$\log F_\pi(s) = \log|F_\pi(s)| + \mathrm{i}\delta(s) = \frac{\sqrt{s - 4m_\pi^2}}{\mathrm{i}\pi}\int_{4m_\pi^2}^{\infty} \mathrm{d}x\frac{\log|F_\pi(x)|}{(x-s)\sqrt{x - 4m_\pi^2}}\,.$$

The absolute value $|F_\pi(x)|$ is measured by $\mathrm{e}^+ + \mathrm{e}^- \to \pi^+ + \pi^-$.

**11.1** The two-pion states (either $\pi^+ + \pi^-$ or $\pi^0 + \pi^0$) have $CP = +1$ eigenvalues by Bose symmetry. Any orbital angular momentum, $\ell$, between any two pions is even by Bose symmetry. Hence the $\ell$ of the remaining pion in the $\pi^0 + \pi^0 + \pi^0$ of the K decay must also be even. So the overall parity of the three neutral pions emitted in K decay has intrinsic $P = -1$ parity. Since the $\pi^0$ has $C = +1$ (from $\pi^0 \to 2\gamma$), the three $\pi^0$ in K decay has $CP = -1$. $\pi^+\pi^-\pi^0$ has an odd $CP$ eigenvalue only if these pions are in a relative ($\ell = 0$) s-state, which is strongly favored by the angular-momentum barrier effects due to the small energy released by $K \to 3\,\pi$.

**11.2** The easiest way to analyze strangeness decays with the $\Delta I = 1/2$ rule is to consider a 'spurion' operator $Sp$ bearing $I = 1/2, I_z = -1/2$. The addition of the isospin $Sp$ with the isospin of the decaying particles must be equal to the overall isospin of the final state. Thus

$$Sp \oplus \Sigma^+ = \sqrt{\frac{1}{3}}\mathcal{M}_{\frac{3}{2}} + \sqrt{\frac{2}{3}}\mathcal{M}_{\frac{1}{2}} \quad , \quad Sp \oplus \Sigma^- = \mathcal{M}_{\frac{3}{2}} \; ,$$

where $\mathcal{M}_{\frac{1}{2}}$ and $\mathcal{M}_{\frac{3}{2}}$ are the $I = \frac{1}{2}$ and $I = \frac{3}{2}$ amplitudes. On the other hand, the isospin decomposition of the nucleon–pion states are

$$\mathrm{n}+\pi^+ = \sqrt{\frac{1}{3}}\mathcal{M}_{\frac{3}{2}} + \sqrt{\frac{2}{3}}\mathcal{M}_{\frac{1}{2}} \; , \; \mathrm{p}+\pi^0 = -\sqrt{\frac{1}{3}}\mathcal{M}_{\frac{1}{2}} + \sqrt{\frac{2}{3}}\mathcal{M}_{\frac{3}{2}} \; , \; \mathrm{n}+\pi^- = \mathcal{M}_{\frac{3}{2}} \; .$$

Hence

$$a_+ = \frac{1}{3}\left(\mathcal{M}_{\frac{3}{2}}\right)^2 + \frac{2}{3}\left(\mathcal{M}_{\frac{1}{2}}\right)^2, \quad a_0 = \frac{\sqrt{2}}{3}\left[\left(\mathcal{M}_{\frac{3}{2}}\right)^2 - \left(\mathcal{M}_{\frac{1}{2}}\right)^2\right], \quad a_- = \left(\mathcal{M}_{\frac{3}{2}}\right)^2 .$$

**11.3, 11.4** The solutions can be found in Chap. 16.

**11.5** Consider the color SU(3) invariant $\sum_{j=1}^{8}(\lambda^j)_{ef}(\lambda^j)_{gh}$. The statement of completeness implies that it can only have the form $A\delta_{ef}\delta_{gh} + B\delta_{eh}\delta_{gf}$. Put $f = g$ and using $\sum_j(\lambda^j)_{eg}(\lambda^j)_{gh} = \frac{16}{3}\delta_{eh}$ to get one relation between $A$ and $B$. Then put $e = f$ and using $\mathrm{Tr}\lambda^j = 0$ to get a second relation. The identity (11.88) follows.

**12.1** The amplitude in the most general form is

$$\mathcal{M} = \epsilon^\mu \left\langle \mathrm{e} \middle| V_\mu^{\mathrm{em}} \middle| \mu \right\rangle = \epsilon^\mu\, \overline{u}\, [q^\nu\, \sigma_{\nu\mu}(a + b\gamma_5)]\, u \, .$$

For on-the-mass-shell photon, the amplitude is a magnetic transition. In the unitarity gauge $\xi = \infty$, there are three Feynman diagrams similar to the penguin diagram (with internal $\nu$ and W in the loop), thus giving the transition $\mu \to \mathrm{e}$ analogous to the $\mathrm{s} \to \mathrm{d}$ transition of penguin. The photon is either emitted from the internal W or from the external fermions $\mu$ and e. Using the leptonic mixing matrix $V_{\mathrm{lep}}$, the coefficients $a, b$ can be estimated to be proportional to $(m_j^2/M_{\mathrm{W}}^2)\; V_{\mu\nu_j} V^*_{\nu_j e}$, where $m_j$ is the neutrino mass.

**12.2** From the solar luminosity $3.86 \times 10^{33}$ erg/s, and 26 MeV $\approx 4.16 \times 10^{-5}$ erg, there are $9.2 \times 10^{37}$ fusions taking place in the sun every second; each fusion produces two neutrinos. So the flux received on earth is

$$\Phi \approx \frac{1.8 \times 10^{37}/\mathrm{s}}{4\pi(1.5 \times 10^{13}\mathrm{cm})^2} \approx 6.4 \times 10^{10}/\mathrm{cm}^2\,/\mathrm{s} \, .$$

**12.3** From (12.10), (12.12), and (12.13), the effective mass of a neutrino in matter is given by $m_\nu^2 = 2E_\nu(\mathcal{V}_{\rm N} + \mathcal{V}_{\rm C}) = 2E_\nu G_{\rm F}(\sqrt{2}N_{\rm e} - N_{\rm n}/\sqrt{2})$. In the solar core, one gets $m_\nu^2 \approx 10^{-4}{\rm eV}^2$, and in a supernova core, $m_\nu^2 \approx 10^8{\rm eV}^2$.

**12.4** It is obvious that by weak decay

$$\frac{\Gamma_{\rm w}(\pi^0 \to {\rm Z}^0 \to {\rm e}^+ + {\rm e}^-)}{\Gamma_{\rm w}(\pi^+ \to {\rm W}^+ \to {\rm e}^+ + \nu_{\rm e})} = \mathcal{O}(1)\,.$$

On the other hand, the electromagnetic decay $\pi^0 \to 2\gamma^* \to {\rm e}^+ + {\rm e}^-$ by conversion of the two virtual photons $\gamma^* + \gamma^*$ into ${\rm e}^+ + {\rm e}^-$, has two diagrams similar to Fig. 4.11 of the Compton scattering. The amplitude $\gamma^* + \gamma^* \to {\rm e}^+ + {\rm e}^-$ is proportional to $\alpha\, m_{\rm e}/M_\pi$. If we assume a pointlike effective coupling $\pi^0\gamma\gamma$, we can draw a triangle loop with two photons and one electron propagators as internal lines, the external lines are $\pi^0$, ${\rm e}^+$ and ${\rm e}^-$. The loop integration would yield a factor $\log(m_{\rm e}/M_\pi)$. All of these considerations yield $\Gamma(\pi^0 \to 2\gamma^* \to {\rm e}^+ + {\rm e}^-) \gg \Gamma_{\rm w}(\pi^0 \to {\rm Z}^0 \to {\rm e}^+ + {\rm e}^-)$.

**12.5** Combine (10.62) with (12.65) and (12.67) to derive this sum rule.

**12.6** For isoscalar targets, the sea contributions cancel in the difference of the $\nu$ and $\overline{\nu}$ cross-sections. Thus

$$\frac{{\rm d}\sigma^{\nu}_{\rm NC} - {\rm d}\sigma^{\overline{\nu}}_{\rm NC}}{{\rm d}x\,{\rm d}y} = \frac{G_{\rm F}^2 ME}{\pi} xQ(x)\left(|u_{\rm L}|^2 + |d_{\rm L}|^2 - |u_{\rm R}|^2 - |d_{\rm R}|^2\right) \times \left[1 - (1-y)^2\right],$$

where $u_{\rm L} = \frac{1}{2} - \frac{2}{3}\sin^2\theta_{\rm W}$, $d_{\rm L} = -\frac{1}{2} + \frac{1}{3}\sin^2\theta_{\rm W}$, $u_{\rm R} = -\frac{2}{3}\sin^2\theta_{\rm W}$, and $d_{\rm R} = \frac{1}{3}\sin^2\theta_{\rm W}$ . This is to be compared with the corresponding CC cross-sections

$$\frac{{\rm d}\sigma^{\nu}_{\rm CC} - {\rm d}\sigma^{\overline{\nu}}_{\rm CC}}{{\rm d}x\,{\rm d}y} = \frac{G_{\rm F}^2 ME}{\pi} xQ(x)\left[1 - (1-y)^2\right].$$

**13.1** The $\tau^- \to \nu_\tau + \eta + \pi^-$ decay amplitude is related to the matrix element $\langle \eta + \pi\,|\,v_\mu\,|\,0\rangle$, where $v_\mu = \overline{u}\gamma_\mu d$. The G-parity of the current $v_\mu$ is opposite to the G-parity of $\eta + \pi$. Also $\eta \to \pi + \pi$ violates CP-invariance.

**13.2** Use the tensor $(T_1)_{\lambda\rho}$ in (13.8) to compute the matrix element squared.

**13.6** Since the momentum transfer $q$ in the n$\to$ p transition is negligible, only two form factors proportional to $\gamma_\mu$ and $\gamma_\mu\gamma_5$ enter the decay amplitude. Use formula (13.62) of the $G(x,y)$ function to obtain the neutron decay rate. A comparison with the data gives $g_1(0) \approx 1.25$.

**13.7** The second term $k^\mu k^\nu/M_{\rm W}^2$ in the numerator of the W propagator, when contracted with the two currents $\overline{\nu}_\tau\gamma_\mu(1-\gamma_5)\tau$ and $\overline{l}\gamma_\nu(1-\gamma_5)\nu_l$, yields the last factor in (13.28) for the rate.

**13.8** The squared matrix element is found to be $|\mathcal{A}|^2 = 64G_{\rm F}^2\,(p_1\cdot p_2)(P\cdot p_3) = 16G_{\rm F}^2(s - m_1^2 - m_2^2)(M^2 + m_3^2 - s)$. Using the three-particle phase space integral of the Appendix, one gets $J(x,y,z)$. For $(V+A)\cdot(V-A)$, we only make the interchange $p_1$ and $p_3$. The integrated width is equal to the $(V-A)\cdot(V-A)$ case.

**14.1** The traceless color matrix $\lambda^j$ in the diagram with two emitted gluons in Fig. 14.3b is at the origin of the noninterference between the tree and bremsstrahlung amplitudes.

**14.2** Analytic expressions for $F_{1,\mathrm{uv}}(q^2)$ and $F_{1,\mathrm{ir}}(q^2)$:

$$\int_0^1 \mathrm{d}v \log\left[1-\frac{q^2}{4m^2}(1-v^2)\right] = -2+\sqrt{\frac{\xi-1}{\xi}}\log\frac{\sqrt{1-\xi}+\sqrt{-\xi}}{\sqrt{1-\xi}-\sqrt{-\xi}}\ ,$$

$$\begin{aligned}\int_0^1 \mathrm{d}v \log\left[1-\frac{q^2}{4m^2}(1-v^2)\right] &= -2+\left[\sqrt{\frac{\eta-1}{\eta}}\log\frac{\sqrt{\eta}-\sqrt{\eta-1}}{\sqrt{\eta}+\sqrt{\eta-1}}+\mathrm{i}\pi\right]\\ &= -2+\sqrt{\frac{1-\eta'}{\eta'}}\tan^{-1}\sqrt{\frac{\eta'}{1-\eta'}}\ ,\end{aligned}$$

where $\xi=-q^2/(4m^2)<0$, $\eta=q^2/(4m^2)>1$, and $0\le\eta'=\frac{q^2}{4m^2}\le 1$.

With a small mass $\zeta$ given to the gluon to regularize the IR divergence, the $F_{1,\mathrm{ir}}(q^2)$ in (14.19) is found to be

$$F_{1,\mathrm{ir}}(q^2)=\frac{4}{3}\frac{g_s^2}{8\pi^2}\left[J(\xi)+K(\xi)\log(\frac{m^2}{\zeta^2})\right]\underset{\xi\to 0}{\longrightarrow}\frac{4}{3}\frac{-g_s^2}{8\pi^2}\log(\frac{m^2}{\zeta^2})\ ,$$

where

$$\begin{aligned}J(\xi)&\equiv-\int_0^1\frac{\mathrm{d}v(1+2\xi)}{1+\xi(1-v^2)}\log[1+\xi(1-v^2)]\ ,\quad 0\le\xi=\frac{q^2}{4m^2}<1\\ &=\frac{-(1+2\xi)}{\sqrt{\xi(1+\xi)}}\left\{\left[\log(\sqrt{1+\xi}+\sqrt{\xi})\right]\left[\log\frac{4(1+\xi)}{\sqrt{1+\xi}+\sqrt{\xi}}\right]\right.\\ &\quad\left.-\frac{\pi^2}{12}-\mathcal{S}p\left(\frac{\sqrt{\xi}-\sqrt{1+\xi}}{\sqrt{\xi}+\sqrt{1+\xi}}\right)\right\}\underset{\xi\to 0}{\longrightarrow}\frac{-2\xi}{3}(1+\frac{4}{5}\xi)\ ,\\ \mathcal{S}p(z)&\equiv-\int_0^z\mathrm{d}t\frac{\log(1-t)}{t}\qquad\text{(Spence function or dilogarithm)}\ ,\\ K(\xi)&\equiv\int_0^1\frac{-\mathrm{d}v(1+2\xi)}{1+\xi(1-v^2)}=\frac{-(1+2\xi)}{2\sqrt{\xi(1+\xi)}}\log\frac{\sqrt{1+\xi}+\sqrt{\xi}}{\sqrt{1+\xi}-\sqrt{\xi}}\underset{\xi\to 0}{\longrightarrow}-(1+\frac{4}{3}\xi)\ .\end{aligned}$$

With dimensional regularization, the IR divergence can also be handled by keeping $\varepsilon\neq 0$ everywhere, even with the UV convergent factor $\Gamma(3-n/2)$ in (14.9). The denominator on the second line of $F_1(q^2)$ in (14.13) is $[m^2-q^2(1-v^2)/4]^{3-n/2}\equiv\Delta^{3-n/2}$. Then we find

$$F_{1,\mathrm{ir}}(q^2)=\frac{4}{3}\frac{g_s^2}{8\pi^2}\left\{J(\xi)-\left[\log\left(\frac{\mu^2}{m^2}\right)+\frac{2}{\varepsilon}\right]K(\xi)\right\}\underset{\xi\to 0}{\longrightarrow}\frac{4}{3}\frac{g_s^2}{4\pi^2}\left[\frac{1}{\epsilon}+\log\left(\frac{\mu}{m}\right)\right]\ .$$

The IR divergence $1/\varepsilon$ pole is identified with $\log(\zeta/\mu)$.

**14.4** We find

$$F_2^{\mathrm{Higgs}}(0)=\frac{f_l^2m_l^2}{8\pi^2M_H^2}\int_0^1\mathrm{d}x\frac{x^2(2-x)}{1-x+x^2m_l^2/M_H^2}\ .$$

With $f_l^2 = \sqrt{2}\, m_l^2\, G_{\mathrm{F}}$, the muon anomalous magnetic moment is more sensitive to the Higgs boson effect than the electron anomalous magnetic moment.

**14.5** We compute $\Sigma(p)$ by using (14.89) and find

$$\Sigma(p) = \frac{4}{3}\frac{g_{\mathrm{s}}^2}{8\pi^2}\int_0^1 \mathrm{d}x\,[2m - \not{p}(1-x)]\log\left[\frac{(1-x)\Lambda^2}{m^2x - p^2x(1-x) + \zeta^2(1-x)}\right].$$

In (14.25) with $m_0 = 0$, we have $m = m\,B(m^2)$ and the gap equation $B(m^2) = 1$, thus

$$1 = \frac{4}{3}\left(\frac{-g_{\mathrm{s}}^2}{8\pi^2}\right)\int_0^1 \mathrm{d}x\,(1-x)\log\left[\frac{(1-x)\Lambda^2}{m^2x^2 + \zeta^2(1-x)}\right]. \tag{1}$$

In principle, the equation can be solved to obtain $m$ in terms of $\Lambda$. However, for the QCD interaction considered here, (1) does not possess a solution since for $\Lambda > m$, the right-hand side of (1) is negative.

**14.6** The quantity $J^{\mu\nu}(q^2)$ as defined by (14.54) has the most general form $J^{\mu\nu}(q^2) = -Aq^2g^{\mu\nu} + Bq^\mu q^\nu$ when $m_2, m_3$ are nonvanishing. To compute the coefficients $A$ and $B$, multiply the left- and the right-hand sides of (14.54) first by $g^{\mu\nu}$, then by $q^\mu q^\nu$. One gets two equations for two unknowns, $A$ and $B$. We find that the two terms $-q^2g^{\mu\nu}$ and $q^\mu q^\nu$ of $J^{\mu\nu}(q^2)$ in (14.58) are now replaced by

$$-q^2g^{\mu\nu} \longrightarrow -q^2g^{\mu\nu}\left[1 - \frac{m_2^2+m_3^2}{2q^2} - \frac{1}{2}\left(\frac{m_2^2-m_3^2}{q^2}\right)^2\right],$$

$$q^\mu q^\nu \longrightarrow q^\mu q^\nu\left[1 + \frac{m_2^2+m_3^2}{q^2} - 2\left(\frac{m_2^2-m_3^2}{q^2}\right)^2\right].$$

The $(E_2, E_3)$ domain $\Delta$, restricted by (14.75) in the massless case, is now replaced by

$$4(E_2^2 - m_2^2)(E_3^2 - m_3^2) \le \left[(\sqrt{q^2} - E_2 - E_3)^2 - m_1^2 - E_2^2 + m_2^2 - E_3^2 + m_3^2\right]^2.$$

**15.1** From the solution $\Sigma(p)$ in Problem 14.5, one gets

$$\begin{aligned} Z_2 - 1 &= \left.\frac{\mathrm{d}\Sigma(p)}{\mathrm{d}\not{p}}\right|_{\not{p}=m} \\ &= \frac{4}{3}\frac{g_{\mathrm{s}}^2}{8\pi^2}\int_0^1 \mathrm{d}x\left[-(1-x)\log\frac{(1-x)\Lambda^2}{m^2x^2+\zeta^2(1-x)} + \frac{2(1-x^2)m^2}{m^2x^2+\zeta^2(1-x)}\right] \\ &\longrightarrow -\frac{4}{3}\frac{g_{\mathrm{s}}^2}{16\pi^2}\log\frac{\Lambda^2}{\mu^2}. \end{aligned}$$

The $\Gamma(\varepsilon/2)$ pole in dimensional regularization is identified with the $\log(\Lambda^2/\mu^2)$ of the Pauli–Villars method.

**15.2** We start with massless fermions QED Lagrangian density $\mathcal{L}(x,t) = -\frac{1}{4}(F_{\mu\nu})^2 + \overline{\psi}(\mathrm{i}\,\not{D})\psi$. Since, by definition, the energy $\int dx \mathcal{L}(x,t)$ has the mass dimension, $\mathcal{L}(x,t)$ must have the dimension (mass)$^2$. The derivative $\partial^\mu$ has the (mass)$^1$ dimension, so from $(F_{\mu\nu})^2$, we deduce that the photon field is dimensionless. From $\overline{\psi}(\mathrm{i}\not{D})\psi$, the fermion field has the dimension (mass)$^{1/2}$, and the coupling constant $e$ has (mass)$^1$ dimension. Using (15.5) and (15.6) with $n=2$, Trace[1] =2 in two dimensions, we get

$$\Pi^{\mu\nu}(q) = \left(q^2 g^{\mu\nu} - q^\mu q^\nu\right)\frac{e^2}{q^2\pi} \Longrightarrow \Pi(q^2) = \frac{1}{q^2}\frac{e^2}{\pi} \Longrightarrow m_\gamma = \frac{e}{\sqrt{\pi}}\,.$$

**15.3** The Coulomb potential derived from the nonrelativistic Fourier transformation of $e^2(q^2)/q^2$ is

$$\begin{aligned} V(\boldsymbol{x}) &= \int \frac{\mathrm{d}^3 q \mathrm{e}^{\mathrm{i}\boldsymbol{q}\cdot\boldsymbol{x}}}{(2\pi)^3}\frac{e^2(q^2)}{q^2} = \int \frac{\mathrm{d}^3 q \mathrm{e}^{\mathrm{i}\boldsymbol{q}\cdot\boldsymbol{x}}}{(2\pi)^3}\frac{-e^2}{|\boldsymbol{q}|^2[1-\widetilde{\Pi}_{\mathrm{ren}}(-|\boldsymbol{q}|^2)]} \\ &\approx -e^2\int \frac{\mathrm{d}^3 q \mathrm{e}^{\mathrm{i}\boldsymbol{q}\cdot\boldsymbol{x}}}{(2\pi)^3}\frac{1}{|\boldsymbol{q}|^2}\left[1+\widetilde{\Pi}_{\mathrm{ren}}(-|\boldsymbol{q}|^2)\right]\,. \end{aligned}$$

For $q^2 \ll m^2$, (15.35) gives $\widetilde{\Pi}_{\mathrm{ren}}(q^2) \approx -\alpha q^2/(15\pi m^2)$ which results in the Coulomb potential being changed into

$$\frac{-\alpha}{|\boldsymbol{x}|} \Longrightarrow V(\boldsymbol{x}) = -\left[\frac{\alpha}{|\boldsymbol{x}|} + \frac{4\alpha^2\delta^3(\boldsymbol{x})}{15m^2}\right]\,.$$

The effect can be measured from the shift of the hydrogen energy levels

$$\Delta E = \int \mathrm{d}^3 x|\psi(\boldsymbol{x})|^2\left(\frac{-4\alpha^2}{15m^2}\right)\delta^3(\boldsymbol{x}) = \frac{-4\alpha^2}{15m^2}|\psi(0)|^2 = \frac{-\alpha^5 m}{30\pi}\,.$$

For $q^2 \gg m^2$, the $4\alpha^2\delta^3(\boldsymbol{x})/15m^2$ term is no longer valid. To perform the Fourier transform (Chap. 10), we may replace $|\boldsymbol{q}|^2$ by $|\boldsymbol{q}|^2+\mu^2$, and take the limit $\mu^2 \to 0$ after the integration has been done. Write $|\boldsymbol{q}| = \mathrm{i}\,(2my)$, $y>0$. The contribution of $\widetilde{\Pi}_{\mathrm{ren}}(4m^2y^2)$ to $V(\boldsymbol{x})$ is due to its imaginary part as given by (15.32). Let us write

$$V(\boldsymbol{x}) = \frac{-e^2(r)}{4\pi r}\,, \quad \text{where } e^2(r) = e^2[1+Q(r)]\,.$$

$$Q(r) = \frac{e^2}{6\pi^2}\int_1^\infty \frac{\mathrm{d}y}{y^2}\mathrm{e}^{-2mry}\left(1+\frac{1}{2y^2}\right)\sqrt{y^2-1}\,.$$

$$\text{For } mr \ll 1\,, \quad Q(r) = \frac{-e^2}{6\pi^2}\left[\log(mr) + \gamma_{\mathrm{E}} + \tfrac{5}{6} + \cdots\right]\,.$$

$$\text{For } mr \gg 1\,, \quad Q(r) = \frac{e^2}{16\pi\sqrt{\pi}}\frac{\mathrm{e}^{-2mr}}{(mr)^{3/2}}\,.$$

As $r$ decreases, $Q(r)$ increases and so does $e^2(r)$.

**15.4** Consider the two-body → two-body scattering amplitude i$\mathcal{M}$ of scalar fields $p_1+p_2 \to p_3+p_4$. To order $\lambda^2$, there are three loop diagrams associated with the variables $s=(p_1+p_2)^2$, $t=(p_3-p_1)^2$, and $u=(p_4-p_1)^2$. For instance, in the s-channel, the amplitude is

$$\mathrm{i}\mathcal{M}_s = (-\mathrm{i}\lambda)^2\left(\frac{1}{2}\right)\int \frac{\mathrm{d}^4k}{(2\pi)^4}\frac{\mathrm{i}}{k^2-m^2}\frac{\mathrm{i}}{(k+P)^2-m^2} \;;\; P^2=s\,,$$

the factor 1/2 arises because of two identical internal $\phi$ fields in the loop. Thus,

$$\begin{aligned}\mathcal{M}_s =& \frac{-\mathrm{i}\lambda^2}{2}\int_0^1 \mathrm{d}x \int \frac{\mathrm{d}^nk}{(2\pi)^n}\frac{1}{[k^2+2xk\cdot p+sx-m^2]^2}\\ =&\Gamma\left(2-\frac{n}{2}\right)\frac{\lambda^2}{2(4\pi)^{n/2}}\int_0^1\frac{\mathrm{d}x}{[m^2-sx(1-x)]^{2-\frac{n}{2}}}\\ =&\Gamma\left(2-(\frac{n}{2})\right)\frac{\lambda^2}{2(4\pi)^{n/2}}\int_0^1\frac{\mathrm{d}x}{[\mu^2]^{2-(n/2)}}+\cdots\,.\end{aligned}$$

The $\cdots$ represent irrelevant finite terms, not necessary for the computation of the $\beta$ function. The renormalization condition for the coupling constant $\lambda$ requires that the $\mathcal{O}(\lambda^2)$ corrections to $\lambda$ vanish at the symmetric point $s=t=u=\mu^2$, where $\mu$ is an arbitrary scale, i.e. the vertex counterterm $\delta_\lambda$ to order $\lambda^2$ exactly cancels the sum $\mathcal{M}_s+\mathcal{M}_t+\mathcal{M}_u$. Thus

$$\delta_\lambda = -\Gamma(2-\tfrac{n}{2})\frac{3\lambda^2}{2(4\pi)^{n/2}}\int_0^1\frac{\mathrm{d}x}{[\mu^2]^{2-\frac{n}{2}}} = -\Gamma(2-\tfrac{n}{2})\frac{3\lambda^2}{2(4\pi)^{n/2}}\frac{1}{[\mu^2]^{2-\frac{n}{2}}};\;.$$

$$\beta(\lambda)=\mu\frac{\partial}{\partial\mu}\left[\frac{-3\lambda^2}{2(4\pi)^{n/2}}\frac{\Gamma(2-\frac{n}{2})}{[\mu^2]^{2-\frac{n}{2}}}\right]=\frac{3\lambda^2}{16\pi^2}\,.$$

On the other hand, at one-loop level there are no corrections to the two-point propagator, i.e. the field-strength corrections to order $\lambda^2$ vanish, implying the CS function $\gamma(\lambda)=0$.

**15.5** The three pure gluon loops of Fig. 15.14 yield the vertex correction

$$Z_1^{\mathrm{glu}} = 1+\frac{g_s^2}{16\pi^2}\left\{\left[\frac{2}{3}+\frac{3}{4}(1-\xi)\right]N_c\right\}\frac{\Gamma(2-\frac{n}{2})}{\mu^{4-n}}\,.$$

This result, combined with $\delta_{\mathrm{glu}}(\mu)$ taken from (15.81) [without the $N_f$ term and with (15.83)], gives $\beta(g_s)=(-g_s^3/16\pi^2)\frac{11}{3}N_c$.

**16.1** The penguin operator $\mathcal{O}_{\mathrm{pen}} = [\overline{d}\,\gamma_\mu(1-\gamma_5)\lambda^b s]\,[\overline{q}\,\gamma^\mu\lambda^b\,q]$ can be rewritten as the product of color-singlet $(V-A)\times V$ currents $2[\overline{d}_j\,\gamma_\mu(1-\gamma_5)\,q_j]\,[\overline{q}_k\,\gamma^\mu\,s_k]$ $-\frac{2}{3}[\overline{d}_i\,\gamma_\mu(1-\gamma_5)\,s_i]\,[\overline{q}_k\,\gamma^\mu\,q_k]$ using (11.88). Writing $V=\frac{1}{2}(V-A)+\frac{1}{2}(V+A)$ and using the Fierz's transformations in the Appendix, one has $\mathcal{O}_{\mathrm{pen}}\sim(S-P)\times(S+P)$.

Taking the derivative of $\langle 0\,|\,\overline{q}\gamma_\mu\gamma_5 s\,|\,\mathrm{K}(p)\rangle \equiv \langle 0\,|\,A_\mu\,|\,\mathrm{K}(p)\rangle = \mathrm{i}f_\mathrm{K}p_\mu$, one gets $\langle 0\,|\,\overline{q}\gamma_5 s\,|\,\mathrm{K}(p)\rangle \equiv \langle 0\,|\,P\,|\,\mathrm{K}(p)\rangle = f_\mathrm{K}m_\mathrm{K}^2/(m_s+m_q)$. We get similar result for

$\langle\pi(p')\,|\,S\,|\,\mathrm{K}(p)\rangle$ in terms of $\langle\pi(p')\,|\,V_\mu\,|\,\mathrm{K}(p)\rangle$. With the light quark masses $m_s, m_q$ (where q = u or d quarks) in the denominators, the matrix elements of $\langle\pi\pi\,|\,\mathcal{O}_{\rm pen}\,|\,\mathrm{K}\rangle$ are enhanced and contribute to the $\Delta I = \frac{1}{2}$ rule in the right direction.

**16.2** There are two charged B mesons, the $\mathrm{B_u^+}$ ($\bar{\mathrm{b}}$u) and the $\mathrm{B_c^+}$ ($\bar{\mathrm{b}}$c). Like $f_\mathrm{K}$ and $f_\pi$, the decay constants $f_{\mathrm{B}_u}, f_{\mathrm{B}_c}$ associated with these $B_\mathrm{u}^+$ and $B_\mathrm{c}^+$ are presumably similar. $f_\mathrm{B}$ can be extracted from the decay rate $B_\mathrm{u}^+ \to \tau^+ + \nu_\tau$:

$$\Gamma(B_\mathrm{u}^+ \to \tau^+ + \nu_\tau) = \frac{G_\mathrm{F}^2 |V_\mathrm{ub}|^2 m_\tau^2 M_\mathrm{B} f_\mathrm{B}^2}{8\pi}\left(1 - \frac{m_\tau^2}{M_\mathrm{B}^2}\right)^2 .$$

This rate is very small due to $|V_\mathrm{ub}|^2 \ll |V_\mathrm{cb}|^2$. The corresponding branching ratio is about $10^{-5}$. The decay $B_\mathrm{c}^+ \to \tau^+ + \nu_\tau$ is a hundred times larger, mainly due to $|V_\mathrm{cb}|^2$.

**16.3** The inclusive rare decay B→ $X_s + \gamma$ comes from the flavor changing b→ s $+\gamma$ which can only occur at the loop level. There are two dominant diagrams similar to the gluonic penguin. In contrast to the gluon, the photon can be emitted either from the internal W or from the internal u, c, t quarks. The amplitude uncorrected by QCD is given by

$$\mathcal{M}(\mathrm{b} \to \mathrm{s} + \gamma) = \frac{G_\mathrm{F}}{\sqrt{2}}\frac{e}{4\pi^2}\sum_{\mathrm{Q=u,c,t}} V_\mathrm{Qb} V_\mathrm{Qs}^* \, F(x_\mathrm{Q})\, q^\mu \varepsilon^\nu \left[\bar{s}\,\sigma_{\mu\nu}(a - b\gamma_5)\, b\right],$$

where $x_\mathrm{Q} = m_\mathrm{Q}^2/M_\mathrm{W}^2$, $a = (m_\mathrm{b} + m_\mathrm{s}), b = (m_\mathrm{b} - m_\mathrm{s})$, $q^\mu$ and $\varepsilon^\nu$ are the photon four-momentum and polarization, and

$$F(x) = \frac{x}{24(x-1)^4}\left[(x-1)(8x^2 + 5x - 7) + 6x(2-3x)\log x\right]$$

is first computed by Inami and Lim (reference in Chap. 11).

The exclusive mode B→ $\gamma$+K (more generally $0^- \to 0^- + \gamma$) cannot occur. The amplitude has the form $\varepsilon^\mu(q)\left\langle \mathrm{K}(p)\,\middle|\,V_\mu^{\rm em}\,\middle|\,\mathrm{B}(P)\right\rangle = \varepsilon^\mu(q)\left[(P+p)_\mu f_+(0) + q_\mu f_-(0)\right] = \varepsilon\cdot(P+p) f_+(0)$, where $q = P - p$ is the photon four-momentum with $q^2 = 0$. Since $q^\mu\left\langle \mathrm{K}(p)\,\middle|\,V_\mu^{\rm em}\,\middle|\,\mathrm{B}(P)\right\rangle = (M_\mathrm{B}^2 - m_\mathrm{K}^2) f_+(0) = 0$ by the conservation of the $V_\mu^{\rm em}$ current, implying that $f_+(0) = 0$.

For $0^- \to 1^- + \gamma$, like $\mathrm{B}(P) \to \mathrm{K}^*(p) + \gamma(q)$, the amplitude is of the type $\mathrm{i}\varepsilon_{\mu\nu\rho\sigma}\varepsilon^\mu(q)\varepsilon^\nu(p)P^\rho p^\sigma C$, the condition $q^\mu\left\langle \mathrm{K}^*(p)\,\middle|\,V_\mu^{\rm em}\,\middle|\,\mathrm{B}(P)\right\rangle = \mathrm{i}\varepsilon_{\mu\nu\rho\sigma}\varepsilon^\nu(p)(P-p)^\mu P^\rho p^\sigma = 0$ is automatically satisfied without forcing $C$ to vanish.

**16.4** In the kinematic region near 'zero recoil' where the final state K is nearly at rest, i.e. the momentum transfer is maximum, it is easy to see that the form factors behave like $f_+ + f_- \sim 1/\sqrt{M_\mathrm{Q}}$, $f_+ - f_- \sim \sqrt{M_\mathrm{Q}}$, where Q stands for b or c quarks, and $f_\pm(t_\mathrm{Q})$ are the usual form factors that enter $\langle \mathrm{K}\,|\,\mathrm{V}^\mu\,|\,\mathrm{B}\rangle$ or $\langle \mathrm{K}\,|\,\mathrm{V}^\mu\,|\,\mathrm{D}\rangle$. Thus

$$f_+^{\mathrm{B\to K}}(t_\mathrm{B}) + f_-^{\mathrm{B\to K}}(t_\mathrm{B}) = \sqrt{\frac{M_\mathrm{B}}{M_\mathrm{D}}}\left[f_+^{\mathrm{D\to K}}(t_\mathrm{D}) + f_-^{\mathrm{D\to K}}(t_\mathrm{D})\right],$$

$$f_+^{\mathrm{B\to K}}(t_\mathrm{B}) - f_-^{\mathrm{B\to K}}(t_\mathrm{B}) = \sqrt{\frac{M_\mathrm{D}}{M_\mathrm{B}}}\left[f_+^{\mathrm{D\to K}}(t_\mathrm{D}) - f_-^{\mathrm{D\to K}}(t_\mathrm{D})\right],$$

where $t_B$ and $t_D$ are given in (16.129).

By factorization and using (16.96), the $B(P) \to \psi(q)+K(p)$ amplitude is

$$a_2 \frac{G_F V_{bc} V_{cs}^*}{\sqrt{2}} M_\psi f_\psi \varepsilon^\mu(q)(P+p)_\mu \, f_+^{B\to K}(M_\psi^2) ,$$

where $f_\psi$ is defined by $\langle 0|\bar{c}\gamma^\mu c|\psi(q,\varepsilon)\rangle = M_\psi f_\psi \varepsilon^\mu(q)$, similar to the decay constant $f_\rho$ of the vector $\rho^0$ meson (Chap. 10). Like $\rho^0 \to e^+ + e^-$, from the $J/\psi \to e^+ + e^-$ rate, one deduces $f_\psi \sim 380$ MeV. The decay width is

$$\Gamma(B\to K+J/\psi) = a_2^2 \frac{G_F^2 |V_{bc}V_{cs}^*|^2}{32\pi M_B^3} f_\psi^2 \left[\lambda(M_B^2, M_\psi^2, m_K^2)\right]^{3/2} |f_+^{B\to K}(M_\psi^2)|^2 .$$

With $t_B = M_\psi^2$, we can use (16.129) to compute the transfer $t_D$ and lastly use (16.128) to express $\Gamma(B \to K + J/\psi)$ in terms of the D→ K form factors measured in D→ K $+e^+ + \nu_e$.

**16.5** The total width $\Gamma_{tot}$ is obtained by summing the three semileptonic widths (e, $\mu$, and $\tau$) with the inclusive hadronic widths $b \to c + q_2 + \bar{q}_3$ where $q_2$ stands for d and s quarks and $q_3$ for u and c quarks. The $b\to u + q_2 + \bar{q}_3$ is negligible. First, if one neglects all the final fermionic masses (including the $\tau$ and the c) and the QCD corrections, the branching ratio is 1/9. Keeping all the fermionic masses and the QCD corrected coefficients $c_\pm$, one finds $Br_{sl} \approx 12\%$, which is to be compared with the measured value $\approx 10\%$.

**16.6** The decay $B^0 \to \pi^+ + \pi^-$ proceeds through the tree diagram (b→ u + $W^-$ followed by $W^- \to d + \bar{u}$) and the loop penguin b→d + g followed by g → $\bar{u}u$, similarly to Fig. 16.13a, b. The tree amplitude is $\sim V_{ub}V_{ud}^* \approx \lambda^3 \approx (0.22)^3$, whereas the penguin amplitude $\sim V_{tb}V_{td}^*$ is also $\approx \lambda^3$ but with an additional smaller coefficient $(\alpha_s/12\pi)\log(m_t^2/m_b^2)$. Hadronic uncertainties in the asymmetry between $B^0 \to \pi^+ + \pi^-$ and $\overline{B}^0 \to \pi^+ + \pi^-$ arise from the mixture of a (small) penguin contribution to the tree diagram. These contributions do not have the same phases, in contrast to the B→ $J/\psi$+ K case. With only the tree diagram, the parameter $\xi$ for the asymmetry defined in (16.118) is

$$\xi = \frac{q}{p}\frac{\overline{A}}{A} = \frac{V_{td}V_{tb}^*}{V_{td}^*V_{tb}} \frac{V_{ub}V_{ud}^*}{V_{ub}^*V_{ud}} = e^{-2i\beta} e^{-2i\gamma} = e^{-2i\alpha} .$$

The asymmetry $\sim \mathrm{Im}\xi = \sin(2\alpha)$ can have either sign, depending on the shape of the unitarity triangle.

The $B_s \equiv \bar{b}$ s meson decay mode $B_s \to \rho^0 + K_S$ proceeds at the quark level through b→ u + $W^-$ followed by $W^- \to d + \bar{u}$ by tree diagram. Like the B→ $J/\psi + K_S$, there is an additional $(q/p)_K$ (besides the $(q/p)_B$) due to the $K^0$ and $\overline{K}^0$ mixing in $K_S$. The corresponding $\xi$ is

$$\xi = \left(\frac{q}{p}\right)_B \left(\frac{q}{p}\right)_K \frac{\overline{A}}{A} = \frac{V_{ts}V_{tb}^*}{V_{ts}^*V_{tb}} \frac{V_{cd}V_{cs}^*}{V_{cd}^*V_{cs}} \frac{V_{ub}V_{ud}^*}{V_{ub}^*V_{ud}} = e^{-2i\gamma} .$$

With an additional minus sign due to the $CP$ eigenvalue of the $\rho^0 + K_S$, the asymmetry is $\sim -\sin(2\gamma)$.

**17.1** There are four tree diagrams, one with a neutrino $\nu_e$ exchanged in the $t$-channel, and three with the photon, the $Z^0$, and the Higgs boson exchanged in the $s$-channel ($e^+ + e^- \to \gamma, Z^0, H \to W^+ + W^-$). For longitudinally polarized W at very high energy, where $\varepsilon_\pm \to k_\pm/M_W$, the $Z^0$-exchange amplitude is

$$\begin{aligned}\mathcal{M} = &\frac{-ig^2}{4(s-M_Z^2)}\,\overline{v}(p_2)\gamma_\mu(g_V^e - g_A^e\gamma_5)u(p_1)\left(g^{\mu\nu} - \frac{P^\mu P^\nu}{M_Z^2}\right)\\ &\times \varepsilon_+^\alpha\varepsilon_-^\beta\,[g_{\alpha\beta}(k_- - k_+)_\nu + g_{\alpha\nu}(k_+ + P)_\beta - g_{\beta\nu}(k_- + P)_\alpha]\\ &\longrightarrow \frac{ig^2}{8M_W^2}\frac{s}{s-M_Z^2}\overline{v}(p_2)(\not{k}_+ - \not{k}_-)(g_V^e - g_A^e\gamma_5)u(p_1)\,,\end{aligned}$$

where $P = k_+ + k_- = p_1 + p_2$, $P^2 = s$. At very high energy, $s \gg (M_Z^2)$, the sum of the contribution to the amplitude from the photon and $Z^0$ exchanges is found to be $\mathcal{M} = -i\sqrt{2}G_F\,\overline{v}(p_2)\;\not{P}(1-\gamma_5)u(p_1)$. This sum is canceled by the amplitude with a neutrino $\nu_e$ exchanged in the $t$-channel, when the electron mass is neglected. However, if $m_e$ is kept nonzero, the sum of the three ($\nu$, $\gamma$, and $Z^0$ exchange) amplitudes is proportional to $m_e\sqrt{s}$ for longitudinally polarized W–W. This residual amplitude, growing like $m_e\sqrt{s}$, is exactly canceled by the Higgs boson exchange, where the coefficient $m_e$ arises from the $He^+e^-$ coupling.

**17.2** At very high energy, $s, M_H^2 \gg M_Z^2$, the amplitude is

$$\mathcal{M}_{LL}(W_L + W_L \to W_L + W_L) = \sqrt{2}G_F M_H^2\left[\frac{s}{s-M_H^2} + \frac{t}{t-M_H^2}\right]\,.$$

Using the partial wave expansion $\mathcal{M}_{LL}(s,t) = 16\pi\sum_J(2J+1)a_{LL}^J(s)P_J(\cos\theta)$ where $P_J(\cos\theta)$ is the Legendre polynomial, one finds $|a_{LL}^{J=0}(s)| = G_F M_H^2/4\pi\sqrt{2}$. The unitarity constraint is $|a_{LL}^J(s)| \le 1 \Longrightarrow M_H < 1.2$ TeV.

**17.3** The amplitude is written as $\varepsilon_\mu(k_1)\varepsilon_\nu(k_2)I^{\mu\nu}(k_1,k_2)$, where $I^{\mu\nu}(k_1,k_2) = T^{\mu\nu}(k_1,k_2) + T^{\nu\mu}(k_2,k_1) = 2T^{\mu\nu}(k_1,k_2)$. The tensor $T^{\mu\nu}(k_1,k_2)$ is calculated from a triangle loop with an internal top quark:

$$\begin{aligned}T^{\mu\nu}(k_1,k_2) = &(-)(-ig_s)^2\left(\frac{-igm_t}{2M_W}\right)\\ &\times\int\frac{d^4p}{(2\pi)^4}\,[t^{\mu\nu}]\,\frac{i}{p^2-m_t^2}\,\frac{i}{(p-k_1)^2-m_t^2}\,\frac{i}{(p+k_2)^2-m_t^2}\,,\end{aligned}$$

where $t^{\mu\nu} = \mathrm{Tr}[(\not{p}+m_t)\gamma^\mu(\frac{1}{2}\lambda_j)(\not{p} - \not{k}_1 + m_t)(\not{p} + \not{k}_2 + m_t)\gamma^\nu(\frac{1}{2}\lambda_j)]$ . The final result comes out to be

$$\begin{aligned}I^{\mu\nu} &= A(k_1,k_2)\,[k_2^\mu k_1^\nu - g^{\mu\nu}k_1\cdot k_2]\,,\\ A(k_1,k_2) &= \frac{4}{3}g_s^2\left(\frac{gm_t}{2M_W}\right)\left(\frac{m_t}{2\pi^2}\right)\int_0^1 dx\int_0^{1-x}dy\,\frac{1-4xy}{M_H^2xy - m_t^2}\,.\end{aligned}$$

# Appendix: Useful Formulas

In this appendix we summarize the notation for Dirac matrices and spinors, and collect together some formulas for the calculation of relativistic cross-sections and decay rates, as well as useful phase space integrals and loop integrals. A summary of the Feynman rules for QCD and electroweak interactions is also given.

## A.1 Relativistic Quantum Mechanics

### Four-vectors

In natural units ($c = \hbar = 1$), the space-time coordinates are denoted by the contravariant four-vector

$$x^\mu = (x^0, x^1, x^2, x^3) = (t,\, x,\, y,\, z) = (t,\, \vec{x})\,, \tag{A.1}$$

to which corresponds the covariant four-vector

$$x_\mu = g_{\mu\nu} x^\nu = (t, -x, -y, -z) = (t, -\vec{x})\,. \tag{A.2}$$

Here the metric tensor is given by

$$g_{\mu\nu} = g^{\mu\nu} = \begin{pmatrix} 1 & 0 & 0 & 0 \\ 0 & -1 & 0 & 0 \\ 0 & 0 & -1 & 0 \\ 0 & 0 & 0 & -1 \end{pmatrix}. \tag{A.3}$$

Other examples of four-vectors are

energy-momentum $\quad p^\mu = (E,\, p_x,\, p_y,\, p_z) = (E, \boldsymbol{p})\,,$

four-gradients $\quad \partial_\mu = \dfrac{\partial}{\partial x^\mu} = \left(\dfrac{\partial}{\partial t}, \nabla\right) = \left(\dfrac{\partial}{\partial t}, \dfrac{\partial}{\partial x}, \dfrac{\partial}{\partial y}, \dfrac{\partial}{\partial z}\right)\,,$

$$\partial^\mu = \frac{\partial}{\partial x_\mu} = \left(\frac{\partial}{\partial t}, -\nabla\right)\,.$$

The scalar product of two four-vectors is given by

$$A\cdot B = g_{\mu\nu} A^\mu B^\nu = A_\mu B^\mu = A_0 B_0 - \boldsymbol{A}\cdot\boldsymbol{B}\,. \tag{A.4}$$

**Dirac Algebra ($n = 4$ dimensions)**

The Dirac matrices satisfy the anticommutation relations

$$\{\gamma^\mu, \gamma^\nu\} = \gamma^\mu\gamma^\nu + \gamma^\nu\gamma^\mu = 2g^{\mu\nu}\boldsymbol{I}_4\,, \qquad \mu,\nu = (0,1,2,3)\,. \tag{A.5}$$

Here and in the following, $\boldsymbol{I}_n$ stands for the $n\times n$ unit matrix. In the standard (Dirac–Pauli) representation, the $\gamma$-matrices have the form

$$\gamma^0 = \begin{pmatrix} \boldsymbol{I}_2 & 0 \\ 0 & -\boldsymbol{I}_2 \end{pmatrix}, \qquad \gamma^i = \begin{pmatrix} 0 & \sigma_i \\ -\sigma_i & 0 \end{pmatrix}. \tag{A.6}$$

Note that $\gamma_\mu = g_{\mu\nu}\gamma^\nu$. The standard Pauli matrices $\sigma_i$ are

$$\sigma_1 = \begin{pmatrix} 0 & 1 \\ 1 & 0 \end{pmatrix}, \qquad \sigma_2 = \begin{pmatrix} 0 & -\mathrm{i} \\ \mathrm{i} & 0 \end{pmatrix}, \qquad \sigma_3 = \begin{pmatrix} 1 & 0 \\ 0 & -1 \end{pmatrix}. \tag{A.7}$$

$$\sigma_i\sigma_j = \delta_{ij} + \mathrm{i}\varepsilon_{ijk}\sigma_k\,, \qquad \varepsilon_{ijk} \text{ totally antisymmetric, } \varepsilon_{123} = +1\,. \tag{A.8}$$

*Hermitian conjugates:*

$$\gamma^0\gamma^{\mu\dagger}\gamma^0 = \gamma^\mu\,, \qquad \gamma^{0\dagger} = \gamma^0, \quad \gamma^{i\dagger} = -\gamma^i\,. \tag{A.9}$$

*Related matrices:*

$$\gamma^5 = \gamma_5 = \mathrm{i}\gamma^0\gamma^1\gamma^2\gamma^3 = -\mathrm{i}\gamma_0\gamma_1\gamma_2\gamma_3\,, \qquad \{\gamma_5, \gamma_\mu\} = 0\,; \tag{A.10}$$

$$\sigma^{\mu\nu} = \frac{\mathrm{i}}{2}[\gamma^\mu, \gamma^\nu] = \frac{\mathrm{i}}{2}(\gamma^\mu\gamma^\nu - \gamma^\nu\gamma^\mu)\,. \tag{A.11}$$

*Identities for the $\gamma$-matrices:*

$$\begin{aligned}
\gamma_\mu\gamma_\nu &= g_{\mu\nu}\boldsymbol{I}_4 - \mathrm{i}\sigma_{\mu\nu}\,, \\
\gamma_\mu\gamma_\nu\gamma_\lambda &= g_{\mu\nu}\gamma_\lambda - g_{\mu\lambda}\gamma_\nu + g_{\nu\lambda}\gamma_\mu - \mathrm{i}\varepsilon_{\mu\nu\lambda\alpha}\gamma^\alpha\gamma_5\,, \\
\gamma_\mu\gamma^\alpha\gamma^\mu &= -2\gamma^\alpha\,, \\
\gamma_\mu\gamma^\alpha\gamma^\beta\gamma^\mu &= 4g^{\alpha\beta}\boldsymbol{I}_4\,, \\
\gamma_\mu\gamma^\alpha\gamma^\beta\gamma^\gamma\gamma^\mu &= -2\gamma^\gamma\gamma^\beta\gamma^\alpha\,, \\
\sigma^{\mu\nu}\sigma_{\mu\nu} &= 12\boldsymbol{I}_4\,.
\end{aligned} \tag{A.12}$$

$$\varepsilon_{\mu\nu\lambda\rho} \text{ is totally antisymmetric;} \qquad \varepsilon_{0123} = -\varepsilon^{0123} = +1; \tag{A.13}$$

$$\varepsilon^{\alpha\beta\mu\nu}\varepsilon_{\alpha\beta\rho\sigma} = -2(\delta^\mu_\rho\delta^\nu_\sigma - \delta^\mu_\sigma\delta^\nu_\rho)\,; \qquad \varepsilon^{\alpha\mu\nu\rho}\varepsilon_{\beta\mu\nu\rho} = -6\,\delta^\alpha_\beta\,. \tag{A.14}$$

*Trace theorems:* The trace of an odd number of $\gamma$-matrices vanishes; also

$$\begin{aligned}
\not{a} \equiv \gamma_\mu a^\mu &= \gamma^0 a^0 - \boldsymbol{\gamma}\cdot\boldsymbol{a}\,, \\
\mathrm{Tr}\,\boldsymbol{I}_4 &= 4\,, \\
\mathrm{Tr}\,\not{a}\not{b} &= 4a\cdot b\,, \\
\mathrm{Tr}\,\not{a}\not{b}\not{c}\not{d} &= 4(a\cdot b\, c\cdot d - a\cdot c\, b\cdot d + a\cdot d\, b\cdot c)\,, \\
\mathrm{Tr}\,\gamma_5\not{a}\not{b} &= 0\,, \\
\mathrm{Tr}\,\gamma_5\not{a}\not{b}\not{c}\not{d} &= 4\mathrm{i}\varepsilon_{\mu\nu\lambda\rho}\, a^\mu b^\nu c^\lambda d^\rho\,.
\end{aligned} \tag{A.15}$$

**Dirac Algebra ($n$ dimensions)**
In $n$-dimensional space-time (1 time, $n-1$ space), the Dirac matrices satisfy

$$\{\gamma^\mu, \gamma^\nu\} = \gamma^\mu\gamma^\nu + \gamma^\nu\gamma^\mu = 2g^{\mu\nu}\boldsymbol{I}_n\,, \qquad \mu,\nu = (0,1,\ldots,n-1)\,. \quad \text{(A.16)}$$

They also have the following properties:

$$\begin{aligned} \gamma_\nu\gamma^\nu &= n\boldsymbol{I}_n\,, \\ \gamma_\mu\gamma^\alpha\gamma^\mu &= (2-n)\gamma^\alpha\,, \\ \gamma_\mu\gamma^\alpha\gamma^\beta\gamma^\mu &= 4g^{\alpha\beta}\boldsymbol{I}_n + (n-4)\gamma^\alpha\gamma^\beta\,, \\ \gamma_\mu\gamma^\alpha\gamma^\beta\gamma^\gamma\gamma^\mu &= -2\gamma^\gamma\gamma^\beta\gamma^\alpha - (n-4)\gamma^\alpha\gamma^\beta\gamma^\gamma\,. \end{aligned} \quad \text{(A.17)}$$

**Dirac Spinors**
A Dirac spinor for a particle of mass $m$, momentum $p$, and polarization $s$ is a four-component complex vector denoted by $u(p,s)$, while for an antiparticle, it is called $v(p,s)$. In each case, $p^\mu = (E, \boldsymbol{p})$ with $E = \sqrt{\boldsymbol{p}^2 + m^2}$. The spinors $u$ and $v$ obey the equations

$$(\not{p} - m)\,u(p,s) = 0\,; \qquad (\not{p} + m)\,v(p,s) = 0\,. \quad \text{(A.18)}$$

The corresponding adjoint spinors, $\bar{u} = u^\dagger\gamma^0$ and $\bar{v} = v^\dagger\gamma^0$, satisfy

$$\bar{u}(p,s)(\not{p} - m) = 0\,; \qquad \bar{v}(p,s)(\not{p} + m) = 0\,. \quad \text{(A.19)}$$

The spinors have the normalization and orthogonality properties

$$\begin{aligned} \bar{u}(p,s)\,u(p,s') &= \;\;2m\delta_{ss'}\,, \qquad & \bar{u}(p,s)\,v(p,s') &= 0\,, \\ \bar{v}(p,s)\,v(p,s') &= -2m\delta_{ss'}\,, \qquad & \bar{v}(p,s)\,u(p,s') &= 0\,. \end{aligned} \quad \text{(A.20)}$$

Their completeness can be expressed in terms of the positive- and negative-energy projection operators $\Lambda_\pm = (\pm\not{p} + m)/(2m)$,

$$\sum_s u(p,s)\,\bar{u}(p,s) = 2m\,\Lambda_+(p)\,, \quad \sum_s v(p,s)\,\bar{v}(p,s) = -2m\,\Lambda_-(p)\,. \quad \text{(A.21)}$$

$\Lambda_\pm$ have the properties

$$\Lambda_\pm^2 = \Lambda_\pm\,; \quad \Lambda_+\Lambda_- = \Lambda_-\Lambda_+ = 0\,; \quad \Lambda_+ + \Lambda_- = 1\,. \quad \text{(A.22)}$$

Projections onto states of well-defined polarizations are performed with the operators $P(\pm s) = \frac{1}{2}\,(1 \pm \gamma_5\not{s})$, where $s^\mu$ is spacelike and defined such that $s^\mu s_\mu = -1$ and $s^\mu p_\mu = 0$,

$$u(p,s)\bar{u}(p,s) = 2m\Lambda_+(p)\,P(+s)\,, \quad v(p,s)\bar{v}(p,s) = -2m\Lambda_-(p)\,P(-s)\,. \text{(A.23)}$$

*Gordon identities:* Any solutions to the Dirac equation satisfy the relations

$$(m_1+m_2)\overline{\psi}_2\gamma_\mu\psi_1 = \overline{\psi}_2\left(-\mathrm{i}\overleftarrow{\partial}_\mu + \mathrm{i}\overrightarrow{\partial}_\mu\right)\psi_1 + \partial^\nu\left(\overline{\psi}_2\sigma_{\mu\nu}\psi_1\right),$$
$$(m_1+m_2)\overline{\psi}_2\gamma_\mu\gamma_5\psi_1 = (-\mathrm{i}\partial_\mu)\left(\overline{\psi}_2\gamma_5\psi_1\right) + \overline{\psi}_2\left(-\mathrm{i}\overleftarrow{\partial}_\nu + \mathrm{i}\overrightarrow{\partial}_\nu\right)\sigma_\mu{}^\nu\gamma_5\psi_1 . \tag{A.24}$$

For positive-energy components, take $\psi = u(p)\mathrm{e}^{-ip\cdot x}$, while for negative-energy components, $\psi = v(p)\mathrm{e}^{ip\cdot x}$. So that in momentum space, (24) gives, for example,

$$(m_1+m_2)\bar{u}(p_2)\gamma_\mu u(p_1) = \bar{u}(p_2)\left[(p_2+p_1)_\mu + \mathrm{i}\sigma_{\mu\nu}(p_2-p_1)^\nu\right]u(p_1),$$
$$(m_1+m_2)\bar{u}(p_2)\gamma_\mu\gamma_5 u(p_1) = \bar{u}(p_2)\left[(p_2-p_1)_\mu + \mathrm{i}\sigma_{\mu\nu}(p_2+p_1)^\nu\right]\gamma_5\, u(p_1),$$
$$(m_1+m_2)\bar{u}(p_2)\gamma_\mu v(p_1) = \bar{u}(p_2)\left[(p_2-p_1)_\mu + \mathrm{i}\sigma_{\mu\nu}(p_2+p_1)^\nu\right]v(p_1). \tag{A.25}$$

*Fierz transformations:* For Lorentz scalar products of bilinear covariants, with all their Lorentz indices contracted,

$$\left(\bar{u}_1\Gamma^i u_2\right)\left(\bar{u}_3\Gamma_i u_4\right) = \sum_j C^i{}_j\left(\bar{u}_1\Gamma^j u_4\right)\left(\bar{u}_3\Gamma_j u_2\right), \tag{A.26}$$

where $\Gamma^i$ denote $\Gamma^{\mathrm{S}} = 1$, $\Gamma^{\mathrm{V}} = \gamma_\mu$, $\Gamma^{\mathrm{T}} = \sigma_{\mu\nu}$, $\Gamma^{\mathrm{P}} = \gamma_5$, and $\Gamma^{\mathrm{A}} = \gamma_\mu\gamma_5$; the coefficients $C^i{}_j$ assume the values given in the following table:

| $i\downarrow \overset{j}{\rightarrow}$ | S | V | T | A | P |
|---|---|---|---|---|---|
| S | 1/4 | 1/4 | 1/8 | −1/4 | 1/4 |
| V | 1 | −1/2 | 0 | −1/2 | −1 |
| T | 3 | 0 | −1/2 | 0 | 3 |
| A | −1 | −1/2 | 0 | −1/2 | 1 |
| P | 1/4 | −1/4 | 1/8 | 1/4 | 1/4 |

Similar to (26) is the relation

$$\left(\bar{u}_1\Gamma^i u_2\right)\left(\bar{u}_3\Gamma_i\gamma_5 u_4\right) = \sum_j C^i{}_j\left(\bar{u}_1\Gamma^j\gamma_5 u_4\right)\left(\bar{u}_3\Gamma_j u_2\right), \tag{A.27}$$

which comes with the same numerical coefficients $C^i{}_j$. The above identities yield the following relations, very useful in the studies of weak processes:

$$\left[\bar{u}_1\gamma_\mu(1\pm\gamma_5)u_2\right]\left[\bar{u}_3\gamma^\mu(1\pm\gamma_5)u_4\right] = -\left[\bar{u}_1\gamma_\mu(1\pm\gamma_5)u_4\right]\left[\bar{u}_3\gamma^\mu(1\pm\gamma_5)u_2\right],$$
$$\left[\bar{u}_1\gamma_\mu(1\pm\gamma_5)u_2\right]\left[\bar{u}_3\gamma^\mu(1\mp\gamma_5)u_4\right] = +2\left[\bar{u}_1(1\mp\gamma_5)u_4\right]\left[\bar{u}_3(1\pm\gamma_5)u_2\right]. \tag{A.28}$$

When anticommuting fields $\psi_i$, rather than c-valued spinors $u_i$, appear in the bilinear products, the right-hand sides of (26)–(28) will carry an additional overall minus sign.

## A.2 Cross-Sections and Decay Rates

The differential cross-section for the reaction $p_1 + p_2 \to p_3 + \ldots + p_n$ is

$$\mathrm{d}\sigma = \frac{|\mathcal{M}|^2}{4F}\,\mathrm{d}\Phi_\mathrm{f}\,\mathcal{S}\,. \tag{A.29}$$

Here, $\mathcal{M}$ is shorthand notation for the invariant amplitude $\langle \mathrm{f}\,|\,\mathcal{M}\,|\,\Phi_\mathrm{i}\rangle$; $F$ is the flux factor

$$F = E_1\,E_2\,|\boldsymbol{v}_{12}| = \left[\,(p_1\cdot p_2)^2 - m_1^2\,m_2^2\,\right]^{1/2}\,; \tag{A.30}$$

and $\mathrm{d}\Phi_\mathrm{f}$ stands for the phase-space volume element of the final state

$$\mathrm{d}\Phi_\mathrm{f} = (2\pi)^4\delta^{(4)}(p_3 + \ldots + p_n - P_\mathrm{i})\,\frac{1}{(2\pi)^{3(n-2)}}\,\frac{\mathrm{d}^3p_3}{2E_3}\cdots\frac{\mathrm{d}^3p_n}{2E_n}\,; \tag{A.31}$$

with $P_\mathrm{i} = p_1 + p_2$; and finally $\mathcal{S}$ is a combinatorial factor needed to avoid overcounting identical configurations whenever there are identical particles in the final state and is given by $\mathcal{S} = \prod_a 1/\ell_a!$, where $\ell_a$ denotes the number of identical particles of type $a$ in the final state.

For the production of two particles, $p_1 + p_2 \to p_3 + p_4$, the differential cross-section in the *center-of-mass* system, where $\boldsymbol{p} = \boldsymbol{p}_1 = -\boldsymbol{p}_2$ and $\boldsymbol{p}' = \boldsymbol{p}_3 = -\boldsymbol{p}_4$, at fixed relative angles $(\varphi, \theta)$ of the final particles, is

$$\left(\frac{\mathrm{d}\sigma}{\mathrm{d}\Omega}\right)_\mathrm{cm} = \frac{|\mathcal{M}|^2}{64\pi^2\,s}\,\frac{|\boldsymbol{p}'|}{|\boldsymbol{p}|}\,\mathcal{S}\,. \tag{A.32}$$

$$|\boldsymbol{p}_\mathrm{cm}|^2 = \frac{1}{4s}\lambda(s, m_1^2, m_2^2)\,, \qquad |\boldsymbol{p}'_\mathrm{cm}|^2 = \frac{1}{4s}\lambda(s, m_3^2, m_4^2)\,; \tag{A.33}$$

where $s = (p_1 + p_2)^2 = (E_1 + E_2)^2$ and $\lambda(a, b, c) \equiv (a - b - c)^2 - 4bc$.

In the *laboratory* system where particle 2 is at rest, $p_2 = (m_2, \boldsymbol{0})$,

$$\left(\frac{\mathrm{d}\sigma}{\mathrm{d}\Omega}\right)_\mathrm{lab} = \frac{|\mathcal{M}|^2}{64\pi^2\,m_2}\,\frac{|\boldsymbol{p}'|}{|\boldsymbol{p}|}\,\frac{1}{E_1 + m_2 - (|\boldsymbol{p}|/|\boldsymbol{p}'|)\,E_3\,\cos\theta_\mathrm{lab}}\,\mathcal{S}\,, \tag{A.34}$$

where $E_1 = \sqrt{\boldsymbol{p}^2 + m_1^2}$, $E_3 = \sqrt{\boldsymbol{p}'^2 + m_3^2}$ and, by energy conservation,

$$E_3(E_1 + m_2) - |\boldsymbol{p}||\boldsymbol{p}'|\cos\theta_\mathrm{lab} = E_1 m_2 + \tfrac{1}{2}\left(m_1^2 + m_2^2 + m_3^2 - m_4^2\right). \tag{A.35}$$

The differential rate of the decay $P \to p_3 + \ldots + p_n$ is

$$\mathrm{d}\Gamma = \frac{|\mathcal{M}|^2}{2E_P}\,\mathrm{d}\Phi_\mathrm{f}\,\mathcal{S}\,, \tag{A.36}$$

where $\mathrm{d}\Phi_\mathrm{f}$ and $\mathcal{S}$ have the same definitions as above. In particular, for a decay from rest into two particles, the rate is

$$\mathrm{d}\Gamma = \frac{|\mathcal{M}|^2\,|\boldsymbol{p}|}{32\pi^2\,M^2}\,\mathrm{d}\Omega\,, \tag{A.37}$$

where $P^2 = M^2$, and $\boldsymbol{p}$ is the momentum of either final particle, given by

$$|\boldsymbol{p}| = \frac{1}{2M}\left\{[M^2 - (m_3 + m_4)^2][M^2 - (m_3 - m_4)^2]\right\}^{1/2}\,.$$

## A.3 Phase Space and Loop Integrals

### Two- and Three-Body Phase Space Integrals

$$\text{(I)} \quad \int \frac{d^3k_1}{2E_1}\frac{d^3k_2}{2E_2}\,\delta^{(4)}(P-k_1-k_2) = \frac{\pi\sqrt{\lambda(P^2,m_1^2,m_2^2)}}{2P^2} \equiv \mathcal{I}(P^2)\,; \qquad \text{(A.38)}$$

a proof of this result is given in Solution 4.1.

$$\text{(II)} \quad \int \frac{d^3k_1}{2E_1}\frac{d^3k_2}{2E_2}\,\delta^{(4)}(P-k_1-k_2)k_1^\mu = \mathcal{I}(P^2)\,\frac{P^2+m_1^2-m_2^2}{2P^2}P^\mu \qquad \text{(A.39)}$$

must have the form $A(P^2)P^\mu$; to compute $A(P^2)$, we multiply both sides of the above equation by $P_\mu$, then use $2k_1\cdot P = P^2+m_1^2-m_2^2$.

$$\text{(III)} \quad \int \frac{d^3k_1}{2E_1}\frac{d^3k_2}{2E_2}\,\delta^{(4)}(P-k_1-k_2)k_1^\mu k_1^\nu = \frac{\mathcal{I}(P^2)}{3P^2}\left\{-\frac{g^{\mu\nu}}{4}\lambda(P^2,m_1^2,m_2^2) + \frac{P^\mu P^\nu}{P^2}\left[(P^2+m_1^2-m_2^2)^2 - P^2m_1^2\right]\right\}; \qquad \text{(A.40)}$$

$$\text{(IV)} \quad \int \frac{d^3k_1}{2E_1}\frac{d^3k_2}{2E_2}\,\delta^{(4)}(P-k_1-k_2)k_1^\mu k_2^\nu = \frac{\mathcal{I}(P^2)}{6P^2}\left\{\frac{g^{\mu\nu}}{2}\lambda(P^2,m_1^2,m_2^2) + \frac{P^\mu P^\nu}{P^2}\left[P^4+P^2(m_1^2+m_2^2)-2(m_1^2-m_2^2)^2\right]\right\}; \qquad \text{(A.41)}$$

when there are terms like $k_1^\mu k_1^\nu$, $k_1^\mu k_2^\nu$, or $k_2^\mu k_2^\nu$ in the integrand, the integral must have the form $a(P^2)g^{\mu\nu} + b(P^2)P^\mu P^\nu$. To compute it, we multiply the integral by the tensors $g_{\mu\nu}$ and $P_\mu P_\nu$ to get two equations for $a(P^2)$, $b(P^2)$ [see also (13.57)].

$$\text{(V)} \quad \int \frac{d^3k_1}{2E_1}\frac{d^3k_2}{2E_2}\frac{d^3k_3}{2E_3}\,\delta^{(4)}(P-k_1-k_2-k_3) = \frac{\pi^2}{4P^2}\int_{(m_1+m_2)^2}^{(\sqrt{P^2}-m_3)^2}\frac{ds}{s}\sqrt{\lambda(s,m_1^2,m_2^2)}\,\sqrt{\lambda(s,m_3^2,P^2)} \qquad \text{(A.42)}$$

(cf. Problem 4.1); put $q = P - k_3$ with $q^2 = s$ and integrate over $\boldsymbol{k}_1$ and $\boldsymbol{k}_2$: $\int \frac{\pi\sqrt{\lambda(s,m_1^2,m_2^2)}}{2s}\frac{d^3k_3}{2E_3}$; then (42) follows from

$$\frac{d^3k_3}{2E_3} = (4\pi)\frac{|\boldsymbol{k}_3|dE_3}{2}\,, \quad |\boldsymbol{k}_3| = \frac{\sqrt{\lambda(s,m_3^2,P^2)}}{2\sqrt{P^2}}\,, \quad \text{and} \quad dE_3 = \frac{ds}{2\sqrt{P^2}}\,.$$

(VI) $$\int \frac{\mathrm{d}^3k_1}{2E_1}\frac{\mathrm{d}^3k_2}{2E_2}\frac{\mathrm{d}^3k_3}{2E_3}\,\delta^{(4)}(P-k_1-k_2-k_3)(k_1\cdot k_3)$$
$$= \frac{\pi^2}{16\,P^2}\int_{(m_1+m_2)^2}^{(\sqrt{P^2}-m_3)^2}\frac{\mathrm{d}s}{s^2}\sqrt{\lambda(s,m_1^2,m_2^2)}\,\sqrt{\lambda(s,m_3^2,P^2)}$$
$$\times\left[(s+m_1^2-m_2^2)(P^2-m_3^2-s)\right]\,; \qquad \text{(A.43)}$$

write $k_1\cdot k_3 = k_1^\mu(P-q)_\mu$ and use (39) and (42).

(VII) $$\int \frac{\mathrm{d}^3k_1}{2E_1}\frac{\mathrm{d}^3k_2}{2E_2}\frac{\mathrm{d}^3k_3}{2E_3}\,\delta^{(4)}(P-k_1-k_2-k_3)(k_1\cdot k_3)^2$$
$$= \frac{\pi^2}{16\,P^2}\int_{(m_1+m_2)^2}^{(\sqrt{P^2}-m_3)^2}\frac{\mathrm{d}s}{s^2}\sqrt{\lambda(s,m_1^2,m_2^2)}\,\sqrt{\lambda(s,m_3^2,P^2)}$$
$$\times\left[\frac{\lambda(s,m_1^2,m_2^2)\lambda(s,m_3^2,P^2)}{3s}+m_1^2\,\lambda(s,m_3^2,P^2)\right.$$
$$\left.+\,m_3^2\,\lambda(s,m_1^2,m_2^2)+4s\,m_1^2\,m_3^2\right]\,; \qquad \text{(A.44)}$$

write $(k_1\cdot k_3)^2 = k_1^\mu k_1^\nu(P-q)_\mu(P-q)_\nu$ and use (40) and (42).

**Loop Integrals**

*Feynman Parameterization.*

$$\frac{1}{a^\alpha b^\beta} = \frac{\Gamma(\alpha+\beta)}{\Gamma(\alpha)\Gamma(\beta)}\int_0^1 \mathrm{d}x\,\frac{x^{\alpha-1}(1-x)^{\beta-1}}{[ax+b(1-x)]^{\alpha+\beta}}\,;$$

$$\frac{1}{a^\alpha b^\beta c^\gamma} = \frac{\Gamma(\alpha+\beta+\gamma)}{\Gamma(\alpha)\Gamma(\beta)\Gamma(\gamma)}\int_0^1 \mathrm{d}x\int_0^{1-x}\mathrm{d}y\,\frac{(1-x-y)^{\alpha-1}x^{\beta-1}y^{\gamma-1}}{[a(1-x-y)+bx+cy]^{\alpha+\beta+\gamma}}\,;$$

$$\frac{1}{abcd} = 6\int_0^1 \mathrm{d}x\int_0^{1-x}\mathrm{d}y\int_0^{1-x-y}\frac{\mathrm{d}z}{[a(1-x-y-z)+bx+cy+dz]^4}\,.$$

The demonstration starts from the identity

$$\frac{1}{a\,b} = \int_0^1 \frac{\mathrm{d}x}{[ax+b(1-x)]^2}.$$

By differentiating the above equation $n$ times with respect to $b$, we get

$$\frac{1}{a\,b^n} = \int_0^1 \mathrm{d}x\,\frac{n(1-x)^{n-1}}{[ax+b(1-x)]^{n+1}}\,,$$

and similarly, by differentiating $(\alpha-1)$ times with respect to $a$ and $(\beta-1)$ times with respect to $b$. To obtain the integral for $1/abc$, we rewrite it as $1/aB$ with $B=bc$ and apply the previous result.

*Gamma Function:*

$$\Gamma(z)=\int_0^\infty \mathrm{d}t\, \mathrm{e}^{-t}t^{z-1}\,, \qquad (\mathrm{Re}\, z>0)\,; \tag{A.45}$$

$$\Gamma(z+1)=z\Gamma(z)=z!\quad,\quad \Gamma(z)\Gamma(1-z)=\frac{\pi}{\sin(\pi z)}\,; \tag{A.46}$$

$$\int_0^\infty \mathrm{d}y\, \frac{y^\alpha}{(1+y^2)^\beta}=\frac{\Gamma\left[\frac{1}{2}(1+\alpha)\right]\Gamma\left[\frac{1}{2}(2\beta-\alpha-1)\right]}{2\,\Gamma(\beta)}\,. \tag{A.47}$$

Formulas (14.49) and (14.56) are also useful.

*Integral over* $\mathrm{d}^4k$*:*

$$\int \mathrm{d}^4k\, \frac{(k^2)^\alpha}{(k^2-A+\mathrm{i}\varepsilon)^\beta}=\frac{\mathrm{i}\,\pi^2\,(-1)^{\alpha-\beta}}{A^{\beta-\alpha-2}}\,\frac{\Gamma(\alpha+2)\Gamma(\beta-\alpha-2)}{\Gamma(\beta)}\,. \tag{A.48}$$

In the above equation, keep in mind that the $+\mathrm{i}\varepsilon$ term is always present in the propagators. We evaluate the $k^0$ integral first, making use of Wick's rotation. The locations of the poles at $k^0=-(\sqrt{|\boldsymbol{k}|^2+A})+\mathrm{i}\varepsilon$ and $k^0=+(\sqrt{|\boldsymbol{k}|^2+A})-\mathrm{i}\varepsilon$ allow us to rotate the contour of the $k^0$ integration in the complex $k^0$ plane by 90° counterclockwise if the integrand falls off sharply enough at large $k^0$, i.e. for $\beta>\alpha+2$. Then we define a *Euclidean* four-momentum variable $k_{\mathrm{E}}$ as $k_{\mathrm{E}}^0=\mathrm{i}k^0$ and $\boldsymbol{k}_{\mathrm{E}}=\boldsymbol{k}$, with the rotated contour going from $k_{\mathrm{E}}^0=-\infty$ to $+\infty$, and we do the integration in four-dimensional spherical coordinates ($k^2=-k_{\mathrm{E}}^2$). Thus,

$$\int \mathrm{d}^4k\, \frac{(k^2)^\alpha}{(k^2-A+\mathrm{i}\varepsilon)^\beta}=\mathrm{i}\int \mathrm{d}\Omega^{(4)}\int_0^\infty \mathrm{d}k_{\mathrm{E}}\, \frac{k_{\mathrm{E}}^3(-k_{\mathrm{E}}^2)^\alpha}{[-k_{\mathrm{E}}^2-A]^\beta}\,, \tag{A.49}$$

and we recover (48) by using (47) and $\int \mathrm{d}\Omega^{(4)}=2\pi^2$ from (14.56).

*Integrals over* $\mathrm{d}^nk$*.* Let us define $D(k,P)\equiv k^2+2k\cdot P-A$ ,

$$C\equiv \frac{\mathrm{i}\,\pi^{\frac{n}{2}}\Gamma\left(\beta-\frac{n}{2}\right)}{\Gamma(\beta)(-P^2-A)^{\beta-(n/2)}} \qquad \text{and} \qquad F\equiv \frac{1}{2}(P^2+A)\frac{\Gamma\left(\beta-1-\frac{n}{2}\right)}{\Gamma\left(\beta-\frac{n}{2}\right)}\,.$$

By making the shift $K=k+P$ and using $\int \mathrm{d}\Omega^{(n)}$ given by (14.56), we get

$$\int \frac{\mathrm{d}^nk}{[D(k,P)]^\beta}=C\,, \tag{A.50}$$

$$\int \frac{\mathrm{d}^nk\, k^\mu}{[D(k,P)]^\beta}=-CP^\mu\,, \tag{A.51}$$

$$\int \frac{\mathrm{d}^nk\, k^\mu k^\nu}{[D(k,P)]^\beta}=C\left[P^\mu P^\nu-g^{\mu\nu}F\right]\,, \tag{A.52}$$

$$\int \frac{\mathrm{d}^nk\, k^\mu k^\nu k^\lambda}{[D(k,P)]^\beta}=C\left[-P^\mu P^\nu P^\lambda+\left(g^{\mu\nu}P^\lambda+g^{\nu\lambda}P^\mu+g^{\lambda\mu}P^\nu\right)F\right]\,. \tag{A.53}$$

## A.4 Feynman Rules

The invariant scattering amplitude $\mathrm{i}\mathcal{M}$ is calculated in the perturbative method by drawing all fully connected Feynman diagrams, excluding self-energy insertions on external lines. Any diagram is either a tree diagram (with no loops) or a loop diagram (with one or more closed lines). Each diagram consists of lines and vertices. In general, the Feynman rules corresponding to lines are model independent, but those associated with vertices depend on the specific interaction model.

### General Rules

External lines (those having at least one free end) represent physical particles in the initial or final states; they have well-defined momenta. The momenta of internal lines (those connected to vertices at both ends) are determined by energy-momentum conservation at the vertices. In a tree diagram, each internal momentum can be fixed by the external momenta. In a loop diagram, to each loop corresponds an internal momentum that cannot be so fixed and must be integrated over.

*External lines* radiate from vertices and receive the following factors:

(a) For spin-$1/2$ fermion of momentum $p$ and spin state $s$
- in initial state: $u(p,s)$ on the right,
- in final state: $\bar{u}(p,s)$ on the left;

(b) For spin-$1/2$ antifermion of momentum $p$ and spin state $s$
- in initial state: $\bar{v}(p,s)$ on the left,
- in final state: $v(p,s)$ on the right;

(c) For spin-0 boson
- in either initial or final state: 1;

(d) For spin-1 boson of helicity $\lambda$ (if massless boson, $\lambda = \pm 1$; if massive boson, $\lambda = 0, \pm 1$)
- in initial state: $\epsilon_\mu(\lambda)$,
- in final state: $\epsilon^*_\mu(\lambda)$.

*Internal lines* represent the propagation of virtual particles between vertices. Each is associated with a propagator which depends on the momentum of the particle and is diagonal in internal labels (isospin, spin, or color):

$$\mathrm{i}\Delta(p) = \frac{\mathrm{i}}{p^2 - m^2 + \mathrm{i}\varepsilon} \qquad \text{spin-0,}$$

$$\mathrm{i}S_{\alpha\beta}(p) = \frac{\mathrm{i}(\not{p} + m)_{\alpha\beta}}{p^2 - m^2 + \mathrm{i}\varepsilon} \qquad \text{spin-}\tfrac{1}{2}\text{ fermion,}$$

$$\mathrm{i}\Delta_{\mu\nu}(p) = \frac{\mathrm{i}}{p^2 + \mathrm{i}\varepsilon}\left[-g_{\mu\nu} + (1-\xi)\frac{p_\mu p_\nu}{p^2}\right] \qquad \text{photon, gluon,}$$

$$\mathrm{i}\Delta_{\mu\nu}(p) = \frac{\mathrm{i}}{p^2 - m^2 + \mathrm{i}\varepsilon}\left[-g_{\mu\nu} + \frac{(1-\xi)p_\mu p_\nu}{p^2 - \xi m^2}\right] \qquad \text{massive vector boson.}$$

Every internal momentum $p$ that is not fixed by momentum conservation at the vertices must be integrated over: $\int \mathrm{d}^4 p/(2\pi)^4$. In addition, identical bosons in the initial or final state must be symmetrized, and identical fermions in the initial or final state, antisymmetrized. Each closed fermion loop receives a factor $-1$, as does each ghost loop. For each closed loop containing $n$ identical bosons, there is a factor $1/n!$.

## QCD Vertex Factors

The indices $a$, $b$ correspond to the quark color and run from 1 to 3; the indices $i, j, \ldots$ correspond to gluon color and run from 1 to 8. $f_{ijk}$ are the SU(3) structure constants. (a) Quark-Gluon vertex

(a) Quark–gluon vertex

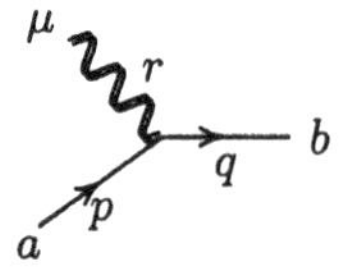

$$-ig_s\gamma_\mu(T_i)_{ba}$$
$$(p+r-q=0;\ T_i=\lambda_i/2)$$

(b) Three-gluon vertex

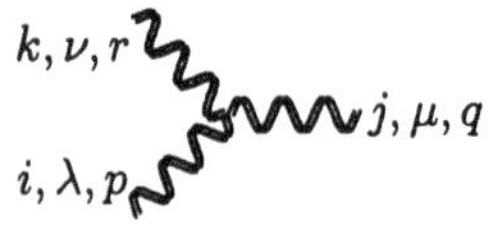

$$-g_s f_{ijk}[g_{\lambda\mu}(p-q)_\nu + g_{\mu\nu}(q-r)_\lambda + g_{\nu\lambda}(r-p)_\mu]$$
$$(p+q+r=0)$$

(c) Four-gluon vertex

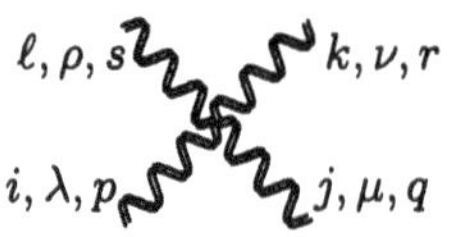

$$-ig_s^2\,[f_{ijm}f_{k\ell m}(g_{\lambda\nu}g_{\mu\rho}-g_{\mu\nu}g_{\lambda\rho}) + f_{ikm}f_{j\ell m}(g_{\lambda\mu}g_{\nu\rho}-g_{\mu\nu}g_{\lambda\rho}) + f_{kjm}f_{i\ell m}(g_{\lambda\nu}g_{\mu\rho}-g_{\lambda\mu}g_{\nu\rho})]$$
$$(p+q+r+s=0)$$

(d) Ghost–gluon vertex

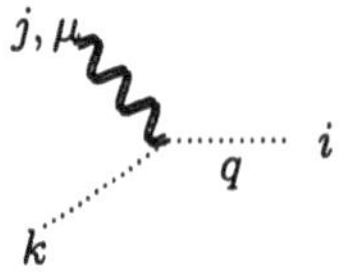

$$g_s f_{ijk}\, q_\mu$$

## Vertex Factors in the Standard Electroweak Model

The indices $A$, $B$ correspond to generations; the indices $a$, $b$ label quark flavors. All couplings are diagonal in color indices. $Q_a$ is the electric charge number of quark $q_a$. The weak coupling $g$ is related to the unit charge $e$ ($e>0$) by $g=e/\sin\theta_W$, where $\theta_W$ is the Weinberg angle, and $g_V^f$ and $g_A^f$ are the weak charges. $V_{AB}$ represents an element of the CKM mixing matrix, and $\tau_-$ is the lowering weak-isospin operator of the quark.

The quark–boson couplings are as follows:

(a) $\gamma$qq vertex

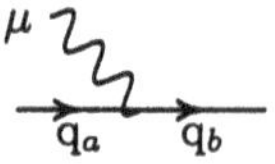

$$-i\,eQ_a\,\gamma_\mu\,\delta_{ba}\,\delta_{BA}$$

(b) Zqq vertex

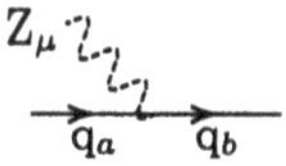

$$\frac{-ig}{2\cos\theta_W}\gamma_\mu\,(g_V^a-g_A^a\gamma_5)\,\delta_{ba}\,\delta_{BA}$$

(c) Wqq vertex

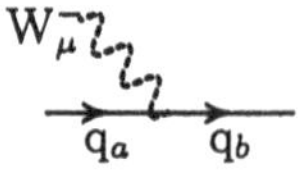

$$\frac{-ig}{2\sqrt{2}}\gamma_\mu\,(1-\gamma_5)\,(\tau_-)_{ba}\,V_{BA}$$

The lepton–boson couplings are:

(a) $\gamma ee$ vertex

$$\mathrm{i}\, e \gamma_\mu \, \delta_{BA}$$

(b) $Z\nu\nu$ vertex

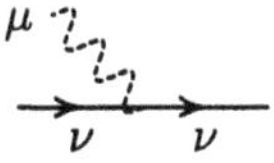

$$\frac{-\mathrm{i}g}{2\cos\theta_{\mathrm{W}}}\gamma_\mu \left(\tfrac{1}{2} - \tfrac{1}{2}\gamma_5\right)\delta_{BA}$$

(c) $Z\ell\ell$ vertex

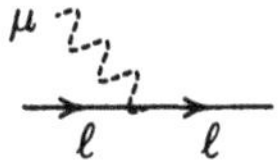

$$\frac{-\mathrm{i}g}{2\cos\theta_{\mathrm{W}}}\gamma_\mu (g_{\mathrm{V}}^{\ell} - g_{\mathrm{A}}^{\ell}\gamma_5)\,\delta_{BA}$$

(d) $W\ell\nu$ vertex

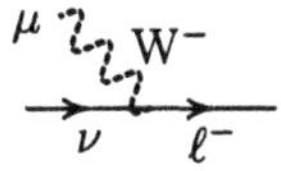

$$\frac{-\mathrm{i}g}{2\sqrt{2}}\gamma_\mu (1-\gamma_5)\,\delta_{BA}$$

The weak charges are $g_{\mathrm{V}} = \frac{1}{2} - \frac{4}{3}\sin^2\theta_{\mathrm{W}}$ and $g_{\mathrm{A}} = \frac{1}{2}$ for u-type quarks; $g_{\mathrm{V}} = -\frac{1}{2} + \frac{2}{3}\sin^2\theta_{\mathrm{W}}$ and $g_{\mathrm{A}} = -\frac{1}{2}$ for d-type quarks; $g_{\mathrm{V}} = \frac{1}{2}$ and $g_{\mathrm{A}} = \frac{1}{2}$ for neutrinos; and $g_{\mathrm{V}} = -\frac{1}{2} + 2\sin^2\theta_{\mathrm{W}}$ and $g_{\mathrm{A}} = -\frac{1}{2}$ for charged leptons. In addition to the quark–boson and quark–lepton couplings given above, there are also the boson–boson and boson–Higgs couplings found in Fig. 9.1b–c.

Furthermore, when dealing with electroweak loop diagrams in the general $R_\xi$ gauge, one must consider both the Faddeev–Popov ghosts $\chi^\pm$ and $\chi_z$ and the Goldstone bosons $w^\pm$, $z$ which come from the spontaneous symmetry breaking of the Higgs mechanism and which are absorbed by the $W^\pm$ and Z bosons to generate mass. The unphysical $\chi^\pm$, $\chi_z$, $w^\pm$, and $z$ appear only in internal lines. Their propagators are

$$\text{for } \chi^\pm, \chi_z: \quad \frac{\mathrm{i}}{p^2 - \xi M_{\mathrm{W,Z}}^2 + \mathrm{i}\varepsilon},$$

$$\text{for } w^\pm, z: \quad \frac{\mathrm{i}}{p^2 - \xi M_{\mathrm{W,Z}}^2 + \mathrm{i}\varepsilon}.$$

Since physical quantities are gauge invariant, the $\xi$-dependence must be canceled among the various diagrams that contribute to a given process.

The Faddeev–Popov ghosts $\chi^\pm$, $\chi_z$ couple only to $W^\pm$, Z, $\gamma$ and the Higgs bosons not to quarks or leptons. The $w^\pm, z$ couplings can either be three-point or four-point vertices, similar to those found in Fig. 9.1b–c. We do not give here the vertices for these unphysical particles, but refer the reader to Appendix B of Cheng and Li's book, where they can be found.

## A.5 Parameters of the Standard Model

In the electroweak sector, the following measured quantities may be taken as inputs for the model:

$$\begin{aligned}\frac{4\pi}{e^2} &= \alpha^{-1} = 137.0359895 \pm 0.0000061\,,\\ G_{\rm F} &= (1.16639 \pm 0.00002) \times 10^{-5}\,\mathrm{GeV}^{-2}\,,\\ \sin^2\theta_{\rm W} &= 0.2315 \pm 0.0004\,,\end{aligned}$$

together with the Higgs mass, $M_{\rm H} > 71$ GeV (LEP-1997), and the fermion masses (Table 7.9). From various experiments, the magnitudes of the CKM matrix elements have been obtained:

$$\begin{array}{lll} |V_{\rm ud}| = 0.9744 \pm 0.001, & |V_{\rm us}| = 0.2205 \pm 0.0011, & |V_{\rm ub}| = 0.0031 \pm 0.0008,\\ |V_{\rm cd}| = 0.204 \pm 0.017, & |V_{\rm cs}| = 1.01 \pm 0.18, & |V_{\rm cb}| = 0.039 \pm 0.0036,\\ |V_{\rm td}| = 0.0092 \pm 0.003, & |V_{\rm ts}| = 0.033 \pm 0.009, & |V_{\rm tb}| = 0.9991 \pm 0.0004\,. \end{array}$$

The running strong coupling strength is given by

$$\alpha_{\rm s}(\mu^2) = \frac{4\pi}{\beta_0 \log(\mu^2/\Lambda^2)}\,,$$

where $\Lambda = 200 \pm 50\,\mathrm{MeV}$, $\beta_0 = 11 - \frac{2}{3}\mathrm{N_f}$, and the running $N_{\rm f}$ is the number of quark species with masses less than the scale $\mu$ of the process considered. In particular, $\alpha_{\rm s}(M_{\rm Z}) \approx 0.118$, corresponding to $N_{\rm f} = 5$.

# Index

# Springer and the environment

At Springer we firmly believe that an international science publisher has a special obligation to the environment, and our corporate policies consistently reflect this conviction.

We also expect our business partners – paper mills, printers, packaging manufacturers, etc. – to commit themselves to using materials and production processes that do not harm the environment. The paper in this book is made from low- or no-chlorine pulp and is acid free, in conformance with international standards for paper permanency.

GPSR Compliance
The European Union's (EU) General Product Safety Regulation (GPSR) is a set of rules that requires consumer products to be safe and our obligations to ensure this.

If you have any concerns about our products, you can contact us on

ProductSafety@springernature.com

In case Publisher is established outside the EU, the EU authorized representative is:

Springer Nature Customer Service Center GmbH
Europaplatz 3
69115 Heidelberg, Germany

www.ingramcontent.com/pod-product-compliance
Ingram Content Group UK Ltd.
Pitfield, Milton Keynes, MK11 3LW, UK
UKHW021903190726
13853UKWH00003B/1389

* 9 7 8 3 6 6 2 0 3 7 1 3 3 *